TEACHER'S EDITION

LIVING BY
CHEMISTRY

SECOND EDITION

Angelica M. Stacy
Professor of Chemistry
University of California at Berkeley

with

Janice A. Coonrod
Bishop O'Dowd High School
Senior Writer and Developer

Jennifer Claesgens
Weber State University
Curriculum Developer

bfw
Freeman

W. H. Freeman and Company / BFW

W. H. Freeman and Company / BFW

Publisher: **Ann Heath**

Sponsoring Editor: **Jeffrey Dowling**

Development Editor: **Donald Gecewicz**

Marketing Manager: **Julie Comforti**

Editorial Assistants: **Matthew Belford, Rachel Chlebowski**

Copyeditor: **Louise B. Ketz**

Director of Editing, Design, and Media Production for the Sciences and Social Sciences: **Tracey Kuehn**

Managing Editor: **Lisa Kinne**

Project Editor: **Robert M. Errera**

Production Manager: **Paul Rohloff**

Photo Editor: **Robin Fadool**

Associate Photo Editor: **Eileen Liang**

Art Director: **Diana Blume**

Text and Cover Designer: **Marsha Cohen, Parallelogram Graphics**

Art Manager: **Matthew McAdams**

Illustrator: **Joe BelBruno**

Technical Artist: **Precision Graphics**

Composition: **Aptara®, Inc.**

Printing and Binding: **BR Printers**

Cover: **Photo by C. Mane/Getty Images**

This material is based upon work supported by the National Science Foundation under award number 9730634. Any opinions, findings, and conclusions or recommendations expressed in this publication are those of the author and do not necessarily reflect the views of the National Science Foundation.

Companion Web site
bcs.whfreeman.com/livingbychemistry2e
The *Living by Chemistry* Companion Web site is available free of charge and provides all access to all media for this program.

Media and e-Book Offerings
For information about additional resources that are available to support *Living by Chemistry*, please visit the catalog page:
highschool.bfwpub.com/LBC2e

Library of Congress Preassigned Control Number: 2015931433

ISBN-10: 1-4641-5639-5

ISBN-13: 978-1-4641-5639-7

2 3 4 5 6 7   23 22 21 20 19

Printed in the United States of America

W. H. Freeman and Company
One New York Plaza
Suite 4600
New York, NY 10004-1562
http://www.highschool.bfwpub.com

Unit 1 | Alchemy
Matter, Atomic Structure, and Bonding

> An engineering project for each unit can be found in the Teacher's Resource Materials for *Living by Chemistry*, 2018 Updated Second Edition.

Unit 2 | Smells
Molecular Structure and Properties

Unit 3 | Weather
Phase Changes and Behavior of Gases

Unit 4 | Toxins
Stoichiometry, Solution Chemistry, and Acids and Bases

Unit 5 | Fire
Energy, Thermodynamics, and Oxidation-Reduction

Unit 6 | Showtime
Reversible Reactions and Chemical Equilibrium

CONTENTS

Dirk Wiersma/Science Source

Unit 1 | Alchemy

Matter, Atomic Structure, and Bonding

Unit 2 | **Smells**

Molecular Structure and Properties

Alex Cao/Getty Images

Stockbyte/Getty Images

Unit 3 | Weather

Phase Changes and Behavior of Gases

Unit 4 | Toxins

Stoichiometry, Solution Chemistry, and Acids and Bases

Charles D. Winters/Science Source

Unit 5 | Fire

Energy, Thermodynamics, and Oxidation-Reduction

Vadym Volodin/iStockphoto

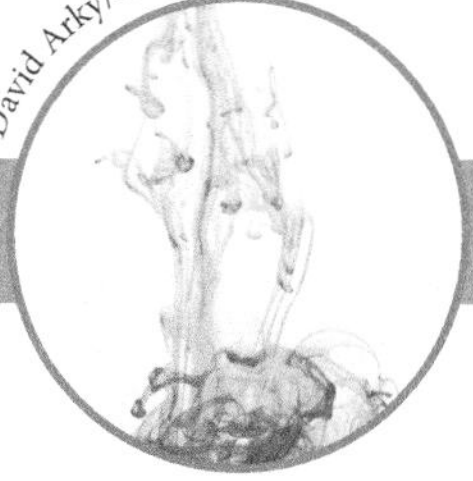

Unit 6 | **Showtime**

Reversible Reactions and Chemical Equilibrium

David Arky/Getty Images

Welcome to the Teacher's Edition of *Living by Chemistry*, second edition. This wraparound Teacher's Edition, along with the full *Living by Chemistry* teaching package, offers complete support for this innovative, full-year, inquiry-based high school chemistry program that meets and exceeds state and national standards.

The Teacher's Edition

New to the second edition, the Teacher's Edition combines all of the material formerly provided in the six separate teacher guides in one comprehensive and easy-to-use resource. Each unit begins with a four-page introduction before the wraparound content. The introductory pages include an explanation of the unit's theme and suggested pacing guides for standard and block schedules. A theme-based curriculum helps to capture student interest and improves retention of concepts. It also serves to ground the study of chemistry in the natural world and everyday life. The margin notes that wrap around the student pages provide teachers with complete, detailed, daily lesson plans for a full year of chemistry.

Teacher's Resource Materials

Look for the green **TRM** button throughout the Teacher's Edition that indicates when Teacher's Resource Materials are available for a unit, chapter, or lesson. Updated for the second edition, the Teacher's Resource Materials include lesson worksheets, handouts, transparencies, card masters, PowerPoint presentations, assessments, and other tools for use throughout the year. Access these carefully developed resources by clicking on the link in the **edAPtext** Teacher's Edition (TE-book), opening the Teacher's Resource Flash Drive (TRFD), or logging onto the Book Companion Site (BCS): bcs.whfreeman.com/livingbychemistry2e (teacher log-in required).

PD Videos

The orange **PD** button appears at the beginning of each chapter. This is to let you know that a "just in time" professional development video presented by the author or an experienced *Living by Chemistry* teacher is available for download. These videos offer tips and advice for teaching the guided-inquiry lessons and help you to make the most of your Teacher's Resource Materials for each chapter and lesson.

Unit Introductions set the stage for teaching the unit.

- Each unit begins with an introduction to the chemistry content and the six overarching themes: Alchemy, Smells, Weather, Toxins, Fire, Showtime.

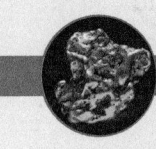

Unit 1 | Alchemy
Matter, Atomic Structure, and Bonding

ALCHEMY AS CONTEXT

Hundreds of years ago, alchemists were trying to turn ordinary substances into gold. The overarching question for this unit—Is it possible to create gold from other substances?—is accompanied by two other important questions: If so, how is it done? and If not, why not? In the course of their investigation, students explore matter and the models scientists have created to explain what matter is and why it behaves the way it does.

ALCHEMY AS THE ROOT OF CHEMISTRY

The study of matter arose partly from practical pursuits. To make food, clothing, and tools, people had to manipulate the substances in their environment. They learned to turn milk into cheese, forge metals, dye wool, and so on. Chemistry as a scientific discipline also has roots in alchemy—the word *chemistry* comes directly from the word *alchemy*. The practice of alchemy can be traced to many cultures, including those of Arabia, China, and ancient Greece. Some alchemists wanted to make gold, while others sought potions that would bring eternal life. A few alchemists renounced these pursuits and sought only to understand the nature of matter.

In light of what we now know about the elements, the pursuit of alchemy seems a bit of a wild goose chase. It is obvious to us now that it is not possible to turn metals into gold by mixing, refining, or fusing substances together. Even so, the discipline of chemistry owes much to the struggles and efforts of the alchemists who developed many practical lab procedures, such as distillation, sublimation, oxidation, and extraction. By manipulating assorted compounds and experimenting with them, much was learned about their properties and passed on to future scientists through extensive bodies of writing.

CONTENT DRIVEN BY CONTEXT

The chemistry content is the top priority in each unit, with enough context provided to engage students and provide a real-world setting. For example, we introduce the periodic table of the elements by challenging students to re-create an early version of the table based on patterns that Dmitri Mendeleyev observed. Next, students consider the merits of historic models of the atom before arriving at a basic atomic model. Throughout the unit we return to the central question: Is it possible to create gold from other substances? The ultimate answer to that question is both yes and no. The gold found here on Earth was created somewhere in the stars through nuclear processes. However, that option is not available to us in a laboratory. So, while it is *theoretically* possible through nuclear chemistry, making gold from other elements is not feasible on Earth.

THE USE OF MODELS TO EXPLAIN MATTER

Another central theme of this unit is the use of models to explain matter. The atomic model is one of the greatest tools chemists have developed to explain the behavior of matter. It is important that students become comfortable working with models and making reasonable deductions and predictions based on evidence. Learning to decipher models is a skill that does not necessarily come overnight. In chemistry, we bombard students with dozens of representations and models for matter, from chemical symbols and formulas to molecules represented as balls bouncing around in a box. Each model has its usefulness, and each has its limitations. There are inevitably inaccuracies in the scale, shape, and size of items shown in models, and these shortcomings may lead to confusion. Remember, it is quite often the model that is limited, not the student.

Unit 1 | **Alchemy**   1-1

- Chapter descriptions and pacing guides help you to plan and prepare.

BUILDING UNDERSTANDING

By starting with the context of alchemy, you can ask simple, fundamental questions—much as early chemists did. You can start at the ground level, without assuming that all students come into the classroom with prior knowledge of chemistry. While students are often familiar with some chemical terminology (such as "atomic" and "nuclear"), most are unclear on the concepts underlying these terms. Using alchemy as a contextual starting point allows you to bring all students along together, wherever they are beginning. It also lets you set the tone for the entire year, modeling an open atmosphere of inquiry and questioning.

Chapter Summary

Chapter	Description	Standard Schedule Days	Block Schedule Days
1	Chapter 1 is an introduction to chemistry, beginning with orientation to lab equipment and safety. Next, students observe the apparent creation of a gold penny. The question arises, "How can you tell if it is gold or not?" To find the answer, students debate the nature of matter, make observations and measurements, and differentiate between types of matter based on properties.	6	3
2	Chapter 2 covers elements and the language of chemistry. Lesson 7: The Copper Cycle provides more food for thought as students consider the question, "What happened to the copper during the copper cycle?"	6	3
3	Chapter 3 provides a chance for students to explore the atomic model, atomic structure, and nuclear chemistry to learn how gold differs from other elements. A central question in this section is "What makes a gold atom a gold atom?" Once students know that the nucleus of an atom holds its elemental identity, and that the nucleus is difficult to change, the focus moves to the parts of an atom that are easier to change—the electrons.	6	3
4	Chapter 4 presents the question, "What happens if the electrons are moved around?" Even though it is not possible to change another element into gold through electron movement, students discover that atoms can be used purposefully to create things as valuable as or more valuable than gold. In an optional lab, students make paint pigments as they learn more about transition metals.	8	4
5	Chapter 5 introduces students to models for ionic bonding, molecular covalent bonding, network covalent bonding, and metallic bonding. The chapter-ending electroplating lab gives students more tangible evidence of electrons and ions. Students explore models of bonding as they ask, "What can I make that is as good as gold?"	6	3

1-2   Unit 1 | **Alchemy**

Pacing Guides
Standard Schedule

Day	Suggested Plan	Day	Suggested Plan
1	Introduction, Chapter 1 Lesson 1	17	Chapter 3 Lesson 16, Chapter 3 Review
2	Safety Quiz, Chapter 1 Lesson 2	18	Chapter 3 Quiz, Chapter 4 Lesson 17
3	Chapter 1 Lesson 3	19	Chapter 4 Lesson 18
4	Chapter 1 Lesson 4	20	Chapter 4 Lesson 19
5	Chapter 1 Lesson 5, Chapter 1 Review	21	Chapter 4 Lesson 20
6	Chapter 1 Quiz, Chapter 2 Lesson 6	22	Chapter 4 Lesson 21
7	Chapter 2 Lesson 7	23	Chapter 4 Lesson 22
8	Chapter 2 Lessons 7 and 8	24	Chapter 4 Lesson 23
9	Chapter 2 Lesson 9	25	Chapter 4 Lesson 24, Chapter 4 Review
10	Chapter 2 Lesson 10	26	Chapter 4 Quiz, Chapter 5 Lesson 25
11	Chapter 2 Review	27	Chapter 5 Lesson 26
12	Chapter 2 Quiz, Chapter 3 Lesson 11	28	Chapter 5 Lesson 27, Chapter 5 Review
13	Chapter 3 Lesson 12	29	Chapter 5 Quiz
14	Chapter 3 Lesson 13	30	Unit 1 Review
15	Chapter 3 Lesson 14	31	Unit 1 Exam
16	Chapter 3 Lesson 15	32	Lab Exam (optional)

Block Schedule

Day	Suggested Plan	Day	Suggested Plan
1	Introduction Chapter 1 Lesson 1	10	Chapter 3 Quiz Chapter 4 Lessons 17 and 18
2	Safety Quiz Chapter 1 Lessons 2 and 3	11	Chapter 4 Lessons 19 and 20
3	Chapter 1 Lessons 4 and 5 Chapter 1 Review	12	Chapter 4 Lessons 21 and 22
4	Chapter 1 Quiz Chapter 2 Lessons 6 and 7	13	Chapter 4 Lessons 23 and 24 Chapter 4 Review
5	Chapter 2 Lessons 7 and 8	14	Chapter 4 Quiz Chapter 5 Lessons 25 and 26
6	Chapter 2 Lessons 9 and 10 Chapter 2 Review	15	Chapter 5 Lesson 27 Chapter 5 Review
7	Chapter 2 Quiz Chapter 3 Lessons 11 and 12	16	Unit Review
8	Chapter 3 Lessons 13 and 14	17	Unit 1 Exam Lab Exam
9	Chapter 3 Lessons 15 and 16 Chapter 3 Review		

Margin content provides complete, step-by-step lesson plans.

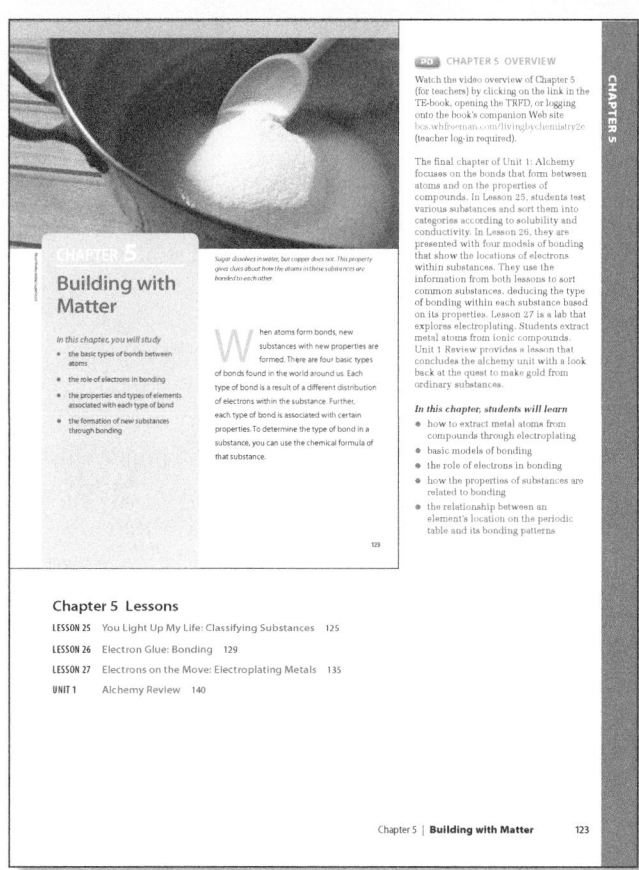

- Each chapter opens with a summary of the concepts covered.

- **PD** **PD Videos** provide "just-in-time" help for each chapter. They offer a brief overview of the chapters with tips and advice for teaching each lesson.

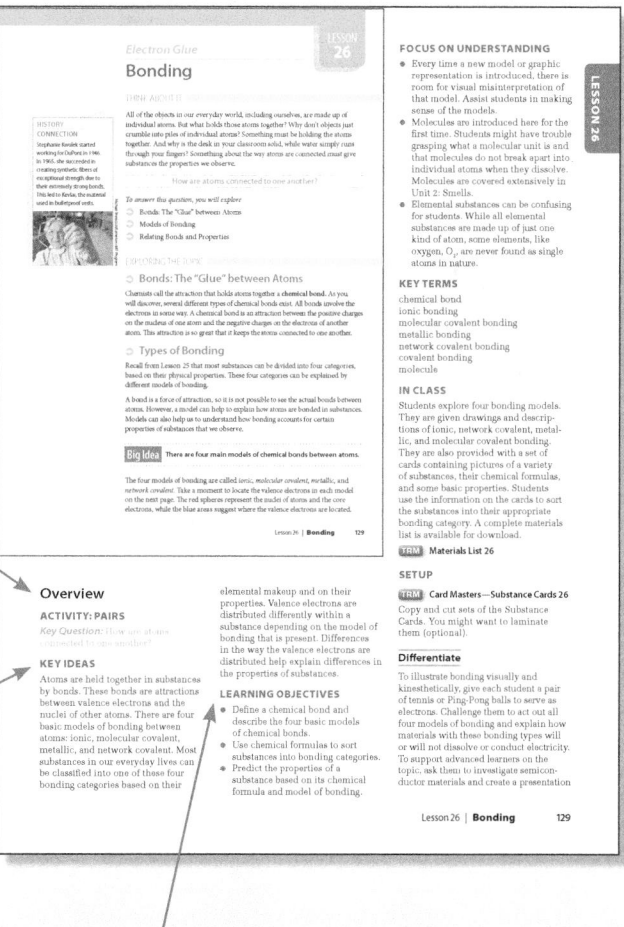

- The lesson **Overview** gives an orientation to the content in the lesson and describes the necessary preparation. Each lesson may be a lab, demo, activity, computer activity, classwork, or a follow-up to the previous lesson.

- **Key Ideas** describes the chemistry content covered in the lesson.

- **Learning Objectives** are stated clearly.

Lessons promote understanding of chemistry concepts.

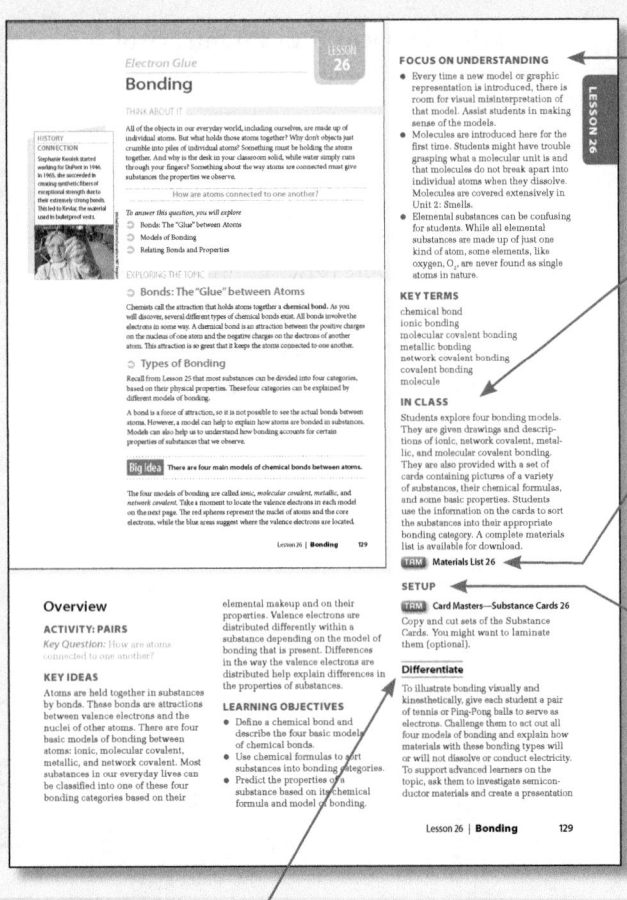

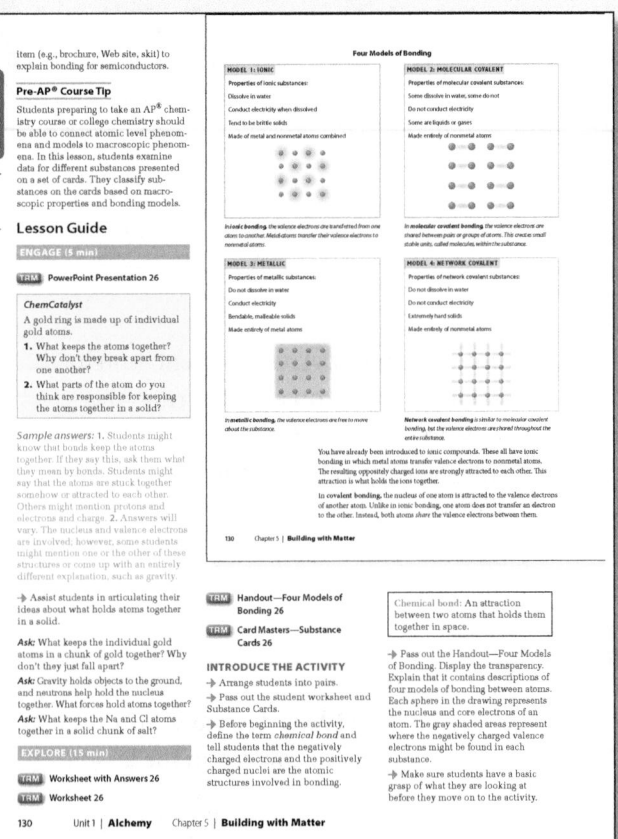

- **Focus on Understanding** points out common misconceptions and pitfalls that students may experience.

- **In Class** describes what will take place during the lesson and alerts you to materials you will need.

- The green **TRM** button indicates that **Teacher's Resource Materials** are available by download from the book's companion Web site, the Teacher's Resource Flash Drive (TRFD), or by clicking on the link in the **edAPtext** Teacher's Edition (TE-book).

- **Setup** and **Cleanup** instructions are spelled out, along with relevant safety reminders.

- **Differentiate** offers tips on how to further adjust the lessons to help struggling students or offer advanced students an additional challenge.

- **Pre-AP® Course Tips** point out science practices and chemistry concepts that help students to prepare for a future AP® or college chemistry course.

- The **Lesson Guide** offers a complete guided-inquiry lesson plan. Each lesson follows a modified version of the 5Es model of teaching and learning. The 5Es describe different stages of a learning sequence: Engage, Explore, Explain, Elaborate, and Evaluate.

Pre-AP® and AP® are trademarks registered and/or owned by the College Board®, which was not involved in the production of, and does not endorse, this product.

Guided inquiry increases retention.

- Every lesson begins with an **Engage** section to get students thinking about chemistry. Lesson PowerPoint presentations are available in the Teacher's Resource Materials that include ChemCatalyst opening questions, Key Points for discussion, and end-of-lesson Check-In questions.

- Students warm up with a **ChemCatalyst,** designed to engage them as well as help the teacher ascertain prior knowledge and misconceptions and set up an interactive classroom by:

  - Writing the ChemCatalyst question on the board or displaying the question using the PowerPoint presentation.
  - Downloading a transparency from the Teacher's Resource Materials for those ChemCatalyst openers with diagrams or illustrations to show in class or printing a copy for each student.

- **Sample questions** help you guide the opening discussion.

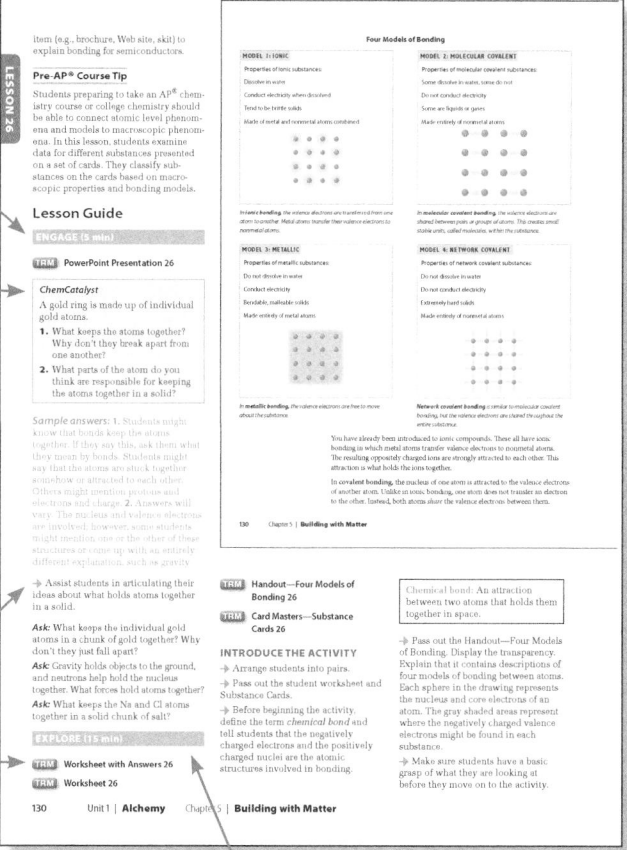

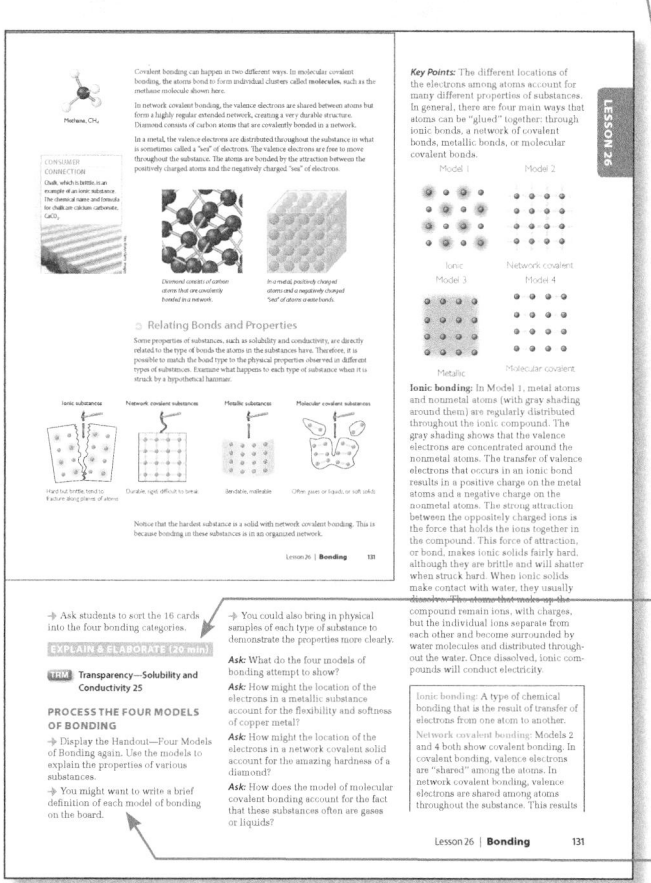

- As part of the **Explore** section of the lesson, students work together to discover and make sense of new chemistry content. Each lesson has a student worksheet to help guide students through the exploration of the chemistry content.

- Download the **Student Worksheet with Answers** and the **Student Worksheet** for each lesson from the Teacher's Resource Materials. Teacher's resources for lessons can also include handouts, card masters, and other printable manipulatives. Everything needed is in the Materials List for each lesson.

- The **Explain & Elaborate** section outlines key points and vocabulary for debriefing the lesson. This is an opportunity for a "making sense" discussion and for students to take notes.

- **Sample questions** help you guide the discussion toward the **key points.**

Exercises provide review and practice.

- A brief **Check-In** question allows you to evaluate how well students understood the main point of the lesson.

  - Write the Check-In question on the board or display the question using the prepared PowerPoint presentation.
  - For Check-In questions with diagrams or illustrations, download a transparency from the Teacher's Resource Materials to show in class or print for students.

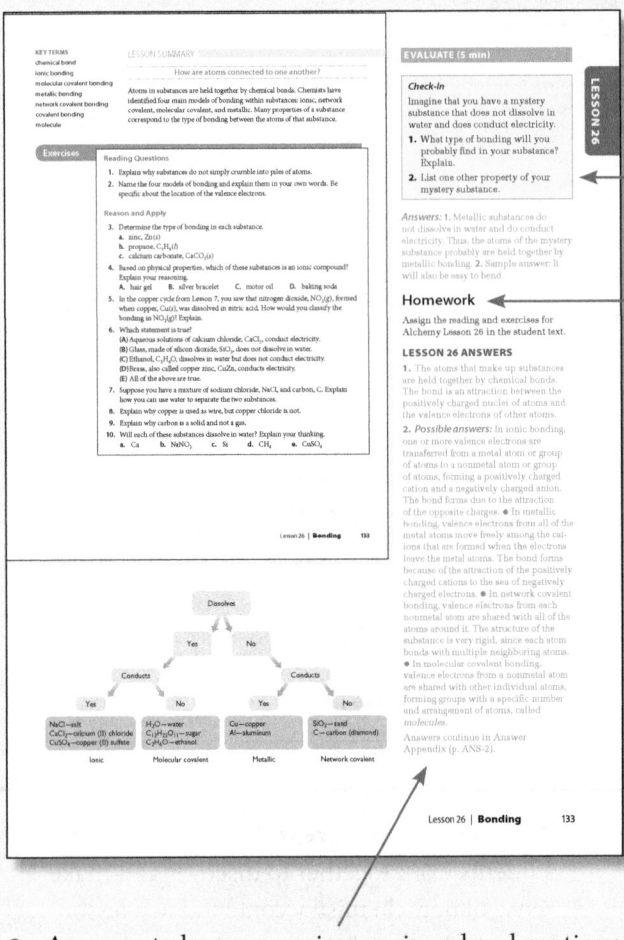

- **Homework,** such as the reading and exercises in the student textbook, or review for an upcoming quiz or test, is listed here.

- **Answers** to lesson exercises are in red and continue in the Answer Appendix at the end of the Teacher's Edition.

- **Answers** to chapter review exercises are provided and continue in the Answer Appendix at the end of the Teacher's Edition.

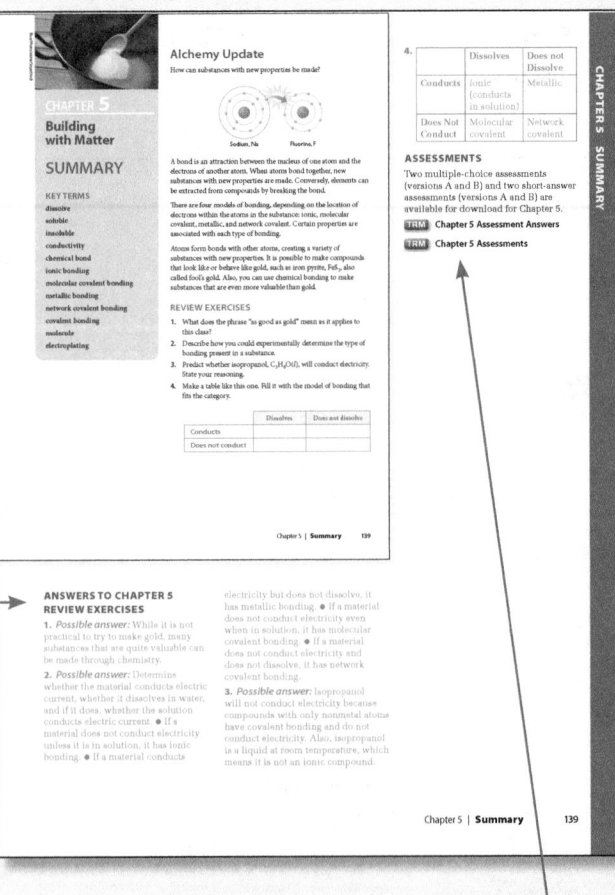

- Prepared chapter **Assessments** are available for download from the Teacher's Resource Materials.

Unit reviews provide additional practice.

- **Unit Reviews** include a lesson plan to review concepts taught throughout the unit.

- **Answers** to unit-review exercises are provided and continue in the Answer Appendix at the end of the Teacher's Edition.

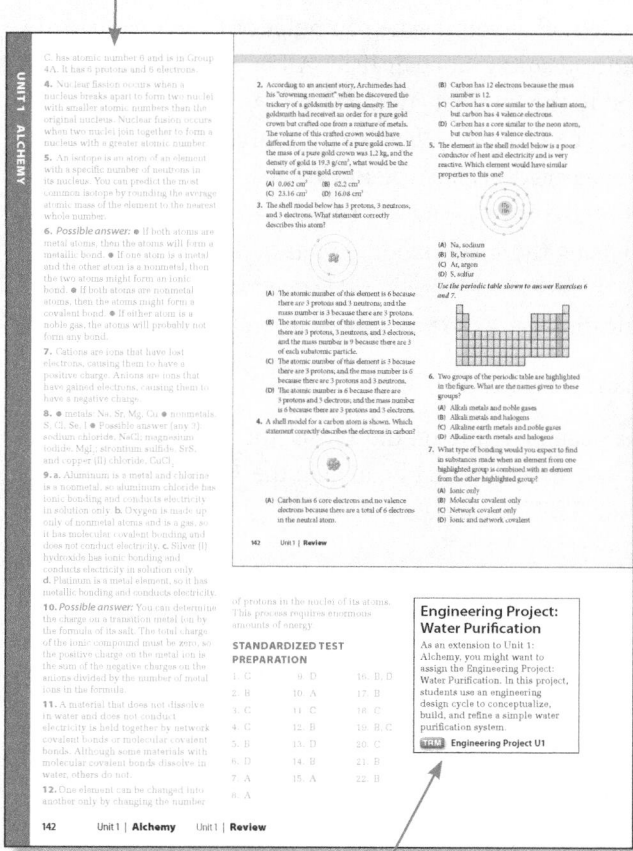

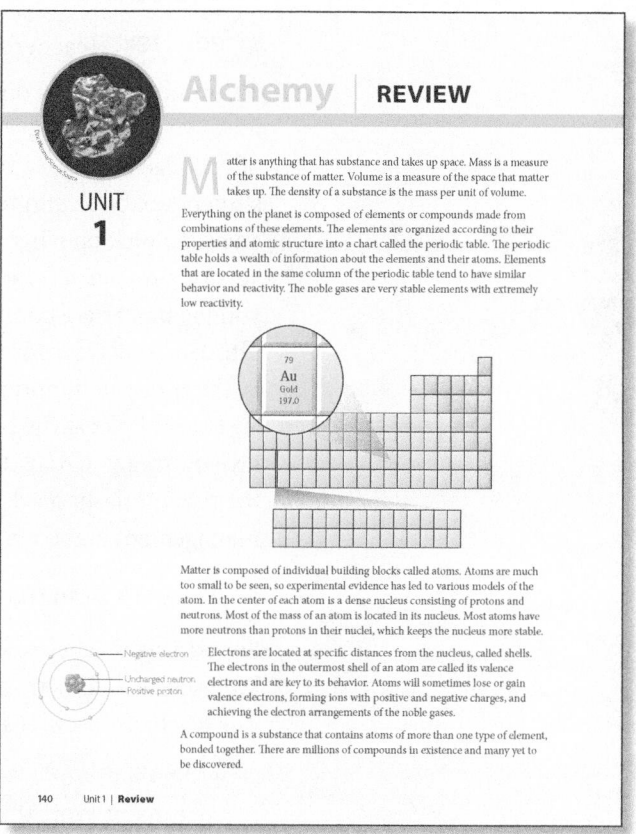

- **Engineering Design Projects** are available at the end of each unit to promote understanding of science and engineering practices related to chemistry.

- Prepared **Unit Assessments** are available for download from the Teacher's Resource Materials.

- **Lab Assessments** are available for each unit to test students' learning of practical laboratory knowledge.

- **Mid-term Assessments** covering Units 1–3, an **End-of-Year Assessment** covering Units 1–6, and a final **Lab Assessment** covering Units 1–6 are available for download from the Teacher's Resource Materials.

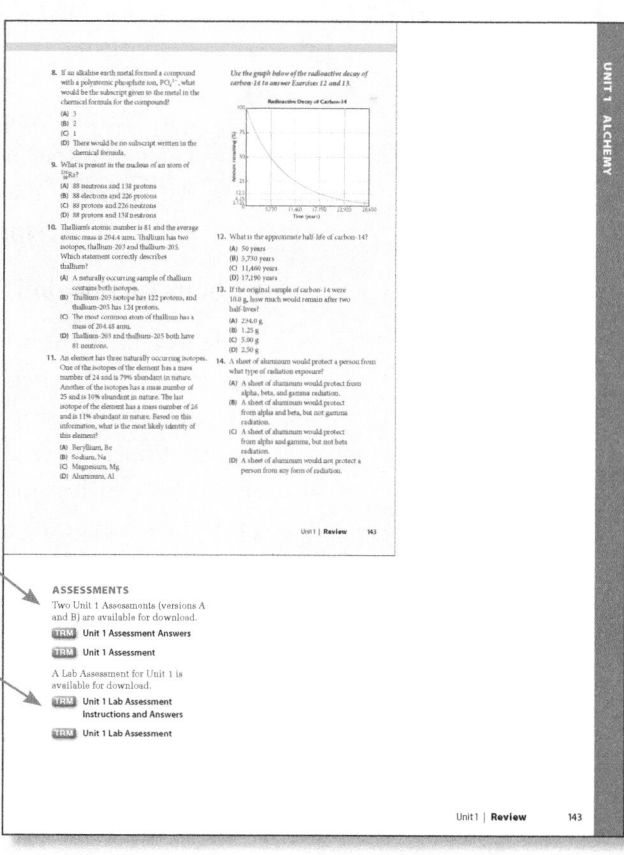

Resources Available for *Living by Chemistry*, 2e

FOR TEACHERS

- **edAPtext** Teacher's Edition (TE-book)

This innovative new digital product gives you full access to the Teacher's Edition for *Living by Chemistry*, second edition, when and where you want it. The TE-book offers page fidelity, so that it matches the print page exactly. You may read the student textbook and teacher content and take notes on the computer at school and then pick up where you left off at your home computer or in a downloaded version on your iPhone/Android smartphone or iPad. All of your notes sync once you log in and are connected to the Internet. Your **edAPtext** TE-book is the ultimate integrator for all of your *Living by Chemistry* resources. While you are in the connected mode, you may access all of the PD Videos, TRM resources, and other links at the point of use. If you want to provide assignments, notes, or give quizzes you may do so using the social-media function of **edAPtext** and the teacher dashboard. The **edAPtext** TE-book is designed to make classroom management and communication easier so you can focus on teaching.

- **Teacher's Resource Flash Drive (TRFD)** 1-4641-5640-9

Supporting teacher resources can be found on the TRFD, the book's companion Web site, as well as through the direct links embedded in the **edAPtext** TE-book. The teacher's resources on the TRFD include:

  - Student Worksheets
  - Student Worksheets with Answers
  - Transparencies, handouts, and other printed hands-on manipulatives, such as card masters, games, and activities to use during class, including specialty items such as Create a Table Cards, Salty 8s Cards, Polyatomic Ions Cards, Nuclear Quest Board Game, Molecules Cards, Structural Formula Cards, and Weather Map transparencies
  - Lesson PowerPoint presentations
  - Assessments, including prepared unit and chapter assessments, a midterm and final assessment, and lab skills assessments

- **Book Companion Site (BCS)**

Students and teachers may access this free password-protected on-line resource 24/7. Teachers have access to all of the Teacher's Resource Materials here, including all student worksheets, handouts, card masters, PowerPoint presentations, and ready-made assessments. Additional support materials are available such as materials lists, pacing guides, and detailed solutions to student edition problem sets. This is a password-protected site found at bcs.whfreeman.com/livingbychemistry2e (login required).

- **ExamView® Assessment Suite** 1-4641-5634-4

The ExamView® Assessment Suite allows teachers to create on-line and paper tests and quizzes quickly and easily. Users may select from the extensive bank of more than 2,000 questions or may write their own. Tests may be printed or administered on-line using the ExamView® Player. Questions can be sorted by unit, chapter, or lesson and include multiple-choice or short-answer/problem-solving questions. The ExamView® Assessment Suite can automatically grade and send results of multiple-choice questions and create reports for teachers.

- **PD** **Chapter PD Videos**

Brief chapter overview videos give teachers a quick guide to what is covered in each chapter. Presented by the author or an experienced *Living by Chemistry* teacher, these videos outline the daily in-class lessons and provide tips on best practices for teaching the chemistry content of the chapter.

- **Lesson PowerPoint Presentations**

There is a PowerPoint presentation for each lesson in *Living by Chemistry,* second edition. The presentations include the ChemCatalyst questions, Key Points for discussion, and Check-In questions. Relevant illustrations and diagrams for the lesson topic are included.

- **Solutions Manual**

This manual includes complete, worked solutions to each exercise in the student textbook.

FOR STUDENTS

- **edAPtext** **Student eBook**

This innovative new digital product gives students full access to *Living by Chemistry,* second edition, when and where they want it. They may read the textbook and take notes on the computer at school and then pick up where they left off in a downloaded version on an iPhone/Android smartphone or iPad. Notes taken off-line will sync once the user logs in when connected to the Internet. In addition to accessing student notes, the social media function allows students to receive group or individual messages from the teacher, to work in study groups, to take quizzes, and to complete other assignments, all at the point of use, within the **edAPtext** program.

About the Author

Angelica M. Stacy is a Professor of Chemistry at the University of California. At Berkeley, she teaches introductory chemistry and does research in materials chemistry and chemistry education. She has received numerous awards and honors for her teaching and held the President's Chair for Teaching at the University of California from 1993 to 1996. In 2005, the National Science Foundation named Dr. Stacy a Distinguished Teacher Scholar. As a part of her work to help more students succeed in science and gain a better understanding of the principles of chemistry, she has participated as a member of the Physical Sciences Design Team that helped develop the Next Generation Science Standards. She has also participated in various AP® chemistry course committees as part of the College Board's efforts to redesign the AP® chemistry course and exam. Dr. Stacy has worked for many years in designing and refining the *Living by Chemistry* curriculum. She developed this curriculum to support students as they learn chemistry by offering them engaging challenges that relate to the world around them.

ACKNOWLEDGMENTS

A number of individuals joined the project as developers for various periods of time to complete this work. Thanks go to these individuals for their contributions to unit development: Karen Chang, David Hodul, Rebecca Krystyniak, Tatiana Lim, Jennifer Loeser, Evy Kavaler, Sari Paikoff, Sally Rupert, Geoff Ruth, Nicci Nunes, Gabriela Waschewski, and Daniel Quach.

David R. Dudley contributed original ideas and sketches for some of the wonderful cartoons interspersed throughout the book. His sketches provided a rich foundation for the art manuscript.

This work would not have been possible without the thoughtful feedback and great ideas from numerous teachers who field-tested early versions of the curriculum. Thanks go to these teachers and their students: Carol de Boer, Wayne Brock, Susan Edgar-Lee, Melissa Getz, David Hodul, Richard Kassissieh, Tatiana Lim, Evy Kavaler, Geoff Ruth, Nicci Nunes, Gabriela Waschewski, and Daniel Quach.

Dr. Truman Schwartz provided a thorough and detailed review of the first edition manuscript. We appreciate his insights and chemistry expertise.

Thank you everyone at Bedford, Freeman & Worth who contributed to the development of this second edition. Special thanks to Jeffrey Dowling for masterfully guiding the development, to Donald Gecewicz for thoughtful development editing, feedback, and advice, and to Karen Misler for keeping us all on track with expert project management. Thank you, Sharon Sikora, for help with assessments and appendix material. Thank you, Kathleen Markiewicz, for help writing tips and advice for the Differentiate feature and Paul Thompson for writing Pre-AP® Course Tips.

Finally, thanks go to the publisher Ann Heath, who believed in this program and helped to assemble and guide the team along the way.

Reviewers and Field Testers

Science Content Advisor

Dr. A. Truman Schwartz,
Macalester College (emeritus), St. Paul, MN

Teaching and Content Reviewers

Scott Balicki
Boston Latin School
Boston, MA

Greg Banks
Urban Science Academy
West Roxbury, MA

Randy Cook
Tri County High School
Howard City, MI

Thomas Holme
University of Wisconsin-Milwaukee
Milwaukee, WI

Mark Klawiter
Deerfield High School
Deerfield, WI

Kathleen Markiewicz
Boston Latin School
Boston, MA

Nicole Nunes
Liberty High School
Benicia, CA

Carri Polizzotti
Marin Catholic High School
Larkspur, CA

Paul Thompson Jr.
High School for Math, Science, and Engineering at CCNY
New York, NY

Matthew Vaughn
Burlingame High School
Burlingame, CA

Rebecca Williams
Richland College
Dallas, TX

Field Testers

Carol de Boer
Amador Valley High School
Pleasanton, CA

Wayne Brock
Life Learning Academy
San Francisco, CA

Janie Burkhalter
Coronado High School
Lubbock, TX

Karen Chang
The Calhoun School
New York, NY

Elizabeth Christopher
El Camino High School
Woodland, CA

Mark Crown
Gateway High School
San Francisco, CA

Susan Edgar-Lee
Hayward High School
Hayward, CA, and
Livermore High School
Livermore, CA

Melissa Getz
Tennyson High School
Hayward, CA

Shannon J. Halkyard
Stuart Hall High School
San Francisco, CA

David Hodul
Bishop O'Dowd High School
Oakland, CA

Kim D. Johnson
Thurgood Marshall Academic High School
San Francisco, CA

Evy Kavaler
Berkeley High School
Berkeley, CA

Bruce Leach
Hill Country Christian School
Austin, TX

Tatiana Lim
Morse High School
San Diego, CA

Kathleen Markiewicz
Boston Latin School
Boston, MA

Steve Maskel
Hillsdale High School
San Mateo, CA

Mardi Mertens
Berkeley High School
Berkeley, CA

Nicole Nunes
Liberty High School
Benicia, CA, and
De La Salle High School
Concord, CA

Tracy A. Ostrom
Skyline High School
Oakland, CA

Pru Phillips
Crawfordsville High School
Crawfordsville, IN

Daniel Quach
Berkeley High School
Berkeley, CA

Carissa Romano
Hayward High School
Hayward, CA

Sally Rupert
Assets High School
Honolulu, HI

Geoff Ruth
Leadership High School
San Francisco, CA

Maureen Wiser
Emery Secondary School
Emeryville, CA

Audrey Yong
Thurgood Marshall Academic High School
San Francisco, CA

INTRODUCTION TO THE *LIVING BY CHEMISTRY* PROGRAM

Living by Chemistry consists of six units, each organized around a specific body of chemistry content and a theme to which students can relate. Most units consist of about 20 lessons of 45-minute duration, which can be combined for 90-minute block periods.

Unit 1	Alchemy	Matter, Atomic Structure, and Bonding	27	Lessons
Unit 2	Smells	Molecular Structure and Properties	21	Lessons
Unit 3	Weather	Phase Changes and Behavior of Gases	19	Lessons
Unit 4	Toxins	Stoichiometry, Solution Chemistry, and Acids and Bases	26	Lessons
Unit 5	Fire	Energy, Thermodynamics, and Oxidation-Reduction	21	Lessons
Unit 6	Showtime	Reversible Reactions and Chemical Equilibrium	7	Lessons

A Thematic Approach

A theme-based curriculum captures students' interest, helps them make connections, and improves retention of concepts. It also serves another purpose—it helps to ground the study of chemistry in the natural world and everyday life. Too often, students view chemistry as an inaccessible discipline centered on synthetic chemicals invented in a laboratory. In reality, chemical processes occur all the time in our bodies and in the world around us. Without most of these processes, life would not be possible. *Living by Chemistry* supports teachers in fostering students' wonder and curiosity about the world around them.

Science as Guided Inquiry

Living by Chemistry is the product of a decade of research and development in high school classrooms, focusing on optimizing student understanding of chemical principles. The curriculum was developed with the belief that science is best learned through firsthand experience and discussion with peers. Guided inquiry allows students to actively participate in, and become adept at, scientific processes and communication. These skills are vital to a student's further success in science as well as beneficial to other future pursuits.

The *Living by Chemistry* curriculum provides you with a student-centered lesson for each day. Students have opportunities to ask questions, make scientific observations, collect evidence, and formulate scientific hypotheses and explanations. In each lesson, students discover concepts and communicate ideas with peers and with the teacher. Formal definitions and formulas often are introduced after students have explored, scrutinized, and developed a concept, providing more effective instruction.

Thinking Like a Scientist

Using guided inquiry as a teaching tool promotes scientific reasoning, critical-thinking skills, and a greater understanding of the concepts. Students develop their own logical conclusions and discover chemistry concepts for themselves, rather than accepting and memorizing facts. The ultimate goal is to foster students who think like scientists and understand the nature of scientific practice. Students learn to study the natural world by asking questions and proposing explanations based on evidence. They learn to reflect on their ideas and review their work and that of their peers, as well as to effectively communicate scientific concepts they have discovered.

Chemistry for All Students

Chemistry is at the core of many aspects of our daily lives. Now, more than ever, the world needs citizens who can make informed decisions about their health, the environment, energy use, nutrition, and safety. Also, chemistry is a required course for myriad different career paths relating to science, engineering, health, and the environment. The *Living by Chemistry* curriculum helps you to promote scientific literacy and support all students in developing valuable skills that extend well beyond the classroom.

State and National Science Standards

While creating the second edition of *Living by Chemistry,* the author referred to recommendations and current trends in national and state science standards, as well as research of best practices of chemical education to guide the revision process. The second edition models current themes in K-12 science standards and frameworks and emphasizes science practices and inquiry-based methods as a means of teaching and learning chemistry. More in-depth correlations of *Living by Chemistry* second edition to current science education standards are available on the Book Companion Site (BCS). Please visit bcs.whfreeman.com/livingbychemistry2e (teacher log-in required).

Preparing Students for AP® Chemistry and College Chemistry Courses

The redesigned AP® Chemistry course centers around six big ideas in chemistry that are linked to science practices. *Living by Chemistry* aligns particularly well to prepare students to take the redesigned AP® Chemistry course. The curriculum meets the prerequisite knowledge requirements and introduces students to concepts that now comprise the AP® Chemistry curriculum. Science practices are emphasized in each daily lesson, and Pre-AP® course tips throughout the Teacher's Edition help teachers to prepare students to take AP® or college chemistry down the road.

Sequenced for Understanding

The sequencing of topics in *Living by Chemistry* is purposeful and well-tested. The topics are ordered and presented in a way that optimizes understanding. Also, the curriculum covers all necessary standards and concepts. (See the Content Coverage Chart on pages xxviii.) You may be tempted to reorder topics or front-load detailed information when a topic is introduced—this is not necessary because *Living by Chemistry* is a spiraling curriculum in which topics are revisited in increasing depth throughout the course. Our experience confirms that by building a solid foundation and by scaffolding all the topics—including the mathematics—*Living by Chemistry* can help you prepare your students for even the most challenging topics.

Unit 1: Alchemy
Matter, Atomic Structure, and Bonding

What is the makeup of the world around us? Students explore atomic theory through a historical lens, learning about matter, elements and compounds, the periodic table, electron configuration, and bonding. Following in the footsteps of scientists, they re-create the modern periodic table and evaluate atomic models in light of experimental evidence.

Unit 2: Smells
Molecular Structure and Properties

How does the nose know? Students investigate the chemistry of smell, learning about molecular formulas, Lewis dot structures, bonding, and the shape and structure of molecules.

Unit 3: Weather
Phase Changes and Behavior of Gases

What's the forecast? Students ask questions related to weather. In the process, they gain an understanding of temperature scales and thermometers, kinetic molecular theory, and the relationships among pressure, temperature, and volume that lead to the gas laws.

Unit 4: Toxins
Stoichiometry, Solution Chemistry, and
Acids and Bases

How much is too much? Students explore the concept of toxicity, learning about the qualitative and quantitative aspects of chemical reactions, including balancing equations, stoichiometry, acid-base chemistry, and precipitation reactions.

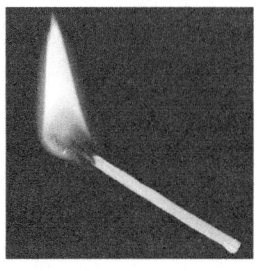

Unit 5: Fire
Energy, Thermodynamics, and
Oxidation-Reduction

What is the nature of fire? Students investigate the intimate connection between chemical and physical change and energy. This unit covers the laws of thermodynamics, combustion, bond energies, heats of reaction, oxidation-reduction, electrochemistry, and electromagnetic radiation.

Unit 6: Showtime
Reversible Reactions and Chemical Equilibrium

Demonstrate what you know! Students investigate reversible reactions and equilibrium concepts while researching and preparing a class demonstration for their final project.

Content	Alchemy					Smells				
	1	2	3	4	5	6	7	8	9	
Atomic and Molecular Structure			I	P	P	P	P	P	M	
Chemical Bonds					I	P	P	P	P	
Conservation of Mass and Stoichiometry	I	P	P							
Gases and Their Properties										
Acids and Bases		I				P				
Solutions		I								
Energy/Chemical Thermodynamics										
Reaction Rates						I				
Chemical Equilibrium										
Organic Chemistry and Biochemistry						I	P	P	P	
Nuclear Processes			I/P							
Electromagnetic Radiation										
Investigation and Experimentation	I	P	P	P	P	P	P	P	P	

The top of the table is headed **Unit and Chapter**.

Content Coverage Chart

Concepts are introduced and then reinforced, often across different units. Coverage usually consists of an introduction (I), then practice (P), and finally teaching to mastery (M).

A Typical Day

START WITH STUDENT UNDERSTANDING

Students come into any class with prior knowledge, assumptions, and misconceptions. When students make sense of new evidence and revise their thinking to accommodate it, they build true understanding. *Living by Chemistry* supports this process by allowing students to build their understanding based on experiences and including discussion questions specifically designed to challenge them to share their reasoning. Lessons are constructed to lead students to a more complete understanding. The *Living by Chemistry* curriculum uses a modified version of the 5Es model of teaching and learning. The 5Es describe different stages of a learning sequence: Engage, Explore, Explain, Elaborate, and Evaluate.

Unit and Chapter

Weather			Toxins					Fire					Showtime	
10	11	12	13	14	15	16	17	18	19	20	21	22	23	24
										M				
			P	P	P	P	P	P	P	P	M			
I	P	M												
						P							P	M
					P	P	P				P		P	M
	I							P	P	P	P		M	
										P			P	P
													I	P
												I		
P	P	P	P	P	P	P	P	P	P	P	P	P	P	M

LESSONS THAT PROMOTE UNDERSTANDING

Engage: At the start of each class, students are immediately engaged in a brief warm-up exercise, called a ChemCatalyst, that focuses on the main goal of the lesson. The purpose of the ChemCatalyst is to determine students' prior knowledge of the subject and encourage participation. In most cases, you can listen to student ideas and ask for explanations without judgment or correction.

Explore: For the next 15 to 20 minutes of class, students explore the key chemistry topics covered that day. Depending on the lesson type, they might solve problems, analyze data, perform a laboratory experiment, build models, or complete a card-sorting activity. They might also watch a brief demonstration or a computer simulation and try to provide explanations for their observations. This is a chance for students to think and build their own understanding, and, most importantly, to support their ideas with evidence. Generally, students work collaboratively in small groups or pairs. During this portion of the lesson, you can circulate from group to group, offering guidance, asking questions, and helping students refine their ideas.

Explain & Elaborate: A teacher-led discussion follows the Explore portion of the class. It allows students to connect their conceptual understanding from the lesson with new chemistry concepts, ideas, tools, and definitions that make up the learning objectives. Sample discussion questions are provided for you, along with summaries of the Key Points to be covered.

Evaluate: At the end of class, a final Check-In question provides both you and students with a quick assessment of their grasp of the day's main concepts.

Homework: Each lesson is accompanied by a reading assignment in the student textbook that reinforces what was learned in class, followed by exercises. The reading includes diagrams, photos, worked examples, and real-world connections. *Living by Chemistry* is designed so that each reading *follows* its corresponding lesson. This reinforces the concepts developed in class and, therefore, optimizes the effectiveness of both the classroom experience and the reading component.

Tips and Classroom-Management Strategies

SHARE THE APPROACH

For some students, the *Living by Chemistry* approach will be new and different from their prior science classroom experiences. Let your students know that they will be taking an active role in their own learning and that seeking the "right" answer will not always be the best strategy. Let them know that you will often be more interested in *why* they think something rather than whether or not what they think is scientifically correct. This will help to establish an open and safe atmosphere where students can share ideas and build their knowledge of chemistry.

BEGIN WITH THE BASICS

Living by Chemistry begins with the basics. We do not assume that students come into your classroom with any exposure to chemistry or its concepts. This scaffolded approach means that students new to chemistry are not at an instant disadvantage. At the same time, students who are acquainted with chemistry vocabulary and principles have an opportunity to clarify and expand their understanding. You will find that while lessons may start simply, they build to embrace sophisticated concepts and principles. Although *Living by Chemistry* may appear to start more slowly than other courses you have seen, all students benefit in the long run by developing a deeper and more comprehensive understanding.

ADJUST PACING TO MATCH THE CLASS

The flexibility of the *Living by Chemistry* curriculum allows you to adjust pacing to meet the needs of each class. For example, for an honors class, the pacing and expectations can be increased. Some lessons can be completed as homework, more projects and Internet research can be assigned, and concepts can be pursued to more sophisticated levels. Other classes may work at a slower pace, with more time spent discussing, practicing, and clarifying concepts. For these classes, you may wish to occasionally spend an extra day on a lesson. For double periods, two lessons can often be easily merged into one. (See Pacing Guides at the beginning of each unit.)

COOPERATIVE LEARNING

In the *Living by Chemistry* curriculum, students work cooperatively in groups, gaining expertise by articulating ideas and communicating concepts. This is one of the best ways to become proficient in a discipline. Students benefit from seeing a variety of approaches to a single problem, and they use their group as a sounding board for their ideas, rather than struggling alone or having to ask you. You might assign new groups at the start of each unit. Some teachers select groups based on students' strengths or other criteria, while some teachers prefer to assign groups randomly.

One strategy that leads to effective group work is a fair division of tasks among the group members. Each member can be assigned a role, such as spokesperson, recorder, equipment and laboratory safety person, or facilitator. In this way, the work is shared, everyone is involved, and everyone has an opportunity to practice communication, leadership, and responsibility.

Differentiated Instruction

A PROGRAM FOR ALL LEARNERS

Living by Chemistry is an inquiry-based, full-year high school chemistry course designed to improve science understanding for all students. The program provides groundbreaking materials for educators who seek a chemistry curriculum that is rigorous, yet accessible to a wide variety of learners.

Research has shown that students learn best when they are actively engaged, investigating questions in an interesting context. *Living by Chemistry* is organized around the big ideas of chemistry to promote subject matter coherence for students. The unit context holds students' interest and provides a real-world foundation for the chemistry concepts, making true learning and retention possible. Pre- and post-tests used in field-test classrooms show that this innovative approach benefits all students and that low-achieving and high-achieving students all make significant gains.

FORMATIVE ASSESSMENT

Formative assessment is built into every *Living by Chemistry* lesson. Each lesson begins with a **ChemCatalyst,** an open-ended question that gives you insight into students' preconceptions. Each lesson also ends with a **Check-In,** a question that helps you gauge if students understood the main point of the lesson. Along the way, **Sample Questions** are provided to help you guide the discussion and gather even more feedback as students develop understanding.

SUPPORTING STUDENTS WITH THE MATHEMATICS

Guided inquiry has been shown to be very effective in helping students understand chemistry. However, many students are still daunted by the mathematics associated with chemistry. To help these students succeed, *Living by Chemistry* scaffolds the mathematics in several important ways.

First, worked examples both in the Teacher's Edition and in the student text model problem-solving techniques, breaking down the solution steps. The solution first models how to approach the problem, often using estimation or by referring to a similar problem. Next, the solution is presented step by step, with an explanation at each step as needed. Finally, the answer is put into context for students and compared to the expected outcome. Second, students have access to Math Spotlight topics in the back of their book. They can refer to this appendix any time they want review or practice with a mathematics concept. In addition, graph interpretation is supported and reinforced throughout the book with the **Using the Graph** feature.

SUPPORTING ENGLISH-LANGUAGE LEARNERS

The activities in *Living by Chemistry* are carefully designed to illustrate and reinforce the ideas that are most difficult for students, and they take different forms, depending on the best vehicle for student understanding of the subject matter: lab experiments, model building, card sorting, card games, group problem solving, classwork, and review. Students often work in groups of two or four. The variety of activities ensures that different learning modalities are addressed and that learning is reinforced in several ways for each student. This approach is particularly helpful to English-language learners, who can rely on visuals, concrete experiences, and the explanations of other students to learn the content—while also acquiring academic language. For English-language learners whose first language is Spanish, an English-Spanish side-by-side glossary is provided in the student text, with page references so students can seek additional information, context, and visuals.

CHALLENGING ADVANCED LEARNERS

Living by Chemistry provides many ways to challenge your advanced students. For students who finish the in-class assignment early, worksheets often end with an **If You Finish Early** question to keep those students engaged and challenged. The student text also includes a variety of projects within each unit. Although these projects are designed to be engaging and educational for all students, you might save some of them as extensions for your advanced students.

Differentiate in the margins offers tips on how to further adjust the lessons to help struggling students or offer advanced students an additional challenge.

Pre-AP® Course Tips in the margins point out science practices and chemistry concepts that help students to prepare for a future AP® or college chemistry course.

CREATING AND MAINTAINING A SAFE LAB ENVIRONMENT

It is widely known that hands-on, inquiry-based science is essential to optimal learning. *Living by Chemistry* emphasizes careful scientific-process skills and science practices to engage students as active participants. Safety is an integral part of every successful experiment, and *Living by Chemistry* emphasizes the requirement for comprehensive and specific safety instruction before beginning any lab activity.

The Teacher's Edition provides general setup, cleanup, and safety instructions for appropriate lessons, but you should also consult your local guidelines and regulations, because these can vary widely.

General safety procedures are covered for students in Lesson 1 of Unit 1: Alchemy and a student Laboratory Safety reference from pp. xx and xxi of the student textbook is reproduced in the walkthrough of the student book that follows. Specific safety instructions should be reviewed before each experiment, as well as during cleanup and disposal of materials. If you modify an activity in some way, check that your changes meet the required safety criteria.

A list of references to research background information on chemical safety, storage, use, and disposal can be found on the Book Companion Site (BCS) bcs.whfreeman.com/livingbychemistry2e (log-in required).

THE STUDENT EDITION

The student edition of *Living by Chemistry,* second edition, is designed to be used for reading and reference after students have completed the engaging in class lessons. This allows students to reflect and review the chemistry taught and to practice problem solving each day. Take a tour through the student edition to see how to use the book to realize success in the course.

Next generation chemistry brought to life.

■ Chapter-opening pages introduce the main ideas and learning goals of the chapter.

■ **THINK ABOUT IT** prepares you to read about chemistry. Take a moment to reflect on the lesson question and topic outline before you read the lesson.

When heated, different metal atoms produce flames of different colors. That is what causes the many brilliant colors of a fireworks display.

CHAPTER 4

Moving Electrons

In this chapter, you will study

• the systematic arrangement of electrons in an atom

• how ionic compounds are formed

• valence electrons and ionic bonding

While the nucleus of an atom can be difficult to change, the electrons are a different story. Electrons can be moved around within an atom or transferred between atoms. When atoms transfer electrons, they become ions with positive and negat

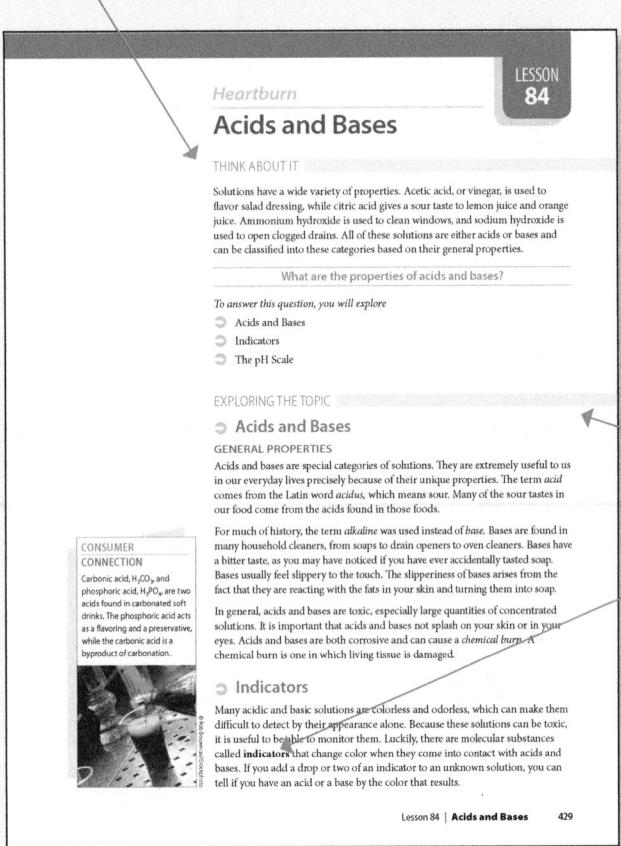

Heartburn

LESSON 84

Acids and Bases

THINK ABOUT IT

Solutions have a wide variety of properties. Acetic acid, or vinegar, is used to flavor salad dressing, while citric acid gives a sour taste to lemon juice and orange juice. Ammonium hydroxide is used to clean windows, and sodium hydroxide is used to open clogged drains. All of these solutions are either acids or bases and can be classified into these categories based on their general properties.

What are the properties of acids and bases?

To answer this question, you will explore
↺ Acids and Bases
↺ Indicators
↺ The pH Scale

EXPLORING THE TOPIC

↺ **Acids and Bases**

GENERAL PROPERTIES

Acids and bases are special categories of solutions. They are extremely useful to us in our everyday lives precisely because of their unique properties. The term *acid* comes from the Latin word *acidus*, which means sour. Many of the sour tastes in our food come from the acids found in those foods.

For much of history, the term *alkaline* was used instead of *base*. Bases are found in many household cleaners, from soaps to drain openers to oven cleaners. Bases have a bitter taste, as you may have noticed if you have ever accidentally tasted soap. Bases usually feel slippery to the touch. The slipperiness of bases arises from the fact that they are reacting with the fats in your skin and turning them into soap.

In general, acids and bases are toxic, especially large quantities of concentrated solutions. It is important that acids and bases not splash on your skin or in your eyes. Acids and bases are both corrosive and can cause a *chemical burn*. A chemical burn is one which living tissue is damaged.

CONSUMER CONNECTION

Carbonic acid, H_2CO_3 and phosphoric acid, H_3PO_4 are two acids found in carbonated soft drinks. The phosphoric acid acts as a flavoring and a preservative, while the carbonic acid is a byproduct of carbonation.

↺ **Indicators**

Many acidic and basic solutions are colorless and odorless, which can make them difficult to detect by their appearance alone. Because these solutions can be toxic, it is useful to be able to monitor them. Luckily, there are molecular substances called **indicators** that change color when they come into contact with acids and bases. If you add a drop or two of an indicator to an unknown solution, you can tell if you have an acid or a base by the color that results.

Lesson 84 | **Acids and Bases**   429

■ **EXPLORING THE TOPIC** allows you to review and practice the chemistry at the right pace. Each lesson follows in step with the daily hands-on, inquiry-based lesson from your class.

■ Look at the **key terms** in bold. Find definitions in English and Spanish at the end of the book.

■ Develop the science practice of using visual representations and models to explain chemistry by studying the **figures, diagrams,** and **illustrations** used throughout the book.

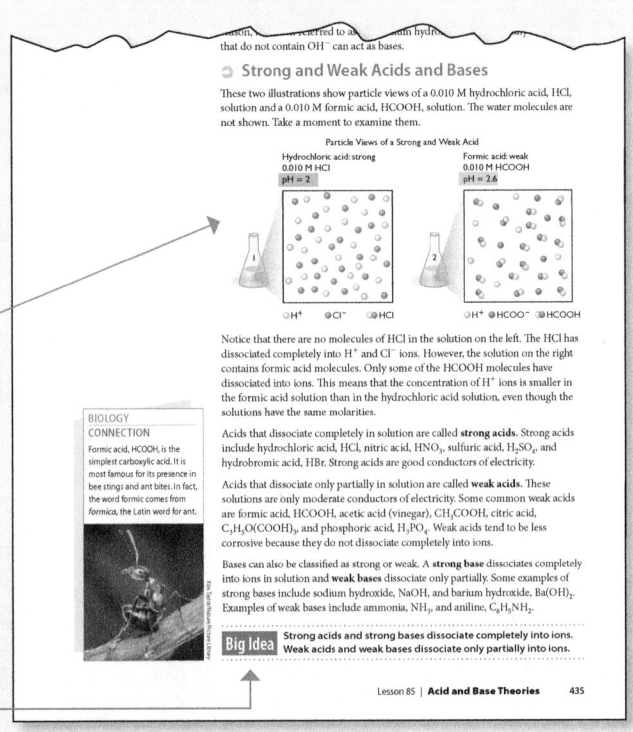

...ason,...referred to a...m hydr...
that do not contain OH^- can act as bases.

↺ **Strong and Weak Acids and Bases**

These two illustrations show particle views of a 0.010 M hydrochloric acid, HCl, solution and a 0.010 M formic acid, HCOOH, solution. The water molecules are not shown. Take a moment to examine them.

Particle Views of a Strong and Weak Acid

Hydrochloric acid: strong
0.010 M HCl
pH = 2

Formic acid: weak
0.010 M HCOOH
pH = 2.6

○ H^+ ● Cl^- ◉ HCl   ○ H^+ ● $HCOO^-$ ◉ HCOOH

Notice that there are no molecules of HCl in the solution on the left. The HCl has dissociated completely into H^+ and Cl^- ions. However, the solution on the right contains formic acid molecules. Only some of the HCOOH molecules have dissociated into ions. This means that the concentration of H^+ ions is smaller in the formic acid solution than in the hydrochloric acid solution, even though the solutions have the same molarities.

BIOLOGY CONNECTION

Formic acid, HCOOH, is the simplest carboxylic acid. It is most famous for its presence in bee stings and ant bites. In fact, the word formic comes from *formica*, the Latin word for ant.

Acids that dissociate completely in solution are called **strong acids**. Strong acids include hydrochloric acid, HCl, nitric acid, HNO_3, sulfuric acid, H_2SO_4, and hydrobromic acid, HBr. Strong acids are good conductors of electricity.

Acids that dissociate only partially in solution are called **weak acids**. These solutions are only moderate conductors of electricity. Some common weak acids are formic acid, HCOOH, acetic acid (vinegar), CH_3COOH, citric acid, $C_3H_5O(COOH)_3$, and phosphoric acid, H_3PO_4. Weak acids tend to be less corrosive because they do not dissociate completely into ions.

Bases can also be classified as strong or weak. A **strong base** dissociates completely into ions in solution and **weak bases** dissociate only partially. Some examples of strong bases include sodium hydroxide, NaOH, and barium hydroxide, $Ba(OH)_2$. Examples of weak bases include ammonia, NH_3, and aniline, $C_6H_5NH_2$.

Big Idea Strong acids and strong bases dissociate completely into ions. Weak acids and weak bases dissociate only partially into ions.

Lesson 85 | **Acid and Base Theories**   435

■ Pay attention to **Big Ideas,** especially when you are reviewing for a quiz or exam. These are fundamental concepts of chemistry.

Important to Know highlights ideas that you should retain for further use throughout the course.

Practice makes perfect.

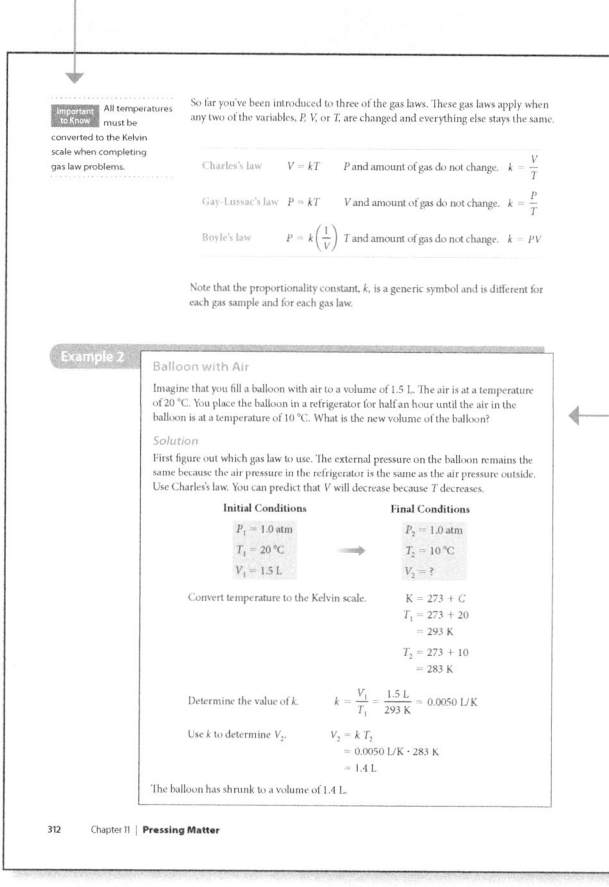

Important to Know All temperatures must be converted to the Kelvin scale when completing gas law problems.

So far you've been introduced to three of the gas laws. These gas laws apply when any two of the variables, P, V, or T, are changed and everything else stays the same.

Charles's law $\quad V = kT \quad$ P and amount of gas do not change. $\quad k = \dfrac{V}{T}$

Gay-Lussac's law $\quad P = kT \quad$ V and amount of gas do not change. $\quad k = \dfrac{P}{T}$

Boyle's law $\quad P = k\left(\dfrac{1}{V}\right) \quad$ T and amount of gas do not change. $\quad k = PV$

Note that the proportionality constant, k, is a generic symbol and is different for each gas sample and for each gas law.

Example 2

Balloon with Air

Imagine that you fill a balloon with air to a volume of 1.5 L. The air is at a temperature of 20 °C. You place the balloon in a refrigerator for half an hour until the air in the balloon is at a temperature of 10 °C. What is the new volume of the balloon?

Solution

First figure out which gas law to use. The external pressure on the balloon remains the same because the air pressure in the refrigerator is the same as the air pressure outside. Use Charles's law. You can predict that V will decrease because T decreases.

Initial Conditions	Final Conditions
$P_1 = 1.0$ atm	$P_2 = 1.0$ atm
$T_1 = 20$ °C	$T_2 = 10$ °C
$V_1 = 1.5$ L	$V_2 = ?$

Convert temperature to the Kelvin scale.
$$K = 273 + C$$
$$T_1 = 273 + 20 = 293 \text{ K}$$
$$T_2 = 273 + 10 = 283 \text{ K}$$

Determine the value of k. $\quad k = \dfrac{V_1}{T_1} = \dfrac{1.5 \text{ L}}{293 \text{ K}} \approx 0.0050 \text{ L/K}$

Use k to determine V_2. $\quad V_2 = k\,T_2 = 0.0050 \text{ L/K} \cdot 283 \text{ K} = 1.4 \text{ L}$

The balloon has shrunk to a volume of 1.4 L.

Lessons offer critical math support. The **examples** show you how to solve problems step by step. Try to answer the question yourself before reading the solution to check your understanding.

The **LESSON SUMMARY** recaps the main ideas of the lesson that answer the opening question. Reading the lesson summaries is a good way to review for a quiz or an exam.

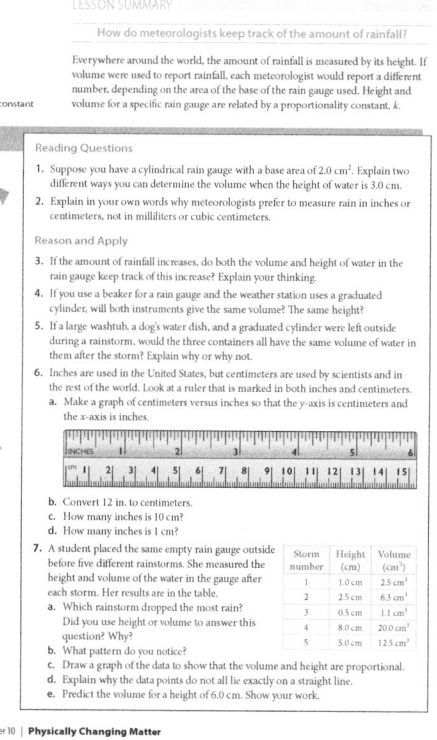

LESSON SUMMARY

How do meteorologists keep track of the amount of rainfall?

KEY TERMS
proportional
proportionality constant

Everywhere around the world, the amount of rainfall is measured by its height. If volume were used to report rainfall, each meteorologist would report a different number, depending on the area of the base of the rain gauge used. Height and volume for a specific rain gauge are related by a proportionality constant, k.

Exercises

Reading Questions

1. Suppose you have a cylindrical rain gauge with a base area of 2.0 cm². Explain two different ways you can determine the volume when the height of water is 3.0 cm.

2. Explain in your own words why meteorologists prefer to measure rain in inches or centimeters, not in milliliters or cubic centimeters.

Reason and Apply

3. If the amount of rainfall increases, do both the volume and height of water in the rain gauge keep track of this increase? Explain your thinking.

4. If you use a beaker for a rain gauge and the weather station uses a graduated cylinder, will both instruments give the same volume? The same height?

5. If a large washtub, a dog's water dish, and a graduated cylinder were left outside during a rainstorm, would the three containers all have the same volume of water in them after the storm? Explain why or why not.

6. Inches are used in the United States, but centimeters are used by scientists and in the rest of the world. Look at a ruler that is marked in both inches and centimeters.
   a. Make a graph of centimeters versus inches so that the y-axis is centimeters and the x-axis is inches.

   b. Convert 12 in. to centimeters.
   c. How many inches is 10 cm?
   d. How many inches is 1 cm?

7. A student placed the same empty rain gauge outside before five different rainstorms. She measured the height and volume of the water in the gauge after each storm. Her results are in the table.

Storm number	Height (cm)	Volume (cm³)
1	1.0 cm	2.5 cm³
2	2.5 cm	6.3 cm³
3	0.5 cm	1.1 cm³
4	8.0 cm	20.0 cm³
5	5.0 cm	12.5 cm³

   a. Which rainstorm dropped the most rain? Did you use height or volume to answer this question? Why?
   b. What pattern do you notice?
   c. Draw a graph of the data to show that the volume and height are proportional.
   d. Explain why the data points do not all lie exactly on a straight line.
   e. Predict the volume for a height of 6.0 cm. Show your work.

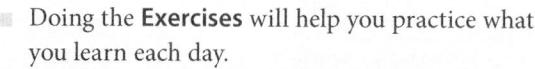

Doing the **Exercises** will help you practice what you learn each day.

Answer the Reading Questions to make sure you understood the main ideas.

Answer the Reason and Apply questions to practice problem solving and using evidence-based reasoning to justify answers.

Features of the Student Edition   xxxiii

Review and prepare for quizzes and exams.

- The **Chapter Summary** and **Review Exercises** help you review after each group of lessons.

- **Projects** may be assigned by your teacher to give you a chance to do some research on your own.

- Two kinds of **Review Exercises** help you to review the unit, starting with **General Review** exercises covering the major concepts in the chapters.

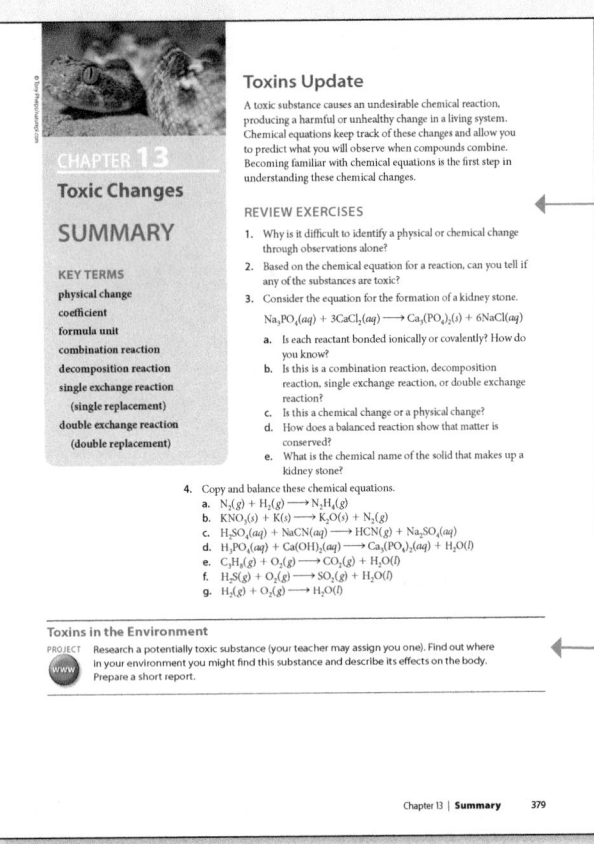

CHAPTER 13
Toxic Changes
SUMMARY

KEY TERMS
physical change
coefficient
formula unit
combination reaction
decomposition reaction
single exchange reaction
(single replacement)
double exchange reaction
(double replacement)

Toxins Update

A toxic substance causes an undesirable chemical reaction, producing a harmful or unhealthy change in a living system. Chemical equations keep track of these changes and allow you to predict what you will observe when compounds combine. Becoming familiar with chemical equations is the first step in understanding these chemical changes.

REVIEW EXERCISES

1. Why is it difficult to identify a physical or chemical change through observations alone?

2. Based on the chemical equation for a reaction, can you tell if any of the substances are toxic?

3. Consider the equation for the formation of a kidney stone.

$$Na_3PO_4(aq) + 3CaCl_2(aq) \longrightarrow Ca_3(PO_4)_2(s) + 6NaCl(aq)$$

   a. Is each reactant bonded ionically or covalently? How do you know?
   b. Is this a combination reaction, decomposition reaction, single exchange reaction, or double exchange reaction?
   c. Is this a chemical change or a physical change?
   d. How does a balanced reaction show that matter is conserved?
   e. What is the chemical name of the solid that makes up a kidney stone?

4. Copy and balance these chemical equations.
   a. $N_2(g) + H_2(g) \longrightarrow N_2H_4(g)$
   b. $KNO_3(s) + K(s) \longrightarrow K_2O(s) + N_2(g)$
   c. $H_2SO_4(aq) + NaCN(aq) \longrightarrow HCN(g) + Na_2SO_4(aq)$
   d. $H_3PO_4(aq) + Ca(OH)_2(aq) \longrightarrow Ca_3(PO_4)_2(aq) + H_2O(l)$
   e. $C_3H_8(g) + O_2(g) \longrightarrow CO_2(g) + H_2O(l)$
   f. $H_2S(g) + O_2(g) \longrightarrow SO_2(g) + H_2O(l)$
   g. $H_2(g) + O_2(g) \longrightarrow H_2O(l)$

Toxins in the Environment

PROJECT Research a potentially toxic substance (your teacher may assign you one). Find out where in your environment you might find this substance and describe its effects on the body. Prepare a short report.

Chapter 13 | Summary   379

- The **Unit Review** summarizes what you learned in that unit.

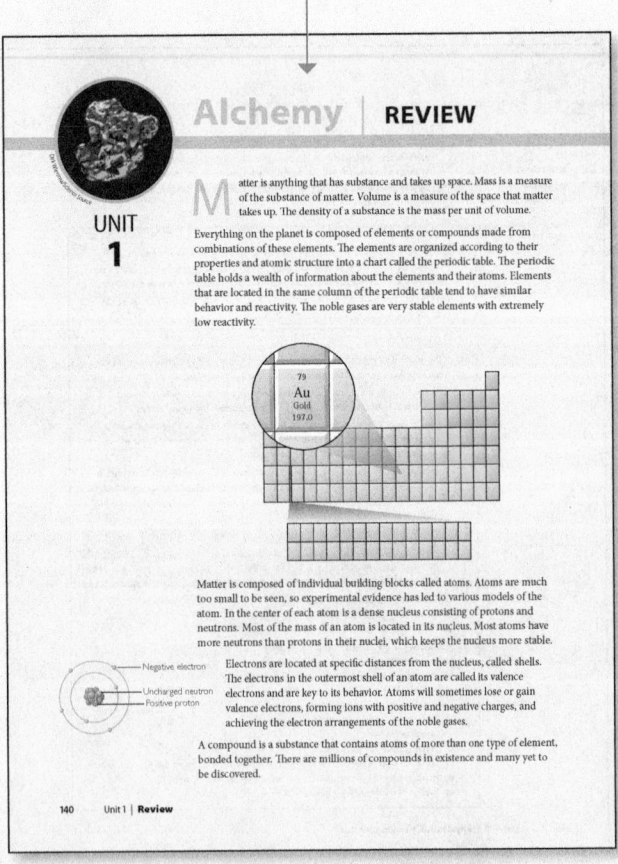

UNIT 1

Alchemy | REVIEW

Matter is anything that has substance and takes up space. Mass is a measure of the substance of matter. Volume is a measure of the space that matter takes up. The density of a substance is the mass per unit of volume.

Everything on the planet is composed of elements or compounds made from combinations of these elements. The elements are organized according to their properties and atomic structure into a chart called the periodic table. The periodic table holds a wealth of information about the elements and their atoms. Elements that are located in the same column of the periodic table tend to have similar behavior and reactivity. The noble gases are very stable elements with extremely low reactivity.

79
Au
Gold
197.0

Matter is composed of individual building blocks called atoms. Atoms are much too small to be seen, so experimental evidence has led to various models of the atom. In the center of each atom is a dense nucleus consisting of protons and neutrons. Most of the mass of an atom is located in its nucleus. Most atoms have more neutrons than protons in their nuclei, which keeps the nucleus more stable.

- Negative electron
- Uncharged neutron
- Positive proton

Electrons are located at specific distances from the nucleus, called shells. The electrons in the outermost shell of an atom are called its valence electrons and are key to its behavior. Atoms will sometimes lose or gain valence electrons, forming ions with positive and negative charges, and achieving the electron arrangements of the noble gases.

A compound is a substance that contains atoms of more than one type of element, bonded together. There are millions of compounds in existence and many yet to be discovered.

140   Unit 1 | Review

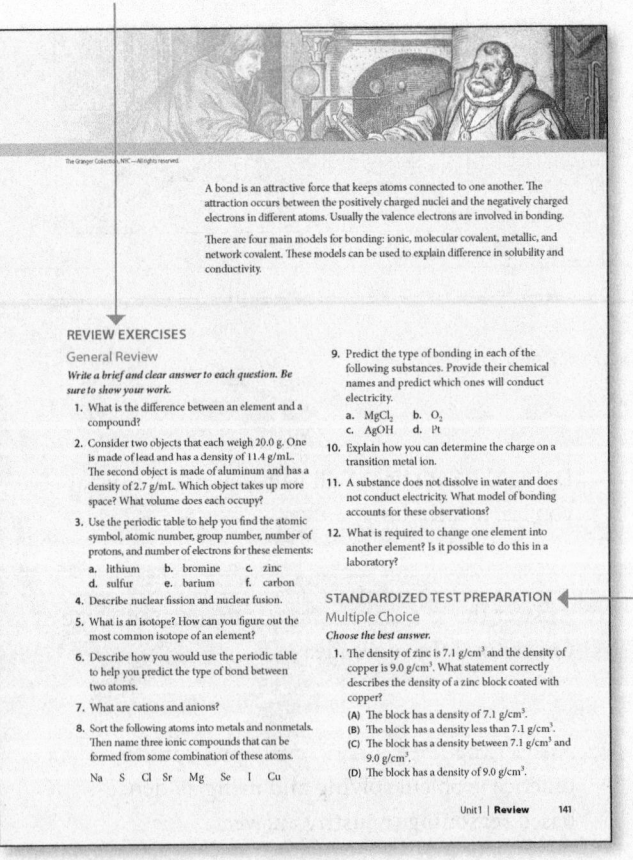

A bond is an attractive force that keeps atoms connected to one another. The attraction occurs between the positively charged nuclei and the negatively charged electrons in different atoms. Usually the valence electrons are involved in bonding.

There are four main models for bonding: ionic, molecular covalent, metallic, and network covalent. These models can be used to explain difference in solubility and conductivity.

REVIEW EXERCISES
General Review

Write a brief and clear answer to each question. Be sure to show your work.

1. What is the difference between an element and a compound?

2. Consider two objects that each weigh 20.0 g. One is made of lead and has a density of 11.4 g/mL. The second object is made of aluminum and has a density of 2.7 g/mL. Which object takes up more space? What volume does each occupy?

3. Use the periodic table to help you find the atomic symbol, atomic number, group number, number of protons, and number of electrons for these elements:
   a. lithium   b. bromine   c. zinc
   d. sulfur   e. barium   f. carbon

4. Describe nuclear fission and nuclear fusion.

5. What is an isotope? How can you figure out the most common isotope of an element?

6. Describe how you would use the periodic table to help you predict the type of bond between two atoms.

7. What are cations and anions?

8. Sort the following atoms into metals and nonmetals. Then name three ionic compounds that can be formed from some combination of these atoms:

   Na   S   Cl   Sr   Mg   Se   I   Cu

9. Predict the type of bonding in each of the following substances. Provide their chemical names and predict which ones will conduct electricity.
   a. MgCl₂   b. O₂
   c. AgOH   d. Pt

10. Explain how you can determine the charge on a transition metal ion.

11. A substance does not dissolve in water and does not conduct electricity. What model of bonding accounts for these observations?

12. What is required to change one element into another element? Is it possible to do this in a laboratory?

STANDARDIZED TEST PREPARATION
Multiple Choice

Choose the best answer.

1. The density of zinc is 7.1 g/cm³ and the density of copper is 9.0 g/cm³. What statement correctly describes the density of a zinc block coated with copper?
   (A) The block has a density of 7.1 g/cm³.
   (B) The block has a density less than 7.1 g/cm³.
   (C) The block has a density between 7.1 g/cm³ and 9.0 g/cm³.
   (D) The block has a density of 9.0 g/cm³.

Unit 1 | Review   141

- The **Standardized Test Preparation** section contains multiple-choice questions. Each Exercise has a stem and four possible answers.

xxxiv   **Features of the Student Edition**

Extra support when you need it.

- When you see a note about **Math Spotlights** in a lesson you can turn to the back of the book for a quick review of that math topic.

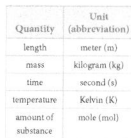

Appendix A: Math Spotlights

SI Units of Measure

Quantity	Unit (abbreviation)
length	meter (m)
mass	kilogram (kg)
time	second (s)
temperature	Kelvin (K)
amount of substance	mole (mol)

Scientists rely on repeatable measurements as they study the physical world. It is important that they use consistent units of measure worldwide. In 1960 an international council standardized the metric system, creating the *Système International d'Unités* (International System of Units), abbreviated as SI.

In the table at left, there are the basic SI units used in chemistry. Other units such as density and volume are combinations of these.

SI units are based on powers of 10. Larger and smaller units get their names by combining standard prefixes with these basic units. For example, the word *centimeter* is a combination of *centi-* and *meter*. A centimeter is one one-hundredth of a meter. These are the prefixes used in the SI, along with their abbreviations and their meanings. (A kilogram is the only basic SI unit that has a prefix as part of its name.)

Prefix	Multiple	Scientific Notation	Prefix	Multiple	Scientific Notation
tera- (T-)	1,000,000,000,000	10^{12}	pico- (p-)	0.000 000 000 001	10^{-12}
giga- (G-)	1,000,000,000	10^{9}	nano- (n-)	0.000 000 001	10^{-9}
mega- (M-)	1,000,000	10^{6}	micro- (μ-)	0.000 001	10^{-6}
kilo- (k-)	1,000	10^{3}	milli- (m-)	0.001	10^{-3}
hecto- (h-)	100	10^{2}	centi- (c-)	0.01	10^{-2}
deka- (da-)	10	10^{1}	deci- (d-)	0.1	10^{-1}

Example 1

Length Conversions

How many meters does each of these lengths represent?
a. 562 centimeters   b. 2.5 kilometers

Solution

a. $562 \text{ cm} \cdot \dfrac{1 \text{ m}}{100 \text{ cm}} = 5.62 \text{ m}$ b. $2.5 \text{ km} \cdot \dfrac{1000 \text{ m}}{1 \text{ km}} = 2500 \text{ m}$

Example 2

The Mass of One Liter

One milliliter, or cubic centimeter, of water at 4 °C weighs 1 g. How much does 1 L of water at 4 °C weigh?

A-0

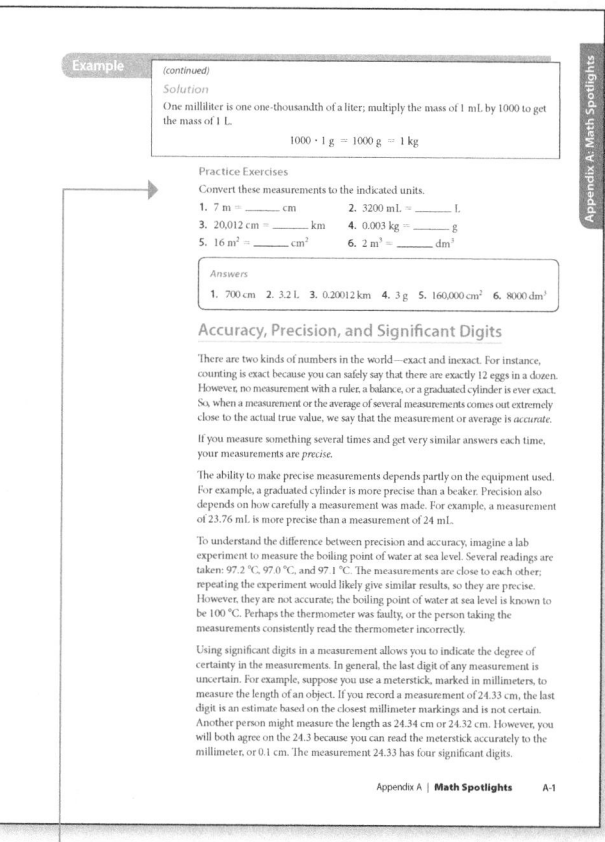

Example (continued)

Solution

One milliliter is one one-thousandth of a liter; multiply the mass of 1 mL by 1000 to get the mass of 1 L.

$$1000 \cdot 1 \text{ g} = 1000 \text{ g} = 1 \text{ kg}$$

Practice Exercises

Convert these measurements to the indicated units.

1. 7 m = _____ cm   2. 3200 mL ≈ _____ L.
3. 20,012 m = _____ km   4. 0.003 kg = _____ g
5. 16 m² = _____ cm²   6. 2 m³ = _____ dm³

Answers

1. 700 cm 2. 3.2 L 3. 0.20012 km 4. 3 g 5. 160,000 cm² 6. 8000 dm³

Accuracy, Precision, and Significant Digits

There are two kinds of numbers in the world—exact and inexact. For instance, counting is exact because you can safely say that there are exactly 12 eggs in a dozen. However, no measurement with a ruler, a balance, or a graduated cylinder is ever exact. So, when a measurement or the average of several measurements comes out extremely close to the actual true value, we say that the measurement or average is *accurate*.

If you measure something several times and get very similar answers each time, your measurements are *precise*.

The ability to make precise measurements depends partly on the equipment used. For example, a graduated cylinder is more precise than a beaker. Precision also depends on how carefully a measurement was made. For example, a measurement of 23.76 mL is more precise than a measurement of 24 mL.

To understand the difference between precision and accuracy, imagine a lab experiment to measure the boiling point of water at sea level. Several readings are taken: 97.2 °C, 97.0 °C, and 97.1 °C. The measurements are close to each other; repeating the experiment would likely give similar results, so they are precise. However, they are not accurate; the boiling point of water at sea level is known to be 100 °C. Perhaps the thermometer was faulty, or the person taking the measurements consistently read the thermometer incorrectly.

Using significant digits in a measurement allows you to indicate the degree of certainty in the measurements. In general, the last digit of any measurement is uncertain. For example, suppose you use a meterstick, marked in millimeters, to measure the length of an object. If you record a measurement of 24.33 cm, the last digit is an estimate based on the closest millimeter markings and is not certain. Another person might measure the length as 24.34 cm or 24.32 cm. However, you will both agree on the 24.3 because you can read the meterstick accurately to the millimeter, or 0.1 cm. The measurement 24.33 has four significant digits.

Appendix A | **Math Spotlights**   A-1

- Do the **Practice Exercises** for a quick refresher. Answers are provided directly below so you can check your understanding.

Glossary

English/Inglés

A

absolute zero The temperature defined as 0 K on the Kelvin scale and − 273.15 °C on the Celsius scale. Considered to be the lowest possible temperature that matter can reach. (p. 279)

absorption Of light, not being transmitted because certain colors are removed by an object. Absorption transfers energy back to the object. (p. 579)

acid A substance that adds hydrogen ions, H^+, to an aqueous solution; a substance that donates a proton to another substance in solution. (p. 430)

actinides A series of elements that follow actinium in Period 7 of the periodic table and that are typically placed separately at the bottom of the periodic table. (p. 44)

activation energy The minimum amount of energy required to initiate a chemical process or reaction. (p. 537)

activity series A table showing elements in order of their chemical activity, with the most easily oxidized at the top of the list. (p. 564)

actual yield The amount of a product obtained when a reaction is run (as opposed to the theoretical yield). (p. 473)

air mass A large volume of air that has consistent temperature and water content. (p. 288)

alkali metals The elements in Group 1A on the periodic table, except for hydrogen. (p. 44)

alkaline earth metals The elements in Group 2A on the periodic table. (p. 44)

Spanish/Español

cero absoluto La temperatura definida como 0 K en la escala Kelvin y − 273.15 °C en la escala Celsius. Se considera como la temperatura más baja que puede alcanzar la materia.

absorción De la luz: cuando esta no se transmite porque ciertos colores han sido eliminados (por un objeto). La absorción transfiere energía al objeto. (p. 579)

ácido Sustancia que cede iones de hidrógeno, H^+, a una solución acuosa; una sustancia que dona un protón a otra sustancia en la solución. (p. 430)

actínidos Serie de elementos que están después del actinio en el séptimo periodo de la tabla periódica y que usualmente, aparecen en la parte de abajo de la tabla. (p. 44)

energía de activación La mínima cantidad de energía que se necesita para iniciar una reacción o un proceso químico. (p. 537)

serie de actividad Una tabla que muestra elementos ordenados de acuerdo con la actividad química de cada uno de dichos elementos, empezando por aquellos que se oxidan con mayor facilidad. (p. 564)

rendimiento real La cantidad de un producto que se obtiene cuando se ejecuta una reacción (al contrario del rendimiento teórico). (p. 473)

masa de aire Gran volumen de aire cuya temperatura y contenido de agua son constantes. (p. 288)

metales alcalinos Los elementos del grupo 1A de la tabla periódica, con excepción del hidrógeno. (p. 44)

metales alcalinotérreos Los elementos del grupo 2A de la tabla periódica. (p. 44)

G-1

- The **Glossary** contains all the key terms, defined in English and Spanish.

Laboratory Safety

Laboratory experiments are an important part of chemistry. Follow these safety precautions to avoid danger.

Before Working in the Lab

- Read and become familiar with the entire procedure before starting.
- Listen to instructions. When you are in doubt, ask your teacher.
- Know the location of emergency exits and escape routes.
- Learn the location and operation of all safety equipment in your laboratory, including the safety shower, eye wash, first aid kit, fire extinguishers, and fire blanket.

Emergencies and Accidents

- Immediately report any accident, however small, to your teacher.
- If you get chemicals on you, rinse the affected area with water.
- In case of chemicals on your face, wash off with plenty of water before removing your goggles. In case of chemicals in your eyes, remove contact lenses and wash eyes with water for at least 15 minutes.
- Minor skin burns should be held under cold, running water.

General Conduct

- Clear your bench top of all unnecessary materials, such as books and jackets, before starting work.
- Do not bring gum, food, or drinks into the laboratory.

Appropriate Apparel

- Always wear protective safety goggles when working in the laboratory.
- Avoid bulky, loose-fitting clothing; roll up long sleeves; and tie back loose hair. Lab coats or aprons may be required.
- Wear long pants and shoes that cover the whole foot (not sandals) so your feet are protected from accidental spills or broken glassware.

Using Glassware and Equipment

- Do not use chipped or cracked glassware or damaged equipment.
- Be careful when handling hot glassware or apparatus. Remember, hot glassware looks like cold glassware.

- Place hot glassware or apparatus (such as a crucible) on an appropriate cooling surface, such as a wire gauze.

- Never point the open end of a test tube toward yourself or anyone else.

- Never fill a pipette using mouth suction. Always use a bulb.

- Keep electrical equipment away from sinks and faucets to minimize the risk of electrical shock.

Using Chemicals

- Never taste substances in the laboratory and avoid touching them if possible.

- Check chemical labels twice to make sure you have the correct substance. Some chemical formulas and names differ by only a letter or a number.

- Read and follow all hazard classifications shown on the label.

- Never pour anything down the drain unless instructed to do so by your teacher.

- When transferring chemicals from a common container to your own test tube or beaker, take only what you need. Do not return any extra material to the original container, because this may contaminate the original.

- Mix all substances together slowly. Add concentrated solutions to dilute solutions. When working with acids and bases, always add concentrated acids and bases to water; never add water to a concentrated acid or base as this can cause dangerous spattering.

- If you are instructed to smell something, do so by wafting (fanning) some of the vapor toward your nose. Do not place your nose near the opening of the container.

Before Leaving the Lab

- Clean your lab station and return equipment to its proper place.

- Make sure that gas lines and water faucets are shut off.

- When discarding used chemicals, carefully follow the instructions provided.

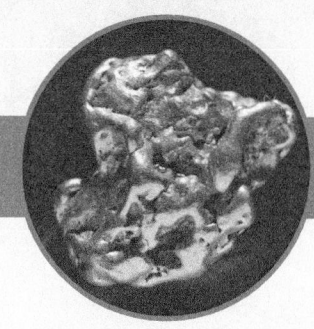

Unit 1 | Alchemy

Matter, Atomic Structure, and Bonding

ALCHEMY AS CONTEXT

Hundreds of years ago, alchemists were trying to turn ordinary substances into gold. The overarching question for this unit—Is it possible to create gold from other substances?—is accompanied by two other important questions: If so, how is it done? and If not, why not? In the course of their investigation, students explore matter and the models scientists have created to explain what matter is and why it behaves the way it does.

ALCHEMY AS THE ROOT OF CHEMISTRY

The study of matter arose partly from practical pursuits. To make food, clothing, and tools, people had to manipulate the substances in their environment. They learned to turn milk into cheese, forge metals, dye wool, and so on. Chemistry as a scientific discipline also has roots in alchemy—the word *chemistry* comes directly from the word *alchemy*. The practice of alchemy can be traced to many cultures, including those of Arabia, China, and ancient Greece. Some alchemists wanted to make gold, while others sought potions that would bring eternal life. A few alchemists renounced these pursuits and sought only to understand the nature of matter.

In light of what we now know about the elements, the pursuit of alchemy seems a bit of a wild goose chase. It is obvious to us now that it is not possible to turn metals into gold by mixing, refining, or fusing substances together. Even so, the discipline of chemistry owes much to the struggles and efforts of the alchemists who developed many practical lab procedures, such as distillation, sublimation, oxidation, and extraction. By manipulating assorted compounds and experimenting with them, much was learned about their properties and passed on to future scientists through extensive bodies of writing.

CONTENT DRIVEN BY CONTEXT

The chemistry content is the top priority in each unit, with enough context provided to engage students and provide a real-world setting. For example, we introduce the periodic table of the elements by challenging students to re-create an early version of the table based on patterns that Dmitri Mendeleyev observed. Next, students consider the merits of historic models of the atom before arriving at a basic atomic model. Throughout the unit we return to the central question: Is it possible to create gold from other substances? The ultimate answer to that question is both yes and no. The gold found here on Earth was created somewhere in the stars through nuclear processes. However, that option is not available to us in a laboratory. So, while it is *theoretically* possible through nuclear chemistry, making gold from other elements is not feasible on Earth.

THE USE OF MODELS TO EXPLAIN MATTER

Another central theme of this unit is the use of models to explain matter. The atomic model is one of the greatest tools chemists have developed to explain the behavior of matter. It is important that students become comfortable working with models and making reasonable deductions and predictions based on evidence. Learning to decipher models is a skill that does not necessarily come overnight. In chemistry, we bombard students with dozens of representations and models for matter, from chemical symbols and formulas to molecules represented as balls bouncing around in a box. Each model has its usefulness, and each has its limitations. There are inevitably inaccuracies in the scale, shape, and size of items shown in models, and these shortcomings may lead to confusion. Remember, it is quite often the model that is limited, not the student.

BUILDING UNDERSTANDING

By starting with the context of alchemy, you can ask simple, fundamental questions—much as early chemists did. You can start at the ground level, without assuming that all students come into the classroom with prior knowledge of chemistry. While students are often familiar with some chemical terminology (such as "atomic" and "nuclear"), most are unclear on the concepts underlying these terms. Using alchemy as a contextual starting point allows you to bring all students along together, wherever they are beginning. It also lets you set the tone for the entire year, modeling an open atmosphere of inquiry and questioning.

Chapter Summary

Chapter	Description	Standard Schedule Days	Block Schedule Days
1	**Chapter 1** is an introduction to chemistry, beginning with orientation to lab equipment and safety. Next, students observe the apparent creation of a gold penny. The question arises, "How can you tell if it is gold or not?" To find the answer, students debate the nature of matter, make observations and measurements, and differentiate between types of matter based on properties.	6	3
2	**Chapter 2** covers elements and the language of chemistry. Lesson 7: The Copper Cycle provides more food for thought as students consider the question, "What happened to the copper during the copper cycle?"	6	3
3	**Chapter 3** provides a chance for students to explore the atomic model, atomic structure, and nuclear chemistry to learn how gold differs from other elements. A central question in this section is "What makes a gold atom a gold atom?" Once students know that the nucleus of an atom holds its elemental identity, and that the nucleus is difficult to change, the focus moves to the parts of an atom that are easier to change—the electrons.	6	3
4	**Chapter 4** presents the question, "What happens if the electrons are moved around?" Even though it is not possible to change another element into gold through electron movement, students discover that atoms can be used purposefully to create things as valuable as or more valuable than gold. In an optional lab, students make paint pigments as they learn more about transition metals.	8	4
5	**Chapter 5** introduces students to models for ionic bonding, molecular covalent bonding, network covalent bonding, and metallic bonding. The chapter-ending electroplating lab gives students more tangible evidence of electrons and ions. Students explore models of bonding as they ask, "What can I make that is as good as gold?"	6	3

Pacing Guides
Standard Schedule

Day	Suggested Plan	Day	Suggested Plan
1	Introduction, Chapter 1 Lesson 1	17	Chapter 3 Lesson 16, Chapter 3 Review
2	Safety Quiz, Chapter 1 Lesson 2	18	Chapter 3 Quiz, Chapter 4 Lesson 17
3	Chapter 1 Lesson 3	19	Chapter 4 Lesson 18
4	Chapter 1 Lesson 4	20	Chapter 4 Lesson 19
5	Chapter 1 Lesson 5, Chapter 1 Review	21	Chapter 4 Lesson 20
6	Chapter 1 Quiz, Chapter 2 Lesson 6	22	Chapter 4 Lesson 21
7	Chapter 2 Lesson 7	23	Chapter 4 Lesson 22
8	Chapter 2 Lessons 7 and 8	24	Chapter 4 Lesson 23
9	Chapter 2 Lesson 9	25	Chapter 4 Lesson 24, Chapter 4 Review
10	Chapter 2 Lesson 10	26	Chapter 4 Quiz, Chapter 5 Lesson 25
11	Chapter 2 Review	27	Chapter 5 Lesson 26
12	Chapter 2 Quiz, Chapter 3 Lesson 11	28	Chapter 5 Lesson 27, Chapter 5 Review
13	Chapter 3 Lesson 12	29	Chapter 5 Quiz
14	Chapter 3 Lesson 13	30	Unit 1 Review
15	Chapter 3 Lesson 14	31	Unit 1 Exam
16	Chapter 3 Lesson 15	32	Lab Exam (optional)

Block Schedule

Day	Suggested Plan	Day	Suggested Plan
1	Introduction Chapter 1 Lesson 1	10	Chapter 3 Quiz Chapter 4 Lessons 17 and 18
2	Safety Quiz Chapter 1 Lessons 2 and 3	11	Chapter 4 Lessons 19 and 20
3	Chapter 1 Lessons 4 and 5 Chapter 1 Review	12	Chapter 4 Lessons 21 and 22
4	Chapter 1 Quiz Chapter 2 Lessons 6 and 7	13	Chapter 4 Lessons 23 and 24 Chapter 4 Review
5	Chapter 2 Lessons 7 and 8	14	Chapter 4 Quiz Chapter 5 Lessons 25 and 26
6	Chapter 2 Lessons 9 and 10 Chapter 2 Review	15	Chapter 5 Lesson 27 Chapter 5 Review
7	Chapter 2 Quiz Chapter 3 Lessons 11 and 12	16	Unit Review
8	Chapter 3 Lessons 13 and 14	17	Unit 1 Exam Lab Exam
9	Chapter 3 Lessons 15 and 16 Chapter 3 Review		

Unit 1 | Alchemy

Die Wennar/Science Source

Emperor Rudolf II (1552–1612) governed the Holy Roman Empire from 1576 to 1608. However, he often neglected his duties as ruler and is better known as an alchemist and patron of the arts.

The Granger Collection, NYC—All rights reserved.

In this unit, you will learn

- what matter is composed of
- to use the language of chemistry
- to decode information contained in the periodic table
- how new substances with new properties are made
- what holds substances together

Why Alchemy?

Chemistry has some of its roots in the ancient practice of alchemy. The alchemists experimented with trying to make gold out of ordinary substances. In the process, they learned a great deal about matter and about chemistry. When you understand the nature of matter and its composition, you will be able to answer the question, "Is it possible to turn ordinary substances into gold?"

1

Watch the video overview of Chapter 1 (for teachers) by clicking on the link in the TE-book, opening the TRFD, or logging onto the book's companion Web site bcs.whfreeman.com/livingbychemistry2e (teacher log-in required).

Chapter 1 introduces the alchemy context and opens the door to the study of chemistry. The focus of this chapter is matter—what it is, how it is measured, and how chemists begin to differentiate one type of matter from another. Lesson 1 covers lab equipment and safety. Lesson 2 is a demonstration of turning a copper penny into what looks like gold. This lesson presents students with the same dilemmas alchemists face: Can ordinary substances be turned into gold, and how do you know if you have been successful in making gold? In Lesson 3, students define matter and learn about some properties of matter. In Lesson 4, they make some measurements of mass and volume, which they then use in Lesson 5 to investigate density.

In this chapter, students will learn

- about lab equipment and lab safety
- a general definition of matter
- to formulate a hypothesis based on observed evidence
- how to solve density problems
- how to identify substances based on their basic properties

De Agostini Picture Library/Getty Images

CHAPTER 1

Defining Matter

In this chapter, you will study

- the tools of chemistry
- how matter is defined
- how to measure mass, volume, and density
- how types of matter differ from one another

This photo shows gold. How can you find out if an object is made of solid gold or simply coated with gold? What makes gold different from other metals? How is it possible to identify a metal based on its properties?

With a simple procedure, you can make a copper penny look like gold. But is it really gold, or has something else been made? A way to tell real gold from other substances is to compare the properties of those substances with the properties of gold. Chemists study various properties of matter and use the results to compare and identify substances.

2

Chapter 1 Lessons

Tools of the Trade

Lab Equipment and Safety

THINK ABOUT IT

A chef depends on a wide variety of gadgets and kitchenware to create delicious meals—from whisks and mixers, to ovens and saucepans. An auto mechanic relies on a toolbox of wrenches. In every profession, it is important to have the right tool for the job. Chemists have their own special tools and equipment that allow them to study the world around them. They also have a set of guidelines for using the tools safely.

What tools and equipment do chemists use?

To answer this question, you will explore

- The Tools Chemists Use
- Laboratory Safety

EXPLORING THE TOPIC

The Tools Chemists Use

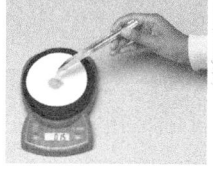

Chemistry often brings to mind a laboratory filled with unusual glassware and bubbling beakers. Chemists depend on a variety of tools in their explorations. In particular, chemists need tools that allow them to measure the mass and volume of substances, mix them, heat and cool them, and observe and separate them. Take a moment to examine these illustrations to see some of the tools that are used for these purposes.

Tools for measuring mass: balance with weighing paper, spatula.

Tools for measuring volume: graduated cylinders, Erlenmeyer flask, burette attached to Erlenmeyer flask.

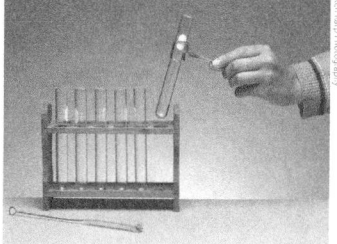

Tools for observing change: test-tube holder, test-tube rack, brush for cleaning.

Tools for mixing: stirring rod and beaker.

Lesson 1 | **Lab Equipment and Safety**   3

location of safety equipment. Then, in groups of four, students search for four different items located somewhere in the laboratory. They bring these items, along with the appropriate equipment card, back to their group. Each group introduces its four pieces of equipment to the rest of the class. A complete materials list is available for download.

TRM Materials List 1

SETUP

The text of the Laboratory Safety handout also appears in the student text, but you may want to hand it out as a contract for the students to sign. Copy one set of the equipment cards from Teaching Resource Materials and cut them out. They can be copied onto heavier card stock or attached to index cards to make them more durable. Place each piece of equipment, with its corresponding card, on a shelf or table close to where it is stored in your lab or classroom. Lock any cupboards that you do not want students opening. *Note:* If your classroom cannot accommodate a scavenger hunt, set up eight stations with the equipment grouped as described in the worksheet.

CLEANUP

Make sure students return equipment to the proper places in the laboratory.

Lesson Guide

ENGAGE (5 min)

TRM PowerPoint Presentation 1

> **ChemCatalyst**
> List at least four tools or pieces of equipment you think a chemist might use in a chemistry laboratory.

Sample answer: Chemists use glass containers, balances, funnels, and thermometers. Students may have heard of beakers or test tubes.

→ Have students name or describe equipment they think chemists use.

Ask: How do you measure the amount of a substance?

Ask: What equipment could you use to combine two substances? To separate two substances?

Ask: How are the tools chemists use like tools you might find in your kitchen or at the grocery store?

Overview

ACTIVITY: GROUPS OF 4

Key Question: What tools and equipment do chemists use?

KEY IDEAS

A number of tools are used in studying matter, especially to keep track of the amount of mass and volume in a sample, to provide containers for mixing substances, and to separate different forms of matter. The purpose of this lesson is to acquaint students with the laboratory, the lab equipment, and key safety considerations.

LEARNING OBJECTIVES

- Recognize common chemistry tools and equipment that they will be using in the course.
- Find all the safety equipment in the laboratory and understand its use.
- Understand the rules of safety in the chemistry laboratory.

IN CLASS

This lesson is a lab orientation adaptable to your school's laboratory and equipment as well as to safety standards in your school district. The activity takes the form of a lab scavenger hunt. First, students create a map of their laboratory, labeling the

TRM Worksheet with Answers 1

TRM Worksheet 1

TRM Card Masters—Equipment Cards 1

INTRODUCE THE ACTIVITY

→ Arrange students into groups of four.

→ Pass out the student worksheet and equipment cards.

→ Assign each group a letter from A through H corresponding to the eight equipment lists on the worksheet. Students must find the correct name, location, and function of each of the four items for their group and share that information first with their group and then with the rest of the class.

→ Give students a blank transparency or poster paper and markers for their presentation (optional).

GUIDE THE ACTIVITY

→ Many of the equipment lists are purposeful groupings and include items that can be used together in a laboratory procedure. You may want to go around the room to discuss the proper use of these items.

EXPLAIN & ELABORATE (20 min)

DISCUSS INFORMATION LEARNED IN THE SCAVENGER HUNT

→ Instruct students to write down the names and locations of the items they found and add them to their maps as each is presented. Encourage students to draw a simple sketch of each item.

→ Instruct each group to present the information they have discovered to the rest of the class in a two- to three-minute presentation. For each presentation, the group should

- Display the names of the items.
- Hold up a sample of each item for all to see. Small items can be passed around.
- Explain the use of each item. If the item is used with other items in the list, demonstrate how the tools are used together.
- Show where each item is located on the map of the classroom.
- Answer any questions to the best of their ability.

Tools for separating: Funnel with filter paper, wash bottle, beaker.

Tools for heating: Hot plate with beaker, boiling chips, and stirring rod, a thermometer.

Tools for heating: Bunsen burner with striker, ring stand with utility clamp, and a triangle holding crucible.

Measuring accurate amounts is important to chemists. They weigh solids on electronic balances and measure volumes of liquids in special glassware. You might notice that many of the containers chemists use are made of glass. Glass is a material of choice because substances in a glass container are visible. Chemists use tempered glass containers, which can be heated over flames without shattering. Also, glass containers are relatively easy to clean and reuse. Finally, notice that chemists use ring stands and special clamps to keep glassware from toppling. Spills can be hazardous.

⤻ Laboratory Safety

The chemistry laboratory is a place for discovery. However, as in any workplace that uses specialized equipment, safety is always important. There are many situations in a lab that can become dangerous. Before participating in any chemistry activities, you should familiarize yourself with the safety equipment in your lab.

Take a moment to examine these illustrations. What safety equipment and precautions do you notice?

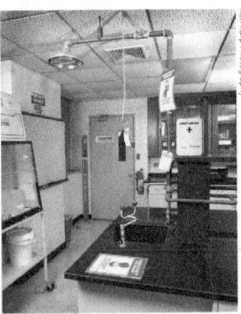

Know the location of the safety equipment and how to use it. Immediately report any laboratory accident, however small, to your teacher.

Never taste or touch chemicals. Never touch hot glassware. If you get chemicals on you, rinse with plenty of water.

DEMONSTRATE THE USE OF THE SAFETY EQUIPMENT

→ Show the class when and how to use all the safety equipment listed here.

- fire blanket
- first-aid kit
- fire extinguisher
- aprons/lab coats
- eyewash
- shower
- goggles
- hood

EVALUATE (5 min)

Check-In

1. Sketch or describe these items:
- graduated cylinder
- Erlenmeyer flask
- test-tube rack
- balance or scale

2. Where are these items located in your classroom?
- eyewash
- fire extinguisher

A few do's and don'ts:

- When working in a lab, dress appropriately. Roll up your sleeves, tie back long hair, and wear closed-toe shoes.
- Be sure that you have read the instructions for the procedure carefully.
- Double-check that you are using the correct chemicals.
- Before you begin working with chemicals or glassware, put on safety goggles.
- Before leaving the lab, clean your lab station and return equipment to its proper place.
- Do not put chemicals back into the original bottle. Doing so might contaminate the chemicals in the bottle.
- Your teacher will provide waste containers; never put chemicals or solutions down the drain unless instructed to do so by your teacher.

LESSON SUMMARY

What tools and equipment do chemists use?

Chemists use their own specialized tools and equipment in the laboratory. These tools and equipment are designed to allow chemists to measure mass and volume, and to mix, heat, cool, observe, and separate substances. It is important to work safely and carefully. When working in a chemistry laboratory, always wear safety goggles. Always wear appropriate clothing and closed-toe shoes. Be prepared to know what to do in case of an accident.

Exercises

Reading Questions

1. Why are most chemistry containers made of glass?
2. Describe the appropriate clothing to wear in a chemistry lab.

Reason and Apply

3. List three things you should do before beginning any laboratory procedure.
4. Describe what you would do in the case of an accidental spill in class.
5. List three things you should do before leaving the laboratory.
6. What is a fire blanket used for? If necessary, do some research to find out.
7. What is a hood used for in the chemistry laboratory? If necessary, do some research to find out.
8. Why do chemists use clamps and ring stands?

goggles, long pants or skirts, and shoes with closed toes. Remove any dangling jewelry.

3. Possible answers (any 3): Know the location of safety equipment. ● Read lab instructions carefully. ● Check to be sure that you are using the right chemicals and equipment. ● Follow directions from the teacher. ● Discuss the steps of the procedure with other members of the group and assign tasks.

4. Possible answer: Immediately report the spill to the teacher and follow instructions for cleaning it up. If you come into contact with any chemicals, rinse the affected area with water for at least 15 minutes.

5. Possible answers (any 3): Clean your work area. ● Make sure all bottles and containers holding chemicals are closed and stored properly. ● Safely dispose of used chemicals as indicated by the teacher. ● Wash your hands.

6. A fire blanket is used to extinguish flames by smothering a fire and depriving it of oxygen.

7. A hood keeps gases and fumes from entering the laboratory by carrying them to a filter or by venting them to the outdoors.

8. Chemists use clamps and ring stands to keep glassware from toppling over and breaking or spilling.

ASSESSMENTS

Two multiple-choice lab safety assessments (versions A and B) and two short-answer lab safety assessments (versions A and B) are available for download.

TRM Lab Safety Assessments Answers

TRM Lab Safety Assessments

LESSON 1

Homework

TRM Handout—Laboratory Safety

Assign the reading and exercises for Alchemy Lesson 1 in the student text. Distribute the Laboratory Safety handout for students to keep in their binders (optional). These rules are also included in the front of the student text and illustrated in the reading for this lesson. For homework, instruct students to review the safety rules. Students should be prepared to answer questions about safety on the next exam or quiz. An optional safety

assessment is included in the Teacher Resource Materials.

LESSON 1 ANSWERS

1. Possible answer: Chemistry containers are made of glass because substances in a glass container are visible and glass containers are relatively easy to clean and reuse. Tempered glass containers can be heated over flames without shattering.

2. Possible answer: Appropriate clothing provides protection and has no loose or dangling parts to interfere with safe lab procedures. While in the lab, tie back long hair and wear safety

Overview

DEMO: WHOLE CLASS

Key Question: What is chemistry?

KEY IDEAS

Centuries ago, alchemists tried to turn ordinary substances into gold. They were not successful. However, they discovered a great deal about matter and its transformations. As a result, the alchemists helped build a foundation for the modern pursuit of chemistry.

LEARNING OBJECTIVES

- Observe a procedure and write observations.
- Define hypothesis and formulate a hypothesis to explain observed phenomena.
- Define chemistry and begin to describe what chemists study.

KEY TERMS

hypothesis
property
chemistry

IN CLASS

This lesson opens with a brief discussion of alchemy and the ancient quest to turn ordinary substances into gold. Students then view a demonstration of a penny being turned into "gold" (the penny is coated with brass). This surprising event leads to a discussion of whether gold was created during the demonstration. This question is the contextual basis for the entire alchemy unit. In the process of conducting the demonstration, you introduce some basic chemistry equipment and experimental procedures. *Note:* If you have sufficient materials and are confident that students are well attuned to lab safety and lab preparedness, you might choose to set up multiple stations to run this lesson as a teacher-guided lab. A complete materials list is available for download.

TRM Materials List 2

Pre-AP® Course Tip

Students preparing to take an AP® chemistry course or college chemistry must be able to substantiate claims based on data collected. In this lesson, students are recording observations to determine gold has actually been created. In the Making Sense part of the worksheet, the students must articulate their claim and cite evidence to support it.

Featured DEMO

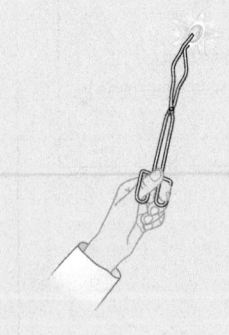

A Penny for Your Thoughts

Purpose

To observe a chemical transformation firsthand.

Materials and Safety

1. List the equipment used in the demonstration.
2. Briefly describe your observations of each substance used in the demonstration.
3. Safety is extremely important in the chemistry lab. Write three important safety considerations for this demonstration.

Procedure and Observations

Record your observations for each step of the demonstration.

1. Place a beaker containing zinc filings and sodium hydroxide on a hot plate set to 4.
2. Use tongs to pick up the penny and place it in the heated beaker.
3. While holding the beaker steady with tongs, remove the penny with the other tongs.
4. Put the hot penny in a beaker of cold water to cool and rinse it.
5. Use tongs to place the penny on the hot plate.
6. When the penny has changed color, use the tongs to place it in the beaker of cold water.

Analysis

Working with the students at your table, spend a few minutes discussing what you observed during the demonstration. Then answer the questions individually on your own paper.

7. Describe what happened to the penny during the demonstration.
8. What do you think turned the penny silver?
9. What do you think turned the penny gold?
10. **Making Sense** Do you think you made real gold? Why or why not? How could you find out?

SETUP

SAFETY

- Safety goggles should be worn at all times.
- Caution: Sodium hydroxide, NaOH, is caustic and will corrode skin, eyes, and clothing. Rinse spills with plenty of water. Have boric acid on hand.
- Use caution with the hot plate and penny. They are very hot.

Before class, add about 125 mL of water to a 250 mL beaker. Place enough zinc filings (~10 g) in a 100 mL beaker to cover the bottom, and carefully add 20 mL of 3 M sodium hydroxide, NaOH. You can use these same beakers for all your classes. Add more sodium hydroxide to the beaker when necessary due to evaporation.

CLEANUP

You will need at least eight gold-colored pennies in total for Lesson 5: All That Glitters. Save those you have made and create extra if needed. Dispose of used sodium hydroxide, NaOH, according to local, state, and federal regulations. Thoroughly rinse the zinc filings and recycle for future use.

A Penny for Your Thoughts

Introduction to Chemistry

THINK ABOUT IT

Gold is worth a lot more than copper. If you could turn pennies into gold, you would be very rich. Beginning in ancient times, people known as alchemists tried to transform substances into other substances. In particular, some of them tried to turn ordinary metals into gold. Today, we recognize these alchemists as early chemists. In fact, the word *chemistry* is derived from Arabic *alkīmiyā'* and from Greek *khēmia*, meaning "the art of transmuting metals."

What is chemistry?

To answer this question, you will explore

- The Roots of Chemistry
- Chemistry: The Study of Matter and Change

EXPLORING THE TOPIC

The Roots of Chemistry

While trying to make gold, alchemists developed some of the first laboratory tools and chemistry techniques. They classified substances into categories and experimented with mixing and heating different substances to create something new. When alchemists succeeded in creating a new substance, they faced the challenge of figuring out whether or not that new substance was really gold. Often, alchemists were fooled into thinking that a substance was gold just because it looked like gold.

In class, you watched a procedure to make a "golden" penny.

During the procedure, when the silver-colored penny was heated, it turned a gold color. You came up with a **hypothesis** to explain what happened. A hypothesis is a possible explanation for an observation. It can be tested by further investigation or experimentation. Suppose your hypothesis is that the penny turned to actual gold during the procedure. To test that hypothesis, you can compare your gold-colored penny to actual gold to see if it has all the **properties,** or characteristics, of gold. You can check the penny's physical properties, such as color, hardness, weight, and the temperature at which it melts. You can also test its chemical properties, such as whether it changes when you pour acid on it or rusts over time.

Over the course of this unit, you will explore whether it is possible to turn copper or any other substance into gold. But unlike the alchemists, you will have the advantage of hundreds of years of chemistry knowledge to help you answer this question.

HISTORY CONNECTION

The ancient art of alchemy has been traced to many different cultures and areas around the world. Some alchemists sought to turn lead into gold or find a potion that would bring eternal life. This painting shows alchemists at work in the late 19th century.

Alchemists, 1893 (oil on canvas), Mehdi (1870–99)/ Golestan Palace Library, Tehran, Iran/Giraudon/ The Bridgeman Art Library

CONSUMER CONNECTION

Why is gold so valuable? It retains its shine and resists change, even after hundreds of years. Gold is soft and easy to fashion into beautiful jewelry. It is a vital component in computers and cellular phones. Gold is also relatively rare.

Lesson 2 | **Introduction to Chemistry**   7

Lesson Guide

ENGAGE (5 min)

TRM PowerPoint Presentation 2

ChemCatalyst

Long ago, early scientists tried to turn ordinary things into gold. This pursuit was called alchemy, and the people who engaged in alchemy were called alchemists. Do you think these early scientists were successful in turning things into gold? Explain your thinking.

Sample answers: No; if you could turn ordinary things into gold, gold would be more common. Or: Sure; you can turn sugar and flour into a cake.

→ Solicit students' ideas on transformations of matter and the practice of alchemy. Allow the discussion to be open-ended.

Ask: Why do you think alchemists wanted to turn ordinary things into gold?

Ask: Do you think the alchemists were successful? Why or why not?

Ask: Do you think modern chemists have found a way to turn cheap metals, such as lead, into gold?

EXPLORE (20 min)

TRM Worksheet with Answers 2

TRM Worksheet 2

INTRODUCE THE DEMONSTRATION

→ Do not pass out the student worksheet right away. Explain to students that they will observe a demonstration as a class twice.

→ Let students know that on the second run, they should record their observations on the worksheet.

→ You may want to have students practice making some observations to help distinguish between observation and inference.

PERFORM THE DEMONSTRATION

→ Run the demonstration once, with minimal explanation. You might want to rinse and cool the penny in a beaker of water and pass it around the room for students to inspect.

→ After the first run, hand out the student worksheet for students to take notes and answer questions.

→ Introduce the equipment: Refer to the equipment (and cards) from Lesson 1 as appropriate. Ask students to explain the use of each piece of equipment used in the demonstration.

→ Introduce the chemicals: Introduce each chemical that is used. Ask students to record their observations of each chemical or mixture of chemicals on the worksheet.

→ Discuss safety: Ask students to list possible safety considerations for this demonstration, and have students record them on the worksheet.

→ Repeat the demonstration slowly, step by step, so students can make observations. Read the directions on the worksheet out loud for each section so students will know where and when to record information. Let students know that they will be recording what they see, hear, and smell. They can also touch the pennies once they have been rinsed and cooled. Remind them that it is not safe to taste anything in a laboratory.

→ Students will probably find this demonstration interesting. If there is time, repeat the demonstration a third time with students assisting you. You will need at least eight gold-colored pennies for Lesson 5: All That Glitters, so this is an opportunity to prepare some of them.

EXPLAIN & ELABORATE (20 min)

DISCUSS STUDENTS' OBSERVATIONS

We suggest that you do not reveal the identity of the alloy formed in this class (brass) so as not to spoil it for students. Let them hypothesize. If students suggest that it is brass, ask them why they think this and how they would prove their assertion.

Ask: What did you observe in the demonstration?

Ask: What do you think happened to change the color of the penny to silver?

Ask: What do you think happened to change the color of the penny to gold?

Ask: Do you think the penny was changed into real gold? Why or why not? How could you find out?

Key Points: The class did not determine if gold was made in the demonstration. To determine if the penny is gold, a scientist would study the properties of the penny. For instance, after the demo, the penny is shiny and yellow like gold, but further comparisons must be made to be certain of its identity. Physical properties, such as the color, hardness, size, and weight, of the penny can be compared with those of real gold. The chemical properties of the penny can be tested, such as what happens when it is placed in various liquids like vinegar or window cleaner. This starts the study of chemistry for the class.

Property: A characteristic of a substance.

Chemistry: The study of substances, their properties, and how they can be transformed. The study of matter and how matter can be changed.

DISCUSS SCIENTIFIC METHODS (OPTIONAL)

Key Points: Scientists use a systematic approach to solve problems in science. Scientific methods usually involve making observations, formulating hypotheses, experimenting, collecting data, analyzing data, and drawing conclusions. Scientific research depends on using a systematic, consistent, and reproducible approach.

Hypothesis: A testable explanation for an observation.

EVALUATE

No Check-In for this lesson.

Homework

Assign the reading and exercises for Alchemy Lesson 2 in the student text.

Chemistry: The Study of Matter and Change

Changes are constantly occurring all around you. Nails rust, colors fade, milk sours, and plants grow. You can mix ingredients and bake them in an oven to make cookies. You can bleach your hair to change its color, and you can freeze water to make ice cubes. Your body can transform cheeseburgers and burritos into muscle, fat, and bone. Chemists seek to understand changes such as these.

Chemistry is the study of what substances are made of, how they behave, and how they can be transformed. It is the study of matter and how matter changes. In this first unit, you will investigate matter and how it can be changed. You will learn to describe and explain what happens when matter is changed and you will begin to understand what changes are possible.

Chemistry at work—an iron train rusting

LESSON SUMMARY

What is chemistry?

KEY TERMS
hypothesis
property
chemistry

Chemistry is the study of the substances in the world around you. It is the study of matter and how matter can be changed. The modern study of chemistry emerged from the experimentation and effort of the alchemists. The alchemists invented useful tools and discovered many valuable laboratory techniques in their efforts to create gold out of ordinary substances.

Exercises

Reading Questions

1. How did the alchemists contribute to the modern study of chemistry?
2. What is chemistry?

Reason and Apply

3. PROJECT Use the library or the Internet to research the development of alchemy in one of these regions: China, India, the Middle East, Greece, Spain, England, or Egypt. Write a two-paragraph essay on the history of alchemy for your chosen region. Be sure to list your sources.

4. PROJECT Use the library or the Internet to research common uses for sodium hydroxide, which is also called *lye*.

5. Write down at least ten changes that you observe in the world around you. Which changes involve chemistry? Explain your reasoning.

LESSON 2 ANSWERS

1. Alchemists developed some of the first laboratory tools and chemistry techniques. They classified substances into categories and experimented with mixing and heating different substances to create something new.

2. Chemistry is the study of what substances are made of, how they behave, and how they can be transformed. It is the study of matter and how matter changes.

Answers continue in Answer Appendix (p. ANS-1).

Defining Matter

THINK ABOUT IT

People tend to value gold over other substances. You don't often see someone wearing aluminum jewelry or putting coal in a high-security bank vault. What is it about gold that makes it unique? Is it possible to create gold from another substance?

This lesson begins to explore the nature of matter as the first step toward proving whether you can or cannot create gold. After all, chemistry is the study of matter and its properties.

What is matter?

To answer this question, you will explore

- Defining Matter
- Is It Matter?
- Measuring Matter

EXPLORING THE TOPIC

Defining Matter

Matter is the word chemists use to refer to all the materials and objects in the world. Your desk, this book, and the paper and ink in the book all are matter. These are all things you can see or feel. However, your senses alone are not always enough to tell you if something is matter. For instance, you cannot see the virus that gave you a cold, but it is matter. Conversely, you can see shadows on the ground cast by the light from the sun, but they are not matter.

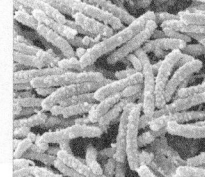

A gold ring, the ink in this book, and a virus each have *substance,* which means they are made out of material, or "stuff." The amount of substance, or material, in an object is called **mass**. Mass is a property of matter that can be measured. So, although the virus has very little substance, it still has mass.

Another property that a gold ring, the ink in this book, and a virus have in common is that they take up space, which means they have dimensions. The amount of space something takes up is called **volume** and is also a property of matter that can be measured.

So, matter is anything that has mass and volume. You can also say matter is anything that has substance and takes up space. This explains why a virus is matter but a shadow is not.

Is It Matter?

It is easy to see that solids and liquids have mass and volume. When water is poured into a container, you can see how much space it takes up. When

Overview

ACTIVITY: GROUPS OF 4

Key Question: What is matter?

KEY IDEAS

Chemistry is the study of matter— what it is composed of and how it can be transformed. Matter can be defined as anything that has substance and takes up space. Scientists refer to substance as *mass,* and to taking up space as *volume.* They define matter as anything that has mass and volume. Almost everything in the universe is matter, with a few exceptions—such as ideas, feelings, and energy, which have

no substance and do not take up space. Chemists might study almost anything in the universe, even things such as energy and pressure, which are not themselves matter but require matter.

LEARNING OBJECTIVES

- Define matter as anything that has mass and volume.
- Classify an item as matter or not matter.

FOCUS ON UNDERSTANDING

- Many students confuse mass with volume. They might also think gases are not matter because they often appear to have no mass and often cannot be seen.

- Students may be unclear on the distinction between matter and energy; e.g., hot air is matter, but heat is not.
- Students may think of chemists as working only with synthetic "chemicals." In fact, chemists study all kinds of matter, both natural and synthetic.

KEY TERMS

matter
mass
volume
meniscus

IN CLASS

Students work in groups to determine if items on a list are matter. Students identify a list of properties common to all the items they classify as matter to construct a definition of matter. The reading in the student text introduces the measurement of mass and volume in preparation for the next lesson. A complete materials list is available for download.

TRM Materials List 3

Differentiate

For English-language learners and other students who struggle with scientific literacy, provide images for the list of items to be categorized as matter, unsure, or not matter in the activity. Encourage advanced students to explain the meanings of these words to students in their groups who need assistance.

Pre-AP® Course Tip

Students preparing to take an AP® chemistry course or college chemistry should be able to analyze data to identify patterns or relationships. In this lesson, students categorize a list of items based on similar properties. After analyzing the patterns they notice, students formulate their own definition of what constitutes matter.

Lesson Guide

ENGAGE (5 min)

TRM PowerPoint Presentation 3

ChemCatalyst

Modern chemistry is defined as the study of matter.

1. What do you think matter is?
2. Name two things that are matter and two things that are not matter.

Sample answers: **1.** Matter is different kinds of substances. **2.** Answers will vary.

Ask: **What do you think chemists study?** (Everything, from food and paper to polymers and pharmaceuticals.)

Ask: **What do you think matter is?**

EXPLORE (15 min)

TRM Worksheet with Answers 3

TRM Worksheet 3

INTRODUCE THE ACTIVITY

→ Arrange students into groups of four.

→ Pass out the student worksheet.

GUIDE THE ACTIVITY

→ Walk around the room and assist students as they discuss the topic. If they seem lost, let them know that matter is another word for "substance."

→ The "correct" answers are not as important as the process of constructing the definition.

→ If one group is struggling, encourage them to discuss with another group.

EXPLAIN & ELABORATE (20 min)

DISCUSS WHAT CONSTITUTES MATTER

→ Write the categories: Matter, Unsure, and Not Matter on the board. As the class reaches consensus on each item, write it on the board in the chosen category.

Ask: **Which items are clearly matter? Not matter? Up for discussion?**

Ask: **If an item is not matter, what is it?** (energy, an idea, a feeling)

Key Points: The items that are clearly matter are all objects or things that are tangible. Gases are matter, too. For example, you do not always see or feel air, but you can inflate a balloon with it, which indicates something is there. Energy and ideas are not matter, but they involve matter. Wind, for instance, could be considered the movement of matter. Experiencing a thought would be impossible without matter in the brain. We experience heat only when matter is around, such as a log burning or warm air from a heater.

DISCUSS THE PROPERTIES OF MATTER

→ As the groups share, complete the list on the board. Assist the class in reaching consensus about the list.

---

filled with water, the container has more mass. You can see this difference if you use a balance. On a two-pan balance, a cup with water will be lower than an identical cup with no water because it has more mass. Thus, water is matter.

When identifying matter, gases can be misleading. For example, most of the time you do not see or feel the air around you and it may seem as if nothing is there. However, when you fill a balloon with carbon dioxide gas, you can see that the carbon dioxide gas inside occupies space. On a two-pan balance, the pan with a balloon filled with carbon dioxide gas will be lower than the pan with an empty balloon because the balloon filled with carbon dioxide gas has more mass. Therefore, gases do have mass and volume even though gases might be harder to detect than solids and liquids.

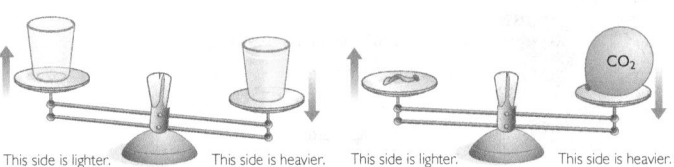

This side is lighter.   This side is heavier.   This side is lighter.   This side is heavier.

What about other things, like heat and sound? Are they matter? When you heat soup, it does not gain mass, and sound may "fill the room," but it doesn't have mass or volume. Sound is the movement of air against your eardrums. Without matter, sound can't exist, but sound itself is not matter. Similarly, heat can't exist without some form of matter, but heat itself is not matter. Both sound and heat are referred to as types of energy, and energy by itself is not matter. Likewise, feelings and thoughts are not matter, although you could argue that they require an interaction of matter and energy.

Example

Is It Matter?

Classify wind, music, and clouds. Are they matter? Explain your answer.

Solution

Wind is the movement of air. Air is matter, but the movement of air is not matter because movement has no mass or volume. Therefore, wind is not matter.

Music is sound, which is the movement of air against your eardrums. Music does not have mass or volume. Therefore, music is not matter.

Clouds are made of water droplets, which have mass and volume. Therefore, clouds are matter.

10   Chapter 1 | **Defining Matter**

Ask: **What is one property that all matter has in common?**

Ask: **Does everything in the Matter column have this property?**

Ask: **Does anything in the Not Matter column have this property?**

Ask: **Can you always see and feel matter?**

Key Points: Matter has some sort of dimension and substance to it. These are two properties of matter. Some might say that most matter can be seen or touched, although there are exceptions. And just because you can feel or sense something does not

always mean it is made of matter. For example, an earthquake is sensed, but it is not matter.

DEFINE MATTER AS A CLASS

→ Have each group share or display its definition of matter on the board.

→ Instruct the class to imagine that they are at a science convention and their task is to create the most accurate definition of matter. They must give evidence to support their claims.

→ Refine the definition until the class reaches consensus. Challenge incomplete definitions with counterexamples.

This electronic balance will measure mass to the nearest thousandth of a gram.

↻ Measuring Matter

MEASURING MASS

To find the mass of something, you weigh it. In everyday life, things are usually weighed in pounds, lb, or kilograms, kg. For example, you've probably weighed yourself and know approximately how many pounds you weigh. And when you go to the grocery store, you might buy a pound of butter or flour. However, chemists measure mass in kilograms, kg, and grams, g (1 kg = 1000 g). [For review of this math topic, see **MATH Spotlight**: SI Units of Measure on page A-0.]

In the chemistry classroom, an electronic balance is used to measure mass. The balance is precise and shows more digits (places) than you may need. You have to decide how many places to pay attention to when using this balance. For example, when you weigh yourself, you measure to the nearest kilogram. You would state your mass as 70 kg, not 69.903 kg. But if you are measuring the mass of a penny, you may want to be more exact. The mass of the copper penny to the nearest hundredth of a gram is 3.11 g. The mass to the nearest gram is 3 g.

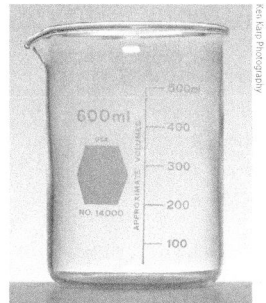

This beaker measures to the nearest tenth of a liter.

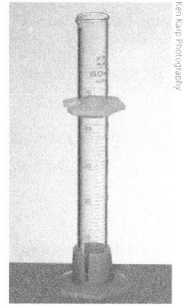

This graduated cylinder measures to the nearest milliliter.

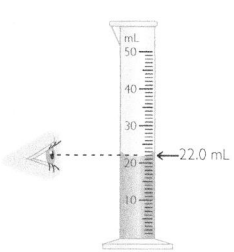

←22.0 mL

Read the measurement at the bottom of the meniscus. Make sure your eye is at the water level.

MEASURING VOLUME

You can measure the volume of a regularly shaped object by measuring its dimensions, such as length, width, and height, and using a mathematical formula. Volume measured this way is reported in cubic units, for example, cubic meters, m^3, or cubic centimeters, cm^3.

Chemists often measure the volume of gases and liquids in liters, L, and milliliters, mL (1 L = 1000 mL). The volume of a large soda bottle is 2 L, or 2000 mL. This is the volume of the bottle, and if the bottle is full, it is also the volume of liquid inside. Chemists use a variety of special containers for measuring volume. For approximate measurements, they use beakers. For more precise measurements, they use graduated cylinders. [For review of this math topic, see **MATH Spotlight**: Accuracy, Precision, and Significant Digits on page A-1.]

Ask: Based on what we have discussed, what is matter?

Ask: Which of these is the best definition of matter? Why? Can it be improved?

Ask: Does everyone agree 100 percent with this definition?

Ask: Does our definition of matter help us sort out items in the Unsure column?

Key Points: A good definition should work 100 percent of the time. Modify and refine the definition until it applies to everything in the Matter column and nothing in the Not Matter column. Incomplete definitions must be fleshed out.

SHARE SOME FORMAL DEFINITIONS OF MATTER

→ Share the definitions of matter given here. Compare the class's definition to the formal ones, looking for similarities and differences. Make sure students do not get the impression that their definition is inferior or wrong.

> Here are some textbook definitions of matter:
>
> **Matter:** Anything that has substance and takes up space.
>
> **Matter:** Anything that has mass and volume.

Ask: How does our class definition compare with these?

Ask: Do you have any criticisms of these other definitions? (They each require that you define other things—mass, volume, substance, space—before you can define matter.)

Ask: What made it difficult to define a term like matter?

BRIEFLY DEFINE ENERGY (OPTIONAL)

Ask: Which items on our list do you think would belong to a category called "energy"?

Ask: What do you think energy is?

Ask: Do coffee and candy contain energy? Explain your thinking.

Ask: Is energy matter?

Ask: Can you detect energy with your senses?

Key Points: Energy is the ability to do work or make reactions happen. There are many different forms of energy, such as heat, light, and electricity. Although we all have an idea of what energy is, it is difficult to define. Students may think of energy as something "contained" in food, caffeine, batteries, and so on. But energy does not take up space or have mass. For instance, a cup of tea does not have more mass when it is hot. However, it may lose mass due to evaporation, which is a loss of matter.

EVALUATE (5 min)

> **Check-In**
>
> Which of the following can be classified as matter? Explain.
>
> **a)** a beam of sunlight
>
> **b)** an automobile
>
> **c)** an idea
>
> **d)** your breath
>
> **e)** rain
>
> **f)** sadness

Sample answer: Parts **b)**, **d)**, and **e)** are matter; **c)** and **f)** are not matter; **a)**, a beam of sunlight, is ambiguous. If we think of a beam of sunlight as light that bounces off dust particles, matter is involved.

Homework

Assign the reading and exercises for Alchemy Lesson 3 in the student text.

LESSON 3 ANSWERS

1. Mass is the amount of material in an object. Volume is the amount of space that the object takes up.

2. *Possible answer:* A bicycle has mass because when you lift it you can feel its weight. You could also prove this by placing the bicycle on a scale. A bicycle has volume because you can see that it fills space that cannot be occupied by anything else. You could also prove this by lowering a bicycle into a bathtub and watching the water level rise. Since it has mass and volume, the bicycle is matter.

3. *Possible answer:* If you use a balance to find the mass of an empty balloon, then inflate the balloon with air and tie off its end, you will see that its mass has increased when you put it back on the balance. The air inside the balloon must have mass. Because the volume of the balloon increases when the balloon is inflated with air, the air inside the balloon also has volume. Because the air inside the balloon has mass and volume, it is matter.

4. *Possible answer:* The Sun is matter because it has mass and volume. Sunlight is not matter because it does not have mass or volume. It is a form of energy that can be detected visually and that makes your skin feel warm.

5. *Possible answer (any 10 of each):* Examples of things that are matter: a car, a tree, a person, a cat, a brick, a desk, a pen, water, a glass, a window, a rock, and the Sun. Each of these objects has mass and volume. ● Examples of things that are not matter: light, sound, movement, gravity, music, heat, time, cell-phone signals, feelings, energy, radio waves, thoughts, memories, and forces. These things do not have mass or volume. They are forms of energy or are intangible.

6. *Possible answers (any 5):* sound, wind, rolling, falling, running, flying, explosion, rain.

You can measure the volume of a sample of water by pouring it into a graduated cylinder and reading the number of milliliters, mL, on the side. However, the surface of the water is curved because of the way it adheres, or clings, to the sides of the cylinder. The curvature of the top of a liquid in a container is called a **meniscus.** It is most accurate to read the water level at its lowest point, the bottom of the meniscus, because most of the liquid is at this level. To get an accurate measurement, you must read the graduated cylinder at eye level.

LESSON SUMMARY

What is matter?

Chemists seek to understand matter and its properties. Matter can be defined as anything that has mass and volume. Mass is the amount of substance or "stuff" in a material or object. Solids, liquids, and gases all have mass and volume and are classified as matter. To measure the mass of something, you use a balance. Volume is the amount of space taken up by matter. To find the volume of something, you can use a container such as a beaker or graduated cylinder, or you can calculate the volume using dimensions, such as length, height, and width.

KEY TERMS
matter
mass
volume
meniscus

Exercises

Reading Questions

1. Explain the difference between mass and volume.
2. Describe how you could prove that a bicycle is matter.

Reason and Apply

3. Someone might claim that air is not matter because you can't see it. Write a paragraph showing how you would prove to this person that air is matter.
4. The Sun is considered matter, but sunlight is not considered matter. Explain why this is so.
5. Sit somewhere and observe your environment. From your observations, make a list of ten things that are matter and a list of ten things that are not matter. Explain your reasoning.
6. Give examples of five words that describe the movement of matter.

Mass Communication

Mass and Volume

THINK ABOUT IT

Suppose you have two samples of gold, a gold ring and a gold nugget. Is there more gold in the ring or in the nugget? They feel similar in weight, and they look fairly similar in volume. While your senses can give you valuable information, they can't tell you exactly how much gold you have in each sample.

How do you determine the masses and volumes of different substances?

To answer this question, you will explore

- Measuring Volume
- Comparing Mass and Volume

EXPLORING THE TOPIC

Measuring Volume

As you learned in Lesson 3, volume is a measure of size, or how much space each sample takes up.

There are two common ways of measuring the volume of solids: (1) by measuring their dimensions and using a geometric formula and (2) by water displacement. The first method is convenient if the object has a regular shape. The second method is more convenient for irregularly shaped objects.

USING GEOMETRIC FORMULAS TO DETERMINE VOLUME

If a solid is rectangular, you can find its volume by measuring its three dimensions—length, width, and height—and multiplying these three values. Volume measured in this way is reported in *cubic* units such as cubic centimeters, cm^3; cubic meters, m^3; or cubic inches, in^3. The formula for volume is

$$V = lwh$$

Consider these two solid blocks. You can use the formula to figure out their volumes.

$V = lwh$
$\quad = 1.0 \text{ cm} \cdot 1.0 \text{ cm} \cdot 1.0 \text{ cm}$
$\quad = 1.0 \text{ cm}^3$

$V = lwh$
$\quad = 1.0 \text{ cm} \cdot 0.5 \text{ cm} \cdot 2.0 \text{ cm}$
$\quad = 1.0 \text{ cm}^3$

1.0 cm

1.0 cm
2.0 cm
0.5 cm

Notice that the two solids have different dimensions, but they have the same volume, 1.0 cubic centimeter.

Overview

LAB: GROUPS OF 4

Key Question: How do you determine the masses and volumes of different substances?

KEY IDEAS

Chemists use a variety of tools to measure the volume and mass of liquids and solids. Volume is measured using geometric formulas, specially calibrated containers, and water displacement. Milliliters, liters, and cubic centimeters are the most common units of volume. Chemists use electronic and triple beam balances calibrated in grams to measure mass. Objects with similar volumes can have vastly different masses, and vice versa. *Note:* The mass and volume of gases will be considered in Unit 3: Weather.

LEARNING OBJECTIVES

- Measure mass using a balance.
- Measure the volume of regularly and irregularly shaped objects.

FOCUS ON UNDERSTANDING

- Students often confuse the concepts of mass, volume, size, weight, heaviness, and density.
- When measuring volume by displacement, students often think incorrectly that the amount of water displaced, or the rise in the water level, is determined by the weight of the object.

KEY TERM

water displacement

IN CLASS

Before the lab, students observe a quick demonstration of the use of electronic or triple beam balances and graduated cylinders. Students work in groups to find the mass and volume of several similarly shaped objects. They answer questions exploring the relationship between mass and volume, setting the foundation for an exploration of density in the next lesson. A complete materials list is available for download.

TRM Materials List 4

SETUP

Make a chart on which groups can record their data for the discussion as a whole class following the lab (optional).

CLEANUP

Return equipment to its appropriate places.

Differentiate

Challenge advanced students to design their own procedure and data tables to measure mass and volume of each object. When they have completed their measurements, provide them with the worksheet to answer the questions.

Pre-AP® Course Tip

Students preparing to take an AP® chemistry course or college chemistry should be able to examine evidence to determine if it supports a hypothesis. In this lesson, students determine how the mass and volume of different objects compare, and they hypothesize about the mass and volume of different objects. Students analyze their collected data to determine if their hypotheses are supported.

Lesson Guide

ENGAGE (5 min)

TRM PowerPoint Presentation 4

ChemCatalyst

1. Which has more mass and weighs more, 5 kilograms of bricks or 5 kilograms of feathers? Explain your thinking.

2. Would it hurt more to be hit with 5 pounds of feathers or 5 pounds of bricks? Explain your thinking.

Sample answers: 1. They have the same mass. (Students may think that feathers are lighter than bricks.) **2.** Bricks are harder and the mass is more concentrated, so they probably would hurt more.

➜ Help students articulate their ideas and distinguish between mass and volume. Students tend to associate mass with volume and are likely to misread the first question.

Ask: Is it easier to lift 5 kilograms of bricks or 5 kilograms of feathers? Explain. (It is the same; if anything, the feathers will be bulkier and more awkward to lift.)

Ask: What tools would you use to measure mass?

Ask: Which has a greater volume, the bricks or the feathers?

Ask: What tools would you use to measure volume?

Ask: Are large objects always heavier than small objects? Explain.

EXPLORE

TRM Worksheet with Answers 4

TRM Worksheet 4

PREPARE FOR THE LAB

➜ Demonstrate how to use the triple beam balance or the electronic balance and how to round to the nearest tenth of a gram.

➜ Draw a graduated cylinder on the board and discuss the meniscus and how to read the correct value.

INTRODUCE THE LAB

➜ Arrange students into groups of four.

➜ Pass out the student worksheet.

➜ Tell students they will be measuring the mass and volume of some simple objects.

GUIDE THE LAB

➜ Do not give away the identity of the metal rods. Students will discover it in the next lesson. They will also calculate density in the next lesson.

➜ While students are working, post the data chart you created or draw data tables on the board for recording student data.

➜ When most groups have completed their data tables, instruct students to post their data values on the class data chart provided.

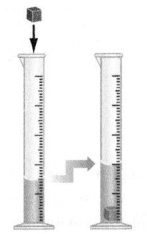

The water level rises by 5.0 mL because the cube has a volume of 5.0 cm³.

USING WATER DISPLACEMENT TO DETERMINE VOLUME

The second method of measuring the volume of a solid object is called **water displacement.** First, pour water into a graduated cylinder. Next, add the object whose volume you are measuring. An object takes up space, so it displaces some of the water when it is placed in the graduated cylinder. This causes the water level to rise by an amount equal to the volume of the object. If you read the volume of water in the graduated cylinder before you submerge the object and again after submerging it, the volume of the object is the difference between these two volumes.

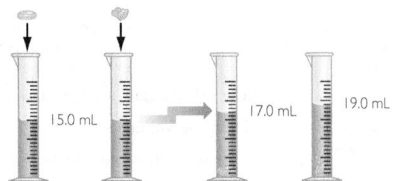

15.0 mL   17.0 mL   19.0 mL

| Important to Know | One milliliter is exactly equal in volume to one cubic centimeter: 1 mL = 1 cm³. |

Imagine you place a gold ring and a gold nugget in graduated cylinders partly filled with water. The graduated cylinder is marked in milliliters, mL. You observe that the water level rises by 2.0 mL in one container and by 4.0 mL in the other. The ring has a volume of 2.0 mL or 2.0 cm³ and the nugget 4.0 mL or 4.0 cm³.

Of course, the water displacement method works only for solids that do not dissolve in water! The solid object also needs to sink so that it is completely submerged. If the object floats, the volume reading will not be accurate.

🔄 Comparing Mass and Volume

Imagine that you have two objects with the same volume that are *not* made from the same material. You can compare their masses using a balance. The masses of 1.0 cm³ of gold and 1.0 cm³ of plastic are given in the table below along with the masses of the same volumes of wood, glass, and copper.

CONSUMER CONNECTION

There are many different types of plastic. Soft-drink bottles are made of polyethylene terephthalate, also called PETE. Plastic pipes and outdoor furniture are made of polyvinyl chloride, also called PVC. Styrofoam cups are made of polystyrene.

1.0 cm³ of gold has much more mass than 1.0 cm³ of plastic.

Material	Volume	Mass
gold	1.0 cm³	19.3 g
plastic	1.0 cm³	0.9–1.5 g
glass	1.0 cm³	2.2–3.1 g
copper	1.0 cm³	9.0 g
wood	1.0 cm³	0.2–1.4 g

The mass of plastic, glass, and wood varies depending on the type.

Notice that 1.0 cm³ of gold is heavier than the same volume of any of the other materials. Thus, two objects can have exactly the same volume but different masses.

EXPLAIN & ELABORATE (15 min)

ANALYZE THE CLASS DATA

➜ Consolidate the class data by finding the mean of group values. You can throw out the highest and lowest values or only outliers.

Ask: Does the largest object you measured have the largest mass? Explain. (No; the 10 cm silver-colored rod or the crayon has the largest volume, but the 5 cm silver-colored rod has the greatest mass.)

Ask: What is the difference between mass and volume? Explain. (Mass is the amount of matter; volume is the amount of space the matter occupies.)

Ask: Which of the four objects do you think contains the greatest amount of matter? (The 5 cm gold-colored rod)

Ask: Do you think any of the objects are the same type of matter? Explain your thinking. (The two silver-colored rods are made of the same type of matter.)

Ask: How might you use mass and volume to decide whether the "golden" penny is real gold? (Answers will vary. Density is covered in the next lesson.)

Important to Know The rise in the water level does not depend on how heavy the cube is. The water-level rise depends only on the volume of the object.

The mass of the plastic depends on the type of plastic being considered. The mass of a 1.0 cm³ sample of plastic varies between 0.9 g and 1.5 g. Wood and glass also show variations. For instance, if you compare a 1.0 cm³ sample of solid wood from an oak tree with a 1.0 cm³ sample of solid wood from a pine tree, they will have different masses. The oak sample will be considerably heavier than the pine. A 1.0 cm³ sample of pure gold will always have a mass of 19.3 g because there is only one type of pure gold.

Example 1

Try to work out the answers yourself before reading the solutions.

Cubes of the Same Volume

Suppose you have two cubes. One cube is made of solid gold and the other cube is made of solid plastic. The sides of the cubes each measure 2.0 cm in length.

a. What is the volume of each cube?

b. How much water will each cube displace?

Solution

a. The volume of a cube is equal to length times width times height.

$$V = lwh$$
$$= 2.0 \text{ cm} \cdot 2.0 \text{ cm} \cdot 2.0 \text{ cm}$$
$$= 8.0 \text{ cm}^3$$

b. Because 1 cm³ = 1 mL, you can reason that 8.0 cm³ = 8.0 mL. So each cube will displace the same amount of water, or 8.0 mL.

Example 2

Cubes of the Same Mass

One cube is made of solid copper and another cube is made of solid glass. They have exactly the same mass. How do their volumes compare?

Solution

If you look at the table of masses on page 14, you will see that 1.0 cm³ of copper has much more mass than 1.0 cm³ of glass. Thus, for the two cubes to have the same mass, the copper cube must be much smaller than the glass cube.

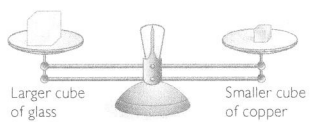

Larger cube of glass   Smaller cube of copper

The small cube of copper exactly balances the mass of the larger cube of glass.

Key Points: Objects with similar volumes can have vastly different masses, and vice versa. For example, a piece of Styrofoam that is the same size as a brick weighs much less. Likewise, objects with similar masses can have vastly different volumes. Styrofoam with the same mass as a brick would take up a lot more volume than a brick. Mass and volume together can help us determine whether the golden penny is real gold.

DISCUSS THE MEASUREMENT OF MASS

→ Write the word *mass* on the board. List the units used to measure mass as the students mention them.

Ask: What types of units are used to measure mass? Weight? (grams, kilograms, pounds, tons, etc.)

Ask: If you have 1 pound of cotton and 1 pound of steel, do they weigh the same? Why or why not? (Yes; they have the same mass.)

Ask: If you have 1 gram of copper and 1 gram of gold, do they weigh the same? Why or why not? (Yes; they have the same mass.)

Ask: How do you think mass and weight are different? (On the Moon, an astronaut weighs less than on Earth, but still has the same mass.)

Key Points: In science, mass is commonly measured in units of grams (g) or kilograms (kg). One gram of any substance has the same mass as one gram of any other substance (1 kg = 1000 g).

The word *weight* is often substituted for *mass*. In your everyday life, you probably are more used to using units of pounds (lb). A kilogram is slightly more than 2 pounds. Pounds measure force or weight, while grams and kilograms measure mass. However, on the surface of the Earth, weight and mass represent the same thing. Mass has to do with how much matter is present, and weight is related to the force of gravity. For example, under water or in space things might weigh less but they still have the same mass (1 lb = 454 g).

DISCUSS THE MEASUREMENT OF VOLUME

→ Write the word *volume* on the board. List the units used to measure volume as the students mention them.

Ask: What types of units are used to measure volume? (Cups, fluid ounces, liters, pints, cubic yards, etc.)

Ask: Where do you use volume in your everyday life? (Gallons of milk or gasoline, measuring cups for baking, etc.)

Ask: How would you measure the volume of a solid using a graduated cylinder and water? Describe your procedure.

Ask: Do heavier objects displace more water than lighter objects when they are submerged in water? Explain. (No; as long as they are submerged completely, masses of objects do not affect volumes.)

Ask: How would you measure the volume of a shoebox? Of sand?

Key Points: Volume is a measure of the amount of space occupied by something. There are several different methods for determining volume, depending on the type of substance or object being measured. In a laboratory, graduated cylinders are used to measure the volume of liquids. The markings on the side are in milliliters (mL) (1000 mL = 1 L).

There are several methods for measuring the volume of solids. Units of cubic centimeters are often used for solids. One milliliter and one cubic centimeter are equal in value (1 mL = 1 cm³).

LESSON 4

The volume of a regularly shaped solid is determined using a geometric formula. For example, the volume of a box that is 10 cm by 20 cm by 15 cm is: 10 cm · 20 cm · 15 cm = 3000 cm³. This is equivalent to 3000 mL.

The volume of a dry powder, such as flour, can be measured using a graduated cup. The volume of an irregularly shaped object often is determined by immersing it in water. The volume of water displaced by the object is equal to the volume of the object. The mass of the object has nothing to do with how much water it displaces. Two substances with very different shapes can have the same volume.

EVALUATE (5 min)

Check-In

A penny has a mass of 2.498 g.

a) What is the mass to the nearest tenth of a gram?

b) How would you determine the volume of a penny?

c) What is the difference between mass and volume?

d) Suppose you find that the golden penny has a mass of 2.6 g. If you compare it with the mass of this penny, what can you conclude?

Answers: **a)** 2.5 g. **b)** By measuring its dimensions and using a geometric formula or by immersing it in water and measuring the change in the volume of the water. **c)** Mass is the amount of matter or "stuff." Volume is the amount of space the matter or "stuff" takes up. **d)** The golden penny has more mass.

Homework

Assign the reading and exercises for Alchemy Lesson 4 in the student text.

LESSON 4 ANSWERS

1. *Possible answer:* To determine the measure of a solid object, measure its dimensions and calculate the volume using a geometric formula or, if it does not float or dissolve, measure the amount of liquid that it displaces when it is submerged.

2. *Possible answer:* Not necessarily. Two objects with the same volume can have very different masses. For example, a cube of iron has much more mass than a cube of wood that has the same volume.

LESSON SUMMARY

How do you determine the masses and volumes of different substances?

The volume of a solid object is determined either by water displacement or by measuring its dimensions and using a geometric formula for volume. The mass of any solid sample is found by weighing it on a balance or scale. The relationship between mass and volume depends on the type of material being considered. Two objects may have the same volume but different masses, or two objects may have the same mass but different volumes.

KEY TERM

water displacement

Exercises

Reading Questions

1. Describe two ways to determine the volume of a solid object.
2. If you have two objects with equal volume, do they have the same mass? Explain your thinking.

Reason and Apply

3. Can you predict the volume of an object just by looking at it? Explain.
4. Can you predict the mass of an object just by looking at it? Explain.
5. If you stretch a rubber band, does it still have the same mass? Explain.
6. You submerge a piece of clay in water and measure the total volume. You change the shape of the clay and put it back into the same amount of water.
   a. The total volume does not change. Explain why.
   b. Does the total mass change? Explain.
7. Describe how you might find the volume of these items:
   a. pancake mix   d. a penny
   b. hair gel   e. lemonade
   c. a shoe box
8. Draw two things with the same mass but different volumes.
9. Draw two things with the same volume but different masses.
10. Use the illustrations to answer these questions.
    a. What is the volume of the liquid inside the container?
    b. What is the volume of the rock in mL? In cm³?
11. Suppose that you have two cubes of exactly the same volume. You weigh them and find a mass of 8.91 g for one cube and 8.88 g for the other cube even though they are made of the same material. How is this possible?

16   Chapter 1 | **Defining Matter**

---

3. Yes, the volume of an object is the amount of space it fills. You can usually see how much space an object fills and estimate its volume based on its dimensions. However, for objects that have an irregular shape, are very thin, or have a surface with lots of holes or pits, determining the volume of the object by sight may be difficult.

4. No, the mass of an object does not depend only on its shape and size, so you cannot predict an object's mass simply by looking at the object. An incorrect answer is that the mass of a known object, such as a piece of metal, can be predicted by looking at its size. In this case, knowledge about density is being used in addition to

visual information, or the assumption is made that an object is made of a certain material just by looking.

5. Yes, the mass of the rubber band is the same because only the shape of the rubber band changes, not the amount of matter in it.

6. a. Volume is the measure of the amount of space the piece of clay occupies. Changing its shape does not change the amount of space it occupies. **b.** No, the mass of the clay does not change, because you did not add or subtract any matter from the clay, and mass is a measure of the amount of matter.

Answers continue in Answer Appendix (p. ANS-1).

All That Glitters

Density

THINK ABOUT IT

Gold and lead are metals that are soft and bendable. They are easy to tell apart at a glance because lead is dull gray. But suppose someone tried to trick you by coating a block of lead with a thin layer of gold. How could you prove that this block was a fake?

How can you use mass and volume to determine the identity of a substance?

To answer this question, you will explore

⟳ The Definition of Density

⟳ Calculating Density

⟳ Identifying Matter Using Density

Gold is a soft, shiny, bendable, metal.

EXPLORING THE TOPIC

⟳ The Definition of Density

Which is heavier, gold or copper? It depends on the amount of each that you have. If you compare a tiny gold earring and a large copper pipe, the copper will be heavier. If you compare the same volume of each, gold is *always* heavier than copper. That is because there is more matter in 1.0 cm³ of gold than in 1.0 cm³ of copper. You could also say that gold is *denser* than copper. The mass of a substance per unit of volume is called its **density**.

This copper has been shaped to form hollow metal pipes.

Imagine comparing 1.0 cm³ cubes of wood, water, and copper from larger samples of each material. All three cubes have the same volume, but each cube has a different mass and a different density.

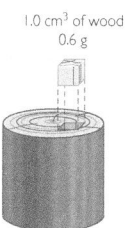
1.0 cm³ of wood
0.6 g

1.0 cm³ of water
1.0 g

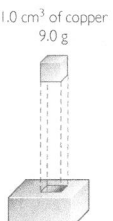
1.0 cm³ of copper
9.0 g

Overview

LAB: PAIRS

Key Question: How can you use mass and volume to determine the identity of a substance?

KEY IDEAS

To differentiate between kinds of matter, we can examine their *intensive* properties, such as density. Density is the measure of the mass of a substance per unit volume. Differences in density explain why two objects may have identical volumes but different masses, or identical masses but different volumes. While the mass and volume of a given substance can change, density does not (at least not at a specified temperature). Density change as a function of temperature will be covered in Unit 3: Weather.

LEARNING OBJECTIVES

- Define density as the amount of mass in a certain space, or mass per unit volume.
- Solve problems for density, mass, or volume using the equation $D = m/V$.
- Explain that density is an intensive property of matter that does not change at a given temperature, regardless of the amount of matter, and therefore can be used to identify a substance.

FOCUS ON UNDERSTANDING

- Students often believe that the density of a substance changes if the size or shape of the substance changes.
- Students might assume that density depends on the size of the sample, growing greater with larger, heavier samples.

KEY TERMS

density
intensive property
extensive property

IN CLASS

This lesson builds on the previous lesson. Students work in pairs to compare the objects they measured in the previous lesson. They use mass and volume data to calculate the density of each item. Density is introduced as an intensive property of matter. Students use density to determine the identities of the various rods and to determine whether the gold-colored penny created in Lesson 2: A Penny for Your Thoughts is real gold. A complete materials list is available for download.

TRM Materials List 5

Differentiate

Students often struggle with chemistry coupled with mathematical concepts. Direct students to **Math Spotlights: Solving Equations** on p. A-3 of the student textbook if they have difficulty with the density calculations. Direct students to **Math Spotlights: SI Units of Measure** on p. A-0 of the student textbook if they need help with units of measurement.

Pre-AP® Course Tip

Students preparing to take an AP® chemistry or college chemistry course should be able to evaluate evidence to answer a scientific question and use that evidence to justify claims. In Part 1 of the activity, students collect data to calculate the density of various objects. They use these data to determine the identity of each object. In Part 2, students use their data to justify the claim that an object's identity can be determined from its density.

Lesson Guide

ENGAGE (5 min)

TRM PowerPoint Presentation 5

ChemCatalyst

In the year 250 B.C.E., King Hiero commissioned a goldsmith to make a crown of pure gold. However, when he received the crown, Hiero suspected that the goldsmith had taken some of the gold and replaced it with a cheaper metal, even though it still weighed the same. He asked Archimedes to determine whether the crown was solid gold. How do you think Archimedes determined whether the crown was solid gold?

Sample answers: He could have examined it to see if it was the right color. He could have measured the mass and the volume of the crown and compared them to the mass and volume of some gold. They both should have the same mass per volume if they are the same substance.

➜ Brainstorm ideas on how to tell similar-looking substances apart. Leave the discussion open-ended.

Ask: How can you tell whether something is gold? (Color, hardness, shininess, etc.; compare it to a pure gold sample.)

Ask: Do you think a crown of gold and a crown of aluminum of identical size have the same volume? Would they have the same mass? (Same volume, but not same mass.)

Ask: According to legend, Archimedes submerged the crown in water. What was Archimedes measuring? (Volume.)

Ask: Could Archimedes determine the identity of the crown's metal only by putting it in water? Explain. (Maybe he would have to compare it to another gold sample with the same mass.)

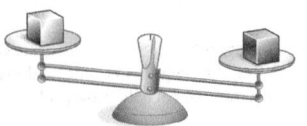

Example 1

Differences in Density

Consider the objects on balances shown in these illustrations. How can differences in density account for what you observe in these pictures?

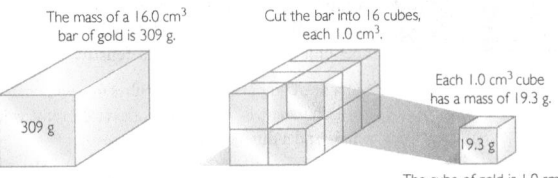

The balance is uneven. The two solid cubes have the same volume but different masses.

The balance is level. The two solid cubes have the same mass but different volumes.

Solution

Because the cubes on the first balance are exactly the same size, for the balance to be uneven one cube must be denser than the other. So the two cubes are made of different materials.

On the second balance, the larger cube must have a lower density than the smaller cube. Again, the two cubes are made of different materials.

⤴ Calculating Density

You can determine the density of a substance without actually cutting out a cubic centimeter sample. Suppose a large chunk of gold has a mass of 309 g and a volume of 16.0 cm³. Divide the mass of the gold by its volume to find its mass *per* cubic centimeter. This is equal to the density.

The mass of a 16.0 cm³ bar of gold is 309 g.

Cut the bar into 16 cubes, each 1.0 cm³.

Each 1.0 cm³ cube has a mass of 19.3 g.

309 g

19.3 g

The cube of gold is 1.0 cm³.

$$309 \text{ g} \div 16.0 \text{ cm}^3 = 19.3 \text{ g/cm}^3$$

The density of gold is 19.3 grams per cubic centimeter, or 19.3 g/cm³. This answer is rounded to three significant digits. [For review of this math topic, see **MATH Spotlight**: Accuracy, Precision, and Significant Digits on page A-1.]

Density

The mathematical formula for density is

$$D = \frac{m}{V}$$

where D is the density, m is the mass, and V is the volume.

TRM Worksheet with Answers 5

TRM Worksheet 5

INTRODUCE THE LAB

➜ Arrange students into pairs.

➜ Pass out the student worksheet.

EXPLAIN & ELABORATE (15 min)

DISCUSS DENSITY AS A PROPERTY OF MATTER

➜ Draw simple diagrams on the board to support the discussion.

Ask: How did the densities of the objects you measured compare to one another?

Ask: How can density be used as evidence to differentiate between substances?

Ask: If two objects have the same volume but different masses, which one is denser? (The object with the greater mass is denser.) What if they have the same mass but different volumes? (The object with the smaller volume is denser.)

Ask: What is the density of a brass rod twice as long as the one you had in class? (The density remains the same regardless of size.)

Key Points: Density can be defined as the amount of substance or mass in a certain space, or the mass per unit volume.

$$D = \frac{m}{V}$$

For purposes of comparison, scientists use 1 cubic centimeter, cm³, as the unit of volume being considered. However, instead of measuring a 1 cm³ cube, you can simply divide the entire mass of an object by its volume in cubic centimeters. For example, if an object has mass 6.0 g and total volume 4.0 cm³, its density can be calculated $D = m/V = 6.0 \text{ g}/4.0 \text{ cm}^3 = 1.5 \text{ g/cm}^3$.

Example 2

Density of a Gold Ring

Suppose you have a gold ring that weighs 7.50 g and has a volume of 0.388 mL. Does it have the same density as a big piece of gold?

Solution

Use the formula to find the density of the ring.

Start with the formula.
$$D = \frac{m}{V}$$

Substitute the values of m and V.
$$D = \frac{7.50 \text{ g}}{0.388 \text{ mL}}$$

Solve for D.
$$= 19.3 \text{ g/mL} = 19.3 \text{ g/cm}^3$$

The ring has the same density as any other sample of gold, big or small. The density of a material does not depend on the shape or size of the sample.

Identifying Matter Using Density

Densities of Some Metals

Metal	Density
copper	9.0 g/cm³
zinc	7.1 g/cm³
gold	19.3 g/cm³
lead	11.4 g/cm³
aluminum	2.7 g/cm³
brass	8.4 g/cm³

Important to Know The density of a substance does not change with its size or its shape. A solid gold bracelet will have the same density as a solid gold brick, and the density of one penny is the same as the density of two pennies.

Every substance has certain properties, such as color, hardness, melting point temperature, and density, that depend on the type of matter, not on the amount or size of the sample. These properties are called **intensive properties.** Intensive properties do not change if the quantity of the substance changes. Therefore, they can be used to help identify that substance.

On the other hand, **extensive properties,** such as mass or volume, do change depending on the amount of matter. Extensive properties alone can't be used to help identify a type of matter. For example, knowing that a metal earring has a mass of 3 g doesn't help you identify what the earring is made of.

Because density is an intensive property, it can be used to help identify the type of matter that an object or sample is made of. First, determine the density of the object and then compare it with known density values in a reference table like this one to help identify the type of matter.

Big Idea Each specific type of substance has a particular density. You can identify a substance by its intensive properties, including density.

THE FAKE BAR OF GOLD

If someone tried to trick you by coating a block of lead with a thin layer of gold, how could you prove the bar is a fake? One approach would be to scratch the bar to reveal that the inside isn't gold. Another method would be to use density to prove that the bar isn't gold.

Fake gold?
Density = ?
Volume = 647.5 cm³
Mass = 7500 g

A block of pure gold will have a density of 19.3 g/cm³.

Lesson 5 | **Density** 19

IDENTIFY SUBSTANCES USING DENSITY VALUES

→ Copy all or part of the density table from the next Key Point on the board.

Ask: Recall the ChemCatalyst question. Was it possible for Archimedes to determine the density of the crown? Why or why not? (Yes; he measured both mass and volume.)

Ask: How could you use density to determine if a crown is solid gold? (Compare its density to the density of pure gold—at least this way you would know if it is not gold.)

Ask: What does the evidence tell us about the gold-colored penny? (It is not pure gold.)

Ask: Do you have a hypothesis about what happened to the penny in Lesson 2: A Penny for Your Thoughts?

Key Points: Density can be used to identify what a substance is made of. The measurements and density calculations you made in class allowed you to identify the gold-colored and silver-colored rods as brass and aluminum. This table shows some common materials and their densities. Gold is the densest metal of those given in the table, even denser than lead.

Material	Density
zinc	7.1 g/cm³
paper	0.9 g/cm³
water	1.0 g/mL
aluminum	2.7 g/cm³
gold	19.3 g/cm³
brass	8.4 g/cm³
copper	9.0 g/cm³
lead	11.4 g/cm³

To determine the density of an object, both the mass and the volume are needed. Archimedes could find the density of the crown if he measured both its volume and its mass. He could then compare the density of the crown to the density of gold.

The density of one of our golden pennies turns out to be around 7.2 g/mL, or 7.2 g/cm³, too low a value to be gold, which has a density of 19.3 g/cm³. However, it appears that the pennies are not made entirely of copper either, because the density value is also lower than that of copper. *Note:* Pennies currently are minted with a thin copper

---

Density is an *intensive* property that does not change with the size, shape, or amount of substance being considered. *Intensive properties,* such as color, hardness, and density, do not depend on the amount, size, or shape of matter and can be used to help identify a substance. *Extensive properties,* such as mass and volume, can change depending on the amount of matter being considered. Mass or volume alone cannot help you identify a substance, but density can. Density changes with temperature. However, if they are at the same temperature, aluminum foil has the same density as an aluminum screw or an aluminum ladder. Some categories of substances

have a range of densities. For example, the densities of plastics or cooking oil vary depending on the type of plastic or oil.

Intensive property: A characteristic that does not depend on the size or the amount of matter.

Density: The mass of a substance per unit volume. $D = m/V$, where D is density, m is mass, and V is volume. Density is usually reported in g/cm³ or g/mL.

Extensive property: A characteristic that is specific to the amount of matter.

---

coating and a core of zinc. This is consistent with the density we measured, which is close to that of zinc, 7.1 g/cm³. We still do not know what the gold-colored coating is. At this point, we might speculate that the coating is brass, which is a combination of zinc and copper, because zinc was in the beaker.

REVIEW A DENSITY CALCULATION (OPTIONAL)

→ You might want to go over the last question from the worksheet to review unit conversion. Students can approach the unit conversions in several ways to arrive at the correct answer. Have students share how they solved the problem on the board or an overhead.

EVALUATE (5 min)

Check-In

In 1999, the U.S. Mint produced a coin called the Golden Dollar. It features an image of Sacagawea, the famous Native American guide for Meriweather Lewis and William Clark. It has a mass of 9.8 g and volume of 1.1 mL. Is this coin truly gold? Explain. (The density of gold is 19.3 g/mL.)

Answer: The density of the Golden Dollar is 9.8 g/1.1 mL or 8.9 g/mL. This is much less than the density of gold, so the Golden Dollar is not solid gold.

Homework

Assign the reading and exercises for Alchemy Lesson 5 and the Chapter 1 Summary in the student text.

LESSON 5 ANSWERS

1. *Possible answer:* Density is the mass of an object divided by its volume.

2. If the density of the penny (its mass divided by its volume) is not equal to the density of gold (19.3 g/cm³), then the penny is not made of solid gold.

3. The density of aluminum is less than the density of gold. More matter is present in a given volume of gold than in the same volume of aluminum.

4. C

5. *Possible answer:* The object that has a density of 2.7 g/cm³ has a larger volume than the object with a density of 8.4 g/cm³. The two objects have the same mass, but the mass is packed into a smaller space in the denser object.

How can you use mass and volume to determine the identity of a substance?

KEY TERMS

density

intensive property

extensive property

To identify substances, you examine their intensive properties, qualities that do not depend on size or amount. Intensive properties include color, hardness, and density. Density is the mass of a substance per unit of volume. If two substances have different densities, then they are probably made of different types of matter.

Exercises

Reading Questions

1. In your own words, define density.
2. Explain how density can be used to determine if the golden penny is made of solid gold.

Reason and Apply

3. How does the density of aluminum compare with the density of gold? What does this tell you about the amount of matter within each?
4. If two objects have the same mass, what must be true? Choose the correct answer(s).
   (A) They have the same volume.
   (B) They are made of the same material.
   (C) They contain the same amount of matter.
   (D) They have the same density.
5. Two objects each have a mass of 5.0 g. One has a density of 2.7 g/cm³ and the other has a density of 8.4 g/cm³. Which object has a larger volume? Explain your thinking.
6. A piece of metal has a volume of 30.0 cm³ and a mass of 252 g. What is its density? What metal do you think this is?
7. A glass marble has a mass of 18.5 g and a volume of 6.45 cm³.
   a. Determine the density of the marble.
   b. What is the mass of six of these marbles? What is the volume? What is the density?
   c. How does the density of one marble compare with the density of six of the marbles?

---

6. 8.40 g/cm³. The metal could be brass because the table shows that the density of brass is 8.4 g/cm³. Brass is a possible identification based on the table, but note that density alone is not adequate to identify a material because other materials could have the same density.

7. a. 2.87 g/cm³. **b.** 38.7 g/cm³. **c.** The density of six glass marbles is the same as the density of one glass marble because density is an intensive property of the glass in the marbles.

CHAPTER 1

Defining Matter

SUMMARY

KEY TERMS

hypothesis
property
chemistry
matter
mass
volume
meniscus
water displacement
density
intensive property
extensive property

Alchemy Update

Can a copper penny be turned into gold through chemical processes?

 ?

The "golden" penny looks like gold and has almost exactly the same volume as a true gold penny. However, the golden penny has a much lower mass than a true gold penny. This means the density of the "golden" penny is less than the density of gold. It cannot be a solid gold penny.

The concept of density has provided the evidence needed to confirm that the gold-colored penny is not actually made of gold. But if it is not gold, then what is it? Many questions remain to be answered. Is it still possible to make gold some other way? And what is it about gold that makes it gold?

REVIEW EXERCISES

1. Explain how you would determine the volume of a powdered solid, a liquid, and a rock.

2. Use your own words to define *matter*.

3. Will an object with a higher density displace more water than an object with a lower density? Explain why or why not.

4. How does the density of one penny compare with the density of two pennies?

1 Penny
Mass = 2.6 g
Volume = 0.36 cm³

2 Pennies
Mass = 5.2 g
Volume = 0.72 cm³

5. A small pebble breaks off of a huge boulder. The pebble has the same density as the boulder. In your own words, explain how this can be true.

6. Archeologists discover a silver crown in an ancient tomb. When they place the crown in a tub of water, it displaces 238.1 cm³ of water. The density of silver is 10.5 g/cm³. If the crown is really silver, what will its mass be?

ASSESSMENTS

Two multiple-choice assessments (versions A and B) and two short-answer assessments (versions A and B) are available for download for Chapter 1.

TRM Chapter 1 Assessment Answers

TRM Chapter 1 Assessments

ANSWERS TO CHAPTER 1 REVIEW EXERCISES

1. *Possible answer:* Determine the volume of a powdered solid or of a liquid by pouring the substance into a graduated cylinder or beaker and reading the markings on the side. Determine the volume of a rock by submerging the rock in a graduated cylinder partially filled with water and then reading how much the water level changes.

2. *Possible answer:* Matter is anything that has mass and takes up space.

3. Density is no help in determining which object will displace more water. A large object will displace more water than a small object no matter how dense the two objects are.

4. 7.2 g/cm³. The density of one penny is the same as the density of two pennies.

5. *Possible answer:* The density is the same because the pebble is made of the same material as the boulder. Although the pebble has a smaller mass, it also has a smaller volume, so the density can be the same.

6. 2500 g.

PD CHAPTER 2 OVERVIEW

Watch the video overview of Chapter 2 (for teachers) by clicking on the link in the TE-book, opening the TRFD, or logging onto the book's companion Web site bcs.whfreeman.com/livingbychemistry2e (teacher log-in required).

Chapter 2 consists of five lessons, one of which is a formal lab. Together, these lessons introduce the language of chemistry and launch a discussion of the elemental building materials of matter. In Lesson 6, students look for patterns in the chemical names and symbols used to represent substances. In Lessons 7 and 8, the class discovers the usefulness of chemical symbols in tracking copper atoms as they are combined with various substances and then recovered again in a lab, which provides evidence of the immutable nature of copper. Lesson 9: Create a Table is a card-sorting activity that introduces students to the periodic table of the elements. In Lesson 10, students learn more about the patterns and information found in the periodic table.

In this chapter, students will learn

- how to decipher basic chemical names and symbols
- how to follow a lab procedure
- to make and record accurate observations
- how to describe a basic chemical reaction
- that matter cannot be created or destroyed
- how the periodic table of the elements is organized
- to recognize patterns related to elemental properties

Images & Volcano/Science Source

CHAPTER 2

Basic Building Materials

In this chapter, you will study

- how chemists use chemical formulas to track changes in matter
- how matter is conserved in chemical processes
- the organization of the elements into the periodic table

Chemists separate substances into their simplest components in order to understand the nature of matter. Shown here is brightly colored, almost pure sulfur found near hot springs.

Matter is composed of components called elements. There are a limited number of elements in existence. Everything in our world, natural and synthetic, is made of individual elements or some combination of them. Chemical formulas indicate what element or elements a substance is made of.

22

Chapter 2 Lessons

A New Language

Chemical Names and Symbols

THINK ABOUT IT

There are two bottles on a shelf in a chemistry lab. Each contains a substance that resembles diamonds. Bottle 1 is labeled C(s), and Bottle 2 is labeled ZrO_2(s). Does either bottle contain diamonds? Do both bottles contain diamonds?

What do chemical names and symbols tell you about matter?

To answer this question, you will explore

- The Language of Chemistry
- Names and Symbols
- Physical Form

EXPLORING THE TOPIC

The Language of Chemistry

Some chemical names are used in daily language, such as aluminum and iron, which cookware and machinery are made of, and ammonia, which is used for cleaning. Other chemical names, such as sodium chloride for salt and calcium carbonate for chalk, are used mainly by chemists.

In the illustration, the symbol C is used to represent carbon and the symbol O is used to represent oxygen.

O_2(g), N_2(g), and CO_2(g) oxygen, nitrogen, and carbon dioxide in air

H_2O(l) water in clouds and puddles

H and He hydrogen and helium in the sun

SiO_2(s) silicon dioxide in windows

$CaCO_3$(s) calcium carbonate in stucco wall

$ZnCrO_4$(s) zinc chromate in paint primer

Fe(s) iron in steel car and sign

CO(g) carbon monoxide in exhaust

Al(s) aluminum in hubcaps

Ni(s) nickel in coin

Overview

ACTIVITY: GROUPS OF 8

Key Question: What do chemical names and symbols tell you about matter?

KEY IDEAS

Learning chemistry can be compared to learning a new language. At first, the chemical names and symbols can be confusing and complex. However, patterns emerge when the names and appearances of substances are examined along with their chemical formulas. Students have to become fluent in this chemical language to succeed in chemistry.

LEARNING OBJECTIVES

- Define the terms *element, compound,* and *aqueous.*
- Determine whether a substance is an element or a compound, based on its chemical formula or symbol.
- Decipher some basic chemical formulas and symbols.

FOCUS ON UNDERSTANDING

- Students might confuse uppercase and lowercase letters in chemical formulas, not knowing where one symbol stops and another begins.
- Even after a lot of exposure to chemical formulas, students may be confused as to whether a subscript

number applies to the element before it or after it in the formula.
- While aqueous is not considered a phase of matter, it is a distinct category of physical form. Aqueous sodium chloride is a dramatically different substance from liquid sodium chloride.

KEY TERMS

element
chemical symbol
compound
chemical formula
phase
aqueous

IN CLASS

Students examine 18 vials containing substances that have been labeled with their chemical name, chemical formula, and phase. The students work in groups to look for patterns in the symbols and substances (e.g., every substance with the symbol (s) after its formula is a solid). Chemical symbols and chemical formulas are defined. Many of these substances will be seen again in Lesson 7: Now You See It. A complete materials list is available for download.

TRM Materials List 6

Differentiate

To help all students build scientific literacy and to support English-language learners, give groups of students the task of creating posters to hang in the classroom that explain several chemical formulas such as $Cu(NO_3)_2$(aq), Zn(s), and H_2O(l). Help students understand the scientific observation of "clear," a term often incorrectly applied by students to mean colorless, by examining vials 9 and 17 together as a class.

SETUP

Prepare a set of vials according to the table on the bottom of page 24. Label each vial with the vial number, chemical name, and chemical formula of substances in the vial. Label vial 18 Unknown. In vial 18, place a mixture of copper (II) sulfate solution, $CuSO_4$(aq) and copper, Cu(s).

Two groups can share each set of vials, or you can prepare one set of vials and divide them among six to eight stations. The former is recommended, because it allows students to examine and compare all the vials together. For eight groups, you will have to prepare four sets of 18 glass vials with screw tops.

Label and fill each vial according to the chart. The exact amounts are not critical; there just has to be enough for students to see. For solids, around 5 g is sufficient for each vial. For an aqueous solution, the concentration is not important. We recommend dilute solutions with just enough compound dissolved for color to be noticed easily. Vial 18 should contain a mixture of copper powder and copper (II) sulfate solution.

CLEANUP

Vials can be used by many classes and stored for future use.

Lesson Guide

ENGAGE (5 min)

TRM PowerPoint Presentation 6

ChemCatalyst

Two bottles are on a shelf in a chemistry lab. Both contain a shiny yellow metal. Bottle A is labeled $Au(s)$. Bottle B is labeled $FeS_2(s)$.

1. What do you think the symbols on the bottles mean?

2. Do you think both bottles contain gold? Why or why not?

Sample answers: **1.** Students may be able to translate Au as gold and FeS_2 as iron sulfide, or fool's gold. **2.** Students may say that the substances are different because the labels are different.

Ask: What do the symbols tell you about the contents of the two bottles?

Ask: Why are the symbols useful to a chemist? (Chemists all around the world can understand the same symbols; the symbols are short and quicker to write than the names.)

↻ Names and Symbols

Elements are the building materials of all matter. In total, there are about 115 known elements. A few are shown here.

Metals

Mercury, Hg *Copper, Cu* *Gold, Au* *Sodium, Na* *Iron, Fe*

Nonmetals

Carbon, C *Chlorine, Cl* *Iodine, I* *Phosphorus, P*

From top left: Harry Taylor/Getty Images, Dmitry Gool/iStockphoto/Thinkstock, Dirk Wiersma/Science Source, Martyn F. Chillmaid/Science Source, Sawayasu Tsuji/iStockphoto, Kae Deezign/Shutterstock, Charles D. Winters/Science Source, Andrew Lambert Photography/Science Source, Charles D. Winters/Science Source

Here are some things you might notice. The **chemical symbols** for the elements consist of one or two letters. The first letter is always capitalized. If there is a second letter, it is lowercase. An element can be a solid, liquid, or gas. Some elements are metals, some are not. Sometimes, the symbol for an element is an abbreviation for its name. Other times, the symbol comes from another source, such as a word in another language. For example, the symbol for iron is Fe from the Latin word *ferrum,* and the symbol for gold is Au from the Latin word *aurum.*

Elements combine in specific ratios to form **compounds.** A compound is represented by a **chemical formula.** For example, sodium chloride, or table salt, is NaCl, which tells you that salt is made of the elements sodium, Na, and chlorine, Cl, in a 1:1 ratio. The chemical formula for carbon dioxide is CO_2. The subscript number "2" in the formula indicates that the elements carbon and oxygen are combined in a 1:2 ratio. (If the subscript is 1, you normally don't write it.)

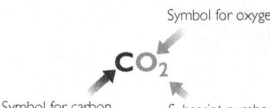

Symbol for oxygen

CO_2 *Chemical formula for carbon dioxide*

Symbol for carbon Subscript number

Compounds can differ greatly in appearance and behavior from the elements that they are composed of. For example, sodium is a shiny metal and chlorine is a gas. They are very different from sodium chloride, table salt.

Big Idea All matter is made up of elements or compounds, or mixtures of these.

CONSUMER CONNECTION

Neon is the gas used in neon signs. The gas is colorless but glows a bright orange-red when an electric current is run through it. Neon gas is put into glass tubes that have been bent and shaped to create colorful signs.

photovideostock/iStockphoto

Contents of Vials

Vial	Chemical name	Chemical formula	Vial	Chemical name	Chemical formula
1	sodium nitrate	$NaNO_3(aq)$	10	copper	$Cu(s)$
2	copper (II) nitrate	$Cu(NO_3)_2(s)$	11	sodium hydroxide	$NaOH(aq)$
3	copper (II) hydroxide	$Cu(OH)_2(s)$	12	copper (II) oxide	$CuO(s)$
4	hydrogen	$H_2(g)$	13	water	$H_2O(l)$
5	sodium nitrate	$NaNO_3(s)$	14	sulfuric acid	$H_2SO_4(aq)$
6	sodium hydroxide	$NaOH(s)$	15	zinc sulfate	$ZnSO_4(aq)$
7	copper (II) sulfate	$CuSO_4(s)$	16	copper (II) sulfate	$CuSO_4(aq)$
8	zinc	$Zn(s)$	17	copper (II) nitrate	$Cu(NO_3)_2(aq)$
9	nitric acid	$HNO_3(aq)$	18	Unknown	Unknown

These minerals contain copper, Cu. Notice that not one of them is copper colored.

Copper oxide,
$Cu_2O(s)$

Copper sulfate,
$CuSO_4(s)$

Copper sulfide,
$CuS(s)$

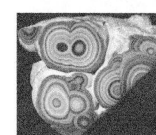
Copper carbonate hydroxide,
$Cu_2(CO_3)(OH)_2(s)$

Water is unusual, because we can encounter it in three phases of matter all at once: liquid, solid ice, and gas in the atmosphere.

⟳ Physical Form

Elements and compounds can exist as solids, liquids, or gases. These forms are called the **phases** of matter, and are represented by using (s), (l), or (g) after the chemical formula. For example, water in the gas phase is written as $H_2O(g)$, water in the liquid phase is written as $H_2O(l)$, and water in the solid phase as ice is written as $H_2O(s)$.

There is another symbol for physical form, the symbol (aq) for **aqueous**. A substance is aqueous when it dissolves and forms a clear mixture with water. Many familiar liquids are actually aqueous solutions, such as grape juice, vinegar, and ocean water.

> **Important to Know** When solid sugar, $C_{12}H_{22}O_{11}(s)$, appears to "melt" in your mouth, it does not become liquid sugar, $C_{12}H_{22}O_{11}(l)$. It dissolves and becomes aqueous sugar, $C_{12}H_{22}O_{11}(aq)$.

An orange is a tasty mixture of compounds.

Most substances are a mixture of compounds. For example, an orange is a mixture of water, H_2O, fructose, $C_6H_{12}O_6$, citric acid, $C_6H_8O_7$, limonene, $C_{10}H_{16}$, and many other compounds. Notice that in mixtures, the components are *not* combined in a specific ratio. Some oranges are sweeter or more sour or more juicy because they have more fructose, citric acid, or water than other oranges. But the ratios of the elements in compounds never change. Water is always water, H_2O.

LESSON SUMMARY

What do chemical names and symbols tell you about matter?

KEY TERMS
element
chemical symbol
compound
chemical formula
phase
aqueous

All matter is composed of elements. Each element has a unique symbol. Elements combine with one another to form compounds. A chemical formula specifies which elements are present in a compound. The letters in chemical formulas are symbols for the elements, and the subscript numbers indicate the amount of each element in that substance. A chemical formula may also include the lowercase letters (s), (g), (l), and (aq), which stand for solid, gas, liquid, and aqueous.

Lesson 6 | **Chemical Names and Symbols**   25

Ask: What elements combine to form water? Nitric acid? Sodium hydroxide?

Ask: What symbols represent sulfate? Nitrate? Hydroxide? (SO_4, NO_3, OH)

Key Points: All matter in the universe either is an element or is made of some combination of elements. Elements can be thought of as the building materials for matter. There are approximately 114 known elements, each represented by a symbol. A substance that consists of two or more elements bound together is called a *compound*.

> **Element:** A unique substance that cannot be broken down into simpler substances through chemical processes. Elements serve as the building materials of all matter.
>
> **Compound:** A pure substance that is a chemical combination of two or more elements in a fixed ratio.

The first letter of an element's symbol is always uppercase, and if there is a second letter, it is always lowercase. The element Co, cobalt, is different from CO, carbon monoxide. The latter is made up of one carbon atom and one oxygen atom. A chemical symbol stands for a single element. Compounds are represented by chemical formulas. For example, copper oxide, CuO, is made up of the elements copper, Cu, and oxygen, O.

> **Chemical formula:** A combination of symbols and subscripts that indicates the number and types of elements in a compound.

Some common groupings of elements have their own specific names. For instance, when O and H appear together as OH, the grouping is named *hydroxide*. Likewise, N and O together as NO_3 are called *nitrate*. The small subscript numbers indicate the proportion of each element. For example, NO_3 has 3 times as much oxygen as it has nitrogen. If there is no subscript, the subscript is understood to be 1.

DISCUSS THE LETTERS IN PARENTHESES: (s), (l), (g), AND (aq)

Ask: What is $H_2O(l)$? $H_2O(g)$? $H_2O(s)$? (water, water vapor, ice)

Ask: What do all the flasks with (aq) on them have in common? What do you think aqueous means? (The root of aqueous is *aqua*. It means that the substance is dissolved in water.)

EXPLORE (20 min)

TRM Worksheet with Answers 6

TRM Worksheet 6

INTRODUCE THE ACTIVITY

→ Arrange students into groups of 8.

→ Pass out the student worksheet and sets of vials.

→ Eight students can share each set of vials, or students can rotate through stations. Do not distribute the eighteenth vial until later in the activity.

→ Students will need to use their completed worksheet later for Lesson 8: What Goes Around Comes Around.

EXPLAIN & ELABORATE (15 min)

DISCUSS THE LANGUAGE OF CHEMISTRY

→ Write several chemical formulas on the board and dissect them together as a class.

Symbol for sulfur   Symbol for oxygen

$$CuSO_4(aq)$$

Symbol for copper   Subscript for oxygen

Ask: Is copper sulfate an element or a compound? How do you know?

Ask: How are NaOH(*s*) and NaOH(*aq*) the same? How are they different?

Key Points: Substances come in different physical forms, called *phases*. Most of us are familiar with the solid, liquid, and gas phases of matter. Ice is what we normally call solid water. Water vapor is gaseous water.

> **Phase:** The physical form a substance is in, such as solid, liquid, or gas. Phase is greatly influenced by temperature.
>
> **Aqueous:** A substance is aqueous when it is dissolved in water.

The ocean is made mostly of water with salt, NaCl(*aq*). Many things we think of as liquids are aqueous solutions, such as filtered apple juice, vinegar, and soda pop. Pure liquids, such as cooking oil, are less common.

COMPARE VIAL ORGANIZING SCHEMES (OPTIONAL)

Ask: How would you organize the 18 vials?

Key Point: There are many ways to group and classify substances. One organizing approach would be to place substances with similar endings together (e.g., sulfate, nitrate, and so on). Another organizing scheme would be to place all the copper substances together and all the zinc substances together (both elements and compounds). Other possibilities are to sort by solid, liquid, gas, and aqueous or to sort by appearance. In stockrooms, chemicals that might react with each other to form dangerous products are kept apart.

EVALUATE (5 min)

> **Check-In**
>
> Imagine that you find a flask labeled $Na_2SO_4(aq)$. What does the label tell you about what is in this flask?

Answer: The substance is sodium sulfate, Na_2SO_4. It is made of sodium, sulfur, and oxygen. The symbol (*aq*) indicates that the sodium sulfate is dissolved in water.

Homework

Assign the reading and exercises for Alchemy Lesson 6 in the student text. Tell students to be sure to complete exercise 6, which is a pre-lab for the next class. Tell them to be prepared to

---

Exercises

Reading Questions

1. Describe the difference between an element and a compound.
2. What is meant by physical form?

Reason and Apply

3. How many elements are included in the chemical formula for sodium nitrate, $NaNO_3$? Name them.
4. What is the difference between NaOH(*s*) and NaOH(*aq*)?
5. You see a ring with a stone that looks like a diamond but wonder why it's so cheap. The jeweler says the stone is a type of diamond called cubic zirconia. How can chemical symbols show that cubic zirconia is not a diamond?
6. You find two containers on a chemical shelf, one labeled $Cu_2O(s)$ and a second labeled CuO(*s*). Are these substances the same or different? Explain.

---

pass a lab readiness quiz at the next class before they begin the lab.

LESSON 6 ANSWERS

1. Elements are the building materials of all matter. A compound is matter that is made up of two or more elements combined in a specific ratio. An element cannot be broken down into simpler substances by chemical means, but a compound can be broken down into elements.

2. The physical form of a substance is its phase: gas, liquid, or solid. An aqueous solution is another physical form.

3. The chemical formula for sodium nitrate, $NaNO_3$, indicates that it has three elements: sodium, Na, nitrogen, N, and oxygen, O.

4. NaOH(*s*) is the symbol for solid sodium hydroxide. NaOH(*aq*) indicates that the sodium hydroxide is dissolved in water.

5. The chemical formula for cubic zirconia is ZiO_2. The chemical formula for diamond is C. The stone cannot be a diamond because it has a different chemical formula than a diamond.

6. Assuming that the contents of the containers are labeled correctly, these substances are two different compounds that are in the solid phase. They each have a different ratio of copper to oxygen in the chemical formulas (Cu_2O is copper (I) oxide and CuO is copper (II) oxide).

Now You See It

The Copper Cycle

THINK ABOUT IT

At this point, you are probably convinced that a copper penny cannot be transformed into real gold in a high-school chemistry classroom. However, the question remains, what happened to the copper penny?

What happens to matter when it is changed?

To answer this question, you will explore

- The Copper Cycle
- Evidence of Chemical Change
- Interpreting Observations

EXPLORING THE TOPIC

The Copper Cycle

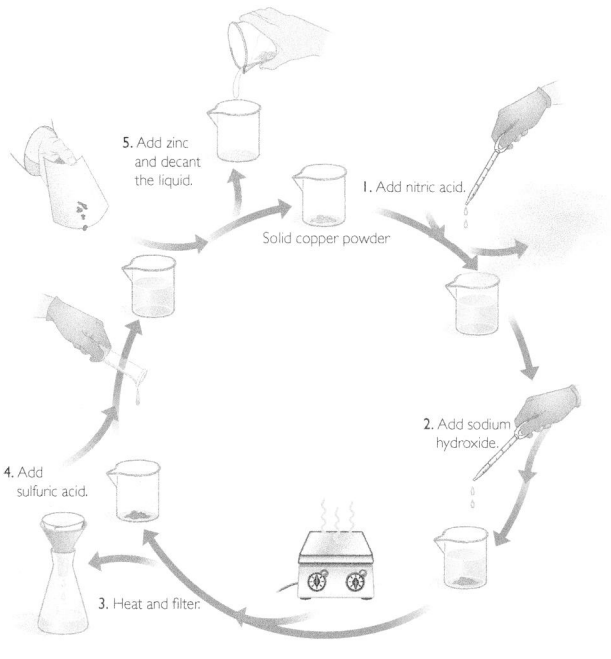

5. Add zinc and decant the liquid.

1. Add nitric acid.

Solid copper powder

2. Add sodium hydroxide.

4. Add sulfuric acid.

3. Heat and filter.

Lesson 7 | **The Copper Cycle**   27

---

tell them that this means "8 molar" and indicates the concentration. Molarity is covered in Unit 4: Toxins.)

KEY TERM

chemical change
chemical reaction

IN CLASS

Elemental copper is taken through a series of chemical reactions and then recovered. New substances with distinct properties are formed as a result of each chemical change along the way. In the discussion that follows the lab, students consider what happened to the copper during the procedure. The full debriefing is left for the next lesson. Students will read about chemical change in the text. This experiment will take two 50-min class periods or one block period to complete. Students will need to use their completed worksheet later in Lesson 8: What Goes Around Comes Around. A complete materials list is available for download.

TRM   Materials List 7

Pre-AP® Course Tip

Students preparing to take an AP® chemistry course or college chemistry should be able to evaluate the degree to which a particular set of observations indicates that the process is chemical versus physical. In this lesson, students conduct an experiment and record their observations for each step. By the end of the lab, the students should be able to articulate their data and discuss what happens to matter when it is changed whether the changes are physical and/or chemical.

SETUP

Have one set of equipment up front to demonstrate safety and measurement, plus a watch glass to show the accumulation of NO_2 in the beaker. For safety reasons, you may want to perform the second step for your students, either in a fume hood or outdoors.

LAB: GROUPS OF 4

Key Question: What happens to matter when it is changed?

KEY IDEAS

Copper, and most other elements, can be observed either in elemental form or as part of different compounds. Although these forms have very different properties, the element can always be recovered from the compound. Chemical changes do not consume elements—they simply rearrange them.

LEARNING OBJECTIVES

- Follow a lab procedure safely.
- Describe a chemical change or chemical reaction.

FOCUS ON UNDERSTANDING

- When a solid shows up in a solution, it is often observed as cloudiness or as a suspension. It might not fit students' more literal definition of a solid.
- Some students might incorrectly think that the copper element leaves the beaker when the brown gas escapes.
- Students might mistake the label 8 M for some form of volume measurement. (If you wish, you can

You can set up enough chemicals for several classes at different stations and have students rotate through. One station should be a fume hood. At the stations put 1 g copper, 1 g zinc, 20 mL of 8 M NaOH, 20 mL of 8 M HNO_3, and 150 mL of 1 M H_2SO_4. Each team will use only 0.1 g copper ($+/- 0.02$ g). To save time, you can preweigh the copper samples for each group.

CLEANUP

Put all the used filter papers into a plastic-lined box to dry. Pour all the liquid into a big jar and allow it to evaporate for a few months until a solid remains. The dry toxic solids and papers can then be disposed of according to local guidelines.

Lesson Guide

ENGAGE (10 min)

TRM PowerPoint Presentation 7

ChemCatalyst

Answer these safety questions about the lab you will be doing today. If you cannot answer these questions, go back and reread the procedure.

1. Name three things that could be dangerous in this lab.

2. Why must part of the lab be done in a fume hood?

3. What do you have to put on before starting the lab? Why?

Sample answers: **1.** The acids, the hot plate, the hot liquid, and the brown gas, NO_2, are all dangerous. **2.** A toxic gas, NO_2, is produced. **3.** Goggles must be worn to protect the eyes from acids, hot liquids, and the possibility of broken glass.

CHECK STUDENTS' LAB READINESS

→ Rather than discussing the answers to the ChemCatalyst questions, spot-check answers to ensure that everyone is prepared for the lab today. If not, the group should reread the procedure and check in with you a second time before doing the lab. More questions here.

Ask: Why is sulfuric acid dangerous? (It burns skin, eyes, clothing.)

Ask: What can you do to protect yourself and others when working with sulfuric acid? (Wear goggles, roll up sleeves, report and clean up spills.)

BIOLOGY CONNECTION

Octopus blood is blue-green in color. It contains the copper-rich protein hemocyanin as opposed to the iron-rich hemoglobin found in vertebrates.

To make sense of the copper penny experiment, it is important to learn more about how copper can be transformed. Copper itself is an element, represented by the chemical symbol Cu(s). As a solid powder, copper has a distinctive orange-brown appearance.

In class, you transformed copper powder through a series of chemical reactions. If all worked well, copper should have reappeared at the end of the experiment. But what happened between the beginning and the end?

The diagram on page 27 shows the steps of the copper cycle. Start at the top and follow the arrows.

Evidence of Chemical Change

The five steps of the copper cycle are shown in these photos. In each step, a substance is changed into a different substance by mixing it with a liquid or by heating it. These changes are called **chemical changes** or **chemical reactions.** You can usually tell when a chemical reaction occurs because there is evidence of new substances forming. For instance, in class you observed some color changes when two of the liquids turned a bright blue color. You also observed the release of a brown gas and the formation of a dark-colored solid. These are all possible signs that new substances are forming. After a chemical reaction, you have new substances with properties that are different from the starting substances.

Copper powder

Step 1: After adding nitric acid

Step 2: After adding aqueous sodium hydroxide

Step 3: After heating

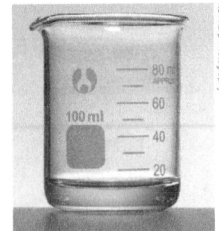
Step 4: After adding sulfuric acid

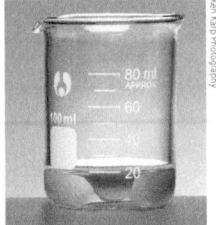
Step 5: After adding zinc

28   Chapter 2 | **Basic Building Materials**

Ask: What is the safest way to pick up a hot beaker? (Use beaker tongs.)

Ask: What do you need to do before you add nitric acid to the copper? Why? (Move to the fume hood; it gives off a toxic gas, NO_2.)

Ask: What should you do if you get acid on your skin or clothes? (Report it, and wash with lots of running water.)

EXPLORE (75 min)

TRM Worksheet with Answers 7

TRM Worksheet 7

TRM Handout—The Copper Cycle 7

INTRODUCE THE LAB

→ Arrange students into groups of four.

→ Pass out the student worksheet and handout.

→ This experiment will take two 50-min class periods to complete.

→ Instruct students to pick up their materials or physically rotate through stations (at your discretion).

→ Let students know what the stopping point should be for today.

→ Review the safety guidelines. You might want to complete the fume hood step for each group while the group observes, to avoid traffic around the hood.

○ Interpreting Observations

Scott P. Orr/Stock Photo

Look again at the first step in the copper-cycle diagram. When nitric acid was added to the copper, a blue liquid formed and a toxic brown gas left the beaker. But where did the copper powder go? You may have wondered if the copper left the beaker with the brown gas. However, copper reappeared at the end of the experiment. This is evidence that the copper doesn't leave the beaker at any point in the cycle.

Was copper put back in again? If copper was not added in again and it did not leave with the brown gas, the only other explanation is that it never left the beaker at all. This would mean that the copper is present in copper-containing compounds at each step of the cycle. Copper must be present somewhere in the blue solutions and the blue solid that were observed. So, the copper was not created or destroyed as a result of its chemical journey.

Over many centuries, chemists have carried out procedures like the copper cycle. In doing so, they have learned about the properties of substances and even discovered new substances. They have also learned how to predict change. Chemists know that pouring nitric acid on copper produces a clear blue liquid and a brown gas. They have even determined the identity of the blue liquid and the brown gas. Perhaps further investigation will allow you to track the changes to the copper penny as well.

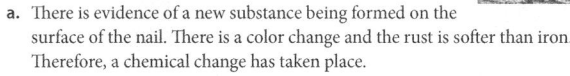

Example

Rusting Nail

An iron nail that stays in contact with water and air starts to form a reddish-brown coating called rust.

   a. Is this a chemical change?

   b. Is rust an element, a compound, or a mixture? Explain your thinking.

   c. How could you gather more evidence to support your answer to part b?

Solution

Rust forms on the surface of the nail, so you can deduce that it is probably caused by exposure to water and air.

   a. There is evidence of a new substance being formed on the surface of the nail. There is a color change and the rust is softer than iron. Therefore, a chemical change has taken place.

   b. Rust is probably a compound, a combination of the iron from the nail plus one or more elements from the air or water.

   c. To gather more evidence for chemical change, you could complete several experiments. For example, you could place a nail in pure water or in pure oxygen to test what the iron is reacting with. (Scientists have done this and found that iron rusts only in the presence of both oxygen and water. They have determined that rust is actually iron oxide hydroxide.)

Lukas Hejtman/Outterstock

Lesson 7 | **The Copper Cycle**   29

SAFETY

- Wear safety goggles at all times.
- Be very careful handling the concentrated nitric acid and concentrated sodium hydroxide, because they will burn any exposed skin. If some gets on your skin, wash the area immediately with water and inform your teacher.
- Use the fume hood when you add the nitric acid to the copper. Nitrogen dioxide, $NO_2(g)$, will be produced. It is a poisonous gas and should not be inhaled.

- When using the hot plate, set it at a medium setting (e.g., setting 4 out of 10). Be careful not to splash when you stir the chemicals.
- Be careful when handling the sulfuric acid. If you get it on your skin, wash it off with plenty of water and inform your teacher.

→ Inform students that when they measure out the copper powder, it is not necessary to have exactly 0.10 g. Plus or minus 0.02 g is fine.

→ Demonstrate the nitric acid procedure in the fume hood. Place a

watch glass over the beaker to show the brown fumes of nitrogen dioxide, $NO_2(g)$.

→ Show students how to remove the beaker from the hot plate with beaker tongs. Remind them to let the beaker cool before going on to the next step.

→ Demonstrate how to set up a funnel and filter paper and how to filter: First, fold the filter paper in half. Fold in half again. Open into a cone. Place filter into funnel.

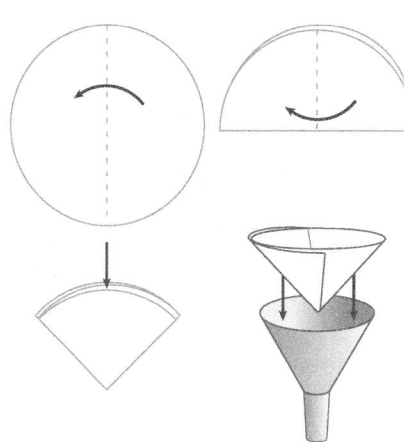

→ Review the process of making scientific observations. Ask students to describe what they see, hear, and smell as clearly, completely, and objectively as possible.

EXPLAIN & ELABORATE (5 min)

DISCUSS STUDENTS' OBSERVATIONS

Ask: What did you observe at each stage of the experiment? (Students can check the consistency of their results with the rest of the class.)

Ask: Did you have any difficulties during the lab?

Ask: How did you end up with copper again? What is your hypothesis? (If they have any doubts that it is copper, the whole experiment can be repeated with the resulting copper to show that what they ended up with is the same as what they started with.)

EVALUATE (5 min)

Check-In

You could see copper only in the first and last steps of today's lab. Where do you think the copper was the rest of the time? Be specific.

Answer: Answers will vary.

Homework

Assign the reading and exercises for Alchemy Lesson 7 in the student text. If you want, you can have students start their lab reports. Instructions are given in the exercises for Alchemy Lesson 8 in the student text.

LESSON 7 ANSWERS

1. *Possible answer:* A chemical reaction is a change leading to the final substance or substances being different from the original substance or substances. Some of the signs that a new substance has formed include color changes, formation of a new solid, formation of a gas, the release of energy as heat or light, an increase or decrease in the mass of material, and a change in the texture of a material.

2. *Possible answer:* The copper would always be present throughout the series of reactions. The copper could be in a different form after each reaction. For example, the copper could become part of a compound, then be dissolved in water, and then be extracted and returned to its pure form.

3. *Possible answer:* A chemical reaction combined the zinc with part of the dissolved copper compound. The resulting zinc compound was also dissolved in water. The zinc was still present, but not as a pure element. In fact, when the zinc was added, it formed a colorless solution of zinc sulfate and the copper came out of the solution as a precipitate.

4. Yes, the ingredients undergo a chemical change. The cookies are a different color and consistency from the cookie dough. These changes are present even after the cookies cool to room temperature. As the cookies bake, the dough absorbs heat energy, which causes a chemical change in the ingredients of the dough. The composition of the cookies is very different from the original ingredients and from the dough.

5. a. The baking soda is a solid, the vinegar is a liquid, the clear colorless liquid is a liquid, and the CO_2 is a gas.
b. Yes, the production of carbon dioxide gas is evidence that a chemical change has occurred after the original liquid and solid substances were mixed. The solution also becomes colder and the odor of the vinegar disappears—other indications of the chemical change.
c. Before the change, the sodium is in the solid baking soda. After the change, the sodium must be in the clear colorless liquid. In fact, when baking soda is mixed with vinegar, the result is sodium acetate, $NaCH_3COO$, dissolved in water.

6. *Possible answer:* Copper ore is crushed into a powder and changed to a copper sulfate solution by a chemical reaction. When an electric current passes through the solution, the copper metal comes out of the solution as copper deposited on a metal plate.

What happens to matter when it is changed?

KEY TERMS
chemical change
chemical reaction

When one substance is transformed, or changed, into another substance, a chemical change has occurred. A chemical change is also called a chemical reaction. When chemical reactions occur, new substances with new properties are produced. Evidence of new substances, such as the production of a gas, a color change, or the formation of a solid in solution, is a sign that a chemical change has taken place. When substances undergo chemical changes, the elements themselves are not created or destroyed.

Exercises

Reading Questions

1. Explain what a chemical reaction is. What are some possible signs that a chemical reaction is taking place?

2. In your own words, explain how it might be possible to start with copper and end up with copper after a series of chemical reactions.

Reason and Apply

3. Look again at the copper-cycle diagram. After adding zinc, the blue solution turns colorless, and copper appears as a solid. What do you think happened to the zinc? Explain your thinking.

4. Suppose you use sugar, butter, eggs, and flour to make some cookie dough. You bake the dough in the oven until the cookies are done. Do the ingredients undergo a chemical change? Give evidence to support your answer.

5. Baking soda is a white powder used for baking or cleaning. When you mix baking soda, $NaHCO_3$, with vinegar, $C_2H_4O_2$, you get a clear colorless liquid and bubbles of CO_2.

Are cookies an example of a chemical change?

  a. Specify the phase of each of the compounds.
  b. Is this a chemical change? Give evidence.
  c. Where is the sodium, Na, before the change? After the change?

6. **PROJECT**

   www   Research how copper is extracted from a compound found in nature. Write a paragraph describing the process.

What Goes Around Comes Around

Conservation of Matter

THINK ABOUT IT

The element copper can be mixed with other substances to make a colorful assortment of compounds. What are these compounds? How can you demonstrate that they all contain copper?

> **What happens to elements in a chemical change?**

To answer this question, you will explore

↻ Translating the Copper Cycle

↻ Tracking an Element

↻ Conservation of Matter

EXPLORING THE TOPIC

↻ Translating the Copper Cycle

Chemical names and chemical formulas are powerful tools you can use to keep track of matter. In fact, you can use them to figure out what was made at various steps in the copper cycle.

There are several approaches you can take to figure out what was made at each step of the cycle. First, you can compare the appearance of the compounds that were obtained in each step with compounds that you have seen before. For example, in an earlier class you examined samples of copper compounds that looked like those shown in the two photos here.

Visual observation of compounds can give you some clues, but it is not enough to make a definite identification. For example, it is hard to tell the difference between copper sulfate and copper nitrate through observation because they are both blue. In fact, several copper compounds are blue.

Another approach is to examine the chemical names and formulas of the substances that were mixed together. The new substances are formed from parts of the starting materials. For example, when sodium *hydroxide*, NaOH, is added to the beaker in the second step of the cycle, a compound called copper *hydroxide*, $Cu(OH)_2$, is produced. Because a blue solid is formed and hydroxide, OH^-, is one of the starting ingredients, you can deduce that copper hydroxide is the product.

ADDING NITRIC ACID TO COPPER

Start at the beginning of the copper cycle. See if you can figure out what was created after the first step by translating each step into chemical symbols and formulas.

Copper nitrate, $Cu(NO_3)_2(s)$

Copper sulfate, $CuSO_4(s)$

Copper hydroxide, $Cu(OH)_2(s)$

Overview

FOLLOW-UP: PAIRS

Key Question: What happens to elements in a chemical change?

KEY IDEAS

Elements combine and recombine but are not destroyed in chemical reactions. Chemists keep track of what is going on in chemical reactions by using symbols and formulas. In this way, specific elements can be tracked through a series of chemical procedures. Tracking the elements provides evidence that new substances are simply rearrangements of the original substances.

LEARNING OBJECTIVES

- Explain that the product of chemical reactions depends on what was present at the time of reaction.
- Explain that matter cannot be created or destroyed in a chemical reaction.

FOCUS ON UNDERSTANDING

- Students might think of substances as each being entirely unique, not aware that there is an underlying structure with elements as building blocks.
- Students might reason that the gases produced during chemical reactions are evidence of matter being lost and not conserved, because gases usually float away.

- This lesson does not introduce chemical equations. Equations for each reaction are included here only to assist you in answering questions students might have.

KEY TERM

law of conservation of mass

IN CLASS

Students process and make sense of the Lab: Copper Cycle from the previous lesson. Students use the symbols for the elements to represent the various reactants and products in the copper cycle. They draw on information from Lesson 6 to identify the various products at each stage of the experiment. They relate the chemical formulas to their observations and determine what happened to the copper at each step of the cycle. The reading in the student text focuses on conservation of mass. The exercises include directions for students to write their first complete lab report. A complete materials list is available for download.

TRM Materials List 8

Lesson Guide

ENGAGE (5 min)

TRM PowerPoint Presentation 8

> *ChemCatalyst*
>
> What do you think happened to the copper powder in the copper cycle experiment when it was mixed with the nitric acid?

Sample answer: The copper powder combined with other elements to create something else, probably a new compound; there was a chemical change. Some students might still think that the copper itself changed into a new substance.

Ask: What were all the different things you observed during the experiment? Describe them all. (Brown gas escaped, blue solution formed, maybe some copper still at bottom of beaker.)

Ask: Where did the copper go during that first step?

Ask: Do you think the copper was destroyed? Why or why not? (Students might guess that the copper is not destroyed because it shows up again at the end of the experiment. Others might think it went away or was in the brown gas after nitric acid was added.)

EXPLORE (20 min)

TRM **Worksheet with Answers 8**

TRM **Worksheet 8**

TRM **Handout—Completed Worksheet from Lesson 6: A New Language**

TRM **Handout—Completed Worksheet from Lesson 7: Now You See It**

INTRODUCE THE FOLLOW-UP

→ Arrange students into pairs.

→ Hand out student worksheet. Students will need their data tables from Lesson 6: A New Language and Lesson 7: Now You See It. They may use their own worksheets or you can give them the prepared handouts.

GUIDE THE FOLLOW-UP

→ Complete Step 1 as a worked example. Write the sentence describing the first step. Then ask students to use their data to help you translate the sentence into a sentence containing chemical formulas, as shown here.

"Colorless nitric acid is added to solid orange-brown copper powder, resulting in a blue-green aqueous solution, a brown gas, and liquid water."

$HNO_3(aq)$ is added to $Cu(s)$, resulting in _____ and _____ and $H_2O(l)$.

Ask: What do you think the blue solution is? (Tell students to compare its appearance with compounds that they saw in Lesson 6.)

Ask: What do you think is the identity of the brown gas? (Students may be able to determine what it is from the incomplete chemical equation.)

Ask: What is the completed chemical sentence? ("Colorless nitric acid is added to solid orange-brown copper powder, resulting in a blue-green aqueous solution, a brown gas, and liquid water.")

$HNO_3(aq)$ is added to $Cu(s)$, resulting in $Cu(NO_3)_2(aq)$ and $NO_2(g)$ and $H_2O(l)$.

EXPLAIN & ELABORATE (15 min)

TRM **Transparency—Copper Cycle Diagram 8**

Step 1: Nitric acid is added to copper powder. A clear blue solution and a brown gas are formed.

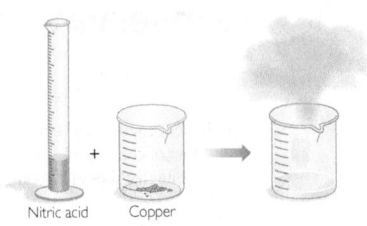

Nitric acid   +   Copper

We know the chemical names that go along with the first half of this step. Nitric acid *is added to* copper powder *resulting in* a blue solution and a brown gas. Using chemical formulas, you can write this as:

$\underline{HNO_3(aq)}$ is added to $\underline{Cu(s)}$ resulting in $\underline{\quad?\quad}$ and $\underline{\quad?\quad}$.
                                                      (blue solution)   (brown gas)

So far in class, you've seen two blue liquids. One was copper sulfate and the other was copper nitrate.

Nitric acid, $HNO_3(aq)$, was added to the copper powder, so you can deduce that the blue solution that formed was copper nitrate, $Cu(NO_3)_2(aq)$, and not copper sulfate, $CuSO_4(aq)$.

$\underline{HNO_3(aq)}$ is added to $\underline{Cu(s)}$ resulting in $\underline{Cu(NO_3)_2(aq)}$ and $\underline{\quad?\quad}$.

How about the brown gas? The brown gas must contain some combination of H, N, O, or Cu because these are the only starting ingredients. Copper, Cu, does not form gaseous compounds, but the other three elements do combine to form several different gases. The chemical formulas and colors of these gases can be found in reference books and are listed in the table below.

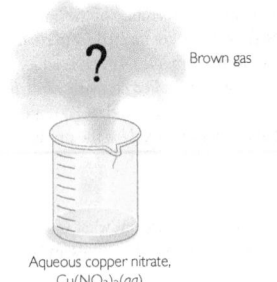

? Brown gas

Aqueous copper nitrate, $Cu(NO_3)_2(aq)$

Gases Containing H, N, or O

Gas	Color
$H_2(g)$, hydrogen gas	colorless
$N_2(g)$, nitrogen gas	colorless
$O_2(g)$, oxygen gas	colorless
$H_2O(g)$, water vapor	colorless
$NH_3(g)$, ammonia gas	colorless
$NO(g)$, nitrogen oxide gas	colorless
$NO_2(g)$, nitrogen dioxide gas	**brown**

DISCUSS WHAT HAPPENED TO THE COPPER THROUGHOUT THE EXPERIMENT

→ Go over the experiment one step at a time with the class. Display the transparency of the Copper Cycle Diagram.

Ask: What happened to the copper after you added the nitric acid?

Ask: What is the chemical formula of the copper compound that was created at this stage? Name this compound. *Note:* We do not recommend that you introduce these balanced chemical equations to the class. They are

provided only for your reference and to assist with any questions that students might raise, such as what gas was produced when the zinc was added? Chemical equations are the focus of Unit 4: Toxins.

Step 1 $Cu(s) + 4HNO_3(aq) \rightarrow Cu(NO_3)_2(aq) + 2H_2O(l) + 2NO_2(g)$

Step 2 $Cu(NO_3)_2(aq) + 2NaOH(aq) \rightarrow Cu(OH)_2(s) + 2NaNO_3(aq)$

Step 3 $Cu(OH)_2(s) + heat \rightarrow CuO(s) + H_2O(l)$

Step 4 $CuO(s) + H_2SO_4(aq) \rightarrow CuSO_4(aq) + H_2O(l)$

Step 5 $CuSO_4(aq) + Zn(s) \rightarrow Cu(s) + ZnSO_4(aq)$

The only *brown* gas in this table is nitrogen dioxide, $NO_2(g)$. The completed chemical sentence for Step 1 is:

$\underline{HNO_3(aq)}$ is added to $\underline{Cu(s)}$ resulting in $\underline{Cu(NO_3)_2(aq)}$ and $\underline{NO_2(g)}$.

This reaction also produces one more compound: water, H_2O. You would not have noticed this because it is clear and colorless. So the final chemical sentence for Step 1 is:

$\underline{HNO_3(aq)}$ is added to $\underline{Cu(s)}$ resulting in $\underline{Cu(NO_3)_2(aq)}$ and $\underline{NO_2(g)}$ and $\underline{H_2O(l)}$.

Notice that all the elements in the starting ingredients also appear in the products. No elements are created or destroyed.

You can deduce the products for the other steps in the copper cycle in a similar way.

Tracking an Element

Once you have figured out the products of Steps 2, 3, and 4 of the copper cycle, you can track the journey of copper through the cycle. The illustration below shows the copper compounds that form at each step of the cycle.

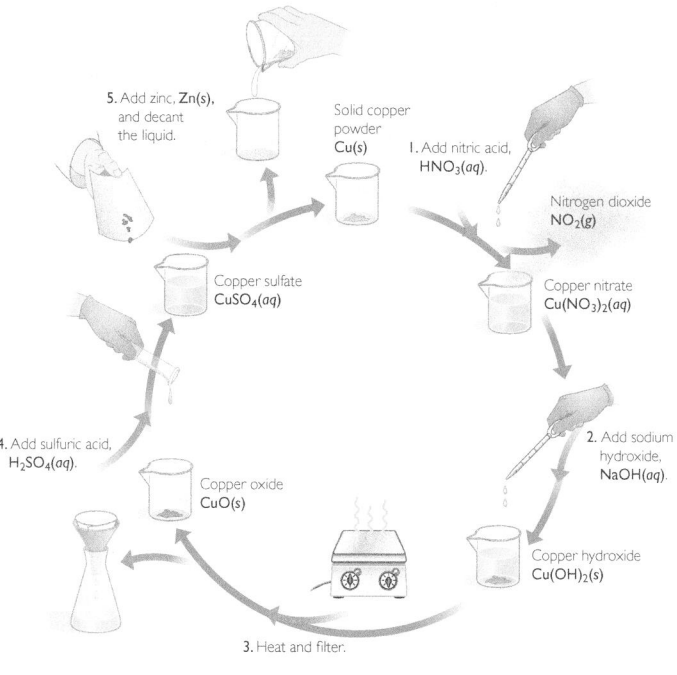

5. Add zinc, $Zn(s)$, and decant the liquid.

Solid copper powder $Cu(s)$

1. Add nitric acid, $HNO_3(aq)$.

Nitrogen dioxide $NO_2(g)$

Copper sulfate $CuSO_4(aq)$

Copper nitrate $Cu(NO_3)_2(aq)$

4. Add sulfuric acid, $H_2SO_4(aq)$.

Copper oxide $CuO(s)$

2. Add sodium hydroxide, $NaOH(aq)$.

Copper hydroxide $Cu(OH)_2(s)$

3. Heat and filter.

Lesson 8 | **Conservation of Matter**   33

focusing on copper. You could prove that copper was not destroyed by measuring the mass of the copper powder at the beginning and at the end of the experiment. If you did the experiment perfectly, you would end up with exactly the same amount of copper powder that you started with. However, in real life, the mass of the copper at the end of this experiment is a bit less than the mass of the copper at the beginning, due to several factors. Little amounts of copper are lost along the way, either because of spills or measurement errors or the copper's staying stuck to the filter paper or beaker. Further, not all the reactions go to completion, so some copper remains in compound form. Nevertheless, mass is still conserved. Even if the original copper did not end up in the beaker at the end, it still exists somewhere.

> **Law of conservation of mass:** The law that states that mass cannot be gained or lost in a chemical reaction—matter cannot be created or destroyed.

REVISIT THE DEFINITION OF AN ELEMENT

Ask: What evidence do you have that copper is an element?

Ask: Why can't the golden penny you made on the first day of class be made of gold, which is an element?

Key Points: No matter what was done to the copper, it was not broken down any further. We were able to track the copper through all its transformations using symbols for the elements. At the end of the experiment, the copper was returned to the form we started the experiment with—solid copper. This experiment provides evidence that elements cannot be broken down or destroyed.

The observations we have made in the copper cycle lab provide further evidence that the golden penny is not gold. Gold and copper are both elements. As elements, each is unique and cannot be broken down into simpler substances or destroyed. If gold is an element, we probably will not be able to convert copper into gold by reactions like those carried out in the copper cycle lab.

Note: Zinc may react with any sulfuric acid remaining from Step 4 to produce $H_2(g)$ bubbles and $ZnSO_4(aq)$.

$$Zn(s) + H_2SO_4(aq) \rightarrow H_2(g) + ZnSO_4(aq)$$

DISCUSS THE IMPLICATIONS OF THE COPPER CYCLE

Ask: Is there evidence of the presence of copper throughout the experiment? Explain.

Ask: What is the advantage of using symbols and formulas to represent the different substances?

Ask: How could you prove that copper was not destroyed during the copper cycle?

Ask: Given that the copper cycle is a circle, do you think we could keep going through it again and again? What might happen?

Key Points: The law of conservation of mass states that matter cannot be created or destroyed. When elemental copper is tracked through a series of reactions, the symbol Cu shows up in the formulas of the new compounds that are made. Thus, elemental copper can be combined with other substances to form new compounds, but it is not destroyed as a result. This is true of all the elements in the copper cycle, even though we are

Check-In

Sodium chloride, NaCl(*aq*), is added to silver nitrate, AgNO$_3$(*aq*), resulting in NaNO$_3$(*aq*) and a white solid. Identify the white solid from the list here. Explain your choice.

(A) AgCl(*s*)

(B) AgCl(*aq*)

(C) AgNO$_3$(*s*)

(D) NaCl(*s*)

Answer: **(A)** If one product contains Na and NO$_3$, then the other probably contains Ag and Cl. Also, the substance is a solid so it cannot be AgCl(*aq*).

Homework

Assign the reading and exercises for Alchemy Lesson 8 in the student text.

HISTORY CONNECTION

Pennies were last made of solid copper in 1836. Pennies made from 1962 to 1982 are 95% copper and 5% zinc. These pennies have a density of 8.6 g/cm^3, which is just slightly less than the density of copper, 9.0 g/cm^3. Since 1982, pennies have been made mostly of zinc with a copper coating. These pennies have a density of 7.2 g/cm^3, which is very close to the density of pure zinc, 7.1 g/cm^3.

The symbol Cu is found at each stage of the cycle. Thus, the element copper is somehow combined in each of these compounds. And of course, the solid that forms at the end of the experiment is elemental copper, Cu(*s*).

What you have observed with copper is true of other elements as well. For example, nickel can be taken through a similar cycle, where various substances are added to nickel powder, Ni(*s*). Just as with the copper cycle, you can recover elemental nickel, Ni(*s*), from compounds that contain nickel.

↻ Conservation of Matter

The copper-cycle experiment brought you full circle, back to where you started. You took a sample of the element copper and added substances to it. After several steps, you ended up with copper powder once again. No matter what was done to the copper, the copper was always there in some form. In other words, it was not created or destroyed by the chemical transformations. Over many centuries, scientists have gathered evidence that matter can never be destroyed or created through chemical transformation. There is so much evidence that this is considered a scientific law.

The **law of conservation of mass** states that mass cannot be gained or lost in a chemical reaction. In other words, matter cannot be created or destroyed.

It is possible to prove that no copper was gained or lost during the copper-cycle experiment by measuring the mass of the copper powder at the beginning and again at the end. If you did the experiment perfectly, you would end up with exactly the same amount of copper powder that you started with. However, in real life, the mass of the copper at the end of this experiment is a bit less than the mass of the copper at the beginning due to several factors. Little amounts of copper are lost along the way, because of spills, measurement errors, and sticking to the filter paper or beaker. These small errors are hard to avoid. In addition, some copper compounds remain in the discarded solutions. Nevertheless, mass is still conserved; even if the copper isn't visible in the beaker at the end, it still exists somewhere.

> **Law of Conservation of Mass**
> Matter cannot be created or destroyed.

LESSON SUMMARY

What happens to elements in a chemical change?

Chemical names and formulas are used to keep track of elements and compounds as they undergo chemical or physical changes. When elemental copper is tracked through a series of reactions, the symbol Cu shows up in the formulas of the new compounds that are made. Elemental copper can be combined with other substances to form new compounds, but it is not destroyed by the chemical transformations. This concept is known as the law of conservation of mass, which states that matter cannot be created or destroyed.

KEY TERM

law of conservation of mass

Reading Questions

1. How can chemical names and symbols help you figure out what copper compound was made in each step of the copper cycle? Give an example.

2. Explain the law conservation of mass in your own words.

Reason and Apply

3. **Lab Report** Write a lab report for the Lab: The Copper Cycle. In your report, include the title of the experiment, purpose, procedure, observations, and conclusions.

(Title)

Purpose: (Explain what you were trying to find out.)

Procedure: (List the steps you followed.)

Observations: (Describe what you observed during the experiment.)

Conclusions: (What can you conclude about what you were trying to find out? Provide evidence for your conclusions.)

Analysis: (Explain what happpened to the copper during the experiment.)

4. Explain how the copper-cycle experiment supports the claim that copper is an element—a basic building block of matter.

5. Nickel sulfate, $NiSO_4(aq)$, is a green solution. Nickel chloride, $NiCl_2(aq)$, is a yellow solution. And hydrochloric acid, $HCl(aq)$, is a clear, colorless solution. If you add nickel, $Ni(s)$, to hydrochloric acid, $HCl(aq)$, what color solution do you expect to form? Explain your reasoning.

6. In the final step of the copper cycle, zinc, $Zn(s)$, is added to copper sulfate, $CuSO_4(aq)$. Elemental copper appears as a solid. Explain what you think happens to the elemental zinc.

7. Matter cannot be created or destroyed. List at least two long-term impacts that this concept has for us on this planet.

LESSON 8 ANSWERS

1. *Possible answer:* The chemical names and symbols indicate what compounds were combined in each step. Because matter is conserved, the products must contain the same elements as the original compounds, which enables you to make a reasonable guess about the products. For example, when sulfuric acid combines with copper oxide, it is likely that copper sulfate and water are the products.

2. *Possible answer:* The law of conservation of mass states that matter cannot be created or destroyed in a chemical reaction.

3. A good lab report will contain
● a title (Lab: The Copper Cycle),
● a statement of purpose (***Possible answer:*** To perform a series of chemical reactions that begin and end with copper powder), ● a procedure (a summary of the steps followed in the experiment), ● results (Check student observations to make sure they noticed the changes that occurred during each step of the copper cycle, including color changes, formation of gas, formation of precipitates, and dissolving of solid materials), ● a conclusion (The conclusion should include a statement related to the conservation of matter. ***Possible answer:*** Although the

copper seemed to disappear, it was still present in the substances in the beaker during the entire series of chemical reactions. This was proved by recovering the copper at the end of the series of reactions.)

4. *Possible answer:* The copper cycle experiment shows that the copper is never lost and implies that it is a component in other compounds formed during the experiment. Elements also cannot be broken down into more fundamental building blocks. The copper cycle experiment does not prove that copper cannot be broken down into simpler building blocks. But the experiment does support the conclusion that copper is an element based upon the definition of element given in Lesson 6.

5. The solution would be yellow because the combination of nickel, Ni, and hydrochloric acid, HCl, can only produce a solution containing compounds with nickel, hydrogen, and chlorine. Nickel chloride, $NiCl_2$, is a possible product and forms a yellow solution. Nickel sulfate, $NiSO_4$, is not a possible product.

6. The zinc reacts with the sulfate part of the copper sulfate to form an aqueous solution of zinc sulfate. The precipitate is pure copper, so the zinc and the sulfate must be present in the dissolved compound.

7. *Possible answers (any 2):* ● Oxygen, carbon dioxide, water, and other important resources can cycle through the environment over and over again. ● Dangerous waste products that cannot break down into safer substances will always be present in the environment. ● The resources that are currently available on the planet are the only resources that will ever be available, unless a method of bringing more material from outer space to Earth is developed.

Overview

DEMO: WHOLE CLASS
ACTIVITY: GROUPS OF 4

Key Question: How is the periodic table organized?

KEY IDEAS

Around 1870, Dmitri Mendeleyev, a Russian chemistry teacher, came up with an ingenious organization of the 63 then-known elements. He organized the elements according to their properties. This organization of elements evolved into our modern periodic table. Mendeleyev was able to predict the existence and properties of as-yet undiscovered elements based on gaps he noticed in his table.

LEARNING OBJECTIVES

● Describe how the organization of the periodic table is based on reactivity and atomic mass.
● Predict the characteristics of a missing element on the periodic table based on its position in the table.

FOCUS ON UNDERSTANDING

● The periodic table is a complex chart with many layers of information. It may take quite a bit of time and exposure for students to grasp its organization.
● The periodic table contains information for both atoms and elements. However, we have not yet introduced the model of the atom. The subtleties of the distinctions between the two usually are lost on students at this point.
● The students probably will not know that the "spokes" on the card-sort cards represent valence electrons. It is not recommended that you share this information yet with the students. Simply allow them to speculate.

KEY TERMS

reactivity
average atomic mass
atomic mass unit, amu
periodic table of the elements

IN CLASS

This lesson introduces students to the periodic table. Students are told about Mendeleyev's efforts to organize the elements into a table. They perform a similar task using cards corresponding to 33 elements (omitting germanium and the transition elements). The cards contain information related to the properties of

the elements. The students re-create the basic organization of the periodic table of the elements based on the data found on the cards, then predict the data that would be on the missing germanium card. Before the card sort, students observe a demonstration to help them understand the concept of reactivity. A complete materials list is available for download.

TRM Materials List 9

Differentiate

Students often struggle at first with the inquiry lesson and may look to their teacher to provide answers. To

help build confidence and support students in self-assessing their card sorts, direct them to move around the classroom in their groups of 4 to observe other groups. Then allow them a few minutes to change their cards if they wish. During the Explain and Elaborate discussion, ask students how or why they changed their sorts.

Pre-AP® Course Tip

Students preparing to take an AP® chemistry course or college chemistry should be able to identify patterns and then use those patterns to refine observations. In this lesson, students organize element cards based on

Featured ACTIVITY

Create a Table

Purpose
To create your own periodic table of the elements from data given on element cards.

Materials
● Create a Table card deck

Instructions
1. Work in groups of four with one set of cards.
2. Find Be, Mg, Ca, and Sr in the deck of cards, and arrange them in a column as shown to the right. These cards are all yellow. Look for similarities and differences in these cards. Find at least one pattern or trend, and describe it to your group.
3. With your group, decide how to organize the rest of the cards into a table. Try to organize them in a way that produces as many patterns as possible.

Questions
1. What characteristics did you use to decide how to sort the cards?
2. What patterns appear in your arrangement? List at least four.
3. Where did you put H and He? What was your reasoning for their placement?
4. Did you notice any cards that didn't quite fit or that seemed out of order? Explain.
5. **Making Sense** To the left are possible cards for the element germanium.
   a. Where does germanium belong in the table?
   b. Which card seems most accurate to you? What is your reasoning?
   c. What would you add to the three empty corners to complete the card?

Create a Table

Properties of the Elements

THINK ABOUT IT

In the late 1860s, a Russian chemist and teacher named Dmitri Mendeleyev was looking for a way to organize the elements known at the time. He wanted to make it easier to remember and understand their chemical behavior. He started by organizing the elements into groups based on similarities in their properties. You can understand Mendeleyev's organization of the elements by examining the information he used to sort them.

How is the periodic table organized?

To answer this question, you will explore

→ Properties of the Elements

→ A Table of Elements

EXPLORING THE TOPIC

⟳ Properties of the Elements

Mendeleyev was intrigued by patterns in the properties of the elements. For example, tin, Sn, sodium, Na, and magnesium, Mg, are all shiny, silvery solids at room temperature.

Tin, Sn   *Sodium, Na*   *Magnesium, Mg*

However, by itself, visual appearance is not a reliable characteristic for sorting the elements into groups. For instance, many elements are shiny, silvery solids at room temperature. And while both oxygen and neon are colorless gases, they are not similar enough in their other properties to be grouped together. Mendeleyev focused on three properties besides appearance to sort the elements: reactivity, the formulas of chemical compounds that form when the element combines chemically, and atomic mass.

REACTIVITY

Mendeleyev focused on the reactivity of elements to sort them. **Reactivity** is a property that describes how easily an element will combine with other substances to form new compounds. An element that is highly reactive combines rapidly

---

patterns and similarities in the properties described on the cards.

SETUP

You might want to cover or remove any periodic tables you have hanging in the room before beginning this activity.

For the reactivity demonstration:
- Put 100 mL of water into each of the two 250 mL beakers.
- Add a few drops of phenolphthalein to each.
- Have 1–2 g magnesium and 1–2 g calcium available to put into the beakers.

CLEANUP

The products of the reactivity demo are $Mg(OH)_2$ and $Ca(OH)_2$. Dispose of these bases according to local guidelines.

Lesson Guide

ENGAGE (5 min)

TRM Transparency—ChemCatalyst 9

TRM PowerPoint Presentation 9

Note: The full-sized image for the ChemCatalyst is available as a transparency in the PowerPoint Presentation.

ChemCatalyst

1. How do you think the elements are organized in this table?

2. What do you think Mendeleyev was trying to do?

Dmitri Mendeleyev's Periodic Table of the Elements

	Group I	Group II	Group III	Group IV	Group V	Group VI	Group VII	Group VIII	
1	H = 1								
2	Li = 7	Be = 9	B = 11	C = 12	N = 14	O = 16	F = 19		
3	Na = 23	Mg = 24	Al = 27	Si = 28	P = 31	S = 32	Cl = 35		
4	K = 39	Ca = 40	— = 44	Ti = 48	V = 51	Cr = 52	Mn = 55	Fe = 56 Co = 59	Ni = 59 Cu = 63
5	Cu = 63	Zn = 65	— = 68	— = 72	As = 75	Se = 78	Br = 80		
6	Rb = 85	Sr = 87	Yt = 88	Zr = 90	Nb = 94	Mo = 96	— = 100	Ru = 104 Rh = 106	Pd = 106 Ag = 108
7	Ag = 108	Cd = 112	In = 113	Sn = 118	Sb = 122	Te = 125	I = 127		
8	Cs = 133	Ba = 137	Di = 138	Ce = 140					
9	—	—							
10			Er = 178	La = 180	Ta = 182	W = 184		Os = 195 Ir = 197	Pt = 198 Au = 199
11	Au = 199	Hg = 200	Tl = 204	Pb = 207	Bi = 208				
12				Th = 231		U = 240			

Note: Mendeleyev's symbol for iodine, "J," has been changed to "I" to match modern symbols.

Sample answers: **1.** Perhaps according to when they were discovered, their properties, or size. Students might say mass, or weight. If students use terms like "atomic mass" or "atomic weight," find out what they mean by those terms. **2.** Answers will vary.

Ask: How do you think this organization of the elements might have been useful to chemists?

Ask: What do you notice about the numbers when you go across the table? Down the table? (Numbers increase going across and down the table.)

Ask: What do you think the blank spaces in the table represent?

EXPLORE (20 min)

TRM Worksheet with Answers 9

TRM Worksheet 9

INTRODUCE THE ACTIVITY

→ Arrange students into groups of 4.

→ Pass out student worksheet. (Wait to pass out cards or students will likely become distracted.)

→ Provide background information on Dimitri Mendeleyev (optional). Mendeleyev was born in Siberia in 1834, the seventeenth child in a very large family. He moved to Saint Petersburg to study medicine, but he was not accepted and instead became a chemistry professor. He noticed patterns in the properties of the elements and is credited with organizing the elements into the first periodic table. He also made many other important contributions to chemistry.

→ Briefly define atomic mass. Atomic mass will be introduced in Lesson 12: Atoms by Numbers. For now, simply tell students that each element has an average atomic mass that is expressed as a decimal number. These are the numbers that appear in the table in the ChemCatalyst and also on the Create a Table cards.

→ Demonstrate the concept of reactivity using magnesium and calcium. You might want to place your beakers on an overhead projector for better visibility.

1. Put 100 mL of water into each of two 250 mL beakers.

2. Add a few drops of phenolphthalein to each.

3. Place 1–2 g magnesium in one beaker and 1–2 g calcium in the other.

4. Solicit students' observations regarding which element appears to be more reactive with water.

Note: The beaker containing the calcium should turn a bright pink color, and some bubbling should be visible. The beaker with the magnesium will not do much at first, but after about 20 minutes, the magnesium will also show some bubbling and a slight pink color.

→ Briefly introduce the concept of reactivity. Reactivity is a property that describes whether an element or compound will chemically combine with other substances to form compounds and also describes the speed of a reaction. Reactivity information is included on the Create a Table cards.

GUIDE THE ACTIVITY

TRM Card Masters–Create a Table 9

→ Pass out sets of Create a Table cards to groups of four students.

→ Ask students to find and arrange the Be, Mg, Ca, and Sr cards in a column as Mendeleyev did (or as they would be on the modern periodic table). The cards should be arranged vertically in this order from top to bottom: Be, Mg, Ca, Sr. These cards are all yellow. Ask students to describe any patterns and trends in this one column.

→ Challenge students to arrange the rest of the cards. Monitor and assist students as needed.

→ There are many approaches to sorting the cards. Students might choose color, size, numbers, reactivity, spokes, etc. All of these approaches are fine. The standard sorting is shown here, but it is best to avoid telling students how to sort the cards. When finished, the sort should look as shown in the illustration.

It took thousands of years for human beings to discover the elements that are the building blocks of matter. Ancient civilizations discovered and used seven metals: gold (Au), copper (Cu), silver (Ag), lead (Pb), tin (Sn), iron (Fe), and mercury (Hg). These seven elements are called the *metals of antiquity*. Many of the early elements were discovered because they are less reactive than other elements and therefore more likely to be found in their pure forms.

with other substances. An explosion, smoke, or a flash of light is a sign that a reaction is proceeding quickly. For instance, when the metal sodium, Na, comes into the slightest contact with water, it reacts vigorously. Sodium even reacts with water vapor in the air. Magnesium, Mg, is another metal that reacts with water. Sodium and magnesium are both shiny, silvery metals that react with water. Based on their properties, these two metals might belong in the same group.

FORMULAS OF COMPOUNDS

Mendeleyev also paid attention to which elements combine with which, and he noted the ratios in which their atoms combine. For example, magnesium, Mg, can combine with chlorine, Cl. When it does, it forms the compound magnesium chloride, $MgCl_2$. This chemical formula indicates that atoms of magnesium and chlorine combine in a 1:2 ratio.

Sodium, Na, also reacts with chlorine, but it combines in a 1:1 ratio, forming sodium chloride, NaCl. Perhaps magnesium and sodium do not belong in the same group after all.

Examine the table, which shows some of the elements that react with chlorine and the compounds that are formed.

You can sort the elements into three groups according to the formulas of compounds with chlorine. In the Activity: Create a Table, you sorted these groups into separate columns, as Mendeleyev did.

Some Elements That React with Chlorine

Element	Symbol	Compound
magnesium	Mg	$MgCl_2$
sodium	Na	NaCl
aluminum	Al	$AlCl_3$
hydrogen	H	HCl
calcium	Ca	$CaCl_2$
indium	In	$InCl_3$
gallium	Ga	$GaCl_3$

ATOMIC MASS

Mendeleyev used another property, called atomic mass, to sort the elements. Each element is made of a different kind of atom. (You will study atoms in Lesson 11: Atomic Pudding.) All atoms of the same element have approximately the same mass. The mass of an atom is called its **atomic mass** and is measured in **atomic mass units,** or **amu.**

Atomic mass will be explained in more detail later in the unit. For now, simply keep in mind that each element has an average atomic mass that is expressed as a decimal number.

You can place the elements in order of their atomic masses. However, sorting the cards just by atomic mass doesn't tell you which elements are similar in their properties.

38   Chapter 2 | **Basic Building Materials**

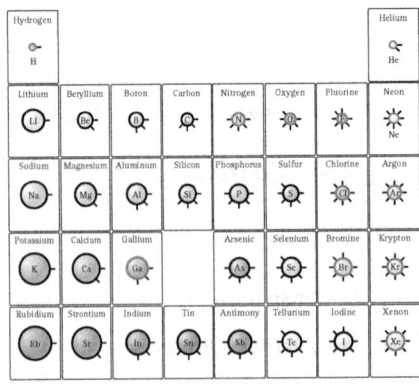

→ If students have trouble deciding where to place H and He, suggest that they save these two element cards for last.

→ The card for germanium, Ge, has been left out intentionally so that students can make conjectures about that element.

EXPLAIN & ELABORATE (15 min)

DISCUSS THE PATTERNS FOUND AND DISCOVERIES MADE DURING THE SORTING ACTIVITY

Ask: What characteristics did you use for sorting?

A Table of Elements

Mendeleyev combined all of these sorting tactics to create his table. He put the elements with similar reactivity and chemical formulas of compounds into columns. Mendeleyev also sorted the elements in order of their atomic masses. He placed the lighter elements at the top of the columns and the heavier elements at the bottom. When he placed the columns next to each other, the atomic masses increased from left to right as well as from top to bottom.

Dmitri Mendeleyev's Periodic Table of the Elements

	Group I	Group II	Group III	Group IV	Group V	Group VI	Group VII	Group VIII	
1	H = 1								
2	Li = 7	Be = 9	B = 11	C = 12	N = 14	O = 16	F = 19		
3	Na = 23	Mg = 24	Al = 27	Si = 28	P = 31	S = 32	Cl = 35		
4	K = 39	Ca = 40	__ = 44	Ti = 48	V = 51	Cr = 52	Mn = 55	Fe = 56	Co = 59
								Ni = 59	Cu = 63
5	Cu = 63	Zn = 65	__ = 68	__ = 72	As = 75	Se = 78	Br = 80		
6	Rb = 85	Sr = 87	Yt = 88	Zr = 90	Nb = 94	Mo = 96	__ = 100	Ru = 104	Rh = 106
								Pd = 106	
7	Ag = 108	Cd = 112	In = 113	Sn = 118	Sb = 122	Te = 125	I = 127		
8	Cs = 133	Ba = 137	Di = 138	Ce = 140	—	—	—	—	
9	—	—			—	—			
10	—	—	Er = 178	La = 180	Ta = 182	W = 184	—	Os = 195	Ir = 197
								Pt = 198	
11	Au = 199	Hg = 200	Tl = 204	Pb = 207	Bi = 208	—	—	—	
12	—	—	—	Th = 231	—	U = 240	—	—	

Note: Mendeleyev's symbol for iodine, "J," has been changed to "I" to match modern symbols.

Mendeleyev organized the 63 elements that were known at the time into a table. In his table, the average atomic masses of the elements increase as you proceed across each row and down each column. The elements in each column have similar physical properties and reactivity, and they tend to form compounds with other elements in the same ratios. The table Mendeleyev created, with elements organized into rows and columns, became known as the **periodic table of the elements.**

The periodic table is an extremely useful organization of the elements. Mendeleyev was even able to predict the existence and properties of as-yet undiscovered elements based on gaps he located in his table.

....

Big Idea Elements are arranged on the periodic table based on similarities in their chemical and physical properties.

....

LESSON SUMMARY

How is the periodic table organized?

Dmitri Mendeleyev created one of the first organized tables of the elements. He sorted the elements based on their properties, specifically reactivity, the formulas of compounds created when elements chemically combine, and atomic mass. He organized the elements into a table so that their atomic masses increased in order

Lesson 9 | **Properties of the Elements**   39

Key Points: Mendeleyev's arrangement of the elements helped predict the existence and properties of as-yet undiscovered elements. Although a few other scientists had the idea of sorting and grouping the elements around the same time as Mendeleyev, he was able to predict as-yet undiscovered elements and their properties. The discovery of these other elements (germanium and scandium) gave his table even more credibility.

EVALUATE (5 min)

TRM Transparency—Check-In 9

Check-In

Which of these elements would be grouped together on the periodic table? Explain your thinking.

cadmium Cd	1. Moderately soft, silvery, solid, metal 2. Reacts very slowly with water 3. Found in $CdCl_2(s)$
zinc Zn	1. Moderately hard, silvery, solid, metal 2. Reacts very slowly with water 3. Found in $ZnCl_2(s)$
iodine I	1. Purple, solid, nonmetal 2. Reacts slowly with metals 3. Found in $ICl(s)$
mercury Hg	1. Silvery, liquid, metal 2. Does not react with water 3. Found in $HgCl_2(s)$

Answer: Zinc, cadmium, and mercury should be grouped together. They are all metals and are found in compounds with the chemical formula—Cl_2. Iodine differs greatly from these three elements.

Homework

Assign the reading and exercises for Alchemy Lesson 9 in the student text.

LESSON 9 ANSWERS

1. Three useful properties for sorting elements are reactivity, formulas of their compounds, and atomic mass.

2. Carbon, C, is most similar to silicon, Si, because both elements are located in the same column of Mendeleyev's periodic table of the elements. Mendeleyev sorted

---

Ask: Did you have to abandon any characteristics and pick different ones?

Ask: Did you discover any new characteristics or patterns during the activity?

Key Points: Mendeleyev organized his periodic table based on the properties of the elements, such as reactivity and atomic mass. Possible sorting characteristics for today's cards are color (representing common properties), size of the circles, the number in the middle of each card (atomic mass), and the number of spokes (valence electrons). The table can be sorted from right to left or from left to right. Similarly, the table may be arranged vertically rather than horizontally. All these are acceptable.

DISCUSS THE MISSING ELEMENT

Ask: Which card did you choose for Ge? Explain your choice. (Card D)

Ask: What other properties do you predict for Ge? Explain. (Metal, shiny and silvery, slow to react with air, forms GeH_4)

Ask: How did Mendeleyev predict new elements before they were discovered? (He used the periodic table and trends in properties to identify missing elements such as germanium.)

the elements, placing elements with similar properties in the same column. Carbon is in the same column as silicon but not in the same column as nitrogen and oxygen.

3. Possible answers:

a.

Si	P	S
Ge		Se
Sn	Sb	

b. (any 2 for each column): ● Column 1: Combines with chlorine in a 1:4 ratio, solid at room temperature, nonreactive. ● Column 2: Combines with chlorine in a 1:3 or a 1:5 ratio, reactive, brittle, solid at room temperature. ● Column 3: Combines with chlorine in a 1:2 ratio, reactive, brittle, solid at room temperature. Multiple correct answers are possible. The answer above uses the same logic as was followed in Lesson 9. The columns are organized using the ratios of the atoms when combined with chlorine. The rows are organized using the atomic mass of the elements. ● silicon: average atomic mass 28.09, gray, solid, metalloid, nonreactive, forms compound $SiCl_4$ ● germanium: average atomic mass 72.61, gray, solid, metalloid, nonreactive, forms compound $GeCl_4$ ● tin: average atomic mass 118.7, silvery-white, solid, metal, nonreactive, forms compound $SnCl_4$ ● phosphorus: average atomic mass 30.97, white or red, solid, nonmetal, brittle, reactive, forms compounds PCl_3 and PCl_5 ● antimony: average atomic mass 121.8, silvery-white, solid, metalloid, brittle, reactive, forms compounds $SbCl_3$ and $SbCl_5$ ● sulfur: average atomic mass 32.07, yellow, solid, nonmetal, brittle, reactive, forms compound $SbCl_2$ ● selenium: average atomic mass 78.96, gray or red, solid, nonmetal, brittle, reactive, forms compound $SeCl_2$

4. Possible answer: Nails are made of iron because iron is hard and fairly stable in air and water. Barium is too soft for a nail, and it reacts easily with water. Phosphorus is a brittle solid that is difficult to shape and would break when struck by a hammer. Phosphorus also reacts too easily with air and water. ● iron: hard, heavy gray metal; reacts slowly with water and oxygen ● barium: soft, silvery metal that reacts easily with water and oxygen ● phosphorus: soft, white or red crystalline solid, brittle, very reactive with water and oxygen

5. a. CaS **b.** The compound with sulfur will have more mass for a given amount of calcium. The two compounds have the same number of atoms, but the atomic

---

across a row from left to right and down a column. The elements in each column of the table have similar properties. The table Mendeleyev created came to be called the periodic table of the elements.

Exercises

Reading Questions

1. List three properties of the elements that are useful in sorting the elements.

2. Do you expect carbon, C, to be more similar to nitrogen, N, oxygen, O, or silicon, Si? Why?

Reason and Apply

3. PROJECT Use a reference book or the Internet to look up the average atomic masses and properties of silicon, Si, germanium, Ge, tin, Sn, phosphorus, P, antimony, Sb, sulfur, S, and selenium, Se.

   a. Organize these elements into rows and columns.
   b. List two properties that the elements in each column have in common.

4. PROJECT Use a reference book or the Internet to look up some of the properties of iron, barium, and phosphorus. Explain why nails are made of iron but they are never made of barium or phosphorus.

5. Suppose you have equal amounts of calcium, Ca, in two beakers. In one beaker, you react the calcium with oxygen, O. In the other beaker, you react the calcium with sulfur, S. The reaction with oxygen forms the compound calcium oxide, CaO.
   a. What do you predict is the chemical formula of the compound formed from the reaction between calcium and sulfur?
   b. Which compound has more mass, the compound containing calcium and oxygen, or the compound containing calcium and sulfur? Explain your thinking.

6. PROJECT Use the Internet to research Dmitri Mendeleyev, the Russian chemist credited with the discovery of the periodic table of the elements. Write a brief paragraph describing Mendeleyev's life and work. Include how he became a chemistry professor and how he came up with the idea for the periodic table.

7. Find at least two different versions of the periodic table and bring a copy of each to class.
   a. Write down what you think makes these two versions similar to each other.
   b. Write down what you think makes these two versions different from each other.

---

mass of sulfur is 32, and the atomic mass of oxygen is only 16.

6. Possible answer: Dmitri Mendeleyev was the youngest child in a large family. He and his mother moved to Saint Petersburg, Russia, when his father died. After finishing school he began teaching science while studying chemistry. He became a professor of chemistry and wrote a textbook called *Principles in Chemistry.* While working on the book, he noticed patterns in the properties of the elements. He made cards with the symbols and properties of the elements and began to arrange them. Based on the patterns he saw in the card

arrangements, Mendeleyev developed his periodic table.

7. Possible answers: a. Both versions arrange the elements in the same order with the same atomic numbers. The columns and rows include the same elements in both versions. Elements are arranged in order of increasing atomic mass. **b.** The specific information about each element, such as average atomic mass, differs between the tables. The names and symbols of elements 110 and higher are different. The placement of the lanthanides and actinides differs between the tables.

Breaking the Code

The Periodic Table

THINK ABOUT IT

The elements copper, Cu, and gold, Au, share many similarities. Both are relatively unreactive elements. They are soft, so it is easy to bend and shape them. They are called *coinage metals* because they have been made into coins by many cultures. Copper and gold have high values as jewelry because they remain shiny for many years. Is the similarity in their properties related to their locations on the periodic table?

What information does the periodic table reveal about the elements?

To answer this question, you will explore
- The Modern Periodic Table
- Trends in Properties

EXPLORING THE TOPIC

The Modern Periodic Table

Scientists have detected around 115 different elements on the planet. Each is unique. Yet groups of elements have similar properties. Recall from Lesson 9: Create a Table that Dmitri Mendeleyev constructed a table based on patterns in the properties of the elements. His table has been replaced over the decades with many updated versions, such as the one shown on pages 42 and 43. The modern periodic table is a storehouse of valuable information about the elements. Over time, you will learn how to make use of the information that is contained there.

ELEMENT SQUARES

Each element has a square on the periodic table. Within each square is information about that element including its name and symbol. The whole number in each square is called the **atomic number.** Hydrogen is the first element in the table and has the atomic number 1. Helium is the second element and has the atomic number 2. Each atomic number corresponds to a different element.

The decimal number in each square on the periodic table is the average atomic mass in amu.

Here is a square from the periodic table.

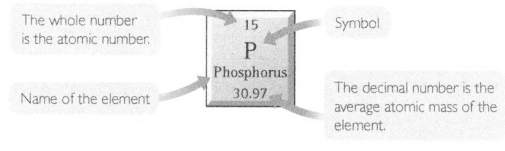

The whole number is the atomic number.

Symbol

15
P
Phosphorus
30.97

Name of the element

The decimal number is the average atomic mass of the element.

Overview

FOLLOW-UP: GROUPS OF 4

Key Question: What information does the periodic table reveal about the elements?

KEY IDEAS

The periodic table contains a tremendous amount of information about the elements. Elements in each column of the periodic table have similar chemical properties. Knowing the precise location of an element on the periodic table can help predict many characteristics of that element, including its reactivity. Chemists have specific names for groups of elements with similar properties.

LEARNING OBJECTIVES

- Use the periodic table to identify elements that are metals, nonmetals, metalloids, alkali metals, alkaline earth metals, transition elements, halogens, noble gases, lanthanides, and actinides.
- Describe the general properties of elements that are periodic.
- Predict the general properties of an element based on its location on the periodic table and identify elements that will exhibit similar chemical behavior.

FOCUS ON UNDERSTANDING

- Many new vocabulary words are associated with the periodic table. These terms are explained more thoroughly in the student edition reading.
- We have chosen to use a specific group numbering system for the periodic table to increase the amount of information students can extract from those designations.

KEY TERMS

atomic number
group
alkali metal
alkaline earth metal
halogen
noble gas
period
main group elements
transition elements
lanthanides
actinides
metal
nonmetal
metalloid

IN CLASS

Students use the Create a Table cards from the previous lesson to look for additional patterns. They learn about physical properties, reactivity, compound formation, and properties of the elements that show up periodically in the table. Students then relate their periodic table card sort to the traditional periodic table, noting how the modern table developed over time. Students read about atomic number and the different parts of the periodic table in the student text. A complete materials list is available for download.

TRM Materials List 10

Differentiate

This lesson is rich in vocabulary. To support all students in building verbal fluency, you could type up and print out the terms associated with the periodic table as well as several long display arrows. During the Explain and Elaborate discussion, guide students to place the terms onto a large, classroom-sized periodic table that you put on a table or on the floor and use the long arrows to identify vertical and horizontal periodic trends from the lesson. Challenge advanced students to research other trends to add on to the list.

Lesson Guide

ENGAGE (5 min)

TRM PowerPoint Presentation 10

TRM Transparency—ChemCatalyst 10

Note: The full-size image for the Chem-Catalyst is available as a transparency or in the PowerPoint Presentation.

ChemCatalyst

The atomic mass of silver is 107.9 amu. The atomic mass of gold is 197.0 amu. Where would you place these elements on the periodic table you created in Lesson 9: Create a Table? Explain your reasoning.

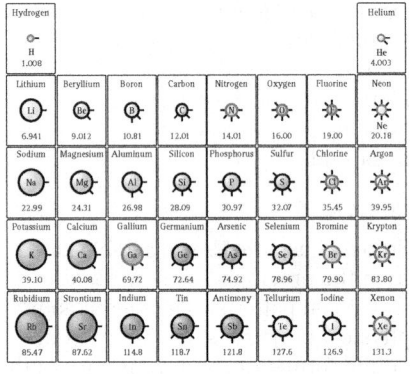

Answers: Students may say that these elements will not fit on the table, or that additional columns and rows would be necessary to fit them on the table.

Ask: Where would you place copper, silver, and gold on your table? Explain your thinking.

Ask: Do these elements fit well into any of the rows or columns?

EXPLORE (20 min)

TRM Worksheet with Answers 10

TRM Worksheet 10

TRM Transparency—Opening Up the Card Sort 10

TRM Transparency—Evolution of the Periodic Table 10

TRM Handout—Periodic Table 10

TRM Card Masters—Create a Table 9

Periodic Table of the Elements

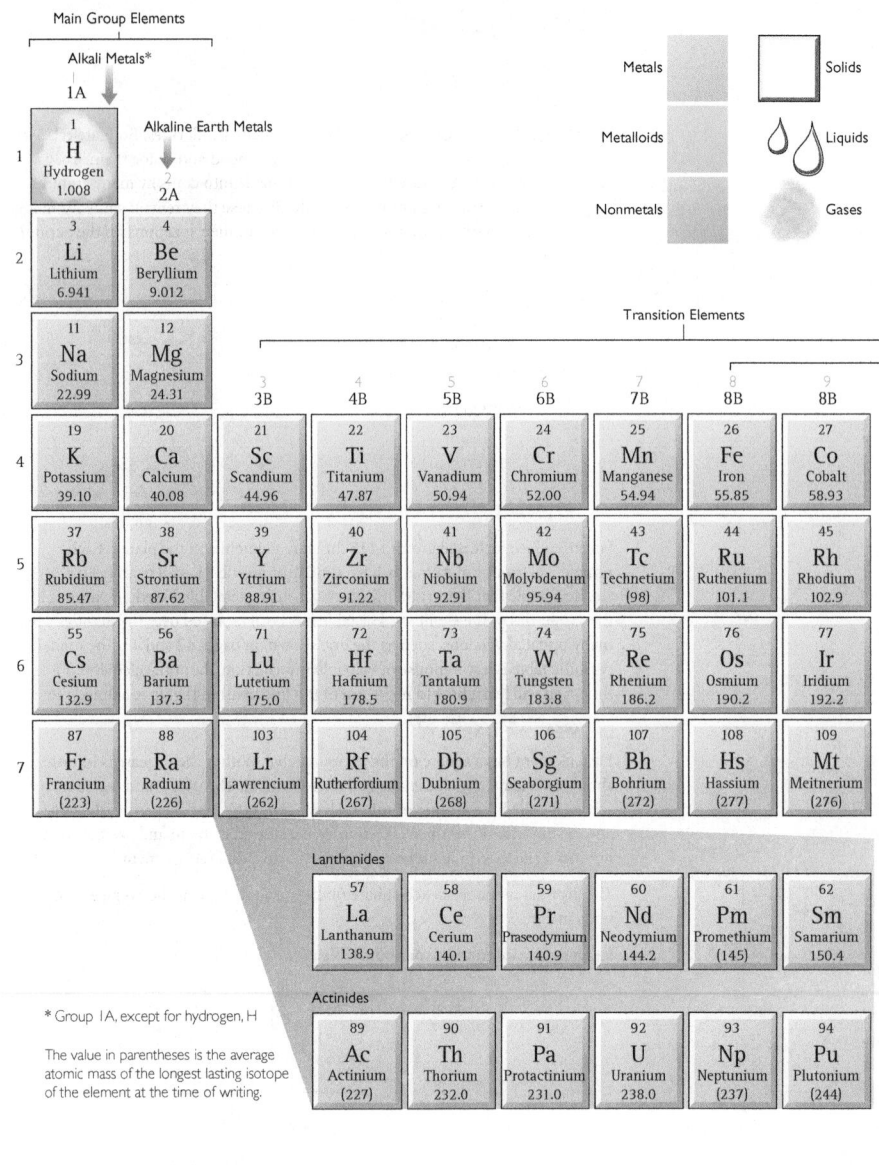

INTRODUCE THE FOLLOW-UP

→ Arrange students into groups of 4.

→ Pass out the student worksheet.

→ Hand out the Create a Table cards from the previous lesson and ask groups to sort them again using the patterns they discovered in Lesson 9: Create a Table.

→ Make a bridge between the card sort table and the modern table. Display the transparency Opening Up the Card Sort. Point out that the card sort table can be opened up to accommodate more elements.

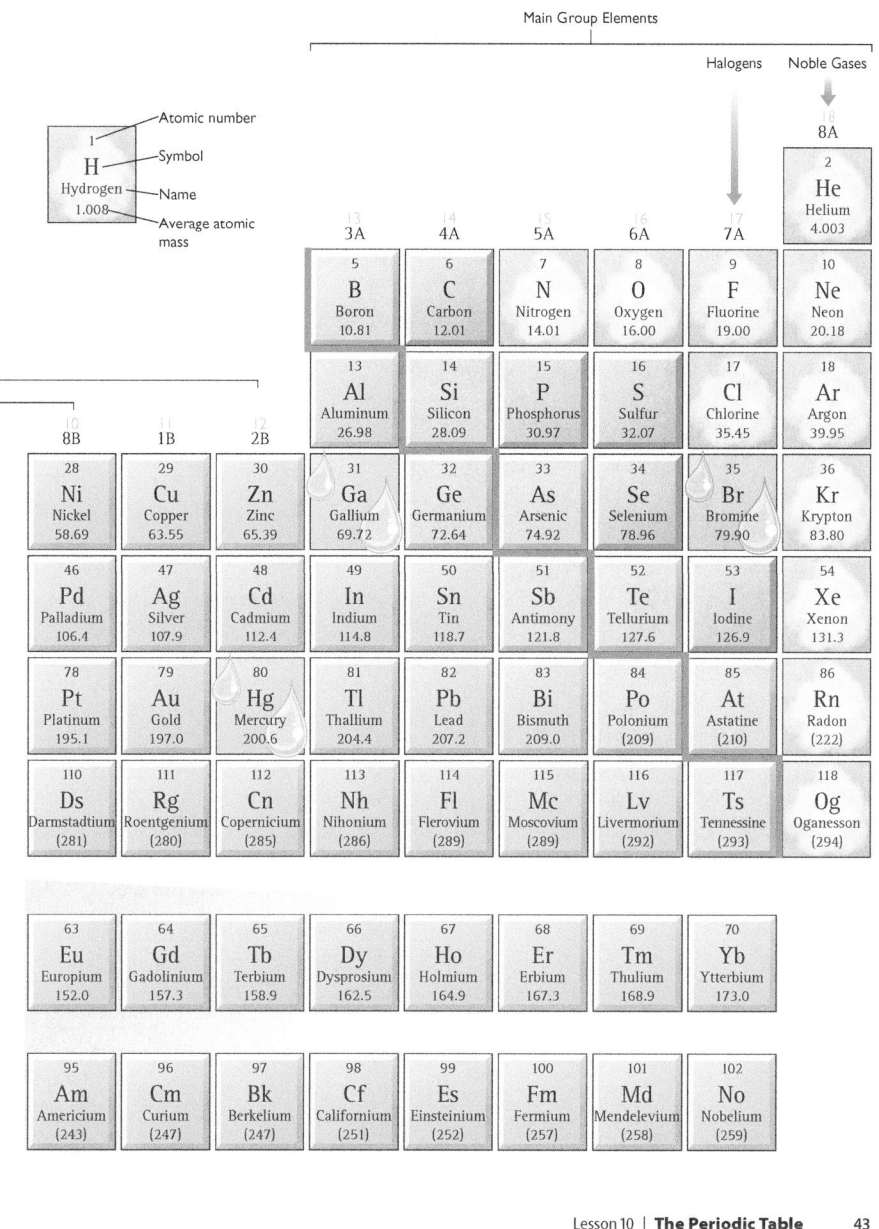

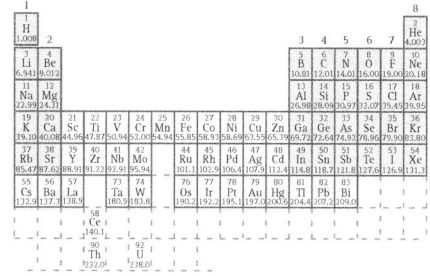

→ Compare to a modern day table.

→ Hand out the Periodic Table of the Elements.

→ Introduce the term *atomic number*.

Atomic number: The consecutive whole numbers associated with the elements on the periodic table.

→ Instruct students to use all of the hands-on items to complete the worksheet.

DISCUSS THE PROPERTIES DESCRIBED ON THE CARDS

→ Hold up a single card or draw one on the board.

Ask: What property is described in each of the four corners of the cards? (upper left: element name, upper right: physical description, lower left: reactivity, lower right: compound formation with hydrogen, H or chlorine, Cl.)

DISCUSS THE VARIOUS PATTERNS DISCOVERED

Ask: Describe a horizontal pattern that you discovered on your table. Are there any exceptions?

Ask: Describe a vertical pattern that you discovered on your table. Are there any exceptions?

Ask: What other color or design patterns do you notice on the cards? What properties do you think these patterns correspond to?

Key Points: There are horizontal patterns on the table (from left to right):

● Atomic mass: The number on the cards increases as you go across (with the exceptions that iodine is smaller than tellurium, and argon is larger than potassium).

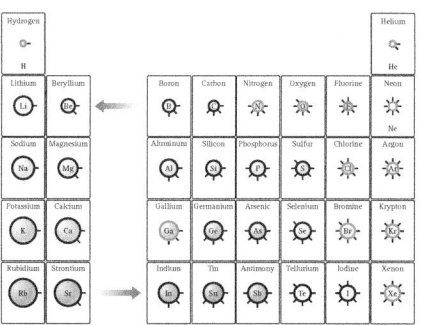

→ Display the transparency Evolution of the Periodic Table. Point out that ten elements have atomic masses that fit between those of calcium, Ca, and gallium, Ga.

→ Add Mendeleyev's remaining elements, keeping them in their columns.

- Radius: The radius of the circles decreases as you go across the rows.
- Spikes: The number of spikes increases by one as you go across the table, then this pattern repeats in the next row.
- Chemical formula: The chemical formula for the compound in the lower right corner increases by one chlorine from column one to column three. Then the chemical formula for the compound decreases by one hydrogen (from four to zero) over the next five columns.

There are vertical patterns on the table (from top to bottom):

- Atomic mass: The number on the cards increases as you go down.
- Color: The elements in each vertical column are color-coded the same.
- Reactivity: As you go down a column, the shading becomes darker if the reactivity increases or lighter if the reactivity decreases.
- Spikes: The elements within each column have the same number of spikes.
- Radius: The radius of the circles increases as you go down the columns.
- Softness: The softness of the metals increases as you go down the columns.

There are other patterns on the table:

- Grainy shading indicates nonmetals and metalloids, solid shading indicates metals.
- Darker hues mean more reactivity.
- A black outline indicates solids, a red outline indicates gases, and a green outline indicates liquids.

INTRODUCE VOCABULARY RELATED TO THE PERIODIC TABLE (OPTIONAL)

→ Display the transparency of the Periodic Table handout and use it to help you introduce the following vocabulary words. These terms are introduced in the reading, so it is not necessary to cover them all.

Group: A vertical column in the periodic table. Elements in a group have similar properties.

Alkali metal: The elements in Group 1A.

Alkaline earth metal: The elements in Group 2A.

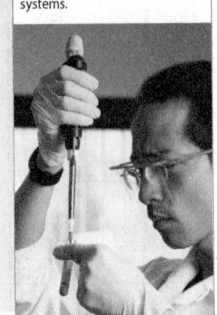

PARTS OF THE PERIODIC TABLE

Most modern periodic tables have 18 vertical columns and 7 horizontal rows. The vertical columns are called **groups**, or families. Hydrogen, H, is in Group 1A, along with lithium, Li, and five other elements in that column. Some of the groups have specific names as shown below.

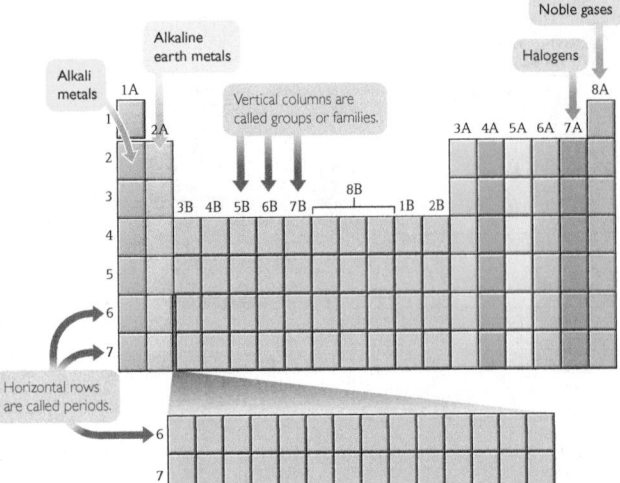

The horizontal rows of the table are called **periods** because patterns repeat periodically, or over and over again, in each row. There are only two elements in Period 1, hydrogen and helium. However, there are eight elements in Periods 2 and 3, and 18 elements in Period 4.

Chemists also have names for sections of the periodic table. Between Group 2A and Group 3A, for example, is where the transition elements fit in.

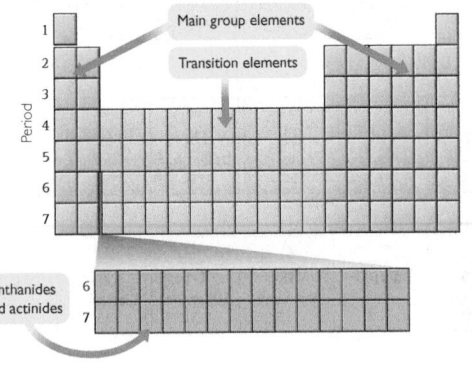

Halogen: The elements in Group 7A.

Noble gas: The elements in Group 8A. They are called noble gases because they are not reactive.

Period: A horizontal row on the periodic table.

Main group elements: The elements in Groups 1A to 8A.

Transition elements: The elements in Groups IB to 8B.

Lanthanides and actinides: The two rows of 14 elements each that are placed separately at the bottom of the periodic table.

Metal: Elements that are excellent conductors of heat and electricity. They generally are shiny and malleable (flexible). They are found to the left of the stair-step line on the periodic table.

Nonmetal: Elements that are poor conductors of heat and electricity. They generally are dull and brittle. They are found to the right of the stair-step line on the periodic table.

Metalloid: The elements between the metals and nonmetals. The metalloids are found along the stair-step line of the periodic table.

In addition, two rows of elements are usually shown at the bottom of the table. These elements are called the lanthanides and actinides. If you examine the atomic numbers of these elements, you'll see that they belong in the sixth and seventh rows. If they were included where they belong, the table would look like this.

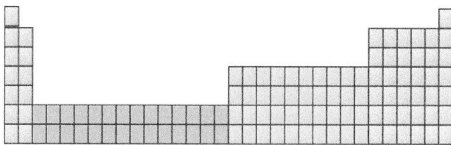

Most periodic tables show these two rows at the bottom so that everything will fit onto one page.

⟳ Trends in Properties

Once the elements are arranged according to their general properties, many other patterns or trends can be found. These three drawings illustrate some of the trends contained within the periodic table.

SOLIDS, LIQUIDS, AND GASES

Most of the elements are solids at room temperature. Several elements are gases at room temperature, but only a few are liquids at or near room temperature.

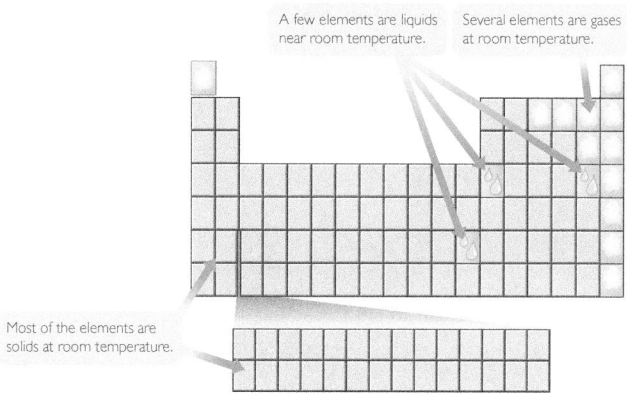

A few elements are liquids near room temperature.

Several elements are gases at room temperature.

Most of the elements are solids at room temperature.

METALS, METALLOIDS, AND NONMETALS

The majority of the elements are **metals**. On most periodic tables, there is a stair-step line that divides the table. Metals are found to the left and **nonmetals**

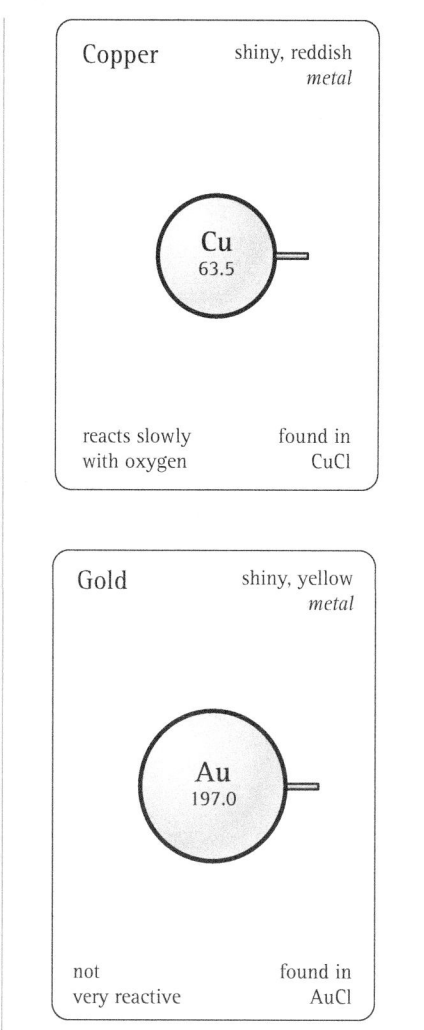

Copper — shiny, reddish *metal*

Cu 63.5

reacts slowly with oxygen — found in CuCl

Gold — shiny, yellow *metal*

Au 197.0

not very reactive — found in AuCl

Sample answer:

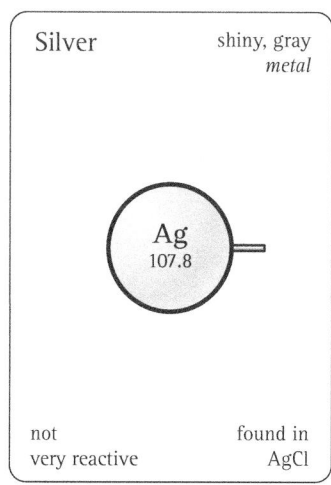

Silver — shiny, gray *metal*

Ag 107.8

not very reactive — found in AgCl

Note: There is debate in the scientific community about where to split the periodic table to show the placement of the lanthanides and actinides. We have chosen to split the table in such a way as to highlight the s-block, p-block, d-block, and f-block of the table. Some tables show lanthanum and actinium in the main body of the table and put lawrencium and lutetium with the lanthanides and actinides. Others place the elements from La–Lu and Ac–Lr all below the table. You may want to share these alternatives with students and use this as a teachable moment to highlight scientific debate and diversity of thought.

EVALUATE (5 min)

TRM Transparency—Check-In 10

Check-In

Look up silver, Ag, on your periodic table. Use these cards for Cu, copper, and Au, gold, to create a card for silver.

Homework

Assign the reading and exercises for Alchemy Lesson 10 and the Chapter 2 Summary in the student text. Students will need a handout of the periodic table for Exercise 3. Assign the project: Element Profile in the student text (optional).

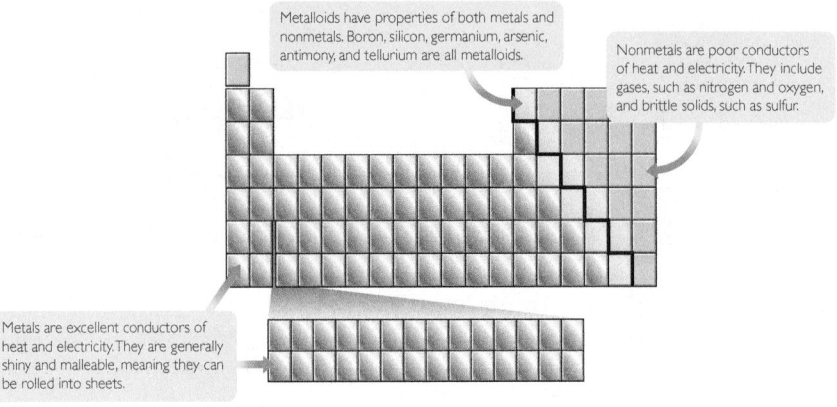

are found to the right of the stair-step line. The elements found along the stair-step line are called **metalloids.** Metalloids have properties like those of both metals and nonmetals.

Metalloids have properties of both metals and nonmetals. Boron, silicon, germanium, arsenic, antimony, and tellurium are all metalloids.

Nonmetals are poor conductors of heat and electricity. They include gases, such as nitrogen and oxygen, and brittle solids, such as sulfur.

Metals are excellent conductors of heat and electricity. They are generally shiny and malleable, meaning they can be rolled into sheets.

REACTIVITY

Elements in the lower left and upper right of the periodic table are the most reactive, with the exception of the noble gases in Group 8A, which are very unreactive. Copper, Cu, silver, Ag, and gold, Au, are metals that are in the middle of the periodic table and are not very reactive.

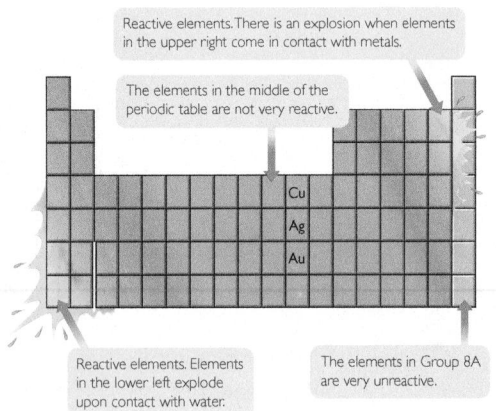

Reactive elements. There is an explosion when elements in the upper right come in contact with metals.

The elements in the middle of the periodic table are not very reactive.

Reactive elements. Elements in the lower left explode upon contact with water.

The elements in Group 8A are very unreactive.

Example 1

Iodine, I

Find iodine, I, on the periodic table.

    a. Find iodine's atomic number, average atomic mass, period, and group.

    b. Would you expect iodine to be a solid, liquid, or gas at room temperature?

    c. Is iodine a metal, metalloid, or nonmetal? How can you tell?

    d. Do you expect iodine to be reactive? Explain.

Solution

Iodine is in the lower-right area of the main group elements.

    a. The atomic number is 53. Average atomic mass is 126.9 amu. Iodine is in Period 5 and Group 7A, halogens.

    b. Iodine is a solid at room temperature.

    c. Iodine is a nonmetal, because it is to the right of the stair-step line.

    d. Yes, you can expect it to be reactive, though not as reactive as elements above it in Group 7A.

Example 2

Coinage Metals

Which element would make the best coin: phosphorus, P, silver, Ag, potassium, K, or xenon, Xe? Explain your thinking.

Solution

Xenon, Xe, is a gas, so it is definitely not a candidate for making a coin. Phosphorus, P, is a nonmetal that is dull and brittle. It would be difficult to shape into a coin. Silver, Ag, is a shiny, malleable metal, so it would make the best coin. Potassium, K, is also a metal, but it is too soft and reactive, so it would not make a good coin. A good coin should not react with other substances.

KEY TERMS

atomic number

group

alkali metal

alkaline earth metal

halogen

noble gas

period

main group elements

transition elements

lanthanides

actinides

metal

nonmetal

metalloid

LESSON SUMMARY

What information does the periodic table reveal about the elements?

The periodic table is an organized chart of the elements. Each element square contains valuable information, including the element name, symbol, atomic number, and average atomic mass. These elements are arranged in vertical columns called groups, or families, and horizontal rows called periods. Most elements are solids and metals, except for those in the upper right of the table. The most reactive elements are located in the lower left and upper right of the table, excluding the noble gases, which are unreactive and located in the last column on the right.

LESSON 10 ANSWERS

1. Within Group 1A, the elements tend to get more reactive as you move from the top of the column to the bottom. Metallic characteristics tend to increase as you move from the top of the table to the bottom. The most reactive metals are located in the lower left portion of the periodic table.

2. *Possible answers (any 2):* ● Reactivity of elements within a period decreases from the left edge to the center and then increases from the center to the right side. However, the element at the far right side of the period is very nonreactive. ● Elements change from metal to metalloid and finally to nonmetal as you move left to right. ● Average atomic mass increases as you move left to right. ● Phase changes from solid on the left to gas on the far right. ● Heat conductivity decreases as you move left to right. ● Conductivity decreases as you move left to right.

3. a. Label groups as follows ● alkali metals: first column on left, excluding hydrogen ● alkaline earth metals: second column from left ● halogens: second column from right ● noble gases: last column on right. **b.** Label sections of table as follows ● main group elements: first two columns on left and last six columns on right (Groups 1A–8A) ● transition elements: the elements in the B groups between the main group elements ● lanthanides: elements 57 through 70 ● actinides: elements 89 through 102

4. Only two elements in Group 2A have average atomic masses greater than 130—barium and radium.

5. B, C, E, and F

6. A, B, and E

7. Copper and mercury are the least reactive. On the periodic table the least reactive elements (aside from the noble gases) are the transition metals that are located in the center of the table. The other elements listed are from more reactive groups near the edge of the table: alkali metals (potassium and rubidium), alkaline earth metals (barium), and halogens (chlorine).

8. *Possible answer:* Copper and platinum, as transition metals, are not very reactive and therefore can be used to make jewelry. Sodium, an alkali metal, is very reactive and not suitable for making jewelry. Neon, a gas, cannot be used to make jewelry.

Exercises

Reading Questions

1. Describe how reactivity changes as you go down Group 1A.
2. Choose two different properties and describe how they vary across a period.

Reason and Apply

3. You will need a handout of the periodic table.
   a. On your periodic table, clearly label the alkali metals, the alkaline earth metals, the halogens, and the noble gases. (If you wish, you may color them and provide a color key at the top.)
   b. Label the main group elements, the transition elements, and the lanthanides and actinides.
4. Name two elements that have properties similar to those of beryllium, Be, and have average atomic masses higher than 130.
5. Which of these elements are solids at room temperature?
   a. fluorine, F   b. titanium, Ti   c. lead, Pb
   d. oxygen, O   e. potassium, K   f. silicon, Si
6. Which of these elements are nonmetals?
   a. bromine, Br   b. carbon, C   c. boron, B
   d. thallium, Tl   e. phosphorus, P   f. aluminum, Al
7. Which two of these elements are the least reactive? Explain your thinking.
   a. chlorine, Cl   b. barium, Ba   c. copper, Cu
   d. rubidium, Rb   e. potassium, K   f. mercury, Hg
8. Can you make jewelry out of each of the elements listed below? Explain your thinking.
   a. copper, Cu   b. neon, Ne
   c. sodium, Na   d. platinum, Pt

CHAPTER 2

Basic Building Materials

SUMMARY

KEY TERMS
element
chemical symbol
compound
chemical formula
phase
aqueous
chemical change (chemical reaction)
law of conservation of mass
reactivity
atomic mass
atomic mass units, amu
periodic table of the elements
atomic number
group
alkali metal
alkaline earth metal
halogen
noble gas
period
main group elements
transition elements
lanthanides
actinides
metal
nonmetal
metalloid
average atomic mass

Alchemy Update

Can an element, such as copper, be transformed into gold through chemical processes? Copper and gold are in the same group on the periodic

 ?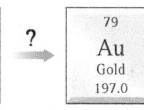

table and have similar properties. However, they are distinct elements with many differences, such as their appearances. While elements can react to form new compounds, from what you have learned so far it does not appear possible to change one element into another element.

REVIEW EXERCISES

1. Make a list of all the information you can extract from the periodic table for the element gold.

2. Explain the law of conservation of mass and how it relates to the copper cycle.

3. What is a chemical formula, and what does it tell you?

4. A filament for a light bulb must conduct electricity. Which of the elements listed below might be useful as a light bulb filament? Explain your thinking.
   A. tungsten, W   **B.** sulfur, S   **C.** bromine, Br

Element Profile

 Research an element. Write a report including

- Your element's name, symbol, and description.
- A list of your element's uses.
- A description of how your element is mined or obtained.

conductor and is not malleable. Also, elements toward the left of the periodic table are better conductors of electricity than sulfur.

Project: Element Profile: A good report would include a definition and explanation of the law of conservation of mass ● a general description of the element and its properties ● a list of all the common uses of an element ● a description of how abundant the element is and how it is mined or extracted ● a description of the role the element plays in the environment, or the environmental impact of the element's extraction and use in industry ● an analysis of how the element's uses and environmental roles relates to the law of conservation of mass ● future projections of the availability and uses of the element

ASSESSMENTS

Two multiple-choice assessments (versions A and B) and two short-answer assessments (versions A and B) are available for download for Chapter 2.

TRM Chapter 2 Assessment Answers

TRM Chapter 2 Assessments

ANSWERS TO CHAPTER 2 REVIEW EXERCISES

1. *Possible answer:* Gold, represented by the symbol Au, is a transition metal that is a solid at room temperature. It has an atomic number equal to 79 and an average atomic mass of 197.0. Gold is nonreactive, a good conductor of heat, and a good conductor of electricity. It has similar properties to copper and silver.

2. *Possible answer:* During a chemical reaction matter changes from one form to another, but the total amount of each element in the matter does not change. The products of a reaction,

such as the reactions in the Lab: The Copper Cycle, can often be predicted based on the materials that react.

3. *Possible answer:* A chemical formula is a symbol that represents a compound. The chemical formula shows what elements are in the compound and the ratio in which the elements combined. It can also show what physical form the compound is in.

4. *Possible answer:* Tungsten might be useful for a lightbulb filament because it is a solid metal. Bromine is unsuitable because it is a liquid. Sulfur is unsuitable because it is not a metal and therefore is a poor

Watch the video overview of Chapter 3 (for teachers) by clicking on the link in the TE-book, opening the TRFD, or logging onto the book's companion Web site bcs.whfreeman.com/ livingbychemistry2e (teacher log-in required).

Chapter 3 focuses on an atomic view of matter. In Lesson 11, students evaluate various atomic models and the evidence that led to their conception. In Lesson 12, students are introduced to the basic structure of an atom and learn how to extract information about atomic structure from the periodic table. They learn how atoms of different elements differ structurally and consider the changes that would be needed to convert one element into another. In Lessons 13 and 14, students explore isotopes and nuclear stability. In Lesson 15, students play Nuclear Quest, a game that introduces nucleosynthesis and radioactivity. Nuclear changes are covered in Lesson 16 as students explore the reactions that result in new elements.

In this chapter, students will learn

- the historical development of the atomic model
- basic atomic structure
- to use the periodic table to extract information about atomic structure
- the composition of stable and radioactive isotopes
- about fission, fusion, and radioactive decay

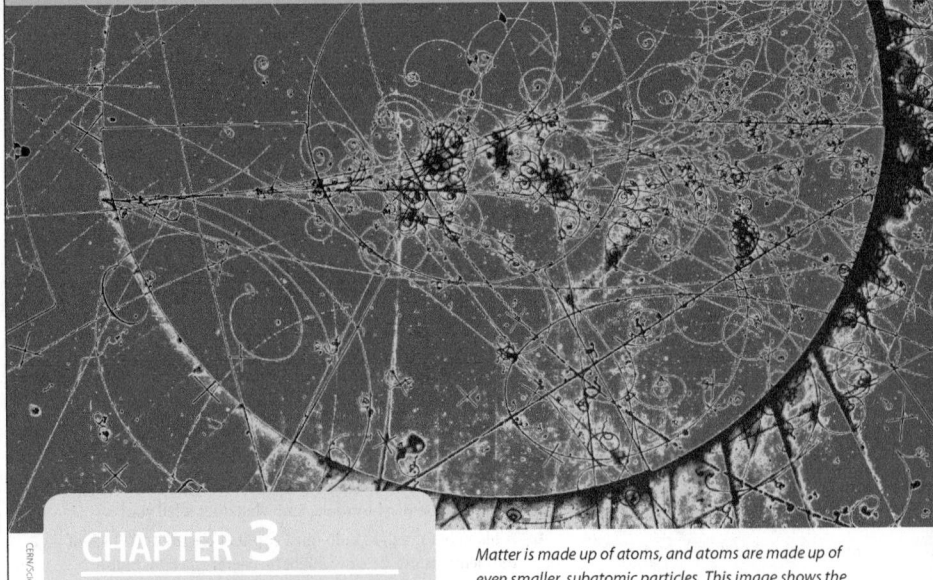

CHAPTER 3

A World of Particles

In this chapter, you will study

- models of the atom
- how atoms differ from one another
- nuclear reactions
- how elements are created

Matter is made up of atoms, and atoms are made up of even smaller, subatomic particles. This image shows the interactions of subatomic particles in a bubble chamber of liquid hydrogen.

All matter is composed of tiny particles called atoms. Based on their observations, chemists agree that atoms themselves are composed of even tinier structures: a nucleus with electrons orbiting around it. So what exactly are atoms, and what does the structure of atoms tell you about matter?

50

Chapter 3 Lessons

Atomic Pudding

Models of the Atom

THINK ABOUT IT

The drawing depicts a very tiny sample of gold taken from a gold ring.

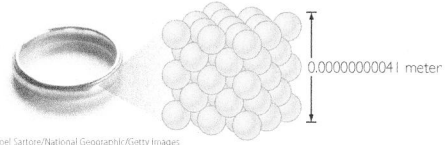

0.00000000041 meter

Joel Sartore/National Geographic/Getty Images

The spheres in the cube of gold are so small that they cannot be seen. What are the spheres, and what does this drawing tell you about the element gold?

How are the smallest bits of matter described?

To answer this question, you will explore

- Atoms: Small Bits of Matter
- Models of the Atom
- Simple Atomic Model

EXPLORING THE TOPIC

Atoms: Small Bits of Matter

Titanium, Ti

+

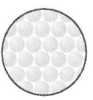

Sulfur, S

↓

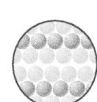

Titanium sulfide, TiS_2

Imagine you break a piece of matter in half, and then break it in half again and again. How many times can you do this? Can you keep going, getting ever smaller? Around 460 B.C.E., the Greek philosopher Democritus wondered the same thing. He thought that if he could just keep breaking matter in half he would eventually end up with the smallest bit of matter possible.

Democritus proposed that all matter was composed of tiny particles that could not be divided further. Today, we use the word **atom** to describe these bits of matter. Of course, atoms are too small to be seen. Democritus' idea was disregarded for the next two thousand years, in part because Democritus did not have evidence to support it.

In 1803, the British scientist John Dalton suggested that the idea of atoms could help explain why elements come together in specific ratios when they form compounds. He imagined atoms of different elements combining to form compounds in the ratios specified by the chemical formulas of the compounds. For example, to form the compound titanium sulfide, TiS_2, titanium and sulfur atoms combine in a 1:2 ratio.

Dalton had more than an idea about atoms. He conducted experiments and made observations to back up his idea. His observations provided strong evidence to support his explanation of how matter behaves.

Lesson 11 | **Models of the Atom**   51

Overview

ACTIVITY: GROUPS OF 2 TO 4

Key Question: How are the smallest bits of matter described?

KEY IDEAS

The atomic theory—the idea that all matter is made up of atoms—is a very old idea dating back to the ancient Greeks. Over time, scientists and philosophers have come up with various models for the atom based on their observations of the world around them. These atomic models help scientists picture and explain the observed behavior of matter. Scientists alter and revise these models as new scientific evidence accumulates. Scientists are now in agreement that the atom consists of a central nucleus with electrons surrounding the nucleus.

LEARNING OBJECTIVES

- Describe the historical development of the current atomic model.
- Describe and draw an atomic model and explain the evidence that supports the existence of atomic structures.
- Describe the dynamic nature of scientific models.

FOCUS ON UNDERSTANDING

- Students might erroneously refer to atoms as molecules and vice versa. Molecules and atoms are not the same.
- Students might confuse models with reality. The scale, size, and shape of atomic structures in an atomic model are usually out of proportion, exaggerated, or even incorrect (e.g., atoms are not flat).
- There is some confusion over what "seeing" something entails. When we state that something cannot be seen, we mean that it is not visible to the unaided eye or even under a microscope. However, scientists can create an image of the overall shape of a molecule.

KEY TERMS

atom
atomic theory
model
nucleus
proton
neutron
electron

IN CLASS

Students are introduced to the term *atom* and to the concept of an atomic model as a tool to help explain the behavior of matter. Then groups of students examine five atomic models. They work together to examine what evidence exists to support each model. Students learn about the nucleus, protons, neutrons, and electrons as tiny parts of the atom. In the discussion, students are introduced to a sixth model, the basic atomic model. A complete materials list is available for download.

TRM Materials List 11

Note: In the student edition for this lesson, students learn about a seventh model, the electron cloud model of the atom.

Differentiate

Students may struggle with conceptualizing scientific models and mistakenly assume that atomic models are pictures of atoms. To support all learners in understanding how scientific models are developed without seeing the object of investigation, try preceding this lesson with a "black box" inquiry activity where students make observations about a mystery object in a sealed box. An online search will reveal many such activities. Ask students to draw a model of what they observe without opening the box to look at the object. After completing Lesson 11, you might have students visit the University of Colorado Boulder's PhET simulation titled "Models of the Hydrogen Atom." You can find a list of URLs for this lesson on the Chapter 3 Web Resources document.

TRM Chapter 3 Web Resources

Pre-AP® Course Tip

Students preparing to take an AP® chemistry course or college chemistry must be able to cite specific evidence that could lead to revising scientific models. In this lesson, students interpret evidence presented by various models of the atom to explain the merits of each model and why they needed to be revised or refined.

Lesson Guide

ENGAGE (5 min)

TRM Transparency—ChemCatalyst 11

TRM PowerPoint Presentation 11

ChemCatalyst

The drawing shown here is a model of a very tiny cube of gold.

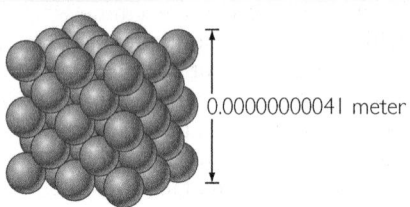

0.00000000041 meter

1. What do you think a scientific model is?
2. The spheres in this model represent atoms. What do you think atoms are?
3. How could you draw a model of the element copper to show that it differs from the element gold?

Sample answers: **1.** Students might mention that models are smaller or larger versions of the real thing. They might give as examples models of the solar system or model cars, airplanes, etc. **2.** Students might say that atoms are the little bits or spheres that make up matter. **3.** Answers will vary.

→ Assist students in sharing their general ideas regarding models and atoms.

In science, the word *theory* indicates that an explanation is supported by overwhelming evidence. The word *theory* allows room for doubt and revision, but indicates a greater degree of certainty than the word does in everyday use. The **atomic theory** states that all matter is made up of atoms. The atomic theory helps us make accurate predictions about the behavior of matter.

⟲ Models of the Atom

Since Dalton's time, scientists have created many **models** to describe atoms and their parts. Models are simplified representations of something you want to explain. For example, a model airplane is a small representation of a larger aircraft. Models take many forms. They can be a plan, a physical structure, a

The Atomic Model through Time

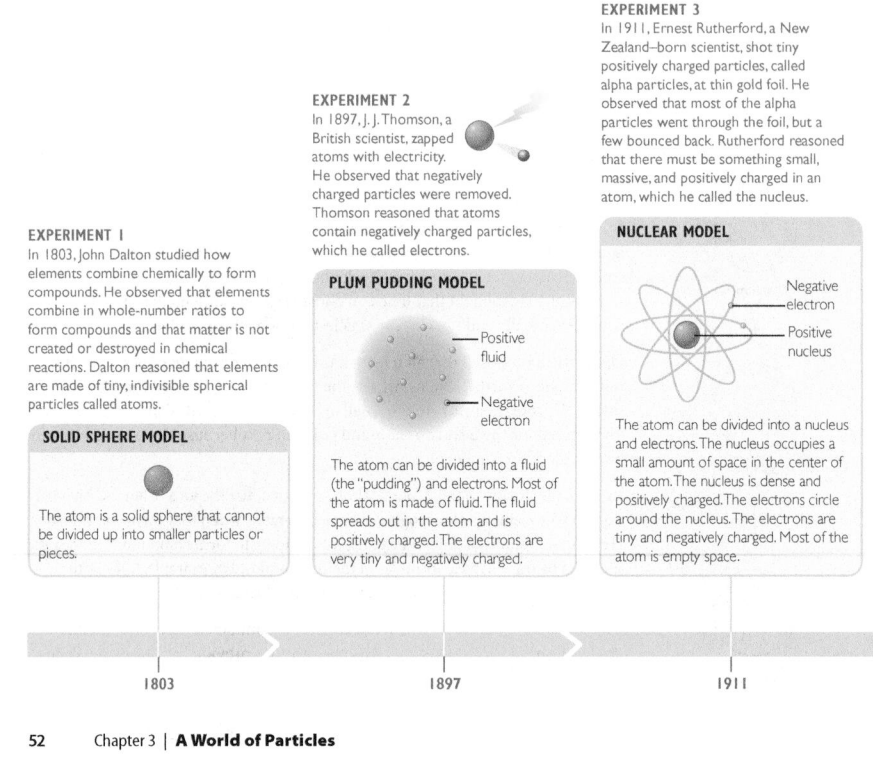

EXPERIMENT 1
In 1803, John Dalton studied how elements combine chemically to form compounds. He observed that elements combine in whole-number ratios to form compounds and that matter is not created or destroyed in chemical reactions. Dalton reasoned that elements are made of tiny, indivisible spherical particles called atoms.

SOLID SPHERE MODEL

The atom is a solid sphere that cannot be divided up into smaller particles or pieces.

EXPERIMENT 2
In 1897, J. J. Thomson, a British scientist, zapped atoms with electricity. He observed that negatively charged particles were removed. Thomson reasoned that atoms contain negatively charged particles, which he called electrons.

PLUM PUDDING MODEL

Positive fluid
Negative electron

The atom can be divided into a fluid (the "pudding") and electrons. Most of the atom is made of fluid. The fluid spreads out in the atom and is positively charged. The electrons are very tiny and negatively charged.

EXPERIMENT 3
In 1911, Ernest Rutherford, a New Zealand–born scientist, shot tiny positively charged particles, called alpha particles, at thin gold foil. He observed that most of the alpha particles went through the foil, but a few bounced back. Rutherford reasoned that there must be something small, massive, and positively charged in an atom, which he called the nucleus.

NUCLEAR MODEL

Negative electron
Positive nucleus

The atom can be divided into a nucleus and electrons. The nucleus occupies a small amount of space in the center of the atom. The nucleus is dense and positively charged. The electrons circle around the nucleus. The electrons are tiny and negatively charged. Most of the atom is empty space.

1803   1897   1911

52   Chapter 3 | **A World of Particles**

Ask: Have you ever created a model of any kind? If so, describe it.

Ask: How do you think scientists come up with models of atoms?

Ask: How do models help scientists (and others)?

Ask: What is an atom? Were atoms invented or discovered?

Ask: If you had a sample of gold measuring 1 cm³, do you think you would have 1000 atoms, 1 million atoms, 1 billion atoms, or more? Explain. (There are many billions of atoms in a cubic centimeter of gold.)

EXPLORE (20 min)

TRM Worksheet with Answers 11

TRM Worksheet 11

TRM Handout—Five Models of the Atom 11

INTRODUCE THE ACTIVITY

→ Tell students to work in groups of two or four.

→ Pass out the student worksheet and the handout Five Models of the Atom. The handout contains descriptions of five atomic models.

drawing, a mathematical equation, or even a mental image. A model that represents the structure of an atom is called an *atomic model*.

Dalton pictured the atom as a hard, solid sphere. Over the next two hundred years, scientists gathered evidence to support and expand on Dalton's model of the atom. It became clear that the atom was more than just a solid sphere.

But how did scientists gather evidence about something too small to be seen? Scientists found they could learn more about atoms and their structure by shooting small pieces of matter at them or by heating them in a flame. Observations from these experiments provided evidence that helped scientists make changes and refine the model of the atom.

EXPERIMENT 4
In 1913, Niels Bohr, a Danish scientist, developed a model of the atom that explained the light given off when elements are exposed to flame or electric fields. He observed that only certain colors of light are given off. For example, hydrogen atoms give off red, blue-green, and blue light. Bohr reasoned that the electrons orbit around the nucleus at different distances like planets orbiting the Sun. The electrons in these orbits have different energies. When an electron falls from an outer to an inner orbit, the color of the light given off depends on the energies of the two orbits.

EXPERIMENT 5
In 1918, Rutherford made a further contribution. He found he could use alpha particles as bullets to knock off small positively charged particles, which he called protons. He reasoned that the nucleus must be a collection of protons.

EXPERIMENT 6
In 1927, Werner Heisenberg, a German scientist, proposed a cloud model of the atom. Heisenberg suggested that the location of an electron could not be specified precisely. Instead, it is only possible to talk about the probability of where an electron might be. This led to a cloud model of the atom. The electron cloud indicates where you will most likely find a single electron.

EXPERIMENT 7
In 1932, a British physicist, James Chadwick, found that the nucleus also included uncharged, or neutral, particles, which he called neutrons. He reasoned that the neutrons were important in holding the positively charged protons together.

SOLAR SYSTEM MODEL

Negative electron
Positive nucleus

The atom can be divided into a nucleus and electrons. The nucleus is in the center of the atom. The nucleus is massive and positively charged. The electrons circle around the nucleus in specified orbits. The electrons are tiny and negatively charged. Different electrons are in orbits at different distances from the nucleus.

PROTON MODEL

The nucleus contains protons. The protons are tiny and positively charged. The electrons circle around the nucleus. The electrons are tiny and negatively charged. Most of the atom is empty space.

ELECTRON CLOUD MODEL

An electron cloud surrounds the nucleus. The cloud is made up of fast-moving electrons. The nucleus is made up of protons: and neutrons:

1913 | 1918 | 1927 | 1932

→ Before beginning the activity, provide students with a basic explanation of the terms model and atom.

Model: A simplified representation of something more complex that facilitates understanding certain aspects of a real object or process.

Atoms: The smallest unit of an element that retains the chemical properties of that element.

EXPLAIN & ELABORATE (15 min)

TRM Transparency—The Atomic Model through Time 11

DISCUSS THE FIVE ATOMIC MODELS AS A CLASS

→ Display the transparency The Atomic Model through Time as you discuss the order in which the models were introduced to the scientific world. *Note:* The simple atomic model and the cloud model also appear on this transparency.

Ask: Can you draw, describe, and explain the model to the rest of the class?

Ask: What evidence does the model explain or not explain?

Ask: In what order did you place the models? Explain your reasoning.

Ask: How could you modify these models to account for Chadwick's discovery of the neutron?

Solid Sphere Model (Dalton, 1803)

Features: a solid sphere

Plum Pudding Model (Thomson, 1897)

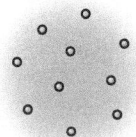

Features: electrons—negative charge; fluid—positive charge

Nuclear Model (Rutherford, 1911)

Features: electrons—negative charge; nucleus—positive charge; electrons circle around the nucleus

Solar System Model (Bohr, 1913)

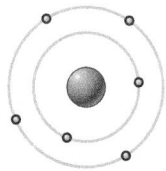

Features: electrons—negative charge; nucleus—positive charge; electrons orbit at different distances from the nucleus

Proton Model (Rutherford, 1918)

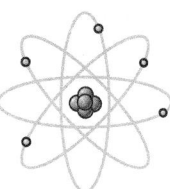

Features: electrons—negative charge; protons—positive charge; electrons circle around the nucleus of protons.

INTRODUCE THE SIMPLE ATOMIC MODEL

→ Point out the last model on the transparency, which includes Rutherford's discovery of protons in the nucleus, Chadwick's discovery of neutrons, and Bohr's discovery of electrons in orbits at specified distances from the nucleus. We refer to this model as the simple atomic model.

SIMPLE ATOMIC MODEL

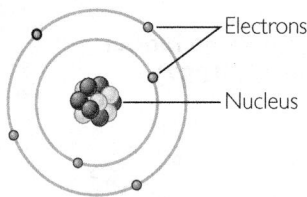
Electrons

Nucleus

Ask: What do we mean when we say that the nucleus is dense?

Ask: If you have six protons and six neutrons, what is the charge on the nucleus?

Ask: If an atom has equal numbers of protons, neutrons, and electrons, what is the total charge on the atom?

Key Points: Contemporary scientists are in agreement that matter is made up of tiny particles called *atoms*. They are also in agreement about the basic structure of an atom. There is a tiny, dense nucleus in the center of the atom. While it occupies a tiny space within the atom, the nucleus accounts for over 99 percent of the mass of the atom. The nucleus contains positively charged particles, called *protons*, and neutrally charged particles, called *neutrons*. Negatively charged electrons are located around the nucleus. The mass of an electron is about 1/1800 the mass of a proton. Although the electrons are very small, they move around the tiny nucleus in a relatively large space outside the nucleus. Thus, the atom consists mainly of empty space.

Nucleus: The dense, positively charged structure found in the center of the atom. It is composed of protons and neutrons.

Proton: A particle with a positive charge, found in the nucleus of atoms.

Electron: A particle with a negative charge. Electrons move very fast around the outside of the nucleus of atoms.

Neutron: A particle that does not have a charge, found in the nucleus of atoms.

TECHNOLOGY CONNECTION

Today, using a scanning tunneling microscope, it is possible to create an image of atoms. This instrument does not magnify a sample of matter like a traditional microscope. Instead, the instrument has a tiny tip that scans the surface of the sample to create a topographic map of the surface.

© Andrew Dunn/Alamy

KEY TERMS
atom
atomic theory
model
nucleus
proton
neutron
electron

54   Chapter 3 | **A World of Particles**

The model of the atom was refined and changed as scientists gathered new evidence. This is what science is all about—a continual process of gathering new knowledge to improve our understanding of the world.

⟳ Simple Atomic Model

The six models shown on the timeline all have something valuable to offer in terms of visualizing matter at an atomic level. At right is a simple atomic model of an atom. In the very center of the atom is the **nucleus**. The nucleus consists of positively charged **protons**, and **neutrons**, which have no charge. The **electrons** are even tinier than the protons and neutrons, and they orbit the nucleus. In this particular atom, the electrons are located at two different distances from the nucleus.

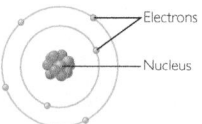
Electrons

Nucleus

Each electron has a charge of −1. The neutrons are neutral and so have no charge. Each proton has a charge of +1. A neutral atom has no overall charge. It has equal numbers of positive protons and negative electrons.

Big Idea   An atom has a nucleus made of protons and neutrons, with electrons orbiting the nucleus.

© Marvin E. Newman/Getty Images

The nucleus occupies a very tiny volume. If an atom were the size of a baseball stadium, the nucleus would be smaller than a baseball.

LESSON SUMMARY

How are the smallest bits of matter described?

Long ago, some philosophers imagined that matter was made up of tiny particles called atoms. Over time, scientists gathered evidence from experimental observations to create models of the atom. Today, we know the atom is made up of protons, neutrons, and electrons. The protons and neutrons are in the center of the atom, in the nucleus. Electrons are outside the nucleus. They are much smaller than the protons and neutrons. In a neutral atom, the positive charges on the protons are equal to the negative charges on the electrons.

A neutral atom has no net charge. This means that the number of protons in the atom is equal to the number of electrons.

neutral atom: positive charge + negative charge = 0

DISCUSS SCIENTIFIC MODELS AND EVIDENCE

Ask: How do models help us understand atoms?

Ask: How does the atomic model further your thinking about matter?

Ask: Do you think that atoms of the same element are alike? Why or why not?

Ask: How do you think you can use the atomic model to explain differences among elements?

Ask: What questions do you still have about how atoms are put together?

Key Points: Scientists have created models to describe atoms. Models are usually simplified imitations of something you want to explain. For example, a model airplane is a small representation of a larger aircraft. A model can be a plan, a physical structure, a drawing, a mathematical equation, or even a mental image. Chemists create models to explain how matter is constructed and how it

Reading Questions

1. What evidence caused Thomson to change Dalton's solid sphere model into the plum pudding model?

2. What evidence caused Rutherford to change Thomson's plum pudding model into the nuclear model?

3. What evidence caused Bohr to change Rutherford's nuclear model into the solar system model?

Reason and Apply

4. Positive and negative charges are attracted to one another. Which of the following are attracted to a negative charge: an electron, a proton, a neutron, a nucleus, an atom? Explain your thinking.

5. Hydrogen and helium are different elements. How can you use the plum pudding model to show how atoms of the two elements might be different from one another?

6. Suppose you discovered protons shortly after Thomson discovered electrons. How would you revise the plum pudding model to include protons? Draw a picture of your revised model of the atom.

7. Draw a solar system model showing one electron, one proton, and one neutron.

8. PROJECT   Use the Internet or another resource to find out how the size of an atom compares with the size of its nucleus. Is the diameter of an atom 10 times, 1,000 times, or 100,000 times the diameter of the nucleus?

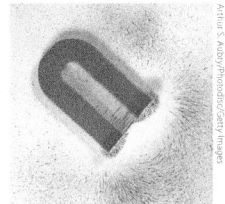

When charged particles are placed near each other, they move toward or away from one another. Similar charges repel, or move away from, one another. Opposite charges attract, or move toward, one another. This photo shows iron filings that have oriented themselves around a magnet's positive and negative ends.

9. The nuclear model and the solar system model both show atoms with electrons circling around the nucleus.
   a. How do these two models differ?
   b. How are these two models similar?
   c. How can you refine the solar system model so that the atoms do not look flat?

10. PROJECT   The ancient Greeks discarded the atomic theory because there was no evidence to support it. Try to provide evidence that atoms do indeed exist. Use the Internet to help you.

11. The ancient Greeks claimed that atoms were the smallest pieces of matter. Were they correct? Explain your thinking.

12. Give an example that shows how science is a process of gathering evidence and refining models.

Lesson 11 | **Models of the Atom**   55

---

behaves. Different models are neither right nor wrong—they simply tell you different things.

Scientific evidence is a collection of observations that everyone agrees on. When a model is supported by a lot of scientific evidence, it is usually accepted by the scientific community. As new evidence is gathered, scientific models are refined and changed. This is what science is all about: a continual process of gathering new knowledge to improve our understanding of the world. The models of the atom we have studied in today's lesson are useful to our thinking about atoms and elements. Later in the unit, we will

learn about additional evidence that has led to further revisions of the atomic model.

EVALUATE (5 min)

TRM Transparency—Check-In 11

Check-In

Here is a model of a carbon atom.

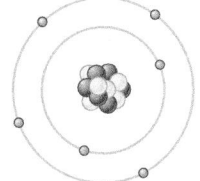

---

1. List two things this model tells us about the carbon atom.

2. List something this model does not tell us about the carbon atom.

Sample answers: 1. The nucleus is at the center of the atom; there are six protons and six electrons; the electrons are at two different distances from the nucleus. 2. There is no indication of scale or relative sizes.

Homework

Assign the reading and exercises for Alchemy Lesson 11 in the student text.

LESSON 11 ANSWERS

1. When Thomson zapped atoms with electricity, he found that a negatively charged particle was removed. Because the solid sphere model does not allow for particles splitting off atoms, he created the plum pudding model. If negatively charged parts of the atom can be removed, a separate part of the atom must be positively charged. Thomson reasoned that an atom consists of a positively charged fluid that contains negatively charged particles called *electrons*.

2. When Rutherford shot alpha particles at a thin layer of atoms, most of them went straight through, but some bounced back. The plum pudding model does not have any solid parts, nor does it have any concentrations of positive charge that would repel the alpha particles. So, Rutherford created the nuclear model. Rutherford reasoned that the positively charged part of the atom must make up a very small part of the total volume. He called this concentration of positive charge the *nucleus* and reasoned that the electrons orbited within the vast empty space surrounding the nucleus.

3. Bohr revised the nuclear model of the atom when he noticed different atoms giving off different colors of light when exposed to flame or electric fields. Because the nuclear model fails to account for this process, he created the solar system model. Bohr suggested a model in which the electrons orbit around the nucleus at different distances, and he proposed that these electrons have different energies. Bohr reasoned that the different levels of light energy emitted correspond to different changes in energy that occur when electrons move from an outer orbit to an inner orbit.

Answers continue in Answer Appendix (p. ANS-1).

Overview

CLASSWORK: INDIVIDUAL

Key Question: How are the atoms of one element different from those of another element?

KEY IDEAS

The periodic table contains valuable information about atomic structure. Each successive element contains atoms with one more proton than the previous element. The atomic number of an element is the number of protons in an atom of that element. In neutral atoms, the number of electrons is equal to the number of protons. The number of protons plus the number of neutrons in an atom provides the approximate mass of a single atom. The decimal number on the periodic table represents the average atomic mass of an element's atoms.

LEARNING OBJECTIVES

- Distinguish between atomic number, mass of an atom, and average atomic mass.
- Describe the structure of an atom and draw a simple atomic model of an atom.
- Extract information from the periodic table related to atomic structure and atomic mass.

FOCUS ON UNDERSTANDING

- The average atomic mass of an element and the mass of an atom are easily confused. Their values are similar, but their meanings are subtly different. The mass of an atom refers to the mass of one atom of a specific isotope. The atomic mass of an element refers to an average mass weighted by isotopic abundance. Isotopes will be covered in Lesson 13: Subatomic Heavyweights and Lesson 14: Isotopia.
- The distinctions between an atom and an element might continue to be troublesome for students.
- It is always useful to remind students of the limitations and exaggerations of the atomic models. For example, the nucleus is much too large relative to the size of the atom, and it is shown flattened out for ease of counting the protons and neutrons.

KEY TERM

atomic number

IN CLASS

The terms *atomic number* and *atomic mass* are explained. Students then use these values, and the periodic table, to figure out the atomic structure of a

variety of elements. They practice extracting information about atomic structure from the periodic table. A complete materials list is available for download.

TRM Materials List 12

Lesson Guide

ENGAGE (10 min)

TRM Transparency—ChemCatalyst 12

TRM PowerPoint Presentation 12

Note: Students should look at a periodic table to complete the ChemCatalyst.

LESSON
12

Atoms by Numbers

Atomic Number and Atomic Mass

THINK ABOUT IT

The element copper is made up of copper atoms. Likewise, gold is made up of gold atoms. We know that copper and gold are different elements. But what makes a copper atom different from a gold atom?

> How are the atoms of one element different from those of another element?

To answer this question, you will explore

- Atomic Number
- Atomic Mass
- The Periodic Table and Atomic Models

EXPLORING THE TOPIC

Atomic Number

In Lesson 9: Create a Table, you learned how Mendeleyev arranged the elements in order of increasing atomic mass. Around 1913, Henry Moseley, a British scientist, discovered an amazingly simple and important property of the elements as well. He determined that the atoms of each element differ by one proton from the atoms of the element before it on the periodic table. The first element, hydrogen, H, has one proton, and the second element, helium, He, has two protons. The third element, lithium, Li, has three protons, beryllium, Be, has four protons, and so on.

The **atomic number** is equal to the number of protons in the nucleus of an element. The elements on the periodic table are arranged in order by their atomic numbers. So the element iron, Fe, with atomic number 26, has 26 protons in its nucleus.

Big Idea If you know the number of protons in an atom, you know its atomic number and what element it is.

For a neutral atom, the atomic number is also equal to the number of electrons. This is because the overall charge of a neutral atom is 0. Protons have a $+1$ charge and electrons have a -1 charge. If the overall charge on an atom is 0, then the number of protons must be equal to the number of electrons.

Atomic Mass

The atomic mass is the mass of a single atom. The protons and neutrons account for almost all of the mass of an atom. Electrons have a much tinier mass by

56   Chapter 3 | **A World of Particles**

ChemCatalyst

Models of a helium atom and a beryllium atom are shown. The nucleus of each contains protons and neutrons. The electrons orbit the nucleus.

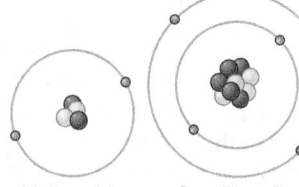

Helium, He   Beryllium, Be

1. Compare the two models. List three similarities and three differences.

comparison. The atomic mass is approximately equal to the total mass of the neutrons and protons because electrons have so little mass.

Every proton in every atom, whether it is an atom of gold or an atom of oxygen, has the same mass. Scientists assign a value of one atomic mass unit, 1 amu, to the mass of a single proton. Neutrons have almost exactly the same mass as protons, so each neutron also has a mass of 1 amu. To determine the mass of a single atom, you add the number of protons and neutrons.

Lithium, Li

Atomic number: 3
Average atomic mass: 6.941 amu

$$\begin{array}{r} 3 \text{ protons} = 3 \text{ amu} \\ + \ 4 \text{ neutrons} = 4 \text{ amu} \\ \hline \text{atomic mass} = 7 \text{ amu} \end{array}$$

Carbon, C

Atomic number: 6
Average atomic mass: 12.01 amu

$$\begin{array}{r} 6 \text{ protons} = 6 \text{ amu} \\ + \ 6 \text{ neutrons} = 6 \text{ amu} \\ \hline \text{atomic mass} = 12 \text{ amu} \end{array}$$

HOW MANY NEUTRONS?

If you look at the atomic models, you will notice that the numbers of protons and electrons are exactly the same as the atomic number, but the number of neutrons is sometimes different from the number of protons. So, how can the periodic table tell you how many neutrons are in an atom? If you know the mass of an atom *and* you know how many protons it has, you can find out how many neutrons it has by subtracting the number of protons from the atomic mass.

AVERAGE ATOMIC MASS

It turns out that not every atom of an element is identical. So, the decimal number in each element square of the periodic table is the average atomic mass of that element in atomic mass units, amu. This number can also be used to estimate the number of neutrons in a nucleus. Simply round the average atomic mass to the nearest whole number and subtract the atomic number (number of protons). For example, lithium has an average atomic mass of 6.941 amu, so a typical lithium atom probably has a mass of 7 amu. The atomic number of lithium is 3, so there are 3 protons. This accounts for 3 amu of the mass. The other 4 amu must be due to 4 neutrons.

In Lesson 13: Subatomic Heavyweights, you will investigate how the average atomic mass is arrived at and how atoms of an element may differ from one another.

....

Important to Know When considering atomic mass, it is necessary to know whether you are focusing on the mass of one particular atom or the average mass of a group of atoms.

....

Lesson 12 | **Atomic Number and Atomic Mass**   57

---

2. Based on the models, why do you think helium is number 2 (the second element) and beryllium number 4 (the fourth element) on the periodic table?

Sample answers: **1.** Both atoms have nuclei, electrons, neutrons, and protons. The beryllium atom is larger than the helium atom; has more protons, neutrons, and electrons; it has two rings, while the helium atom has only one. **2.** These numbers correspond to the number of protons and electrons. For He, the number also corresponds to the number of neutrons.

Ask: Do helium and beryllium have the same number of protons? Neutrons? Electrons?

Ask: How do the locations of He and Be on the periodic table relate to their number of protons, neutrons, and electrons?

EXPLORE (15 min)

TRM Worksheet with Answers 12

TRM Worksheet 12

INTRODUCE THE CLASSWORK

→ Let students know they will be working individually.

→ Pass out the student worksheet.

→ Tell students that the whole number in an element square on the periodic table is called the atomic number and the decimal number is called the average atomic mass.

EXPLAIN & ELABORATE (15 min)

DISCUSS ATOMIC NUMBER

Ask: What information can you find out from the atomic number of an atom?

Ask: How do you find the atomic number on the periodic table?

Ask: Can you draw an atomic model for krypton, Kr, from only the atomic number? Explain your process.

Key Points: The atomic number of an element is the same as the number of protons in the nucleus of an atom of that element. In a neutral atom, the number of protons is the same as the number of electrons. This is why a neutral atom has no net charge.

> **Atomic number:** The number of protons in the nucleus of an atom of an element. In the periodic table, the elements are arranged in order by atomic number.

Each successive element has one more proton than the element before it. It is the number of protons that makes atoms of one element different from atoms of another element. Likewise, assuming the atoms are neutral, each successive element has one more electron than the element before it. The atomic number does not tell you how many neutrons an atom of an element has. However, for smaller atoms, the number of protons is close to the number of neutrons.

DISCUSS ATOMIC MASS

→ Demonstrate how to find the approximate mass of an atom, where to find the average atomic mass of an element, and how to estimate the number of neutrons in an atom from information in the periodic table.

Ask: Which atomic particles account for most of the mass of an atom? (protons and neutrons)

Ask: How can you estimate the mass of an atom in atomic mass units (amu)?

Ask: What is the approximate mass of an atom of chlorine with 18 neutrons? (35 amu)

Ask: How does this atomic mass compare to the average atomic mass of chlorine given in the periodic table? (35 amu versus 35.45 amu)

Ask: What information would you need to draw a basic atomic model of neon, Ne? Where would you find that information?

Key Points: Protons and neutrons account for most of the mass of an atom. A proton and a neutron each have a mass of 1 amu (atomic mass unit). By comparison, the mass of an electron is 0.0005 amu, or about 1/2000 the mass of a proton or neutron. Therefore, the mass of an atom in atomic mass units is approximately equal to the sum of the number of protons and neutrons, because the mass of the electrons is so small. Average atomic mass is covered in Lesson 13: Subatomic Heavyweights.

Mass of one proton: 1 amu
Mass of one neutron: 1 amu
Mass of one electron: ~0.0005 amu

Note: The official definition of amu is 1/12 the mass of a carbon atom. Because carbon has six neutrons and six protons, 1 amu is equal to the mass of one proton or one neutron. However, the simplified definition given above is probably more useful to students at this point.

You can estimate the number of neutrons in an atom by subtracting the number of protons from the average atomic mass of the element (rounded to the nearest whole number). For example, neon has atomic number 10 and an average atomic mass of 20.1797 amu, which means we can expect a typical neon atom to have a mass of 20 amu. The atomic number indicates ten protons in the nucleus, each with mass 1 amu, giving a total mass of 10 amu. So there must be ten neutrons to account for the remaining 10 amu.

number of neutrons = average atomic mass rounded to the nearest whole number − atomic number

The atomic mass of an atom determined by summing the number of protons and neutrons is not identical to the average atomic mass of the element given in the periodic table. This difference will be discussed further in the next lesson.

RELATE TODAY'S FINDINGS TO THE ALCHEMY CONTEXT

Ask: What differences did you discover between an atom of gold and an atom of copper?

The Periodic Table and Atomic Models

To draw an atomic model of a specific element, you must know the numbers of protons, neutrons, and electrons. You find this information on the periodic table. This illustration shows you how to get information about atomic structure from each square of the periodic table to build a basic atomic model of an element.

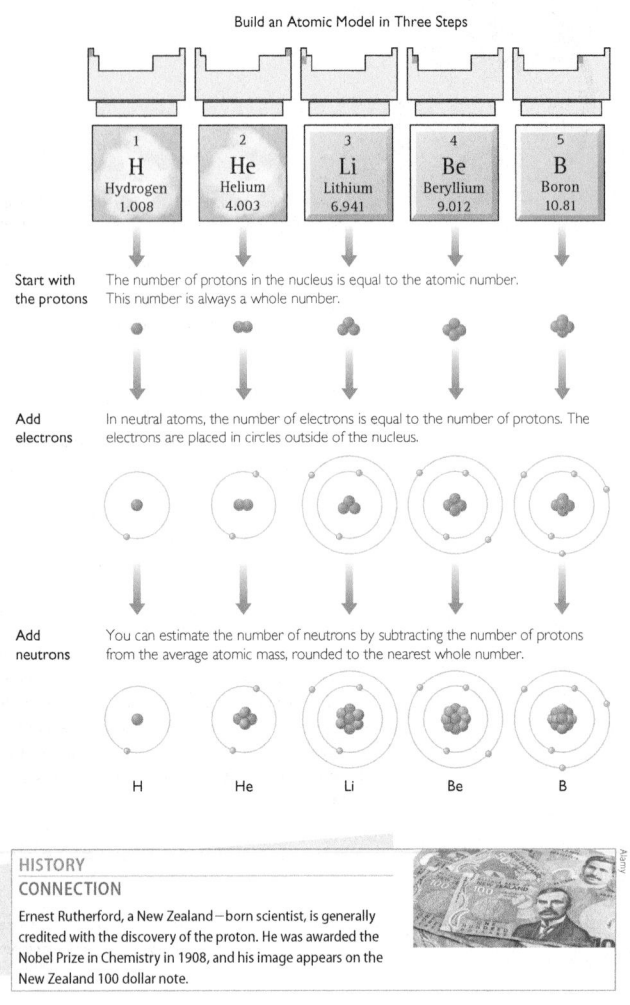

Build an Atomic Model in Three Steps

Start with the protons The number of protons in the nucleus is equal to the atomic number. This number is always a whole number.

Add electrons In neutral atoms, the number of electrons is equal to the number of protons. The electrons are placed in circles outside of the nucleus.

Add neutrons You can estimate the number of neutrons by subtracting the number of protons from the average atomic mass, rounded to the nearest whole number.

HISTORY CONNECTION

Ernest Rutherford, a New Zealand–born scientist, is generally credited with the discovery of the proton. He was awarded the Nobel Prize in Chemistry in 1908, and his image appears on the New Zealand 100 dollar note.

Ask: What would you have to change in an atom of copper to make it like an atom of gold? (You would have to add 50 protons, 82 neutrons, and 50 electrons.)

Ask: Do you think it is possible to add protons, neutrons, and electrons to an atom of copper? Explain.

Ask: If you are trying to make gold from other substances, do you think there is a better element to start with than copper?

Key Point: If you change the number of protons in an atom, you also change the elemental identity of that atom. For example, if you could somehow add a proton to the nucleus of a carbon atom so that it went from six protons to seven protons, you would no longer have a carbon atom, but a nitrogen atom. Thus, if you know the number of protons in an atom, you can also figure out its identity.

EVALUATE (5 min)

Check-In

Use your periodic table to identify these elements.

a. Atomic number is 18.

b. Has three electrons when atoms are neutral.

c. Atomic mass is 16.0.

Copper and Gold Atoms

How is an atom of gold, Au, different from an atom of copper, Cu?

Solution

You can find the information you need on the periodic table. The atomic number of copper is 29. So neutral copper atoms have 29 protons and 29 electrons.

To estimate the number of neutrons in the atom, round the atomic mass to 64 and subtract the atomic number.

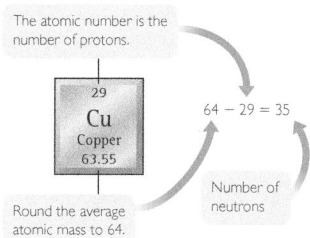

The atomic number is the number of protons.

29
Cu
Copper
63.55

$64 - 29 = 35$

Number of neutrons

Round the average atomic mass to 64.

Number of protons = 29
Number of electrons = 29
Number of neutrons ≈ 35

You can follow the same steps for gold atoms.

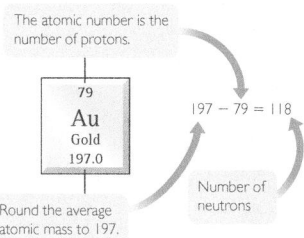

The atomic number is the number of protons.

79
Au
Gold
197.0

$197 - 79 = 118$

Number of neutrons

Round the average atomic mass to 197.

Number of protons = 79
Number of electrons = 79
Number of neutrons ≈ 118

So a gold atom has 50 more protons, 50 more electrons, and about 83 more neutrons than a copper atom.

Answers: **a.** argon, Ar **b.** lithium, Li
c. oxygen, O

Homework

Assign the reading and exercises for Alchemy Lesson 12 in the student text.

LESSON 12 ANSWERS

1. The atomic number indicates the number of protons in the nucleus of an atom.

2. The atomic mass is the mass of an atom. If the mass is given in amu, this number is equal to the number of protons and neutrons in the nucleus of the atom.

3. An atom that has 12 protons in its nucleus has an atomic number of 12. The periodic table indicates that element number 12 is magnesium.

4. The piece of information that identifies an element is its atomic number, or the number of protons in each atom of the element. The number of protons in the nucleus is a constant characteristic of all atoms of an element.

5. The atomic mass is the sum of the protons and the neutrons in the nucleus of an atom. Although boron and carbon each have six neutrons, carbon has six protons while boron only has five protons.

6.

Element	Sym.	at.#	#p	#e	#n	amu
nickel	Ni	28	28	28	31	58.69
neon	Ne	10	10	10	10	20.18
magnesium	Mg	12	12	12	12	24.31
phosphorus	P	15	15	15	16	30.97
zinc	Zn	30	30	30	35	65.38

7.

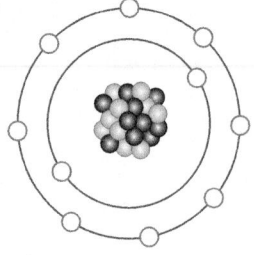

8. From lowest to highest number of protons: ● N: nitrogen, atomic number 7, 7 protons, 7 electrons, Group 5A ● Na: sodium, atomic number 11, 11 protons, 11 electrons, Group 1A ● Mg: magnesium, atomic number 12, 12 protons, 12 electrons, Group 2A ● S: sulfur, atomic number 16, 16 protons, 16 electrons, Group 6A ● Se: selenium, atomic number 34, 34 protons, 34 electrons, Group 6A ● Sr: strontium, atomic number 38, 38 protons, 38 electrons, Group 2A

How are the atoms of one element different from those of another element?

The periodic table reveals information about atomic structure. The atomic number of an element is equal to the number of protons in each of its atoms. The atomic number is also equal to the number of electrons in a neutral atom of an element. You can identify an element by the number of protons in the nucleus of an atom of the element. The protons and neutrons account for almost all the mass of an atom. Therefore, the mass of an atom in amu is approximately equal to the number of protons plus the number of neutrons. You can estimate the number of neutrons in an atom by subtracting the number of protons from the average atomic mass.

KEY TERM

atomic number

Exercises

Reading Questions

1. What does the atomic number tell you?

2. What does the atomic mass tell you?

Reason and Apply

3. If you have a sample of atoms and each atom has 12 protons in its nucleus, which element do you have?

4. If you want to identify an element, what one piece of information would you ask for? Explain your thinking.

5. Why does carbon, C, have a larger atomic mass than boron, B, even though they each have six neutrons?

6. Make a table like the one below. Use a periodic table to fill in the missing information.

Element	Chemical symbol	Atomic number	Number of protons	Number of electrons	Number of neutrons	Average atomic mass
nickel						
	Ne					
						24.31
		15				
				30		

7. Draw a simple atomic model for an atom of neon, Ne.

8. Place the following elements in order from lowest number of protons to highest number of protons: S, Mg, N, Na, Se, Sr. Then give the following information about a neutral atom of each: name, atomic number, number of protons, number of electrons, group number.

Subatomic Heavyweights

Isotopes

THINK ABOUT IT

It might surprise you that all atoms of the element copper are not exact replicas of one another. Although all copper atoms have many similarities, their masses can differ by 1 or 2 amu. What is the same and what is different about these atoms?

How can atoms of the same element be different?

To answer this question, you will explore

⤳ Isotopes

⤳ Average Atomic Mass

EXPLORING THE TOPIC

⤳ Isotopes

VARIATIONS IN THE NUCLEUS

Carbon is the sixth element on the periodic table. Its average atomic mass is listed as 12.01 amu. Why is the atomic mass listed as 12.01 rather than simply as 12?

Most of the carbon atoms that are found in nature do have a mass of 12 amu. However, for every 100 carbon atoms, it is typical to find one carbon atom with a mass of 13 amu and even more rarely a carbon atom with a mass of 14 amu. This makes the *average* atomic mass 12.01 amu. Consider the structures of these three varieties of carbon atom. A model of each is shown here.

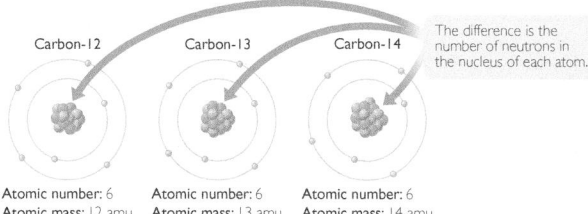

Carbon-12	Carbon-13	Carbon-14	The difference is the number of neutrons in the nucleus of each atom.
Atomic number: 6	Atomic number: 6	Atomic number: 6	
Atomic mass: 12 amu	Atomic mass: 13 amu	Atomic mass: 14 amu	

Compare the nucleus of each atom. Each has 6 protons. That is what makes them all carbon atoms. Recall that each atom of an element has the same number of protons as every other atom of that element.

However, one carbon atom has 6 neutrons, another carbon atom has 7 neutrons, and the third has 8 neutrons. The mass of an atom is the sum of the number of protons and neutrons, so one carbon atom has a mass of 12 amu and the others have masses of 13 and 14 amu. These atoms are referred to as carbon-12, carbon-13, and carbon-14 to show that they are all atoms of carbon but have different masses.

FOCUS ON UNDERSTANDING

- The idea that different atoms of the same element can be slightly different from one another and still be the same element can be confusing to students.
- There may be continued confusion over the term *atomic mass* as students refine their understanding. To assist students, we consistently use "average atomic mass" when talking about the value found in the periodic table.
- Instructors might refer to the average atomic mass value as the atomic weight. However, atomic weight is associated with moles of atoms. The term *atomic weight* is being phased out in favor of the term *relative atomic mass* or *molar mass*.

KEY TERMS

isotope
mass number
average atomic mass

IN CLASS

Students explore isotopes and their effect on the value of the average atomic mass of an element. A complete materials list is available for download.

TRM   Materials List 13

Differentiate

To assist English-language learners and struggling readers with new vocabulary terms, such as *mass number* or *atomic mass,* create a word wall in your classroom that displays the definition of the term as well as a contextualized example. To support learners who struggle with mathematical concepts or students with a visual/kinesthetic learning preference, try the University of Colorado Boulder's PhET simulation "Isotopes and Atomic Mass." You can find a list of URLs for this lesson on the Chapter 3 Web Resources document.

TRM   Chapter 3 Web Resources

Pre-AP® Course Tip

Students preparing to take an AP® chemistry course or college chemistry should be able to use mathematics to solve problems that describe the physical world, such as determining the average atomic mass of an element. In this lesson, students first learn what an isotope is and then mathematically determine the average atomic mass of an element based on the natural abundances of the isotope of that element.

Overview

CLASSWORK: PAIRS

Key Question: How can atoms of the same element be different?

KEY IDEAS

Not all atoms of an element are identical to one another. While the number of protons in the nucleus of an element is the same for all atoms of a given element, there is some variation in the number of neutrons. Atoms of an element that have different numbers of neutrons are called *isotopes*. The average atomic mass of an element, provided in the periodic table, is a weighted average of the atomic masses of the isotopes of that element.

LEARNING OBJECTIVES

- Define isotope and write and interpret the symbol for a specific isotope.
- Determine the average atomic mass of an element based on the natural abundance of isotopes of that element.
- Predict the number of protons, neutrons, and electrons in the most abundant isotope of an atom, based on average atomic mass.

<table>
<tr><td>

Lesson Guide

ENGAGE (5 min)

TRM Transparency—ChemCatalyst 13

TRM PowerPoint Presentation 13

ChemCatalyst

A chemist investigating a sample of lithium found that some lithium atoms have a lower mass than other lithium atoms. The chemist drew models of the two different types of lithium atoms, as shown here.

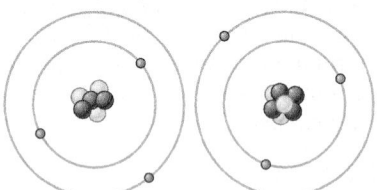

1. What is different about the two atoms?
2. What is the atomic number of each atom?
3. What is the atomic mass of each atom?

Sample answers: **1.** Their nuclei are different. One has more particles in it than the other, either more protons or more neutrons. **2.** According to the periodic table, the atomic number of lithium is 3. Both have this atomic number. **3.** The atomic mass is equal to the number of protons plus the number of neutrons, so the atomic masses are 6 amu and 7 amu.

→ Solicit students' ideas about similarities and differences between atoms of the same element.

Ask: How did you arrive at your answers for atomic number? Atomic mass?

Ask: What parts of an atom account for most of its mass?

Ask: The average atomic mass of lithium listed in the periodic table is 6.94. Where do you think this number comes from? Why isn't it a whole number?

EXPLORE (20 min)

TRM Worksheet with Answers 13

TRM Worksheet 13

INTRODUCE THE CLASSWORK

→ Arrange students into pairs.
→ Pass out the student worksheet.

</td><td>

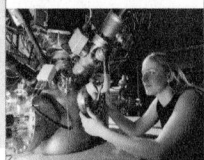

TECHNOLOGY CONNECTION

A mass spectrometer is used to determine the isotopic composition of an element. It makes an extremely accurate measurement of the mass of an individual atom, using the principle that a heavier particle will travel a straighter path through a magnetic field than a lighter particle.

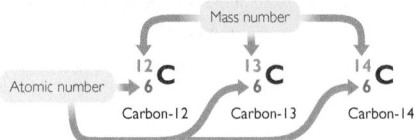

Atoms of an element that have different numbers of neutrons are called **isotopes.** Carbon has three isotopes. In nature, almost all the elements have at least two isotopes. A few elements have as many as ten isotopes.

ISOTOPE SYMBOLS

Symbols for the three isotopes of carbon are shown below. Notice that the mass of each isotope is shown as a superscript number (on top). This number is a whole number. It is the sum of the numbers of protons and neutrons in the atom and is sometimes called the **mass number.** The subscript number (on the bottom) is the atomic number of the element, in this case, 6. The number of neutrons in each atom is equal to the top number minus the bottom number.

Isotope Symbols for Carbon

$^{12}_{6}C$ $^{13}_{6}C$ $^{14}_{6}C$
Carbon-12 Carbon-13 Carbon-14

These three isotopes are nearly identical in their properties. For example, they all form the same compounds.

Big Idea Isotopes are atoms of the same element but with different numbers of neutrons.

Average Atomic Mass

You may have noticed that the **average atomic mass** values in each square of the periodic table are decimal numbers, usually to the nearest hundredth of a unit. These values are averages of the masses of the isotopes in a sample. For example, neon has an average atomic mass of 20.18 amu. How is this average calculated?

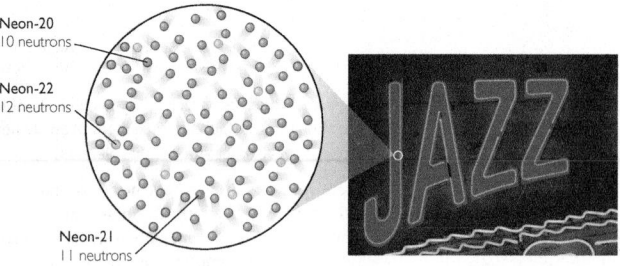

Neon-20 10 neutrons
Neon-22 12 neutrons
Neon-21 11 neutrons

The circle contains a sample of 100 atoms of neon.

</td></tr>
</table>

→ Define isotopes as atoms of an element that have different numbers of neutrons.

EXPLAIN & ELABORATE (15 min)

TRM Transparency—Sample of 100 Neon Isotopes 13

DISCUSS ISOTOPES

Ask: What is an isotope? Give an example.

Ask: How are the atoms of two isotopes different? How are they the same?

Key Points: Atoms of the same element that have different numbers of neutrons are called *isotopes.* Most elements consist of more than one type of isotope, differing only by the number of neutrons in their nuclei. Everything else about them is identical. If two atoms have different numbers of neutrons it means that their atomic masses of two atoms will not be the same even though they are atoms of the same element. The percentage of each isotope of an element that occurs in nature is called the *natural percent abundance* of the isotope. For example, carbon exists in nature as three isotopes. A sample of carbon would be 98.89% carbon-12, 1.11% carbon-13, and less than 1% carbon-14.

About 90% of all neon atoms have an atomic mass of 20 amu, 9% have an atomic mass of 22 amu, and 1% have an atomic mass of 21 amu. By considering a random sample of 100 neon atoms, you can calculate their average atomic mass like this:

$$\text{average atomic mass} = \frac{\text{total mass}}{\text{number of atoms}}$$
$$= \frac{(90)(20\,\text{amu}) + (9)(22\,\text{amu}) + (1)(21\,\text{amu})}{100}$$
$$= 20.19\,\text{amu}$$

As you can see, this number is nearly identical to 20.18, the average atomic mass listed for neon on the periodic table.

The percentage of each isotope of an element that occurs in nature is called the *natural abundance* of the isotope. For example, the natural abundance of neon-20 is about 90.48%.

Important to Know The mass of an isotope refers to the mass of a single specific atom of an element. The average atomic mass given on the periodic table is the average of the masses of all the isotopes in a large sample of that element.

Example

Isotopes of Copper

There are two different isotopes of copper. The isotope names and symbols are given here.

$${}^{63}_{29}\text{Cu} \qquad {}^{65}_{29}\text{Cu}$$
copper-63   copper-65

a. Explain why both symbols have 29 as the bottom number.
b. Explain how the two isotopes are different from each other.
c. Scientists have found the natural abundances of each isotope: 69% copper-63 and 31% copper-65. Explain why the average atomic mass listed on the periodic table for copper is 63.55.

Solution

a. Copper's atomic number is 29, so both isotope symbols have a subscript 29 indicating 29 protons.
b. The two isotopes have different atomic masses. Both isotopes have 29 protons, so copper-63 has 34 neutrons and copper-65 has 36 neutrons.
c. You could consider a sample of 100 atoms of copper and calculate their average mass. The average mass of the 100 atoms is determined by adding the masses of the 100 atoms and dividing this total by 100.

$$\frac{69(63\,\text{amu}) + 31(65\,\text{amu})}{100} = \frac{6362\,\text{amu}}{100} = 63.6\,\text{amu}$$

This is very close to the value found on the periodic table.

INTRODUCE SYMBOLS ASSOCIATED WITH ISOTOPES

→ Write the symbols for the two isotopes of boron on the board. Ask students to help decipher the information contained in the symbols.

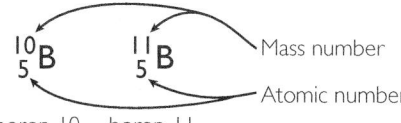

$${}^{10}_{5}\text{B} \qquad {}^{11}_{5}\text{B}$$
Mass number
Atomic number
boron-10   boron-11

Ask: What information do you think is contained in these two symbols representing isotopes of boron?

Ask: How would you write the symbol for the isotope carbon-12? $\left({}^{12}_{6}\text{C}\right)$

Ask: How would you write the symbol for the isotope carbon-13? $\left({}^{13}_{6}\text{C}\right)$

Ask: What do you think uranium-235 is?

Key Points: Chemists use a special notation to symbolize an isotope. The superscript number (on top) in this symbol is the mass number, or the mass of a single atom. It is always a whole number and is equal to the number of protons plus the number of neutrons. The bottom number is the atomic number of the element, which is equal to the number of protons. One way to refer to isotopes is to use their mass numbers. The symbols above represent boron-10 and boron-11. You might have heard of uranium-235. It

is a specific isotope of uranium, with an atomic mass of 235 amu, used in nuclear reactors.

CALCULATE THE ATOMIC MASS OF NEON

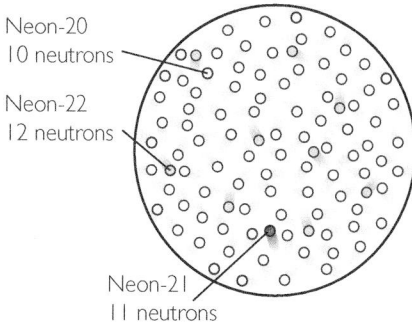

Neon-20
10 neutrons

Neon-22
12 neutrons

Neon-21
11 neutrons

→ Show the transparency, Sample of 100 Neon Atoms. Explain to the students what the drawing represents.

→ Complete the calculations for the atomic mass of neon on the board.

Ask: Can you explain in your own words why the atomic mass of an element is an average?

Ask: Can you explain how you could use the illustration to figure out the average atomic mass of the element neon?

Ask: How can you predict the most common isotope of an element?

Ask: How many protons does element 57, lanthanum, have in its nucleus? How many neutrons could it have? How did you figure that out? (57 protons— from the atomic number; around 82 neutrons—subtracted the number of protons from the average atomic mass)

Key Points: The average atomic mass of an element is the weighted average of the masses of the isotopes in a sample of the element. Scientists determine the average atomic mass by taking a weighted average of the masses of all the isotopes in a sample. For example, the drawing shows that a sample of neon atoms contains three different isotopes. Most of the neon atoms (90%) have 10 neutrons in the nucleus, while some (9%) have 12 neutrons in the nucleus. Fewer than 1% have 11 neutrons in the nucleus.

Calculations:

Neon atoms have 10 protons.
Mass of atoms with 10 neutrons:
$$10 + 10 = 20\,\text{amu}$$
Mass of atoms with 11 neutrons:
$$10 + 11 = 21\,\text{amu}$$
Mass of atoms with 12 neutrons:
$$10 + 12 = 22\,\text{amu}$$

Assume a sample of 100 atoms.
90 atoms have mass 20 amu:
$$90(20) = 1800 \text{ amu}$$
1 atom has mass 21 amu: $1(21) = 21$ amu
9 atoms have mass 22 amu:
$$9(22) = 198 \text{ amu}$$

The total mass of 100 atoms:
$$1800 + 21 + 198 = 2019 \text{ amu}$$

The average atomic mass of the neon in this sample is 2019/100, or 20.19 amu. This is very close to the value 20.1797 listed in the periodic table.

The most common isotope of an element often has a mass that is close to the average atomic mass given in the periodic table. To find the number of neutrons in the most common isotope, round the average atomic mass to the nearest whole number, then subtract the number of protons.

EVALUATE (5 min)

Check-In

Nitrogen, N, has two naturally occurring isotopes. Predict the number of neutrons in the two isotopes of nitrogen. Which isotope do you predict to be more abundant? How do you know?

Answer: Nitrogen's average atomic mass is 14.01 amu, so we can expect two isotopes: nitrogen-14 and nitrogen-15. The first has seven neutrons, and the second has eight neutrons. Nitrogen-14 must be more abundant, because the average atomic mass is 14.01, barely greater than 14. Indeed, 99.63% of the nitrogen found naturally is nitrogen-14. Only 0.37% of the nitrogen found is nitrogen-15.

Homework

Assign the reading and exercises for Alchemy Lesson 13 in the student text.

LESSON 13 ANSWERS

1. Atomic number refers to the number of protons in the nucleus of an atom. Atomic mass, when expressed in amu, is the sum of the number of protons and the number of neutrons.

2. *Possible answer:* The mass of an atom in amu can be found by adding the number of protons to the number of neutrons in that particular atom. This number is always a whole number. The

---

Example

(continued)

Otherwise, you can convert each isotope's percent natural abundance to a decimal number, then multiply this by the isotope's mass number. Do this for each isotope, then add the products:

$$69\% = \frac{69}{100} = 0.69 \qquad 31\% = \frac{31}{100} = 0.31$$

$$(0.69)(63 \text{ amu}) + (0.31)(65 \text{ amu}) = 63.6 \text{ amu}$$

LESSON SUMMARY

How can atoms of the same element be different?

KEY TERMS
isotope
mass number
average atomic mass

Elements are composed of nearly identical atoms, each with the same number of protons. However, not every atom of an element has the same number of neutrons in its nucleus. Atoms of an element with different numbers of neutrons are called isotopes. Because neutrons account for part of the mass of an atom, isotopes have different masses. The average atomic mass is an average of the masses of all the different isotopes, taking natural abundance into account.

Exercises

Reading Questions

1. Explain the differences between atomic number and atomic mass.
2. Explain the difference between the average atomic mass given on the periodic table and the mass of an atom.

Reason and Apply

3. How are potassium-39, potassium-40, and potassium-41 different from each other? Write the isotope symbols for the three isotopes of potassium.
4. How many protons, neutrons, and electrons are in each?
   a. fluorine-23   **b.** $^{59}_{27}\text{Co}$ **c.** molybdenum-96
5. An isotope of iron, Fe, has 26 protons and 32 neutrons.
   a. What is the approximate mass of this isotope?
   b. How would you write the symbol for this isotope?
6. Find the element phosphorus, P, on the periodic table.
   a. What is the average atomic mass of phosphorus?
   b. What is its atomic number?
   c. Predict which isotope you would find in greatest abundance for phosphorus.
7. Chlorine, Cl, is 76% chlorine-35 and 24% chlorine-37. Determine the average atomic mass of chlorine.
8. Lithium, Li, is 7.6% lithium-6 and 92.4% lithium-7. Determine the average atomic mass of lithium.
9. Which isotope of nitrogen is found in nature? Explain your reasoning.
   a. $^{7}_{14}\text{N}$ **b.** $^{14}_{7}\text{N}$ **c.** $^{15}_{6}\text{N}$

---

average atomic mass given on the periodic table is the average of the masses of all the isotopes in a large sample of that element. The periodic table in the textbook lists the average atomic mass to four significant digits.

3. *Possible answer:* The isotopes differ from one another in the number of neutrons in their nuclei. Each potassium isotope has 19 protons, but potassium-39 has 20 neutrons, potassium-40 has 21 neutrons, and potassium-41 has 22 neutrons. The isotopes are $^{39}_{19}\text{K}$, $^{40}_{19}\text{K}$, and $^{41}_{19}\text{K}$.

4. a. Fluorine-23 has 9 protons, 14 neutrons, and 9 electrons.

b. $^{59}_{27}\text{Co}$ has 27 protons, 32 neutrons, and 27 electrons. **c.** Molybdenum-96 has 42 protons, 54 neutrons, and 42 electrons.

5. a. 58 amu **b.** $^{58}_{26}\text{Fe}$

6. a. 30.97 amu **b.** 15 **c.** Phosphorus-31.

7. 35.48 amu

8. 6.920 amu

9. Nitrogen has an atomic number of 7. The only isotope symbol of the three choices that has an atomic number of 7 is $^{14}_{7}\text{N}$. The correct answer is B.

Isotopia

Stable and Radioactive Isotopes

THINK ABOUT IT

Some isotopes are found in nature and others exist only because they were created in the laboratory or under unusual conditions. If you went digging in a copper mine and analyzed the samples, you would find copper atoms with either 34 or 36 neutrons. But do any other isotopes of copper exist?

What types of isotopes do the various elements have?

To answer this question, you will explore
- Naturally Occurring Isotopes
- Stable and Radioactive Isotopes

EXPLORING THE TOPIC

○ Naturally Occurring Isotopes

When an atom or element can be found somewhere on Earth, it is called *naturally occurring*. Most chemists agree that there are about 92 naturally occurring elements on the periodic table. The rest of the elements on the table have existed only as a result of human activity or unusual circumstances.

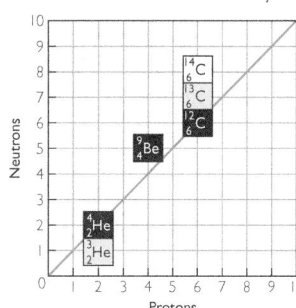

The element beryllium, Be, has only one naturally occurring isotope. This means all beryllium atoms have four protons and five neutrons. The element helium, He, has two naturally occurring isotopes. Its atoms have two protons and either one or two neutrons. Helium and beryllium are shown plotted on a graph of neutrons versus protons. Notice that helium-4 lies on the diagonal line, because its neutrons and protons are equal in number.

Now consider an element like carbon, C, with more than one isotope. Its isotopes, carbon-12, carbon-13, and carbon-14, are also on the graph. Notice that carbon's isotopes line up vertically. This is because they all have 6 protons. One of carbon's isotopes, with 6 protons and 6 neutrons, lies on the diagonal. The other two isotopes have more neutrons than protons in their nuclei.

ISOTOPES OF THE FIRST 95 ELEMENTS

The graph on page 66 has been expanded to include *all* of the isotopes of the first 95 elements. As before, each square plotted on the graph represents a different isotope. Take a few minutes to study the graph.

Lesson 14 | **Stable and Radioactive Isotopes**   65

Overview

ACTIVITY: PAIRS

Key Question: What types of isotopes do the various elements have?

KEY IDEAS

Radioactive isotopes have unstable nuclei that emit particles in a process called *radioactive decay*. The neutron-to-proton ratio in the nucleus is key to the stability of an isotope. In atoms with a low atomic mass, the ratio of neutrons to protons is close to 1:1. In larger nuclei, the ratio is larger.

LEARNING OBJECTIVES

- Interpret a graph of naturally occurring isotopes.
- Describe the general nuclear composition of a stable nucleus.
- Differentiate between a stable isotope and a radioactive isotope

FOCUS ON UNDERSTANDING

- Many more isotopes than are listed in the graph of neutrons versus protons can be created in the laboratory. Many very rare isotopes have been left off the graph to keep it from being cluttered.

- Radioactivity is a random and spontaneous process. We cannot predict when a specific atom will decay. Also, radioactive isotopes vary greatly in their rate of decay, so samples of what we call "unstable" isotopes can exist for millions of years.

KEY TERM

radioactive isotope

IN CLASS

This lesson focuses on a graph that plots the number of neutrons in a nucleus against the number of protons for elements in the periodic table up to atomic number 95. Radioactive isotopes and the most abundant isotopes of each of these elements are also indicated. Students work with the graph, extracting information and discovering trends. They learn about the composition of the nucleus of isotopes that are considered naturally occurring. Students are introduced to the concept of radioactivity, which is covered in greater depth in the next two lessons. Half-life is covered in Lesson 16: Old Gold. A complete materials list is available for download.

TRM Materials List 14

Differentiate

Help students develop scientific literacy with the concepts of stable versus unstable isotopes using a demonstration from the American Nuclear Society titled "Modeling Radioactive & Stable Atoms." To challenge your advanced learners, have them research and create a classroom brochure on useful radioisotopes in medicine and technology. You can find a list of URLs for this lesson on the Chapter 3 Web Resources document.

TRM Chapter 3 Web Resources

Pre-AP® Course Tip

Students preparing to take an AP® chemistry course or college chemistry should be able to use representations, such as this lesson's graphical analysis of naturally occurring isotopes, to discern between stable and unstable isotopes.

Lesson Guide

ENGAGE (5 min)

TRM PowerPoint Presentation 14

Lesson 14 | **Stable and Radioactive Isotopes**   65

ChemCatalyst

Which of the following are isotopes of copper, CU? Explain your reasoning.

A. $^{63}_{29}Cu$

B. $^{197}_{79}Au$

C. $^{63}_{28}Cu$

D. $^{87}_{29}Cu$

E. $^{34}_{29}Cu$

F. $^{65}_{29}Cu$

Sample answer: **A** and **F** are the only isotopes of copper. **B** is gold (Au). **C** has atomic number 28, so it is not copper. **D** has the correct atomic number but far too many neutrons to be an isotope of copper. **E** also has the correct atomic number, but it has far too few neutrons to be an isotope of copper. Students might say that **D** and **E** are isotopes of copper because they have learned that atomic number indicates what element you have. Today's lesson will help clarify why these are not isotopes of copper.

→ Sample students' understanding of isotope symbols.

Ask: Which isotope symbols represent atoms of copper? Why?

Ask: Which number indicates that you have an atom of copper? (the atomic number: 29)

Ask: Why can atoms of copper have different superscript numbers in the isotope symbol?

Ask: Are atomic masses of 87 and 34 reasonable for copper? Why or why not?

Ask: What are the names of the isotopes of copper? (copper-63 and copper-65)

EXPLORE (20 min)

TRM Worksheet with Answers 14

TRM Worksheet 14

TRM Handout—Chart of Naturally Occurring Isotopes 14

INTRODUCE THE ACTIVITY

→ Have students work in pairs.

→ Pass out student worksheets and the Handout—Chart of Naturally Occurring Isotopes 14. Make sure students also have access to a periodic table.

TECHNOLOGY CONNECTION

Technetium, Tc, and promethium, Pm, are not found in nature. They are human-made isotopes that are very unstable. Two forms of technetium, though, can be used as radioactive tracers in living tissues for research and for diagnosis.

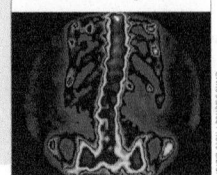

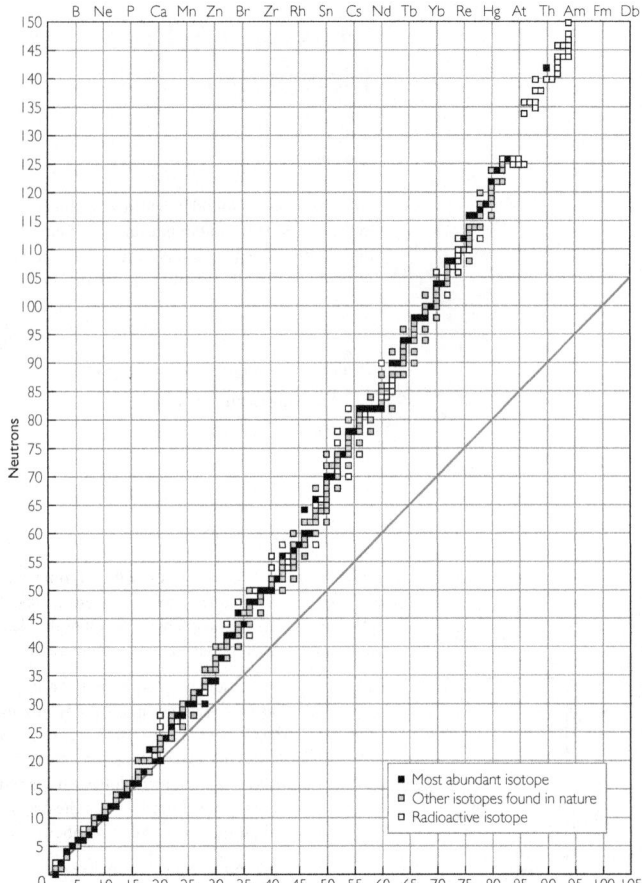

Isotopes of the Elements

Here are some things you might notice:

- Some elements have only one isotope. However, *most* elements have more than one isotope.
- The squares that lie on the diagonal line represent isotopes with equal numbers of protons and neutrons.
- Except in hydrogen-1 and helium-3 atoms, the number of neutrons in an atom's nucleus is always equal to or greater than the number of protons. So most of the points lie on or above the diagonal line.

EXPLAIN & ELABORATE (15 min)

DISCUSS NATURALLY OCCURRING ISOTOPES

Ask: What are three things you learned from your examination of the isotope chart?

Ask: What do you think we mean when we say that an isotope is naturally occurring?

Ask: How does the isotope chart compare to the periodic table? When would you use each?

Ask: What information would have to be on the isotope chart so that you can determine the average atomic mass of an element?

Key Points: The graph of naturally occurring isotopes gives us an idea of how many different isotopes of the elements are found in nature. Some elements (e.g., fluorine or phosphorus) have only one naturally occurring isotope, while other elements (e.g., tin or calcium) have several naturally occurring isotopes. Isotopes that can only be made in a laboratory (e.g., technetium or promethium) are not considered naturally occurring.

The words *atom*, *isotope*, and *element* are interrelated. To review: All matter is made up of unique building materials called *elements*. Elements

TECHNOLOGY CONNECTION

Rubidium-strontium dating is a method used to determine the age of geological and lunar rock samples. It is based on the fact that rubidium-87 decays over time to become strontium-87.

- Radioactive isotopes are not common. (You will learn more about radioactivity later in this lesson.)
- No element has more than 10 isotopes. Most elements have between 1 and 6 isotopes.
- The number of neutrons is roughly equal to the number of protons for atoms up to atomic number 20.
- The collection of plotted points curves up, away from the diagonal line. Beyond atomic number 20, elements have considerably more neutrons than protons.
- The majority of radioactive isotopes are elements with atomic numbers above 80.
- The elements with even numbers of protons have more isotopes than elements with odd numbers of protons. Even numbers of neutrons are also more common than odd numbers.

Stable and Radioactive Isotopes

When atoms are not stable, they emit small particles. This means that small bits of the nucleus come flying out of the atom. These less stable isotopes are called **radioactive isotopes.** They decay over time as particles are spontaneously emitted from the nucleus in a process called *radioactive decay.* You will learn more about radioactive decay in upcoming lessons.

As you saw in the graph, hydrogen, H, argon, Ar, and cerium, Ce, are examples of elements that have radioactive isotopes shown with white squares. There are only a few white squares scattered throughout the graph until you reach polonium, Po, element number 84. However, technetium, Tc, promethium, Pm, and every element from polonium and beyond, have *only* radioactive isotopes.

STABILITY STARTS IN THE NUCLEUS

A stable isotope has a stable nucleus. Stability is related to the balance between the number of protons and the number of neutrons in the nucleus. An atom that doesn't have enough neutrons will disintegrate. So will an atom with too many neutrons. The larger the atom, the more neutrons it takes to make a stable nucleus. For example, isotopes of element number 74, tungsten, W, require at least 110 neutrons to be stable.

Example

Hafnium-144

Do you expect the isotope hafnium-144 to exist in nature? Explain your thinking.

Solution

The task here is to find out if there is a square on the isotope graph that corresponds to hafnium-144. The way to do this is to figure out how many protons and neutrons are in hafnium-144. Use the periodic table to find the atomic number of hafnium, Hf. The atomic number is 72, so a hafnium atom has 72 protons. An isotope of hafnium with a mass of 144 has $144 - 72 = 72$ neutrons. There is no square on the graph corresponding to 72 protons and 72 neutrons. So a hafnium atom with only 72 neutrons does not exist in nature.

Lesson 14 | **Stable and Radioactive Isotopes**   67

are made up of individual particles called *atoms*. All the atoms of an element are identical, with one exception: Some atoms of an element may have different numbers of neutrons. An isotope is one type of atom of an element with a particular number of neutrons.

DISCUSS THE NUCLEAR COMPOSITION OF ISOTOPES

Ask: What patterns did you notice in the numbers of neutrons and protons?

Ask: What does the slanted line on the isotope graph indicate? (1:1 ratio of neutrons to protons)

Ask: What is the ratio of neutrons to protons for small atoms? (about 1:1) Large atoms? (about 1.4:1)

Key Points: Nearly all atoms have at least one neutron for every proton in the nucleus. Points that lie on the diagonal line on the graph indicate isotopes with equal numbers of neutrons and protons. Many of the most abundant isotopes with atomic numbers less than or equal to 40 have equal numbers of protons and neutrons.

Many isotopes have more neutrons than protons. The curve described by the data points gets farther from the line as the atomic number increases,

indicating greater numbers of neutrons than protons. The more protons an atom has (the larger it is), the more neutrons it needs in the nucleus to be stable. For the largest atoms, this ratio is close to three neutrons to every two protons. One explanation is that the neutrons serve as a neutral packing material in the nucleus to help hold the positively charged protons together.

BRIEFLY INTRODUCE THE CONCEPT OF RADIOACTIVITY

Ask: How common are radioactive isotopes on the chart? Explain.

Ask: Do any of the elements on the chart have both radioactive and nonradioactive isotopes?

Ask: Do any of the elements on the chart have only radioactive isotopes? If so, which ones?

Ask: When an isotope is referred to as unstable, what do you think is meant?

Key Points: A handful of the isotopes on the chart are unstable. This means that over time they emit particles from the nucleus; they decay. The white squares on the graph designate these unstable isotopes, which are called *radioactive isotopes*. This means that some atoms are lost over time because they undergo radioactive decay or spontaneously emit particles from their nuclei. Radioactive decay is covered in the next two lessons.

Some elements have a naturally occurring radioactive isotope. For example, hydrogen-3 (also called *tritium*) is hydrogen's radioactive isotope. Each tritium isotope has one proton and two neutrons in its nucleus. Tritium isotopes are rare, possibly only one per billion or so hydrogen atoms. Argon, technetium, cerium, and promethium also have radioactive isotopes. Technetium and promethium are special cases—they are not found in nature at all but are human-made and very unstable.

The isotopes of the elements after bismuth (atomic number 84 and up) are all radioactive. Moreover, all the elements after uranium (atomic number 93 and up) on the periodic table were discovered in the laboratory and are radioactive. Most of the human-made elements disintegrate rapidly, with some existing only for microseconds.

Radioactive isotope: Any isotope that has an unstable nucleus and decays over time.

EVALUATE (5 min)

Check-In

1. Use the chart to determine how many neutrons you would need to make a stable element with 79 protons.

2. What element is this? Write its isotope symbol.

Answers: 1. A stable element with 79 protons would have about 118 neutrons in the nucleus. **2.** This element is gold-79, $^{197}_{79}$Au.

Homework

Assign the reading and exercises for Alchemy Lesson 14 in the student text.

LESSON 14 ANSWERS

1. Possible answer: An element is a fundamental building block of matter. An atom is the smallest possible unit of an element. An atom is the smallest unit of an element that still has the same characteristics as the element.

2. Possible answer: An isotope is an atom with a specific number of protons and neutrons. The word *atoms* can refer to a group of atoms with different mass numbers.

3. Oxygen has 3 stable isotopes.
● Neodymium has 5 stable isotopes.
● Copper has 2 stable isotopes. ● Tin has 10 stable isotopes.

4. A black box at (12, 12) represents $^{24}_{12}$Mg. ● A pink box at (35, 46) represents $^{81}_{35}$Br. ● No box is present at (60, 92) for $^{152}_{60}$Nd. ● A black box at (78, 117) represents $^{195}_{78}$Pt. ● A white box at (92,146) represents $^{238}_{92}$U. Uranium-238 is naturally occurring even though it is not stable.

5. The diagonal line on the graph represents isotopes that have equal numbers of protons and neutrons, because the line passes through points that have the same *x*- and *y*-coordinate.

6. a. Possible answer: helium-4, $^{4}_{2}$He; boron-10, $^{10}_{5}$B; and oxygen-16, $^{16}_{8}$O. Any three isotopes that fall on the diagonal line are acceptable. The mass number will be twice the atomic number. **b. Possible answer:** calcium-40, $^{40}_{20}$Ca, and calcium-42, $^{42}_{20}$Ca. Any two isotopes that have the same atomic number, representing the same element, are acceptable. **c. Possible answer:** fluorine-19, $^{19}_{9}$F, and neon-20, $^{20}_{10}$Ne.

Example (continued)

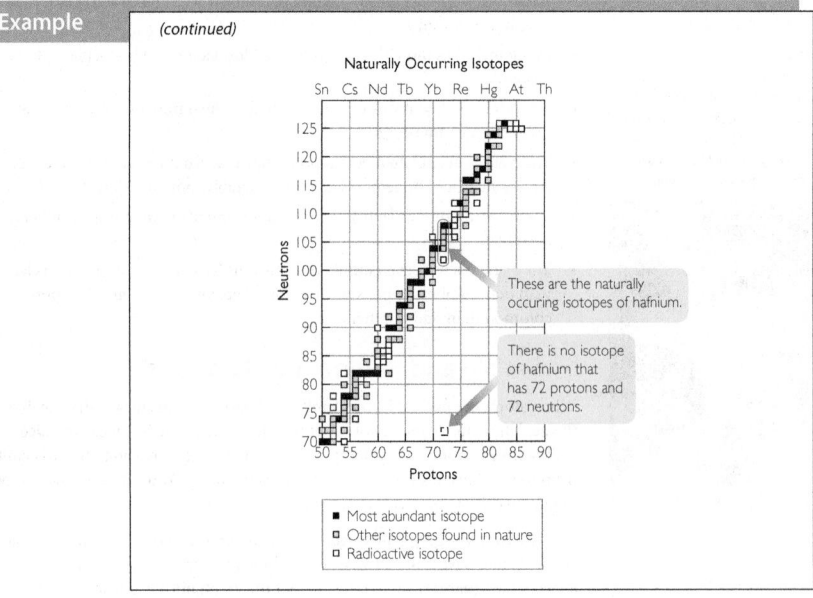

Naturally Occurring Isotopes

These are the naturally occurring isotopes of hafnium.

There is no isotope of hafnium that has 72 protons and 72 neutrons.

■ Most abundant isotope
□ Other isotopes found in nature
▫ Radioactive isotope

LESSON SUMMARY

What types of isotopes do the various elements have?

KEY TERM

radioactive isotope

Most elements have more than one naturally occurring isotope. Most of these isotopes are stable. Stable isotopes have stable nuclei, with just the right balance of protons and neutrons. In atoms with atomic numbers up to 20, the number of neutrons is roughly equal to the number of protons. Atoms with atomic numbers between 20 and 84 require progressively more neutrons in the nucleus to attain stability. Unstable isotopes are called radioactive isotopes. The nucleus of an unstable isotope will decay and emit radioactive particles. All elements beyond atomic number 83 are unstable, and therefore, all their isotopes are radioactive.

Exercises

Reading Questions

1. Explain the relationship between the words *atom* and *element*.
2. Explain the relationship between the words *atom* and *isotope*.

Reason and Apply

3. Find these elements on the isotope graph on page 66. How many stable isotopes does each element have?

    oxygen, O   neodymium, Nd   copper, Cu   tin, Sn

68   Chapter 3 | **A World of Particles**

Any two isotopes with the same variation between their mass number and atomic number are acceptable. **d. Possible answer:** zinc-70, $^{70}_{30}$Zn, and germanium-70, $^{70}_{32}$Ge. Any two isotopes with the same mass number are acceptable.

7. No stable isotopes exist beyond $^{40}_{20}$Ca that have equal numbers of protons and neutrons. The assumption must be made that a stable isotope would occur naturally if it existed (and therefore be plotted on the graph of the isotopes of the first 95 elements).

8. Each drawing should have 8 protons and 8 electrons, 2 in the inner orbit and 6 in the outer orbit. The only difference should be in the number of neutrons represented in the nucleus. Oxygen-16 should have 8 neutrons, oxygen-17 should have nine neutrons, and oxygen-18 should have 10 neutrons.

9. Possible answer: polonium: polonium-209 (1.49) and polonium-210 (1.50) ● astatine: astatine-210 (1.47) and astatine-211 (1.48) ● radon: radon-211(1.45), radon-220 (1.56), and radon-222 (1.58) ● francium: francium-223 (1.56) ● radium: radium-223 (1.53), radium-224 (1.55), radium-226 (1.57), and radium-228 (1.59) The possible elements are technetium (atomic number 43), promethium (atomic

(continued)

4. Use the isotope graph to determine which of these isotopes would be found in nature.

$$^{24}_{12}\text{Mg} \qquad ^{81}_{35}\text{Br} \qquad ^{152}_{60}\text{Nd} \qquad ^{195}_{78}\text{Pt} \qquad ^{238}_{92}\text{U}$$

5. What does the diagonal line on the graph represent?

6. Use the isotope graph to find the isotopes described below. In each case, give the isotope name and the isotope symbol.
   a. Find three isotopes with equal numbers of neutrons and protons.
   b. Find two isotopes with the same number of protons.
   c. Find two isotopes with the same number of neutrons.
   d. Find two isotopes with the same atomic mass units.

7. Is an atom with a nucleus of 31 protons and 31 neutrons a stable isotope? Why or why not?

8. Draw basic atomic models for all the stable isotopes of oxygen, O.

9. Name five elements on the periodic table that have only radioactive isotopes. Determine the neutron-to-proton ratio for each of their isotopes, in decimal form.

10. How many protons could a stable atom with 90 neutrons have? Which elements would these be?

11. Would an atom with 60 protons and a mass number of 155 be stable?

12. Which of the following isotopes are likely to be found in nature? Identify the element by name if it is an isotope that might be found in nature.

$$^{162}_{63}? \qquad ^{75}_{33}? \qquad ^{112}_{56}? \qquad ^{260}_{88}?$$

13. Choose the best answer. The diagonal line on the isotope graph on page 66 represents nuclei with
   (A) the same number of protons.
   (B) the same number of neutrons.
   (C) the same number of protons as neutrons.
   (D) the same number of protons plus neutrons.

14. Choose the best answer. In general, elements with an even atomic number
   (A) have only one isotope.
   (B) have more isotopes than those with an odd atomic number.
   (C) have the same number of protons as neutrons.
   (D) are halogens.

Lesson 14 | **Stable and Radioactive Isotopes**   69

number 61), and any elements with an atomic number greater than 83 except thorium (atomic number 90). Each ratio is determined by dividing the number of neutrons by the number of protons.

10. Stable elements with 90 neutrons can have 62, 63, 64, or 66 protons. These elements are samarium, europium, gadolinium, and dysprosium. The assumption must be made that a stable isotope would occur naturally if it existed.

11. No, an atom with 60 protons and a mass of 155 would have 95 neutrons. The graph of isotopes of the first 95 elements indicates that no stable isotope has 60 protons and 95 neutrons.

12. According to the graph of isotopes of the first 95 elements $^{75}_{33}\text{Ar}$ is an isotope of arsenic that occurs in nature. $^{162}_{63}?$, $^{112}_{52}?$, and $^{260}_{88}?$ are not likely to occur in nature.

13. C

14. B

Overview

ACTIVITY: GROUPS OF 4

Key Question: What are nuclear reactions?

KEY IDEAS

Nuclear chemistry is the chemistry of the nucleus of the atom. Changes to the nucleus come in three main forms: radioactive decay, fission, and fusion. Radioactive decay includes alpha decay, beta decay, and gamma radiation. Fission is a process by which the nucleus breaks into smaller pieces, transferring large amounts of energy. The reverse, nuclear fusion, is the process by which nuclei combine to make a larger nucleus. Most of these nuclear processes involve changes in atomic number, which mean changes in atomic identity.

LEARNING OBJECTIVES

- Explain the different processes involved in nuclear changes and the conditions required for those processes.
- Explain the connection between nuclear changes and changes in atomic identity.

FOCUS ON UNDERSTANDING

- Students might be a bit overwhelmed by all the new terminology in this lesson. However, the next lesson is also devoted to nuclear chemistry and will help consolidate student learning.
- Some nuclear processes—such as positron emission and neutron capture—are not covered in this lesson.
- There is some natural overlap in these concepts. For instance, although fission is not considered a form of radioactive decay, radioactive decay may accompany a fission process. Also, alpha decay produces two new nuclei but is not usually considered fission. You can downplay these details. What is important is that students grasp the larger concepts.

KEY TERMS

nuclear reaction	half-life
radioactive decay	radiation
alpha decay	gamma ray
alpha particle	fission
beta decay	fusion
beta particle	

Featured ACTIVITY

Sort cards into two piles.

Nuclear Quest

Purpose

To explore nuclear chemistry.

Instructions

Play Nuclear Quest. You will need a game board, a pair of dice, Nuclear Quest cards, Gamma Radiation cards, and a game piece for each player.

Goal

Each player rolls the dice to move along the periodic table, then draws a card from the Nuclear Quest pile and does what the card says. The goal of the game is to discover element 117, give it your own name, and find out if scientists have given it an "official" name. This is accomplished by using nuclear chemistry to proceed through the periodic table.

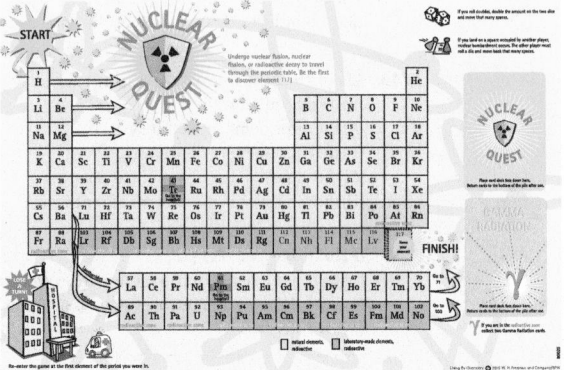

Making Sense

Write a paragraph describing what you learned about nuclear chemistry from this game.

IN CLASS

Students are exposed to the vocabulary and general concepts of nuclear science through a game called Nuclear Quest. The goal of the game is to journey through the periodic table using nuclear processes. The first student to reach element 117 wins the game. The team then works together to name the new element. After all the teams have finished the game, they complete a short student worksheet to find out what information they discovered about nuclear changes. A complete materials list is available for download.

TRM Materials List 15

SETUP

Before class, inventory the game materials. If you like, read the instructions and play a practice round.

Nuclear Reactions

THINK ABOUT IT

When the nucleus of an atom is changed in some way, a nuclear reaction is taking place. Nuclear reactions take place in the Sun and sometimes on Earth. Scientists can also make nuclear reactions happen in laboratories and in nuclear power plants.

What are nuclear reactions?

To answer this question, you will explore

- Nuclear Changes
- Radioactive Decay
- Fission and Fusion

EXPLORING THE TOPIC

Nuclear Changes

Changes in the nucleus are called **nuclear reactions.** Nuclear reactions differ greatly from chemical reactions. Nuclear reactions involve changes to the nucleus and can change one element into another element. The nucleus of an atom is not so easy to change.

Some nuclear reactions don't require any help to get started. But changing the nucleus of an atom *on purpose* can require massive amounts of energy, so you would not be able to do this in your chemistry lab. There are several ways the nucleus of an atom can change. The nucleus can lose particles, it can be split into smaller nuclei, or it can combine with another nucleus or particle. Nuclear changes can happen only under certain circumstances. They happen without intervention in radioactive elements, scientists make them happen, with great difficulty, in a nuclear reactor, and they happen all the time in the cores of stars.

Nuclear chemistry occurs in the fiery cores of stars.

Radioactive Decay

Recall that radioactive isotopes are unstable. By ejecting a particle, the nucleus of a radioactive atom can become stable. The process of ejecting or emitting pieces from the nucleus of an atom is called **radioactive decay.**

Scientists have identified the various *subatomic particles,* particles smaller than the atoms themselves, that are emitted from or shot out of the nucleus during radioactive events. These include alpha particles and beta particles.

Sample answer: The first four cards in the first row all involve changes to the nucleus, as do the first and last cards in the second row. These cards move players around on the game board, which is a periodic table.

→ Discuss the cards in the Nuclear Quest game.

Ask: Which cards change the nucleus of one element into the nucleus of an element with a larger atomic number? How can you tell? (cards 1, 3, 4, and 6; because you move ahead in the periodic table)

Ask: Which cards change the nucleus of one element into the nucleus of an element with a smaller atomic number? (cards 2 and 10; because the nucleus gets smaller and you move back)

Ask: What is alpha decay? Beta decay? (loss of two protons and two neutrons; the neutron turning into a proton and shooting off an electron)

Ask: What should you do when you get radiation sickness during this game? (Go to the hospital.)

EXPLORE (30 min)

TRM Worksheet with Answers 15

TRM Worksheet 15

INTRODUCE THE ACTIVITY

→ Arrange students into groups of four.

→ Pass out the student worksheet.

→ Inform students that the goal of the game is to discover element 117, give it your own name, and find out if scientists have given it an "official" name. This is accomplished by moving the nucleus through the entire periodic table.

→ Briefly review the game instructions. Playing the game is the best way to learn its nuances, even if all the rules are not understood before playing. There may be time to play the game more than once.

Lesson Guide

ENGAGE (5 min)

TRM Handout—Nuclear Quest 15

TRM Transparency—ChemCatalyst 15

TRM PowerPoint Presentation 15

TRM Nuclear Quest game board 15

TRM Card Masters—Nuclear Quest 15

TRM Card Masters—Gamma Radiation 15

ChemCatalyst

Hand out one Nuclear Quest game to each group. Ask students to find one of each of the ten kinds of cards in the decks of cards. Which cards cause the nucleus of one element to change into the nucleus of a different element in the game?

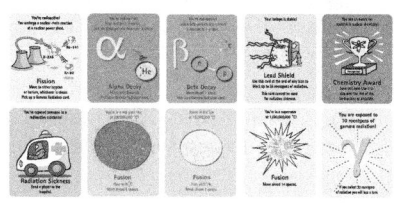

EXPLAIN & ELABORATE (10 min)

PROCESS THE NUCLEAR QUEST GAME

→ You might want to draw diagrams like those below to illustrate alpha decay and beta decay.

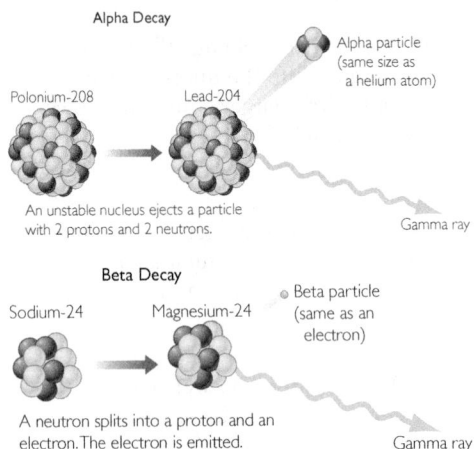

Alpha Decay

Polonium-208 → Lead-204

Alpha particle (same size as a helium atom)

An unstable nucleus ejects a particle with 2 protons and 2 neutrons.

Gamma ray

Beta Decay

Sodium-24 → Magnesium-24

Beta particle (same as an electron)

A neutron splits into a proton and an electron. The electron is emitted.

Gamma ray

Ask: What do you think the game is trying to teach you? Describe it in your own words.

Ask: What happened to your nucleus that allowed you to move it around the board?

Ask: Are nuclear changes dangerous? Explain your thinking.

Ask: What do you think fission is? Fusion?

Ask: What did your team choose to name element 117 in the game?

Ask: What name did scientists decide on for new element 117?

Key Points: Nuclear chemistry is the study of changes to the nucleus. There are several ways the nucleus of an atom can change: radioactive decay, fission, and fusion. When a nucleus emits particles, it is called *radioactive decay*. Alpha decay and beta decay are ways in which the nucleus loses particles. (*Note:* Radioactive decay will be defined more fully in the next lesson.) The splitting of a nucleus into two smaller nuclei is called *fission*. The joining of nuclei is called *fusion*. Fission, alpha decay, and beta decay are often accompanied by gamma radiation, which is the release of a high-energy photon. Radiation is potentially harmful to living things because it damages tissue. However, a thin sheet of lead stops most forms of radiation, though gamma radiation requires lead shields several centimeters thick for effective protection. (*Note:* The vocabulary below is covered extensively in the student edition and in the next lesson. It is not necessary to go over each term.)

Important to Know Many elements have both stable and radioactive isotopes. For example, carbon-12 and carbon-13 are stable while carbon-14 is radioactive.

ALPHA DECAY

Alpha decay involves the ejection of an alpha particle from the nucleus of an atom. An **alpha particle** consists of two protons and two neutrons and can be represented by the Greek letter α, alpha. Because an alpha particle has two protons, it is the same as the nucleus of a helium atom.

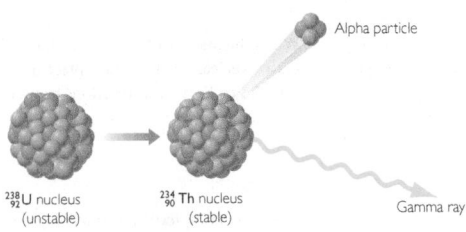

Alpha particle

$^{238}_{92}$U nucleus (unstable) → $^{234}_{90}$Th nucleus (stable)

Gamma ray

An alpha particle carries two protons away from the nucleus of an atom. This changes the identity of the element. So, when an atom of uranium, U, atomic number 92, undergoes alpha decay, an atom of thorium, Th, atomic number 90, is formed. This is why alpha decay causes you to move two spaces backward on the Nuclear Quest game board. Notice that a gamma ray is often released as a result of alpha decay.

HISTORY CONNECTION

Marie Curie received the Nobel Prize for her groundbreaking work on radioactive substances. She was the first woman to win the prize, and she actually won two Nobel Prizes—one in Physics (1903) and one in Chemistry (1911). Like other pioneers, she was unaware of the dangers of radioactive samples and sometimes carried them around in her pockets, which probably cut short her life.

Periodic Table

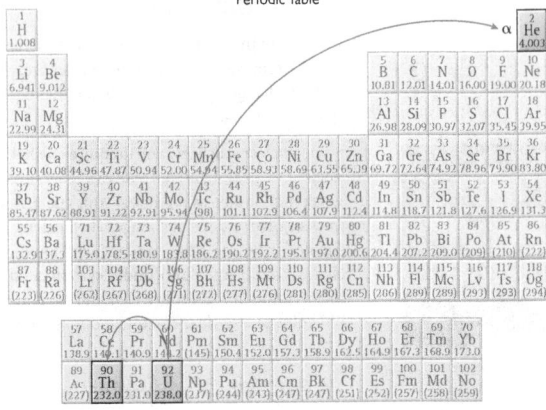

Big Idea The elemental identity of an atom is determined by the number of protons it has.

BETA DECAY

Beta decay involves the ejection of an electron from the nucleus of an atom. The ejected electron is called a **beta particle**. Beta particles can be represented by the Greek letter β, beta. You may wonder how an electron got into the nucleus of an

Nuclear reaction: A process that involves changes to the nucleus of an atom.

Radioactive decay: A spontaneous process by which an atom emits radiation or a particle from its nucleus to become more stable.

Fusion: The joining of two nuclei to form a larger nucleus accompanied by a release of energy.

Fission: The splitting apart of an atomic nucleus into two smaller nuclei accompanied by a release of energy.

Alpha particle: A particle composed of two protons and two neutrons, equivalent to the nucleus of a helium atom.

Beta particle: An electron emitted from the nucleus of an atom during beta decay.

Gamma ray: A form of high-energy electromagnetic radiation emitted during nuclear reactions.

Alpha decay: A nuclear reaction in which an atom emits an alpha particle. Its atomic number decreases by two, and its mass number decreases by four.

atom. In fact, electrons *do not* exist by themselves within the nucleus. However, in beta decay a neutron can split into two parts, becoming an electron and a proton. The electron is ejected and the proton stays behind in the nucleus.

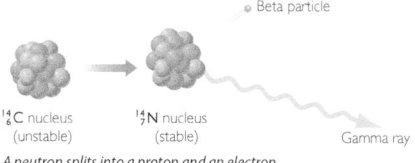

$^{14}_{6}$C nucleus
(unstable)

$^{14}_{7}$N nucleus
(stable)

Beta particle

Gamma ray

A neutron splits into a proton and an electron.

Removal of an electron from the nucleus of an atom through this unique process leaves that atom with one less neutron, one more proton, and a new identity. This is why beta decay causes you to move one space forward in the Nuclear Quest game. Beta decay can also result in the release of gamma rays.

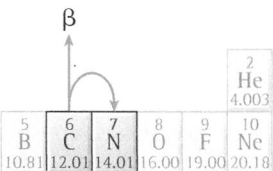

Example

Plutonium-241

The radioactive isotope plutonium-241, $^{241}_{94}$Pu, decays by emitting a beta particle. What isotope is formed?

Solution

The atomic number of plutonium is 94, so it has 94 protons. In beta decay, a neutron becomes a proton. This increases the number of protons in the nucleus by one, but decreases the number of neutrons by one. The atomic mass stays the same because protons and neutrons have the same mass. The atomic number of the new atom is 95, and the new mass is still 241 amu. Element 95 is americium, Am.

$$^{241}_{94}\text{Pu} \rightarrow \beta + ^{241}_{95}\text{Am}$$

The new isotope is americium-241.

HALF-LIFE

Carbon-14 has a **half-life** of 5730 years. This means that it will take 5730 years for half of a carbon-14 sample to decay to nitrogen-14. It will take another 5730 years for half of the remaining carbon-14 to decay, and so on. By measuring the remaining amount of carbon-14 compared to the amount of stable carbon, the age of the sample can be determined.

Beta decay: A nuclear reaction in which a neutron changes into a proton, and the atom emits a beta particle. The atom's atomic number increases by one.

Changes in the nucleus of an atom can change the identity of an element. Alpha decay results in a decrease in the atomic number by two. Beta decay results in an increase in the atomic number by one. Nuclear fission involves one nucleus splitting apart to form two new nuclei, each with smaller atomic numbers. Nuclear fusion involves two nuclei combining to form a nucleus with a larger atomic number. Nuclear fusion takes place in extremely hot environments, such as the cores of stars. Larger elements are formed in the hottest stars.

Nuclear changes often involve the transfer of large amounts of energy. This is why nuclear reactions can provide energy to operate power plants.

EVALUATE

No Check-In for this lesson.

Homework

Assign the reading and exercises for Alchemy Lesson 15 in the student text.

CAREER CONNECTION

Carbon-14 is a rare radioactive isotope of carbon that builds up to a steady level in living things as long as they are alive. When they die, the carbon-14 begins to decay at a predictable rate. Scientists have learned how to determine the age of a carbon fossil or artifact by measuring the amount of carbon-14 left in a sample. This process is called *carbon dating.*

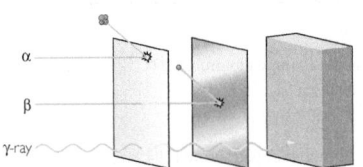

RADIATION

The particles and rays that are emitted during nuclear reactions are forms of **radiation.** So alpha and beta particles are both forms of radiation. Gamma rays are often emitted during radioactive events. **Gamma rays** are a kind of radiation similar to light, microwaves, and x-rays except they are much higher in energy, so they can be very dangerous. Gamma rays are represented by the Greek letter γ, gamma. When gamma rays are emitted, the identity of the emitting atom does not change.

α

β

γ-ray

A sheet of paper, or human skin, stops alpha particles, and a sheet of aluminum foil stops beta particles. Gamma rays can penetrate several inches of lead before stopping.

Naturally occurring radiation does exist, with low levels coming from sources in the Earth, as well as from the Sun and beyond. Humans have learned how to use certain types of radiation for a variety of scientific purposes. However, radiation from nuclear reactions can be harmful to your health.

Even though radiation can be dangerous, it can also be used to help people. For example, some cancer patients receive radiation treatments to kill cancerous cells so that it kills those cells but doesn't damage the rest of the body. Radiation can also be used to kill bacteria and viruses on food, particularly meat and vegetables. This helps reduce the risk of food poisoning.

Fission and Fusion

The splitting apart of a nucleus is called **fission.** The result is two atoms with smaller nuclei. Fission can occur spontaneously when the nucleus of an atom is unstable. Scientists can also make nuclear fission happen by shooting neutrons at a nucleus. Fission reactions are accompanied by radiation.

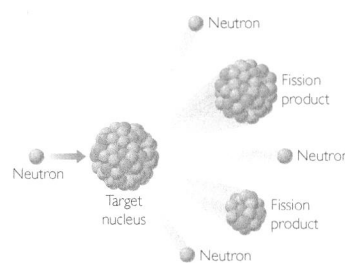

Another type of nuclear reaction is **fusion.** During fusion, nuclei join to form a larger nucleus, resulting in an atom of a different element. Temperatures of around 100,000,000 °C are required before nuclei can successfully fuse together. So the natural conditions on Earth do not support fusion.

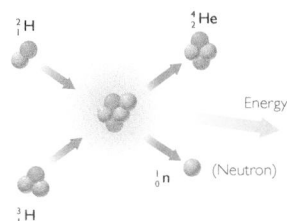

Two isotopes of hydrogen fuse to become helium.

CAREER CONNECTION

X-rays are similar to gamma rays except that they are not generated by radioactive samples. Using quick exposures, doctors and dentists can image your bones. However, long-term exposure to radiation can lead to illness and even death. So when patients are having x-rays taken, they wear a lead shield to protect their organs and health professionals leave the room.

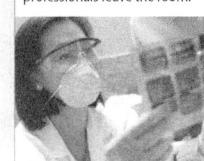

KEY TERMS

nuclear reaction

radioactive decay

alpha decay

alpha particle

beta decay

beta particle

half-life

radiation

gamma ray

fission

fusion

LESSON SUMMARY

What are nuclear reactions?

Radioactive decay, fission, and fusion are all forms of nuclear reactions. When radioactive decay is taking place, particles are emitted from the nucleus. Fission is the process by which the nucleus breaks into smaller nuclei. Fusion is the process by which nuclei combine to make a larger nucleus. Nuclear reactions can involve changes in atomic number, which mean changes in atomic identity. Radiation, such as the emission of alpha particles, beta particles, and gamma rays, accompanies decay and fission. All forms of radiation can cause harm to humans, but some forms are more damaging than others.

LESSON 15 ANSWERS

1. A nuclear reaction is a change in the nucleus of an atom.

2. *Possible answer:* In alpha decay, a nucleus emits a particle made up of two protons and two neutrons. In beta decay, a nucleus emits an electron. In alpha decay, the atomic number decreases by two and the atomic mass decreases by four. In beta decay, the atomic number increases by one and the atomic mass remains unchanged.

3. *Possible answer:* Gamma radiation is the most harmful because it has the most power to penetrate into living tissues and cause damage. This answer can be inferred from the diagram that shows the penetrating power of different kinds of radioactive decay.

4. The mass of the atom changes during alpha decay because the nucleus loses two electrons and two protons, which have a combined mass of 4 amu.

5. The mass of the atom does not change during beta decay. The nucleus loses one electron, which has a mass that is a tiny fraction of the mass of the nucleus. The conversion of a neutron into a proton does not change the mass because neutrons and protons have nearly identical masses.

6. The alpha particle is emitted from the nucleus and consists only of protons and neutrons. Because it has two positively charged particles but no negatively charged particles, the alpha particle has a net positive charge of +2 and is not a neutral atom.

7. a. calcium-42, $^{40}_{20}$Ca ✓ **b.** xenon-131 $^{131}_{54}$Xe ✓, **c.** magnesium-24, $^{24}_{12}$Mg ✓ **d.** cobalt-52, $^{52}_{27}$Co

8. a. osmium-171 $^{171}_{76}$Os ✓ **b.** samarium-145, $^{145}_{62}$Sm ✓ **c.** neptunium-237, $^{237}_{93}$Np ✓ **d.** radium-228, $^{228}_{88}$Ra

9. The age of an object can be determined by the amount of carbon-14 found in a sample of the object relative to the amount of stable carbon in the sample. Objects that can be dated using carbon dating are composed partially of carbon, which means that they are usually the remains of living organisms. The wooden handle of the axe, ashes, and jawbone originated in living organisms. The stone blade, clay pot, and arrowhead are composed mainly of other elements.

10. *Possible answer:* According to the graph of the radioactive decay of carbon-14, 67% of the original amount of carbon-14 will remain after about

3500 years. Sources of error in this determination are the ability to read the data value on the graph, the measurement of the amount of carbon-14 remaining in the sunken ship, and the contamination of the sample due to being under water for thousands of years.

11. C

12. C

Exercises

Reading Questions

1. What is a nuclear reaction?
2. What is the difference between alpha decay and beta decay?
3. What type of radiation is most harmful to living things: alpha radiation, beta radiation, or gamma radiation? Why?

Reason and Apply

4. Explain why the mass of an atom changes when an alpha particle is emitted.
5. Explain why the mass number of an atom does not change when a beta particle is emitted.
6. An alpha particle is not a neutral atom. It has a charge of +2. Why is this the case?
7. Suppose each of these isotopes emits a beta particle. Give the isotope name and symbol for the isotope that is produced. Place a checkmark next to the symbol for isotopes that are stable. (Consult the isotope graph on page 66 of Lesson 14: Isotopia.)
   a. potassium-42
   b. iodine-131
   c. iron-52
   d. sodium-24
8. Suppose each of these isotopes emits an alpha particle. Give the isotope name and symbol for the isotope that is produced. Place a checkmark next to the isotopes produced that are stable. (Consult the isotope graph on page 66 of Lesson 14: Isotopia.)
   a. platinum-175
   b. gadolinium-149
   c. americium-241
   d. thorium-232
9. The following items were found at an ancient campsite: an ax with a stone blade and a wooden handle, a clay pot, ashes of a campfire, a jawbone of a deer, and an arrowhead. Which objects could be used to determine the age of the site? Explain.
10. Scientists determine that the wooden beams of a sunken ship have 67% of the concentration of carbon-14 that is found in the leaves of a tree alive today. Find the age of the ship and explain how you determined it. List three possible sources of error in your determination.
11. What fraction of the original $^{14}_{6}$C would be in a wooden ax handle that was 17,190 years old?
    (A) $\frac{1}{4}$　　(B) $\frac{3}{4}$　　(C) $\frac{1}{8}$　　(D) $\frac{1}{2}$
12. Suppose $^{197}_{119}$Pt undergoes beta decay. The isotope that forms then also undergoes beta decay. What is the final product?
    (A) $^{195}_{120}$Ir　　(B) $^{197}_{118}$Au　　(C) $^{195}_{119}$Hg　　(D) $^{197}_{121}$Tl

Formation of Elements

THINK ABOUT IT

The element gold is essentially a collection of identical atoms. Gold has only one stable isotope. So, every gold atom has 79 protons, 79 electrons, and 118 neutrons. Perhaps it is possible to make gold atoms by using nuclear processes to add or subtract protons, neutrons, and electrons. To understand if this is possible, it is useful to consider what processes lead to the creation of new elements.

How are new elements formed?

To answer this question, you will explore

- Making New Elements
- Writing Nuclear Equations
- Nuclear Chain Reactions

EXPLORING THE TOPIC

Making New Elements

Some nuclear reactions occur without intervention on Earth when radioactive isotopes decay. It is much more difficult to change the nucleus of a stable atom. However, there are places with the right conditions and enough available energy to change these nuclei. Nuclear changes take place continuously in the Sun and other stars. And on Earth, scientists can carry out certain nuclear reactions in specially designed facilities like nuclear reactors or particle accelerators.

New elements can be created through nuclear fission or nuclear fusion. Whether a nucleus is split apart or joined with another nucleus, both processes result in the formation of different elements. And both processes require an energetic push to get them started.

Big Idea | The only way to change one element into another is to change the number of protons in the nucleus.

MAKING ELEMENTS IN THE STARS

The creation of new elements through nuclear chemistry is called *nucleosynthesis*. Most nucleosynthesis takes place far from Earth, inside stars.

The Sun is a giant ball of hydrogen, H, and helium, He. The amazing light and heat energy that radiates from the Sun is the result of continuous fusion reactions. However, the Sun is not hot enough to produce elements beyond helium on the periodic table. Higher temperatures are needed for the formation of larger atoms.

Overview

CLASSWORK: GROUPS OF 4

Key Question: How are new elements formed?

KEY IDEAS

The nucleus of the atom is difficult to change. Most changes to the nucleus require lots of energy. Some nuclear reactions occur spontaneously on Earth when radioactive isotopes, which are unstable, decay. Nuclear fission can be controlled and harnessed. Nuclear fusion occurs only in the cores of stars, in hydrogen bombs, and in laboratories using tremendously powerful lasers or strong magnets. These processes result in the production of new elements. Nuclear equations associated with nuclear reactions track the number of protons and neutrons before and after the reactions.

LEARNING OBJECTIVES

- Explain how different elements are formed through nuclear reactions.
- Write a balanced nuclear equation.
- Describe the mechanism behind a nuclear chain reaction.

FOCUS ON UNDERSTANDING

The term *isotope* refers to an atom of a particular element.

KEY TERMS

nuclear equation
parent isotope
daughter isotope
chain reaction

IN CLASS

Students complete a student worksheet dealing with specific nuclear reactions. Students learn to write equations to describe nuclear changes. They gain proficiency in determining how the nucleus changes in response to each type of nuclear reaction, as well as in determining what isotopes are produced. Finally, they consider how nuclear chemistry might be used to create gold. A complete materials list is available for download.

TRM **Materials List 16**

Differentiate

Nuclear reactions often present challenges to students because of the combination of mathematical and chemical concepts along with symbolic notation. For example, many students find it difficult to predict the parent isotope when given the daughter isotope. During the Explain & Elaborate discussion, have students work in pairs or small groups with whiteboards and dry erase markers (or some equivalent). Ask them to write a variety of nuclear equations on the whiteboard after you have provided them with only the daughter or parent isotope and the type of decay. Allow students to present their work to their classmates for feedback, making sure to ask them to explain their reasoning in approaching the problem.

Lesson Guide

ENGAGE (5 min)

TRM Transparency—ChemCatalyst 16

TRM PowerPoint Presentation 16

ChemCatalyst

These equations represent fusion reactions:

$$_2^4\text{He} + {_2^4}\text{He} \rightarrow {_4^8}\text{Be} + \text{energy}$$
$$_4^8\text{Be} + {_2^4}\text{He} \rightarrow {_6^{12}}\text{C} + \text{energy}$$
$$_6^{12}\text{C} + {_2^4}\text{He} \rightarrow {_8^{16}}\text{O} + \text{energy}$$

1. What patterns do you notice in the fusion reactions?

2. Do you think that gold can be created on Earth by a fusion reaction? Explain your thinking.

Sample answers: **1.** All the reactions release energy and involve fusing with a helium nucleus, creating larger nuclei. No protons or neutrons are lost in the process. **2.** Some students will say yes, you can fuse the two appropriate nuclei to create an atom of gold. Others will reason that if this were possible, scientists would be doing it now.

→ Solicit students' general ideas on how fusion could be used to create specific elements.

Ask: What patterns do you notice in the fusion reactions?

Ask: Can you obtain oxygen-16 by fusing beryllium-8 and helium-4? Explain. (No; there are not enough protons to make oxygen.)

Ask: Can you obtain nitrogen-14 by fusing boron-10 with helium-4? Explain. (yes)

Ask: Do you think that gold can be created on Earth by a fusion reaction? Explain.

EXPLORE (15 min)

TRM Worksheet with Answers 16

TRM Worksheet 16

TRM Handout—Chart of Naturally Occurring Isotopes 14

INTRODUCE THE CLASSWORK

→ Arrange students in groups of four.

→ Pass out the student worksheets.

→ Explain to students that nuclear equations are used to help track the changes that occur during nuclear reactions. Students will need a periodic table and the handout—Chart of Naturally Occurring Isotopes.

EXPLAIN & ELABORATE (20 min)

TRM Transparency—Nuclear Chain Reaction 16

DISCUSS NUCLEAR EQUATIONS

→ Call students up to the board to write out nuclear equations. Make sure the equations are numerically balanced. Label each equation with its appropriate nuclear process.

Ask: Can you write the nuclear equation showing technetium-100 undergoing alpha decay? ($^{100}_{43}$Tc → $^{96}_{41}$Nb + α)

Ask: What do you think is happening in this equation? $^{14}_{6}$C → $^{0}_{-1}$e $^{14}_{7}$N (beta decay)

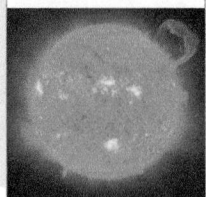

© NASA/JPL-Caltech/Corbis

Stars do not exist forever. Some stars collapse in on themselves after millions of years. Other stars explode and become what astronomers call *supernovas*. Where does gold come from? Small stars like our own can produce only helium. Larger, hotter stars can produce heavier elements up to iron. Only a supernova explosion can produce elements heavier than iron, including gold.

⮌ Writing Nuclear Equations

Nuclear reactions can be expressed as **nuclear equations.** You can use nuclear equations to track the number of neutrons and protons in the atoms, along with the identity of the products. Nuclear equations use isotope symbols to represent each particle. The alpha decay of an isotope of barium-140 to form xenon-136 can be written as a nuclear equation using α for the alpha particle:

$$^{140}_{56}\text{Ba} \rightarrow \alpha + ^{136}_{54}\text{Xe}$$

An alpha particle is the same as a helium nucleus, so it can also be written this way:

$$^{140}_{56}\text{Ba} \rightarrow ^{4}_{2}\text{He} + ^{136}_{54}\text{Xe}$$

In the second equation, you can track the protons and the mass that is removed from the barium nucleus. Notice that the equation is balanced numerically. The mass numbers on both sides are equal: $140 = 4 + 136$. The numbers of protons are also equal on both sides: $56 = 2 + 54$. In radioactive decay, the starting isotope is called the **parent isotope** and the resulting isotope is called the **daughter isotope.** In this equation, $^{140}_{56}$Ba is the parent isotope and $^{136}_{54}$Xe is the daughter isotope.

Beta decay can be shown in two ways as well, because a beta particle is the same as an electron and is sometimes shown as e⁻, or in isotope form as $^{0}_{-1}$e.

$$^{140}_{56}\text{Ba} \rightarrow \beta + ^{140}_{57}\text{La} \quad \text{or} \quad ^{140}_{56}\text{Ba} \rightarrow ^{0}_{-1}\text{e} + ^{140}_{57}\text{La}$$

Recall that the beta particle, β, comes from a neutron in the nucleus splitting into a proton and an electron. The new element $^{140}_{57}$La has a higher atomic number because it has one more proton in its nucleus than barium.

Fusion equations have two starting isotopes coming together to form a new product. When a carbon-12 isotope fuses with a nitrogen-14 isotope to form an isotope of aluminum, the equation looks like this:

$$^{12}_{6}\text{C} + ^{14}_{7}\text{N} \rightarrow ^{26}_{13}\text{Al}$$

Notice once again that the equation is balanced numerically on both sides.

Example

Radium-222

Write the equation representing the alpha decay of radium-222. What is the daughter isotope?

Solution

Find radium, Ra, on the periodic table. Its atomic number is 88. Now you can write the isotope symbol for radium, $^{222}_{88}$Ra.

Ask: Can you write the nuclear equation showing barium undergoing beta decay? ($^{140}_{56}$Ba → β + $^{140}_{57}$La or $^{140}_{56}$Ba → $^{0}_{-1}$e + $^{140}_{57}$La)

Ask: What element fuses with copper to make gold? Explain your thinking. (possibly tin, Sn, because it has 50 protons)

Key Points: Nuclear processes can be written as nuclear equations. Such equations must be numerically balanced, both in mass (upper numbers) and charge (lower numbers).

DISCUSS TYPES OF NUCLEAR REACTIONS AND THEIR EQUATIONS

Key Points: During alpha decay, the nucleus of an atom emits a helium nucleus, transforming the element into an element with a smaller nucleus. Chemists use the symbol α (the Greek letter alpha) to represent the helium nucleus. To show the alpha decay of a uranium-238 atom, a chemist might write this equation:

Alpha decay: $^{238}_{92}$U → α + $^{234}_{90}$Th or

$$^{238}_{92}\text{U} \rightarrow ^{4}_{2}\text{He} + ^{234}_{90}\text{Th}$$

Example

(continued)

An alpha particle is the same as a helium nucleus. Because the alpha particle is emitted, it goes on the right side of the equation. Balancing the equation gives the resulting nucleus, which has atomic number 86, and which is radon, Rn.

$$^{222}_{88}\text{Ra} \rightarrow {}^{4}_{2}\text{He} + {}^{218}_{86}\text{Rn}$$

The daughter isotope is radon-218.

Nuclear Chain Reactions

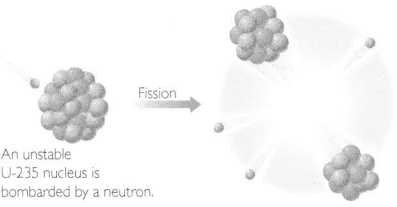

An unstable U-235 nucleus is bombarded by a neutron.

A barium nucleus, a krypton nucleus, three neutrons, and energy are produced.

Nuclear fission can be provoked by striking nuclei with fast-moving particles, such as neutrons. And, sometimes, one nuclear reaction can lead to others. A nuclear **chain reaction** can take place when the neutrons emitted strike surrounding nuclei, causing them to split apart as well.

For example, natural uranium consists of three isotopes: uranium-238, uranium-235, and uranium-234. All of these uranium isotopes are mildly radioactive but are still fairly stable. Uranium isotopes can be provoked into undergoing fission reactions. Scientists do this by firing a neutron at a uranium atom. This results in the formation of two smaller isotopes and the emission of three other neutrons, which in turn cause other uranium isotopes to undergo fission. The result is a chain reaction releasing enormous amounts of energy.

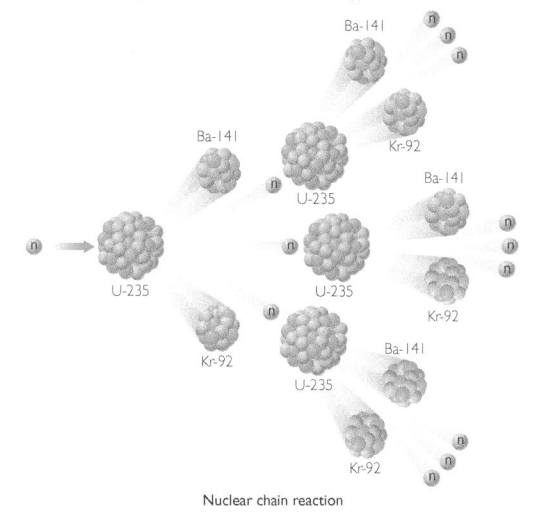

Nuclear chain reaction

During beta decay, a neutron inside the nucleus of an atom emits an electron. Electrons do not exist by themselves within the nucleus. However, in beta decay, a neutron splits into an electron and a proton. This explains why an atom gains a proton and loses a neutron during beta decay, keeping the atom's mass essentially the same. Nuclear chemists represent the neutron-splitting step with this equation:

Formation of beta particle:
$${}^{1}_{0}\text{n} \rightarrow {}^{0}_{-1}\text{e} + {}^{1}_{1}\text{p}$$

Notice that the equation is numerically balanced, showing that both mass (upper number) and charge (lower number) are conserved.

Chemists use the symbol β to represent a beta particle. Here is the equation for the beta decay of barium-140:

Beta decay: ${}^{140}_{56}\text{Ba} \rightarrow \beta + {}^{140}_{57}\text{La}$ or
${}^{140}_{56}\text{Ba} \rightarrow {}^{0}_{-1}\text{e} + {}^{140}_{57}\text{La}$

Nuclear fusion involves the joining together of nuclei. Nuclear fusion requires an extraordinary amount of energy.

A fusion reaction: ${}^{52}_{24}\text{Cr} + {}^{4}_{2}\text{He} \rightarrow {}^{56}_{26}\text{Fe}$

Fission involves a nucleus breaking up into smaller nuclei. Often, a few neutrons are also emitted.

A fission reaction:
${}^{235}_{92}\text{U} \rightarrow {}^{141}_{56}\text{Ba} + {}^{92}_{36}\text{Kr} + {}^{1}_{0}\text{n} + {}^{1}_{0}\text{n}$

DISCUSS THE FORMATION OF NEW ELEMENTS

Ask: Why does the identity of the isotope change when an alpha particle is emitted?

Ask: What isotope might emit an alpha particle to yield gold-197? (thallium-201)

Ask: What isotope might emit a beta particle to yield gold-197? (platinum-197)

Ask: Do you think you can make large amounts of gold by radioactive decay? Explain your thinking.

Key Points: Nuclear reactions change the identity of an element. The starting isotope is called the *parent* isotope, and the resulting isotope is called the *daughter* isotope. A beta particle, β, is smaller than an alpha particle and does not have any significant mass. It carries a negative charge. Its removal from the nucleus results in the conversion of a neutron to a proton, and the atomic number of the isotope increases by one. Once again, a different element is formed.

Nuclear fusion produces bigger (heavier) elements from smaller (lighter) ones. Nuclear fusion requires extraordinarily high temperatures. Most fusion reactions also release lots of energy, and fusion is responsible for the production of energy in stars, including our Sun. If we could harness fusion on Earth, we might have a very clean and powerful energy source on the planet. Generating temperatures close to those on the Sun, though, is currently out of the question. Thus, making gold using the process of fusion is also out of the question.

DISCUSS NUCLEAR CHAIN REACTIONS (OPTIONAL)

→ Show the transparency, Nuclear Chain Reaction.

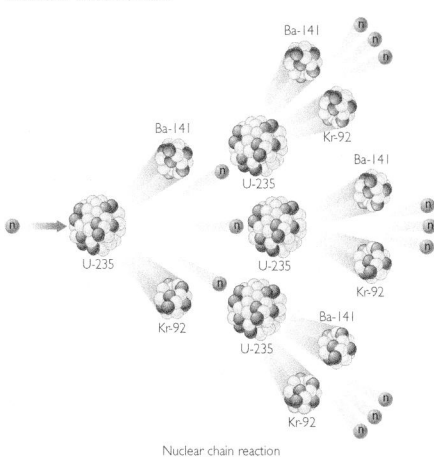

Nuclear chain reaction

Key Points: Nuclear fission is a process that releases enormous amounts of energy. A nucleus must be unstable before it will undergo fission. For heavy nuclei that are stable, nuclear chemists sometimes provoke nuclear instability by bombarding them with neutrons.

Nuclear fission can result in a nuclear chain reaction that produces a great deal of energy. The transparency shows how the neutrons released from one nucleus create instability in surrounding nuclei, resulting in a chain reaction. This process is potentially dangerous, but when controlled, it is a huge source of power. It is the process responsible for powering nuclear-powered submarines and providing many communities with electricity.

EVALUATE (5 min)

Check-In

In a paragraph, defend this statement: If you want to find gold, your best bet is to dig "old" gold out of the ground. Your chances of making "new" gold are slim.

Answer: The gold that is on Earth has been here since the planet formed. Thus, it is "old" gold. It is not feasible to make "new" gold from other elements, as the alchemists dreamed, because nuclear fusion requires temperatures of billions of degrees. And nuclear fission requires parent isotopes with very specific numbers of protons and neutrons: thallium-201 and platinum-119. These isotopes do not exist naturally.

Homework

Assign the reading and exercises for Alchemy Lesson 16 and the Chapter 3 Summary in the student text. Optional: Assign the project Nuclear Power in the student text, or give students a choice of projects from the unit.

LESSON 16 ANSWERS

1. *Possible answers (any 4):* In alpha decay, a nucleus emits a particle consisting of two protons and two neutrons. The atomic number decreases by two and the atomic mass decreases by four. ● In beta decay, a nucleus emits an electron. The atomic number increases by one as one of the neutrons becomes a proton. The atomic mass does not change. ● In nuclear fission, a nucleus splits apart to form the nuclei of two or more lighter elements. ● In nuclear fusion, two nuclei combine to form the nucleus of a heavier element. ● Nucleosynthesis is the process

Scientists have learned how to control and harness nuclear fission here on Earth. Nuclear fission is used to produce electricity, power submarines, and propel a variety of large ships. Uranium-235 is the isotope most commonly used for nuclear power.

If a nuclear chain reaction is controlled and the energy is released slowly, it can be used to generate electricity. If the energy is released all at once, the result is a nuclear explosion.

MAKING NEW GOLD

Most of the elements found on Earth are the products of fusion reactions in stars that exploded billions of years ago. The elements have been here for a very, very long time. There has been little change in the amounts of these elements. Small changes to the amounts of each element occur over time due to radioactive decay and fission. The gold we have on Earth came from supernova explosions that may have taken place billions of years ago. The prospect of making large quantities of new gold through nuclear chemistry appears slim.

LESSON SUMMARY

How are new elements formed?

Elements are formed through nuclear reactions. Radioactive decay, nuclear fission, and nuclear fusion are all possible sources of new elements. The elements we have on Earth came from fusion reactions in stars and supernova explosions long ago. It is difficult to cause nuclear fusion to occur on Earth because large amounts of energy are required to fuse nuclei. Nuclear fission, on the other hand, has been harnessed as a source of power. The power that is generated by nuclear plants is a result of a controlled fission chain reaction.

KEY TERMS
nuclear equation
parent isotope
daughter isotope
chain reaction

Exercises

Reading Questions

1. Describe four processes that result in the formation of new elements.
2. What is a nuclear chain reaction?

Reason and Apply

3. Write the nuclear equation for the beta decay of cerium-141.
4. Write the nuclear equation for the alpha decay of platinum-191.
5. Write a nuclear equation for the formation of iron-54 through fusion.
6. Consider the fission of uranium-235.
   a. The fission of uranium-235 begins with the addition of a neutron to the nucleus. What isotope is formed? Give the isotope symbol.
   b. After the neutron is added, the uranium atom is more unstable and undergoes fission. A possible set of products is krypton-94, barium-139, and 3 neutrons. How many protons are in these products?
   c. Were any protons lost?
7. **PROJECT** Write a paragraph explaining how a nuclear reactor works. Be sure to explain the purpose of the control rods.

of elements forming inside stars through nuclear chemistry. ● A nuclear chain reactions is a series of nuclear reactions when particles emitted by one fission reaction strike other nuclei.

2. A nuclear chain reaction is a sequence of nuclear fission reactions occurring when a particle strikes a nucleus causing it to break apart and form more particles that strike other nuclei, breaking them apart.

3. $^{141}_{58}\text{Ce} \rightarrow\ ^{0}_{-1}\text{e} +\ ^{141}_{59}\text{Pr}$

4. $^{50}_{24}\text{Pt} \rightarrow\ ^{4}_{2}\text{He} +\ ^{54}_{26}\text{Fe}$

5. Possible Answer $^{50}_{24}\text{Cr} +\ ^{4}_{2}\text{He} \rightarrow\ ^{54}_{26}\text{Fe}$

6. a. Uranium-236; $^{236}_{92}\text{U}$ **b.** Krypton-94 has 36 protons and barium-139 has 56 protons. **c.** A uranium atom has

92 protons and the reaction products have 92 protons, so no protons were lost.

7. *Possible answer:* Inside a nuclear power plant, uranium-235 is concentrated into rods and submerged in water. The atoms break apart when a neutron strikes them. More neutrons are produced during the reaction, and they strike other uranium-235 atoms. These fission reactions produce heat that is used to convert water into steam. The steam turns a turbine to generate electric power. Control rods are made of material that absorbs the neutrons without breaking apart. Raising and lowering the rods in the water controls the rate of the nuclear reactions by changing the amount of neutrons absorbed.

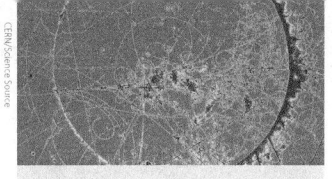

CHAPTER 3

A World of Particles

SUMMARY

KEY TERMS

atom
atomic theory
model
nucleus
proton
neutron
electron
atomic number
isotope
mass number
average atomic mass
radioactive isotope
nuclear reaction
radioactive decay
alpha decay
alpha particle
beta decay
beta particle
half-life
radiation
gamma ray
fission
fusion
nuclear equation
parent isotope
daughter isotope
chain reaction

Alchemy Update

Can a copper atom be transformed into a gold atom through nuclear processes?

Elements are collections of similar atoms. The number of protons in the nucleus determines the identity of an atom. For instance, all atoms of gold have 79 protons. However, atoms of an element may have different numbers of neutrons. The number of neutrons is related to the stability of the atom. Stable gold atoms have 118 neutrons.

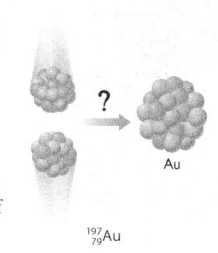

$^{197}_{79}$Au

Nuclear fission, nuclear fusion, and radioactive decay can result in new elements. But nuclear fusion and nuclear fission are difficult processes to control and often involve large amounts of energy. Nuclear reactions do take place in the stars where atoms are created. So far it is not yet possible or practical to create gold atoms through nuclear reactions here on Earth.

REVIEW EXERCISES

1. Describe the different processes that might result in a change in atomic identity.

2. Give an example of evidence used by scientists to revise the atomic model.

3. Explain how to use the periodic table to deduce the number of protons, neutrons, and electrons of an atom of a specified element.

4. Write a nuclear equation for the hypothetical fusion of a copper atom with another nucleus to make gold. Can this happen? Why or why not?

Nuclear Power

PROJECT
www

Research nuclear power. Find out how a nuclear power plant works. Write a report including

- an explanation (with a simple drawing) of how electric power is produced from a nuclear power plant.

- the major benefits and risks of nuclear power.

4. $^{63}_{29}$Cu + $^{134}_{50}$Sn → $^{197}_{79}$Au. The equation is not plausible because tin-134 is not a naturally occurring isotope of tin. Copper cannot be fused directly with another atom to make gold because there is no combination of isotopes of copper and tin that have enough neutrons to form a stable isotope of gold.

Project: Nuclear Power: A good report would include description of how a nuclear reaction produces energy ● a description of how the heat product in nuclear reactions is converted into electricity ● a drawing of the inside of a reactor core and how the steam turns a turbine inside a generator ● a list of the advantages and disadvantages of nuclear power when compared to fossil fuels and alternative energy sources ● an opinion on whether nuclear power will be and should be an important source of energy in the future

ASSESSMENTS

Two multiple-choice assessments (versions A and B) and two short-answer assessments (versions A and B) are available for download for Chapter 3.

TRM Chapter 3 Assessments

TRM Chapter 3 Assessment Answers

ANSWERS TO CHAPTER 3 REVIEW EXERCISES

1. *Possible answer:* An atom changes identity when the number of protons in its nucleus changes. Processes in which this can occur are radioactive decay, fission, and fusion. In radioactive decay, emission of an alpha particle decreases the atomic number by two, while emission of a beta particle increases the atomic number by one. Nuclear fission is the process in which a nucleus breaks apart, forming two or more smaller nuclei. In nuclear fusion, two nuclei join together to form one larger nucleus.

2. *Possible answer:* Elements are formed in nature by nuclear fusion. Most of the elements found on Earth are the products of fusion reactions in stars that exploded billions of years ago. After forming, these elements can decay into other elements through radioactive decay.

3. The atomic number of an element is the same as the number of protons and electrons in a neutral atom of that element. You can estimate the number of neutrons by rounding the average atomic mass to a whole number and subtracting the number of protons.

Watch the video overview of Chapter 4 (for teachers) by clicking on the link in the TE-book, opening the TRFD, or logging onto the book's companion Web site bcs.whfreeman.com/livingbychemistry2e (teacher log-in required).

Chapter 4 focuses on electrons and the formation of ions. Students begin by performing flame tests in Lesson 17, which allows them to verify the presence of specific atoms within ionic compounds. This serves two purposes: It provides evidence that atoms are not destroyed when they are within compounds, and it demonstrates that electrons are a fairly mobile part of the atom. Lessons 18 and 19 introduce students to electron shells, electron arrangement, and the formation of ions. Lessons 20 and 21 introduce ionic compounds and reinforce the concept with the card game Salty Eights. Lessons 22 and 23 present polyatomic ions and the chemistry of the transition metals. The section concludes with an exploration of electron configurations in Lesson 24.

In this chapter, students will learn

- about electron shells and the arrangement of electrons within them
- how to identify valence and core electrons
- how to use the periodic table to determine electron arrangement
- ionic bonding patterns
- about transition metals and polyatomic ions

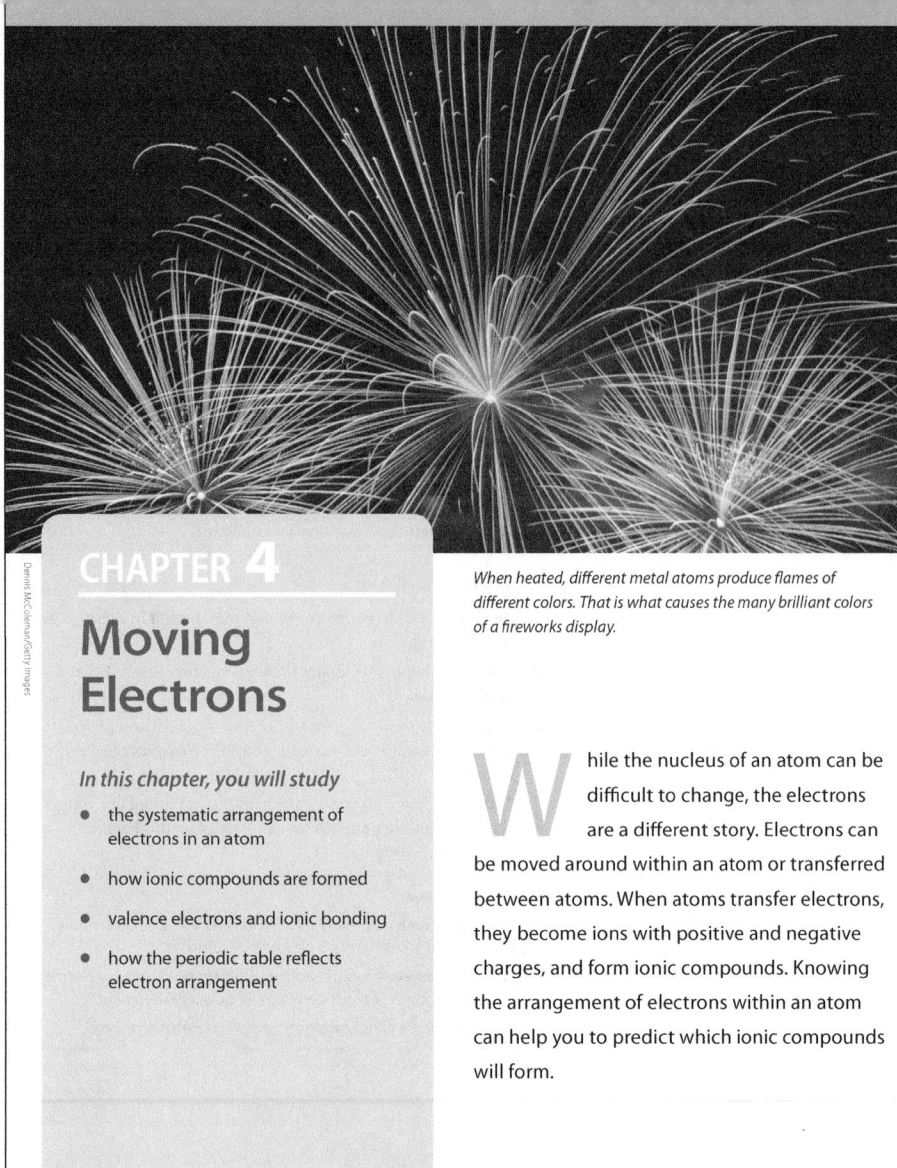

CHAPTER 4
Moving Electrons

In this chapter, you will study
- the systematic arrangement of electrons in an atom
- how ionic compounds are formed
- valence electrons and ionic bonding
- how the periodic table reflects electron arrangement

82

When heated, different metal atoms produce flames of different colors. That is what causes the many brilliant colors of a fireworks display.

While the nucleus of an atom can be difficult to change, the electrons are a different story. Electrons can be moved around within an atom or transferred between atoms. When atoms transfer electrons, they become ions with positive and negative charges, and form ionic compounds. Knowing the arrangement of electrons within an atom can help you to predict which ionic compounds will form.

Chapter 4 Lessons

Flame Tests

Purpose

To identify metal atoms in a variety of compounds by using a flame test and to provide evidence for the presence of certain atoms within compounds.

Materials

- Bunsen burner
- set of tongs
- copper wire
- penny
- set of 11 solutions: sodium carbonate, Na_2CO_3; potassium nitrate, KNO_3; copper (II) nitrate, $Cu(NO_3)_2$; strontium nitrate, $Sr(NO_3)_2$; potassium chloride, KCl; sodium chloride, NaCl; copper (II) sulfate, $CuSO_4$; strontium chloride, $SrCl_2$; sodium nitrate, $NaNO_3$; copper (II) chloride, $CuCl_2$; potassium sulfate, K_2SO_4
- 11 pieces of nichrome wire, each with a loop at one end

Procedure and Observations

1. For each solution, follow these steps and record your observations.
   - Place the loop end of the nichrome wire into the solution.
   - Place the loop end with the solution on it into the flame.
   - Record the color of the flame in your table.
   - Place the nichrome wire back in the correct solution.
2. Using tongs, place the penny and the copper wire into the flame one at a time and observe the results.

Analysis

1. Group the substances based on the color of flame produced. What patterns do you notice in the groupings?
2. **Making Sense** Can a flame test be used to identify a metal atom in a compound? Why or why not? What about a nonmetal atom?
3. **If You Finish Early** Copper oxide, CuO, is a black solid. It doesn't look at all like the element copper. What color flame would it produce? Draw a model of copper oxide to explain the color of the flame that it would produce.

IN CLASS

Students conduct flame tests on a set of 11 solutions and 2 metallic solids. They collect their results and group the substances according to flame colors. Students will look for patterns involving the chemical names of the substances and the colors of the flames. A complete materials list is available for download.

TRM Materials List 17

SETUP

Prepare solutions: Create (~0.5 M) solutions of each of the 11 metal salts listed by dissolving the amounts given in 100 mL of distilled water. (The exact concentration does not matter—keep adding the salt until no more dissolves.) Use distilled water, because tap water might be contaminated and affect your results. Divide each solution into four labeled 50 mL beakers.

Prepare testing wires: Cut 44 pieces of nichrome wire ahead of time, each about 6 inches long. Bend to form a very small loop at one end of each wire. Use labeling tape and permanent marker to label each wire with its appropriate solution formula so students will not cross-contaminate samples. Place one nichrome wire in each 50 mL beaker of solution, with the loop in the liquid. An alternative to nichrome wires are wooden splints, soaked in the solutions overnight.

Prepare copper samples: Clean the copper wire and the pennies in hydrochloric acid before class.

Set up four stations with each solution, clean copper wire, a clean copper penny, tongs, and Bunsen burners.

SAFETY

Students will be using chemicals and fire today. They should wear goggles, follow basic fire safety precautions, and learn where the eyewash, fire blanket, and fire extinguisher are located.

CLEANUP

Reuse: The same solutions can be used for several classes. Be sure to instruct students to place the nichrome wires back in the correct solutions to prevent cross-contamination.

Dispose of properly: Combine the remaining amounts of each type of solution. Place the beakers in a safe

Overview

LAB: GROUPS OF 8

Key Question: What evidence is there that certain atoms are present in a compound?

KEY IDEAS

Flame tests provide evidence of the presence of certain atoms in compounds. Different metal atoms produce different-colored flames when heated. Flame tests give some evidence that atoms are not destroyed when they combine to form compounds.

LEARNING OBJECTIVES

- Conduct a flame test and use the results to determine the identity of a compound.
- Interpret evidence of the presence of certain atoms within compounds.

FOCUS ON UNDERSTANDING

- The compounds tested in flame tests are being heated and vaporized, not burned. In other words, the compounds break apart into atoms. Thus, it is important to avoid using the term "burn."
- The most common contaminant in flame testing is sodium, which gives a bright yellow flame. Students should be forewarned about this.

KEY TERM

flame test

location and allow the liquid to evaporate. Dispose of the remaining toxic solids according to state and local regulations.

Pre-AP® Course Tip

Students preparing to take an AP® chemistry course or college chemistry should be able to analyze data to identify certain patterns and relationships among chemical phenomena. In this lesson, students perform a flame test on various compounds. After analyzing their results, they identify patterns in their groupings and are able to conclude which element is responsible for the characteristic color of the flame.

Lesson Guide

ENGAGE (5 min)

TRM PowerPoint Presentation 17

TRM Transparency—ChemCatalyst 17

ChemCatalyst

These drawings are models that show solid copper, solid copper chloride, and aqueous copper chloride as collections of atoms.

1. Describe each model.
2. What is similar about each model? What is different?

Cu(s) solid copper

CuCl₂ (s) solid copper (II) chloride

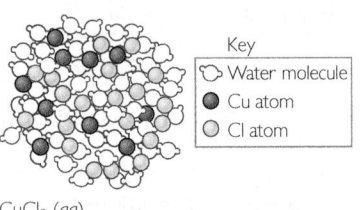

Key
- ◌ Water molecule
- ● Cu atom
- ○ Cl atom

CuCl₂ (aq)
aqueous copper (II) chloride

Technicolor Atoms

Flame Tests

THINK ABOUT IT

On the Fourth of July, you can watch colorful fireworks without considering the chemistry behind them. But each color that bursts forth in the sky is associated with a particular chemical compound. For example, any green sparkles you might see are probably due to a compound such as barium sulfate. But what's responsible for the color? Is it the entire barium sulfate compound, or is it one of the atoms in the compound?

> What evidence is there that certain atoms are present in a compound?

To answer this question, you will explore

⟳ Flame Tests

⟳ Evidence for Atoms in Compounds

⟳ Excited Electrons

EXPLORING THE TOPIC

⟳ **Flame Tests**

Fireworks originated in China about 2000 years ago. The legend surrounding the discovery suggests that fireworks were discovered by a Chinese alchemist who mixed charcoal, sulfur, and potassium nitrate and accidentally produced a colorful gunpowder. Since then, the noise and the bright colors associated with fireworks have been used in celebrations all over the world.

In a chemistry lab, it is relatively easy to obtain the colorful flames associated with fireworks. You can do so by heating certain compounds in a flame, such as the flames of a Bunsen burner. The flame colors produced by heating four different compounds are shown here.

   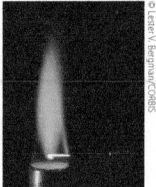

Different compounds produce different colors in a flame test. The metal in the compound may determine color.

HISTORY CONNECTION

The original firecracker, called *baozhu*, was created in China around 200 B.C.E. It was a segment of green bamboo that was thrown onto a fire. Upon heating, the trapped gases inside the bamboo expanded, causing the bamboo to explode.

Sample answers: **2.** Each model has the copper spheres, because they all contain copper. The first two models are solid, and the third is aqueous. The first is a single element, while the others are compounds that include chlorine.

→ Assist students in sharing their insights about and interpretations of the models.

Ask: What information about the three substances is depicted in the models?

Ask: Is the model for copper (II) chloride consistent with the chemical formula? Explain.

Ask: What is similar about the models of CuCl₂(s) and CuCl₂(aq)? What is different?

EXPLORE (20 min)

TRM Worksheet with Answers 17

TRM Worksheet 17

Compound Colors

Red	Orange	Yellow-orange	Green	Blue-green	Pink-lilac
lithium nitrate $LiNO_3$	calcium nitrate $Ca(NO_3)_2$	sodium nitrate $NaNO_3$	barium nitrate $Ba(NO_3)_2$	copper nitrate $Cu(NO_3)_2$	potassium nitrate KNO_3
lithium chloride $LiCl$	calcium chloride $CaCl_2$	sodium chloride $NaCl$	barium chloride $BaCl_2$	copper chloride $CuCl_2$	potassium chloride KCl
lithium sulfate Li_2SO_4	calcium sulfate $CaSO_4$	sodium sulfate Na_2SO_4	barium sulfate $BaSO_4$	copper sulfate $CuSO_4$	potassium sulfate K_2SO_4
lithium Li	calcium Ca	sodium Na	barium Ba	copper wire Cu	potassium K

The colors for several more compounds are provided in the table. Take a moment to examine the data. What patterns do you notice?

Notice that each metal atom, Li, Ca, Na, and so on, is associated with a specific flame color. Lithium compounds all make a red flame, while the barium compounds all make a green flame. The nonmetal atoms in these compounds do not seem to affect the color of the compound. So $CaSO_4$ and $CaCl_2$ and $Ca(NO_3)_2$ all produce the same color flame, but $CaCl_2$ and $NaCl$ and $CuCl_2$ do not. The metal atom must somehow be responsible for the color of the flame.

Evidence for Atoms in Compounds

Chemists have found these flame color patterns to be quite helpful. A **flame test** can be used to quickly confirm the presence of certain metal atoms in an unknown sample. So, a potassium compound can quickly be distinguished from a calcium compound by heating samples of the compounds.

Also, the color patterns associated with the flame tests clearly indicate whether certain metal atoms are present within the compounds. For example, copper metal turns the flame blue-green and so does copper chloride. This is evidence that there are copper atoms in the copper chloride.

Excited Electrons

Heating metal compounds in a Bunsen burner does not destroy the atoms in them or create new elements. This is because the temperature of a flame is not high enough to change the nuclei of metal atoms. However, the heat does have an impact on the electrons of these metal atoms.

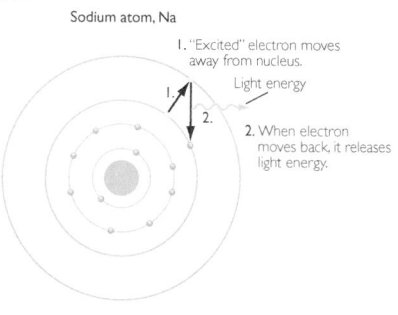

Sodium atom, Na

1. "Excited" electron moves away from nucleus.
Light energy
2. When electron moves back, it releases light energy.

Lesson 17 | **Flame Tests**   85

INTRODUCE THE LAB

→ Arrange students in groups of eight at each station.

→ Remind students about fire and chemical safety.

SAFETY

- Wear safety goggles.
- Tie back long hair and remove dangling jewelry.
- Roll up long sleeves and keep clothing away from flames.
- Know the location of the eyewash, fire blanket, and fire extinguisher.

→ Tell students that they should work as two groups of four at each station, with one group testing half the solutions and then switching to test the other half.

→ Demonstrate how to do a flame test. Dip the looped end of the nichrome wire into one of the solutions and then place it directly in the flame. You might want to dim the lights in the room for effect.

→ Remind students to return the wires to the appropriate beakers to prevent cross-contamination.

→ Pass out the student worksheet and let students begin.

EXPLAIN & ELABORATE (15 min)

TRM Transparency—Sodium Atom 17

DISCUSS CLASS RESULTS OF THE FLAME TEST

→ Ask students to share their findings.

→ Write the flame colors on the board and group the substances by flame color.

RED

strontium nitrate
strontium chloride

BLUE-GREEN

copper (II) nitrate
copper (II) sulfate
copper (II) chloride
solid copper

YELLOW-ORANGE

sodium carbonate
sodium chloride
sodium nitrate

PINK-LILAC

potassium nitrate
potassium chloride
potassium sulfate

Ask: What patterns do you notice in the data?

Ask: What evidence do you have from the flame tests that copper is responsible for producing a blue-green flame?

Ask: Does nitrate produce a specific-colored flame? Explain your thinking.

Ask: How does the flame test provide evidence of individual atoms?

Key Points: The metal element in each chemical formula appears to be responsible for the flame colors. So, the flame test can be used to detect the presence of certain metal atoms. No matter what form the copper atom is in, whether it is an element (copper wire) or in a compound like copper sulfate or copper nitrate, it always produces a characteristic blue-green flame. The potassium in potassium nitrate, potassium chloride, and potassium sulfate always produces a pinkish flame.

Only certain elements produce colorful flames. Every substance containing the same metal gives the same flame color. At the same time, all the substances that are called nitrates do not produce the same flame color, nor do all the substances that are called chlorides or sulfates. The nonmetal elements—nitrogen, oxygen, sulfur, and chlorine—do not give rise to colors when placed in a flame.

> **Flame test:** A test used in the laboratory to look for the presence of certain metal atoms. A sample of a compound is heated in a flame, and the resulting color is noted.

DISCUSS THE EVIDENCE THAT COMPOUNDS ARE COMPOSED OF ATOMS

→ Display the ChemCatalyst transparency again.

Ask: What evidence do you have that individual atoms continue to exist in compounds? (flame test, chemical formula, copper cycle)

Ask: Why do the three models all contain dark-gray spheres?

Ask: When you heat the solutions in the flame, is one atom converted into another? (no)

Ask: What do you think would happen if you heated the substance shown in the third model? (Water would evaporate, leaving behind copper and chlorine atoms in the form of copper chloride.)

Key Points: Elements and compounds are collections of atoms. All matter is made up of atoms of the elements listed in the periodic table. The chemical formula tells you what atoms are present in a compound. For example, copper (II) oxide, CuO, is a collection of copper and oxygen atoms. Because copper atoms are present in copper (II) oxide, a blue-green flame is observed. So, even though copper (II) oxide is black and not shiny like copper metal, the copper atoms are still there. Compounds are simply mixtures of atoms of different elements.

The only way to change one atom into another is to change the nucleus through a nuclear reaction. The number of protons in the nucleus determines the identity of an atom. Changing the number of protons does change the identity of the atom. However, a nuclear change requires either a radioactive nucleus or specialized conditions. The temperature of the flame of the Bunsen burner is simply not hot enough to change one atom into another. (The temperature of the flame is hundreds of degrees, not millions of degrees, as would be required.)

Electrons are located at different average distances from the nucleus of an atom. When heated, some electrons of metal atoms get "excited" and move to distances farther from the nucleus. This move is only temporary. When the electrons move back to their original distance from the nucleus, they release energy in the form of colored light. This is what you see during a flame test.

The fact that it is relatively easy to affect the electrons in atoms opens up new possibilities for exploration. Perhaps the solution to creating substances with the properties of gold rests in altering the electrons in some way. The nuclei of atoms cannot be changed easily, so it is time to explore ways in which the electrons in atoms can be changed. The rest of this unit explores the role of the electrons in the chemistry of atoms and compounds.

> **Example**
>
> ### Flame Colors
>
> Which of these compounds will give similar flame colors when heated?
>
> $NaCl$ $CaCl_2$ $SrCl_2$ $Sr(NO_3)_2$ $Cu(NO_3)_2$
>
> *Solution*
> The color of the flame depends on the metal. The only two compounds with the same metal are strontium chloride, $SrCl_2$, and strontium nitrate, $Sr(NO_3)_2$.

KEY TERM
flame test

LESSON SUMMARY

What evidence is there that certain atoms are present in a compound?

Many metal atoms produce a characteristic colored flame when compounds containing those atoms are heated. The colors are a result of light energy produced when excited electrons return to their original distance from the nucleus in the metal atoms. Flame tests provide evidence that certain atoms are present in compounds and that they are not destroyed during chemical changes. The flame test also demonstrates that it is easier to add or remove electrons in atoms, which can be done with a small amount of energy, than to alter the protons and neutrons in the nucleus, which requires a great deal of energy.

EXPLAIN THE SOURCE OF THE COLORED LIGHT

→ Display the transparency Sodium Atom.

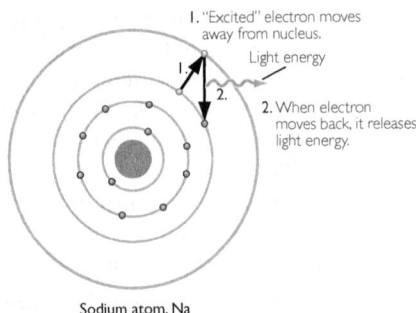

1. "Excited" electron moves away from nucleus.
   Light energy
2. When electron moves back, it releases light energy.

Sodium atom, Na

Ask: What does this model tell you about the flame test?

Ask: What do you think the different rings represent?

Ask: Does the movement of electrons change the identity of an atom? Explain.

Key Points: The illustration indicates that the flame colors are associated with movements of the electrons within the sodium atom. The heat from the flame causes electrons to move farther from the nucleus. When the electrons move back to their original distance from the nucleus, energy is released as colored light. The electrons, not the nucleus, are affected by the heat from the flame.

Exercises

Reading Questions

1. How did the flame test provide evidence that specific atoms are present in compounds?

2. Explain what is responsible for the colors during a flame test.

Reason and Apply

3. PROJECT

Find out why fireworks are so colorful. What substances are used to produce the colors?

4. Predict the color of the flame for the compound sodium hydroxide, NaOH. Explain your reasoning.

5. Imagine that you were in charge of creating a red and purple fireworks display. Name two combinations of compounds you could use.

6. What evidence do you have from flame tests that copper is responsible for producing a flame with a blue-green color?

7. Does nitrate produce a colored flame? Explain your thinking.

8. Would it matter whether you did a flame test with sodium chloride, NaCl, in solid form or sodium chloride as an aqueous solution? Explain.

9. What flame colors would be produced by these compounds? Explain your choices.
   a. Na_2CO_3 b. $Ba(OH)_2$ c. KOH   d. K_2CO_3 e. BaO

10. If two chemical samples both produce an orange flame upon testing, which statement is true?
    (A) The two samples contain identical compounds.
    (B) The samples both contain chlorine atoms.
    (C) The samples both contain calcium atoms.
    (D) The samples both contain potassium atoms.
    (E) The samples contain different compounds.

11. What evidence supports the claim that chloride, Cl^-, does not cause the flame to have a color?
    (A) Lithium chloride, LiCl, and sodium chloride, NaCl, have different colors.
    (B) Ammonium chloride, NH_4Cl, does not cause the flame to have a color.
    (C) Sodium chloride, NaCl, and sodium nitrate, $NaNO_3$, both produce flames with a yellow-orange color.
    (D) All of the above.

Lesson 17 | **Flame Tests**   87

Bohr's model of the atom came directly from evidence like that produced in class today. He noted that atoms emit light of certain colors when they are heated. He hypothesized that electrons are located in specific shells (represented by circles in the model). The electrons can energetically "jump" between these shells but cannot occupy the space between the shells. When the electrons fall back to their original positions, they release energy in the form of light energy.

EVALUATE (5 min)

Check-In

Predict the flame colors produced when heating these substances. Explain your thinking.
- copper (II) carbonate
- calcium chloride

Answers: The copper compound will emit a blue-green color. There is not enough information to determine the color of the calcium chloride flame test, but it probably would be different from the colors observed today. (It is brick red or deep orange.)

Homework

Assign the reading and exercises for Alchemy Lesson 17 in the student text.

LESSON 17 ANSWERS

1. The color of the flame produced during a flame test is a characteristic of particular metallic elements. When a compound containing one of the metallic elements is heated, its atoms emit light of a specific color.

2. When a compound is heated, some of the electrons in the metallic atoms gain energy and move farther away from the nucleus. When the electrons move back to their original distance, they release energy in the form of colored light.

3. *Possible answer:* Fireworks are colorful because they are made using metallic compounds that emit visible light during the explosions. Electrons of the metal atoms become excited when they are heated, and the atoms release energy as light when the electrons return to their original state. Salts of various metals produce the range of fireworks colors: ● red: lithium or strontium ● orange: calcium ● yellow: sodium ● green: barium ● blue: copper

4. The flame for sodium hydroxide will be yellow-orange because the metal atom determines the color. Sodium atoms emit a yellow-orange color when placed in a flame.

5. *Possible answer:* A red color in fireworks could result from using a lithium salt. A purple color could result by mixing compounds that would produce red and blue colors separately. Any mixture of lithium and copper salts would produce purple fireworks.

6. *Possible answer:* Today's lab involved the performance of a flame test on five different compounds containing copper: copper nitrate, copper sulfate, copper chloride, copper wire, and a copper-coated penny. Each compound produced a blue-green flame when heated by the Bunsen burner. Since the presence of copper is the only thing these compounds have in common, the copper is responsible for the blue-green flame.

7. No, the flame color of each nitrate compound is different and matches the flame color of the metal in the compound. This indicates that the nitrate is not responsible for the color of the flame.

Answers continue in Answer Appendix (p. ANS-2).

Lesson 17 | **Flame Tests**   87

Overview

CLASSWORK: PAIRS

Key Question: Why do elements in the same group in the periodic table have similar properties?

KEY IDEAS

In the shell model, electrons occupy distinct areas around the nucleus of an atom. The electrons in the outermost electron shell of the atom are called *valence electrons*. All other electrons are called *core electrons*. The periodic table reflects the predictable patterns found in electron arrangements. Among main group elements, elements in the same group have the same number of valence electrons.

LEARNING OBJECTIVES

- Create a shell model diagram of an atom, placing the correct number of electrons in the correct shells.
- Explain the difference between a valence electron and a core electron.
- Describe the patterns in the periodic table associated with electron arrangements.

FOCUS ON UNDERSTANDING

- Students might be confused by all the different models of the atom. For example, it is difficult for them to reconcile an electron cloud and a shell model.
- Only a portion of the periodic table is covered in this lesson (the main group elements, Groups 1A–8A) so that students can better grasp the overall patterns without worrying about exceptions at this point.

KEY TERMS

valence shell
valence electron
core electron

IN CLASS

At the beginning of class, students are introduced to the shell model of the atom. They examine a handout showing the arrangement of electrons in shells around the nucleus of a number of atoms. Students use the data on the handout as a guide to construct shell models for other atoms. Finally, they identify patterns in the outermost, or valence, electrons. A complete materials list is available for download.

TRM Materials List 18

Life on the Edge

Valence and Core Electrons

THINK ABOUT IT

Lithium, Li, and sodium, Na, are both located in Group 1A on the periodic table because they have very similar properties. Both are soft, silvery metals that react with water. Both form compounds with chlorine in a 1:1 ratio—lithium chloride, LiCl, and sodium chloride, NaCl. What do lithium atoms have in common with sodium atoms that make them behave similarly?

Why do elements in the same group in the periodic table have similar properties?

To answer this question, you will explore

- Electron Shells
- Patterns in Atomic Structure
- Valence and Core Electrons

EXPLORING THE TOPIC

Electron Shells

Recall that elements that are grouped together in columns on the periodic table share similar properties.

Consider lithium, Li, and sodium, Na, from Group 1A on the periodic table. You can get the following information from these two element squares on the periodic table:

3
Li
Lithium
6.941

3 protons, 3 electrons
Average atomic mass = 6.941 amu

11
Na
Sodium
22.99

11 protons, 11 electrons
Average atomic mass = 22.99 amu

This basic information doesn't provide any evidence of similarities between atoms of lithium and atoms of sodium. However, examining the structures of these atoms in more detail reveals some patterns.

Lithium, Li   Sodium, Na

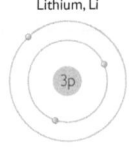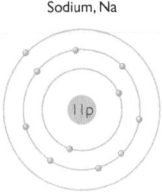

Pre-AP® Course Tip

Students preparing to take an AP® chemistry course or college chemistry should be able to draw representations of accepted atomic models. In this lesson, students draw shell-model diagrams for various atoms. From their models, students are able to determine how many valence electrons are in an atom and the relationship between the period number and number of shells that an atom has.

Lesson Guide

ENGAGE (5 min)

TRM Transparency— ChemCatalyst 18

TRM PowerPoint Presentation 18

ChemCatalyst

Hydrogen							Helium
H							He
Lithium	Beryllium	Boron	Carbon	Nitrogen	Oxygen	Fluorine	Neon
Li	Be	B	C	N	O	F	Ne
Sodium	Magnesium	Aluminum	Silicon	Phosphorus	Sulfur	Chlorine	Argon
Na	Mg	Al	Si	P	S	Cl	Ar
Potassium	Calcium	Gallium	Germanium	Arsenic	Selenium	Bromine	Krypton
K	Ca	Ga	Ge	As	Se	Br	Kr
Rubidium	Strontium	Indium	Tin	Antimony	Tellurium	Iodine	Xenon
Rb	Sr	In	Sn	Sb	Te	I	Xe

Although electron configurations are easier to draw in two dimensions, keep in mind that atoms are not flat. Electron shells are really surfaces of spheres. Each model is useful in understanding different aspects of the atom.

In these models, the electrons are shown orbiting in concentric circles around the nucleus. Chemists call these circles *electron shells*, and these atomic models are called *shell models*.

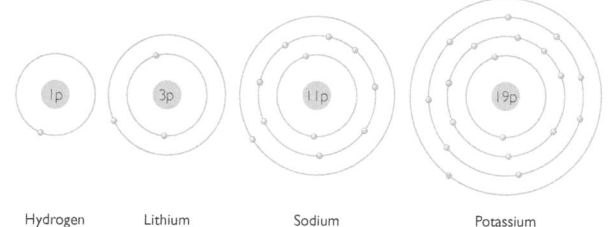

| Hydrogen | Lithium | Sodium | Potassium |

Lithium has a total of three electrons in two shells. Sodium, on the other hand, has a total of 11 electrons in three shells. But the atoms of both elements do have one thing in common: a single electron in their outer shells. If this feature is responsible for the similar properties of lithium and sodium, you would expect the atoms of the other elements in the same group to also have a single electron in their outer shells.

The atoms of two more Group 1A elements are shown in the illustration. Compare the models of hydrogen and potassium with those of lithium and sodium. The pattern holds: Each of the Group 1A elements has one electron in the outer shell of its atoms. This pattern also extends to other Group 1A elements not shown here, such as rubidium and cesium.

↻ Patterns in Atomic Structure

Examining the shell models of elements beyond lithium on the periodic table reveals further patterns in the outermost electrons. Shell models for the first 14 elements are shown in this illustration. A repeating pattern emerges from the diagram.

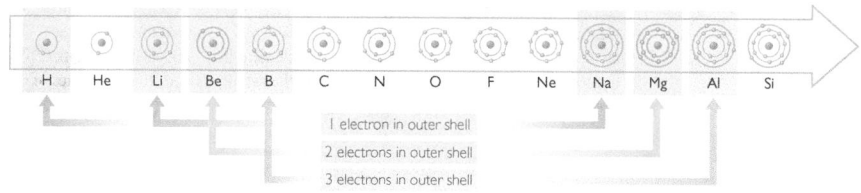

| H | He | Li | Be | B | C | N | O | F | Ne | Na | Mg | Al | Si |

1 electron in outer shell
2 electrons in outer shell
3 electrons in outer shell

Comparing electron arrangements reveals patterns in atomic structure.

Starting with lithium, notice how the number of electrons in the outer shell repeats in a regular pattern, going from one to eight and then starting over again.

TRM Worksheet with Answers 18

TRM Worksheet 18

TRM Handout—Table of Electron Shells (with answers) 18

TRM Handout—Table of Electron Shells 18

TRM Transparency—The Shell Model 18

TRM Transparency—Table of Valence and Core Electrons 18

PREPARE FOR THE CLASSWORK

→ Display the Transparency—The Shell Model 18, which shows a 2-D and a 3-D representation of electron shells.

→ Make sure students understand that these models of the atom are simplified and are not to scale. Remind them that atoms take up space in three dimensions but flat models are easier to draw.

→ Provide a basic definition of electron shells: Electron shells are the levels around the nucleus where electrons can be found. These levels are represented on models by spheres or circles.

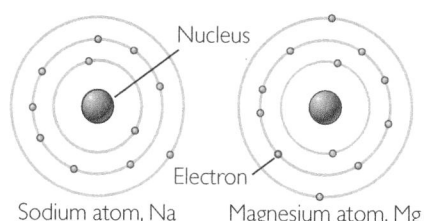

Nucleus

Electron

Sodium atom, Na   Magnesium atom, Mg

INTRODUCE THE CLASSWORK

→ Arrange students into pairs.

→ Give students both the student worksheet and the Handout—Table of Electron Shells that shows the arrangements of electrons for the elements in the first four rows of the periodic table. Students will also need a periodic table.

→ Explain to students that they will be looking for patterns in the arrangements of the electrons and creating shell models based on these patterns.

---

1. What do you notice about the number of spokes on the circles?

2. The spokes represent electrons. Do the spokes represent the total number of electrons? Explain your thinking.

Sample answers: **1.** The number of spokes increases by one as you go across the table from left to right in each period. The number of spokes remains the same as you go down a group. **2.** Except for helium and hydrogen, the spokes do not represent all the electrons. For example, sodium and potassium have 11 and 19 electrons, respectively, not 1.

→ Solicit students' ideas regarding the patterns they see in the spokes.

Ask: What patterns do you find in the number of spokes?

Ask: Do the spokes represent the total number of electrons? Why or why not?

Ask: The pattern in the number of spokes is periodic. What do you think this means? (The pattern changes in a regular, repeating way. Or, it repeats in each period of the table.)

EXPLAIN & ELABORATE (15 min)

DISCUSS GENERAL PATTERNS OF ELECTRONS

Ask: Based on your handout, what do lithium, sodium, and potassium all have in common? (Each has two electrons in the first shell and one electron in the outermost shell.)

Ask: If you know what group an element is in, what can you predict about its electron arrangement? (the number of electrons in the outermost shell)

Ask: What does the period an element is in tell you about its electrons? (the number of electron shells)

Ask: How can you determine the arrangement of an element's electrons from its position in the periodic table?

Key Points: The atomic number of an element is the same as the total number of electrons. As you proceed from one element to the next in the periodic table, the number of electrons increases by one. The number of electrons corresponds to the atomic number of each element.

The period (row) number of the element is the same as the number of electron shells. As you go down the periodic table, from one period to the next, the number of electron shells increases by one. In the first period, the atoms have electrons in only one shell. In the fourth period, the atoms have electrons in four shells.

For main-group elements, the group number of the element is the same as the number of electrons in the outermost shell. For example, all the Group 4A elements have four electrons in their outermost shell, while all the Group 8A elements have eight electrons in their outermost shell.

DISCUSS PATTERNS IN VALENCE AND CORE ELECTRONS

→ Define valence and core electrons.

→ Display the Transparency—Table of Valence and Core Electrons 18.

→ With help from the students, determine the number of core and valence electrons from the handout and fill in the Transparency—Table of Valence and Core Electrons.

When these electron shell models are organized according to the periodic table, even more is revealed. Examine this table. The atoms in the first row, or Period 1, have only one shell. The atoms in Period 2 have two shells. The atoms in each new period of the periodic table have an additional electron shell.

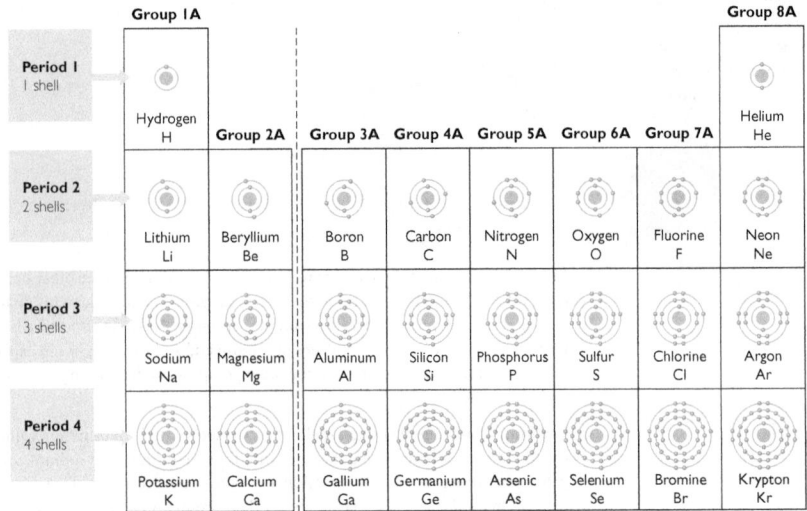

It turns out that for main group elements, Groups 1A through 8A, everything you need to know about the placement of an atom's electrons can be decoded from the periodic table.

> **Big Idea** The arrangement of elements on the periodic table reflects the arrangement of electrons in the atoms of those elements.

In the first column of the table, every element has one electron in its outermost electron shell. In the second column, each has two electrons in the outermost shell, and so on. In the eighth column, each element has eight electrons in its outermost shell. This feature turns out to be the key to many of the properties of the atoms.

Valence and Core Electrons

The outermost shell that electrons occupy is very important in chemistry and is referred to as the **valence shell**. The electrons in that shell are called the **valence electrons**. The electrons that are located in all of the inner shells are known as **core electrons**.

Valence electrons

90   Chapter 4 | **Moving Electrons**

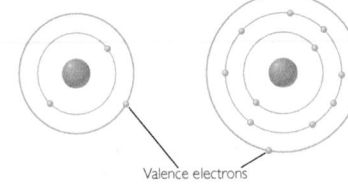

Ask: What patterns do you see in the numbers of valence electrons in the table?

Ask: How do the numbers of core electrons vary across the second row? (They remain at 2.)

Ask: Why is there a change in the number of core electrons across the fourth row? (There are ten transition elements between Ca and Ga, so ten more electrons get filled in—but not in the valence shell, for reasons given later in Lesson 24: Shell Game).

Key Points: The outermost electron shell is referred to as the *valence shell*. The electrons in this shell are called *valence electrons*. The electrons located in all the inner shells combined are referred to as *core electrons*. For reasons we will discover later, chemists pay a lot of attention to the number of electrons that are in the outermost electron shell.

For main group elements, the number of electrons in the valence shell corresponds to the group number on the periodic table. So, every Group 4A element has four electrons in its outermost shell, while every Group 8A element has eight electrons in its outermost shell.

Elements with the same number of electrons in their outermost shells have similar properties. In other words, elements with the same number of valence electrons have similar properties.

Example

Sulfur Atoms

Find the element sulfur, S, on the periodic table.

  a. What group is sulfur in?

  b. How many valence electrons does a sulfur atom have?

  c. Explain why selenium, Se, has properties similar to those of sulfur.

Solution

Sulfur is not a metal, so it is located toward the right side of the periodic table.

  a. Sulfur is in Group 6A of the periodic table.

  b. The number of valence electrons is the same as the group number, so sulfur has six valence electrons.

  c. Selenium, Se, is also in Group 6A. Even though it has many more total electrons than sulfur has, selenium also has six valence electrons. This is why sulfur and selenium have similar properties.

LESSON SUMMARY

Why do elements in the same group in the periodic table have similar properties?

Elements that are grouped together in columns on the periodic table share similar properties. For main group elements, atoms in the same group have the same number of electrons in their outermost shells. The outermost shell is called the valence shell, and the electrons in that shell are called valence electrons. Electrons in the inner shells are called core electrons. The number of valence electrons in an atom of a main group element corresponds to the element's group number. So, the periodic table is a direct reflection of the atomic structures of the atoms.

KEY TERMS

valence shell
valence electron
core electron

Exercises

Reading Questions

1. How can you determine the arrangement of an element's electrons from its position on the periodic table?

2. If you know what group an element is in, what can you predict about its electron arrangement?

Valence shell: The outermost electron shell in an atom.

Valence electrons: The electrons located in the outermost electron shell of an atom.

Core electrons: All other electrons in an atom besides the valence electrons.

The arrangement of electrons in their shells is highly predictable. The table of valence and core electrons shows that for main group electrons the number of valence electrons increases by one as you go across the table, from Group 1A to Group 8A. This pattern repeats in each row. Thus, for main group elements the number of valence electrons in an atom corresponds to the group number of its element.

The numbers of core electrons also exhibit patterns across each row of the periodic table. The elements in the first row have only one shell, so there are no core electrons. All the elements in the second row have 2 core electrons. All the elements in the third row have 10 core electrons. All the elements in the fourth row before the transition elements have 18 core electrons, while the elements in the fourth row after the transition elements have 28 core electrons.

Check-In

Provide each piece of information for element 34.

  a. The element's name and symbol.

  b. The total number of electrons in an atom of this element.

  c. The number of core electrons in an atom of this element.

  d. The number of valence electrons.

  e. The group number for this element.

  f. The names of other elements with the same number of valence electrons.

Answers: **a.** Selenium, Se. **b.** It has a total of 34 electrons. **c.** It has 28 core electrons. **d.** It has 6 valence electrons. **e.** Group 6A. **f.** Oxygen, sulfur, tellurium, and polonium.

Homework

Assign the reading and exercises for Alchemy Lesson 18 in the student text.

LESSON 18 ANSWERS

1. For main group elements, the number of shells containing electrons is equal to the period number. All of the shells, except the highest, are completely filled. The group number determines the number of electrons in the outermost shell.

2. For elements in the main group, the number of valence electrons is equal to the group number.

3. *Possible answer:* Beryllium, magnesium, and calcium are all alkaline earth metals located in Group 2A. These elements have similar chemical and physical properties because they have two valence electrons.

4. Boron has two core electrons in the first shell and three valence electrons in the second shell.

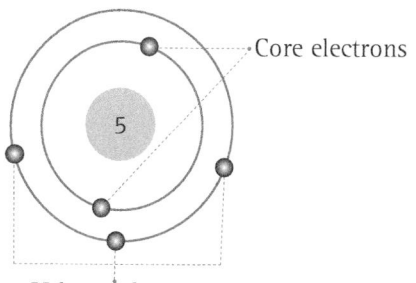

5. The number of core electrons does not change across a period for main group elements.

6. a. Carbon has 6 electrons and silicon has 14 electrons.

b.

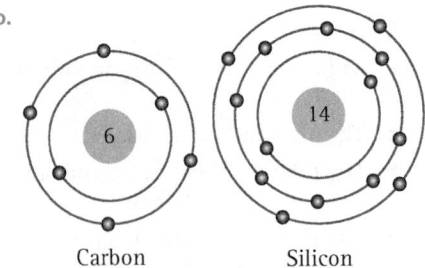

Carbon   Silicon

c. Carbon and silicon both have 4 valence electrons.

d. Carbon has 2 core electrons and silicon has 10 core electrons.

e. Carbon and silicon have similar properties because atoms of the two elements have the same number of valence electrons.

7. a. Element number 17 is chlorine. It has the chemical symbol Cl and is located in Group 7A. This information comes directly from square 17 on the periodic table.

b. The nucleus contains 17 protons. The number of protons is equal to the atomic number.

c. *Possible answer:* The average atomic mass of chlorine to the nearest whole number is 35 amu. Therefore, the nucleus will have 35 − 17 = 18 neutrons. According to the graph of isotopes of the first 95 elements, chlorine-35 is a naturally occurring isotope.

d. The number of electrons in a neutral atom of chlorine is 17. In a neutral atom, the number of electrons equals the number of protons.

e. Chlorine has 7 valence electrons. The number of valence electrons is equal to the group number.

f. Chlorine has 10 core electrons. The number of core electrons equals the difference between the total number of electrons and the number of valence electrons, or 17 − 7.

g. *Possible answer (any 3 elements):* Fluorine, bromine, iodine, and astatine all have the same number of valence electrons. All of these elements are in Group 7A along with chlorine.

8. a. Element number 50 is tin. It has the chemical symbol Sn and is located in Group 4A. This information comes directly from square 50 on the periodic table.

b. The nucleus contains 50 protons. The number of protons is equal to the atomic number.

Exercises *(continued)*

Reason and Apply

3. What do beryllium, Be, magnesium, Mg, and calcium, Ca, all have in common?

4. Draw a shell model for boron, B. Identify the core and valence electrons.

5. For main group elements, how does the number of core electrons vary across a period?

6. Consider the elements carbon, C, and silicon, Si.
   a. How many electrons does an atom of each of these elements have?
   b. Draw shell models for atoms of each of these elements.
   c. How many valence electrons do atoms of each of these elements have?
   d. How many core electrons do atoms of each of these elements have?
   e. Why are the properties of carbon and silicon similar?

7. Provide the following information for element number 17. For each answer, explain how you know.
   a. the element's name, symbol, and group number
   b. the number of protons in the nucleus
   c. the number of neutrons in the nucleus
   d. the number of electrons in a neutral atom of the element
   e. the number of electrons that are valence electrons in a neutral atom
   f. the number of electrons that are core electrons in a neutral atom
   g. the names of three other elements with the same number of valence electrons

8. Answer Exercise 7, parts a through g, for element number 50.

9. If an element is located in Group 4A of the periodic table, what can you conclude about atoms of this element?
   (A) It has four valence electrons.
   (B) It has four electron shells.
   (C) It has four core electrons.
   (D) It has properties similar to carbon.
   (E) Both A and D.

10. What do the elements in Period 3 have in common?
   (A) Their atoms have the same number of valence electrons.
   (B) Their atoms have the same number of electron shells.
   (C) Their atoms have the same properties.
   (D) None of the above.

92   Chapter 4 | **Moving Electrons**

---

c. *Possible answer:* The average atomic mass of tin to the nearest whole number is 119 amu. Therefore, the nucleus will have 119 − 50 = 69 neutrons. According to the graph of isotopes of the first 95 elements, tin-119 is a naturally occurring isotope.

d. The number of electrons in a neutral atom of tin is 50. In a neutral atom, the number of electrons equals the number of protons.

e. Tin has 4 valence electrons. The number of valence electrons is equal to the group number.

f. Tin has 46 core electrons. The number of core electrons equals the difference between the total number of electrons and the number of valence electrons, or 50 − 4.

g. *Possible answer (any 3 elements):* Carbon, silicon, germanium, and lead all have the same number of valence electrons. All of these elements are in Group 4A along with tin.

9. E

10. B

LESSON 19

Ions

THINK ABOUT IT

Some atoms are more chemically stable than others. In other words, they don't readily combine with other atoms to form new compounds. The atoms of the noble gas elements are considered the most chemically stable atoms on the periodic table. The most reactive elements are located just before and just after the noble gases, in Groups 1A and 7A. Perhaps the electron arrangements of these atoms are related to the atoms' reactivity.

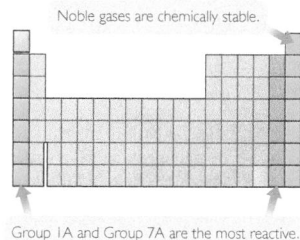

Noble gases are chemically stable.

Group 1A and Group 7A are the most reactive.

How is chemical stability related to the arrangements of electrons in atoms?

To answer this question, you will explore

↪ Noble Gases

↪ Noble Gas Envy

↪ Patterns in Ion Charges

EXPLORING THE TOPIC

↪ Noble Gases

Some elements are chemically stable, or rarely react with other elements. Others are reactive and combine readily with other elements to form compounds. What makes some elements chemically stable and others reactive?

INDUSTRY CONNECTION

Noble gases are so stable that they rarely form compounds. At one time it was thought that they *never* formed compounds with other atoms. But in 1962 a British scientist, Neil Bartlett, created xenon hexafluoroplatinate, a yellow solid, by accident. Chemists have not yet been able to create any compounds with either helium or neon. Shown here is the "light signature" of xenon.

Lesson 19 | **Ions**   93

Overview

ACTIVITY: GROUPS OF 8

Key Question: How is chemical stability related to the arrangements of electrons in atoms?

KEY IDEAS

Some atoms gain or lose electrons when they interact. Atoms that have lost or gained electrons are called *ions*. When an electron is gained or lost the resulting atom has a charge on it. The positively charged ions are called *cations*, and the negatively charged ions are called *anions*. Ions tend to have an electron arrangement identical to that of a noble gas.

LEARNING OBJECTIVES

● Explain that an ion is formed when an atom loses or gains electrons and state the difference between a cation and an anion.
● Determine the charge on an ion based on an atom's placement in the periodic table.
● Explain the relationship between ion charge and valence electrons.

FOCUS ON UNDERSTANDING

● The last few lessons have emphasized the notion that atoms have a set number of electrons. The concept of atoms losing and gaining electrons might now be confusing to students.

● Atoms that lose electrons have a positive charge, represented by a plus sign (+). This can seem counterintuitive to students, who might reason that something with a charge of +2 must have gained two particles.
● Students are familiar with stable nuclei versus radioactive nuclei. The term *stability* is used differently here: The focus is on chemical stability, not nuclear stability. These different applications of the concept of stability might confuse students.

KEY TERMS

ion
cation
anion

IN CLASS

In this activity, students create a table of ions for a portion of the periodic table. They do so by first creating individual ion cards for elements 1 through 20 and 31 through 36. Teams then organize the cards according to the periodic table and look for patterns. You might want to have students create a more permanent table by gluing their ion cards to a piece of butcher paper after they have arranged them. Students read about "noble gas envy" and ionic charge in the student text. A complete materials list is available for download.

TRM Materials List 19

SETUP

Before class, prepare blank cards. Each group of eight students will need 28 cards. Use index cards or cut standard 8½-by-11 inch paper into quarters.

Differentiate

Students may struggle with understanding how taking away electrons results in a positively charged cation or how adding electrons results in a negatively charged anion. You might want to have students visit the University of Colorado Boulder's PhET simulation titled "Build an Atom" during the making-sense discussion to provide a visual to support students in identifying an ion's overall charge. You can find a list of URLs for this lesson on the Chapter 4 Web Resources document.

TRM Chapter 4 Web Resources

Lesson Guide

ENGAGE (5 min)

TRM Transparency—ChemCatalyst 19

TRM PowerPoint Presentation 19

ChemCatalyst

Chemists have found that metal atoms transfer electrons to nonmetal atoms when they form compounds. Examine the shell model showing how a lithium atom might transfer an electron to a fluorine atom.

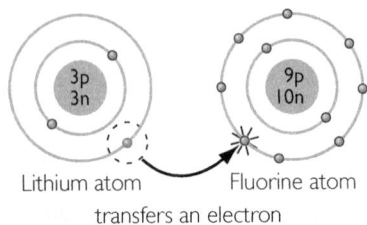

Lithium atom   Fluorine atom
transfers an electron

1. What effect does this electron transfer have on the charge of each atom? Explain.
2. What element does each atom resemble after the electron has been transferred?

Sample answers: **1.** The transfer of an electron gives the fluorine atom a −1 charge and the lithium atom a +1 charge. **2.** The lithium atom now resembles a helium atom in its electron arrangement. The fluorine atom now resembles a neon atom in its electron arrangement.

DISCUSS THE CHEMCATALYST

→ Discuss what happens when atoms lose or gain electrons.

Ask: What is happening in the model?

Ask: How many protons and electrons does each atom have after the transfer? What does this do to the charge on each atom?

Ask: Have any other parts of the lithium and fluorine atoms been changed during this process? (no)

EXPLORE (20 min)

TRM Worksheet with Answers 19

TRM Worksheet 19

Noble gases are the most chemically stable elements. Shell models for the noble gas elements helium, He, neon, Ne, argon, Ar, and krypton, Kr, are shown below. Take a moment to compare the valence electrons of each.

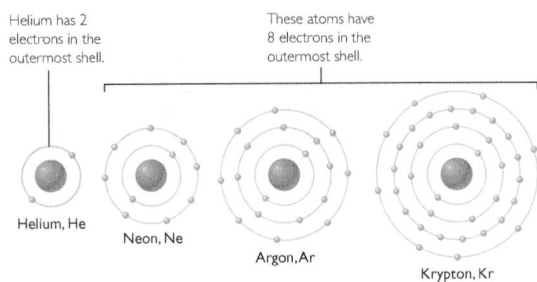

Helium has 2 electrons in the outermost shell.

These atoms have 8 electrons in the outermost shell.

Helium, He

Neon, Ne

Argon, Ar

Krypton, Kr

Helium has two valence electrons, which is the maximum for the first shell. The remaining noble gases each have eight valence electrons. The stability of the noble gases is associated with the number of valence electrons they have.

⤺ Noble Gas Envy

Consider two highly reactive elements, sodium, Na, and fluorine, F. The outer shell of a fluorine atom, F, has seven electrons. This is just one short of the eight electrons that neon, Ne, has in its outer shell. The outer shell of a sodium atom, Na, has one electron. This is just one more electron than neon has in total.

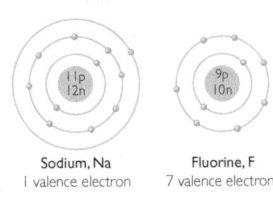

Sodium, Na
1 valence electron

Fluorine, F
7 valence electrons

Now examine what happens to these two atoms when they combine to form a compound. Sodium, Na, gives one electron to fluorine, F.

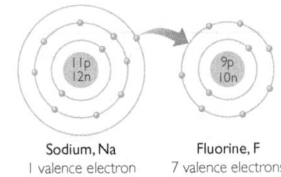

Sodium, Na
1 valence electron

Fluorine, F
7 valence electrons

INDUSTRY CONNECTION

Sodium and chlorine are both highly reactive. In elemental form, sodium is a soft, silvery metal, and chlorine is a greenish poisonous gas. However, they are never found in elemental form in nature. Yet they are often found as sodium chloride, which is used as table salt.

PREPARE FOR THE ACTIVITY

→ Explain the formation of ions. Draw models like these on the board to illustrate the transfer of an electron. Focus students on taking an inventory of protons and electrons in each atom.

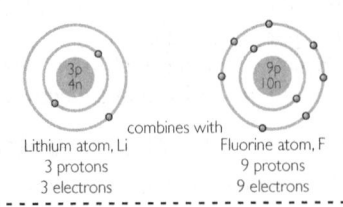

combines with

Lithium atom, Li
3 protons
3 electrons

Fluorine atom, F
9 protons
9 electrons

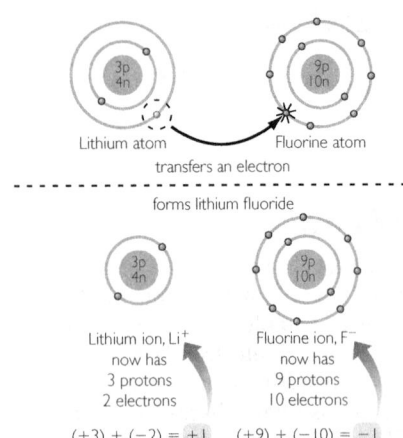

Lithium atom   Fluorine atom
transfers an electron

forms lithium fluoride

Lithium ion, Li⁺ now has
3 protons
2 electrons

Fluorine ion, F⁻ now has
9 protons
10 electrons

$(+3) + (-2) = +1$ $(+9) + (-10) = -1$

Now both atoms have an electron arrangement like that of neon, Ne.

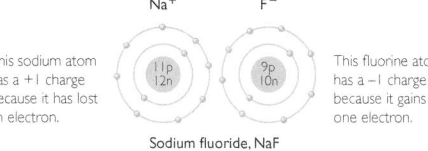

This sodium atom has a +1 charge because it has lost an electron.

This fluorine atom has a −1 charge because it gains one electron.

Sodium fluoride, NaF

Both ions now have an electron arrangement like the noble gas neon. They form an ionic compound, NaF.

Important to Know When an atom loses or gains electrons, the result is a charge on the atom. The rest of the atom stays the same, and the identity of the element does not change.

The movement of an electron from one atom to another atom alters the balance of charges on both atoms. Fluorine now has more electrons than protons. Sodium has more protons than electrons. The atoms are now called **ions** because they possess a charge. The sodium ion has a charge of +1 and its symbol is written as Na^+. The fluorine ion has a charge of −1 and its symbol is written as F^-. Because Na^+ and F^- have opposite charges, they attract one another. So the movement of an electron from one atom to the other forms a new compound, sodium fluoride, NaF. As a result of this electron transfer, sodium atoms are now bonded to fluorine atoms.

Example

Calcium Oxide, CaO

Consider the compound calcium oxide, CaO.

a. Draw shell models for calcium, Ca, and oxygen, O.

b. With arrows, show how electrons can be transferred so that each atom has the same electron arrangement as that of a noble gas.

c. What are the charges on the calcium, Ca, and oxygen, O, ions after electrons have been transferred?

Solution

a. You can draw a shell model for a neutral atom, an atom that has no charge, using the element number and the element's position on the periodic table. Calcium is in Period 4 and Group 2A, so it has four shells and two valence electrons. Oxygen is in Period 2 and Group 6A, so it has two electron shells and six valence electrons.

Calcium, Ca   Oxygen, O

The calcium atom loses 2 electrons.   The oxygen atom gains 2 electrons.

Calcium, Ca   Oxygen, O

Lesson 19 | **Ions**   95

Ion: An atom (or group of atoms) that has a positive or negative charge because it has lost or gained electrons.

INTRODUCE THE ACTIVITY

→ Divide the class into groups of eight.

→ Pass out the student worksheet.

→ Draw examples of two ion cards on the board and explain the information on them. Let students know that the atoms in Groups 1A, 2A, and 3A lose electrons, that atoms in Group 4A can either gain or lose electrons, and that

atoms in Groups 5A, 6A, and 7A gain electrons. In each case, the atom will take on an electron arrangement like that of a noble gas. Tell students to use both sides of the index card for elements in Group 4A.

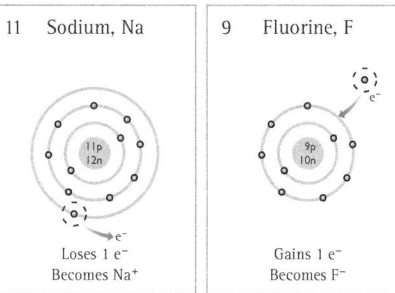

11   Sodium, Na   9   Fluorine, F

Loses 1 e^-
Becomes Na^+

Gains 1 e^-
Becomes F^-

→ Remind students how to estimate the number of neutrons from the average atomic mass.

→ At the end of the activity, have students glue their ion cards to a piece of butcher paper as they have arranged them. (optional)

EXPLAIN & ELABORATE (15 min)

TRM Transparency—Ion Cards Arranged 19

TRM Transparency—Noble Gas Envy 19

DISCUSS PERIODIC PATTERNS OF ION CHARGES

→ Display the transparency Ion Cards Arranged and point out the patterns that are evident.

Ask: What kinds of periodic patterns do you notice for the charges on the ions?

Ask: Why do ions have a charge?

Ask: Does the identity of an atom stay the same when electrons are transferred? Explain.

Key Points: The table of arranged ion cards shows that the charges on ions are predictable. Ions in each group have the same ion charge.

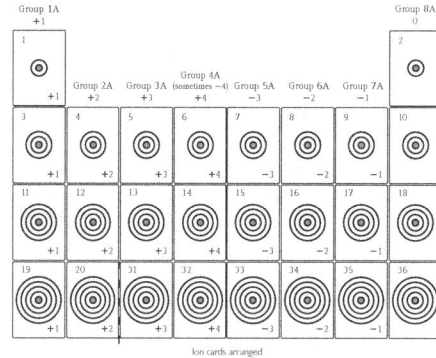

When electrons are removed from or added to an atom, the rest of the atom stays the same. The number of protons in the nucleus is not changed, so the removal or addition of electrons results in a charge on the atom, making it an ion, but does not change the identity of the element.

The charge on an ion is noted with a superscript. For example, if one electron is removed from lithium, Li, its ion is written as Li^+. Removing three electrons from Al results in Al^{3+}. If one electron is added to Cl, its ion is written as Cl. Adding two electrons to O results in a negative charge, written as O^{2-}. Atoms with a

Lesson 19 | **Ions**   95

positive charge are called *cations*. Atoms with a negative charge are called *anions*.

> **Cation:** An ion with a net positive charge. Usually, these are formed from metal atoms.
>
> **Anion:** An ion with a net negative charge. Usually, these are formed from nonmetal atoms.

The ions formed from transition elements are exceptions to these simple rules. The charges on transition elements are not as predictable as those on main-group elements. We will address these elements later in the chapter.

INTRODUCE "NOBLE GAS ENVY"

➡ Show the transparency Noble Gas Envy.

Ask: What connection can you find between the shell models of ions and the shell models of noble gases?

Ask: Why do you think this lesson is titled *Noble Gas Envy*?

Key Points: After an electron transfer occurs, the electron arrangements of the resulting ions look like the electron arrangements of noble gases. For example, when the compound lithium fluoride is formed, lithium gives up an electron to fluorine. Each atom now has the same electron arrangement as one of the noble gases: The lithium cation has the same electron arrangement as helium, He, and the fluorine anion has the same electron arrangement as neon, Ne.

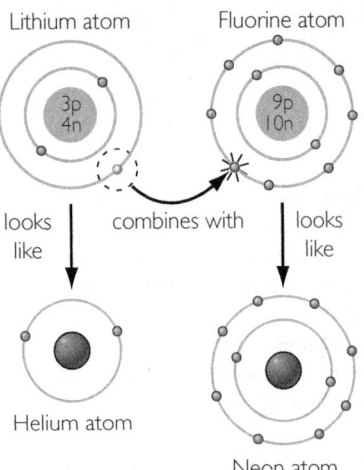

Atoms tend to lose or gain electrons to attain the electron arrangement of a noble gas. This tendency is what we are calling "noble gas envy." It can be a powerful tool in helping predict the charges on ions. Atoms seem to "want" to have electron arrangements like

---

Example *(continued)*

b. If calcium transfers two electrons to oxygen, they will both have noble gas electron configurations. The electron arrangement of oxygen now resembles that of neon, Ne. The electron arrangement of calcium now resembles that of argon, Ar.

c. Because calcium has lost two electrons, it now has more protons than electrons, giving it a $+2$ charge. The ion can be written as Ca^{2+}. Oxygen has gained 2 electrons, so its charge is -2. Its ion can be written as O^{2-}.

In the calcium and oxygen example, each ion has an electron arrangement that is identical to that of a noble gas. It is reasonable to assume that there is some advantage to having this sort of electron arrangement.

Noble gases are very stable the way they are, without reacting or exchanging any electrons with other elements. Apparently, other atoms can achieve some of the stability of the noble gases by exchanging electrons and becoming ions.

➲ Patterns in Ion Charges

When atoms lose or gain electrons, they become ions, with negative or positive charges. Below is a portion of the periodic table with the various charges on the ions filled in. In every case, the ion that forms has an electron arrangement identical to the noble gas it is nearest to on the table.

Take a moment to examine the table. What patterns do you notice?

Ion Charges on Main Group Elements

1A	2A	3A	4A	5A	6A	7A	8A
1+							
1 H^+	2+	3+	4+ (or 4−)	3−	2−	1−	2 He
3 Li^+	4 Be^{2+}	5 B^{3+}	6 C^{4+}	7 N^{3-}	8 O^{2-}	9 F^-	10 Ne
11 Na^+	12 Mg^{2+}	13 Al^{3+}	14 Si^{4+}	15 P^{3-}	16 S^{2-}	17 Cl^-	18 Ar
19 K^+	20 Ca^{2+}	31 Ga^{3+}	32 Ge^{4+}	33 As^{3-}	34 Se^{2-}	35 Br^-	36 Kr

Notice that the elements on the left side of the table, which are mostly metals, tend to form ions with a positive charge. Ions with a positive charge are called **cations**. The valence shells of these atoms have fewer electrons in them than those of the elements on the right. So, it is easier for these elements to form compounds by giving up these few electrons.

On the other hand, the elements on the right side of the table, which are mostly nonmetals, tend to form ions with a negative charge. Ions with a negative charge are called **anions**.

For the first four groups of the periodic table, the ion charge is the same as the group number. For Groups 5A through 7A, the ion charge is negative and goes from -3 to -1.

---

those of the noble gases because of the chemical stability of the noble gases. It seems that atoms will combine with each other to form compounds only if the result is something fairly stable.

EVALUATE (5 min)

Check-In

1. Draw a shell for calcium, Ca, showing the arrangement of its electrons.

2. What would have to happen for an atom of calcium to have an electron arrangement like that of a noble gas? Explain.

Answers. 1. Calcium, Ca, is located in the second column with two valence electrons in the fourth shell.

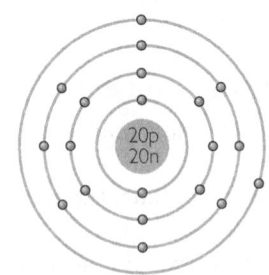

LESSON SUMMARY

How is chemical stability related to the arrangements
of electrons in atoms?

Noble gases have particularly stable atoms. This is attributed to their electron
arrangements. Helium has two electrons, which is the maximum for the first shell.
The remaining noble gases all have eight electrons in their outermost shells. Other
atoms on the periodic table are more reactive than the noble gases. However,
these reactive atoms reach greater chemical stability by gaining or losing valence
electrons. Atoms tend to lose or gain electrons to attain an electron arrangement
similar to that of a noble gas. Atoms do this by combining with other atoms to
form new compounds. When an atom loses or gains electrons, it becomes an ion,
with a charge. Ions with a positive charge are called cations. Ions with a negative
charge are called anions.

KEY TERMS

ion

cation

anion

Exercises

Reading Questions

1. Explain the difference between an anion and a cation.

2. Explain what is meant by noble gas envy.

Reason and Apply

3. How many electrons, protons, and neutrons does Li^+ have?

4. Give two similarities and two differences between Cl and Cl^-.

5. Give two similarities and two differences between Be and Be^{2+}.

6. Which noble gas is closest to magnesium, Mg, on the periodic table? What must
happen to a magnesium atom for it to have an electron arrangement similar to that
of a noble gas?

7. Which noble gas is closest to sulfur, S, on the periodic table? What must happen to
a sulfur atom for it to have an electron arrangement similar to that of a noble gas?

8. List four ions that have the same number of electrons as neon, Ne.

9. List four ions that have the same number of electrons as argon, Ar.

10. What charge would an arsenic, As, ion have?

11. What is the symbol of an ion with 22 protons, 24 neutrons, and 18 electrons?

12. When chlorine gains an electron to become a chloride ion with a −1 charge,
it ends up with the electron arrangement of argon. Why doesn't it become an
argon atom?

13. Explain why the elements on the right side of the periodic table gain electrons
instead of losing them.

14. What periodic patterns do you notice for the charges on the ions?

15. Which of these ions have the correct charge? Choose all that apply.
   a. Na^{2+} **b.** Li^+ **c.** Al^{4+} **d.** Ca^{2+} **e.** Ga^{3+}

16. Which of these ions have the same number of electrons as S^{2-}? Choose all that apply.
   a. Cl^- **b.** Ca^{2+} **c.** Na^+ **d.** O^{2-} **e.** P^{3-}

---

2. If an atom of calcium loses two
electrons, it will have an electron
arrangement resembling that of argon,
Ar. A calcium ion would be formed
with a charge of +2, written as Ca^{2+}.

Homework

Assign the reading and exercises for
Alchemy Lesson 19 in the student text.

LESSON 19 ANSWERS

1. A cation is an ion that has a positive
charge. An anion is an ion that has a
negative charge.

2. "Noble gas envy" refers to the
tendency of atoms to gain or lose
electrons to obtain an electron
configuration in which the outer
shell is completely filled.

3. Li^+ has 3 protons, 2 electrons. The
average atomic mass of lithium to the
nearest whole number is 7 amu.
Therefore, the nucleus will likely
have 4 neutrons.

4. *Possible answer:* Cl and Cl^- are
similar in that they both have the
same number of protons and the same
number of neutrons. But Cl has 17
electrons and Cl^- has 18 electrons. Cl
is more reactive because it is missing
an electron in its valence shell, while
Cl^- is more stable because it has a full
valence shell.

5. *Possible answer:* Be and Be^{2+} are
alike in having the same number of
protons and the same number of
neutrons. But Be has four electrons and
Be^{2+} has two electrons. Be is more
reactive because it has only two
electrons in its valence shell, while
Be^{2+} is more stable because it has a full
valence shell.

6. The noble gas closest to magnesium
is neon. A magnesium atom loses two
electrons to have an electron arrangement like that of neon.

7. The noble gas closest to sulfur is
argon. A sulfur atom gains two electrons
to have an electron arrangement like
that of argon.

8. *Possible answers (any 4):* N^{3-}, O^{2-},
F^-, Na^+, Mg^{2+}, Al^{3+}

9. *Possible answers (any 4):* P^{3-}, S^{2-},
Cl^-, K^+, Ca^{2+}

10. It has a charge of 3^-.

11. The element that has 22 protons
is identified on the periodic table as
titanium, Ti. Because the ion has 22
protons and 18 electrons, it has a net
positive charge of 4. The number of
neutrons is not needed to determine
the ion.

12. When chlorine gains an electron, it
does not become an argon atom because
the ion still has 17 protons. The number
of protons determines the identity of
the element.

13. Elements on the right side of the
table gain electrons to have a noble gas
arrangement. They tend not to lose
electrons because the charge would be
too large.

14. In each period, ions on the left side
have a positive charge that increases
toward the center of the table. On the
right side, ions have a negative charge,
which increases from the second row
from the right toward the center. The
element at the far right of a period does
not form ions.

15. B, D, and E

16. A, B, and E

Overview

CLASSWORK: PAIRS

Key Question: How can valence electrons be used to predict chemical formulas?

KEY IDEAS

Metal and nonmetal atoms come together to form ionic compounds. The atoms in ionic compounds combine in specific ratios. We can predict the formulas of the compounds that form by examining their valence shells. The metal ions are cations, and the nonmetal atoms are anions. Both have electron arrangements resembling those of noble gas atoms. The rule of zero charge requires that the charges on the cations and anions in an ionic compound sum to 0. Further, the number of valence electrons in the atoms of a compound after bonding usually totals 8 or a multiple of 8.

LEARNING OBJECTIVES

- Predict the chemical formulas of compounds that will form between metal and nonmetal atoms.
- Explain how an ionic compound forms and determine if it follows the rule of zero charge.

FOCUS ON UNDERSTANDING

Students might think that metal atoms "want" to give away electrons. However, a metal atom will give up electrons only if there is a nonmetal atom that can accept the electrons. The electrons do not float away freely from a metal under normal conditions.

KEY TERMS

ionic compound
rule of zero charge

IN CLASS

Students examine the patterns in the formulas of ionic compounds formed between metal and nonmetal elements to come up with a generalization about which atoms will come together and in what proportions. Students complete a data table in which they combine atomic models to create compounds. They look for patterns in the table and arrive at rules for the formulas of compounds formed between metal and nonmetal atoms. They practice this skill in the next lesson, in which they play the Salty Eights game. A complete materials list is available for download.

TRM Materials List 20

Getting Connected

Ionic Compounds

THINK ABOUT IT

Sodium and chlorine atoms combine to form sodium chloride, NaCl, when sodium transfers a valence electron to chlorine. There are no other compounds that form between sodium and chlorine. For example, $NaCl_4$ and Na_3Cl are not possible compounds. Why is NaCl the compound that forms, rather than $NaCl_4$ or Na_3Cl?

How can valence electrons be used to predict chemical formulas?

To answer this question, you will explore

- Ionic Compounds
- The Rule of Zero Charge
- More Complex Ionic Compounds

EXPLORING THE TOPIC

Ionic Compounds

There are millions of different compounds on the planet. Each compound is a result of a combination of atoms and has a chemical formula. However, some combinations of atoms are simply not possible. Valence electrons are the key to figuring out chemical formulas and to determining which compounds are possible.

COMPOUNDS BETWEEN METALS AND NONMETALS

Atoms combine with other atoms to achieve the stability of the noble gases. One way that atoms accomplish this is to transfer valence electrons to other atoms. When atoms gain or lose electrons, the resulting atoms are no longer neutral.

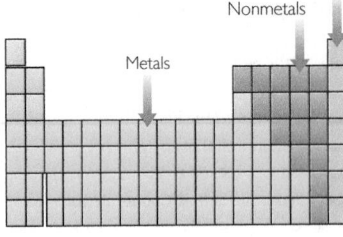

They become ions with charges. The compounds that form in this way are called **ionic compounds.** Ionic compounds, like the noble gases, are very stable.

Ionic compounds form between metal atoms and nonmetal atoms. The metal atoms, on the left side of the periodic table, tend to give up electrons and become cations, which have positive charges. The nonmetal atoms, on the right side of the periodic table, tend to accept electrons and become anions, which have negative charges. Because the cation and anion have opposite charges, they are strongly attracted to one another.

Pre-AP® Course Tip

Students preparing to take an AP® chemistry course or college chemistry should be able to analyze data to identify patterns and relationships among chemical phenomena. In this lesson, students use patterns found on the periodic table to determine the correct chemical formula of ionic compound.

Lesson Guide

ENGAGE (5 min)

TRM PowerPoint Presentation 20

ChemCatalyst

Metal elements combine with the nonmetal element chlorine, Cl, to form compounds. The formulas are given in the tables.

Table 1

Element	Compound
Na	NaCl
K	KCl

Table 2

Element	Compound
Mg	$MgCl_2$
Ca	$CaCl_2$

Big Idea Ionic compounds form when valence electrons are transferred between atoms.

FORMULAS OF SIMPLE IONIC COMPOUNDS

The table below contains data on three simple ionic compounds. These compounds are all composed of a metal atom and a nonmetal atom in a 1:1 ratio. Pay particular attention to the total charge and the number of valence electrons on each of the atoms that combine.

Simple Ionic Compounds

Compound	Metal	Cation	Number of valence electrons	Nonmetal	Anion	Number of valence electrons
NaCl	Na	Na^+	1	Cl	Cl^-	7
CaO	Ca	Ca^{2+}	2	O	O^{2-}	6
GaAs	Ga	Ga^{3+}	3	As	As^{3-}	5

Note that in each case, the charge on the cation is equal to and opposite the charge on the anion.

◌ The Rule of Zero Charge

You can see a distinct pattern to the ions that form in ionic compounds. Every time a metal atom and a nonmetal atom bond, they form a compound with an overall zero charge. This is known as the **rule of zero charge.**

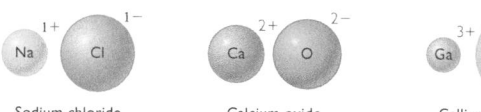

| Na $1+$ Cl $1-$ | Ca $2+$ O $2-$ | Ga $3+$ As $3-$ |
| Sodium chloride | Calcium oxide | Gallium arsenide |

Notice how the ions in each compound add up to zero.

The charges on the metal cations and nonmetal anions add up to zero. The rule of zero charge is useful in predicting the chemical formulas for ionic compounds, which form between a metal and a nonmetal.

PREDICTING IONIC BONDING USING THE PERIODIC TABLE

You may have noticed that the atoms that combine to form ionic compounds come from very predictable places on the periodic table. Sodium, with one valence electron, can combine with chlorine, fluorine, or bromine, each with seven valence electrons. In fact, any Group 1A element can combine with any Group 7A element to form an ionic compound.

CONSUMER CONNECTION

All ionic compounds form crystals. If you look at grains of table salt with a magnifying glass, you will see that the pieces are shaped like tiny boxes.

Lesson 20 | **Ionic Compounds** 99

Table 3

Element	Compound
Ne	none
Ar	none

1. Compare the three tables. What do you notice?

2. Predict the formula of a compound formed between lithium, Li, and chlorine, Cl. Which table would you put it in?

Sample answers: 1. The noble gases do not form compounds with chlorine. The elements in Table 1 form compounds in a 1:1 ratio with chlorine.

The elements in Table 2 form compounds in a 1:2 ratio with chlorine. 2. The probable compound is LiCl and belongs in Table 1.

Ask: How are the formulas related to an element's position on the periodic table?

Ask: Why do you think that neon, Ne, and argon, Ar, do not form compounds?

Ask: How can chlorine obtain an electron arrangement similar to that of a noble gas?

Ask: Why do you think sodium, Na, combines with chlorine, Cl, in a 1:1 ratio but magnesium, Mg, combines with chorine, Cl, in a 1:2 ratio?

EXPLORE (20 min)

TRM Worksheet with Answers 20

TRM Worksheet 20

INTRODUCE THE CLASSWORK

→ Arrange students into pairs.

→ Pass out the student worksheet.

→ Define ionic compound for the class.

Ionic compound: An ionic compound is a compound composed of positive and negative ions, formed when metal and nonmetal atoms combine.

EXPLAIN & ELABORATE (15 min)

TRM Transparency—Magnesium Chloride 20

TRM Transparency—Chemical Formulas of Ionic Compounds 20

DISCUSS THE TABLE OF IONIC COMPOUNDS

→ Use the transparency Magnesium Chloride to show how the electrons are transferred.

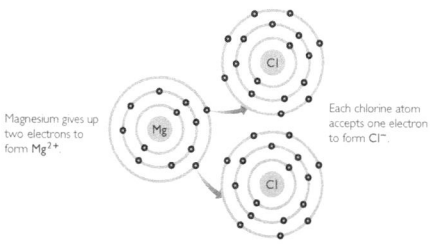

Magnesium gives up two electrons to form Mg^{2+}

Each chlorine atom accepts one electron to form Cl^-.

$Mg^{2+} + Cl^- + Cl^-$ produces $MgCl_2$ with zero charge.

Ask: Why does magnesium, Mg, combine with chlorine, Cl, in a 1:2 ratio?

Ask: Can magnesium form a cation with a charge other than +2? Explain your thinking.

Ask: Why does magnesium combine with oxygen, O, in a 1:1 ratio?

Ask: Why must the total charge on ionic compounds always sum to 0?

Key Points: Metal and nonmetal elements combine to form ionic compounds. The metal cations and nonmetal anions combine so that the total charge is 0. For example, Mg^{2+} and Cl combine to form $MgCl_2$ so that the +2 charge on the magnesium cation is balanced by the −1 charge on two chloride anions. Ionic compounds are sometimes called *salts*. You have undoubtedly heard of table salt, or sodium chloride, NaCl. Table salt is by far the most common ionic compound on our planet.

The electron arrangements of the cations and anions resemble the arrangements of a noble gas atom. For this to be true, the metal atoms can give up only valence electrons, and the nonmetal atoms can accept electrons only into their valence shell.

DISCUSS RULES TO DETERMINE FORMULAS OF IONIC COMPOUNDS

→ Introduce the rule of zero charge.

Ask: What is meant by the rule of zero charge?

Ask: How can you use the rule of zero charge to determine the formulas of ionic compounds?

Ask: What do you notice about the total numbers of valence electrons in ionic compounds?

Ask: Does Li_3N satisfy the rule of zero charge?

Key Points: The rule of zero charge can be used to determine the chemical formulas of ionic compounds. The chemical formula is correct if the charges on all the metal cations and nonmetal anions in the formula sum to 0. For example, one magnesium atom cannot combine with three chlorine atoms to form $MgCl_3$, because the resulting compound would have a charge of -1, not 0. This rule is useful in predicting the chemical formulas of ionic (metal-nonmetal) compounds. The rule does not apply to compounds formed between two metal elements or between two nonmetal elements.

> **Rule of zero charge:** In an ionic compound, the positive charges on the metal cations and the negative charges on the nonmetal anions sum to 0.

EXAMINE CHEMICAL FORMULAS OF IONIC COMPOUNDS

→ Use the transparency Chemical Formulas of Ionic Compounds to show patterns in the chemical formulas.

Valence Electrons in Main Group Elements

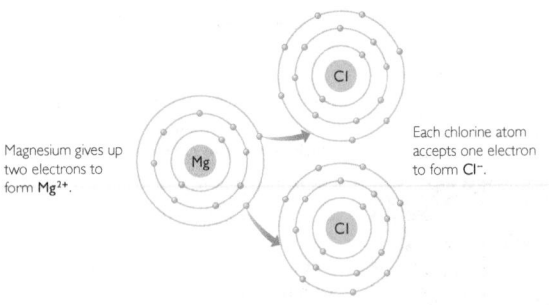

The drawing illustrates how you can use the periodic table to decide which atoms will combine in a 1:1 ratio to form an ionic compound. There are dozens of possible combinations. For example, an atom in Group 3A can form a compound with an atom in Group 5A.

Important to Know The atoms in Group 4A can either transfer or accept electrons.

More Complex Ionic Compounds

All of the compounds discussed in this lesson so far have a metal-to-nonmetal ratio of 1:1. However, it is also possible to have compounds with different ratios. For example, one magnesium atom can combine with two chlorine atoms to form magnesium chloride, $MgCl_2$. Before bonding, each magnesium atom has two valence electrons. These two electrons are transferred to two different chlorine atoms, each of which needs one more electron to have the same number of electrons as the noble gas argon, Ar.

CONSUMER CONNECTION

Energy drinks help replace necessary ions, known as electrolytes, in your body.

Magnesium gives up two electrons to form Mg^{2+}.

Each chlorine atom accepts one electron to form Cl^-.

$Mg^{2+} + Cl^- + Cl^-$ produces $MgCl_2$ with zero charge.

100   Chapter 4 | **Moving Electrons**

Chemical Formulas of Ionic Compounds

Example	Number of valence electrons for the metal	Number of valence electrons for the nonmetal	Total number of valence electrons	Total positive charge	Total negative charge	Total charge
NaF	1	7	8	+1	−1	0
MgO	2	6	8	+2	−2	0
AlN	3	5	8	+3	−3	0
K_2Se	1	6	8	2(+1)	−2	0
$MgCl_2$	2	7	16	+2	2(−1)	0
AlF_3	3	7	24	+3	3(−1)	0
Al_2O_3	3	6	24	2(+3)	3(−2)	0

Notice in this table that the atoms combine in such a way that the resulting compound has a zero charge.

More Complex Ionic Compounds

Compound	Metal	Cations	Nonmetal	Anions
$MgCl_2$	Mg	Mg^{2+}	Cl	Cl^- Cl^-
Na_2O	Na	Na^+ Na^+	O	O^{2-}
Al_2O_3	Al	Al^{3+} Al^{3+}	O	O^{2-} O^{2-} O^{2-}

Note that the charges on cations are equal to and opposite the charges on anions. For example, for aluminum oxide, $+3 + 3 - 2 - 2 - 2 = 0$. The total charge on the five atoms in the compound adds up to zero.

You can see that sometimes more than one atom of a particular element must combine to result in a compound with a zero charge. So, ionic compounds in ratios of 1:2, 2:1, or even 2:3 may result.

Example

Sodium Sulfide

The chemical formula for sodium sulfide is Na_2S.

a. What is the charge on each sodium ion, Na? On each sulfur ion, S?

b. Show that the charges on the ions add up to zero.

c. Is Na_2Cl a possible ionic compound? Why or why not?

Solution

a. Sodium, Na, is in Group 1A and has one valence electron. Sodium atoms give up one electron to have a charge of $+1$. Sulfur, S, is in Group 6A and has six valence electrons. Sulfur atoms gain two electrons to have a charge of -2.

b. Check the charge on the compound:

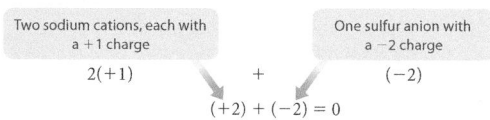

Two sodium cations, each with a $+1$ charge

$2(+1)$

One sulfur anion with a -2 charge

(-2)

$(+2) + (-2) = 0$

The charges on the ions add up to zero, as they should.

c. Each sodium atom has one valence electron and each chlorine atom has seven valence electrons. If two sodium atoms gave up electrons to one chlorine atom, the chlorine atom would have nine valence electrons, which is not a stable number resembling a noble gas. So Na_2Cl is not a compound that is likely to form. Also, such a compound would not have a neutral charge.

Lesson 20 | **Ionic Compounds** 101

Ask: What patterns of metal and nonmetal elements result in a ratio of 1:1? 1:2? 2:1?

Key Points: The number of electrons associated with the atoms of an ionic compound generally totals 8 or a multiple of 8. The atoms of the noble gases already have eight electrons in their valence shell. They tend not to combine with other elements.

strontium cations have a charge of $+2$, and Group 6A anions have a charge of -2, for a net charge of 0.

Homework

Assign the reading and exercises for Alchemy Lesson 20 in the student text.

EVALUATE (5 min)

Check-In

What elements will combine with strontium, Sr, in a 1:1 ratio? Explain your thinking.

Answers: Strontium has two valence electrons, so it will combine in a 1:1 ratio with the Group 6A elements, which have six valence electrons, for a total of eight valence electrons. Also,

LESSON 20 ANSWERS

1. The number of valence electrons can be used to predict whether an atom will form a cation or an anion, as well as the charge on the ion. Ionic compounds form between cations and anions in a ratio where the charges are balanced.

2. The rule of zero charge can be used to predict the ratio of ions that will form a compound that has no overall charge.

3. a. $+1$

b. -3

c. Lithium nitride has three lithium ions with a charge of $+1$ and one nitride ion with a charge of -3. $3(+1) + (-3) = 3 + (-3) = 0$

d. 8

4. a. $+3$

b. -3

c. Aluminum arsenide has one aluminum ion with a charge of $+3$ and one arsenide ion with a charge of -3. $(+3) + (-3) = 0$

d. 8

5. a. KBr has one potassium ion with a charge of $+1$ and one bromide ion with a charge of -1. $(+1) + (-1) = 0$

b. CaO has one calcium ion with a charge of $+2$ and one oxide ion with a charge of -2. $(+2) + (-2) = 0$

c. Li_2O has two lithium ions with a charge of $+1$ and one oxide ion with a charge of -2. $2(+1) + (-2) = 2 + (-2) = 0$

d. $CaCl_2$ has one calcium ion with a charge of $+2$ and two chloride ions with a charge of -1. $(+2) + 2(-1) = 2 + (-2) = 0$

e. $AlCl_3$ has one aluminum ion with a charge of $+3$ and three chloride ions with a charge of -1. $(+3) + 3(-1) = 3 + (-3) = 0$

6. a. 8

b. 8

c. 8

d. 16

e. 24

7. All three compounds violate the rule of zero charge.

LESSON SUMMARY

How can valence electrons be used to predict chemical formulas?

Metal atoms and nonmetal atoms combine to form ionic compounds through electron transfer. The metal atoms give up electrons. They form ions with a positive charge, called cations. The nonmetal atoms accept electrons and form ions with a negative charge, called anions. You can use the positions of the elements on the periodic table to predict the ionic compounds they can form. When an ionic compound forms, the total charge on the atoms adds up to zero. This is known as the rule of zero charge.

Exercises

Reading Questions

1. How can you use valence electrons to predict which ionic compounds will form?

2. How does the rule of zero charge help you predict the formula of an ionic compound?

Reason and Apply

3. Lithium nitride has the formula Li_3N.
   a. What is the charge on the lithium ion?
   b. What is the charge on the nitrogen ion?
   c. Show that the charges on the ions add up to zero.
   d. What is the total number of valence electrons in all the atoms in Li_3N?

4. Aluminum arsenide has the formula AlAs.
   a. What is the charge on the aluminum ion?
   b. What is the charge on the arsenic ion?
   c. Show that the charges on the ions add up to zero.
   d. What is the total number of valence electrons in all the atoms in AlAs?

5. For each of these compounds, show that the charges on the ions add up to zero.
   a. KBr   **b.** CaO   **c.** Li_2O **d.** $CaCl_2$ **e.** $AlCl_3$

6. What are the total numbers of valence electrons in all the atoms in each of these compounds?
   a. KBr   **b.** CaO   **c.** Li_2O **d.** $CaCl_2$ **e.** $AlCl_3$

7. Explain why the following compounds do not form.
   a. $NaCl_2$ **b.** CaCl   **c.** AlO

Salty Eights

Purpose

To make and name as many ionic compounds as possible.

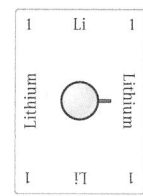 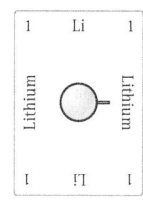 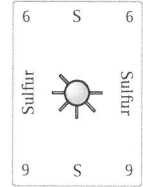

Materials

• Salty Eights cards

Instructions

1. Deal eight cards to each player and place the rest in a draw pile. On your turn, you must put down an ionic compound or a noble gas. If you cannot put one down, draw cards until you can. Play until one player has used all of his or her cards.

2. Keep score with a data table like this one. Point values are on the Rules card. List your compounds in the table. Write the metal first and the nonmetal second. If you use more than one atom of an element, don't forget to put the right number in the formula (for example, write Li_2S for lithium sulfide).

Cation	Anion	Compound formula	Compound name	Point value
			Total:	

Bonus Question Determine the compound with the highest point value you could make with the cards. Name that compound.

Lesson 21 | **Formulas for Ionic Compounds** 103

each compound, with more points given for more complex compounds. The game ends when a student plays out all the cards from his or her hand. The activity will help students become familiar with common ionic compounds and their naming schemes. A complete materials list is available for download.

TRM Materials List 21

Lesson Guide

ENGAGE (5 min)

TRM PowerPoint Presentation 21

ChemCatalyst

Find these cards in your Salty Eights card deck: Li, F, Na, Mg, O

1. List the ionic compounds you can make with pairs of cards, using two different elements.

2. List the ionic compounds you can make with three cards and only two different elements.

3. What rule must all these compounds satisfy?

Sample answers: **1.** You can make LiF, NaF, and MgO with two cards. **2.** You can make MgF_2, Na_2O, and Li_2O with three cards. **3.** They must satisfy the rule of zero change.

→ Discuss the possible compounds that can be formed with the cards.

Ask: What compounds can you make with pairs of cards?

Ask: What compounds can you make with three cards?

Ask: Can you use these cards to make a compound with four cards? (not if we keep it to two elements)

EXPLORE (25 min)

TRM Worksheet with Answers 21

TRM Worksheet 21

TRM Handout—Rules for Salty Eights 21

TRM Card Masters—Salty Eights 21

INTRODUCE THE ACTIVITY

→ Arrange students into groups of four.

→ Pass out the student worksheet, Salty Eights Cards, and the Handout—Rules for Salty Eights.

→ Briefly explain how to play the Salty Eights game.

Overview

ACTIVITY: GROUPS OF 4

Key Question: How can you predict chemical formulas and name ionic compounds?

KEY IDEAS

The number of valence electrons in an atom can be used to predict the chemical formulas of ionic compounds formed between metal and nonmetal main-group elements.

LEARNING OBJECTIVES

• Use valence electrons to predict ionic compounds.

• Develop proficiency at naming binary ionic compounds and writing their chemical formulas.

FOCUS ON UNDERSTANDING

This activity uses imagery from the periodic table card sort in Lesson 9 to represent elements. "Spokes" on each atom represent valence electrons. These visual clues can be very helpful to students.

IN CLASS

In this lesson, students are divided into teams of four. They play a card game that requires them to form ionic compounds. Points are awarded for

- The green cards are metals, the pink cards are nonmetals, the blue cards are noble gases, and there are several wild cards.
- The numbers on each card indicate the number of valence electrons.
- Each player starts with eight cards. The rest of the cards are placed face down in the draw pile.
- Players take turns putting down a compound. The compounds can have two, three, or more cards, but they cannot have more than two different elements (e.g., $MgCl_2$, but not MgFCl). Noble gases are played alone. The number of valence electrons must total 8 or a multiple of 8.
- When a player cannot put down a compound, the player must draw cards until he or she can put down a compound or a noble gas.
- The game ends when one player has no cards left. This player receives 20 bonus points.
- A player receives points based on how many cards he or she used to make the compound. Points are deducted for compounds and noble gases left in each player's hand at the end of the game.
- You can add a rule that if a player does not name a compound correctly upon playing it, the first player to name it correctly can earn the same number of points for it. (optional)
- Students should keep track of their compounds and their points on the worksheet.

EXPLAIN & ELABORATE (10 min)

→ Summarize what was learned from the game.

Ask: What difficulties did you encounter in writing chemical formulas or in naming the compounds?

Ask: What was the technique you used to create compounds using the cards?

Ask: What compound had the highest point value in the game you played?

Key Points: In general, atoms come together to form an ionic compound if the number of valence electrons totals 8 or a multiple of 8. The noble gases have eight electrons in their valence shells already, so they tend not to combine with other elements. The spokes on the cards show how many valence electrons are missing from a noble-gas arrangement. For example, oxygen can bond with calcium to form calcium oxide,

---

LESSON
21

Salty Eights

Formulas for Ionic Compounds

THINK ABOUT IT

There are at least 50 common metal elements. These combine with over 15 nonmetal elements to make a wide variety of ionic compounds. The task of predicting compounds might seem complex given the number of possible ways to combine a metal and a nonmetal. However, if you know how many valence electrons each atom has, you can reliably predict which ionic compounds can be made.

How can you predict chemical formulas and name ionic compounds?

To answer this question, you will explore

⮕ Predicting Chemical Formulas

⮕ Naming Ionic Compounds

EXPLORING THE TOPIC

⮕ Predicting Chemical Formulas

An ionic compound that is made up of only one kind of metal atom and one kind of nonmetal atom is known as a *salt*. Table salt, NaCl, is the most familiar example. To predict formulas for salt compounds, follow these guidelines:

- Make sure you are combining two different elements—a metal and a nonmetal.
- No matter how many atoms you use, the total number of valence electrons must add up to 8, 16, 24, or another multiple of 8.
- Make sure that the charges on the metal cations and nonmetal anions in your ionic compound add up to zero (rule of zero charge).

This process applies only to the main group elements. Ionic compounds using transition elements are covered in Lesson 23: Alchemy of Paint.

> **HISTORY CONNECTION**
>
> Salt has been a valuable trade item since ancient times. For thousands of years—before refrigeration—humans depended largely on salt as a preservative for foods, especially meat and fish. The wealth of a city, country, or individual could often be linked to the amount of salt possessed or stored. People who are "worth their salt" are considered worthwhile and valuable. Salzburg, Austria, shown here, is named for its many salt mines, which contributed to its great wealth.

Example 1

Practice Writing Chemical Formulas

Each of these cards represents an atom.

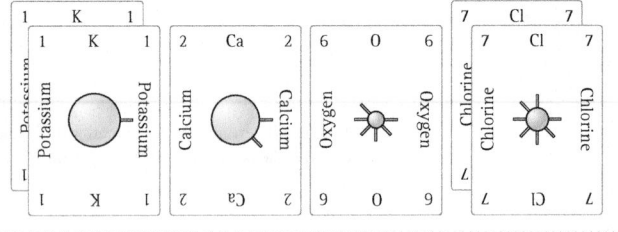

CaO. In some cases, the missing electrons can be gained by bonding with more than one atom.

Needs 1 electron — Chlorine

Needs 2 electrons — Oxygen

Has 2 electrons to transfer — Calcium

Writing correct chemical formulas is a matter of keeping track of exactly how many atoms come together to make a compound. Each ionic compound consists of at least one metal atom and at least one nonmetal atom. Some

examples are $AlCl_3$, Li_3N, Na_2O, Mg_3N_2, and Al_2O_3.

Creating correct chemical names is a matter of remembering some basic guidelines. A binary ionic compound is named by taking the name of the first element (the metal) and combining it with the name of the second element (the nonmetal). However, the ending of the second element is replaced with "ide." Aluminum chloride, magnesium nitride, and sodium oxide are all correct names. Notice that you cannot tell how many atoms of each element are in the ionic compound from the name alone. Prefixes such as "mono-,"

Example

(continued)

a. Which atoms are metals and which are nonmetals? How can you tell?

b. What ionic compounds can you make from two of these cards?

c. What ionic compounds could you make from three cards but only two different elements?

Solution

a. The atoms from the left side of the periodic table, with fewer valence electrons, are the metals. So, potassium, K, and calcium, Ca, are metal atoms. The atoms from the right side of the periodic table, which have closer to eight valence electrons, are the nonmetals. These include oxygen, O, and chlorine, Cl.

b. Look at the number of spokes on each card. The number of spokes represents the number of valence electrons. To create ionic compounds, combine a metal with a nonmetal. The number of spokes must add up to 8. Two ionic compounds are possible with these cards. These are shown here.

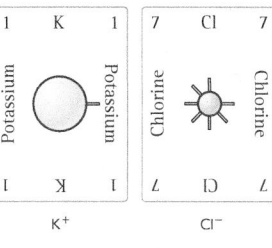

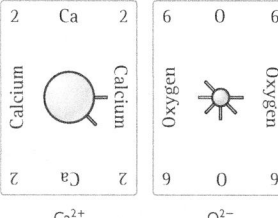

c. If two atoms with one valence electron combine with one atom with six valence electrons, the result is eight valence electrons. One possible compound is shown.

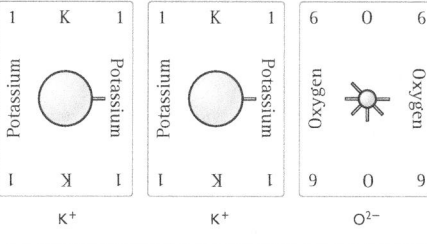

"di-," and "tri-" are only used for molecular compounds, which will be introduced later.

EVALUATE (5 min)

Check-In

Which of these compounds are likely to form? Explain.

a. Na_2S

b. K_2Mg

c. $AlBr_2$

d. Na_3N

e. OCl

f. CaO

Homework

Assign the reading and exercises for Alchemy Lesson 21 in the student text.

(continued)

You can also make $CaCl_2$. Notice that the valence electrons add up to 16, which is a multiple of 8. The charges add up to zero.

2 spokes + 7 spokes + 7 spokes = 16 spokes

Ca^{2+} Cl^- Cl^-

⟳ Naming Ionic Compounds

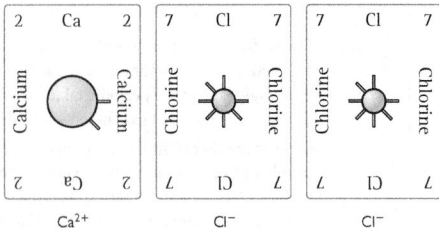

INDUSTRY CONNECTION

Potassium bromide is currently in use as an anti-seizure medication for dogs and cats with epilepsy.

In the names of ionic compounds, the name of the metal atom comes first, followed by the name of the nonmetal atom. However, the name of the nonmetal atom is altered slightly by replacing the last part of the word with the letters "-ide." Here is a list of ionic compounds with their names. Take a moment to examine them.

Compound	Name	Compound	Name
NaCl	sodium chloride	Al_2O_3	aluminum oxide
MgF_2	magnesium fluoride	GaP	gallium phosphide
Li_2S	lithium sulfide		

So in ionic compounds, sulfur becomes sulfide, nitrogen becomes nitride, and bromine becomes bromide.

Notice that the chemical name for ionic compounds does not have anything to do with the subscript numbers in the chemical formula. For example, $MgCl_2$ is simply magnesium chloride, *not* magnesium dichloride.

Example 2

Naming Compounds

Write the chemical formula and name for the compound created from each pair of elements.

    a. potassium and bromine   **b.** oxygen and calcium

    c. oxygen and potassium   **d.** sodium and chlorine

    e. sodium and oxygen

Example

(continued)

Solution

To write the formula for the compound, make sure the atoms are in the correct ratio so that the compound follows the rule of zero charge. To name the compound, write the metal atom first, followed by the nonmetal atom. Change the ending of the nonmetal atom's name to "-ide."

a. KBr, potassium bromide   b. CaO, calcium oxide

c. K_2O, potassium oxide   d. NaCl, sodium chloride

e. Na_2O, sodium oxide

LESSON SUMMARY

How can you predict chemical formulas and name ionic compounds?

There are several important guidelines to follow in creating ionic compounds. Metal atoms are combined with nonmetal atoms. Next, the total number of valence electrons adds up to eight or a multiple of eight. Finally, the charges on the metal cations and nonmetal anions in ionic compounds add up to zero. When naming ionic compounds made from two different types of elements, the name of the metal atom comes first, followed by the name of the nonmetal atom. In addition, the ending of the name of the nonmetal atom is changed to "-ide."

Exercises

Reading Questions

1. Explain how to use the periodic table to determine the charges on ions.
2. Explain how to use the periodic table to determine the correct formulas for ionic compounds.

Reason and Apply

3. Which chemical formulas are consistent with the guidelines for creating ionic compounds? Explain your thinking.
   a. LiCl   b. $LiCl_2$ c. MgCl   d. $MgCl_2$ e. $AlCl_3$
4. Give examples of six ionic compounds with a metal-to-nonmetal ratio of 1:1. Specify the total number of valence electrons for each compound. Name each compound.
5. Give examples of three ionic compounds with a metal-to-nonmetal ratio of 2:1. Specify the total number of valence electrons for each compound. Name each compound.
6. Give examples of three ionic compounds with a metal-to-nonmetal ratio of 1:2. Specify the total number of valence electrons for each compound. Name each compound.

Lesson 21 | **Formulas for Ionic Compounds**   107

LESSON 21 ANSWERS

1. For main group elements, the group number shows the number of valence electrons. Metals lose all their valence electrons to form positive ions adding a positive charge for each ion lost. Nonmetal atoms gain electrons to have eight valence electrons, adding a negative charge for each electron gained.

2. For main group elements, the periodic table is used to determine the charge on each ion. Ions combine to form ionic compounds in ratios that have a total charge equal to zero.

3. a. LiCl is possible because the total of the charges on the ions equals zero. The lithium ion has a charge of +1 and the chloride ion has a charge of −1. $(+1) + (−1) = 0$

b. $LiCl_2$ is not possible because the total of the charges on the ions equals −1. $(+1) + 2(−1) = 1 + (+2) = 1$

c. MgCl is not possible because the total of the charges on the ions equals +1. Magnesium ions have a charge of +2. $(+2) + (−1) = +1$

d. $MgCl_2$ is possible because the total of the charges on the ions equals zero. $(+2) + 2(−1) = 2 + (−2) = 0$

e. $AlCl_3$ is possible because the total of the charges on the ions equals zero. Aluminum ions have a charge of +3. $(+3) + 3(−1) = 3 + (−3) = 0$

4. *Possible answer (any 6):* Sodium chloride, NaCl, has 8 valence electrons; calcium oxide, CaO, has 8 valence electrons; rubidium iodide, RbI, has 8 valence electrons; potassium bromide, KBr, has 8 valence electrons; sodium fluoride, NaF, has 8 valence electrons; magnesium oxide, MgO, has 8 valence electrons; strontium sulfide, SrS, has 8 valence electrons; calcium sulfide, CaS, has 8 valence electrons. Any combination of a Group 1A metal with a Group 7A nonmetal, or a Group 2A metal with a Group 6A nonmetal, or a Group 3A metal with a Group 5A nonmetal is correct.

5. *Possible answer (any 3):* Sodium oxide, Na_2O, has 8 valence electrons; lithium oxide, Li_2O, has 8 valence electrons; rubidium sulfide, Rb_2S, has 8 valence electrons; potassium sulfide, K_2S, has 8 valence electrons. Any combination of a Group 1A metal with a Group 6A nonmetal is correct.

6. *Possible answer (any 3):* Magnesium chloride, $MgCl_2$, has 16 valence electrons; calcium fluoride, CaF_2, has 16 valence electrons; barium iodide, BaI_2, has 16 valence electrons; strontium bromide, $SrBr_2$, has 16 valence electrons. Any combination of a Group 2A metal with a Group 7A nonmetal is correct.

7. a. $AlBr_3$, aluminum bromide **b.** Al_2S_3, aluminum sulfide **c.** AlAs, aluminum arsenide **d.** Na_2S, sodium sulfide **e.** CaS, calcium sulfide **f.** Ga_2S_3, gallium sulfide

8. a. Mg^{2+}, $O^{2−}$, MgO

b. Rb^+, $Br^−$, RbBr

c. Sr^{2+}, $I^−$, SrI_2

d. Be^{2+}, $F^−$, BeF_2

e. Al^{3+}, $Cl^−$, $AlCl_3$

f. Pb^{4+}, $S^{2−}$, PbS_2

(continued)

7. Predict the formulas for the ionic compounds that are formed when these metal and nonmetal elements are combined. Name each compound.
   a. Al and Br
   b. Al and S
   c. Al and As
   d. Na and S
   e. Ca and S
   f. Ga and S

8. For each compound, write the cation and anion with the appropriate charge. Then write the chemical formula for each compound.

   Example: sodium fluoride, Na^+, F^-, NaF
   a. magnesium oxide
   b. rubidium bromide
   c. strontium iodide
   d. beryllium fluoride
   e. aluminum chloride
   f. lead sulfide

Polyatomic Ions

THINK ABOUT IT

So far, we have considered ionic compounds made of only two elements, a metal and a nonmetal. However, there are some ionic compounds that consist of more than two elements. For example, sodium hydroxide, NaOH, which is commonly found in drain cleaner, consists of one metal element and *two* nonmetal elements. You might wonder about the O and the H. Are they both ions?

What is a polyatomic ion?

To answer this question, you will explore

↻ Polyatomic Ions

↻ Predicting Chemical Formulas for Polyatomic Ions

EXPLORING THE TOPIC

↻ Polyatomic Ions

Polyatomic ion	Name
OH^-	hydroxide
NO_3^-	nitrate
CO_3^{2-}	carbonate
SO_4^{2-}	sulfate
PO_4^{3-}	phosphate
BrO_3^-	bromate
NH_4^+	ammonium

Some ionic compounds contain ions that consist of two or more elements. These ions are called **polyatomic ions**. In contrast, ions that have only one element are called **monatomic ions.** *Mono* means "one" and *poly* means "many." Calcium sulfate, $CaSO_4$, is an example of a compound made up of a monatomic ion and a polyatomic ion. The calcium ion is monatomic and the sulfate ion is polyatomic.

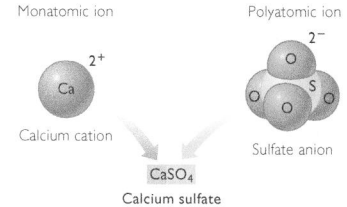

Monatomic ion — Calcium cation Ca^{2+}

Polyatomic ion — Sulfate anion SO^{2-}

$CaSO_4$
Calcium sulfate

CONSUMER CONNECTION

Sodium hydroxide, or lye, has many uses. It is used to straighten or curl hair, but if left in too long, it can damage hair and skin.

It is important to keep in mind that polyatomic ions are a *group* of atoms that stay together. In calcium sulfate, the entire group of atoms in the sulfate ion has a negative charge.

NAMES OF POLYATOMIC IONS

Ionic compounds with polyatomic ions follow specific naming rules. Each polyatomic ion has its own name, as shown in the table. When naming a compound, you simply insert the polyatomic ion name at either the beginning or ending of the chemical name. The cation is first and the anion is second. For example, a

Overview

ACTIVITY: GROUPS OF 4

Key Question: What is a polyatomic ion?

KEY IDEAS

Polyatomic ions are composed of more than one atom. The charge on a polyatomic ion is shared by the entire group of atoms. Most polyatomic ions are anions, with negative charges. If a chemical formula contains more than one polyatomic ion, parentheses are placed around the polyatomic ion and a subscript indicates the number. The rule of zero charge can be used to

predict the formulas of compounds that contain polyatomic ions. Compounds that contain polyatomic ions have their own naming guidelines.

LEARNING OBJECTIVES

- Recognize and name polyatomic ions.
- Write names and chemical formulas of compounds with polyatomic ions.

FOCUS ON UNDERSTANDING

- Students have just learned to write formulas of binary ionic compounds. Polyatomic ions present a new level of complexity.
- The optimal approach to polyatomic ions is for students to learn

to recognize these ions and to familiarize themselves with their names and charges.

KEY TERMS

polyatomic ion
monatomic ion

IN CLASS

Students play two games using the Polyatomic Ions cards. In the game Ionic Grid, students lay out the cards face up and take turns creating compounds with neutral charges. In the game Three-Minute Bonding, each student gets one card and must form and write down as many correct compounds as possible by combining with cards from other students in three minutes. A complete materials list is available for download.

TRM **Materials List 22**

Differentiate

To develop scientific literacy and support English-language learners, you might have your students create a set of hint cards. On the front of an index card have students write a term such as *mono-atomic ion* or *polyatomic ion* and then define the term, provide an example, and draw a model on the back of the card. Punch a hole in each card and place them on a metal ring to make them easy to flip through. Ask students to create hint cards for frequently confused ions such as sulfate and sulfide. They can also create hint cards to explain processes such as how to apply rule of zero charge to write ionic formulas or when to include parentheses in an ionic formula.

Lesson Guide

TRM **Transparency—ChemCatalyst 22**

TRM **PowerPoint Presentation 22**

ChemCatalyst

These cards show a sodium ion and three polyatomic ions.

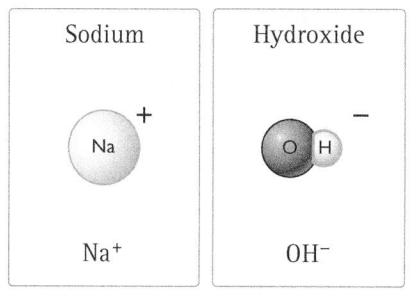

Sodium	Hydroxide
Na $^+$	O H $^-$
Na^+	OH^-

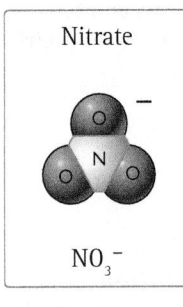
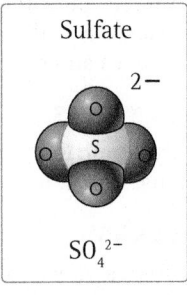

Nitrate	Sulfate
NO_3^-	SO_4^{2-}

1. What do you think a polyatomic ion is?

2. Name three compounds formed between sodium ions and each of the three polyatomic ions. Use the rule of zero charge to write the compounds' formulas.

Sample answers: 1. A polyatomic ion is an ion with more than one atom. 2. Sodium hydroxide, sodium nitrate, and sodium sulfate. Sodium ions have a +1 charge. Therefore, you can make NaOH, $NaNO_3$, and Na_2SO_4. Students might not write the formulas correctly yet.

→ Assist students in sharing their ideas about polyatomic ions.

Ask: What do the ions on the cards have in common?

Ask: What does the term *nitrate* refer to in the compound sodium nitrate?

Ask: Why is the formula of sodium sulfate Na_2SO_4?

EXPLORE (25 min)

TRM Worksheet with Answers 22

TRM Worksheet 22

TRM Card Masters—Polyatomic Ions 22

TRM Handout—Rules for Ionic Grid and Three-Minute Bonding 22

INTRODUCE THE ACTIVITY

→ Arrange students into groups of four.
→ Pass out handouts and worksheets to students and let them know that they will be playing two games today, Ionic Grid and Three-Minute Bonding. The goal of both games is to create ionic compounds with polyatomic ions.

GUIDE THE ACTIVITY

→ Briefly go over the instructions for Ionic Grid. In this game, four students share one deck of cards that are laid out face up in a grid. Students take turns creating ionic compounds using the cards. (See the handout.)

→ Monitor students as they play, and settle any challenges. After students play Ionic Grid for 10 minutes, collect

the card decks and move on to Three-Minute Bonding.

→ Briefly go over the instructions for Three-Minute Bonding. This game uses one deck of cards for the whole class. Remove the wild cards, shuffle the deck, and deal out one card to each student. Students have three minutes to mingle and form as many compounds as they can using their cards. (See the handout)

→ At some point during Three-Minute Bonding, yell "Switch!" so students will know to trade cards with whoever is closest.

→ At the end of Three-Minute Bonding, yell, "Stop bonding!" Process the game as a class. Ask the three students with

the most compounds to write their lists of chemical formulas on the board. Other students can compare their formulas and make any needed corrections to them. Any student with a compound not already listed should share it.

EXPLAIN & ELABORATE (10 min)

TRM Transparency—Polyatomic Ions 22

INTRODUCE POLYATOMIC IONS

→ Use the transparency Polyatomic Ions during the discussion. Stress the fact that for polyatomic ions the entire group of atoms shares the charge.

→ Go over the notation. For example, $Mg(OH)_2$ means there are two hydroxide ions.

---

compound that ends with NO_3 is a nitrate. $NaNO_3$ is called sodium *nitrate*, and $Ca(NO_3)_2$ is called calcium *nitrate*.

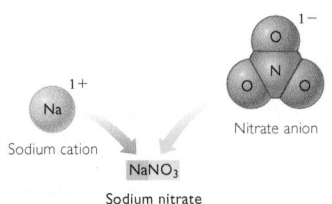

Sodium cation → $NaNO_3$ ← Nitrate anion

Sodium nitrate

Compound	Chemical formula
calcium sulfate	$CaSO_4$
magnesium hydroxide	$Mg(OH)_2$
sodium nitrate	$NaNO_3$
calcium nitrate	$Ca(NO_3)_2$
ammonium carbonate	$(NH_4)_2CO_3$

Important to Know A polyatomic ion is treated as a unit. For example, the formula for calcium nitrate is written as $Ca(NO_3)_2$ with parentheses around NO_3. Notice that it is not written as CaN_2O_6.

⊃ Predicting Chemical Formulas for Polyatomic Ions

Several chemical formulas with polyatomic ions are listed in the table. Notice that in some cases there are parentheses around the polyatomic ion, with a subscript after the second parenthesis. Why are the formulas written this way? How do you determine the subscript?

Recall that you can use the rule of zero charge to determine the chemical formulas associated with ionic compounds. For example, a single calcium ion combines with a single sulfate ion because the charges add up to zero:

$$Ca^{2+} + SO_4^{2-} \text{ becomes } CaSO_4, \text{ calcium sulfate}$$

However, like simple ionic compounds, compounds with polyatomic ions do not always combine in a 1:1 ratio. When there is more than one of the same polyatomic ion in a formula, the ion is enclosed in parentheses and a subscript number indicates how many ions are in the compound. For example, a single calcium ion combines with *two* nitrate ions:

$$Ca^{2+} + NO_3^- + NO_3^- \text{ becomes } Ca(NO_3)_2, \text{ calcium nitrate}$$

Below are two more examples of ionic compounds containing polyatomic ions. Each of these compounds consists of three ions. One compound has monatomic and polyatomic ions, while the other has only polyatomic ions. The rule of zero charge applies to both.

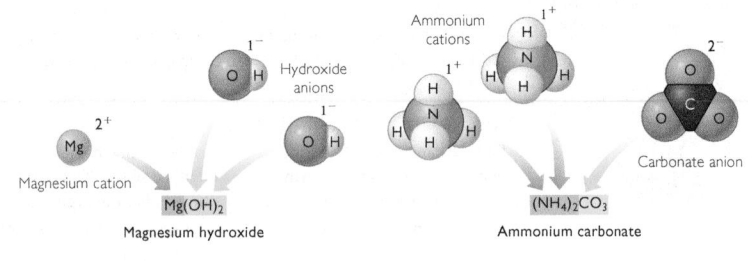

Magnesium cation — Hydroxide anions — Ammonium cations — Carbonate anion

$Mg(OH)_2$ — Magnesium hydroxide — $(NH_4)_2CO_3$ — Ammonium carbonate

LESSON SUMMARY

What is a polyatomic ion?

Some ions are composed of more than one atom. These are called polyatomic ions. The entire cluster of atoms shares the charge on a polyatomic ion. If there are two or more of the same polyatomic ion in a chemical formula, parentheses are placed around that ion and a subscript number indicates how many ions are present. To determine the formula of a compound with polyatomic ions, you can use the rule of zero charge. Each polyatomic ion has a unique name that is used when naming the compound it is in.

KEY TERMS
polyatomic ion
monatomic ion

Exercises

Reading Questions

1. What is a polyatomic ion?
2. How can you tell from a chemical formula if there is a polyatomic ion in a compound?

Reason and Apply

3. Write the name for each ionic compound listed here.
   a. NH_4Cl
   b. K_2SO_4
   c. $Al(OH)_3$
   d. $MgCO_3$
4. Write the chemical formula for each compound listed here.
   a. lithium sulfate
   b. potassium hydroxide
   c. magnesium nitrate
   d. ammonium sulfate
5. Sodium cyanide, NaCN, contains a cyanide ion. What is the charge on the cyanide ion?
6. Calcium phosphate, $Ca_3(PO_4)_2$, contains phosphate ions. What is the charge on a phosphate ion?
7. Which chemical formula does not represent a possible compound with sulfate, SO_4^{2-}? Explain your answer.
   a. Na_2SO_4 b. KSO_4 c. $Al_2(SO_4)_3$ d. $CaSO_4$

Ask: What is a polyatomic ion?

Ask: How can you tell if an ionic compound contains a polyatomic ion?

Ask: What were the most challenging parts of the two games?

Key Points: Polyatomic ions contain more than one atom. The word *poly-* means "many." Polyatomic ions may contain two, three, or more atoms. The main thing to remember is that these atoms stay together as a group and the entire group of atoms shares the negative or positive charge. Some common polyatomic ions are listed here.

hydroxide	OH^-
nitrate	NO_3^-

carbonate	CO_3^{2-}
sulfate	SO_4^{2-}
ammonium	NH_4^+
chromate	CrO_4^{2-}
phosphate	PO_4^{3-}
cyanide	CN^-

Most polyatomic ions are anions, with negative charges. The only significant polyatomic cation to remember is ammonium, NH_4^+. Students will encounter polyatomic ions frequently, so it is helpful for them to learn the names and charges of the most common ones.

Polyatomic ion: An ion composed of a group of atoms with an overall

positive or negative charge. Most polyatomic ions are anions.

DISCUSS FORMULAS AND NAMES OF POLYATOMIC IONS

Ask: How can you determine the formula of an ionic compound containing a polyatomic ion?

Ask: If you are given a chemical formula such as calcium phosphate, $Ca_3(PO_4)_2$, how can you determine the charge on the phosphate ion?

Ask: What naming patterns did you notice during the games?

Key Points: The rule of zero charge can be used to predict the formulas of compounds that contain polyatomic ions. Remember that ionic compounds always have a total charge of 0. The charges on the cations and anions sum to 0. If there is more than one of the same polyatomic ion in a formula, that polyatomic ion is placed in parentheses and a subscript number shows how many ions are needed. For example, when a calcium ion, Ca^{2+}, combines with a hydroxide ion, OH^-, the chemical formula is $Ca(OH)_2$. Two hydroxide ions are needed to combine with this cation because each hydroxide ion has a charge of only -1.

Compounds containing polyatomic ions have their own unique naming guidelines. As with other ionic compounds, the cation is named first and the anion second. For any compound containing a polyatomic ion, simply insert the name of the polyatomic ion where it belongs. For example, NaOH becomes sodium hydroxide.

EVALUATE (5 min)

Check-In

1. What is the name of the compound $Be(NO_3)_2$?
2. What ions are present in this compound, and what are the charges on the ions?

Answers: **1.** The compound is called *beryllium nitrate.* **2.** The ions present are Be^{2+} and NO_3^-, with charges of $+2$ and -1, respectively.

Homework

Assign the reading and exercises for Alchemy Lesson 22 in the student text.

LESSON 22 ANSWERS

Answers continue in Answer Appendix (p. ANS-2).

Overview

LAB: GROUPS OF 4

Key Question: What types of compounds are made from transition metals? *Note:* Students have to look at a periodic table to complete the ChemCatalyst.

KEY IDEAS

Transition metals bond with nonmetal atoms and with polyatomic anions to form ionic compounds. Unlike the main-group atoms, most transition metals can have more than one ion charge. The rule of zero charge can help you figure out the charge on a transition metal ion. Colorful paint pigments frequently are composed of transition metal compounds.

LEARNING OBJECTIVES

- Recognize transition metal compounds and their names.
- Determine the charge on transition metal ions given their chemical formula.

FOCUS ON UNDERSTANDING

Students often have difficulty keeping the rules for ions straight. Main-group atoms and polyatomic ions have fixed ion charges. The charges on transition metal ions can change from one compound to another.

IN CLASS

Students spend this class period in the laboratory creating compounds similar to those created by the ancient alchemists. They are provided with recipes for three paint pigments and one paint binder. You might want to assign one, two, or three pigments to each team depending on the amount of time you have. If time allows, students can add a binder to their pigments and paint in class. This activity gives students firsthand experience with the chemistry of transition metals. Students read about transition metal ions in the student text. *Note:* This lab is optional. If you are short on time, you could assign the alternate student Worksheet 23.1 that directs students to look for patterns in transition metal compound names and formulas and then have students move on to the reading in the text. A complete materials list is available for download.

TRM Materials List 23

SETUP

If time is an issue, you might want to weigh out the metal salts before class.

Alchemy of Paint

Transition Metal Chemistry

THINK ABOUT IT

For thousands of years, human beings have expressed themselves through painting. Over time, people have discovered pigments with a wide variety of brilliant colors. Almost all paint pigments contain a transition metal cation.

What types of compounds are made from transition metals?

To answer this question, you will explore

- Transition Metal Compounds
- Charges on Transition Metal Cations

EXPLORING THE TOPIC

Transition Metal Compounds

The transition metals are named for their location in the middle of the periodic table. Many of the metals that we use in our daily life, such as copper, iron, nickel, silver, and gold, are located in this part of the periodic table.

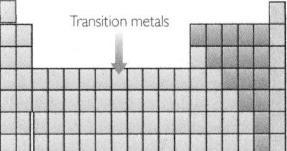
Transition metals

Transition metal compounds tend to be brightly colored. Hence, they are commonly found as *pigments* used to color paints. For example, red ochre is a pigment made from iron (III) oxide, Fe_2O_3. It is thought to be the first pigment ever used by human beings for creating art.

The ancient alchemists worked with the transition metals a great deal, mostly because these metals were closer in their properties to gold than to other metals. As a result, alchemists occasionally discovered paint pigments while trying to create gold.

The table on the next page lists various paint pigments that you can buy at an art store. Examine the data. Pay particular attention to the different charges on the transition metal cations.

Notice that the chemical names are a bit unusual. In the middle of each chemical name for a transition metal compound is a Roman numeral: I, II, III, or IV (meaning 1, 2, 3, or 4). This Roman numeral indicates what the charge is on the transition metal cation. Thus, cobalt (II) oxide has a +2 charge on the cobalt ion, and manganese (IV) carbonate has a +4 charge on the manganese ion.

HISTORY CONNECTION

Ionic compounds containing transition metals were used to create ancient cave paintings. This cave painting is from Lascaux, France, and is approximately 17,000 years old.

© The Gallery Collection/Corbis

You might also want to crack the eggs for students and separate the yolks. You can assign each group to make a different pigment or let groups choose. If you have enough time and materials, groups can make all three. To make paint, they will also need to make the binder.

SAFETY

- Students will be working with transition metal compounds and acid. Safety goggles must be worn at all times.
- Students will heat a compound in a test tube. Remind them that they must keep this test tube pointed away from their face or anyone else's.
- If students will be painting, remind them that pigments containing transition metal compounds are moderately toxic. They should avoid prolonged skin contact and be careful not to get any in their mouth. In the event of contact, they should wash immediately with plenty of water.
- Students must keep hair tied back and secure loose clothing. No dangling jewelry is permitted. Lab aprons and lab gloves are recommended.

Some Pigments Containing Ionic Compounds with Transition Metals

Color	Pigment name	Chemical name	Chemical formula	Cation	Anion
blue	cobalt blue	cobalt (II) oxide	CoO	Co^{2+}	O^{2-}
earth tone	red ochre	iron (III) oxide	Fe_2O_3	Fe^{3+}	O^{2-}
dark green	viridian	chromium (III) oxide	Cr_2O_3	Cr^{3+}	O^{2-}
brown	umber	manganese (IV) dioxide	MnO_2	Mn^{4+}	O^{2-}
blue-green	malachite	copper (II) carbonate	$CuCO_3$	Cu^{2+}	CO_3^{2-}
white	titanium white	titanium (IV) dioxide	TiO_2	Ti^{4+}	O^{2-}
red	cuprite	copper (I) oxide	Cu_2O	Cu^+	O^{2-}

Note that copper, Cu, has a charge of +1 in copper (I) oxide, Cu_2O, and a charge of +2 in copper (II) carbonate, $CuCO_3$.

There is an older naming system for the transition metal ionic compounds. In this system, Fe_2O_3 and FeO were called ferric oxide and ferrous oxide. The "-ic" ending indicated the higher ion charge and the "-ous" ending referred to the lower ion charge. So Co^{2+} was cobaltous and Co^{3+} was cobaltic.

↻ Charges on Transition Metal Cations

Recall that you can determine the charges on the main group metals and nonmetals from their positions on the periodic table. For example, all of the alkali metals, Group 1A, form cations with +1 charges. All of the halogens, Group 7A, form anions with −1 charges. However, you cannot simply determine the charge of a transition metal ion from its location on the periodic table. Instead, it is necessary to use chemical formulas to determine the charge of the ion.

To determine the charges on transition metal cations, you use the charges on ions that you do know and apply the rule of zero charge. For example, iron combines with oxygen to form both Fe_2O_3 and FeO. Oxygen is a main group anion, so you can use the periodic table to determine that oxygen atoms form ions with a −2 charge. The rule of zero charge states that the charges on the ions in the compound should add up to zero. Working backward, you can determine the charge on the iron cations in each compound.

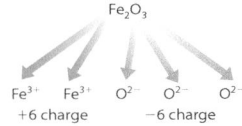

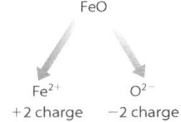

The charges on the three oxygen anions add up to −6. Each iron cation must have a charge of +3.

The charge on the single oxygen anion is −2. The iron cation must have a charge of +2.

Important to Know Unlike the main group metals, most transition metals can form several ions with different charges.

These two different iron compounds are formally named iron (III) oxide and iron (II) oxide.

Lesson 23 | **Transition Metal Chemistry**   113

CLEANUP

Put filter papers in a plastic-lined box to dry. Pour liquid filtration into a large container and allow it to evaporate. Dispose of dry toxic solids according to local guidelines.

Differentiate

Students may need additional scaffolding to understand ionic nomenclature and writing ionic formulas involving transition metals, which have variable ionic charges. Use the supplemental Worksheet 23.1 to provide guided practice in furthering their science literacy and the language of chemistry.

Lesson Guide

ENGAGE (10 min)

TRM PowerPoint Presentation 23

ChemCatalyst

1. What types of substances will you be creating in today's lab?

2. What safety precautions are especially important for today's lab?

3. Describe one of the procedures you will be completing in class today.

Sample answers: **1.** Many different answers will satisfy. Students might say that they will be creating pigments from transition metal compounds. **2.** It is important that students wear safety goggles and tie back their hair. **3.** One procedure used in today's class is filtering.

CHECK STUDENTS' LAB READINESS

Check to see that students are familiar with the lab procedure. If a student cannot answer the ChemCatalyst questions, he or she should reread the lab handout and check in with you a second time before the group begins.

EXPLORE (30 min)

TRM Worksheet with Answers 23

TRM Worksheet 23

TRM Handout—Making Paint Pigments 23

TRM Worksheet with Answers 23.1

TRM Worksheet 23.1

INTRODUCE THE LAB

→ Arrange students into groups of four.

→ Pass out the student worksheet.

→ Tell students that they will be making paint pigments. You can assign different groups to make different pigments or let groups choose. If you have enough materials and time, all groups can make all the pigments.

→ Go over the safety instructions. Goggles must be worn at all times. Hair should be pulled back. No dangling jewelry or loose clothing is permitted. When a compound is being heated in a test tube, the open end of the test tube should never point toward anyone.

EXPLAIN & ELABORATE (10 min)

PROCESS THE LABORATORY ACTIVITY

→ Show students an image of Van Gogh's *Fifteen Sunflowers* that you find online.

Ask: What types of properties are necessary for a good paint? (color that lasts, dries, mixes with water or oil, has the right consistency to spread well, covers surfaces well, etc.)

Ask: What types of chemical bonds are present in the substances you made today in class? (ionic bonds, metal-nonmetal salts)

Lesson 23 | **Transition Metal Chemistry**   113

Ask: What are the anions and cations in the pigments you created? (cobalt (II) chloride: Co^{2+}, Cl^-; zinc oxide: Zn^{2+}, O^{2-}; potassium nitrite: K^+, NO_2^-; copper (II) sulfate: Cu^{2+}, SO_4^{2-}; sodium carbonate: Na^+, CO_3^{2-})

Key Points: Many of the paint pigments that artists historically used and continue to use are ionic metal-nonmetal compounds. The cobalt yellow that you made in class today is a color used extensively in Van Gogh's *Fifteen Sunflowers*, a painting he produced in 1888. Today, this painting is worth at least $50 million. That's a lot of gold!

Transition metals bond with nonmetal atoms to form ionic compounds. Unlike the main-group elements, most transition metals can have more than one ion charge. When naming ionic compounds, a Roman numeral is used to indicate the charge on the transition metal cation. To determine the charges on transition metal ions in ionic compounds, use the known charges of other ions in the compound and apply the rule of zero charge. *Note:* Worksheet 23.1 and the student reading provide direction and practice for working with transition metal ions in compounds.

Check-In

There is no Check-In for this lesson.

Homework

Assign the reading and exercises for Alchemy Lesson 23 in the student text.

Example

Transition Metal Compounds

Determine the charge on the transition metal cation in each of the compounds given. Then name the compound.

   a. Ag_2S **b.** $Fe(NO_3)_3$

Solution

You can determine the charge on each transition metal cation from the charges on the anions using the rule of zero charge.

   a. Sulfur anions have a charge of -2. So each silver cation must have a charge of $+1$, Ag^+. The compound name is silver (I) sulfide.

   b. The polyatomic ion is nitrate with a -1 charge. There are three nitrate ions, so the iron cation must have a charge of $+3$, Fe^{3+}. The compound name is iron (III) nitrate.

CONSUMER CONNECTION

Transition metal ions are responsible for colors in different-colored gem stones. For example, chromium ions, Cr^{3+}, are present in red ruby, green emerald, and pink topaz. Iron ions, Fe^{3+}, are present in citrine and yellow sapphire, while Fe^{2+} ions make sapphires blue. The transition metals in crystals absorb certain colors of light while allowing other colors to pass through and be seen.

LESSON SUMMARY

What types of compounds are made from transition metals?

Transition metals bond with nonmetal atoms to form ionic compounds. Unlike the main group atoms, most transition metals can have more than one ion charge. The best way to determine the charge on a transition metal cation is to work backward from the anion, whose charge is known. When naming ionic compounds, a Roman numeral is used to indicate the charge on the transition metal cation. Colorful paint pigments are frequently composed of transition metal compounds.

Exercises

Reading Questions

1. What does the Roman numeral in a chemical name indicate?
2. Explain how you determine the charge on a transition metal cation from the chemical formula.

Reason and Apply

3. Determine the charge on the transition metal cation in each of the compounds listed. Then name each compound.
   a. HgS
   b. $CuCO_3$
   c. $NiCl_2$
   d. $Co(NO_3)_3$
   e. $Cu(OH)_2$
   f. $FeSO_4$

4. Write the cation and anion in each compound, then determine the correct chemical formula.
   a. copper (II) sulfide
   b. nickel (II) nitrate
   c. iron (II) carbonate
   d. cobalt (II) sulfate
   e. iron (III) carbonate
   f. chromium (VI) oxide

5. Cobalt violet is a paint pigment discovered in 1859. If the cation for this compound is Co^{2+} and the anion is PO_4^{3-}, what is the chemical formula?

LESSON 23 ANSWERS

1. The Roman numeral indicates the charge on the transition metal cation in the compound.

2. Determine the total charge on the anions in the compound. The total charge of the cations is the opposite of the charge on the anions. Divide by the number of cations in the chemical formula to find the charge of each cation.

3. a. +2, mercury (II) sulfide **b.** +2, copper (II) carbonate **c.** +2, nickel (II) chloride **d.** +3, cobalt (III) nitrate **e.** +2, copper (II) hydroxide **f.** +2, iron (II) sulfate

4. a. Cu^{2+}, S^{2-}, CuS **b.** Ni^{2+}, NO_3^-, $Ni(NO_3)_2$ **c.** Fe^{2+}, CO_3^{2-}, $FeCO_3$ **d.** Co^{2+}, SO_4^{2-}, $CoSO_4$ **e.** Fe^{3+}, CO_3^{2+}, $Fe_2(CO_2)_3$ **f.** Cr^{6+}, O^{2-}, CrO_3

5. $Co_3(PO_4)_2$

Overview

CLASSWORK: PAIRS

Key Question: What does the periodic table indicate about the arrangements of electrons?

KEY IDEAS

Scientific evidence provides substantial support for the idea that electron shells in an atom are further divided into subshells. These subshells are designated s, p, d, and f. Like the shells, each electron subshell can hold a specific maximum number of electrons. The s subshells can hold 2 electrons, the p subshells can hold 6, the d subshells can hold 10, and the f subshells can hold 14. Chemists use a shorthand notation called an *electron configuration* to keep track of electrons and the subshells they inhabit.

LEARNING OBJECTIVES

- Describe the structure of an atom in terms of electron shells and subshells.
- Use the periodic table to determine the electron arrangement in an atom and to write electron configurations.
- Explain the organization of the periodic table in terms of the arrangements of electrons in subshells.

FOCUS ON UNDERSTANDING

- The 4s subshell fills with electrons before the 3d subshell. This is confusing to students because they expect the opposite to be true.
- Electron configurations can be written with the 4s subshell before the 3d or vice versa. It does not matter as long as the electrons are placed in the correct subshell. We have chosen to highlight the location of the last electron into the shells. This allows students to locate the element on the periodic table quickly.
- Some exceptional electron configurations, especially in the transition elements, do not follow the general pattern. This lesson does not cover these exceptions.

KEY TERM

electron configuration

IN CLASS

Students work in pairs to understand the relationship between electron subshells and the structure of the periodic table. They learn that the periodic table is a valuable tool for determining the

electron arrangement of an atom and for writing electron configurations. *Note:* This lesson is optional. If you are short on class time, you can continue directly to Lesson 25: You Light Up My Life. A complete materials list is available for download.

TRM Materials List 24

Differentiate

To build fluency with writing electron configurations and to reinforce the relationship between the properties in the periodic table and electronic structure, model and guide students through a finger-

tracing method for deriving an electron configuration, such as for gallium, Ga. Using a periodic table, point at hydrogen (first electron in the 1s subshell), trace across to helium (1s subshell is now filled), down to lithium, across to neon (filling the 2s and 2p subshells), and so on until reaching gallium. If students are able to locate the s, p, d, and f blocks in the periodic table, this method can help them recognize that electrons fill the 3d subshell before 4p. To challenge advanced learners, ask them to determine the electron configuration of an f-block element.

Shell Game

Electron Configurations

THINK ABOUT IT

Recall that the chemistry of the elements is closely related to the number of valence electrons in their atoms. The valence electrons are found in the outermost electron shell of an atom.

> **What does the periodic table indicate about the arrangements of electrons?**

To answer this question, you will explore

- Subshells in Atoms
- Electron Configurations
- Connecting the Periodic Table to Electron Arrangements
- Noble Gas Shorthand

EXPLORING THE TOPIC

⟳ Subshells in Atoms

Electrons are arranged into shells numbered $n = 1, 2, 3$, and so on. The number of electron shells in an atom is the same as the number of the period where the element is located on the periodic table. Each shell has a maximum number of electrons. For instance, the $n = 2$ shell cannot have more than 8 electrons.

Scientific evidence has led chemists to propose that electron shells are further divided into electron subshells. Imagine magnifying the basic atomic model and finding that each shell is composed of subshells. Notice that the number of subshells that a shell has is equal to n.

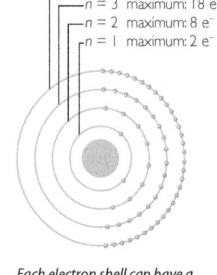

$n = 4$ maximum: 32 e⁻
$n = 3$ maximum: 18 e⁻
$n = 2$ maximum: 8 e⁻
$n = 1$ maximum: 2 e⁻

Each electron shell can have a maximum number of electrons.

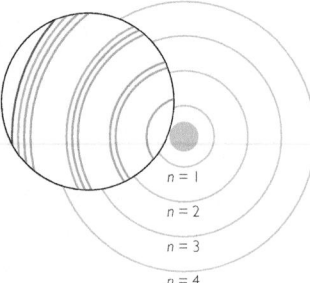

Electron shells are further divided into subshells.

$n = 1$
$n = 2$
$n = 3$
$n = 4$

THE s, p, d, AND f SUBSHELLS

The subshells have special names. They are called the s, p, d, and f subshells. Just like the basic shells, each subshell has a maximum capacity of electrons. The s subshells can have a maximum of 2 electrons, p subshells can have a maximum of 6 electrons, and d subshells can have a maximum of 10 electrons. Finally, f subshells can have a maximum of 14 electrons.

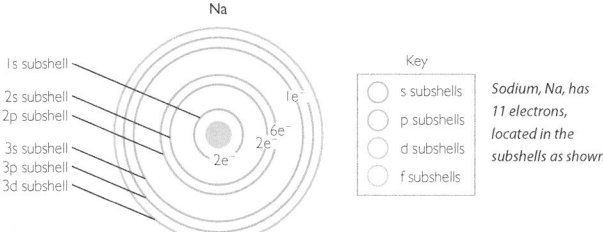

Na

1s subshell
2s subshell
2p subshell
3s subshell
3p subshell
3d subshell

1e⁻
6e⁻
2e⁻
2e⁻

Key

○ s subshells
○ p subshells
○ d subshells
○ f subshells

Sodium, Na, has 11 electrons, located in the subshells as shown.

Notice that the name of each subshell is labeled using both the basic shell number and the subshell letter (1s, 2s, 2p, and so on).

Example 1

Electron Arrangements

Use the illustration of the subshells in a sodium atom above to help you answer these questions:

  a. How many total electrons are there in a sodium, Na, atom? Which shells are they in?

  b. How many valence electrons does sodium have? Which subshell are they in?

  c. How many electrons are there in the 3s subshell of sodium? In the 3p subshell?

Solution

The atomic number of sodium is 11.

  a. There are a total of 11 electrons in a neutral sodium atom. The electrons are in shells, $n = 1$, $n = 2$, $n = 3$.

  b. Sodium has one valence electron, located in the 3s subshell.

  c. There is one electron in the 3s subshell of sodium, and none in the 3p subshell.

⟳ Electron Configurations

It can be time-consuming to draw subshell models of the atoms to show the arrangements of the electrons, especially for atoms with large atomic numbers. Chemists have developed a shorthand notation called an **electron configuration** to keep track of the electrons in an atom. The electron configurations for the first ten elements are shown here.

H $1s^1$	C $1s^2 2s^2 2p^2$
He $1s^2$	N $1s^2 2s^2 2p^3$
Li $1s^2 2s^1$	O $1s^2 2s^2 2p^4$
Be $1s^2 2s^2$	F $1s^2 2s^2 2p^5$
B $1s^2 2s^2 2p^1$	Ne $1s^2 2s^2 2p^6$

← Nitrogen has a total of 7 electrons.

Lesson Guide

ENGAGE (5 min)

TRM Transparency—ChemCatalyst 24

TRM PowerPoint Presentation 24

ChemCatalyst

These drawings show two different ways to represent the arrangement of the electrons in atoms of the element calcium, Ca.

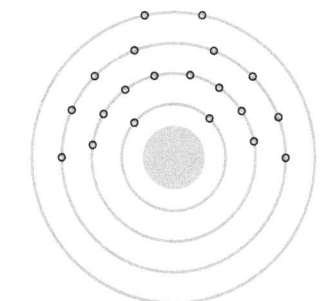

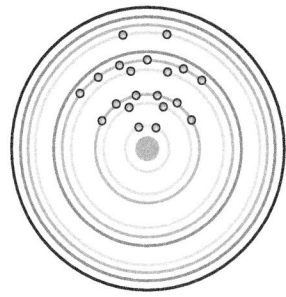

1. Name at least two differences between the drawings.

2. Name at least two similarities between the drawings.

Sample answers: **1.** There are more circles in the drawing on the right. The electrons are distributed differently in each model. **2.** Both drawings show a total of 20 electrons. The electrons are on circles around the nucleus. The innermost circle is identical in both drawings.

→ Assist students in examining this new way to represent electron arrangements.

Ask: What does the drawing on the right imply about electron shells?

Ask: Based on the drawings, what do you think a subshell is?

Ask: What are the names given to the subshells? (2s, 3s, 3p, 3d, 4s, etc.)

Ask: How many valence electrons does calcium have? (2)

Ask: Where are the valence electrons located? (in the outermost shell)

EXPLORE (15 min)

TRM Worksheet with Answers 24

TRM Worksheet 24

TRM Transparency—Electron Shells and Subshells 24

TRM Transparency—Electron Configurations 24

TRM Transparency—Subshell Blocks 24

INTRODUCE THE CLASSWORK

→ Arrange students into pairs.

→ Pass out the student worksheet.

→ Tell students that they will be using the periodic table to determine the arrangements of electrons into subshells.

EXPLAIN & ELABORATE (20 min)

DISCUSS ELECTRON SHELLS AND SUBSHELLS

→ Display the Transparency—Electron Shells and Subshells. Have students help you fill in the numbers in the table.

Ask: How many subshells are in each shell?

Ask: How many electrons are in each subshell? In each shell?

Electron Shells and Subshells

Shell	Number of electrons in the shell	Subshell	Number of electrons in the subshell
$n = 1$	2	1s	2
$n = 2$	8	2s	2
		2p	6
$n = 3$	18	3s	2
		3p	6
		3d	10
$n = 4$	32	4s	2
		4p	6
		4d	10
		4f	14

Key Points: The electron shells in the shell model of an atom (except for $n = 1$) are divided into subshells. The letter n represents the main shell or energy level. The maximum numbers of electrons that can occupy the main shells, $n = 1, 2, 3,$ and 4, are 2, 8, 18, and 32, respectively. Each subshell is indicated by its main shell number and a letter, either s, p, d, or f. The number of subshells in each shell is the same as the shell number. The maximum numbers of electrons that can occupy s, p, d, and f subshells are 2, 6, 10, and 14, respectively.

DISCUSS ELECTRON CONFIGURATIONS

→ Use the Transparency—Electron Configurations to display the electron configurations of elements up to oxygen so students can see where each electron goes.

Ask: How do you write an electron configuration?

Ask: How is the outermost subshell of an atom of an element related to the element's position on the periodic table?

Ask: How do you think electron configurations may be valuable to chemists?

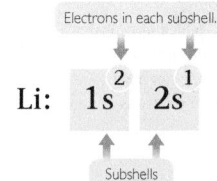

Each subshell is written using the shell number and the subshell letter. In addition, the number of electrons in each subshell is indicated with a superscript number.

Notice that the superscript numbers add up to the total number of electrons for that atom.

The sequence in which electrons fill up the subshells is 1s, 2s, 2p, 3s, 3p. After the element argon, the pattern changes slightly.

Example 2

Electron Configuration of Sulfur

Write the electron configuration for sulfur, S.

Solution

Sulfur is located in the third row in Group 6A. The atomic number of sulfur is 16, so there are 16 electrons that need to be distributed in subshells, beginning with the 1s subshell.

The electron configuration of sulfur is $1s^2 2s^2 2p^6 3s^2 3p^4$.

⟳ Connecting the Periodic Table to Electron Arrangements

An outline of the periodic table appears here with color-coding to show the subshell for the outermost electron of each element. For example, any element located in the green area will have its outermost electron(s) in a p subshell.

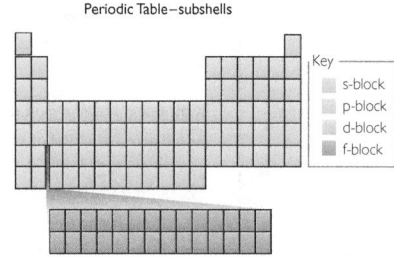

Periodic Table—subshells

Key
- s-block
- p-block
- d-block
- f-block

As you proceed across the periodic table from one element to the next, one additional proton and one additional electron are added, along with one or more neutrons. Each additional electron goes into a specific subshell. If an atom is located in the orange areas of the table, the last electron is placed into an s subshell. If an atom is located in the blue area of the table, the last electron is placed into a d subshell. And so on.

The elements in each block have related properties. The elements in the s-block are reactive metals. The elements in the d-block tend to form colorful compounds that are used as pigments. The elements in the p-block tend to form colorless compounds.

DECODING THE TABLE

To write out an electron configuration for a specific element, you can simply "read" from the periodic table, moving across from left to right and then down to the

---

Electron configuration: A shorthand way to keep track of all the electrons in an atom of an element for all the subshells that have electrons. The number of electrons in each subshell is shown as a superscript.

RELATE ELECTRON CONFIGURATIONS TO THE PERIODIC TABLE

→ Display the Transparency—Subshell Blocks, which shows the periodic table.

→ Show students how they can use the periodic table to predict the

electron arrangement of the atoms of an element.

Ask: What is the relationship between the structure of the periodic table and electron subshells?

Ask: If you know where an element is located on the periodic table, what do you know about the electron structure of its atoms?

Ask: How you can figure out an element's electron configuration from the periodic table?

Key Points: The periodic table is organized in subshell blocks. As you proceed across the periodic table from one element to the next, the number of

next row. For example, the sequence of subshells for argon, Ar, is 1s, 2s, 2p, 3s, 3p. The electron configuration for argon is $1s^22s^22p^63s^23p^6$.

Periodic Table—subshells

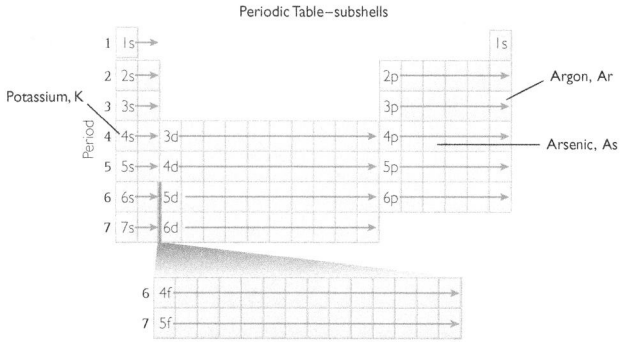

Everything runs smoothly until you reach the fourth row of the periodic table. After argon, you might expect the next electron to be in the 3d subshell. However, this does not happen. The next element is potassium, K. Like the other elements in Group 1A, potassium has one electron in the s subshell. So, you place an electron in 4s before 3d. The electron configuration for potassium is $1s^22s^22p^63s^23p^64s^1$. The electron configuration for arsenic is $1s^22s^22p^63s^23p^64s^23d^{10}4p^3$.

You may have noticed that you only have to look at the ending of each electron configuration to figure out the identity of the element associated with it. The ending provides you with the exact spot on the periodic table where you can find the element.

> **Important to Know** The s subshells fill with electrons before the d subshells from the previous shell. For example, the 4s subshell fills before the 3d subshell, the 5s subshell fills before the 4d subshell, and so on.

Example 3

Electron Configuration of Cobalt

Write the electron configuration for cobalt, Co.

Solution

Locate cobalt on the periodic table. It is element number 27 and is located in the fourth period of the periodic table.

Simply trace your finger across the periodic table of subshells, writing the subshells as you go. When you get to cobalt, stop writing. Every subshell up to the 4s subshell is completely filled. In addition, cobalt has seven electrons in the 3d subshell. The answer is $1s^22s^22p^63s^23p^64s^23d^7$.

You can check that you have the correct electron configuration by adding the superscript numbers to make sure there are 27 electrons.

protons and electrons increases by one. Each added electron goes into a specific subshell as indicated on the Subshell Blocks table. Notice that the s-, p-, and d-blocks are 2, 6, and 10 elements wide, respectively. This corresponds to the maximum number of electrons in each of these subshells.

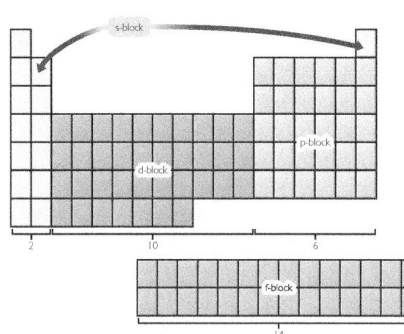

The order of filling of subshells does not always correspond to the numerical order of the subshells. Chemists have found that electrons fill the subshells in the order 1s, 2s, 2p, 3s, 3p, 4s, 3d, 4p. Notice that the 4s subshell fills before the 3d subshell. In spite of this, you can still determine the correct electron configuration by simply following from left to right on the Subshell Blocks table.

The elements in the different subshell blocks have related properties. The elements in the s-block are reactive metals. The elements in the d-block tend to form colorful compounds that

are used as pigments. The elements in the p-block tend to form white compounds.

INTRODUCE THE NOBLE-GAS SHORTHAND NOTATION (OPTIONAL)

→ On the Transparency—Electron Configurations, circle the noble gases helium, He, neon, Ne, and argon, Ar, and write some noble-gas shorthand notations. You might want to underline the noble-gas equivalent in each long version as well.

Ask: How do you identify the core electrons in an atom?

Ask: What do you think is meant by the noble-gas core?

Ask: Write the electron configuration of the elements in the third row of the periodic table, beginning with sodium, using the noble-gas shorthand.

Ask: If you know the location of the last electron placed into a subshell of an atom, can you figure out the identity of that atom? Explain.

Key Points: There is shorthand to write electron configurations using the noble gases. Each noble gas has its s subshell and p subshell entirely filled. You can use the noble gases as placeholders to symbolize all the filled subshells up to that point. For example, the shorthand to write the electron configurations of lithium and beryllium is Li: $1s^2s^1 = $ [He]$2s^1$ and Be: $1s^22s^2 = $ [He]$2s^2$.

The noble-gas shorthand notation can provide information about the number of valence electrons in an atom. Consider the electron configuration of the Group 2A elements:

beryllium	Be	[He]$2s^2$
magnesium	Mg	[Ne]$3s^2$
calcium	Ca	[Ar]$4s^2$
strontium	Sr	[Kr]$5s^2$
barium	Ba	[Xe]$6s^2$

The similarity in valence electrons is consistent with the similarities in the properties of elements in this group.

The noble-gas shorthand also emphasizes how the number of valence electrons changes as you move from element to element across a period. Consider the electron configurations of elements in the third period.

sodium	Na	$[Ne]3s^1$
magnesium	Mg	$[Ne]3s^2$
aluminum	Al	$[Ne]3s^23p^1$
silicon	Si	$[Ne]3s^23p^2$
phosphorus	P	$[Ne]3s^23p^3$
sulfur	S	$[Ne]3s^23p^4$
chlorine	Cl	$[Ne]3s^23p^5$
argon	Ar	$[Ne]3s^23p^6 = Ar$

You can easily identify an element from its shorthand configuration. The ending provides you with the exact position on the periodic table where you can find the element. For example, in $[Ne]3s^23p^5$, the [Ne] indicates that the element should be in the period after neon. The last subshell term, $3p^5$, indicates that this element is in the p-block. The number in front of the p indicates the exact subshell, which in turn indicates the row on the periodic table. In this case, the element is in the third row. The superscript 5 indicates five p electrons. Counting five spaces across the p-block, you arrive at chlorine, Cl. Chlorine has seven more electrons than neon, which is also apparent from the shorthand configuration.

EVALUATE (5 min)

Check-In

Identify the element with this electron configuration:
$1s^22s^22p^63s^23p^64s^23d^{10}4p^3$

Answer: The electron configuration is for the element arsenic, As. The ending of the configuration indicates that this atom is in the fourth period, three spaces into the p-block. This places arsenic in Group 5A.

Homework

Assign the reading and exercises for Alchemy Lesson 24 and the Chapter 4 Summary in the student text.

⟳ Noble Gas Shorthand

Depending on the element, the electron configuration can be lengthy to write. Plus, each element just repeats the electron configuration of the previous element but adds one more electron.

Rather than repeat the same thing every time, chemists have devised a quicker way to write out electron configurations. They use the noble gas at the end of each period as a placeholder to symbolize all of the filled subshells before that place on the table. Using this "shorthand" method, the electron configuration of cobalt is $[Ar]4s^23d^7$.

Shorthand notation allows you to make some interesting comparisons. Notice that the noble gas shorthand notation emphasizes the valence electrons. Using this method, it is easy to see that each element in Group 2A has two valence electrons, both located in an s subshell.

Group 2A Elements

Element	Symbol	Electron configuration
beryllium	Be	$[He]2s^2$
magnesium	Mg	$[Ne]3s^2$
calcium	Ca	$[Ar]4s^2$
strontium	Sr	$[Kr]5s^2$
barium	Ba	$[Xe]6s^2$
radium	Ra	$[Rn]7s^2$

Example 4

Electron Configuration of Selenium

Find the element Selenium, Se, element number 34, on the periodic table.

a. What is the electron configuration of selenium?

b. Write the electron configuration using noble gas shorthand.

c. In what subshells are selenium's valence electrons?

Solution

a. Selenium is in the p-block, in Period 4. The electron configuration of selenium is $1s^22s^22p^63s^23p^64s^23d^{10}4p^4$.

b. The noble gas that comes before selenium is argon, Ar. So, the noble gas shorthand for this configuration is $[Ar]4s^23d^{10}4p^4$.

c. Selenium's valence electrons are in subshells 4s and 4p.

LESSON SUMMARY

What does the periodic table indicate about the arrangements of electrons?

Electrons in atoms are arranged into basic shells labeled $n = 1, 2, 3$, and so on. These shells are divided into subshells. The number of subshells in each shell is equal to n. The subshells are referred to as s, p, d, and f subshells. The s, p, d, and f subshells can have a maximum of 2, 6, 10, and 14 electrons, respectively. Chemists use electron configurations to specify the arrangements of electrons in subshells. The periodic table provides the information needed to write electron configurations.

KEY TERM
electron configuration

Exercises

Reading Questions

1. What are electron subshells?
2. What is an electron configuration?
3. How is the arrangement of electrons in an atom related to the location of the atom on the periodic table?

Reason and Apply

4. How many subshells are in each shell: $n = 1$, $n = 2$, $n = 3$, $n = 4$?
5. What is the total number of subshells for elements in Period 5 of the periodic table?
6. Draw a subshell model for each of these elements, putting the electrons in their appropriate places.
   a. sodium, Na   b. neon, Ne   c. carbon, C   d. vanadium, V
7. What is the outermost subshell for bromine, Br?
8. Name an element with electrons in the f subshell.
9. Consider the element with the atomic number 13.
   a. What is the electron configuration for the element with atomic number 13?
   b. How many valence electrons does element number 13 have? How do you know?
   c. How many core electrons does element number 13 have? How did you figure that out?
10. Explain why the chemical properties of argon, krypton, and xenon are similar, even though there are 18 elements between argon and krypton and 32 elements between krypton and xenon.
11. Write the electron configuration for each of these atoms. Then write it using the noble gas shorthand method.
    a. oxygen   b. chlorine   c. iron   d. calcium
    e. magnesium   f. silver   g. silicon   h. mercury
12. You should be able to figure out the identity of an atom from its electron configuration alone. Describe at least two ways you could do this.
13. Which elements are described by these electron configurations?
    a. $1s^2 2s^2 2p^6 3s^2 3p^6 4s^2 3d^4$
    b. $1s^2 2s^2 2p^6 3s^2 3p^2$
    c. $1s^2 2s^2 2p^3$
    d. $1s^2 2s^2 2p^6 3s^2 3p^6 4s^2 3d^{10} 4p^6 5s^2 4d^{10} 5p^6 6s^1$
    e. $1s^2 2s^2 2p^6 3s^2 3p^6 4s^2 3d^{10} 4p^6 5s^2 4d^{10} 5p^6 6s^2 4f^{14} 5d^{10} 6p^2$
    f. $[Kr] 5s^2 4d^9$

Lesson 24 | **Electron Configurations**   121

c.

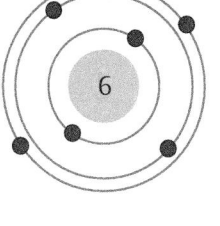

d.

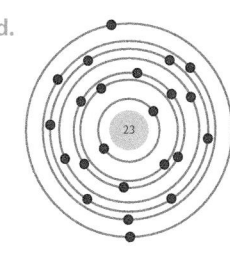

7. $4p$

8. *Possible answer:* gold. You can accept any element that has an atomic number of 57 or greater as a correct answer.

9. a. $1s^2 2s^2 2p^6 3s^2 3p^1$

b. Element 13 has 3 valence electrons because it has three electrons in the outer shell, $n = 3$.

c. Element 13 has 10 core electrons because it has two shells filled completely: $n = 1$ with 2 electrons and $n = 2$ with 8 electrons.

10. Argon, krypton, and xenon have similar properties because they all have the same number of valence electrons in their outermost shells.

11. a. $1s^2 2s^2 2p^4$, [He] $2s^2 2p^4$

b. $1s^2 2s^2 2p^6 3s^2 3p^5$, [Ne] $3s^2 3p^5$

c. $1s^2 2s^2 2p^6 3s^2 3p^6 4s^2 3d^6$, [Ar] $4s^2 3d^6$

d. $1s^2 2s^2 2p^6 3s^2 3p^6 4s^2$, [Ar] $4s^2$

e. $1s^2 2s^2 2p^6 3s^2$, [Ne] $3s^2$

f. $1s^2 2s^2 2p^6 3s^2 3p^6 4s^2 3d^{10} 4p^6 5s^2 4d^9$, [Kr] $5s^2 4d^9$

g. $1s^2 2s^2 2p^6 3s^2 3p^2$, [Ne] $3s^2 3p^2$

h. $1s^2 2s^2 2p^6 3s^2 3p^6 4s^2 3d^{10}$ $4p^6 5s^2 4d^{10} 5p^6 6s^2 4f^{14} 5d^{10}$, [Xe] $6s^2 4f^{14} 5d^{10}$

12. *Possible answer:* Add the total number of electrons in the configuration and look up the atomic number. Find the location on the periodic table that corresponds to the end of the electron configuration.

13. a. chromium
b. silicon
c. nitrogen
d. cesium
e. lead
f. silver

LESSON 24 ANSWERS

1. Electron subshells are divisions within a specific electron shell of an atom.

2. An electron configuration is a shorthand notation used to show the position of electrons within shells and subshells of an atom.

3. As the number of electrons in an element increases, they are added in a specific sequence that is illustrated by the position of the element on the periodic table. Each section of the table corresponds to a particular subshell of electrons.

4. The shell $n = 1$ has one subshell. The shell $n = 2$ has two subshells. The shell $n = 3$ has three subshells. The shell $n = 4$ has four subshells.

5. The sum of the subshells of the shells is 15.

6. a.

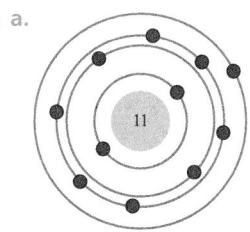

b.

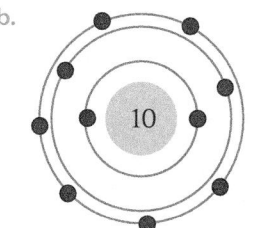

ANSWERS TO CHAPTER 4 REVIEW EXERCISES

1. Valence electrons are important because they are the outermost part of an atom and will interact with other atoms.

2. Ionic compounds are combinations of cations and anions that are bound together by the electrical attraction of their opposite charges. They form when electrons are transferred between atoms leaving each resulting ion with a noble gas arrangement.

3. As the number of electrons in an element increases, they are added in a specific sequence that is illustrated by the position of the element on the periodic table. Each section of the table corresponds to a particular subshell of electrons.

4. $1s^2 2s^2 2p^6 3s^2 3p^1$, [Ne] $3s^2 3p^1$

5. a. The anion is Mg^{2+} and has a charge of $+2$. The cation is Cl^- and has a charge of -1.

b. The anion is Ca^{2+} and has a charge of $+2$. The cation is NO_2^- and has a charge of -1.

6. a. Na_2SiO_3, $NaClO_2$, $NaHCO_3$

b. $CaSiO_3$, $Ca(ClO_2)_2$, $Ca(HCO_3)_2$

ASSESSMENTS

Two multiple-choice assessments (versions A and B) and two short-answer assessments (versions A and B) are available for download for Chapter 4.

TRM **Chapter 4 Assessment Answers**

TRM **Chapter 4 Assessments**

CHAPTER 4
Moving Electrons

SUMMARY

KEY TERMS

flame test
valence shell
valence electron
core electron
ion
cation
anion
ionic compound
rule of zero charge
polyatomic ion
monatomic ion
electron configuration

Alchemy Update

What do the electrons of an atom have to do with its properties?

The outermost electron shell contains an atom's valence electrons. The number of valence electrons in main group elements is the same as their group number. Valence electrons are responsible for many of an element's properties, so elements in the same group have similar properties. While you cannot change copper atoms into gold atoms, you can combine elements into compounds with new properties.

The electrons in an atom are fairly easy to change, unlike the parts of a nucleus. Atoms become more stable when they transfer electrons to other atoms, forming ionic compounds.

REVIEW EXERCISES

1. Explain the importance of valence electrons.

2. What are ionic compounds, and how do they form?

3. How is the arrangement of electrons within atoms connected to the periodic table?

4. Write the electron configuration for aluminum. Then write it using noble gas shorthand.

5. For each compound listed, identify the anion and cation and the charge on each.
   a. magnesium chloride, $MgCl_2$
   b. calcium nitrite, $Ca(NO_2)_2$

6. Three polyatomic ions are listed here.
   silicate, SiO_3^{2-} chlorite, ClO_2^- bicarbonate, HCO_3^-
   a. Write the chemical formulas for sodium silicate, sodium chlorite, and sodium bicarbonate.
   b. Write the chemical formulas for calcium silicate, calcium chlorite, and calcium bicarbonate.

CONSUMER CONNECTION

Sodium bicarbonate, $NaHCO_3$, is commonly known as baking soda. When it is used in baking, it reacts with other ingredients to release carbon dioxide. For example, baking soda is what causes the batter to rise and bubble when you make pancakes.

CHAPTER 5

Building with Matter

In this chapter, you will study

- the basic types of bonds between atoms
- the role of electrons in bonding
- the properties and types of elements associated with each type of bond
- the formation of new substances through bonding

Sugar dissolves in water, but copper does not. This property gives clues about how the atoms in these substances are bonded to each other.

W hen atoms form bonds, new substances with new properties are formed. There are four basic types of bonds found in the world around us. Each type of bond is a result of a different distribution of electrons within the substance. Further, each type of bond is associated with certain properties. To determine the type of bond in a substance, you can use the chemical formula of that substance.

123

PD CHAPTER 5 OVERVIEW

Watch the video overview of Chapter 5 (for teachers) by clicking on the link in the TE-book, opening the TRFD, or logging onto the book's companion Web site bcs.whfreeman.com/livingbychemistry2e (teacher log-in required).

The final chapter of Unit 1: Alchemy focuses on the bonds that form between atoms and on the properties of compounds. In Lesson 25, students test various substances and sort them into categories according to solubility and conductivity. In Lesson 26, they are presented with four models of bonding that show the locations of electrons within substances. They use the information from both lessons to sort common substances, deducing the type of bonding within each substance based on its properties. Lesson 27 is a lab that explores electroplating. Students extract metal atoms from ionic compounds. Unit 1 Review provides a lesson that concludes the alchemy unit with a look back at the quest to make gold from ordinary substances.

In this chapter, students will learn

- how to extract metal atoms from compounds through electroplating
- basic models of bonding
- the role of electrons in bonding
- how the properties of substances are related to bonding
- the relationship between an element's location on the periodic table and its bonding patterns

Chapter 5 Lessons

Overview

LAB: PAIRS

Key Question: How can substances be sorted into general categories?

KEY IDEAS

Most substances can be sorted into four categories according to whether or not they dissolve in water and whether or not they conduct electricity. These categories reflect differences in the types of bonds between the atoms of the substances. This lesson sets up the exploration of bonding that makes up this chapter, the last of Unit 1: Alchemy.

LEARNING OBJECTIVES

- Classify substances into four categories based on solubility and conductivity.
- Explain the difference between the terms *soluble* and *insoluble.*
- Begin to describe the atomic makeup of substances based on whether they are soluble and/or conduct electricity.

FOCUS ON UNDERSTANDING

- Students will be introduced to the concept of solubility later in the lesson. Use the term *dissolving* until after the activity.
- Both dissolving and conductivity are invisible processes and therefore subject to many misconceptions. For instance, the fact that substances disappear visually when they dissolve suggests that the substance is gone.

KEY TERMS

dissolve
soluble
insoluble
conductivity

IN CLASS

The lesson begins with a brief demonstration of the dissolving and conductivity of a sports drink. This demo allows the teacher to demonstrate the proper use of the conductivity apparatus. During the lab, teams of students first predict the solubility and conductivity of ten substances and then test them at ten stations. Finally, they look for patterns in their data. Solubility is covered in more depth in Unit 4: Toxins. A complete materials list is available for download.

TRM Materials List 25

SETUP

Use a string of holiday lights as the source of bulbs and wire. Cut 48 pieces of 6 in wire from the light string. Strip approximately 1 in on both ends of each wire. Assemble one conductivity apparatus for the demonstration. Have distilled water on hand for the dissolving portion of the demo. Test your sports drink to make sure it conducts.

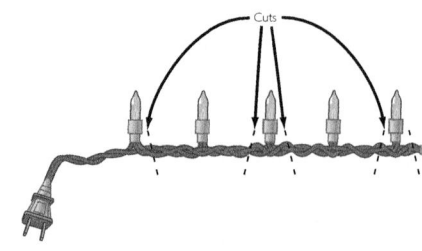

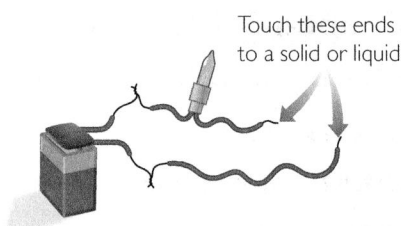
Touch these ends to a solid or liquid.

Set up ten stations around the classroom with materials for testing. At each station, set up two beakers: one for the pure substance and one for the pure substance mixed with distilled water. At the station for water, place two beakers filled with distilled water. To prevent contamination, set

Materials

- bulb with wires
- 9-volt battery with snap connector
- wire with stripped ends
- 100 mL beakers
- paper clips
- salt, NaCl
- sand, SiO_2
- paraffin wax, $C_{20}H_{42}$
- calcium chloride, $CaCl_2$
- copper, Cu
- copper (II) sulfate, $CuSO_4$
- aluminum foil, Al
- sucrose, $C_{12}H_{22}O_{11}$
- distilled water, H_2O

Classifying Substances

Purpose

To investigate the properties of substances.

Predictions and Data

Predict whether each substance will conduct electricity and whether it will dissolve in water. Make a table for your predictions and data.

Procedure

1. Assemble your conductivity tester as shown.
2. Take your conductivity tester to each station. Test the pure substance for conductivity.
3. Next, observe the substance in water. Did the substance dissolve?
4. Finally, test the substance mixed with the distilled water in the second beaker for conductivity. Record your results in a data table.

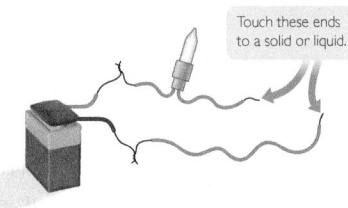

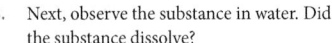
Touch these ends to a solid or liquid.

Analysis

5. Group the substances according to their properties. How do your results compare with your predictions? What do the substances in each group have in common? Summarize your findings in a paragraph.

You Light Up My Life

Classifying Substances

LESSON 25

THINK ABOUT IT

If you accidentally drop a piece of jewelry into a tub of water, you should have no trouble getting it back again. However, if you drop a cube of sugar into the water, it will dissolve. Some substances dissolve in water and others do not. So, dissolving is one property of matter that you can use to sort substances into general groups.

> How can substances be sorted into general categories?

To answer this question, you will explore

- Dissolving and Conductivity
- Testing and Sorting Substances

EXPLORING THE TOPIC

Dissolving and Conductivity

There are millions of different substances in our world. Even chemists do not know the identities of all the possible compounds. However, many of the known substances can be sorted into general categories by looking at two properties: dissolving and conductivity.

DISSOLVING SUBSTANCES IN WATER

We live on a watery planet. Most substances come into contact with water at some point. Some substances, like sugar, **dissolve** in water to make an aqueous solution. Other substances, like gold, do not dissolve in water and will remain unchanged when placed in water. A substance that dissolves in water is **soluble** in water. A substance that does not dissolve is **insoluble** in water.

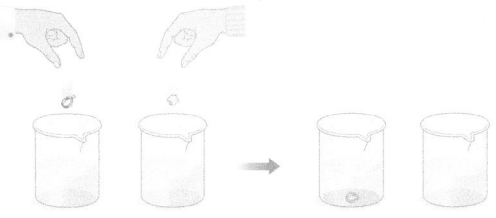

Many substances can be sorted into one of two categories based on whether they dissolve in water.

CONDUCTIVITY

By examining a second property, it is possible to sort matter even further. The property of **conductivity** has to do with how well electricity flows through a substance. Electrical conductivity requires the movement of charge as either ions or

> **BIOLOGY CONNECTION**
>
> Not only solid substances are soluble. Gases can dissolve in liquids too. This is why fish are able to survive underwater. The ocean is full of dissolved oxygen that fish can extract from the water using their gills.

---

up paper clips in each beaker so that when students visit that station they simply need to touch the ends of the wires of their conductivity tester to each clip. Be sure that the clips will complete the circuit—especially for the metals in the dry beaker.

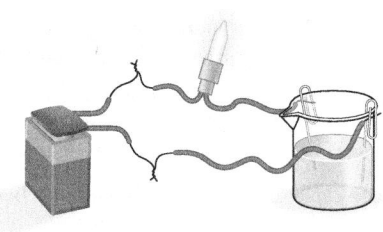

Testing materials: water, aluminum foil, sugar, salt, sand, paraffin, ethanol, copper, calcium chloride, copper (II) sulfate. Safety goggles must be worn at all times.

CLEANUP

Save and reuse: Evaporate water from test materials so they can be stored for future use. Disassemble, dry, and store the conductivity apparatus.

Lesson Guide

ENGAGE (5 min)

TRM PowerPoint Presentation 25

ChemCatalyst

1. If you were to drop a spoonful of salt, NaCl, into a glass of water, what would happen?
2. If you were to drop a gold ring into a glass of water, what would happen?
3. What do you think is different about the atoms of these two substances? Why do you suppose the gold atoms do not break apart?

Sample answers: **1.** The salt would sink to the bottom of the glass and would begin to dissolve. Eventually, it would all mix into the water. Some students might say the salt will "disappear." **2.** The gold ring also will sink to the bottom of the glass, but then nothing will happen to it. **3.** Students might hypothesize that the salt can break apart because it is made from two different elements, or that gold atoms have a stronger connection with one another.

→ Assist students in verbalizing their thoughts about the properties of gold and about dissolving in particular.

Ask: What do you think is happening to the atoms when salt dissolves?

Ask: Why do you think gold does not dissolve?

Ask: How are the atoms of these two substances different?

Ask: Why do you think some solids dissolve while others do not?

EXPLORE (20 min)

TRM Worksheet with Answers 25

TRM Worksheet 25

INTRODUCE THE LAB

→ Arrange students into pairs.

→ Pass out the student worksheet.

→ Briefly define the terms *dissolving* and *conductivity*.

> **Dissolve:** To disperse evenly into another substance. For example, a solid can dissolve in a liquid.
>
> **Conductivity:** A property that describes how well a substance transmits electricity.

→ Demonstrate dissolving and conductivity with a powdered sports drink. You might model the assembly for the students and touch the two wires together to show that it works.

- Ask the students to predict whether or not the dry sports drink powder will conduct electricity and light the bulb.
- Test the powder for conductivity. The bulb should not light.
- Ask the students to predict whether sports drink powder will dissolve in water or not; then complete the test.
- Show the students how you plan to run an electric current through the sports drink solution. Ask them to predict whether or not the bulb will light up.
- Test the sports drink solution for conductivity. The solution should light the bulb.
- You might want to have students speculate on what is happening, or you could simply go directly to the lab portion of the lesson.

→ Review safety considerations and tell students that they will be predicting test results for ten substances and then completing the tests themselves. Ask students to work in pairs to make a conductivity apparatus. Then one or two pairs of students can go to each station to test the different substances.

EXPLAIN & ELABORATE (15 min)

TRM Transparency—Solubility and Conductivity 25

SHARE STUDENTS' GENERALIZATIONS ABOUT CONDUCTIVITY

→ Summarize students' generalizations and write them on the board as students share them. Ask for a consensus on the statements as you accept them.

Ask: What generalizations can you make about the substances that did not light up the bulb?

Ask: What generalizations can you make about the substances that did light up the bulb?

Ask: Based on your data, why do you think the sports drink lit up the bulb when dissolved?

Key Points: Generalizations about substances that do not light up the bulb:

- Compounds made up of C, H, and O atoms do not conduct electricity.
- Compounds made up entirely of nonmetals do not light up the bulb.
- Compounds made up of a combination of metals and nonmetals do

electrons. Copper wire is a great conductor of electricity. Your entire home is wired with hundreds of feet of copper or aluminum wiring that carries electricity from the power lines outside to the electrical outlets in your home. The human body is another good conductor of electricity. However, in the body, electricity moves as ions, not as electrons, as is the case in wiring.

Some substances do not conduct electricity. The electrical wires in your house are covered with a coating that does not conduct electricity. That way, you can plug in a stereo or a lamp and not get an electrical shock.

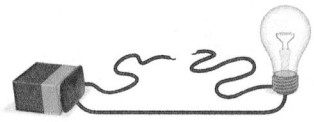 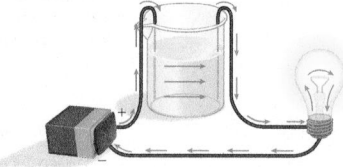

The light bulb doesn't light up because there's a break in the wire.

Even though there is a break in the wire, the bulb lights up because the liquid in the beaker conducts electricity.

Electrical conductivity can be tested by setting up a simple electrical circuit. Wires connect the terminals of a battery to a light bulb. When the circuit is complete, and an electrical current is flowing, the bulb will light up. If the flow of current is interrupted by the presence of a substance that does not conduct electricity, the light bulb will not light up. In the figure above, the red arrows show the direction of flow of positive charge. The electrons flow in the opposite direction.

⟳ Testing and Sorting Substances

You can use a two-step test to sort substances by both properties. First, drop a substance into water to find out if it dissolves. Next, test it for conductivity. This allows you to sort all matter into four basic categories as shown in the illustration below.

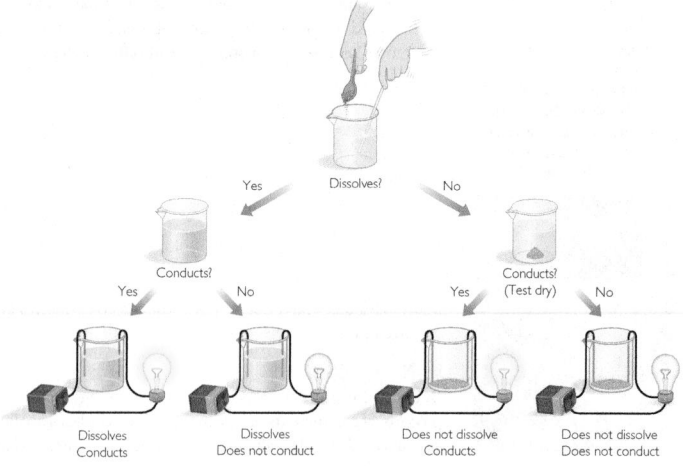

not light up the bulb when they are in their solid form.

Generalizations about substances that do light up the bulb:

- Everything that lights up the bulb has a metal atom in it.
- Compounds made of metal and nonmetal atoms, such as salts, light up the bulb when they are dissolved in water. (The sports drink is a solution of water, various salts, sugar, and a dye.)
- Metal solids light up the bulb.

ANALYZE THE RESULTS

→ Use the Transparency—Solubility and Conductivity to summarize the

results. Ask students to assist you with filling in the substances at the bottom of the chart.

→ After sorting the substances, introduce the terms *soluble* and *insoluble*. Label the appropriate boxes with these terms.

Ask: What statement can you make about ionic compounds and dissolving?

Ask: What generalizations can you make about substances made up entirely of metal atoms?

Ask: What generalizations can you make about substances made up entirely of nonmetal atoms?

Examples of substances in each of the four categories are given in the table. Take a moment to look for patterns in the chemical formulas of the substances in each category. Note whether substances are made from metal atoms, nonmetal atoms, or some combination.

Types of Substances

Properties	dissolves: yes conducts: yes	dissolves: yes conducts: no	dissolves: no conducts: yes	dissolves: no conducts: no
Examples	salt, NaCl; calcium chloride, $CaCl_2$; copper sulfate, $CuSO_4$	water, H_2O; sugar, $C_{12}H_{22}O_{11}$; ethanol, C_2H_6O	gold, Au; copper, Cu; aluminum, Al	sand, SiO_2; paraffin, $C_{20}H_{42}$
Types of atoms	metal and nonmetal atoms	nonmetal atoms only	metal atoms only	nonmetal atoms only

Here are some patterns you might notice:

- The substances that are *soluble* in water and *conduct* electricity are made up of a metal element and one or more nonmetal elements. Remember, these are called ionic compounds.
- The substances that are *soluble* in water but *do not conduct* electricity are often made up of carbon, hydrogen, and oxygen atoms joined together. These are all nonmetal elements.
- The substances that are *insoluble* in water but *do conduct* electricity are made of metallic elements.
- The substances that are *insoluble* in water and *do not conduct* electricity are made up of nonmetal atoms.

Notice that only substances that contain metal atoms will conduct electricity. Also, substances that are made entirely of metal atoms will not dissolve in water. Many ionic compounds dissolve in water.

In Lesson 26: Electron Glue, you will discover that these two properties, solubility and conductivity, are directly related to the manner in which the individual atoms in these substances are linked.

Important to Know Ionic compounds conduct electricity only when they are dissolved in water. Dry ionic solids do not conduct electricity.

Example

Predicting Properties

Predict whether the following substances dissolve in water and whether they conduct electricity.

a. lead, Pb
b. potassium bromide, KBr

Key Points: We can place all the substances tested into one of the four categories. A substance that dissolves in another substance is said to be *soluble* in that substance. A substance that does not dissolve in another substance is said to be *insoluble* in that substance. Once we find out whether or not a substance dissolves in water, we can further sort according to whether or not the dissolved substance conducts electricity. In the next lesson, we will take a closer look at these categories to figure out what else the substances have in common or what might account for their common properties.

Soluble: Describes a substance that is capable of being dissolved in another substance.

Insoluble: Describes a substance that is incapable of being dissolved in another substance.

Notice that substances that conduct electricity are either solid metals or ionic compounds dissolved in water. Substances made entirely of nonmetal atoms, such as sugar, do not conduct electricity. *Note:* Some substances are considered slightly soluble. More discussion of solubility will take place in Unit 4: Toxins.

EVALUATE (5 min)

Check-In

Predict whether $MgSO_4(aq)$, commonly known as Epsom salts, will conduct electricity or not. State your reasoning.

Answer: This salt contains both a metal and a nonmetal and the symbol (*aq*) next to the formula indicates that it dissolves in water. This substance will conduct electricity.

Homework

Assign the reading and exercises for Alchemy Lesson 25 in the student text.

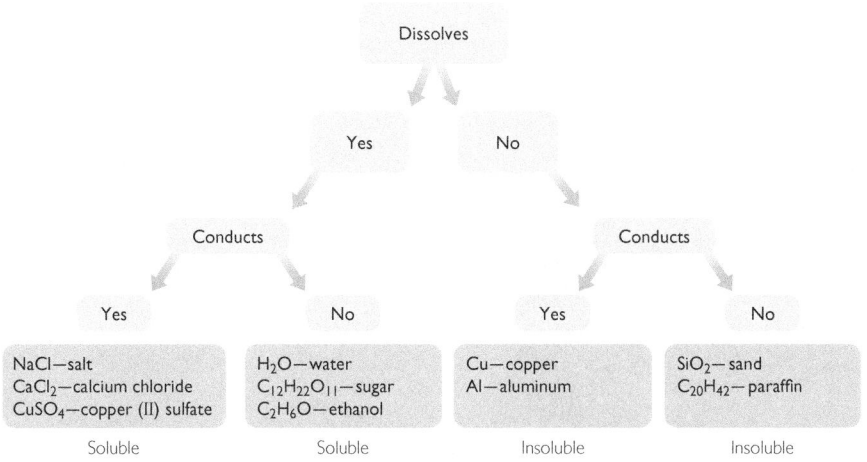

LESSON 25 ANSWERS

1. A substance is insoluble if it fails to dissolve in a particular solvent.

2. *Possible answer:* Electrical conductivity can be tested by setting up a simple electrical circuit. When the circuit is complete, a bulb that is part of the circuit will light up. If the circuit is interrupted by a substance that does not conduct electricity, the bulb will not light up.

3. *Possible answer:* The substance is most likely an ionic compound. Many compounds dissolve in water, but electrical conductivity in solution is a characteristic of compounds that separate into ions, such as ionic compounds.

4. No, though ionic compounds generally contain metals, they do not conduct electricity in their solid form. Substances that are composed only of metals always conduct electricity. However, in a solid ionic compound, where a metal is bonded with a nonmetal, the charged ions are locked into position so they cannot move freely and carry electricity.

5. No, though ionic compounds generally do not conduct electricity as solids, they do conduct electricity as aqueous solutions. In solution, the ions of an ionic compound are able to move around, so they can carry electric charges from one place to another.

6. *Possible answers:* **a.** Acetone is not likely to conduct electricity either in solution or in its pure form because it does not contain any metallic elements. Acetone will probably dissolve in water because compounds composed only of carbon, hydrogen, and nitrogen usually dissolve in water. **b.** Titanium is a metallic element, so it conducts electricity and does not dissolve in water. **c.** Lithium nitrate is an ionic compound, so it dissolves in water and does not conduct electricity in its solid form. Once dissolved, it will conduct electricity. **d.** Bronze is a mixture that is composed only of metallic elements. Therefore, it conducts electricity and does not dissolve in water.

Example

(continued)

Solution

  a. Lead is a metal element. Because it is made only of metal atoms, it does not dissolve in water, but it does conduct electricity.

  b. Potassium bromide consists of a metal and a nonmetal. It is an ionic compound. Therefore, it does dissolve in water, and the mixture does conduct electricity. The dry compound does not conduct electricity.

LESSON SUMMARY

KEY TERMS
dissolve
soluble
insoluble
conductivity

How can substances be sorted into general categories?

Most substances on the planet can be sorted into four categories based on two properties: whether they are soluble in water and whether they conduct electricity. As you will discover in coming lessons, the properties of conductivity and solubility are directly related to the way in which atoms are connected to each other.

Exercises

Reading Questions

1. What does insoluble mean?
2. Describe a way to determine whether a substance conducts electricity.

Reason and Apply

3. What generalization can you make about a substance that is soluble in water and conducts electricity once it has dissolved? Explain your thinking.

4. Do all solid substances containing metals conduct electricity? Explain your reasoning.

5. If a substance does not conduct electricity as a solid, does this mean that it will not conduct electricity if it dissolves? Explain your reasoning.

6. Predict whether each substance listed will conduct electricity, dissolve in water, and/or conduct electricity once it has dissolved. Explain your thinking in each case.
  a. $C_3H_6O(l)$ acetone
  b. $Ti(s)$ titanium
  c. $LiNO_3(s)$ lithium nitrate
  d. $CuZn(s)$ bronze

Electron Glue

Bonding

THINK ABOUT IT

All of the objects in our everyday world, including ourselves, are made up of
individual atoms. But what holds those atoms together? Why don't objects just
crumble into piles of individual atoms? Something must be holding the atoms
together. And why is the desk in your classroom solid, while water simply runs
through your fingers? Something about the way atoms are connected must give
substances the properties we observe.

How are atoms connected to one another?

To answer this question, you will explore

○ Bonds: The "Glue" between Atoms

○ Models of Bonding

○ Relating Bonds and Properties

EXPLORING THE TOPIC

○ Bonds: The "Glue" between Atoms

Chemists call the attraction that holds atoms together a **chemical bond**. As you
will discover, several different types of chemical bonds exist. All bonds involve the
electrons in some way. A chemical bond is an attraction between the positive charges
on the nucleus of one atom and the negative charges on the electrons of another
atom. This attraction is so great that it keeps the atoms connected to one another.

○ Types of Bonding

Recall from Lesson 25 that most substances can be divided into four categories,
based on their physical properties. These four categories can be explained by
different models of bonding.

A bond is a force of attraction, so it is not possible to see the actual bonds between
atoms. However, a model can help to explain how atoms are bonded in substances.
Models can also help us to understand how bonding accounts for certain
properties of substances that we observe.

> **Big Idea** There are four main models of chemical bonds between atoms.

The four models of bonding are called *ionic, molecular covalent, metallic,* and
network covalent. Take a moment to locate the valence electrons in each model
on the next page. The red spheres represent the nuclei of atoms and the core
electrons, while the blue areas suggest where the valence electrons are located.

Overview

ACTIVITY: PAIRS

Key Question: How are atoms
connected to one another?

KEY IDEAS

Atoms are held together in substances
by bonds. These bonds are attractions
between valence electrons and the
nuclei of other atoms. There are four
basic models of bonding between
atoms: ionic, molecular covalent,
metallic, and network covalent. Most
substances in our everyday lives can
be classified into one of these four
bonding categories based on their

elemental makeup and on their
properties. Valence electrons are
distributed differently within a
substance depending on the model of
bonding that is present. Differences
in the way the valence electrons are
distributed help explain differences in
the properties of substances.

LEARNING OBJECTIVES

● Define a chemical bond and
describe the four basic models
of chemical bonds.

● Use chemical formulas to sort
substances into bonding categories.

● Predict the properties of a
substance based on its chemical
formula and model of bonding.

FOCUS ON UNDERSTANDING

● Every time a new model or graphic
representation is introduced, there is
room for visual misinterpretation of
that model. Assist students in making
sense of the models.

● Molecules are introduced here for the
first time. Students might have trouble
grasping what a molecular unit is and
that molecules do not break apart into
individual atoms when they dissolve.
Molecules are covered extensively in
Unit 2: Smells.

● Elemental substances can be confusing
for students. While all elemental
substances are made up of just one
kind of atom, some elements, like
oxygen, O_2, are never found as single
atoms in nature.

KEY TERMS

chemical bond
ionic bonding
molecular covalent bonding
metallic bonding
network covalent bonding
covalent bonding
molecule

IN CLASS

Students explore four bonding models.
They are given drawings and descriptions of ionic, network covalent, metallic, and molecular covalent bonding.
They are also provided with a set of
cards containing pictures of a variety
of substances, their chemical formulas,
and some basic properties. Students
use the information on the cards to sort
the substances into their appropriate
bonding category. A complete materials
list is available for download.

TRM Materials List 26

SETUP

TRM Card Masters—Substance Cards 26
Copy and cut sets of the Substance
Cards. You might want to laminate
them (optional).

Differentiate

To illustrate bonding visually and
kinesthetically, give each student a pair
of tennis or Ping-Pong balls to serve as
electrons. Challenge them to act out all
four models of bonding and explain how
materials with these bonding types will
or will not dissolve or conduct electricity.
To support advanced learners on the
topic, ask them to investigate semicon-
ductor materials and create a presentation

item (e.g., brochure, Web site, skit) to explain bonding for semiconductors.

Pre-AP® Course Tip

Students preparing to take an AP® chemistry course or college chemistry should be able to connect atomic level phenomena and models to macroscopic phenomena. In this lesson, students examine data for different substances presented on a set of cards. They classify substances on the cards based on macroscopic properties and bonding models.

Lesson Guide

ENGAGE (5 min)

TRM PowerPoint Presentation 26

ChemCatalyst

A gold ring is made up of individual gold atoms.

1. What keeps the atoms together? Why don't they break apart from one another?

2. What parts of the atom do you think are responsible for keeping the atoms together in a solid?

Sample answers: **1.** Students might know that bonds keep the atoms together. If they say this, ask them what they mean by bonds. Students might say that the atoms are stuck together somehow or attracted to each other. Others might mention protons and electrons and charge. **2.** Answers will vary. The nucleus and valence electrons are involved; however, some students might mention one or the other of these structures or come up with an entirely different explanation, such as gravity.

→ Assist students in articulating their ideas about what holds atoms together in a solid.

Ask: What keeps the individual gold atoms in a chunk of gold together? Why don't they just fall apart?

Ask: Gravity holds objects to the ground, and neutrons help hold the nucleus together. What forces hold atoms together?

Ask: What keeps the Na and Cl atoms together in a solid chunk of salt?

EXPLORE (15 min)

TRM Worksheet with Answers 26

TRM Worksheet 26

Four Models of Bonding

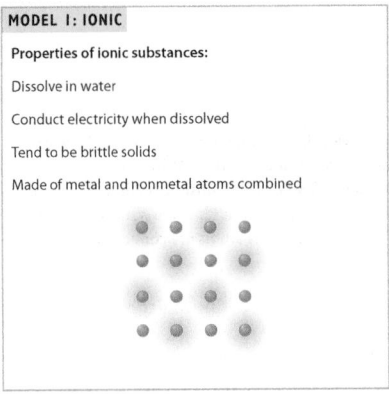

MODEL 1: IONIC

Properties of ionic substances:

Dissolve in water

Conduct electricity when dissolved

Tend to be brittle solids

Made of metal and nonmetal atoms combined

In **ionic bonding**, the valence electrons are transferred from one atom to another. Metal atoms transfer their valence electrons to nonmetal atoms.

MODEL 2: MOLECULAR COVALENT

Properties of molecular covalent substances:

Some dissolve in water, some do not

Do not conduct electricity

Some are liquids or gases

Made entirely of nonmetal atoms

In **molecular covalent bonding**, the valence electrons are shared between pairs or groups of atoms. This creates small stable units, called molecules, within the substance.

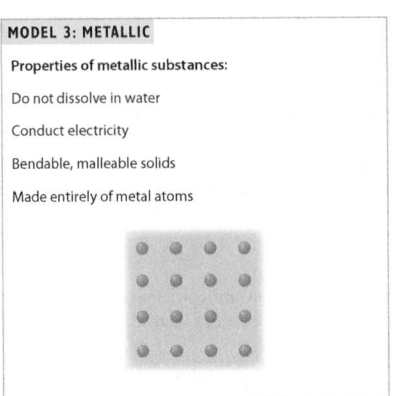

MODEL 3: METALLIC

Properties of metallic substances:

Do not dissolve in water

Conduct electricity

Bendable, malleable solids

Made entirely of metal atoms

In **metallic bonding**, the valence electrons are free to move about the substance.

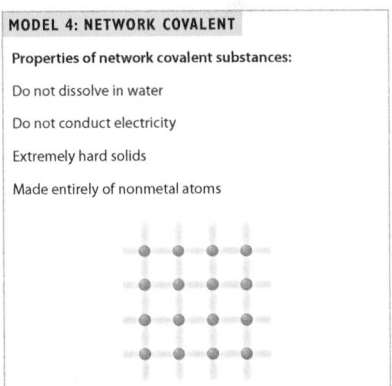

MODEL 4: NETWORK COVALENT

Properties of network covalent substances:

Do not dissolve in water

Do not conduct electricity

Extremely hard solids

Made entirely of nonmetal atoms

Network covalent bonding is similar to molecular covalent bonding, but the valence electrons are shared throughout the entire substance.

You have already been introduced to ionic compounds. These all have ionic bonding in which metal atoms transfer valence electrons to nonmetal atoms. The resulting oppositely charged ions are strongly attracted to each other. This attraction is what holds the ions together.

In **covalent bonding**, the nucleus of one atom is attracted to the valence electrons of another atom. Unlike in ionic bonding, one atom does not transfer an electron to the other. Instead, both atoms *share* the valence electrons between them.

130   Chapter 5 | **Building with Matter**

TRM Handout—Four Models of Bonding 26

TRM Card Masters—Substance Cards 26

INTRODUCE THE ACTIVITY

→ Arrange students into pairs.

→ Pass out the student worksheet and Substance Cards.

→ Before beginning the activity, define the term *chemical bond* and tell students that the negatively charged electrons and the positively charged nuclei are the atomic structures involved in bonding.

Chemical bond: An attraction between two atoms that holds them together in space.

→ Pass out the Handout—Four Models of Bonding. Display the transparency. Explain that it contains descriptions of four models of bonding between atoms. Each sphere in the drawing represents the nucleus and core electrons of an atom. The gray shaded areas represent where the negatively charged valence electrons might be found in each substance.

→ Make sure students have a basic grasp of what they are looking at before they move on to the activity.

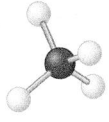

Methane, CH₄

Covalent bonding can happen in two different ways. In molecular covalent bonding, the atoms bond to form individual clusters called **molecules,** such as the methane molecule shown here.

In network covalent bonding, the valence electrons are shared between atoms but form a highly regular extended network, creating a very durable structure. Diamond consists of carbon atoms that are covalently bonded in a network.

In a metal, the valence electrons are distributed throughout the substance in what is sometimes called a "sea" of electrons. The valence electrons are free to move throughout the substance. The atoms are bonded by the attraction between the positively charged atoms and the negatively charged "sea" of electrons.

Diamond consists of carbon atoms that are covalently bonded in a network.

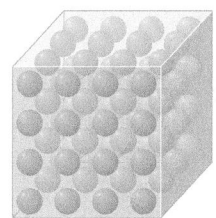

In a metal, positively charged atoms and a negatively charged "sea" of atoms create bonds.

⮌ Relating Bonds and Properties

Some properties of substances, such as solubility and conductivity, are directly related to the type of bonds the atoms in the substances have. Therefore, it is possible to match the bond type to the physical properties observed in different types of substances. Examine what happens to each type of substance when it is struck by a hypothetical hammer.

Ionic substances
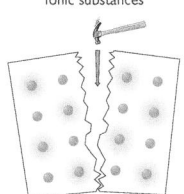
Hard but brittle, tend to fracture along planes of atoms

Network covalent substances
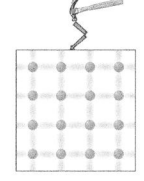
Durable, rigid, difficult to break

Metallic substances
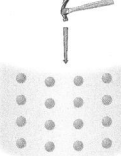
Bendable, malleable

Molecular covalent substances
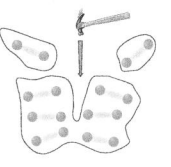
Often gases or liquids, or soft solids

Notice that the hardest substance is a solid with network covalent bonding. This is because bonding in these substances is in an organized network.

Lesson 26 | **Bonding**   131

Key Points: The different locations of the electrons among atoms account for many different properties of substances. In general, there are four main ways that atoms can be "glued" together: through ionic bonds, a network of covalent bonds, metallic bonds, or molecular covalent bonds.

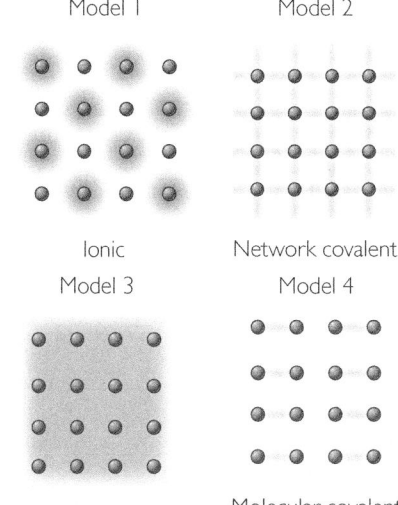

Model 1 — Ionic

Model 2 — Network covalent

Model 3 — Metallic

Model 4 — Molecular covalent

Ionic bonding: In Model 1, metal atoms and nonmetal atoms (with gray shading around them) are regularly distributed throughout the ionic compound. The gray shading shows that the valence electrons are concentrated around the nonmetal atoms. The transfer of valence electrons that occurs in an ionic bond results in a positive charge on the metal atoms and a negative charge on the nonmetal atoms. The strong attraction between the oppositely charged ions is the force that holds the ions together in the compound. This force of attraction, or bond, makes ionic solids fairly hard, although they are brittle and will shatter when struck hard. When ionic solids make contact with water, they usually dissolve. The atoms that make up the compound remain ions, with charges, but the individual ions separate from each other and become surrounded by water molecules and distributed throughout the water. Once dissolved, ionic compounds will conduct electricity.

Ionic bonding: A type of chemical bonding that is the result of transfer of electrons from one atom to another.

Network covalent bonding: Models 2 and 4 both show covalent bonding. In covalent bonding, valence electrons are "shared" among the atoms. In network covalent bonding, valence electrons are shared among atoms throughout the substance. This results

---

→ Ask students to sort the 16 cards into the four bonding categories.

EXPLAIN & ELABORATE (20 min)

TRM Transparency—Solubility and Conductivity 25

PROCESS THE FOUR MODELS OF BONDING

→ Display the Handout—Four Models of Bonding again. Use the models to explain the properties of various substances.

→ You might want to write a brief definition of each model of bonding on the board.

→ You could also bring in physical samples of each type of substance to demonstrate the properties more clearly.

Ask: What do the four models of bonding attempt to show?

Ask: How might the location of the electrons in a metallic substance account for the flexibility and softness of copper metal?

Ask: How might the location of the electrons in a network covalent solid account for the amazing hardness of a diamond?

Ask: How does the model of molecular covalent bonding account for the fact that these substances often are gases or liquids?

Lesson 26 | **Bonding**   131

in a very uniform and durable structure. Network covalent substances typically are solids at room temperature. They are very hard and do not dissolve in water. The valence electrons are not free enough to move about the substance and conduct electricity. A diamond is a classic example of a network covalent solid. It is made up solely of carbon atoms bonded to other carbon atoms in all directions.

Covalent bonding: A type of chemical bonding in which one or more pairs of valence electrons are shared between the atoms. Covalent bonding can be molecular covalent or network covalent.

Metallic bonding: Model 3 shows metal atoms surrounded by a "sea" of valence electrons. The atoms can have a momentary positive charge because the valence electrons move around in this "sea." Metals are hard but bendable and malleable. They usually dent when you strike them with a hammer. They also are ductile, meaning they can be drawn into wires. Finally, metallic substances conduct electricity. The distribution of the valence electrons throughout the substance allows the electrons to move freely.

Metallic bond: A bond between metal atoms in which the valence electrons are free to move throughout the substance.

Molecular covalent bonding: Models 2 and 4 show examples of covalent bonding. In covalent bonding, the valence electrons are "shared" among the atoms. Unlike the atoms in network covalent substances, atoms in molecular covalent substances are not connected covalently throughout the substance. Instead, the atoms are covalently connected in many identical units called *molecules*. These units may be composed of only two atoms, as in chlorine gas, Cl_2, or they may be composed of dozens of atoms, as in polyethylene plastic. These substances often are gases or liquids at room temperature. Olive oil, gasoline, butter, propane gas, and plastic bags are all made up of molecules.

Molecule: A group of atoms covalently bonded together.

Bonding can also help to explain the properties of dissolving and conductivity. Examine the illustration below representing dissolving. Water is represented by the lighter blue areas. Ionic solids and molecular covalent substances dissolve in water. Metallic solids and network covalent solids do not.

Ionic substances	Network covalent substances	Metallic substances	Molecular covalent substances

Dissolve into metal and nonmetal ions	Do not dissolve	Do not dissolve	Dissolve and molecules scatter in water

Conduction requires the movement of a charged particle, either an ion or an electron. Metals conduct electricity because the valence electrons are free to move throughout the solid. Ionic compounds that have been dissolved in water conduct electricity because the cations and anions are free to move in the solution. Network covalent solids and molecular covalent substances do not conduct electricity. The charge cannot move in these substances because the electrons are "stuck" between the atoms and are not available to move.

The periodic table is a valuable tool in figuring out bonding. You can use the table to determine if the elements in a compound are metals, nonmetals, or both.

- Ionic compounds, such as salts, are made from metal and nonmetal elements.
- Metallic compounds, such as brass, are made only of metal atoms.
- Network covalent compounds, such as diamonds, and molecular covalent compounds, such as methane, are made from nonmetals.

Example

Identifying the Type of Bonding

Determine the bonding in each of the following substances. What general physical properties can you expect of each substance?

    a. magnesium chloride, $MgCl_2$

    b. rubbing alcohol, C_3H_8O

Solution

    a. Magnesium chloride, $MgCl_2$, is an ionic compound with ionic bonding because it is made of a metal and a nonmetal element. It is probably brittle, dissolves in water, and conducts electricity when dissolved.

    b. Rubbing alcohol, C_3H_8O, is a molecular covalent substance with molecular covalent bonding because it is made entirely of nonmetal atoms and is a liquid.

RELATE BONDING TO SOLUBILITY AND CONDUCTIVITY

→ Use the Transparency—Solubility and Conductivity from Lesson 25 to show the relationship between bonding and properties such as solubility and conductivity.

Ask: If you know the chemical formula of a substance, what can you figure out about its properties?

Key Points: The chart created in the previous lesson can now be labeled with the four types of bonds.

BONDING ALSO RELATES TO THE TYPE OF ATOM IN THE SUBSTANCE—METAL OR NONMETAL

metal atoms only: metallic bonding
nonmetal atoms only: network covalent bonding, molecular covalent bonding
metal and nonmetal atoms: ionic bonding

Most substances can be classified this way, with a few exceptions. Also, once the type of bonding in a substance is known, many properties of that substance are known, too.

KEY TERMS

chemical bond

ionic bonding

molecular covalent bonding

metallic bonding

network covalent bonding

covalent bonding

molecule

LESSON SUMMARY

How are atoms connected to one another?

Atoms in substances are held together by chemical bonds. Chemists have identified four main models of bonding within substances: ionic, network covalent, molecular covalent, and metallic. Many properties of a substance correspond to the type of bonding between the atoms of that substance.

Exercises

Reading Questions

1. Explain why substances do not simply crumble into piles of atoms.

2. Name the four models of bonding and explain them in your own words. Be specific about the location of the valence electrons.

Reason and Apply

3. Determine the type of bonding in each substance.
   a. zinc, $Zn(s)$
   b. propane, $C_3H_8(l)$
   c. calcium carbonate, $CaCO_3(s)$

4. Based on physical properties, which of these substances is an ionic compound? Explain your reasoning.
   A. hair gel   B. silver bracelet   C. motor oil   D. baking soda

5. In the copper cycle from Lesson 7, you saw that nitrogen dioxide, $NO_2(g)$, formed when copper, $Cu(s)$, was dissolved in nitric acid. How would you classify the bonding in $NO_2(g)$? Explain.

6. Which statement is true?
   (A) Aqueous solutions of calcium chloride, $CaCl_2$, conduct electricity.
   (B) Glass, made of silicon dioxide, SiO_2, does not dissolve in water.
   (C) Ethanol, C_2H_6O, dissolves in water but does not conduct electricity.
   (D) Brass, also called copper zinc, $CuZn$, conducts electricity.
   (E) All of the above are true.

7. Suppose you have a mixture of sodium chloride, $NaCl$, and carbon, C. Explain how you can use water to separate the two substances.

8. Explain why copper is used as wire, but copper chloride is not.

9. Explain why carbon is a solid and not a gas.

10. Will each of these substances dissolve in water? Explain your thinking.
    a. Ca b. $NaNO_3$ c. Si d. CH_4 e. $CuSO_4$

Lesson 26 | **Bonding**   133

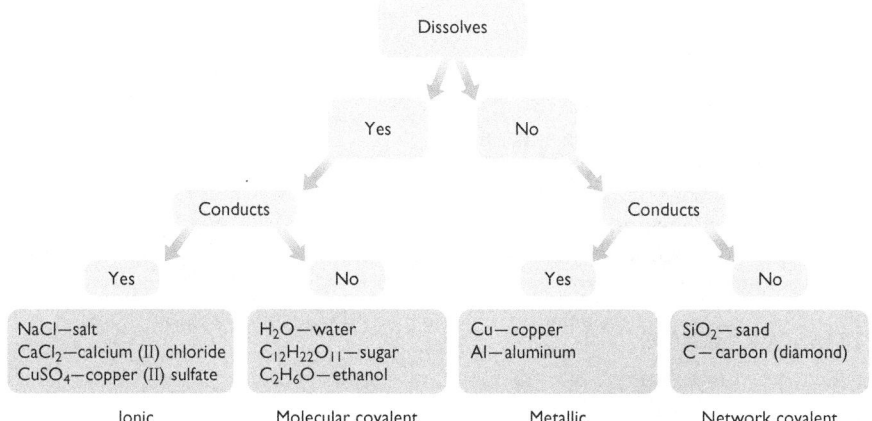

LESSON 26

EVALUATE (5 min)

Check-In

Imagine that you have a mystery substance that does not dissolve in water and does conduct electricity.

1. What type of bonding will you probably find in your substance? Explain.

2. List one other property of your mystery substance.

Answers: 1. Metallic substances do not dissolve in water and do conduct electricity. Thus, the atoms of the mystery substance probably are held together by metallic bonding. 2. Sample answer: It will also be easy to bend.

Homework

Assign the reading and exercises for Alchemy Lesson 26 in the student text.

LESSON 26 ANSWERS

1. The atoms that make up substances are held together by chemical bonds. The bond is an attraction between the positively charged nuclei of atoms and the valence electrons of other atoms.

2. *Possible answers:* In ionic bonding, one or more valence electrons are transferred from a metal atom or group of atoms to a nonmetal atom or group of atoms, forming a positively charged cation and a negatively charged anion. The bond forms due to the attraction of the opposite charges. ● In metallic bonding, valence electrons from all of the metal atoms move freely among the cations that are formed when the electrons leave the metal atoms. The bond forms because of the attraction of the positively charged cations to the sea of negatively charged electrons. ● In network covalent bonding, valence electrons from each nonmetal atom are shared with all of the atoms around it. The structure of the substance is very rigid, since each atom bonds with multiple neighboring atoms. ● In molecular covalent bonding, valence electrons from a nonmetal atom are shared with other individual atoms, forming groups with a specific number and arrangement of atoms, called *molecules*.

Answers continue in Answer Appendix (p. ANS-2).

Lesson 26 | **Bonding**   133

Overview

LAB: PAIRS

Key Question: How can you extract an element from a compound?

KEY IDEAS

Atoms are not destroyed when they combine with other atoms to form compounds. Elements can be extracted from compounds. Cations are simply metal atoms with charges on them. Using electrochemistry, you can "add" electrons to cations and retrieve the elemental metal from the compound in solution.

LEARNING OBJECTIVES

● Assemble an electroplating apparatus.
● Explain how to extract elemental metal from an ionic compound through electroplating.

KEY TERM

electroplating

IN CLASS

Students will use an electroplating setup to observe the transfer of electrons in copper atoms. The purpose of this procedure is to extract an elemental metal from an ionic solution. Nickel strips serve as the plating surface for the copper metal. The electrical leads are reversed several times so students can observe the copper atoms going in and out of solution as the charge on the atoms changes. (Optional: You can have students use the clear nail polish to make a design on the nickel electrode. Students then can use their apparatus to create a small copper-plated art piece. Alternatively, you can have students copper-plate a nickel or a dime.)

Note: This lesson is optional. If you are short on class time, you can simply assign the reading in the student edition. A complete materials list is available for download.

TRM Materials List 27

SETUP

Cut enough nickel strips for each class. Standard first-aid or paramedic scissors can be used. Prepare the copper (II) sulfate solution by dissolving 840 g of copper (II) sulfate in 3 L of water. Add 240 mL of 6 M H_2SO_4 and 104 mL of 0.1 M HCl. Finally, add enough water to bring the volume to 4 L. This will make enough solution for a class of 32 students. The solution can be reused for multiple classes.

Electroplating Metals

Purpose

To use electrochemistry to extract metals from ionic compounds in a solution.

Procedure

1. Set up the electroplating apparatus as shown. Observe what happens.

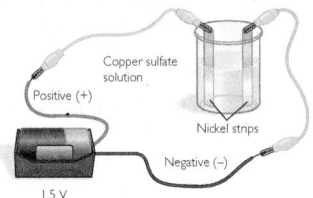

2. Switch the sides of the battery to which the two alligator clips are attached. Wait at least one minute or until you notice a change.

3. Reverse the wiring back to its original position.

Observations

1. What did you observe when you hooked up the nickel strips to the battery?

2. What happened when you reversed the flow of electricity?

3. Where does the copper come from that ends up on the nickel strip?

4. What is in the copper sulfate solution?

5. Write a short paragraph explaining your observations.

6. **Making Sense** Are copper atoms and copper ions the same element? Explain your thinking.

7. **If You Finish Early** Consider a sample of gold chloride, $AuCl_3$. Explain what procedure you might follow in order to extract solid gold from the compound.

◆ SAFETY Instructions

Wear safety goggles at all times.

The solution contains acid. Handle carefully. Rinse the nickel strips after they have been in the copper solution.

Materials

- copper sulfate plating solution
- 250 mL beaker
- 2 nickel strips (cut into about 1-by-3-in. strips)
- 1.5-volt D-cell battery with holder
- 2 insulated wires with alligator clips

CLEANUP

Reuse: Save the copper (II) sulfate solution for reuse. Use a copper strip as one electrode, and reverse the current on the electroplating apparatus to clean off any nickel electrodes. This process takes about five minutes.
Dispose: Let other solutions evaporate, and dispose of toxic solids according to local guidelines.

Lesson Guide

ENGAGE (5 min)

TRM Transparency—ChemCatalyst 27

TRM PowerPoint Presentation 27

ChemCatalyst

1. What is the charge on the copper ions in this copper (II) chloride compound?

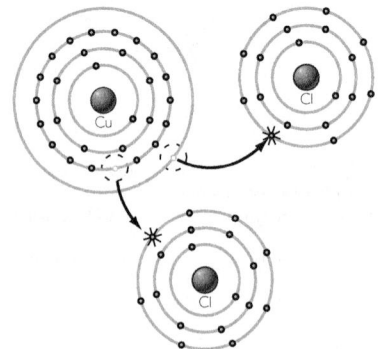

Electrons on the Move

Electroplating Metals

THINK ABOUT IT

Metals are important in our lives. They are used for everything from electrical wiring, jewelry, and soda cans, to cars, airplanes, and bridges. Most metals are dug out of the ground as ionic compounds. In ionic compounds, the metal atoms are bonded to nonmetal atoms. Through the ages, people have struggled to extract the pure metals from these compounds by figuring out ways to separate the metal atoms from the nonmetal atoms.

How can you extract an element from a compound?

To answer this question, you will explore

- The Quest for Precious Metals
- Electroplating Elemental Metals

EXPLORING THE TOPIC

The Quest for Precious Metals

Sometime around 6000 B.C.E., metals came into widespread use. Before that, most tools and implements were made from earth, bone, wood, and stone. The first metals to be discovered were gold, copper, and silver. Later, lead, tin, iron, and mercury were added to the list. Most of these metals are not readily found in nature in pure form.

The first metals that were used were the ones people stumbled upon in their pure form in the earth. Then, people discovered that, using heat, they could extract metals from other compounds dug out of the earth. For example, if iron (III) oxide, Fe_2O_3, is heated in the presence of charcoal, the oxygen in the compound is removed as CO_2 gas. This leaves behind solid iron, Fe. This process requires very high temperatures.

Molten iron being poured out to solidify.

For nearly 7000 years, only a handful of metals was in widespread use. Other metals, such as aluminum, Al, could not be extracted easily by heating and remained unavailable.

Electroplating Elemental Metals

USING ELECTRICITY TO EXTRACT METALS

Over time, the construction of better furnaces has resulted in the extraction of a variety of metals. More recently, it was discovered that electricity could be used to

HISTORY CONNECTION

The first metal tools, implements, and weapons were made from copper. Gold, copper, silver, tin, lead, iron, and mercury are often referred to as the Metals of Antiquity. For some 7000 years, these seven metals were the only ones in widespread use. These metals were known to the Mesopotamians, Greeks, Egyptians, and Romans.

INTRODUCE THE LAB

→ Ask students to work in pairs.

→ Pass out the student worksheet.

→ Go over safety guidelines. Students must wear safety goggles at all times. The solution contains acid, which is corrosive. Before handling the nickel strip, students must rinse it with water.

→ Demonstrate the assembly of the electroplating apparatus. Informally introduce the term *electrode* by telling students that in this apparatus the nickel strips are considered electrodes. They allow electricity to enter and leave the solution.

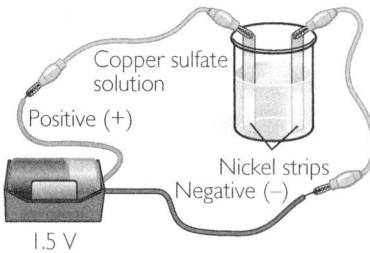

→ If you have time and enough nickel strips, you can let the students create a copper-plated art piece. Give students nail polish to make a design on the nickel strip. Copper will electroplate only where there is no nail polish. Give the students nail polish remover to remove the nail polish after the copper is plated. The result is a small piece of metal art.

TRM Transparency—Electroplating 27

DISCUSS STUDENTS' OBSERVATIONS

Ask: What did you observe when you attached the wires to the battery and the metal strips?

Ask: What happened when the wires were reversed?

Ask: Where did the copper metal come from that ended up on the nickel strip?

Key Points: It is possible to transform metal cations in solution into neutral metal atoms, using electricity. This is what you did in class. The metal cations in a copper sulfate solution gained electrons and were deposited as neutral copper metal onto one of the nickel strips. This process is called *electroplating*. When the nickel strips are connected to the opposite battery terminals, the plating process is reversed. The nickel strip that was plated with copper loses its copper,

---

2. How do you think we could get solid copper from a sample of copper (II) sulfate, $CuSO_4$?

Sample answers: **1.** The copper ions in copper (II) chloride have a +2 charge. **2.** Students might speculate that the way to get solid copper is to figure out how to give back electrons to the copper. Others might say to use chemical reactions similar to those in the copper cycle experiment.

→ Assist students in verbalizing their ideas about extracting copper from a copper compound.

Ask: What is happening in the copper (II) chloride illustration?

Ask: What is the chemical formula of copper (II) chloride? How do you know? ($CuCl_2$)

Ask: What happens when you remove electrons from an atom of copper?

Ask: What do you think you would have to do to extract elemental copper from ionic copper (II) chloride? Copper (II) sulfate?

TRM Worksheet with Answers 27

TRM Worksheet 27

and the other nickel strip becomes plated with copper. The only source of copper is in the solution in the form of copper ions.

DISCUSS THE PROCESS AT A PARTICLE LEVEL

→ Display the Transparency—Electroplating as you discuss what is going on at a particle level before and during the activity. Focus on the copper ions and atoms.

Ask: What is in the beaker?

Ask: Each nickel strip has either a positive or a negative charge after the battery has been hooked up. How can you figure out which charge is on which nickel strip?

Ask: How does this experiment provide evidence that electrons are transferred when ionic compounds are created?

Ask: What is the main difference between copper atoms and copper ions?

Ask: Is there a setup that might help you extract gold metal from gold chloride, $AuCl_3$? Describe it.

Key Points: Aqueous copper (II) sulfate, $CuSO_4\,(aq)$, is really copper cations, Cu^{2+}, and sulfate anions, SO_4^{2-}. There are molecules of water, H_2O, surrounding the cations and anions. (Acid is also added to the solution to facilitate the reaction.) In reality, there are more than a billion billion ions and water molecules in the beaker.

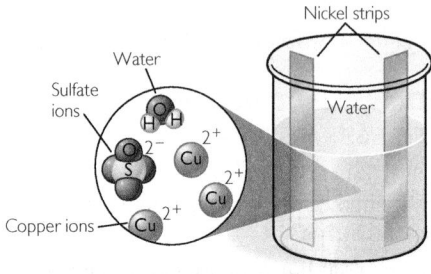

Before connecting to the battery

Once the battery is hooked up, one nickel strip has a positive charge, and the other has a negative charge. The current runs from the positive terminal of the battery, through the wires, to the positively charged nickel strip. The current then runs through the solution to the negatively charged nickel strip and back through the wire to the negative terminal of the battery. The copper cations, Cu^{2+}, are attracted to the nickel strip with the negative charge. Electrons are transferred from the nickel strip to the copper ions, Cu^{2+}, causing neutral atoms of copper to be deposited on the

extract metals from compounds. Recall that electrons are transferred from metal atoms to nonmetal atoms in ionic compounds. Scientists discovered they could "give" electrons back to the metal ions, re-forming the neutral metal atoms once again. For example, copper metal can be extracted from a solution of copper sulfate by running an electrical current through the solution. A simple setup is shown here.

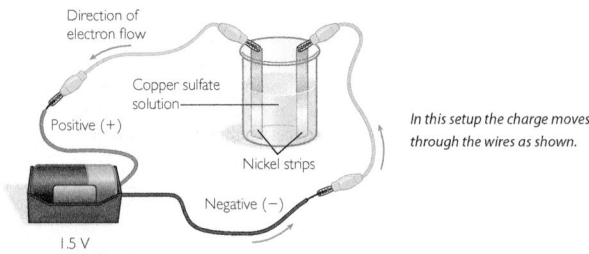

In this setup the charge moves through the wires as shown.

Once the battery is hooked up, one nickel strip has a positive charge and the other has a negative charge. In this electrical circuit, the electrons move from the negative terminal of the battery to the negatively charged nickel strip. Electrons also move from the other nickel strip toward the positive terminal of the battery. Ions moving in solution complete the circuit.

The positive copper ions, Cu^{2+}, are attracted to the nickel strip with the negative charge. Electrons are added to the copper cations in solution. Each copper ion that gains enough electrons becomes elemental copper. When the copper cations become elemental copper atoms, they come out of solution and coat the negatively charged nickel strip. This process is called **electroplating.**

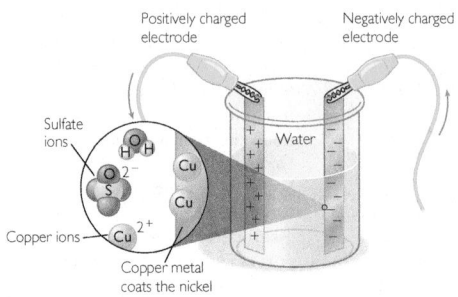

When the experiment shown is completed, the negatively charged nickel becomes coated with the pure copper metal.

You can reverse the process and produce metal cations from metal atoms. The battery has a positive and a negative terminal. By switching the wires in the test setup, you can change the direction the electrical charge moves. This causes the reaction to reverse. The copper coating comes off of the first nickel strip and reenters the solution. Copper from the solution plates onto the other nickel strip.

strip. This is visible as a coating that looks like copper metal. Matter is conserved in this process. No new atoms or new elements are made.

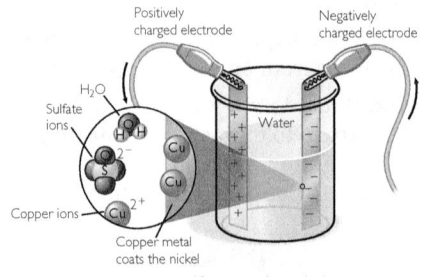

After connecting to the battery

Many elements are found in nature only in combination with other atoms

in compounds. For example, while aluminum is the most abundant metallic element in Earth's crust, it is never found in a pure state in nature. All of Earth's aluminum is found as aluminum ore in the form of compounds such as aluminum oxide, Al_2O_3. To obtain a pure sample of aluminum metal, the aluminum atoms have to be extracted from aluminum ore. Electroplating methods commonly are used to extract metals from ores. While you cannot make gold by moving electrons, you can plate thin layers of gold on jewelry. If you dissolve a gold compound, such as $AuCl_3$, and do an experiment

ELECTROPLATING GOLD

Is it possible to design an apparatus similar to the one in the Lab: Electroplating Metals to extract gold atoms from ionic compounds containing gold ions? Yes, by all means. However, gold is one of those few elements that *is* commonly found in nature in its elemental form. Gold *compounds* are seldom found in nature. This is because gold is fairly stable as it is. So harvesting gold atoms from gold ions is not the best way to make a pile of gold bars.

However, electroplating techniques can be used to make a little bit of gold go a long way. Using electroplating or other plating techniques, less expensive metals can be covered with thin layers of gold atoms.

Today, electroplating is an extremely important industrial process. For example, aluminum metal has been in widespread use only recently because chemists finally found a way to extract it from ionic compounds using electrochemical methods. Electroplating is also used to coat electronic parts in computers.

Copper bowl plated with 24 karat gold

Example

Plating Gold from AuCl₃

Suppose that you have a sample of $AuCl_3(s)$.

  a. What is the charge on the Au ions in $AuCl_3$?
  b. What is the chemical name of $AuCl_3$?
  c. Do you expect $AuCl_3$ to conduct electricity as a solid? Explain your thinking.
  d. Do you expect $AuCl_3$ to dissolve in water? Explain your thinking.
  e. Describe how you can extract gold from $AuCl_3$.

Solution

  a. You need to find the charge on the main group element first. Cl is in Group 7B in the periodic table, which means that it has seven valence electrons. So, Cl needs to gain an electron, which will make it Cl^-. The Au becomes Au^{3+} because the charges in an ionic compound must add up to zero.

  b. The gold has a $+3$ charge. So, the name is gold (III) chloride.

  c. $AuCl_3$ is an ionic solid. Ionic solids do not conduct electricity. However, if you dissolve $AuCl_3$ in water, the solution will conduct electricity.

  d. Because $AuCl_3$ is ionic, you can assume it will dissolve in water, forming Au^{3+} and Cl^- ions.

  e. Gold can be plated onto a piece of nickel by using a battery to move electrons through a solution containing $AuCl_3$.

similar to the copper plating experiment, it is possible to coat metals with gold. Thus, even though you cannot create new gold, you can make other objects look like gold.

EVALUATE (5 min)

Check-In

1. What is required to transform $CuCl_2(aq)$ into $Cu(s)$?

2. What is required to transform $CuCl_2(aq)$ into $Au(s)$?

Answer: **1.** The Cu^{2+} cations in $CuCl_2(aq)$ can be transformed into $Cu(s)$ by adding two electrons to the copper cations. This can be done with a battery and electroplating setup. **2.** Transforming $CuCl_2$ into $Au(s)$ requires changing the nucleus of either Cu or Cl by nuclear chemistry. The correct number of protons (and neutrons) needs to be added. This is not feasible.

Homework

Assign the reading and exercises for Alchemy Lesson 27 and the Chapter 5 Summary in the student text.

LESSON 27 ANSWERS

1. *Possible answers (any 3):* ● finding pure metals in nature ● heating ionic compounds to separate the metal ● extracting metals with electricity ● reusing and recycling discarded metals

2. *Possible answer:* Place two nickel strips in a beaker of copper chloride solution. Connect the strips to the two terminals of a battery. Allow the battery current to flow until the strip connected to the negative terminal of the battery has become plated with copper.

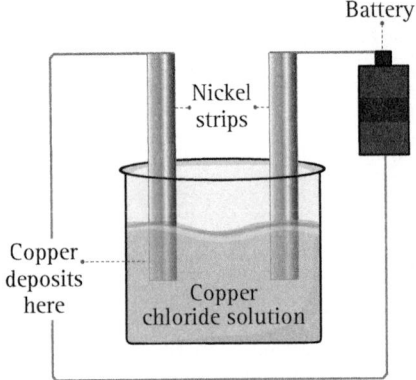

3. *Possible answer:* Attach the coated object to the positive terminal of a battery in an electroplating circuit. When an object is attached to the positive terminal, electrons flow from the object, forming cations. The metal ions are no longer attracted to the coated object, and they enter the solution.

4. A good lab report will contain: ● a title (Lab: Electroplating) ● a statement of purpose (*Possible answer:* To demonstrate how metals can be extracted from an ionic compound dissolved in water) ● a procedure (a summary of the steps followed in the experiment) results (check student observations to make sure they address all of the observation questions in the lab) ● a conclusion (that includes a statement about the attraction of the metal cations to the metal strip. *Possible answer:* When an ionic compound is dissolved in water, running an electric current through two metal strips and the solution will cause the metal cations to deposit on the negatively charged metal strip.).

5. The copper sulfate solution is composed of cations, Cu^{2+}, and anions, SO_4^-, dissolved in water. The only thing that is added to the solution during the experiment is a stream of electrons. When the electrons are added to the copper ions, copper atoms, Cu, are formed on the metal strip. This indicates that the Cu^{2+} ions are simply copper ions that are missing electrons.

6. *Possible answer:* The mass of the negatively charged nickel strip should increase as it is plated with copper. There is no change to the nickel itself, but as a layer of copper forms on top of the nickel, the mass will increase by the mass of the copper plating.

7. *Possible answer:* Nickel cannot change into copper unless the number of protons in the nucleus changes. The plating apparatus only adds electrons to the nickel strip, causing the plating to occur. Adding electrons does not change the nucleus, so it cannot change the identity of the atoms. You cannot be sure simply from observation that the strip is copper-coated nickel and not pure copper. Reversing the experiment only proves that the copper appears and disappears on the strip. Observations must be taken into account along with knowledge about the structure of atoms. Taking the copper-plated strip and cutting it or measuring its density to prove that the nickel is still there is not a valid answer because it does not explain why the nickel had to remain intact.

How can you extract an element from a compound?

KEY TERM

electroplating

Most metal atoms are found in nature combined with other atoms, in the form of ionic compounds. Some metals can be extracted from ionic compounds through heating. Elements can also be extracted from ionic compounds by using electricity to move electrons between atoms. Using electroplating, you can create a variety of substances that could be considered "as good as gold."

Exercises

Reading Questions

1. Describe three ways in which people have managed to obtain pure metals to use for making tools and other objects.

2. Explain how to plate copper metal onto a nickel strip. Write a simple procedure and draw a labeled sketch showing how to carry it out.

3. Explain how you can *remove* a metal coating from an object.

Reason and Apply

4. **Lab Report** Complete a lab report for the copper plating experiment. In your report, give the title of the experiment, purpose, procedure, observations, and conclusions.

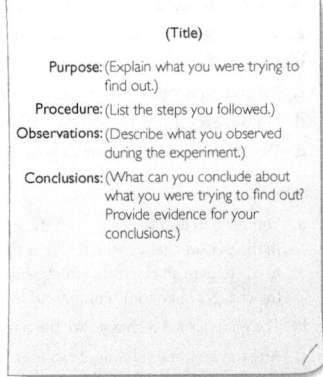

5. Explain how the copper plating experiment supports the claim that Cu^{2+} is just a Cu atom that is missing two electrons.

6. If you measure the mass of the nickel strip before and after plating copper, do you expect the mass to change? Explain your thinking.

7. Suppose that a classmate claimed that the nickel in the electroplating lab changed into copper. Provide an argument to prove that this is not true.

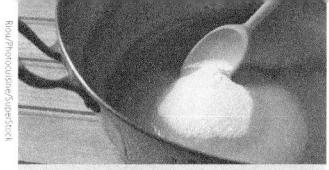

CHAPTER 5

Building with Matter

SUMMARY

KEY TERMS

dissolve

soluble

insoluble

conductivity

chemical bond

ionic bonding

molecular covalent bonding

metallic bonding

network covalent bonding

covalent bonding

molecule

electroplating

Alchemy Update

How can substances with new properties be made?

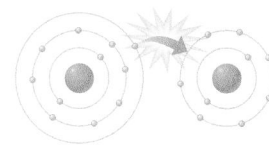

Sodium, Na   Fluorine, F

A bond is an attraction between the nucleus of one atom and the electrons of another atom. When atoms bond together, new substances with new properties are made. Conversely, elements can be extracted from compounds by breaking the bond.

There are four models of bonding, depending on the location of electrons within the atoms in the substance: ionic, molecular covalent, metallic, and network covalent. Certain properties are associated with each type of bonding.

Atoms form bonds with other atoms, creating a variety of substances with new properties. It is possible to make compounds that look like or behave like gold, such as iron pyrite, FeS_2, also called fool's gold. Also, you can use chemical bonding to make substances that are even more valuable than gold.

REVIEW EXERCISES

1. What does the phrase "as good as gold" mean as it applies to this class?

2. Describe how you could experimentally determine the type of bonding present in a substance.

3. Predict whether isopropanol, $C_3H_8O(l)$, will conduct electricity. State your reasoning.

4. Make a table like this one. Fill it with the model of bonding that fits the category.

	Dissolves	Does not dissolve
Conducts		
Does not conduct		

4.

	Dissolves	Does not Dissolve
Conducts	Ionic (conducts in solution)	Metallic
Does Not Conduct	Molecular covalent	Network covalent

ASSESSMENTS

Two multiple-choice assessments (versions A and B) and two short-answer assessments (versions A and B) are available for download for Chapter 5.

TRM **Chapter 5 Assessment Answers**

TRM **Chapter 5 Assessments**

ANSWERS TO CHAPTER 5 REVIEW EXERCISES

1. *Possible answer:* While it is not practical to try to make gold, many substances that are quite valuable can be made through chemistry.

2. *Possible answer:* Determine whether the material conducts electric current, whether it dissolves in water, and if it does, whether the solution conducts electric current. ● If a material does not conduct electricity unless it is in solution, it has ionic bonding. ● If a material conducts electricity but does not dissolve, it has metallic bonding. ● If a material does not conduct electricity even when in solution, it has molecular covalent bonding. ● If a material does not conduct electricity and does not dissolve, it has network covalent bonding.

3. *Possible answer:* Isopropanol will not conduct electricity because compounds with only nonmetal atoms have covalent bonding and do not conduct electricity. Also, isopropanol is a liquid at room temperature, which means it is not an ionic compound.

Overview

CLASSWORK: PAIRS

Key Question: Why were the alchemists unable to make gold from other substances?

KEY IDEAS

Students review what they learned in Unit 1: Alchemy.

LEARNING OBJECTIVE

Create a list of topics and concepts to study for an upcoming exam.

FOCUS ON UNDERSTANDING

Provide clarity for students regarding the alchemists' quest. It is possible to create gold using nuclear reactions. This happens in the stars. But it is very difficult to reproduce this outcome on Earth.

IN CLASS

Students review what they have learned in the unit about atoms, elements, compounds, and the periodic table. An extensive discussion follows to clear up any conceptual misunderstandings and to clarify the scope of the unit exam. A complete materials list is available for download.

> **TRM** Materials List R1

GUIDE TO UNIT REVIEW

ENGAGE (5 min)

ChemCatalyst

1. Name three topics that might be on an exam covering the entire alchemy unit.

2. Write a problem or question that could be included on the exam.

Sample answers: **1.** Atomic structure, nuclear reactions, ion formation, the periodic table, valence electrons, and so on. **2.** Use the periodic table to figure out the most common isotope of bromine.

→ Write general topics on the board as students share their answers. This becomes a review list.

EXPLORE (20 min)

> **TRM** Worksheet with Answers R1

> **TRM** Worksheet R1

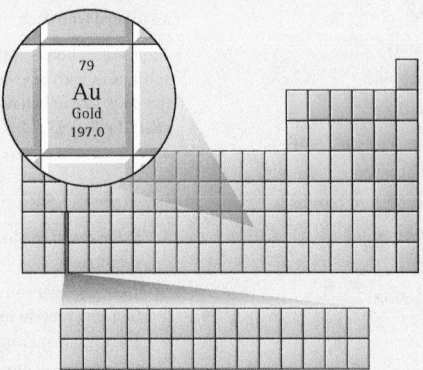

Alchemy | REVIEW

UNIT 1

Matter is anything that has substance and takes up space. Mass is a measure of the substance of matter. Volume is a measure of the space that matter takes up. The density of a substance is the mass per unit of volume.

Everything on the planet is composed of elements or compounds made from combinations of these elements. The elements are organized according to their properties and atomic structure into a chart called the periodic table. The periodic table holds a wealth of information about the elements and their atoms. Elements that are located in the same column of the periodic table tend to have similar behavior and reactivity. The noble gases are very stable elements with extremely low reactivity.

Matter is composed of individual building blocks called atoms. Atoms are much too small to be seen, so experimental evidence has led to various models of the atom. In the center of each atom is a dense nucleus consisting of protons and neutrons. Most of the mass of an atom is located in its nucleus. Most atoms have more neutrons than protons in their nuclei, which keeps the nucleus more stable.

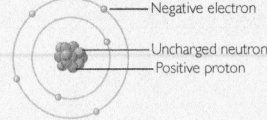

— Negative electron
— Uncharged neutron
— Positive proton

Electrons are located at specific distances from the nucleus, called shells. The electrons in the outermost shell of an atom are called its valence electrons and are key to its behavior. Atoms will sometimes lose or gain valence electrons, forming ions with positive and negative charges, and achieving the electron arrangements of the noble gases.

A compound is a substance that contains atoms of more than one type of element, bonded together. There are millions of compounds in existence and many yet to be discovered.

INTRODUCE THE CLASSWORK

→ Ask students to work in pairs.

→ Pass out student worksheets.

EXPLAIN & ELABORATE (20 min)

REVIEW THE KEY CHEMISTRY CONCEPTS

→ Create a list of topics and terms on the board as a study guide.

→ Address any questions or sources of confusion that came up during the completion of the worksheet.

Ask: How is an atom different from an element?

Ask: What are structures found in an atom?

Ask: What is a compound?

Ask: What types of bonds can form between atoms?

Ask: What is happening with the electrons in each type of bond?

Ask: If you know an atom's position in the periodic table, what else do you know?

Ask: What is the nuclear equation for the beta decay of nitrogen-16?
$(^{16}_{7}N \rightarrow {}^{0}_{-1}e^{-} + {}^{16}_{8}O)$

A bond is an attractive force that keeps atoms connected to one another. The attraction occurs between the positively charged nuclei and the negatively charged electrons in different atoms. Usually the valence electrons are involved in bonding.

There are four main models for bonding: ionic, molecular covalent, metallic, and network covalent. These models can be used to explain difference in solubility and conductivity.

REVIEW EXERCISES

General Review

Write a brief and clear answer to each question. Be sure to show your work.

1. What is the difference between an element and a compound?

2. Consider two objects that each weigh 20.0 g. One is made of lead and has a density of 11.4 g/mL. The second object is made of aluminum and has a density of 2.7 g/mL. Which object takes up more space? What volume does each occupy?

3. Use the periodic table to help you find the atomic symbol, atomic number, group number, number of protons, and number of electrons for these elements:

   a. lithium   **b.** bromine   **c.** zinc
   d. sulfur   **e.** barium   **f.** carbon

4. Describe nuclear fission and nuclear fusion.

5. What is an isotope? How can you figure out the most common isotope of an element?

6. Describe how you would use the periodic table to help you predict the type of bond between two atoms.

7. What are cations and anions?

8. Sort the following atoms into metals and nonmetals. Then name three ionic compounds that can be formed from some combination of these atoms.

   Na   S   Cl   Sr   Mg   Se   I   Cu

9. Predict the type of bonding in each of the following substances. Provide their chemical names and predict which ones will conduct electricity.

   a. $MgCl_2$ **b.** O_2
   c. $AgOH$ **d.** Pt

10. Explain how you can determine the charge on a transition metal ion.

11. A substance does not dissolve in water and does not conduct electricity. What model of bonding accounts for these observations?

12. What is required to change one element into another element? Is it possible to do this in a laboratory?

STANDARDIZED TEST PREPARATION

Multiple Choice

Choose the best answer.

1. The density of zinc is 7.1 g/cm³ and the density of copper is 9.0 g/cm³. What statement correctly describes the density of a zinc block coated with copper?

   (A) The block has a density of 7.1 g/cm³.
   (B) The block has a density less than 7.1 g/cm³.
   (C) The block has a density between 7.1 g/cm³ and 9.0 g/cm³.
   (D) The block has a density of 9.0 g/cm³.

Ask: What properties might make gold a valuable substance?

List of Possible Topics for Review

- Defining matter
- Measuring matter—mass, volume, and density
- Chemical symbols
- Conservation of matter
- Organization of the periodic table
- Atomic structure
- Isotopes
- Nuclear changes
- Properties of substances—conductivity and solubility
- Chemical bonds—ionic, metallic, molecular covalent, network covalent

- Ionic compounds—charges, valence electrons, polyatomic ions
- Electron arrangements—shells, subshells

REVIEW THE PURSUIT OF MAKING GOLD

Ask: Why do you think the alchemists were unsuccessful in creating gold from other substances?

Ask: Can you convert copper into gold? Why or why not?

Ask: What is required to change one element into another?

Ask: Would you say that chemistry is more about moving protons around

or moving electrons around? Explain your thinking.

Key Points: An atom of gold has exactly 79 protons in its nucleus, which is what makes it gold and not some other element. It is not possible to convert copper into gold in chemical reactions. It can be done only by nuclear reactions, which result in changes to the nuclei of atoms. Nuclear reactions can be challenging to control and require an enormous amount of energy and very specific conditions.

Chemistry can still be used to create things that are as good as gold. Atoms of one element can be combined with atoms of other elements to make new substances with new properties. In chemical reactions, the nucleus is not changed, and atoms retain their identities. The manner in which atoms combine is dictated by how many electrons they have and where those electrons are located. Electrons are the glue that holds atoms together in bonds. Thus, most of the changes we observe in chemistry are associated with electrons.

EVALUATE

There is no Check-In for this review.

Homework

Assign the Unit 1: Alchemy Review in the student text in preparation for the unit exam.

ANSWERS

GENERAL REVIEW

1. *Possible answer:* An element is the basic building block of compounds. An element has only one type of atom, while a compound has at least two types of atoms held together by chemical bonds.

2. lead: $V = 1.75$ mL; aluminum: $V = 7.41$ mL; the aluminum object takes up more space.

3. **a.** Lithium, Li, has atomic number 2 and is in Group 2A. It has 2 protons and 2 electrons. **b.** Bromine, Br, has atomic number 35 and is in Group 7A. It has 35 protons and 35 electrons. **c.** Zinc, Zn, has atomic number 30 and is in Group 2B. It has 30 protons and 30 electrons. **d.** Sulfur, S, has atomic number 16 and is in Group 6A. It has 16 protons and 16 electrons. **e.** Barium, Ba, has atomic number 56 and is in Group 2A. It has 56 protons and 56 electrons. **f.** Carbon,

C, has atomic number 6 and is in Group 4A. It has 6 protons and 6 electrons.

4. Nuclear fission occurs when a nucleus breaks apart to form two nuclei with smaller atomic numbers than the original nucleus. Nuclear fusion occurs when two nuclei join together to form a nucleus with a greater atomic number.

5. An isotope is an atom of an element with a specific number of neutrons in its nucleus. You can predict the most common isotope by rounding the average atomic mass of the element to the nearest whole number.

6. *Possible answer:* ● If both atoms are metal atoms, then the atoms will form a metallic bond. ● If one atom is a metal and the other atom is a nonmetal, then the two atoms might form an ionic bond. ● If both atoms are nonmetal atoms, then the atoms might form a covalent bond. ● If either atom is a noble gas, the atoms will probably not form any bond.

7. Cations are ions that have lost electrons, causing them to have a positive charge. Anions are ions that have gained electrons, causing them to have a negative charge.

8. ● metals: Na, Sr, Mg, Cu ● nonmetals: S, Cl, Se, I ● Possible answer (any 3): sodium chloride, NaCl; magnesium iodide, MgI_2; strontium sulfide, SrS; and copper (II) chloride, $CuCl_2$.

9. a. Aluminum is a metal and chlorine is a nonmetal, so aluminum chloride has ionic bonding and conducts electricity in solution only. **b.** Oxygen is made up only of nonmetal atoms and is a gas, so it has molecular covalent bonding and does not conduct electricity. **c.** Silver (I) hydroxide has ionic bonding and conducts electricity in solution only. **d.** Platinum is a metal element, so it has metallic bonding and conducts electricity.

10. *Possible answer:* You can determine the charge on a transition metal ion by the formula of its salt. The total charge of the ionic compound must be zero, so the positive charge on the metal ion is the sum of the negative charges on the anions divided by the number of metal ions in the formula.

11. A material that does not dissolve in water and does not conduct electricity is held together by network covalent bonds or molecular covalent bonds. Although some materials with molecular covalent bonds dissolve in water, others do not.

12. One element can be changed into another only by changing the number of protons in the nuclei of its atoms. This process requires enormous amounts of energy.

---

2. According to an ancient story, Archimedes had his "crowning moment" when he discovered the trickery of a goldsmith by using density. The goldsmith had received an order for a pure gold crown but crafted one from a mixture of metals. The volume of this crafted crown would have differed from the volume of a pure gold crown. If the mass of a pure gold crown was 1.2 kg, and the density of gold is 19.3 g/cm^3, what would be the volume of a pure gold crown?

(A) 0.062 cm^3 (B) 62.2 cm^3
(C) 23.16 cm^3 (D) 16.08 cm^3

3. The shell model below has 3 protons, 3 neutrons, and 3 electrons. What statement correctly describes this atom?

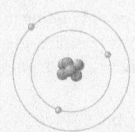

(A) The atomic number of this element is 6 because there are 3 protons and 3 neutrons; and the mass number is 3 because there are 3 protons.

(B) The atomic number of this element is 3 because there are 3 protons, 3 neutrons, and 3 electrons; and the mass number is 9 because there are 3 of each subatomic particle.

(C) The atomic number of this element is 3 because there are 3 protons; and the mass number is 6 because there are 3 protons and 3 neutrons.

(D) The atomic number is 6 because there are 3 protons and 3 electrons; and the mass number is 6 because there are 3 protons and 3 electrons.

4. A shell model for a carbon atom is shown. Which statement correctly describes the electrons in carbon?

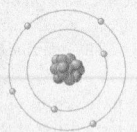

(A) Carbon has 6 core electrons and no valence electrons because there are a total of 6 electrons in the neutral atom.

(B) Carbon has 12 electrons because the mass number is 12.

(C) Carbon has a core similar to the helium atom, but carbon has 4 valence electrons.

(D) Carbon has a core similar to the neon atom, but carbon has 4 valence electrons.

5. The element in the shell model below is a poor conductor of heat and electricity and is very reactive. Which element would have similar properties to this one?

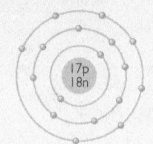

(A) Na, sodium
(B) Br, bromine
(C) Ar, argon
(D) S, sulfur

Use the periodic table shown to answer Exercises 6 and 7.

6. Two groups of the periodic table are highlighted in the figure. What are the names given to these groups?

(A) Alkali metals and noble gases
(B) Alkali metals and halogens
(C) Alkaline earth metals and noble gases
(D) Alkaline earth metals and halogens

7. What type of bonding would you expect to find in substances made when an element from one highlighted group is combined with an element from the other highlighted group?

(A) Ionic only
(B) Molecular covalent only
(C) Network covalent only
(D) Ionic and network covalent

STANDARDIZED TEST PREPARATION

1. C	9. D	16. B, D
2. B	10. A	17. B
3. C	11. C	18. C
4. C	12. B	19. B, C
5. B	13. D	20. C
6. D	14. B	21. B
7. A	15. A	22. B
8. A		

Engineering Project: Water Purification

As an extension to Unit 1: Alchemy, you might want to assign the Engineering Project: Water Purification. In this project, students use an engineering design cycle to conceptualize, build, and refine a simple water purification system.

TRM Engineering Project U1

8. If an alkaline earth metal formed a compound with a polyatomic phosphate ion, PO_4^{3-}, what would be the subscript given to the metal in the chemical formula for the compound?

    (A) 3

    (B) 2

    (C) 1

    (D) There would be no subscript written in the chemical formula.

9. What is present in the nucleus of an atom of $^{226}_{88}Ra$?

    (A) 88 neutrons and 138 protons

    (B) 88 electrons and 226 protons

    (C) 88 protons and 226 neutrons

    (D) 88 protons and 138 neutrons

10. Thallium's atomic number is 81 and the average atomic mass is 204.4 amu. Thallium has two isotopes, thallium-203 and thallium-205. Which statement correctly describes thallium?

    (A) A naturally occurring sample of thallium contains both isotopes.

    (B) Thallium-203 isotope has 122 protons, and thallium-205 has 124 protons.

    (C) The most common atom of thallium has a mass of 204.48 amu.

    (D) Thallium-203 and thallium-205 both have 81 neutrons.

11. An element has three naturally occurring isotopes. One of the isotopes of the element has a mass number of 24 and is 79% abundant in nature. Another of the isotopes has a mass number of 25 and is 10% abundant in nature. The last isotope of the element has a mass number of 26 and is 11% abundant in nature. Based on this information, what is the most likely identity of this element?

    (A) Beryllium, Be

    (B) Sodium, Na

    (C) Magnesium, Mg

    (D) Aluminum, Al

Use the graph below of the radioactive decay of carbon-14 to answer Exercises 12 and 13.

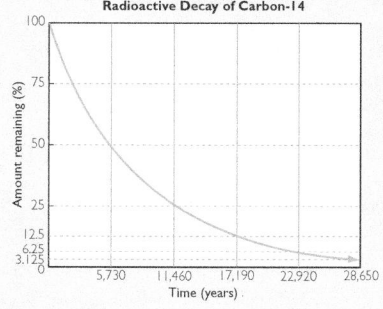

12. What is the approximate half-life of carbon-14?

    (A) 50 years

    (B) 5,730 years

    (C) 11,460 years

    (D) 17,190 years

13. If the original sample of carbon-14 were 10.0 g, how much would remain after two half-lives?

    (A) 234.0 g

    (B) 1.25 g

    (C) 5.00 g

    (D) 2.50 g

14. A sheet of aluminum would protect a person from what type of radiation exposure?

    (A) A sheet of aluminum would protect from alpha, beta, and gamma radiation.

    (B) A sheet of aluminum would protect from alpha and beta, but not gamma radiation.

    (C) A sheet of aluminum would protect from alpha and gamma, but not beta radiation.

    (D) A sheet of aluminum would not protect a person from any form of radiation.

Unit 1 | **Review**   143

ASSESSMENTS

Two Unit 1 Assessments (versions A and B) are available for download.

TRM **Unit 1 Assessment Answers**

TRM **Unit 1 Assessment**

A Lab Assessment for Unit 1 is available for download.

TRM **Unit 1 Lab Assessment Instructions and Answers**

TRM **Unit 1 Lab Assessment**

Use the diagram of the copper cycle below to answer Exercises 15–19.

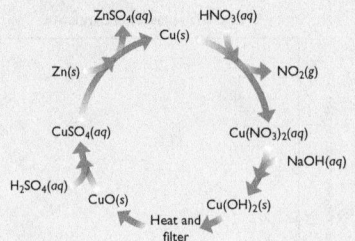

15. Which of these statements correctly describes the law of conservation of mass?

(A) The law of conservation of mass would be supported if the amount of copper at the beginning and end of a copper-cycle experiment were accurately measured.

(B) The law of conservation of mass states that good laboratory techniques will help prevent the accidental loss of lab materials.

(C) The law of conservation of mass states that small amounts of mass are lost in the course of an experiment because of spills, measurement errors, and other factors.

(D) The law of conservation of mass does not apply to gases or aqueous solutions.

16. Which of the following are solids produced in the copper cycle? (Choose all that apply.)

(A) Nitrogen dioxide
(B) Copper hydroxide
(C) Sulfuric acid
(D) Copper oxide

17. What is true about the copper ions present in the copper cycle?

(A) Only copper (I) is present in the ionic compounds.

(B) Only copper (II) is present in the ionic compounds.

(C) Both copper (I) and copper (II) are present in the ionic compounds.

(D) Only elemental copper is present in the copper cycle.

18. What is the charge of the copper atoms in the copper solid, $Cu(s)$?

(A) $+2$ (B) $+1$
(C) 0 (D) -1

19. Which substances, as written below, would most likely conduct electricity? (Choose all that apply.)

(A) $NO_2(g)$
(B) $Cu(NO_3)_2(aq)$
(C) $NaOH(aq)$
(D) $Cu(OH)_2(s)$

20. The acetate ion $C_2H_3O_2{}^-$ has a charge of -1. What is the correct formula for calcium acetate?

(A) $CaC_2H_3O_2$
(B) $Ca_2C_2H_3O_2$
(C) $Ca(C_2H_3O_2)_2$
(D) $C_2H_3O_2Ca$

21. Which of the following compounds will generally *not* form?

(A) KBr
(B) $NeBr_2$
(C) $AlBr_3$
(D) All of these compounds will form.

22. Four models of chemical bonding are illustrated in the diagram. The red spheres represent the nuclei and core electrons of atoms and blue areas represent where valence electrons are located.

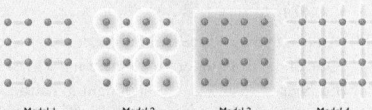

Model 1   Model 2   Model 3   Model 4

Which statement below correctly identifies the type of bonding for the compound listed?

(A) $MgCl_2(s)$ bonds according to Model 1 because valence electrons are shared between the atoms.

(B) $CO_2(g)$ bonds according to Model 1 because valence electrons are shared between the atoms.

(C) $Li(s)$ atoms are bonded according to Model 4 because valence electrons are free to move about the substance.

(D) $N_2(g)$ bonds according to Model 4 because valence electrons are shared throughout the entire substance.

Unit 2 | Smells

Molecular Structure and Properties

SMELL AS CONTEXT

Smell is a highly intriguing and familiar context that activates your students' curiosity. There is plenty of room in this unit to share information, hypotheses, memories, and experiences. The students get to be the experts. As the teacher, you can choose how long or involved the discussions become. The goal is to spark an interest that will be sustained throughout the unit.

Students become more actively involved in learning when they can relate to the subject matter taught. When the immediate relevance of chemistry concepts is demonstrated, students find the chemistry easier to grasp and apply.

SMELL CLASSIFICATION AS A FIELD OF RESEARCH

The mechanism by which we detect the variety of odors in our environment is not yet completely understood. For decades, scientists have been trying to identify what they consider to be the "primary odors" and to agree on a naming system for these smells. Most studies provide evidence that scents can be categorized into more or less discrete classes.

For this unit, we have chosen to focus on five terms that have been consistently chosen by researchers to describe groups of common odors: minty, fishy, sweet, putrid, and camphor (pinelike). The category we call "sweet" includes both fruity and floral smells, for convenience's sake. While not every odor observed in daily life will fit into these five categories, they do account for a large number of the smells one might encounter. These groups provide us with a common language and a structure to use in our pursuit of smell chemistry. For a summary of smell classifications covered in this unit and the chemistry behind them, see Lesson 46.

CONTENT DRIVEN BY CONTEXT

This unit establishes students' understanding of molecules through tangible experience. Students can draw on their own experience with scents to analyze molecular properties and behavior. They begin by examining smells and molecular data for several compounds. Students are then asked to formulate a hypothesis based on patterns in the data that relates smell and molecular structure. The lessons are designed so that students continue to test their hypotheses, refining their conclusions as new evidence is discovered. Throughout the unit, students learn the chemistry related to smell. They explore molecular formulas, Lewis dot symbols, structural formulas, polarity, molecular size and shape, and bonding patterns. They learn to identify particular functional groups within molecules and are introduced to organic chemistry and the chemistry of living things.

VARIATIONS IN EXPERIENCE

Somewhere between 0.5% and 1% of the population have a condition called *anosmia* and cannot perceive smell. Students with anosmia or respiratory difficulties can allow others to smell for them during the smelling activities. There are also variations in the ability to smell. These inconsistencies in smell perception may come up in the classroom. These discrepant events represent a more specialized area of inquiry that students can pursue in the project at the end of the unit. The majority experience will guide this particular unit. We are trying to teach students the basics of smell chemistry. Encourage those with dissenting opinions to continue to speak up, but rely on consensus to direct the activities.

THE USE OF MODELS TO EXPLAIN MOLECULES

This unit continues to use models to explain matter. Students explore two-dimensional structural formulas and three-dimensional ball-and-stick and space-filling models. They learn to make predictions about the smell of molecules based on evidence that they discover using models. Current theories propose that molecules of different shapes fit into matching receptor sites located on the surface of tiny hairlike cilia, which line the nasal passages. Once the molecule has docked, it triggers a neural impulse to the portion of the brain responsible for smell. Students explore this theory with models, and they learn to classify and categorize different compounds.

BUILDING UNDERSTANDING

In Unit 1: Alchemy, students began their study of matter and atomic properties. In this unit, students gradually construct understanding about molecules and their properties as they explore ideas about smell. This foundation in molecular structure and behavior serves to prepare students for a better understanding of the gas laws and chemical reactions, the subjects of later units.

Chapter Summary

Chapter	Description	Standard Schedule Days	Block Schedule Days
6	**Chapter 6** is an introduction to the smells context. Students first compare smell categories and consider the question, "What is the connection between smell and molecular composition?" They develop a hypothesis, which they have the opportunity to revise as they explore Lewis dot structures, the octet rule, and then functional groups. The chapter culminates with a lab in which they synthesize an ester from an alcohol and an organic acid.	9	4.5
7	In **Chapter 7,** students work with ball-and-stick models to study the role of electron pairs and molecular shape. They rethink and refine the hypotheses they proposed in the first chapter as new smell compounds are introduced. They continue their study using space-filling models to generate links between molecular shape and smell. In this chapter, they consider the receptor site theory for smell.	7	3.5
8	**Chapter 8** focuses on interactions between molecules. Students observe evidence of polarity in the lab and learn about electronegativity. Electronegativity values are used to determine the polarity of molecules and the direction of bond dipoles. In this chapter, students answer the question, "How are polarity, phase, shape, and bonding patterns related to smell?"	6	3
9	**Chapter 9** focuses on biological molecules. Students study the "handedness" of molecules and are challenged to distinguish between mirror image isomers. In this chapter, students are introduced to amino acids and build models to see how amino acids link and fold to form proteins. Students use this information as they revisit receptor site theory and review the chemistry related to smell.	6	3

Pacing Guides

Standard Schedule

Day	Suggested Plan	Day	Suggested Plan
1	Chapter 6 Lesson 28	15	Chapter 7 Lesson 41, Chapter 7 Review
2	Chapter 6 Lesson 29	16	Chapter 7 Quiz
3	Chapter 6 Lesson 30	17	Chapter 8 Lesson 42
4	Chapter 6 Lesson 31	18	Chapter 8 Lesson 43
5	Chapter 6 Lesson 32	19	Chapter 8 Lesson 44
6	Chapter 6 Lesson 33	20	Chapter 8 Lesson 45
7	Chapter 6 Lesson 34	21	Chapter 8 Lesson 46, Chapter 8 Review
8	Chapter 6 Lesson 35, Chapter 6 Review	22	Chapter 8 Quiz
9	Chapter 6 Quiz	23	Chapter 9 Lesson 47
10	Chapter 7 Lesson 36	24	Chapter 9 Lesson 48, Chapter 9 Review
11	Chapter 7 Lesson 37	25	Chapter 9 Quiz
12	Chapter 7 Lesson 38	26	Unit Review
13	Chapter 7 Lesson 39	27	Unit Exam
14	Chapter 7 Lesson 40	28	Lab Exam (Optional)

Block Schedule

Day	Suggested Plan	Day	Suggested Plan
1	Chapter 6 Lessons 28 and 29	9	Chapter 8 Lessons 43 and 44
2	Chapter 6 Lessons 30 and 31	10	Chapter 8 Lessons 45 and 46 Chapter 8 Review
3	Chapter 6 Lessons 32 and 33	11	Chapter 8 Quiz Chapter 9 Lesson 47
4	Chapter 6 Lessons 34 and 35 Chapter 6 Review	12	Chapter 9 Lesson 48 Chapter 9 Review
5	Chapter 6 Quiz Chapter 7 Lessons 36 and 37	13	Chapter 9 Quiz
6	Chapter 7 Lessons 38 and 39	14	Unit Review
7	Chapter 7 Lessons 40 and 41 Chapter 7 Review	15	Unit 2 Exam, Lab Exam (Optional)
8	Chapter 7 Quiz Chapter 8 Lesson 42		

Unit 2 | Smells

Each flower puts out a scent that attracts specific insects. Not all flowers smell pleasant. The corpse flower, which blooms once every three years, smells like decaying flesh.

In this unit, you will learn

- how atoms form molecules
- to predict the smell of a compound
- to interpret molecular models
- how the nose detects different molecules
- about amino acids and proteins

Why Smells?

The sense of smell is a familiar and important part of our lives. It helps us detect pleasant smells as well as unpleasant smells that alert us to possible dangers. Sometimes, our sense of smell helps us to detect things that we can't see. Sometimes, we detect something all the way across the room or even in another room. But how does it work? The way atoms are connected in molecules, and the structures of the molecules, have a great deal to do with the properties of those molecules. When you understand the chemistry of molecules, you will begin to answer the question, "What is the chemistry of smell?"

145

PD CHAPTER 6 OVERVIEW

Watch the video overview of Chapter 6 (for teachers) by clicking on the link in the TE-book, opening the TRFD, or logging onto the book's companion Web site bcs.whfreeman.com/livingbychemistry2e (teacher log-in required).

Chapter 6 introduces students to the smell context. In Lesson 28, students sample several smells and examine molecular data for these compounds. They discover patterns in the data that relate smell to molecular composition. Lessons 29 and 30 focus on structural formulas and the covalent bonding tendencies of hydrogen, oxygen, nitrogen, and carbon. Lessons 31 and 32 introduce students to Lewis dot symbols and the octet rule. In Lesson 33, students explore the relationship between functional groups and smell. As students explore and integrate the topics in this chapter, they refine their hypotheses about how smell is related to a molecule's composition and structure. Finally, in Lessons 34 and 35, students participate in a lab to synthesize a sweet-smelling ester from an alcohol and an organic acid and analyze the results.

In this chapter, students will learn

- to interpret molecular formulas
- to create molecular structures from molecular formulas
- to use Lewis dot symbols to predict molecular bonding and structure
- to identify and name functional groups within molecules
- about relationships among molecular formulas, molecular structures, functional groups, and smell

Fotosearch/Getty Images

CHAPTER 6

Speaking of Molecules

In this chapter, you will study

- different representations for molecules
- bonding patterns in molecular covalent compounds
- how molecular structure relates to smell

Our sense of taste is closely related to our sense of smell. For this reason, many chefs have very sophisticated senses of smell.

The nose is able to identify hundreds of different smells. Most substances that smell are made of molecules. So, how does the nose differentiate one molecule from another? When you analyze the molecular composition and structure of compounds that smell, a variety of patterns emerges. Knowing the number and types of atoms in molecules, and the way atoms are connected in molecules, is vital to understanding the chemistry of smell.

146

Chapter 6 Lessons

Sniffing Around

Molecular Formulas

THINK ABOUT IT

The topic of smell is both friendly and familiar. Most of us have had many experiences with smells: smells that attract, smells that disgust, and smells that evoke strong memories of vacations, holidays, or other experiences. It turns out that our noses are fantastic chemical detectors. Learning more about your nose and how it works will help you learn more about chemistry.

___

What does chemistry have to do with smell?

___

To answer this question, you will explore

- The Sense of Smell
- Classifying Smells
- Molecular Formula, Name, and Smell

EXPLORING THE TOPIC

The Sense of Smell

Someone opens a perfume bottle across the room. After a few seconds, you notice the sweet smell of the perfume. How did the smell get to your nose and how did your nose detect it?

The smells that come from a vial of perfume and end up in your nose are made of molecules. Molecules are groups of nonmetal atoms that are covalently bonded together. In this unit, you will examine mechanisms that your nose uses to detect and distinguish various smells.

To be consistent, you will always work with the consensus smell. In other words, you will work with the smell reported by the majority of your classmates.

Classifying Smells

To compare smells, we need a common language for describing them. How would you describe the smells of the items in these pictures?

> **BIOLOGY CONNECTION**
>
> Not every human being experiences smell in exactly the same way. Some people may smell one odor completely differently from other people or may not smell that one odor at all. These exceptions are of particular interest to smell researchers, who use this information to figure out how many distinct smells the nose can detect.

---

molecular formula

IN CLASS

Working in groups, students sample the smells of five substances. Then they examine data for the molecular formula and chemical name of each of the five compounds. Students work in groups to look for patterns in the data. These patterns are discussed and summarized on the board. The class forms a hypothesis about how the smell of a compound can be predicted from certain molecular information. This hypothesis will be expanded and revised several times during the unit as students consider more information. A complete materials list is available for download.

TRM **Materials List 28**

SETUP

Prepare sets of vials as described in the following table. Label the vials with only the letters A, B, C, D, and E. Place a cotton ball in each vial, then use the plastic pipettes to place three to five drops of each essence in the appropriately lettered vial. Use a different pipette for each essence. Place sets of the five vials in reusable plastic bags to make it easy to distribute them to groups of students.

Label	Contents	Smell
A	spearmint oil	minty
B	fish oil	fishy
C	pineapple extract	sweet
D	banana extract	sweet
E	peppermint oil	minty

CLEANUP

Save the vials for reuse in other classes and over the course of this unit. Within a few weeks, however, you will need to remove the cotton balls and air out the vials. Otherwise, the different smells will mix with one another and begin to smell putrid.

When you are finished using vials A–E in all your classes, remove the cotton balls from the vials, place them in a plastic bag, and dispose of them. Let the vials air out in a hood or rinse them with acetone for reuse the next time you teach this unit.

Overview

ACTIVITY: GROUPS OF 4

Key Question: What does chemistry have to do with smell?

KEY IDEAS

We can talk about molecules by using molecular formulas and chemical names. A preliminary look at the connection between smell and chemistry shows relationships among the molecular formula, chemical name, and smell of a compound.

LEARNING OBJECTIVES

- Detect patterns in chemical formulas and relate these patterns to a molecular property.
- Create a hypothesis based on analysis of data.

FOCUS ON UNDERSTANDING

- Students with very strong allergies or asthma should allow others to do the smelling for them and then work from their classmates' data.
- In rare instances, a student might have no sense of smell. Some students' sense of smell will differ from the norm. These cases can make a good discussion topic.

Lesson Guide

ENGAGE (5 min)

TRM PowerPoint Presentation 28

ChemCatalyst

1. What do you think is happening when you smell something?

2. Why do you think we have a sense of smell?

Sample answers: **1.** Students might say the nose is detecting a "gas" or different "molecules." Ask them to clarify and to give evidence. Keep the discussion open-ended. **2.** Students might say the sense of smell is to help us find food, detect danger (skunks, toxins), enjoy life (flowers), find a mate, and so on.

Ask: Why do some things smell good and other things smell bad?

Ask: How does a smell get from one place to another?

Ask: Describe what you think your nose is detecting when you smell something. What do you think chemistry has to do with smell?

EXPLORE (15 min)

TRM Worksheet with Answers 28

TRM Worksheet 28

INTRODUCE THE ACTIVITY

→ Tell students to work in groups of four.

→ Tell students they will examine molecular covalent substances in this unit. These substances consist of molecules. Each molecule stays together as a unit and is represented by a molecular formula.

→ Review the concept of a molecule and introduce molecular formulas.

> **Molecular formula:** The chemical formula of a molecular substance, showing the types of atoms in each molecule and the ratios of those atoms to one another.

→ Instruct students on how to smell safely. Chemicals may have very strong odors or be caustic. Remind students to use wafting when they smell anything in the laboratory. They should use their hand to draw air toward them, not sniff directly from the container. The essences are food grade, and the smells are no

stronger than those we encounter daily. However, any students with asthma or allergies should not do the smelling part of the activity.

→ Pass out the student worksheet.

→ Pass out vials A–E. Emphasize to students that they have to put the caps back on the vials immediately after smelling and not exchange caps among the vials. Before collecting the vials, tell students to tighten all the caps.

GUIDE THE ACTIVITY

→ Encourage students to use the convention shown in the table when writing their molecular formulas: carbon, then hydrogen, then oxygen (e.g., C_3H_8O).

BIOLOGY CONNECTION

A moth uses its antennae to detect smells, while a snake uses its tongue.

PureStock/Getty Images

Important to Know A molecular formula is a chemical formula for a molecular covalent compound. Recall that chemical formulas can describe other compounds that are not molecules.

HEALTH CONNECTION

Dogs are trained to sniff out everything from termites and drugs to bombs and natural gas leaks. There is evidence that some dogs may even be able to detect when a person has developed certain cancers.

Electrofood/SuperStock

Scientists classify smells by placing similar-smelling substances in the same category. For example, apples, bananas, strawberries, and pineapples might all fit into a category called "fruity" smelling. Roses, geraniums, and lilies might all fit into a category called "floral" smelling. Further, fruity- and floral-smelling substances might be grouped together in a larger category called "sweet" smelling.

Not all scientists agree on the specific names for smell categories or on the number of different categories that should exist. As you begin studying smells, you will start with three categories: sweet, minty, and fishy.

↻ Molecular Formula, Name, and Smell

A closer investigation reveals connections among a substance's chemical name, its chemical formula, and its smell. For a molecular covalent compound, the chemical formula is also called the **molecular formula**, because it describes the makeup of each molecule.

Examine the drawings of the five vials. Each vial contains a cotton ball soaked in a different molecular compound. See if you can find any connections between the smells and their chemical names and molecular formulas.

Minty	Sweet	Sweet	Minty	Fishy
1	2	3	4	5
$C_{10}H_{14}O$	$C_7H_{14}O_2$	$C_8H_{16}O_2$	$C_{10}H_{16}O$	$C_8H_{19}N$
L-carvone	Ethyl pentanoate	Hexyl acetate	D-pulegone	Diisobutylamine

Here are some patterns that you might notice.

ALL THE MOLECULES

- contain carbon and hydrogen atoms
- have more hydrogen atoms than carbon atoms
- are composed of three different types of atoms

SWEET-SMELLING MOLECULES

- contain two oxygen atoms
- have twice as many hydrogen atoms as they do carbon atoms
- have two-word names ending in "-ate"

MINTY-SMELLING MOLECULES

- contain one oxygen atom
- contain ten carbon atoms

EXPLAIN & ELABORATE (20 min)

ASSIST STUDENTS IN SHARING THEIR EXPERIENCE OF SAMPLING SMELLS

→ Draw this table on the board. Fill in as groups share their smell classifications.

Vial A	Vial B	Vial C	Vial D	Vial E
minty	fishy	sweet	sweet	minty

Ask: Which smell classification did you use for each of the five mystery smells?

Ask: Which vials would you put in the same category? (A and E, C and D)

- have even numbers of carbon and hydrogen atoms
- have names ending in "-one"

FISHY-SMELLING MOLECULES

- contain a nitrogen atom
- do not contain oxygen atoms

Perhaps *every* molecule with two oxygen atoms smells sweet. Perhaps *every* molecule that smells fishy contains a nitrogen atom. These are reasonable hypotheses based on very little data. You can test these hypotheses by gathering more data.

Example

Predicting Smell

How would you expect a molecule with the molecular formula $C_9H_{18}O_2$ to smell? Explain.

Solution

The molecule has two oxygen atoms. Based on data given so far, this suggests that the molecule smells sweet. It would be useful to know the name to confirm this. So far, we have seen that the names of sweet-smelling molecules end in "-ate."

LESSON SUMMARY

What does chemistry have to do with smell?

KEY TERM

molecular formula

The smells that your nose detects are made of molecules. The molecular formulas and chemical names of molecular substances are directly related to the way these substances smell. You may be able to predict the smell of a substance by examining its chemical formula, its chemical name, or both.

Exercises

Reading Questions

1. How do scientists classify smells? What purpose might this serve?
2. From the evidence so far, how are molecular formulas related to smell? How are chemical names related to smell?

Reason and Apply

3. Name five substances from home that do not fit into the smell categories of sweet, minty, or fishy. What new categories would you put them in?
4. Make a list of five things from home that have no smell.
5. In a paragraph or two, write about *one* of the following. You may use a cartoon format instead, if you wish.
   a. Describe a memory that you have that relates to smell. Describe the smell itself and what it reminds you of. (Avoid using adjectives that are not informative, such as "weird.")

Ask: Why do you think there are sometimes disagreements over how to classify the smells of different substances?

COLLECT A LIST OF PATTERNS GENERATED BY THE STUDENTS

→ Write the patterns on the board as the students share them.

Ask: What patterns did you discover?

Ask: According to the data, if a molecule has one oxygen atom, how would you expect it to smell? (minty) What if it has a nitrogen atom? (fishy) What if it has two oxygen atoms? (sweet)

Ask: What other patterns help you predict the smell of a molecule?

PATTERNS STUDENTS MIGHT NOTICE

All the molecules:

- All the molecules have H and C atoms.
- There are more H atoms than C atoms.
- The names of molecules that have similar smells have similar endings.
- Molecules that smell good all have even numbers of H atoms.
- Smell seems to be related to the atoms other than C and H atoms.

Sweet-smelling molecules:

- Molecules that smell sweet have two O atoms.

- Molecules that end with "-ate" smell sweet.

Minty-smelling molecules:

- Molecules that smell minty have one O atom and no N atoms.
- Minty-smelling substances have ten C atoms and end in "-one."

Fishy-smelling molecules:

- Molecules that smell fishy have one N atom.
- The molecules with names that end with "-ine" smell fishy.

Note: While many of these patterns are useful in predicting smells, there will be exceptions, and some statements need refinement. The goal is to refine these ideas so that we improve our accuracy in predicting smells.

COME UP WITH A HYPOTHESIS ABOUT PREDICTING SMELLS

→ At the appropriate point in the discussion, write a hypothesis about predicting smells on the board. Allow students to refine it.

Ask: If I gave you a new smell vial, what would you want to know about the substance inside to predict its smell?

Ask: Do you think you can predict the smell of a substance simply from knowing its chemical name and molecular formula? Why or why not?

Ask: What should we do to test this hypothesis?

Key Points: A possible hypothesis is that the smell of a substance can be predicted if you know its name and/or its chemical formula. For instance, molecules with two oxygen atoms may all smell sweet, molecules containing nitrogen may all smell fishy, molecules containing one oxygen atom may all smell minty, molecules ending in "-ate" may all smell sweet, and so on. As students learn more about molecules and smell properties throughout the unit, they will have a chance to refine their hypotheses.

EVALUATE (5 min)

Check-In

1. How would you expect a compound with the molecular formula $C_8H_{16}O_2$ to smell? Explain.
2. How sure are you of your prediction?

Answers: 1. From the preliminary data, you can predict it will smell sweet because it has two oxygen atoms. **2.** You cannot be all that sure, because you have smelled only five molecules. You have to collect more data.

Homework

Assign the reading and exercises for Smells Lesson 28 in the student text.

LESSON 28 ANSWERS

1. Scientists classify smells by placing similar types of smells in a category. This allows scientists to talk about smells in a consistent way.

2. *Possible answer:* Substances with similar smells have similarities in the type and number of specific atoms in their molecular formula. The suffix at the end of a chemical name and the structure of the name of a substance seem to be related to the type of smell that the substance has.

3. *Possible answers (any 5):* vinegar, sour; cocoa powder, chocolate; perfume, flowery; damp paper, musty; window cleaner, ammonia; coffee, roasted; onion, pungent.

Category names should be general enough to include multiple smells and also be descriptive of the smells within the category.

4. *Possible answers (any 5):* salt, silverware, glass, aluminum foil, CD, telephone, table

5. a. Answer should include a descriptive term related to the smell. **b. *Possible answer:*** A super-smelling ability would allow the superhero to detect very small traces of certain substances or to detect substances from a long distance away. Smells could help pinpoint the location of people or dangerous substances. **c.** Answer should include some explanation of why a sense of smell is important in daily life.

6. *Possible answer:* a. The sense of taste and the sense of smell are closely related. When the sense of smell is taken away, relying on taste alone makes it hard to identify foods. The taste of the food also changes, because it is no longer enhanced by the smell. **b.** During the process of tasting, some molecules enter the nose. Nasal congestion can reduce or eliminate the sense of taste along with the sense of smell. The two senses are strongly related.

7. *Possible answer:* ● Methyl octenoate probably smells sweet because

Exercises

(continued)

  b. If you were a smell superhero, how would you use your smell superpowers or your enhanced sense of smell to assist the world?

  c. What would your life be like if you couldn't smell?

6. Try tasting something while having your nose plugged and your eyes closed. You may wish to work with a partner who can provide you with a variety of different-flavored jelly beans or other food items to sample.

  a. Describe your experience. How difficult is it to taste something when you can't smell it and you don't know what it is?

  b. How do you think taste is related to smell?

7. Predict the smells of these molecules. Explain your reasoning.

  a. methyl octenoate, $C_9H_{18}O_2$

  b. monoethylamine, C_2H_7N

  c. ethyl acetate, $C_4H_8O_2$

8. What do you think 1,5-pentanedithiol ($C_5H_{12}S_2$) would smell like? How confident of your answer are you? Explain why.

9. What makes something smell the way it does? Discuss your own hypothesis about why different things smell different. Include questions you still have.

150   Chapter 6 | **Speaking of Molecules**

it has two oxygen atoms and twice as many hydrogen atoms as carbon atoms in each molecule. The chemical name also ends in "-ate." ● Monoethylamine probably smells fishy because it has one nitrogen atom and no oxygen atoms in each molecule. ● Ethyl acetate probably smells sweet because it has two oxygen atoms and twice as many hydrogen atoms as carbon atoms in each molecule. The chemical name also ends in "-ate."

8. *Possible answer:* Because sulfur is in the same group as oxygen, a molecule with two sulfur atoms might have a sweet smell. I am not confident

in this answer because the chemical formula for 1,5-pentanedithiol does not fit any of the patterns I have learned about so far.

9. *Possible answer:* Something has a smell when some of its molecules travel through air and enter the nose. The smell occurs when the molecules interact with molecules inside the nose. The way the molecules interact with the smell receptors in the nose determines their smell. We still have to learn exactly how the nose detects smell, and what part or parts of a molecule determine how it smells.

Molecules in Two Dimensions

Structural Formulas

THINK ABOUT IT

Would you rather smell sweaty gym socks or fresh pineapple? The choice is pretty clear. But what if you were asked if you would rather smell ethyl butyrate or hexanoic acid? How would you know which compound is safe to sniff? Both molecules have the same molecular formula: $C_6H_{12}O_2$.

Alberto Pérez Veiga/Shutterstock

© RedHelga/iStockphoto

How can molecules with the same molecular formula be different?

To answer this question, you will explore

- Predicting Smells
- Structural Formula
- Orientations of Structural Formulas

EXPLORING THE TOPIC

Predicting Smells

In Lesson 28: Sniffing Around, you arrived at some generalizations that relate smell to molecular formula and chemical name. What happens if you use these generalizations to make predictions about some new compounds? Consider the five new compounds listed in the table.

Vial	Chemical name	Molecular formula	Predicted smell
1	pentyl acetate	$C_7H_{14}O_2$	sweet
2	pentanoic acid	$C_5H_{10}O_2$	sweet
3	propyl acetate	$C_5H_{10}O_2$	sweet
4	ethylthiol	C_2H_6S	a new smell category
5	hexylamine	$C_6H_{15}N$	fishy

All three have two oxygen atoms. Two end in "-ate" and one in "acid": sweet smelling?

Contains a sulfur atom: new smell category?

Contains a nitrogen atom: fishy?

The predictions based on molecular formula and chemical name alone are reasonably accurate. Four out of five are correct. Ethylthiol has a skunk smell, so it belongs in a new smell category. The only prediction that is incorrect is the pentanoic acid in vial 2. It has a molecular formula *identical* to the molecular formula of a sweet-smelling molecule, but it has a very putrid smell. Apparently,

Lesson 29 | **Structural Formulas** 151

Overview

ACTIVITY: GROUPS OF 4

Key Question: How can molecules with the same molecular formula be different?

KEY IDEAS

Structural formulas are models that chemists use to show how the atoms in a molecule are connected to one another. Two molecules might have the same molecular formula but different structures. Such molecules are called isomers.

LEARNING OBJECTIVES

- Describe the difference between structural formulas and molecular formulas.
- Recognize isomers.

FOCUS ON UNDERSTANDING

- It might take some time for students to be able to visually distinguish the details of structural formulas and to realize that the same molecule can be represented in different ways on the page.
- Functional groups will be introduced in a later lesson. For now, students only have to understand that atoms can be

connected in different ways to form different overall structures.

KEY TERMS

structural formula
isomer

IN CLASS

At the beginning of class, students predict how three new mystery substances will smell based on their molecular formulas and chemical names. Students then test their predictions by smelling these new substances. They examine the structural formulas of these same three molecules, looking for similarities and differences. Students explore the orientation of structural formulas and look at ball-and-stick models to become more proficient at differentiating between different molecules and different orientations. Students are introduced to isomers. A complete materials list is available for download.

TRM Materials List 29

SETUP

Prepare sets of vials F, G, H according to the following table. Label the vials with the letter only. Place a cotton ball in each vial, then use the plastic pipettes to deliver three to five drops of each stock smell solution to the appropriately lettered vial. Use a new pipette for each essence. You should prepare the vials in a hood or outdoors, especially vial G.

Label	Contents	Smell
F	apricot perfume oil (ethyl pentanoate)	sweet
G	butyric acid	putrid
H	rum flavor extract (ethyl acetate)	sweet

Build one ball-and-stick molecular model each of butyric acid, ethyl acetate, and 1-propanol (molecule 1 from the worksheet). Here are the structural formulas and the pieces you will need. Use 4-hole carbons.

Butyric acid

Ethyl acetate

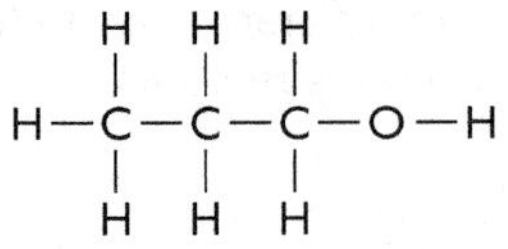

1-propanol

	Carbons (black)	Oxygens (red)	Hydrogens (white)	Single bonds (straight sticks)	Double bonds (curved sticks)
butyric acid	4	2	8	12	2
ethyl acetate	4	2	8	12	2
1-propanol	3	1	8	11	0

CLEANUP

Save the vials for reuse in other classes and over the course of this unit. Within a few weeks, however, you should remove the cotton balls and air out the vials. Otherwise, the substances' smells will mix and begin to smell putrid. When you are finished using vials F, G, H in all your classes, remove the cotton balls from the vials, place them in a plastic bag, and dispose of them. Let the vials air out in a hood or rinse them with acetone for reuse the next time you teach this unit.

Lesson Guide

ENGAGE (5 min)

TRM PowerPoint Presentation 29

Check-ChemCatalyst

Predict the smells of these three new molecules. Provide evidence to support your prediction.

Vial F:	ethyl pentanoate	$C_7H_{14}O_2$
Vial G:	butyric acid	$C_4H_8O_2$
Vial H:	ethyl acetate	$C_4H_8O_2$

Sample answer: Some students might predict that all three vials will smell sweet, based on their chemical formulas and the presence of two oxygen atoms. However, vial G may smell different, based on the chemical name.

Ask: What smells did you predict for each vial? What was your reasoning?

Ask: What do these molecules have in common? How do they differ? Do you think vial G and vial H will smell identical? Why or why not?

Ask: What information do you need to find out if your predictions are correct?

---

two compounds can have identical formulas, different names, and very different smells. Perhaps this is why its name does not end in "-ate" like the other sweet-smelling compound.

As you continue to study molecules, some of the patterns discovered here will hold up, while others will be revised or abandoned. You can continue to refine your ideas about smell so that you can better understand the chemistry of smell.

Big Idea Science involves coming up with ideas based on observations and then refining these ideas based on further observation.

↺ Structural Formula

As it turns out, two compounds can have identical molecular formulas but smell much different. This is because having the same molecular formula does not guarantee that the molecules themselves are identical.

To show how the individual atoms in a molecule are bonded to each other, chemists use what is called a structural formula. A **structural formula** is a diagram or drawing that shows how the atoms in a molecule are arranged and where they are connected. The lines between the atoms represent covalent bonds holding the atoms together. Take a moment to examine the structural formulas for ethyl butyrate and hexanoic acid below. What differences do you see?

These compounds have the same molecular formula, but different smells.

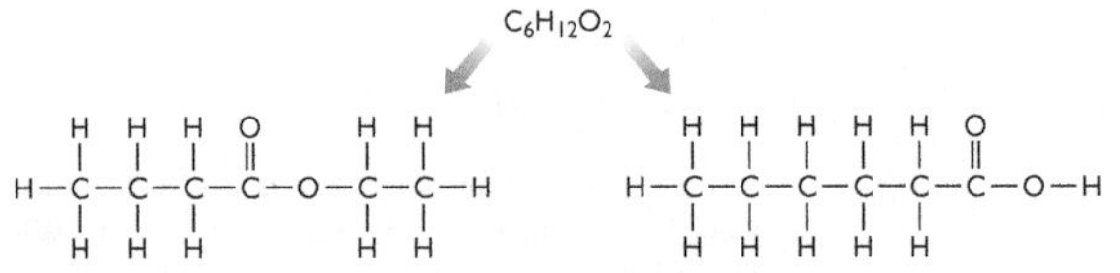

$C_6H_{12}O_2$

Ethyl butyrate, pineapple smell

Hexanoic acid, sweaty gym sock smell

Notice that the oxygen atoms are located in different places within the two molecules. In one molecule, the oxygen atoms are in the middle of the molecule, whereas in the other molecule, the oxygen atoms are at one end.

ISOMERS

In some compounds, the atoms can be connected only a certain way. For instance, the formula C_2H_6 has only one structural formula.

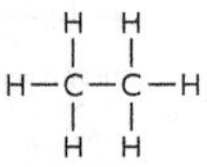

C_2H_6

As the number and types of atoms in a molecule increase, there are more possible ways to connect the atoms. Molecules with the same molecular formula and different structural formulas are called **isomers**. The formula C_2H_6O has two isomers, shown here.

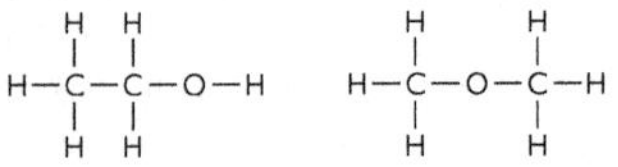

C_2H_6O

BIOLOGY CONNECTION

The spray of a skunk is a defensive adaptation. The spray contains the compound ethylthiol. Predators will usually leave a skunk alone once they have been blasted with this stinky substance that burns the eyes and nose and is difficult to get rid of.

© W. Perry Conway/CORBIS

EXPLORE (15 min)

TRM Worksheet with Answers 29

TRM Worksheet 29

INTRODUCE THE ACTIVITY

→ Tell students to work in groups of four.

→ Pass out the worksheet to each group.

→ Ask students to write their predictions in the table before you pass out the vials with the new molecules.

→ Pass out vials F, G, H to groups of four students.

GUIDE THE ACTIVITY

→ After smell sampling is complete, ask students to tighten the caps on the vials. Collect the vials.

→ When students get to question 6, hold up the model of 1-propanol. Show them how different spheres represent different kinds of atoms while the sticks represent bonds. Show them how turning the model or rotating the bond can make it look different, as in the third or fifth drawing, but does not change the structure of the molecule (i.e., it was not taken apart and put back together).

Ethyl butyrate and hexanoic acid are considered isomers of each other. The formula $C_6H_{12}O_2$ has many more isomers. In fact, there are more than 25 different molecules with the molecular formula $C_6H_{12}O_2$, each with unique properties.

Drawings of structural formulas are flat and two-dimensional. However, molecules are not flat. They take up three dimensions in space.

Important to Know Structural formulas are used to represent molecules that are covalently bonded. They are not used to represent ionic compounds.

Ethanol

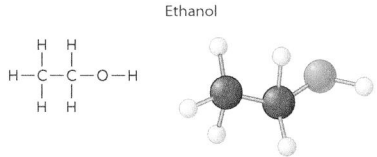

⟳ Orientations of Structural Formulas

There are several ways to draw the *same* structural formula. Consider these two drawings.

These two molecules have the same molecular formula and each atom has the same connections. The structural formula has just been turned 90° in space. These molecules are *not* isomers; they have the same structural formula.

Here are four structural formulas. How many different compounds are represented?

Molecule 1   Molecule 2   Molecule 3   Molecule 4

If you build three-dimensional models of these molecules, you will find that there are only three different compounds here. The first and the third structural formulas are identical. All the atoms in the first and third structures are connected in the same way.

The illustration on the next page shows how molecule 1 can be turned and rotated to match molecule 3 without taking the model apart.

→ Show the models of butyric acid, $C_4H_8O_2$ and ethyl acetate, $C_4H_8O_2$. Show that butyric acid cannot be turned into ethyl acetate without taking the molecule apart.

Ask: What information is contained in a structural formula?

Ask: How is the information in a structural formula different from the information in a molecular formula?

Ask: When a molecular structure is different, does that mean the molecule is different? Explain.

Key Points: A structural formula is a two-dimensional drawing of a molecule showing how the atoms in a molecule are connected. Each line represents a covalent bond. A double line represents a double bond. Thus, every molecule in a particular substance has the same very specific structure that does not change.

> **Structural formula:** A drawing or diagram that a chemist uses to show how the atoms in a molecule are connected. Each line represents a covalent bond.

There are several ways to draw the same structural formula without changing the identity of the molecule. A molecule can be drawn forward, backward, or vertically, and it will still be the same molecule. This is because flipping the molecule in space, or turning it, does not change the way the atoms are connected. It is still the same substance. When two molecules have the same molecular formula but different structural formulas, they are called *isomers* of each other. Isomers have distinct properties. Therefore, we would expect molecules with different structures to have different smells.

> **Isomers:** Molecules with the same molecular formula but different structural formulas.

Butyric acid

Ethyl acetate

DISCUSS THE THREE SMELLS—F, G, AND H

TRM Transparency—Structural Formulas 29

→ You might want to redisplay the ChemCatalyst.

Ask: How did your experience with the vials match your predictions?

Ask: How do you explain your results?

Ask: How did vial G smell? What smell classification might you use for vial G and other similar smells?

Key Points: Even though the molecules in vials G and H have identical molecular formulas, they have different smells and therefore must be different somehow. Even though the molecules in vial G have two oxygen atoms, like the sweet-smelling molecules, vial G smells disgusting. Smell chemists call substances that smell like vial G *putrid*. We will use the classification "putrid" in the remainder of the unit. We now have four smell classifications: sweet, minty, fishy, and putrid.

→ Show the model of 1-propanol, C_3H_8O, from the worksheet. Show how you can rotate the model or individual bonds to match the third and fifth drawings but not the second or fourth.

EXPLORE THE RELATIONSHIP BETWEEN STRUCTURAL FORMULA AND SMELL

→ Display the Transparency—Structural Formulas.

Ask: What do the structural formulas tell you about the smells of molecules G and H?

Ask: How are these two molecules different?

Ask: Do you think you can predict the smell of a compound from only its chemical name and molecular formula? Why or why not?

Ask: If you were given another putrid-smelling compound, what would you expect it to look like structurally?

Ask: If you were given a new compound to smell, what information would you want to have in order to predict its smell?

Key Points: Molecules can smell different even if they have the same molecular formula. There is more to predicting smell than simply looking at the molecular formula. The structural formulas of molecules G and H show us that two substances can have identical molecular formulas but different structures. In particular, one of the oxygen atoms in molecule G is connected to a carbon atom and a hydrogen atom, whereas one of the oxygen atoms in molecule H is connected to two carbon atoms.

Molecules can smell similar even if they have different molecular formulas. Molecules F and H both smell sweet, even though their molecular formulas are different. However, both have a similar structural feature in the middle of the molecules: One oxygen atom is connected by two lines to a carbon atom, and another oxygen atom is between two carbon atoms.

EVALUATE (5 min)

Check-In

For each compound, predict the smell or describe what information you would want in order to predict the smell.

a. $C_6H_{12}O_2$

b. $C_6H_{15}N$

Answers: **a.** Either sweet or putrid. Having the chemical name of the molecule would help: "-ate" indicates a

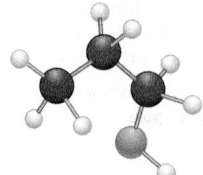 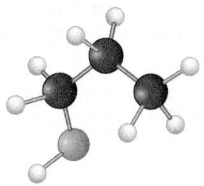 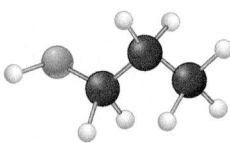

Start with molecule 1.   *Turn it around.*   *Rotate the carbon atom on the left. This is the same as molecule 3.*

Structural formulas show only the sequence in which one atom is connected to the next. The way atoms are connected in a molecule is like a charm bracelet. The middle charm is always between the same two charms regardless of how you hold the bracelet. It does not matter if the oxygen atom is drawn on the left or right side of the molecule. It also does not matter if the O—H is drawn horizontally or vertically. It is only the number and type of connections that matter.

Example

Isomers of C_3H_9N

How many different isomers are represented below?

Molecule 1   Molecule 2   Molecule 3   Molecule 4

Solution

There are three isomers. Molecules 2 and 3 are identical. So there are three different compounds represented here.

LESSON SUMMARY

How can molecules with the same molecular formula be different?

Compounds with identical molecular formulas may have different structures. Chemists use drawings called structural formulas to show how the atoms in molecules are connected to each other. Each line in a structural formula represents a covalent bond between atoms. When molecules have the same molecular formula but different structural formulas, they are called isomers. Isomers represent different compounds with different properties. Because smell is a property, compounds with the same molecular formula may smell different from each other because they are different substances.

KEY TERMS
structural formula
isomer

sweet smell; "acid" indicates a putrid smell. The structural formula would also help, because you could look for structural features. **b.** Fishy. We might want to smell a few more compounds containing nitrogen atoms to be sure they all smell fishy.

Homework

Assign the reading and exercises for Smells Lesson 29 in the student text.

Reading Questions

1. What information does a structural formula provide?

2. What are isomers?

Reason and Apply

3. If you are given a structural formula of a molecule, do you also know its molecular formula? Explain your thinking.

4. The words *isotope* and *isosceles* also have the prefix "iso-." How are their meanings similar to that of the word *isomer*?

5. Structural formulas for six molecules are shown below. Write the molecular formula for each of the molecules.

a.
```
        O  H  H
        ||  |  |
 H—O—C—C—C—H
            |  |
            H  H
```

b.
```
    H  O  H  H  H
    |  ||  |  |  |
 H—C—C—C—C—C—H
    |     |  |  |
    H     H  H  H
```

c.
```
    H  O  H  H
    |  ||  |  |
 H—C—C—C—C—H
    |     |
    H   H—C—H
            |
            H
```

d.
```
    H  H  O        H
    |  |  ||       |
 H—C—C—C—O—C—H
    |  |          |
    H  H          H
```

e.
```
    H  H  H  H  H
    |  |  |  |  |
 H—C—C—C—N—C—H
    |  |  |     |
    H  H  H     H
```

f.
```
       H
       |
 H—C—O—C—H
    ||      |
    O       H
```

6. Two molecules have the same molecular formula, yet one smells sweet and the other smells putrid. Explain how you think this might be possible.

7. For each of the molecules in Exercise 5, draw an isomer of the compound.

7. a.
```
       H  O  H
       |  ||  |
 H—C—C—C—O—H
       |     |
       H     H
```

b.
```
    H  H  O  H  H
    |  |  ||  |  |
 H—C—C—C—C—C—H
    |  |     |  |
    H  H     H  H
```

c.
```
                H
                |
       H  H—C—H  H
       |     |     |
 H—C—C————C————C—H
    ||  |     |     |
    O  H     H     H
```

d.
```
       H  O  H  H
       |  ||  |  |
 H—O—C—C—C—C—H
          |     |  |
          H     H  H
```

e.
```
    H  H  H  H  H
    |  |  |  |  |
 H—C—C—N—C—C—H
    |  |     |  |
    H  H     H  H
```

f.
```
    H
    |
 H—C—H
    |
    C=O
    |
    O
    |
    H
```

LESSON 29 ANSWERS

1. Structural formulas show what atoms are present in a molecule and how the atoms are bonded to one another.

2. Isomers are molecules that have the same molecular formula but different structural formulas.

3. Yes, the structural formula shows all of the atoms in a molecule. Therefore, the molecular formula can be determined from the structural formula

4. *Isotope* and *isomer* both define objects that are equal in some ways and unequal in other ways.

5. a. $C_3H_6O_2$ b. $C_5H_{10}O$ c. $C_5H_{10}O$
d. $C_4H_8O_2$ e. $C_4H_{11}N$ f. $C_2H_4O_2$

6. *Possible answer:* The two molecules are different compounds with different properties. Although they have the same molecular formula, the atoms are connected in different ways, so they have different structural formulas. Based on the information presented in this lesson, the molecule with a sweet smell has two oxygen atoms in the middle of the molecule, and the molecule with the putrid smell has two oxygen atoms at one end of the molecule.

Overview

CLASSWORK: PAIRS

Key Question: What are the rules for drawing structural formulas?

KEY IDEAS

Within most molecules, hydrogen forms one bond, oxygen forms two bonds, nitrogen forms three bonds, and carbon forms four bonds. These bonding tendencies can be summarized with the mnemonic device HONC 1234.

LEARNING OBJECTIVES

- Create accurate structural formulas from molecular formulas.
- Identify and differentiate between isomers and molecules oriented differently in space.
- Explain and use the HONC 1234 rule.

FOCUS ON UNDERSTANDING

The orientation of a molecule in space might continue to be a source of confusion for students, making the accurate identification of isomers challenging.

KEY TERM

HONC 1234 rule

IN CLASS

Students are introduced to the HONC 1234 rule and use it to construct accurate molecular structures. Students work from molecular formulas to make two or three isomers for each formula. The sometimes subtle differences between different structural orientations versus different isomers are reinforced in this activity. A complete materials list is available for download.

TRM Materials List 30

SETUP

Build models of one or two C_3H_9N isomers from Question 3 on the worksheet for the Explain & Elaborate discussion.

Differentiate

Understanding how two-dimensional structural formulas can be oriented in space can be challenging for students. To help develop this concept and to support students in learning the meaning of the term *isomer*, ask students to work in small groups of 3–4 to draw as many isomers as they can for C_3H_9N on a whiteboard or piece of paper. Have one student from each

group present the group's isomers. Students will likely have drawn some identical isomers on their whiteboard or two groups will have drawn an identical isomer differently. Engage students in an open discussion around similarities and differences between the whiteboards and ask students as a class to determine how many total isomers they have come up with for this molecule.

Pre-AP® Course Tip

Students preparing to take an AP® chemistry course or college chemistry should be able to analyze data to identify certain types of patterns

and relationships among chemical phenomena. In this lesson, students first identify patterns in the types of bonding among hydrogen, oxygen, nitrogen, and carbon in organic compounds. Based on the established patterns, students are able to draw structural formulas of various compounds.

Lesson Guide

ENGAGE (5 min)

TRM PowerPoint Presentation 30

TRM Transparency—ChemCatalyst 30

HONC If You Like Molecules

Bonding Tendencies

THINK ABOUT IT

To understand the chemistry of smell, it is necessary to understand how molecules are put together. If you examine the structural formulas of molecules, you can see patterns in the way the atoms are connected.

For example, hydrogen atoms are always arranged around the outside of the molecule. Plus, hydrogen atoms always have only one line connecting them to other atoms, while carbon atoms always have more than one line connecting them to other atoms.

What are the rules for drawing structural formulas?

To answer this question, you will explore

↻ The HONC 1234 Rule

↻ Drawing Structural Formulas

EXPLORING THE TOPIC

↻ **The HONC 1234 Rule**

The structural formulas for three molecules are shown here. Take a moment to count the number of times each type of atom—hydrogen, oxygen, nitrogen, and carbon—is connected to other atoms.

$C_6H_{12}O_2$ $C_{10}H_{18}O$ $C_4H_{11}N$

- Every **Hydrogen** atom has **one** line connecting it to other atoms.
- Every **Oxygen** atom has **two** lines connecting it to other atoms.
- Every **Nitrogen** atom has **three** lines connecting it to other atoms.
- Every **Carbon** atom has **four** lines connecting it to other atoms.

This information is sometimes referred to as the **HONC 1234 rule.** Within most molecules, hydrogen makes one bond, oxygen makes two bonds, nitrogen makes three bonds, and carbon makes four bonds. The bonds in a structural formula are represented by lines. One line connecting two atoms is called a *single bond*. A pair of lines connecting the same two atoms, as in C=O, is called a *double bond*.

The HONC 1234 rule tells you how four of the most common nonmetal atoms will bond. You will learn about the bonding of other nonmetal atoms like sulfur, S, and chlorine, Cl, in later lessons.

Example 1

HONC 1234

Are the following molecules correct according to the HONC 1234 rule? If not, what is wrong with them?

Molecule 1   Molecule 2

Solution

Both molecules are incorrect according to the HONC 1234 rule. In molecule 1, there is an oxygen atom with four bonds. It should have only two bonds. In molecule 2, there is a hydrogen atom with two bonds. It should have only one. The hydrogen atom cannot go in the middle of the molecule, between two carbon atoms.

Drawing Structural Formulas

The HONC 1234 rule is all you need to draw the structural formulas for thousands of different molecules correctly. Some examples are given here and on top of the next page.

STRUCTURAL FORMULA FOR C$_2$H$_6$

Consider a molecule with two carbon atoms and six hydrogen atoms. Its molecular formula is C$_2$H$_6$. What structural formula would this molecule have? Start by connecting the two carbon atoms with a single bond. The "carbon backbone" is generally a good place to start when drawing molecules.

Step 1: Connect the carbon atoms: C—C

Step 2: Add hydrogen atoms. Make sure each hydrogen atom has just one bond and each carbon atom has a total of four bonds:

The structural formula shown above is consistent with the HONC 1234 rule. This is the only structure you can make out of this molecular formula. It represents a substance called *ethane*.

STRUCTURAL FORMULA FOR C$_2$H$_6$O

Consider a molecule with two carbon atoms, six hydrogen atoms, and one oxygen atom. Its molecular formula is C$_2$H$_6$O. What structural formula could this molecule have?

ASTRONOMY CONNECTION

At room temperature, ethane, C$_2$H$_6$, is a flammable gas. On Earth, ethane is one of the components of natural gas. It has also been detected in the atmospheres of Jupiter, Saturn, Uranus, and Neptune.

Lesson 30 | **Bonding Tendencies**   157

Ask: What patterns do you notice in the way atoms are connected?

Ask: What do you think HONC 1234 means?

Ask: How many connections do atoms of each element make with other atoms?

EXPLORE (20 min)

TRM Worksheet with Answers 30

TRM Worksheet 30

INTRODUCE THE ACTIVITY

→ Tell students to work in pairs.

→ Pass out the student worksheet.

→ Introduce the HONC 1234 rule. Write the letters H, O, N, and C on the board in a vertical column. Fill in the number of bonds as students respond.

hydrogen	H	1	Hydrogen atoms form one bond.
oxygen	O	2	Oxygen atoms form two bonds.
nitrogen	N	3	Nitrogen atoms form three bonds.
carbon	C	4	Carbon atoms form four bonds.

→ Tell students they will use the HONC 1234 rule to construct structural formulas from molecular formulas.

EXPLAIN & ELABORATE (15 min)

PROCESS THE DIFFERENT STRUCTURES FROM THE CLASSWORK

→ Ask students to come to the board and draw their structural formulas.

Ask: Does the drawing follow the HONC 1234 rule? If not, how can the drawing be altered?

Ask: Can anyone draw this molecule in a different way?

Ask: Can anyone draw a different molecule with the same molecular formula?

Ask: How does knowing the HONC 1234 rule limit the number of possible structural formulas that you can draw from a particular molecular formula?

Key Points: The HONC 1234 rule is a way to remember the bonding tendencies of hydrogen, oxygen, nitrogen, and carbon atoms in molecules. Hydrogen tends to form one bond, oxygen two, nitrogen three, and carbon four.

ChemCatalyst

Examine these molecules. What patterns do you see in the bonding of atoms of hydrogen, oxygen, carbon, and nitrogen?

Molecule K
Diisobutylamine; fishy

Molecule E
Menthone; minty

Sample answer: Students will notice that hydrogen atoms bond only once. Carbon always has four bonds connecting it to other atoms. Oxygen has two bonds, and nitrogen has three.

LESSON 30

DISCUSS DIFFERENT WAYS TO REPRESENT THE SAME MOLECULAR FORMULA

→ You might want to pick one molecular formula and stick with it, clarifying the difference between different isomers and different orientations. You could also use three-dimensional models to clarify.

Ask: How many isomers does C_3H_9N have? How did you find out?

Ask: How can you tell whether a drawing represents a new molecule?

Ask: How can you tell how many structural formulas a molecular formula represents?

Key Points: When trying to decide if two structures represent the same molecule, you must check how the atoms are connected. Try mentally turning the molecule in space, paying attention to how and where in the molecule different atoms are placed. All these structures represent the same molecule. Only their orientations in space are different.

CONSUMER CONNECTION

Ethanol, C_2H_6O, can be made from plants such as sugarcane, corn, and switchgrass, and can be used as an alternative fuel for automobiles. It can be used alone or blended with gasoline. However, it currently takes about a liter of fuel to produce one liter of ethanol, which may offset its environmental benefit.

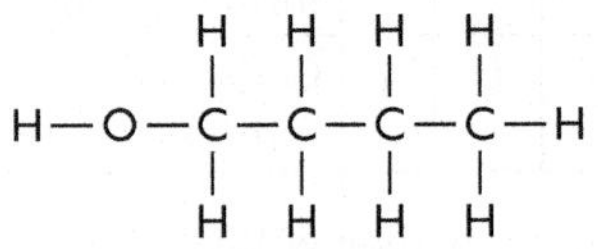

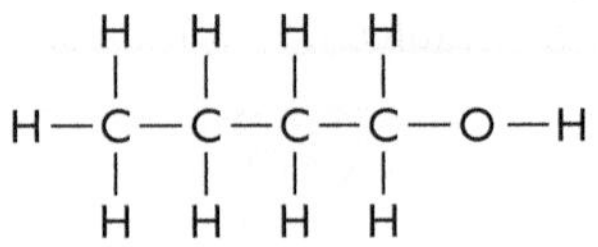

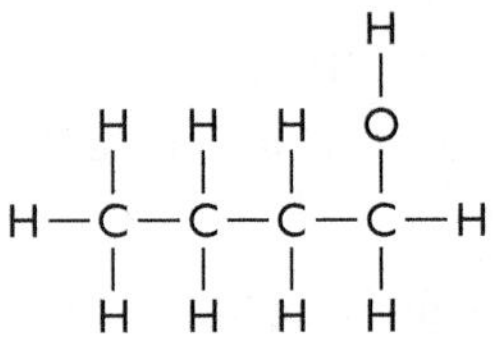

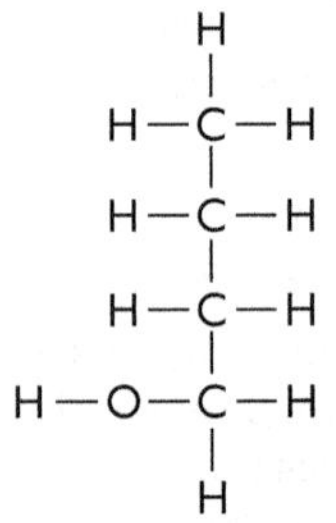

Start by connecting the carbon atoms. Then add the oxygen atom.

Step 1: Connect the carbon atoms: C—C

Step 2: Add the oxygen atom to the carbon chain: C—C—O

Step 3: Add hydrogen atoms last:

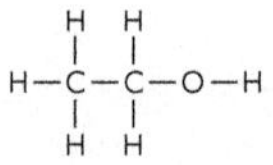

Make sure your structural formula follows the HONC 1234 rule. Notice that hydrogen atoms are generally the last atoms you add when creating a structural formula.

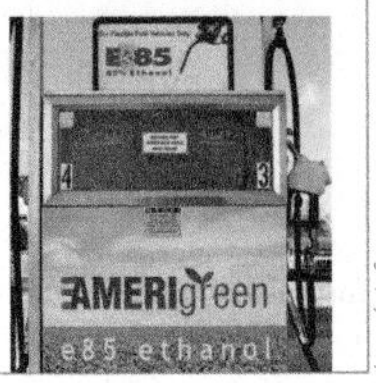

This molecule is called *ethanol* or *ethyl alcohol*.

A SECOND STRUCTURE—AN ISOMER

Ethanol is not the only possible structure for this particular molecular formula. The atoms in C_2H_6O can also be arranged a different way, to form a completely different molecule. This new structure is created by placing the oxygen atom between the two carbon atoms.

Step 1: Connect the carbon and oxygen atoms: C—O—C

Step 2: Add hydrogen atoms:

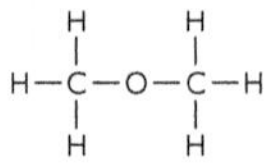

This structure *also* follows the HONC 1234 rule and is a correct structural formula for C_2H_6O. This molecule is called *dimethyl ether*. Ethanol and dimethyl ether are isomers. They have the same molecular formula but different structural formulas. Dimethyl ether has different properties from ethanol, including smell.

Not all structural formulas are simple to figure out. As the number and type of atoms increase, there are more possible ways to connect the atoms.

Example 2

Structural Formula for $C_4H_{11}N$

Draw a structural formula for the molecular formula $C_4H_{11}N$.

Solution

Start by making a chain of the carbon atoms. Place the nitrogen atom anywhere in the chain. Finish by adding hydrogen atoms so that every carbon atom has four bonds and the nitrogen atom has three bonds. Two possible solutions are shown here, but many more are possible.

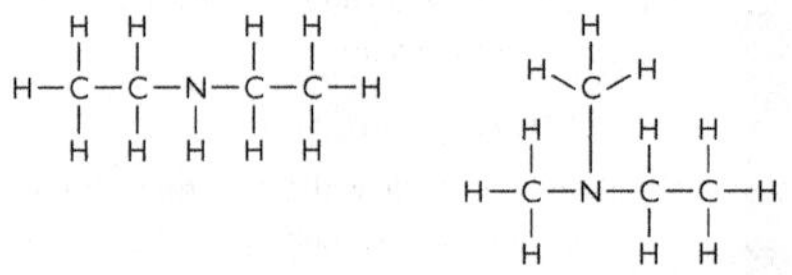

EVALUATE (5 min)

Check-In

Will any of the molecules shown here have similar smells? Explain your thinking.

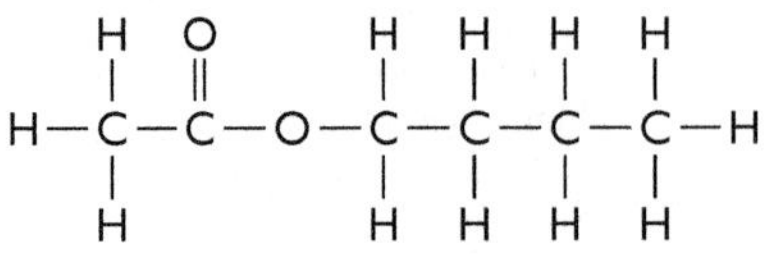

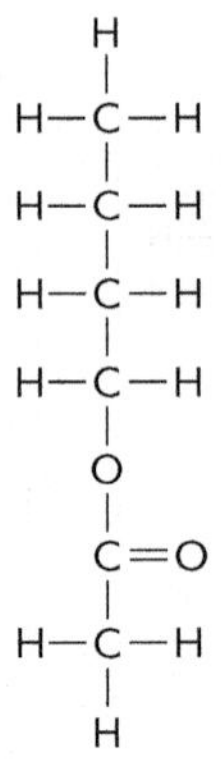

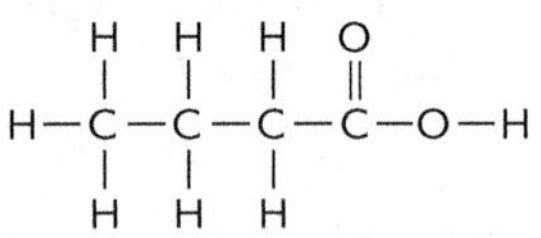

LESSON SUMMARY

What are the rules for drawing structural formulas?

The individual atoms in molecules are not connected randomly. Each atom in a molecule has a tendency to bond a specific number of times. The HONC 1234 rule describes the bonding patterns of hydrogen, oxygen, nitrogen, and carbon atoms. The HONC 1234 rule can be a valuable tool for creating structural formulas from molecular formulas.

KEY TERM

HONC 1234 rule

Exercises

Reading Questions

1. What is the HONC 1234 rule?

2. Explain why one molecular formula can represent more than one structural formula.

Reason and Apply

3. Use the HONC 1234 rule to create possible structural formulas for molecules with these molecular formulas. Remember that it is easiest to start with the carbon atoms.
   a. $C_3H_8O_2$ b. $C_4H_{11}N$ c. C_4H_{10} d. $C_5H_{12}O$

4. Is each structural formula correct according to the HONC 1234 rule? For any molecules that don't follow the HONC 1234 rule, repair the incorrect structural formula. (*Note:* There may be more than one correct way to repair a formula.)

a.
```
    H
    |
H — C — H
    |
H — C — H
    |
H — C — H
    |
    O
    |
    H
```

b.
```
    H
    |
H — O — H
    |
H — C — H
    |
H — C — H
    |
    H
```

c.
```
  H   H
  |   |
H—C — C—H
  |   |
  O   H
  |
  H
```

d.
```
  H   H
  |   |
H—C — C — N—H
  |   |
  O   H
  |
  H
```

e.
```
    H
    |
H — C — O — H
    |
H — C — H
    |
    H
```

5. Think about how molecules might interact with your nose. Why do you think molecules with different structural formulas have different smells?

b.
```
          H   H
           \ /
  H   H    N    H
  |   |    |    |
H—C — C — C — C—H
  |   |    |    |
  H   H    H    H
```

c.
```
  H   H   H   H
  |   |   |   |
H—C — C — C — C—H
  |   |   |   |
  H   H   H   H
```

d.
```
  H   H   H   H   H
  |   |   |   |   |
H—C — C — C — C — C—O—H
  |   |   |   |   |
  H   H   H   H   H
```

Other structural formulas are acceptable if they follow the HONC 1234 rule.

4. a. correct

b. One oxygen atom has three bonds, which is one too many. Possible diagram:

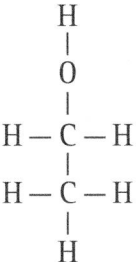

c. correct

d. One nitrogen atom has two bonds, which is one too few. Possible diagram:

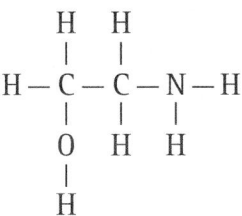

e. correct

5. *Possible answer:* Atoms on the molecules might react with smell receptors inside the nose. Different receptors may detect different types of molecular structures, leading to different sensations of smell.

Answer: The first and second molecules will smell the same, because they are the same molecule.

Homework

Assign the reading and exercises for Smells Lesson 30 in the student text.

LESSON 30 ANSWERS

1. The HONC 1234 rule describes the bonding patterns in molecules. Hydrogen forms one bond with other atoms. Oxygen forms two bonds with other atoms. Nitrogen forms three bonds with other atoms. Carbon forms four bonds with other atoms.

2. The atoms can form connections with different atoms or in a different order, making two or more different molecules with the same number of each type of atom.

3. *Possible answers:*

a.
```
        H   H   H
        |   |   |
H—O — C — C — C—O—H
        |   |   |
        H   H   H
```

Overview

ACTIVITY: GROUPS OF 4

Key Question: How does one atom bond to another in a molecule?

KEY IDEAS

Nonmetal atoms bond covalently. The bonding tendencies of nonmetal atoms are directly related to the number of valence electrons they have. Lewis dot symbols keep track of the valence electrons of different atoms. In covalent bonds, atoms share pairs of electrons, so only unpaired valence electrons are available for bonding. We can use Lewis dot symbols and the HONC 1234 rule to construct accurate structural formulas.

LEARNING OBJECTIVES

- Create accurate structural formulas using Lewis dot symbols.
- Describe the type of bonding found in molecular substances.
- Explain the chemistry behind the HONC 1234 rule.

FOCUS ON UNDERSTANDING

- This latest model—of electrons being paired within atoms—appears to contradict earlier models of the atom, in which electrons are found in a cloud or distributed around a shell. This might confuse students.
- Students might wonder why electrons pair, given that particles with like charges repel each other. Acknowledge that this is strange but is consistent with observations.
- In this lesson, students learn by doing. It is not necessary to thoroughly explain Lewis dot symbols before the activity.

KEY TERMS

Lewis dot symbol
Lewis dot structure
bonded pair
lone pair

IN CLASS

In this lesson, students use Lewis dot symbols and the HONC 1234 rule to create simple molecular structures from molecular formulas. Students are reintroduced to the concept of covalent bonding at the beginning of class. They are provided with Lewis dot manipulative materials to use in creating a variety of molecular structures. A complete materials list is available for download.

TRM Materials List 31

Featured ACTIVITY

Materials

- Lewis dot puzzle pieces

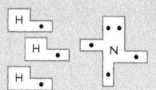

Three hydrogen atoms combine with one nitrogen atom to form

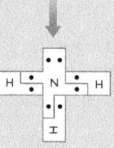

...a molecule of NH₃.

H:N:H
 ̈H

Lewis dot structure for NH₃

Connect the Dots

Purpose

To investigate the role of electrons in covalent bonding.

Instructions

The puzzle pieces you have been given represent Lewis dot symbols. The puzzle pieces allow you to pair up electrons and create molecules.

1. Use the puzzle pieces to construct the molecules given below. Then draw the Lewis dot structure for each molecule.

PH_3 HOCl F_2 CH_3Cl

2. Use the puzzle pieces to create more molecules, following the directions given below. For each molecule, draw the Lewis dot structure and write the molecular formula.
   a. Use one S atom and as many H atoms as you need.
   b. Use one Si atom and as many F atoms as you need.
   c. Use two O atoms and as many H atoms as you need.

3. Use the puzzle pieces to construct a molecule with the molecular formula C_2H_6. Draw its Lewis dot structure and its structural formula.

4. Use the puzzle pieces to construct all the possible isomers of C_3H_8O. Draw Lewis dot structures for each isomer. Do the molecules follow the HONC 1234 rule?

5. Use the puzzle pieces to design your own molecule with at least five carbon atoms. Draw its Lewis dot structure. What is the molecular formula of your designer molecule? Does it obey the HONC 1234 rule?

6. Find a puzzle piece for each type of atom. Put hydrogen and helium aside. Sort the rest of the puzzle pieces according to the periodic table. Record your sort by copying it into a grid. Include the symbols and the dots.

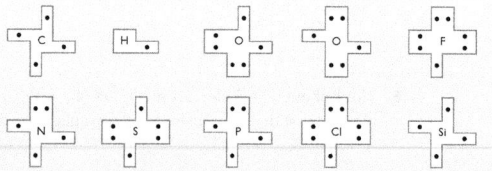

7. List two patterns that you notice.

8. **Making Sense** Using what you've learned, explain why the HONC 1234 rule works.

9. **If You Finish Early** Draw the Lewis dot structures of two different molecules with the molecular formula C_2H_7N.

SETUP

Direct students to the Lewis Dot Puzzle Pieces Web application on the book's companion Web site. Alternatively, you can download a set of pieces to print and cut out from the Teacher's Resource Materials. If you are using printed pieces, place a set of puzzle pieces in a resealable bag for each group of four.

Lesson Guide

ENGAGE (5 min)

TRM PowerPoint Presentation 31

TRM Transparency—ChemCatalyst 31

ChemCatalyst

These diagrams are called Lewis dot symbols.

·Ċ·   ·N̈·   :Ö·   H·

1. What is the relationship between the number of dots, the number of valence electrons, and the HONC 1234 rule?

2. Create a Lewis dot symbol for fluorine, F. How many bonds will fluorine make?

Sample answers: 1. Carbon has four electrons and makes four bonds. Nitrogen has five electrons and makes

Connect the Dots

Lewis Dot Symbols

THINK ABOUT IT

The HONC 1234 rule is a great trick to help you figure out the structures of molecules. But why does it work? Why does carbon connect with four atoms, while hydrogen can connect with only one atom? To answer these questions, let's take a closer look at bonds.

How does one atom bond to another in a molecule?

To answer this question, you will explore

- The Covalent Bond
- Lewis Dot Symbols and Structures
- Bonded Pairs and Lone Pairs

EXPLORING THE TOPIC

The Covalent Bond

All of the smelly compounds we have studied so far have been molecules. In fact, most of the substances that smell are made up of molecules, not ionic, metallic, or network covalent compounds.

The bonds that are found in molecules are called covalent bonds. Covalent bonds are formed between the atoms of nonmetallic elements. There are only about 15 nonmetallic elements in the periodic table. That is not many compared to the number of metallic elements.

The atoms involved in a covalent bond *share* a pair of valence electrons between them. The drawing below shows the difference between covalent bonds and ionic bonds.

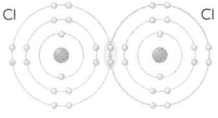
Covalent bonding: electron sharing

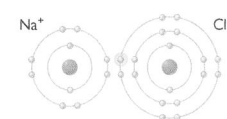
Ionic bonding: electron transfer

In covalent bonds, the nonmetal atoms are *sharing* electrons. As a result, the nonmetal atoms are tightly bound together, but the atoms do not become ions with charges.

 Big Idea A covalent bond is one in which nonmetal atoms *share* one or more pairs of electrons with one another. An ionic bond is one in which a metal atom *gives up* electrons to a nonmetal atom.

---

three bonds. Oxygen has six electrons and makes two bonds. Hydrogen has one electron and makes one bond. 2. Fluorine has seven valence electrons, so its Lewis dot symbol should show three paired electrons and one unpaired electron. Fluorine will make one bond.

Ask: How are the Lewis dot symbols related to the number of valence electrons?

Ask: How can you use the Lewis dot symbol to determine the number of bonds an atom will make?

Ask: What symbol did you draw for fluorine, F? How many bonds does fluorine make?

According to what you have learned so far, what is a covalent bond?

EXPLORE (20 min)

TRM Worksheet with Answers 31

TRM Worksheet 31

TRM Lewis Dot Puzzle Pieces (print version) 31

INTRODUCE THE ACTIVITY

→ Tell students to work in groups of four.

→ Pass out the student worksheet.

→ Introduce Lewis dot symbols. Draw one or two on the board. Tell students

that Lewis dot symbols are a way to keep track of valence electrons. These symbols are extremely useful tools for figuring out bonding and for creating correct molecular structures.

> **Lewis dot symbol:** A diagram that uses dots to show the valence electrons of a single atom.

→ Introduce students to the Lewis dot puzzle pieces. Show them that each puzzle piece contains the correct number of valence electrons for that atom. Also show them that each puzzle piece contains the appropriate number of tabs for bonding.

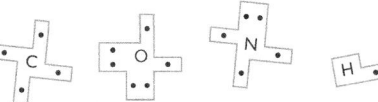

→ Tell students that they will be using the puzzle pieces to create molecules according to the worksheet directions.

EXPLAIN & ELABORATE (15 min)

TRM Transparency—Electron Pairs 31

DISCUSS THE USE OF LEWIS DOT SYMBOLS

→ Ask students to come to the board to draw the structural formulas of some of the molecules they created.

→ Use the HONC 1234 rule to check each molecule. Have students correct them as needed.

Ask: How do the puzzle pieces keep track of how many bonds each atom makes?

Ask: How many possible isomers does the molecular formula C_3H_8O have? Explain how you figured this out. (It has three.)

Key Points: You can use Lewis dot symbols to create Lewis dot structures. When you draw an atom using dots to represent the valence electrons, the drawing is called a *Lewis dot symbol*. A Lewis dot structure is a structural formula of a molecule in which the lines (covalent bonds) have each been replaced with a pair of dots. All the valence electrons on each atom in the molecule are shown as dots. Lewis dot structures make it clear that covalent bonds involve the sharing of a pair of electrons.

> **Lewis dot structure:** A diagram that uses dots to show the valence electrons of a molecule.

DISCUSS COVALENT BONDS AND THE PAIRING UP OF VALENCE ELECTRONS

→ Display the Transparency—Electron Pairs to introduce unpaired electrons, lone pairs, and so on.

Ask: Oxygen has six valence electrons and forms two bonds. Which electrons are involved in the bonding? (the solitary, or unpaired, ones)

Ask: How is the number of bonds related to the numbers of paired and unpaired electrons?

Ask: Why does the HONC 1234 rule work?

Key Points: A covalent bond is the sharing of a pair of electrons between two nonmetal atoms. This pair of electrons is referred to as a bonded pair. The Lewis dot symbols show us that the number of covalent bonds an atom makes is dependent on how many single, unpaired electrons it has.

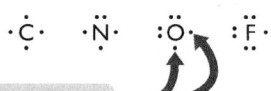

Oxygen makes two bonds. It has two unpaired valence electrons.

> **Bonded pair:** A pair of electrons that are shared in a covalent bond between two atoms.

Some valence electrons are not involved in bonding. The transparency shows the group of atoms that come together to form the structure methanol, CH_4O. The bonded pairs can be seen shared between two atoms. However, notice the pair of electrons not shared between two atoms in the molecule. These are referred to as lone pairs of electrons. They are not involved in bonding within the molecule.

> **Lone pair:** A pair of valence electrons not involved in bonding within a molecule. The two electrons belong to one atom.

EVALUATE (5 min)

Check-In

The molecular formula $C_4H_{10}O$ has seven different isomers. Draw the structural formula of one of them. You can use your puzzle pieces to assist you.

Answer: Isomers can have a string of four carbon atoms, or CH_3 groups that branch

---

CONSUMER CONNECTION

The smell of ammonia is powerful and irritating. Ammonia is toxic and can damage the interior of your nose and lungs. However, solutions with dissolved ammonia make good household cleansers. And small amounts of ammonia are used in smelling salts to revive a person who has fainted.

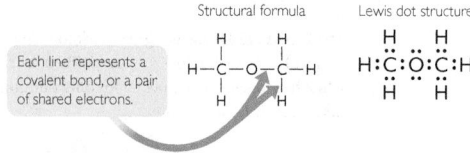

© Jonathan Nourok/PhotoEdit

◔ Lewis Dot Symbols and Structures

Structural formulas are one way to show the structure of a molecule on paper. Each line in a structural formula can be replaced with a pair of dots to represent the electrons that are being shared.

Each line represents a covalent bond, or a pair of shared electrons.

Structural formula | Lewis dot structure

A drawing of a molecule that uses dots to represent the valence electrons is called a **Lewis dot structure.** Above is a molecule of dimethyl ether, C_2H_6O. Both the structural formula and the Lewis dot structure represent the same molecule.

Lewis dot structures keep track of every valence electron in every atom of a molecule. If you break the molecule apart into its individual atoms, you can see where each valence electron came from.

A **Lewis dot symbol** is the symbol of an element with dots to show the number of valence electrons that a single atom of that element has. Examine these Lewis dot symbols for hydrogen, oxygen, nitrogen, and carbon. Notice that some electrons are not paired up. These electrons are referred to as *unpaired electrons* and are available for bonding. The unpaired electrons form covalent bonds with other atoms. Lewis dot symbols can help to explain the HONC 1234 rule and the bonding of nonmetal atoms.

1 valence electron	6 valence electrons	5 valence electrons	4 valence electrons
H·	·Ö:	·Ṅ:	·Ċ·
1 electron available for sharing	2 electrons available for sharing	3 electrons available for sharing	4 electrons available for sharing

Recall that you can determine the number of valence electrons in an atom by locating its element on the periodic table. Hydrogen is in Group 1A and has one valence electron. Carbon is in Group 4A and has four valence electrons. Once you know the number of valence electrons an atom has, it is easy to draw a Lewis dot symbol for it.

The Lewis dot symbol is named after Gilbert Newton Lewis, a chemist who was a professor at the University of California at Berkeley. Lewis first introduced the idea of shared electrons and covalent bonds in 1916. He was the first scientist to use a system of dots to represent valence electrons in atoms.

◔ Bonded Pairs and Lone Pairs

Lewis dot symbols can be used to draw Lewis dot structures and structural formulas for molecules. Imagine bringing together the Lewis dot symbols for hydrogen and carbon to form methane, CH_4.

H:C:H (with H above and below)

---

off the main carbon chain, or an oxygen atom between two carbon atoms. Two possible isomers are shown.

(structural formula diagram of C4H10O isomers)

Homework

Assign the reading and exercises for Smells Lesson 31 in the student text.

Because carbon bonds four times, it will bond with four hydrogen atoms. Once they are bonded, every valence electron in all of the atoms is paired up. The pairs of electrons between atoms are called **bonded pairs.** There are four bonded pairs in the methane molecule shown at the bottom of the previous page.

A pair of electrons in a molecule that are not shared between atoms is called a **lone pair** of electrons. In a molecule of ammonia, the nitrogen atom has one lone pair of electrons.

Example

Lewis Dot Structure of PCl_3

Examine the Lewis dot structure of PCl_3, phosphorus trichloride.

a. Draw the Lewis dot symbols for phosphorus and chlorine.

b. How many bonds does a phosphorus atom make? How many does each chlorine atom make?

c. How many bonded pairs does this molecule have? How many lone pairs?

d. How many covalent bonds does this molecule have?

Solution

a. The periodic table can help you figure out how many valence electrons phosphorus and chlorine have. Phosphorus is in Group 5A and chlorine is in Group 7A, so they have five and seven valence electrons, respectively.

b. Phosphorus has three unpaired electrons, so it makes three bonds. Chlorine has only one unpaired electron, so it makes one bond.

The three bonded pairs of electrons are circled.
All of the other pairs of electrons are lone pairs.

c. The molecule has three bonded pairs and ten lone pairs.

d. Each bonded pair represents one covalent bond. There are three covalent bonds in the molecule.

LESSON SUMMARY

How does one atom bond to another in a molecule?

Molecules are made of nonmetal atoms that are covalently bonded. A covalent bond is a bond in which a pair of electrons is shared by two atoms. Lewis dot symbols keep track of the number of valence electrons in each atom. They can

Lesson 31 | **Lewis Dot Symbols**   163

LESSON 31 ANSWERS

1. *Possible answer:* In an ionic bond, a valence electron is transferred from one metal atom to a nonmetal atom. In a covalent bond, two nonmetal atoms share a pair of electrons. Ionic and covalent bonds are similar because both involve the valence electrons of the atoms that are bonded together. The result of the bond is that the two atoms have an outer shell that is filled. In both types of bonds, electromagnetic forces hold the atoms together.

2. *Possible answer:* A Lewis dot structure is a diagram used to keep track of the valence electrons in an element or compound. To draw a Lewis dot structure, start by surrounding the symbol for an element with a dot for each valence electron. Then combine the elements according to the chemical or structural formula such that all unpaired electrons form bonded pairs with other atoms.

3. a.

K•	In•	•Pb•	•Bi•	•Te:	:I:
Group 1	3	4	5	6	7

b. Only nonmetal atoms form covalent bonds. Metal atoms form ionic bonds. Therefore, potassium, K, indium, In, lead, Pb, and bismuth, Bi, do not form covalent bonds. Tellurium, Te, forms two covalent bonds. Iodine, I, forms one covalent bond.

4. a. ••Ge•   •Sb•   •Se: :Br:

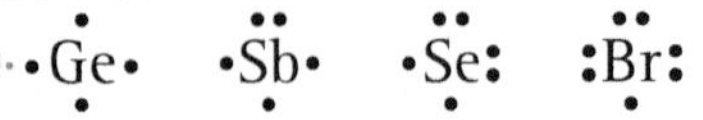

b.

```
    H              ••            ••           ••
H:Ge:H      H:Sb:H       H:Se:        :Br:
    H              H             H            H
```

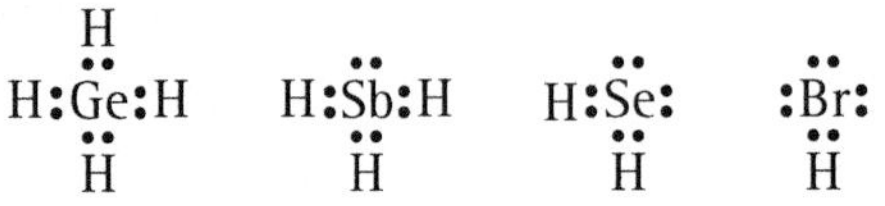

c. One hydrogen atom bonds with each unpaired electron. The total number of hydrogen atoms is equal to the number of unpaired electrons in the original Lewis dot structure.

5. a.

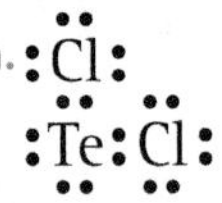

```
  ••
:Cl:
:Te:Cl:
  ••
```

b.

```
:I:H
```

c.

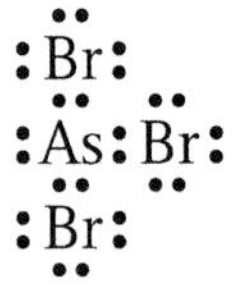

```
:Br:
:As:Br:
:Br:
```

KEY TERMS

Lewis dot structure
Lewis dot symbol
bonded pair
lone pair

help you to predict covalent bonding between atoms. A Lewis dot structure shows how the atoms in an entire molecule are bonded together. The valence electrons that are involved in bonding are called a bonded pair. The valence electrons that are paired up in a molecule, but do not take part in bonding, are called lone pairs.

Exercises

Reading Questions

1. How are an ionic bond and a covalent bond different? Similar?
2. What is a Lewis dot structure? Explain how you would create one.

Reason and Apply

3. Draw the Lewis dot symbols for these elements:

    Te   I   K   Bi   In   Pb
    a. Arrange them in order of their group numbers on the periodic table.
    b. Determine how many covalent bonds each element would make.

4. Germanium, antimony, selenium, and bromine each bond to a different number of hydrogen atoms.

    GeH_4 SbH_3 H_2Se HBr
    a. Draw Lewis dot symbols for Ge, Sb, Se, and Br.
    b. Draw a Lewis dot structure for each molecule.
    c. Explain the pattern in the number of hydrogen atoms.

5. Draw Lewis dot structures for the molecules listed here.
    a. $TeCl_2$ **b.** HI   **c.** $AsBr_3$ **d.** SiF_4 **e.** F_2

6. How many lone pairs does each of the molecules in Exercise 5 have?

7. In your own words, explain why the HONC 1234 rule works.

8. Draw Lewis dot puzzle pieces for Si, P, S, and Cl. What rule would you make for Si, P, S, and Cl? What would be the name of this bonding rule?

---

d.

```
   :F:
:F:Si:F:
   :F:
```

e. :F:F:

6. a. 8 lone pairs **b.** 3 lone pairs
c. 10 lone pairs **d.** 12 lone pairs
e. 6 lone pairs

7. *Possible answer:* The HONC 1234 rule works because it is based on the number of unpaired electrons in the atoms of each element. An atom can form one covalent bond for each unpaired valence electron.

8.

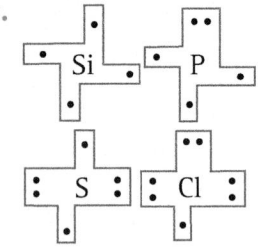

Silicon forms four bonds, phosphorus forms three bonds, sulfur forms two bonds, and chlorine forms one bond. The rule name could be ClSPSi 1234.

Eight Is Enough

Octet Rule

THINK ABOUT IT

When atoms bond covalently to form molecules, they share electrons to obtain an electron arrangement similar to that of a noble gas atom. Lewis dot structures can help you to discover how atoms share electrons to form molecules.

How do atoms bond to form molecules?

To answer this question, you will explore

↻ The Octet Rule

↻ Double and Triple Bonds

EXPLORING THE TOPIC

↻ The Octet Rule

When carbon, nitrogen, oxygen, and fluorine combine with hydrogen, they form these molecules:

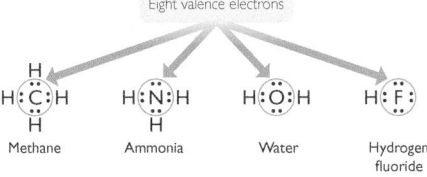

All of the above compounds are extraordinarily different. However, their Lewis dot structures reveal a striking similarity. Once carbon, nitrogen, oxygen, and fluorine are bonded, they are each surrounded by eight valence electrons.

Eight valence electrons

H:C:H   H:N:H   H:O:H   H:F:
Methane   Ammonia   Water   Hydrogen
                             fluoride

This tendency to bond until eight valence electrons surround an atom is called the **octet rule.** (The word *octet* comes from *octo*, which is Latin for "eight.") After these atoms are bonded, they all resemble atoms of a noble gas in their electron arrangements. Recall that the electron arrangements of the noble gases are very stable.

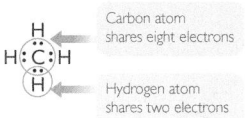

Carbon atom shares eight electrons

Hydrogen atom shares two electrons

Lesson 32 | **Octet Rule**   165

Overview

ACTIVITY: GROUPS OF 4

Key Question: How do atoms bond to form molecules?

KEY IDEAS

In molecular substances, atoms—with the exception of hydrogen—share electrons in such a way as to obtain a total of eight valence electrons. Thus, the electron arrangement of a covalently bonded atom resembles that of a noble gas. This tendency to achieve eight valence electrons through bonding is referred to as the *octet rule.* The octet rule also allows for double and triple bonding in covalent compounds.

LEARNING OBJECTIVES

● Apply the octet rule to predict bonding in molecules.
● Draw Lewis dot structures and structural formulas for molecules that contain double and triple bonds.

KEY TERMS

octet rule
double bond
triple bond

IN CLASS

Students are introduced to the octet rule at the opening of class. While completing a worksheet, students

practice creating structural formulas. They problem-solve structures with double and triple bonds. A complete materials list is available for download.

TRM Materials List 32

Differentiate

Students may need extra support in understanding how certain atoms can form multiple bonds. One way to help students build on this concept is to create a wall chart that illustrates how structural formulas, Lewis dot structures, and the number of valence electrons relate to each other. For students who quickly master Lewis dot structures, challenge them to develop structures for molecules such as CO and SO_3 and polyatomic ions such as NH_4^+ and CO_3^{2-}.

Lesson Guide

ENGAGE (5 min)

TRM PowerPoint Presentation 32

> *ChemCatalyst*
>
> Draw the Lewis dot structure for the two covalently bonded molecules shown here. Explain how you arrived at your answer.
>
> **a.** Cl_2
>
> **b.** O_2

Sample answers: **a.** The Lewis dot symbol for chlorine has seven dots, so it has only one electron available for bonding. Two chlorine atoms must share a pair of electrons. **b.** Oxygen has six valence electrons, so it has two electrons available for bonding. Perhaps two oxygen atoms share two pairs of electrons.

:Cl· + ·Cl: ⟶ :Cl:Cl:

:O· + ·O: ⟶ :O::O:

→ Guide students toward the octet rule as they discuss their various solutions to the ChemCatalyst task.

→ Ask a student to draw the Lewis dot structure for Cl_2 on the board. Work with it until it is correct. Circle the eight valence electrons associated with each chlorine atom in Cl_2.

:Cl· + ·Cl: ⟶ (:Cl:Cl:)

→ Ask students to draw the Lewis dot structure for O_2 on the board. Work with it until it is correct. Circle the eight valence electrons associated with each oxygen atom in O_2.

Ask: How did you arrive at your drawing for Cl_2?

Ask: What other molecules have similar Lewis dot structures? (all the halogens—F_2, Br_2, I_2, At_2)

Ask: How did you arrive at your drawing for O_2?

> **Octet rule:** Nonmetal atoms combine so that each atom has a total of eight valence electrons by sharing electrons.

EXPLORE (20 min)

TRM **Worksheet with Answers 32**

TRM **Worksheet 32**

TRM **Lewis Dot Puzzle Pieces (print version) 31**

INTRODUCE THE ACTIVITY

→ Tell students to work in groups of four.

→ Pass out the student worksheet.

→ Introduce the octet rule. Draw one or two other Lewis dot structures to show that the octet rule applies to most molecules.

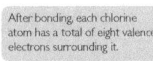
Cl_2 PCl_3 H_2S

→ Tell the class that they will use the octet rule to create models of molecules.

→ Remind students that each line in a structural formula represents a pair of bonded electrons.

→ Direct students to the Lewis dot puzzle pieces Web application or hand out a printed set to each group.

EXPLAIN & ELABORATE (15 min)

DISCUSS THE MOLECULAR STRUCTURES FROM THE CLASSWORK

→ Ask students to come to the board to draw the various structures from their worksheet.

Ask: What process did you use in completing the Lewis dot structures?

Note that hydrogen is an exception to the octet rule. Each hydrogen atom shares two electrons, not eight. A covalently bonded hydrogen atom resembles the noble gas helium, He, which has only two valence electrons.

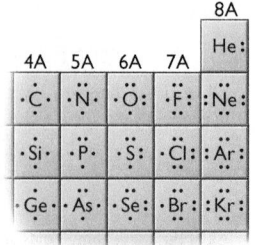
F_2 SCl_2 CH_4O

PERIODIC TRENDS

To predict covalent bonding in molecules, you can use the periodic table. In the illustration, Lewis dot symbols are shown for nonmetal elements in the second, third, and fourth rows. Notice that the elements in the same group have similar Lewis dot symbols and bond in a similar way. For example, selenium, Se, which is below sulfur, S, and oxygen, O, has two unpaired electrons and forms two bonds, just like oxygen.

Elements in the same group have the same number of valence electrons and therefore have similar Lewis dot symbols. Helium is an exception.

↻ Double and Triple Bonds

There is more than one way to satisfy the octet rule through bonding. Quite a few of the structural formulas you have examined so far have a bond with two lines. This type of bond is called a **double bond**. In a Lewis dot structure, a double bond is represented by four dots instead of the usual two. A double bond contains four electrons that are shared between atoms. Methyl methanoate, $C_2H_4O_2$, is an example of a molecule that contains a double bond. Its structural formula and Lewis dot structure are shown here.

> A double bond means four shared electrons.

The carbon and oxygen atoms with double bonds are surrounded by a total of eight valence electrons each, just like the atoms with single bonds.

Ask: How can you use the HONC 1234 rule to check your Lewis dot structures?

Ask: How did you know where to put lone pairs of electrons on the structural formulas?

Key Points: The HONC 1234 rule and the octet rule both help you figure out Lewis dot structures and structural formulas. The HONC 1234 rule allows you to figure out how many bonds there are for each H, O, N, and C atom. For example, nitrogen forms three bonds. Lewis dot structures show that the bonds are pairs of electrons shared between two atoms. In addition, the Lewis dot structure shows lone pairs of electrons. For example, nitrogen

has three bonding pairs of electrons and one lone pair.

DISCUSS DOUBLE AND TRIPLE BONDING

→ Draw the Lewis dot structure and structural formula of C_2H_2, C_2H_4, and C_2H_6 on the board.

$H:C:::C:H$

$H-C\equiv C-H$

Example

Carbon Dioxide

Draw the Lewis dot structure and structural formula for carbon dioxide, CO_2.

Solution

Step 1: Start with the Lewis dot symbols. Bring the atoms together.

:O: •C• •O: → :O:C:O: Incorrect

Step 2: Check to see if the octet rule is satisfied. This Lewis dot structure isn't correct. Move the remaining unpaired electrons to create double bonds.

Incorrect   Correct
:O:C:O: → :O::C::O:

Step 3: Check your molecule again to see that each atom in it satisfies the octet rule and still has the correct number of valence electrons.

 Each atom is now surrounded by a total of eight valence electrons.

Step 4: To make the structural formula, replace each pair of dots with a line.

$$O=C=O$$

If two atoms can share four electrons, can they share six electrons, or even more? The answer is yes.

N≡N   O=O   H—H

:N:::N: :O::O: H:H

Triple bond   Double bond   Single bond

Nitrogen gas, N_2, has a **triple bond** between the two nitrogen atoms, so each nitrogen atom has six shared valence electrons. Notice that there are also two lone pairs in this molecule. Oxygen gas, O_2, has a double bond between the oxygen atoms. The four lone pairs in the oxygen molecule have been adjusted slightly in this Lewis dot structure to space the electrons more evenly around the atoms. The placement of the dots in a Lewis dot structure can vary as long as the number of pairs of electrons is correct. Quadruple bonds are also possible, but quite rare.

Some of the most common substances in the world around you are molecules with double and triple bonds. The air you breathe is composed mainly of nitrogen gas and oxygen gas. In fact, the air is 78% nitrogen gas.

LESSON SUMMARY

How do atoms bond to form molecules?

KEY TERMS
octet rule
double bond
triple bond

When nonmetal atoms bond, they share electrons to obtain the same electron arrangement as a noble gas atom. Nonmetal atoms will share electrons with other atoms so that both atoms share a total of eight valence electrons each. This is called the octet rule. Hydrogen still fits the sharing pattern, but it ends up with only two

---

H   H
H:C:C:H
H   H

H   H
H—C—C—H
H   H

Ask: When do you need double bonds to satisfy the octet rule? Explain.

Ask: Do molecules with double bonds satisfy the HONC 1234 rule? Explain.

Ask: Is it possible to make a triple-bonded oxygen compound? Explain.

Ask: In theory, are quadruple-bonded carbon compounds possible? Explain.

Key Points: Both the HONC 1234 rule and the octet rule can be satisfied by using double and triple bonds appropriately. Double and triple bonds are sometimes referred to as multiple bonds. It is not possible to create a triple-bonded oxygen compound, according to the HONC rule. Oxygen has only two unpaired electrons and thus forms only two bonds. A quadruple-bonded carbon compound or silicon compound is possible, but these compounds are not very stable and do not last long.

There are exceptions to the bonding rules laid out here. One notable exception is carbon monoxide, a

dangerous gas that has almost no smell. It has a triple bond between the carbon and the oxygen atoms, meaning that it does not satisfy the HONC 1234 rule. However, it does satisfy the octet rule in an unusual fashion. The oxygen atom contributes four electrons to the triple bond, while the carbon atom contributes two.

•C• and •O: form :C:::O:

SPECULATE ON THE SMELLS ASSOCIATED WITH THE STRUCTURES IN THE CLASSWORK

Ask: What smells did you predict for the various molecules? What was your reasoning?

Ask: Have you smelled any of these compounds? (Carbon dioxide and carbon monoxide are both odorless gases. HCN, hydrogen cyanide, smells like almonds.)

Key Points: Here are a few smells that students might identify. Molecules with a nitrogen atom might smell fishy, molecules with a single oxygen atom might smell minty, and molecules with two oxygen atoms might smell sweet. Carbon monoxide and carbon dioxide do not have a smell. CH_4O_2 has no double bonds, as do most sweet-smelling molecules, so perhaps it does not smell sweet.

EVALUATE (5 min)

Check-In

1. Which one of these compounds has multiple bonds in it? Explain.

C_4H_{10}	C_4H_6

2. Draw one possible structural formula for C_4H_6.

Answers: 1. C_4H_6. The compound with fewer hydrogen atoms must have multiple bonds. **2.** Two possible structural formulas for C_4H_6 are shown (there are others).

H   H
H—C=C—C=C—H
        H   H

H   H
H—C≡C—C—C—H
        H   H

Homework

Assign the reading and exercises for Smells Lesson 32 in the student text.

LESSON 32 ANSWERS

1. *Possible answer:* Nitrogen has three unpaired electrons, as shown in the Lewis dot structure. Hydrogen atoms have one unpaired electron, so three hydrogen atoms can form bonds with one nitrogen, giving the nitrogen an octet of electrons and each hydrogen two electrons. In NH_2, nitrogen would have only seven electrons, and in NH_4, an extra hydrogen atom is left over after all the unpaired electrons in the nitrogen atom have formed bonds.

2. The octet rule states that atoms combine so that each atom (except hydrogen and helium) is surrounded by eight valence electrons. The rule is used to determine the number of bonds each atom forms and to show the structure of connections between atoms in a molecule.

3. *Possible answers (any 3):* hydrogen, chlorine, bromine, iodine, astatine, or another fluorine.

4. Because hydrogen has one unpaired electron to share, the elements that will combine with three hydrogen atoms are those that have three unpaired valence electrons. Nitrogen and phosphorus are the only nonmetal elements in Group 5A.

5.

6.

valence electrons like the noble gas helium. Atoms can also satisfy the octet rule by forming double and triple bonds in which they share four or six valence electrons.

Reading Questions

1. Explain why nitrogen bonds with hydrogen to form NH_3, but not NH_2 or NH_4. Use Lewis dot structures to support your argument.

2. What is the octet rule, and how can you use it to create a molecular structure?

Reason and Apply

3. List three nonmetal elements that combine with only one fluorine atom to satisfy the octet rule.

4. List two nonmetal elements that combine with three hydrogen atoms to satisfy the octet rule.

5. Draw Lewis dot structures for these molecules. Notice that in part d and part f, the formulas are written in a way that emphasizes the structure of the molecule.
   a. CF_4 b. CH_3Cl c. $SiCl_2H_2$
   d. CH_3OH e. $HOCl$ f. CH_3NH_2

6. Use the octet rule to draw Lewis dot structures for all the stable molecules with the molecular formula C_3H_8O. (*Hint:* There are three total molecules.)

7. Consider the molecules C_2H_2, N_2H_2, and H_2O_2.
   a. Draw a Lewis dot structure for each one. What pattern do you notice?
   b. What can you do to check that your Lewis dot structures are correct? Name at least two ways.

8. Consider the molecules C_2H_4 and N_2H_4. Draw a Lewis dot structure for each of the molecules.

9. Which is more likely to exist in nature, a molecule of CH_3 or a molecule of CH_4? Explain your reasoning.

7. a. *Possible answer:* The number of bonds between the central atoms decreases as the number of valence electrons increases.

b. *Possible answers:* Make sure that each atom other than hydrogen is surrounded by eight electrons, and each hydrogen is surrounded by two electrons. Make sure that the number of connections follows the HONC 1234 rule. Compare the Lewis dot structure to a structural formula. Each bond should be replaced be a bonding pair, with double bonds having two bonding pairs and triple bonds having three bonding pairs.

8.

9. *Possible answer:* CH_4 would form a stable compound because all of the atoms in the molecule are surrounded by the most stable number of valence electrons—eight for carbon, two tor hydrogen. In CH_3, the carbon atom would only have seven electrons in its outer shell.

Where's the Fun?

Functional Groups

THINK ABOUT IT

It makes sense that the structure of a molecule would affect its properties. The atoms in ethyl butyrate (pineapple smell) and hexanoic acid (dirty sock smell) are connected differently, so they behave differently when they enter your nose. But what about molecules that smell similar? What is it about these structures that causes them to have similar properties?

What does the structure of a molecule have to do with smell?

To answer this question, you will explore

- Relating Smell to Molecular Structure
- Functional Groups
- Classifying Molecules

EXPLORING THE TOPIC

Relating Smell to Molecular Structure

PUTRID-SMELLING MOLECULES

Consider two molecules in the same smell category. Both hexanoic acid and butyric acid smell putrid. The first smells like stinky feet while the second smells like a carton of spoiled milk. Take a moment to compare the structural formulas of hexanoic acid and butyric acid.

Hexanoic acid, $C_6H_{12}O_2$ Butyric acid, $C_4H_8O_2$

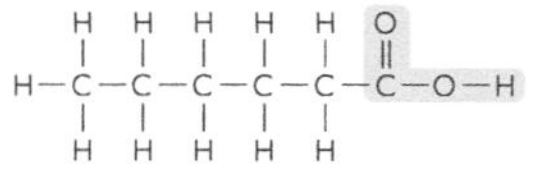

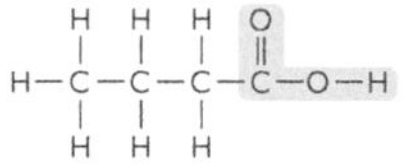

Notice that both molecules have two oxygen atoms bonded on the end in an identical fashion. They both contain the structural feature highlighted.

Perhaps all putrid-smelling molecules have this same structural feature. Two more compounds that smell putrid are shown here.

Isopentanoic acid, $C_5H_{10}O_2$ Isobutyric acid, $C_4H_8O_2$

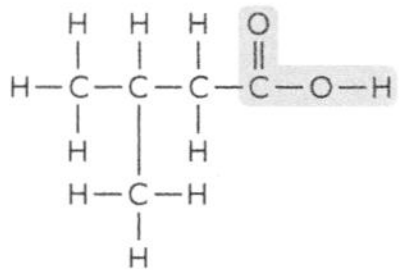

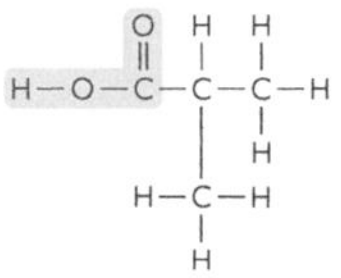

Lesson 33 | **Functional Groups**   169

Overview

ACTIVITY: PAIRS

Key Question: What does the structure of a molecule have to do with smell?

KEY IDEAS

The properties of a molecule are intimately related to common structural features of those molecules, in particular, to functional groups. Smell is one property directly related to functional groups. The presence of functional groups in molecules is also the key to naming molecular compounds.

LEARNING OBJECTIVES

- Identify and name basic functional groups within molecules.
- Relate certain functional groups to certain smell categories.
- Describe the naming patterns found among molecules associated with specific functional groups.
- Deduce the probable smell of a compound from its name or structural formula.

FOCUS ON UNDERSTANDING

It takes practice to visually distinguish functional groups. The differences between functional groups can be subtle. Also, a molecule's orientation

in space can cause confusion in identifying functional groups.

KEY TERM

functional group

IN CLASS

This activity introduces functional groups to the class. Pairs of students sort cards containing information on the smell, molecular formula, chemical name, and structural formula of 21 different molecules. They look for patterns in the data that might relate the structure of the molecules of a compound to the compound's smell. After results are shared, the instructor introduces the term *functional group* and clarifies each functional group's specific structure. Finally, a Smell Summary chart is compiled based on the evidence acquired so far. *Note:* The Teacher Edition and Student Edition use a combination of IUPAC and common names for molecules. The IUPAC names are more systematic, but in some cases the common name is much shorter or more familiar (e.g., L-carvone rather than its IUPAC name, 2-methyl-5 prop-l-en-2-yl-cyclohex-2-en-1-one.) A complete materials list is available for download.

TRM Materials List 33

SETUP

Prepare card sets to pass out to pairs of students. There should be 21 cards in each set.

Differentiate

This lesson provides a good opportunity for students to summarize their learning. One way is to work as a class to generate a Smell Summary chart as described in the lesson. Another way is to make the chart an individual assignment for students to work on throughout the unit. Ask students to create their own visual aid to summarize the chemistry learned by studying smells. The visual aid could be similar to the Smell Summary chart suggested in the lesson but could also incorporate drawings or images students find that relate to the chemistry of smell. The assignment can be ongoing throughout the unit as more chemistry is learned.

Pre-AP® Course Tip

Students preparing to take an AP® chemistry course or college chemistry should be able to analyze data to identify certain types of patterns and relationships among chemical phenomena. In this lesson, students study patterns on a set of structural formula cards to find common functional groups in different organic compounds. Based on the type of functional group, they can hypothesize the type of smell a compound will have.

Lesson Guide

ENGAGE (5 min)

TRM PowerPoint Presentation 33

TRM Transparency—ChemCatalyst 33

TRM Card Masters—Structural Formulas 33

ChemCatalyst

Consider the following compounds. List at least three differences and three similarities between the two molecules.

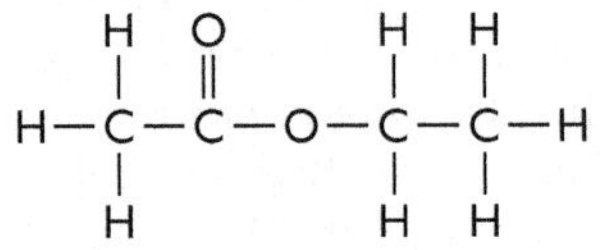

Molecule 1

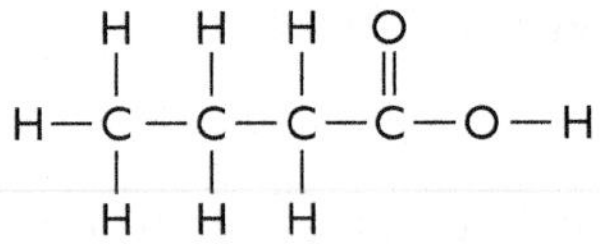

Molecule 2

Sample answers: Differences: The structures are different. The double-bonded oxygen atom is connected to the second carbon atom in one molecule and to the fourth carbon atom in the other. The oxygen atoms are in the middle of one molecule and near the end of the other molecule. Similarities: They are composed of the same elements. Both have the same molecular formula, $C_4H_8O_2$. Both have two oxygen atoms and a double-bonded oxygen atom. They are similar in shape.

Ask: What are the major differences in the structures of these molecules?

Ask: How do you think these two molecules smell? Would you be surprised to learn that they have different smells? Why or why not?

Ask: Do you think there is a relationship between the structure of a molecule and its properties? Explain your thinking.

CONSUMER CONNECTION

The ingredients list on most of your household products reads like a chemical inventory. Examine the labels of your shampoos, cleansers, hair gels, mouthwash, and so on, to see which chemical names you recognize.

Catherine Bausinger

You can see that this same structural feature is present in isopentanoic acid and isobutyric acid as well.

SWEET-SMELLING MOLECULES

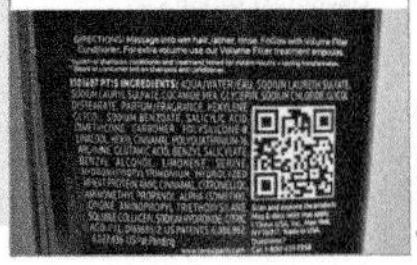

Now look at the structural formulas for two sweet-smelling molecules.

Ethyl butyrate, $C_6H_{12}O_2$

Ethyl acetate, $C_4H_8O_2$

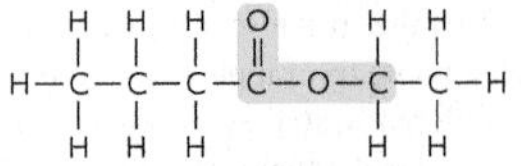

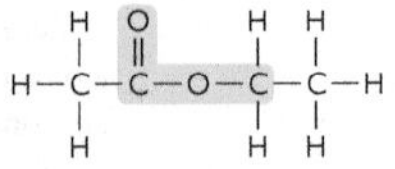

Just like the putrid-smelling molecules, these sweet-smelling molecules have an identical structural feature, which is highlighted in yellow. This feature is similar to the one found in putrid-smelling molecules, but it is slightly different.

Functional Groups

The structural features that groups of molecules have in common are called **functional groups.** A functional group often stands out as an unusual or unique portion of a molecule. The functional group found in the putrid-smelling molecules is called a *carboxyl* functional group. The functional group found in the sweet-smelling molecules is called an *ester* functional group.

Two other functional groups are associated with fishy smells and minty smells.

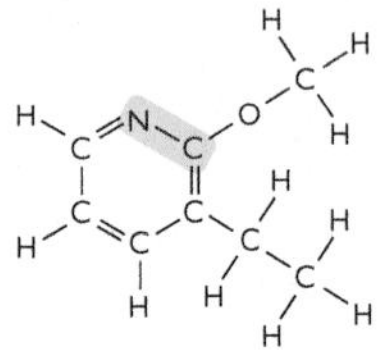

Amine group, fishy

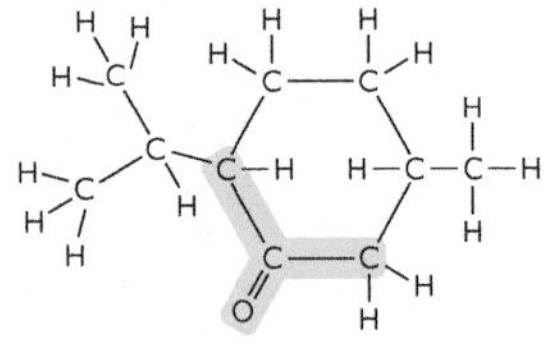

Ketone group, minty

Example 1

Functional Group and Smell

The structural formulas of two molecules are shown here. Predict how the molecules will smell. What is your reasoning?

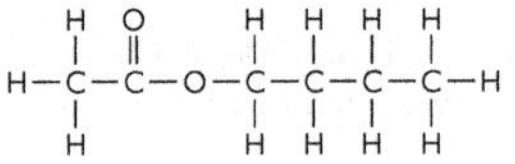

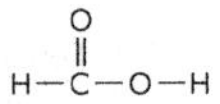

Solution

This molecule contains an ester functional group. It is probably a sweet-smelling molecule.

This molecule contains a carboxyl functional group. It is probably a putrid-smelling molecule.

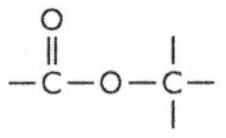

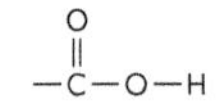

EXPLORE (15 min)

TRM Worksheet with Answers 33

TRM Worksheet 33

INTRODUCE THE ACTIVITY

→ Tell students to work in pairs.

→ Pass out the worksheet to each group.

→ Hand out Structural Formula cards to each pair to sort according to the directions on the worksheet.

EXPLAIN & ELABORATE (20 min)

TRM Transparency—Ketones 33

TRM Transparency—Amines 33

TRM Transparency—Carboxylic Acids 33

TRM Transparency—Esters 33

TRM Transparency—Alkanes 33

DISCUSS THE PATTERNS FOUND IN THE MOLECULAR STRUCTURES

→ Make a list of students' generalizations.

→ Allow students to assist you in identifying the common structural features in molecules that smell the same. Have students draw these features on the board.

OTHER FUNCTIONAL GROUPS

There are other functional groups besides the four discussed so far. For instance, a molecule may have a hydroxyl group (−OH). Molecules containing this feature are called *alcohols*. A molecule containing an aldehyde group is similar to a ketone, except that the carbon that is double bonded to the oxygen is between a carbon and a hydrogen.

Hydroxyl group   Aldehyde group

Another type of molecule is one containing a single oxygen located between two carbon atoms. Molecules containing this feature are called *ethers*.

Ether group

Some molecules have no functional groups. For example, *alkanes* consist of only carbon and hydrogen atoms connected with single bonds.

Alkane

⟳ Classifying Molecules

The easiest way to classify molecules is by functional group. All the molecules that have ester functional groups are referred to as *esters*. All the sweet-smelling compounds you have encountered so far contain an ester functional group and have two-word names that end in "-yl" and "-ate." Eth**yl** butyr**ate,** eth**yl** acet**ate,** and hex**yl** acet**ate** all smell sweet.

The putrid-smelling compounds have a carboxyl group and names that end with "-ic acid." Butyr**ic acid** and hexano**ic acid** both smell putrid.

You may be able to identify the smell of a molecule by paying attention to its chemical name. Examine the compounds in the table. Each of these molecules has a unique smell even though they all have three carbon atoms and at least six hydrogen atoms. Their names are also very different.

Name	Compound type	Functional group	Smell	Common use or source
propane C_3H_8	alkane	no functional group	no smell or gasoline-like	fuel in camp stoves
propionic acid $C_3H_6O_2$	carboxylic acid	carboxyl	putrid	in sweat

Lesson 33 | **Functional Groups**   171

atom connected to two carbon atoms, one of which is bonded to a second oxygen atom. The other parts of the sweet-smelling molecules may be quite different from each other, but every sweet-smelling molecule in our sample contains this same feature.

You can see that minty-smelling molecules all have an identical structural feature as well: a carbon atom connected to a single oxygen atom by a double bond. These specific features are referred to as functional groups.

> Functional group: A cluster of atoms in a molecule that is responsible for many of its properties.

INTRODUCE FUNCTIONAL GROUPS

→ Display the transparencies showing the molecules sorted according to smell.

→ You might draw and label each functional group separately on the board for clarity.

Ask: What functional group is found in all the fishy-smelling molecules? The putrid-smelling molecules?

Ask: How are the putrid-smelling molecules and the sweet-smelling molecules similar?

Ask: How are they different?

Ask: What generalization can you make about the molecules that smell like gasoline?

Ask: How are the names of the molecules related to their functional groups?

Key Points: The functional groups have names, and molecules frequently are named according to the functional groups they contain.

Carboxyl group

Ask: What were some of the discoveries you made as a result of sorting the cards?

Ask: Do your best to describe the structural features that all the sweet-smelling molecules have in common.

Ask: What structural features do you find in all the fishy-smelling molecules?

Ask: What do the molecules ending in "-ane" have in common?

Ask: What do all the molecules that smell like gasoline have in common?

Key Points: A number of patterns emerge from the card sort. For example, the smell and the name of a

molecule are directly connected. Also, both the sweet-smelling molecules and the putrid-smelling molecules have two oxygen atoms. The minty-smelling molecules all contain a ring and have a double-bonded oxygen atom. If the cards are sorted by smell, you will find five groups: fishy, minty, sweet, putrid, and gasoline smell. Sorting by name should produce the same outcome.

Each group of molecules with a similar smell has something identical in its structure. All the sweet-smelling molecules, for example, contain an identical piece in them: an oxygen

LESSON 33

Ester group

Ketone group

Amine group

Functional groups can be shown without the rest of the molecule.

Carboxyl group

Ketone group

Ester group

Amine group

All the molecules containing ester functional groups end in "-ate." All the ketones end in "-one." All the amines end in "-ine."

There are other functional groups not covered here. Alcohols are molecules that all contain a hydroxyl group (–OH). The names of all alcohols end in "-ol," for example, ethanol or propanol. Students will discover more about the smells of alcohols in later lessons.

Hydroxyl group

Molecules that smell like gasoline do not seem to have any particular functional group. These molecules are called *alkanes*, and they are made entirely of carbon and hydrogen atoms.

Name	Compound type	Functional group	Smell	Common use or source
ethyl formate $C_3H_6O_2$	ester	ester	sweet	in raspberries
acetone C_3H_6O	ketone	ketone	sweet	nail polish remover
isopropanol C_3H_8O	alcohol	hydroxyl	medicinal	rubbing alcohol
timethylamine C_3H_9N	amine	amine	fishy	in bad breath

Compounds that have the same functional group tend to have similar properties. For example, esters smell sweet, dissolve in water, and change phase from liquid to gas fairly easily. There are exceptions. For example, larger alkanes tend to have a gasoline smell, while smaller alkanes like propane in the table have almost no smell.

LESSON SUMMARY

What does the structure of a molecule have to do with smell?

Based on what the lessons have covered so far, compounds that have similar smells also appear to have similar structural features. These features are called functional groups. Functional groups have names. Compounds are frequently named according to the functional groups they contain. If you identify the functional group in a compound, you may also be able to predict how that compound will smell.

KEY TERM
functional group

172   Chapter 6 | **Speaking of Molecules**

Alkane

The structure of an aldehyde is like that of a ketone, except that the carbon atom that is double bonded to the oxygen atom is attached directly to at least one hydrogen atom. In the structure of formaldehyde, the carbon atom with the double-bonded oxygen is attached directly to two hydrogen atoms.

Aldehyde group

An ether has a single oxygen atom located between two carbon atoms.

Ether group

There are exceptions to the patterns related to smell. For example, small alkanes such as methane, CH_4, or propane, C_3H_8, have almost no smell or no smell at all.

CREATE A SMELLS SUMMARY CHART

→ Tape up a large piece of paper (about 3 ft by 3 ft). Create a poster that summarizes the class's thinking so far.

Reading Questions

1. What is a functional group?

2. What information would you want to have to predict the smell of a compound? Explain why.

Reason and Apply

3. Explain why C_2H_4 has fewer hydrogen atoms than C_2H_6.

4. Create a four-carbon molecule that is
   a. an alkane
   b. a carboxylic acid
   c. an alcohol
   d. an ester

5. Consider a compound called hexanol. Chemists can tell from its name that it has six carbons (hex-) and it is an alcohol (-ol).
   a. What is the molecular formula of hexanol?
   b. Draw a possible structural formula for hexanol.
   c. Is the molecular formula or the structural formula more useful in determining the smell of hexanol? Explain.

6. **Research** Examine the ingredients lists on some household products, such as shampoo, lotion, or cleanser. Write down the names of compounds and functional groups that you were able to identify.

7. If you were a chemist and you wanted to invent a new smell, what would you think about doing or creating?

8. If you were a chemist and you wanted to change the smell of a molecule, what might you try to do to that molecule?

9. Explain what you think is going on in this cartoon.

Pentyl propionate   Octanoic acid

→ Solicit information about how molecular formula, chemical name, and functional group are related to smell.

Ask: Is the molecular formula of a molecule ever helpful in predicting smell? If so, when?

Ask: What specific information in the chemical name helps you predict smell?

Ask: How can you use a functional group to predict smell?

Ask: Do any other features or properties help you predict smell?

Ask: Do we have to go further in our investigation to understand the chemistry of smell? Explain your thinking.

EVALUATE (5 min)

Check-In

If a molecule is sweet-smelling, what other things do you know about it? List at least three things that are probably true.

Answer: It probably contains two oxygen atoms and an ester functional group. One of the oxygen atoms is double bonded. The chemical name probably ends in "-ate."

Homework

Assign the reading and exercises for Smells Lesson 33 in the student text.

LESSON 33 ANSWERS

1. A functional group is a portion of a molecular structure that is the same in all molecules of a certain type.

2. *Possible answer:* The smell of a compound is related to the functional groups of its molecules. To figure out the smell, you have to know what atoms are in the molecule and how they are arranged.

3. A C_2H_4 has fewer hydrogen atoms than C_2H_6 because the two carbon atoms are held together by a double bond. There are only two unpaired electrons on each carbon atom available to form bonds with hydrogen.

4. *Possible answers:*

a.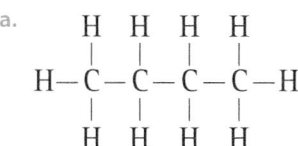

Smells Summary Chart

Molecular Formula
1 N atom = fishy
2 O's = sweet or putrid
1 O atom = minty

Some other feature?

Smell

Functional Group
amine = fishy
ester = sweet
ketone = minty
carboxyl = putrid
alkane = gasoline
alcohol = ?

Chemical Name
"-ine" = fishy
"-ate" = sweet
"-one" = minty
"-ic acid" = putrid
"-ane" = gasoline
"-ol" = ?

b.

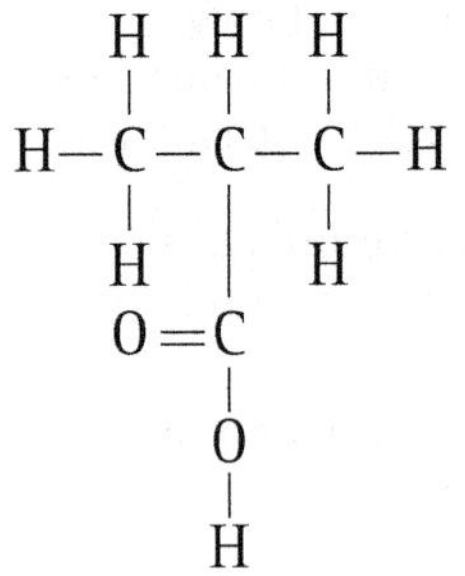

```
      H   H   H
      |   |   |
  H — C — C — C — H
      |   |   |
      H   |   H
          O = C
              |
              O
              |
              H
```

c.

```
      H   H   H   H
      |   |   |   |
  H — C — C — C — C — O — H
      |   |   |   |
      H   H   H   H
```

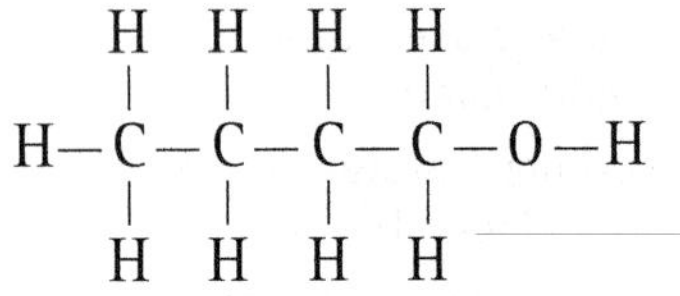

d.

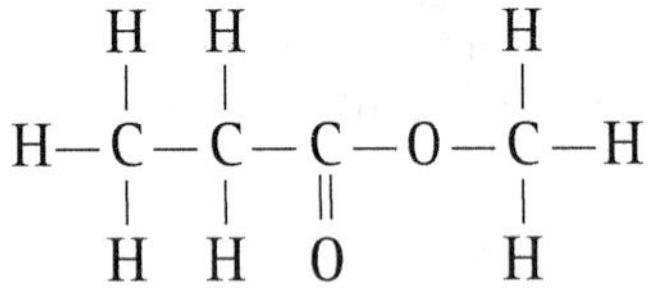

```
      H   H           H
      |   |           |
  H — C — C — C — O — C — H
      |   |   ‖       |
      H   H   O       H
```

5. a. $C_6H_{14}O$

b.

```
      H   H   H   H   H   H
      |   |   |   |   |   |
  H — C — C — C — C — C — C — O — H
      |   |   |   |   |   |
      H   H   H   H   H   H
```

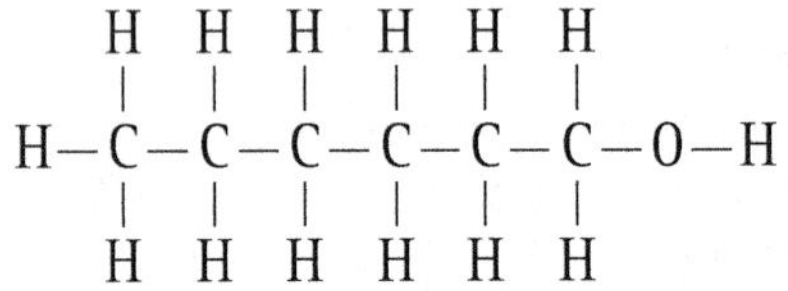

c. The structural formula is more useful because it shows the hydroxyl functional group (–OH) that is characteristic of alcohols, and alcohols have a characteristic smell.

6. Possible answer: glycol distearate, ester; sodium benzoate, ester; cetyl alcohol, alcohol; polythethylene glycol, alcohol; triethanolamine, alcohol and amine; dimethicone, ketone.

7. Possible answer: To create a new smell, decide on the type of smell wanted. Then pick the functional group that is present in similar smells. Then try making small changes in the rest of the molecule and see how that affects the smell.

8. Possible answer: To make large changes, change the functional group of the molecule. To make smaller changes in the smell, keep the same functional group and change other parts of the molecule.

9. Possible answer: The witch and the wizard are each creating a concoction that has the same molecular formula, so they are isomers of each other. However, the substance in the left cauldron, pentyl propionate, smells good, while the substance in the right cauldron, octanoic acid, smells bad. It is possible that they have different functional groups.

10. a. carboxyl **b.** ketone **c.** ketone
d. ester **e.** amine **f.** carboxyl

11. a. $C_3H_6O_2$ **b.** $C_5H_{10}O$ **c.** $C_5H_{10}O$
d. $C_4H_8O_2$ **e.** $C_4H_{11}N$ **f.** $C_2H_4O_2$

(continued)

10. Structural formulas for six molecules are shown here. Identify (by name) the functional group in each molecule.

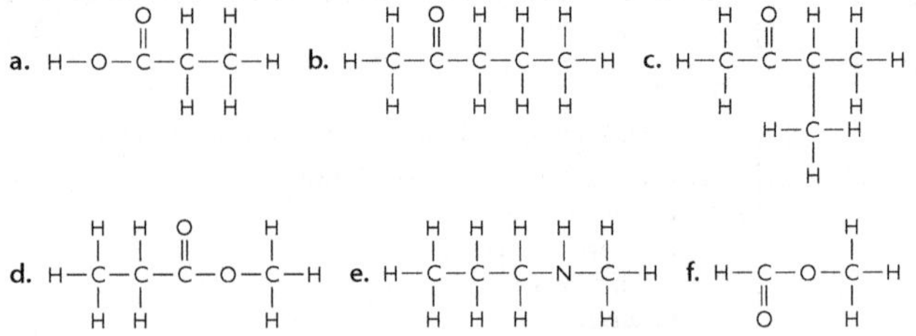

11. Write the molecular formula for each of the molecules in Exercise 10.

Create a Smell

Ester Synthesis

THINK ABOUT IT

Body odor, stinky shoes, moldy carpets, dog odors, musty basements, bad breath—these are just a few of the smells that we consider unpleasant. But the structures of these molecules are not so different from the structures of sweet-smelling substances. Molecules that have a carboxyl functional group are very similar to molecules that have an ester functional group. Perhaps there is a way to change one type of molecule into the other.

> How can a molecule be changed into a different molecule by using chemistry?

To answer this question, you will explore

 Transforming Smells

Chemical Reactions

Chemical Synthesis

EXPLORING THE TOPIC

Transforming Smells

Every year, consumers spend millions of dollars to purchase products that will deal with unwanted odors. Often, these products do nothing more than cover up a foul smell with a pleasing smell. But some products use chemistry to change the molecular structure of the smelly molecules.

Butyric acid, $C_4H_8O_2$, is a carboxylic acid with a putrid smell. However, the difference between this molecule and a sweet-smelling molecule is minimal. Butyric acid can be chemically changed into a sweet-smelling ester.

CONSUMER CONNECTION

The putrid-smelling compound butyric acid is formed when butter goes bad. In fact, *butyric* means "from butter." Butyric acid is also found in Parmesan cheese and vomit.

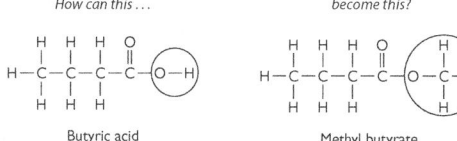

How can this ...   *become this?*

Butyric acid   Methyl butyrate

You can see that to accomplish this task, you would have to somehow remove the hydrogen atom that is attached to an oxygen atom in the butyric acid molecule and attach in its place a carbon atom with three hydrogen atoms. This procedure is possible in a chemistry lab.

Lesson 34 | **Ester Synthesis**   175

Overview

LAB: PAIRS

Key Question: How can a molecule be changed into a different molecule by using chemistry?

KEY IDEAS

It is possible for chemists to create different smells in the laboratory by creating molecules with specific functional groups. For instance, an alcohol and a carboxylic acid can be induced to react to create a new molecule that smells sweet. The different smells can be attributed to specific sequences of atoms, or

functional groups. *Note:* The terms *chemical reaction* and *synthesis* are covered more thoroughly in Lesson 35 of the Student Edition.

LEARNING OBJECTIVE

Complete a lab procedure to produce sweet-smelling esters.

KEY TERMS

chemical reaction
synthesis

IN CLASS

This class is a formal lab experiment that results in the formation of three esters (sweet-smelling) from various

organic acids and alcohols. The ChemCatalyst is designed to check for lab readiness. Before the lab, address the class as a whole on the basic steps of the experiment and on safety concerns. A worksheet provides lab instructions and questions to answer. Students work in pairs to set up their equipment and complete the procedure. The lab ends with cleanup and a sharing of preliminary results. Results of the lab are analyzed more thoroughly in Lesson 35. A complete materials list is available for download.

TRM Materials List 34

SETUP

Arrange the chemicals in reagent bottles at stations. Set up waste containers for the esters at each station. Have baking soda available to neutralize acid spills. You may choose to distribute the more concentrated acids yourself rather than have the students handle them. Note the safety precautions for the lab. Addtional safety reminders for students are found in the Explore section.

SAFETY

- There should be **no open flames** in this lab, because the organic chemicals are flammable.
- All bottles or containers should be closed when not in use to lessen fire danger and to minimize odors in the room.
- Extreme care should be taken when handling 18 M sulfuric acid. Have baking soda available to neutralize spills.
- Heating must be done slowly and carefully.
- Butyric acid is extremely pungent. Take extra precautions to avoid spills and drips.

CLEANUP

Put boiling stones and used pipettes in the trash. Have students pour ester products into a common waste receptacle. Place leftover esters in a safe place, allow the liquid to evaporate, and dispose of the solids at an appropriate waste disposal site. Stopper and save remaining alcohol and acid reagents.

Lesson Guide

ENGAGE (5 min)

TRM PowerPoint Presentation 34

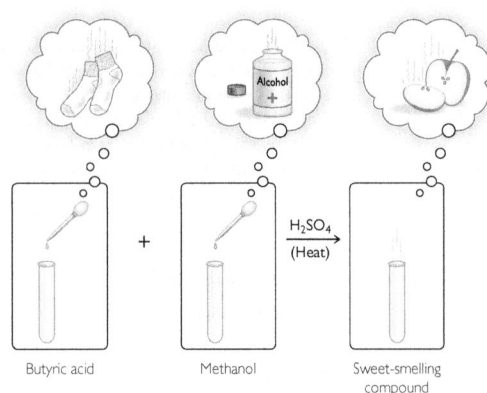

ChemCatalyst

1. What are some of the starting ingredients you will be using in this lab?
2. Name something you will be doing to the chemicals in this experiment.

Answers will vary.

CHECK STUDENTS FOR LAB READINESS

→ If answers are incorrect, have students reread the lab until they can provide correct answers.

EXPLORE (25 min)

TRM **Worksheet with Answers 34**

TRM **Worksheet 34**

INTRODUCE THE LAB

→ Tell students to work in pairs.

→ Pass out the worksheet to each group.

→ Go over safety guidelines with the class.

SAFETY

- Everyone must wear safety goggles.
- You will be using a hot plate (medium heat) to heat the ingredients. There should be **no open flames,** because several of the chemicals are flammable.
- Remember to waft the chemicals when you want to smell them. Some of the chemicals you are using smell very bad.
- All uncovered bottles of alcohols and esters should be kept far away from flames, because they are extremely flammable.
- Concentrated sulfuric acid will also be used in the reactions. It is very caustic and can burn the skin. Baking soda is available to neutralize any spills.

→ Tell students where they can pick up the equipment and chemicals in your classroom.

→ Emphasize the importance of exact measurements. If the reactants are not mixed in the appropriate proportions, leftover carboxylic acid might mask any ester that has been created, producing a stinky outcome instead of a sweet one.

CONSUMER CONNECTION

An entire industry is devoted to the creation of fragrances and flavors for use in our foods and commercial products. Some companies even manufacture room odorizers to inject a pleasant aroma into public environments or businesses. These aromas include the smell of freshly brewed coffee, baking bread, or an evergreen forest. The idea is to create environments that evoke pleasant memories or experiences for the user—and to stimulate the appetite or the pocketbook.

Pablo Rivera/SuperStock

Chemical Reactions

In the ester synthesis lab, you mixed some substances together with a catalyst and heated them, following a specific set of instructions. The result was a molecule that no longer smelled bad.

Butyric acid + Methanol $\xrightarrow[\text{(Heat)}]{H_2SO_4}$ Sweet-smelling compound

When butyric acid and methanol are mixed and heated, no change is visible to the eye. However, your nose tells you that a dramatic change in properties has occurred. The foul smell of sweaty socks disappears and the sweet smell of apples fills the room.

The change in smell is clear evidence that a chemical reaction has occurred. A **chemical reaction** is the process by which matter is changed so that new substances are formed. For new substances to form, the atoms must rearrange. Some bonds between atoms must be broken and new bonds must be formed. The result of a chemical reaction is a new compound with entirely different properties. In this case, one of the new properties is a different smell.

The detection of a sweet-smelling compound is evidence that you have made an ester. In fact, methyl butyrate smells like apples. In Lesson 35: Making Scents, you will examine this transformation in greater detail.

Chemical Synthesis

Chemists often work in the laboratory with the goal of producing a specific compound. This process is called synthesis. **Synthesis** is the process of producing a chemical compound, usually by combining two simpler compounds. Many of the products that we have come to depend on are products that chemists have synthesized, like plastics, fabrics, cosmetics, deodorants, cleaning products, vitamins, and medications.

Sometimes chemists work to synthesize a molecule that is very rare or hard to find. For instance, many lifesaving, anti-cancer compounds have been discovered in plant and animal life in the rainforests of South America. It may be difficult to harvest

→ Show students an example of the apparatus they will be using. Go over the important points of the lab as outlined here.

- The water should be at a gentle boil.
- The three test tubes must be labeled 1, 2, and 3.
- Make certain you carefully waft the starting chemicals to smell them.
- Smell the mixtures again before heating.
- Record any errors that crop up during the experiment (e.g., putting too many drops of sulfuric acid into the tube). These may be important in processing the lab later.

- If the mixture still smells putrid after heating, resume heating for another 5 minutes or until the smell changes.

EXPLAIN & ELABORATE (10 min)

GATHER THE CLASS DATA

→ Draw the data table on the board and collect the results of the experiment as a class.

→ Make sure students have all the information they need to complete a formal lab write-up as homework.

these compounds or find enough of a substance to treat the many patients who are in need. Chemists study the structure of a compound like this and work to create it in the laboratory. Most of the medicines in use today are synthesized in laboratories.

The U.S. National Cancer Institute has identified 3000 plants containing compounds with anti-cancer properties. Most are found in the rainforest. Here, a guide points out a medicinal plant.

Sometimes the point of synthesis is to create a new molecule that is *not* found in nature. All of the plastics that we have in our lives are synthesized from small molecules that have been strung together through chemical reactions. These compounds are valued for their amazing properties. Many plastics are durable, flexible, and easily molded into a variety of useful shapes—everything from shoes to computer keyboards.

LESSON SUMMARY

How can a molecule be changed into a different molecule by using chemistry?

To transform a putrid-smelling molecule into a sweet-smelling molecule, it is necessary to transform a carboxylic acid molecule into an ester molecule. This transformation requires a chemical reaction in which bonds between certain atoms are broken and new bonds are made. The result of any chemical reaction is a different compound or compounds than the compound you started with. And when the functional group changes, the smell changes. Chemists control chemical reactions in the laboratory to synthesize specific molecules with valuable properties.

KEY TERMS
chemical reaction
synthesis

Exercises

Reading Questions

1. What would you do to cause an alcohol and an acid to react?
2. What is chemical synthesis?

Reason and Apply

3. In class, why do you think the foul smell of butyric acid disappeared when you mixed it with methanol and heated the mixture in the Lab: Ester Synthesis?

Lesson 34 | **Ester Synthesis**   177

LESSON 34 ANSWERS

1. Mix the alcohol and acid together with a strong acid and heat them.

2. Chemical synthesis is the process of producing a chemical compound by combining two or more simpler compounds.

3. *Possible answer:* When butyric acid is heated with methanol, the two compounds react to form a new compound, which is a sweet-smelling ester. The butyric acid, which has a foul smell, is no longer present.

4. The first two molecules are acids, so they have carboxyl groups (–COOH). The next three molecules are alcohols, and have hydroxol groups (–OH.)

5. *Possible answer:* Combine the acetic acid with an alcohol and sulfuric acid and then heat the mixture. The hydrogen atom of the acid is replaced by the part of the alcohol molecule that is attached to the –OH functional group. The product is an ester, which is similar in structure to a carboxylic acid, but with the oxygen atom bonded to two carbon atoms instead of to one carbon atom and one hydrogen atom.

6. A good lab report will contain: ● a title (Lab: Ester Synthesis) ● a statement of purpose (Possible answer: To create new smells and analyze the products) ● a procedure (a summary of the steps followed in the experiment) ● results (Check student observations to make sure they properly recorded observations of the smells of the original ingredients and the smells of the final products.) ● a conclusion (Possible answer: A new compound was formed in each test tube having different properties from the original materials.)

Ask: How did the smells after heating compare to those before heating? Did you notice any patterns?

Ask: What functional group do you think is present in the final molecules? Explain.

Ask: What happened to the molecules to change the smell?

Key Points: In this lab, you used chemistry to produce new molecules with new properties. A chemical change took place as the result of a chemical reaction. When chemists purposely produce a specific compound, the process is called chemical synthesis.

Synthesis: The creation of specific compounds by chemists through controlled chemical reactions.

EVALUATE (5 min)

There is no Check-In for this class. Results of this lab will be discussed in detail in the next lesson.

Homework

Assign the reading and exercises for Smells Lesson 34 in the student text.

Exercises

(continued)

4. The structural formulas for the molecules used in the Lab: Ester Synthesis are given in the table. Identify the functional groups.

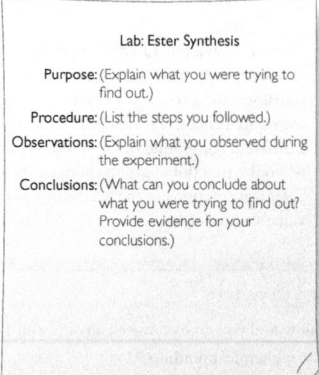

Test tube	Organic acid	Structural formula	Alcohol	Structural formula
1	acetic acid		isopentanol	
2	acetic acid		butanol	
3	butyric acid		ethanol	

5. Explain how you can convert acetic acid into an ester. Be specific about how the molecule needs to change.

6. **Lab Report** Write a lab report for the Lab: Ester Synthesis. In your report, include the title of the experiment, purpose, procedure, observations, and conclusions.

Lab: Ester Synthesis

Purpose: (Explain what you were trying to find out.)

Procedure: (List the steps you followed.)

Observations: (Explain what you observed during the experiment.)

Conclusions: (What can you conclude about what you were trying to find out? Provide evidence for your conclusions.)

Making Scents

Analyzing Ester Synthesis

THINK ABOUT IT

When you heat a mixture of a carboxylic acid and an alcohol, a chemical reaction takes place. You can smell the outcome. A really putrid-smelling substance is transformed into one that smells sweet. An ester has been produced. But what has happened to the original substances that were mixed together?

> What happened to the molecules during the creation
> of a new smell?

To answer this question, you will explore

⟳ Chemical Equations

⟳ The Role of the Catalyst

⟳ Ester Synthesis

EXPLORING THE TOPIC

⟳ Chemical Equations

Chemists use **chemical equations** to keep track of changes in matter. These can be certain physical changes or changes due to chemical reactions. A chemical equation usually uses chemical formulas to describe what happens when substances are mixed together and new substances with new properties are formed. In class, you used structural formulas to describe what happened to the molecules that were combined.

Consider what happens if you add butyric acid to ethanol and create a pineapple smell. A description of the reaction is given here. Each of these substances consists of a collection of molecules that are structurally different. The structural formulas for the molecules in each of the substances are also shown along with their chemical formulas.

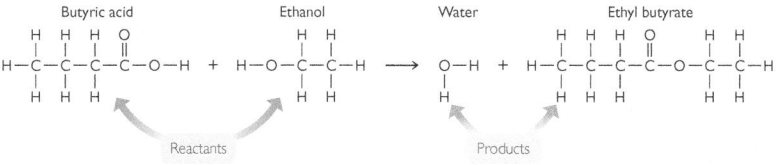

Reactants   Products

Butyric acid, $C_4H_8O_2$, is added to ethanol, C_2H_6O, to produce water, H_2O, and ethyl butyrate, $C_6H_{12}O_2$.

IN CLASS

This lesson processes the previous lab in detail and begins to develop the concept of a chemical reaction. The worksheet provides students with the structural formulas of the reactants. Students are guided to come up with the molecular structures of the ester products they produced in the previous class. Students gain a conceptual understanding of what happened to the specific molecules they mixed and heated in the lab. They are introduced to a rudimentary chemical equation. The formal chemical equations have been purposely left out so students can focus on how the molecular structures change during the reaction. Chemical reactions are handled formally in Unit 4: Toxins after students have had experience with moles in Unit 3: Weather. The purpose of this lesson is to provide a conceptual foundation for understanding chemical reactions. A complete materials list is available for download.

TRM **Materials List 35**

Pre-AP® Course Tip

Students preparing to take an AP® chemistry course or college chemistry should be able to analyze data to identify certain types of patterns and relationships among chemical phenomena. In this lesson, students analyze their results from Lesson 34 where they made different esters. Based on their results, they can predict products of an esterification reaction and draw accurate structural formulas of an ester.

Lesson Guide

ENGAGE (5 min)

TRM **PowerPoint Presentation 35**

ChemCatalyst

What do you think happened in the experiment in the previous class to transform an acid molecule and an alcohol molecule into a sweet-smelling molecule?

Sample answers: Students might say that the molecules in the test tube exchanged pieces or that the molecules came apart entirely and rearranged.

Overview

FOLLOW-UP: INDIVIDUAL

Key Question: What happened to the molecules during the creation of a new smell?

KEY IDEAS

In the ester lab, a chemical change, or chemical reaction, took place. The starting substances of a chemical reaction are called the *reactants*. The ending substances are called the *products*. During chemical reactions, new compounds with new properties are formed. On a molecular level, a chemical change is a rearrangement of atoms that involves the breaking and making of chemical bonds.

LEARNING OBJECTIVES

● Explain what happened at a molecular level during the ester synthesis lab.
● Predict the product of a reaction between an alcohol and a carboxylic acid.
● Give a general definition of a chemical reaction.
● Define what a catalyst is.

KEY TERMS

chemical equation   product
reactant   catalyst

Ask: What do you think took place in the test tube when you mixed and heated the initial substances?

Ask: What evidence do you have to support your answer?

TRM Worksheet with Answers 35

TRM Worksheet 35

TRM Transparency—Data Table 35

INTRODUCE THE FOLLOW-UP

→ Let students know they will be working individually.

→ Pass out the worksheet to each student.

→ Introduce students to a rudimentary chemical equation. It is not necessary to use chemical formulas yet.

● Chemists show what happens during a chemical reaction with a chemical sentence written as a chemical equation. It might look like this, where A and B are the starting ingredients and the new substances produced are C and D, or just E.

  A + B → C + D   or   A + B → E

> **Chemical equation:** A chemical sentence that tracks what happens during a change in matter. Chemical equations are written with chemical formulas and keep track of the atoms involved in the changes.

→ Use the Data Table transparency to display the results of the lab experiment. This will serve as a reference as students complete the worksheet.

Data Table

Test tube	Organic acid	Alcohol	Smell of mixture before heating	Smell of mixture after heating
1	acetic acid	isopentanol	putrid	fruity, banana smell
2	acetic acid	butanol	strongly putrid	fruity, pear smell
3	butyric acid	ethanol	putrid	fruity, pineapple smell

PROCESS THE STRUCTURES OF THE ESTER PRODUCTS

→ Ask students to draw the structures from Exercises 3, 5, 7, and 8 on the board. If a structure is incorrect, guide the students to fix it using the HONC 1234 rule or conservation of matter.

In a chemical reaction, the substances that are mixed together are called the **reactants.** The new substances that are produced are called the **products.**

The reactants in this chemical reaction are butyric acid and ethanol. One product that is easily detected by smell is ethyl butyrate. However, this is not the only product. Although it is difficult to observe with the five senses, water is also a product of this chemical reaction. The chemical equation helps to make this clear. The chemical equation shows that in a chemical reaction, no matter is created or destroyed.

Big Idea Matter cannot be created or destroyed. Matter is conserved.

The next illustration highlights the areas of the reactant molecules that change during the chemical reaction. You can see that a hydroxyl, $-OH$ group, breaks off from the butyric acid molecule. In addition, the $H-O$ bond in the ethanol molecule must also break. A new bond forms as the larger molecular pieces come together to form ethyl butyrate.

Finally, you can see that the $H-$ and $-OH$ pieces that are left over combine to form H_2O, water.

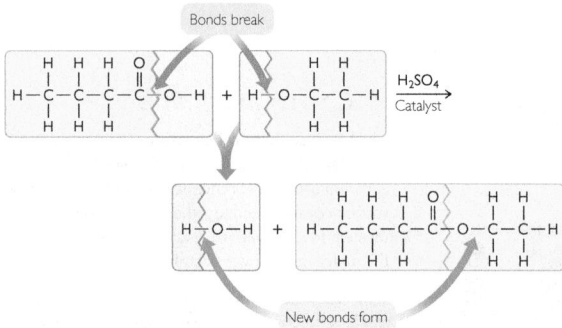

> **CONSUMER CONNECTION**
> Esters are found in a number of household products, from cleansers and air fresheners to hand lotions and shampoos. Their main role is to make a product smell a certain way, although some are detergents as well.

For every butyric acid molecule that reacts with an ethanol molecule, one molecule of water is produced along with one molecule of ethyl butyrate.

The complete chemical equation for this chemical reaction is

$$C_4H_8O_2 + C_2H_6O \longrightarrow H_2O + C_6H_{12}O_2$$

There are the same number of each type of atom on both sides of the equation. All of the matter is accounted for.

↻ The Role of the Catalyst

There's one more ingredient in the chemical reaction that we have not yet discussed. When you try to carry out this reaction in a lab, you will find that simply mixing butyric acid and ethanol does not produce ethyl butyrate. In fact,

---

Ask: How did you decide what the product molecules should look like?

Ask: Are the same atoms present before and after the chemical reaction?

Key Points: The products of these reactions smell sweet, so they must all contain an ester functional group. The atoms are not destroyed and must all be accounted for. The final structure is arrived at by combining the two original molecules and making sure an ester functional group is present. Three atoms are left over, two H atoms and an O atom, which form water. Many different acids and alcohols can be brought together to form an ester

and water. The general description of this reaction is:

  acid + alcohol → water + ester

ASSIST STUDENTS IN MAKING SENSE OF THE CHEMICAL REACTIONS

→ Draw one of the chemical reactions in its entirety on the board. During your discussion, draw a box around the atoms that break off to form water.

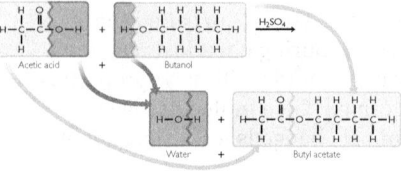

Clive Streeter/Getty Images

nothing happens. To make the chemical reaction happen, it is necessary to add sulfuric acid, H_2SO_4, and heat the mixture.

$$C_4H_8O_2 + C_2H_6O \xrightarrow{H_2SO_4} H_2O + C_6H_{12}O_2$$

The sulfuric acid does not get used up during the reaction. It is still present at the end. In this reaction, the sulfuric acid is simply helping the reaction along. Chemists refer to a substance that assists a chemical reaction as a **catalyst.** A catalyst speeds up a reaction, but the catalyst itself is not consumed by the reaction. In chemical equations, a catalyst is written above the arrow.

Ester Synthesis

Many different acids and alcohols can be brought together to form an ester and water. The general description of this reaction is

acid + alcohol $\longrightarrow$ water + ester

Take another look at the illustration on page 180 showing the bonds breaking. It is not necessary for every single bond in the acid and alcohol molecules to break apart to form the ester. All of the $C-H$ bonds remain connected during this reaction.

During chemical reactions, the major changes in the molecules often occur where the functional groups are located. The part of the molecule that is easiest to change is the functional group. This makes sense with what you already know: The functional group is directly related to the properties of a molecule.

LESSON SUMMARY

What happened to the molecules during the creation of a new smell?

When chemical reactions occur, new compounds with different properties and structures are produced. The substances that are combined are called the reactants. The substances that are produced are called the products. When chemical reactions occur, bonds break and new bonds form. Chemical equations use chemical formulas to track the changes that occur during chemical reactions. Because matter is conserved, all atoms are accounted for in a chemical equation. Some chemical reactions require a catalyst to help them get started or proceed more rapidly. A catalyst is not consumed by the chemical reaction it is assisting.

KEY TERMS

chemical equation

reactant

product

catalyst

Exercises

Reading Questions

1. Explain why converting a carboxylic acid to an ester might be useful.

2. Describe what happens during a chemical reaction.

3. What is a catalyst?

Lesson 35 | **Analyzing Ester Synthesis**   181

Reactant: An element or compound that is a starting substance in a chemical reaction. Reactants are written to the left of the arrow in a chemical equation.

Product: An element or compound that results from a chemical reaction. Products are written to the right of the arrow in a chemical equation.

Catalyst: A substance that accelerates a chemical reaction but is itself not permanently consumed or altered by the reaction. A catalyst is written above the arrow in a chemical equation.

When atoms are rearranged during chemical reactions, not all of the bonds must break. Notice that most of the molecule stays together. In this reaction, only one bond breaks in each molecule. Two new bonds form. Usually, chemical reactions happen in the area of a functional group on a molecule. A bond may break within or next to a functional group, and new bonds then form with other pieces of molecules.

DISCUSS THE PATTERNS FOUND IN THE CHEMICAL NAMES (OPTIONAL)

→ Write out several of the chemical reactions as sentences. Show students the naming patterns.

Ask: What do the chemical names of the reactants in Question 5 have in common with the chemical names of the products?

Ask: What patterns do you see in the naming of the ester products?

Key Points: The naming of chemical compounds is not random. There is a pattern to it. When butyric acid reacts with ethanol, it forms ethyl butyrate. When formic acid reacts with octanol, it forms octyl formate. The first part of the name of the acid becomes the second part of the name of the ester that is formed. The first part of the name of the alcohol becomes the first part of the name of the ester.

Formic acid reacts with octanol to form octyl formate.

From this pattern, we can correctly predict that the product of acetic acid and propanol will be propyl acetate.

Ask: What are the products in this equation?

Ask: What part of each molecule changed during the chemical reaction? Which atoms created the water molecule?

Ask: Why do you think H_2SO_4 is shown above the arrow?

Ask: According to our drawing, how many bonds were broken in this reaction? What new bonds were formed?

Key Points: The lab procedure you completed resulted in a chemical reaction. When chemical reactions take place, new compounds with new properties are produced.

During chemical reactions, bonds are broken and the atoms rearrange themselves to form new compounds.

It is possible to track the changes to the structure of the molecules through chemical equations. In a chemical equation, the substances to the left of the arrow are the substances that are mixed and are called the *reactants.* The new substances that are produced are shown to the right side of the arrow and are called the *products.* Sometimes, a catalyst is added to assist a chemical reaction. This substance speeds up or facilitates the reaction without being in any way changed. A catalyst is not considered a reactant.

EVALUATE (5 min)

Check-In

1. Predict the structural formula of the product of this reaction.

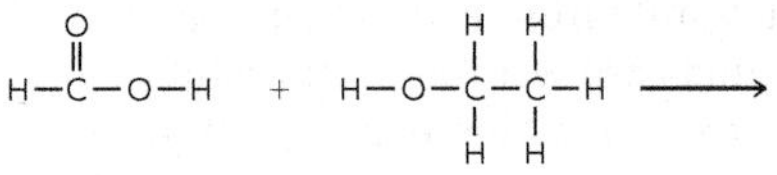

Formic acid   +   Ethanol

2. What smell would you expect the product to have?

Answers: **1.** The structure of this molecule is shown below. Water is also a product.

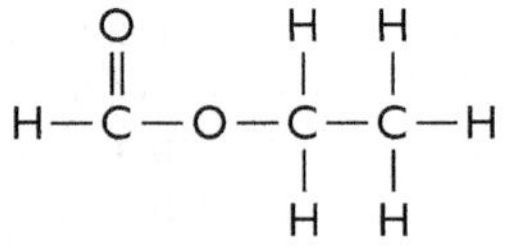

Ethyl formate

2. Because ethyl formate is an ester, you would expect it to smell sweet.

Homework

Assign the reading and exercises for Smells Lesson 35 in the student text. Students complete their Ester Synthesis lab report in Exercise 1. Assign the Chapter 6 Summary.

LESSON 35 ANSWERS

1. *Possible answer:* Converting a carboxylic acid into an ester is a way to make a new compound with more desirable properties. For example, eliminating a putrid odor is one possible goal of converting a carboxylic acid into an ester.

2. During a reaction, chemical bonds break and new chemical bonds form. This process changes the reactants into different chemical substances, called products.

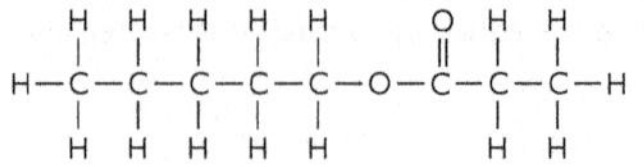

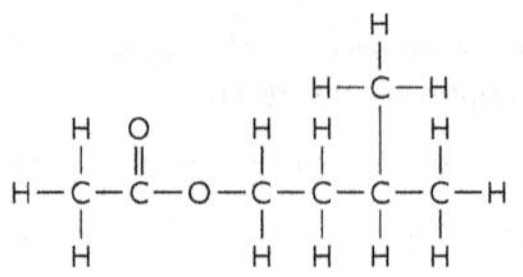

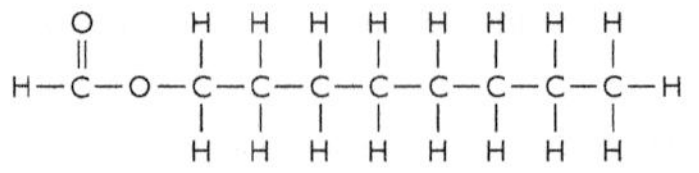

3. A catalyst is a chemical that is added to a reaction mixture to help get the reaction started, but is not consumed by the reaction.

4. *Possible answer:* Scientists can create the smells they want in the laboratory by identifying the type of smell they desire and then creating a molecule with functional groups that will facilitate that type of smell. For example, for a sweet-smelling molecule, start with an alcohol and react it with an acid to create an ester. In the Lab: Create a Smell, we discovered that these kinds

Exercises

(continued)

Reason and Apply

4. Write a paragraph answering the question, "How can scientists use chemistry to create compounds with specific smells?" Include evidence to support your answer.

5. Below are the structural formulas for four esters. Write the correct molecular formula for each one.

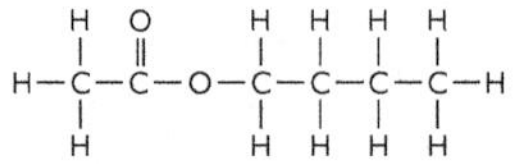

6. Ester molecules are named for the alcohol and acid that form them using the convention

(alcohol name)yl (acid name)ate

For example, methanol + ethanoic acid combine to make methyl ethanoate. Name the esters formed from

a. isopropanol + methanoic acid
b. caproic acid + butanol
c. salicylic acid + ethanol

of reactions are possible. Then perform other procedures, through trial and error and knowledge of functional groups, to alter the structure and the functional groups further to create a new smell.

5. a. $C_8H_{16}O_2$ **b.** $C_7H_{14}O_2$ **c.** $C_9H_{18}O_2$ **d.** $C_6H_{12}O_2$

6. a. isopropyl methanoate **b.** caproyl butanoate **c.** salicyl ethanoate

CHAPTER 6

Speaking of Molecules

SUMMARY

KEY TERMS

molecular formula

structural formula

isomer

HONC 1234 rule

Lewis dot structure

Lewis dot symbol

bonded pair

lone pair

octet rule

double bond

triple bond

functional group

chemical reaction

synthesis

chemical equation

reactant

product

catalyst

Smells Update

The investigation into the chemistry of smell has challenged you to explore how molecules are put together. So far you know that the smell of a substance is related to its molecular formula, chemical name, and any functional groups present.

$$-\overset{\overset{O}{\|}}{C}-O-\overset{|}{C}-$$

This is based on the data investigated so far. Further investigation may lead you in new directions.

REVIEW EXERCISES

1. What functional groups are present in each of these molecules?

   a. H—C—C—O—C—C—C—C—C—C—H

   b. H—C—C—C—C—C—H

   c. H—C—C—C—C—C—O—H

   d. H—C—C—C—N—H

2. Draw the Lewis dot structure and the structural formula for each of these molecules.
   a. SiF_4 b. CO_2 c. CH_4 d. SF_2
   e. C_2H_4 f. C_2H_2 g. C_2H_6

3. How many lone pairs are in each of these molecules?
   a. CO_2 b. SiF_4 c. CH_4

4. Draw at least two structural formulas for the molecular formula C_3H_6O.

5. From what you've learned so far, how is molecular structure related to smell?

Functional Groups

PROJECT Research a functional group. Choose a functional group and find out as much as you can about molecules that contain it. Create a poster that has these details:

- A large drawing of the functional group along with its name
- The structural formulas for at least five molecules that possess your functional group, along with their chemical names
- A brief description of the properties associated with this group of compounds

Chapter 6 | **Summary**   183

4. *Possible answer:*

$$\underset{\underset{H}{|}}{\overset{\overset{H}{|}}{C}}=\underset{\underset{H}{|}}{\overset{\overset{H}{|}}{C}}-\underset{\underset{H}{|}}{\overset{\overset{H}{|}}{C}}-O-H$$

$$H-\underset{\underset{H}{|}}{\overset{\overset{H}{|}}{C}}-\underset{\underset{H}{|}}{\overset{\overset{H}{|}}{C}}-\underset{\underset{H}{|}}{\overset{\overset{H}{|}}{C}}=O$$

$$H-\underset{\underset{H}{|}}{\overset{\overset{H}{|}}{C}}-\underset{\underset{O}{\|}}{C}-\underset{\underset{H}{|}}{\overset{\overset{H}{|}}{C}}-H$$

5. *Possible answer:* The smell of a compound is strongly related to the functional group of the compound's structural formula.

Project: Functional Groups

A good poster would include:

- A diagram of the structure of the functional group clearly showing bonds and atom types
- Five or more molecules with the functional group highlighted in the molecule and the name of each molecule clearly labeled
- Information on how molecules containing the functional group are named
- A list of common properties that molecules with the functional group possess
- A description of common uses for compounds with the functional group

ASSESSMENTS

Two multiple-choice assessments (versions A and B) and two short-answer assessments (versions A and B) are available for download for Chapter 6.

TRM Chapter 6 Assessment Answers

TRM Chapter 6 Assessments

ANSWERS TO CHAPTER 6 REVIEW EXERCISES

1. a. ester **b.** ketone **c.** carboxyl **d.** amine

2. a.

:F:Si:F: F—Si—F

b. Ö::C::Ö   O=C=O

c.

H:C:H   H—C—H

d. :F:S:F: F—S—F

e.

H H   H H
C::C   C=C
H H   H H

f. H:C:::C:H   H—C≡C—H

g.

H H   H H
H:C:C:H   H—C—C—H
H H   H H

3. a. 4 lone pairs **b.** 12 lone pairs **c.** no lone pairs

PD CHAPTER 7 OVERVIEW

Watch the video overview of Chapter 7 (for teachers) by clicking on the link in the TE-book, opening the TRFD, or logging onto the book's companion Web site bcs.whfreeman.com/livingbychemistry2e (teacher log-in required).

Chapter 7 explores molecular shape and the role of electron pairs. Lesson 36 introduces three-dimensional ball-and-stick models. Some new smells force students to rethink their previous hypothesis linking smell to functional groups. To help students understand overall molecular shape, they are introduced to electron domain theory and the geometries of small molecules in Lessons 37 and 38. These lessons highlight the role of electrons—both bonded pairs and lone pairs—in determining the shapes of molecules. After using what they have learned to construct some molecules, students move to space-filling models in Lesson 39. In Lesson 40, they come up with generalizations that link molecular shape and smell. Finally, in Lesson 41, they integrate their learning into a model of how the nose works, and the receptor site model is introduced.

In this chapter, students will learn

- how to interpret three-dimensional models of molecules
- how electron pairs determine molecular geometry
- how to use Lewis dot structures to predict possible molecular structures from molecular formulas
- general relationships between molecular shape and smell
- the receptor site or "lock and key" theory of molecular interaction

Molecular models help chemists to picture something too tiny to see.

CHAPTER 7

Building Molecules

In this chapter, you will study

- three-dimensional molecular models
- the role of valence electrons in determining molecular shape
- the receptor site theory and have a chance to develop your own model of how a nose works

Two compounds with the same functional group may have different smells. In addition to molecular structure, the *shape* of a molecule seems to be related to its smell. Molecular shape is determined by the bonds between atoms and the locations of electrons. Because we experience different molecular shapes as different smells, our experience tells us that the nose can somehow detect these differences.

184

Chapter 7 Lessons

New Smells, New Ideas

Ball-and-Stick Models

THINK ABOUT IT

The structural formula is a valuable source of information about a molecule. You can use the structural formula to identify functional groups in a molecule. Knowing the functional group also helps you to predict molecular properties, including smell. However, sometimes only knowing the functional group is not enough to predict the smell of a compound.

What three-dimensional features of a molecule are important in predicting smell?

To answer this question, you will explore

- New Smell Molecules
- Three-Dimensional Models

EXPLORING THE TOPIC

New Smell Molecules

Below are three compounds, each with a distinctive smell. In spite of their wide range of smells, each of the three compounds contains the same functional group.

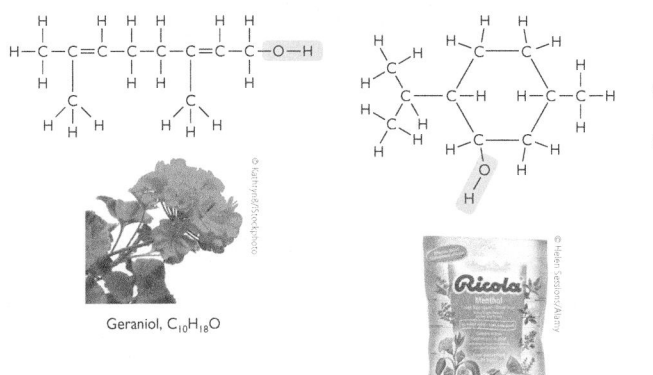

Geraniol, $C_{10}H_{18}O$

Menthol, $C_{10}H_{20}O$

Fenchol, $C_{10}H_{18}O$

Take a moment to locate their functional groups. Each of these molecules has a hydroxyl functional group, and each name ends in "-ol." All three molecules are alcohols. So, why are the three smells so different? Something besides just functional group affects the smell of a compound. The answer lies in the three-dimensional shape of the molecules of each compound.

Lesson 36 | **Ball-and-Stick Models**   185

- Students may or may not notice the specific structural features mentioned here, or they may describe them differently. It is all right to allow them their current theories.

KEY TERM

ball-and-stick model

IN CLASS

Five new smells are sampled in this class. These compounds are all alcohols, but they belong in three different smell categories (sweet, camphor, and minty). Students examine the molecular and structural formulas of these molecules and attempt to explain how molecules with the same functional group might belong to completely different smell categories. They discover that the overall shape of a molecular compound also affects the way it smells. A complete materials list is available for download.

TRM Materials List 36

SETUP

Prepare sets of vials I, J, K, L, M by first placing a cotton ball in each vial and then using the plastic pipettes to deliver three to five drops of the stock smell solutions to the appropriately lettered vials. Use a new pipette for each essence. You should prepare the vials in a hood or outdoors. Place each set of the five vials in a plastic sandwich bag to make it easy to distribute them to groups of students.

Label	Contents	Smell
I	rose perfume	sweet
J	pine cleaner	camphor
K	jasmine perfume oil	sweet
L	mint flower extract	minty
M	pine oil	camphor

Build ball-and-stick models of citronellol, menthol, and fenchol. Label the models 1, 2, and 3. Do not label the molecules with their chemical names or smells. Build one set for the class to view, or if you have enough modeling materials, build a set for each group or enough for groups to share.

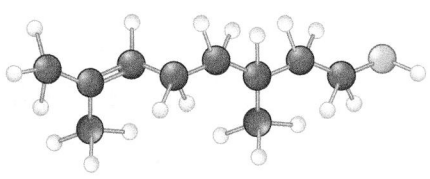

1

Overview

ACTIVITY: GROUPS OF 4

Key Question: What three-dimensional features of a molecule are important in predicting smell?

KEY IDEAS

Structural formula and functional group are related to the properties of a molecule, such as smell. However, the overall shape of a molecule can also account for differences in smell. Chemists use ball-and-stick models as three-dimensional representations of molecules to allow them to explore the overall shape of a molecule.

LEARNING OBJECTIVES

- Interpret three-dimensional ball-and-stick molecular representations.
- Translate between molecular models, molecular formulas, and structural formulas.
- Describe connections between molecular properties and molecular structure.

FOCUS ON UNDERSTANDING

- Students gain valuable conceptual understanding from firsthand contact with the molecular models. The more models available, the better.

2

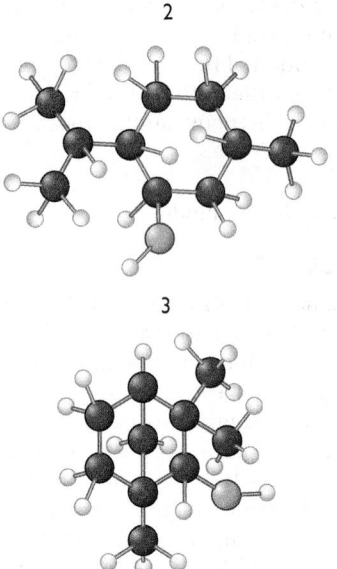

3

CLEANUP

Save the vials for reuse in other classes and over the course of this unit. Within a few weeks, however, you will have to remove the cotton balls and air out the vials. Otherwise, the substances' smells will mix with one another and begin to smell putrid. When you are finished using vials I, J, K, L, M in all your classes, remove the cotton balls from the vials, place them in a plastic bag, and dispose of them. Let the vials air out in a hood or rinse them with acetone for reuse the next time you do this unit.

Lesson Guide

ENGAGE (5 min)

TRM Transparency—ChemCatalyst 36

TRM PowerPoint Presentation 36

ChemCatalyst

Do you think any of these molecules will smell similar? What evidence do you have to support your prediction?

citronellol, $C_{10}H_{20}O$

geraniol $C_{10}H_{18}O$

---

menthol $C_{10}H_{20}O$

Sample answer: Students might predict that these molecules will all smell minty because they have one oxygen atom. Or that the ones that look similar have similar smells. *Note:* Do not give away the smell categories in advance. Students will discover these in today's activity.

⟳ Three-Dimensional Models

It is difficult to show a three-dimensional drawing of a molecule on a flat piece of paper. With a molecular model set, you can build a three-dimensional structure.

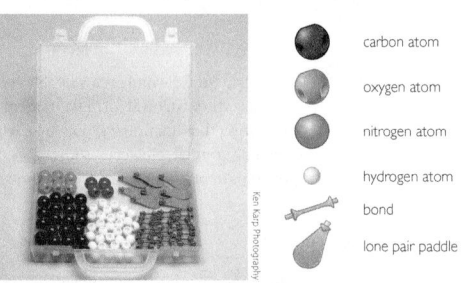

carbon atom

oxygen atom

nitrogen atom

hydrogen atom

bond

lone pair paddle

Notice that the bonds in a model kit look like little sticks. And the atoms are small spheres. These models are called **ball-and-stick models.**

A picture of a ball-and-stick model for ethyl acetate, $C_4H_8O_2$, is shown here. The carbon atoms are shown in black, the hydrogen atoms are white, and the oxygen atoms are red.

The carbon atoms are not in a straight line.

Some atoms are pointing away from you.

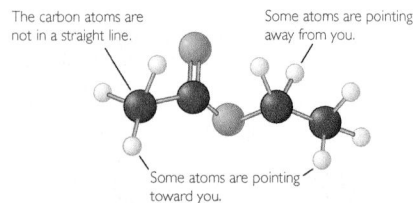

Some atoms are pointing toward you.

Illustrations of ball-and-stick models have some drawbacks, in that some atoms may be partially visible or entirely hidden. However, the illustrations convey more information about molecular shape than structural formulas do. Using a real model is the best way to examine the three-dimensional shape of a molecule. The two representations for a molecule of citral, $C_{10}H_{16}O$, are shown for comparison. Look for similarities and differences between them.

Structural formula

Ball-and-stick model

Citral, $C_{10}H_{16}O$

→ Discuss the possible smells of the three compounds.

Ask: What do the molecules have in common?

Ask: Do you think any of these compounds will smell similar? Give your reasoning.

Ask: Suppose you are told that two of the compounds smell sweet. Which two would you predict are the sweet-smelling ones? Explain your reasoning.

EXPLORE (20 min)

TRM Worksheet with Answers 36

TRM Worksheet 36

Notice that the structural formula shows the citral molecule as flat with all of the carbon atoms in a line. The ball-and-stick representation, on the other hand, shows that the carbon atoms are not arranged in a line, but are connected in a zigzag fashion. Both representations contain much of the same information, but the ball-and-stick model adds information about the way the atoms are arranged in space.

Everett Collection/SuperStock

Example

Ball-and-Stick Model

Examine the drawing of this ball-and-stick model of isopentylacetate, $C_7H_{14}O_2$.

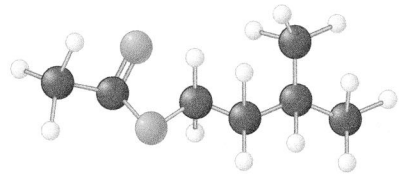

a. Draw the structural formula of this compound.
b. What functional group is in the compound?
c. Predict a smell for this compound.

Solution

a. The structural formula of this compound is

$$H-\overset{\overset{\displaystyle H}{|}}{\underset{\underset{\displaystyle H}{|}}{C}}-\overset{\overset{\displaystyle O}{||}}{C}-O-\overset{\overset{\displaystyle H}{|}}{\underset{\underset{\displaystyle H}{|}}{C}}-\overset{\overset{\displaystyle H}{|}}{\underset{\underset{\displaystyle H}{|}}{C}}-\overset{\overset{\overset{\displaystyle H}{|}}{\underset{\underset{\displaystyle H}{|}}{C}}}{\underset{}{C}}-\overset{\overset{\displaystyle H}{|}}{\underset{\underset{\displaystyle H}{|}}{C}}-H$$

b. This compound has an ester functional group.
c. Because it has an ester group and its name ends in "-ate," it probably smells sweet. (In fact, this molecule smells like bananas.)

INTRODUCE THE ACTIVITY

→ Arrange students into groups of four.

→ Pass out the student worksheet.

→ Pass out vials I, J, K, L, M to groups of four students. Remind students of the correct procedure for smelling, and tell them to replace and tighten the caps when they finish. After sampling is complete, collect the vials.

→ When students have completed Part 1 individually, have students work with the ball-and-stick models to complete Part 2.

EXPLAIN & ELABORATE (15 min)

DISCUSS THE RESULTS OF THE SMELLING

→ Create a table on the board for vials I, J, K, L, M. Solicit students' smell categories. Guide them to the categories listed, and introduce the new smell classification "camphor."

Ask: What information did you discover about the molecules in vials I, J, K, L, M?

Ask: There is a new smell category in two of the vials. How did you identify it?

Ask: What similarities did you find among the structural formulas of the molecules that smelled similar?

Ask: Are molecular formula and functional group enough information to predict smell? Why or why not?

Vial I	Vial J	Vial K	Vial L	Vial M
sweet	camphor	sweet	minty	camphor

Key Points: Each molecule in the activity has a hydroxyl group (–OH) and is an alcohol, but the molecules do not all smell the same. Three distinctly different smells were present: sweet, minty, and a new smell we call "camphor." Camphor is a strong smell usually associated with mothballs and certain ointments. Apparently, something besides functional group is responsible for the smell of a compound.

There are a number of similarities among those alcohols that smell alike. The sweet-smelling alcohols in this group all have one oxygen atom and ten carbon atoms. Further, the carbon atoms in the sweet-smelling alcohols do not form a ring. The camphor-smelling alcohols all have a ring structure with some sort of crosspiece or bridge. The minty-smelling alcohol has a ring but no bridge.

DISCUSS BALL-AND-STICK MODELS

→ Have the ball-and-stick models of citronellol, menthol, and fenchol at hand so you can discuss the various features.

Ask: What new information do you gain about molecules from looking at the ball-and-stick models?

Ask: How is a ball-and-stick model similar to other representations you have seen?

Ask: If you know the molecular formula of a molecule, can you build a ball-and-stick model of it? Why or why not?

Ask: How would you describe the process used to identify the three ball-and-stick models?

Ask: If someone handed you a new model, do you think you could determine how the molecule it represents might smell just by looking at it?

Key Points: A ball-and-stick model shows the three-dimensional shape of a molecule. The molecules are not flat, and the carbon atoms are not connected in a straight line. Rather, the atoms are arranged in three dimensions in bent, crooked, or branched chains. The molecular formula and the structural formula of a molecule can both be determined from the ball-and-stick model. However, you typically cannot build a ball-and-stick model from just a molecular formula. The models in today's lesson are color-coded. All the hydrogen atoms are white, the carbon atoms are black, and the oxygen atoms are red. Sticks represent bonds between atoms, much like the lines in structural formulas.

> **Ball-and-stick model:** A three-dimensional representation of a molecule that uses sticks to represent bonds and color-coded balls to represent atoms.

It is hard to tell at this point if the shape of a molecule is related to its smell. There are some general structural similarities between the two sweet-smelling molecules and between the two camphor-smelling molecules. More information is needed to be able to generalize about molecular shape.

EVALUATE (5 min)

Check-In

Predict the smells of these molecules:

1. Propyl butyrate

2. $C_6H_{14}O$

3.

```
      H   H   H   H   H   O
      |   |   |   |   |   ||
  H — C — C — C — C — C — C — O — H
      |   |   |   |   |
      H   H   H   H   H
```

Answers: **1.** The first molecule, propyl butyrate, will probably smell sweet. You can tell it is an ester by its name, which ends in "-ate." So far, we have lots of evidence that esters smell sweet. **2.** The second molecule could have several different smells because it could have any number of different structures. We have just seen that alcohols can smell minty, sweet, or even like camphor. It could also be an ether. **3.** The third molecule has a carboxylic acid functional group. It will probably smell putrid.

Homework

Assign the reading and exercises for Smells Lesson 36 in the student text.

LESSON 36 ANSWERS

1. *Possible answer:* A structural formula shows the types of bonds within a molecule and the arrangement of atoms in two dimensions. A ball-and-stick

---

Ball-and-stick models for the three compounds introduced at the beginning of this lesson are below.

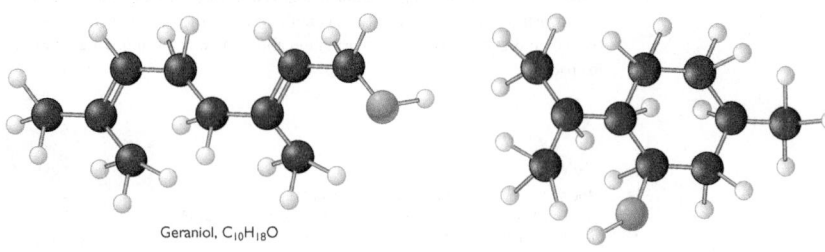

Geraniol, $C_{10}H_{18}O$

Menthol, $C_{10}H_{20}O$

Fenchol, $C_{10}H_{18}O$

The overall shapes of the molecules differ greatly. The carbon atoms in geraniol are connected in a long chain, while the menthol and fenchol molecules both contain ring structures. Perhaps these overall shapes are related to the different smells of these three alcohols.

LESSON SUMMARY

> What three-dimensional features of a molecule are important in predicting smell?

KEY TERM

ball-and-stick model

Molecules are three-dimensional. A ball-and-stick model kit is a tool that you can use to construct models of molecules. Ball-and-stick models allow you to see how the various atoms in molecules are arranged in space, as well as the overall shape of each molecule. Molecular shape may have something to do with smell.

Exercises

Reading Questions

1. What are the differences between a structural formula and a ball-and-stick model?

2. What is your hypothesis about why the three alcohols smell different even though they have the same functional group?

---

model adds information about the arrangement in three dimensions.

2. *Possible answer:* Although the functional group explains the general type of smell, the shape and size of the rest of the molecule also affect the smell of a compound.

3. Ten black balls representing carbon atoms, 18 gray balls representing hydrogen atoms, 1 red ball representing an oxygen atom, and 30 connectors representing chemical bonds.

4. a.

```
  H  H  H  H  H  H  ..O..
  .. .. .. .. .. ..  : :
H:C:C:C:C:C:C:C:C:O:H
  .. .. .. .. .. ..  ..
  H  H  H  H  H  H
```

b.

```
      H   H   H   H   H   H   O
      |   |   |   |   |   |   ||
  H — C — C — C — C — C — C — C — O — H
      |   |   |   |   |   |
      H   H   H   H   H   H
```

c. carboxylic acid

d. The name is probably two words and ends in "acid." The compound probably has a putrid smell.

5. a.

```
      H ..O..   H
      .. : : ..
  H:C:C:O:C:H
      ..     ..
      H       H
```

b.

```
      H   O       H
      |   ||      |
  H — C — C — O — C — H
      |           |
      H           H
```

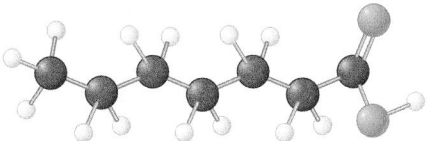

Exercises

(continued)

Reason and Apply

3. What model pieces do you need to build a ball-and-stick model of geraniol?
4. Consider this model. Its molecular formula is $C_7H_{14}O_2$.

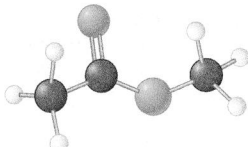

    a. Draw the Lewis dot structure.
    b. Draw the structural formula.
    c. What is the functional group in the molecule?
    d. What can you predict about the name and smell of this compound?

5. Consider this model. Its molecular formula is $C_3H_6O_2$.

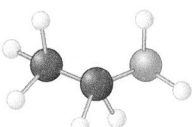

    a. Draw the Lewis dot structure.
    b. Draw the structural formula.
    c. What is the functional group in the molecule?
    d. What can you predict about the name and smell of this compound?

6. Consider this model. Its molecular formula is C_2H_7N.

    a. Draw the Lewis dot structure.
    b. Draw the structural formula.
    c. What is the functional group in the molecule?
    d. What can you predict about the name and smell of this compound?

7. What evidence do you have that the structural formula may not always be useful in predicting smell?

c. ester

d. The name probably ends in "-ate." The compound probably has a sweet smell.

6. a.

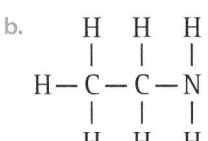

b.

```
      H   H   H
      |   |   |
  H — C — C — N
      |   |   |
      H   H   H
```

c. amine

d. The name probably ends in "-amine." The compound likely has a fishy smell.

7. Possible answer: The three molecules geraniol, menthol, and fenchol each have ester functional groups in their structural formulas. However, each of the three molecules has a distinct smell.

Overview

ACTIVITY: PAIRS

Key Question: How do electrons affect the shape of a molecule?

KEY IDEAS

Lone pairs of electrons have an effect on the shape of a molecule. The space occupied by a pair of electrons, whether a bonded pair or a lone pair, is called an *electron domain*. In a molecule, electron domains are located as far apart from one another as is physically possible. This distribution of electron domains results in the tetrahedral, pyramidal, and bent shapes of CH_4, NH_3, and H_2O, respectively.

LEARNING OBJECTIVES

- Determine the shapes of small molecules.
- Explain how lone pairs of electrons influence molecular shape.
- Describe electron domain theory and how it relates to molecular shape.

FOCUS ON UNDERSTANDING

- Many textbooks refer to the ideas presented in this lesson as *valence shell electron pair repulsion (VSEPR) theory.* We have opted instead to refer to these ideas as *electron domain theory,* because the phrase "electron pair repulsion" can confuse students in light of the existence of double and triple bonds in molecules and in light of the confounding fact that two particles with the same charge should repel each other rather than pair up in a bond.
- Students might come away from this lesson thinking that all small molecules have an underlying tetrahedral shape. There are alternatives, such as BF_3, which is trigonal planar. These are covered in Lesson 38: Let's Build It: Molecular Shape.

KEY TERMS

tetrahedral shape
electron domain
electron domain theory
pyramidal shape
bent shape

IN CLASS

After being introduced to the concept of electron domains, students are given gumdrops, marshmallows, and toothpicks from which to create ball-and-stick models of methane, CH_4. They are challenged to manipulate and refine their

---

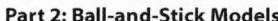

Two's Company

Purpose
To use three-dimensional models to visualize small molecules.

Materials
- gumdrops, marshmallows, and toothpicks
- ruler
- ball-and-stick molecular model set

Part 1: Gumdrop Molecules

1. Create a methane molecule using gumdrops, marshmallows, and toothpicks.
2. Make sure every pair of electrons in the molecule is as far away as possible from every other pair of electrons. Use a ruler to check the distances.
3. Draw Lewis dot structures for the following molecules:
   a. CH_4 **b.** NH_3 **c.** H_2O
4. How many pairs of electrons are located around the central atom of each molecule?
5. Besides the identity of the central atom, what is different about these three molecules?
6. Using gumdrops and toothpicks, create ball-and-stick models of NH_3 and H_2O.
7. Did you remember to include lone pairs? Fix your models if you have to so that lone pairs are represented. Do the lone pairs affect the shape of the molecule?
8. Compare your three gumdrop models. Describe any similarities.

Part 2: Ball-and-Stick Models

1. Use the molecular model sets to create models of CH_4, NH_3, H_2O, and HF. Use black for carbon, white for hydrogen, red for oxygen, and blue for nitrogen.
2. Add the appropriate lone pair paddles to your models.
3. How many lone pair paddles would you need for an atom of neon? Explain your answer.
4. Draw sketches of your three-dimensional models. What is the shape of each molecule if you ignore the lone pair paddles?
5. **Making Sense** Explain how the lone pairs affect the shapes of these molecules.

190   Chapter 7 | **Building Molecules**

---

model so that the electron domains end up as far apart as possible. This should result in a tetrahedral shape. When students successfully complete this step, they move on to build models of ammonia and water, with a worksheet as a guide. Finally, groups of four create ball-and-stick models of four small molecules to further explore molecular geometry. A complete materials list is available for download.

 **Materials List 37**

SETUP

Use plastic sandwich bags as containers for sets of gumdrops, marshmallows,

and toothpicks for each pair of students. Slightly stale gumdrops and marshmallows work best and discourage consumption. You may want to remind students about why it is important never to taste or eat anything in the class or laboratory. You could also bag ball-and-stick model materials for the four models—methane, ammonia, water, and hydrogen fluoride.

CLEANUP

The supplies can be saved for multiple class periods, although you will want to have surplus supplies to replace any that are damaged or lost. The candies and toothpicks

---

Two's Company
Electron Domains

THINK ABOUT IT

You may have noticed that the atoms in the molecular model kits have a certain number of holes and that the sticks only fit in certain places. So, when you connect the atoms, the molecules automatically end up with the correct three-dimensional shape. But what is it that determines the shape of a molecule?

How do electrons affect the shape of a molecule?

To answer this question, you will explore

⟳ Shapes of Molecules

⟳ Electron Domains

EXPLORING THE TOPIC

⟳ Shapes of Molecules

These illustrations show models of ethanol, C_2H_6O.

Ethanol

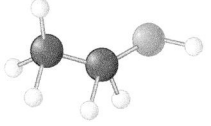

Ball-and-stick model   *Structural formula*

What accounts for the three-dimensional structure of ethanol? The answer lies in the valence electrons. They determine the ultimate shape of a molecule.

To understand molecular shape, it is useful to begin by examining a simple molecule such as methane, CH_4. While the structural formula is flat and cross-shaped, the ball-and-stick model shows that methane has a three-dimensional structure. Both models show that there are four bonds.

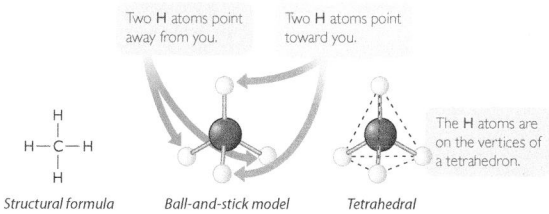

Two **H** atoms point away from you.   Two **H** atoms point toward you.

The **H** atoms are on the vertices of a tetrahedron.

Structural formula   *Ball-and-stick model*   *Tetrahedral*

Lesson 37 | **Electron Domains**   191

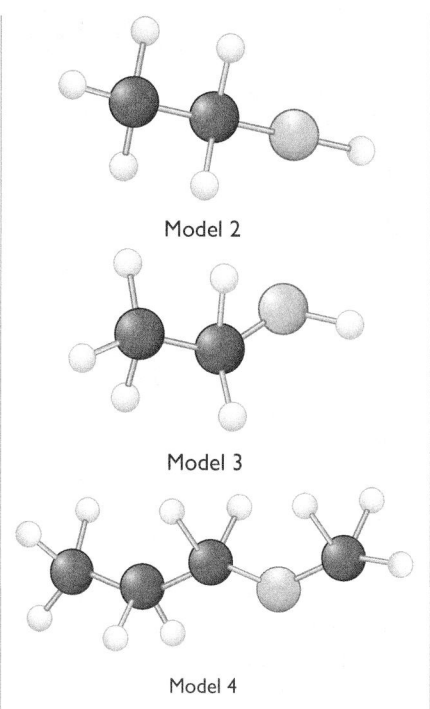

Model 2

Model 3

Model 4

Sample answer: Most students will easily eliminate model 4 because it has the wrong number of atoms. They may also eliminate model 2 because it is flat and not three-dimensional. However, they probably will not know whether model 1 or model 3 is correct. Listen to their reasoning. Model 3 is correct.

Ask: Why is each drawing a correct or an incorrect model for ethanol?

Ask: Why isn't model 2 correct? Model 4?

Ask: Why do you think ball-and-stick models are crooked?

Ask: How are models 1 and 3 different from each other?

EXPLORE (20 min)

TRM Worksheet with Answers 37

TRM Worksheet 37

INTRODUCE THE ACTIVITY

→ Arrange students into pairs.

→ Pass out the student worksheet.

→ Go over the general instructions for building molecules from the materials provided.

● The toothpicks represent bonded electrons.

● The marshmallows represent hydrogen atoms.

● The gumdrops represent other atoms, such as carbon.

● The goal is to first create a correct model for CH_4, methane.

---

are common household supplies and can be disposed of as such.

SAFETY
Remind students that they should not taste or eat anything in the science classroom or laboratory.

Lesson Guide

ENGAGE (5 min)

TRM Transparency—ChemCatalyst 37

TRM PowerPoint Presentation 37

ChemCatalyst

Examine the structural formula of ethanol. Which is the correct ball-and-stick model for ethanol? Explain your reasoning.

H   H
| |
H—C—C—O—H
| |
H   H

Model I

- The only rule is that all the electron pairs must be equally distant from each other.
- Students must show the teacher their methane model before completing the worksheet.

GUIDE THE ACTIVITY

→ Challenge students to keep working if they show you incorrect methane models. If necessary, repeat the rule that all the electron pairs must be equally distant from each other. Point out which electrons in their model do not satisfy this rule.

→ Students should not continue further on their worksheet until their tetrahedral methane model has been given the okay by you. Instruct students to continue to Part 2 when their model is complete.

→ For Part 3, students can work in groups of four with one set of molecular models.

EXPLAIN & ELABORATE (15 min)

DISCUSS THE GUMDROP MODEL OF CH₄

→ Have on hand some of your own examples of incorrect models of CH₄ to discuss their relative merits and flaws.

Ask: Are all the C–H bonds identical in your model? How can you tell?

Ask: If you were to measure the distances between each pair of atoms or electron pairs in your model, would they be the same? Are all the angles the same?

Ask: The structural formula shows CH₄ as a two-dimensional cross. Are all the hydrogen atoms in a cross equidistant from one another?

Ask: How would you show that each hydrogen atom in a tetrahedral molecule is equidistant from the other three hydrogen atoms?

Key Points: The overall geometric shape of a methane model is tetrahedral. In the tetrahedron, each bonding pair is equidistant from the other three pairs. All the H–C–H bond angles are identical (109°). The tetrahedral shape is identical any way it is turned in space. In contrast, the bonding pairs are not equidistant from one another in a cross arrangement. In a cross, each bonding pair is 90° from two other bonding pairs but 180° from a third bonding pair. The bonding pairs can get farther apart

The word used to describe the shape of the methane molecule is **tetrahedral.** A tetrahedral molecule has a symmetrical shape, with one atom exactly in the center. The distance between any other two atoms bonded to the central atom is the same.

You can prove to yourself that a ball-and-stick model of methane is symmetrical in every direction by spinning the molecule as shown in the illustrations. The molecule looks exactly the same no matter what direction it is spun.

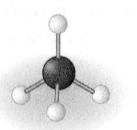 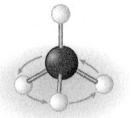 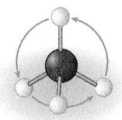

*One **H** atom sticks straight up. Three **H** atoms rest on the table.* *You can spin the molecule around and it looks identical.* *You can grab one of the **H** atoms on the table and spin the molecule around and it still looks identical.*

↻ Electron Domains

METHANE, CH₄

The illustration here shows a methane molecule with the shared valence electrons superimposed, or laid, over the ball-and-stick model. The bonded electron pairs occupy space between the carbon atom and each of the four hydrogen atoms. The space occupied by the electrons is called an **electron domain.** An electron domain can consist of a bonded pair of electrons, a lone pair of electrons, or multiple bonded pairs of electrons.

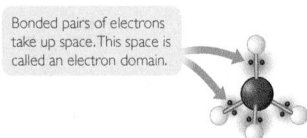

Bonded pairs of electrons take up space. This space is called an electron domain.

A methane molecule has four electron domains.

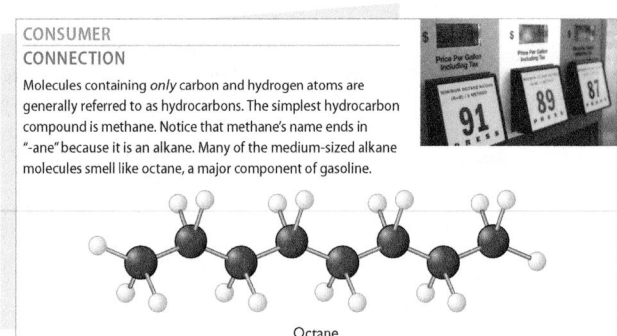

CONSUMER CONNECTION

Molecules containing *only* carbon and hydrogen atoms are generally referred to as hydrocarbons. The simplest hydrocarbon compound is methane. Notice that methane's name ends in "-ane" because it is an alkane. Many of the medium-sized alkane molecules smell like octane, a major component of gasoline.

Octane

from one another by moving out of the plane of the cross.

This area is an electron domain.

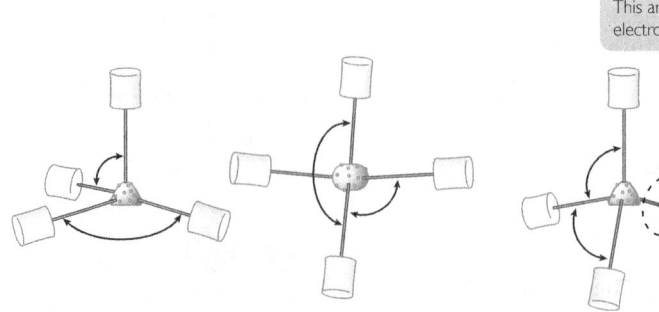

This area is an electron domain.

Incorrect models–electron pairs are not equally distant.   Correct models–all angles between bonds are the same.

Each electron domain in the tetrahedral methane molecule is the same distance from the other three electron domains. They are as far apart from one another as possible. Because of this arrangement, the hydrogen atoms are as far apart from one another as possible and the distance between any two hydrogen atoms is the same. Additionally, in the tetrahedral molecule, if you measure the angle formed between any two hydrogen atoms and the carbon atom, you'll find that all these bond angles have equal measures. This is not accurately shown in the structural formula.

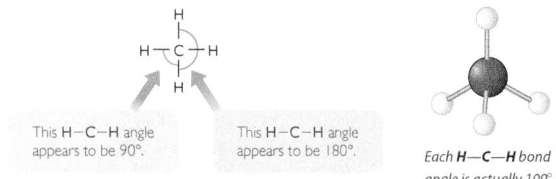

This H—C—H angle appears to be 90°.   This H—C—H angle appears to be 180°.   Each **H—C—H** bond angle is actually 109°.

> **Important to Know** Electrons are negatively charged, so they repel one another. However, in molecules, electrons bond in pairs. It is the electron pairs that repel one another.

This tendency for electron pairs to be as far apart from one another as possible is called **electron domain theory**. This theory is also referred to as *valence shell electron pair repulsion theory*, or *VSEPR theory*.

AMMONIA, NH₃

The structural formula and ball-and-stick model for ammonia, NH₃, are shown below. The ball-and-stick model shows that the hydrogen atoms are all located on one side of the nitrogen atom.

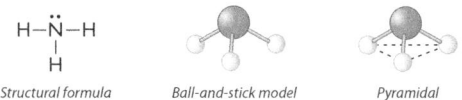

Structural formula   Ball-and-stick model   Pyramidal

The word used to describe the shape of the ammonia molecule is **pyramidal.**

The nitrogen atom has a lone pair of electrons. Therefore, while there are only three bonds, there are four electron domains. The four electron domains are arranged in a tetrahedral shape to get as far apart as possible, similar to what happens in methane.

An ammonia molecule has four electron domains, just like methane.

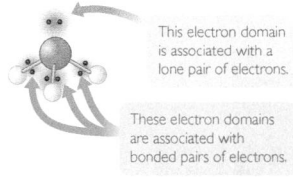

This electron domain is associated with a lone pair of electrons.

These electron domains are associated with bonded pairs of electrons.

Lesson 37 | **Electron Domains**   193

> **Tetrahedral shape:** The shape around an atom with four bonded pairs of electrons. This is the shape of a methane molecule.

INTRODUCE THE ELECTRON DOMAINS

Key Points: An electron domain describes the area occupied by a set of electrons in a bond or a lone pair. Electron pairs tend to remain as far apart as possible from other pairs of electrons within a molecule. You can use the idea of electron domains to help determine the shape of a molecule.

> **Electron domain:** The space occupied by valence electrons in a molecule, either a bonded pair(s) or a lone pair. Electron domains affect the overall shape of a molecule.
>
> **Electron domain theory:** The idea that every electron domain in a molecule is as far as possible from every other electron domain in that molecule.

DISCUSS THE TETRAHEDRAL SHAPE OF MOLECULES

→ Use the models of water and ammonia built by the students as examples.

→ At the appropriate time in the discussion, introduce the ball-and-stick models of methane, ammonia, water, and hydrogen fluoride.

→ Ask students to draw the structural formula of each molecule on the board.

Ask: How did the Lewis dot structures assist you with your gumdrop models?

Ask: What gumdrop model did you create for NH₃? How did you arrive at that shape?

Ask: What is the total number of electron domains in each of the three molecules in Part 2 of the worksheet?

Ask: How do lone pairs influence the location of the hydrogen atoms?

Key Points: Even though the molecules you created today have different numbers of atoms, they all have a similar underlying shape. The underlying shape is called a *tetrahedron*. If you examine molecular models that include a special piece representing a lone pair (called a paddle), you can see the common geometries in these molecules (and even in a neon atom). The images below also appear on p. 194 of the student text.

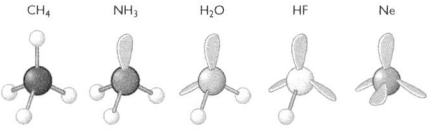

CH₄   NH₃   H₂O   HF   Ne

Most molecular models do not include lone-pair paddles in their representations. Chemists simply remember that lone pairs are present in certain atoms. Lone pairs of electrons "take up space" and affect the ultimate shape of a molecule. The molecular model kits are designed so that the appropriate angles are created when the pieces are placed together. If you remove the lone-pair paddles from the models, you see the more familiar representations of these molecules. In general, these shapes are referred to by the terms *tetrahedral, pyramidal, bent, linear,* and *point*. (*Note:* The pyramidal shape is trigonal pyramidal.) The images below also appear on p. 194 of the student text.

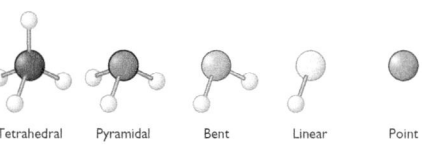

Tetrahedral   Pyramidal   Bent   Linear   Point

Lesson 37 | **Electron Domains**   193

Pyramidal shape: The shape around an atom with one lone pair of electrons. This is the shape of an ammonia molecule.

Bent shape: The shape around an atom with two lone pairs of electrons. This is the characteristic shape of a water molecule.

EVALUATE (5 min)

Check-In

Use your model kit to build a model for ethanol. Be sure to use lone pairs to help you with your overall structure.

$$
\begin{array}{ccccc}
& \text{H} & & \text{H} & \\
& | & & | & \\
\text{H}- & \text{C} & - & \text{C} & -\text{O}-\text{H} \\
& | & & | & \\
& \text{H} & & \text{H} &
\end{array}
$$

Answer: The model should look like model 3 from the ChemCatalyst. The geometry around each of the carbon atoms is tetrahedral. There are two lone pairs in this molecule, both on the oxygen atom. Because the lone pairs take up space, the C–O–H angle is less than 180°.

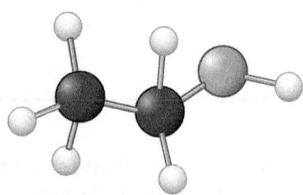

Homework

Assign the reading and exercises for Smells Lesson 37 in the student text.

WATER, H₂O

Take a look at one more molecule—water, H_2O. The word used to describe the shape of a water molecule is **bent.**

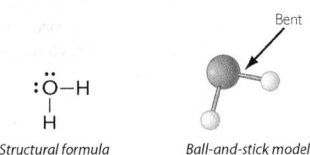

Structural formula   Ball-and-stick model

A water molecule has four electron domains, two bonding pairs and two lone pairs. The four electron domains of water are arranged in a tetrahedral shape, as in methane and ammonia. This results in a bent shape for the water molecule. In all three molecules, the bond angles are 109° or close to this.

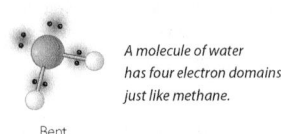

A molecule of water has four electron domains, just like methane.

Bent

BALL-AND-STICK MODELS WITH LONE PAIR PADDLES

In a molecular model kit, the lone pairs are sometimes represented as plastic paddles. This helps you to visualize where the final shape of a molecule comes from.

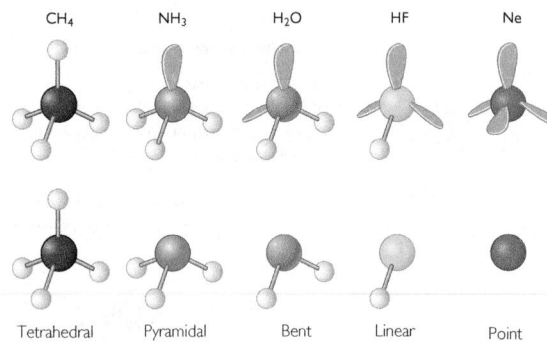

CH₄	NH₃	H₂O	HF	Ne
Tetrahedral	Pyramidal	Bent	Linear	Point

Notice that you name the shape of the molecule based only on the arrangement of atoms. The lone pairs are not considered.

194   Chapter 7 | **Building Molecules**

Example

Phosphine, PH₃

What is the shape of phosphine, PH_3? Explain your thinking.

Solution

Begin by drawing the Lewis dot structure and a structural formula of phosphine. A phosphorus atom has five valence electrons, and each hydrogen atom contributes one electron, for a total of eight electrons. There are three bonding pairs and one lone pair. The four atoms in phosphine are arranged in a pyramidal shape.

LESSON SUMMARY

How do electrons affect the shape of a molecule?

KEY TERMS

tetrahedral shape
electron domain
electron domain theory
pyramidal shape
bent shape

The three-dimensional shape of a molecule is determined by the various electron pairs in the molecule. *Both* bonded pairs of electrons *and* lone pairs of electrons affect the shape of a molecule. Each electron pair takes up space, called an electron domain. The final shape of a molecule is determined by the fact that electron domains in a molecule are located as far apart from one another as possible.

Exercises

Reading Questions

1. In your own words, describe a tetrahedral shape.

2. What is meant by electron domain theory?

Reason and Apply

3. If you were going to predict the three-dimensional structure of a small molecule, what would you want to know?

4. Predict the three-dimensional structure of H_2S. Explain your thinking.

5. List three molecules that have a tetrahedral shape.

6. List three molecules that have a bent shape.

7. What is the shape of arsine, AsH_3? Explain your thinking.

8. Predict the three-dimensional shape of HOCl. Explain your thinking.

9. Draw a methane, CH_4, molecule and show how it fits inside a tetrahedron. Do the same for ammonia, NH_3, and water, H_2O.

8. *Possible answer:* A molecule of HOCl will have a bent shape with the oxygen atom between the hydrogen and chlorine atoms. The oxygen atom has two lone pairs and two single bonds. The single bonds will probably form a bent shape.

9.

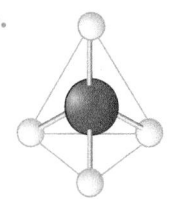
Methane

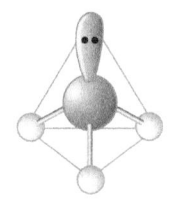
Ammonia

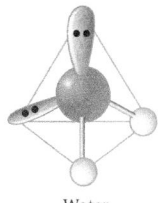
Water

LESSON 37 ANSWERS

1. *Possible answer:* A tetrahedral shape has four single bonds spaced equally around one central atom.

2. Electron domain theory states that electron pairs in a molecule will be as far from one another as possible.

3. *Possible answer:* I would want to know the number of lone pairs and bonded pairs for the central atom. Then I could determine the arrangement of the atoms in three dimensions.

4. *Possible answer:* An H_2S molecule would have a bent shape like a water molecule. The sulfur atom has two bonds and two lone pairs. The four electron domains are arranged in a tetrahedral shape, causing them to be spaced as far apart as possible. The two bonds form a bent shape. Any molecule with only three atoms will form either a linear or a bent shape.

5. *Possible answers (any 3):* CH_4, $SiCl_4$, CF_4, $SiBr_4$

6. *Possible answers (any 3):* H_2O, H_2S, SCl_2 OBr_2

7. AsH_3 will have a pyramidal shape because the three bonded pairs and the one lone pair on the arsenic atom form a tetrahedron. Therefore, the three single bonds form a pyramid with the arsenic atom at its top.

Overview

ACTIVITY: GROUPS OF 4

Key Question: How can you predict the shape of a molecule?

KEY IDEAS

Lewis dot structures are a useful tool in the three-dimensional modeling of molecules. They help us determine how many electron domains are in a molecule. Electron domain theory can be applied to molecules with double and triple bonds, resulting in other shapes besides tetrahedral. Although lone pairs influence the ultimate shape of the molecule, they are not included in the actual description of the shape of the molecule.

LEARNING OBJECTIVE

● Predict and explain molecular shape, including of molecules with multiple bonds.

FOCUS ON UNDERSTANDING

● Lewis dot structures may lead students to think that bond angles must be 90° or 180°. Building ball-and-stick models gives students a better grasp of molecular shape.
● In the next lesson, smaller "shapes" come together to form larger molecules, describing another level of overall shape.

KEY TERMS

trigonal planar shape
linear shape

IN CLASS

Students use candy and toothpicks to build a model of formaldehyde, CH_2O, to learn about the effect of double bonds on the shape of a molecule. Groups of four then use ball-and-stick model-building materials to construct models of 13 different molecules, using Lewis dot structures as a guide. Students then draw and describe the different shapes of these molecules. A complete materials list is available for download.

TRM Materials List 38

SETUP

Place any gumdrops, marshmallows, and toothpicks in bags or trays. Inventory the model-building sets. You may want to remove the space-filling modeling parts from the set (disk connectors, white hemispheres, 3-hole carbons) so students have only ball-and-stick modeling parts to work with today.

LESSON 38

Let's Build It

Molecular Shape

THINK ABOUT IT

So far you have considered the shapes of small molecules containing single bonds. All of the molecules you've studied have had four electron domains. However, small molecules that have only three or even two electron domains also exist. These molecules have double and triple bonds. So, there are more molecular shapes to consider.

How can you predict the shape of a molecule?

To answer this question, you will explore

⟳ More Shapes
⟳ Larger Molecules

EXPLORING THE TOPIC

⟳ More Shapes

DOUBLE BONDS

Formaldehyde, CH_2O, is a very simple molecule that contains one double bond. The double bond counts as only one electron domain even though it contains two pairs of bonded electrons. This means that the central atom in the formaldehyde molecule (the carbon atom) has only three electron domains surrounding it, spread out into a flat triangle. The phrase used to describe the underlying shape of the formaldehyde molecule is **trigonal planar.**

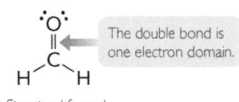
The double bond is one electron domain.

Structural formula

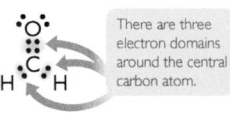
There are three electron domains around the central carbon atom.

Lewis dot structure

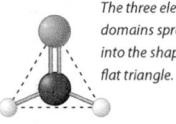
The three electron domains spread out into the shape of a flat triangle.

Trigonal planar

Both ammonia, NH_3, and formaldehyde, CH_2O, have four atoms but have very different shapes. The difference in shape is due to the number of electron domains around the central atom.

The three hydrogen atoms are below the nitrogen atom to make room for the lone pair on nitrogen.

Ammonia

Formaldehyde

There are no lone pairs, so the three electron domains spread out into a triangle.

Pyramidal   Trigonal planar

196   Chapter 7 | **Building Molecules**

Build one ball-and-stick model of citronellol for the Explain & Elaborate discussion. See the image in the margin on p. 185 as a reference.

Differentiate

Students may benefit from a wall chart that lays out the steps to predicting electron domain geometry. Make a simple poster that shows how to (1) create a Lewis dot structure for the molecule, (2) count the number of electron domains around each central atom(s), (3) use the number of electron domains to determine the electron domain geometry, (4) predict the bonding geometry based on the electron domain geometry and presence of any lone pairs. You might have students visit the University of Colorado Boulder's PhET simulations titled "Molecular Shapes Basics" and "Molecular Shapes." A list of URLs is in the Chapter 7 Web Resources.

TRM Chapter 7 Web Resources

Pre-AP® Course Tip

Students preparing to take an AP® chemistry course or college chemistry must be able to use representations and models to make predictions. In this lesson, students work with ball-and-stick models and Lewis dot structures to predict and describe the shape of various molecules.

TWO DOUBLE BONDS

In a molecule with a double bond, the double bond counts as one electron domain. So a molecule like carbon dioxide, with two double bonds, has only two electron domains around the central atom that have to be as far apart as possible. This shape is described as **linear.**

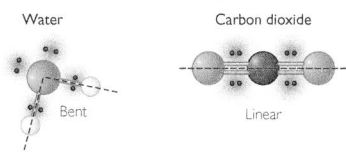

O=C=O

Each double bond is one electron domain.

There are two electron domains around the central carbon atom.

The electron domains are as far apart as possible.

Linear

:Ö::C::Ö:

Both water, H_2O, and carbon dioxide, CO_2, have three atoms. However, the shape of water is bent while carbon dioxide is linear. The difference in shape is due to the number of electron domains around the central atom.

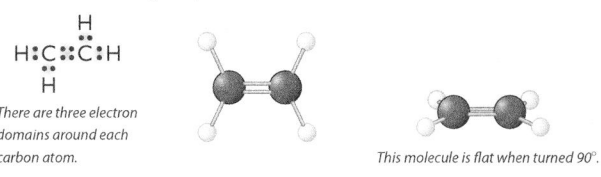

Water

Bent

Carbon dioxide

Linear

Example 1

Ethene, C_2H_4

Determine the molecular shape of ethene, C_2H_4.

Solution

Begin by drawing the Lewis dot structure for ethene to determine the number of electron domains around each carbon atom. Next, translate that into a three-dimensional representation of the molecule. In this case, there are no lone pairs. The molecule is flat. Each half of ethene is trigonal planar.

H:C::C:H

There are three electron domains around each carbon atom.

This molecule is flat when turned 90°.

INDUSTRY CONNECTION

Ethene, C_2H_4, is also called ethylene. It is a colorless gas with a slightly sweet odor. Ethene is used in the manufacture of many common products like polyethylene plastic and polystyrene. Ethene also can be used as a plant hormone to control the ripening of fruit or the opening of flowers.

Lesson 38 | **Molecular Shape**   197

Lesson Guide

ChemCatalyst

1. What is the Lewis dot structure of formaldehyde, CH_2O?

2. Draw formaldehyde's structural formula.

3. How many electron domains do you think this molecule has? Explain your reasoning.

Sample answers:

1.

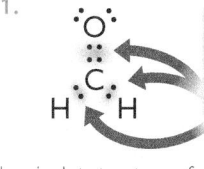

There are three electron domains around the central carbon atom.

Lewis dot structure of formaldehyde, CH_2O

2.

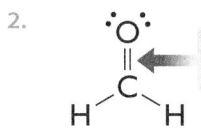

The double bond is one electron domain.

Structural formula of formaldehyde, CH_2O

---

3. There are three electron domains around the carbon atom: the two single bonds with hydrogen and the double bond with oxygen. Students might think that the double bond counts for two electron domains or that this molecule has five electron domains, because oxygen has two lone pairs. However, tell them you are most interested in the electron domains around the central carbon atom.

DISCUSS THE CHEMCATALYST

➔ Have students draw their Lewis dot structure and structural formula of formaldehyde on the board.

Ask: How many electron domains are there around the carbon atom? How do you account for the double bond?

Ask: What overall shape do you think this molecule has? Is it tetrahedral?

Ask: What effect does the double bond have on the overall shape of the molecule?

PREPARE FOR THE ACTIVITY

➔ Pass out the gumdrop, marshmallow, and toothpick sets to groups of students to build a model of formaldehyde, CH_2O.

➔ Allow students to work out the shape of this molecule and describe it. The oxygen atom and hydrogen atoms have to be as far apart as possible.

➔ Tell students the shape of formaldehyde is referred to as *trigonal planar*. Point out that the overall shape is flat and triangular.

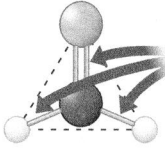

There are three electron domains around the central carbon atom.

The shape of formaldehyde is trigonal planar.

➔ Point out the double bond and the three electron domains around the central atom. Explain that the four electrons in the double bond are all in the same electron domain. Collect the gumdrops and marshmallows from the students before beginning the activity.

Lesson 38 | **Molecular Shape**   197

INTRODUCE THE ACTIVITY

→ Arrange students into groups of four.

→ Pass out the student worksheet.

→ Pass out sets of ball-and-stick models. Tell students they will be drawing Lewis dot structures and using them as guides to create ball-and-stick models of the compounds in the table. Remind them that an electron pair is a bond shown with a line in a structural formula and a stick in a ball-and-stick model.

→ Students can use black spheres for carbon, white spheres for hydrogen, red spheres for oxygen, and blue for nitrogen. Pairs of curved connectors can be used to make double bonds.

→ Advise students that some molecules may be large enough to combine more than one shape. They should describe the shape around each central atom when necessary.

→ Groups can divide up the work of building the models so that each group has one set of models to work with.

EXPLAIN & ELABORATE (15 min)

DISCUSS THE MODEL-BUILDING ACTIVITY

→ Ask students to share their models. Hold up each model in turn.

→ Construct a table containing the main shapes as you solicit answers from students. A complete table is available for download.

TRM Model Building Table 38

Ask: Identify the central atom or atoms in this molecule. How many electron domains are around the central atom?

Ask: Which molecules have three electron domains around a central atom? What shape do these molecules take? (trigonal planar) What shape do molecules with two electron domains take? (linear)

Ask: What do double bonds do to the shape of a molecule?

Ask: Both formaldehyde, CH_2O, and ammonia, NH_3, have four atoms. Why don't they have the same shape?

Key Points: Double or triple bonding changes the number of electron domains around an atom, affecting the overall shape of a molecule. Atoms tend to arrange themselves in tetrahedral shapes around atoms unless multiple bonds are present. When a carbon atom has a double bond and two single bonds,

TRIPLE BONDS

Another example of a linear molecule is hydrogen cyanide, HCN. It also has two electron domains around the central atom, like carbon dioxide. One electron domain is a single bond between carbon and hydrogen, and the other is a triple bond between carbon and nitrogen.

 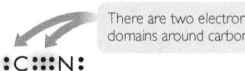 

Hydrogen cyanide has one single bond and one triple bond. This results in two electron domains around the central atom.

Example 2

Ethyne, C_2H_2

Determine the molecular shape of ethyne, C_2H_2.

Solution

Begin by drawing the Lewis dot structure of ethyne to determine the number of electron domains around the carbon atoms. To satisfy the octet rule and the HONC 1234 rule, the carbon atoms must have a triple bond between them. Each carbon atom has two electron domains surrounding it. Ethyne is a linear molecule.

 There are two electron domains around each carbon atom.

⟳ Larger Molecules

The shapes of all of these small molecules help to explain why chains of carbon atoms are crooked rather than straight. The illustration shows a ball-and-stick model for heptanoic acid, $C_7H_{14}O_2$, with seven carbon atoms in a chain. The arrangement of atoms around six of the carbon atoms is tetrahedral. Most of the molecule is a series of overlapping tetrahedral shapes.

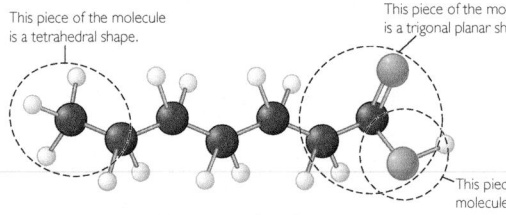

This piece of the molecule is a tetrahedral shape.

This piece of the molecule is a trigonal planar shape.

This piece of the molecule is a bent shape.

Heptanoic acid, $C_7H_{14}O_2$

The various electron domains in this molecule cause its overall shape to be crooked, or zigzag.

Number of domains around central atom	4	4	Download table for more examples.
Number of lone pairs	0	1	
Shape	tetrahedral	pyramidal	
Example	CH_4 CH_3Cl, CH_2Cl_2	NH_3	
Sample Sketch			

LESSON SUMMARY

How can you predict the shape of a molecule?

The shape of a molecule is affected by the location and number of its electron domains. Lewis dot structures help you to determine the number of electron domains in a molecule. An atom involved in a double bond or triple bond will have fewer electron domains surrounding it. This results in molecules that have trigonal planar and linear shapes. Large molecules can have various geometric shapes within them. Areas around double bonds are flat, and those around triple bonds are linear.

KEY TERMS
trigonal planar shape
linear shape

Exercises

Reading Questions

1. What shapes are possible if a molecule has three electron domains?

2. What shapes are possible for a molecule with three atoms? Explain your thinking.

Reason and Apply

3. Describe the shape of each of these molecules. (Use Lewis dot structures, the periodic table, and the HONC 1234 rule to assist you.)

Cl_2 CO_2 H_2O

4. Which of the molecules in Exercise 3 have the same shape?

5. What shape or shapes do you predict for a molecule with two atoms? A molecule with three atoms? Four atoms? Five atoms? You can draw Lewis dot structures or electron domains to assist you in answering the questions.

6. Predict the shapes of these molecules:

CF_4 NF_3 H_2Se H_2CS

7. Consider the butane, C_4H_{10}, molecule.
   a. Draw the Lewis dot structure for butane.
   b. How many electron domains does the molecule have?
   c. What shape would you predict for C_4H_{10}?
   d. Explain why the carbon atom chain is not straight.

Lesson 38 | **Molecular Shape**   199

there are only three electron domains. The atoms arrange themselves in a flat triangle around the carbon atom. When a central atom has only two electron domains, as in carbon dioxide or hydrogen cyanide, the molecule ends up in a linear shape.

Trigonal planar shape: A flat triangular shape found in small molecules with three electron domains surrounding the central atom.

Linear shape: A geometric shape found in small molecules with two electron domains surrounding the central atom.

The number of electron domains is more important in determining the structure of a molecule than is the number of atoms. Both water and carbon dioxide have three atoms, but water has a bent shape due to the four electron domains around the oxygen atom. Carbon dioxide, on the other hand, has two double bonds, resulting in a linear shape.

DISCUSS THE GEOMETRIES OF A LARGER MOLECULE

→ Use the ball-and-stick model of citronellol to show the overall effect of small shapes on the shape of a larger molecule.

Ask: What shapes are present within this large molecule of citronellol? (tetrahedral, trigonal planar, and bent)

Ask: What causes shapes other than tetrahedral ones to show up in a molecule?

Can you explain why the overall molecule is crooked instead of straight like the structural formula?

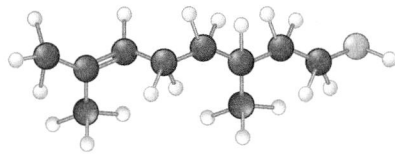

Citronellol

Key Points: The more atoms in a molecule, the more combinations of shapes you might see together. Ethane has two carbon atoms. You can consider these both central atoms, each with four electron domains. The result is two tetrahedra stuck together at a vertex. A very large molecule, like citronellol, is simply a combination of many tetrahedra, with a trigonal planar area involving a double bond and a bent area next to the oxygen atom. The result is a molecule in which the electron domains are all as far apart as possible from one another. In general, shapes other than tetrahedral ones show up in a molecule when there are multiple (double or triple) bonds.

EVALUATE (5 min)

Check-In

What is the shape of this molecule?
H_2S

Answer: Bent, similar to water. Sulfur has six valence electrons, so it has two lone pairs, like oxygen. This gives sulfur four electron domains, which arrange themselves tetrahedrally. The result is a bent molecule.

Homework

Assign the reading and exercises for Smells Lesson 38 in the student text.

LESSON 38 ANSWERS

Answers continue in Answer Appendix (p. ANS-2).

Overview

ACTIVITY: GROUPS OF 4

Key Question: How is the shape of a molecular compound related to its smell?

KEY IDEAS

The overall shape of a molecular compound is directly related to its smell. Space-filling models provide another way to examine molecules three-dimensionally. Smell is related to the overall shape of the whole molecule.

LEARNING OBJECTIVES

- Build a space-filling molecular model given the structural formula.
- Begin to relate the overall shapes of molecules to their smell categories.

FOCUS ON UNDERSTANDING

- Here we focus on the overall shapes of molecules, as opposed to the geometric shapes surrounding the central atoms, as we did in Lessons 37 and 38.
- We have carefully chosen examples for simplicity and consistency in the smells categories. In reality, for instance, not every minty-smelling molecule is flat and planar.

KEY TERM

space-filling model

IN CLASS

Students use space-filling molecular models to examine the overall shape of six smell molecules. They compare and contrast similar-smelling compounds and different-smelling compounds, looking for shape trends. They find that there are three different overall shapes of these six molecules: stringy, ball-shaped, and frying-pan shaped. The Check-In involves an optional mini-drama that offers practice in forming a hypothesis as well introducing a new smell molecule. A complete materials list is available for download.

TRM Materials List 39

SETUP

To save class time, you could build all the space-filling models yourself in advance. Label them with the appropriate number (see worksheet). However, this is very time consuming—you will probably want help. Refer to the worksheet for the structural formulas. If you have enough materials and want students to build the models during class, you might make

available only the appropriate parts to build space-filling models today. If you decide to perform the mini-drama, have the protective gear and vial Y hidden away but easily accessible.

CLEANUP

Collect and inventory model-building sets or models.

Lesson Guide

ENGAGE (5 min)

TRM Transparency—ChemCatalyst 39

TRM PowerPoint Presentation 39

What Shape Is That Smell?

Space-Filling Models

THINK ABOUT IT

Methyl octenoate, $C_9H_{16}O_2$, is a compound that smells like violets. A ball-and-stick model of methyl octenoate shows that it is a series of overlapping tetrahedral shapes stuck together. There is a trigonal planar segment in the area of the double bonds. But how would you describe the shape of the *whole* molecule? And does the shape of the whole molecule have anything to do with its smell?

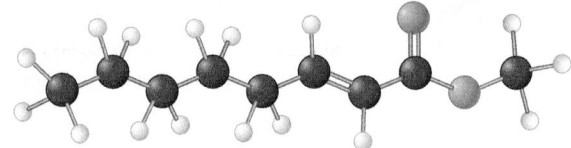

How is the shape of a molecular compound related to its smell?

To answer this question, you will examine

➲ Space-Filling Models

➲ Relating Shape and Smell

EXPLORING THE TOPIC

➲ Space-Filling Models

Most of the smell molecules you have encountered are considerably larger than three, four, or five atoms. The best way to look at the overall shape of these molecules is with a different type of model, called a **space-filling model.**

Take a moment to compare the ball-and-stick model of methyl octenoate with the space-filling model. In a space-filling model, the sticks between atoms have been eliminated. There is no space between atoms. Instead, bonded atoms are shown slightly overlapping.

Methyl octenoate, $C_9H_{16}O_2$

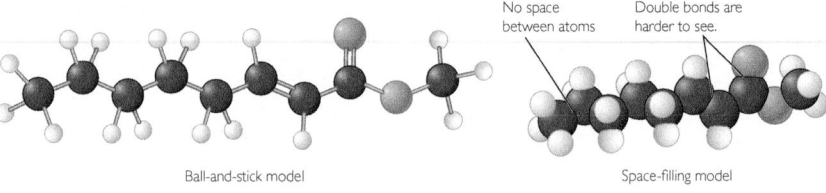

No space between atoms

Double bonds are harder to see.

Ball-and-stick model

Space-filling model

200   Chapter 7 | **Building Molecules**

ChemCatalyst

What similarities and differences do you see between these two different types of models?

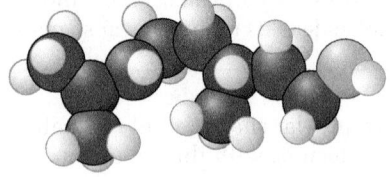

Space-filling model of citronellol

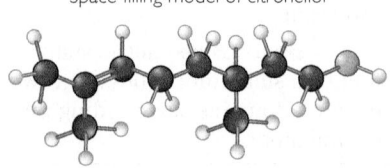

Ball-and-stick model of citronellol

In some ways, a space-filling model could be considered more accurate than a ball-and-stick model. In reality, a stick has no resemblance to a bond. In a molecule, the atoms *share* electrons with neighboring atoms. This would suggest that the atoms are located very close to, or overlapping, one another. However, in an illustration of a space-filling model, you can't see multiple bonds, and some atoms may be hidden behind others.

⟳ Relating Shape and Smell

Space-filling models for molecules that smell sweet, minty, and like camphor (piney) are shown here. Take a moment to look for patterns that may indicate a connection between the shapes of these molecules and their smells.

SWEET-SMELLING MOLECULES

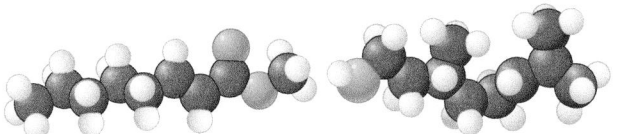

Methyl octenoate, $C_9H_{16}O_2$ Citronellol, $C_{10}H_{20}O$

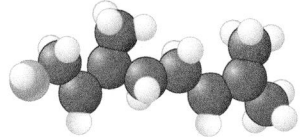

Geraniol, $C_{10}H_{18}O$

MINTY-SMELLING MOLECULES

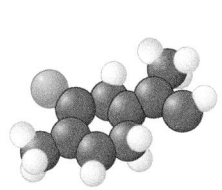   

Carvone, $C_{10}H_{14}O$ Pulegone, $C_{10}H_{16}O$

Menthone, $C_{10}H_{18}O$

ENVIRONMENTAL CONNECTION

Geraniol is a sweet-smelling molecule. It is one of the molecules that make roses smell like roses.

Sample answer: One model has sticks, and the other does not. One model is compact and chunky, and the spheres are partially hidden. Both models represent the same molecule, with the same number and color of each type of atom connected in the same order.

Ask: How are the atoms represented in the space-filling model?

Ask: How are the bonds represented in each model?

Ask: Which model do you think is more accurate, or more like real life? What is your reasoning?

Ask: Why do you think the atoms are not all perfectly round in a space-filling model?

EXPLORE (15 min)

TRM Worksheet with Answers 39

TRM Worksheet 39

INTRODUCE THE ACTIVITY

→ Arrange students into groups of four.

→ Pass out the student worksheet.

→ Tell students they will be examining another kind of three-dimensional model for molecules, called a space-filling model.

→ Distribute a set of space-filling models or model-building sets to each group of four students. If students are building the models, assign each pair a different model to build. Then they can share their models to answer the worksheet questions.

→ Go over what the pieces represent.

EXPLAIN & ELABORATE (15 min)

ASSIST STUDENTS IN SHARING THEIR OBSERVATIONS

→ Write the three smell classifications—sweet, minty, and camphor—on the board. At the appropriate time, make a list of some of the students' shape names under the smell headings.

Ask: What is the process you used to identify the molecules represented by the models?

Ask: Which do you think looks more like a real molecule, a ball-and-stick model or a space-filling model? Explain your thinking.

Ask: Which models represent similar smells? (1 and 6, 3 and 4, 2 and 5)

Ask: What structural similarities did you find in the molecules that smell similar?

Ask: How would you describe the overall shape of the sweet-smelling molecules? The minty-smelling molecules? The camphor-smelling molecules?

Key Points: A space-filling model is a three-dimensional model that a chemist uses to show how the atoms in a molecule are arranged in space and how they fill this space. Although we cannot say precisely what a molecule looks like, space-filling models are considered slightly more accurate than ball-and-stick models. For one thing, covalent bonds between atoms cause atoms to be attracted to each other. Thus, atoms in a molecule probably are right next to each other rather than a stick's length apart.

To identify the space-filling models, students might have found it useful to start by figuring out the molecular formulas of the models and of the illustrations. They could also narrow down the answer by counting the number of oxygen atoms, looking for rings, or figuring out exactly where certain double bonds and oxygen atoms are located. Molecules 1 and 6 both smell sweet. They also resemble each other in overall shape—especially compared to the minty-smelling and camphor-smelling molecules. The same can be said of the minty-smelling molecules and the camphor-smelling molecules.

INTRODUCE SHAPE NAMES FOR LARGER MOLECULES

→ Starting from the students' suggestions, introduce the three shape names we will be using from now on.

Ask: What overall shape do the molecules in each smell category have?

Ask: What did you predict for the smells of pulegone and isopentyl acetate? What was your reasoning?

Ask: Do you have enough evidence to link smell to shape? (More data would be useful.)

Key Points: The shape of a molecular compound seems to be directly related to its smell. To talk about molecular shape, it is necessary to agree on some simple shape classifications. The sweet-smelling molecules in today's activity were all long and stringy molecules. The camphor-smelling molecules were also similar in shape— they looked bunched up. The minty-smelling molecules have a ring with a handle, sort of like a frying pan. For consistency, we will call these shapes stringy, ball-shaped, and frying pan. We can summarize our observations:

Possible student answers:

Sweet: Long, stringy, snakelike, flat
Minty: ring-shaped, ball-shaped, frying pan
Camphor: bunched up, cluster, ball-shaped

Note that these are hypotheses, not rules that apply to all molecules. While ball-shaped molecules typically are spherical, molecules shaped like a frying pan do not always smell minty. Remind students that many more experiments are necessary to clarify the relationship between shape and smell.

EVALUATE (10 min)

TRM Transparency—Check-In 39

COMPLETE A MINI-DRAMA DEMONSTRATION (OPTIONAL)

→ Dress up in protective garb and use protective equipment to dramatize the introduction of the molecule for the Check-In. Using tongs, remove the newest vial from a box, place it gingerly

in a jar or beaker, and cover the container with something like a watch glass. Act as if you are handling a very dangerous substance.

→ Write "Smell Y" on the board. Make sure the students see your actions, but do not give away anything.

→ Students are likely to ask, "What are you doing?" or "Is that stuff dangerous?" or "Why are you wearing all that stuff?" or "What's in the bottle?" Point out that they are beginning an inquiry.

→ Ask students what they think you are doing. They will say things like "We're smelling something dangerous,"

and "You're going to do a demonstration with a toxic substance." Point out that they have made a hypothesis about what you are doing. A hypothesis is an educated or best guess.

→ Ask students how they arrived at their hypothesis and what evidence they have that supports their hypothesis. Suggest to students that they can make a hypothesis about the smell of the substance in vial Y. Tell them you are willing to give them some evidence to help them in making their hypothesis. (Transition into the Check-In. If you want, reveal the information one piece at a time.)

CAMPHOR-SMELLING MOLECULES

 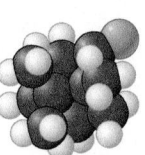

Fenchol, $C_{10}H_{18}O$ Camphor, $C_{10}H_{16}O$

The sweet-smelling molecules are all long and stringy. The minty-smelling molecules all have a six-carbon ring structure. They have a shape that resembles a frying pan. The camphor-smelling molecules are a tight cluster of atoms in the shape of a ball. So far, there appear to be three smell categories that are directly related to the molecular shape.

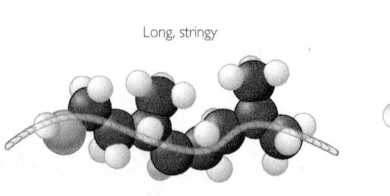

Long, stringy

Sweet smell

Frying pan–shaped

Minty smell

Ball-shaped

Camphor smell

LESSON SUMMARY

How is the shape of a molecular compound related to its smell?

A space-filling model is a three-dimensional model that shows how the atoms in a molecule are arranged in space and how they fill this space. The shape of a molecule appears to be related to its smell. For example, the sweet-smelling molecules explored here are all long and stringy in shape. The minty-smelling molecules are frying pan–shaped, and the camphor-smelling molecules are ball-shaped. More data will certainly be helpful to confirm a connection between a molecule's shape and its smell.

KEY TERM
space-filling model

Exercises

Reading Questions

1. How is a space-filling model useful?
2. How is a space-filling model different from a ball-and-stick model?

Reason and Apply

3. Draw a possible structural formula for a molecule that smells sweet and has the molecular formula $C_9H_{20}O$. What general shape does this molecule have?
4. If someone told you a molecule had a six-carbon ring, what smell would you predict for that compound?
5. Draw a possible structural formula for a molecule that smells minty and has the molecular formula $C_{10}H_{16}O$. What general shape does this molecule have?
6. If someone told you a molecule is ball-shaped, what else would you want to know to predict the smell of the molecule?
7. Which do you think has more influence on smell: the functional group that is present or the shape of a molecule? Explain why you think so.

Check-In

What smell do you predict for the substance in vial Y? Explain your reasoning.

Vial Y

Molecular formula:	$C_{12}H_{20}O_2$
Chemical name:	bornyl acetate

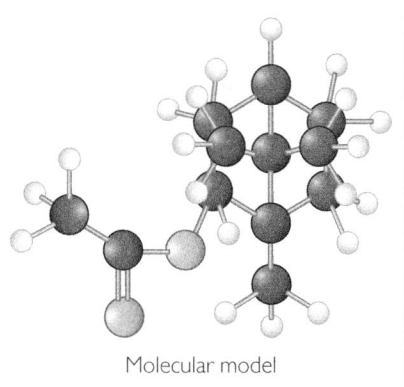

Structural formula

Molecular model

Answer: The two oxygen atoms indicate either a sweet or putrid smell. The chemical name indicates that it is an ester, with a sweet smell. The structural formula confirms the ester but also

has features of a minty-smelling or a camphor-smelling molecule. The molecular model confirms a ball-shaped camphor-smelling molecule. In reality, this molecule, bornyl acetate, is an insect pheromone. It smells both sweet and medicinal. It draws large numbers of insects to one place. (This is why we are not going to smell it and why the mini-drama emphasized caution!)

Homework

Assign the reading and exercises for Smells Lesson 39 in the student text.

LESSON 39 ANSWERS

1. A space-filling model is a more accurate model of how the atoms of the molecule are arranged in space. It can give a better picture of the shape of the molecule than a ball-and-stick model.

2. A space-filling model does not show the bonds between atoms as a ball-and-stick model does. Instead, the atoms overlap to show their positions more accurately.

3. *Possible answer:*

$C_9H_{20}O$ is long and stringy, with a zigzag shape.

4. Minty.

5.

$C_{10}H_{16}O$ has a ring-shaped structure of carbon atoms with a handle. The overall shape of the molecule resembles a frying pan.

6. *Possible answer:* I would want to know what functional groups are on the molecule.

7. *Possible answer:* Because the types of smells we have learned about contain molecules that share the same functional group, the functional group seems to be more important in determining the smell. Shape is also a factor, but functional group seems to matter more.

Overview

ACTIVITY: GROUPS OF 4

Key Question: What chemical information is most useful in predicting the smell of a compound?

KEY IDEAS

Patterns in the chemical information of molecular compounds suggest general rules for some smell categories. Functional groups can be used to predict the smells of amines and carboxylic acids. Molecular shape is a predictor for minty-, sweet-, and camphor-smelling compounds. Even the chemical formula and name can be enough information to narrow down a smell.

LEARNING OBJECTIVES

- Summarize the connections explored so far between molecular structure and smell.
- Predict smells of a wide variety of compounds by examining molecular formulas, chemical names, molecular structures, and molecular shapes.

IN CLASS

Groups of students are given a set of 24 molecule cards with the chemical name, molecular formula, structural formula, and three-dimensional structure of these molecules. Students are challenged to sort the cards according to smell based on the evidence they have seen so far. Students use a worksheet to help them organize their discoveries. During the discussion, the class creates general rules for each smell category. A complete materials list is available for download.

TRM Materials List 40

Differentiate

This lesson provides an opportunity for students to learn about the process of making predictions in science. One characteristic, such as shape or functional group, may not be the sole factor in predicting smell for all molecules. Many students may find this hard to accept. To help them during the whole-class discussion, solicit ideas from students about how to create a "smell prediction" flow chart that incorporates both overall shape and functional group.

Sorting It Out

Shape and Smell

THINK ABOUT IT

So far the investigation into the chemistry of smell has turned up evidence that molecular formula, functional group, chemical name, and the overall shape of molecules can help predict smell. Which pieces of information are most useful? What should you consider first?

> **What chemical information is most useful in predicting the smell of a compound?**

To answer this question, you will explore

↻ A Summary of the Smell Investigation So Far
↻ General Rules for Predicting Smell

EXPLORING THE TOPIC

↻ A Summary of the Smell Investigation So Far

Clearly, the relationship between molecules and smell is quite complex. Often, when you smell and examine new molecules, you find out new information and refine your hypotheses. Here is what you might conclude about smell so far.

MOLECULAR FORMULA

Molecular formulas can give some indication of the way compounds smell.

2 O atoms	sweet or putrid
1 N atom	fishy
Only C and H	gasoline or no smell
1 O atom	sweet, minty, or camphor

Molecular formula is most useful in predicting fishy smells. It does not work so well for sweet, putrid, minty, or camphor smells.

FUNCTIONAL GROUP

Functional groups are even better than molecular formulas in narrowing down a smell. For example, compounds with a carboxyl functional group smell putrid. Compounds with an ester functional group smell sweet.

Carboxylic acids, putrid   Esters, sweet

PSYCHOLOGY CONNECTION

Scientists have found that memories evoked by specific smells seem to be stronger and more emotion-based than memories brought up by visual or auditory cues. The exact connection between smell and memory is unclear, but sensations from the smell receptors in the nose travel to areas of the brain that deal with memory.

204   Chapter 7 | **Building Molecules**

Pre-AP® Course Tip

Students preparing to take an AP® chemistry course or college chemistry must be able to draw connections to topics outside chemistry, such as the connection between molecular structure and smell. In this lesson, students predict the smells of a wide variety of compounds by examining their structure, shape, and functional groups.

Lesson Guide

ENGAGE (5 min)

TRM PowerPoint Presentation 40

ChemCatalyst

What smell or smells do you predict for a compound made of molecules that are long and stringy in shape? What is your reasoning?

Sample answer: Most students would predict a sweet smell for a long, stringy molecule. All the evidence so far has pointed to this shape for sweet-smelling molecules.

Ask: Based on the evidence collected in the previous class, how do you think a compound with stringy molecules would smell? Provide evidence to support your answer.

Because alcohols and ketones might smell sweet, minty, or like camphor, functional group is not always correct in predicting smells.

CHEMICAL NAME

Molecules are named according to their functional groups. A carboxylic acid molecule, for instance, will have a chemical name ending in "-ic acid."

Molecular Names and Smells

Compound type	Name ending	Examples	Smell
carboxylic acid	-ic acid	butyric acid, $C_4H_8O_2$ pentanoic acid, $C_5H_{10}O_2$	putrid
ester	-yl -ate	ethyl acetate, $C_4H_8O_2$ propyl acetate, $C_5H_{10}O_2$	sweet
amine	-ine	hexylamine, $C_6H_{15}N$ diisobutylamine, $C_8H_{19}N$	fishy
alkane	-ane	methane, CH_4 pentane, C_5H_{12} hexane, C_6H_{14}	no smell or gasoline
alcohol	-ol	citronellol, $C_{10}H_{20}O$ menthol, $C_{10}H_{20}O$	sweet, minty, camphor
ketone	-one	menthone, $C_{10}H_{18}O$ pulegone, $C_{10}H_{16}O$	sweet, minty, camphor

Because the functional group is not useful in predicting smells for alcohols and ketones, the chemical name is also not very useful.

MOLECULAR SHAPE

The overall shape of a molecule can be an important link to smell. Camphor-smelling compounds are made of ball-shaped molecules. Minty-smelling compounds are flat in overall shape. And compounds made of long and stringy molecules often smell sweet.

Long, stringy

Flat /frying pan–shaped

Ball-shaped

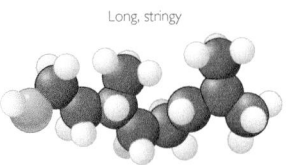

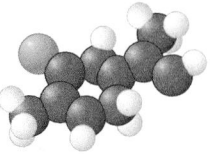

Citronellol, sweet smell

Carvone, minty smell

Fenchol, camphor smell

⟳ General Rules for Predicting Smell

Each piece of chemical information appears to be valuable at different times, depending on the smell category. It is helpful to look for general guidelines you can use to predict smell.

Lesson 40 | **Shape and Smell** 205

→ Tell students they will work in groups of four to sort the cards into groups of molecules that will smell similar.

→ Suggest to students that they put any cards they are unsure about into a separate pile.

TRM Transparency—Smell Classification and Molecular Characteristics 40

DISCUSS STUDENT OBSERVATIONS

→ Put the five smell classifications on the board. Ask groups to share their results.

→ Reach consensus on the sorting of the cards. The molecules cards are placed in their correct smell classifications in the table below.

Ask: What process did you use to sort your cards?

Ask: Which cards were you not 100% sure of? Why? Where did you decide to put them?

Ask: What did this activity help you discover?

GENERALIZE CHARACTERISTICS FOR DETERMINING A COMPOUND'S SMELL CATEGORY

→ Display the transparency Smell Classification and Molecular Characteristics. Fill in the table with class input. Circle the important characteristic for each smell category.

Ask: What chemical information would you like to have to determine if a compound was made up of putrid or fishy-smelling molecules?

TRM Worksheet with Answers 40

TRM Worksheet 40

TRM Card Masters—Molecules Cards 40

INTRODUCE THE ACTIVITY

→ Arrange students into groups of four.

→ Pass out the student worksheet.

Smell classification	Cards	Molecular formula info	Shape(s)	Functional group(s)
sweet	C, D, F, H, L, M, Q, U	2 O atoms in some, 1 O atom in others	stringy	alcohol, ester, and aldehyde (students should say it is a ketone)
minty	A, E, J, R	rings, 1 O atom in all	frying pan	ketones and alcohols
camphor	P, S, T, V	double rings, 1 O atom in all	ball-shaped	ketones and alcohols
putrid	G, I, N, W	2 O atoms in all	stringy and frying pan	carboxyl
fishy	B, K, X	1 N atom in all	stringy and frying pan	amines

Ask: What do all the sweet-smelling compounds have in common? (stringy shape)

Ask: What general statement can you make about carboxylic acids? (They are always putrid, no matter what their shape.)

Ask: How can you tell a camphor-smelling compound from the others? (Its molecules will be ball-shaped.)

Ask: What do you know about compounds that contain a nitrogen atom? (So far, they are all fishy-smelling.)

Ask: When is a ring-shaped compound not minty-smelling? (when it is an amine or a carboxylic acid)

Ask: Are generalizations still useful even if we find some exceptions? Explain.

Key Points: In each smell category, it is possible to find one distinctive feature that sets that group apart from the other smell categories. For sweet, minty, and camphor compounds, the shape will always be consistent. Sweet compounds are stringy, minty compounds are frying-pan–shaped with a six-carbon ring, and camphor compounds are ball-shaped. Putrid and fishy compounds, on the other hand, may take a variety of shapes, but their functional groups are always consistent. So far in our investigation, all the putrid molecules have been carboxylic acids. All the fishy-smelling molecules have been amines.

REVISE THE SMELLS SUMMARY CHART

→ Tape up the Smells Summary chart you used in Lesson 33: Where's the Fun? Have the students help you revise the information. (Changes are in bold.)

Ask: How should we revise the Smells Summary chart to bring it up to date?

This table summarizes the chemical information of each of the five smell categories you have encountered. In each category, the most useful information for predicting that smell has been highlighted.

Summary of Chemical Information

Smell classification	Molecular formula info	Shape(s)	Functional group(s)
sweet	2 O atoms in some, 1 O atom in others	stringy	hydroxyl and ester
minty	rings, 1 O atom in all	frying pan–shaped	ketone and hydroxyl
camphor	double rings, 1 O atom in all	ball-shaped	ketone and hydroxyl
putrid	2 O atoms in all	stringy	carboxyl
fishy	1 N atom in all	stringy	amine

The only category that causes a little trouble is the sweet-smelling category. Not every compound made of stringy molecules is sweet smelling. But, if the compound has stringy molecules and is not a carboxylic acid or an amine, then it is sweet smelling.

Example

Predict a Smell

Examine the following molecular compound and predict its smell. What information is most useful in predicting the smell? What type of compound is this?

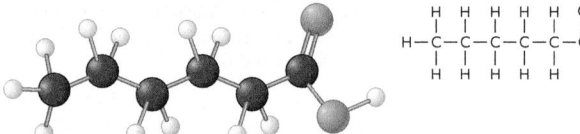

Solution

Here is what you can deduce:

- Molecular shape: Because the molecule is stringy, it probably does not have a minty or camphor smell. Its smell may be sweet, putrid, or fishy.
- Molecular formula: There is no N atom, so its smell can't be fishy. It has two O atoms, so it might have a sweet or putrid smell.
- Functional group: The carboxyl functional group indicates that the molecule smells putrid.
- The putrid smell and carboxyl group indicate that this is a carboxylic acid.

The functional group was most useful in predicting the smell of this compound.

Smells Summary Chart

Molecular Shape
frying pan = minty
ball = camphor
stringy = sweet

Molecular Formula
1 N atom = fishy
2 O's = sweet or putrid
1 O atom = minty, **camphor, sweet**

Smell

Functional Group
amine = fishy
ester = sweet
ketone = minty, **camphor, sweet**
carboxyl = putrid
alkane = gasoline
alcohol = **minty, camphor, sweet**

Chemical Name
"-ine" = fishy
"-ate" = sweet
"-one" = minty
"-ic acid" = putrid
"-ane" = gasoline
"-ol" = **minty, camphor, sweet**

LESSON SUMMARY

What chemical information is most useful in predicting the smell of a compound?

By looking for patterns in the chemical information of molecules, it is possible to come up with general guidelines for certain smell categories. Some chemical information is more useful than other information depending on the smell category. Molecular shape is an accurate predictor for minty and camphor compounds. Functional group can be used to predict the smells of amines and carboxylic acids. Each piece of chemical information, including chemical formula and name, allows you at least to narrow down the possible smells of a compound.

Exercises

Reading Questions

1. Can the molecular formula of a compound help you to predict its smell? Explain your reasoning.

2. What one piece of chemical information would you want in order to predict the smell of a molecule? Explain your choice.

Reason and Apply

3. What is the minimum amount of information you need to know to determine if a compound smells sweet?

4. If someone tells you a compound smells sweet, what can you assume about its molecules? What can't you assume about its molecules?

5. If someone told you a compound was made of molecules that had a stringy three-dimensional shape, what else would you want to know to determine that compound's smell?

3. *Possible answer:* The minimum required information to determine that a molecule smells sweet is the shape and the functional group. If the molecule is an ester, just the functional group is necessary.

4. *Possible answer:* I can assume that the molecule is stringy. I can assume that the molecule has an ester or alcohol functional group, but I cannot assume which it has. I can also assume that the molecule has at least one oxygen atom in its chemical formula, but I cannot assume exactly how many oxygen atoms it has.

5. You would need to know what functional groups are on the molecule to determine the compound's smell.

EVALUATE (5 min)

Check-In

If a compound is sweet smelling, what other things do you know about it? List at least three things that are probably true.

Answer: A compound that is sweet smelling probably contains two oxygen atoms and an ester functional group. One of the oxygen atoms is double-bonded. The chemical name of the molecule probably ends in "-ate." It has stringy molecules.

Homework

Assign the reading and exercises for Smells Lesson 40 in the student text.

LESSON 40 ANSWERS

1. The molecular formula of a compound can provide some information about its smell, but is not very useful in many cases because it fails to give information about functional groups.

2. The one piece of chemical information that is most useful in predicting the smell of a molecule is the functional group. Most molecules with the same functional group have similar smells, although there is variation based on other factors.

Overview

ACTIVITY: GROUPS OF 4

Key Question: How does the nose detect and identify different smells?

KEY IDEAS

Scientists have come up with many models to explain the chemistry of smell. Most agree that the shape of a molecular compound is important to the way the compound smells. The receptor site model is one of the most widely accepted models of smell. In this model, molecules with different shapes are like keys that fit into locks, or receptors, in the nose. Molecules with different shapes trigger different smells.

LEARNING OBJECTIVES

- Come up with a plausible model to explain how smell works in the nose, based on the evidence thus far.
- Describe the receptor site model.

FOCUS ON UNDERSTANDING

- The receptor site model, sometimes referred to as *lock-and-key,* is considered one of the main ways molecules trigger responses in the body. For example, pain medication is detected in receptor sites. Thus, it is an extremely useful concept for students.
- During the presentations, students may ask clarifying questions but should not attempt to either disprove or validate anyone's model.
- Students might incorrectly assume that molecules break apart when they become airborne. The fact that molecular shape is important in determining smell provides evidence that molecules actually stay together. This concept is fleshed out in the reading for this lesson.

KEY TERM

receptor site theory

IN CLASS

Groups of students speculate as to what takes place in the interior of the nose so that we can detect different smells of different compounds. Each group comes up with a model to explain the mechanics of smell and creates a poster showing how the model works. Each group then presents its poster to the class, explaining the intricacies of the model. After the presentation, students are introduced to the receptor site model. *Note:* This lesson may take more

than one standard class period. If your time is limited, direct students to keep their presentations to one minute or select only a few groups to present. A complete materials list is available for download.

TRM Materials List 41

SETUP

Make sure you have some way to post or display the students' posters during the groups' presentations.

Differentiate

Students often have different preferences for how they represent their

learning. Provide students with options of creating the smells poster as outlined, which supports visual learners, writing up an explanation for how they think smell works, or using everyday materials as well as the molecular model kits to make three-dimensional representations to explain how smell works. Allow students to choose based on each student's preference, or assign students to a mode of expression and support them in building their ability to reason in alternate ways. If time permits, other ways to summarize might be to allow students to build a Web site or make a short movie. By letting students use a variety of methods for

---

How Does the Nose Know?

Receptor Site Theory

THINK ABOUT IT

Imagine you are sipping a cool glass of lemonade on a hot summer day. The lemonade contains some fresh mint leaves for extra flavor. You detect both the lemon smell and the minty smell. The molecules associated with these two smells are quite different. What is happening inside your nose that allows you to detect both smells at the same time and to tell them apart? How does the nose know these molecules are different?

How does the nose detect and identify different smells?

To answer this question, you will explore

- Receptor Site Theory
- Phase Change and Molecular Stability

EXPLORING THE TOPIC

Receptor Site Theory

The chemistry of smell is a fairly new science, and much remains to be discovered about how the sense of smell works. A number of theories exist to explain what happens in the nose when a smell is detected. While scientists are still uncertain about all of the details about how smell works, one theory seems to fit the evidence better than the others. This theory is called the **receptor site theory.** It is a widely accepted theory of how the interior of the nose detects different smells.

The receptor site theory uses the "lock and key" model. In this model, molecules are like "keys" that fit only into certain "locks." In other words, molecules have specific shapes that will fit only certain receptor sites, and the lining of the nose is covered with receptor sites.

HEALTH CONNECTION

Many medications work by using a lock and key mechanism. The molecules found in a medication may be designed to *stimulate* a receptor site, resulting in the desired physical response. Or a medication may have been designed to *block* receptor sites, preventing a certain unwanted physical response.

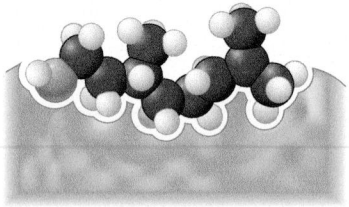

Molecules have specific shapes that fit into specific receptor sites in the nose, the way a key fits into a specific keyhole of a lock.

The lemony-smelling and minty-smelling molecules in the compounds of the lemonade might fit into different receptor sites as shown in the illustration below.

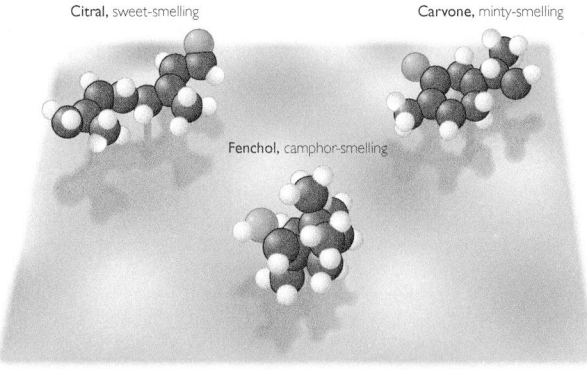

Citral, sweet-smelling   Carvone, minty-smelling

Fenchol, camphor-smelling

In this model, each type of molecule fits only into its own receptor sites, not into the others. According to receptor site theory, this is how the nose distinguishes between two molecules.

Scientists are unsure how many receptor sites are in the nose or how many different shapes the sites accommodate. However, they do know that receptor sites are made of large, very intricate protein molecules. These protein molecules consist of hundreds of atoms bonded together. They are much more elaborate and complex than the simple drawings of receptor sites shown in the illustration.

Scientists think that once a smell molecule has "docked" in a receptor site, nerves are stimulated and send a message to the brain. If minty receptor sites are stimulated, then the brain registers a minty smell.

It is possible that smell is much more complex than just described. A specific smell might be the result of a combination of different molecules stimulating a combination of receptor sites.

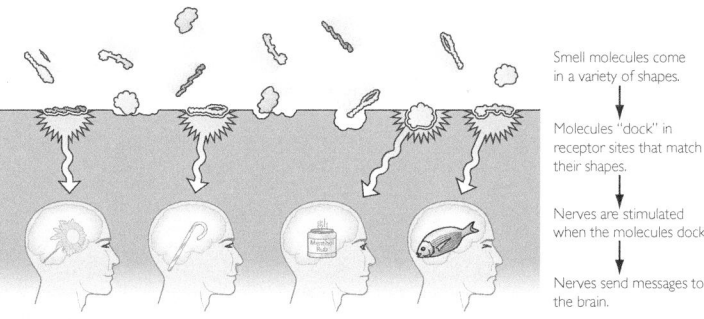

Smell molecules come in a variety of shapes.

Molecules "dock" in receptor sites that match their shapes.

Nerves are stimulated when the molecules dock.

Nerves send messages to the brain.

The brain interprets the messages as smells.

Lesson 41 | **Receptor Site Theory**   209

---

presenting models, you are supporting the development of scientific literacy around scientific models, which are not strictly visual.

Pre-AP® Course Tip

Students preparing to take an AP® chemistry course or college chemistry must be able to draw pictures that represent particles that cannot be macroscopically observed but that coincide with accepted models. In this lesson, students work in groups to design their own model to explain how smell works in the nose based on the evidence they have gathered from explorations in Unit 2.

Lesson Guide

ENGAGE (5 min)

TRM PowerPoint Presentation 41

ChemCatalyst

1. Suppose that you needed to separate coins but could not see them. Explain how you would make a machine that detects and sorts coins.

2. How do you think your nose detects a smell?

Sample answers: **1.** Students might suggest sorting them by size. If you shake the coins over a series of holes, those that are smaller than the hole will fall through. **2.** Some students might have heard of the receptor site model, and they may say that molecules fit into spaces in the nose, that the nose can detect the shapes of functional groups, or that the molecules travel up into the brain where they are detected.

Ask: What is a smell?

Ask: Can you only smell things that are in the air? Explain.

Ask: What do you think is going on in the nose when you smell something?

Ask: When you have a cold, you cannot smell things as well. What do you think is happening that affects your sense of smell?

EXPLORE (20 min)

TRM Worksheet with Answers 41

TRM Worksheet 41

INTRODUCE THE ACTIVITY

→ Arrange students into groups of four.

→ Pass out the student worksheet.

→ Tell the students they will be designing a model for how smell works and illustrating it on a poster. Their mechanism must explain how the smell reaches the nose, what happens to smell molecules when they are in the nose, and how we perceive smell and identify it with other experiences.

→ Let students know they will present their models to the class.

→ Tell students they will have limited time to complete their posters (perhaps ten minutes of group discussion and ten minutes of poster creation).

→ Let students know when half the time has passed so they can get started drawing if they are not already doing so.

EXPLAIN & ELABORATE (20 min)

GROUP PRESENTATIONS OF SMELL MODELS

→ Allow a minute or two for each presentation. You might choose one student from each group to be the spokesperson. Provide positive feedback for ideas.

→ Check whether the model proposed by each group explains the evidence we have about the sense of smell. Evidence about the sense of smell:

- You can detect a smell at a distance from the source. Wafting works.
- The shape of the molecule is related to how it smells.
- Some people cannot detect specific smells.
- You cannot detect smells as well when you have a cold.
- The sense of smell deteriorates as you get older.
- We all react similarly to certain bad smells like rotten eggs.
- Smells are connected with memories.
- For sweet, minty, and camphor molecules, the shape will always be consistent. Sweet molecules are stringy, minty molecules are frying-pan–shaped, and camphor molecules are ball-shaped.
- Putrid and fishy molecules may take a variety of shapes, but their functional groups are always consistent.
- So far in our investigation, all the putrid molecules have been carboxylic acids, and vice versa. All the fishy-smelling molecules have been amines, and vice versa.

→ Get the class to assist in making suggestions for improvement of each model to support the evidence. (*Note:* Remind students that this is what scientists do. They test hypotheses and try to improve their models based on evidence. Remind the class that scientists to this day do not know the details of how the sense of smell works.)

Ask: How do molecules get from a substance into your nose?

Ask: How does the nose distinguish among molecules with different shapes and functional groups?

Ask: What evidence do you have that molecules do not fall apart when traveling through the air?

Ask: What evidence is there that our brain is involved?

Ask: What experiments could you do to test each model?

Key Points: A good model should show that the nose can distinguish between molecules with different shapes and functional groups. Students might propose that molecules with the same functional groups react chemically inside the nose or that molecules with similar shapes fit into receptor sites. They might suggest that nose hairs collect smells. A

good model should explain why some people cannot detect certain smells and why you cannot smell as well when you have a cold.

A model might show a connection to the brain. This explains why memories are associated with smell. It can also account for the deterioration in the sense of smell as you get older. The connection to the brain may also be the reason we react similarly to bad smells such as rotten eggs and to good smells such as flowers.

A model might demonstrate that molecules remain intact. We have

found that the shape of a molecule of a compound is important to the way it smells. This means that the molecule must enter the nose intact, not as a cluster of atoms.

INTRODUCE CURRENT RECEPTOR SITE THEORIES (OPTIONAL)

→ Point out any student models that happen to match the scientists' models.

→ Highlight the receptor site model. Draw several rudimentary receptor sites on the board to demonstrate.

Key Points: Scientists have proposed many theories about how smell works

---

Important to Know A substance does not have to boil in order for some of its molecules to become a gas. At the surface of molecular liquids and solids, some molecules escape and become a gas.

KEY TERM

receptor site theory

Phase Change and Molecular Stability

SMELLS ARE GASES

When you bring an onion home from the supermarket, it doesn't give off much of a smell. But when you slice or chop an onion, the odor can become quite overpowering. What's happening? When you cut up the onion, your knife is cutting through some of the cell walls of the onion and releasing a lot of liquid.

One of the molecules that is in this liquid and is responsible for the distinctive smell of onions is allyl propyl disulfide. Its structural formula is shown here. Its molecular formula is $C_6H_{12}S_2$. This is the molecule your nose detects when you smell onions.

Allyl propyl disulfide, $C_6H_{12}S_2$

But something else has to happen for you to smell the onion. Molecules of allyl propyl disulfide have to get from the onion into your nose. The only way that can happen is if some of the molecules travel through the air to your nose. To do this, the compound changes phase to become a gas or vapor that floats through the air and enters your nostrils. There is no other possible explanation for your ability to smell an onion several feet away.

You cannot smell something unless some portion of it is in the form of a gas. This means that everything you are able to smell is a substance that has become a gas.

MOLECULES ARE STABLE UNITS

Because your nose is detecting the shape and functional group of a molecule, this means that the molecule must enter your nose in one piece. When molecules change phase, they do not break apart into pieces or into individual atoms. Molecules go from the solid phase, to the liquid phase, to the gas phase without breaking apart into individual atoms. This is because molecules are stable units, with strong covalent bonds between the atoms. For example, water is still water whether it is in the form of water vapor, liquid water, or ice.

Big Idea Molecules are stable units. They do not break apart when they change phase.

LESSON SUMMARY

How does the nose detect and identify different smells?

The receptor site theory explains how the nose detects different smells. This theory suggests that smell molecules travel to receptor sites in the lining of the nose. Once a smell molecule docks in a receptor site, it stimulates a nervous-system response, sending a signal to the brain. Everything you smell is in the form of molecules in the gas phase.

Exercises

Reading Questions

1. Explain the receptor site theory and how it applies to smell detection.

2. What does phase change have to do with smell?

Reason and Apply

3. Here is a structural formula for an active ingredient in muscle ointment. How does this compound smell? How does your nose detect the smell?

4. A chemist creates a new molecule that has a completely different three-dimensional shape from other molecules humans have ever encountered. Would you be able smell it? Select the best answer.
   (A) Probably not, because synthetic compounds do not have a smell.
   (B) Definitely, because all esters smell sweet.
   (C) Probably not, because our noses would not have developed receptor sites to detect it.
   (D) Definitely, because parts of the molecule can break off to fit into receptor sites in our noses.

5. Some smells are similar, such as popcorn and freshly baked bread. Explain how you might account for similar smells using the receptor site theory.

6. How might the receptor site theory explain why a dog has a better sense of smell than a person?

7. Why do you think you might "get used to" a smell and hardly detect it? Use the receptor site model in your explanation.

8. Are some smells "faster" than others? Explain.

Lesson 41 | **Receptor Site Theory**   211

and created models corresponding to these theories. The last model described here (model 4) is called the receptor site model; it reflects the most widely accepted theory about how the interior of the nose detects different smells.

- Model 1: Molecules vibrate inside the nose, with each distinct molecule vibrating differently. Our nerves and brain determine smell by detecting these different molecular vibrations.

- Model 2: Smell molecules react chemically with the inside of the nose in such a way that our nerves detect and our brain deduces smells from the chemical reactions.

- Model 3: Molecules of compounds press against the inside of the nose or even punch through it. This causes chemical or physical changes inside the nose that allow us to determine the smell of the compounds.

- Model 4: Smell molecules of compounds fit into receptor sites, which are like molds that mirror the shape and size of the molecules. Each receptor site is specific to one shape. When there is a molecule sitting in a particular receptor site, it generates a signal to our brain and we register this as a specific smell.

LESSON 41

> **Receptor site theory:** The currently accepted model explaining how smells are detected in the nose. Molecules fit into receptor sites that correspond to the overall shape of the molecule. This stimulates a response in the body.

EVALUATE (5 min)

TRM Transparency—Check-In 41

> #### Check-In
>
> One of the molecules that gives coffee its smell is 2-furylmethanethiol.
>
> 1. Write down everything you know about how this molecule is detected by the nose.
>
> 2. Draw a possible receptor site for this molecule

Answers: **1.** We know that a molecule must first enter the nose. It must be airborne, and it must be inhaled. It lands on the lining of the nose and fits into a specific receptor site that is shaped similarly to the molecule. The sulfur atom in this molecule may also affect its smell. **2.** A receptor site for this molecule might look something like a pentagon with a bump on it.

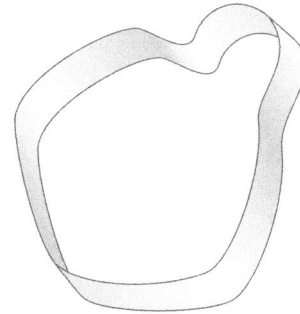

Homework

Assign the reading and exercises for Smells Lesson 41 and the Chapter 7 Summary in the student text.

LESSON 41 ANSWERS

Answers continue in Answer Appendix (p. ANS-3).

Lesson 41 | **Receptor Site Theory**   211

ANSWERS TO CHAPTER 7 REVIEW EXERCISES

1. a. $C_5H_{10}O_2$

b.

H H H   :O: H
            ‥
H:C:C:C:O:C:C:H
  ‥ ‥ ‥   ‥
H H H   H

c. ester **d.** The molecule probably has a sweet smell.

2. a. Ammonia has a pyramidal shape because it has four atoms and a central nitrogen atom that has four electron domains (it has one lone pair). **b.** Silicon chloride has a tetrahedral shape because it has five atoms and a central silicon atom that has four electron domains, each of which is shared with a chlorine atom. **c.** Hydrogen sulfide has a bent shape because it has three atoms and the sulfur atom has four electron domains (it has two lone pairs). The four domains form a tetrahedron, so the two bonds are bent. **d.** Hydrogen cyanide has a linear shape because it has three atoms and the carbon atom has only two electron domains. The carbon and nitrogen atoms form a triple bond. **e.** Formaldehyde has a trigonal planar shape because it has four atoms and a central carbon atom that has three electron domains. The carbon is single-bonded to the hydrogen atoms and double-bonded to the oxygen atom.

3. The shape of a molecule is determined by the number of electron domains around its atoms. The electron domains are all negatively charged, so they are positioned as far apart as possible.

4. *Possible answer:*

H H H H
     ‥
H:C:C:C:C:O:H
     ‥
H H H H

5. *Possible answer:* Although the functional group appears to be a main factor that determines the smell of a molecule, its shape also affects the smell because it helps determine which receptors will be stimulated to send a signal to the brain.

Project: Other Smell Classifications

A good poster would include:

● a general description of the smell
● a list of substances that give off this particular smell, as well as where they occur and their uses
● a diagram of the functional group(s) associated with the smell
● a diagram of the structure of the functional group clearly showing bonds and atom types
● five or more structural formulas, with the functional group(s) highlighted in the diagram and the names of molecules clearly labeled
● information on how molecules containing the smell are named
● a list of common properties that molecules with the particular smell possess

CHAPTER 7

Building Molecules

SUMMARY

KEY TERMS

ball-and-stick model

tetrahedral shape

electron domain

electron domain theory

pyramidal shape

bent shape

trigonal planar shape

linear shape

space-filling model

receptor site theory

Smells Update

The smell of a substance sometimes has more to do with molecular shape than with the functional groups that are present. In this chapter, you learned about three-dimensional molecular structure and related it to smell. Ball-and-stick and space-filling models show how atoms are arranged within a molecule. The actual shape of a molecule is determined by the locations of electron domains.

No matter what shape a molecule has, it must fit into an appropriate receptor site in order for the nose to detect its smell.

REVIEW EXERCISES

1. Examine the ball-and-stick model for propyl acetate.
   a. What is the molecular formula for this molecule?
   b. Draw the Lewis dot structure and the structural formula.
   c. What functional group is in the molecule?
   d. Predict the smell of the compound.

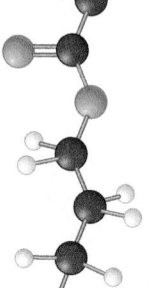

2. Predict the shape of each molecule. Support your answer.
   a. ammonia, NH_3
   b. silicon chloride, $SiCl_4$
   c. hydrogen sulfide, H_2S
   d. hydrogen cyanide, HCN
   e. formaldehyde, CH_2O

3. Explain what determines a molecule's shape.

4. Draw a possible Lewis dot structure for $C_4H_{10}O$.

5. From what you've learned so far, how is molecular shape related to smell?

Other Smell Classifications

 PROJECT

Research a smell classification not discussed in class, such as musk, woody, spicy, nutty, leather, or tobacco. Create a poster that includes

● A drawing describing the smell classification and sources of the smell

● The structural formulas and ball-and-stick models of at least five molecules in the smell classification that you investigated, along with their chemical names

● A brief description of the properties associated with this group of molecules

212   Chapter 7 | **Summary**

ASSESSMENTS

Two multiple-choice assessments (versions A and B) and two short-answer assessments (versions A and B) are available for download for Chapter 7.

TRM Chapter 7 Assessment Answers

TRM Chapter 7 Assessments

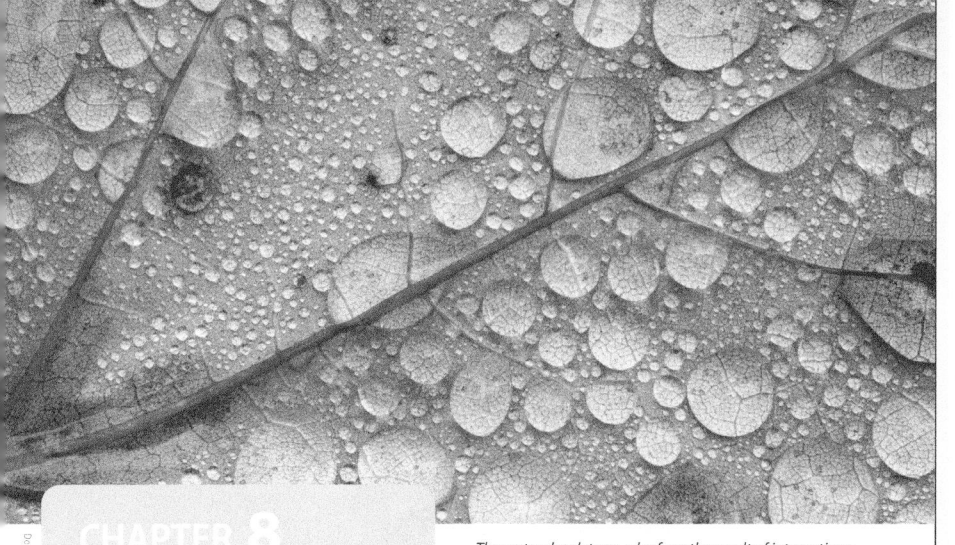

CHAPTER 8

Molecules in Action

In this chapter, you will study

- polarity
- intermolecular interactions
- why some substances do not smell
- how phase, molecular size, polarity, shape, and bonding patterns relate to smell

The water droplets on a leaf are the result of interactions between molecules in the water and the surface of the leaf.

Molecules interact with each other in a variety of ways. The process of smelling involves interactions between the molecules that enter the nose and the molecules that make up the lining of the nose. One key to molecular interactions is the fact that atoms in a molecule often do not share electrons equally. This unequal sharing creates partial positive and negative charges on different atoms in the molecule. These partial charges influence how molecules interact with each other and help to determine properties like phase, smell, and whether the substance will dissolve in another substance. Molecular size, shape, and bonding patterns all contribute to smell.

213

PD CHAPTER 8 OVERVIEW

Watch the video overview of Chapter 8 (for teachers) by clicking on the link in the TE-book, opening the TRFD, or logging onto the book's companion Web site bcs.whfreeman.com/livingbychemistry2e (teacher log-in required).

Chapter 8 focuses on interactions between molecules. Lesson 42 is a charged wand demonstration and lab that provides evidence of polarity in molecules. Lesson 43 uses an information-packed comic strip to develop the concept of polarity and dipoles. Lesson 44 introduces the electronegativity scale. Students use electronegativity values to compare atoms and bonds and to figure out the direction of bond dipoles. In Lesson 45, students go one step further, using electronegativity values to determine the direction of polarity of entire molecules. They discover that perfectly symmetrical molecules are nonpolar and often do not smell. Lesson 46 explores how molecules, polarity, phase, shape, and bonding patterns are related to smell.

In this chapter, students will learn

- the role of polarity in smell chemistry
- about electronegativity and bond dipoles
- about the continuum of bonding from nonpolar covalent bonds to ionic bonds
- to predict whether a substance will have a smell based on its composition, bonding, phase, size, and polarity
- how polarity relates to intermolecular interactions

Chapter 8 Lessons

Overview

LAB AND DEMO: GROUPS OF 4

Key Question: Why do some molecules smell while others do not?

KEY IDEAS

Molecules have an equal number of protons and electrons and no net charge. However, many molecules have regions of positive and negative partial charge. These molecules are called polar molecules. Polarity in molecules contributes to intermolecular attractions. Polarity affects the overall behavior of a substance and may play a role in smell properties. Water is one of the best-known polar compounds.

LEARNING OBJECTIVES

- Describe the behavior of polar molecules.
- Explain the general difference between a polar and a nonpolar molecule.
- Describe basic intermolecular attractions.
- Define a partial charge.

FOCUS ON UNDERSTANDING

- Students might not realize that "static" is really static electricity and occurs when an object has gained or lost some electrons. Depending on the rod and the fabric used, the rod may develop a positive or a negative charge, but it will attract the polar liquids either way.
- Water is a small polar molecule, yet for humans, it does not have a smell. Some students might notice this discrepancy. The subject is handled in Lesson 45. Put simply, there are so many H_2O molecules in the nose that we do not smell the "additional" water molecules.
- Students might think of a force as something imposed on an object from an outside source. Emphasize the "attraction" aspect of intermolecular forces: The molecules are attracting each other.

KEY TERMS

polar molecule
nonpolar molecules
partial charge
intermolecular force

IN CLASS

Students first consider why air does not smell. Then the instructor demonstrates the attraction of a stream of water to a charged wand. Students consider possible interactions between water molecules and the charged wand and then work in groups of four trying the charged wand on three new liquids located at different stations, all but one of which are attracted to the wand. Students also examine a droplet of each of the four liquids placed on waxed paper: Three bead up, but one does not. Polarity is introduced, and intermolecular interactions are put forth as a possible factor in the smells of compounds. Optional: You could leave out a set of drops overnight and have students observe the rate of evaporation, another indication of the intermolecular forces present.

A complete materials list is available for download.

TRM Materials List 42

SETUP

Before class, set up burets containing three different liquids at stations around the room (one or more stations per liquid, depending on your class size). Set up a 500 mL beaker under each buret to catch the stream of liquid. A wand and a soft cloth should also be placed at each station. Include an index card or sticker with the name of the liquid. Also have pieces of waxed paper and droppers available at each station, or set up another

---

Attractions between Molecules

THINK ABOUT IT

Not everything has a smell. In fact, there are more substances on the planet that we *don't* smell than substances that we *do* smell. If everything around us had a smell, our noses would probably be overwhelmed most of the time.

> Why do some molecules smell while others do not?

To answer this question, you will explore

- Compounds That Do Not Have a Smell
- Polar Molecules
- Intermolecular Force

EXPLORING THE TOPIC

Compounds That Do Not Have a Smell

Take a big whiff of air. Perhaps you are able to detect some odors—a freshly mowed lawn, somebody's food cooking. But the air itself does not smell. Small molecules, such as nitrogen, oxygen, carbon dioxide, and water vapor, obviously fit inside the receptor sites. So, why can't we smell them?

Perhaps these molecules are too small to stay docked in a receptor site. But some small molecules do have a smell. Hydrogen sulfide, H_2S, smells like rotting eggs and ammonia, NH_3, smells pungent.

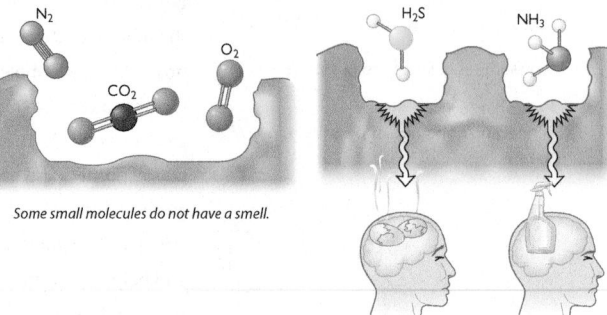

Some small molecules do not have a smell.

But H_2S and NH_3 both have strong smells.

So the molecular size of a compound does not explain whether we can smell it. It turns out that a molecular property called *polarity* is involved.

◌ Polar Molecules

When two different materials are rubbed together, some electrons can transfer
from one material to the other. The result is an imbalance of positive and negative
charges. One material has an excess of positive charge and the other an excess of
negative charge. This is known as static electricity.

Suppose that you rub a plastic wand on a piece of cloth. The plastic wand will
end up with a charge. In class you tested a number of liquids by holding a
charged wand near them and observed that some of the streams bend toward
the wand.

In this table, all of the liquids in the first column are attracted to a charged wand.
They are made up of **polar molecules.** The liquids in the second column are not
attracted to a charged wand. They consist of **nonpolar molecules.**

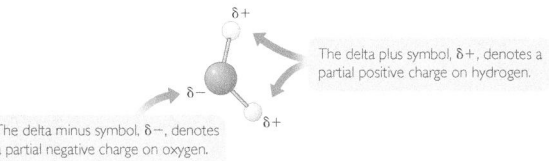

	Polar molecules	Nonpolar molecules	
These molecules are all **polar.** They all have a smell.	water vinegar nail polish remover rubbing alcohol antifreeze	mineral oil hexane motor oil	These molecules are all **nonpolar.** The two oils do not have a smell.

Water is a polar molecule. As this illustration shows, there are partial charges
at different locations on the water molecule. These **partial charges** are much
smaller than the charge on an individual electron or proton. However, the
charges are large enough to cause a stream of water to be attracted to a
charged wand.

$\delta+$

The delta plus symbol, $\delta+$, denotes a
partial positive charge on hydrogen.

$\delta-$

$\delta+$

The delta minus symbol, $\delta-$, denotes
a partial negative charge on oxygen.

The oxygen atom has a partial negative charge, and the hydrogen atoms each have
a partial positive charge.

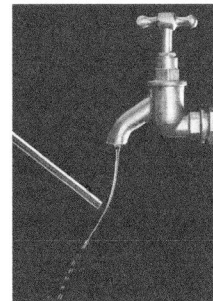

◌ Intermolecular Force

EXPLAINING THE CHARGED WAND

When a wand with a negative charge is placed next to a stream of water, the water
molecules orient themselves in space so that their positive ends are lined up in the
direction of the wand. As a result, the whole stream of water is attracted to, or
pulled toward, the charged wand. The illustration on the next page shows a
possible way to depict what happens to the molecules.

Lesson 42 | **Attractions between Molecules**   215

---

station with the same liquids and
waxed paper. Set up one buret at the
front of the room for a demonstration
with water.

CLEANUP

Store liquids for use with future
classes.

Lesson Guide

TRM Transparency—ChemCatalyst 42

TRM PowerPoint Presentation 42

ChemCatalyst

If a molecule fits into a receptor site
in the nose, it seems as if it should
smell. Yet most of the molecules in
air—O_2 (oxygen), N_2 (nitrogen), CO_2
(carbon dioxide), and Ar (argon)—
do not have a smell. What do you
think is going on?

Answers: Students might say that
small molecules simply "fall out" of
receptor sites or are too small to fit the
site or that air smells but we are just
used to it, so we do not notice it. They
might also say that the smell of air is

---

too faint for humans to detect. All are
reasonable answers.

➡ Stimulate discussion about why
some substances do not smell.

Ask: Why do you think that clean air
does not smell?

Ask: If the small molecules in air fit into
receptor sites, why wouldn't they smell?

Ask: What do you think it would be like
if we could smell the air around us all
the time?

Ask: Do you think molecules in the air
interact with one another? Explain.

Ask: Do you think molecules in the air
interact with the nose? Explain.

TRM Worksheet with Answers 42

TRM Worksheet 42

PREPARE FOR THE LAB

➡ Demonstrate the effect of a charged
wand on water. Use the charged wand
and a fine flow of water from a buret
to provide macroscopic evidence that
some molecules are attracted to a charge.
Rub a plastic wand with a cloth to create
a charge on the wand. Hold the charged
wand close to, but not touching, the
stream of liquid. Repeat, moving the
wand close to and away from the stream
of water.

➡ Use the demonstration as an
opportunity to introduce students
to the use of a buret.

INTRODUCE THE LAB

➡ Arrange students into groups of four.

➡ Pass out the student worksheet.

➡ Tell students that they will be using
this same technique to test three more
liquids in class. The liquid should not
touch the charged wand.

➡ Direct students how to rotate through
the stations.

➡ Tell students that if the burets are
close to empty they should be refilled
carefully with the contents of the catch
beaker. The valve on the buret should be
closed during refilling.

TRM Transparency—Water Molecule
Interacting 42

TRM Transparency—Charged Wand 42

Lesson 42 | **Attractions between Molecules**   215

DISCUSS RESULTS OF THE CHARGED WAND ACTIVITY

→ Display the transparency of water interacting with the charged wand.

→ Draw a large water molecule, with its partial charges, on the board, or refer to the illustration on the transparency. Introduce the delta symbol.

Ask: What did you observe when you tested the various liquids with the charged wand?

Ask: What do you think is going on at a molecular level to attract the liquids to the wand?

Ask: Why do you think the charged wand does not affect some liquids?

Ask: Do you think the same liquids would be attracted to a positive charge? Why or why not?

Key Points: The charged wand experiment provides evidence that some molecules are attracted to a charge. The most obvious explanation for this observation is that those molecules must have some sort of charge on them. This is true. Some molecules have a slight charge on opposite ends of the molecule. These molecules are called *polar molecules*. According to our results today, water, acetic acid, and isopropanol are all polar molecules. Other molecules, such as hexane, are not attracted to the charged wand. Such molecules are called *nonpolar molecules*.

> **Polar molecules:** Molecules that are attracted to a charge because they have partial charges on them.
>
> **Nonpolar molecules:** Molecules that are not attracted to a charge.

One end of a polar molecule has a partial negative charge, and the other end of the molecule has a partial positive charge. It is important to stress that this charge is "partial," as compared to the type of charge on an ion. To differentiate between full charges and partial charges, chemists use the symbols δ+ and δ− (delta plus and delta minus) to indicate partial charges. The hydrogen atoms in a water molecule have partial positive charges, while the oxygen atom has a partial negative charge (as shown in the drawing).

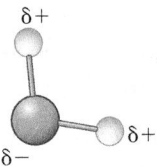

The individual molecules in polar liquids will respond when another charged substance comes near. In the figure, you can see that the water molecules are all pointed in the same direction. This is because the positively charged ends of the water molecules are attracted to the negatively charged wand. The overall attraction between the water and the wand causes the liquid to move in the direction of the wand.

DISCUSS INTERMOLECULAR INTERACTIONS

→ You might have students draw their interpretations of the Making Sense question on the board. Ask them to explain their drawings.

→ For the first key point, display the transparency that shows water molecules interacting.

Ask: What did you observe during the droplet exercise?

Ask: What is happening in your Making Sense drawing?

Ask: How are the water molecules interacting on the transparency? Did any of your drawings come close to matching this illustration?

Ask: What effect do you think these interactions might have on the properties of water?

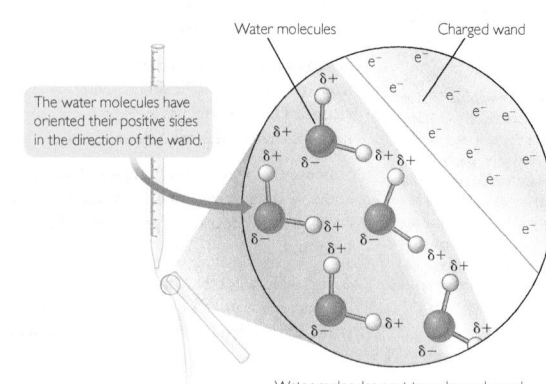

The water molecules have oriented their positive sides in the direction of the wand.

Water molecules

Charged wand

Water molecules next to a charged wand

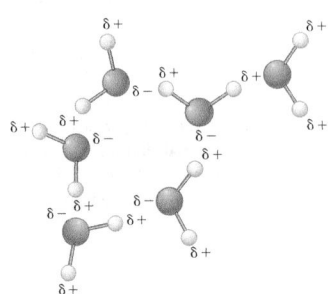

Water molecules interacting

LANGUAGE CONNECTION

The saying "Oil and water don't mix" is one way to remember that a nonpolar substance and a polar substance tend not to dissolve in each other.

INTERMOLECULAR BEHAVIOR

The partial charges on polar molecules cause more than an attraction to a charged wand. They also cause attractions between molecules. As the molecules in a polar liquid tumble around, they tend to align with each other because partial negative charges are attracted to partial positive charges. This attraction between individual molecules is an **intermolecular force.** The prefix *inter-* means "between" and the force is a force of attraction.

Intermolecular attractions can be used to explain many observable properties.

Observation: Water beads up on waxed paper. Oil spreads out.

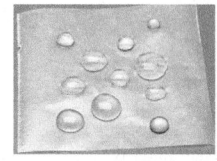 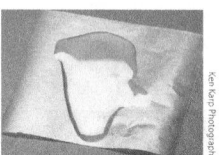

Explanation: The water molecules are polar, while the oil molecules are nonpolar. The individual water molecules are attracted to each other and "cling together." Individual oil molecules are not attracted to one another to the same extent, and they spread out.

Observation: Water is a liquid at room temperature, but methane is a gas at room temperature. They have roughly the same molecular mass.

Explanation: Water is polar and methane is nonpolar. The individual water molecules are attracted to each other and stay together as a liquid. The attractions between methane molecules are much weaker. The methane molecules spread throughout the room as a gas.

Observation: Methanol dissolves easily in water. Oil floats on top of water but does not dissolve.

Explanation: The polar methanol molecules are attracted to the polar water molecules and go into solution. The nonpolar oil molecules are not attracted to the polar water molecules.

LESSON SUMMARY

Why do some molecules smell while others do not?

KEY TERMS

polar molecule
nonpolar molecule
partial charge
intermolecular force

Molecules can be divided into two classes: polar and nonpolar. Polar molecules are attracted to a charged wand because they have partial charges distributed within the molecule. Nonpolar molecules are not attracted to a charged wand. Polarity is responsible for intermolecular attractions that affect many properties of molecules, possibly including smell properties.

Exercises

Reading Questions

1. Explain in your own words what a polar molecule is.
2. What are intermolecular attractions?

Reason and Apply

3. **Lab Report** Write a lab report for the Lab: Attractions Between Molecules. In your conclusion, explain why some liquids were attracted to the wand and others were not, and why some liquids beaded up on waxed paper while others did not.

(Title)

Purpose: (Explain what you were trying to find out.)

Procedure: (List the steps you followed.)

Results: (Explain what you observed during the experiment.)

Conclusions: (What can you conclude about what you were trying to find out? Provide evidence for your conclusions.)

Lesson 42 | **Attractions between Molecules** 217

Ask: How might molecular interactions account for the beading behavior observed in some of the liquid droplets?

Key Points: The partial charges on polar molecules cause individual molecules to be attracted to each other. As the molecules in a polar liquid tumble around, they tend to align with each other, such that the side that is partially negative is closer to the partially positive side of another molecule. The attraction that happens between individual polar molecules is called an *intermolecular*

force or an *intermolecular attraction.* (The prefix *inter-* means "between.")

Intermolecular forces: The forces of attraction that occur between molecules.

These interactions are responsible for many observable properties of polar substances. For instance, we observed in this lesson that polar liquids tend to bead up because the individual molecules are attracted to each other and cling together. All molecules interact with each other, but the attractions between polar molecules tend to be stronger than those between nonpolar molecules. In general, polar molecules are more likely to be liquids and not gases at room temperature. This is because the attractions between individual molecules cause them to form a liquid rather than to disperse as a gas. However, even with nonpolar molecules, the random motion of electrons can cause momentary imbalances of charge, resulting in a momentary partial charge. This force, called the *London dispersion force,* draws even nonpolar molecules together to some extent.

EVALUATE (5 min)

Check-In

Acetone is polar. Name two other things that are probably true about acetone.

Answer: Because acetone is polar, it probably is a liquid at room temperature. Also, it should bead up when placed on a piece of waxed paper. Students might also say that acetone will have a smell and that it is attracted to a charged wand.

Homework

Assign the reading and exercises for Smells Lesson 42 in the student text. Exercise 3 is a lab report for today's lab. Optional: Assign students to read "The Bare Essentials of Polarity" comic strip in the student text in preparation for the next lesson.

LESSON 42 ANSWERS

1. *Possible answer:* A polar molecule is a molecule in which the charge is not evenly distributed around the molecule. This means that different portions of the molecule will have partial electric charges.

Lesson 42 | **Attractions between Molecules** 217

2. Possible answer: Intermolecular attractions are forces that cause charged sections of polar molecules to move closer together. The negative part of one molecule is attracted to the positive part of another molecule and vice versa.

3. A good lab report will contain:
● a title (Lab: Attractions between Molecules) ● a statement of purpose (**Possible answer:** To observe the behavior of certain liquids near a charged wand and as droplets on waxed paper.) ● a procedure (a summary of the steps followed in the experiment) ● results (Check student observations to make sure they properly recorded observations in their data table. The expected results are that the water, acetic acid, and isopropanol will be attracted to the wand and form round drops on the waxed paper, while the hexane will not be attracted to the wand and will spread out across the waxed paper.) ● a conclusion (Students should synthesize material from Part 2 of the lab in their conclusion. **Possible answer:** Liquids that were attracted to the wand also formed droplets on the waxed paper. The molecules in those liquids must have some electrical charge on them so that they are attracted to the wand and to each other.)

4. a.

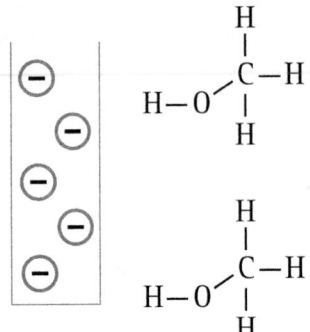

b. The methanol will bead up on the paper because it is polar. The hydrogen atom on one molecule is attracted to the oxygen atom of another molecule. This mutual attraction causes the liquid to retain a shape on waxed paper.

c. Yes, methanol should dissolve in water because the partial charges on the polar methanol molecules are attracted to the partial charges on the polar water molecules.

5. Hexane would not be expected to dissolve in water because the information given indicates that it is a nonpolar compound. Compounds with nonpolar molecules do not tend to dissolve in water because the water molecules are more attracted to each other than they are to the molecules of the compound.

6. For a liquid to be repelled by a charged wand, the molecules of a liquid must have the same charge as the wand. However, molecules in a liquid are either nonpolar or polar and have both partial positive charges and partial negative charges. If the molecules are nonpolar, the liquid is neither attracted to the wand nor repelled by the wand. If the molecules are polar, the end with a different charge than the wand is attracted to the wand.

(continued)

4. Methanol, CH_3OH, is attracted to a charged wand. The hydrogen atoms have a partial positive charge and the oxygen atom has a partial negative charge.
  a. Draw a picture showing how you predict the methanol molecules are oriented when they are attracted to a wand with negative charges.
  b. Do you expect methanol to bead up or spread out on waxed paper? Explain your thinking.
  c. Do you expect methanol to dissolve in water? Explain your thinking.

5. Hexane, C_6H_{14}, is not attracted to a charged wand and it spreads out on waxed paper. Do you expect hexane to dissolve in water? Explain your thinking.

6. Explain why no liquids are repelled from a charged wand.

The Overview for lesson 43, which references The Bare Essentials of Polarity comic strip, begins on p. 223.

Lesson 42 | **Attractions between Molecules**   219

The Overview for lesson 43, which references The Bare Essentials of Polarity comic strip, begins on p. 223.

Because the elements have such varying electronegativities and can bond in many different combinations, there is really a continuum of polarity in bonding. We can break the continuum down into three categories.

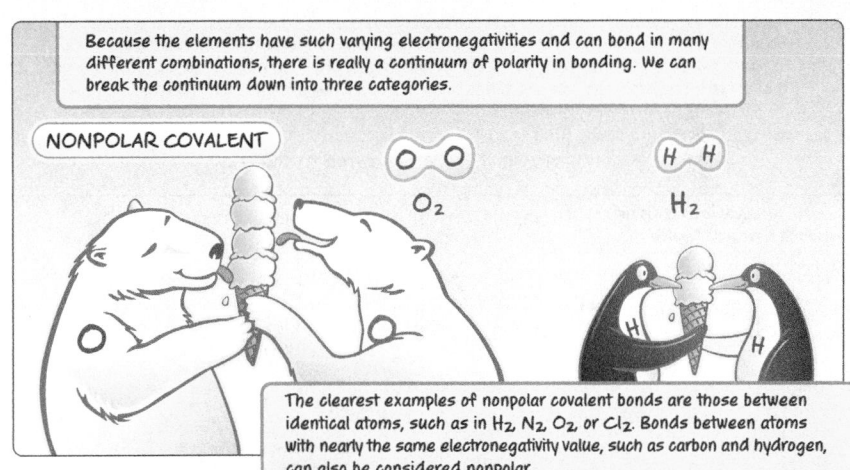

NONPOLAR COVALENT

O_2

H_2

The clearest examples of nonpolar covalent bonds are those between identical atoms, such as in H_2, N_2, O_2, or Cl_2. Bonds between atoms with nearly the same electronegativity value, such as carbon and hydrogen, can also be considered nonpolar.

POLAR COVALENT

HF

Partial negative charge

Partial positive charge

In a polar covalent bond, two atoms share bonded pairs of electrons somewhat unequally. The electrons are more attracted to one atom than the other. Examples include bonds between carbon and oxygen atoms, or between hydrogen and fluorine atoms.

IONIC

Negative charge

Come back!

NaCl

Positive charge

A large difference in electronegativity results in the winner-take-all situation of ionic bonding. The more electronegative atom takes the bonding electrons and becomes a negative ion, while the other atom becomes a positive ion. The opposite charges on the ions attract each other.

Lesson 42 | **Attractions between Molecules**   221

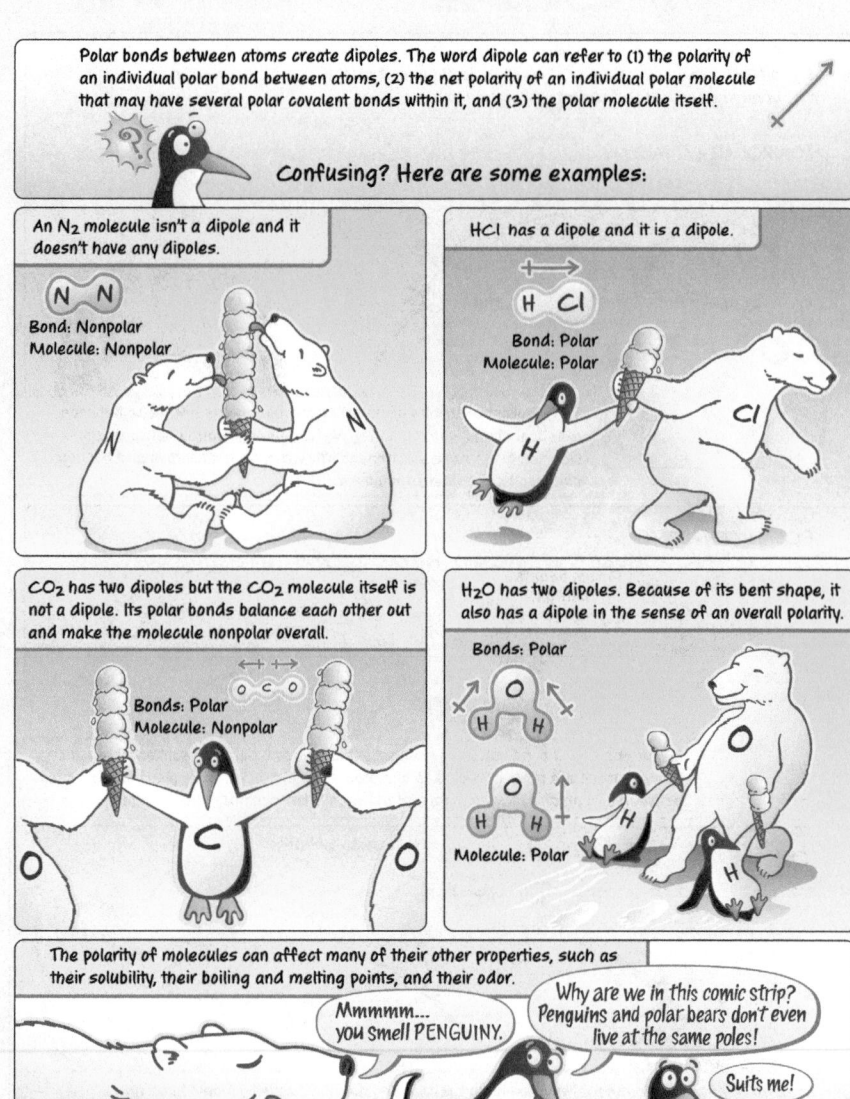

Electronegativity and Polarity

THINK ABOUT IT

Hydrogen chloride, HCl, is a colorless but very toxic gas. Its smell is described as a suffocating, acrid odor. Like most other small molecules that smell, HCl molecules are polar. But what makes an HCl molecule polar? Where do the partial charges come from on the atoms in a polar molecule?

What makes a molecule polar?

To answer this question, you will examine

- Electronegativity
- Nonpolar Molecules
- Electronegativity and Bonding

EXPLORING THE TOPIC

Electronegativity

The hydrogen atom and the chlorine atom in hydrogen chloride, HCl, form a covalent bond by sharing a pair of electrons. This cartoon represents HCl as a penguin and a polar bear. The bonded pair of electrons is represented as two scoops of ice cream. Although the penguin and the polar bear are sharing the ice cream, they are not sharing it equally.

 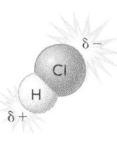

Differences in atoms' electronegativity lead to differences in sharing electrons.

Similarly, the hydrogen atom and the chlorine atom in a hydrogen chloride molecule do not share the bonded pair of electrons equally. The chlorine atom attracts the shared electrons much more strongly than the hydrogen atom does. As a result, the shared electrons spend more time near the chlorine atom than they do near the hydrogen atom. Because of this displacement of the electrons, the hydrogen atom has a partial positive charge and the chlorine atom has a partial negative charge.

Lesson 43 | **Electronegativity and Polarity**   223

Overview

ACTIVITY: PAIRS

Key Question: What makes a molecule polar?

KEY IDEAS

The tendency of an atom to attract shared electrons is called *electronegativity*. When two different atoms bond together, the atom that is more electronegative will attract the shared electrons, causing a partial negative charge on that atom. A molecule is nonpolar if it has no partial charges or if the partial charges cancel. A molecule is polar if it has a positive end and a negative end. There is a continuum of bonding from equal sharing to unequal sharing of shared pairs of electrons due to differences in electronegativity. Electronegativity is covered more quantitatively in the next lesson.

LEARNING OBJECTIVES

- Explain what causes polarity and polar molecules.
- Describe the different types of bonding that correspond to different combinations of electronegative atoms.

- Predict the general direction and strength of a dipole for any two atoms, using the periodic table.

FOCUS ON UNDERSTANDING

- The term *electronegative* can seem contrary to some students. It is not the tendency of an atom to be negative. Rather, it is the tendency of an atom to attract negatively charged electrons.
- The word *dipole* can be a source of confusion because it has several subtly different meanings. Chemists refer to polar molecules as dipoles, and they also say that individual bonds have dipoles (which are numeric and measurable). Finally, molecules with polar bonds can have net dipoles, also called a *dipole moment*.

KEY TERMS

electronegativity
dipole

IN CLASS

Students read the multipage comic strip, "The Bare Essentials of Polarity," which focuses on polarity, electronegativity, and bonding. They answer questions on a worksheet to analyze the illustrations in the comic strip. The concept of electronegativity is introduced, along with three different categories of bonds. A complete materials list is available for download.

TRM Materials List 43

Pre-AP® Course Tip

Students preparing to take an AP® chemistry course or college chemistry must be able to use representations and models to make predictions. In this lesson, students encounter a comic strip that introduces concepts of electronegativity. Based on the comic strip, the students explain and make predictions about polarity and polar molecules.

Lesson Guide

ENGAGE (5 min)

TRM Transparency—ChemCatalyst 43

TRM PowerPoint Presentation 43

Consider this illustration:

1. If the penguin represents a hydrogen atom and the polar bear represents a chlorine atom, what does the ice cream represent in the drawing? What do you think the picture is trying to illustrate?

2. Would HCl be attracted to a charged wand? Explain your thinking.

Sample answers: 1. Most students will figure out that the ice cream represents the electrons shared between the atoms. Other answers are possible. The picture is trying to illustrate that the chlorine atom is somehow stronger than the hydrogen atom and pulls on the electrons more. **2.** Students might say that HCl will be attracted to a charged wand because it has a lopsided molecule that is more negative on one side than on the other. Some might say they cannot tell from the evidence.

➜ You might have a student draw the Lewis dot structure of HCl on the board to remind students of the shared electrons.

Ask: What do you think the polar bear and penguin drawing is trying to illustrate?

Ask: Why is the penguin being swept off its feet by the polar bear?

Ask: How successfully are the polar bear and penguin sharing the ice cream cone?

Ask: What explanation of polarity does the ChemCatalyst illustration provide?

EXPLORE (20 min)

TRM Worksheet with Answers 43

TRM Worksheet 43

TRM Handout—The Bare Essentials of Polarity 43

INTRODUCE THE ACTIVITY

➜ Arrange students into pairs.

➜ Have students individually read the multipage comic strip, "The Bare Essentials of Polarity," before passing out the worksheets. Students can work in pairs on the worksheet.

The tendency of an atom to attract shared electrons is called **electronegativity.** An atom that has a large electronegativity value strongly attracts shared electrons. In this case, the chlorine atom, like the polar bear, is stronger in attracting electrons, so it is more electronegative than the hydrogen atom. The atoms that are more electronegative are the ones that end up with a partial negative charge. The atoms that are less electronegative end up with a partial positive charge. The result is a polar bond.

A polar molecule is called a **dipole,** because it has two poles: a positive end and a negative end. A dipole can be shown with an arrow starting at the positive end and pointing to the negative end of the molecule. The polar bond itself is also called a dipole.

➲ Nonpolar Molecules

When two atoms with identical electronegativities bond together, the attraction of the shared electrons is identical. As a result, the molecule is nonpolar. For example, H_2 and Cl_2 are both nonpolar molecules. They have no partial charges.

A contest between two polar bears of equal strength or two penguins of equal strength would result in a tie.

There is another way to end up with a nonpolar molecule. Examine the next illustration. Why is carbon dioxide, CO_2, a nonpolar molecule?

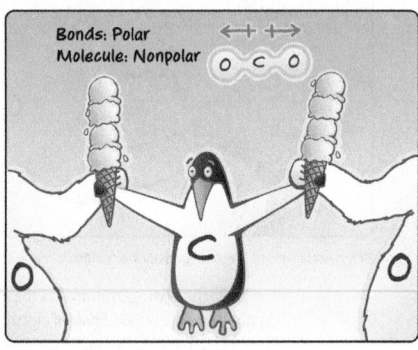

CO_2 is a nonpolar molecule.

The two dipoles in CO_2 balance each other, and there is no partial positive end to the molecule. So the overall molecule is nonpolar.

EXPLAIN & ELABORATE (15 min)

DEFINE ELECTRONEGATIVITY

➜ Draw a molecule of HCl on the board. At the appropriate time in the discussion, add the partial charges and the dipole.

Ask: What are some of the things you learned about polarity from the comic strip?

Ask: How would you use polar bears and penguins to illustrate a polar molecule? A nonpolar molecule?

Ask: According to the comic strip, which elements tend to attract shared electrons to the greatest degree?

Ask: According to the comic strip, what is electronegativity?

Key Points: The tendency of an atom to attract shared electrons is called *electronegativity.* When two atoms with different electronegativities bond, they attract the bonding electrons to different degrees. The bonding electrons spend more time around one of the atoms, resulting in a partial negative charge on that atom. An atom that strongly attracts the shared electrons is considered highly electronegative. The atom with lower electronegativity will end up with a partial positive charge on it. The

↻ Electronegativity and Bonding

There are two types of covalent bonds: polar covalent bonds and nonpolar covalent bonds. When two atoms with different electronegativities bond together, the result is a polar covalent bond.

Different electronegativities result in polar bonds.

When the electronegativities of the atoms that bond are identical, the result is a nonpolar covalent bond.

Identical electronegativities result in nonpolar bonds.

If the electronegativities of the two atoms differ greatly, it is possible for electrons to be pulled entirely toward one of the atoms in a bond. The result is an ionic bond.

Extremely different electronegativities result in ionic bonds.

result is a polar bond, and the polar molecule is a dipole. Hydrogen chloride, HCl, is shown here as an example.

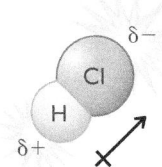

Hydrogen chloride, HCl
Electrons pulled in the direction of the dipole arrow.

Electronegativity: The tendency of an atom to attract the electrons that are involved in bonding.

Dipole: A polar molecule or a polar bond between atoms. A crossed arrow is used to show the direction of a dipole. The crossed end of the arrow indicates the partial positive (+) end of the polar bond, and the arrow points in the direction of the partial negative (−) end.

RELATE ELECTRONEGATIVITY TO BONDING

→ Provide students with a general description of the three categories of bonds summarized at the end of the comic strip.

Ask: How does the comic strip illustrate the range of possible ways of sharing electrons?

Ask: What are the names of the types of bonds that form between atoms? How are these bonds different from one another?

Ask: Why do some elements in molecules have partial charges?

Ask: What is an ion? How does an ion form?

Key Points: Bonds that involve sharing or transferring electrons fall into three categories. In nonpolar covalent bonds, the electrons are shared equally. (For instance, in diatomic molecules such as H_2, the atoms are identical, so there is no difference in the degree to which each atom in the molecule attracts the shared electrons.) In polar covalent bonds, the electrons are shared unequally. The electrons are more attracted to atoms with higher electronegativity. Ionic bonds involve the transfer of one or more electrons from one atom to another. When the electronegativities of two atoms differ greatly, the bond between them is considered ionic. The dividing line between polar covalent bonding and ionic bonding is not clear.

EVALUATE (5 min)

Check-In

Consider hydrogen iodide, HI.

1. Is HI a polar molecule? Explain your reasoning.
2. How would the atoms be portrayed in the comic strip—as polar bears, penguins, or both? Explain.

Answers: **1.** Hydrogen iodide is a polar molecule. Its atoms do not share the electrons equally. Iodine is located on the right side of the periodic table, where the electronegativities are greater. Hydrogen is on the left side. **2.** In a drawing, hydrogen would be a penguin and iodine would be a polar bear. The iodine atom would have a partial negative charge on it.

Homework

Assign the reading and exercises for Smells Lesson 43 in the student text.

LESSON 43 ANSWERS

1. Possible answer: In a polar covalent bond, the electron is attracted more by one atom than by the other, so one of the atoms has a partial negative charge and the other has a partial positive charge. In a nonpolar covalent bond, in which the two atoms share the electrons equally, there are no partial charges.

2. In a polar covalent bond, electrons are shared by two atoms, but are more strongly attracted to one atom than the other. In an ionic bond, the electron is transferred completely from one atom to the other.

3. Although the two carbon-oxygen bonds of carbon dioxide are polar, the molecule itself has a linear shape, so the partial negative charges are on opposite sides of the molecule. These charges balance one another, so there is no net dipole.

4. a. nonpolar covalent **b.** polar covalent **c.** nonpolar covalent **d.** polar covalent **e.** polar covalent

5. Possible answer: The drawing should contain three penguins and one polar bear. Each penguin should be sharing two scoops of ice cream with the polar bear. The polar bear should be labeled "N" for nitrogen and the penguins should be labeled "H" for hydrogen.

6. Possible answer:

As shown in the Lewis dot structure, HOCl will have a bent shape similar to that of a water molecule. Although the electronegativity of various atoms has not yet been taught, it can be inferred that bent molecules formed from multiple elements should be polar because the bonds between different types of atoms will be polar. For three-atom molecules, only a linear shape can have no dipole.

7. Possible answer: Because of the difference in electronegativity of atoms in a molecule, some molecules have dipoles in which the electrons are attracted more to one end of the bond than the other. The molecule then acts like a miniature magnet, and the positive end of the molecule is attracted to a negatively charged wand.

Ionic compounds represent the extreme of polar bonds, in which electrons are transferred to the more electronegative atom in the pair.

LESSON SUMMARY

What makes a molecule polar?

Electronegativity is a measure of the ability of an atom to attract the electrons that are involved in a bond. Different atoms have different electronegativities. When two different kinds of atoms bond together, they do not attract the bonding electrons equally. The electrons are attracted to the atom with greater electronegativity. The result is a polar covalent bond. When atoms with identical electronegativities bond together, the electrons are shared equally, and the result is a nonpolar covalent bond. If the two atoms involved in a bond differ greatly in electronegativity, the result is electron transfer and an ionic bond.

KEY TERMS
electronegativity
dipole

Exercises

Reading Questions

1. What is the difference between a polar covalent bond and a nonpolar covalent bond?
2. What is the difference between a polar covalent bond and an ionic bond?
3. Explain why carbon dioxide is a nonpolar molecule even though its bonds are polar.

Reason and Apply

4. Are these molecules polar covalent or nonpolar covalent?
   a. N_2 **b.** HF   **c.** F_2 **d.** NO   **e.** FCl
5. Using polar bears and penguins, create an illustration showing an ammonia, NH_3, molecule. (*Hint:* You may wish to start with a Lewis dot structure.)
6. Is the molecule HOCl polar or nonpolar? Use a Lewis dot structure to explain your thinking.
7. Use electronegativity to explain why some molecules are attracted to a charged wand.

Thinking (Electro) Negatively

Electronegativity Scale

THINK ABOUT IT

Some atoms are "greedier" than other atoms when it comes to sharing the bonding electrons between them. A bond between two atoms with very different electronegativities is more polar than a bond between atoms with similar electronegativities. How can the polarity of different bonds be compared?

How can electronegativity be used to compare bonds?

To answer this question, you will explore

⟳ Electronegativity Scale

⟳ Diatomic Molecules

EXPLORING THE TOPIC

⟳ Electronegativity Scale

Chemists have assigned each atom a number, called an *electronegativity value*. This number corresponds to the tendency of an atom to attract bonding electrons. By using these numbers, it is possible to compare the polarity of different bonds. The table here shows the electronegativity value for an atom of each element in the periodic table.

Electronegativity scale

H 2.10																	He
Li 0.98	Be 1.57											B 2.04	C 2.55	N 3.04	O 3.44	F 3.98	Ne
Na 0.93	Mg 1.31											Al 1.61	Si 1.90	P 2.19	S 2.58	Cl 3.16	Ar
K 0.82	Ca 1.00	Sc 1.36	Ti 1.54	V 1.63	Cr 1.66	Mn 1.55	Fe 1.83	Co 1.88	Ni 1.91	Cu 1.90	Zn 1.65	Ga 1.81	Ge 2.01	As 2.18	Se 2.55	Br 2.96	Kr
Rb 0.82	Sr 0.95	Y 1.22	Zr 1.33	Nb 1.60	Mo 2.16	Tc 1.90	Ru 2.2	Rh 2.28	Pd 2.20	Ag 1.93	Cd 1.69	In 1.78	Sn 1.96	Sb 2.05	Te 2.1	I 2.66	Xe
Cs 0.79	Ba 0.89	La* 1.10	Hf 1.30	Ta 1.50	W 2.36	Re 1.90	Os 2.20	Ir 2.20	Pt 2.28	Au 2.54	Hg 2.00	Tl 1.62	Pb 2.33	Bi 2.02	Po 2.00	At 2.20	Rn
Fr 0.70	Ra 0.89	Ac* 1.10															

* Electronegativity values for the lanthanides and actinides range from about 1.10 to 1.50.

Notice that the values generally increase from left to right and bottom to top. Noble gases are often not assigned values because they generally do not form compounds.

Lesson 44 | **Electronegativity Scale**   227

Overview

CLASSWORK: INDIVIDUAL

Key Question: How can electro-negativity be used to compare bonds?

KEY IDEAS

The electronegativity scale is an arbitrary scale, invented by Linus Pauling in 1932, that assigns a numerical value to the electronegativity of each atom in the periodic table. By comparing the electronegativities of two atoms, chemists can determine if a resulting bond between them will be highly polar. The numbers that Pauling generated allow us to quantify the continuum of bonding—nonpolar covalent, polar covalent, and ionic bonds.

LEARNING OBJECTIVES

- Use the electronegativity scale to compare atoms and to compare (calculate) the polarity of different bonds.
- Use the electronegativity scale to predict bond dipoles and bond type.
- Describe the continuum of nonpolar, polar, and ionic bonding in terms of electronegativity.

FOCUS ON UNDERSTANDING

- While the noble gases are considered unreactive, there are occasional exceptions, and sometimes they are assigned electronegativity values. To keep from confusing students, we have not included this information.

- It can be difficult for students to grasp the idea of a unitless scale.

KEY TERM

diatomic molecule

IN CLASS

The concept of electronegativity that was introduced in the polarity comic strip in Lesson 43 is expanded on. Students learn to read and use the electronegativity scale. Students use electronegativity values to compare atoms and bonds and to figure out the direction of bond dipoles. They also apply electronegativity values to the continuum of bonding from nonpolar covalent to polar covalent to ionic. A complete materials list is available for download.

TRM Materials List 44

Differentiate

To develop scientific literacy and support English-language learners, engage students in creating a "word wall" with the vocabulary from the past few lessons. In particular, have students explain the various meanings for the word *dipole*. Include symbols, such as the partial charge symbols, in the word wall, too, because many students are unfamiliar with the Greek alphabet. To extend this lesson for students interested in history, find resources on the various electronegativity scales (Pauling, Mulliken, Allred-Rochow, and so on), including how they were developed and why scientists favor particular scales. This can be a good resource for teaching students about the human factor in how science is conducted.

Lesson Guide

ENGAGE

TRM Transparency—ChemCatalyst 44

TRM PowerPoint Presentation 44

ChemCatalyst

1. Explain how the illustration and the table below it might be related to each other.

2. What patterns do you see in the numbers in the table?

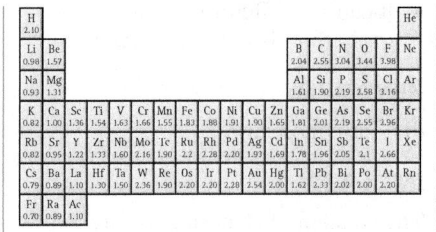

Sample answers: 1. Some students will say that the numbers in the table must stand for the electronegativity of each atom. The sizes of the polar bears and penguins represent the numbers. The thickness of the ice also is related to the magnitude of the numbers. **2.** The numbers in the table increase across a row and decrease down a group.

Ask: How is the illustration related to the table shown below it?

Ask: What do you think the numbers on the table represent?

Ask: What parts of the table do the penguins and polar bears represent? What does the ice represent?

Ask: What do the polar bears and penguins look like where the elements are most chemically reactive?

Ask: Which animals represent nonmetals? Why are there some bears in the metals area of the periodic table?

EXPLORE (15 MIN)

TRM Worksheet with Answers 44

TRM Worksheet 44

INTRODUCE THE CLASSWORK

➡ Let students know they will be working individually.

➡ Pass out the student worksheet and the handout Electronegativity Scale.

➡ Tell students they will be exploring electronegativity in a quantitative way.

EXPLAIN & ELABORATE (20 min)

TRM Transparency—Electronegativity Scale 44

TRM Transparency—Bonding Continuum 44

DISCUSS THE TABLE OF ELECTRONEGATIVITIES

➡ Display the transparency Electronegativity Scale.

Ask: What general trends in electro-negativity do you notice from the table?

HISTORY CONNECTION

Linus Pauling, inventor of the electronegativity scale, is one of only two people (the other is Marie Curie) to receive Nobel Prizes in two different fields. In 1954, he received the Nobel Prize in Chemistry, and in 1962 he received the Nobel Peace Prize for his work in campaigning against above-ground nuclear testing.

Joe McNally/Hulton Archive/Getty Images

A scale for electronegativity was first proposed by Linus Pauling in 1932. The scale chemists use today ranges in value from 4.0 down to 0. The electronegativity scale allows you to compare individual atoms. For example, fluorine, with an electronegativity value of 3.98, attracts shared electrons more strongly than hydrogen, with an electronegativity value of 2.10.

➲ Diatomic Molecules

The electronegativity scale also allows you to compare the polarity of bonds. In a polar covalent bond, the electrons tend to spend more time around the more electronegative atom. Several substances with only two atoms are shown below. Molecules with two atoms are called **diatomic molecules.** By looking at the numerical difference between the electronegativities of the two atoms, it is possible to compare the polarity of one bond to another. Bonds that have a greater difference in electronegativity between the two atoms are more polar.

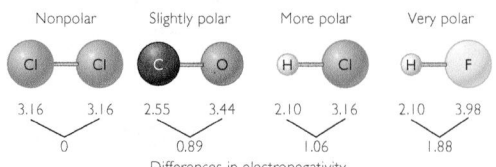

Nonpolar	Slightly polar	More polar	Very polar
Cl—Cl	C—O	H—Cl	H—F
3.16 3.16	2.55 3.44	2.10 3.16	2.10 3.98
0	0.89	1.06	1.88

Differences in electronegativity

When the difference in electronegativity between the two atoms is very large, the bond is no longer considered covalent. In that case, the electrons are transferred from the less electronegative atom to the more electronegative atom. A cation and an anion form that are attracted to one another in an ionic bond. As shown on this electronegativity scale, when the difference between the electronegativities of two bonded atoms is greater than about 2.1, the bond is considered ionic:

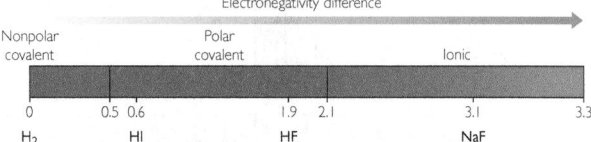

Electronegativity difference

Nonpolar covalent	Polar covalent	Ionic
0 0.5 0.6	1.9 2.1	3.1 3.3
H_2 HI	HF	NaF

Example

Potassium and Chlorine

Predict the type of bond you would find between potassium and chlorine.

Solution

Use the electronegativity scale to find the electronegativities of potassium, K, and chlorine, Cl. It shows that the values are 0.82 for K and 3.16 for Cl. The difference is 2.34, so the bond is ionic.

Ask: Why are metals referred to as electropositive?

Key Points: In 1932, Linus Pauling created a scale for electronegativity and assigned numerical values for the electronegativities of the elements. This scale can be found on page 227 of the student textbook. The scale ranges from 4.0 down to 0, with 4.0 the highest electronegativity. The lowest electronegativities can be found in the lower-left portion of the periodic table and in the noble gas family, and the highest electronegativities are found in the upper-right portion of the table. The scale was created by comparing the polarities of a variety of molecules. This scale is unitless.

DISCUSS BONDING BETWEEN PAIRS OF ATOMS

➡ Display the transparency Bonding Continuum.

Ask: How would you determine which end of a molecule with two atoms has a partial negative charge and which end has a partial positive charge? Explain.

Ask: Which pairs of atoms do you expect to be most polar?

Ask: When two atoms have a very large difference in electronegativity, the bond between them is considered ionic. Explain why.

Ask: What is a polar covalent bond?

LESSON SUMMARY

How can electronegativity be used to compare bonds?

Electronegativity values can help you to compare and classify different bonds. If there is no difference in electronegativity between the two atoms, the bond is considered nonpolar covalent. The greater the electronegativity difference, the more polar the bond. The electrons in a polar bond tend to spend more time around the more electronegative atom. When the difference between the electronegativities of two bonded atoms is greater than about 2.1, the bond is considered ionic and one atom gives up an electron to the other atom, forming two ions.

KEY TERM

diatomic molecule

Exercises

Reading Questions

1. Explain how electronegativity values help you to determine the polarity of a bond between two atoms.
2. How can you determine which atom in a covalent bond is partially positive?

Reason and Apply

3. Consider the following pairs of atoms. Place each set in order of increasing bond polarity. Describe the trend.
   a. Li−F   Na−F   K−F   Rb−F   Cs−F
   b. Mg−O   P−S   N−F   K−Cl   Al−N
4. Place a partial positive or a partial negative charge on each atom in the following pairs of atoms. Describe the trend.
   a. H−B   b. H−C   c. H−N   d. H−O   e. H−F
5. Is hydrogen always partially positive when bonded to another atom? Explain.
6. Name three pairs of atoms with ionic bonds. For each pair, show the difference in electronegativity between the two atoms.
7. Name three pairs of atoms with polar covalent bonds. For each pair, show the difference in electronegativity between the two atoms.
8. Describe or draw what happens to the electrons in a polar covalent bond, a nonpolar bond, and an ionic bond.
9. What do we mean when we say that bonding is on a continuum?

from one atom to the other. The more electronegative atom gets the electron(s) and becomes a negative ion. The bond is called *ionic,* because ions are formed.

Note: When the difference in electronegativity between two atoms is very small (less than 0.5), the bond is so slightly polar that it is often considered nonpolar. For example, the electronegativity difference between carbon and hydrogen is less than 0.5, so it is often considered nonpolar.

The dividing line between polar covalent bonding and ionic bonding is not clear-cut. The value 2.1 given for the difference in electronegativity as the dividing line is only a guide. Bonding between atoms is on a continuum, as shown in the diagram Electronegativity Difference, found on page 228 of the student edition.

EVALUATE (5 min)

Check-In

1. Is the bond in potassium chloride, KCl, nonpolar, polar, or ionic? Explain.
2. To what degree do the K and Cl atoms in KCl, potassium chloride, share electrons?

Answers: 1. The electronegativity of K is 0.82, and the electronegativity of Cl is 3.16. The difference is 2.34. In fact, the difference in electronegativity is so great that we can think about the bond as one in which one electron is given up by the potassium atom and is transferred to the chlorine atom. The bond is ionic. 2. The potassium and chlorine atoms do not share the electron.

Homework

Assign the reading and exercises for Smells Lesson 44 in the student text.

LESSON 44 ANSWERS

1. Electronegativity values help determine the polarity of the bond between two atoms because they can be used to determine the tendency of an electron to be attracted to one atom rather than another atom.

2. In a polar covalent bond, the atom with the lower electronegativity value will have a partial positive charge.

Answers continue in Answer Appendix (p. ANS-3).

Ask: Where would Cl_2 be on the continuum of bonding? (nonpolar covalent) HCl? (polar covalent)

Key Points: By determining the numerical difference between electronegativities in a bond, you can compare the polarities of bonds. For instance, hydrogen has an electronegativity of 2.10. Chlorine has an electronegativity of 3.16. The difference in their electronegativities is 1.06, making HCl a polar molecule.

Numerical differences in electronegativity can also help predict the type of bond that will be found.

- No difference in electronegativity: When the two atoms bonded together are identical, the electrons are shared equally. The bond is called *nonpolar covalent.*
- Small differences in electronegativity (below 2.1): When there is a small difference in electronegativity between two atoms bonded together, there is unequal sharing of electrons. The electrons are attracted toward the atom with the greater electronegativity. Partial charges are set up on each atom, and the bond is called *polar covalent.*
- Large differences in electronegativity (above 2.1): When there is a large difference in electronegativity between two atoms bonded together, the electron essentially is transferred

Overview

ACTIVITY: PAIRS

Key Question: What does polarity have to do with smell?

KEY IDEAS

The shape and symmetry of a molecule can affect its polarity. In general, polar molecules smell, and nonpolar molecules do not smell. Polarity affects smell in two ways. First, it affects whether a molecule will dissolve in the watery mucous lining of the nose. Second, the fact that polar molecules are attracted to each other may assist them in docking with the polar portions of the large protein molecules that make up the receptor sites.

LEARNING OBJECTIVES

- Assess a molecule for symmetry and determine if it is likely to be polar.
- Use electronegativity values to locate the partial negative and partial positive portions of a molecule.
- Explain the connection between polarity and smell.

FOCUS ON UNDERSTANDING

Students often assume that the location of a lone pair will be the location of a partial negative charge. This is not always the case.

IN CLASS

Using the electronegativity scale, students figure out the location of partial charges on several molecules. This assists them with determining the direction of polarity. Students are given paper representations of eight molecules that they must cut out and orient properly in relationship to a receptor site and a molecule of water. Students are challenged to determine which molecules do not smell and why. Students learn about the importance of polarity in smell chemistry and are introduced to current theories about receptor sites, polarity, and smell. A complete materials list is available for download.

TRM Materials List 45

SETUP

You might want to have a few ball-and-stick models available at the front of the class to demonstrate the symmetry of small nonpolar molecules and the asymmetry of small polar molecules. Suggested nonpolar models: CF_4, CO_2, N_2, CH_4. Suggested polar molecules: CF_3Cl, H_2O. You might also want to have a medium-sized molecule such as citronellol available.

LESSON
45

Polar Molecules and Smell

THINK ABOUT IT

Polar molecules have a potent smell. The smell of ammonia, NH_3, in some window cleaners is quite strong. In contrast, the methane gas, CH_4 does not have a smell even though its C—H bonds are all polar. So, why can't you smell CH_4 gas?

What does polarity have to do with smell?

To answer this question, you will examine

- Polarity of Molecules
- Nonpolar Molecules
- Polarity and Smell

EXPLORING THE TOPIC

➔ Polarity of Molecules

If the *bonds* between atoms of a molecule are polar, what about the molecule as a whole unit? One way to tell whether a molecule is polar is to examine its overall shape. Notice that each of the three molecules shown here has an irregular shape, or some sort of asymmetry. Asymmetric molecules are polar.

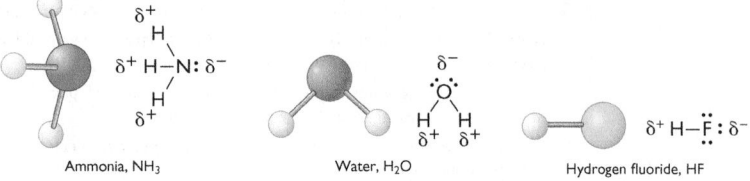

Ammonia, NH_3 Water, H_2O Hydrogen fluoride, HF

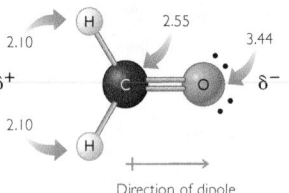

Direction of dipole
Formaldehyde, CH_2O

Ammonia, water, and hydrogen fluoride are all polar. In these molecules, there is a partial negative charge on the nitrogen, oxygen, and fluorine atoms because these atoms are more electronegative than hydrogen atoms. The hydrogen atoms have partial positive charges. To figure out which end of a molecule has a partial negative charge and which end has a partial positive charge, check the electronegativities of the atoms and the overall shape of the molecule.

Consider formaldehyde, CH_2O, shown here. The numbers next to the atoms are their electronegativities.

230   Chapter 8 | **Molecules in Action**

Differentiate

This lesson can be a challenge for students who struggle with conceptualizing three-dimensional molecules from two-dimensional representations. Some students find it challenging to determine overall dipoles for molecules such as CF_3Cl. You might have students visit the University of Colorado Boulder's PhET simulation called Molecule Polarity. Use one of the teacher-support activities that allow students to explore how electronegativity differences affect bond polarity, how individual bond dipoles and molecular geometry affect overall polarity, and

how real molecules behave in terms of polarity. You can find a list of URLs for this lesson on the Chapter 8 Web Resources document.

TRM Chapter 8 Web Resources

Pre-AP® Course Tip

Students preparing to take an AP® chemistry course or college chemistry must be able to draw connections to topics outside of chemistry. In this lesson, students assign partial positive and partial negative signs to molecules to determine polarity. By doing so, the students begin to explain the connection between polarity (chemistry) and smell (biology/biochemistry).

The oxygen end of the formaldehyde molecule has a partial negative charge because the oxygen atom is the most electronegative atom in the formaldehyde molecule. Similarly, the hydrogen atoms have partial positive charges because they are the least electronegative atoms in the molecule.

LESSON 45

Example 1

Phosphine, PH₃

Is phosphine, PH_3, polar? What is the direction of the dipole?

Solution

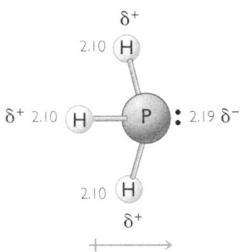

A Lewis dot structure of PH_3 shows four pairs of electrons, three bonding pairs and one lone pair. These four pairs of electrons are arranged around the phosphorus, P, atom in a tetrahedral shape.

Because the molecule is asymmetrical, it is polar. You can find the electronegativity values for phosphorus and hydrogen in the electronegativity scale in Lesson 44: Thinking (Electro)Negatively. The electronegativity value for P is 2.19 and for H is 2.10. Because the electronegativity of the phosphorus atom is greater, the P atom attracts the electrons more strongly and has a partial negative charge. The H atoms have partial positive charges. The overall dipole is shown.

⟳ Nonpolar Molecules

Diatomic molecules with two identical atoms are nonpolar. The electrons between the two identical atoms are shared equally. However, these are not the only kind of nonpolar molecules. Some molecules are symmetrical. The symmetry in these molecules can make them nonpolar even though they have polar bonds within them. The polarities of the individual polar bonds balance each other out due to the shape of the molecule.

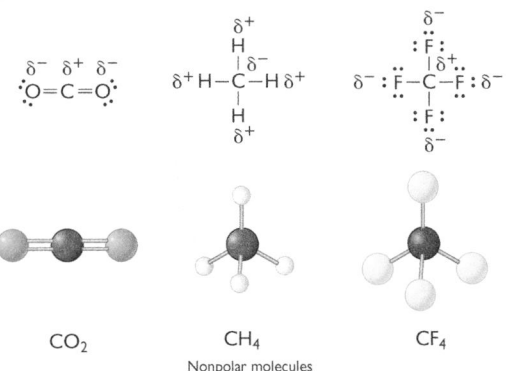

CO₂   CH₄   CF₄

Nonpolar molecules

Lesson 45 | **Polar Molecules and Smell**   231

Lesson Guide

ENGAGE (5 min)

TRM PowerPoint Presentation 45

ChemCatalyst

Hydrogen chloride, HCl, and ammonia, NH_3, have a smell, and large amounts of each dissolve in water. Oxygen, O_2, nitrogen, N_2, and methane, CH_4, do not have a smell, and only a small amount of each dissolves in water. How can you explain these differences?

Sample answer: Students might say that the molecules that dissolve in water are more polar than those that

do not dissolve much in water. They may or may not add that polarity must have something to do with smell.

→ You might want to pursue the ChemCatalyst question in depth by having students draw structural formulas (including lone pairs) of these molecules on the board.

Ask: How are the molecules different? (HCl and NH_3 are asymmetrical, polar molecules, whereas O_2, N_2, and CH_4 are symmetrical, nonpolar molecules.)

Ask: How are the molecules in the ChemCatalyst similar? (They are all relatively small.)

Ask: Why do you think HCl and NH_3 dissolve easily in water? Why do you think they have a smell?

Ask: How do you think polarity is related to smell?

EXPLORE (20 min)

TRM Worksheet with Answers 45

TRM Worksheet 45

TRM Transparency—Electronegativity Scale 44

TRM Handout—Molecules Cutouts

INTRODUCE THE ACTIVITY

→ Arrange students into pairs.

→ Pass out the student worksheet, Handout—Molecules Cutouts, a printout of the Transparency—Electronegativity Scale 44, glue sticks, and scissors. To save time and cutting, you could have each pair share a worksheet.

EXPLAIN & ELABORATE (15 min)

DISCUSS THE POLARITY OF MOLECULES WITH MORE THAN TWO ATOMS

→ Have your own set of cut-out molecules available for demonstration. You might want to use tape to stick them on the board or on a piece of butcher paper.

→ Ask students to come up and draw the location and identity of the partial charges on each molecule.

Ask: What process did you use to figure out the polarity of a molecule?

Ask: Do the lone pairs tell you where the more electronegative part of the molecule is? Explain.

Ask: What is the best way to determine which part of a molecule has a partial negative charge and which part has a partial positive charge?

Key Points: Polarity of diatomic molecules is fairly easy to determine. If the two atoms are identical (such as two hydrogen atoms), the molecule is not polar. If the two atoms are different (such as HCl), you must figure out which atom is more electronegative. That is the end with the partial negative charge.

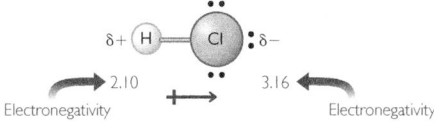

Both electronegativity and the overall symmetry of the molecules help determine the polarity of molecules with more than two atoms. For example, to figure out the polarity of formaldehyde, CH_2O, all you

231

need to know is that oxygen is more electronegative than hydrogen. Likewise, the nitrogen atom in ammonia, NH_3, is more electronegative than the hydrogen atoms. If you cannot remember electronegativity trends, you can check the electronegativity chart.

CH₂O

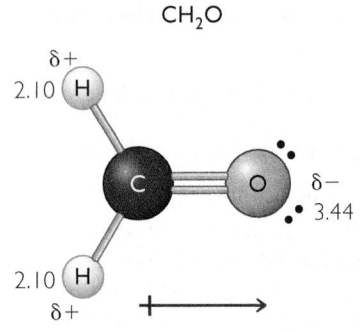

NH₃

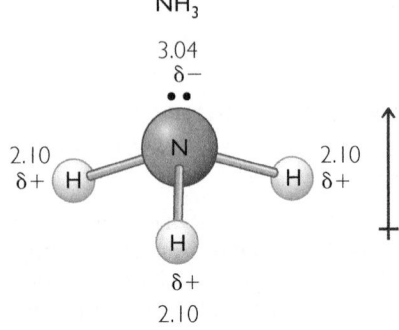

RELATE POLARITY TO SMELL

→ You might want to have ball-and-stick models on hand to demonstrate the symmetry of small nonpolar molecules. Suggested models: CF_4, CO_2, N_2, CH_4.

Ask: Which molecules did you pick as the ones that did not have a smell? What was your reasoning?

Ask: Why doesn't carbon dioxide have a smell? What about methane?

Ask: What does the overall symmetry of a molecule have to do with its smell?

Ask: Can some molecules be symmetrical in shape and still be polar? Explain.

Key Points: Small nonpolar molecules do not have a smell. The most obvious examples of small nonpolar molecules are molecules with two identical atoms, such as oxygen, O_2, and nitrogen, N_2. Molecules with more than two atoms can also be nonpolar if they have a symmetrical shape. For example, tetrafluoromethane, CF_4, is nonpolar. Although all the fluorine atoms are highly electronegative, the entire molecule is a perfectly symmetrical tetrahedron. No portion of the molecule is any more electronegative than any other portion.

232

Chlorotrifluoromethane, also known as Freon 13, was developed early in the 20th century and used widely as a refrigerant. However, use of Freon 13 and other chlorofluorocarbons (CFCs) was significantly reduced in the late 1980s due to their effects on the ozone layer of the atmosphere.

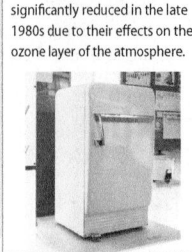

DAJ/Getty Images

→ Polarity and Smell

Small molecules that are polar have a smell. Small molecules that are nonpolar do not have a smell. What does polarity have to do with the smell of small molecules?

Polar: Have a smell	Nonpolar: Do not have a smell	Exceptions
HF, HCl, H_2S, CH_3F, CH_2F_2, CHF_3, NH_3, PH_3, HOCl	N_2, O_2, CO_2, CH_4, CF_4	F_2, Cl_2, Br_2, and CCl_4 are nonpolar but have a smell. H_2O is polar but does not have a smell.

Polar molecules have properties that distinguish them from nonpolar molecules. Polar molecules dissolve in water and are attracted to other polar molecules. One well-accepted theory is that polar molecules *dissolve* in the mucous membrane of the nose and then attach to receptor sites. The mucus is composed largely of water.

RECEPTOR SITES IN THE NOSE

Receptor sites in the nose contain large polar protein molecules. Scientists believe that small polar molecules are attracted to the polar receptor sites after they enter the nose. You could think of polarity as working somewhat like a magnet, with small polar molecules "sticking" to the polar part of a receptor site. Nonpolar molecules are not attracted to the same extent as polar molecules, so they may not be detected by the nose.

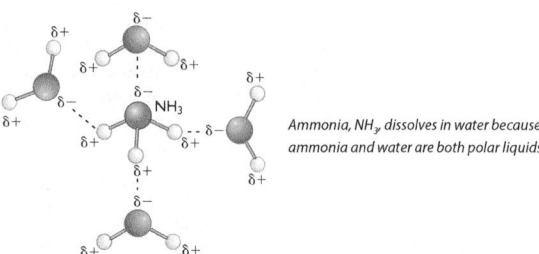

Ammonia, NH₃, dissolves in water because ammonia and water are both polar liquids.

Example 2 Chlorotrifluoromethane, CClF₃, and Tetrafluoromethane, CF₄

Compare chlorotrifluoromethane with tetrafluoromethane. Explain why one of these molecules has a smell and the other does not.

Solution

Both molecules have a tetrahedral shape. In tetrafluoromethane, each C—F bond is polar. However, the four C—F bonds are distributed symmetrically so that the molecule as a whole is nonpolar. Therefore, it does not have a smell.

CF₄

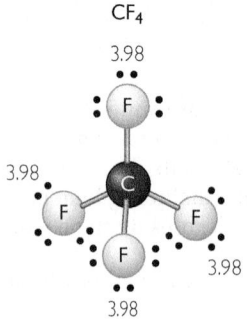

Tetrafluoromethane is symmetrical and nonpolar.

If the overall shape of a molecule is asymmetrical and the molecule is made from more than one kind of atom, chances are it is a polar molecule. If the overall shape of a molecule is highly symmetrical, it is necessary to look

closer to see whether any imbalance in electronegativity is present. The presence of different atoms on different sides of the molecule, as in chlorotrifluoromethane, is a sign of polarity.

CClF₃

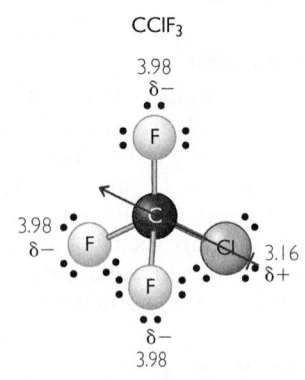

Chlorotrifluoromethane is polar because of the chlorine atom on one side.

Example 2

(continued)

Chlorotrifluoromethane is not as symmetrical. This is because one of the F atoms is replaced by a Cl atom. The molecule is polar because F and Cl have different electronegativities. Therefore, it has a smell.

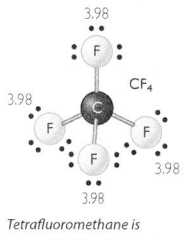

Tetrafluoromethane is symmetrical and nonpolar.

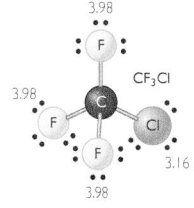

Chlorotrifluoromethane is asymmetrical and polar.

NATURE CONNECTION

Many animals, such as elephants, seem to be able to smell water at great distances.

LESSON SUMMARY

What does polarity have to do with smell?

Small asymmetrical molecules with polar bonds are polar. Molecules that are symmetrical in every way are nonpolar. Theories suggest that the polarity of a molecule may help it to "stick" in a receptor site in the nose. Further, polar substances are more soluable in water. This may allow polar substances to dissolve in the mucous membrane of the nose, whereas nonpolar substances are simply exhaled undissolved and undetected.

Exercises

Reading Questions

1. How can you determine if a molecule is polar?
2. Describe one theory of why small nonpolar molecules do not have a smell.

Reason and Apply

3. For each of the molecules listed, draw a Lewis dot structure and indicate the shape of each molecule. Decide whether these substances smell. Explain your reasoning for each.
   a. H_2Se **b.** H_2 **c.** Ar   **d.** HOF   **e.** $CHClF_2$ **f.** CH_2O
4. For each of the polar molecules in Exercise 3, draw the dipole.
5. Water is an exception to our rule about small molecules. It is a polar molecule, yet humans don't smell it. What do you think is going on?
6. Do you think it might be useful if you could smell the air? Explain your thinking.
7. Methane, CH_4, gas leaks can be very dangerous and can be explosive. Why do you think dimethyl sulfide, C_2H_6S, is added to natural gas that is used in homes and buildings?

DISCUSS WHAT MIGHT HAPPEN IN THE NOSE

→ Ask students to show the class how a molecule of ammonia might interact with molecules of water.

→ You might have available a model of a molecule such as citronellol to discuss the polarity of molecules of this size.

Ask: Why do you think polarity is important to smelling? What does it have to do with the nose? With receptor sites?

Ask: What is the inside of the nose like? Do you think the inside of the nose is polar? Explain.

Ask: Do you think a larger molecule like citronellol is considered a polar molecule? Does it dissolve in water? Explain.

Key Points: Inside the nose is a watery mucous lining. One theory is that polar molecules dissolve in the mucous lining and then attach to receptor sites in the nose. Because of intermolecular attractions, polar molecules dissolve very easily in other polar molecules. Nonpolar molecules, on the other hand, do not dissolve in water or in other polar molecules. This might explain why we cannot smell small nonpolar

molecules: They cannot move through the mucous lining to the receptor sites. Ammonia dissolves in water.

Molecules have to be attracted to receptor sites to be detected. Another smell theory is that polar molecules "dock" in polar receptor sites after they enter the nose. Receptor-site molecules have polar aspects to them, allowing a smell molecule to "attach" to the site. Nonpolar molecules, such as methane and carbon dioxide, would not dock in receptor sites. Medium-sized molecules that smell, like citronellol, usually are asymmetrical in shape and often have lone pairs somewhere on the molecule. They are polar enough to dock in receptor sites.

The small molecules that constitute our air do not have a smell. They can all be shown to be nonpolar. Presumably, they simply move into our nostrils and breathing passages and then move out again. One exception is water, which is a small polar molecule that does not have a smell. Scientists think that because so much water is present at all times in our nose the signal to our brain never changes, and hence, we do not smell it. We have become so used to it that it is like background noise. Many animals, including elephants, seem to be able to smell water at great distances.

EVALUATE (5 min)

Check-In

Is hydrogen cyanide, HCN, a polar molecule? Will it smell? Why or why not?

Answer: Even though HCN is a linear molecule, you can tell at a glance that it is an asymmetrical molecule. It has a nitrogen atom on one end and a hydrogen atom on the other end. The nitrogen atom is more electronegative than the hydrogen atom. Thus, the nitrogen end has a partial negative charge on it. This molecule smells a little like almonds.

Homework

Assign the reading and exercises for Smells Lesson 45 in the student text.

LESSON 45 ANSWERS

Answers continue in Answer Appendix (p. ANS-3).

Overview

CLASSWORK: PAIRS

Key Question: What generalizations can you make about smell and molecules?

KEY IDEAS

Generalizations can be made relating smell to polarity, molecular size, phase, and type of bonding. For instance, most medium-sized molecules—those with about 5 to 19 carbon atoms—have a smell, including nonpolar ones; small molecules tend to have a smell only if they are polar; and large molecules generally do not have a smell. Ionic and metallic solids do not vaporize and therefore cannot be smelled.

LEARNING OBJECTIVES

- Explain the connections between smell and polarity, molecular size, phase, and type of bonding.
- Predict if a molecule will have a smell based on its structure, composition, and phase.

FOCUS ON UNDERSTANDING

- Some substances seem to have a smell, but what we smell are contaminants. For example, students might claim that metals have a smell. Smells coming from metal objects usually are oils or other molecular substances. Plastics seem to smell, but what we smell are compounds used in their manufacture.
- Carbon atoms can form four bonds allowing for large molecules. The number of carbon atoms is often used to indicate molecular size.

IN CLASS

Students are given a worksheet with information about how polarity, size, phase, and type of bonding are related to whether a substance will have a smell or not. Students refine their hypotheses about smell and put smell into a broader context. A complete materials list is available for download.

TRM Materials List 46

SETUP

Post the Smells Summary Chart and blank paper for the No Smell Summary Chart.

Lesson Guide

ENGAGE (5 min)

TRM PowerPoint Presentation 46

Sniffing It Out

Phase, Size, Polarity, and Smell

THINK ABOUT IT

You can smell chicken frying or someone's perfume from across a room, but you can't smell a piece of paper or a binder on your desk. And cookies right out of the oven are much easier to smell than cold ones. How can you use what you have learned about molecules to understand these observations?

> What generalizations can you make about smell and molecules?

To answer this question, you will explore

⟳ Bonding and Smell

⟳ Molecular Size and Smell

EXPLORING THE TOPIC

⟳ Bonding and Smell

In Unit 1: Alchemy, you sorted matter into four classes based on whether the substance dissolved in water and whether the substance conducted electricity. This sorting led to four models of bonding: ionic, molecular covalent, metallic, and network covalent. How does the property of smell relate to bonding?

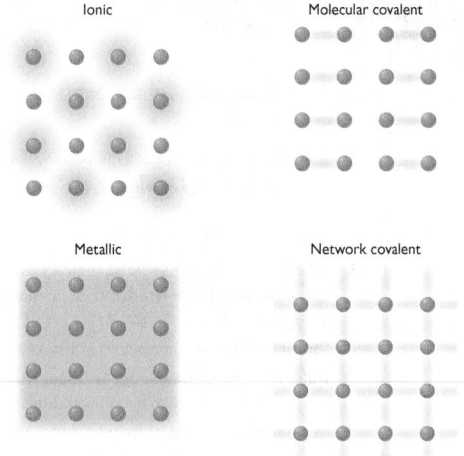

Red spheres represent nuclei and core electrons. Blue areas represent bonding electrons.

ChemCatalyst

1. If you place an open perfume bottle and a piece of paper in a sunny window, the aroma of the perfume will soon fill the air, but you will not smell the paper at all. Explain what is going on.

2. What is the heat from the Sun doing to the perfume to increase the smell?

Sample answers: **1.** Students should relate heating to phase change and offer explanations based on the movement of molecules and molecules becoming airborne. Students might say that to smell a substance, the substance must become a gas. **2.** The heat from the Sun provides enough energy to overcome the intermolecular forces that are keeping the perfume in liquid form.

Ask: Why can you smell a solid substance, like a cookie, when it is being cooked in the oven?

Ask: If a plastic bag, a piece of paper, or a book warms up in the Sun, will you suddenly be able to smell it? Why or why not?

EXPLORE (15 min)

TRM Worksheet with Answers 46

TRM Worksheet 46

Recall that all the substances you smelled in this unit were molecules with nonmetal atoms. This evidence suggests that substances that have a smell fit into the molecular covalent category.

Chemical Bonding and Smell

Substance type and bonding	Smell?	Phase	Examples	
			Name	Formula
ionic metals bonded to nonmetals	no	solid	sodium chloride (table salt)	NaCl
			calcium oxide (lime)	CaO
			calcium carbonate (chalk)	CaCO₃
molecular covalent nonmetals bonded in molecules	yes, with some exceptions	gas, liquid, or solid	nitrogen	N₂
			ammonia	NH₃
			menthol	C₁₀H₁₈O
metallic elemental metals	no	solid	gold	Au
			copper	Cu
			aluminum	Al
network covalent nonmetals bonded in a network	no	solid	carbon (diamond)	C
			silicon dioxide	SiO₂

Molecular covalent substances exist as gases, liquids, or solids at room temperature. Substances in the other three bonding categories are strictly solids (with a few exceptions, such as liquid mercury). It is reasonable to propose that substances that are ionic, metallic, or network covalent do not have a smell because they do not form gases easily at room temperature. This means that they cannot get into the receptor sites in our noses.

Big Idea Only molecular covalent substances form gases at room temperature.

Example 1

Bonding and Smell

Predict whether you would be able to smell the following substances. Explain your reasoning.

    a. propanol, C₃H₈O   **b.** iron, Fe   **c.** copper sulfate, CuSO₄

Solution

    a. Propanol consists of nonmetal atoms, so it is a molecular covalent substance. It probably has a smell.

    b. Iron is a metallic solid. It does not have a smell.

    c. Copper sulfate consists of both metal and nonmetal atoms, so it is an ionic solid. It does not have a smell.

Lesson 46 | **Phase, Size, Polarity, and Smell**   235

INTRODUCE THE CLASSWORK

→ Arrange students into pairs.

→ Pass out the student worksheet.

→ Point out that the range for molecular size is small: fewer than 5 carbon atoms; medium: 5 to 19 carbon atoms; large: 20 or more carbon atoms.

EXPLAIN & ELABORATE (20 min)

TRM Transparency—Three Phases 46

PROPOSE A COMPREHENSIVE MODEL FOR SMELL

→ There are many good ways to summarize this information. You could write the headings "polarity," "phase," and "molecular size" on the board and write student generalizations beneath the appropriate headings.

Ask: When determining smell, when is it important to consider the polarity of a molecule? (when small molecules are being considered)

Ask: What is a medium-sized molecule according to the data table? (a molecule with around eight to ten carbon atoms)

Ask: What did you discover about phase and molecular size?

Ask: Now that you have seen all of this information, what determines whether a substance smells?

Key Points: Phase and molecular size both play a role in smell properties. Medium-sized molecules all seem to have a smell, and they are liquids or gases. Their smell is determined by shape and functional group. Large molecules do not have a smell. They are too big and bulky to become gases and move into the nose.

Polarity determines the smell of small molecules. Small polar molecules have a smell. Small nonpolar molecules do not have a smell.

Many solids do not evaporate into gases and therefore do not have a smell. Solids tend not to smell unless they can become gaseous (e.g., components of a chocolate bar). Nonmolecular solids (ionic and metallic solids) do not have a smell.

RELATE PHASE, BONDING, AND SMELL

→ Display the Transparency—Three Phases showing methane, octane, and polystyrene to assist you in guiding the discussion.

Ask: What does phase have to do with the process of smelling? (The substance has to be in the gas phase.)

Ask: How is the type of bonding related to phase? (Mainly molecular substances vaporize at room temperature.)

Ask: How is the size of a molecule related to its phase? (As molecules get bigger, they tend to be solids at room temperature.)

Ask: What are the attributes of molecules that have a smell? (Fewer than about 20 carbon atoms; mainly molecules with carbon; small molecules must be polar.)

Ask: Why has our study of smells focused on molecules? (Ionic, network covalent, and metallic solids generally are not gases at 25 °C, and thus, they have no smell because they cannot enter the nose.)

Key Points: Molecular substances tend to have a smell because it is easy for them to become airborne. A substance has to be airborne to get into the nostrils and be detected by the nose. If a substance is a gas at 25 °C, it probably is composed of molecules. Methane is a good example of a molecule that is a gas at 25 °C. However, methane does not have a smell because it is not a polar molecule. Note that a substance that smells, such as dimethyl sulfide or butanethiol, is usually added to methane to help to detect leaks.

H—C—H (with H above and below the C)

Methane, CH$_4$(g)

Substances that are liquids at ordinary temperatures tend to have a smell. Gasoline is an example of a medium-sized molecule that is a liquid and vaporizes easily and, therefore, can be smelled. Even though gasoline is nonpolar, it does have a smell. Most molecules with 5 to 20 carbon atoms are liquids at 25 °C and have a smell.

H—C—C—C—C—C—C—C—C—H (chain of 8 carbons each with H above and below)

Gasoline, C$_8$H$_{18}$(l)

Molecular solids are volatile and have a smell. Some solids sublime; that is, a few molecules go directly from the solid to the gas phase without passing through the liquid phase. These substances consist of molecules, and they do have a smell. This is how you are able to smell a piece of chocolate. Ionic, network, and metallic substances do not smell. They do not consist of molecules, so there are no small units that can enter the gas phase readily. Polystyrene is an example of a solid that does not have a smell. You can think of it as consisting of long-chain molecules or as a network covalent solid.

Polystyrene, (C$_8$H$_8$)$_n$(s)

UPDATE THE SMELLS SUMMARY CHART

➜ Add the new information to the existing Smells Summary Chart.

➔ Molecular Size and Smell

In the previous lesson, you learned that small molecules have a smell only if they are polar. What about medium-sized and large molecules? Take a moment to examine the data in this table.

Molecular Size, Polarity, and Smell

Molecular size and polarity	Smell?	Phase	Examples	
			Name	Formula
small nonpolar molecules	no	gas	nitrogen	N$_2$
			carbon dioxide	CO$_2$
			methane	CH$_4$
small polar molecules	yes	gas	hydrogen sulfide	H$_2$S
			ammonia	NH$_3$
			fluoromethane	CH$_3$F
medium-sized polar and nonpolar molecules	yes	liquid	octane	C$_8$H$_{18}$
			geraniol	C$_{10}$H$_{18}$O
			carvone	C$_{10}$H$_{14}$O
			pentylpropionate	C$_8$H$_{16}$O$_2$
large polar and nonpolar molecules	no	solid	1-triacontyl palmitate (beeswax)	C$_{46}$H$_{92}$O$_2$
			polystyrene	C$_{8000}$H$_{8000}$
			cellulose	C$_6$H$_{10}$O$_5$

Important to Know Some substances smell only because they are contaminated with small molecules. For example, plastic itself does not have a smell, but it is often contaminated with small molecules that are used to make the plastic.

Many pieces of information can be gathered from the table. First, small molecules are gases at room temperature, medium-sized molecules are liquids, and large molecules are solids. One reason for this sorting by size is that large molecules tend to have more intermolecular attractions. This means that larger molecules are less likely to be gaseous and are less likely to get to your nose.

Note that a molecule is considered medium-sized if it has about 5 to 19 carbon atoms and large if it has 20 or more carbon atoms. Carbon is a unique element because it bonds easily to other carbon atoms and forms four bonds. This makes it ideal for forming long and complex molecules.

In general, all medium-sized molecules have a smell, whether they are polar or nonpolar. The medium-sized molecules form a liquid because they are attracted to one another, but they are volatile, meaning they can go easily into the gas phase. Once these molecules get inside the nose, the intermolecular attractions between the molecules and the receptor sites allow you to detect a smell.

HEATING AND COOLING

What happens when you warm or cool molecular substances? Small molecules are already gases at room temperature. But larger molecules are liquids or solids. These molecules must undergo a phase change in order to be smelled. Warming

BIOLOGY CONNECTION

The grassy odor that is emitted when the lawn is mowed is cis-3-hexenal. This molecule is unstable and tends to rearrange to form trans-2-hexenal. Scientific evidence connects trans-2-hexenal with healing from stress.

© Jim Craigmyle/Corbis

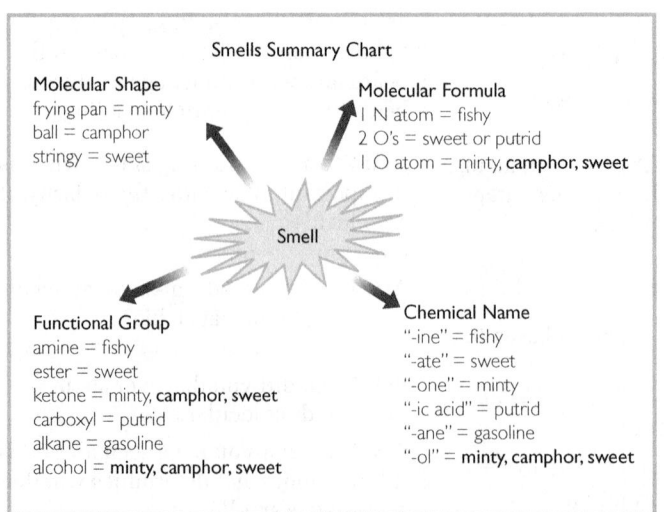

Smells Summary Chart

Molecular Shape
frying pan = minty
ball = camphor
stringy = sweet

Molecular Formula
I N atom = fishy
2 O's = sweet or putrid
I O atom = minty, **camphor, sweet**

Smell

Functional Group
amine = fishy
ester = sweet
ketone = minty, **camphor, sweet**
carboxyl = putrid
alkane = gasoline
alcohol = **minty, camphor, sweet**

Chemical Name
"-ine" = fishy
"-ate" = sweet
"-one" = minty
"-ic acid" = putrid
"-ane" = gasoline
"-ol" = **minty, camphor, sweet**

a molecular liquid or solid will generally give it a more intense aroma. This is because heating a substance increases the rate at which a substance becomes a gas. Cooling a substance has the opposite effect.

When there are more molecules in the air, you are more likely to detect them. So the temperature of a substance will usually affect the strength of its smell. For example, you are more likely to smell warm food than cold leftovers from the fridge.

Example 2

Predict the Smell

Below is the structural formula of a substance called 2-acetylthiazole. Predict whether this substance will have a smell at room temperature. Give evidence to support your answer.

2-acetylthiazole

Solution

This molecule's formula is C_5H_5NSO. It has five carbon atoms. This is a medium-sized molecule, so it probably has some sort of smell. (In fact, it does. It is described as having a hazelnut or popcorn smell.)

LESSON SUMMARY

What generalizations can you make about smell and molecules?

Humans can smell some molecular substances. We cannot smell ionic, metallic, or network covalent substances. This is because the first requirement for a substance to be smelled is that it be in a gaseous form. The size of the molecule seems to be another important factor in predicting smell. Most medium-sized molecules will smell. Most large-sized molecules will not smell. Much of this is due to the effect of size on phase change. For small molecules, polarity also plays a role. Small nonpolar molecules do not have a smell. Small polar molecules do have a smell.

BIOLOGY CONNECTION

There is scientific evidence that some fish are better at smelling and tasting underwater than dogs are, out in the air. This explains why fish will often turn away from or avoid a fisherman's lure. It is said that a salmon's sense of smell can detect when a person puts their hand in the water 100 ft upstream.

EVALUATE (5 min)

TRM Transparency—Check-In 46

Check-In

Which of these will have a smell? Explain your reasoning.

Substance	Structure	Phase
CaCl₂, calcium chloride	Cl Ca²⁺ Cl⁻ (repeating throughout the solid in three dimensions)	solid
C₈H₈O₃ vanillin	*(structural formula)*	liquid
HCN, hydrogen cyanide	H—C≡N	gas

Answer: Calcium chloride will not have a smell. It is an ionic compound and is also a solid. Vanillin probably will have a smell. It is a medium-sized molecule (that smells like vanilla). Hydrogen cyanide also has a smell. It is a small polar molecule.

Homework

Assign the reading and exercises for Smells Lesson 46 and the Chapter 46 Summary in the student text.

→ Create a new, "No Smell" summary chart.

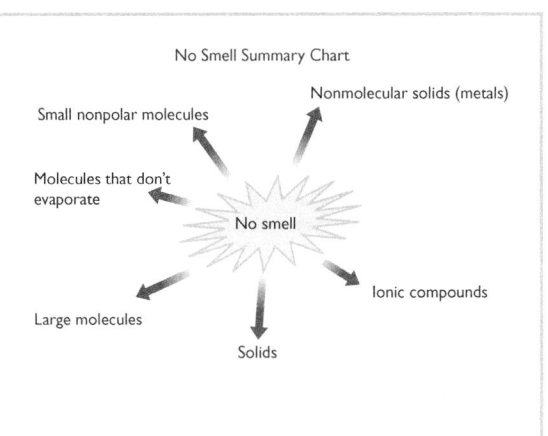

No Smell Summary Chart

Small nonpolar molecules

Nonmolecular solids (metals)

Molecules that don't evaporate

No smell

Large molecules

Solids

Ionic compounds

LESSON 46 ANSWERS

1. *Possible answer:* Substances that are most likely to have a smell are molecular covalent compounds that form small to medium-sized polar molecules. Small and medium-sized molecules are most likely to travel to the nose because they are more likely to become gases. Polar compounds are more likely to have a smell because they interact better with receptor sites in the nose.

2. *Possible answer:* Substances that do not have a smell include those that are made up of small nonpolar molecules, very large molecules, metals, and compounds that have ionic bonds. Nonpolar molecules are not attracted to receptor sites in the nose, and large molecules do not enter the gas phase so they do not get into the nose. The atoms of metals and ionic compounds are held in place by strong bonds, so they do not enter the gas phase.

3. Decanol has a smell because it is a medium-sized molecular compound and it is polar. Lead does not have a smell because it is a metal. Iron oxide does not have a smell because it is an ionic compound. Potassium chloride does not have a smell because it is an ionic compound.

4. Neon does not have a smell because it is a single atom and, therefore, is not polar. Ethane does not have a smell because it is a small, nonpolar molecule. The polar bonds are arranged symmetrically around the carbon atoms. Decane does have a smell because it is a medium-sized molecule. The molecule is nonpolar, but medium-sized nonpolar molecules still have a smell. Also, the molecule is an alkane because of its *-ane* ending, which indicates that it has a gasoline smell. Methylamine has a smell because it is a small, polar molecule. Methylamine is an amine with a fishy smell from the suffix and from the presence of nitrogen.

5. *Possible answer:* If a substance can become a gas under normal conditions, you should be able to smell it as long as receptor sites in the nose can detect it. In general, small nonpolar molecules would be odorless.

6. Molecules containing more than about 20 carbon atoms do not have a smell because they do not easily enter the gas phase, so they do not reach the receptors in the nose.

Exercises

Reading Questions

1. Explain what types of substances you would expect to have a smell and why.
2. Explain what types of substances you would expect *not* to have a smell and why.

Reason and Apply

3. Predict whether you would be able to smell these substances. Explain your reasoning.

   decanol, $C_{10}H_{22}O$ lead, Pb   iron oxide, Fe_2O_3 potassium chloride, KCl

4. Predict whether you would be able to smell these substances. Explain your reasoning.

   neon, Ne   ethane, C_2H_6 decane, $C_{10}H_{22}$ methylamine, CH_5N

5. If a substance is capable of becoming a gas under normal conditions, should you be able to smell it? Explain.

6. Molecules that have a structure containing more than about 20 carbon atoms rarely smell. Explain why this is so.

7. Your cotton T-shirt smells when you take it out of the clothes dryer. Explain why.

8. Occasionally you might go into a restaurant or a room that smells and after a while you stop noticing the odor. Explain what you think is going on.

9. From a biological point of view, why do you think it is useful for humans to be able to smell certain substances and not others?

7. *Possible answer:* When the T-shirt comes out of the clothes dryer, it is so warm that molecules from the detergent and fabric softener are likely to have changed phase and become gases. When they travel to your nose, you are able to smell them. The cotton itself is not producing a smell, even though it is warmed.

8. *Possible answer:* When you first enter the room, molecules in the air enter your nose and attach to receptor sites. After you have been in the room awhile, all of the receptor sites are filled and you no longer smell the molecules in the air.

9. *Possible answer:* It is useful to be able to smell substances that humans need to find, such as food, or that they need to avoid, such as toxic substances. It is not useful to smell such common materials as air and water, because they are constantly around us.

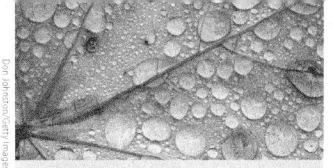

CHAPTER 8

Molecules in Action

SUMMARY

KEY TERMS

polar molecule

nonpolar molecule

partial charge

intermolecular force

electronegativity

dipole

diatomic molecule

Smells Update

Some substances, like hot apple cider, have a smell. Other substances, like a granite tabletop, do not. The process of smelling is an interaction between molecules. To be smelled, a substance must become airborne and enter the nose. Once it is in the nose, it must dissolve and be detected. Receptor sites are composed of large, folded protein molecules with a polar area that attracts polar molecules.

Summary of Investigation into Smell

	Has a smell	Does not have a smell
Bonding in compound	molecular covalent	ionic, metallic, network covalent
Size of compound	medium-sized molecules and small polar molecules (see polarity)	large molecules
Polarity of compound	small polar molecules	small nonpolar molecules
Phase of compound	gas phase	no molecules in the gas phase

REVIEW EXERCISES

1. Which element is more electronegative?
   a. Zn or Br   b. Li or Cs   c. Au or Al

2. Place the following pairs of atoms in order of increasing bond polarity. Explain your reasoning.
   H–I   H–Cl   H–F

3. Draw the structural formula and place a partial positive and a partial negative charge on each atom in these molecules. Which molecules are polar? Explain your reasoning.
   a. HCl   b. CH_4 c. H_2O

4. At cold temperatures, hydrogen bromide, HBr, is a liquid.
   a. Draw the Lewis dot structure for HBr. Label the dipole.
   b. Would HBr be attracted to a charged wand? Would it form a round drop on waxed paper? Explain your reasoning.
   c. In HBr, what type of bond exists between H and Br?
   d. Do you expect HBr to have a smell? Explain.

5. Use what you know about molecules to explain why you can't smell perfume while the bottle is closed but can smell it once the bottle is opened.

6. Which can you smell better, cold bread or warm bread right out of the oven? Explain why.

attracted to the negatively charged wand. Polar molecules form round drops on the waxed paper because the opposite partial charges attract one another. **c.** The bond between hydrogen and bromine in HBr is a polar covalent bond, because the difference in electronegativity between the two atoms is $2.96 - 2.10 = 0.86$. **d.** I would expect hydrogen bromide to have a smell because it is a small polar molecule.

5. *Possible answer:* To be smelled, the molecule must reach your nose in the gas phase. When the bottle is open, some of the molecules enter the gas phase, and you are able to smell them.

6. You can smell the warm bread better because warm molecules are more likely to enter the gas phase and travel through the air to your nose.

Project: Sense of Smell Study

A good experiment and its associated write-up would include ● a title that describes the experiment ● a goal statement that describes the purpose ● a description of the subjects of the study ● a clear, reproducible procedure that involves gathering data in an objective manner ● a summary of the raw data and analysis of the data through tables and graphs ● a conclusion or conclusions that can be drawn from the data, as well as the limitations of and assumptions made for each conclusion. An exceptional write-up would include ● ideas for further study or inquiry.

ASSESSMENTS

Two multiple-choice assessments (versions A and B) and two short-answer assessments (versions A and B) are available for download for Chapter 8.

TRM Chapter 8 Assessment Answers

TRM Chapter 8 Assessments

ANSWERS TO CHAPTER 8 REVIEW EXERCISES

1. a. Br **b.** Li **c.** Au

2. In order of increasing polarity from left to right: H—I, H—Cl, H—F. The electronegativity of the halogen atoms decreases from the top to the bottom of the group. Therefore, the difference in electronegativity between hydrogen and halogen atoms, which is the bond polarity, increases from bottom to top.

3. a. $^{\delta+}H - Cl^{\delta-}$ **b.** $^{\delta+}H - C^{\delta-}_{} - H^{\delta+}$ with $H^{\delta+}$ above and $H^{\delta+}$ below

c. $^{\delta-}O$ with $H^{\delta+}$ and $H^{\delta+}$

HCl and H_2O are polar molecules because they are asymmetrical and have polar bonds. CH_4 is not a polar molecule because its polar bonds are symmetrically arranged around the carbon atom.

4. a. H:Br: **b.** Hydrogen bromide would be attracted to a charged wand and would form a round drop on waxed paper because it is a polar molecule. The positive end of the molecule is

Sense of Smell Study

PROJECT

Plan and conduct an experiment to compare how the sense of smell differs among different groups of people.

- **Create a goal statement.** Write a sentence or two stating the goal of your study. What do you hope to discover or explore?

- **Choose subjects to study.** Pick two categories of people to study and compare. These categories should be clear and easy to determine (for example, children under the age of 12 versus adults, vegetarians versus meat-eaters, women versus men).

- **Write a proposal.** Write several sentences stating how you propose to accomplish your goal, including how you will conduct your study, how you will randomly choose your sample of participants, and how you will set up the control variables. Clear your proposal with your teacher before conducting your study.

- **Conduct your study.** Keep your data organized in a table or chart. Keep careful notes of everything that you do and what you observe.

- **Write up your results and conclusions.** Write up the results of your study and any conclusions you have come to based on your collected data. Include your data table. If possible, create a graph using your data.

L-Carvone

D-Carvone

Archie McPhee and Company

CHAPTER 9

Molecules in the Body

In this chapter, you will study

- mirror-image isomers
- amino acid molecules
- protein molecules

These molecules are mirror images of each other and have different properties. D-Carvone produces the scent of dill, the other the fragrance of mint. The two unite in this unusual treat.

Compounds that you smell are sensed by receptor sites in your nose. Receptor sites are made of large molecules called proteins. The protein molecules consist of amino acid molecules that are linked together to form long chains. There are a great variety of protein molecules, made of various combinations of different amino acids. Protein molecules are essential to nearly all body functions, including the sense of smell.

241

PD CHAPTER 9 OVERVIEW

Watch the video overview of Chapter 9 (for teachers) by clicking on the link in the TE-book, opening the TRFD, or logging onto the book's companion Web site bcs.whfreeman.com/livingbychemistry2e (teacher log-in required).

The final section of the Smells unit focuses on biological molecules. Lesson 47 covers the "handedness" of molecules and challenges students to visually and structurally distinguish between mirror-image isomers. Lesson 48 introduces amino acids and shows students how to connect and fold them to form a protein. This lesson also revisits the receptor site, or "lock-and-key," mechanism that characterizes the sense of smell. The unit review summarizes the context of smell and all that has been learned about molecules.

In this chapter, students will learn

- how to identify mirror-image isomers
- how "handedness" affects molecular properties
- about the formation of proteins from amino acids
- about basic "lock-and-key" mechanisms in biological processes

Chapter 9 Lessons

Overview

ACTIVITY: GROUPS OF 4

Key Question: What are mirror-image isomers?

KEY IDEAS

Mirror-image isomers are a pair of molecules that differ only in that they are the mirror image of each other. These molecules are said to have a "handedness" because they differ from one another the way a right hand and left hand differ. Mirror-image isomers can each have different properties. For example, they can have different smells.

LEARNING OBJECTIVES

● Distinguish mirror-image molecular structures.
● Explain what it means for molecules to be superimposable.
● Explain why mirror-image isomers have different properties.

FOCUS ON UNDERSTANDING

● It is sometimes difficult to visualize how mirror-image isomers differ from each other. Demonstrate with models as much as possible.
● Students might need help in understanding what it means that two molecules are superimposable. Show them that your hands are not superimposable. No matter how you turn and twist your left hand, it will never have the same orientation as your right hand.
● Be aware that some individuals cannot detect the difference in smell between D-carvone and L-carvone. This "teaching moment" is an opportunity to point out that there often are exceptions to the rule. These exceptions do not necessarily invalidate the hypothesis, but they do indicate that there is more to learn about smell receptors.

KEY TERM

mirror-image isomer

IN CLASS

Students make models of several tetrahedral structures and view their images in a mirror. Students then build the mirror-image model of each molecule and explore whether the pair of molecules are superimposable on each other. They compare the smell of L-carvone (which they smelled in

---

Mirror-Image Isomers

THINK ABOUT IT

Limonene is a molecule that comes in two forms. The structural formulas for both forms are identical. Yet one smells like pine needles, while the other has the lemon-orange fragrance of citrus fruits. How can two molecules with the same molecular formula and the same structural formula have such distinct smells?

What are mirror-image isomers?

To answer this question, you will explore

↪ Mirror Images of Molecules
↪ Properties of Mirror-Image Isomers

EXPLORING THE TOPIC

↪ Mirror Images of Molecules

Many things in nature are mirror images of each other: your ears, for example, or your eyebrows. While your ears may look physically identical, your left ear would not work well on the right side of your head. Your hands and feet are other good examples. Consider your left and right hands. They are made of the same parts and have the same overall structures. But if you try to place your left hand on your right hand with your palms down, the thumbs will point in opposite directions.

Your hands are mirror images of each other. That is, when you observe your left hand in a mirror, it will look identical to your right hand and vice versa. However, you cannot *superimpose* them on each other. No matter how you position your left hand, it is never identical to your right hand.

MOLECULES WITH A HANDEDNESS

Just as your right hand cannot be superimposed on your left hand, some molecules have a "handedness."

The two limonene molecules described earlier have identical structures that are mirror images. The mirror images are not superimposable on one another. They are actually different molecules. You could say that one molecule is left-handed and the other is right-handed. They are isomers of one another.

Determining if two molecules are mirror images or the same molecule can be visually challenging. A three-dimensional model of the molecule and a mirror can help.

242   Chapter 9 | **Molecules in the Body**

Lesson 28) with that of D-carvone and explain the nature of the mirror images of these molecules. A complete materials list is available for download.

TRM Materials List 47

SETUP

Vial A from Lesson 28 contains L-carvone. Prepare vial Z by first placing a cotton ball in the vial. Use a plastic pipette to deliver three to five drops of caraway seed oil. Ground-up or crushed caraway seeds can be substituted if they have a strong enough smell. Place sets of the two

vials in plastic sandwich bags for ease of distribution.

Build one or more sets of space-filling models of L-carvone and D-carvone with the molecular model sets for students to examine as they are working on the worksheet. Use the disk connectors to represent bonds (double disks for double bonds) and white hemispheres for hydrogen. Put a tag on each with the correct name. Ball-and-stick models and structural formulas are shown here for reference. You can build one isomer and use a mirror to construct the other.

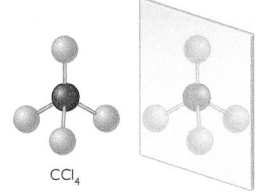

CCl₄

Consider the mirror image of a simple tetrahedral molecule, tetrachloromethane, CCl_4. It is fairly easy to see that the mirror image can be superimposed on the original molecule. (In other words, you can imagine picking it up and turning or sliding it so that it matches the original.) So, tetrachloromethane does not have a mirror-image isomer. The molecule and its mirror image are identical.

Now consider dichloromethane, CCl_2H_2, and chlorofluoromethanol, $CHClFOH$. Take a moment to decide if either of the mirror images can be superimposed on its original molecule.

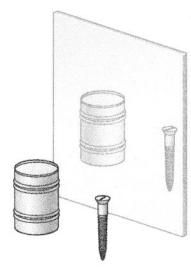

A can is identical to its mirror image, but a screw is not. The screw's thread spirals in the opposite direction in the mirror image.

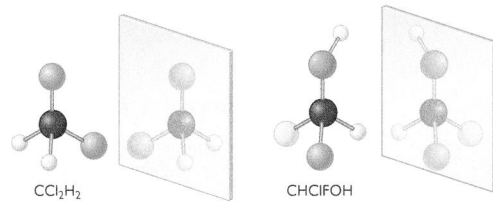

CCl₂H₂   CHClFOH

The mirror image of dichloromethane, CCl_2H_2, can be superimposed on the original molecule. However, when there are four different groups attached to a carbon atom as in the second molecule, the mirror image cannot be superimposed on the original molecule.

The two molecules of chlorofluoromethanol are **mirror-image isomers.** They have different properties.

Example

Objects in a Mirror

Which of the objects listed here look identical in a mirror? Which would look different? Explain any differences.

a. a can

b. a shoe

c. fluoromethane, CH_3F

d. 2-chlorobutane, C_4H_9Cl

© D4Fish/iStockphoto (left). Pritmova Svetana/Shutterstock (right)

CLEANUP

Save the vials for reuse in other classes. When you are finished using all the vials in all of your classes, remove the cotton balls from the vials, place the cotton balls in a plastic bag, and dispose of them. Let the vials air out in a hood or rinse them with acetone for reuse the next time you do this unit. If you do not remove the cotton balls, the substances will mix with each other and begin to smell putrid.

Differentiate

This lesson provides an opportunity for students to build literacy and learn about careers in chemistry such as chemical engineering or pharmaceutical chemistry. Have students research and write a report on topics related to this lesson, such as medicines that exist as a single mirror-image isomer or a mixture of mirror-image isomers. If possible, have students create molecular models of the isomers they research.

Lesson Guide

ENGAGE (5 min)

TRM PowerPoint Presentation 47

> *ChemCatalyst*
>
> Which of these objects look identical in a mirror? Which look different? Explain any differences.
>
> **1.** glove   **2.** barbell
>
> **3.** spring   **4.** tetrahedron

Sample answers: **1.** A glove does not look identical in a mirror. When both thumbs point in the same direction, the palms are opposite. **2.** A barbell looks identical in a mirror. Both "sides" are the same (but any writing will be reversed). **3.** A spring does not look identical in a mirror. The mirror image of a spring has the opposite rotation. If one twists clockwise, the other twists counterclockwise. **4.** Some students might be uncertain about a tetrahedron. Accept their guesses. They will have an opportunity to test their ideas in the activity.

DISCUSS THE CHEMCATALYST

→ Assist students in describing mirror images.

Ask: What other objects look different from their mirror image? (feet, ears, shoes, car doors, side-view mirrors, baseball mitts, golf clubs)

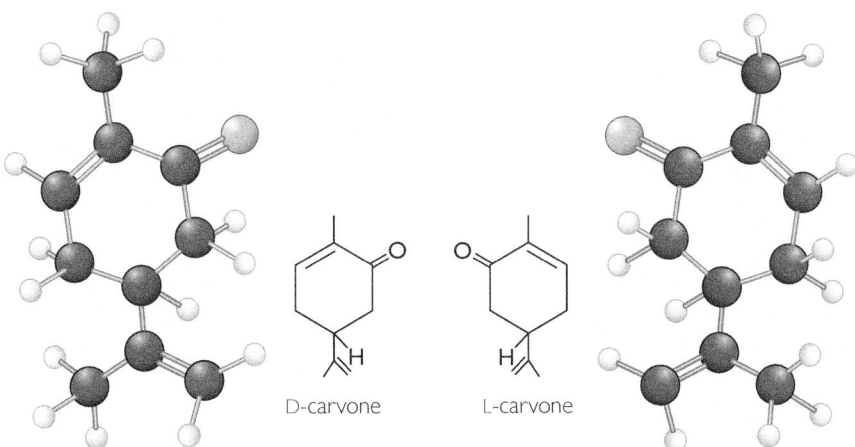

D-carvone   L-carvone

Ask: What objects do not look different from their mirror images? (symmetrical objects such as cowboy hats and rearview mirrors)

Ask: What does it mean that something has "handedness"?

Ask: If a molecule and its mirror image are different, do you think the two molecules will have the same smell? Why or why not?

EXPLORE (20 min)

TRM Worksheet with Answers 47

TRM Worksheet 47

INTRODUCE THE ACTIVITY

→ Arrange students into groups of four.

→ Pass out the student worksheets, mirrors, and molecular-model kits.

→ Introduce the concept of handedness through a few quick demonstrations.

Hide both hands behind a binder or a large piece of paper. Extend one hand. Ask students to identify it as left or right. Ask them how they could tell. Ask students what a receptor site might look like for each hand and how the sites might be different. (A glove would be a receptor-site metaphor.)

GUIDE THE ACTIVITY

→ Suggest that students view the mirror from behind the model, not from the side. It will be more obvious if the images are identical.

→ Be sure students clearly understand that superimposable means you can place two objects on top of each other so that they look identical in every way. You cannot superimpose your left hand on your right hand. Either one hand will be palm up and the other palm down, or your thumbs will be pointing in opposite directions.

EXPLAIN & ELABORATE (15 min)

DISCUSS MIRROR-IMAGE MOLECULES

→ Have space-filling models of the molecules from the worksheet available to demonstrate the concepts. A large mirror is also helpful.

Ask: How can you determine whether a molecule has a mirror-image isomer?

Ask: How does the mirror help you with the activity? Explain what it tells you.

Ask: What did you discover when you made models of the mirror images?

Example *(continued)*

Solution

A and C look identical in a mirror. B and D do not.

In the 2-chlorobutane molecule, the second carbon atom from the left has four different groups attached to it: CH_3, H, Cl, and CH_2CH_3. Therefore, this molecule has a mirror-image isomer. When the H and Cl atoms of the two mirror-image isomers are in the same location, the CH_3 and CH_2CH_3 groups are in opposite locations.

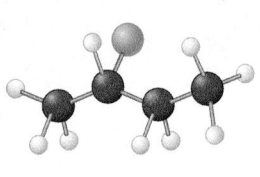

Properties of Mirror-Image Isomers

One way to figure out if a molecule has a mirror-image isomer is to check to see if any of the carbon atoms are attached to four different groups.

A structural formula for limonene, $C_{10}H_{16}$, is shown below (left). The illustration below (right) shows that nine of the carbon atoms in a limonene molecule are not attached to four *different* groups.

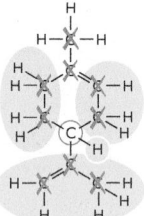

The remaining carbon atom, circled in the diagram, is attached to a hydrogen atom and three carbon "groups." The two sides of the carbon ring are different from one another and considered different groups. You can conclude that limonene has a mirror-image isomer because it contains a carbon atom attached to four different groups of atoms. The two isomers are referred to as D-limonene and L-limonene.

The mirror-image isomers of limonene are shown on top of the next page. The main difference between them has to do with the position of the CCH_2CH_3 group highlighted in yellow. If you flip the L-limonene molecule around and try to superimpose it onto the D-limonene molecule, the CCH_2CH_3 group will point toward you in one isomer and away from you in the other isomer. It is as if you have aligned your thumbs on your right and left hands, only to find one palm down and the other palm up.

244 Chapter 9 | **Molecules in the Body**

Ask: Look at the two carvone molecules on the worksheet. In each, the carbon atom at the bottom of the carbon ring is attached to four different "things." What happens to these four things in the mirror image?

Key Points: Mirror-image molecules that cannot be superimposed on each other are called mirror-image isomers. Your left hand and your right hand are mirror images of each other. They cannot be superimposed on each other. Similarly, CHFClBr and C(CH₃)HFCl have mirror-image isomers. The mirror images are not superimposable; thus, they are distinct molecules. Molecules also can have mirror images that can

be superimposed on each other. These are not considered mirror-image isomers because they are both identical. Methane, CH_4, and fluoro-methane, CH_3F, are examples.

Tetrahedral molecules in which four different atoms or groups are attached to a carbon atom always have mirror-image isomers. Indeed, any molecule that contains even one carbon atom with four different atoms or groups attached to it will have a mirror-image isomer. The carbon atom at the bottom of the carbon ring in the carvone molecule on the worksheet is an example.

$$H \quad H$$
$$\backslash \quad /$$
$$C = C$$
$$/ \quad \backslash$$
$$CH_3 \quad CH_3$$

Cis isomer

$$H \quad CH_3$$
$$\backslash \quad /$$
$$C = C$$
$$/ \quad \backslash$$
$$CH_3 \quad H$$

Trans isomer

KEY TERM

mirror-image isomer

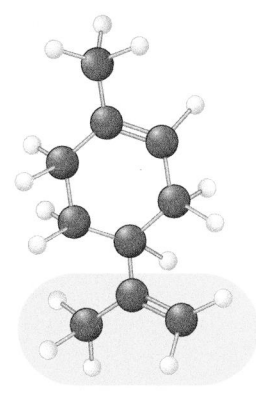

D-limonene

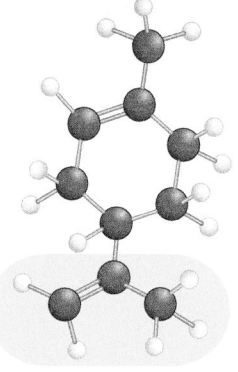

L-limonene

DIFFERENT SMELLS

Just as your left foot does not fit into your right footprint, D-limonene does not fit into the smell receptor site for L-limonene.

Because the mirror-image isomers do not fit into the same receptor site, you would expect them to have distinct smells. Indeed, this is the case: D-limonene smells like citrus while L-limonene smells like pine. The two molecules have the same molecular formula, the same structural formula, and the same shape, but they are mirror images of one another.

LESSON SUMMARY

What are mirror-image isomers?

Nature is full of mirror images that are not identical. Some objects and some molecules can be superimposed on their mirror images. For other objects, the mirror image is not superimposable. The mirror image cannot be positioned such that it is identical to the original object or molecule. Molecules with mirror images that cannot be superimposed are called mirror-image isomers. These molecules fit into different smell receptor sites and therefore have different smells.

Exercises

Reading Questions

1. What do you have to do to the mirror image of the letter "D" to superimpose the mirror image onto the original letter?

2. What types of molecules have mirror-image isomers?

DISCUSS PROPERTIES OF MIRROR-IMAGE ISOMERS

Ask: What did you conclude about D-carvone and L-carvone?

Ask: How are D-carvone and L-carvone similar? How are they different?

Ask: Do you expect D-carvone and L-carvone to fit differently into receptor sites? Why or why not?

Key Points: The mirror-image isomers D-carvone and L-carvone have different smells. Thus, they probably do not fit into the same receptor site. Just as a left hand does not fit into a right-hand glove, mirror-image isomers do not fit into the same lock-and-key sites. This

means that they can have different properties, including smell.

The mirror-image isomers have a "handedness." Human hands are symmetrical but different from each other. No matter how hard you try, you cannot superimpose a right hand onto a left hand so that the two are identical. The thumbs will always be on opposite sides, or the palms will be on opposite sides. Mirror-image isomers are molecules that have a handedness. Just as your right hand and left hand require different gloves, mirror-image isomers may have different receptor sites.

Check-In

1. Which of these molecules will have a mirror-image isomer? Explain your reasoning.

   a. CF_4

   b. CHF_3

   c. $C(CH_3)_4$

Answer: None of these molecules will have a mirror-image isomer. These molecules are superimposable on their mirror images, and the mirror images represent the same molecule. They do not have a carbon atom with four different things attached to it.

Homework

Assign the reading and exercises for Smells Lesson 47 in the student text.

LESSON 47 ANSWERS

1. The mirror image of the letter D can be superimposed on the original as long as the mirror is placed horizontally with respect to the letter. Then the mirror image and the original image are identical. The letter D has one line of symmetry—a horizontal line halfway between the top and the bottom of the letter.

2. *Possible answer:* Molecules that have a carbon atom that is attached to four different atoms or groups of atoms will have mirror-image isomers. Other types of molecules have mirror-image isomers. This example is the one mentioned in the textbook.

3. *Possible answers (any 5 of each):*
● Objects that look different in a mirror include a glove, a written sentence, a pair of scissors, a clock face, a globe of Earth, and a book. ● Objects that look the same in a mirror include a blank cube, a table, a spoon, an empty glass, a pencil with no writing on it, and a wrapped package.

4. *Possible answer:*

A B C D E F G H I J K L M N

O P Q R S T U V W X Y Z

Objects that are symmetrical do not have mirror-image isomers because flipping the object about its center, which is how a mirror image is formed, does not change the object.

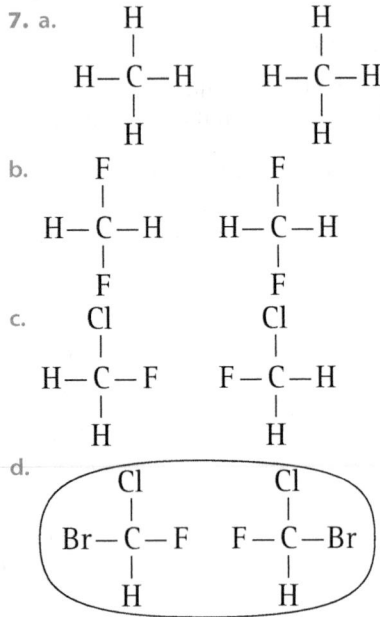

LESSON 47

5. *Possible answer:*

n m l k j i h g f e d c b a
z y x w v u t s r q p o

The letters b and d are mirror images of one another, as are the letters p and q. Answers will vary depending on handwriting styles

6.

$$\underset{H}{\overset{H}{\diagdown}}C{=}O \qquad \underset{Cl}{\overset{H}{\diagdown}}C{=}O$$

CH_2O does not have a mirror-image isomer because it is the same on both sides of the carbon-oxygen bond. ClHCO does not have a mirror-image isomer because its mirror image can be rotated in three dimensions around the carbon-oxygen bond such that it can be superimposed on the original.

7. a.

H—C—H (with H top and bottom)   H—C—H (with H top and bottom)

b.

H—C—H (with F top and bottom)   H—C—H (with F top and bottom)

c.

H—C—F (with Cl top, H bottom)   F—C—H (with Cl top, H bottom)

d.

Br—C—F (with Cl top, H bottom)   F—C—Br (with Cl top, H bottom) *(circled)*

8. a. Butyric acid does not have mirror-image isomers because there are no carbon atoms that are attached to four different atoms or groups. **b.** Ethyl formate does not have mirror-image isomers because there are no carbon atoms that are attached to four different atoms or groups. **c.** 2-butanol does have mirror-image isomers because the second carbon atom from the left in the structural formula is bonded to four different groups.

9. a. citronellol: $C_{10}H_{20}O$, geraniol: $C_{10}H_{18}O$ **b.** Citronellol has two distinct smells because the structural formulas have mirror-image isomers that interact with different receptors in the nose. Citronellol has mirror-image isomers because the third carbon atom from the

right in the structural formula is bonded to four different groups. **c.** The geraniol molecule has only one smell because the molecule and its mirror image are identical. None of the carbon atoms in the structure is bonded to four different groups.

Exercises (continued)

Reason and Apply

3. List five objects that look different in a mirror and five that do not.

4. Draw mirror images of the capital letters in the English alphabet. Explain why symmetrical objects do not have mirror-image isomers.

5. Draw mirror images of the lowercase letters in the English alphabet. Which lowercase letters are mirror images of one another?

6. Draw the structural formulas for CH_2O and ClHCO. Explain why these two molecules do not have mirror-image isomers.

7. Draw structural formulas for the molecules listed below. Draw the mirror image of each. Circle the molecules with mirror-image isomers.
   a. CH_4
   b. CFH_3
   c. $CClFH_2$
   d. $CBrClFH$

8. Examine the molecules here. Which have mirror-image isomers? Explain your reasoning.
   a. butyric acid   b. ethyl formate   c. 2-butanol

9. Two structural formulas are shown.

   Citronellol

   Geraniol

   a. Write the molecular formulas for citronellol and geraniol.
   b. Molecules with the structural formula for citronellol can smell either like insect repellant or like flowers. Explain why there are two distinct smells.
   c. All molecules with the structural formula for geraniol smell like roses. Explain why there is only one smell for this molecule.

246   Chapter 9 | **Molecules in the Body**

Amino Acids and Proteins

THINK ABOUT IT

Smelly molecules are small or medium-sized molecules that travel as a gas into your nose. Once in your nose, these molecules interact with receptor sites, which send a signal to your brain. Receptor sites are also built of atoms bonded together to create a "pocket" into which a specific smell molecule can fit.

What is a receptor site made of?

To answer this question, you will explore

⟳ Amino Acids

⟳ Proteins

EXPLORING THE TOPIC

⟳ Amino Acids

Amino acids are the building materials for many of the structures in the body, especially those related to the functioning of cells. They are important molecules in all living organisms. As the name implies, amino acids have two functional groups: an amine group and a carboxyl group. The general formula for an amino acid is $C_2H_4NO_2R$, where R is simply a placeholder in the formula for a variety of possible groups of atoms. The groups consist of carbon and hydrogen, and sometimes nitrogen and/or oxygen.

The structure of an amino acid molecule

Here are structural formulas for three amino acids with the R groups highlighted. Notice that everything else about them is identical.

Glycine   Alanine   Valine

Lesson 48 | **Amino Acids and Proteins**   247

Overview

ACTIVITY: GROUPS OF 4

Key Question: What is a receptor site made of?

KEY IDEAS

Smell receptor sites are composed of proteins, which are made from a chain of amino acids. One end of an amino acid is an amine functional group, and the other end is a carboxyl group. The amine group from one amino acid links with the carboxyl group of another amino acid to form proteins. There are 20 amino acids used by the body. Of the two mirror-image isomers

of each amino acid, only the left-handed one is usable in the human body.

LEARNING OBJECTIVES

● Explain that protein molecules are chains of amino acid molecules.
● Describe how the smell receptor sites are protein chains folded to form a receptor of a specific shape.
● Explain the handedness of a smell receptor site.

FOCUS ON UNDERSTANDING

● It is useful to point out that receptor sites are not mirror images of molecules but rather "prints" of molecules themselves. For example,

your right foot is not a receptor site for your left foot. A left footprint is.
● A protein is a chain of more than 100 amino acids. A shorter chain typically is referred to as a polypeptide. However, for the sake of simplicity, we do not make a clear distinction in this lesson.

KEY TERMS

amino acid
peptide bond (amide bond)
protein

IN CLASS

Students examine the structures of amino acid molecules. They build the mirror-image isomer of the amino acid molecule alanine used by the body. Students then build two other amino acid molecules and link them to form a dipeptide. In the Explain & Elaborate discussion, the class examines a model of a protein with six linked amino acid molecules. The protein is folded to form a smell receptor site. A complete materials list is available for download.

TRM Materials List 48

SETUP

If you have enough modeling sets, you might build a few ball-and-stick models from the handout ahead of time. Some of these can be the same. All should be the left-handed mirror-image isomer. You can tell that you have the left-handed isomer by viewing the molecule with the H atom pointing toward you. The other three groups should be positioned with COOH, the R group, and NH_2 in clockwise order. You can remember this with the acronym CORN, where CO stands for the COOH group, R for the R group, and N for the NH_2 group. You can refer to the diagram on page 248 of the student edition that shows how to identify left-handed versus right-handed amino acids.

Here is an example of linking amino acids to form a protein molecule.

Peptide bond

$+ H_2O$

You will use this model after the activity to demonstrate how a protein molecule folds to form a receptor site. You can practice ahead of time twisting it so it coils like a helix and then into a fairly compact bundle with an indentation. If you do not have enough modeling materials for each group to build two amino-acid molecules, you can join pairs or groups to link their amino acids together for Question 3 on the worksheet. You can do the If You Finish Early at the front of the room, with groups linking their amino acids together to make a protein and fold it into a receptor site.

CLEANUP

Allow time for students to take apart the protein molecule in preparation for the next class.

Pre-AP® Course Tip

Students preparing to take an AP® chemistry or college chemistry course should be able to evaluate data and use the given information to address a certain question. In this lesson, students use molecular-model sets to build different mirror-image isomers and explain why a molecule is or is not superimposable on its mirror image. After analyzing different structural features of different isomers, students should be able to use these data to answer the question: Why might mirror-image isomers have different smells?

Lesson Guide

ENGAGE (5 min)

TRM PowerPoint Presentation 48

TRM Transparency—ChemCatalyst 48

ChemCatalyst

The mirror-image isomers of carvone are shown.

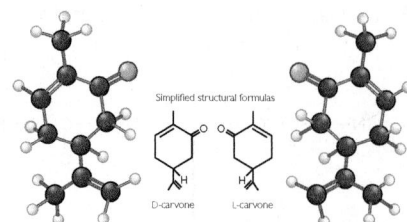

1. Explain how the receptor sites for D-carvone and L-carvone might be different from each other.

2. Sketch receptor sites for D-carvone and L-carvone. Assume that the polar side (the side with the oxygen atom) attaches to the receptor site.

Sample answers: **1.** Because the molecules are mirror images of each other, the receptor sites for these molecules probably are mirror images of each other as well. **2.** The receptor site probably is narrow, because the molecule enters with the flatter part perpendicular to the receptor site. For D-carvone, the bottom of the receptor site bends to the right to fit the C(CH₂)(CH₃) group. For L-carvone, the bottom bends to the left.

→ Ask one or two students to draw receptor sites for D-carvone and L-carvone on the board.

→ (Optional) Use a tray with sand or soft clay to make "molecule prints" of D-carvone and L-carvone. Show that each isomer fits only into its own molecule print, just as your right foot fits only into its right footprint.

Ask: What would receptor sites for your left hand and right hand look like?

You might notice that the carbon atom in the center of the amino acid molecule has four different groups attached to it. Therefore, amino acids have mirror-image isomers and can be left-handed or right-handed molecules. There's a trick to figuring out the handedness of an amino acid. The illustration shows you how.

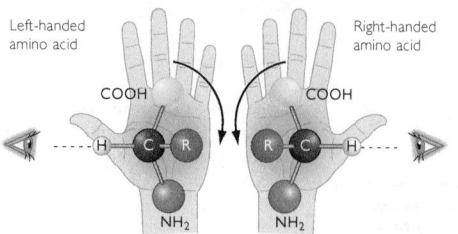

Left-handed amino acids: When you point the H atom toward your eye, the COOH group, the R group, and the NH₂ group are arranged clockwise. This is sometimes referred to as the CORN rule for COOH, R, and NH₂.

The human body uses only the left-handed amino acids, but both mirror images are available. How did life begin selecting only one of the two mirror images? This is a question of heated debate. Currently, no one has a good explanation.

There are 20 amino acids, with different R groups, that your body needs. Your body produces about half of these amino acids. You get the others as part of your diet. Six of the amino acids are listed in this illustration. These amino acids are placed into two categories: either hydrophobic ("water fearing") or hydrophilic ("water loving"). Take a moment to determine the difference.

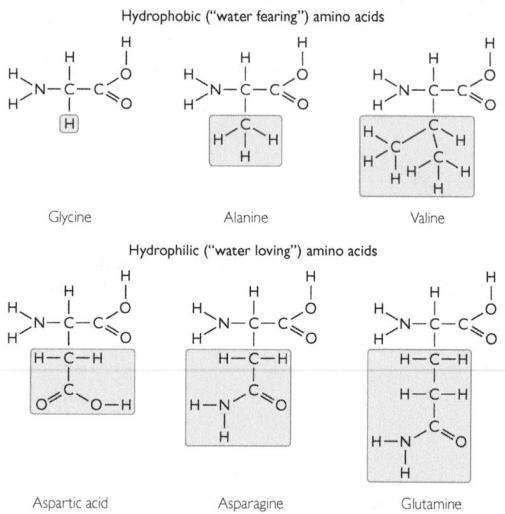

The R groups for the amino acids classified as hydrophobic contain only carbon and hydrogen atoms. These R groups are nonpolar. They are not attracted to water molecules. The R groups for the amino acids classified as hydrophilic contain atoms such as oxygen and nitrogen. These R groups are polar and are attracted to water molecules. As you will see, the nature of the R group is important when amino acids are linked to form proteins.

Proteins

As shown in the next illustration, the carboxyl group of one amino acid can link to the amine group on another amino acid. The bond between two amino acids is called an **amide bond**, or a **peptide bond**. In the process, two hydrogen atoms and an oxygen atom break off from the amino acids and form a water molecule. In this way, it is possible to form long chains of amino acids called **proteins.** Because they are very large molecules, proteins are often called *macromolecules.*

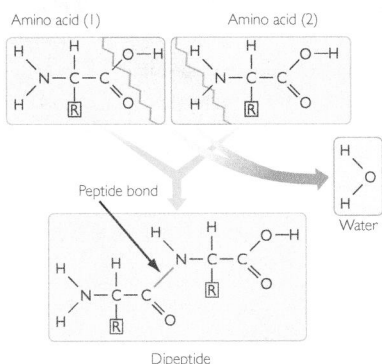

Dipeptide

Proteins are essential in all living organisms. Protein molecules send messages throughout your body, they help the immune system fight off infections, and they are building blocks of many parts of your body, including organs, hair, and muscles. Proteins have a wide variety of structures and functions because many different sequences of amino acids are possible.

As the chain of amino acids forms, it twists and bends. In the photograph, each bead represents an amino acid. The entire chain of beads represents a protein molecule. Notice that amino acids can repeat.

The R groups on each amino acid play a major role in determining the resulting structure of the protein. The hydrophobic R groups tend to cluster in the center of the molecule. The hydrophilic R groups tend to be on the outside of the molecule, near the watery environment.

Smell receptor sites are protein molecules that have a pocket that fits a particular smell molecule. When the smell molecule attaches to the pocket, it changes the structure of the protein just a bit. This small change is enough to affect a neighboring protein, which triggers a signal to the brain.

Lesson 48 | **Amino Acids and Proteins** 249

Ask: How is a handprint of your left hand different from one of your right hand?

Ask: What would you observe if you made molecule prints of D-carvone and L-carvone?

Ask: What might explain why some people cannot smell L-carvone at all?

EXPLORE (20 min)

TRM Worksheet with Answers 48

TRM Transparency—Amino Acids 48

TRM Worksheet 48

TRM Handout—Twenty Amino Acids 48

INTRODUCE THE ACTIVITY

➜ Arrange students into groups of four.

➜ Pass out the student worksheet, models, and the handout Twenty Amino Acids.

➜ Explain that amino acids are molecules the body uses to make proteins. Proteins are long chains of amino acids. The human body uses 20 amino acid molecules. Each has a mirror-image isomer. The human body can use only the left-handed mirror-image isomer.

> **Amino acid:** A molecule with a carboxyl functional group and an amine functional group.
>
> **Protein:** A large molecule consisting of amino acids bonded together.

➜ Explain R groups. Aside from the carboxyl functional group, the amine functional group, and the hydrogen atom, an amino acid has a fourth group that identifies it, called the R group. On the handout, the R groups are shown at the bottom of each molecule (everything below the second carbon atom).

➜ Explain how to tell that an amino acid is left-handed. You can tell that you have the left-handed isomer by viewing the molecule with the H atom pointing toward you. The other three groups should be positioned with COOH, the R group, and NH_2 in clockwise order. You can remember this with the acronym CORN, where CO stands for the COOH group, R for the R group, and N for the NH_2 group.

➜ Tell students that they will be working with amino acids to build protein molecules that fold into smell receptor sites.

EXPLAIN & ELABORATE (15 min)

DISCUSS AMINO ACIDS

Ask: What are the main features of amino acid molecules?

Ask: Why do amino acid molecules all have mirror-image isomers?

Ask: How do amino acid molecules link together?

Ask: How can you show that the HONC 1234 rule applies to a dipeptide?

Key Points: Amino acids are molecules with a carboxyl group, COOH, and an amine group, NH_2. There are hundreds of naturally occurring amino acids. Each amino acid has a mirror-image isomer because four different groups are attached to a central carbon atom. Your body uses 20 different amino acids, and it uses the left-handed isomers exclusively. Amino acids can link together to form long chains called *proteins*. The carboxyl group of one amino acid can link to the amine group of another amino acid with a peptide bond. Water is lost in the process. It is possible to make a wide variety of proteins by linking amino acids in different sequences. Your hair, your skin, enzymes, and smell receptor sites are all made of proteins with different sequences of amino acids.

DISCUSS FOLDING OF PROTEINS

→ Use your ball-and-stick models of about six amino acid molecules linked together. Demonstrate how this chain of amino acids can be twisted to form a smell receptor site.

Ask: Why do you think protein molecules tend to fold into globs rather than remain stretched out?

Ask: Why are amino acids in different parts of the protein attracted to one another?

Ask: Which amino acids are likely to be attracted to polar R groups?

Ask: How can protein molecules account for the variety of smell receptor sites in the nose?

Key Points: Protein molecules fold to form pockets because of attractions between the R groups on different amino acids in the chain. The polar R groups tend to attract other polar R groups. These hydrophilic (water-loving) R groups tend to be on the outside of the molecule, near the watery environment. The hydrophobic (water-hating), nonpolar R groups tend to cluster in the center of the molecule. The lining of the nose is made of protein molecules. Each kind of protein folds in a different way to create a smell receptor site for a different smell molecule. When a smell molecule attaches to the site, a nerve carries a signal from that site to the brain.

EVALUATE (5 min)

Check-In

Name two concepts from the Smells Unit that were used to help you understand proteins.

Answer: Possible answers are the HONC 1234 rule, structural formulas, polarity, attractions between polar molecules, and so on.

Homework

Assign the reading and exercises for Smells Lesson 48 and the Chapter 9 Review in the student text.

LESSON SUMMARY

What is a receptor site made of?

There are 20 different amino acid molecules that are essential building blocks of the human body. Amino acids have both a carboxyl functional group and an amine functional group. Amino acids all have mirror-image isomers, but only left-handed amino acid molecules are used by the human body. The carboxyl group on one amino acid can link with the amine group on another amino acid to form a peptide bond. In this way, a long chain of amino acids called a protein forms. A large variety of protein molecules exists because the amino acids can be linked in a variety of sequences. Some amino acids in a protein chain are polar and others are nonpolar. This affects the folding of a chain of amino acids in the protein. Smell receptor sites are composed of proteins.

KEY TERMS
amino acid
peptide bond (amide bond)
protein

Exercises

Reading Questions

1. What is an amino acid?
2. What is a protein?

Reason and Apply

3. Draw a diagram to show how glycine and alanine combine to form a peptide bond.
4. Imagine a protein molecule composed of 200 glycine molecules and 200 aspartic acid molecules. Describe how the protein is folded. What is on the inside and what is on the outside?
5. PROJECT Use the Internet to research amino acids. Describe three different sources of amino acids in a person's diet.
6. Describe how forming a peptide bond is like forming an ester from an acid and an alcohol.

LESSON 48 ANSWERS

1. Amino acids are the building materials for many of the structures in the body, especially those related to the functioning of cells. An amino acid has a carbon backbone and two functional groups, an amine group and a carboxylic acid group.

2. A protein is a biological molecule consisting of a long chain of amino acids linked together by peptide bonds.

3. *Possible answer:*

Answers continue in Answer Appendix (p. ANS-4).

CHAPTER 9

Molecules in the Body

SUMMARY

KEY TERMS

mirror-image isomer

amino acid

peptide bond (amide bond)

protein

Smells Update

Receptor sites exist throughout living systems. They serve as the mechanism for triggering many biological responses and processes, including smell. Receptor sites can be very specific, allowing only certain molecules to dock. So, the molecular structure of the receptor site is important. Some molecules have mirror-image isomers that have different properties, including smell. Amino acid molecules exist as mirror-image isomers. Amino acids combine to form a vast array of proteins, including those that make up receptor sites.

REVIEW EXERCISES

1. How do you determine if two molecular models represent mirror-image isomers?

2. Draw the structural formula for these molecules. Which have mirror-image isomers? Explain your reasoning.
   a. CBrClFH   b. CH_4 c. CH_2Cl_2

3. Explain how you could use your feet to show that the "handedness" of a molecule can determine if it fits into a receptor site.

4. What are amino acids and what purpose do they serve in your body?

5. Using the receptor site model, explain how a person can smell L-carvone but not its mirror-image isomer D-carvone.

Modeling a Receptor Site

PROJECT

Build a space-filling model of a smell molecule from a compound in one of the five smell classifications you have studied. Use clay, plaster, sand, or other materials to build a receptor site for this smell molecule. Describe other molecules that might fit into the receptor site you built. Then create a space-filling model of a molecule that will not fit into the receptor site you built.

Your project should include

- the receptor site model that you built, labeled with the smell that it detects and a description of molecules that fit into it

- a space-filling model of a molecule that fits into the receptor site and one that does not (label which is which)

ANSWERS TO CHAPTER 9 REVIEW EXERCISES

1. *Possible answer:* Two molecular models represent mirror-image isomers if they are mirror images of each other and one cannot be rotated in space to be superimposed on top of the other. When the models have a carbon atom that is attached to four different groups, they will be mirror-image isomers.

2. a.

```
      Cl
      |
Br —  C — F
      |
      H
```

b.

```
      H
      |
H —   C — H
      |
      H
```

c.

```
      H
      |
Cl —  C — H
      |
      Cl
```

The only molecule that has mirror-image isomers is CBrClFH. In this molecule, the carbon atom is attached to four different atoms. The other two molecules have at least two identical atoms attached to the carbon atom, so the mirror image can be superimposed on the original molecule.

3. *Possible answer:* A molecule fits into its receptor site only when the receptor has the same "handedness" as the molecule. Likewise, each of your feet will fit into only one shoe of a pair of shoes, even though the shoes are mirror images of one another.

4. *Possible answer:* Amino acids are chemical compounds that have a carbon backbone and two functional groups, an amine group and a carboxyl group. In the body, amino acids are the building blocks of proteins, which are long molecules that form when a chain of amino acid molecules is linked together.

5. *Possible answer:* According to the receptor site model, a molecule only causes a smell receptor to send a message to the brain when it fits perfectly into the receptor. If a molecule, such as carvone, has mirror-image isomers, the isomers will not fit into the same three-dimensional space. If a person has a receptor that matches L-carvone but no receptor that matches D-carvone, that person's nose will only detect L-carvone.

Project: Modeling a Receptor Site

A good project would include ● two space-filling models of different molecules, with distinct colors used for carbon, hydrogen, oxygen, and any other atoms present in the molecules ● a receptor site that fits one of the models precisely (like a cast and mold) ● a labeled functional group on the model that fits into the receptor site ● a description of the smell that the receptor site detects, which should be specific to the molecule that fits into it ● a description of other molecules that may fit into the receptor site, given its shape.

ASSESSMENTS

Two multiple-choice assessments (versions A and B) and two short-answer assessments (versions A and B) are available for download for Chapter 9.

TRM Chapter 9 Assessment Answers

TRM Chapter 9 Assessments

Overview

CLASSWORK: PAIRS

Key Question: How is smell related to molecular structure and properties?

KEY IDEAS

Students review what they have learned in Unit 2: Smells regarding molecular structure and its relationship to the property of smell. Molecular formula, structural formula, functional group, three-dimensional shape, polarity, molecular size, molecular phase, and handedness all provide relevant information that can be taken into account in determining the smell of a molecule.

LEARNING OBJECTIVES

● Ask clarifying questions regarding concepts covered in this unit.
● Create a list of topics and concepts to study for an upcoming exam.

FOCUS ON UNDERSTANDING

It is important for students to understand that there probably are exceptions to all our generalizations about smells. Nevertheless, the broad generalizations still hold true and still are valuable.

IN CLASS

Using a worksheet, students review the unit. Students apply and integrate what they have learned about smells and chemistry. They draw structural formulas from molecular formulas, identify functional groups, and predict the smells and properties of compounds from assorted chemical information. Last, they predict the smell of a compound and then test the accuracy of their prediction. You might provide students with model sets to complete the review if it is assigned as a class activity. The worksheet can be completed as a homework assignment. A complete materials list is available for download.

TRM Materials List R2

GUIDE TO UNIT REVIEW

ENGAGE (5 min)

ChemCatalyst

Name three items that might be on an exam covering the entire Smells Unit. Compose a question that could be included on the exam.

UNIT 2

Smells | REVIEW

Alex Cao/Getty Images

S mell chemistry is an active area of research. Scientists are still refining theories about how smell works. However, one thing is clear: Smell is all about molecules. To understand smell, it is important to understand what molecules are and how they are put together.

The smell of a compound is greatly dependent on its composition and structure. This is true of other molecular properties besides smell. You do not smell some molecules because they are too large or heavy to become airborne, or because they are nonpolar and do not stick in receptor sites. Other molecules have a type of smell that depends on what atoms or functional groups are present. The smell of a molecule may also depend on the shape of the entire molecule.

By exploring the topic of smell, you have studied covalent bonding and the general rules governing how atoms come together to form the amazing structures called molecules.

REVIEW EXERCISES

General Review

Write a brief and clear answer to each question. Be sure to show your work.

1. What are isomers?

2. Draw two isomers with the molecular formula C_2H_4O.

3. A molecule can be described using a molecular formula, a structural formula, a ball-and-stick model, or a space-filling model. What information does each of these provide?

4. Consider the structural formula for methyl salicylate. Name two functional groups that are in this molecule.

5. Here is a ball-and-stick model of methyl salicylate.

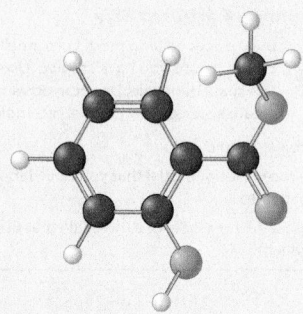

a. Does it follow the HONC 1234 rule?
b. Write the molecular formula for methyl salicylate.
c. Describe the overall shape of this molecule.
d. Find the oxygen atom that is bonded to two carbon atoms. Explain why this bond is bent.

Sample answers: Students might mention structural formulas, functional groups, the octet rule, Lewis dot symbols, electronegativity, polarity, electron domain theory, the receptor site model, and so on.

DISCUSS THE CHEMCATALYST

→ Begin to review what was learned in Unit 2: Smells.

→ Write general topics on the board as students share their answers. This becomes a review list.

→ Ask students to save their exam questions for later in the class.

Ask: What kinds of things might be on an exam for this unit?

EXPLORE (20 min)

TRM Worksheet with Answers R2

TRM Worksheet R2

INTRODUCE THE CLASSWORK

→ Arrange students into pairs.
→ Pass out the student worksheet.

EXPLAIN & ELABORATE (15 min)

GO OVER ANSWERS TO REVIEW EXERCISES

→ Ask students to come to the board to draw their structural formulas as time allows.

6. A polar molecule has the molecular formula C_3H_6O.
   a. Draw the structural formula.
   b. Draw the Lewis dot structure and label the shared electrons and the lone pairs.
   c. Identify the dipoles.
   d. Name the geometric shape around each carbon atom.
   e. Discuss what the name of this compound might be. Explain your reasoning.

7. What is a dipole? Explain why methane, CH_4, has four dipoles but has no overall dipole.

8. Table salt, NaCl, has no smell because
   A. It is not made up of molecules.
   B. It is an ionic compound.
   C. No part of it enters the gas phase at room temperature.
   D. All of the above.
   E. None of the above.

STANDARDIZED TEST PREPARATION
Multiple Choice
Choose the best answer.

1. Based on its name and chemical formula, predict the smell of formic acid, CH_2O_2.
   (A) Sweet   (B) Minty
   (C) Fishy   (D) Putrid

2. Which of the following compounds are covalent molecules? (Choose all that apply.)
   (A) $NH_4C_2H_3O_2$ (B) $C(CH_3)_4$
   (C) CH_2BrF (D) Na_2CO_3

3. Which statement best describes what happens to the valence electrons in a covalent bond?
   (A) Valence electrons are shared between two metal atoms.
   (B) Valence electrons are shared between two nonmetal atoms.
   (C) Valence electrons are transferred from a metal atom to a nonmetal atom.
   (D) Valence electrons are transferred from a nonmetal atom to a metal atom.

4. Which of the following best describes how two different compounds made up of molecules that are isomers of each other might smell?
   (A) The two compounds will most likely smell different from each other because isomers have the same molecular formula but different structural formulas.
   (B) The two compounds will most likely smell the same because isomers have the same molecular formula but different structural formulas.
   (C) The two compounds will most likely smell the same because isomers have the same molecular formula and the same structural formulas.
   (D) The two compounds will most likely smell different from each other because isomers have different molecular formulas and the same structural formulas.

REVIEW THE KEY CHEMISTRY IDEAS

→ Continue to create a list of topics and terms on the board as a study guide.

→ Ask students to share the exam questions they wrote for the ChemCatalyst question. Use these as review questions. Some sample questions are provided.

Ask: What information does a molecular formula contain?

Ask: What does a molecular formula tell you about smell?

Ask: Why do smell chemists use smell classifications?

Ask: What is a structural formula? How do you draw a structural formula?

Ask: What is a functional group? Name some. What can functional groups tell you about a molecule?

Ask: What is the HONC 1234 rule, and where does it come from?

Ask: What is a Lewis dot structure? How do you draw one?

Ask: What are lone pairs and bonded pairs?

Ask: Why are some molecules with three atoms bent while others are linear?

Ask: How is three-dimensional molecular structure related to smell?

Ask: What is a covalent bond? An ionic bond? A polar covalent bond?

Ask: What is a dipole? How do you determine the direction of a dipole?

Ask: What does polarity have to do with smell?

Ask: What does the size of a molecule have to do with smell?

Ask: What types of substances do not have a smell?

Ask: Why does the Smells Unit deal mostly with molecules?

EVALUATE (5 min)

TRM Transparency—Check-In R2

→ Test students' predictions for methyl salicylate.

→ Record the smell predictions of the class for methyl salicylate. You might have a ball-and-stick model of methyl salicylate available for inspection.

→ Pass out one candy to each student for students to smell. *Note:* Remind students never to taste any substances in the laboratory.

Ask: What do you predict methyl salicylate will smell like? What is your reasoning?

Ask: How does it actually smell?

Ask: What smell categories does it fit into?

Ask: Does the overall shape of the molecule fit with the smell? (both stringy and planar)

Key Points: The methyl salicylate molecule has structural features of both a minty molecule and a sweet molecule. Methyl salicylate is the compound in wintergreen candy that smells minty fresh and sweet. It is shaped roughly like a frying pan and also has a long, stringy tail. It also has an ester functional group, like many sweet-smelling molecules.

© GEORGIOS KEFALAS/epa/Corbis

Check-In

Would compounds made of either of these molecules have a smell? If so, try to predict what the smell would be. Explain your reasoning.

Answer: Both molecules would have a smell. The first molecule (decanoic acid) is a medium-sized carboxylic acid. Both its name and its structural formula indicate that it should smell putrid, which it does. The second molecule (dimethyl ether) is a small polar molecule, so we predict that it should have a smell. (Students have not smelled ether, so they probably will not be able to predict a smell for this compound.)

Homework

Assign the Unit 2: Smells Review in the student text in preparation for the unit exam.

ANSWERS

GENERAL REVIEW

1. Isomers are molecules that have the same molecular formula but different structural formulas.

2. *Possible answer:*

A third isomer of C_2H_4O is possible, in which the oxygen and carbon atoms form a triangle with three bonds.

3. A molecular formula provides the elements and number of atoms per element in a compound. ● A structural formula shows the way in which these atoms are bound to each other. ● A ball-and-stick model provides a three-dimensional image of the bonds. ● A space-filling model shows the amount of space each atom takes up, with atoms that are sharing electrons overlapped in space.

4. Ester group, hydroxyl (alcohol) group

5. a. Yes, each hydrogen atom is part of one covalent bond, each oxygen atom is part of two bonds (counting double bonds twice), and each carbon atom

5. Which of the following statements best describes the molecule below?

(A) The molecular formula for the molecule is $C_9H_{15}O$.
(B) The functional group is an ester.
(C) The shape around all of the carbon atoms is linear.
(D) The name of the compound will most likely end with "-one."

6. Which statement about the Lewis dot symbol of an element is true?

(A) The total number of dots surrounding the symbol is the number of valence electrons for that element.
(B) The number of dots surrounding the symbol will be the same for all elements of the same group on the periodic table.
(C) For elements in the same period on the periodic table, the number of dots surrounding the element symbol will increase as you move from left to right on the periodic table.
(D) All of the above statements are true.

7. Which statement correctly describes the Lewis dot structure of CF_4?

(A) The Lewis structure for CF_4 has a total of 40 dots in the structure because there are 5 atoms.
(B) In the Lewis structure for CF_4, carbon is the central atom with four bonded pairs of electrons, and each fluorine atom has a total of three bonded pairs of electrons.
(C) In the Lewis structure for CF_4, carbon is the central atom with four bonded pairs of electrons, and each fluorine atom has three lone pairs of electrons.
(D) In the Lewis structure for CF_4, the fluorine atom shares seven electrons with each carbon atom.

8. Which statement best describes water, H_2O?

(A) Water is a bent molecule that has two lone pairs around the central oxygen atom.
(B) Water is a linear molecule that has two lone pairs around the central oxygen atom.
(C) Water is a bent molecule with no lone pairs around the central oxygen atom.
(D) Water is a bent molecule with no lone pairs around the central hydrogen atom.

9. Which of the molecules below have a double bond in their structural formula?

$H_2, N_2, O_2, Cl_2, C_2H_2, C_2H_4, C_2H_6, H_2CO$

(A) H_2, N_2, O_2, Cl_2
(B) $C_2H_2, C_2H_4, C_2H_6, H_2CO$
(C) O_2, C_2H_4, H_2CO
(D) H_2, N_2, Cl_2, C_2H_6

10. How many electrons are shared between the carbon and nitrogen in hydrogen cyanide, HCN?

(A) Six electrons
(B) Five electrons
(C) Four electrons
(D) Three electrons

11. Based on the functional group, what is the predicted smell for the compound below?

(A) The molecule has a carboxyl group, so it likely smells putrid.
(B) The molecule has a carboxyl group, so it likely smells minty.
(C) The molecule has an ester group, so it likely smells sweet.
(D) The molecule has a ketone group, so it likely smells minty.

is part of four bonds. **b.** $C_8H_8O_3$ **c.** The molecule has a flat section with the ester functional group attached to its side, similar to the shape of a frying pan. **d.** The bond is bent because there are four electron domains. The oxygen atom has two bonded pairs and two lone pairs. The shape that separates the four domains as much as possible is tetrahedral, which causes the two bonds to be bent.

6. a. *Possible answer:*

b.

Other pairs of electrons are shared electrons. **c.** The dipoles are the carbon-oxygen bonds (the arrow pointing to the carbon atom), the carbon-hydrogen bonds (the arrow pointing toward the carbon atom) and the hydrogen-oxygen bond (the arrow pointing toward the oxygen). **d.** Each carbon atom has four electron domains. The shape that allows them to be as far apart as possible is a tetrahedron. **e.** The name of the molecule might end in *-ol* because it has a hydroxyl function group and is therefore an alcohol.

12. Which of the following correctly describes electron domains and molecular shape?

    (A) Only bonded pairs of electrons affect the shape of a molecule.
    (B) Only lone pairs of electrons affect the shape of a molecule.
    (C) Electron pairs have a tendency to repel one another due to their negative charge.
    (D) Electron pairs have a tendency to be as close to one another as possible.

13. What molecular shapes are represented in this picture of ethanol, C_2H_5OH?

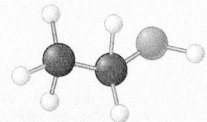

    (A) Pyramidal and bent
    (B) Tetrahedral and pyramidal
    (C) Tetrahedral and bent
    (D) Pyramidal and linear

14. Aspects of the receptor site theory are given below. What can you conclude about smell molecules from the receptor site theory?

    I: Smell molecules come in a variety of molecular shapes.
    II: Smell molecules change phase to become a gas or vapor that floats through the air to enter your nose and be detected.
    III: Smell molecules "dock" in receptor sites that match their molecular shapes.
    IV: Nerves are stimulated when the molecules dock and send messages to the brain, which are interpreted as a particular smell.

    (A) Smell molecules must break apart into individual atoms when they become a gas.
    (B) Smell molecules must lose their functional groups when they become a gas.
    (C) Smell molecules must change their shape to fit the "dock" in the receptor sites.
    (D) Smell molecules must keep their same structural formula when they become a gas.

15. What is the *bond* polarity of the C—H bonds in methane, CH_4? Refer to the electronegativity scale at the beginning of Lesson 44 to answer.

    (A) The C—H bonds are nonpolar covalent bonds because the electronegativity difference between carbon and hydrogen is large.
    (B) The C—H bonds are polar covalent bonds because the electronegativity difference between carbon and hydrogen is large.
    (C) The C—H bonds are ionic bonds because the electronegativity difference between carbon and hydrogen is large.
    (D) The polarity of C—H bonds is small because the electronegativity difference between carbon and hydrogen is small.

16. Which of these molecules are polar? (Choose all that apply.)

    (A) NH_3
    (B) CCl_4
    (C) C_2H_6
    (D) CH_3OH

17. Which of the choices below illustrates the concept "like dissolves like?"

    (A) Iodine, I_2, is more soluble in water than in oil.
    (B) Iodine, I_2, is more soluble in oil than in water.
    (C) Vinegar, CH_3COOH, is more soluble in oil than in water.
    (D) Sodium chloride, NaCl, is more soluble in oil than in water.

Engineering Design Project: Scent-Detection Device

As an extension to Unit 2: Smells, you can assign the Engineering Design Project: Scent-Detection Device. In this project, students use an engineering design cycle to conceptualize how they might build and refine a simple scent-detection device.

TRM Engineering Project U2

ASSESSMENTS

Two Unit 2 Assessments (versions A and B) are available for download.

TRM Unit 2 Assessment Answers

TRM Unit 2 Assessments

ASSESSMENTS

A Lab Assessment for Unit 2 is available for download.

TRM Unit 2 Lab Assessment Instructions and Answers

TRM Unit 2 Lab Assessment

---

7. A dipole has two ends, one that is positively charged and one that is negatively charged. In a methane molecule, each of the bonds is a polar bond, because the electronegativities of carbon and hydrogen are different. The hydrogen has a partial positive charge and the carbon has a partial negative charge. However, methane has a tetrahedral shape to keep the four electron domains as far apart as possible. This means that the outside of the molecule has four identical partial positive charges. Therefore, the entire methane molecule does not have two poles and is not a dipole.

8. D

STANDARDIZED TEST PREPARATION

1. D	8. A	15. D
2. B, C	9. C	16. A, D
3. B	10. A	17. B
4. A	11. C	18. C
5. D	12. C	19. B
6. D	13. C	20. A
7. C	14. D	

18. Which of the following cannot be explained by describing the intermolecular forces between molecules?

(A) Water molecules form drops rather than spreading out on wax paper.

(B) Water is a liquid at room temperature while methane is a gas at room temperature, even though both water and methane have similar molecular mass.

(C) The O–H bond in water is polar because of differences in electronegativity of the atoms.

(D) Oil tends to float on water while ethanol tends to dissolve in water.

19. Which group of elements has the highest electronegativity values?

(A) C, H, O

(B) O, Cl, F

(C) Ne, H, C

(D) Cl, Br, I

20. Which of the following molecules has a mirror-image isomer?

(A) $CH(CH_3)FBr$

(B) CHF_3

(C) CH_2F_2

(D) CF_4

HISTORY CONNECTION

In 2004, Richard Axel and Linda Buck shared a Nobel prize for their advances in understanding the mechanism of smell. Axel and Buck found that mammals have approximately 1000 different genes for olfactory receptors, which led to a better understanding of smell.

© Jennifer Altman/epa/Corbis (left), © Dan Lamont/Corbis (right)

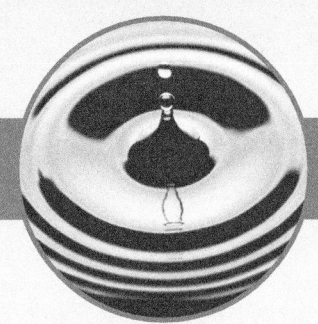

Unit 3 | Weather

Phase Changes and Behavior of Gases

WEATHER AS CONTEXT

Most of the weather phenomena we experience are a result of physical changes in matter. Differences in density cause the movement of air that we experience as wind. Heat from the Sun causes water to evaporate and rise into the atmosphere. Water vapor is moved about by temperature and pressure differentials in the air, only to change phase again and become some form of precipitation. To understand weather, we have to understand phase changes and the interplays of variables such as pressure, temperature, relative humidity, and density.

CONTENT DRIVEN BY CONTEXT

There are notable advantages to using weather as a context. Students have probably constructed personal explanations for the causes of various weather phenomena, from fog to hurricanes, that they experience. Building on these first-hand experiences motivates study of a topic as invisible and abstract as gas behavior. Throughout the unit, students search for patterns among the weather variables. Understanding and predicting the weather is an intricate science, but students relish this opportunity to become amateur meteorologists and quickly gain some expertise. Chemists study gas behavior by examining samples in closed containers. In contrast, our atmosphere is a constantly changing open system. To address this problem, in this unit we focus on air masses, which have uniform conditions of temperature and pressure. We also relate the study of specific volumes of gas or amounts of gas under controlled conditions to understanding what is happening in less controlled conditions. Ultimately, understanding the gas laws helps students to explain weather phenomena.

MATHEMATICAL RELATIONSHIPS

The gas laws involve mathematics that can be challenging for some students. Besides algebra involving multiple variables, students will see equations that may seem unfamiliar if they are used to solving for only x and y. Luckily, most of the mathematical relationships in chemistry are proportional ones, and they become more accessible to students when this is pointed out. Lessons in Unit 1: Alchemy laid the groundwork by introducing proportional relationships and proportionality constants such as density. As they work with these proportional relationships, students begin to truly grasp the big picture and become more confident with algebraic relationships.

In this unit, we have taken special care with how the mole is introduced. Students regularly have difficulty with this concept, usually with grasping its magnitude and what it concretely represents. For this reason, we wait to use scientific notation until Unit 4: Toxins. The lessons in this unit make the mole as concrete as possible for students so that they are fully prepared for the stoichiometry in the next three units.

References to Math Spotlights on solving equations, dimensional analysis, significant digits, and other math topics appear where appropriate in the readings and examples in the student textbook. Math Spotlights at the back of the student textbook provide review and practice of common math topics for your students.

THE USE OF MODELS TO EXPLAIN MOLECULAR BEHAVIOR

Understanding pressure, temperature, and volume changes of a gas hinges on understanding kinetic molecular theory. To succeed at the concepts in this unit, students have to visualize gas molecules. To that end, the activities and the student textbook regularly help and encourage students to visualize gas behavior at the molecular level. Students also explore molecular motion with computer simulations that help them visualize pressure, volume, and temperature changes.

BUILDING UNDERSTANDING

So far, students have constructed a solid foundation in chemistry by studying atomic and molecular structure and properties. In this unit, they build on this foundation by studying gas laws, phase changes, temperature scales, and the concept of the mole. Topics in this unit are introduced in a deliberate sequence with the content and math carefully scaffolded.

Chapter Summary

Chapter	Description	Standard Schedule Days	Block Schedule Days
10	**Chapter 10** focuses on physical changes to matter, more specifically changes in phase, volume, and density of matter in response to changes in temperature. The proportional relationship between volume and temperature of a gas is introduced along with the Kelvin scale.	6	3
11	In **Chapter 11,** students explore gas pressure qualitatively, quantitatively, and from a particulate viewpoint. By focusing on high- and low-pressure systems in the atmosphere, students learn to explain macroscopic observations of matter in terms of the behavior of particles. Lessons explore relationships among volume, temperature, and pressure of gases.	6	3
12	**Chapter 12** explores the relationship between the number of particles, n, and all the previously mentioned variables. This chapter introduces the mole, setting up a firm foundation for stoichiometry in the next unit. Exploring the number of particles allows us to investigate humidity, providing explanations for weather phenomena such as fog, dew, rain, and snow. Finally, the many variables are brought together in a lesson on hurricanes.	6	3

Pacing Guides

Standard Schedule

Day	Suggested Plan	Day	Suggested Plan
1	Chapter 10 Lesson 49	13	Chapter 11 Lesson 60
2	Chapter 10 Lesson 50	14	Chapter 11 Lesson 61
3	Chapter 10 Lesson 51	15	Chapter 11 Lesson 62, Chapter 11 Review, Project (optional)
4	Chapter 10 Lesson 52	16	Chapter 11 Quiz, Chapter 12 Lesson 63
5	Chapter 10 Lesson 53	17	Chapter 12 Lesson 64
6	Chapter 10 Lesson 54	18	Chapter 12 Lesson 65
7	Chapter 10 Lesson 55, Chapter 10 Review, Project (Optional)	19	Chapter 12 Lesson 66
8	Chapter 10 Quiz, Chapter 11 Lesson 56	20	Chapter 12 Lesson 67
9	Chapter 11 Lesson 57	21	Chapter 12 Review, Project (optional)
10	Chapter 11 Lesson 58	22	Chapter 12 Quiz
11	Chapter 11 Lesson 59	23	Unit Review, Lab Exam (optional)
12	Chapter 11 Lesson 59 (cont.)	24	Unit Exam

Block Schedule

Day	Suggested Plan	Day	Suggested Plan
1	Chapter 10 Lessons 49 and 50	8	Chapter 11 Lessons 60 and 61
2	Chapter 10 Lessons 51 and 52	9	Chapter 11 Lesson 62, Chapter 11 Review, Project (optional)
3	Chapter 10 Lessons 53 and 54	10	Chapter 11 Quiz Chapter 12 Lessons 63 and 64
4	Chapter 10 Lesson 55, Chapter 10 Review, Project (optional)	11	Chapter 12 Lessons 65 and 66
5	Chapter 10 Quiz, Chapter 11 Lessons 56 and 57	12	Chapter 12 Lesson 67, Chapter 12 Review, Project (optional)
6	Chapter 11 Lessons 57 (cont.) and 58	13	Chapter 12 Quiz, Unit Review
7	Chapter 11 Lesson 59	14	Unit Exam, Lab Exam (optional)

Unit 3 | Weather

StockbyteGetty Images

Water changes phase and is transported around the planet. It may then show up as snow or rain.

Why Weather?

Thunderstorms dump great quantities of rain, fog seeps into a bay at nightfall, warm temperatures entice us to the beach, and hurricanes devastate coastal communities. For better or worse, the weather is a part of our everyday lives. Physical change is at the core of weather. Weather occurs when matter undergoes changes in location, density, phase, temperature, volume, and pressure. When you understand the relationships between these changes in matter, you will begin to answer the question, "What is the chemistry of weather?"

In this unit, you will learn

- about proportional relationships
- about temperature scales and how thermometers work
- the effects of changing temperature, pressure, and volume on matter
- about the behavior of gases
- how to read weather maps and make weather predictions

Wes Walker/Lonely Planet Images/Getty Images

257

Watch the video overview of Chapter 10 (for teachers) by clicking on the link in the TE-book, opening the TRFD, or logging onto the book's companion Web site bcs.whfreeman.com/ livingbychemistry2e (teacher log-in required).

Chapter 10 of the Weather unit introduces students to the physical changes of matter and explores how those changes are monitored, measured, and tracked. In Lesson 49, students explore the information typically found on weather maps. They look for correlations among the locations of the jet stream, clouds, precipitation, temperature, high and low pressure, and warm and cold fronts. In Lesson 50, students are introduced to the first of many proportional relationships when they investigate why rainfall is measured in units of height rather than volume. Students explore phase change behavior in Lesson 51, one of several lessons in the unit that highlight the interactions among phase, density, temperature, and volume. In Lesson 52, students construct simple liquid and gas thermometers. Lesson 53 introduces students to the Kelvin scale and the kinetic theory of gases. Lesson 54 explores the effect of temperature changes on gases and quantifies the relationship between gas volume and temperature. In Lesson 55, students apply what they have learned about gas behavior to the topics of weather fronts, cold and warm air masses, and the formation of storms and precipitation.

In this chapter, students will learn

- how to interpret basic weather maps
- about basic proportional relationships
- about thermometers and temperature scales
- the kinetic theory of gases
- phase change behavior
- Charles's law relating the volume and temperature of a gas

Image courtesy of the Earth Science and Remote Sensing Unit, NASA Johnson Space Center

CHAPTER 10
Physically Changing Matter

In this chapter, you will study

- how to read basic weather maps
- proportional relationships
- density and phase changes in matter
- temperature scales and thermometers
- the relationship between volume and temperature of gases

258

A view of Earth's atmosphere from space shows weather in action.

Weather is the result of interactions between Earth, the atmosphere, water, and the Sun. It is about the movement of matter around the planet. Water, for example, changes phase from liquid to gas and then back to liquid as it moves from oceans to the atmosphere, to clouds, and down to Earth as rain. The atmosphere is composed of different gases, so the study of the behavior of gases is important in understanding weather. The gas laws are mathematical equations that describe the relationship between changing temperature, volume, and pressure of gases. Understanding the behavior of matter in relation to temperature allows weather experts to make predictions about the weather.

Chapter 10 Lessons

Weather or Not

Weather Science

THINK ABOUT IT

You have probably seen a weather forecast, either on television or in the newspapers. You may have wondered—what are highs and lows? What are fronts? What does the jet stream have to do with the weather?

What causes the weather?

To begin to answer this question, you will explore

↻ Earth, Air, Water, and Sun

↻ Weather Maps

EXPLORING THE TOPIC

↻ Earth, Air, Water, and Sun

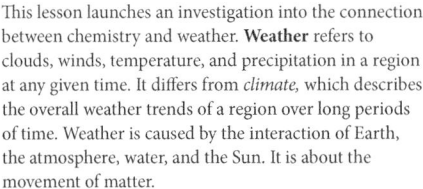

From outer space, you can see clouds above the surface of the planet. A satellite picture of Earth also clearly shows that our planet's surface is mostly water.

This lesson launches an investigation into the connection between chemistry and weather. **Weather** refers to clouds, winds, temperature, and precipitation in a region at any given time. It differs from *climate,* which describes the overall weather trends of a region over long periods of time. Weather is caused by the interaction of Earth, the atmosphere, water, and the Sun. It is about the movement of matter.

A planet must have an *atmosphere* to have weather. An atmosphere is a layer of gases surrounding the planet. In contrast, the Moon has no weather because it barely has any atmosphere. The atmosphere on Earth is often referred to as *air.* Air is composed of a mixture of gases. Earth's atmosphere consists of nitrogen (78%), oxygen (21%), argon (1%), carbon dioxide (0.04%), and traces of other gases. There is some water vapor, $H_2O(g)$, in the atmosphere, but the amount varies from place to place, from day to day, and even from hour to hour. On average, it makes up about 1% of the atmosphere.

The weather we experience is caused by physical changes in the atmosphere. **Physical changes** alter the form a substance takes but do not change the identity of the substance. Examples of physical changes are changes in volume, temperature, shape, size, and pressure. These are different from chemical changes, which produce new substances. The heating of Earth by the Sun is the root cause of most of the physical changes in Earth's atmosphere.

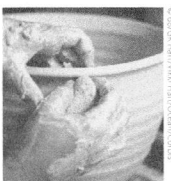

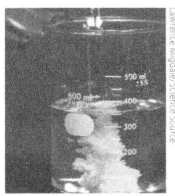

Heating and reshaping a substance are both physical changes. Mixing chemicals to form a new compound is a chemical change.

Lesson 49 | **Weather Science**   259

Overview

ACTIVITY: WHOLE CLASS OR GROUPS OF 4

Key Question: What causes the weather?

KEY IDEAS

Weather on our planet is a result of the dynamic interplay among Earth, the atmosphere, water, and the Sun. Meteorologists monitor moisture, air pressure, temperature, storm fronts, and the jet stream to predict storms and precipitation. Weather science is closely related to the chemistry of phase changes and gas laws.

LEARNING OBJECTIVES

● Explain the phenomenon of weather in general terms.
● List the variables meteorologists study or measure to predict the weather.
● Describe the basic components of a weather map.

KEY TERMS

weather
physical change
phase change

IN CLASS

This lesson introduces the Weather unit. Students examine data on weather maps to begin to understand

weather forecasting. Data on the jet stream, temperature, cloud cover, weather fronts, precipitation, and air pressure are provided on individual maps, each on a separate color transparency. These transparencies can be overlaid during the class discussion so that students can look for relationships among the weather variables. Students also are introduced to some basic weather vocabulary they will use throughout this unit. (*Note:* We recommend that you print enough color transparencies so that students can work in groups on this activity, each with their own set, or photocopy the handout onto a transparency for each group and cut it apart so they can have their own black-and-white set. Alternatively, the lesson can be run using one set of color transparencies and an overhead projector. To do this, be sure to give students the handout Weather Variables, so they can hypothesize about the relationships between variables. A complete materials list is available for download.

TRM Materials List 49

Lesson Guide

ENGAGE (5 min)

TRM Transparency—ChemCatalyst 49

TRM PowerPoint Presentation 49

ChemCatalyst

The table gives the weather conditions on a day in late October in Miami, Florida (shown on the map with a star). Predict the weather for later today. Indicate whether you think each of the conditions will increase, decrease, or stay the same. Explain your reasoning.

Current Conditions at 1:30 P.M. in Miami, Florida

Temperature	82 °F/27.8 °C
Pressure	29.95 in Hg, falling
Fronts	Cold front to pass through today
Conditions	Mostly cloudy, wind gusts up to 30 mph NW
Humidity	71%

Sample answer: Mainly cloudy and humid. High around 86 °F with showers and a thunderstorm today. Showers, thunderstorms, and wind tonight with a low of 72 °F. Clearing and sunny on the following day after morning showers. Reasons that students offer will vary.

Ask: What do you think is the most useful information in the weather data? Explain your reasoning.

Ask: What was your forecast? What did you base it on?

Ask: What things do you have to know about to forecast the weather?

Ask: What do you think causes weather?

EXPLORE (20 min)

TRM Worksheet with Answers 49

TRM Worksheet 49

TRM Weather Map Transparencies 49

TRM Handout—Weather Variables 49

INTRODUCE THE ACTIVITY

➜ Arrange students into groups of four.

➜ Pass out the student worksheet.

Define weather. Point out that it is different from climate, which describes the average weather for a region over long periods of time.

> **Weather:** The state of the atmosphere in a region over a short period of time. Weather is the result of the interaction among Earth, the atmosphere, water, and the Sun. It refers to clouds, winds, temperature, and rainfall or snowfall.

Tell students that a weather forecast is a prediction made by a meteorologist about the weather.

➜ Pass out a set of weather map transparencies to each group of four students. Tell students that these transparencies show weather data for the same day. They will look for patterns in these weather maps to better understand the variables that affect the weather. They can overlay the maps on a white sheet of paper, two or three at a time, to compare different variables. If you are conducting a whole-class activity, have students refer to the handout Weather Variables to decide which variables should be compared.

➜ You may remind students that a variable is a quantity or characteristic that can change or vary.

GUIDE THE ACTIVITY

➜ If you are conducting a whole-class activity with one set of transparencies,

The **phase changes** of water between solid, liquid, and gas are types of physical changes that play an important role in weather. For example, rain is evidence that a phase change from gas to liquid has taken place.

Big Idea Weather is the result of physical changes to matter.

⟳ Weather Maps

Meteorologists use maps to keep track of atmospheric conditions and to predict the weather. Weather information for a specific day is shown on the maps here. What patterns do you observe as you compare the six maps?

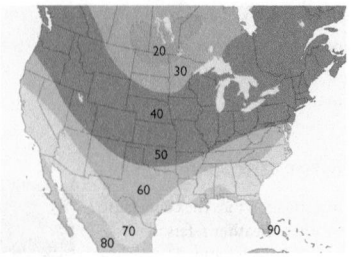
Temperature highs in degrees Fahrenheit

Air pressure highs and lows

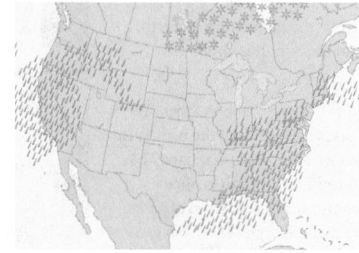
Precipitation: rain and snow

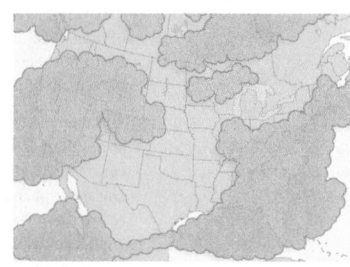
Cloud cover

Fronts

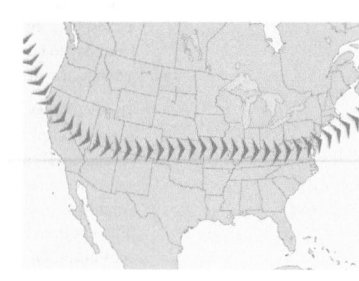
Jet stream

start by displaying each transparency over the base map so students can answer Question 1. For Question 2, overlay Precipitation, Cloud Cover, and Temperature highs. Next, ask students to hypothesize about which weather variables may be related. Overlay the appropriate transparencies so students can check their hypotheses. You might also do part 2 of the worksheet as a class discussion.

EXPLAIN & ELABORATE (15 min)

DISCUSS THE TOOLS OF WEATHER PREDICTION

➜ Display the appropriate transparencies as needed.

➜ Draw weather symbols on the board and label them.

Ask: What weather variables were on the weather maps you studied? (jet stream, temperature, cloud cover, precipitation, fronts, high and low pressure)

Ask: What are some of the symbols found on the maps? What do you think they stand for?

Ask: Why is it necessary for a meteorologist to use more than one map to predict the weather?

Key Points: Several factors affect the weather in North America:

Jet stream. The jet stream map shows the location of high-level winds.

In addition to familiar features such as temperature and precipitation, weather maps often show a curve called the *jet stream*. The jet stream shows the location of high-altitude winds. These winds are at least 57 miles per hour (up to 190 mi/h) and are in the upper atmosphere above 20,000 feet in altitude (that's 4 miles up). These winds have a great effect on what happens in the air below them and generally "steer" storms.

Here are some features to take note of:

- In North America, temperatures are colder in the north and warmer in the south.
- The shape of the jet stream coincides with the contour lines on the temperature map.
- If it is raining or snowing, then clouds must be present, but clouds do not necessarily mean it will rain or snow.
- Fronts are located in areas of low pressure.
- Clouds are more likely to form near fronts, so it is more likely to rain and snow there.
- Warm fronts and cold fronts are associated with low-pressure systems.
- There is no precipitation for regions with high pressure. High pressure is associated with sunny days. A region with high pressure can be either hot or cold.

Example

Weather Maps

Describe the weather in these areas using the six weather maps in this lesson. Include the temperature, air pressure, cloud cover, precipitation, and storm fronts.

   a. West Coast   **b.** middle of the United States

Solution

   a. West Coast: The temperature is about 60 °F, slightly colder to the north. The air pressure is low. There are storm fronts and it is cloudy, with rain to the south and clearing to the north.

   b. Middle of the United States: It is about 40 °F and sunny. The air pressure is high, there are no clouds, and there is no precipitation.

LESSON SUMMARY

What causes the weather?

KEY TERMS
weather
physical change
phase change

Weather is about the movement of air and water. The gases in our atmosphere and the water on Earth's surface undergo physical changes and move around the planet. Water exists in all three phases (solid, liquid, and gas) on Earth. Water shows up as snow, ice, rain, clouds, fog, and water vapor. Meteorologists use detailed maps to monitor the physical changes taking place in our atmosphere and to keep track of the movement of moisture and air around the planet. Information on weather maps includes temperature, air pressure, cloud cover, precipitation, warm and cold fronts, and the location of the jet stream.

Lesson 49 | **Weather Science**   261

high pressure. Ls show where large air masses with consistently low air pressure are located.

DISCUSS GENERAL RELATIONSHIPS AMONG WEATHER VARIABLES

→ Use the map transparencies as needed to show the general relationships between weather variables.

Ask: What connections did you find between the different maps when they were placed on top of each other?

Ask: If you were going to try to predict whether it will rain in a certain city, which weather maps would you want to have? Explain.

Key Points: There are connections between the different maps:

- The precipitation map is most closely associated with low-pressure areas and with weather fronts of all kinds.
- The curves of the jet stream map generally match the curves of the temperature maps.
- The lows on the air pressure map are associated with the warm and cold fronts.
- Rain requires clouds overhead. Yet the presence of clouds does not necessarily mean precipitation.
- The jet stream divides cold arctic air from warmer air in the south. As winter sets in, the jet stream moves lower down over the continent. Broadly speaking, you can predict the general location of the jet stream by examining the pattern created by the temperature bands and placing the jet stream between the warmer and colder temperatures.

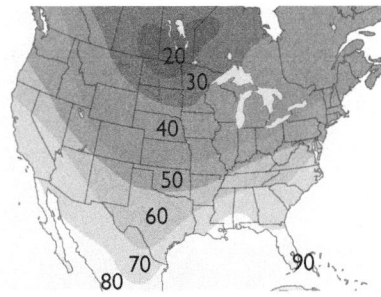

Temperature highs in degrees Fahrenheit

In the jet stream, air moves swiftly from west to east across the United States. Storms located in the atmosphere below the jet stream generally move from west to east across the continent. If you want to predict the weather in the United States, it often helps to look to the west.

These winds are at least 57 mi/h (up to 190 mi/h) and are in the upper atmosphere above 20,000 ft (4 mi). These winds generally "steer" storms around the planet.

Temperature. Bands of color are used to display variations in temperature, with reds being hotter temperatures and blues colder temperatures. Note that temperatures are not random but form a clear pattern across the continent.

Cloud cover. Gray areas designate the location of clouds.

Fronts. Two main types of fronts are shown, cold fronts and warm fronts. In a warm front, warm air is moving into a region. It is depicted by a series of red scallops pointing in the direction in which the warm air is moving. In a cold front, cold air is moving into a region. It is depicted by a series of blue triangles pointing in the direction in which the cold air is moving.

Cold front   Warm front

Precipitation. Raindrops and snowflakes show where there is rain or snow.

Pressure. Hs show the locations of large air masses with consistently

Lesson 49 | **Weather Science**   261

EVALUATE (5 min)

Check-In

What do weather maps keep track of? How do they help meteorologists?

Answer: Weather maps keep track of the jet stream, temperature, cloud cover, weather fronts, precipitation, air pressure, and other variables. Meteorologists can examine these maps and reach conclusions about what kind of weather likely will happen.

Homework

Assign the reading and exercises for Weather Lesson 49 in the student text.

LESSON 49 ANSWERS

1. *Possible answer:* To have weather, a planet must have a layer of gases surrounding its surface.

2. *Possible answer (any 3):*
● Temperature refers to the amount of heat in the atmosphere. ● Air pressure refers to the pressure air exerts on Earth's surface. ● Precipitation is the amount of rain or snow that falls to Earth's surface. ● Cloud cover is the type and location of clouds in an area. ● Forecasters study the speed and direction of winds and wind gusts. ● Forecasters also determine the location of the jet stream.

3. A physical change is a change in the form of a substance that does not change the identity of the substance.

4. *Possible answer (any 3):*
● Temperature maps show the temperature in different areas. Temperature maps can show actual values or different-colored areas based on the range the temperature falls in. ● Air pressure maps indicate areas of high or low air pressure using the symbols H and L. ● Precipitation maps show where water is currently falling using different symbols or colors for rain and snow. ● Cloud cover maps show the location of clouds. Darker areas indicate thicker clouds. ● Maps of atmospheric fronts use shapes of symbols on a line to show warm fronts, cold fronts, and other types of fronts. ● Jet stream maps show the path of the high-speed, high altitude winds in the upper atmosphere.

5. Answers will vary. Make sure answers identify the map or maps used

Exercises

Reading Questions

1. What substances are necessary for a planet to have weather?
2. Describe three variables that meteorologists study to make weather predictions.
3. What is a physical change?

Reason and Apply

4. Name three types of maps that meteorologists use. Describe the maps and the information that they contain.
5. Consult a newspaper, the Internet, or a weather report on television.
   a. Name three states that are currently experiencing a cold front (or warm front) moving through.
   b. Which states are experiencing high-pressure systems? What is the weather forecast in those states?
   c. If you were traveling to New York tomorrow, would you pack a raincoat? Explain.
6. Nitrogen, $N_2(g)$, oxygen, $O_2(g)$, carbon dioxide, $CO_2(g)$, and water vapor, $H_2O(g)$, are all gases found in the atmosphere.
   a. Draw the structural formula for each and include lone pair electrons.
   b. Draw the Lewis dot structure for water.
   c. Which molecules are polar and which molecules are nonpolar?
   d. Which one of the four substances is naturally found as a liquid, a solid, and a gas on Earth?

ENVIRONMENTAL CONNECTION

On other planets, the compounds that are involved in the weather may be quite different from those found on Earth. The atmosphere of Mars is mostly carbon dioxide. The polar ice caps on Mars are made of frozen carbon dioxide.

Science Source

and the date and time that the map was produced for Web-based variable maps. **a.** The states identified as having fronts passing through should have a scalloped line drawn within the state indicating a warm front or a line with triangles indicating a cold front. **b.** A capital H denotes high pressure on weather maps. A high-pressure system usually indicates clear weather. **c.** Determine whether there are symbols for rain or snow over New York on the map.

6. a.

$$:N \equiv N:\qquad \ddot{\underset{..}{O}}=\ddot{\underset{..}{O}}\qquad \ddot{\underset{..}{O}}=C=\ddot{\underset{..}{O}}\qquad \ddot{\underset{..}{O}}-H$$

b.

H
$$\ddot{\underset{..}{O}}:H$$

c. Nitrogen and oxygen both have nonpolar bonds because the atoms are identical. Carbon dioxide has polar bonds, but the molecule is symmetric so the molecule itself is nonpolar. Water vapor has polar bonds and a bent shape, so it is a polar molecule.

d. Of these substances, only water is naturally found in all three phases on Earth. Nitrogen, oxygen, and carbon dioxide are gases unless the temperature falls well below 0 °C.

Measuring Liquids

THINK ABOUT IT

After a rainstorm, rain in the gutter or in a puddle can be several inches deep, yet on the sidewalk and grass, it may not be that deep. The water runs off the sidewalks, runs down hills, collects in depressions, and soaks into grass. So, how do you determine how much rain fell?

> How do meteorologists keep track of the amount of rainfall?

To answer this question, you will explore

↻ Rain Gauges

↻ Volume Versus Height

↻ Proportional Relationships

EXPLORING THE TOPIC

↻ Rain Gauges

Suppose you have three rain gauges that are different sizes. You place them next to one another out in the rain. After a few hours, these containers might look like this:

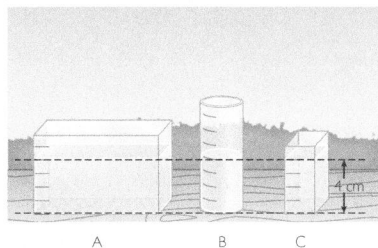

A B C

Sample data for these three containers are given in the table below.

Container	Area of the base (cm²)	Height of rain (cm)	Volume of rain (cm³)
A	10.0 cm²	4.0 cm	40.0 cm³
B	2.5 cm²	4.0 cm	10.0 cm³
C	2.0 cm²	4.0 cm	8.0 cm³

You can see that the volume of rain in each container is quite different. However, the *height* of rain is the same. This is why meteorologists report height of rainfall, in centimeters or inches, not volume of rainfall. The height of rainfall is the same for these containers, but the volume is not.

Overview

LAB: PAIRS

Key Question: How do meteorologists keep track of the amount of rainfall?

KEY IDEAS

This lesson focuses on proportional relationships, partly in preparation for gas laws, introduced later in the unit. Meteorologists measure rainfall by recording the height of rain that falls, in inches or millimeters. This is a valid reflection of rain amount because the volume of a container is proportional to its height. Height is a more useful measurement than volume for rainfall because it is not dependent on the overall size of the container used.

LEARNING OBJECTIVES

- Identify a proportional relationship.
- Describe several methods for solving a problem involving proportional variables.
- Explain why rainfall is measured in terms of height rather than volume.

FOCUS ON UNDERSTANDING

- Many rulers do not start at zero. Students should correct for the offset. If they do not, they will not obtain a straight line for the relationship between volume and height.

- This is a good opportunity to review accuracy, precision, and significant digits.
- For simplicity, we refer to directly proportional relationships in this unit as proportional relationships.

KEY TERMS

proportional
proportionality constant

IN CLASS

In this activity, students explore the relationship between volume and height as they investigate the optimal way to measure rainfall. The lesson begins with a quick demonstration to simulate rain falling over different surface areas. During the activity, students measure the volume of rain in two containers to find specific rainfall amounts. They then plot volume versus height on a graph. The graph allows students to see the proportional relationship between the height and the volume of rain that falls into a container with parallel sides. Proportional relationships are discussed further in the next lesson. A complete materials list is available for download.

TRM Materials List 50

SETUP

For the optional wet demonstration, you will need a large transparent container (e.g., a fish tank); a small beaker, a larger beaker, and a Florence flask to fit inside; and a foil roasting pan or plastic tub to place on top. With a thumbtack, poke many tiny, evenly spaced holes in the tub (e.g., a 0.5 in² array). To watch it rain inside the large transparent container, place the plastic tub on top and fill it with water. Graduated beakers, if available, may be used instead of graduated cylinders to speed up the volume measurements for the lab.

Lesson Guide

ENGAGE (5 min)

TRM Transparency—ChemCatalyst 50

TRM PowerPoint Presentation 50

> *ChemCatalyst*
>
> **1.** How is rainfall usually measured? Describe the type of instrument you think is usually used.
>
> **2.** Which of these containers would make the best rain gauge? Explain your reasoning.

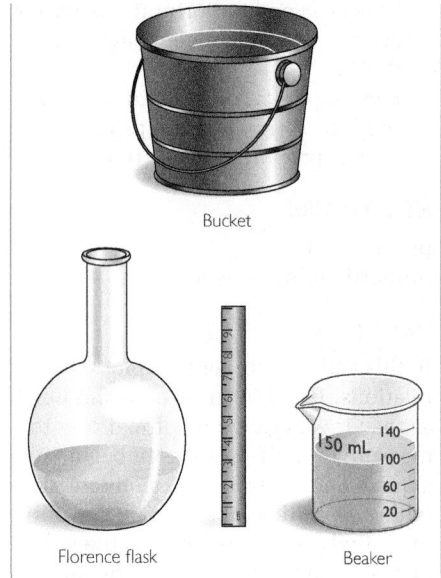

Bucket

Florence flask

Beaker

Sample answers: **1.** Rain is measured with a rain gauge, which looks a little like a graduated cylinder. **2.** Each container shown has flaws. The narrow neck of the flask restricts how rain can enter the container. The bucket lacks graduations and has sloped sides. The beaker has straight sides with somewhat imprecise graduations that measure volume, not height.

Ask: What units and instruments are used to track rainfall?

Ask: What are some of the issues with the containers shown in the ChemCatalyst?

Ask: Describe how you would measure rainfall. What kind of container would you use?

Ask: How is volume different from height?

EXPLORE (20 min)

TRM Worksheet with Answers 50

TRM Worksheet 50

COMPLETE THE DEMONSTRATION

→ Simulate rainfall over a region, with either a wet or a dry demonstration (optional). Show students a commercial rain gauge (optional).

DRY DEMONSTRATION

→ Show students several flat objects with different areas, such as a piece of binder paper, a circular coaster, and a sticky note. Place these objects on a table (or in a plastic tub). Ask students to imagine a rainstorm passing through the classroom with rain falling consistently on all these areas.

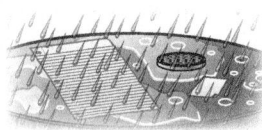

→ Ask students if the same amount of rain falls on each area. Ask them how they might measure the amount of rain that falls on the objects.

WET DEMONSTRATION (OPTIONAL)

→ Place a small beaker, a larger beaker, and a Florence flask in a large transparent container. Place the tub with holes on top of the container.

⟳ Volume Versus Height

Imagine that you place a rain gauge with a square base of 2.0 cm² in a steady rain for several hours. Each hour, you go outside to measure how much rain has collected in the rain gauge.

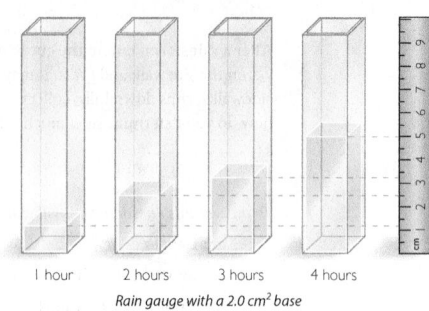

| 1 hour | 2 hours | 3 hours | 4 hours |

Rain gauge with a 2.0 cm² base

The graph shows how volume changes as height changes, or volume versus height. *Versus* means *compared with* and is abbreviated *vs.* Both the volume and the height of the rain in the gauge increase as more rain is collected. Take a moment to examine the data.

The graph shows that volume increases in a steady and predictable way in relationship to height. And once some data points are known, others can be predicted.

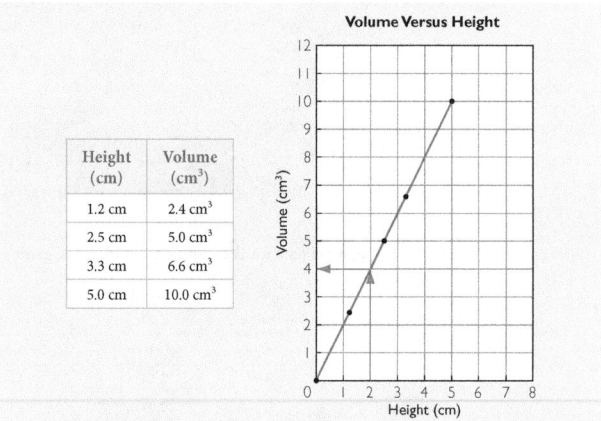

Height (cm)	Volume (cm³)
1.2 cm	2.4 cm³
2.5 cm	5.0 cm³
3.3 cm	6.6 cm³
5.0 cm	10.0 cm³

Volume Versus Height

Using the Graph

You can use this graph to predict the volume of water corresponding to various heights of water in a rain gauge. Each point on the slanted line corresponds to a different volume and height.

→ Ask students to predict the height of the water in the three small containers when you pour water into the tub.

→ Pour the water into the tub and let the students observe.

INTRODUCE THE LAB

→ Arrange students into pairs.

→ Pass out the student worksheet.

→ Tell students they will measure different amounts of rain in one rain gauge and then repeat the procedure with a second rain gauge.

Suppose you want to predict the volume of water when the height is 2.0 cm. You can either use the graph to find the answer or use math to calculate it. [For review of this math topic, see **Math Spotlight**: Graphing on page A-8.]

Method 1: First, find the height, 2.0 cm, on the *x*-axis. Move vertically from this spot until you reach the slanted line. Now move horizontally until you reach the *y*-axis. This point on the line corresponds to a volume of 4.0 cm³.

Method 2: Notice that for every data point in the table, the volume is always 2 times the height. The volume for a height of 2.0 cm is 4.0 cm³. Similarly, the volume for a height of 6 cm would be 6 · 2 = 12 cm³.

Proportional Relationships

The relationship between a container's volume and its height is a **proportional** one. This means that volume and height are related to one another by a single number called the **proportionality constant.** When the height is multiplied by this number, you get the volume. In the example we just used, the height and volume of rain in the rain gauge were related by the number 2.0. In other words, for this container, the volume is always 2.0 times the height, and 2.0 is the proportionality constant.

Volume of rain = (proportionality constant) · (height of rain)

A proportionality constant can be represented by the letter *k* in a math formula.

Volume = *k* · height

In the case of the rain gauge, the proportionality constant is also equal to the area of the base.

Volume = 2.0 cm² · height

↑
Proportionality constant, *k*

The graph provides another way to look at proportional relationships. Whenever two variables are proportional to each other, the graph of the data points is *always* a straight line passing through the origin (0, 0). [For review of this math topic, see **MATH Spotlight**: Ratios and Proportions on page A-11.]

Volume and height are proportional for containers with a uniform shape and parallel walls.

Volume *is proportional to height.*   Volume *is not proportional to height.*

Amount of rainfall is not proportional to time. This is because rainfall is not necessarily steady. It can rain a lot and then a little.

Important to Know In the lab, volume is measured in milliliters, mL, in a graduated cylinder. Conveniently, 1 mL = 1 cm³.

AGRICULTURE CONNECTION

Rain gauges often have a wide funnel to collect water into a narrow tube. Because the collection area is large, the height corresponding to 1 inch of rain in the rain gauge is much larger than an actual inch. Rain gauges allow a farmer to see these "large inches" of water from inside his or her home without having to walk out to the field in the rain.

© Matt Dutty/Alamy

EXPLAIN & ELABORATE (15 min)

TRM Transparency—Volume Versus Height of Water 50

DISCUSS THE GRAPH

Display the Transparency—Volume Versus Height of Water. At the appropriate time in the discussion, draw a line or curve through each set of points.

Ask: Why do the data points lie more or less along a straight line for the beaker but not for the Florence flask?

Ask: Which would make a better rain gauge, the beaker or the Florence flask?

Explain your thinking.

Ask: How did you made your predictions for the volume of 10.0 cm of water in the 400 mL beaker? Explain.

Key Points: The data points show that volume increases in a steady and predictable way in relation to the height of the beaker. There are several good ways to make volume predictions for the beaker from the data. They are outlined here:

• Use the data in the data table to estimate the volume at a different height.

• Use the straight line on the graph to predict other data points.

• Calculate new volume values by using proportions. For example, in the same beaker, 50 cm of rain will produce 10 times as much volume as 5 cm of rain.

All these methods are possible because the height and volume of a container are proportional to each other for the container. The data points for the beaker lie on a straight line going through zero. Whenever a graph of two variables results in a straight line that passes through the origin, (0, 0), the two variables are proportional to each other. The data points for the Florence flask do not lie on a straight line. Thus, volume is not proportional to height for the flask.

> **Proportional:** Two variables are directly proportional when you can multiply the value of one by a constant to obtain the value of the other.

The graphed line and the math associated with the line represent an ideal. This is why the data points measured in the experiment for the beaker are not all exactly on the line drawn. There is always some error in measuring data. The line describes what would happen if your data points were the result of perfect measurements. Math and measurement do not always match exactly. Nevertheless, the math allows you to make accurate predictions.

EXPLORE PROPORTIONAL RELATIONSHIPS

Ask: How are the volume and height of a container with parallel walls related to each other mathematically?

Ask: If the amount of rainfall increases, do both the volume and the height of water in the rain gauge keep track of this increase? Explain your thinking.

Ask: Why do meteorologists report the height of rainfall instead of the volume of rainfall?

Ask: What is another proportional relationship?

Key Points: The volume of rainfall increases regularly in relation to the height of the rainfall. This is because height is proportional to volume. The height of rain collected in a rain gauge does not depend on the diameter of the container. A rainstorm that drops 1 in of rain in one container drops 1 in of rain in any other container (provided the walls of the containers are parallel). The volume of rain collected does depend on the diameter of the container. After a storm, a container with a larger diameter will have more water in it than one with a smaller diameter.

If meteorologists reported rainfall in terms of volume, they all would have to use a rain gauge with the exact same size and shape to be able to compare measurements. Thus, meteorologists report rainfall in inches, centimeters, or millimeters.

There are countless examples of proportional relationships in the world around you. For example:

- The number of steps in a staircase is proportional to the total height of the staircase.
- The distance traveled by an automobile going a certain speed is proportional to the amount of time it is driven.
- The number of pages in a book is proportional to the thickness of the book.
- The number of eggs you have is proportional to the number of cartons of eggs you have.
- The relationship between centimeters and inches is proportional.

All these proportional relationships can be expressed mathematically.

- The relationship between volume and height of rain gauges can be expressed mathematically by this formula:

  Volume = (area of base) · (height)

- The relationship between distance and time traveled by an object can be expressed by this formula:

  Distance = (speed) · (time)

- The relationship between centimeters and inches is expressed as

  Inches = (2.5) · (centimeters)

In each case, one number in the equation does not change. If the area of the base does not change, that becomes a constant in the first equation. Similarly, if the speed is steady (e.g., 55 mi/h on the highway), speed becomes the constant in the second equation. This unchanging number is often referred to as the *proportionality constant*. You can see that if the variable on one side of the equation is doubled, or tripled, or quadrupled, the variable on the other side of the equation must also be doubled, tripled, or quadrupled to satisfy the equality.

KEY TERMS
proportional
proportionality constant

Everywhere around the world, the amount of rainfall is measured by its height. If volume were used to report rainfall, each meteorologist would report a different number, depending on the area of the base of the rain gauge used. Height and volume for a specific rain gauge are related by a proportionality constant, k.

Exercises

Reading Questions

1. Suppose you have a cylindrical rain gauge with a base area of 2.0 cm². Explain two different ways you can determine the volume when the height of water is 3.0 cm.

2. Explain in your own words why meteorologists prefer to measure rain in inches or centimeters, not in milliliters or cubic centimeters.

Reason and Apply

3. If the amount of rainfall increases, do both the volume and height of water in the rain gauge keep track of this increase? Explain your thinking.

4. If you use a beaker for a rain gauge and the weather station uses a graduated cylinder, will both instruments give the same volume? The same height?

5. If a large washtub, a dog's water dish, and a graduated cylinder were left outside during a rainstorm, would the three containers all have the same volume of water in them after the storm? Explain why or why not.

6. Inches are used in the United States, but centimeters are used by scientists and in the rest of the world. Look at a ruler that is marked in both inches and centimeters.
   a. Make a graph of centimeters versus inches so that the y-axis is centimeters and the x-axis is inches.

   b. Convert 12 in. to centimeters.
   c. How many inches is 10 cm?
   d. How many inches is 1 cm?

7. A student placed the same empty rain gauge outside before five different rainstorms. She measured the height and volume of the water in the gauge after each storm. Her results are in the table.

Storm number	Height (cm)	Volume (cm³)
1	1.0 cm	2.5 cm³
2	2.5 cm	6.3 cm³
3	0.5 cm	1.1 cm³
4	8.0 cm	20.0 cm³
5	5.0 cm	12.5 cm³

   a. Which rainstorm dropped the most rain? Did you use height or volume to answer this question? Why?
   b. What pattern do you notice?
   c. Draw a graph of the data to show that the volume and height are proportional.
   d. Explain why the data points do not all lie exactly on a straight line.
   e. Predict the volume for a height of 6.0 cm. Show your work.

Proportionality constant: The number that relates two variables that are proportional to each other. It is represented by a lowercase k.

EVALUATE (5 min)

Check-In

Suppose you find that a cylindrical rain gauge contains a volume of 8 mL of rain for a height of 2 cm of rain. Describe how you might figure out the volume of rain for a height of 10 cm of rain in this same container.

Answer: Graph the data point and draw a line through this point and the origin, (0, 0). Then find the volume for 10 cm. Or use proportionality. Because the new height, 10 cm, is 5 times 2 cm, the new volume will be 5 times 8 mL, or 40 mL.

Homework

Assign the reading and exercises for Weather Lesson 50 in the student text.

LESSON 50 ANSWERS

Answers continue in Answer Appendix (p. ANS-4).

Having a Meltdown

Density of Liquids and Solids

THINK ABOUT IT

Much of the country depends on snowfall for the year's water supply. The mountains get many feet of snow every winter, and this snow eventually melts and travels from creek to stream to river to reservoir. Scientists measure the snowpack, or the total amount of snow that has accumulated on the ground, to predict the amount of water that will be available for consumption the rest of the year. But is the amount of water in a foot of snow in the mountains equal to the amount in a foot of rainfall?

How much water is present in equal volumes of snow and rain?

To answer this question, you will explore

- Density and Phase
- Converting Snowfall to Rainfall
- The Density of Ice

EXPLORING THE TOPIC

Density and Phase

Just as rainfall is measured in inches, snowfall can also be measured in inches. All you need is a ruler. However, scientists who study water distribution are more interested in how much liquid water the snow represents.

When snow melts, its volume decreases. However, its mass is still the same. This is because the same water molecules are present in the frozen snow and in the melted snow.

If the volume of the sample changes but its mass stays the same, then mathematically, the density of the sample has to change. The density of snow must be lower than the density of water.

Recall that density is the mass per unit volume.

$$D = \frac{m}{V}$$

Unlike water, snow has a wide range of densities. Snow can be fluffy or packed. So, the volume of liquid water contained in any snow sample depends on the density of the snow being considered.

Snow · · · · · · · Water

Mass is not lost, but the volume decreases.

Lesson 51 | **Density of Liquids and Solids** · · · · 267

Overview

LAB: GROUPS OF 4

Key Question: How much water is present in equal volumes of snow and rain?

KEY IDEAS

When a substance changes phase, its density also changes. For instance, water has different densities depending on whether it is in the form of snow, ice, or rain. Density relates mass and volume. If you know the density of snow, you can calculate the volume of water obtained when a given volume of snow melts.

LEARNING OBJECTIVES

- Make density calculations, converting volumes of liquids and solids.
- Explain how phase changes affect the density of a substance.
- Use density equations to calculate the volume of water in a sample of snow or ice.

FOCUS ON UNDERSTANDING

- For the relationship between mass and volume, the proportionality constant, k, is also the density, D. The proportionality constant does not always have a name.
- Many students have trouble rearranging equations to solve for the value they need.

IN CLASS

Students begin class by exploring the idea that snow and liquid water have different densities. They then design a procedure to measure the volume and mass of different amounts of water to determine the density of water. Finally, they use their data to calculate the volume of water obtained when a given volume of snow melts. Students practice working with proportional relationships. The worksheet gives them the opportunity to use the formula $D = m/V$, which they learned in Unit 1: Alchemy. This time they also see the relationship graphically, with a graph relating mass to volume. They can use a Math Card, which summarizes calculations and conversions for proportional relationships. The fact that the slope of the line is the proportionality constant is covered in the reading for this lesson in the student text. A complete materials list is available for download.

TRM Materials List 51

SETUP

If snow is available where you live, this class could easily be modified to allow students to collect, measure, and melt snow and calculate the densities of both water and snow.

The Math Cards are optional. They give students a scaffold until they become more familiar with the proportional relationships in this unit. All the Math Cards for this unit have been included in the Teaching Resource Materials for this lesson. Only the first one pertains to this lesson, but you might copy and cut them all at the same time, saving the rest for future lessons. Students will also need the Triangle Instructions card, which describes how to use the Math Cards.

ENGAGE (5 min)

TRM PowerPoint Presentation 51

ChemCatalyst

Water resource engineers measure the depth of the snowpack in the mountains during the winter months to predict the amount of water that will fill lakes and reservoirs the following spring.

1. Do you think 3 inches of snow is the same as 3 inches of rain? Explain your reasoning.
2. How could you figure out the volume of water that will be produced by a particular depth of snow?

Sample answers: 1. Students might say that 3 inches of snow does not contain as much water as 3 inches of rain because the snow is fluffier or less dense than rain. **2.** Students might suggest melting the snow to measure the volume of water contained in it.

→ Discuss how to compare amounts of snow and rain.

Ask: How is snow different from rain? How is it the same?

Ask: Does 10 g of snow have the same mass as 10 g of water? Explain.

Ask: Which do you think will be denser, snow, ice, or rain? Explain your reasoning.

EXPLORE (15 min)

TRM Worksheet with Answers 51

TRM Worksheet 51

INTRODUCE THE LAB

→ Arrange students into groups of four.

→ Pass out the student worksheet.

→ Tell students that they will measure the mass of four different volumes of water and use their data to determine the density of water.

Leave plenty of time for the Explain & Elaborate portion of the class.

EXPLAIN & ELABORATE (20 min)

TRM Card Masters—Math Cards 51

TRM Transparency—Mass Versus Volume of Water 51

DISCUSS DENSITY CHANGES ASSOCIATED WITH PHASE CHANGES

→ Display the transparency of the graph showing mass versus volume for water in different phases.

Ask: How did you figure out the density of liquid water? What value did you obtain?

Ask: How do the densities of different volumes of liquid water compare?

Ask: How do the densities of snow and ice compare to the density of liquid water?

Ask: What happens to the density of a substance when it changes phase? What evidence do you have to support this idea?

Ask: How do the graphs differ for ice, water, and snow? How are they the

same? (They have different slopes, or different steepness; each is a straight line going through the origin.)

Key Points: You can determine the density of water by measuring the mass of a certain volume of water. The density, *D*, is the mass, *m*, divided by the volume, *V*:

$$D = \frac{m}{V}$$

The ratio of mass to volume for liquid water is 1.0 g/mL (at room temperature and sea level). The value is the same for any amount of water. The density of a small glass of water and a large

pool of water is the same, 1.0 g/mL, as long as both are the same temperature. In both cases, 1 mL of water will have a mass of 1 g.

The densities of snow and ice are less than the density of water. Ice has a density of 0.92 g/mL. In other words, 1 mL of ice has a mass of only 0.92 g, so it floats in liquid water. Snow is even less dense, approximately 0.50 g/mL, depending on the type of snow. The graph you created allows you to compare the densities of the different phases of water. The differences in density for the three phases are reflected in the steepness of the lines. The steeper the line made by

---

ENVIRONMENTAL CONNECTION

Snowpack density can vary from about 0.1 g/mL to 0.5 g/mL, depending on conditions. Dry, fluffy snow is less dense than wet, slushy snow, and therefore easier to shovel.

MASS AND VOLUME ARE PROPORTIONAL

The equation for density relates mass and volume.

As long as the density of the substance doesn't change, its mass and volume are proportional. The proportionality constant here is the density, *D*.

$$m = DV$$

Because mass and volume are proportional to each other for a specific substance, the data points lie on a straight line that goes through the origin. Examine the graphs of mass versus volume for rain, snow, and ice. How does the slope of the line relate to the density?

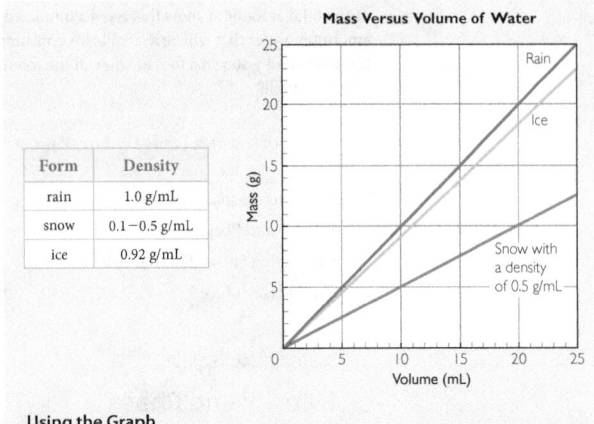

Form	Density
rain	1.0 g/mL
snow	0.1–0.5 g/mL
ice	0.92 g/mL

Using the Graph

The line for rain is steepest because liquid water has the greatest density. For every 1.0 mL increase in rain volume, the mass of the water increases by 1.0 g. The line for snow is the least steep because the density of snow is less than that of both rain and ice. For snow, a 1 mL increase in volume corresponds to only a 0.5 g increase in mass.

Big Idea Matter changes density when it changes phase.

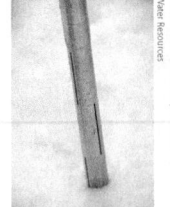

Scientists called hydrologists measure snowpack by weighing an aluminum tube filled with snow.

↻ Converting Snowfall to Rainfall

CALCULATING THE DENSITY OF SNOW

One way to determine the volume of water in a snowpack is to melt the snow. But melting a large amount of snow is inconvenient.

Another way to study the water volume of the snowpack is to determine the density of the snow and calculate the volume of liquid water it represents. When new snow has fallen in the mountains, hydrologists take samples of the snow with a large aluminum tube. They push the empty tube all the way down into the snow. Then they weigh the tube with the snow still inside. The density of the snow is the mass divided by the volume.

COMPLETE WORKED EXAMPLES

➡ Write the following sample problems on the board and have students solve them before discussing the solutions. You can also revisit the ChemCatalyst questions at this time.

➡ Display the transparency of the graph for student reference.

➡ Hand out a Density Math Card and a Triangle Instructions Card to each student (optional). Explain that these cards act as guides in using the density equation to solve problems for any of the three variables. Show students how to use the cards.

Example

Volume of Water from Snow Melt

Suppose you have a volume of 20 mL of snow with a density of 0.25 g/mL. What volume of liquid water do you get when you melt this amount of snow?

Solution

When the snow melts, its mass will not change. Use the density equation to convert between mass and volume.

Use the volume and density of snow to find its mass.

$$m_{snow} = D_{snow} V_{snow}$$
$$= (0.25 \text{ g/mL}) \cdot (20 \text{ mL}) = 5 \text{ g}$$

The mass of snow is equal to the mass of water.

$$m_{snow} = m_{water} = 5 \text{ g}$$

Rearrange the density equation to find the volume of water.

$$V_{water} = \frac{m_{water}}{D_{water}} = \frac{5 \text{ g}}{1.0 \text{ g/mL}} = 5 \text{ mL}$$

When 20 mL of snow with a density of 0.25 g/mL melts, you get 5 mL of water.

You can also solve this problem using a graph. The graph here shows mass versus volume for liquid water, as well as snow with a density of 0.25 g/mL.

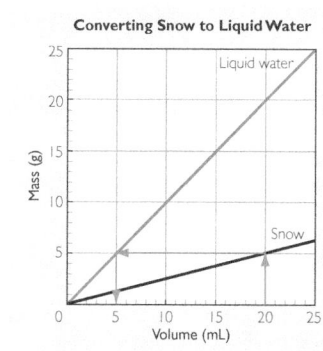

Converting Snow to Liquid Water

Using the Graph
The arrows on the graph show how to convert from snow volume to liquid water volume. You begin with 20 mL of snow, which is equivalent to 5 g of snow. When snow melts, the mass remains the same. 5 g of liquid water is equivalent to 5 mL of snow.

SPORTS CONNECTION

Skiers and snowboarders enjoy fresh "powder." This is the term for untouched, freshly fallen snow with a particularly low density. Powder is perfect for landing on because it forms a natural cushion. This means that it does not hurt as much when you land on it compared with snow that is compacted.

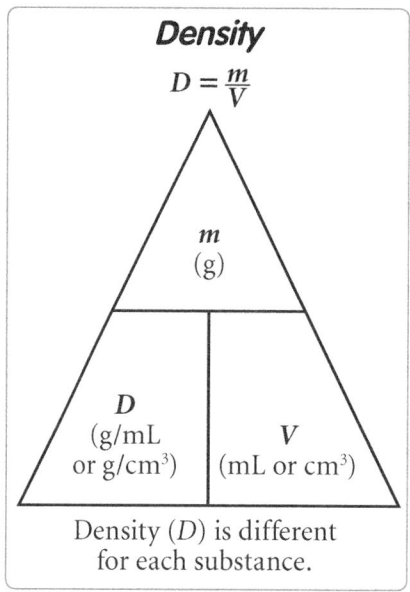

Density

$$D = \frac{m}{V}$$

m (g)

D (g/mL or g/cm³) V (mL or cm³)

Density (D) is different for each substance.

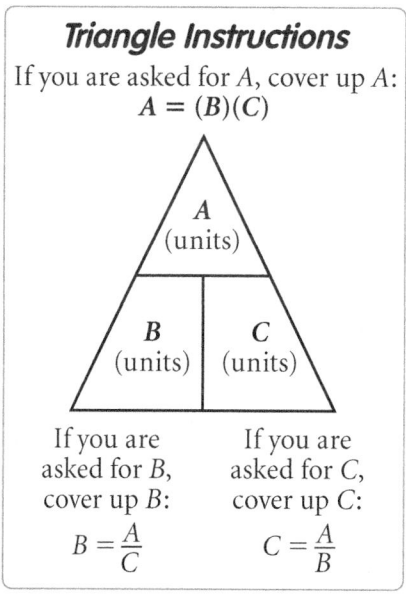

Triangle Instructions

If you are asked for A, cover up A:
$$A = (B)(C)$$

A (units)

B (units) C (units)

If you are asked for B, cover up B:
$$B = \frac{A}{C}$$

If you are asked for C, cover up C:
$$C = \frac{A}{B}$$

---

connecting the data points, the greater the density of the substance.

DISCUSS THE PROPORTIONAL RELATIONSHIP m = DV

Ask: If you know the volume of a sample of snow, how can you find its mass without using a scale?

Ask: What evidence do you have that mass and volume are proportional to each other?

Ask: How are density units related to the units for mass and volume?

Key Points: The relationship $D = \frac{m}{V}$ can also be written as $m = DV$. When density

is constant, the mass and volume are proportional. The proportionality constant is the density. Notice that the unit of mass is grams, the unit of volume is milliliters or cubic centimeters, and the unit of density is g/mL or g/cm³.

For any proportional relationship, the graph is a straight line that passes through the origin, (0, 0). The three lines students graphed on the worksheet represent proportional relationships. The slope of each line is equal to the proportionality constant.

Ask: If you have the same mass of rain and snow, will the volumes occupied by the rain and the snow be the same or different? Explain.

Ask: If you have 100 mL of snow, how would you determine the volume of the same mass of rain?

Key Points: Scientists measure snowpack in terms of depth—meters or feet—and then make conversions to obtain the volume of water. This is another application of the law of conservation of mass.

EXAMPLE 1

Imagine that you have a box with volume 14.5 mL. What mass of ice will just fill this box?

SOLUTION

The mass is equal to the density times the volume.

$$m = DV$$
$$m = 0.92 \text{ g/mL} \cdot 14.5 \text{ mL} = 13 \text{ g of ice}$$

EXAMPLE 2

You have 12 g of snow with density 0.50 g/mL. What volume does this snow occupy in milliliters?

SOLUTION

Solve the mathematical equation for volume, then substitute known values.

$$m = DV$$
$$V = \frac{m}{D}$$
$$V = \frac{12 \text{ g}}{0.050 \text{ g/mL}} = 24 \text{ mL}$$

EXAMPLE 3

If you have 100 mL of snow, what volume of water do you have?

SOLUTION

When the snow melts, the volume will change but the mass will stay the same. First, determine the mass of snow. Assume the snow has a density of 0.5 g/mL.

$$m = DV = 0.5 \text{ g/mL} \cdot 100 \text{ mL}$$
$$= 50 \text{ g of snow}$$

When the snow melts, you get the same mass of water.

$$50 \text{ g of snow} = 50 \text{ g of water}$$

Finally, determine the volume of liquid water. The density of water is 1.0 g/mL.

$$V = \frac{m}{D} = \frac{50 \text{ g}}{1.0 \text{ g/mL}}$$
$$= 50 \text{ mL}$$

↻ The Density of Ice

The density of ice is 0.92 g/mL. It is less than the density of liquid water, which is 1.0 g/mL. Because water becomes less dense when it freezes, its volume increases. This unique property of water is due to a special kind of intermolecular force called *hydrogen bonding.* In both liquid and solid water, the hydrogen atoms in one water molecule attract the highly electronegative oxygen atoms in another. When water freezes, the molecular motion slows down and the molecules lock into a hexagonal structure due to hydrogen bonding.

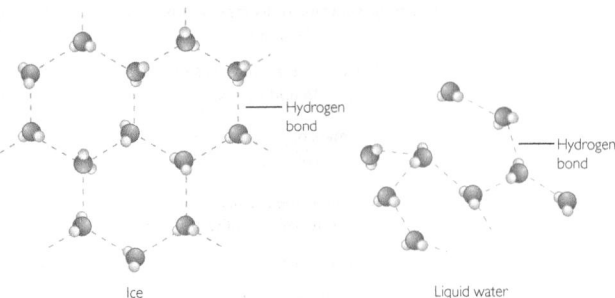

Ice   Liquid water

LESSON SUMMARY

How much water is present in equal volumes of snow and rain?

The density of snow is less than the density of liquid water. You can use mathematical equations or graphs to convert the volume of snow to the volume of liquid water. To use graphs, you plot two lines of mass versus volume, one for snow and another for liquid water. The slope of each line is the density of that substance. To use mathematical equations, you have to know the density of the snow and the density of liquid water.

Exercises

Reading Questions

1. Explain how to use mathematical equations to convert the volume of snow in a snowpack to the volume of water in the snowpack.

2. Explain how to use graphs to convert the volume of snow in a snowpack to the volume of water in the snowpack.

Reason and Apply

3. How are snow and ice different?

4. When ice melts, will the liquid water occupy more or less volume than the ice? Explain your thinking.

5. Suppose that you melt 24 mL of ice. What is the volume of liquid water that results?

Because water is denser, the volume of liquid water is less than the volume of snow.

EVALUATE (5 min)

Check-In

1. Imagine that you have equal masses of snow and rain. Which has a greater volume? Explain your thinking.

2. What is the mass of 14 mL of rainwater?

Answers: **1.** Snow is less dense than rain and will occupy a larger volume

for a given mass. **2.** The mass of 14 mL of rainwater is 14 g.

Homework

Assign the reading and exercises for Weather Lesson 51 in the student text.

LESSON 51 ANSWERS

1. You can convert the volume of snow in a snowpack to a volume of water by using the mathematical equation for density, $D = \frac{m}{V}$. If you know the volume and density of snow, you can determine the mass of the snow in the snowpack. If the snow melts, the mass of the water will be

Exercises

(continued)

6. Which has more water for equal volumes of snow: snow with a density of 0.5 g/mL or snow with a density of 0.25 g/mL? Explain your thinking.

7. Suppose you have a box with a volume of 17.5 mL.
   a. If you fill this box with ice, what mass of ice do you have? (The density of ice is 0.92 g/mL.)
   b. If you fill this box with liquid water, what mass of liquid water do you have? (The density of liquid water is 1.0 g/mL.)

8. Suppose that you have a box that is full and contains 500 grams of a substance.
   a. What is the volume of the box if the substance inside is corn oil? (The density of corn oil is 0.92 g/mL.)
   b. What is the volume of the box if the substance inside is lead? (The density of lead is 11.35 g/mL.)

9. Lead, Pb, is more dense than iron, Fe.
   a. Which occupies a larger volume: 4.3 g of Pb or 4.3 g of Fe? Explain your thinking.
   b. Which has a larger mass: 2.6 mL of Pb or 2.6 mL of Fe? Explain your thinking.

7. a. 16.1 g. b. 17.5 g.

8. a. 543 mL. b. 44 mL.

9. a. 4.3 g of iron, Fe, occupies a greater volume because volume increases as density decreases, and iron is less dense than lead. b. 2.6 mL of lead, Pb, has a larger mass because mass increases as density increases, and lead is more dense than iron.

Lesson 51 | **Density of Liquids and Solids**   271

the same as the mass of the snow. Then you can calculate the volume of water by dividing the mass by the density of water, which is equal to 1 g/mL.

2. You can use a graph of mass versus volume to determine the mass of water in the snowpack given the volume, as long as you have the graph for the right density of snow. Read the mass of snow that corresponds to the volume of snow. Then use a graph that shows the mass versus the volume of water to read the corresponding volume of water for the same mass.

3. *Possible answer:* Snow and ice are different because there is air between the crystals of ice in a snowpack. This gives the snow a different density from that of pure ice. Ice has a solid structure, whereas snow is composed of individual crystals.

4. When ice melts, the liquid water occupies less volume than the ice because the density of water is greater than the density of ice. Because the density is greater, the volume is less.

5. 22.1 mL

6. Snow that has a density of 0.5 g/mL contains twice as much water as an equal volume of snow that has a density of 0.25 g/mL because a higher density means there is a greater mass of material in the same volume.

Overview

LAB: GROUPS OF 4

Key Question: How is temperature measured?

KEY IDEAS

The volume of matter changes in response to changes in temperature. Almost all substances expand on heating and contract on cooling. A thermometer is constructed to measure the volume change of a substance in response to temperature changes. Then it is calibrated to reflect temperature. Physical phenomena that are reproducible, such as the boiling point and freezing point of water, are used to construct temperature scales.

LEARNING OBJECTIVES

- Create a thermometer and a temperature scale.
- Describe how a thermometer works.
- Explain the Fahrenheit and Celsius temperature scales.

FOCUS ON UNDERSTANDING

- The idea that thermometers do not measure temperature directly but instead measure the expansion of a liquid or a gas may not have occurred to students.
- Students may have trouble grasping the idea that a temperature scale is fairly arbitrary and that they can create any temperature scale they want provided it accurately reflects at least two standard repeatable temperatures.
- The gas thermometer can be difficult to understand, because the trapped gas is not visible.

KEY TERMS

melting point (melting temperature)
boiling point (boiling temperature)

IN CLASS

Students use a rudimentary thermometer made from glycol, a plastic straw, a vial, and a rubber stopper. This instrument is placed in water of varying temperatures, and the height of the liquid is marked. Students calibrate their thermometer and figure out the current temperature in the room. They also make a rudimentary gas thermometer and explore the expansion and contraction of air in response to temperature. Students are introduced to the Fahrenheit and Celsius temperature scales.

Thermometers

Purpose

To examine how the volume of a liquid and a gas change in response to temperature.

Part 1: Liquid Thermometer

Procedure

Follow the handout instructions to build a thermometer like the one shown here. Then mark the liquid level in the straw for these conditions.

- room temperature
- vial warmed by your hand
- ice water
- ice water with 1 tablespoon of salt per 200 mL water
- boiling water (do not allow the thermometer to touch the bottom of the beaker)

Observations and Analysis

1. What did you observe? What is happening to the liquid in the vial to make it move up and down in the straw?

2. Create a scale for the thermometer. Assign numbers for the places you marked on the straw for boiling water and ice water. What numbers did you choose and why?

3. Based on your newly created temperature scale, estimate the temperature in the room. How did you arrive at your answer?

Part 2: Gas Thermometer

Materials

- 250 mL beakers (3)
- ice
- 10 mL graduated cylinder
- food coloring
- test-tube holder—wire
- hot plate

Procedure

1. Set up three beakers: one with ice water, one with room-temperature water, and one with hot water (cooled enough so that it is no longer steaming). Add 1–2 drops of food coloring to each beaker.

2. Hold the 10 mL graduated cylinder with a test-tube holder. Invert the graduated cylinder and immerse it in each beaker for one minute.

Making Sense

Describe how a thermometer works.

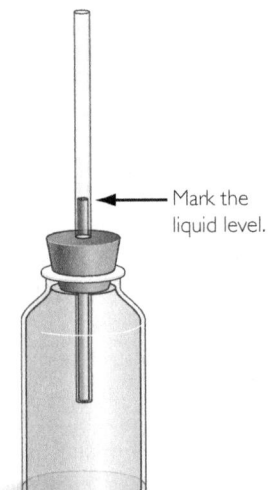

← Mark the liquid level.

A complete materials list is available for download.

TRM   Materials List 52

SETUP

You will need about 2 lb of crushed ice per class. Assemble eight ethylene glycol thermometers. Use a cork borer to put a hole through each rubber septum or stopper. Put the hard plastic straw through the hole. (Petroleum jelly helps the straw slip through. Take care not to get petroleum jelly inside the straw.) Fill the vials with ethylene glycol so that the liquid level is all the way

---

Hot Enough

Thermometers

THINK ABOUT IT

On a single day, a thermometer in Anchorage, Alaska, might read −15 degrees, while a thermometer in Sydney, Australia, reads 35 degrees, and a thermometer in your classroom reads 68 degrees. It is interesting to consider how a small glass tube with some red liquid inside can indicate how hot or cold it is.

This thermometer shows the two most widely used temperature scales, Fahrenheit and Celsius.

How is temperature measured?

To answer this question, you will explore

- Changes in Volume
- Creating Temperature Scales
- Fahrenheit Versus Celsius

EXPLORING THE TOPIC

Changes in Volume

Matter responds to increasing temperature in a variety of ways. A wad of bubble gum gets softer, a "mood" ring changes color, and the air in a balloon expands. By systematically tracking any one of these changes, you could make a thermometer.

However, to make a *good* thermometer, it is better to track a physical change that is reproducible and easily measurable.

The liquid crystals in a mood ring change color in response to changes in temperature.

One way to make a thermometer is to fill a thin glass or plastic tube with a liquid. As the liquid in the thermometer is heated, its volume and height increase. As the liquid is cooled, its volume and height decrease. Because volume and height are proportional, you can detect the increase in volume as an increase in height.

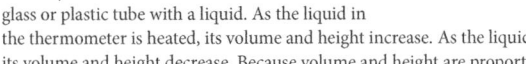
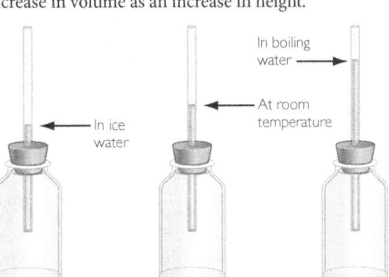

In ice water

At room temperature

In boiling water

Lesson 52 | **Thermometers**   273

---

to the top. You might want to add a few drops of food coloring to the ethylene glycol so it is easier to see in the straw. Put the septum with the straw on the vial so that liquid rises in the straw. You can have students place 250 mL beakers with water on hot plates before the ChemCatalyst so they do not waste time waiting for the water to boil.

SAFETY

- Students should wear safety goggles at all times.

CLEANUP

You can use the previously assembled thermometers for another class. Use alcohol to wipe off any marks left on the straws.

Differentiate

To support scientific literacy, you might assign students to research the history of thermometers and temperature scales using a "jigsaw" activity. Assign small groups of students a temperature scale to become an expert on. There are many available—Fahrenheit, Celsius, Kelvin, Rankine, Rømer,

---

Newton, Delisle, Réaumur. Students in these groups become "experts" on their scale, knowing why it was developed, who uses it, why it is popular or unpopular, etc. Then mix students so that there is an "expert" for each scale in a new group and allow students to confer and share their learning. You might give them an open-ended topic to discuss as a mixed group, such as "How does society decide which temperature scale to use?"

Lesson Guide

ENGAGE (5 min)

TRM PowerPoint Presentation 52

ChemCatalyst

The weather forecast in Moscow, Russia, calls for a 60% chance of precipitation with highs reaching 30 °C, while in Washington, D.C., the weather forecast calls for a 70% chance of precipitation with highs reaching 50 °F.

1. Which city will be warmer? Explain your thinking.
2. Do you think it will rain or snow in either of the two cities? Explain your reasoning.

Sample answers: 1. Students might confuse the two temperature scales. Moscow will be warmer because 30 °C (equivalent to 86 °F) is warmer than 50 °F. 2. Some students will confuse the two scales and say it will snow in Moscow. However, it will not snow in either place because it is too warm.

→ Solicit students' ideas about how a thermometer works.

Ask: What weather did you predict for the two cities? Why?

Ask: What is the difference between degrees Celsius and degrees Fahrenheit?

Ask: What is temperature? How is it measured?

Ask: Why is a liquid used in a thermometer? (Students may be more familiar with digital thermometers, which rely on the fact that electron resistance changes with temperature.)

EXPLORE (20 min)

TRM Worksheet with Answers 52

TRM Worksheet 52

INTRODUCE THE ACTIVITY

➡ Arrange students into groups of four.

➡ Pass out the student worksheet.

➡ Briefly go over the procedure for the liquid thermometer. Show students the preassembled ethylene glycol thermometer. Show students how to hold the thermometer with a test-tube holder in water at different temperatures. The thermometer should not touch the sides or bottom of the beaker. Note that the bottom of the beaker on the hot plate can get hotter than boiling water and that prolonged contact with it can damage the thermometer. Briefly go over the procedure for the gas thermometer. Show students how to measure the gas volume using the markings on the graduated cylinder.

EXPLAIN & ELABORATE (15 min)

DISCUSS THE HEATING AND COOLING OF MATTER

➡ You might sketch the illustrations below for the gas thermometer discussion.

Ask: How does a thermometer work?

Ask: Explain the construction of the homemade liquid thermometer. Why is a straw needed?

Ask: How can you provide evidence that air is trapped inside the graduated cylinder when it is held upside down in the water?

Ask: How can you explain how you can use the sample of trapped air to construct a thermometer?

Key Points: The volume of matter changes in response to changes in temperature. Matter generally expands when heated and contracts when cooled. This property can be used to measure temperature, because the volume of liquid in a thermometer changes as the temperature changes.

Many thermometers contain a liquid, such as alcohol or mercury. The thermometer has a reservoir of liquid connected to a thin glass tube. The glass tube is needed because the volume change is relatively small. Narrowing the area makes the height change easier to detect.

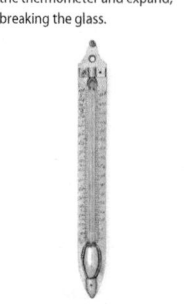

GAS THERMOMETERS

Gases also expand and contract upon heating and cooling. A gas thermometer requires a container with a flexible volume. Suppose you have a sample of air in hot water trapped in an inverted graduated cylinder. If you heat the trapped air sample, some of the air will bubble out of the cylinder. This is evidence that the air is expanding. If you cool the trapped air sample in ice water, water will rise into the cylinder. This is evidence that the air sample contracts as it cools, leaving room for some water to move up into the cylinder. The expansion and contraction of air is important in understanding the weather.

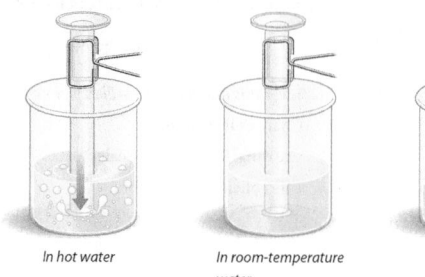

In hot water | In room-temperature water | In ice water

Big Idea Most matter expands in volume as it is heated and contracts as it is cooled. In solids, the change is usually very small. In gases, the change is dramatic.

↻ Creating Temperature Scales

To make a thermometer, you could put a colored liquid in a glass or plastic tube. Then you need to construct a scale, or number line, on the tube to assign numbers to different temperatures. You can construct a scale by measuring two temperatures, such as the boiling point and freezing point of water. It is typical to measure water because water is easily accessible and always boils and freezes at the same temperatures at sea level.

The temperature at which a substance melts or freezes is called the **melting point**, or the **melting temperature**. The temperature at which it boils or condenses to a liquid is called the **boiling point**, or the **boiling temperature**. Each time you measure the height of the liquid at the melting point and boiling point of water, you will get the same two heights on your thermometer.

Once you have two marks, you can divide the length between the marks into a round number of intervals. Each interval is one unit of temperature. The intervals

These two thermometers are different sizes but have the same temperature scale. Both show the same temperature.

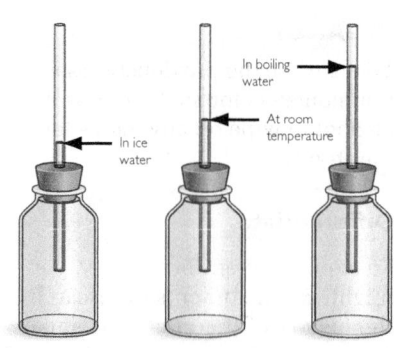

Gases expand and contract as the temperature changes. Water does not fill the inverted graduated cylinder when you push it into the water,

evidence that air is trapped inside. As the sample of trapped air is warmed in the hot water, air bubbles escape. The air trapped inside expands to fill the entire volume. When the graduated cylinder with hot air is placed in water at room temperature, the air sample cools and water rises in the graduated cylinder. This is evidence that the air sample contracts as it cools, leaving room for some water to move up into the graduated cylinder. The volume occupied by the air decreases even more when the air is cooled further with ice.

provide numerical values for other temperatures besides the melting and boiling points of water. For instance, you can use your thermometer to measure the room's temperature. Someone else who makes a thermometer using the same points and intervals that you used will be using the same scale, and you can accurately compare temperatures.

↻ Fahrenheit Versus Celsius

One commonly used temperature scale is the Celsius scale. This scale sets the melting temperature of ice at 0° and the boiling temperature of water at 100° as measured near sea level as opposed to in the mountains (water boils at a lower temperature when you are at higher altitudes). This scale uses degrees Celsius, °C, as units. The Swedish astronomer Anders Celsius created this temperature scale in 1747. Most of the world and most scientists use the Celsius temperature scale.

In contrast, a weather report in the United States gives temperature in degrees Fahrenheit, °F. German physicist Daniel Gabriel Fahrenheit had proposed his temperature scale in 1724. On the Fahrenheit scale, the melting temperature of ice is 32 °F and the boiling temperature of water is 212 °F.

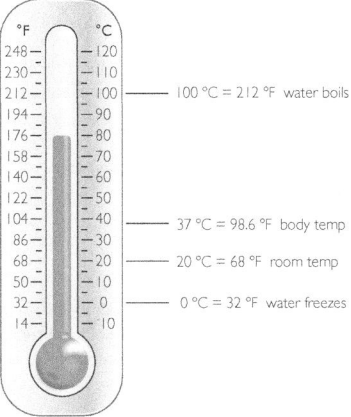

There are 180 Fahrenheit degrees between 0 °C and 100 °C. This means that a Fahrenheit degree unit is a smaller change in temperature than a Celsius degree unit.

In this course, you will use the Celsius scale. It is useful to know how to convert between the two scales. The formula given here allows you to convert from degrees Celsius, C, to degrees Fahrenheit, F.

$$F = \frac{9}{5}C + 32$$

[For review of this math topic, see **MATH Spotlight**: SI Units of Measure on page A-0.]

In hot water   In room-temperature water   In ice water

DISCUSS HOW TEMPERATURE IS MEASURED

Ask: If you know the volume or height of a liquid at two different temperatures, can you predict the temperature for a third volume or height? Explain your thinking.

Ask: How can you use a trapped gas to measure temperature?

Ask: How reliable is boiling water in setting a temperature scale? What about ice water?

Key Points: To set a temperature scale, you need at least two measurements. It is best to choose temperatures that are reproducible to set the scale. For example, ice water and boiling water are at very specific temperatures at sea level (the temperatures vary at different altitudes). In contrast, body temperature and room temperature are not good choices because both can vary from one moment (or one person) to the next.

Melting point or melting temperature: The temperature at which a substance melts or freezes. At this temperature, both solid and liquid phases of the substance are present.

Boiling point or boiling temperature: The temperature at which a substance boils or condenses. At this temperature, both liquid and gas phases of the substance are present.

Note: These definitions of melting point and boiling point refer to phases at equilibrium. Equilibrium is covered in Unit 6: Showtime.

Once you have set a scale with two points, you can determine other temperatures. Imagine that the temperature of ice water is called 0° and that of boiling water is called 60°. You can then make a mark halfway between these points and call it 30°. You can continue to divide up the scale and label it. On this particular scale, room temperature would be about 12°. Once two temperature points are noted on a scale, it is possible to figure out where any other temperature would be on that scale.

RELATE THE HISTORY OF THE THERMOMETER

→ Share all or part of the following information with your class. Students can name their own temperature scales, just as Fahrenheit and Celsius did.

Key Points: In 1724, German physicist Daniel G. Fahrenheit invented the first modern thermometer—the mercury thermometer. To set his scale, he called the temperature of an ice-salt mixture 0 °F and called his own body temperature 96 °F. Then he divided the scale into single degrees. (The scale has since been recalibrated so that normal body temperature is now 98.6 °F.) On his scale, the freezing point of pure water happens to occur at 32 °F (and the boiling point at 212 °F).

Degree: The increment of temperature that corresponds to one unit on a thermometer. The size of a degree depends on the temperature scale used.

In 1747, Anders Celsius, a Swedish astronomer, created a thermometer with a different scale. Celsius used 0° and 100° for the melting point of snow and the boiling point of water, respectively. The Celsius temperature scale is now part of the metric system of measurement (SI). Most of the world and most scientists measure temperature in degrees Celsius, °C. It was formerly called the centigrade scale.

DISCUSS THE RELATIONSHIP BETWEEN °C AND °F

Ask: Which units are larger, Celsius degrees or Fahrenheit degrees? (Celsius degrees)

Ask: Would you be colder at 0 °F or 0 °C? (0 °F)

Ask: When a thermometer reads 100 °C, what is the temperature in degrees Fahrenheit? (212 °F)

Ask: Is there any advantage of one scale or the other?

Key Points: Degrees Celsius and degrees Fahrenheit both measure the same thing (temperature). Celsius units are larger than Fahrenheit units. Thus, a change of one degree Celsius represents a greater change in temperature than does a change of one degree Fahrenheit. To convert from degrees Celsius to degrees Fahrenheit, use this formula:

$$F = \frac{9}{5}(C) + 32$$

or

$$F = 1.8(C) + 32$$

EVALUATE (5 min)

Check-In

The temperature is 37 °C in Spain in July. How does this compare with body temperature, which is 98.6 °F?

Answer: Converting 37 °C to Fahrenheit: $F = \frac{9}{5}(32 \text{ °C}) + 32 = 98.6$ °F. This is identical to body temperature and well above room temperature, so it will feel rather hot.

Homework

Assign the reading and exercises for Weather Lesson 52 in the student text.

LESSON 52 ANSWERS

1. *Possible answer:* Liquids usually expand when they are heated and contract when they are cooled. At any given temperature, the volume of a

---

Example

Weather Forecast

The weather forecast in Tokyo, Japan, calls for a 60% chance of precipitation with temperatures reaching 30 °C, while in Washington, D.C., the weather forecast calls for a 70% chance of precipitation with temperatures reaching 50 °F.

  a. Which city will be warmer? Justify your answer by converting the temperatures to the same scale.

  b. Assuming that there is precipitation in each of these cities, will rain or snow be seen in each place? Explain your thinking.

Solution

To compare temperatures, you convert them to the same scale.

  a. You can convert 30 °C to °F:

$$F = \frac{9}{5}C + 32$$

    Substitute the known value. $F = \frac{9}{5}(30) + 32$

    Solve for the unknown value. $F = 86$

  So, 30 °C = 86 °F. You could also convert 50 °F to °C:

  Tokyo is warmer because 86 °F (30 °C) is higher than 50 °F (10 °C).

  b. Because ice and snow melt at 32 °F (0 °C), you would expect rain rather than snow in both cities.

LESSON SUMMARY

How is temperature measured?

KEY TERMS

melting point (melting temperature)

boiling point (boiling temperature)

Matter changes volume in response to changes in temperature. Matter generally expands when heated and contracts when cooled. This property can be used to measure temperature: The volume of liquid in a thermometer changes as the temperature changes. To set a temperature scale, you need to start with at least two points. Once these two temperature points are noted on a scale, it is possible to figure out where any other temperature would be on that scale.

Exercises

Reading Questions

1. Explain how the height of a liquid can be used to measure temperature.

2. Describe in your own words how to construct a temperature scale.

Reason and Apply

3. What are the advantages of the Celsius temperature scale over the Fahrenheit temperature scale?

---

fixed amount of liquid will always be the same. If the liquid is in a tube, the height of the column of liquid will change predictably with changes in temperature. The key point is that liquids will change volume with temperature and this change can be measured.

2. *Possible answer:* Construct a temperature scale by finding the height of a liquid in a tube at two known temperatures, such as the melting and boiling points of water. Mark these two values on the tube. Then subdivide the length between the marks into equal intervals.

3. *Possible answer:* One advantage of the Celsius temperature scale over the Fahrenheit scale is that it is simpler to use because it is divided into exactly 100 degrees between the reference points 0 °C and 100 °C. Remembering the melting and boiling points of water is easy on this scale. Also, because scientists use the Celsius scale so as to have a common reference, using the Celsius scale makes it easier to interpret scientific data.

4. No, 0 °C refers to the melting point of water. On the Fahrenheit scale, the melting point of water is 32 °F, so 0 °F refers to a temperature colder than the melting point of ice.

Exercises

(continued)

4. When the temperature is 0 °C, is it also 0 °F? Explain.

5. Convert −40 °C to °F. Show your work.

6. Which is larger: one Celsius degree or one Fahrenheit degree? Explain.

7. The doctor tells you that your body temperature is 40 °C. Are you sick? Show your work and explain your answer.

8. You will be traveling to Hawaii where the forecast is for a temperature of 31 °C during the day, dropping to 28 °C overnight. Your friend recommends that you bring clothing for warm weather. Is this a good recommendation? Why or why not?

9. Create a graph comparing the Celsius and Fahrenheit scales.
   a. Plot the freezing point and the boiling point of water on your graph. Draw a straight line connecting the two points.
   b. Use the graph to determine the temperature in °F when it is 10 °C.
   c. If the weather forecast says it is 55 °F in Washington, D.C., what is the temperature in °C?

9. a.

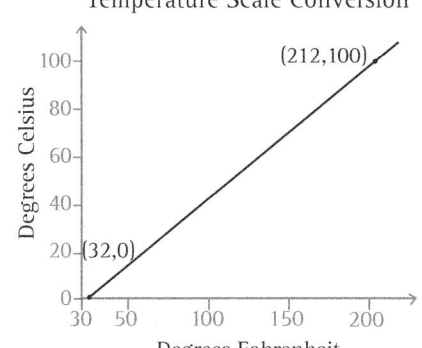

Temperature Scale Conversion

b. 50 °F

c. 13 °C

5. $F = \frac{9}{5}C + 32$

$\quad = \frac{9}{5}(-40) + 32$

$\quad = -72 + 32$

$\quad = -40$

6. One Celsius degree is larger than one Fahrenheit degree because there are 100 Celsius degrees between the melting point and the boiling point of water but there are 180 Fahrenheit degrees between these two points.

7. $F = \frac{9}{5}C + 32$

$\quad = \frac{9}{5}(40) + 32$

$\quad = 72 + 32$

$\quad = 104$

Normal body temperature is about 98.6 °F. A body temperature of 40 °C is the same as 104 °F. Any body temperature over 100 °F indicates fever and probable illness, so you are sick if you have a temperature of 40 °C.

8. *Possible answer:* The recommendation to take warm weather clothes was a good recommendation because the temperatures in the forecast correspond to a temperature range of about 82 °F to 88 °F.

Overview

COMPUTER ACTIVITY: INDIVIDUAL

Key Question: How cold can substances become?

KEY IDEAS

Celsius and Fahrenheit are the two most commonly used temperature scales. The Kelvin scale is a third temperature scale widely used by scientists. The Kelvin scale is based on the idea that the temperature of a gas should be assigned a value of zero when the volume of the gas is zero. Zero Kelvin, referred to as absolute zero, corresponds to −273.15 °C. The average speed of gas particles decreases as temperature decreases.

LEARNING OBJECTIVES

- Describe the relationship between the Celsius and Kelvin temperature scales.
- Explain the concept of absolute zero.
- Describe the motion of gas molecules according to the kinetic theory of gases.

FOCUS ON UNDERSTANDING

- Negative temperatures can be confusing to students—especially the idea that, for negative temperatures, as the temperature decreases, the absolute value of the temperature increases.
- Students might find it baffling that Kelvin units and Celsius units are equivalent in size, given their dramatic difference in values.

KEY TERMS

absolute zero
Kelvin scale
kinetic theory of gases
temperature

IN CLASS

Students complete a worksheet that introduces the Kelvin temperature scale. First, they explore the Celsius scale and contemplate what happens when the volume of a gas drops to zero. The new scale is introduced and compared to other temperature scales. Students observe a simulation depicting the motions of gas particles. Then they explain the expansion and contraction of gases with changing temperature as changes in the average speeds of the particles. A complete materials list is available for download.

TRM Materials List 53

Kelvin Scale

THINK ABOUT IT

The coldest temperature recorded on Earth was in Antarctica: a chilling −89 °C. The temperature is even lower on other planets that are farther away from the Sun. Researchers have recorded temperatures as low as −235 °C on the surface of Triton, a moon of Neptune!

How cold can substances become?

To answer this question, you will explore

↻ Absolute Zero

↻ Kelvin Scale

↻ Molecules in Motion

HISTORY CONNECTION

The French inventor and physicist Guillaume Amontons first proposed the existence of absolute zero in 1702. Although he became deaf in his early teens, he did not allow this to stop him from having a productive scientific career. In addition to his work on temperature and pressure, Amontons developed the gas thermometer.

EXPLORING THE TOPIC

↻ Absolute Zero

The volumes of most substances decrease as the temperature decreases. But there is a limit to how much a substance can shrink. It does not make sense that matter would have a negative volume. Hypothetically, the lowest temperature possible would correspond to a volume of *zero*.

Consider an example. The volume of a gas inside a flexible container is measured as the gas is cooled. Several data points are then plotted on a graph.

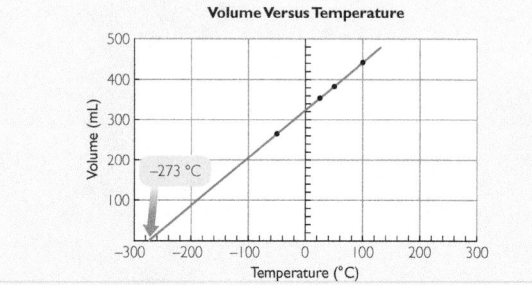

Using the Graph
This graph shows that as the temperature decreases, the volume decreases in a predictable way. The data points lie more or less on a straight line. If you extend this line to the *x*-axis, you can determine the theoretical temperature of the gas at zero volume.

SETUP

Search online for a gas properties simulation, such as the University of Colorado Boulder's PhET simulation: Gas Properties. Set up the projector to display the simulation and run through the simulation before class. You can find a list of URLs for this lesson on the Chapter 10 Web Resources document.

Pre-AP® Course Tip

Students preparing to take an AP® chemistry or college chemistry course should be able to use models and theories to generate predictions or claims occurring in nearly all facets of chemistry. In the second part of this lesson, students use a computer simulation to explain the behavior of gas particles based on the kinetic molecular theory.

Lesson Guide

ENGAGE (5 min)

TRM PowerPoint Presentation 53

More precise measurements reveal −273.15 °C as the temperature at zero volume. Scientists hypothesize that this value corresponds to the lowest temperature possible, and call it **absolute zero.** In actuality, as the temperature is lowered, a gas would condense to a liquid and then to a solid well above this temperature, so zero volume is a property that is still a hypothesis.

Kelvin Scale

On the Celsius scale, the temperature at which the volume of gas is theoretically zero is −273 °C. If you want to create a scale where the temperature is 0 when the volume is 0, you must add 273 to each temperature. This new temperature scale is called the **Kelvin scale.** One kelvin is the same size as one Celsius degree.

$$K = C + 273$$

$$C = K − 273$$

For example, to convert 20 °C to kelvins, add 273. This temperature corresponds to 293 K.

Note that the word *degree,* or the symbol for degree, °, is not used with the Kelvin scale. The unit of temperature on the Kelvin scale is the kelvin, K.

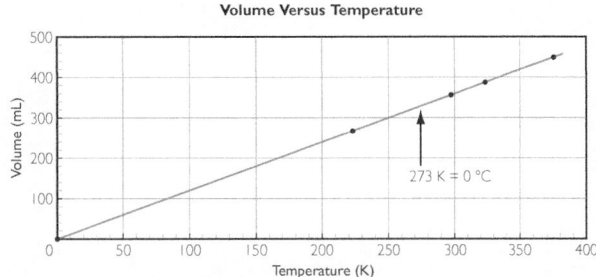

Volume Versus Temperature

273 K = 0 °C

Using the Graph
The Kelvin scale sets the zero point at absolute zero. As a result, all temperatures are positive when they are expressed in kelvins. Notice that this graph is the same as the previous graph, but the *x*-axis has been shifted by 273. The value for 0 °C is 273 K.

Room temperature is often considered to be 20 °C, or 68 °F, so 293 K is a temperature you will often find in problems. If you subtract 273 from a temperature in kelvins, you will be back at the Celsius temperature.

Molecules in Motion

Gases are quite different from liquids and solids. The molecules and atoms in solids and liquids are held close together by forces between molecules and atoms. Liquids lie at the bottom of any container they are in, and solids generally maintain their own shape without a container. In contrast, the molecules in gases

Lesson 53 | **Kelvin Scale**   279

ChemCatalyst

Researchers have recorded the temperature on Triton, a moon of Neptune, as −235 °C.

1. Do you think carbon dioxide, CO_2, would be a solid, a liquid, or a gas at this temperature? Explain your reasoning.

2. What do you think is the coldest temperature something can get to? What limits how cold something can get?

Sample answers: **1.** Carbon dioxide would be a solid at this temperature. Everything becomes a solid if it gets cold enough. **2.** Some students may say zero degrees, while others may note that there are negative Celsius and Fahrenheit temperatures. They might say there is no limit to how cold something can get. Some might say that the lowest temperature occurs when all motion in a sample of matter stops.

➡ Assist students in sharing their initial ideas on cold temperatures.

Ask: What do you think will happen to gaseous carbon dioxide molecules when the temperature gets as low as −235 °C? Explain your thinking.

Ask: Do you think the molecules move closer together or farther apart as they are cooled?

Ask: What is the lowest temperature you think a substance can reach? Explain your thinking.

Ask: What is the smallest volume you think a gas could occupy? Explain your thinking.

EXPLORE (15 min)

TRM Worksheet with Answers 53

TRM Worksheet 53

INTRODUCE THE ACTIVITY

➡ Let students know they will be working individually.

➡ Pass out the student worksheet.

➡ Tell students that they will be introduced to a new temperature scale and shown a computer simulation of the motions of gas particles. Students will work together as a class with the teacher on the worksheet.

GUIDE THE ACTIVITY

For Part 2: Computer Activity, follow these steps:

➡ Select constant volume.

➡ Add about 20 gas molecules.

➡ Ask students to observe the motions of the molecules.

➡ Ask students to observe the motions of the molecules as you raise and lower the temperature.

EXPLAIN & ELABORATE (15 min)

TRM Transparency—Absolute Zero 53

DISCUSS THE KELVIN SCALE

➡ Display the transparency comparing the Celsius scale and the Kelvin scale.

➡ During the discussion, draw vertical lines showing where 0 °C and 0 K each intersect the opposite scale.

Ask: Can the volume of a gas be negative? Why or why not?

Ask: What is the lowest temperature that you think can be reached on the Celsius scale? Explain your thinking.

Ask: Do you think it is possible for a substance to reach absolute zero? Why or why not?

Ask: What might be the advantages of the Kelvin scale?

Key Points: On the Celsius scale, the temperature at which the volume of a gas is theoretically equal to 0 is −273 °C. (The actual value is −273.15 °C.) If you shift the Celsius temperature scale to the left by 273° by adding 273 to each

Lesson 53 | **Kelvin Scale**   279

temperature, the result is a scale on which the temperature will be zero when the volume is zero. This new temperature scale is called the Kelvin scale. The unit of temperature on the Kelvin scale is the kelvin (K). The symbol ° and the word degree are not used with the Kelvin scale. For example, it is correct to say 5 kelvins, not 5 degrees Kelvin.

$$K = °C + 273$$
$$°C = K - 273$$

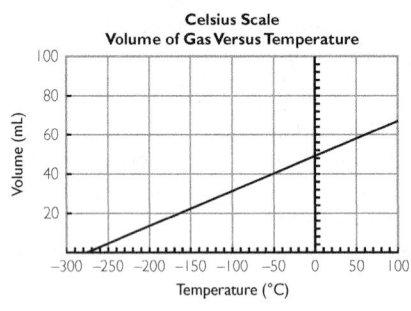

Celsius Scale
Volume of Gas Versus Temperature

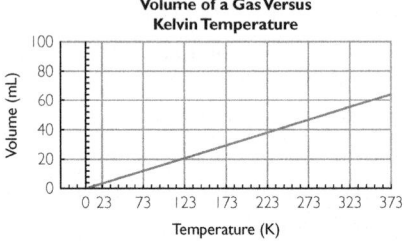

Volume of a Gas Versus Kelvin Temperature

A temperature of 0 K is referred to as *absolute zero*. It is considered to be the lowest temperature that can hypothetically be reached. (In January 2013, though, an experiment by German scientists that achieved temperatures below absolute zero was reported.) The lowest temperature ever recorded on Earth is −89 °C, in Antarctica. This is equivalent to 184 K. The lowest temperature recorded in the solar system was −235 °C, or 38 K, on Triton, a moon of Neptune. At this temperature, the surface of Triton likely consists of oceans of nitrogen and glaciers of methane.

DISCUSS THE COMPUTER ACTIVITY

→ Run the computer simulation again during the discussion portion of the class.

→ To help students see the distribution of speeds, run the simulation at a low temperature.

Ask: How would you describe the motions of the gas particles?

Ask: What happens to the gas particles as they are heated?

Ask: If the container walls were flexible, what would happen?

Ask: How can you explain the change in volume of a gas in terms of the motions of particles?

ASTRONOMY CONNECTION

The Boomerang Nebula is the coldest place known in the universe. It is 5000 light-years from Earth in the constellation Centaurus and has a temperature of 1 K.

KEY TERMS
absolute zero
Kelvin scale
kinetic theory of gases
temperature

disperse to fill whatever space they are in. This indicates that there are no forces between the molecules in a gas.

So, why is it that gas molecules are found throughout their container and not just at the bottom of it? And why is it that the space a gas occupies expands so dramatically as temperature increases? To explain many of the properties of gases, scientists rely on a model called the **kinetic theory of gases.** This model proposes that gas particles—atoms or molecules—are in constant motion. The word *kinetic* comes from a Greek word, *kinetikos,* meaning "moving."

Big Idea On a particle level, all matter is in constant motion.

These illustrations show the motion of three gas particles at three different times, based on the kinetic theory of gases.

Three gas particles are moving in a container.

The gas particles travel in straight lines. The directions are random.

The gas particles bounce off walls and each other.

The gas particles move at different speeds.

One characteristic of gases described by the kinetic theory is that not all gas particles are moving at the same speed. For example, the gas particle shown in blue moves more from frame to frame than the one shown in gray. However, the kinetic theory focuses on the average speed for all the particles at a given temperature. Indeed, the **temperature** of a gas can be defined as a measure of the *average* kinetic energy of the gas particles. When the temperature increases, the average speed of the gas particles increases. Scientists hypothesize that if you could cool matter to absolute zero, the atoms in the substance would stop moving.

Big Idea Temperature is a measure of the average speed of the atoms or molecules in a sample.

LESSON SUMMARY

How cold can substances become?

The Kelvin scale assigns a value of 0 K to the hypothetical temperature of a gas with zero volume. This point is called absolute zero and is at −273 °C. Scientists consider this to be the lowest hypothetical temperature that matter can reach. They hypothesize that all motion stops at absolute zero. The kinetic theory of gases describes the motions of the gas particles. The atoms or molecules in a gas are constantly moving at an average speed that increases with increasing temperature.

Ask: What do you think would happen to the motions of the particles that could cool a gas to 0 K?

Key Points: The model displayed in the simulation is the kinetic theory of gases. This theory maintains that gas particles exhibit these features:

- Gas particles are constantly moving.
- The motion of gas particles is random.
- Gas particles move in straight lines.
- The speeds of the particles are not all the same.
- Gas particles have a lot of space to move around. (They are tiny compared to the space they are found in.)

- Gas particles change directions when they hit each other or the walls of the container.

As the temperature increases, the average speed of the particles increases. At higher temperatures, the gas particles move faster, and you can imagine that they bounce harder off the walls. At lower temperatures, the particles move slower, and you can imagine that they hit the walls with less force. Indeed, if the container were flexible, like a balloon, the hot gas particles would expand the balloon. The balloon would shrink if the gas particles were cooled.

Reading Questions

1. What is absolute zero? Why is it considered a hypothetical temperature?
2. What advantages does the Kelvin scale have over the Celsius scale?
3. How does the kinetic theory of gases explain temperature?

Reason and Apply

4. What are the freezing and boiling temperatures of water in degrees Celsius, kelvins, and degrees Fahrenheit?
5. Which unit is the smallest: one Celsius degree, one kelvin, or one Fahrenheit degree? Explain your thinking.
6. Would you describe each of these temperatures as warm, hot, or cold?
   a. 100 K   b. 60 °C   c. 250 K   d. 25 °C   e. 300 K
   f. −100 °C   g. 400 K
7. Convert each of the Kelvin temperatures in Exercise 6 to degrees Celsius, and vice versa.
8. What do you think is the highest temperature that can be reached by a substance? What is your reasoning?
9. Here are a few common temperatures on the Fahrenheit scale. Convert each of these to the Kelvin scale.
   a. 95 °F (a hot day)   b. 350 °F (oven temperature)   c. 5 °F (freezer)
10. The temperature on the surface of Venus is 736 K. Convert this temperature into degrees Fahrenheit and degrees Celsius. Describe what the atmosphere of Venus might be like.
11. Choose the best answer. According to the kinetic theory of gases, particles in a sample of gas
    (A) Move randomly, in curved paths, at different speeds, and bounce off each other and the walls of their container.
    (B) Move randomly, in straight-line paths, at different speeds, and bounce off each other and the walls of their container.
    (C) Move randomly, in straight-line paths, at the same speed, and bounce off each other and the walls of their container.
    (D) Move randomly, in curved paths, at the same speed, and bounce off each other and the walls of their container.
12. Use the kinetic theory of gases to explain why gases expand upon heating and shrink upon cooling.

The temperature of a gas is a measure of the average energy of motion of the gas particles. Scientists hypothesize that if it were possible to reach absolute zero, the motions of atoms and particles would stop.

EVALUATE (5 min)

Check-In

1. Describe three features of the motions of gas particles.
2. Use the motions of gas particles to explain why gases expand when they are heated.

Answers: 1. *Sample answers:* Gas particles are in constant, random, straight-line motion. Gas particles move at different speeds and generally do not interact with one another because they have a lot of space to move around in. Gas particles change directions when they hit the walls of the container or each other. 2. When gases are heated, the gas particles move faster. They push harder on the walls of the container. If the container has a flexible size, its size will increase.

Homework

Assign the reading and exercises for Weather Lesson 53 in the student text.

LESSON 53 ANSWERS

1. *Possible answer:* Absolute zero is the temperature at which the volume of a gas would be equal to zero. It is considered a theoretical temperature because a real gas would condense into a liquid and then a solid before reaching absolute zero, making a volume of zero impossible.

2. *Possible answer:* One advantage that the Kelvin scale has over the Celsius scale is that its zero point is the lowest possible temperature. This means that there can never be any negative temperatures in the Kelvin scale.

3. *Possible answer:* The kinetic theory of gases defines the temperature of a gas as the average kinetic energy of the particles of the gas.

4. The freezing temperatures of water are 0 °C, 273 K, and 32 °F. The boiling temperatures of water are 100 °C, 373 K, and 212 °F.

5. The smallest unit is 1 °F because there are 180 Fahrenheit degrees between the freezing temperature of water and boiling temperature of water, while there are only 100 Celsius degrees or kelvins between the freezing temperature of water and boiling temperature of water.

6. a. cold, 100 K is below the freezing point of water. b. hot, 60 °C is equivalent to 140 °F. c. cold, 250 K is below the freezing point of water. d. warm, 25 °C is equivalent to 75 °F, the temperature of a warm room. e. warm, 300 K is equal to about 81 °F. f. cold, −100 °C is below the freezing point of water. g. hot, 400 K is above the boiling point of water.

7. K = C + 273; C + K − 273
a. −173 °C b. 333 K c. −23 °C d. 298 K
e. 27 °C f. 173 K g. 127 °C

8. *Possible answer:* There is no highest possible temperature because particles can keep moving faster and faster, causing the temperature to increase indefinitely.

9. a. 308 K b. 450 K c. 258 K

10. 865 °F. The atmosphere of Venus would be filled with hot, energetic gases. Water clouds could not exist, because at such high temperatures water could only be in the gas phase.

11. B

12. *Possible answer:* According to the kinetic theory of gases, the particles of a gas move faster when they are heated and more slowly when they are cooled. When the particles move faster, they will collide more often or the gas will expand to take up more room. When the gas is cooled the particles do not move as fast, so they remain closer together and the gas contracts.

Overview

CLASSWORK: PAIRS

DEMO (OPTIONAL): PAIRS

Key Question: How can you predict the volume of a gas sample?

KEY IDEAS

The mathematical relationship between volume and temperature for a certain amount of gas at constant pressure is known formally as Charles's law. It is expressed mathematically as $V = kT$ or $k = V/T$, where k is the proportionality constant. This proportionality holds true only if the Kelvin scale is used. If one set of values for volume (V) and temperature (T) is known for a specific gas sample, it is possible to solve for any value of T for that gas sample. The value of the proportionality constant, k, depends on the amount of gas and the pressure of the gas. If the pressure or the amount of gas changes, the proportionality constant changes.

LEARNING OBJECTIVES

- Explain Charles's law and use it to solve simple gas law problems involving volume and temperature.
- Explain two methods for determining the volume of a gas if its temperature is known.

FOCUS ON UNDERSTANDING

- The conditions under which the proportionality constant, k, changes sometimes confuse students. The value of k stays the same for a given sample of gas. The value of k changes if the amount of gas or its pressure is changed.
- Students have not yet been introduced to the concept of pressure, so we avoid direct mention of it in the classwork. When defining Charles's law in the Explain & Elaborate discussion, we say that the pressure does not change rather than say that the pressure is constant.
- Throughout the unit, we avoid using the term *constant* to mean unchanging or fixed to avoid confusion with the proportionality constant.

KEY TERM

Charles's law

IN CLASS

In this lesson, students investigate the mathematical relationship between the volume and temperature of a gas. You may want to find an online video

showing balloons being placed in liquid nitrogen, to show the change in the volume of a gas in response to a change in temperature. Students then complete a worksheet that supports them in reasoning and in performing calculations using Charles's law. Finally, students solve gas law problems by finding volume and temperature using the value of the proportionality constant, $k = V/T$. A complete materials list is available for download.

 TRM Materials List 54

Differentiate

This lesson is mathematics-intensive. Some students may feel comfortable

making the calculations, and other students may struggle. To assess students' needs, allow students to work on the first worksheet question in their groups. After they finish, use a simple show of hands to gauge readiness. Ask students to raise their hands and vote on a scale of 1 to 5, with 5 fingers indicating that they understand how to solve the problems and can teach someone else and 1 finger indicating that they are unsure of how to begin solving these types of problems. Once you check student readiness on these problems, you might break students into groups of similar readiness. Students who voted with 5 fingers may benefit from more

---

Charles's Law

THINK ABOUT IT

The vent for a heater in your house is typically placed near the floor. This is because hot air rises and cooler air descends. If the vent were near the ceiling, the room would still be cold near the floor because the hot air would remain near the ceiling. So hot air must be less dense than cold air. For a similar mass, it occupies a larger volume than cooler air.

How can you predict the volume of a gas sample?

To answer this question, you will explore

➲ Predicting Gas Volume
➲ Charles's Law

EXPLORING THE TOPIC

➲ Predicting Gas Volume

A *piston* is a movable part that traps a sample of gas within a cylinder. It can move up and down if the volume of the gas changes. The illustration shows how the volume of nitrogen gas changes as the temperature changes. At first, the volume of the gas is 600 mL at a temperature of 300 K. When the cylinder is cooled, the gas contracts and the piston moves down. Then when the cylinder is heated, the gas expands and the piston moves up.

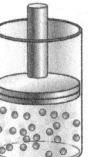 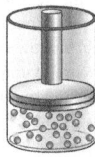 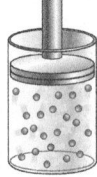

$V_1 = 600$ mL $V_2 = 400$ mL $V_3 = 800$ mL
$T_1 = 300$ K $T_2 = 200$ K $T_3 = 400$ K

Subscripts are used to indicate corresponding volumes and temperatures for the same gas under different conditions. For example, V_1 is the volume at the first temperature, T_1. Notice that if you determine the ratio of the volume of the gas to the temperature of the gas, the ratio is the same at all three of these temperatures.

$$\frac{V_1}{T_1} = 2.0 \text{ mL/K} \qquad \frac{V_2}{T_2} = 2.0 \text{ mL/K} \qquad \frac{V_3}{T_3} = 2.0 \text{ mL/K}$$

In fact, if nothing else changes, the ratio of volume to temperature for this gas will be 2.0 mL/K at any temperature. The relationship between the volume and

Important to Know Each gas sample has a unique value for the proportionality constant, $k = V/T$, depending on the amount of gas in the sample.

temperature for this sample of gas is proportional. So the volume and temperature for this gas are related by the proportionality constant, $k = V/T$, or 2.0 mL/K. With this proportionality constant, you can determine the volume of this gas at any temperature. [For review of this math topic, see **MATH Spotlight**: Ratios and Proportions on page A-11.]

CHANGING THE AMOUNT OF GAS

Suppose you start with a *different* amount of gas in the same cylinder. For this new sample of gas, the proportionality constant, $k = V/T$, is 1.5 mL/K.

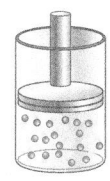 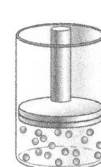 

$V_1 = 450$ mL $\quad$ $V_2 = 300$ mL $\quad$ $V_3 = 600$ mL

$T_1 = 300$ K $\quad$ $T_2 = 200$ K $\quad$ $T_3 = 400$ K

$\dfrac{V_1}{T_1} = 1.5$ mL/K $\quad$ $\dfrac{V_2}{T_2} = 1.5$ mL/K $\quad$ $\dfrac{V_3}{T_3} = 1.5$ mL/K

Again, you can use $k = V/T$ for this gas sample to calculate the volume at any temperature.

If you graph volume versus temperature for the two gas samples described previously, the result will be two different lines with different slopes.

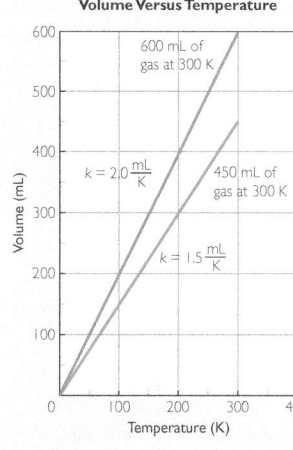

Volume Versus Temperature

600 mL of gas at 300 K

$k = 2.0 \dfrac{mL}{K}$

450 mL of gas at 300 K

$k = 1.5 \dfrac{mL}{K}$

Volume (mL) / Temperature (K)

Using the Graph

Because the relationship between temperature and volume is proportional, the lines pass through the origin. The slope of each line is equal to the proportionality constant for that sample. You can use the graph to predict the volume of either gas sample at any temperature.

Lesson 54 | **Charles's Law** $\quad$ 283

---

complex Charles's law problems because they likely will finish early. Students who voted with 1 or 2 fingers may need extra support from you while they work or other scaffolds such as Math Cards.

Pre-AP® Course Tip

Students preparing to take an AP® chemistry or college chemistry course should be able to cite reasons for using a particular mathematical approach such as graphical analysis for the direct relationship between V and T to justify the use of direct proportionality in mathematical problem solving. In this lesson, students explain Charles's

law and two methods for determining the volume of a gas at a constant temperature.

TRM PowerPoint Presentation 54

ChemCatalyst

A lava lamp contains a waxy substance and water, which do not mix, and a light bulb at the base. As the bulb heats the waxy substance, it rises. Near the top of the lamp, the waxy substance cools and falls. Explain why this happens.

Sample answer: Students may not have a complete answer at this point. They may say that the wax expands as it is heated and contracts as it cools. The hotter wax occupies a larger volume and therefore is less dense. Substances that are less dense float on top of denser substances.

Ask: How does the density of a substance change as it is heated?

Ask: Why does the wax in the lava lamp rise and fall?

Ask: Why are heating vents in a room near the floor and not on the ceiling?

Ask: What do you think happens to air warmed by the Sun during the day?

Ask: How do you think the expansion and contraction of gases affect the weather?

TRM Worksheet with Answers 54

TRM Worksheet 54

COMPLETE THE DEMONSTRATION (OPTIONAL)

➜ If possible, show a video that demonstrates the change in volume of a balloon as it is heated or cooled.

INTRODUCE THE CLASSWORK

➜ Arrange students into pairs.

➜ Pass out the student worksheet.

➜ Tell them they will be exploring how the volume of a gas is related to its temperature.

TRM Transparency—Balloon Volume Versus Temperature 54

INTRODUCE CHARLES'S LAW

➜ Display the transparency—Balloon Volume Versus Temperature.

Ask: How can you use the value of k to determine the volume of a gas sample at different temperatures?

Ask: How does the value of k change with the size of the gas sample?

Ask: How does the value of k relate to the steepness of the line on the graph?

Ask: Why is it necessary to measure only one volume and temperature to draw a graph of volume versus temperature for a gas?

Key Points: Charles's law states that volume, V, is proportional to the Kelvin temperature, T. This means that the volume is equal to the proportionality constant, k, times the temperature: $V = kT$. This relationship was described first by the French scientist Jacques Charles in 1802. It holds true only if the amount of gas and the gas pressure do not change.

> **Charles's law:** For a given sample of gas at a certain pressure, the volume of gas is directly proportional to its Kelvin temperature.

The proportionality constant, k, indicates how much the volume of a gas changes per kelvin. You can calculate the value of k from one measurement of volume and temperature for a gas sample: $k = V/T$. Then you can multiply a different temperature by k to find the volume at this temperature. This is true only if temperatures are given in the Kelvin scale.

Because volume is proportional to temperature, the graph of volume versus temperature for a gas sample is a straight line that goes through the origin, (0, 0). Each point on the line is a volume-temperature pair that satisfies the equation $V = kT$. You can use the graph to determine the volume at any temperature. The value of k is equal to the slope of the line and is different for different quantities of gas. This means that the lines will vary in steepness for balloons with different quantities of gas inside.

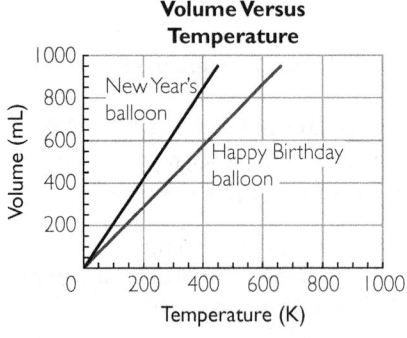

Volume Versus Temperature

COMPLETE AN EXAMPLE USING CHARLES'S LAW

→ Hand out Math Card: Charles's Law (optional) and help students relate V, T, and k.

⊃ Charles's Law

Jacques Charles was a French inventor, scientist, mathematician, and very active balloonist. Because of his interest in hot air balloons, he gave a lot of thought to the relationship between the volume and temperature of a gas. In 1787, he described the proportionality of gas volume and temperature mathematically. This description is now known as **Charles's law.**

> **Charles's Law**
> If pressure and the number of particles of a gas stay the same, then volume is proportional to the Kelvin temperature.
> $$V = kT \text{ or } k = \frac{V}{T}$$

Example 1

Calculate Gas Volume

Imagine that a cylinder with a piston contains 10 mL of air at 295 K. What will the volume of gas in the cylinder be at 550 K?

Solution

Calculate the volume by using Charles's law or by drawing a graph. You can predict that the volume will increase when the temperature increases. So the answer should be greater than 10 mL.

Using the Formula

Find the value of k.
$$k = \frac{V}{T} = \frac{10 \text{ mL}}{295 \text{ K}} = 0.034 \text{ mL/K}$$

Solve for the volume at 550 K.
$$V = kT = 0.034 \text{ mL/K} \cdot 550 \text{ K} = 18.5 \text{ mL}$$

Graphical Analysis

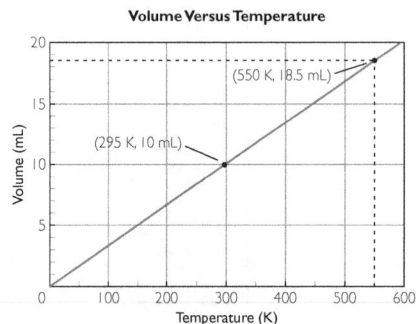

First, plot the data point for the initial conditions. Draw a line through this point and the origin, since at 0 K the *y*-value is theoretically 0 mL. Use the line to find the volume when the *x*-value is 550 K. Both methods give the same answer. At 550 K, the volume is 18.5 mL.

284   Chapter 10 | **Physically Changing Matter**

Charles's Law

$$k = \frac{V}{T}$$

(triangle diagram)
- V (L)
- k (L/K)
- T (K)

The proportionality constant, k, is different for each gas sample.

EXAMPLE

The first thing in the morning, you fill a balloon with air to a volume of 180 mL at 50 °C. After several hours out in the Sun, the air inside the balloon has warmed to 85 °C. Calculate the new volume of the balloon.

SOLUTION

Step 1: Predict whether the volume will increase or decrease. The temperature increases, so the volume of the balloon must increase.

HISTORY
CONNECTION

On May 6, 1937, the airship *Hindenburg* caught fire as it landed at an airfield in New Jersey. The *Hindenburg* was kept afloat with massive chambers filled with hydrogen, which is less dense than air but is also highly flammable. Thirty-six people died when a fire on board spread rapidly.

Example 2

Calculate Gas Temperature

A balloon contains 500 mL of helium at 20 °C. You want to increase the volume to 550 mL. What temperature in °C will give you a volume of 550 mL for this sample of gas?

Solution

You can predict that the temperature will be higher for a larger volume.

Convert the temperature to kelvins.	$K = C + 273$
	$= 20 + 273$
	$= 293$ K
Determine the proportionality constant, k.	$k = \dfrac{V}{T} = \dfrac{500 \text{ mL}}{293 \text{ K}} = 1.7$ mL/K
Rearrange the equation in terms of T and solve for temperature at 550 mL.	$T = \dfrac{V}{k} = \dfrac{550 \text{ mL}}{1.7 \text{ mL/K}} = 324$ K
Convert back to degrees Celsius.	$C = K - 273 = 324 - 273 = 51$ °C

A temperature of 51 °C is needed to give a volume of 550 mL. The temperature is higher for a larger volume, as predicted.

Important to Know To solve problems using Charles's law, temperature values must first be converted to the Kelvin scale.

KEY TERM
Charles's law

[For review of this topic, see **Math Spotlight**: Solving Equations on page •••.]

LESSON SUMMARY

How can you predict the volume of a gas sample?

The volume and temperature of a sample of gas are proportional to each other if nothing else changes (such as pressure or the amount of gas). This relationship between the volume and temperature of a gas in kelvins is described by the equation $V = kT$ and is known as Charles's law. The proportionality constant, k, is different for each different amount of gas.

DISCUSS WHY HOT AIR RISES

Ask: What happens to the density of a gas as the temperature increases?

Ask: Suppose a balloon is heated such that the density is lower than that of the surrounding air. Do you expect the balloon to float or sink? Explain.

Ask: What are two ways to decrease the density of the gas inside a balloon.

Key Points: Hot air rises because it is less dense than cooler air. When the temperature of a mass of air increases, the molecules move faster and the volume increases. Because density is given by m/V, the density of the air decreases when V gets larger. Less dense substances float on denser substances, so a larger warm air mass floats above a larger cooler air mass.

EVALUATE (5 min)

Check-In

A sample of gas has a volume of 120 L at a temperature of 40 °C. The temperature drops to −10 °C. If nothing else changes, what is the new volume of the gas?

Answer: The temperature decreases, so the volume decreases. Convert the temperatures to the Kelvin scale. Then determine the proportionality constant, $k = V/T = 120$ L/313 K = 0.38 L/K. Now use the value of A: to determine the new volume at −10 °C (263 K). $V = kT = (0.38 \text{ L/K})(263 \text{ K}) = 100$ L.

Homework

Assign the reading and exercises for Weather Lesson 54 in the student text.

Step 2: Convert all temperatures to the Kelvin scale:

$$T_1 = 50 \text{ °C} + 273 = 323 \text{ K}$$
$$T_2 = 85 \text{ °C} + 273 = 358 \text{ K}$$

Step 3: Find the proportionality constant, k. Use the volume and temperature when you first inflated the balloon to determine the value of k:

$$k = \frac{V}{T} = \frac{180 \text{ mL}}{323 \text{ K}} = 0.56 \text{ mL/K}$$

Step 4: Apply Charles's law. Use the value of k to find V at the new temperature:

$$V = kT = (0.56 \text{ mL/K})(358 \text{ K})$$
$$= 200 \text{ mL}$$

Step 5: Check that the answer makes sense and matches prediction.

The volume has increased from 180 mL at 50 °C to 200 mL at 85 °C. Thus, the volume has increased slightly, as expected for a small increase in temperature.

LESSON 54 ANSWERS

1. Determine the proportionality constant, *k*, for a gas by dividing the volume of the gas by its temperature measured on the Kelvin temperature scale.

2. Determine the volume of a gas by multiplying the proportionality constant, *k*, and the temperature of the gas measured on the Kelvin temperature scale. The solution assumes constant pressure and that a specific mass of gas was used in determining the proportionality constant.

3. 689 mL

Volume vs. Temperature

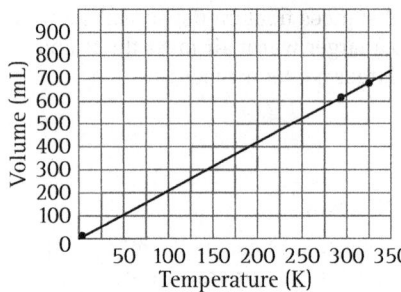

4. 1320 mL

5. 186 K

6. Possible answers: ● Calculate the proportionality constant using $k = V/T$ and then use it to calculate the new temperature, $T = V/k = 147$ K. ● Make a graph using the proportionality constant. Find the value that corresponds to 1 L by reading right from 1 L to the graph and down to the temperature in degrees Kelvin. From reading the graph, the gas should be cooled to about 145 K.

Volume vs. Temperature

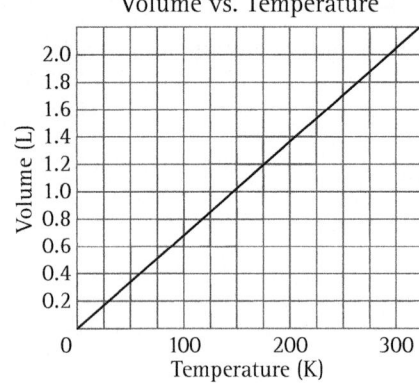

Exercises

Reading Questions

1. Explain how to determine the proportionality constant, *k*, for a sample of gas.

2. Explain how to determine the volume of a gas at a certain temperature using the proportionality constant, *k*.

Reason and Apply

3. A gas sample in a cylinder has a volume of 620 mL at 293 K. If you allow the piston to move while you heat the gas to 325 K, what will the volume of the gas be? Check your answer by drawing a graph.

4. A gas sample in a cylinder has a volume of 980 mL at a temperature of 27 °C. If you allow the piston to move while you heat the gas to 325 K, what will the volume of the gas be at 325 K?

5. A gas sample in a cylinder with a piston has a volume of 330 mL at 280 K. What temperature will you need to heat it to in order to change the volume to 220 mL?

6. A 2.0 L gas sample at 20 °C must be cooled to what temperature for the volume to change to 1.0 L? Show at least two different ways to solve this problem.

7. Imagine that you have a huge helium balloon for a parade. Around noon, it is 27 °C when you fill the balloon with helium gas to a volume of 25,000 L. Later in the day, the temperature drops to 15 °C.
   a. What is the proportionality constant, $k = V/T$, at the beginning of the day?
   b. Calculate the volume of the balloon when the temperature has dropped to 22 °C.
   c. What will the proportionality constant, $k = V/T$, be at the end of the day when the temperature is 15 °C? Explain your answer.

8. Would this cup make a good rain gauge? Explain your thinking.

7. a. 83.3 L/K **b.** 24,600 L **c.** 83.3 L/K; The proportionality constant for a particular gas does not change as the temperature changes as long as the amount of gas is the same.

8. Possible answer: The cup would not make a good rain gauge because the sides of the cup slope inward from top to bottom so no proportionality constant applies to the whole cup. The cross-sectional area of the cup is not constant.

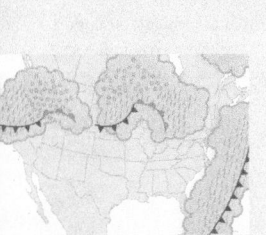

Density, Temperature, and Fronts

Purpose
To investigate the movement of air masses and their role in determining the weather.

Part 1: Weather Maps
Reexamine the weather maps from Lesson 49: Weather or Not to answer these questions.

1. Examine the first map at the left. What relationships do you see between fronts, clouds, and precipitation?
2. Examine the second map. What relationship do you see between fronts and lows? Between fronts and highs?
3. Describe any connections you find between fronts and the jet stream.
4. Where would you expect to see warm and cold air masses on each map?

Part 2: Warm and Cold Fronts

1. Why is a cold air mass denser than a warm air mass?
2. Explain why clouds might form when a warm air mass collides with a cold air mass.
3. **Making Sense** What does air density have to do with weather fronts?
4. **If You Finish Early** Eighty percent of the air in our atmosphere is made up of nitrogen gas, N_2, while only 1% of the air is made up of water vapor. Why doesn't it rain liquid nitrogen instead of rainwater?

Lesson 55 | **Density, Temperature, and Fronts**   287

Overview

ACTIVITY: WHOLE CLASS OR GROUPS OF FOUR

DEMO: WHOLE CLASS
Key Question: How do weather fronts affect the weather?

KEY IDEAS
Weather fronts, which occur where two air masses meet, are responsible for the majority of storms that occur. A cold front forms when a cold air mass catches up with a warm air mass, and a warm front forms when a warm air mass catches up with a cold air mass. When warm and cold air meet, density differences force the warm air upward, and clouds form.

LEARNING OBJECTIVES
- Explain the roles of temperature and density in the movement of cold and warm air masses.
- Describe the weather patterns associated with warm fronts and cold fronts.

FOCUS ON UNDERSTANDING
Air mass may seem like odd terminology to students. It refers to a large mass of air that is consistent with respect to temperature and moisture.

KEY TERM
air mass

IN CLASS
This lesson summarizes what has been learned about the weather so far and integrates the chemistry learning with the context. Students observe a demonstration at the beginning of class that shows water masses of two different temperatures and densities mixing, to simulate the meeting of air masses. Weather variables are re-examined using the color transparencies from Lesson 49: Weather or Not. Students complete a worksheet to explore the interaction between weather fronts and precipitation. A complete materials list is available for download.

TRM Materials List 55

SETUP
We recommend that you complete at least one practice run with the demonstration apparatus before class. The apparatus is a clear, flat acrylic tank divided into two equal volumes with a removable partition. The key to success is to remove the partition at a speed that facilitates the slow movement of one body of colored water over the other.

While students are working on the ChemCatalyst, add cold water (20 °C) and blue food coloring to one chamber in the tank. Add hot water (30 °C) and red food coloring to the other chamber. If you have several sets of color transparencies, students can do Part 1 on the worksheet in groups. If you have one set, you can use an overhead projector and do Part 1 as a whole-class discussion. Superimpose the transparencies as needed for each question.

CLEANUP
Water with food coloring can be disposed of in the sink. Rinse out the tank in preparation for the next class.

Lesson Guide

TRM PowerPoint Presentation 55

TRM Transparency—ChemCatalyst 55

ENGAGE (5 min)

ChemCatalyst

Large air masses form over different regions of land and ocean. These air masses have a consistent temperature and moisture content.

1. What patterns do you notice in the temperatures and moisture content of the air masses shown on the map?

2. Why do you think clouds form when the continental polar air mass collides with the maritime tropical air mass?

3. Use the concept of density to explain why warm air in the maritime tropical air mass rises, while cold air in the continental polar air mass descends.

Sample answers: **1.** Cold air masses are in the north. Warm air masses are near the equator. Air masses over water have higher moisture content than those over land. **2.** A tropical air mass has a lot of moisture. When it meets a polar air mass, the temperature decreases. The water condenses to form clouds. **3.** The denser, cool air descends. The less dense, warm air rises.

Ask: What is an air mass?

Ask: Why is warm air less dense than cold air?

Ask: What do you predict will happen when a warm air mass collides with a cold air mass?

Ask: Why does warm air rise and cold air descend?

EXPLORE (20 min)

TRM Worksheet with Answers 55

TRM Worksheet 55

COMPLETE THE DEMONSTRATION

→ Add cold water (20 °C) and blue food coloring to one chamber in the tank. Add hot water (30 °C) and red food coloring to the other.

→ Use the plastic chamber and colored water to simulate what happens when two air masses of differing temperatures meet each other. Done properly, the mixing water masses resemble a weather front that forms at the boundaries of two air masses.

- Tell students that the red water represents a tropical air mass and the blue water represents a polar air mass.
- Ask students to predict what will happen if you remove the partition dividing the two chambers.
- Hold a white piece of paper behind the container so that the interaction between the two solutions can be seen more clearly.
- Carefully remove the center dividing wall, allowing the two solutions to interact.
- Ask students to describe what they observe and to offer explanations for their observations.

- Bring the class's attention back to the apparatus when the warm and cold water samples have layered vertically.

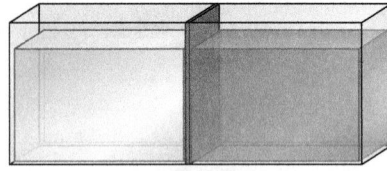

Before mixing

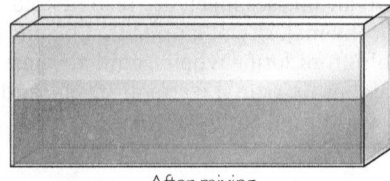

After mixing

Front and Center

Density, Temperature, and Fronts

THINK ABOUT IT

Meteorologists have identified large masses of air that have a consistent temperature and water content. These **air masses** form in particular locations on Earth's surface and cover thousands of square miles. They have a great influence on the weather. Weather fronts occur at the boundaries of these air masses, where warm and cold air meet.

How do weather fronts affect the weather?

To answer this question, you will explore

⤴ Air Density

⤴ Weather Fronts

⤴ Relationships between Weather Variables

EXPLORING THE TOPIC

⤴ Air Density

SMALL AIR MASSES

To understand a large mass of air on Earth's surface, it is helpful to examine the behavior of a small sample of gas. When a gas sample gets warmer, it expands to fill a larger volume. Notice that as the temperature increases, there are fewer molecules per milliliter of volume in the cylinders. The number of molecules does not change, but the spacing between those molecules does change. The density, $D = m/V$, decreases because the same mass, m, is divided by a larger volume, V, resulting in a smaller number.

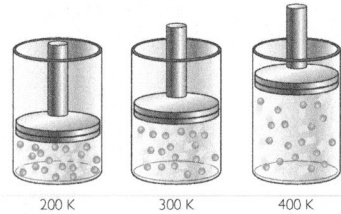

| 200 K | 300 K | 400 K |

Air expands and becomes less dense as it is heated.

LARGE AIR MASSES

In Unit 1: Alchemy, you found that every substance has a specific density. For example, the density of gold is 19.3 g/mL, and the density of water is 1.0 g/mL.

This is true only at a specified temperature. The density of a substance changes as the temperature changes. For liquids and solids, this change in density is very small. For gases, the change is much larger.

When air gets warmer, it expands as well as rises, because warm air is less dense than colder air. This is what happens when a warm air mass meets a cold air mass. This fact is key to understanding weather fronts and storms.

⟳ Weather Fronts

WARM AND COLD AIR MEET

When a cold air mass meets a warm air mass, two different weather situations can arise. These two situations are shown in the illustrations and are described here.

Cold front. Cold air overtakes warm air.

Warm front. Warm air overtakes cold air.

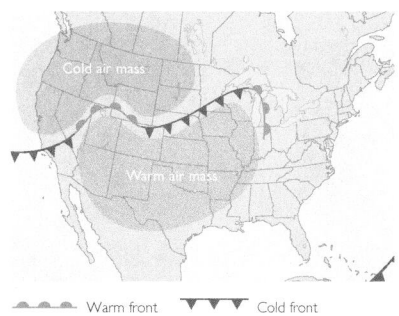

Warm front Cold front

→ Review the definition of a weather front. Draw the symbols for the weather fronts on the board, by way of review. A cold front forms when a cold air mass catches up with a warm air mass, while a warm front forms when a warm air mass catches up with a cold air mass.

▲▲▲ Cold front   ⌒⌒⌒ Warm front

INTRODUCE THE ACTIVITY

→ Arrange students into groups of four.

→ Pass out the student worksheet

→ Place the first set of transparencies on the overhead, or hand out a set to each group.

EXPLAIN & ELABORATE (15 min)

TRM Transparency—Fronts 55

DISCUSS WEATHER FRONTS

→ Use the transparencies Fronts and base map to assist the class in processing what they have learned.

Ask: What differences do you see between cold fronts and warm fronts?

Ask: What is the relationship between weather fronts and precipitation?

Ask: How are fronts and air masses related?

Ask: How does density influence weather fronts? How does temperature influence weather fronts?

Key Points: Fronts occur between the boundaries of warm and cold air masses. In general, warm (tropical) air masses move up across the North American continent from the south, and cold (polar) air masses move down across the continent from the north. The shape of a front tells you something about where the cold or warm air masses lie. You can see that the cold fronts are bowed down, and the warm fronts are bowed up on the map of North America on page 289 of the student text.

Warm and cold air masses have different densities. Warm air masses rise above cold air masses because they are less dense. When this happens, clouds form. The cold temperatures cause water in the warm air mass to change phase from a vapor to a liquid. This is why fronts are associated with storms and rain. See the images on page 289 of the student text.

The weather associated with cold and warm fronts differs. The clouds that form with cold fronts tend to be thicker, puffy clouds like those seen with thunderstorms. The clouds form quickly, directly in the area of the cold front. The clouds associated with warm fronts are thinner and not as puffy. In advance of warm fronts, you may have days of clouds before the rain arrives. Precipitation tends to occur at or just behind cold fronts and comes in advance of warm fronts.

SUMMARIZE THE WEATHER UNIT SO FAR

→ Use the base map transparency to review the different weather concepts learned so far.

→ Draw one warm front and one cold front on the map, as shown on page 290 of the student text. Ask students where to draw clouds, precipitation, highs, lows, and the jet stream in the appropriate places.

→ Ask students to provide their reasoning for what is placed on the map.

Ask: What are some generalizations you can make about weather and the different variables that affect weather?

Ask: If you were looking at a map that showed only weather fronts, where would you predict precipitation?

- What would you know about air masses from such a map?
- What would you know about cloud cover?
- What would you know about the location of lows and highs?

Key Points: Interactions among the temperature, volume, and density of air masses contribute significantly to the formation of weather. Here are some generalizations about weather:

- Fronts occur at the boundaries between warm and cold air masses. Warm air overtaking cold air is called a *warm front*. Cold air overtaking warm air is called a *cold front*.
- Warm air, which is less dense, layers over the denser cold air.
- Clouds and steady light rain form ahead of a warm front. Clouds and heavy showers form at and behind a cold front.
- On weather maps, Ls (lows) are closely associated with fronts, while Hs (highs) appear away from the fronts. Highs are associated with clear skies. Lows are associated with storms and cloudy skies.

EVALUATE (5 min)

Check-In

A warm front is approaching your hometown. It is one day away. What type of weather would you expect to observe?

Answer: Clouds and rain form in front of an advancing warm front. There will be increasing clouds and steady and light rainfall.

Homework

Assign the reading and exercises for Weather Lesson 55 and the Chapter 10 Summary in the student text.

Clouds at the leading edge of a cold front

Clouds at the leading edge of a warm front

NATURE CONNECTION

Cold fronts are responsible for squalls, tornadoes, strong winds, and other destructive weather.

COLD FRONTS

A cold front occurs when cold air overtakes warm air. The warm air is suddenly pushed up as the advancing cold air forces its way underneath it. The warm air cools at higher altitudes, which causes water vapor in the air to condense into tiny liquid cloud droplets. The clouds associated with cold fronts can be seen directly in the area of the advancing front. They form quickly and are often thick, puffy clouds, like the ones seen before thunderstorms. Heavy rains are often associated with cold fronts.

WARM FRONTS

A warm front occurs when warm air overtakes cold air. The warm air gradually flows up and over the cold air because the warm air is less dense. The warm air cools at higher altitudes and water vapor condenses to form clouds. The clouds associated with warm fronts usually form ahead of the place where the air masses meet. You will see high, wispy clouds when a warm front is still hundreds of miles away. As the warm front gets closer, the clouds get thicker and lower, and precipitation falls. Steady, light rain over a large area is associated with warm fronts.

⊃ Relationships between Weather Variables

The clouds, high- and low-pressure systems, precipitation, and the jet stream can be added to a weather map that contains some fronts. Take a moment to draw some conclusions about how these weather variables relate to one another.

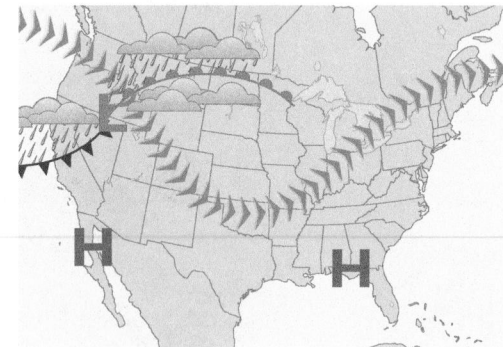

- Fronts occur at the boundaries between warm and cold air masses.
- Clouds and steady, light rain form ahead of a warm front. Clouds and heavy showers form at and behind a cold front.
- On weather maps, Ls are closely associated with fronts while Hs appear to be away from the fronts. Hs represent areas of high pressure and are associated with clear skies. Ls represent areas of low pressure and are associated with storms and cloudy skies.

When air masses form over Earth, they get their properties from the area beneath them. In other words, if an air mass forms over water, it will tend to be full of moisture; if it forms over land, it will tend to be dry.

LESSON SUMMARY

How do weather fronts affect the weather?

KEY TERM

air mass

When a cold air mass and a warm air mass meet, the warm air rises and layers on top of the cold air. This is because warm air is less dense than cold air. A cold front occurs when a cold air mass overtakes a warm air mass, and a warm front occurs when a warm air mass overtakes a cold air mass. Clouds and precipitation occur with both warm and cold fronts. Warm fronts produce steady, light rain, and cold fronts produce sudden, heavy showers.

Exercises

Reading Questions

1. Explain why hot air rises.
2. Explain the difference between a warm front and a cold front.

Reason and Apply

3. Suppose you have two gas samples in flexible containers with the same outside pressure and the *same amount of gas* in each. Sample A is at a temperature of 25 °C and Sample B is at 5 °C. Which of these statements is true?
   (A) Sample A occupies a larger volume and has lower density.
   (B) Sample A has higher density and a smaller volume.
   (C) Sample B has molecules moving at a greater average speed.

4. A cold front is approaching your hometown and is due to arrive tomorrow. What kind of weather would you expect to observe?

5. A warm front is approaching your hometown and is due to arrive tomorrow. What kind of weather would you expect to observe?

6. The continental polar air mass overtakes the maritime tropical air mass.
   a. What kind of front develops?
   b. What happens to the air masses when they meet?
   c. What sort of weather would you expect and where?

continental polar air mass pushes the warmer maritime tropical air mass upward. **c.** Stormy weather at the front is expected where the two air masses meet.

7. *Possible answers:* **a.** The temperature will be warmer than on the previous day. **b.** Cloudy weather is predicted ahead of the warm front, with a chance of showers. **c.** Stormy weather is predicted ahead of the cold front. **d.** Stormy weather and decreasing temperatures are predicted as the cold front passes. **e.** Warm and humid weather is predicted ahead of the cold front.

LESSON 55 ANSWERS

1. When air gets warmer, its volume increases and its density decreases. Air that is less dense than the surrounding air will rise.

2. A warm front occurs when warm air overtakes cold air and is gradually pushed upward. A cold front occurs when cold air overtakes warm air and pushes it rapidly upward.

3. A

4. *Possible answer:* The temperature would be lower tomorrow than today. Also, heavy rain and possibly thunderstorms are expected tomorrow.

5. *Possible answer:* High clouds are expected to start forming in the sky today, and slow, steady precipitation will start tomorrow.

6. a. When the colder continental polar air mass overtakes the warmer maritime tropical air mass, a cold front is formed. **b.** The colder

(continued)

7. PROJECT Look in the newspaper or on the Internet. Find a recent weather map with at least one warm front and one cold front.

    a. Find a warm front on the map. What weather is predicted for tomorrow in the region of the warm front?

b. What weather is predicted in the direction in which the warm front is moving?

c. Find a cold front on the map. Describe the weather forecast given for the region near the cold front.

d. What weather is predicted for tomorrow in the region close to the cold front?

e. What weather is predicted in the direction in which the cold front is moving?

CHAPTER 10

Physically Changing Matter

SUMMARY

KEY TERMS

weather

physical change

phase change

proportional

proportionality constant

melting point
  (melting temperature)

boiling point
  (boiling temperature)

absolute zero

Kelvin scale

kinetic theory of gases

temperature

Charles's law

air mass

Weather Update

The water and gases that make up Earth's atmosphere change density and temperature as they move about the planet. Physical changes of matter cause the weather we experience. Most matter expands when heated and contracts when cooled. When heated enough, matter changes phase, from solid to liquid to gas. The density of gases is dramatically lower than the density of either liquids or solids. Changes in temperature and density play a role in the weather.

REVIEW EXERCISES

1. What is 50 °F in kelvins?

2. Use the kinetic theory of gases to explain temperature.

3. A gas sample in a flexible container has a volume of 650 mL at a temperature of 27 °C. If the pressure stays the same when the sample is heated to 80 °C, what will be the new volume?

4. You fill a balloon with helium to a volume of 3 L in the morning, when the temperature is 23 °C. At the end of the day, the temperature drops to 10 °C.
   a. What is the proportionality constant, $k = V/T$, at the start of the day?
   b. What is the proportionality constant at the end of the day? Explain your thinking.
   c. What will the volume of the balloon be when the temperature is 10 °C?

5. a. What is the volume of 5.2 g of solid CO_2 if the density of solid CO_2 is 1.56 g?
   b. What will the volume of the 5.2 g(s) be if all of the solid changes to a gas? The density of $CO_2(g)$ is 0.0019 g/mL. Explain your reasoning.

Different Thermometers

 PROJECT Different thermometers use different properties of matter to measure temperature. Research how one of these thermometers works. Write a description and include a diagram.

Liquid crystal thermometer

Infrared ear thermometer

Digital resistance thermometer

Cooking thermometer

Galileo density thermometer

to an appropriate scale ● a diagram of the thermometer and the components that measure and record the temperature ● a list of the different applications for the thermometer and reasons why the thermometer is better suited than others for the particular application.

ASSESSMENTS

Two multiple-choice assessments (versions A and B) and two short-answer assessments (versions A and B) are available for download for Chapter 10.

TRM Chapter 10 Assessment Answers

TRM Chapter 10 Assessments

ANSWERS TO CHAPTER 10 REVIEW EXERCISES

1. 283 K

2. According to the kinetic theory of gases, molecules in a gas are in constant motion. As the gas becomes warmer, its particles have more energy and they move faster. Temperature is a measure of the average kinetic energy of the particles in a gas.

3. 760 mL (answer given to two significant figures)

4. a. 0.0101 L/K. **b.** 0.0101 L/K; according to Charles's law, if the amount of a gas at constant pressure remains the same, then the proportionality constant for the gas will remain the same. **c.** 2.86 L.

5. a. 3.3 mL. **b.** 2700 mL (answer given to two significant figures). The volume of gas can be calculated using the density equation. The mass of the carbon dioxide does not change when it changes phase and becomes a gas.

Project: Different Thermometers

A good project would include: ● a description of the type of thermometer chosen for the project ● a general description of how the thermometer measures the temperature of its environment and the mechanism of how the temperature is converted

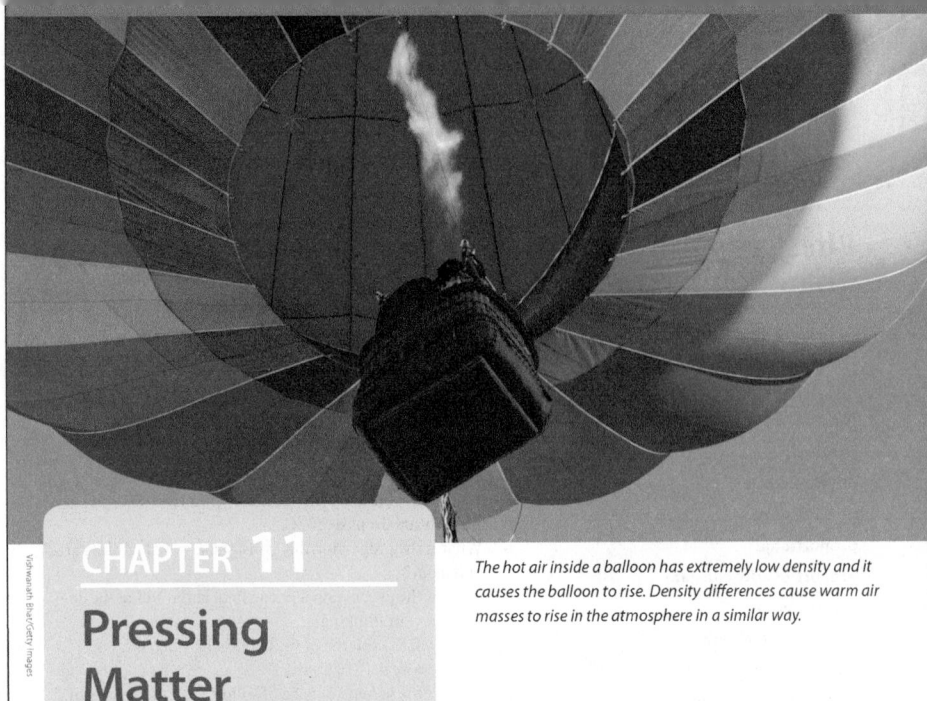

CHAPTER 11

PD **CHAPTER 11 OVERVIEW**

Watch the video overview of Chapter 11 (for teachers) by clicking on the link in the TE-book, opening the TRFD, or logging onto the book's companion Web site bcs.whfreeman.com/ livingbychemistry2e (teacher log-in required).

Chapter 11 of the Weather unit consists of seven lessons focusing on gas pressure. Lesson 56 is a lab that provides students firsthand experience with gas pressure and density. In Lesson 57, students observe a series of demonstrations featuring the effects of gas pressure on the world around them. Lesson 58 introduces Boyle's law. In Lesson 59, students investigate flexible and rigid containers and their effects on gas samples. Computer simulations allow students to explore the kinetics of gas behavior under different conditions in Lesson 60. The combined gas law is introduced in Lesson 61 as students follow the progress of a weather balloon through a changing atmosphere. In Lesson 62, students simulate the creation of a cloud inside a bottle and connect their learning to the weather associated with high-pressure and low-pressure systems.

In this chapter, students will learn

- about the density of gases
- a particulate view of gas pressure
- the relationships among the pressure, volume, and temperature of gases
- Boyle's law, Gay-Lussac's law, and the combined gas law
- weather associated with high-pressure and low-pressure systems

Vishwanath Bhat/Getty Images

CHAPTER 11

Pressing Matter

In this chapter, you will study

- the density of gases
- the behavior of gas particles
- the relationships between gas pressure, temperature, and volume
- high- and low-pressure weather systems

The hot air inside a balloon has extremely low density and it causes the balloon to rise. Density differences cause warm air masses to rise in the atmosphere in a similar way.

The atoms and molecules in solids, liquids, and gases are always in motion. However, gas molecules are free to move to fill the entire space of their container. Gas molecules collide with one another and with anything else they come in contact with, causing pressure. When samples of gases are placed in containers, it is possible to measure the relationships between gas pressure, temperature, and volume. The air pressure of the atmosphere has a great deal of influence on the weather.

294

Chapter 11 Lessons

It's Sublime

Gas Density

THINK ABOUT IT

Earth is unique in that it is the only planet in the solar system with so much liquid water on its surface. The weather and the life on Earth depend on the movement of moisture around the planet. The phase changes of water are mostly responsible for this movement. When water changes phase from liquid to gas, the airborne water molecules spread out in the atmosphere and the density changes dramatically.

How do the densities of a solid and a gas compare?

To answer this question, you will explore

↺ Density of Gases

↺ The Molecular View

↺ Comparing Density and Phase of Water

EXPLORING THE TOPIC

↻ Density of Gases

An airplane glides easily through the air at around 500 miles per hour. The density of air must be very low for objects to pass so easily through it. Determining the density of a gas is not as straightforward as determining the densities of liquids and solids. This is because it is not easy to measure the mass of a gas. It is hard to put a gas on a balance, and many gases, like helium, rise up in the air around them.

One approach is to measure the mass of a substance while it is a liquid or a solid, and then turn it into a gas. Provided you don't let any escape, the substance will have the same mass as a gas as it did as a liquid or a solid. You can then measure the volume of the gas to determine its density.

SUBLIMATION OF DRY ICE

One substance that you can use to study phase changes is carbon dioxide, CO_2. Carbon dioxide turns *directly* from a solid to a gas at the extremely low temperature of $-78\,°C$ ($-108\,°F$). This means the temperature in your classroom is warm enough to turn solid carbon dioxide into a gas.

When a solid changes directly into a gas without forming a liquid, the phase change is called **sublimation**. In class, you gathered data on the density of $CO_2(g)$ by finding the mass of a chunk of $CO_2(s)$ and then placing it into a plastic bag and allowing it to sublime. The volume of the gas is huge compared to the volume of the same mass of solid.

$CO_2(s)$ is commonly referred to as "dry ice" because it does not go through a liquid phase when it changes to a gas.

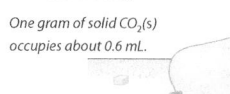

One gram of solid $CO_2(s)$ occupies about 0.6 mL.

One gram of gaseous $CO_2(g)$ occupies about 500 mL.

Lesson 56 | **Gas Density**   295

Overview

LAB: GROUPS OF 4

Key Question: How do the densities of a solid and a gas compare?

KEY IDEAS

When a substance changes phase from a gas to either a liquid or a solid, there is a dramatic change in density. This is because the particles in that substance are much farther apart in the gas phase than are those same particles in the solid or liquid phase. When solid carbon dioxide, $CO_2(s)$, changes phase, it goes directly from a solid to a gas. This type of phase change is called *sublimation*. Solid CO_2 is called "dry ice," because it sublimes directly to a gas without passing through the liquid phase.

LEARNING OBJECTIVES

● Describe the density differences that occur during phase changes.

● Explain how moisture gets into the atmosphere.

● Calculate the density of a gas from mass and volume measurements.

FOCUS ON UNDERSTANDING

● Students often believe that they can see water vapor because they confuse it with fog and what most people refer to as steam. Fog and the mist above boiling water (what most people call steam) consist of visible liquid water droplets. We cannot see water vapor because it is a colorless gas.

● Students may have difficulty extrapolating from dry ice, which

---

sublimes, to the evaporation of water; however, the magnitudes of the density changes are similar.

● Students might struggle when asked to provide evidence to support molecular views of gases. Encourage them to use their experiences to provide evidence.

KEY TERMS

sublimation
evaporation

IN CLASS

In this class, students observe a demonstration of the properties of dry ice. After the demonstration, they perform a lab in which they observe the dramatic change in density associated with a phase change from a solid to a gas. At the beginning of class, teams of students quickly measure the mass of a sample of dry ice and place it in a sealed garbage bag to sublime. This step is completed before the ChemCatalyst so the dry ice will have enough time to sublime completely before the end of class. Students complete worksheet questions before measuring the volume of their bags of CO_2 gas. They then calculate the density of carbon dioxide gas from their data. Sketches of molecular views assist students in conceptualizing gases. A complete materials list is available for download.

TRM Materials List 56

SETUP

Before class, acquire 5–7 lb of dry ice. Use a hammer to crush the dry ice into a powder.

> ### SAFETY
> ● Do not touch dry ice with your bare hands. Use gloves or tongs.
> ● Keep the dry ice in a polystyrene foam cooler or ice chest.

CLEANUP

You can ask students to "release" the carbon dioxide gas after the lesson and to fold the garbage bags for reuse by the next class.

Lesson Guide

INTRODUCE THE LAB (5 min)

TRM Transparency—Dry Ice Setup 56

→ Arrange students into groups of four.

→ Pass out the student worksheet.

→ Tell students they will be setting up the first part of their experiment before the ChemCatalyst. Emphasize to students that they must work quickly during this setup.

→ Distribute a range of amounts of crushed dry ice in polystyrene foam cups to the groups—each between 5 g and 20 g of dry ice.

→ Distribute the larger amounts of dry ice first so they will have more time to sublime.

→ Show the transparency Dry Ice Setup to quickly explain the procedure. The mass of solid dry ice changes continuously as it sublimes, so it is important to get the dry ice into the garbage bags quickly.

1. One student from each group should get a polystyrene foam cup filled with 5–10 g of dry ice from the teacher and a 5-gallon plastic garbage bag.

2. Remove all the air from the 5-gallon garbage bag.

3. Find the mass of the cup and the dry ice together.

4. Quickly pour the dry ice into the deflated bag and close it tightly so the bag does not leak. Be careful to keep air out.

5. Weigh the empty cup. Subtract this weight from the mass of the cup containing the dry ice to determine the mass of the dry ice.

ENGAGE (5 min)

TRM PowerPoint Presentation 56

ChemCatalyst

Water exists in many forms, including water vapor, fog, clouds, and liquid water.

1. Why do you think you cannot see water vapor in the air?

2. How are fog and clouds different from water vapor?

3. Why do you think airplanes can fly through clouds?

Sample answers: **1.** Some students will know that water vapor cannot be seen because it is a clear colorless gas, with the water molecules very far apart from one another. **2.** Many students will know that clouds and fog consist of tiny droplets of water collected together, while water vapor consists of gaseous water molecules distributed throughout other air molecules. **3.** Students might say a cloud is only water droplets and air, or that a cloud is not very dense.

→ Assist students in sharing their ideas about water vapor, clouds, and fog.

The density of the carbon dioxide changes dramatically as the phase changes. $CO_2(s)$ has a density of 1.56 g/mL. In contrast, $CO_2(g)$ has a very low density of 0.0019 g/mL. The volume of the gas is nearly 800 times as large as the volume of the solid!

Big Idea The densities of gases are extremely low compared to the densities of liquids and solids.

The Molecular View

Why did the bag puff up so much as the $CO_2(s)$ sublimed? Because you cannot see gases, you must depend on hypotheses about how the molecules in a gas behave. Consider four possible hypotheses and the corresponding models for the sublimation of solid carbon dioxide.

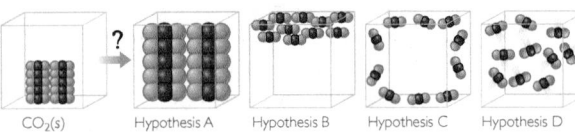

CO₂(s)   Hypothesis A   Hypothesis B   Hypothesis C   Hypothesis D

Hypothesis A: The CO_2 molecules become larger in the gas phase compared with the solid phase.

Evidence against Hypothesis A: You can move freely through a gas, so the molecules must not take up much space. The mass stays the same, so it is difficult to imagine how the molecules would get bigger. The evidence doesn't support Hypothesis A.

Hypothesis B: The CO_2 molecules rise to the top of the container.

Evidence against Hypothesis B: The bag filled with $CO_2(g)$ was pushed out in all directions. This indicates that the molecules are spread throughout, not just at the top. For example, the air in a room is everywhere, not just near the ceiling. The evidence doesn't support Hypothesis B.

Hypothesis C: The molecules go to the outer edges of the bag, pushing the sides out.

Evidence against Hypothesis C: If gases went to the outer edges of their containers, you would walk into a room and find all the air on the walls and none in the middle. Since you can breathe everywhere in the room, this indicates that gas molecules are spread throughout. The evidence doesn't support Hypothesis C.

Hypothesis D: The CO_2 molecules spread apart from one another and bounce around inside the bag.

Evidence supporting Hypothesis D: This model is consistent with the large volume and the low density of the gas. This model is also consistent with the bag staying puffed out in all directions and offering resistance when you push on it. This evidence supports Hypothesis D.

Model D can still be improved upon. It shows ten molecules in a small cubic space. In actuality, the gas molecules are much tinier, much farther apart, much more numerous, and moving very fast. When a gas is trapped in a bag, the moving molecules bang into the bag and push it out. If you were to remove the bag, the

Hypothesis A

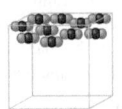

Hypothesis B

Hypothesis C

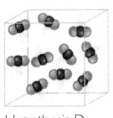

Hypothesis D

Ask: Can you see gases in the air? Explain your thinking.

Ask: Why can solid objects move easily through gases?

Ask: Which phase of water is responsible for the composition of clouds?

Ask: What are the differences between water vapor, fog, and liquid water? Explain.

Ask: How do the densities of water vapor and clouds compare to the density of liquid water?

EXPLORE (15 min)

TRM Worksheet with Answers 56

TRM Worksheet 56

DEMONSTRATE SUBLIMATION

→ The purpose of this demonstration is to introduce the concept of sublimation and to discuss the identity of the "fog" that appears around the dry ice.

→ This demonstration requires a hot plate set to medium heat, a small beaker of water, a small beaker of mineral or clear cooking oil, an ice cube, and some dry ice.

1. Show students a beaker containing a piece of dry ice and another beaker containing an ice cube (frozen water). Ask students to describe any differences they observe between the dry ice and the water ice.

2. Ask students to predict what will happen if you put the beaker with

molecules would quickly spread throughout the room. This is what happens to water when it **evaporates**, or changes into a vapor, and spreads out in the atmosphere.

| Big Idea | A sample of gas expands to fill whatever space it is in.

↻ Comparing Density and Phase of Water

The table shows the density for some different forms of water. Water has three phases: solid, liquid, and gas. But these phases can take a few different forms. For example, snow and ice both represent solid forms of water.

Different Forms of Water

Substance	Chemical formula	Density (g/mL)	Volume of 1000 g (mL)
rain	$H_2O(l)$	1.0 g/mL*	1,000 mL
water vapor	$H_2O(g)$	0.0008 g/mL*	1,250,000 mL
ice	$H_2O(s)$	0.92 g/mL†	1,087 mL
snow	$H_2O(s)$	0.1–0.5 g/mL	2,000–10,000 mL
clouds	$H_2O(l)$	0.001–0.002 g/mL	500,000–1,000,000 mL

*The density is for 25 °C. †The density is for 0 °C.

Notice what happens to the *volume* of 1000 grams of water as it goes from liquid to water vapor. The volume increases more than a thousand-fold when evaporation occurs.

The density landscape illustration shows the relationship between the location of matter and its density.

Clouds could be considered a mixture of water vapor and liquid water. It takes 15 million cloud droplets to form one drop of rain. The extremely low density of clouds explains why airplanes can move easily through them.

ASTRONOMY CONNECTION

The lowest temperature ever recorded on Earth was −129 °F in 1983 in Antarctica. At this temperature, $CO_2(g)$ condenses to form solid dry ice. The southern polar ice cap on Mars contains dry ice because the temperature there is below the sublimation temperature of CO_2.

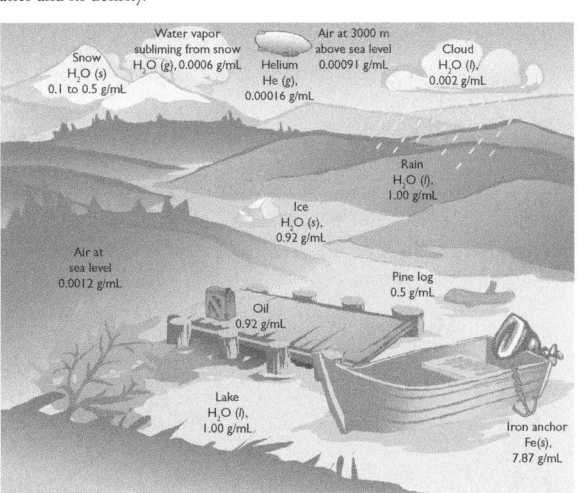

Lesson 56 | **Gas Density**   297

CONTINUE THE LAB ACTIVITY

→ Pass out worksheets to individual students. Ask them to complete Part 1 of the worksheet.

→ When 10–15 min remain in the class period, ask groups to complete Part 2 of the worksheet, finding the volume of their sublimed gas.

EXPLAIN & ELABORATE (15 min)

DISCUSS THE DRY ICE DEMONSTRATION

Ask: What happens when beakers of dry ice and water ice are put on the hot plate?

Ask: Why is a fog not visible when you put the dry ice in oil?

Ask: What do you think the "fog" that you see is?

Ask: Can you explain the difference between fog and water vapor?

Key Points: Heating water ice and dry ice causes phase changes in both. When ice made from water is heated, it melts into a liquid. If heating continues, the liquid water will become water vapor, a gas. Dry ice is solid carbon dioxide. When dry ice is heated, it becomes a gas, skipping the liquid phase. This is why it is called "dry" ice. When a solid is heated to become a liquid, the process is called *melting*. When a liquid is heated to become a gas, it is called *evaporation*. When a solid substance becomes a gas without forming a liquid, it is called *sublimation*. The "fog" you see around dry ice is tiny droplets of liquid water. The fog forms in the vicinity of the dry ice because the temperature is low enough to cause water vapor in the air to become a liquid. Notice that you cannot see the water vapor in the air until it is cold enough to condense. Gaseous carbon dioxide and gaseous water are not visible.

DISCUSS THE MASS AND DENSITY OF CO_2 GAS

Ask: How many times larger is the volume of $CO_2(g)$ than that of $CO_2(s)$? (~800 times larger)

Ask: You determined the mass of the solid CO_2. Is it the same as the mass of the gaseous CO_2 in the bag? Why or why not?

Ask: What density did you calculate for $CO_2(g)$ using your own data?

Ask: Why does the gas feel as if it weighs very little compared to the solid?

Key Points: The space that gaseous CO_2 can occupy is dramatically larger than the space that solid CO_2 occupies. The volume of gaseous carbon dioxide is

the water ice on the hot plate. Complete this step. (The water ice will melt, and there will be a hissing sound as the liquid evaporates. The water ice moves around as it "floats" on the liquid water beneath it. Fog is visible.)

3. Ask students to predict what will happen if you put the beaker with dry ice on the hot plate. Complete this step. (The dry ice will slowly disappear. A liquid is not visible, and the dry ice does not move around. You do see some fog, which is from water vapor in the air that condenses.)

4. Ask students what they think the fog is.

5. Put a small piece of dry ice in the beaker with water. Put a small piece

of dry ice in the beaker with oil. Ask students to describe their observations. (Dry ice releases bubbles, evidence that a gas forms. There is a fog over the water but only a very small amount of fog over the oil, evidence that the fog is made up of water droplets. Water has condensed out of the air because it is cooled by the dry ice. Nothing is visible inside the gas bubbles in the oil, indicating that carbon dioxide gas is not visible.)

→ Provide students with a basic definition of sublimation.

> **Sublimation:** The phase change from a solid to a gas.

about 800 times larger than the volume of the solid. The mass of the gas is the same as the mass of the solid, because no molecules are lost or added. Thus, by finding the mass of the solid CO_2 beforehand and measuring the volume of the gaseous CO_2 afterward, it is possible to calculate the density of gaseous CO_2. The actual density of $CO_2(g)$ is 0.0019 g/mL. When any substance changes into a gas, it expands dramatically. The main substance that condenses and evaporates in large quantities on our planet is water. The large volume change associated with water evaporation and condensation is one key to the dynamics of weather.

> **Evaporation:** The phase change from a liquid to a gas.

Discuss accuracy and possible sources of error. The accepted value for the density of $CO_2(g)$ is 0.0019 g/cm³. Students' measurements usually lead to a higher calculated value. Usually, some CO_2 sublimes before the bag is sealed, resulting in a lower volume measurement, which leads to a higher calculated value for density. You can have students calculate their percent error.

> Percent error
> $$\frac{|\text{experimental value} - \text{accepted value}|}{\text{accepted value}} \cdot 100$$

DISCUSS PARTICLE VIEWS OF GASES

→ Draw pictures on the board similar to those on the worksheet.

Ask: What happens to the solid carbon dioxide molecules when dry ice sublimes?

Ask: How can you explain the large increase in volume when carbon dioxide sublimes?

Ask: It is much easier to compress gases than solids. Is this consistent with the models? Explain.

Ask: Does each model explain why airplanes can fly through clouds? Explain.

Key Points: When solid carbon dioxide sublimes, the individual molecules move farther and farther apart, as shown in Hypothesis D on page 296 of the student edition. Gas molecules do not individually expand, as shown in Hypothesis A, nor do they all float to the top of a container or the surfaces of the container, as shown in Hypotheses B and C. Instead, the container inflates because the gas molecules move around. They are relatively far apart from one another, so much of the total volume is empty

Substances that are less dense float in substances that are denser. For example, ice floats in liquid water. When the temperature of a mass of air increases, the air expands, resulting in a lower density. This change in density causes the warmer air to rise above cooler air. These changes are related to the movement of matter on the planet and are essential to causing different weather.

LESSON SUMMARY

> ### How do the densities of a solid and a gas compare?

Gases form when solids sublime and liquids evaporate. During both sublimation and evaporation, molecules spread far apart from each other. This increases the volume of the substance by about 1000 times. It also decreases the density of the substance dramatically. Solids are generally denser than liquids. Gases are much less dense than both solids and liquids. Gases expand to fill whatever container they are in.

KEY TERMS
sublimation
evaporation

Exercises

Reading Questions

1. How does the density of a gas compare with the density of a solid?
2. Draw a molecular view of carbon dioxide gas.

Reason and Apply

3. In your own words, define sublimation and evaporation.
4. Describe three ways to show that gases exist.
5. When water freezes, the water molecules move apart very slightly. What evidence can you provide to support this claim?
6. Draw a molecular view for water vapor, liquid water, and ice.
7. What is the volume of 6.4 g of $CO_2(s)$? (density = 1.56 g/mL)
8. What is the mass of 3.5 L of $CO_2(g)$? (density = 0.0019 g/mL)
9. How many grams of $CO_2(s)$ would you need to fill a 6.5 L bag with $CO_2(g)$?
10. A person has 15 g of dry ice and wants to completely fill a bag that has a volume of 8 L with carbon dioxide gas. Is this enough dry ice to fill the bag with $CO_2(g)$? Explain.

space. The molecules of a gas are about 1000 times more dispersed (less dense) than the molecules of a solid. In fact, you could say that this is essentially what a gas is—molecules (or atoms) of a substance extremely far apart from one another. The drawing in Hypothesis D is not completely accurate. One improvement that could be made is to place much more distance between the gas molecules to reflect a density 800 times lower than that of the solid. However, to show molecules large enough to be seen would mean having only one CO_2 molecule in the box, with the next one several feet away.

EVALUATE (5 min)

Check-In

A sample of oxygen gas has a mass of 1.43 g and occupies a volume of almost exactly 1000 mL. What is the density of this oxygen gas? Is it more or less dense than carbon dioxide gas?

Answer: Density is equal to mass divided by volume: 1.43 g divided by 1000 mL equals 0.00143 g/mL or g/cm³. This gas is less dense than carbon dioxide gas: 0.00143 < 0.0019.

Homework

Assign the reading and exercises for Weather Lesson 56 in the student text.

LESSON 56 ANSWERS

Answers continue in Answer Appendix (p. ANS-4).

Air Force

Air Pressure

THINK ABOUT IT

Earth is surrounded by the *atmosphere,* a layer of air nearly 80 miles thick. The atmosphere is made up predominantly of colorless, odorless gases, which you can move through. The air molecules are tiny, and they are spaced far apart, so most of the time it feels as if nothing is there. However, the weight of all that air above you exerts a great deal of pressure on you.

> **What evidence do we have that gases exert pressure?**

To answer this question, you will explore

↻ Evidence of Air Pressure

↻ Explanation for Air Pressure

↻ The Atmosphere

EXPLORING THE TOPIC

↻ Evidence of Air Pressure

When you fill a car's tires with air, the rubber is stretched quite tightly. The pressure exerted by the air in the tires is high enough to push the entire car up off the ground. Indeed, the air pressure is holding up more than 2000 pounds. How can air be so powerful?

The ability of car tires to remain inflated while holding up so much weight provides evidence that air is pushing on the inside of the tire in all directions. This "pushing" property of a gas is called **pressure.** Pressure is defined as a force over a specific surface area. A gas exerts pressure on all surfaces it comes in contact with.

There are many ways to demonstrate that gases exert pressure on the things around them. Here are two examples.

SUBMERGED PAPER

When a plastic cup is turned upside down and submerged in water, the paper inside the cup stays dry. There is air trapped in the cup that pushes out in all directions. The air pushes on the water with enough pressure to keep the water out of the cup.

BALLOON IN A BOTTLE

Suppose you try to blow up a balloon inside a bottle. Even with a great deal of effort, the balloon will inflate only a tiny bit. The balloon does not inflate because there is air inside the bottle. Even though it looks like

Overview

DEMO: WHOLE CLASS

Key Question: What evidence do we have that gases exert pressure?

KEY IDEAS

Gas pressure can be defined as the force per unit area that results from gas molecules striking the walls of whatever container or object they come in contact with. There is air pressure from the gases that naturally surround us all the time. This air pressure is called *atmospheric pressure.*

LEARNING OBJECTIVES

● Describe and define gas pressure.
● Explain what causes air pressure.
● Complete simple air pressure calculations.

FOCUS ON UNDERSTANDING

Students often find the idea of the pressure of the air around them baffling, because they cannot feel it. They have to see evidence of its presence, because pressure differentials affect so many situations involving gases. Chapter 11 is devoted to the concept of pressure.

KEY TERMS

pressure
atmospheric pressure
atmosphere (atm)

IN CLASS

Students view a series of demonstrations involving air pressure. They write down observations, make drawings to show the various pressures, and discuss with a partner what they think happened. Then each demonstration is discussed as a class. Much discussion is integrated into the activity. Finally, the concepts of gas pressure, atmosphere, and atmospheric pressure are generally defined. *Note:* Atmospheric composition and properties are covered in more detail in the student text. A complete materials list is available for download.

TRM Materials List 57

SETUP

Instructions for seven demos are included. Practice the demos. You can set up to do all or some, depending on the time and equipment available. Consumer hand-pump vacuum chambers for preserving food are available in stores or online.

ENGAGE (5 min)

TRM PowerPoint Presentation 57

(Optional) Use a balloon to demonstrate the ChemCatalyst.

> *ChemCatalyst*
>
> If you blow up a balloon and let it go, it flies around the room.
>
> **1.** Why does the gas inside the balloon come out?
>
> **2.** How can you change how fast the balloon moves?
>
> **3.** How does this demonstration provide evidence of air pressure?

Sample answers: **1.** Some students may say that the gas comes out because there is an opening in the container. Others may say it is because there is more pressure in the balloon. **2.** If you put less air in the balloon, it will move slower. **3.** This demonstration shows that air molecules have mass and can exert pressure, like the wind.

Ask: Why doesn't the balloon stay in one place when you let go of it?

Ask: Why is the balloon moving so fast? How can you change that?

Ask: What do you think air pressure is? What does it have to do with weather?

EXPLORE (20–25 min)

TRM Worksheet with Answers 57

TRM Worksheet 57

PERFORM THE DEMONSTRATIONS

→ Tell students they will view the demo as a class.

→ Pass out the student worksheet

→ You will conduct demonstrations with some assistance from the class. Students can work individually or in pairs on their worksheets.

→ You can complete all or some of them depending on materials and time constraints.

→ After each demo, ask for possible explanations. Provide time for students to record their observations and answers in the chart on the worksheet. You may want students to draw diagrams to explain what they observed in each demo. Ask students to consider air pressure inside and outside each container. They should use arrows to indicate where there is air pressure. More arrows means greater air pressure.

1. Balloon in a Bottle *Note:* For health reasons, you will need one balloon for each student who tries to do the demonstration. Put an uninflated balloon inside a 2 L plastic bottle. Fold the opening of the balloon back over the mouth of the bottle so it stays in place. See the photograph on p. 300 of the student text. Try to blow air into the balloon inside the bottle. Do this as a race, with one student using the balloon inside the bottle setup and another blowing up a balloon in the usual way (no bottle). Explain: Why was the balloon in the bottle so difficult to inflate?

2. Soft Drink Can Put 5 mL of water in the bottom of an empty aluminum soft drink can. Heat the can on a hot plate until you see steam coming out of the opening. Holding the can with a pair of tongs, quickly invert the can into a dishpan filled halfway with cold water. The can will collapse suddenly and dramatically. Explain: What causes the can to collapse?

3. Submerged Cup Fill a large beaker about two-thirds full with water. Crumple a dry piece of paper and squeeze it into the bottom of a plastic cup. Invert the cup,

PHYSICS CONNECTION

A gas can even push against another gas. In fact, gas with a high pressure pushes a space shuttle into the air.

"nothing is there," the bottle is full of air, and when you try to inflate the balloon, you are pushing on this air, which takes a lot of work!

⟳ Explanation for Air Pressure

Collisions of gas molecules with surrounding objects are what we experience as air pressure. There are huge numbers of molecules in an automobile tire, many more than are shown in the illustration. The tiny push from each one adds up to a lot, enough to hold a car up.

The pressure of the air inside your tires is expressed in pounds per square inch, or lb/in^2. This unit is sometimes written as *psi*, for pounds per square inch. That is, each square inch of surface area on the tire experiences a certain number of pounds of force from the air inside. A tire on a racing bicycle might be inflated to a pressure of 100 lb/in^2.

A tire filled with air is firm because the air molecules are colliding with the inside walls of the tire, thereby pushing the walls out.

Example

Air Pressure in Car Tires

Suppose a car weighs 2000 pounds. Each of its four tires touches the road over an area that is about a 4-by-4-inch square. What air pressure do you need in each tire to hold up the car?

Solution

Each of the four tires needs to push up ¼ of the 2000 pounds, or 500 pounds. A 4-by-4-inch square has an area of 16 in^2, so each tire is in contact with a 16 in^2 area of the ground. Pressure is force per unit area.

$$\text{Pressure} = \text{force/area}$$
$$= 500 \text{ lb/16 in}^2$$
$$= 31 \text{ lb/in}^2$$

The pressure in each tire needs to be at least 31 lb/in^2.

300   Chapter 11 | **Pressing Matter**

making sure that the paper stays up in the cup. Immerse the inverted cup with the paper completely underwater in the beaker, holding it as vertical as possible. See the photograph on p. 299 of the student text. Take the cup out of the water and then remove the crumpled paper to show that it remained dry. Explain: Why didn't the paper in the cup get wet?

4. Air-Pressure Mat Place an air-pressure mat on a smooth, flat surface. See image on p. 301 of the student text. Show that it is easy to pick up the mat if you grab one of the corners and peel it up. Then try to lift the

mat by the hook in the middle. It is impossible to pick it up this way. Allow students to try. Explain: Why is it so difficult to lift the mat by the hook?

5. Cup and Card Fill a clear plastic cup with water. Place a laminated card over the top of the cup. Hold the card to the mouth of the cup and invert over a plastic tub. You can now let go of the card, it remains suspended, and the water does not spill out. Explain: Why doesn't the card fall?

6. Balloon in a Vacuum Inflate a balloon to about 2 or 3 in in diameter. Tie it off. Place the balloon inside a

The Atmosphere

Earth's atmosphere is a mixture of gases, mostly nitrogen, with some oxygen, carbon dioxide, water vapor, and argon. This mixture of gases is what we call air. The density of Earth's atmosphere changes as you travel up in altitude—the air becomes thinner at higher elevations. This concept will be covered more extensively in later lessons.

Big Idea The atmosphere is a mixture of gases, including gaseous water.

Although we barely notice it, the air around us exerts pressure on us all the time. **Atmospheric pressure** is air pressure that is always present on Earth as a result of air molecules colliding with objects on the planet. The pressure due to the atmosphere is the equivalent of a 14.7-pound weight pushing on every square inch of surface. We say that the pressure due to the atmosphere is 14.7 pounds per square inch, or 14.7 lb/in². This pressure is measured at sea level.

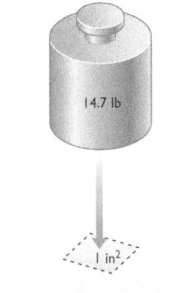

Scientists also use a unit called an **atmosphere,** or **atm,** to measure air pressure.

Atmospheric Pressure At sea level and 25 °C, there is 1 atm of pressure.

$$1 \text{ atm} = 14.7 \text{ lb/in}^2$$

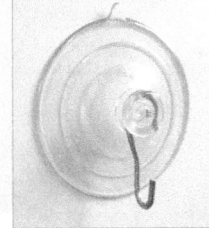
Atmospheric pressure played a role in all of the demonstrations you completed in class. The air pressure mat is particularly good at demonstrating how much pressure the air around us exerts on objects. If the mat has a surface area of 100 square inches and standard air pressure from the atmosphere is 14.7 lb/in², the total pressure on the mat from the air is 1470 pounds! No noder no one can lift it.

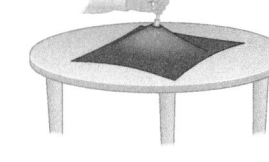

LESSON SUMMARY

What evidence do we have that gases exert pressure?

Earth is surrounded by an atmosphere of gases. We call this mixture of gases air. Although we cannot see it, the air around us exerts pressure. Gas molecules are

DISCUSS EACH DEMONSTRATION

→ Go over the demonstrations. Ask students to explain what they think happened in each demonstration.

→ Ask groups of students to put their drawings on the board, with arrows showing air pressures. You might want to complete this during the demonstrations.

Ask: How is air pressure involved in each demonstration?

Key Points: In each demonstration, air is trapped somewhere. In each demonstration, either the pressure of the trapped air is changed or the pressure of the air outside the container with the trapped air is changed.

● **Balloon in a Bottle:** The bottle already has air in it. When you try to put air into the balloon inside the bottle, you are pushing against the pressure of the air already in the container.

● **Soft Drink Can:** Heating the can produces water vapor inside it. When the can is placed upside down in cold water, the opening is sealed off. The water vapor inside the can cools quickly and turns into liquid water. The result is a dramatic decrease in air pressure inside the can. The can collapses because the air pressure on the outside of the can is suddenly so much greater than the pressure on the inside.

● **Submerged Cup:** When the cup is inverted in water, there is air inside the cup. This air takes up space. As a result, the water goes only partway up the inside of the cup. The paper stays dry.

● **Air Pressure Mat:** The mat is held down by atmospheric pressure. Because it is flat against the table, there is no air underneath pushing up on it. (*Note:* Atmospheric pressure is approximately 15 lb/in². The mat is about 100 in². Thus, the total force pushing down on the mat is about 1500 lb!)

● **Cup and Card:** The air pressure pushing up on the card from the atmosphere is larger than the weight

hand-pump vacuum chamber. Put on the lid, and pump to remove the air. This will result in an increase in the size of the balloon. Allow the air back in to decrease the size of the balloon.

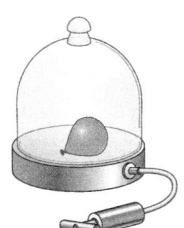

Explain: Why does the balloon increase in size in the vacuum chamber?

7. Marshmallows Place several marshmallows in the hand-pump vacuum chamber (or put mini-marshmallows into a large syringe

and put the cap on). Pump to remove the air. The marshmallows will puff up and increase in size. Then allow the air back in to decrease the marshmallows' size. The marshmallows will be smaller than they were when you started.

Explain: Why do the marshmallows increase in size inside the vacuum chamber? Why is the final size of the marshmallows smaller than their original size?

of the water and the card pushing down. The result is that the card is held in place.

- **Balloon in a Vacuum:** The pressure from the air outside the balloon decreases as the air is pumped out of the chamber. Thus, the balloon increases in size until the air pressure outside the balloon and the air pressure inside the balloon are equal.

- **Marshmallows:** Marshmallows have tiny pockets of trapped air inside them. When the air outside the marshmallows is removed, the air trapped inside the marshmallows expands and the marshmallows puff up. The air pressure in the pockets inside the marshmallows is equalizing with the air pressure outside the marshmallows. Some of the air pockets within the marshmallows will burst, so when air is put back into the chamber the marshmallows may be smaller than they were initially.

DEFINE AIR PRESSURE

Ask: What observations in the demonstrations provide evidence that gases take up space?

Ask: Can you provide evidence from the demonstrations to support the claim that gases are compressible? Explain.

Ask: What observations provide evidence that gases exert pressure?

Ask: What is air pressure?

Key Points: Air pressure is the force per unit area exerted on objects as a result of gas molecules colliding with those objects. When gas molecules are trapped in a container, those gas molecules collide with and push on the walls of the container. Thus, gas pressure often is expressed in pounds per square inch, or lb/in^2. Even though gas molecules are very far apart from one another compared to the molecules in a solid or a liquid, they can account for a great deal of force.

> **Pressure:** Force applied over a specific area. Force per unit area. Gas pressure is caused by gas molecules striking objects or the walls of a container.

In examples like those in this lesson, there are generally two types of air pressure to consider: air trapped inside a container and air from the atmosphere outside a container. All gases exert gas pressure, no matter what their identity. All the things we learn about air pressure apply to other gases as well, such as helium, oxygen, and fluorine.

KEY TERMS

pressure
atmospheric pressure
atmosphere (atm)

constantly moving and colliding with anything they come in contact with. Air pressure is defined as the force over a surface area. This force is caused by the collisions of the gas molecules. Gas pressure is measured in pounds per square inch, lb/in^2, or in atmospheres, atm. At sea level and 25 °C, the atmospheric pressure is 14.7 lb/in^2, or 1 atm.

Exercises

Reading Questions

1. Give three pieces of evidence that air exerts pressure.

2. Explain what causes air pressure.

Reason and Apply

3. High up in space, there are no molecules to collide with the outside walls of a balloon. Therefore, the air pressure inside the balloon is much greater than the air pressure outside of the balloon. Describe what would happen to an inflated balloon if it were suddenly put in space.

4. Calculate the weight in pounds of the atmosphere pushing down on the back of your hand.

5. When you fly in an airplane, it is common to feel painful pressure in your ears. Explain what might be happening in terms of air pressure.

6. What is the minimum pressure you need in bike tires to hold up a person who weighs 150 lb? Assume that the tires touch the ground in a square that is 1.5 in by 1.5 in.

7. The air pressure from the atmosphere measures 0.5 atm at an altitude of 18,000 ft. How much pressure is this in pounds per square inch? Calculate the weight in pounds of the atmosphere pushing down on the back of your hand at this altitude.

8. PROJECT Go to a library or do a Web search to find out how gases, such as helium or oxygen, are transported for industrial use.

9. If a car runs over the tip of your foot, it doesn't break your toes. Explain why.

The mixture of gases that surrounds you at all times is called the *atmosphere*. We commonly refer to it simply as "air." There is constant pressure on you from this air. While you might not feel the air pushing against you, the gases in the air exert a significant amount of pressure, called *atmospheric pressure*.

> **Atmospheric pressure:** Air pressure that is always present on Earth as a result of air molecules colliding with the surfaces of objects on the planet. At sea level and 25 °C, there is 14.7 lb/in^2 of air pressure from the air around us. This is referred to as one atmosphere of pressure, or 1 atm.

EVALUATE (5 min)

Check-In

Give evidence that gas molecules exert pressure on the walls of whatever container they are in.

Sample answer: A balloon stretched out tight when it is blown up is one example of the air pressure exerted by gas molecules.

Homework

Assign the reading and exercises for Weather Lesson 57 in the student text.

LESSON 57 ANSWERS

Answers continue in Answer Appendix (p. ANS-5).

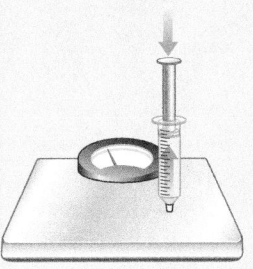

Boyle's Law

Purpose

To observe and quantify the relationship between gas pressure and gas volume.

Materials

- 50 mL plastic syringe with cap screwed on tight
- bathroom scale

Procedure

1. Start with the syringe at 50 mL. Make sure the cap is on tight.
2. Hold the syringe so that the tip is on top of a bathroom scale.
3. Repeat these steps for at least five different volumes.
   - One person depresses the plunger by a few milliliters.
   - A second person reads the exact volume.
   - A third person reads the number of pounds that is exerted on the bathroom scale.
   - Everyone records the volume and weight data in a table like this one.

Data

Trial	Volume (mL)	Weight that you apply (lb)	Pressure that you apply (lb/in²)	Atmospheric pressure (lb/in²)	Total pressure (lb/in²)
1				14.7 lb/in²	
2				14.7 lb/in²	

SAFETY Instructions

The cap on the tip of the syringe should always be pointed down, away from eyes. Wear safety goggles.

Analysis

1. Estimate the cross-sectional area inside of the syringe in square inches. Determine the pressure applied in pounds per square inch. Calculate the rest of the values needed to complete your table.
2. Graph the total pressure versus volume. What happens to the pressure of the gas as the volume of the gas decreases?
3. **Making Sense** Using today's observations, explain how the pressure and volume of a gas change in relation to each other.

Lesson 58 | **Boyle's Law**   303

Overview

LAB: PAIRS

DEMO: WHOLE CLASS

Key Question: How does gas volume affect gas pressure?

KEY IDEAS

When an amount of gas is compressed into a smaller volume, its pressure increases. The pressure of a gas is inversely proportional to the volume the gas occupies (if the temperature and amount of gas do not change).

Thus, when one variable gets larger, the other variable must get smaller, and vice versa. Mathematically, the relationship between pressure and volume is expressed as $P_1V_1 = P_2V_2 = k$, or $P = k/V$. This mathematical relationship is known as *Boyle's law.*

LEARNING OBJECTIVES

- Explain the relationship between gas pressure and gas volume.
- Define an inversely proportional relationship.
- State Boyle's law.

FOCUS ON UNDERSTANDING

- Among the weather variables, this is the only inversely proportional relationship students will encounter. Thus, it may give them some difficulty initially.
- So far we have avoided using the term *constant* to mean fixed or unchanging to prevent confusion with the proportionality constant. However, in this lesson, you may want to point out this meaning of the word to students.

KEY TERMS

inversely proportional
Boyle's law

IN CLASS

Students explore the relationship between gas pressure and gas volume experientially using large syringes. Students work in pairs for the first portion of the lab before gathering in larger groups for Part 2. The larger groups gather pressure and volume data using a syringe with air trapped inside the chamber. A bathroom scale measures the force required to change the volume inside the syringe. See figure on p. 303 of the student text. (Instead of the scale and syringe, you can use a platform apparatus like the one shown—sometimes called an *elasticity of gases apparatus*—and have students place a book or other objects with known mass on top.)

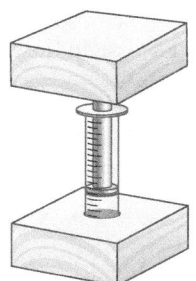

Students graph the data and are guided to recognize that the relationship between pressure and volume is an inverse relationship. A complete materials list is available for download.

TRM Materials List 58

SETUP

You will need two types of syringes for this lab: uncapped syringes for Part 1 of the lab and capped syringes for Part 2. It is important to check the capped syringes to make sure the caps are tightly screwed on or firmly secured with superglue. Instruct students to wear safety glasses and to always point the syringe with cap down as they push on it. The caps can fly off at very high pressures. The number of groups for Part 2 will depend on the number of bathroom scales you have. Thus, you could do Part 2 in larger groups or as a student-assisted demo.

Differentiate

Students may not know or understand the roles of independent and dependent variables as well as experimental controls. The experiment in this lesson can help illustrate how the variable of pressure is dependent on volume. Students can "see" how this takes place, and you can help students grasp the importance of experimental design by asking what might happen if the syringe developed a leak (changing number of particles) or if you repeated the experiment at the beach or in a blizzard (changing temperature). Have students label their graph on the student worksheet with these terms in the appropriate places.

Pre-AP® Course Tip

Students preparing to take an AP® chemistry or college chemistry course should be able to cite reasons for using a particular mathematical method such as graphical analysis for the inverse relationship between P and V to justify the use of direct proportionality in mathematical problem solving. In this lesson, students explain the relationship between gas pressure and gas volume and state Boyle's law.

Lesson Guide

ENGAGE (5 min)

TRM PowerPoint Presentation 58

ChemCatalyst

An empty plastic water bottle has a cork fitted into the opening.

1. Predict what would happen if you stepped on the plastic bottle.
2. Explain your answer in terms of pressure and volume.

Sample answers: 1. The cork should fly out of it. **2.** The bottle contains trapped air. When you step on the bottle, you decrease the amount of space this air has to fit into, causing air pressure to increase inside the bottle. The cork flies

out because it is the weakest place in the system.

→ You might want to demonstrate with a bottle and cork. Be sure to point the cork away from anyone and from anything breakable.

Ask: Is the bottle truly empty? Explain.

Ask: What is happening in terms of pressure and volume?

Ask: If you squeeze some gas into a smaller volume, what happens to the pressure inside the container?

Feeling under Pressure

Boyle's Law

THINK ABOUT IT

A flat basketball does not bounce very well. However, you can push air into the ball with a pump. The air you trap inside pushes on the skin of the ball with a lot of pressure. The amount of air you push into the ball determines how firm and bouncy the ball is. What would happen if you pushed the same amount of air into a smaller ball?

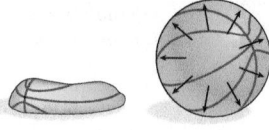

How does gas volume affect gas pressure?

To answer this question, you will explore

↻ The Syringe and Scale

↻ Graphing Pressure–Volume Data

↻ Boyle's Law

EXPLORING THE TOPIC

↻ **The Syringe and Scale**

In class, you examined how the pressure of a gas trapped inside a container is related to the volume the gas occupies. A syringe makes a good container for such an investigation, because it can be sealed off and the volume of air trapped inside the syringe can be measured.

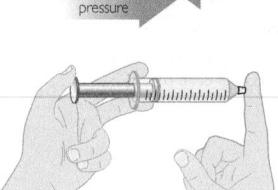

If you push down on the plunger of the syringe, you can feel the pressure of the gas inside the syringe. Air molecules take up space and as you squeeze them into a smaller space, they push back more and more. This is because the same number of molecules occupies a smaller volume, resulting in more collisions between the molecules and the container. More collisions mean greater gas pressure.

GATHERING DATA

You can learn more about the relationship between the volume of a gas and its pressure by using a capped syringe and a bathroom scale. As you push down on the plunger of the syringe, the gas pressure inside the syringe increases, and the

---

EXPLORE (20 min)

TRM Worksheet with Answers 58

TRM Worksheet 58

INTRODUCE THE LAB

→ Arrange students into pairs.

→ Pass out the student worksheet.

→ Discuss limiting the number of variables. Write the terms *temperature, volume, pressure,* and *number of molecules* on the board. In this lab, the number of molecules does not change, and the gas is assumed to remain at room

---

weight on the scale increases. You can record this quantitative, or numerical, data. To calculate the pressure of the gas, divide the measured weight on the scale by the surface area of the plunger in pounds per square inch, or lb/in². Weight is a measure of force.

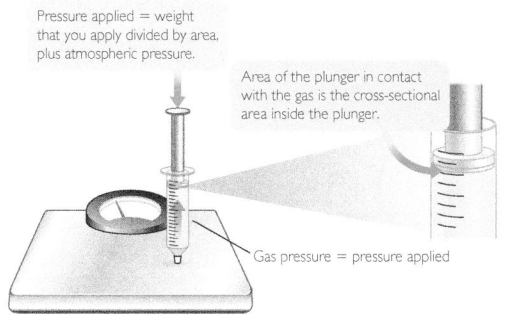

Pressure applied = weight that you apply divided by area, plus atmospheric pressure.

Area of the plunger in contact with the gas is the cross-sectional area inside the plunger.

Gas pressure = pressure applied

Important to Know Because the atmosphere is also pushing on the syringe, you need to add 14.7 lb/in² to the pressure that you apply.

Note that the pressure you apply plus atmospheric pressure is equal to the gas pressure inside the syringe. When you push down, the gas pushes back with the same pressure that is being applied to it.

Data for this experiment are given in the table. Notice that as the volume decreases, the pressure of the gas inside the syringe increases.

Pressure and Volume Data

Trial	Volume (mL)	Total pressure (lb/in²)
1	100 mL	15 lb/in²
2	80 mL	19 lb/in²
3	60 mL	25 lb/in²
4	40 mL	38 lb/in²
5	30 mL	50 lb/in²
6	20 mL	75 lb/in²

Big Idea If you squeeze a sample of gas into a smaller container, its pressure will increase.

↻ Graphing Pressure–Volume Data

Many of the relationships you have explored so far, like mass versus volume and volume versus temperature of a gas, are proportional relationships. The relationship between pressure and volume is not directly proportional. A graph

the gas in the syringe, which you measured in pounds on the scale. Pounds can be converted into pressure by dividing by the area on which you are pushing. In this case, the area is the cross-sectional area of the chamber in the syringe. In today's class, you used units of pounds per square inch, or lb/in². It is necessary to add 14.7 lb/in² to all the measurements to account for the force of the air pressure from the atmosphere.

When the volume of a gas is decreased, its pressure goes up. This is because the same number of molecules must occupy a smaller space. A smaller space containing the same number of molecules means more collisions and, therefore, higher pressure. Conversely, when the volume of a gas is increased, its pressure goes down. This is because the same number of gas molecules must occupy a larger space. There are fewer collisions, and therefore, the gas molecules exert a lower pressure.

PROCESS THE GRAPH OF PRESSURE VERSUS VOLUME

→ Draw x- and y-axes on the board with Volume on the x-axis and Pressure on the y-axis. Have students help you create a simple graph using the data from the syringe-and-scale experiment.

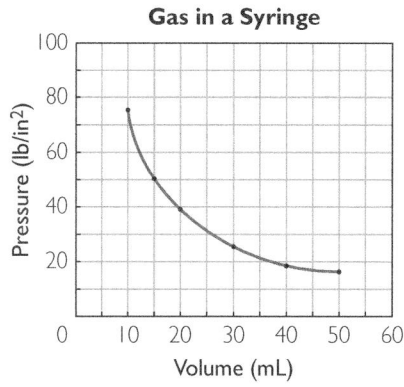

Gas in a Syringe

Ask: What does the graph show about the relationship between pressure and volume?

Ask: Why isn't the graph a straight line?

Ask: Does the curve ever touch the axis? Explain.

Ask: Does the graph of pressure versus volume describe a proportional relationship? Why or why not?

Key Points: The relationship between gas pressure and gas volume is called an *inverse proportion*. This means that when one variable increases, the other

temperature. Today's class focuses on pressure and volume and how these two variables interact.

GUIDE THE LAB

→ Students can work in pairs on Part 1. Depending on the number of bathroom scales you have, you can do Part 2 as a whole-class demo or in groups. Students are asked to estimate the cross-sectional area inside the syringe. They might need help visualizing or calculating this area. It probably will be about 1–2 in², depending on the syringe.

DISCUSS AIR PRESSURE AS IT RELATES TO THE ACTIVITY

Ask: What did you observe when you explored the syringe with no cap on it?

Ask: Does the amount of air change when you push in the plunger?

Ask: Can you decrease the volume of the syringe to zero? Why or why not?

Ask: Why does it get harder and harder to depress the plunger as the volume gets smaller and smaller?

Key Points: In the experiment with syringe and scale, you put weight on

decreases, and vice versa. The graph of gas pressure versus gas volume is a curve, not a straight line going through the origin. In contrast, with the rain gauge and with Charles's law, the variables you encountered are directly proportional. Both variables increase together.

> **Inversely proportional:** Two variables are inversely proportional to each other if one variable increases when the other decreases.

The curve on the graph is nearly vertical for small volumes and nearly horizontal for large volumes. For small volumes, the gas pressure in the syringe is extremely high. After a certain point, it is difficult to make the volume any smaller, no matter how hard you push. In other words, you hit a limit as to how small the gas volume can be. At large volumes, the pressure is extremely low in the syringe. Once the volume is large, the pressure does not change much. Between the two extremes, changes in volume and pressure affect each other more.

INTRODUCE BOYLE'S LAW

➡ You might complete a graph of the data for *P* versus *1/V* on the board.

➡ Hand out the Math Card for Boyle's law to students (optional).

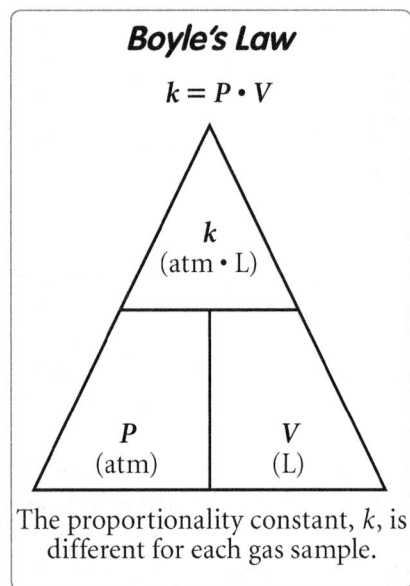

Boyle's Law

$$k = P \cdot V$$

k
(atm · L)

P
(atm)

V
(L)

The proportionality constant, *k*, is different for each gas sample.

Ask: What happens if you create a graph of *P* versus *1/V*?

Ask: What is the approximate value of *P* times *V* for your data? (It should be somewhere around 600–700.)

---

of the data for this experiment results in a curved line that does not go through the origin.

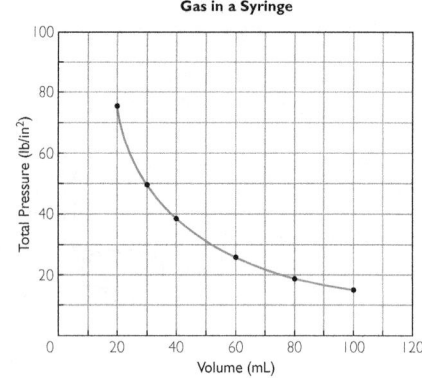

Gas pressure and gas volume are **inversely proportional** to one another. In other words, when one variable increases, the other decreases, and vice versa.

When the volume of the gas in the syringe is very small, the gas pressure is quite high. When the volume of the gas is large, the gas pressure is quite low. However, as long as there are gas molecules in a container, neither the gas pressure nor the gas volume can ever reach zero.

⟳ Boyle's Law

In 1662, a British scientist named Robert Boyle discovered that gas pressure and volume are inversely proportional to each other. This relationship is known as **Boyle's law.**

> **Boyle's Law**
> The pressure P of a given amount of gas is inversely proportional to its volume V, if the temperature and amount of gas are not changed. The relationship is
> $$PV = k \text{ or } P = k\left(\frac{1}{V}\right), \text{ where } k \text{ is the proportionality constant.}$$

Examine the Pressure and Volume Data table on page 305. If you multiply pressure by volume, the product will always be around 1500. (It will vary slightly due to rounding and slight errors in measurement.) Boyle's law expresses this relationship as $PV = k$. In this experiment, the proportionality constant, k, is 1500 mL · lb/in². [For a review of this math topic, see **MATH Spotlight**: Ratio and Proportions on page A-11.]

Boyle's law can be used to solve problems involving gas pressure and gas volume. Once you have one set of pressure and volume measurements for a gas, you can determine the proportionality constant. With the proportionality constant, you can determine the pressure of the gas at any volume.

INDUSTRY CONNECTION

Gases are often stored in tanks at high pressure or even at low temperatures so that they're liquid, because otherwise the containers needed for them would be huge. These tanks are usually made of thick metal so that they don't change volume.

Ask: What would happen to your data if you heated the air in the syringe and then measured pressure and volume? (*k* is different at a different *T*)

Key Points: The mathematical relationship between pressure and volume is described by the equation

> $$PV = k \quad \text{or} \quad P = k/V$$

This relationship is known as *Boyle's law,* after Robert Boyle, the British scientist who discovered the relationship. If you were able to make very accurate measurements in your syringe-and-scale experiment, the proportionality constant, *k,* would be the same number for every value of *P* times *V*. You can have students calculate the product *PV* for their data. The products will be nearly identical, but not exactly, due to experimental error.

A graph of *P* versus *1/V* is a straight line going through the origin. This means that there is an inversely proportional relationship between *P* and *V*. It is important to emphasize that the proportionality constant, *k,* is a different number if the amount of gas or the temperature is changed.

Example

Gas in a Syringe

Determine what the pressure of the gas inside this syringe would be if the volume were reduced to 10 mL.

Solution

Start with Boyle's law.

$$P = k\left(\dfrac{1}{V}\right)$$

Substitute values for k and V.

$$= 1500 \text{ mL} \cdot \text{lb/in}^2\left(\dfrac{1}{10 \text{ mL}}\right)$$

Solve.

$$= 150 \text{ lb/in}^2$$

You can check this answer: $PV = 150 \text{ lb/in}^2 \cdot 10 \text{ mL} = 1500 \text{ mL} \cdot \text{lb/in}^2$.

LESSON SUMMARY

How does gas volume affect gas pressure?

If you squeeze a gas into a smaller volume, it will exert more pressure on its container because a smaller space means more collisions between the molecules and the container. If you let a gas spread out into a huge volume, the gas pressure will get very low because there will be fewer collisions between the molecules and the container. Gas pressure and gas volume are inversely proportional. This means that as one variable increases, the other decreases, and vice versa. This relationship is described by Boyle's law: $P = k(1/V)$.

KEY TERMS

inversely proportional
Boyle's law

Exercises

Reading Questions

1. Explain how you can use a scale to measure the pressure of the gas in a syringe.
2. Describe how pressure varies as the volume of a trapped gas changes.

Reason and Apply

3. **Lab Report** Write a lab report for the Lab: Boyle's Law. In the results section, provide two graphs of the data. Plot pressure versus volume and plot pressure versus the inverse of the volume, $1/V$. Also discuss how accurate your results are and what factors might contribute to experimental error in this lab. In the

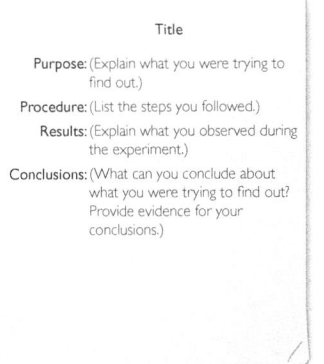

Title

Purpose: (Explain what you were trying to find out.)

Procedure: (List the steps you followed.)

Results: (Explain what you observed during the experiment.)

Conclusions: (What can you conclude about what you were trying to find out? Provide evidence for your conclusions.)

Thus, for each new situation, k has a different value, equal to PV.

> **Boyle's law:** The pressure of a given amount of gas is inversely proportional to its volume, if the temperature and amount of gas are not changed. The relationship between pressure and volume can be expressed as $PV = k$, or $P = k(1/V)$, where k is the proportionality constant.

Boyle's law enables you to solve problems involving gas pressure and gas volume.

EVALUATE (5 min)

Check-In

A balloon full of gas occupies 7.5 L, and the pressure on the outside of the balloon is 1.0 atm. What do you predict will happen to the pressure inside the balloon if the balloon is placed underwater to a depth at which its new volume is 2.5 L?

Answer: The pressure will increase, because the volume decreases.

Homework

Assign the reading and exercises for Weather Lesson 58 in the student text.

LESSON 58 ANSWERS

1. *Possible answer:* As you push down on the plunger of the syringe, the gas pressure inside the syringe increases and the pressure on the scale increases. The pressure inside the syringe is equal to the pressure you are exerting on the plunger, which can be read from the scale. To calculate the pressure of the gas, divide the weight on the scale by the surface area of the plunger. Then add atmospheric pressure.

2. *Possible answer:* As the volume of the gas decreases, the pressure of a trapped gas increases. As the volume of the gas increases, the pressure of the trapped gas decreases.

3. A good lab report will contain ● a title (Lab: Boyle's Law) ● a statement of purpose (***Possible answer:*** to discover the relationship between gas pressure and gas volume) ● a procedure (a summary of the steps followed in the experiment) ● results (a list of observations made during each step in the procedure, including two graphs and a list of sources of experimental error ● The first graph should show a curve that decreases asymptotically toward the volume axis as volume increases and toward the pressure axis as pressure increases. ● The second graph should show a curve that is nearly linear, which will allow students to connect their data with Boyle's law. ● Possible sources of experimental error that could be listed include the accuracy of the weight given by the scale, the accuracy of the scale on the side of the syringe, and the difficulty of maintaining a constant pressure on the plunger to get a steady reading on the scale. ● a conclusion (***Possible answer:*** Pressure is inversely proportional to volume. This is reflected in Boyle's law, because k is the proportionality constant between pressure, P, and the inverse of volume, $1/V$.)

4. Pumping the tire increases the amount of air in the tire and causes the pressure inside the tire to increase. This causes the air inside the tire to push harder against the pump, making it more difficult to force additional air into the tire.

5. *Possible answer:* The relationship between gas pressure and gas volume is different from the relationship between gas volume and gas temperature because an increase in pressure causes the volume to decrease, while an increase in temperature causes the volume to increase.

6. a. Because pressure and volume have an inverse relationship, doubling the pressure outside of the balloon halves the balloon's volume. The product of pressure and volume is a constant, assuming no other parameters change. **b.** The pressure of the air inside the balloon will double. Because the balloon's surface is flexible, the air pressure inside the balloon equilibrates with the water pressure outside the balloon.

7. a. *Possible answer:* Show that the pressure is proportional to the inverse of the volume. First, compute the inverse of each volume and create a table that shows the pressure and the inverse of volume for each trial. Then divide the second column into the first column and see if the result is a constant.

Trial	P (lb/in^2)	$1/V$ (1/mL)	k (mL lb/in^2)
1	10	1/60	600
2	15	1/40	600
3	20	1/30	600
4	40	1/15	600
5	60	1/10	600

b. *Possible answer:* Show that pressure and volume always multiply to the same constant. Use the original table of data and multiply the pressure and volume for each trial.

$k = PV$

Trial 1: 60 mL $\cdot$ 10 lb/in^2 = 600 mL $\cdot$ lb/in^2

Trial 2: 40 mL $\cdot$ 15 lb/in^2 = 600 mL $\cdot$ lb/in^2

Trial 3: 30 mL $\cdot$ 20 lb/in^2 = 600 mL $\cdot$ lb/in^2

Trial 4: 15 mL $\cdot$ 40 lb/in^2 = 600 mL $\cdot$ lb/in^2

Trial 5: 10 mL $\cdot$ 60 lb/in^2 = 600 mL $\cdot$ lb/in^2

c. The graph shows that gas pressure and volume are inversely proportional.

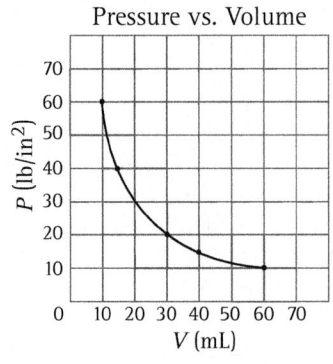

Pressure vs. Volume

d. The graph shows that the values are directly proportional.

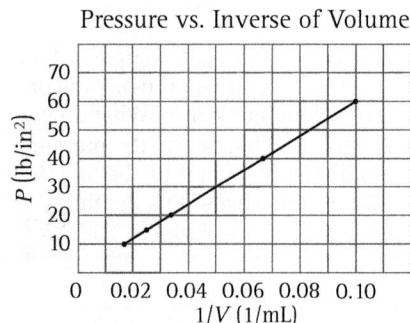

Pressure vs. Inverse of Volume

e. 20 mL

f. 12 lb/in^2

Exercises (continued)

conclusions section, explain the relationship between pressure and volume. Explain how your data relate to Boyle's law.

4. You are using a bicycle pump to fill a bicycle tire with air. It gets harder and harder to push the plunger on the pump the more air is in the tire. Explain what is going on.

5. How is the relationship between gas pressure and gas volume different from the relationship between gas volume and gas temperature?

6. Imagine you blow up a balloon before going scuba diving. You put on your gear and descend to 30 ft with the balloon. The total pressure from the ocean at that depth is 2 atm. Assume temperature is constant.
   a. What happens to the volume of the balloon?
   b. What happens to the pressure of the air on the inside of the balloon?

7. This table shows experimental data for gas volume and pressure.

Trial	Volume (mL)	Pressure applied (lb/in^2)
1	60 mL	10 lb/in^2
2	40 mL	15 lb/in^2
3	30 mL	20 lb/in^2
4	15 mL	40 lb/in^2
5	10 mL	60 lb/in^2

   a. Show that $P = k(1/V)$ for each trial. Remember to add 14.7 lb/in^2 to the values in the last column to account for atmospheric pressure.
   b. Show that $PV = k$ for each trial.
   c. Create a graph of P versus V. What does the graph tell you about the relationship between gas pressure and volume?
   d. Create a graph of P versus $1/V$. What does the graph tell you about the relationship between gas pressure and inverse volume?
   e. Use the graphs from parts c and d to estimate the volume when the pressure is 30 lb/in^2.
   f. Use the graphs from parts c and d to estimate the pressure when the volume is 50 mL.

Egg in a Bottle

Gay-Lussac's Law

THINK ABOUT IT

Gases are trapped in balloons, soda cans, pickle jars, bicycle tires, and in your lungs if you hold your breath. Some of these containers are flexible and can vary in size, while others are rigid. A gas sample can be described by pressure, volume, and temperature. The gas might be heated, cooled, compressed, taken underwater, left outside overnight, or taken up in an airplane.

> **How does gas pressure change in flexible and rigid containers?**

To answer this question, you will explore

⟳ Gay-Lussac's Law

⟳ Gas Law Problems

EXPLORING THE TOPIC

⟳ Gay-Lussac's Law

Gases are sometimes stored in rigid gas containers. For example, underwater divers wear a metal gas tank strapped to their backs. To supply air to the diver for a sufficient amount of time, a lot of air is pushed or compressed into the tank. The result is that the gas inside the tank is at a high pressure.

With rigid containers the pressure of the air inside the container changes if the temperature changes. Indeed, there is a risk of an explosion if you heat the gas inside a pressurized tank.

A scuba diver carries a tank of compressed gas.

PRESSURE VERSUS TEMPERATURE

Consider a tank containing helium. If this tank is placed in the sun and the temperature increases, the pressure of the gas inside the tank also increases.

Helium Tank Data

Temperature (K)	Volume (L)	Pressure inside tank (atm)	P/T (atm/K)
293 K	55 L	130 atm	0.44
313 K	55 L	138 atm	0.44
283 K	55 L	125 atm	0.44
240 K	55 L	106 atm	0.44

Lesson 59 | **Gay-Lussac's Law**   309

Overview

CLASSWORK: INDIVIDUAL

DEMO: INDIVIDUAL

Key Question: How does gas pressure change in flexible and in rigid containers?

KEY IDEAS

Changing the temperature of a gas will change its pressure if the volume is kept the same. The pressure of a gas is proportional to the temperature of a gas if the volume and amount of gas do not change and if the temperature of the gas is expressed in kelvins. Mathematically, the relationship

between pressure and temperature is $P = kT$, where k is the proportionality constant. This mathematical relationship was first investigated by Guillaume Amontons in the late 1600s but is often called Gay-Lussac's law. Boyle's law and Gay-Lussac's law are two gas laws that apply to situations in which the pressure changes in response to changes in either volume or temperature. The type of container a gas occupies—rigid or flexible—affects which of these gas laws applies.

LEARNING OBJECTIVES

● Describe the qualitative and quantitative relationships between the pressure and temperature of a gas.

● Explain how flexible and rigid containers affect the pressure, volume, and temperature of a gas sample.

● Complete gas law problems involving changes in pressure.

FOCUS ON UNDERSTANDING

● Flexible and rigid containers present some conceptual challenges. For a flexible container, the pressure differences both outside and inside the container must be considered.

● Students often forget about the effects of atmospheric pressure.

● The pressure change in a rigid container can be observed only with a pressure gauge or if the container bursts.

● Several mathematical approaches to solving gas law problems exist, including using proportional reasoning and dimensional analysis. Students will use whichever method works best for them based on the math strategies they have already learned. Our approach presents the gas laws in such a way as to best assist students with lower math abilities while not getting in the way of students with higher abilities.

KEY TERM

Gay-Lussac's law

IN CLASS

Students observe a demonstration of Gay-Lussac's law. A hard-cooked egg is forced into an Erlenmeyer flask by cooling the air inside to alter the air pressure inside the flask. Students complete a worksheet that allows them to gain some proficiency at completing problems involving Gay-Lussac's law and Boyle's law to determine pressure. Students consider the effects of flexible and rigid containers on the pressure, volume, and temperature of a gas sample. Additional worked examples reinforce the gas laws. *Note:* This lesson can be completed in a block period or two standard class periods. A complete materials list is available for download.

TRM Materials List 59

SETUP

Prepare hard-cooked eggs the night before class. You need only one egg per class if all goes well, but it is useful to have a few extra. Set up the flask with a small amount of water inside and the egg on top of the flask so students will see the setup when they answer the ChemCatalyst.

Pre-AP® Course Tip

Students preparing to take an AP® chemistry or college chemistry course should be able to use mathematics to solve problems that describe the physical world. In this lesson, students use mathematics to complete gas law problems involving changes in pressure.

Lesson Guide

ENGAGE (5 min)

TRM PowerPoint Presentation 59

→ Assemble the egg-and-bottle demonstration at the beginning of class so students can look at the setup and answer the ChemCatalyst.

ChemCatalyst

Examine the setup. How could you use gas pressure to get the egg into the bottle? What variables would you change: pressure, volume, and/or temperature?

Sample answer: Some students may suggest increasing air pressure outside to push on the egg, while others will suggest that the air pressure inside the bottle be reduced. Some may suggest warming the outside air and cooling the air inside the flask. Others will suggest increasing the pressure outside the flask and reducing it inside.

→ Discuss ideas about how to get the egg into the bottle.

Ask: How would you get the egg into the bottle using gas pressure?

Ask: What happens to the gas inside the bottle when it is cooled?

Ask: What force might push the egg into the bottle?

EXPLORE (50 min)

TRM Worksheet with Answers 59

TRM Worksheet 59

COMPLETE THE DEMONSTRATION

→ Demonstrate how differences in gas pressure can be used to force an egg into a bottle.

1. With a small amount of water in the bottom of the flask, heat the flask on a preheated hot plate for several minutes. (There should be enough water so it does not all boil away.) Remove the flask from the hot plate with an oven mitt when you see steam coming out of the flask.

COOKING CONNECTION

A pressure cooker can cook food in a fraction of the time it takes with a regular pot. Because the lid fits tightly, a higher pressure is achieved, resulting in a higher temperature.

Pressure Versus Temperature: Helium Gas in a Tank

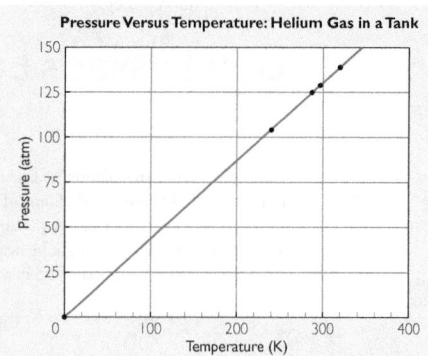

Using the Graph
Because the pressure and the Kelvin temperature are proportional, the graph is a straight line through the origin. If you used the Celsius temperature, the graph would still be a straight line, but it would not go through the origin.

Big Idea If you heat a sample of gas in a closed rigid container, its pressure will increase.

If you graph the helium tank data, the points lie on a straight line. The straight line through the data points and the point $(0, 0)$ indicate that there is a proportional relationship between pressure and temperature. Joseph Louis Gay-Lussac, a French scientist, described this proportional relationship in the early 1800s.

Gay-Lussac's Law The pressure P and the Kelvin temperature T of a gas are proportional when the volume and the amount of gas do not change. The relationship is $P = kT$. The value of the proportionality constant, k, depends on the specific gas sample.

Example 1

Propane Gas Tank

A 5.0 L propane gas tank on a camping stove contains propane gas at a pressure of 2.0 atm when the temperature is 14 °C (about 57 °F) in the morning. During the day, the tank warms up to a temperature of 34 °C (about 93 °F). What is the pressure of the propane gas inside the tank when the gas temperature is 34 °C?

Solution
The volume of the tank does not change. The temperature of the gas increases, causing the pressure of the gas to increase.

310   Chapter 11 | **Pressing Matter**

---

2. Place the hard-cooked egg on the opening of the flask so it makes a seal. Observe what happens as the air inside the flask cools (the egg gets pushed into the flask).

3. You can speed up the process by placing the flask in cool water or ice water.

4. Ask students to explain what they think happened.

5. Ask students how they could use gas pressure to get the egg back out of the flask. Listen to their suggestions. Then turn the flask upside down so that the egg falls into the opening. Hold the flask so that it is tipped sideways, and reheat the flask on the

bottom until the egg is pushed back out of the flask. A Bunsen burner works best for this last procedure, but you can use a hot plate or a blow dryer. Be sure to use an oven mitt when holding the flask.

INTRODUCE THE CLASSWORK

→ Let students know they will be working individually.

→ Pass out the student worksheet

→ Tell students they will be exploring a number of situations in which the pressure changes in response to either a change in volume or a change in temperature.

Example 1

(continued)

You can use the subscript 1 for the first set of conditions, and subscript 2 for the second set of conditions. These are sometimes called initial conditions and final conditions. You need to find P_2, the final pressure.

Initial Conditions	Final Conditions
$P_1 = 2.0$ atm	$P_2 = ?$
$T_1 = 14\,°C$	$T_2 = 34\,°C$
$V_1 = 5.0$ L	$V_2 = 5.0$ L

The variables that are changing are P and T. Use Gay-Lussac's law: $P = kT$. Recall that the temperature must be in kelvins.

Convert the temperature to the Kelvin scale.

$$K = 273 + C$$
$$T_1 = 273 + 14 = 287 \text{ K (in the morning)}$$
$$T_2 = 273 + 34 = 307 \text{ K (during the day)}$$

Determine the value of k using P_1 and T_1.

$$k = \frac{P_1}{T_1} = \frac{2.0 \text{ atm}}{287 \text{ K}} = 0.0070 \text{ atm/K}$$

Use k to solve for P_2.

$$P_2 = k\,T_2$$
$$= 0.0070 \text{ atm/K} \cdot 307 \text{ K}$$
$$= 2.15 \text{ atm}$$

At 34 °C, the pressure of the propane gas in the tank has increased to 2.15 atm.

Gas Law Problems

Gas pressure, gas volume, and gas temperature can change depending on the conditions. But no matter what, the gases must obey the gas laws!

There are two general types of containers: rigid and flexible. A rigid container has a volume that doesn't change. A flexible container can vary in size.

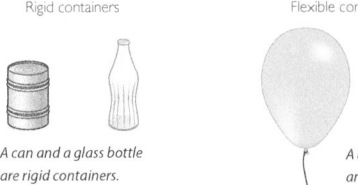

Rigid containers

A can and a glass bottle are rigid containers.

Flexible containers

A balloon and a syringe are flexible containers.

The type of container has an effect on what happens to the gas inside when the conditions change. For example, Boyle's law and Charles's law apply when a gas sample is in a flexible container, which can change volume. Gay-Lussac's law applies for gas samples in rigid containers.

SPORTS CONNECTION

The tire pressure of race cars is extremely important. If the tire pressure is too high, the tires won't have enough contact with the track. If the tire pressure is too low, the cars won't handle well. But the tires heat up considerably during a race, increasing the pressure. The average temperature of a tire being removed during a pit stop is about 100 °C.

© David Madison/Corbis

INTRODUCE GAY-LUSSAC'S LAW

Ask: Describe a situation in which you would use Gay-Lussac's law.

Ask: Why is it dangerous to heat a sealed container?

Ask: How does the type of gas container (rigid or flexible) relate to this gas law?

Key Points: Gas pressure is proportional to temperature if the volume and amount of gas are not changed and if the temperature is expressed in kelvins. This relationship is known as Gay-Lussac's law. It is named after French scientist Joseph Louis Gay-Lussac, who described the relationship in 1802. It is expressed as $P = kT$. The proportionality constant, k, describes how much the pressure increases per degree of temperature in kelvins. The value of k for a set of conditions is P/T. Gay-Lussac's law applies only to rigid containers for which the volume does not change.

> **Gay-Lussac's law:** The pressure of a given amount of gas is directly proportional to temperature if the gas volume and amount of gas do not change and if the temperature is expressed in kelvins.

COMPARE GASES IN RIGID AND FLEXIBLE CONTAINERS

→ Tell students to take out the Math Cards for Charles's law, Boyle's law, and Gay-Lussac's law (optional).

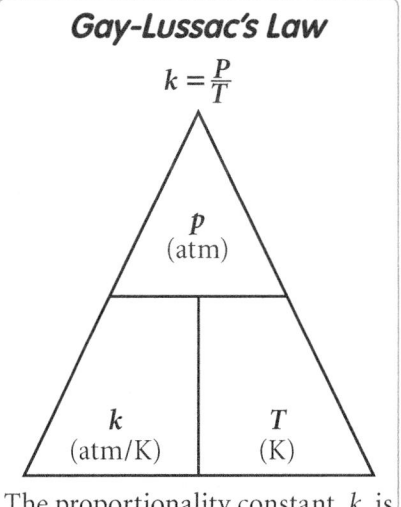

Gay-Lussac's Law

$$k = \frac{P}{T}$$

The proportionality constant, k, is different for each gas sample.

Ask: What are some examples of flexible containers? (balloon, piston and cylinder, syringe, bag)

Ask: What are some examples of rigid containers? (bottle, can, gas tank, pressure cooker)

BRIEFLY DISCUSS THE DEMONSTRATION

Ask: What you think caused the egg to go into the flask? Explain.

Ask: The egg was pushed, not sucked, into the flask. What does this mean?

Key Points: As the gas inside the flask cools, the pressure of the gas inside the flask decreases. Because the air pressure decreases inside the flask, the force on the egg exerted by the air outside the flask is greater than the force exerted on the egg by the air inside the flask. When the difference becomes great enough, the egg is pushed into the flask. The change in gas pressure happens both

because gas pressure is proportional to temperature and because some of the water vapor in the flask becomes a liquid. Notice that the egg is not *sucked* into the flask. It is *pushed* into the flask by the higher air pressure outside the flask. Gaseous molecules cannot *suck*.

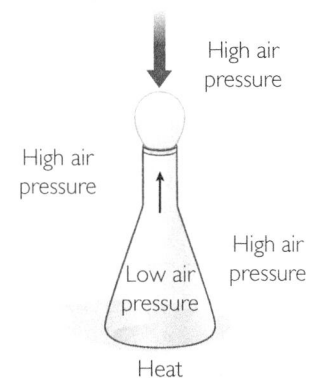

High air pressure

High air pressure

High air pressure

Low air pressure

Heat

Ask: How does the type of gas container (rigid or flexible) relate to Charles's law, Boyle's law, and Gay-Lussac's law?

Key Points: The three gas laws help predict gas temperature, pressure, and volume when two of these variables change and the third remains fixed. There are two types of containers for the gas: one that can vary in volume, such as a balloon or a syringe, and one with a fixed volume, such as a sealed glass bottle.

- Flexible container: When the gas is in a flexible container, changing the temperature or the pressure causes the volume to change. Charles's law applies when volume and temperature vary but pressure does not change. Boyle's law applies when volume and pressure vary but temperature does not change.
- Rigid container: When a gas is in a rigid container, increasing the temperature causes the pressure to increase. Gay-Lussac's law applies when the pressure and temperature vary but volume does not change.

It is important to notice that in all these cases gas cannot enter or escape the containers. The amount of gas is fixed. We will explore the effects of changing the number of gas molecules in Chapter 12.

SOLVE SAMPLE GAS-LAW PROBLEMS

KEYS TO SOLVING GAS-LAW PROBLEMS

1. Identify which variable is not changing: P, V, or T.

2. Identify the two variables that are changing: P and V, P and T, or V and T.

3. Identify the gas law formula that should be used to solve the problem.

4. Determine the proportionality constant k from P_1, V_1, or T_1.

5. Use k to find P_2, V_2, or T_2.

6. Remember that all temperatures must be in kelvins.

EXAMPLE 1

Imagine that you have a syringe with a plunger that slides easily. Air is trapped inside. The gas pressure is 1.0 atm, the gas temperature is 20 °C, and the volume of the gas is 52 mL. You place the syringe in the freezer when the gas pressure is 1.0 atm. The volume of the air trapped in the syringe changes to 45 mL, and the pressure is still 1.0 atm. What is the temperature of the air in the freezer?

Important to Know All temperatures must be converted to the Kelvin scale when completing gas law problems.

So far you've been introduced to three of the gas laws. These gas laws apply when any two of the variables, P, V, or T, are changed and everything else stays the same.

Charles's law	$V = kT$	P and amount of gas do not change.	$k = \dfrac{V}{T}$
Gay-Lussac's law	$P = kT$	V and amount of gas do not change.	$k = \dfrac{P}{T}$
Boyle's law	$P = k\left(\dfrac{1}{V}\right)$	T and amount of gas do not change.	$k = PV$

Note that the proportionality constant, k, is a generic symbol and is different for each gas sample and for each gas law.

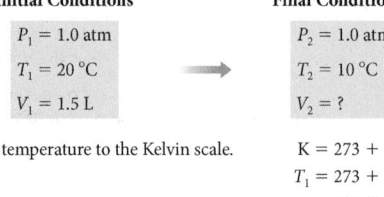

Example 2

Balloon with Air

Imagine that you fill a balloon with air to a volume of 1.5 L. The air is at a temperature of 20 °C. You place the balloon in a refrigerator for half an hour until the air in the balloon is at a temperature of 10 °C. What is the new volume of the balloon?

Solution

First figure out which gas law to use. The external pressure on the balloon remains the same because the air pressure in the refrigerator is the same as the air pressure outside. Use Charles's law. You can predict that V will decrease because T decreases.

Initial Conditions	Final Conditions
$P_1 = 1.0$ atm	$P_2 = 1.0$ atm
$T_1 = 20$ °C	$T_2 = 10$ °C
$V_1 = 1.5$ L	$V_2 = ?$

Convert temperature to the Kelvin scale.

$$K = 273 + C$$
$$T_1 = 273 + 20$$
$$= 293 \text{ K}$$
$$T_2 = 273 + 10$$
$$= 283 \text{ K}$$

Determine the value of k. $\quad k = \dfrac{V_1}{T_1} = \dfrac{1.5 \text{ L}}{293 \text{ K}} = 0.0050$ L/K

Use k to determine V_2.
$$V_2 = k\,T_2$$
$$= 0.0050 \text{ L/K} \cdot 283 \text{ K}$$
$$= 1.4 \text{ L}$$

The balloon has shrunk to a volume of 1.4 L.

312   Chapter 11 | **Pressing Matter**

SOLUTION

Note that the pressure of the air in the syringe is the same as the pressure of the air outside, because the container is flexible.

P_1	1.0 atm		P_2	1.0 atm
T_1	20 °C		T_2	—?—
V_1	52 mL		V_2	45 mL

You use the relationship between V and T (Charles's law). Always convert the temperature to kelvins:

$$T_1 = 20 \text{ °C} + 273 = 293 \text{ K}$$

Then calculate k. Use the volume, V_1, and the temperature, T_1, of the air trapped in the syringe:

$$k = \frac{V_1}{T_1}$$
$$= \frac{52 \text{ mL}}{293 \text{ K}}$$
$$= 0.18 \text{ mL/K}$$

Next, use k to determine T_2, the temperature inside the freezer:

$$T_2 = \frac{V_2}{k} = \frac{45 \text{ mL}}{0.18 \text{ mL/K}}$$
$$= 250 \text{ K} = -23 \text{ °C}$$

LESSON SUMMARY

How does gas pressure change in flexible and rigid containers?

Gases can be trapped in an assortment of containers and subjected to various conditions of pressure, temperature, and volume. The gas laws allow you to calculate new values for gas temperature, gas pressure, and gas volume when two of these three variables change while the other remains the same. Charles's law and Gay-Lussac's law are proportional relationships: $V = kT$ and $P = kT$. Boyle's law is different because P and V are inversely proportional: $P = k(1/V)$. The proportionality constant, k, is the key to solving gas law problems.

KEY TERM

Gay-Lussac's law

Exercises

Reading Questions

1. Describe three ways in which you can change a gas sample.
2. Explain the difference between flexible and rigid containers.

Reason and Apply

3. A scuba-diving tank holds 18 L of air at a pressure of 40 atm. If the temperature does not change, what volume would this same air occupy if it were allowed to expand until it reached a pressure of 1.0 atm?

4. Imagine you fill a balloon with air to a volume of 240 mL. Initially, the air temperature is 25 °C and the air pressure is 1.0 atm. You carry the balloon with you up a mountain where the air pressure is 0.75 atm and the temperature is 25 °C.
   a. When the balloon is carried up the mountain, what changes? What stays the same?
   b. The air pressure on the outside of the balloon has decreased. Can the air pressure on the inside decrease so that the pressures are equal? Why or why not?
   c. What happens to the volume occupied by the air inside the balloon? Explain your thinking.
   d. Solve for the new volume of the balloon.

5. The air inside a 180 mL glass bottle is at 1.0 atm and 25 °C when you close it. You carry the glass bottle with you up a mountain where the air pressure is 0.75 atm and the temperature is 5 °C.
   a. The air pressure on the outside of the glass bottle has decreased. What happens to the volume of air inside of the bottle? Explain your thinking.
   b. Do you expect the temperature of the air inside the bottle to cool to 5 °C when you're at the top of the mountain? Explain your thinking.
   c. What happens to the pressure inside the glass bottle? Explain your thinking.
   d. Solve for the new pressure of the gas.

Lesson 59 | **Gay-Lussac's Law**   313

The temperature of the air in the syringe decreased. This is why the volume decreased.

EXAMPLE 2

A gas is sealed in a 35 mL metal can at a pressure of 1.0 atm and a temperature of 25 °C. The can is rated for a maximum pressure of 2.0 atm. You heat the can until the pressure relief valve bursts. What is the temperature of the gas when the pressure relief valve bursts?

P_1	1.0 atm
T_1	25 °C
V_1	35 mL

P_2	2.0 atm
T_2	—?—
V_2	35 mL

SOLUTION

The volume does not change in this example. You need to use the relationship between P and T (Gay-Lussac's law). Convert the temperature to kelvins:

$$T_1 = 25 \text{ °C} + 273$$
$$= 298 \text{ K}$$

Calculate k. Use the initial pressure, P_1, and the initial temperature, T_1, of the air trapped in the can:

$$k = \frac{P_1}{T_1} = \frac{1.0 \text{ atm}}{298 \text{ K}} = 0.0034 \text{ atm/K}$$

Next, use k to determine T_2, the temperature at which the can bursts:

$$T_2 = \frac{P_2}{k}$$
$$= \frac{2.0 \text{ atm}}{0.0034 \text{ atm/K}}$$
$$= 588 \text{ K}$$
$$588 \text{ K} - 273 = 315 \text{ °C}$$

So, the valve would burst at a temperature between 310 °C and 320 °C. The external pressure is not the same as the internal pressure until the can bursts.

EVALUATE (10 min)

Check-In

In the factory, a potato chip bag is filled with 50.0 mL of air. The pressure of the air is 1.0 atm, and the temperature is 25 °C. Imagine that you take the potato chips with you on an airplane. At higher altitudes, the air pressure in the cabin is 0.85 atm, and the temperature is 25 °C. The bag puffs up.

1. Which gas law applies?
2. Explain why the bag puffs up in the airplane.
3. What is the volume of the gas in the bag when it is at a higher altitude? Show your work.

Answers: **1.** The temperature remains unchanged, but both the pressure and the volume change. You have to use the relationship between P and V (Boyle's law). **2.** The bag puffs up because the air pressure outside the bag is lower than the air pressure inside the bag. **3.** First, calculate k using P_1 and V_1. $k = P_1V_1 = 1.0 \text{ atm} \cdot 50 \text{ mL} = 50 \text{ mL} \cdot \text{atm}$. Second, use k to determine V_2. $V_2 = k/P_2 = 50 \text{ mL} \cdot \text{atm}/0.85 \text{ atm} = 59 \text{ mL}$.

Homework

Assign the reading and exercises for Weather Lesson 59 in the student text.

LESSON 59 ANSWERS

1. *Possible answer:* You can change the pressure, volume, or temperature of a gas sample.

Answers continue in Answer Appendix (p. ANS-5).

Lesson 59 | **Gay-Lussac's Law**   313

Overview

COMPUTER ACTIVITY: WHOLE CLASS

Key Question: How do molecules cause gas pressure? Show students balloons filled with sand, water, and air.

KEY IDEAS

Gas molecules are constantly in motion. They exert pressure on any object they collide with. Changing conditions, such as temperature and volume, can change the amount of pressure exerted by a gas. When gas molecules are squeezed into a smaller volume, they hit the walls of the container more often and thus exert more pressure. This relationship is expressed mathematically by the equation $k = PV$ (Boyle's law). When gas molecules are heated, they move faster and collide with the walls of their container more vigorously and more often. If the container is rigid, the volume cannot change, and the molecules exert more pressure. This relationship is expressed mathematically by the equation $k = P/T$ (Gay-Lussac's law). If the container is flexible, the volume changes so that the pressure remains the same. This relationship is expressed mathematically by the equation $k = V/T$ (Charles's law).

LEARNING OBJECTIVES

- Describe the motions of gas particles under changing conditions.
- Explain changes in pressure, volume, and temperature based on the motions of molecules.

KEY TERM

kinetic theory of gases

IN CLASS

In this lesson, students examine a computer simulation that focuses on the effects of changing conditions on the pressure of a gas. Students discuss how to explain Charles's law, Boyle's law, and Gay-Lussac's law in terms of molecular speeds and collisions with the walls of the container. In an optional simulation, they can "be the molecule." This lesson reviews and reinforces the gas laws learned so far. A complete materials list is available for download.

TRM Materials List 60

TRM Chapter 11 Web Resources

SETUP

Search online for a gas properties simulation, such as the University of

---

LESSON 60

Be the Molecule

Molecular View of Pressure

THINK ABOUT IT

Imagine you have three balloons: one filled with sand, a second filled with water, and a third filled with air. The air balloon has the least mass but it has the largest volume. It is also the most uniform and spherical, indicating that the gas molecules are pushing out in all directions.

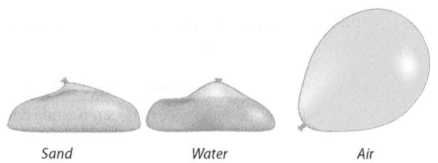

| Sand | Water | Air |

How do molecules cause gas pressure?

To answer this question, you will explore

- Kinetic View of Pressure
- Changing Gas Pressure

EXPLORING THE TOPIC

🗢 Kinetic View of Pressure

Inside the sand balloon, the atoms that make up sand are tightly packed together. Inside the water balloon, the forces between the water molecules keep them close to each other but not quite as close as the molecules in solids. Inside the air balloon, however, there is a lot of movement and activity. The gas molecules are moving rapidly and bouncing off each other and the walls of the balloon.

> **Big Idea** Gas pressure is caused by the collisions of molecules or atoms with all surfaces such as the walls of a container or your skin.

COLLISIONS CAUSE PRESSURE

According to the **kinetic theory of gases**, gas molecules are in constant motion. The gas molecules in a sample move at many hundreds of miles per hour in straight line paths. Even though they are very tiny, the molecules inside the container hit each other and the walls of their container. There are so many molecules hitting the walls that the collisions add up to a measurable pressure.

Note that gas pressure is caused by molecules hitting the walls of the container only. When gas molecules collide with each other, this may change their speed or direction, but it does not affect the overall gas pressure.

314 Chapter 11 | **Pressing Matter**

Colorado Boulder's PhET simulation, Gas Properties, which allows you to explore the gas laws by varying pressure, temperature, volume, and the number of particles. You can find a list of URLs for this lesson on the Chapter 11 Web Resources document. Set up the projector to display the simulation and run through the simulation before class. Prepare three balloons as a visual for the ChemCatalyst exercise. Fill one balloon with sand, one with water, and one with air. Mark an area on the floor for the "Be the Molecule" simulation at least 10 ft by 15 ft.

Differentiate

Students may need support to develop scientific literacy about the kinetic theory of gases because it is abstract. Try the resources that go with PhET's Gas Properties simulation to give students more opportunities to experiment with how changing variables such as temperature or volume affect gas pressure. Other students may benefit by making connections to how KMT can explain everyday pressure-change phenomena. Either let students come up with their own or provide students with an event

BIOLOGY CONNECTION

The diaphragm is a muscle that separates the chest from the abdomen. When the diaphragm moves up, air flows out of your nose and mouth. When the diaphragm moves down, the volume of the chest cavity increases, allowing air to flow in.

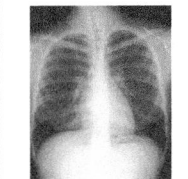

The illustration shows a gas trapped in a container with a piston that can move up and down. The temperature and pressure of the gas inside the container are indicated on a thermometer and pressure gauge. Notice that the gas inside the container exerts enough pressure on the piston to hold it up. The piston does not fall downward, even though the gas molecules are very far apart.

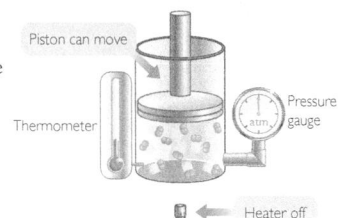

Changing Gas Pressure

Look again at the illustration. Suppose you keep the temperature the same but decrease the volume of the cylinder. What happens to the gas molecules inside the container? There is less space for the same number of molecules. The pressure increases. If you keep the volume the same but heat the gas inside the container, the gas molecules move faster. The pressure increases again. The drawings here illustrate these two changes.

CHANGING VOLUME

Action: The piston is pushed downward to decrease the volume. The temperature is held constant by allowing heat to exchange with the surroundings.

Observation: The gas pressure increases.

Explanation: When the gas molecules are forced into a smaller space, they hit the walls of the container more often. The gas pressure is higher. This change is associated with Boyle's law.

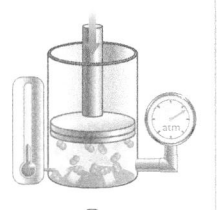

CHANGING TEMPERATURE

Action: The gas is heated. The volume is held constant by clamping the piston so that it cannot move.

Observation: The gas pressure increases.

Explanation: At higher temperatures, the gas molecules move faster. They hit the walls of the container harder and more often, which increases the gas pressure. This change is associated with Gay-Lussac's law.

Lesson 60 | **Molecular View of Pressure**   315

---

to explain, and have students work in pairs to develop a storyboard (snapshots in time) to illustrate what is happening. For students who might need a challenge, this lesson may be a good time to introduce the concept of buoyancy.

Lesson Guide

TRM PowerPoint Presentation 60

ChemCatalyst

Consider three balloons: one filled with sand, a second filled with water, and a third filled with air.

1. Describe at least three differences between the balloon containing air and the balloons containing sand and water.

2. What are the individual gas molecules doing inside the balloon containing air to make it big and round?

Sample answers: **1.** The balloons containing sand and water are much heavier. They are not round, and they

---

do not maintain their shape. **2.** The individual gas molecules must be hitting the inside of the balloon and pushing it out equally in all directions.

→ Discuss differences among the three balloons.

Ask: Why don't the sand and water balloons maintain their shape?

Ask: What makes it so hard to deform the air balloon?

Ask: Why do gas molecules exert pressure in all directions?

Ask: What is keeping the molecules in the liquid water from exerting pressure in all directions? (The motion of the molecules is restricted because the molecules are attracted to one another.)

→ Tell students that they will view computer simulations of gas molecules and discuss their observations as a class.

TRM Worksheet with Answers 60

TRM Worksheet 60

INTRODUCE THE COMPUTER SIMULATION

→ Tell students they will work as a class.

→ Pass out the student worksheet.

→ Tell students that they will view computer simulations of gas molecules and discuss their observations as a class.

→ Download the directions to run the computer simulations for constant volume and constant temperature.

TRM Computer Simulations 60

→ Download directions for the student simulation of gas molecules.

TRM Be the Molecule Simulation 60

TRM Transparency—Be the Molecule 60

SUMMARIZE THE GAS LAWS STUDIED SO FAR

→ You might display the computer simulations as you discuss Boyle's law, Charles's law, and Gay-Lussac's law.

→ Write the three gas laws on the board.

Ask: Can you describe a situation in which you would use Charles's law?

Lesson 60 | **Molecular View of Pressure**   315

Ask: Do any of the gas laws apply to a gas in an open container? Explain your thinking.

Ask: What variable does not change for each gas law? (For Boyle's law, *T* does not change; for Charles's law, *P* does not change; for Gay-Lussac's law, *V* does not change.)

Ask: What variable stays the same for all three gas laws? (amount of gas, or number of molecules)

Key Points: The gas laws allow you to calculate new values for gas temperature, pressure, and volume when two of these conditions change. With the gas laws discussed so far, it is understood that the gas is trapped or sealed off so that the number of molecules remains unchanged.

> Charles's law:
> $V = kT, k = V/T$

Pressure and amount of gas do not change.

> Gay-Lussac's law:
> $P = kT, k = P/T$

Volume and amount of gas do not change.

> Boyle's law:
> $P = k \cdot (1/V), k = PV$

Temperature and amount of gas do not change.

Keep in mind that Charles's law and Gay-Lussac's law are proportional relationships: $V = kT$ and $P = kT$. Once you determine the proportionality constant, *k*, for one set of conditions for a specific gas sample, you can determine *V* and *T,* or *P* and *T,* for all other conditions for that gas provided the number of molecules does not change. Boyle's law is different, because *P* and *V* are inversely proportional: $P = k \cdot (1/V)$. But the mathematical manipulations are similar. Note that all temperatures must be converted to the Kelvin scale when you are solving gas law problems.

In the kinetic theory of gases, the gas molecules are in constant motion. They collide with the walls of the container. These collisions are the direct cause of the pressure on the walls of the container. When the temperature increases, the average speed of the molecules increases. When the molecules move faster, they collide with the walls more often and hit the walls harder. The result is higher gas pressure at higher temperature. Similarly, when the volume of a gas is reduced, the

same number of molecules must occupy a smaller space. This means that they also collide with the walls of the container more often, resulting in higher pressure.

EVALUATE (5 min)

Check-In

A family went for a drive in the desert. In the morning, the air pressure in the tires of their car was around 28 lb/in². In the afternoon, the tire pressure was around 32 lb/in². Provide an explanation on the molecular level for why this happened.

LESSON SUMMARY

How do molecules cause gas pressure?

The collisions of gas molecules with the walls of a container cause gas pressure. If the volume decreases while the temperature stays the same, the molecules collide with the walls more frequently. This causes an increase in gas pressure. If the volume stays the same and the temperature increases, the molecules move faster. They collide with the walls more frequently and they hit the walls harder. This causes an increase in gas pressure.

KEY TERM
kinetic theory of gases

Exercises

Reading Questions

1. Use the kinetic theory of gases to explain why increasing the gas volume decreases the gas pressure.

2. Use the kinetic theory of gases to explain why decreasing the gas temperature decreases the gas pressure.

Reason and Apply

3. Suppose that the pressure of a gas in a cylinder has increased. What might have changed to cause this? Explain your thinking.

4. Why is it dangerous to heat a gas in a sealed container?

5. Can you decrease the volume of a gas to zero? Why or why not?

6. When you fly in a commercial airplane, you often feel the change in air pressure in your ears. It feels painful. Use the kinetic theory of gases to explain what you think is happening.

7. At sea level, the pressure of trapped air inside your body is 1 atm. It is equal to the pressure of air outside your body at sea level. Imagine that you do a deep sea dive. You descend slowly to a depth where the pressure outside your body is 3.5 atm.
   a. How does the volume of your lungs compare at 3.5 atm with the volume at sea level?
   b. Why is it dangerous to hold your breath and ascend quickly?

8. As a helium balloon floats up into the sky, the pressure of the atmosphere around it decreases as the balloon's altitude increases. If the gas temperature does not change, what do you expect will happen to the volume of the balloon? Explain your thinking.

9. A gas is trapped in a cylinder with a piston as shown. The gas pressure is 1.0 atm, the volume is 500 mL, and the temperature is 300 K.
   a. Draw the cylinder after the volume has been changed to 1000 mL. The gas temperature is 300 K.
   b. Does the pressure inside the cylinder increase or decrease? Explain your thinking.
   c. Which gas law applies?

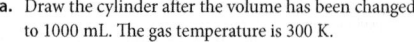

10. PROJECT
    Look up information on steam engines in a book or on the Web. Explain how a steam engine works.

Answer: It is likely that the pressure increased because the daytime temperature increased and the tires got hot moving along the hot pavement. The gas molecules move faster as they heat up, increasing the pressure they exert. (*Note:* The volume of the tires may have increased a little, which would tend to decrease the pressure caused by the increase in temperature.)

Homework

Assign the reading and exercises for Weather Lesson 60 in the student text.

LESSON 60 ANSWERS

Answers continue in Answer Appendix (p. ANS-5).

What Goes Up

Combined Gas Law

THINK ABOUT IT

In some circumstances, more than two gas variables change at once. For example, if you released a balloon at sea level and it rose into the atmosphere, you would be dealing with three changing gas variables: pressure, temperature, and volume. What effect does this have on the balloon?

> **What are the relationships among pressure, volume, and temperature for a sample of gas?**

To answer this question, you will explore

↻ Changing Pressure, Volume, and Temperature

↻ Combined Gas Law

EXPLORING THE TOPIC

↻ Changing Pressure, Volume, and Temperature

VARIATIONS IN PRESSURE AND TEMPERATURE WITH ALTITUDE

Both the temperature and the pressure of the atmosphere decrease with altitude. These two conditions (temperature and pressure) will naturally vary as a balloon rises and will affect the overall volume of the balloon. The sample of gas inside the balloon responds to the conditions outside the balloon.

The illustration indicates the approximate average pressure and temperature of the atmosphere at various altitudes. At an altitude of 40,000 ft, the air pressure is only 0.2 atm and the air temperature is around a chilly $-57\,°C$ ($-70\,°F$).

WEATHER BALLOONS

Weather balloons are large balloons filled with helium or hydrogen gas that carry instruments aloft to study the atmosphere at high altitudes. As the weather balloon rises, the instruments measure atmospheric conditions like the temperature, relative humidity, and pressure. This information is relayed back to a

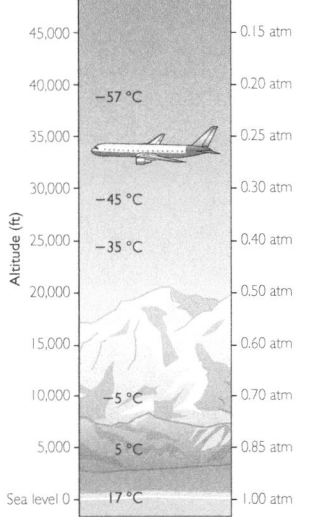

Altitude (ft)		
45,000		0.15 atm
40,000	$-57\,°C$	0.20 atm
35,000		0.25 atm
30,000		0.30 atm
25,000	$-45\,°C$	0.40 atm
	$-35\,°C$	
20,000		0.50 atm
15,000		0.60 atm
10,000	$-5\,°C$	0.70 atm
5,000	$5\,°C$	0.85 atm
Sea level 0	$17\,°C$	1.00 atm

CONSUMER CONNECTION

Have you ever wondered why boxes of cake mix sometimes include high-altitude directions? At higher altitudes, water boils at a lower temperature, and the air bubbles from the leavening expand more, causing cakes to expand too much while they're cooking. To adjust for high altitudes, bakers usually add more water (so the cake doesn't dry out too much), decrease the amount of leavening, and increase the baking temperature (so the cake won't have as much time to rise before it sets).

Overview

CLASSWORK: GROUPS OF 4

Key Question: What is the relationship among pressure, volume, and temperature for a sample of gas?

KEY IDEAS

A weather balloon rising into the atmosphere is affected by two variables, pressure and temperature, changing simultaneously. The temperature of the gas sample decreases as the balloon rises in altitude. The pressure of the gas also decreases as the balloon rises. The mathematical equation that describes the interactions of pressure, volume, and

temperature together while the number of molecules remains the same is called the *combined gas law*. It is described mathematically by the equation $k = PV/T$. This equation can be used to predict the volume of the weather balloon under various conditions.

LEARNING OBJECTIVES

● Define the combined gas law.
● Solve gas law problems that involve changes in all three of the variables, *P*, *V*, and *T*.

FOCUS ON UNDERSTANDING

● The math that accompanies this gas law is a bit trickier than the math

associated with the three gas laws discussed previously. We have taken an approach that focuses on the proportionality constant, *k*.
● Students may have difficulty grasping the concept that the temperature and pressure inside the balloon will match the temperature and pressure outside the balloon.

KEY TERM

combined gas law

IN CLASS

Students are introduced to weather balloons and the combined gas law. They complete a worksheet focusing on the ascent of a weather balloon through the changing atmosphere. They calculate how the volume of the gas sample changes with increasing altitude, using the equation $k = PV/T$. They wrestle with the question of which has a greater effect on the volume of the balloon, the decreasing air temperature or the decreasing air pressure. A complete materials list is available for download.

TRM Materials List 61

Differentiate

Student groups may work at varying speeds during this lesson. Provide each small group with 3 cups. Green means they are doing well. Yellow means they need some assistance. Red means they are stuck and need a lot of help. This signaling can help you identify who needs help with a quick visual scan of the classroom. Another way to use the cups is to ask groups finishing early to disperse and offer help to groups with red or yellow cups.

Pre-AP® Course Tip

Students preparing to take an AP® chemistry or college chemistry course should be able to use mathematics to solve problems that describe the physical world. In this lesson, students use mathematics to solve gas law problems that involve changes in pressure, volume, and temperature.

Lesson Guide

ENGAGE (5 min)

TRM PowerPoint Presentation 61

ChemCatalyst

A weather balloon is inflated with helium to a volume of 125,000 L. When it is released, it rises high into the atmosphere, where both the pressure and the temperature are lower.

1. Explain why the balloon rises.

2. Will the balloon pop at a high altitude? Explain your thinking.

Sample answers: **1.** The gas in the helium balloon is less dense than the surrounding air, because the mass of the balloon (He) is less per unit volume than the mass of the air molecules (mostly O_2 and N_2). **2.** When a sample of gas decreases in temperature, it contracts (Charles's law). When a sample of gas decreases in pressure, it expands (Boyle's law). Because these are opposing effects, it is difficult to predict what will happen to the balloon from what we know at this point. It depends on whether temperature changes or pressure changes are more important.

→ Assist students in sharing their ideas about how temperature and pressure changes affect a weather balloon.

Ask: How do pressure and temperature affect the volume of the balloon?

Ask: How would you explain the observation that a weather balloon has burst at some altitude and fallen back to Earth?

Ask: How do you think a weather balloon can be used to probe the atmosphere?

EXPLORE (20 min)

TRM Worksheet with Answers 61

TRM Worksheet 61

INTRODUCE THE CLASSWORK

→ Arrange students into groups of four.

→ Pass out the student worksheet

→ Tell students that they will be working with pressure, temperature, and volume data for a weather balloon. Briefly introduce weather balloons and atmospheric changes with altitude to the class.

→ Tell students that weather balloons are large balloons filled with helium or hydrogen gas that carry instruments for studying the atmosphere at high altitudes. As the balloon rises, the instruments take measurements. This

METEOROLOGY CONNECTION

A weather balloon has three parts: the elastic balloon, the instrument package, called a *radiosonde,* and the parachute. Weather balloons are released twice a day from sites all around the world. There are about 900 release sites worldwide; 95 of these sites are in the United States. Data from weather balloons are fed into National Weather Service supercomputers across the country. This information is used to help predict weather around the country.

weather station by a transmitter. Wind speed and wind direction can be calculated from the collected data.

To follow the changes to the gas inside a weather balloon, you must first know the starting conditions of the balloon. Imagine a clear cool morning, where the air temperature is 17 °C and the atmospheric pressure is 1 atm. A team of meteorologists inflates a weather balloon with helium to a volume of 8000 L. If released, this weather balloon will rise no matter what the outside conditions. This is because the balloon is full of helium and helium is less dense than air.

CONFLICTING OUTCOMES

Consider the effect of changing conditions on the volume of the rising weather balloon. Charles's law indicates that the volume of the balloon will shrink with decreasing temperature. However, Boyle's law indicates that as the air pressure outside the balloon decreases, the balloon will expand in volume.

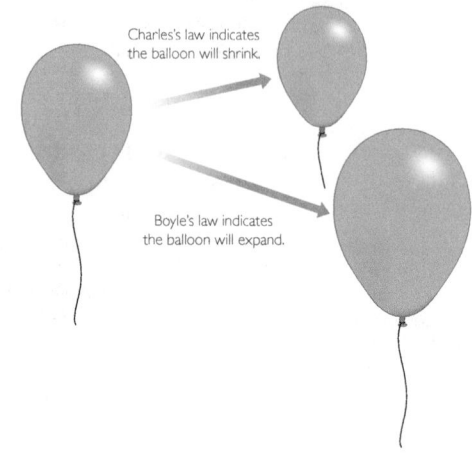

Charles's law indicates the balloon will shrink.

Boyle's law indicates the balloon will expand.

There are two conflicting outcomes. Will the balloon expand or shrink?

↻ Combined Gas Law

For a sample of gas, the relationship among gas pressure, temperature, and volume can be expressed in a mathematical equation called the **combined gas law.**

Combined Gas Law

If you know the temperature, T, pressure, P, and volume, V, of a gas, you can determine k. Then you can use k to determine the volume of the gas for other pressures and temperatures.

$$k = \frac{PV}{T}$$

information is relayed back to a station via a transmitter. Unlike a hot-air balloon, a weather balloon is a closed container.

→ Remind students that the atmosphere changes in pressure and temperature with altitude. In general, both these variables decrease with altitude. A gas sample in the air will respond to these changes.

→ Introduce the combined gas law, $k = \frac{PV}{T}$, and complete a sample problem.

EXAMPLE

A balloon has a volume of 12,000 L at sea level where the pressure is 1.0 atm

and the temperature is 285 K. The balloon is released and travels to an altitude of 5,000 feet where the temperature is 278 K and the pressure is 0.80 atm. What is the new volume of the balloon?

SOLUTION

$$k = \frac{P_1 V_1}{T_1}$$

$$= \frac{1.0 \text{ atm} \cdot 12,000 \text{ L}}{285 \text{ K}}$$

$$= 42.1 \frac{\text{atm} \cdot \text{L}}{\text{K}}$$

Example

Weather Balloon at 25,000 Ft

A weather balloon is filled at sea level with 8000 L of helium. The pressure at sea level is 1 atm and the temperature is 17 °C. Calculate the volume of the balloon at 25,000 ft when the atmospheric pressure has decreased to 0.4 atm.

Solution

Because pressure, volume, and temperature are all changing, you need to use the combined gas law. The first step in using the combined gas law is to make note of all the values for P, V, and T. Notice that we already know five out of the six values. Remember to convert °C to K. Then it is a matter of solving for V_2.

P_1 = 1.0 atm	P_2 = 0.4 atm
V_1 = 8000 L	V_2 = ?
T_1 = 290 K	T_2 = 238 K

Use the combined gas law.
$$k = \frac{P_1 V_1}{T_1} = \frac{P_2 V_2}{T_2}$$

Determine the proportionality constant, k
$$k = \frac{P_1 V_1}{T_1}$$
$$= \frac{1.0 \text{ atm} \cdot 8000 \text{ L}}{290 \text{ K}}$$
$$= 27.6 \frac{\text{atm} \cdot \text{L}}{\text{K}}$$

Use k to solve for V_2.
$$V_2 = \frac{kT_2}{P_2}$$
$$V_2 = \frac{\left(27.6 \frac{\text{atm} \cdot \text{L}}{\text{K}}\right)(238 \text{ K})}{0.4 \text{ atm}}$$
$$V_2 = 16{,}422 \text{ L}$$

The balloon has expanded to about 16,000 L at this altitude.

[For a review of this math topic, see MATH Spotlight: Accuracy, Precision, and Significant Digits on page A-1.]

PRESSURE VERSUS TEMPERATURE

The example shows that decreasing air pressure has a greater effect on a weather balloon than decreasing temperature does. The volume of the balloon continues to expand as the balloon goes up in altitude, in spite of the decreasing temperature. The balloon eventually ruptures. There is a parachute on every weather balloon so that the weather instruments can be carried safely to the ground.

Lesson 61 | **Combined Gas Law**   319

$$k = \frac{P_2 V_2}{T_2}$$
$$42.1 \frac{\text{atm} \cdot \text{L}}{\text{K}} = \frac{(0.80 \text{ atm})(V_2)}{278 \text{ K}}$$
$$V_2 \approx 15{,}000 \text{ L}$$

EXPLAIN & ELABORATE (15 min)

TRM Transparency—Altitude Table 61

INTRODUCE THE COMBINED GAS LAW

→ Write the combined gas law on the board.

Combined gas law:
$$k = \frac{PV}{T} \quad \text{or} \quad \frac{P_1 V_1}{T_1} = \frac{P_2 V_2}{T_2}$$

Ask: If the temperature and volume of a gas change, can you determine the pressure?

Ask: If the volume of a gas increases and the pressure decreases, can you determine the temperature?

Ask: If you change the temperature of the gas in a balloon, will the value of PV/T remain the same? Explain.

Ask: Suppose you have two balloons filled with different amounts of gas. Will the value of PV/T be the same? Explain.

Key Points: The relationship among the pressure, temperature, and volume of a gas is described by the combined gas law. It states that the value of PV/T stays the same for a set of conditions for a given sample of gas. This law allows you to calculate one variable when the other two change.

EXPLORE HOW CHANGING T AND P AFFECTS THE WEATHER BALLOON

→ Use the Transparency—Altitude Table to assist in the discussion.

Ask: Does a decrease in air temperature have an effect on the gas pressure in the balloon? Explain.

Ask: The weather balloon continues to rise no matter what conditions we impose on it. Why? (It is full of helium.)

Ask: Why do you think the balloon eventually pops?

Ask: Which change has a greater effect on the volume of the balloon when you increase in altitude, the air temperature or the air pressure?

Key Points: The weather balloon rises no matter what the outside conditions because the balloon is full of helium, and helium is less dense than air. With a decrease in temperature, we would expect the balloon to decrease in volume (Charles's law). However, with a decrease in pressure, we would expect the balloon to increase in volume (Boyle's law). It appears that the changing air pressure has a greater effect on the volume of the balloon than does the changing air temperature. The balloon increases in volume until the pressure inside is too great for the material of which the balloon is made, and the balloon pops.

EVALUATE (5 min)

Check-In

A sample of neon gas occupies a volume of 1.0 L at 300 K and 1.0 atm.

1. Calculate the value of the proportionality constant, k.

2. Suppose you increase the temperature to 600 K and decrease the pressure to 0.50 atm. Does the volume of the gas increase or decrease? Explain your answer.

Answers:

1.

$$k = \frac{P_1 V_1}{T_1} = \frac{1.0 \text{ L} \cdot 1.0 \text{ atm}}{300 \text{ K}}$$

$$= 0.0033 \frac{\text{atm} \cdot \text{L}}{\text{K}}$$

2. The volume definitely will be larger under the new conditions.

$$k = \frac{P_2 V_2}{T_2}$$

$$0.0033 \frac{\text{atm} \cdot \text{L}}{\text{K}} = \frac{0.50 \text{ atm} \cdot V}{600 \text{ K}}$$

$$= 4.0 \text{ L}$$

Homework

Assign the reading and exercises for Weather Lesson 61 in the student text.

LESSON 61 ANSWERS

1. The combined gas law is a mathematical equation that relates gas pressure, temperature, and volume. This law applies when all three variables can change at the same time.

2. *Possible answer:* According to the combined gas law, temperature and volume are directly proportional. The temperature of the atmosphere does affect the volume of a weather balloon as it rises. Because the weather balloon is a flexible container, an increase in temperature will cause the volume to increase, and a decrease in temperature will cause the volume to decrease.

3. a. Because the pressure increases, the volume of the gas will decrease because pressure and volume are inversely proportional when the temperature and the amount of gas remain unchanged.
b. $PV = k$ **c.** 0.49 L.

4. a. Because the temperature increases, the pressure of the air in the tire will increase because pressure and temperature are directly proportional when the volume and the amount of gas are constant. **b.** $P = kT$ **c.** 1.1 atm

5. a. The volume of the balloon will increase because, based on the values given, the change in pressure has a greater effect on the final volume than the change in temperature.
b. $\frac{PV}{T} = k$ **c.** 1.6 L
6. 164 atm

What are the relationships among pressure, volume, and temperature for a sample of gas?

If there is a situation involving a gas in which volume, temperature, and pressure are all changing, then you can use the combined gas law to determine the effects of changing two variables on the third variable. The value of the proportionality constant, k, is equal to PV/T. Therefore, $P_1 V_1/T_1 = P_2 V_2/T_2$. Remember that the combined gas law works only in situations where the amount of gas involved remains the same.

KEY TERM

combined gas law

Exercises

Reading Questions

1. What is the combined gas law? When do you apply it?
2. Does the temperature of the atmosphere have any effect on the volume of a weather balloon as it rises? Explain.

Reason and Apply

3. A gas collected in a flexible container at a pressure of 0.97 atm has a volume of 0.500 L. The pressure is changed to 1.0 atm. The amount of gas and the temperature of the gas do not change.
   a. Will the volume of the gas increase or decrease? Explain.
   b. Which equation should you use to calculate the new volume of the gas?
   c. Calculate the volume of the gas at a pressure of 1.0 atm.
4. The gas pressure in an automobile tire was 1.0 atm at 21 °C. After an hour of driving, the tire heated up to 55 °C. Assume that the tire remained at constant volume and no gas escaped from the tire.
   a. Did the gas pressure inside the tire increase or decrease? Explain.
   b. Which equation should you use to calculate the new gas pressure of the tire?
   c. Calculate the gas pressure of the tire at 55 °C.
5. The helium inside a balloon has a volume of 1.5 L, a pressure of 1.0 atm, and a temperature of 25 °C. The balloon floats up into the sky where the air pressure is 0.95 atm and the temperature is 20 °C.
   a. Will the volume of the balloon increase or decrease? Explain.
   b. Which equation should you use to calculate the new volume of the balloon?
   c. Calculate the volume of the balloon at 20 °C and 0.95 atm.
6. Someone leaves a steel tank of nitrogen gas in the sun. The gas pressure inside the tank was 150 atm at 27 °C to begin with. After several hours, the internal temperature rises to 55 °C. What is the gas pressure in the tank now?
7. A car airbag is an example of a trapped gas. The airbag is designed to inflate to 65 L if you have an accident in a location where the temperature is 25 °C and the air pressure is 1.0 atm.
   a. Determine the size of the airbag if you have an accident in the mountains where the temperature is −5 °C and the air pressure is 0.8 atm.
   b. Suppose that you have an accident in a location in which the airbag inflates to 60 L. What has to be true about the temperature and air pressure in this location?

320   Chapter 11 | **Pressing Matter**

---

7. a. 73 L **b.** The pressure divided by the temperature must be 0.0036 atm/K. Because the volume of the airbag is slightly smaller, this could mean that the temperature is slightly colder or the air pressure is slightly higher.

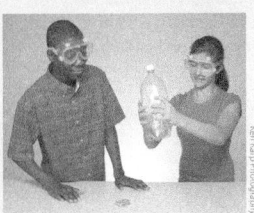

SAFETY Instructions

Safety goggles must be worn at all times.

High and Low Air Pressure

Purpose
To find the connection between air pressure and the weather forecast.

Materials
- 2 L plastic bottle, or vacuum chamber and vacuum pump
- warm tap water
- long safety matches

Procedure
1. Put a small amount (about 15 mL) of warm water into the plastic bottle.
2. Light a match. Blow it out and then hold it inside the bottle to collect some smoke. Quickly remove the match and put the cap tightly on the bottle. Shake the chamber to add moisture to the air inside.
3. Squeeze the bottle and release. Pump a few times while observing the air inside the bottle.
4. Repeat the experiment, but this time use 10 mL of cold water. Next, repeat the experiment with a dry bottle. Do not add water. Just create smoke, close the bottle, and squeeze and release.

Observations
1. What did you observe inside the bottle when you squeezed and released the bottle?
2. What happens to the pressure, volume, and temperature of the air inside the bottle when you squeeze it? When you release it?
3. What gas law best explains how the properties of the gas inside the bottle change during this experiment? Explain your choice.
4. How does this gas law explain what's going on in the bottle?
5. What did you observe when you used a dry bottle?

Lesson 62 | **High and Low Air Pressure**   321

Overview

DEMO: WHOLE CLASS

LAB: GROUPS OF 4

Key Question: How are areas of high and low air pressure related to the weather?

KEY IDEAS

Meteorologists regularly track the pressure of the air around us to better forecast the weather. From air pressure measurements, high-pressure and low-pressure areas are charted on weather maps. Low-pressure areas are associated with clouds and storms, and high-pressure areas are associated with the Sun and clear skies.

LEARNING OBJECTIVE

Explain the influence of high-pressure and low-pressure systems on the weather.

IN CLASS

This lesson applies what has been learned about gas behavior to the weather context. Students watch a demo of volume change involving temperature and phase changes. Then, working in groups, students make a "cloud in a bottle." During the discussion, students make connections between high and low pressure and weather. A complete materials list is available for download.

TRM Materials List 62

SETUP

Practice the demo as well as making a cloud in a bottle. For making a cloud, a vacuum chamber and vacuum pump work best, but an empty plastic soda bottle also works.

Lesson Guide

ENGAGE (5 min)

TRM PowerPoint Presentation 62

> ### ChemCatalyst
> Clouds are tiny water droplets suspended in the air.
> 1. Are pressure, temperature, or volume changes involved in the formation of clouds? Explain your thinking.
> 2. Cloud formation is related to low pressure. Explain why.

Sample answers: 1. Students may speculate that clouds form when moisture in the air gets cold enough to undergo a phase change. 2. If water condenses, there are fewer gas molecules in the air. Perhaps fewer gas molecules means lower pressure.

Ask: How does water vapor get into the atmosphere? Why does water vapor rise?

Ask: What causes water vapor to condense to form clouds?

EXPLORE (20 min)

TRM Worksheet with Answers 62

TRM Worksheet 62

COMPLETE THE DEMONSTRATION

→ Demonstrate how evaporation is associated with high pressure and condensation is associated with low pressure.

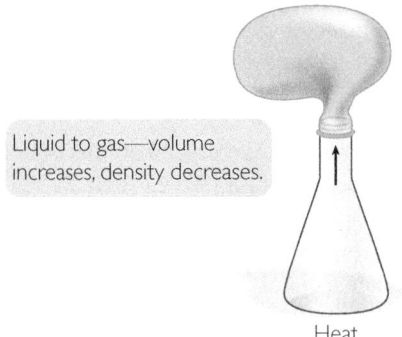

Liquid to gas—volume increases, density decreases.

Heat

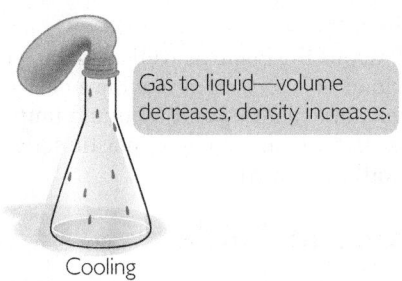

Gas to liquid—volume decreases, density increases.

Cooling

1. Draw students' attention to a 250 mL Erlenmeyer flask with 5 mL of water in it. Place a balloon over the mouth of the flask.
2. Heat the flask on a hot plate. Do not let all the water boil away.
3. After several minutes, use an oven mitt to remove the flask from the hot plate.
4. Solicit students' observations.
5. Ask why the balloon gets so large. The number of gas molecules increases as water evaporates.

➔ Repeat the procedure, but this time, attach the balloon after removing the flask from the hot plate.

1. Place the flask upright in ice water.

Gas to liquid—a dramatic change in density can have a dramatic outcome.

Cooling

2. Solicit students' observations.
3. Ask why the balloon goes inside the flask. (Before the balloon was placed on the flask, the air in the flask was mainly water vapor. When this water vapor condenses, the pressure decreases inside the flask. The air on the outside pushes the balloon into the flask.)

INTRODUCE THE LAB

➔ Arrange students into groups of four.

➔ Pass out the student worksheet.

➔ Tell students they will be using the pressure-temperature relationships to create a cloud inside a bottle. (If the cloud in the bottle is too difficult to see, you can demonstrate the cloud in the bottle with a vacuum pump and vacuum

---

Cloud in a Bottle

High and Low Air Pressure

THINK ABOUT IT

Weather maps typically contain information about high air pressure and low air pressure. You often hear meteorologists refer to high- and low-pressure systems. But how does this information help you figure out whether it will be rainy or sunny?

How are areas of high and low air pressure related to the weather?

To answer this question, you will explore

↻ High and Low Air Pressure
↻ Cloud Formation

EXPLORING THE TOPIC

↻ High and Low Air Pressure

Meteorologists measure air pressure at different locations on Earth's surface. These readings of air pressure are summarized on an air-pressure map like this one.

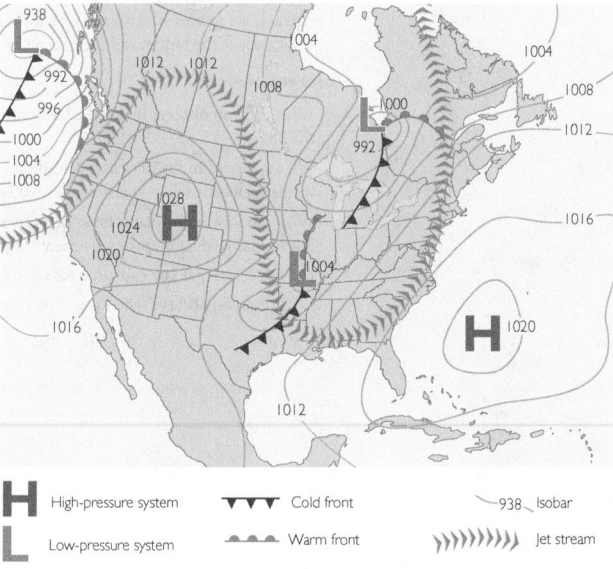

| **H** | High-pressure system | ▼▼▼ Cold front | 938 Isobar |
| **L** | Low-pressure system | ⌒⌒⌒ Warm front |)))))) Jet stream |

---

chamber, if you have the apparatus. Note that this will change the situation, as *V* will remain the same while *n* changes. Changes in *n* are covered in Chapter 12.

EXPLAIN & ELABORATE (15 min)

PROCESS THE LAB

Ask: What variables did you change when you pumped the air out of the bottle?

Ask: What caused the cloud to form?

Ask: Do you think the water vapor in the air in the room had any effect on the cloud formation?

Key Points: In the plastic bottle, pressure and temperature were changing, but the volume stayed (nearly) the same. This situation is covered by Gay-Lussac's law, $P = kT$. When the air pressure inside the bottle is lowered, the temperature inside the bottle also decreases. This decrease in temperature causes water vapor in the bottle to change phase and form droplets. The smoke provides some particles in the air for the water droplets to cling to or form around, which can be seen as a faint cloud inside the bottle. Gay-Lussac's law also tells us that when the pressure inside the bottle is increased, the

The lines labeled with numbers on the map are called *isobars*. They mark areas with the same air pressure. The numbers correspond to air pressure measurements in millibars. One atmosphere is equal to 1013.25 millibars. Notice that each new line represents an air pressure difference of 4 millibars.

Low-pressure areas are associated with lower numbers, high-pressure areas with higher numbers. However, high and low air pressure are measured relative to the surrounding air. In other words, a high-pressure air system is not associated with a certain number but simply shows an area that stands out as having higher air pressure than the areas around it. If you examine the map, you can also see that the isobars encircle areas with high- and low-pressure systems.

The map also shows the jet stream as a series of blue arrows traveling from west to east across the continent. Warm and cold fronts are also included. Take a moment to look for patterns.

- The jet stream weaves between the Hs and Ls instead of going through the middle of them.
- The jet stream appears to follow the general curvature of the isobar lines. For this particular map, what this means is that the jet stream will guide western storms up into Canada.
- The location of the weather fronts is also associated with air pressure. Notice that the weather fronts appear to be centered on the low-pressure areas.
- There are no weather fronts near the high-pressure areas, so you wouldn't expect to see any storms in these areas.

WEATHER ASSOCIATED WITH HIGH AND LOW PRESSURE

In general, low-pressure areas are associated with fronts, clouds, and precipitation. This is because when warm and cold air masses meet, the warm air rises, leaving behind an area of low pressure. Areas of lower air pressure coincide with areas where fronts are located.

The centers of all storms are areas of low air pressure. When forecasters say a low-pressure area is moving toward your region, cloudy weather and precipitation usually are on the way.

In contrast, high-pressure areas are places where skies are clear and pleasant. When forecasters say that a high-pressure system has moved into your area, you can expect clear skies and sunny days. High-pressure centers mean that denser air must be sinking, which inhibits precipitation and cloud formation.

Lesson 62 | **High and Low Air Pressure**   323

This causes the water vapor in the air to condense into droplets, forming clouds. A low-pressure system is created near Earth's surface when warm air rises up into the atmosphere. A high-pressure system is created near Earth's surface when cool air sinks from high in the atmosphere.

Enough water vapor must be present in the air in order for clouds to form. This is why more clouds form over rainforest than over desert. Fog is common near bodies of water. On warm days, when the air has the potential to hold more moisture, large puffy clouds can be observed at lower altitudes. On cold days, when there is less moisture in the air, the clouds tend to be wispy and high in the sky.

EVALUATE (5 min)

Check-In

On a camping trip you take a sealed plastic water bottle to a higher elevation. When you arrive on the mountain you notice the bottle is slightly larger and condensation has formed on the inside of the bottle. Explain what happened in terms of P, V, T, and the quantity PV.

Answer: The volume has increased slightly because the atmospheric pressure is lower. So the pressure inside the bottle has increased. Because water condensed, the temperature must have decreased. The quantity of PV must have decreased because the ratio PV/T is a constant.

Homework

Assign the reading and exercises for Weather Lesson 62 and the Chapter 11 Summary in the student text.

temperature also must increase. This is why the cloud evaporates again when you squeeze the bottle.

DISCUSS CLOUD FORMATION IN EARTH'S ATMOSPHERE

Ask: Is the air in our atmosphere in a container? Explain.

Ask: What are the first steps in the formation of clouds? (evaporation of water, followed by decreases in pressure and temperature)

Ask: Why does water vapor rise?

Ask: What happens to the temperature and pressure of the water vapor as it rises?

Key Points: Clouds form when water vapor in the atmosphere changes phase and forms water droplets. Water vapor enters the atmosphere due to evaporation from bodies of water. When the water is warmer, more water evaporates. This is one reason why warm, tropical areas have much rainfall. Because no "container" is associated with Earth's atmosphere, it is easier to explain weather by focusing on pressure and temperature changes of air masses.

Warm air rises because it is less dense than cold air. As warm air rises, the temperature and pressure decrease.

LESSON 62 ANSWERS

1. *Possible answer:* High-pressure areas are generally associated with clear skies. Low-pressure areas are generally associated with cloudy weather and precipitation.

2. *Possible answer:* As the water of a lake heats in the Sun, it evaporates into the atmosphere as water vapor. As the air containing the water vapor becomes warmer, its density decreases and it rises. When the air mass rises, it cools and the water vapor condenses into droplets of liquid water that form clouds.

3. Assume that the pressure at each high- or low-pressure center is equal to the value of the last isobar labeled that surrounds the center. The actual pressure could be up to 3 millibars higher or lower. The pressures at the two highs are 1028 and 1020 mbar. The pressures at the three lows are 984 mbar, 1004 mbar, and 992 mbar. To convert the measures to atmospheres, divide each value by 1013.25 mbar/atm.

High pressure centers:
1028 mbar = 1.015 atm
1020 mbar = 1.007 atm

Low pressure centers:
984 mbar = 0.971 atm
1004 mbar = 0.991 atm
992 mbar = 0.979 atm.

The values at the two highs are both greater than 1 atm. The values at the three lows are all less than 1 atm.

4. A good answer will include ● a list of several types of clouds and how to tell types of cloud apart ● under what conditions each type of cloud forms, why each type of cloud forms, and at what altitude each type of cloud forms.

North of the equator, winds around high-pressure areas move in a clockwise direction. The winds around low-pressure areas move in a counterclockwise direction. This is one reason why the jet stream loops up and over highs and down and under lows as it travels across the United States. South of the equator, winds around high- and low-pressure systems move in the opposite directions.

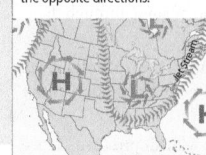

↻ Cloud Formation

When you look up at the sky and see big, puffy, white clouds, it's hard to believe that they are just a bunch of condensed water droplets suspended in the air. Because they are a mixture of water droplets and air, our eyes see them as white or sometimes as gray. But what causes clouds to form in the first place?

Because there is no volume or container associated with Earth's atmosphere, it makes sense to explain weather by focusing on pressure and temperature changes in air masses.

Warm air rises because it is less dense than cold air. As it rises, its temperature and pressure decrease. As the water vapor cools, it condenses into droplets, forming clouds.

A cloud is made up of millions of suspended water droplets that condense from atmospheric water vapor.

Enough water vapor must be present in the air for clouds to form. This is why more clouds form over rainforests than over deserts. On warm days, large puffy clouds are observed at lower altitudes because there are many water molecules in the air. On cold days, if there are clouds at all, they tend to be wispy and high in the sky because there are so few water molecules in the air.

LESSON SUMMARY

How are areas of high and low air pressure related to the weather?

Air pressure varies across the surface of Earth. On air pressure maps, areas of high and low air pressure are indicated with Hs and Ls. Areas of high pressure indicate clear skies and sunny days. Areas of low pressure are associated with storms. For clouds to form, there must be moisture in the air. If this moisture is in a warm air mass, it travels upward, cools off, and undergoes a phase change, forming clouds composed of tiny water droplets. Clouds form under conditions of low temperature and low pressure.

Exercises

Reading Questions

1. What weather is associated with high air pressure? Low air pressure?
2. Describe how a cloud forms. Start with water in a lake.

Reason and Apply

3. One atmosphere of pressure, 1 atm, is equivalent to 1013.25 millibars. Convert the pressure from millibars to atmospheres for the two highs and the two lows on the air pressure map on page 322. What do you notice about these values?

4. PROJECT Find out about different types of clouds and the conditions under which they form. Try to explain how the type of cloud relates to the conditions.

CHAPTER 11

Pressing Matter

SUMMARY

KEY TERMS

sublimation

evaporation

pressure

atmospheric pressure

atmosphere (atm)

inversely proportional

Boyle's law

Gay-Lussac's law

kinetic theory of gases

combined gas law

Weather Update

Air pressure at Earth's surface is constantly changing because air molecules are moving from place to place. Areas of high pressure are associated with clear skies and warm temperatures. The gas molecules in regions of high pressure are moving rapidly and expand into areas of lower pressure. Areas of low pressure are associated with clouds and storms. The gas molecules in regions of low pressure are moving more slowly and water molecules are condensing. Air pressure keeps changing as water evaporates and condenses, which changes the number of air molecules.

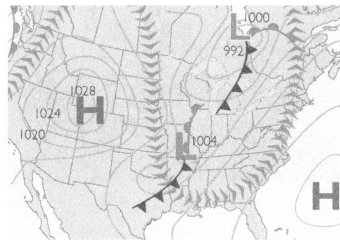

Studying gas behavior in small containers helps us to explain the behavior of gases in the atmosphere. Three gas laws relate gas pressure to temperature and volume—Charles's law, Gay-Lussac's law, and Boyle's law. The combined gas law relates pressure, volume, and temperature.

REVIEW EXERCISES

1. Use the kinetic theory of gases to explain how temperature affects the pressure of a gas.

2. Explain why a weather balloon pops when it reaches a sufficiently high altitude.

3. A cylinder with a movable piston contains 26.5 L of air at a pressure of 1.5 atm. If the temperature is kept the same but the volume of the container is expanded to 50.0 L, what is the new pressure of the gas?

4. A gas sample in a rigid container has a pressure of 3.0 atm at 350 K. If the temperature is raised to 450 K, what will the new gas pressure be?

5. A sample of gas in a flexible container has a volume of 7.5 L, a pressure of 2.5 atm, and a temperature of 293 K. What will the new pressure be if the volume expands to 10.0 L and the temperature is raised to 315 K?

High and Low Pressure

 PROJECT In a newspaper or on the Web, find a weather map of the United States.

1. Find the areas of high and low pressure.

2. Explain where the air is moving up and where the air is moving down.

3. Where do you expect to find stormy weather? Explain your thinking.

---

3. Stormy weather can occur in areas of low pressure where the map indicates a cold front. Advancing cold air pushing warm air upward causes stormy weather.

ASSESSMENTS

Two multiple-choice assessments (versions A and B) and two short-answer assessments (versions A and B) are available for download for Chapter 11.

TRM Chapter 11 Assessment Answers

TRM Chapter 11 Assessments

ANSWERS TO CHAPTER 11 REVIEW EXERCISES

1. The pressure of a gas is proportional to the temperature of the gas. As a gas becomes warmer, its molecules move faster and collide with each other and other objects more often, so the pressure increases.

2. *Possible answer:* As a weather balloon rises in the atmosphere, the pressure outside the balloon decreases. As the pressure decreases, the volume increases because pressure and volume are inversely proportional and the balloon is a flexible container. When the balloon expands too far, it will pop.

3. 0.80 atm

4. 3.85 atm

5. 2.0 atm (to two significant figures)

Project: High and Low Pressure

A good project would refer to the specific weather map of the United States used to answer the numbered exercises. Check student answers against the map used.

1. A capital H indicates high pressure, while a capital L indicates low pressure.

2. The air is moving up in areas of low pressure. The air is moving down in areas of high pressure.

CHAPTER 12

Watch the video overview of Chapter 12 (for teachers) by clicking on the link in the TE-book, opening the TRFD, or logging onto the book's companion Web site bcs.whfreeman.com/livingbychemistry2e (teacher log-in required).

Chapter 12 of the Weather unit focuses on the number of gas particles in a sample and how this is related to pressure, temperature, and volume. Lesson 63 explores the atmosphere and how the density of gas molecules varies with altitude. In Lesson 64, students learn about standard temperature and pressure and how the number of particles is related to the volume of a gas. Lesson 65 is a lab in which students measure the volume of an average breath and then calculate the number of gas particles in that volume using the ideal gas law. Lesson 66 extends the concept of number density to water vapor as students learn about relative humidity. Extreme weather conditions are the subject of Lesson 67.

In this chapter, students will learn

- about pressure and temperature variations in the atmosphere
- the relationship between the pressure and the number density of a gas
- about the unit called the *mole*
- Avogadro's hypothesis and the ideal gas law
- about water vapor density and humidity
- about the formation of hurricanes

K. Kent/Science Source

CHAPTER **12**
Concentrating Matter

In this chapter, you will study

- variations in the atmosphere with changing altitude
- how to determine the numbers of gas molecules in a sample
- the relationships between pressure, temperature, volume, and number of molecules in a gas
- humidity
- conditions that lead to hurricane formation

326

Lightning is an extreme weather phenomenon that typically occurs during thunderstorms.

Gas molecules are too small and numerous to count. However, the mathematical relationships between gas temperature, pressure, volume, and number of molecules allow scientists to determine the exact number of molecules in a gas sample. Knowing the number of water molecules in a particular volume of air helps scientists to predict when precipitation will occur. Air pressure and air temperature greatly affect the amount of water vapor in the air. When moist air cools, the water vapor in the air changes from a gas to a liquid. When extreme conditions of temperature and pressure occur, hurricanes can form.

Chapter 12 Lessons

$\frac{n}{s}$ *for Number*

Pressure and Number Density

THINK ABOUT IT

Airplanes fly at about 35,000 ft (~6.6 mi) above Earth. At this altitude, the air pressure is less than 0.25 atm and the air temperature is less than −45 °C. It is very cold, and the air is very thin up there! You might wonder how many oxygen molecules are available at this air pressure to breathe.

How is the number of gas molecules in a sample related to pressure?

To begin to answer this question, you will explore

↻ Pressure and Number Density

↻ Measuring Air Pressure

EXPLORING THE TOPIC

↻ Pressure and Number Density

The illustration provides information on the *troposphere,* or lower atmosphere, where clouds form and weather takes place. The containers with spheres show how the density of air molecules decreases with altitude.

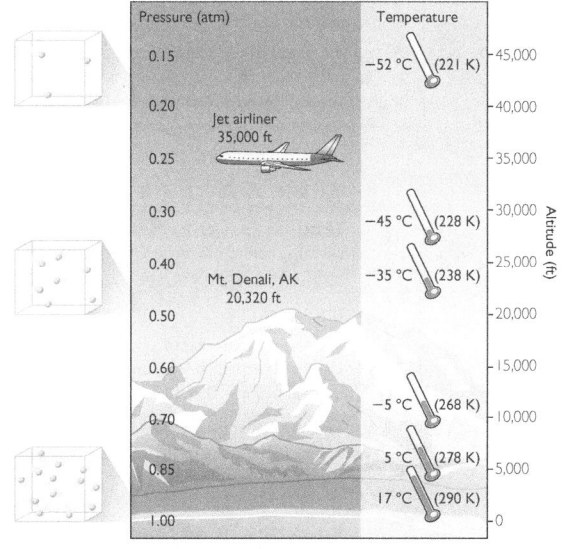

Lesson 63 | **Pressure and Number Density**   327

Overview

LAB: GROUPS OF 4

Key Question: How is the number of gas molecules in a sample related to pressure?

KEY IDEAS

The density of molecules in the atmosphere can be expressed as n/V, where n is the number of gas particles and V is the unit volume. This is known as the *number density* (as opposed to the *mass* density). As n/V increases, the pressure of the gas also increases, and vice versa. When the temperature remains the same, the pressure of a gas is directly proportional to the number

density of the gas: $P = k \cdot (n/V)$. In our atmosphere, the value of n/V decreases as you go up in altitude, and therefore the pressure, P, decreases.

LEARNING OBJECTIVES

● Define the number density of a gas.
● Describe the number density of the atmosphere as it relates to altitude.
● Explain the relationship between number density and gas pressure.
● Describe one way to measure gas pressure.

FOCUS ON UNDERSTANDING

Students might wonder why this particular mathematical relationship

lacks a gas-law name. You might provide them with an easy way to refer to it, such as the "number density equation" or the "number/pressure relationship."

KEY TERM

number density

IN CLASS

This lesson examines the effect of the number of air molecules or atoms on air pressure. Students are provided with a handout that summarizes the changes in Earth's atmosphere with altitude. Part 1 of the worksheet focuses on how changing the number of particles, *n*, affects gas pressure in our atmosphere. Part 2 contains an activity in which students are challenged to change the water level in a U-tube by manipulating pressure, volume, or the number of gas particles. This leads to the idea that the difference in height of the liquid in the U-tube reflects the atmospheric pressure. The discussion focuses on how number density, *n/V,* is related to gas pressure. A complete materials list is available for download.

TRM Materials List 63

SETUP

Students will put about 30 mL of water in the tubing. There may be an overflow of water as students manipulate the U-tubes. Set up catch basins or have students work over a sink. Note that it is very easy to make water squirt some distance out of the U-tube using the syringe. Plan accordingly for this possibility. Have one setup available for the Explain & Elaborate discussion following the lab.

Lesson Guide

ENGAGE (5 min)

TRM PowerPoint Presentation 63

TRM Transparency—ChemCatalyst 63

ChemCatalyst

Earth's Atmosphere
Compare the atmosphere at sea level and at 34,000 ft, the altitude at which airplanes fly.

1. Describe at least three differences.

2. Explain why it is difficult to breathe at 34,000 ft.

Sample answers: 1. The temperature and air pressure are lower. There are fewer molecules per unit of volume at 34,000 ft. **2.** Each breath you take at 34,000 ft contains only one-quarter as many air molecules as air at sea level.

➡ Discuss air pressure in the atmosphere.

Ask: How does air change as altitude increases?

Ask: How does the amount of oxygen in one breath at 34,000 ft compare to the amount in one breath at sea level? Explain your thinking.

Ask: Use the kinetic view of gases to explain how the number of gas molecules is related to pressure.

EXPLORE (15 min)

TRM Worksheet with Answers 63

TRM Worksheet 63

TRM Handout—Earth's Atmosphere 63

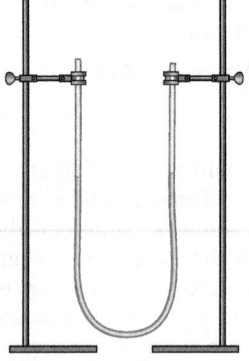

INTRODUCE THE LAB

➡ Arrange students into groups of four.

➡ Pass out the student worksheet and the handout Earth's Atmosphere.

➡ Tell students that they will answer a set of questions for Part 1. Then briefly explain the procedure for Part 2. Ask students to work in groups of four.

➡ Tell students to wear safety goggles.

➡ Hold up the U-tube equipment for students to see.

➡ Tell students they will be challenged to change the water level inside the tube using only the materials provided.

➡ Remind students that the air pressure of the atmosphere is pressing down on the water in both sides of the open tube.

➡ Tell students that the letter *n* refers to the number of atoms or molecules being considered in a sample of gas.

HISTORY CONNECTION

The "weather glass" is an old-fashioned instrument for predicting the weather. The liquid is higher in the spout than in the bottle. This indicates that the atmospheric pressure is lower than average and a storm is approaching.

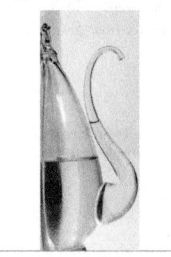

Chemists use the variable *n* to refer to the number of molecules, atoms, or particles they are considering. The number of air molecules per unit of volume is referred to as the **number density,** *n/V.*

The number density of air molecules is relatively low at high altitudes. This is why it gets more difficult to breathe at higher altitudes. There are fewer air molecules per volume of air.

If there are fewer gas molecules in a certain volume, then the air pressure will be lower. Air pressure is proportional to number density. This relationship can be expressed by the equation

$$P = k\left(\frac{n}{V}\right)$$

The decrease in air pressure with increasing altitude is due mainly to a decrease in the number density of gas molecules.

Big Idea The more gas molecules you have in a space, the greater the gas pressure will be.

Example

Bicycle Tire

Use the kinetic theory of gases to explain why the air pressure increases as you pump up a bicycle tire.

Solution

As you pump up a tire, the number of gas molecules per unit of volume becomes greater and greater. This increases the number of collisions between molecules and the walls of the tire. The pressure inside the tire increases until the tire is almost rigid.

⟳ Measuring Air Pressure

There are a number of different ways to measure air pressure. Meteorologists and chemists commonly use either a manometer or a barometer. Both instruments measure the height of a liquid in a tube.

MANOMETER

A *manometer* consists of a liquid in a U-tube as shown in the illustration. If the U-tube is partially filled with water, the water levels on both sides of the U will be at equal heights. This is because the air pressure is the same on both sides.

Imagine that you seal off one side of the U-tube with a stopper so that the number of molecules in the sealed-off area cannot change. The heights of the water will be the same, at least at first. However, if the outside air pressure increases, it will push on the water in the open end causing the water to move up on the side with the stopper, compressing the air on that side. The air pressure of the trapped gas increases

EXPLAIN & ELABORATE (15–20 min)

DISCUSS THE RELATIONSHIP BETWEEN PRESSURE AND NUMBER DENSITY

➡ Write the following on the board: The number of gas molecules per unit of volume, *n/V,* is called the *number density.*

➡ At the appropriate point in the discussion, write the proportional relationship *P = k · (n/V)* on the board.

Ask: How does the kinetic theory of gases explain why *P* increases as *n* increases?

Ask: How does the kinetic theory of gases explain why *P* increases as *n/V* increases?

Ask: Why is it useful to talk in terms of changing number density, *n/V,* rather than just changing number, *n*?

Key Points: The gas pressure increases as the number of gas molecules per unit of volume increases. If you add more gas molecules to a container that does not change size, the pressure of the gas increases because there are more molecules colliding with the walls of the container. The relationship between *P* and *n/V* is proportional.

$$P = k\left(\frac{n}{V}\right)$$

In general, it is more useful to consider how *n/V* changes rather than simply *n,* especially in a flexible container with

because n/V increases. If the outside air pressure decreases, the water in the open end will rise. The air pressure of the trapped gas also decreases because n/V is smaller.

HISTORY CONNECTION

Evangelista Torricelli invented the barometer. Pressure is still sometimes reported in torrs. A *torr* is equivalent to a millimeter of mercury, or mm Hg. 1 atm = 760 mm Hg = 760 torr

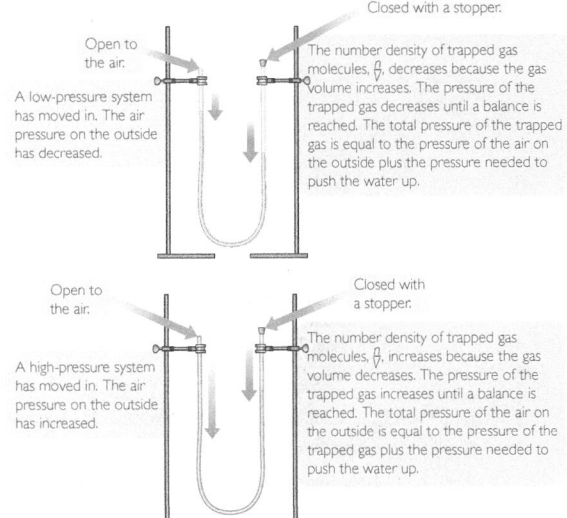

Open to the air.
A low-pressure system has moved in. The air pressure on the outside has decreased.

Closed with a stopper.
The number density of trapped gas molecules, n, decreases because the gas volume increases. The pressure of the trapped gas decreases until a balance is reached. The total pressure of the trapped gas is equal to the pressure of the air on the outside plus the pressure needed to push the water up.

Open to the air.
A high-pressure system has moved in. The air pressure on the outside has increased.

Closed with a stopper.
The number density of trapped gas molecules, n, increases because the gas volume decreases. The pressure of the trapped gas increases until a balance is reached. The total pressure of the air on the outside is equal to the pressure of the trapped gas plus the pressure needed to push the water up.

In a U-tube manometer, the water level on the closed side *rises* when the air pressure in the atmosphere increases. The water level on the closed side *falls* when the air pressure in the atmosphere decreases.

BAROMETER

A *barometer* consists of a tube filled with liquid, which is inverted in a dish of the same liquid. The surface of the pool of liquid is exposed to the atmosphere. There is almost no gas pressure inside the tube.

The barometer shown in the illustration measures the difference in pressure on two surfaces of a liquid. The pressure on the liquid inside the tube is close to zero, so it is the mass of liquid in the tube that balances the air pressure on the outside.

Important to Know The liquid does not spill out of the inverted tube because the force of the air pushing on the surface of the liquid is greater than the weight of the liquid pushing down in the inverted tube.

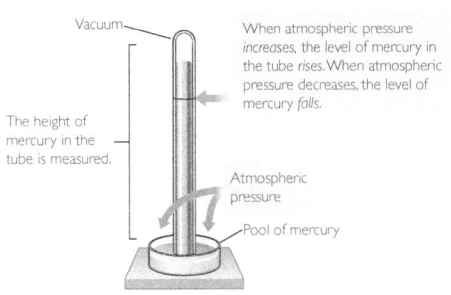

Vacuum

When atmospheric pressure *increases*, the level of mercury in the tube *rises*. When atmospheric pressure decreases, the level of mercury *falls*.

The height of mercury in the tube is measured.

Atmospheric pressure

Pool of mercury

When the U-tube is open to the atmosphere at both ends, the water levels are always at the same height, even if one end of the tubing is higher than the other. This is because the gas pressure from the atmosphere is the same on both sides.

To make the water levels uneven, you have to seal off one end with a finger, a syringe, a cork, or a balloon. This allows you to manipulate the volume of the gas molecules on the side that is sealed off and thereby change the number density, n/V, relative to the number density of the atmosphere. The side with the lower water level has the higher pressure.

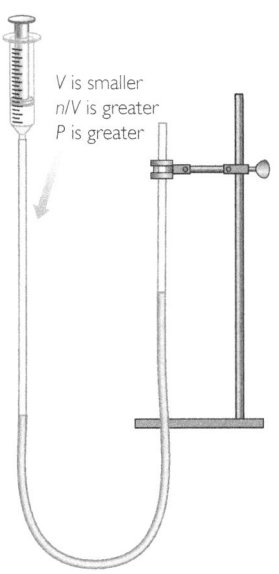

V is smaller
n/V is greater
P is greater

Consider the syringe as part of the system. Depressing the plunger reduces the volume while n stays the same. This increases n/V, which increases the pressure P on that side. You can tell that the pressure is greater on the side with the syringe, because the trapped air is pushing up the water on the other side.

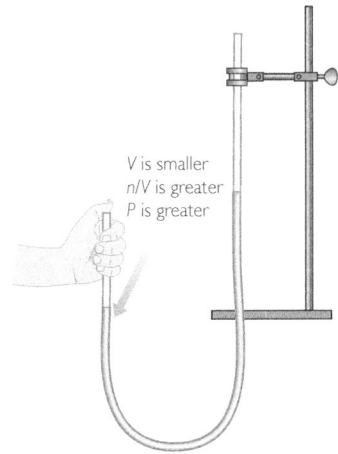

V is smaller
n/V is greater
P is greater

Sealing off one end means n will stay the same on that side. When you lower that end, the weight of the water compresses the air on that side, causing the gas volume to decrease; n/V increases, so P increases.

---

variable volume. Pumping up a bicycle tire is an example. The bicycle pump puts more and more air molecules into the tire, increasing the number density and the air pressure inside the tire.

Number density: The number of gas particles per unit volume.

$$\text{Number density} = \left(\frac{n}{V}\right)$$

DISCUSS STUDENTS' SOLUTIONS TO THE CHALLENGE

→ Have a U-tube with water available at the front of the room for demonstration.

→ Ask student groups to share, draw a picture of, or demonstrate how they

managed to change the water levels in the U-tube.

Ask: What happens to the water levels when the U-tube is open to the atmosphere on both ends?

Ask: How did your group manage to make the water levels inside the tube uneven?

Ask: What you did you do in terms of changing pressure and number density? Explain.

Key Points: The height of the water levels in a U-tube indicates differences in pressure.

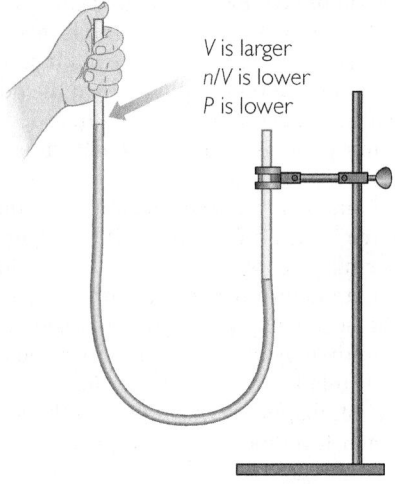

V is larger
n/*V* is lower
P is lower

If you seal off one end and then raise that end, the number of molecules, *n*, stays the same on the side that is sealed off. The weight of the water makes it move down in the tubing, resulting in a slightly larger gas volume. This decreases *n*/*V*, and therefore *P*, on the sealed-off side.

DISCUSS MEASUREMENT OF ATMOSPHERIC PRESSURE

Ask: Suppose you have a U-tube partially filled with water. One end has a stopper on it, and the other end is open. What happens to the water levels if you take the U-tube up a mountain?

Ask: What happens to the water levels in this same U-tube if a high-pressure system moves into a region?

Key Points: Air pressure can be determined by measuring the difference in height of a liquid. The U-tube can be used to measure the air pressure of the atmosphere. Scientists call instruments that measure air pressure *barometers*. In a U-tube barometer, one end is capped, and the other is open to the atmosphere. When the air pressure in the atmosphere increases, the water level on the open side moves down. When the air pressure in the atmosphere decreases, the water level on the open side moves up. In the atmosphere, the number density of the air decreases with increasing altitude. This causes the air pressure to decrease with increasing altitude.

EVALUATE (5 min)

Check-In

A balloon is filled with helium, tied off, and then released. As it climbs into the air, its volume slowly increases. Explain what is going on with the helium atoms inside the balloon and the air molecules outside the balloon in terms of number density and pressure.

Mercury is often the liquid of choice for barometers due to its high density. Air pressure is sometimes expressed in inches of mercury or millimeters of mercury.

1 atm = 29.9 in Hg = 760 mmHg

Meteorologists will sometimes say that the mercury is *rising* when a high-pressure system arrives, or that the mercury is *falling* when a low-pressure system moves in. They are referring to the motion of the liquid mercury in the inverted tube.

LESSON SUMMARY

How is the number of gas molecules in a sample related to pressure?

KEY TERM
number density

Number density is a measure of the number of molecules, atoms, or particles per unit of volume, *n*/*V*. The number density of air molecules in the atmosphere decreases with increasing altitude resulting in lower gas pressure. It is possible to monitor atmospheric pressure by monitoring the level of a liquid in a tube that is sealed at one end. Manometers and barometers are two instruments that measure gas pressure.

Exercises

Reading Questions

1. How can you account for the decrease in air pressure with increasing altitude?
2. Use the kinetic theory of gases to explain the relationship between number density and gas pressure.
3. Explain the difference between a manometer and a barometer.

Reason and Apply

4. What effect does changing the number of gas particles in a container have on the gas pressure in the container? Use the kinetic theory of gases in your explanation.
5. Name three ways that you can increase the pressure of a tire.
6. The boxes in the illustrations show tiny samples of air. Assume they are at the same temperature.

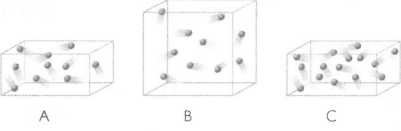

A   B   C

  a. List the samples in order of increasing gas pressure. Explain your reasoning.
  b. Sketch a volume of air that has a pressure in between the pressures in boxes A and B.
7. Explain how the mercury level in a barometer changes when a high-pressure system moves into a region.
8. Explain why the liquid in the barometer does not spill out.

Answer: As the balloon ascends, the number density of the air molecules outside the balloon decreases, so the outside air pressure pushing in on the balloon decreases. The balloon expands until the pressure inside matches the pressure outside. The number of helium atoms does not change, but the number density inside the balloon decreases because the volume increases.

Homework

Assign the reading and exercises for Weather Lesson 63 in the student text.

LESSON 63 ANSWERS

1. *Possible answer:* Air pressure decreases with increasing altitude because the number of molecules of air in a given volume decreases. This means that there are fewer collisions of air molecules with one another and with other objects, and therefore, the air exerts less pressure.

2. According to the kinetic theory of gases, gas pressure is the result of collisions of moving molecules with a surface. Gas pressure increases as the number density of gas molecules increases, because the higher number of molecules will lead to more collisions with the surface.

Answers continue in Answer Appendix (p. ANS-6).

LESSON 64
The Mole and Avogadro's Law

THINK ABOUT IT

Suppose you have two balloons, one filled with helium and the other with carbon dioxide. The pressure, temperature, and volume of the two gases are identical. However, the masses of the balloons are different. While the helium balloon floats, the carbon dioxide balloon sinks. How can you figure out the number of gas particles in each balloon?

How do chemists keep track of the number of gas particles?

To begin to answer this question, you will explore

- Counting Gas Particles
- Avogadro's Law

EXPLORING THE TOPIC

Counting Gas Particles

If you take two air samples of identical volume in the same room, you might predict that both samples will contain the same number of air molecules. A smaller air sample from this room will contain fewer air molecules than a larger air sample.

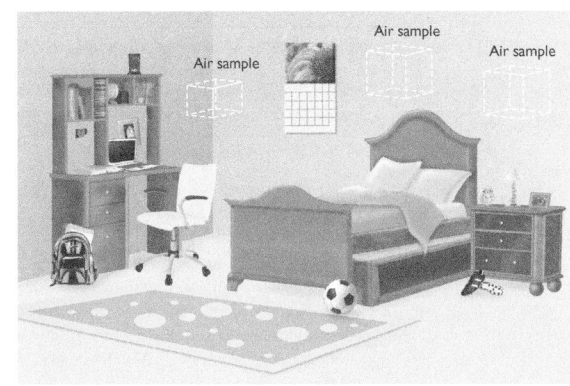

Air sample
Air sample
Air sample

Because the air molecules are spread uniformly over the space the gas occupies, the number of air molecules, n, is proportional to the volume, V, the gas occupies.

Lesson 64 | **The Mole and Avogadro's Law**   331

SPORTS CONNECTION

A hot air balloon is not a closed container. The flame from a propane burner heats the air inside the balloon. As the gas is heated, it expands and some moles of gas molecules escape through the opening at the base. This decreases the number density, n/V, and the mass density, m/V, of the gas inside the balloon. The lower mass density of the captured gas causes the balloon to rise.

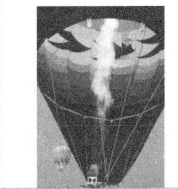

Overview

CLASSWORK: INDIVIDUAL

DEMO: INDIVIDUAL

Key Question: How do chemists keep track of the number of gas particles?

KEY IDEAS

Avogadro's law states that equal volumes of gases contain equal numbers of gas molecules if they are at the same temperature and pressure. At a standard pressure of 1 atm and a standard temperature of 273 K, 22.4 L of any gas will contain exactly one mole of gas molecules. One mole represents 602,000,000,000,000,000,000,000 molecules.

LEARNING OBJECTIVES

- Define a mole.
- Explain Avogadro's law.
- Define standard temperature and pressure.

FOCUS ON UNDERSTANDING

- The curriculum introduces the mole without overemphasizing the exact number the mole represents. At this point, it is enough for students to understand that one mole represents an extraordinarily large number of atoms or

molecules. The mole is covered more thoroughly in the next unit, Toxins.

- Students have a better grasp of the mole if it is not presented right away in scientific notation. They will use Avogadro's number and scientific notation extensively in the next unit, Toxins.

KEY TERMS

mole
Avogadro's number
standard temperature and pressure (STP)
Avogadro's law

IN CLASS

Students observe a simple demonstration with balloons filled with different gases. Students then complete a worksheet that expands on the concept of number density as it relates to gas pressure. Students are introduced to the notion that equal volumes of gas at the same temperature and pressure have equal numbers of gas molecules. Avogadro's hypothesis, STP, and the unit of a mole are introduced. A complete materials list is available for download.

TRM   Materials List 64

SETUP

The demonstration requires that you have two balloons of the same volume containing different gases. Bring a helium balloon to class. Blow up a second balloon so that it matches the helium balloon in size. This is your CO_2 balloon. It is important that the two balloons look as if they have identical volumes. Clearly label each balloon with its identity.

Differentiate

The *mole* presents a variety of challenges for students, particularly where the unit and number itself came from. Students may benefit from researching online to learn more about the history of Avogadro's hypothesis and the concept of the mole.

Lesson Guide

ENGAGE (5 min)

TRM   PowerPoint Presentation 64

ChemCatalyst

There are two balloons. One is filled with helium, He, and the other with carbon dioxide, CO_2.

1. Describe what happens when the balloons are released.

2. For the two balloons, state whether these properties are the same or different, and explain why:

● pressure, P
● temperature, T
● volume, V
● mass, m
● number density, n/V
● number, n
● density, m/V

Sample answers: **1.** The helium, He, balloon rises, while the carbon dioxide, CO_2, balloon falls. **2.** All quantities except m and m/V are probably the same. P and T are the same because the pressure and temperature in the room are the same. The same pressure means the same number density, n/V. The volumes appear similar. If V and n/V are the same, then n must be the same. Because helium floats and carbon dioxide sinks, the densities and hence the masses must be different.

➜ Discuss the two balloons.

Ask: Why do the two balloons have the same number of particles?

Ask: How can two gases have the same number of particles and the same volume but different densities?

Ask: What do you think accounts for the different floating and falling behavior?

Ask: How you could measure out equal numbers of helium atoms and carbon dioxide molecules? Explain.

EXPLORE (15 min)

TRM Worksheet with Answers 64

TRM Worksheet 64

INTRODUCE THE CLASSWORK

➜ Let students know they will be working individually.

➜ Pass out the student worksheet.

➜ Write the following on the board and introduce the unit of a mole:

> 1 mole =
> 602,000,000,000,000,000,000,000

● Tell students that gas particles are far too small and too numerous to count individually. So a special counting unit is used called a *mole*. One mole represents a very large number of gas particles, equal to 602,000,000,000,000,000,000,000, or 602 sextillion, particles.

● Explain that chemists use a unit called a *mole* to describe the number of gas particles in a sample.

● Explain that what may appear to be small samples, (such as 1 mole, 2 moles, half a mole, or 0.1 moles) refer to very large numbers of particles. For example, 2 moles is 2 times the number on the board. Half a mole is half that number.

● Pass out worksheets. Students will work individually on the worksheet to explore the relationship between gas pressure, P, and number of molecules, n.

A MOLE OF GAS PARTICLES

Gas particles are far too numerous to count individually. So chemists have defined a counting unit for small particles called a **mole**. The abbreviation for mole is mol. A mole represents a very large number of gas particles. A mole is equal to 602,000,000,000,000,000,000,000 particles. This is 602 sextillion! This number is also called **Avogadro's number**, after Amedeo Avogadro, a 19th century Italian scientist.

The number of gas particles in a sample is referred to in terms of moles of gas. Keep in mind that when referring to 1 mole, 2 moles, or half a mole of gas particles, these quantities all refer to very large numbers. For example, 2 moles of gas particles is 1,204,000,000,000,000,000,000,000 particles (1,204 sextillion). Half a mole of gas particles is 301,000,000,000,000,000,000,000 (301 sextillion).

STANDARD TEMPERATURE AND PRESSURE, STP

Chemists have figured out how to calculate the number of gas particles in a sample of gas if they know the temperature, volume, and pressure of the sample. However, when comparing gases, it is convenient to define just one set of conditions. Chemists have chosen a gas pressure of 1 atm and a gas temperature of 273 K as the conditions for comparing gases. These conditions are called **standard temperature and pressure,** or **STP**. At STP, 1 mol of gas particles occupies a volume of 22.4 L.

Example 1

Two Moles of Air Molecules

Suppose you want to collect 2.0 mol of air molecules at STP. What size box do you need?

Solution

Because 1 mol of air molecules has a volume of 22.4 L at STP, 2 mol has a volume of 44.8 L. So, you need a box with a volume of 44.8 L.

Big Idea The mole is a counting unit. One mole of gas particles at standard temperature and pressure occupies a volume of 22.4 liters.

➲ Avogadro's Law

Several cylinders each contain different gas samples at STP. The data for each cylinder is shown in the table on the next page. Notice that the volumes of the samples vary.

A few molecules (or atoms, in the case of helium and neon) are drawn in the table to help you visualize the gas. Obviously, only a few gas particles are drawn because it is impossible to draw 602 sextillion. However, the numbers of molecules and atoms are in correct proportion to each other so that you can make comparisons. Take a moment to look for patterns.

Notice that the mass of the gas sample depends on the identity of the gas. A hydrogen molecule has a mass of 2.0 amu, while each neon atom has a mass of 20 amu. So 0.25 mol of neon gas has ten times the mass of 0.25 mol of hydrogen gas.

EXPLAIN & ELABORATE (20 min)

INTRODUCE STANDARD TEMPERATURE AND PRESSURE, STP

Key Points: Standard temperature and pressure, or STP, is 1 atm of pressure at a temperature of 273 K. To be consistent in comparing gases, scientists have agreed on a standard set of conditions under which gases can be measured and compared: 1 atm of pressure and a temperature of 273 K. Note that this is 0 °C.

> **Standard temperature and pressure, STP:** One atmosphere of pressure and a temperature of 273 Kelvin.

Various Gases at STP

Sample	1	2	3	4	5
Gas	H_2	Ne	CO_2	O_2	He
n (moles)	0.25	0.25	0.50	0.50	1.0
Volume	5.6 L	5.6 L	11.2 L	11.2 L	22.4 L
Mass	0.50 g	5.0 g	22.0 g	16.0 g	4.0 g

The volume of each cylinder is determined by the number of gas particles. For example, a 22.4-liter sample at STP always has twice as many gas particles as an 11.2-liter sample, regardless of the identity of the gas.

If two gases have the same pressure, volume, and temperature, then they have the same number of gas particles *independent of the identity of the gases.* This generalization is known as **Avogadro's law.** Avogadro's law is extremely useful. Amedeo Avogadro first proposed this hypothesis in 1811. The reverse is also true: If two gas samples with the same pressure and temperature have the same number of particles, then they occupy the same volume.

Big Idea If two gas samples have the same pressure, temperature, and volume, they contain the same number of particles, even if you are comparing two different types of gases.

Example 2

Helium and Carbon Dioxide Balloons

Suppose you have two balloons, one filled with helium and the other with carbon dioxide. The pressure, temperature, and volume of the two gases are identical.

a. Why is the mass of the carbon dioxide balloon greater?

b. What do you know about the number of atoms in the balloons?

Solution

a. Even though the number of CO_2 molecules is identical to the number of He atoms, the mass of the carbon dioxide balloon is greater because individual molecules of carbon dioxide, CO_2, have a greater mass than atoms of helium, He.

Lesson 64 | **The Mole and Avogadro's Law**   333

Mole: A unit invented by chemists to count large numbers of gas particles. There are 602,000,000,000,000,000,000,000 particles in 1 mole. This number is 602 sextillion.

EVALUATE (5 min)

Check-In

One balloon contains 22.4 L of Ar, argon gas, and another contains 22.4 L of Ne, neon gas. Both balloons are at 273 K and 1 atm.

1. Do the balloons contain the same number of atoms? Why or why not?

2. Will the balloons have the same mass? Why or why not?

Answers: **1.** The balloons contain the same number of atoms because they are at the same temperature and pressure, STP. **2.** The balloons will not have the same mass. Argon atoms are heavier than neon atoms.

Homework

Assign the reading and exercises for Weather Lesson 64 in the student text.

INTRODUCE AVOGADRO'S LAW

→ Optional: Use the two balloons as a visual aid to help Avogadro's law.

Ask: Assume that the gases in the two balloons occupy the same volume and are at the same temperature and pressure. What else must be true about the two gas samples?

Ask: Does it make sense that the numbers of gas particles in each of the two balloons are the same?

Ask: How can two balloons hold identical numbers of gases but have different weights?

Key Points: Equal volumes of gases contain equal numbers of gas particles if the temperature and pressure are the same. This is true regardless of what gas is sampled. This generalization is known as Avogadro's law and was first proposed by Italian scientist Amedeo Avogadro in 1811. Thus, if two gases have the same temperature, pressure, and volume, they also have the same number of gas particles.

There are exactly 602,000,000,000,000,000,000,000 particles in 22.4 L at STP. This is true for *any* gas. Chemists call this enormous number 1 mole. The abbreviation for this unit is mol.

LESSON 64 ANSWERS

1. *Possible answer:* Chemists invented the mole to have a sufficiently large unit with which to count the enormous number of particles in a sample of gas.

2. *Possible answer:* Avogadro's law relates the number of particles in a gas to its volume, temperature, and pressure. According to Avogadro's law, the number of particles in two gases is the same if the temperature, pressure, and volume are the same, even if the gases are of two different types.

3. a. Each gas sample contains 1 mol of atoms. **b.** 0.0446 mol/L **c.** The xenon sample has the largest mass because the mass of each xenon atom is greater than the mass of each neon atom and argon atom, according to the periodic table. **d.** Because all of the samples have the same volume, the sample with the largest mass, xenon, also has the greatest mass density.

4. a. By the definition of a mole, 22.4 L of any gas contains 1 mol of particles. **b.** 0.0446 mol/L **c.** hydrogen: $2 \cdot 1.008$ amu = 2.016 amu nitrogen: $2 \cdot 14.01$ amu = 28.02 amu carbon dioxide: 12.01 amu + $2 \cdot 16.00$ amu = 43.01 amu. Carbon dioxide has the largest mass, because even though there are equal numbers of each type of molecule, the mass of a carbon dioxide molecule is greater than the mass of either a hydrogen or a nitrogen molecule. **d.** Carbon dioxide has the largest number of atoms because each molecule consists of three atoms, meaning there are 3 mol of carbon and oxygen atoms in 1 mol of carbon dioxide molecules. **e.** All of the samples have the same volume, so the sample with the largest mass, carbon dioxide, also has the greatest mass density.

5. helium: $8.0 \text{ g} \cdot 1 \text{ mol}/4.0 \text{ g} = 2.0$ mol argon: $40.0 \text{ g} \cdot 1 \text{ mol}/40.0 \text{ g} = 1.0$ mol. 8.0 g of helium has more atoms than 40.0 g of argon.

6. The balloon filled with hydrogen has more particles than the balloon filled with oxygen because it has a greater volume. The number density of both gases is the same if the pressure and temperature are identical. Therefore, the balloon with greater volume holds more particles.

Example

(continued)

b. There are more atoms in the CO_2 balloon. The number of molecules of CO_2 is equal to the number of atoms of He according to Avogadro's law. Because each molecule of CO_2 consists of three atoms, there are three times as many atoms in the CO_2 balloon than in the He balloon.

LESSON SUMMARY

KEY TERMS
mole
Avogadro's number
standard temperature and pressure (STP)
Avogadro's law

How do chemists keep track of the number of gas particles?

Gas particles move randomly and are distributed uniformly in the space they occupy. At standard temperature and pressure, STP, the gas pressure is 1.0 atm and the temperature is 273 K. At STP in a volume of 22.4 L, there is 1 mol of gas particles. One mole represents 602,000,000,000,000,000,000,000 particles. This is 602 sextillion, also called Avogadro's number. Avogadro's law states that equal volumes of any gas at the same pressure and temperature have the same number of gas particles, independent of the identity of the gas.

Exercises

Reading Questions

1. Explain why chemists invented the unit called a mole.
2. Explain Avogadro's law.

Reason and Apply

3. Suppose you have 22.4 L of the following gases at STP: neon, Ne, argon, Ar, and xenon, Xe.
   a. How many atoms are there in each gas sample?
   b. What is the number density of atoms, n/V, for each sample?
   c. Which sample has the largest mass? Explain your reasoning.
   d. Which sample has the largest mass density, m/V?

4. Suppose you have 22.4 L of the following gases at STP: hydrogen, H_2, nitrogen, N_2, and carbon dioxide, CO_2.
   a. How many molecules are there in each gas sample?
   b. What is the number density of molecules, n/V, for each sample?
   c. Which sample has the largest mass? Explain your reasoning.
   d. Which sample has the largest number of atoms? Explain your reasoning.
   e. Which sample has the largest mass density, m/V?

5. Which has more atoms: 8.0 g of helium, He, or 40.0 g of argon, Ar? Explain your reasoning.

6. Which has more particles: a balloon filled with 10 L of oxygen, O_2, gas, or a balloon filled with 15 L of hydrogen, H_2, gas? Explain your reasoning. Assume STP.

7. At 25 °C, which balloon has the greater volume: an oxygen, O_2, balloon at 1.2 atm with a mass of 16.0 g, or a helium, He, balloon at 1.2 atm with a mass of 2.0 g? Explain your reasoning.

7. oxygen: $16 \text{ g} \cdot 1 \text{ mol}/32 \text{ g} = 0.50$ mol helium: $2.0 \text{ g} \cdot 1 \text{ mol}/4 \text{ g} = 0.50$ mol According to Avogadro's law, if the temperature and pressure are the same, 0.50 mol of one gas has the same volume as 0.50 mol of another gas. So, the balloons have the same volume.

Take a Breath

Ideal Gas Law

THINK ABOUT IT

Climbing Mount Everest is extremely difficult without bringing along a tank of oxygen to help you breathe. As you climb higher and higher, the air pressure gets lower and lower. At 29,000 ft, on top of Mount Everest, the air pressure is only 0.33 atm. You might wonder how many oxygen molecules are available at this pressure.

> How can you calculate the number of moles of a gas
> if you know *P, V,* and *T* ?

To begin to answer this question, you will explore

⟳ Pressure, Volume, Temperature, and Number

⟳ The Ideal Gas Law

EXPLORING THE TOPIC

⟳ Pressure, Volume, Temperature, and Number

Four variables describe gas samples: pressure, volume, temperature, and number of moles of atoms or molecules. Any two variables can be related to each other mathematically if the other two are kept the same. You might begin to wonder if you can relate all four variables in a single equation. The combined gas law relates three of the variables: pressure, volume, and temperature.

So far you've learned that at STP, there is 1.0 mol of gas molecules in a volume of 22.4 liters for *any* gas sample.

INCREASING THE PRESSURE

Suppose you want to adjust this sample of gas so that it has a pressure of 2.0 atm. In what ways can you accomplish this?

You can change the pressure to 2.0 atm in different ways.

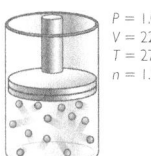
$P = 1.0$ atm
$V = 22.4$ L
$T = 273$ K
$n = 1.0$ mol

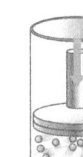
$P = 2.0$ atm
$V = 11.2$ L
$T = 273$ K
$n = 1.0$ mol

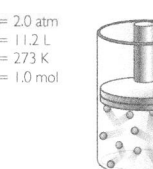
$P = 2.0$ atm
$V = 22.4$ L
$T = 546$ K
$n = 1.0$ mol

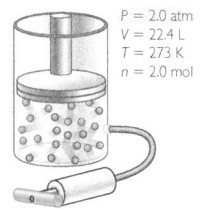
$P = 2.0$ atm
$V = 22.4$ L
$T = 273$ K
$n = 2.0$ mol

Original gas sample

Decrease V. *Squeeze the gas at 273 K into a volume of 11.2 L.*

Increase T without changing V. *Heat the gas in 22.4 L to a temperature of 546 K.*

Increase n without changing V. *Add 1.0 mol of gas molecules to the container at 273 K at a volume of 22.4 L.*

Lesson 65 | **Ideal Gas Law**   335

Overview

LAB: GROUPS OF 4

Key Question: How can you calculate the number of moles of a gas if you know *P, V,* and *T*?

KEY IDEAS

The ideal gas law, $PV = nRT$, allows scientists to relate gas pressure, volume, moles of particles, and temperature. *R* is a constant and does not change; however, its value depends on the units that are used. If you know pressure, volume, and temperature, you can calculate the number of moles of gas in a sample by using this equation.

LEARNING OBJECTIVES

● Define the ideal gas law.
● Define the universal gas constant, *R*.
● Complete calculations for finding *n*, using the ideal gas law.

FOCUS ON UNDERSTANDING

Note that we still are not using scientific notation for the number of molecules in a mole. This lets students to become familiar with this number and grasp its magnitude.

KEY TERMS

ideal gas law
universal gas constant (*R*)

IN CLASS

Students determine the volume of one average breath of air from their lungs. They blow air through a tube into a 2 L plastic soft drink bottle, displacing an amount of water. Students use their measurements to calculate the volume of a single breath. They then use the ideal gas law to calculate the number of moles of air in one breath at sea level and the number of moles of air in one breath at a higher elevation. A complete materials list is available for download.

TRM Materials List 65

SETUP

Cut drinking straws into three or four pieces to serve as mouthpieces for the tubing. Each student should use a different straw. You might have your own equipment set up at the front of the room to demonstrate the procedure.

Pre-AP® Course Tip

Students preparing to take an AP® chemistry or college chemistry course should be able to use mathematics to solve problems that describe the physical world. In this lesson, students use mathematics to complete calculations relating the number of moles of gas, *n*, to the pressure, *P*, temperature, *T*, and volume, *V*, of the gas using the ideal gas law.

Lesson Guide

ENGAGE (10 min)

TRM PowerPoint Presentation 65

ChemCatalyst

1. Describe how you can determine the volume of a breath of air.

2. Name four factors that might affect the volume you measure.

3. What do you have to know to determine the number of molecules in a breath of air?

Sample Answers: **1.** You could blow into a bag and measure the volume of the bag. **2.** The volume you measure will depend on how deep a breath you take, how large a person you are, the pressure in the atmosphere (or your altitude), and the air temperature. **3.** You have to know the relationship between the number of molecules and volume.

→ Discuss measuring a breath of air.

Ask: What type of container would you use to measure the volume of a breath? (a flexible container)

Ask: How large a container should you use? (several liters)

Ask: Do you think it is enough to take only one measurement? Why or why not?

Ask: What variables do you think affect the number of molecules in a breath of air? (the volume of a breath, air pressure, air temperature, your size, how deeply you breathe)

TRM Worksheet with Answers 65

TRM Worksheet 65

INTRODUCE THE LAB

→ Arrange students into groups of four.

→ Pass out the student worksheet.

→ Tell students that the ideal gas law allows us to determine the number of moles of gas molecules in a gas sample if temperature, pressure, and volume are known. The equation for the ideal gas law is

$$PV = nRT$$

where R is equivalent to the proportionality constant, k, for this equation.

$$R = \frac{PV}{nT} \qquad R = 0.082 \text{ L} \cdot \text{atm/mol} \cdot \text{K}$$

→ Ask students to work in groups of four. Let them know they must wear their safety goggles.

→ Go over the general procedure for Part 1 with the class. Emphasize to students that the goal is to find the volume of a normal breath. (This is not a competition to determine who can exhale the largest volume of air.)

→ In Part 2, students complete calculations using the volume of one breath and the ideal gas law to determine how many moles of air particles they would breathe at different altitudes.

So while the pressure is the same in each case, you can see how the other variables have changed.

Target pressure	Volume	Temperature	Moles
2.0 atm	11.2 L	273 K	1.0 mol
2.0 atm	22.4 L	546 K	1.0 mol
2.0 atm	22.4 L	273 K	2.0 mol

INCREASING THE VOLUME

Suppose you want to adjust the original sample of gas so that it has a volume of 44.8 L. How can you accomplish this?

You can produce a sample of gas with a volume of 44.8 L in several ways.

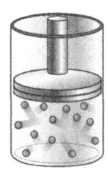

P = 1.0 atm
V = 22.4 L
T = 273 K
n = 1.0 mol

Original gas sample.

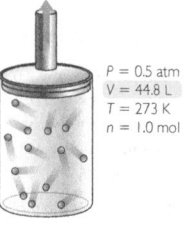

P = 0.5 atm
V = 44.8 L
T = 273 K
n = 1.0 mol

Increase V. *Expand 1.0 mol of gas molecules at 273 K until the pressure is 0.5 atm.*

P = 1.0 atm
V = 44.8 L
T = 546 K
n = 1.0 mol

Increase T. *Heat 1.0 mol of gas molecules at 1.0 atm to a temperature of 546 K.*

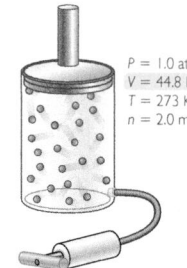

P = 1.0 atm
V = 44.8 L
T = 273 K
n = 2.0 mol

Increase n. *Add 1.0 mol of gas molecules to the container at STP, to arrive at 2.0 total moles.*

Pressure	Target volume	Temperature	Moles
0.5 atm	44.8 L	273 K	1.0 mol
1.0 atm	44.8 L	546 K	1.0 mol
1.0 atm	44.8 L	273 K	2.0 mol

It is apparent that these variables are mathematically related. In fact, if you combine the equation for the combined gas law with the relationship described by Avogadro's law, the result is a mathematical relationship known as the ideal gas law.

DISCUSS HOW GROUPS DETERMINED THE VOLUME OF ONE BREATH OF AIR

You might have each group write its results for the volume of one breath of air on the board. (Note that the average total lung capacity for an adult is around 6 L of air. However, the amount of air breathed in or out during normal respiration is closer to 500 mL.)

Ask: What are some different ways you can determine the volume of one breath of air?

Ask: What are some reasons why the calculated volumes differ from group to group?

Ask: What are some possible causes of error in this procedure?

Ask: How could you improve the experiment you did? (measure the volume of one breath several times, measure several breaths and divide by the number of breaths, collect the breath in a larger bottle, try to breathe normally, etc.)

Key Points: We should see differences in the volume of one breath of air from group to group. Each person's lung capacity is slightly different. Further,

The Ideal Gas Law

The **ideal gas law** relates the pressure, volume, temperature, and number of moles for a gas sample.

Ideal Gas Law

The product of the pressure, P, and volume, V, of a gas is proportional to the product of the number of moles, n, and the temperature, T.

$$PV = nRT$$

The two products are related by the constant, R.

$$R = 0.082 \frac{L \cdot atm}{mol \cdot K}$$

The **universal gas constant**, R, is the same for all gases.

Because R has units of $L \cdot atm/mol \cdot K$, you can use this value of R only if volume is measured in *liters*, pressure in *atmospheres*, temperature in *kelvins*, and number of gas molecules in *moles*. When these units are used, R is always equal to $0.082 \, L \cdot atm/mol \cdot K$.

If you know three of the four variables for any gas sample, you can use the ideal gas law to calculate the fourth variable. This is most useful in determining the moles of gas molecules in a gas sample because the pressure, volume, and temperature can be easily measured.

Example

Moles of Air on Mount Everest

How many moles of air are in a 0.5 L breath on top of Mount Everest? The pressure is 0.33 atm and the temperature is 254 K.

Solution

The ideal gas law relates all these quantities. Insert the values for P, V, T, and R into the equation and solve for n.

$$PV = nRT$$

$$(0.33 \text{ atm})(0.5 \text{ L}) = n\left(0.082 \frac{L \cdot atm}{mol \cdot K}\right)(254 \text{ K})$$

$$n = \frac{(0.33 \text{ atm})(0.5 \text{ L})}{\left(0.082 \frac{L \cdot atm}{mol \cdot K}\right)(254 \text{ K})}$$

$$n = 0.008 \text{ mol of air}$$

So there is only 0.008 mol of air molecules in a 0.5 L breath of air atop Mount Everest.

Lesson 65 | **Ideal Gas Law**   337

---

the water is displaced. Then you can divide the total volume of the bottle by the number of breaths.

DISCUSS THE IDEAL GAS LAW

→ You might want to guide the class through the last question on the worksheet.

Ask student volunteers to assist you at the board.

Ask: How did you use the ideal gas law to determine the number of moles of gas particles in one breath?

Ask: How did you convert moles to number of gas particles?

Ask: What did you find out about a breath of air at sea level compared to a breath of air on a 10,000-ft-high mountain?

Ask: How would you figure out the number of moles of air in a breath at the top of Mount Everest or at an elevation 30,000 ft? Explain.

Key Points: The ideal gas law allows scientists to relate gas pressure, volume, moles of particles, and temperature. The ideal gas law is a result of combining the ideas in the combined gas law with Avogadro's law. The ideal gas law allows you to calculate the number of moles of gas particles in any given volume. You must know the pressure and temperature of the gas, and the values must be expressed in the appropriate units (in this case, pressure in atmospheres, volume in liters, temperature in Kelvin, and number in moles).

> **Ideal gas law:** The ideal gas law states that $PV = nRT$, where R, the universal gas constant, is equivalent to the proportionality constant, k, for this equation.
>
> $$R = \frac{PV}{nT} \qquad R = 0.082 \, L \cdot atm/mol \cdot K$$

Note that R is the same for all gases but the value of R does change, depending on if the units change. For example, $R = 62.4 \, L \cdot mmHg/mol \cdot K$ and $R = 8.314 \, L \cdot kPa/mol \cdot K$. We have opted to round off R to 3 significant digits for simplicity. Note that the value for R is often reported to more significant digits than this.

In this activity, the results for the value of n vary from group to group. However, if a breath volume is 0.50 L, then n is equal to 0.020 mol at sea level and 0.016 mol at 10,000 ft.

---

some students may take a deep breath, while others may take a shallow breath. Because there is variation from one breath to the next, it may be useful to measure the volumes of several breaths and average the outcomes.

There is more than one way to figure out the volume of air that was exhaled into the bottle.

- One way is to begin the procedure by calibrating the bottle. You can do so by putting known quantities of water (say, 250 mL or 500 mL) into the bottle and marking the side with a marker.
- Another method for measuring breath volume is to (1) breathe into

the bottle, (2) cap off the bottle underwater, (3) remove the bottle from the tub and invert it, and (4) refill the bottle with water using measured amounts.

- A third method for figuring out the breath volume is to measure the amount of water that is left behind in the 2 L bottle and then subtract that amount from 2 L. This gives you the amount of volume occupied by the air. It might be more precise to measure an exact volume of the bottle before starting, in case it is slightly over or under 2 L.
- A fourth method is to keep breathing into the bottle until all

The number of moles can be converted to the total number of gas molecules by multiplying by 602 sextillion. The difference in number of moles between sea level and 10,000 ft is about 0.004 mol. This small difference in moles is about 2.4 sextillion total molecules. So, there are many fewer molecules in each breath at 10,000 ft than at sea level.

The ideal gas law can be used to solve for variables other than n. For example, you can use the ideal gas law to solve for P if n, V, and T are known.

EVALUATE (5 min)

Check-In

You cap a 1.0 L plastic bottle on a mountaintop where the air pressure is 50 atm and the temperature is 298 K.

1. How many moles of gas are in the bottle?

2. What is the number density, n/V, of the gas inside the bottle on the mountaintop?

3. At sea level, the volume of the bottle becomes 0.50 L. What is the number density of the gas inside the bottle at sea level?

Answers: **1.** Using the ideal gas law to solve for n: $n = PV/RT = 0.020$ mol.
2. Number density on the mountaintop: $n/V = 0.020$ mol/1.0 L = 0.020 mol/L.
3. Number density at sea level: $n/V = 0.020$ mol/0.50 L = 0.040 mol/L.

Homework

Assign the reading and exercises for Weather Lesson 65 in the student text.

LESSON 65 ANSWERS

1. The ideal gas law is an equation that relates the variables pressure, volume, number of moles, and temperature for any sample of any gas. The equation is $PV = nRT$.

2. *Possible answer:* Use the ideal gas law whenever you know three of the variables for a gas (pressure, volume, number density, temperature), and you want to find the value of the fourth variable.

3. 0.13 mol

4. 34 L

5. 3.90 atm (to three significant figures)

6. 430 mol (to two significant figures)

7. *Possible answers (any 3):*
● Reduce the number of particles to 0.5 mol. ● Reduce the temperature to 137 K. ● Increase the volume to 44.8 L.

● Reduce the number of particles to 0.1 mol, but increase the volume to 112 L.

8. Yes; according to the ideal gas law, if the number of particles, volume, and temperature of two different gases is the same, then the pressure will be the same.

How can you calculate the number of moles of a gas if you know P, V, and T?

Any sample of gas can be described by four variables: pressure, volume, temperature, and moles. The ideal gas law relates these four variables to each other mathematically: $PV = nRT$, where R is the universal gas constant. The value of R is 0.082 L · atm/mol · K for all gas samples if units of atmospheres, liters, moles, and kelvins are used. If you know three of the four variables for any gas sample, you can use the ideal gas law to calculate the fourth variable.

KEY TERMS
ideal gas law
universal gas constant, R

Exercises

Reading Questions

1. What is the ideal gas law?
2. Describe when you might want to use the ideal gas law.

Reason and Apply

3. How many moles of hydrogen, H_2, gas are contained in a volume of 2 L at 280 K and 1.5 atm?

4. What volume would 1.5 mol of nitrogen, N_2, gas occupy at standard temperature and pressure?

5. Find the pressure of 3.40 mol of gas if the gas temperature is 40.0 °C and the gas volume is 22.4 L.

6. How many moles of helium, He, gas are contained in a 10,000 L weather balloon at 1 atm and 10 °C?

7. Suppose you have 1.0 mol of gas molecules in 22.4 L at STP. Describe three ways you can get a gas pressure of 0.50 atm.

8. Will the pressure of helium, He, gas be the same as the pressure of oxygen, O_2, if you have 1 mol of each gas, each at a volume of 22.4 L and each at 273 K? Explain your thinking.

Humidity, Condensation

THINK ABOUT IT

Exercising can make you work up a sweat. Sweating helps to regulate your body temperature. When a breeze blows across your sweaty forehead, you may notice that your skin feels cooler. However, on a humid day, with a lot of moisture in the air, it seems as if no amount of sweating helps you to cool down.

What is humidity and how is it measured?

To answer this question, you will explore

- Evaporation and Condensation
- Humidity
- Relative Humidity

EXPLORING THE TOPIC

Evaporation and Condensation

After a summer rainstorm, puddles of water on the ground often disappear quickly. The rainwater is evaporating. Recall that evaporation is a phase change from a liquid to a gas.

Evaporation is the reverse process of condensation, when water vapor becomes a liquid. These two processes, evaporation and condensation, are both occurring wherever water is present. In fact, there is a competition between the two processes that can result in net evaporation or net condensation.

The rates of both evaporation and condensation depend mainly on temperature and the amount of water vapor already in the air.

Important to Know The water in rain puddles does not have to boil to go into the gas phase. Evaporation takes place on the surface of a liquid all the time. Try leaving a glass of water on a table. The water will gradually disappear, leaving the glass empty.

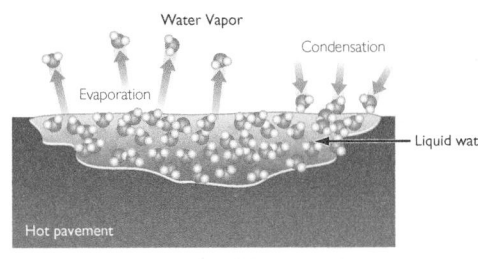

Cloud formation, rain and snowfall, fog, frost, and the appearance of dew are all events associated with the condensation of water vapor out of the atmosphere.

Lesson 66 | **Humidity, Condensation**   339

Overview

LAB: GROUPS OF 4

Key Question: What is humidity and how is it measured?

KEY IDEAS

Humidity is the amount of water vapor in a volume of air. Thus, it is a measure of water vapor density, n/V. Humidity can be expressed as mass density in grams per cubic centimeter or as number density, in moles per 1000 liters. Air temperature affects how much water vapor potentially can be in the air. Humidity generally is expressed as relative humidity, the percentage of water vapor in the air relative to the maximum amount possible at that temperature. At maximum humidity, the air is saturated with water. Maximum humidity increases with increasing air temperature.

LEARNING OBJECTIVES

- Define humidity and relative humidity.
- Explain the relationship between humidity and phenomena such as cloud formation, fog, rainfall, and dew.
- Explain the relationship between water vapor density and air temperature.

FOCUS ON UNDERSTANDING

What meteorologists refer to as humidity is not water vapor density but actually relative humidity, which is expressed as a percent.

KEY TERMS

humidity
partial pressure
relative humidity

IN CLASS

Students investigate the amount of water vapor in the air. They also explore the relationships among condensation, humidity, and air temperature. They complete two different procedures that allow them to determine the humidity of the air in their classroom. One procedure involves discovering the temperature at which condensation occurs. The other involves a wet-bulb thermometer. If you are pressed for time, Part II is optional or can be done as a demo. A complete materials list is available for download.

TRM Materials List 66

SETUP

You can set up half the stations for Part 1 and half for Part 2. For Part 1, set up a 250 mL beaker, water, ice, thermometer, and stirring rod. For Part 2, set up a beaker containing 200 mL of room-temperature water, two thermometers, small rubber bands, and a small piece of cheesecloth—approximately 2 in by 6 in—for students to wrap around the bulb of the thermometer.

Differentiate

Students struggling to master the language of chemistry may benefit from using a word splash to make sense of vocabulary from Unit 3. A word splash is a collection of terms that you put on display for students to examine before they begin a lesson or reading. You can find different word splash applications online. Create a word splash of key vocabulary for Unit 3 and share it with students. Ask them to take turns in small groups explaining how the chosen terms relate to each other and to the unit's theme of weather. Have one student in each group keep a record of key points to report back to the rest of the class.

Lesson Guide

ENGAGE (5 min)

TRM PowerPoint Presentation 66

1. Is there water vapor in the air right now? What evidence do you have to support your answer?

2. What do you think humidity means? How does humidity depend on temperature?

Sample answers: **1.** Most students will say that some water vapor is in the air at all times, because water is evaporating all around us or that clouds can be seen overhead. **2.** Humidity is a measure of the amount of water vapor in the air. Students may say that warm air can "hold" more water vapor than cold air.

→ Assist the class in sharing ideas about humidity.

Ask: Does the amount of water vapor in the air change? Explain.

Ask: What do you think affects the amount of water vapor in the air?

Ask: How do meteorologists determine the moisture content of the air? How do they express how much water vapor is in the air?

Ask: What is humidity? Do you feel warmer or colder when there is humidity?

EXPLORE (15 min)

TRM Worksheet with Answers 66

TRM Worksheet 66

TRM Handout—Relative Humidity 66

INTRODUCE THE LAB

→ Arrange students into groups of four.

→ Pass out the student worksheet.

→ Introduce the term *humidity*. Tell students that humidity describes the amount of water vapor in the air.

→ Explain to students that they will be working in groups of four to complete two procedures to determine the amount of water vapor in the air.

Humidity: The density of the water vapor in the air at any given time. Humidity is dependent on air temperature and pressure.

EXPLAIN & ELABORATE (20 min)

TRM Transparency—Water Vapor Density Versus Temperature 66

GEOLOGY CONNECTION

It is estimated that Earth has about 326 million cubic miles of water. This includes all of the water in the oceans, underground, and locked up as ice. The U.S. Geological Survey estimates that about 3100 mi³ of this water is in the air—mostly as water vapor, but also as clouds or precipitation—at any one time.

METEOROLOGY CONNECTION

Weather forecasters use a sling psychrometer to determine the relative humidity. It consists of two thermometers, one dry and one wrapped in a wet cloth. The thermometers are spun through the air and then read.

During spinning, water evaporates from the cloth, cooling the wet-bulb thermometer. The amount of cooling that occurs is directly related to how much moisture is in the air. When there is more moisture in the air, less evaporation and less cooling occur.

You might have noticed a tendency for fog to form on cool nights after a warm, humid day. This is because a lot of water has evaporated during the day. When the temperature decreases, the amount of water vapor in the air exceeds the maximum allowable for the cooler temperature and fog forms.

↻ Humidity

Water vapor is present in the air around us all the time. There is more or less water vapor in the air depending on the weather conditions. Warmer air masses can have more water vapor than colder air masses.

Humidity is the number density of water vapor in the air in moles per unit of volume, or n/V. Sometimes, it is expressed as grams per unit of volume (such as cm³ or liters). For a given temperature, there is a limit to the number density of water molecules that can be in the air. At a certain point, the air becomes saturated, and no further net evaporation takes place. The graph shows the maximum number density for air temperatures between $-1\,°C$ and $40\,°C$. You can see that the maximum humidity depends on temperature.

On a cool day, around $10\,°C$ ($50\,°F$), for instance, the maximum number density of water molecules in the air is approximately 0.5 mol per 1000 liters of air. This amount represents the maximum humidity for this temperature. The actual humidity may be at or below this value.

Water Vapor Density Versus Temperature

Points on the curve correspond to 100% relative humidity.

Water vapor density (mol/1000 L) vs *Temperature (°C)*

Using the Graph

This graph represents an inequality. The actual value for humidity can be any point on or below the curve. The graph shows that the total amount of water vapor that can be in the air increases with increasing temperature.

If the temperature of the air drops or the humidity rises, precipitation can occur. If these two values, T and n/V, result in a data point above the curve, there will be precipitation or condensation of some sort. The condensation of water on the

DISCUSS WATER VAPOR DENSITY AND HUMIDITY

Ask: What does water vapor density describe?

Ask: What does the condensation procedure indicate?

Ask: Under what conditions would the water vapor condense out of the air at a lower temperature? At a higher temperature?

Ask: What is the water vapor density of the air in your classroom? Explain how you know.

Key Points: Humidity is a measure of the amount of water vapor in the air. Humidity can be expressed as number

density, n/V, or as mass density, g/cm³. Temperature affects maximum humidity. Warm air can have a higher maximum water vapor density than cold air.

The condensation procedure provides evidence that water vapor is present in the air. When the temperature drops low enough, the water vapor in contact with the beaker changes phase and condenses on the outside of the glass. There is no other place this moisture could come from. Water will condense from air if the air cools sufficiently.

The temperature at which water vapor condenses indicates how much water vapor is in the air. Suppose water condenses at a temperature of 15 °C in your classroom. The graph shows this

outside of a glass of ice water is a sign that the air next to the glass has cooled enough for the water vapor to change phase to a liquid.

Like all gases, water vapor exerts a pressure. This pressure is part of the total pressure exerted by all the gases in the atmosphere. Each gas exerts a **partial pressure**, and these partial pressures add up to the atmospheric pressure.

Relative Humidity

When meteorologists consider humidity, they focus on the relative humidity. **Relative humidity** is the percent of the maximum humidity for a specified temperature. When the air contains the maximum amount of water vapor for the temperature of the air, it is at 100% relative humidity.

The terms *humidity* and *relative humidity* are often used interchangeably in weather reports. Meteorologists might say the humidity today is at 35% when they are really talking about the relative humidity.

When air comes in contact with a cold surface, its temperature drops, causing some of the water vapor in the air to condense on the surface.

Example

Relative Humidity

Imagine that on a certain day the water vapor density is 1.6 mol/1000 L and the temperature is 30 °C.

a. What is the relative humidity?

b. At night, the temperature drops to 20 °C. Do you expect there will be precipitation?

Solution

a. You can read the maximum water vapor density at 30 °C on the graph Water Vapor Density Versus Temperature. It is 1.7 mol/1000 L. This corresponds to 100% humidity. The relative humidity is the measured water vapor density divided by the maximum possible water vapor density.

$$\frac{1.6}{1.7} = 0.94 = 94\%$$

b. The point (20 °C, 1.6 mol/1000 L) on the graph is above the 100% humidity curve, so you can expect precipitation.

Matter that is in contact with an evaporating liquid will cool off. Sweating helps cool you because the evaporation of water on your skin transfers heat away from the body. When there is more water vapor in the air, it becomes harder for sweat to evaporate. This is why you often feel much hotter in humid air than in dry air.

Lesson 66 | **Humidity, Condensation**   341

---

that can be present at 30 °C is approximately 1.69 moles of water vapor per 1000 L of air sampled. The maximum water vapor density is called *100% relative humidity*. At 100% relative humidity, the air is saturated with water vapor, and no further evaporation can take place. The water vapor density corresponding to 100% humidity depends on temperature. Thus, conditions of 100% humidity at 40 °C mean more water vapor in the air than 100% humidity at 10 °C.

Humidity is sometimes expressed as relative humidity. Relative humidity is a percentage of the maximum humidity at a specified temperature. If the relative humidity is 50%, the amount of water vapor in the air is half the maximum amount possible for that temperature. At 90% humidity, you would expect it to be very close to raining or very foggy. You would almost feel the moisture in the air. Summers in the Midwest can have very high humidities. People tend to feel most comfortable at a relative humidity of about 45%.

> **Relative humidity:** The amount of water vapor in the air compared to the maximum amount of water vapor possible for a specific temperature, expressed as a percent.

PROCESS THE WET- AND DRY-BULB PROCEDURE (OPTIONAL)

→ Use the handout showing relative humidity.

Ask: What did the wet- and dry-bulb procedure allow you to determine?

Ask: Examine the table of relative humidity. What does a large temperature difference between wet and dry bulbs suggest? a small temperature difference?

Ask: Suppose your classroom has a relative humidity of 60%. What does that mean?

Key Points: The wet- and dry-bulb procedure allows you to determine the relative humidity of the air. While the thermometer is being waved around, the water evaporates from the wick, cooling the wet-bulb thermometer. When water evaporates, it transfers heat away from whatever it is in contact with. That is why sweating cools your skin.

The larger the temperature difference between the wet and dry bulbs, the drier the air. Likewise, the smaller the

---

as equivalent to approximately 0.70 mol of water vapor per 1000 L of air.

INTRODUCE RELATIVE HUMIDITY

→ Display the transparency Water Vapor Density Versus Temperature.

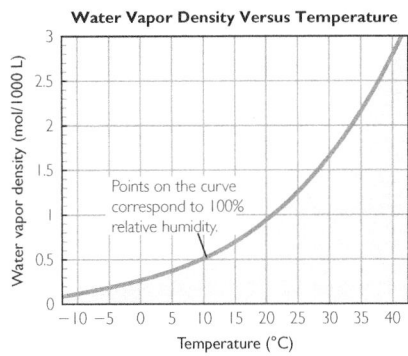

Ask: What does the curve on the graph of water vapor density versus temperature represent? What do points in the shaded area represent?

Ask: Is it possible for there to be any values above the curve, say 3 moles of water vapor per 1000 L of air at 30 °C? Explain.

Ask: Once the air has reached a maximum water vapor density for the temperature of the air, what happens?

Key Points: There is an upper limit to the amount of water vapor that can be present in the atmosphere at a given temperature. If you look at the graph, the maximum amount of water vapor

temperature difference between the bulbs, the moister the air. If the surrounding air is dry, more moisture evaporates from the wick, cooling the wet-bulb thermometer more. This creates a greater difference between the temperatures of the two thermometers. If the surrounding air is at a relative humidity of 100%, there is no difference between the two temperatures.

Meteorologists have worked out charts of these differences for each degree of temperature so that observers can find relative humidity easily.

EVALUATE (5 min)

Check-In

On a hot summer day, a firefighter records a dry-bulb temperature of 30 °C and a wet-bulb temperature of 12 °C. What does this tell you about the relative humidity?

Answer: The difference in the two bulb temperatures is 18°. This number is not even on the table, but it is apparent that the air is quite dry. There is very little water vapor in the air, perhaps as low as 4% relative humidity.

Homework

Assign the reading and exercises for Weather Lesson 66 in the student text.

LESSON 66 ANSWERS

1. Humidity is a measure of the number density of molecules of water vapor in the air.

2. Relative humidity is a percentage comparison of the actual number density of water molecules in the air to the maximum possible number density of water molecules in the air at a certain temperature.

3. About 1.7 mol per 1000 L

4. No, according to the graph of water vapor density versus temperature, the maximum possible number density of water molecules in air at 40 °C is about 2.8 mol per 1000 L. Water will condense if the vapor density exceeds that value.

5. 2.2 mol per 1000 L

6. 18%

7. 0.43 mol per 1000 L

8. Both vapor densities can be 100% relative humidity because the maximum vapor density is a function of temperature. The value for 100% relative humidity is

LESSON SUMMARY

What is humidity and how is it measured?

Humidity refers to the amount of water vapor in the air. It is a measure of the number density of water molecules. Air temperature affects how much water vapor can be in the air. Warmer air can contain more water vapor than colder air. There is an upper limit to the amount of water vapor that can be in the air at any given temperature. This upper limit is called maximum humidity, or 100% relative humidity. At 100% relative humidity, no more water can evaporate into because the air is saturated. You might expect rain or fog when the relative humidity is close to 100%.

KEY TERMS

humidity
partial pressure
relative humidity

Exercises

Reading Questions

1. What does humidity measure?
2. Explain what is meant by relative humidity.

Reason and Apply

3. What is the maximum vapor density possible at 30 °C?
4. Is it possible for the water vapor density in the air to reach 10 mol per 1000 L at a temperature of 40 °C? Explain why or why not.
5. Use the graph Water Vapor Density Versus Temperature to predict the water vapor density at 100% humidity for 35 °C.
6. What is the relative humidity if there is 0.5 mol of water vapor per 1000 L of air at 40 °C?
7. When the humidity is 25% and the temperature is 30 °C, what is the water vapor density?
8. On one day, 100% humidity corresponds to 1.9 mol of water vapor per 1000 L of air. On another day, 100% humidity corresponds to only 1.0 mol of water vapor per 1000 L of air. How can two different water vapor densities both be at 100% humidity?
9. Explain why it is easy to get dehydrated when you are exercising at a temperature of 0 °C and a relative humidity of 20%.
10. Do you feel cooler at 100% humidity at 30 °C, or at 50% humidity at 30 °C? Explain your thinking.
11. Suppose the humidity is 65% during the day when the temperature is 30 °C.
    a. If the temperature drops to 25 °C at night, do you expect fog? Explain.
    b. If the temperature drops to 20 °C at night, do you expect fog? Explain.
12. On a cold winter day the relative humidity is 50% outdoors but only 5% indoors. Which answer best explains what is going on?
    (A) It must be raining outside.
    (B) It must be snowing outside.
    (C) The heater is on inside, so the same humidity is a lower relative humidity.
    (D) The heater is on inside, and it is evaporating some of the humidity in the air.

equal to the maximum vapor density for the particular temperature. The temperature on the first day is about 33 °C and the temperature on the other day is about 20 °C.

9. Possible answer: Dehydration can occur quickly during exercise when the relative humidity is very low because sweat evaporates quickly at a low relative humidity. At low relative humidity, liquid water evaporates more rapidly than it does at a higher relative humidity.

10. Possible answer: People feel cooler at 50% humidity because more water evaporates from the skin, absorbing energy as it evaporates. At

100% humidity, the rate of evaporation is slower, equal to the rate of condensation, and heat energy is transferred away from the skin more slowly.

11. The water vapor density in the air during the day is 1.1 mol/1000 L
a. The maximum water vapor density at 25 °C is about 1.3 mol per 1000 L, so fog will not form. **b.** The maximum water vapor density at 20 °C is about 0.9 mol per 1000 L, so fog will form as water vapor condenses out of the gas phase.

12. C

Hurricane!

Extreme Physical Change

THINK ABOUT IT

Water and air provide our planet with its weather and support plant and animal life. Weather itself is a result of a dynamic interplay of physical change between water, Earth's atmosphere, and energy from the Sun. There are times when these physical changes create conditions that can be catastrophic. A hurricane is one example of an extreme weather phenomenon that can be quite destructive.

What are hurricanes and what causes them?

To answer this question, you will explore

- The Properties of Hurricanes
- Hurricane Formation
- Global Warming

EXPLORING THE TOPIC

The Properties of Hurricanes

A *hurricane* is an enormous tropical rainstorm with powerful winds that forms over the ocean. It has a distinctive shape to it, resembling a giant rotating pinwheel. The wind speeds in a hurricane are over 75 mi/h and may even exceed 150 mi/h. The storm may bring torrential rains, destructive winds, and flooding to areas of land that it passes over. The sheer size of these storms is shown in this photo of a hurricane in the Gulf of Mexico.

These giant storms may be from 125 to 1000 mi across and 15 mi high. Further, they are characterized by very low air pressure at their centers. In general, the lower the air pressure, the larger the storm. Consider the data in the table for three hurricanes in the same year.

Hurricane name	Year	Wind speed (mi/h)	Pressure (atm)	Category
Hurricane Rita	2005	173	0.885	5
Hurricane Ike	2008	145	0.923	4
Hurricane Sandy	2012	115	0.928	3

Overview

CLASSWORK: INDIVIDUAL

Key Question: What are hurricanes and what causes them?

KEY IDEAS

Hurricanes form as a result of specific weather conditions: very warm oceans, humid air, and converging winds. A continuing cycle of evaporation and condensation drives the hurricane by setting up a large air pressure differential. The gas-to-liquid phase change fuels hurricanes. This phase change is directly related to air and ocean temperatures. Concerns have been raised that global warming may

be increasing the frequency and/or intensity of hurricanes.

LEARNING OBJECTIVES

- Describe the meteorological conditions that result in a hurricane.
- Explain the role of phase change, air pressure, and temperature in hurricane formation.
- Define climate and global warming.

IN CLASS

Students complete a worksheet about the anatomy and physical characteristics of a hurricane. The conditions that affect the formation and intensity of hurricanes are also explored, with a focus on changes in temperature. The discussion

and the reading in the student text make connections between hurricanes and global warming. If you are pressed for time, Part 1 is optional. A complete materials list is available for download.

TRM Materials List 67

SETUP

If you have the necessary technology in your classroom, a wealth of satellite images and footage of hurricanes is available from NASA and other organizations.

Pre-AP® Course Tip

Students preparing to take an AP® chemistry or college chemistry course should be able to draw connections to topics outside chemistry. In this lesson, students discuss the meteorological conditions that result in a hurricane and the role of phase change, air pressure, and temperature in hurricane formation.

Lesson Guide

ENGAGE (5 min)

TRM PowerPoint Presentation 67

ChemCatalyst
1. What is a hurricane? What characteristics does it have?
2. Where do hurricanes form?

Sample answers: 1. Hurricanes are large, spinning storms that bring high winds and lots of rain. When they move over landmasses, they can do tremendous damage. 2. Hurricanes form over the ocean.

Ask: Where and when do hurricanes form?

Ask: What makes a hurricane destructive?

Ask: In what direction do hurricanes turn? What explains this observation?

Ask: What is the difference between a hurricane and a large rainstorm? A tornado?

Ask: What is the "eye" of the hurricane?

EXPLORE (15 min)

TRM Worksheet with Answers 67

TRM Worksheet 67

INTRODUCE THE CLASSWORK

→ Let students know they will be working individually.

→ Pass out the student worksheet.

EXPLAIN & ELABORATE (20 min)

INTRODUCE HURRICANES AND THEIR GENERAL CHARACTERISTICS

Ask: What is a hurricane?

Ask: How are hurricanes ranked?

Key Points: Hurricanes are destructive storms characterized by strong winds and large amounts of rainfall. Hurricanes form only over very warm ocean waters—at least 80 °F. For this reason, hurricane season in the United States tends to be at the end of the hot summer months, from June to November, when oceans near the equator are warmest.

Tropical depressions can build to tropical storms, which can build to hurricanes. A future hurricane first develops as a tropical depression, a clearly defined low-pressure system with winds below 38 mi/h. Some tropical depressions continue to build and become tropical storms, with lower air pressure and higher winds. A tropical storm that builds to wind speeds greater than 75 mi/h is considered a hurricane.

There are five categories of hurricane, with category 1 the least intense and category 5 the most intense. Hurricanes are ranked by their wind speeds. Any storm with winds greater than 150 mi/h is considered a category 5 hurricane. Because hurricanes often move from water to land, they are accompanied by a surge in the ocean waters at the coastline. This wave of water, called a *storm surge,* can reach 20 ft or higher. Hurricanes are not to be confused with tornadoes. Tornadoes are violent rotating windstorms that form over land and affect a much smaller area.

DISCUSS THE FORMATION OF HURRICANES

→ Display the transparency Anatomy of a Hurricane.

→ As you discuss the formation of hurricanes, show some of the online satellite images and footage (optional).

Hurricanes are categorized according to their strength, as indicated in the table. They are accompanied by a surge in coastal waters called a *storm surge.* A storm surge can reach 20 ft or more in height, causing major coastal flooding.

⟳ Hurricane Formation

Hurricanes form only over very warm ocean waters of at least 80 °F. In addition, they require a great deal of moisture in the air. For these two reasons, hurricanes originate southeast of the United States, in the tropical ocean waters near the equator.

You have probably heard people refer to "hurricane season." Most hurricanes form during summer and fall when the waters are warmest in tropical zones. For the United States, hurricane season stretches from June through November.

A storm must grow through several stages before it is considered a hurricane. A future hurricane starts out as a tropical depression, which is a clearly defined low-pressure system with winds below 38 mi/h. Some tropical depressions continue to build and become tropical storms. In tropical storms, the air circulates around a low-pressure system with winds between 38 and 74 mi/h. Finally, a tropical storm that builds to wind speeds beyond 75 mi/h is considered a hurricane.

It can take several days for a tropical storm to develop into a hurricane. It begins with moist warm air from the ocean surface rising rapidly. The water vapor in this warm air condenses and forms storm clouds. The heat released by this condensation warms the cooler air above it, causing the air to rise even more. This process continues, with more and more warm air moving up into the developing storm. A spiral wind pattern begins to develop, and the hurricane takes on its characteristic shape. In the very center of the storm, an "eye" develops and cool air descends, creating a calm storm-free area. The eye of a hurricane may

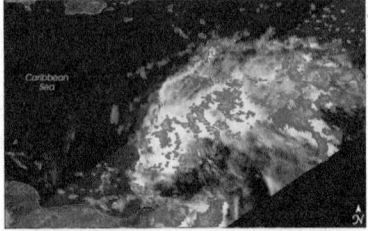
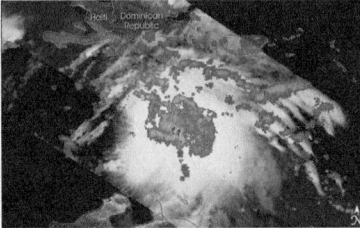

These satellite images show the stages of hurricane formation.

Ask: How does a hurricane form? What conditions are necessary?

Ask: What do the movies of hurricanes tell you about them?

Key Points: Tropical storms begin forming when a great deal of warm water evaporates into the atmosphere. This warm, moist air rises into the atmosphere, where a small change in temperature can cause condensation and cloud formation. Rapid condensation has several outcomes, the most significant of which is to dramatically lower the air pressure in the area. As all these water-vapor molecules become liquid cloud droplets, the value of n drops dramatically, affecting the partial pressure of water vapor in the atmosphere. So a hurricane ends up with very low pressure below and high pressure above. The pressure differential feeds the hurricane further, drawing more air into the storm. This cycle feeds on itself and is responsible for moving massive numbers of molecules, resulting in extraordinarily strong winds. As the storm moves over areas of warmer water, evaporation increases. This also feeds the storm. A hurricane picks up wind speed as it moves over warmer water and loses wind speed over cooler water.

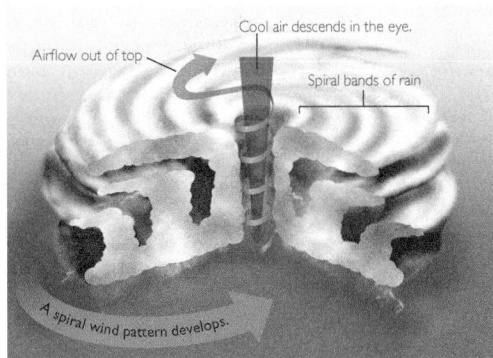

A spiral wind pattern develops.

Airflow out of top

Cool air descends in the eye.

Spiral bands of rain

be between 12 and 60 mi in diameter. When the eye of the storm passes over an area, people will experience calm, clear conditions, as if there is no hurricane at all. The eye is surrounded by the eyewall, which is the most violent part of the storm. Near the ocean surface, spiral bands of rain stretch out for miles.

Global Warming

Earth is currently in a warming phase. Several recent hurricane seasons in the United States were unusually destructive. The number and severity of storms caused scientists to speculate that there is a connection between the increasing number and severity of hurricanes and the warming of the entire planet. In other words, Earth may be undergoing long-term climate change.

Experts think that the warming of the planet is worsened by our practice of using petroleum products for our energy needs. Burning gasoline, coal, and natural gas releases enormous amounts of carbon dioxide into our atmosphere. Increases in carbon dioxide have jumped up rapidly in recent years. This increase causes a greater amount of heat to be trapped in the Earth's atmosphere and is linked to an increase in average air temperature.

The effects of global warming can be seen in the loss of habitat for polar animals.

Dramatic effects of climate change are shown in these two photos of Glacier Bay National Park, taken 60 years apart.
Field, W. O. 1941. Muir Glacier. From the Glacier Photograph Collection. Boulder, Colorado USA: National Snow and Ice Data Center/World Data Center for Glaciology. Digital media. (Left)

Water vapor density is related to temperature. As students saw in the previous lesson, the graph of these two variables is a curve that gets quite steep as the temperature increases. The steepness of this curve means that at warmer temperatures, small increases in temperature can have a dramatic effect on water vapor density. Thus, the warmer the temperature of the air and the ocean, the greater the chance that optimal conditions will exist for hurricane formation.

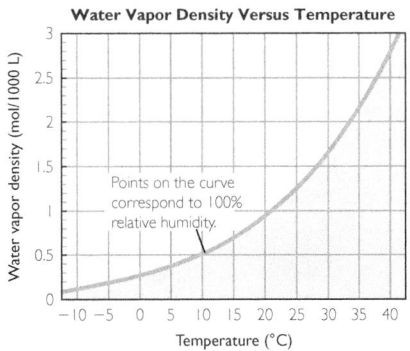

Water Vapor Density Versus Temperature

Points on the curve correspond to 100% relative humidity.

DISCUSS GLOBAL WARMING (OPTIONAL)

→ Display the transparency Global Warming Trends showing temperature changes on the planet since 1880.

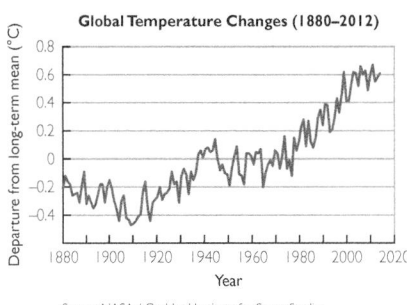

Global Temperature Changes (1880–2012)

Source: NASA / Goddard Institute for Space Studies.

Ask: What is global warming?

Ask: Why is global warming a potentially controversial subject?

Key Points: There is a great deal of scientific evidence supporting the idea that our planet is in a warming cycle, often referred to as *global warming.* Many experts also hypothesize that the burning of so many petroleum products on our planet is a huge contributor to this warming. Using gasoline, coal, heating oil, and so on, as fuels releases enormous amounts of carbon dioxide gas into the atmosphere. Increases in atmospheric carbon dioxide have jumped rapidly in recent years. This increase causes a greater amount of heat to be trapped in Earth's atmosphere and is linked to an increase in average air temperature.

Meteorologists hypothesize that increases in ocean temperatures of only a few degrees may have dramatic effects on the planet's weather. A change of 1 °F may seem insignificant, but it has a dramatic effect on the water on the planet by raising ocean levels by a few inches and causing more rainfall and stronger storms. Polar ice is melting at an accelerated rate. Sea level has risen 4–8 in over the past century. Both the number and the intensity of hurricanes have increased.

Scientists hope energy conservation will help to reverse or slow down this warming trend. An example of a step in this direction is the active involvement of automobile makers in creating vehicles that run on alternative energy sources.

EVALUATE (5 min)

Check-In

Why do most hurricanes have their origins near the equator?

Answer: Hurricanes form as a result of warm oceans and moist air, conditions prevalent in the ocean waters near the equator.

Homework

Assign the reading and exercises for Weather Lesson 67 in the student text. Assign the Chapter 12 Summary to help students review for the quiz.

LESSON 67 ANSWERS

1. *Possible answer:* For a hurricane to form, the temperature at the ocean's surface must be at least 80 °F and the air must contain a lot of moisture.

2. A good paragraph should include: ● a description of three or more plausible changes to Earth's environment that could result from global warming ● a description of how life on Earth is affected by the changing environment and how people have adapted to the new environment ● a writing style that invokes vivid imagery in the reader's mind, instead of merely presenting technical facts and observations. Paragraphs may focus on human life or on other forms of life. The paragraphs will vary based on assumptions made about the adaptability of human life, other forms of life, and the technologies that will become available. However, realistic scenarios that could result from global warming should be presented. Possible scenarios include: ● more extreme weather conditions, such as larger storms and more frequent hurricane formation, ● extensive drought and desertification, ● flooding of coastal cities and a global rise in sea level, ● melting of polar ice caps, ● changes in wind patterns and ocean currents that lead to dramatic changes in local climate, ● changes in wildlife habitat and in biodiversity, ● changes in crop yields and available land for farming a particular crop or farming in general.

3. Answers will vary according to the data that are used. Data should indicate that wind speed in a hurricane increases as the surface water temperature increases. Data can take the form of a series of satellite images, a case study of one or more hurricanes, or a table or chart of wind speed versus temperature. Data should come from a reliable source such as the National Weather Service, a weather forecasting service, or a university Web site.

4. A complete answer will list three pieces of evidence and include for each piece of evidence a description of the evidence as well as the way in which the evidence is an indicator of global warming. Possible evidence of global warming includes: ● changes in ocean surface temperature, ● changes in atmospheric temperature, ● a loss of polar ice and glacier ice, ● rapidly changing climate patterns and habitat losses (both terrestrial and oceanic), ● the poleward migration of ecosystems, ● increases in greenhouse-gas levels in the atmosphere.

GEOLOGY CONNECTION

This drawing is a depiction of the projected new coastline of the eastern United States if the polar ice caps melt completely to water.

Meteorologists hypothesize that increases in ocean temperatures of only a few degrees may have dramatic effects on the weather of our planet, including the possibility of increased frequency and intensity of hurricanes. One degree may not sound like much, but it is sufficient to raise ocean levels by a few inches and to cause more rainfall and stronger storms.

Studies show a steady increase in global temperature over the past 120 years. Scientists hope energy conservation will help to reverse or slow down this warming trend. Automobile makers are actively involved in creating vehicles that run on alternative energy sources.

LESSON SUMMARY

What are hurricanes and what causes them?

A hurricane is a large and powerful tropical storm with intense spiraling winds. Hurricanes form over the warm tropical waters near the equator. At the very center of a hurricane is an area of extremely low air pressure. The heat released when moist warm air rises and condenses into storm clouds powers hurricane formation. Also, the low pressure caused by condensation draws more air into the storm from areas of relatively higher pressure. This cycle of evaporation and condensation of very large amounts of water is a key feature of hurricanes. Experts studying climate changes due to global warming are monitoring data about increases in the frequency and intensity of hurricanes.

Exercises

Reading Questions

1. What conditions are necessary for hurricane formation?
2. Write a creative paragraph describing what life on the planet may be like if global warming continues at its present pace.

Reason and Apply

3. PROJECT   Find data showing that ocean temperature is related to the wind speed of a hurricane.

4. PROJECT   Discuss three pieces of evidence that global warming is occurring.

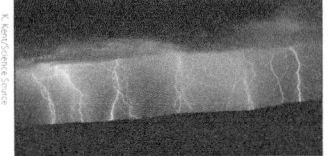

CHAPTER 12

Concentrating Matter

SUMMARY

KEY TERMS

number density

mole

Avogadro's number

standard temperature and pressure (STP)

Avogadro's law

ideal gas law

universal gas constant, R

humidity

partial pressure

relative humidity

Weather Update

Air pressure is directly related to the number density, or number of particles per unit of volume, of the gas molecules in a sample of air. Avogadro's law states that two gas samples have the same number of molecules if temperature volume and pressure are the same. Avogadro's law and the combined gas law together form the ideal gas law. The ideal gas law relates pressure, volume, temperature, and the number of moles by a proportionality constant. This constant, R, is called the universal gas constant.

Extreme weather can occur if conditions support rapid evaporation of water molecules followed by rapid condensation of these molecules.

The amount of water vapor in the air plays a big role in determining the weather. Water vapor density, or humidity, is dependent on air temperature and air pressure. When air is at 100% humidity, chances are it is raining, snowing, or densely foggy because the air is saturated with water molecules.

REVIEW EXERCISES

1. Explain how you can determine the number of molecules in a breath of air.

2. What type of weather do you predict if the relative humidity is 100%?

3. Suppose that you have 22.4 L of helium, He, gas and 22.4 L of neon, Ne, gas at STP. Which of these statements is *false*?
    (A) Both samples contain the same number of atoms.
    (B) Both samples have the same number density, n/V.
    (C) The two samples have different masses.
    (D) The two samples have the same mass density, D.

4. How many moles of nitrogen gas, $N_2(g)$, are contained in 2 liters at 350 K and 1.5 atm?

5. Suppose the water vapor density is 1.5 mol/1000 L at a temperature of 35 °C, and the maximum water vapor density at 35 °C is 2.2 mol/1000 L. What is the relative humidity?

Global Climate Change

 PROJECT   Earth's overall climate is currently in a warming phase. Research some of the causes and effects of global warming on Earth.

• What evidence do scientists have that average global temperatures are increasing?

• What are some possible causes of global warming?

• How might global warming affect the severity and frequency of storms in a region?

• What are some possible effects of global warming on different living things and their environments?

ASSESSMENTS

Two multiple-choice assessments (versions A and B) and two short-answer assessments (versions A and B) are available for download for Chapter 12.

TRM Chapter 12 Assessment Answers

TRM Chapter 12 Assessments

ANSWERS TO CHAPTER 12 REVIEW EXERCISES

1. *Possible answer:* Determine the number of molecules in a breath of air by capturing the air in a flexible container, such as a spherical balloon or plastic bag, so that its volume can be measured. In addition to the volume of air, record the temperature. The pressure of the air sample will be one atmosphere (1 atm). You can then calculate the number of molecules, n, using the ideal gas law, $PV = nRT$.

2. *Possible answer:* Some type of precipitation or fog is likely. If the relative humidity is 100%, the maximum amount of water vapor is in the air for the current temperature.

3. D

4. 0.10 mol

5. 68%

Project: Global Climate Change

A good report will include: ● scientific evidence that global temperatures are increasing, ● three or more possible contributing factors to global warming, ● possible effects of global warming on storms and climate, ● possible effect of global warming on living things and their environments, ● citations for all the sources used for the information, and wherever possible, verification of facts from multiple sources.

Overview

CLASSWORK: INDIVIDUAL

Key Question: What does chemistry have to do with weather?

KEY IDEAS

The weather is driven by the chemistry of phase changes and gas laws and how these relate to the weather.

IN CLASS

Students review what they have learned about phase changes and gas laws. Students work in pairs on a review worksheet, then create a list of review topics for a unit exam. A complete materials list is available for download.

TRM Materials List R3

GUIDE TO UNIT REVIEW

ENGAGE (5 min)

> **ChemCatalyst**
>
> **1.** Describe three different physical changes involved in weather.
>
> **2.** What physical changes of matter affect the weather?

Sample answers: **1.** Phase changes, changing temperature, changing density with temperature, changes in gas pressure, etc. **2.** The phase changes of water cause water to cycle from the land to the air and back to the land. The expansion and contraction of air with changing temperature cause the air to move around the planet.

→ Assist students in summing up how the topics they learned about relate to weather forecasting.

Ask: What does chemistry have to do with the weather?

Ask: What causes matter to move on a global scale?

Ask: How does changing temperature affect water? Gases?

EXPLORE (20 min)

TRM Worksheet with Answers R3

TRM Worksheet R3

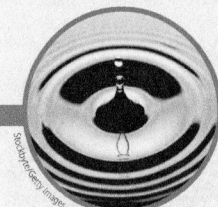

Weather | REVIEW

UNIT
3

The atmosphere surrounding our planet is a mixture of gases, including water vapor. The weather that we experience every day is almost entirely caused by physical changes to these gases. Energy from the Sun causes water molecules to change phase and evaporate into the air. This water vapor moves around, sometimes changing phase again and forming clouds, rain, ice, snow, dew, fog, and so on.

By studying small samples of gas in containers, we can learn more about how gases in the atmosphere behave. The kinetic theory of gases explains the behavior of individual gas particles under varying conditions. For example, when a gas in a rigid container is heated, its pressure increases because the molecules move faster, increasing the force and number of collisions with whatever they contact. When a sample of gas is squeezed into a smaller volume, it also exerts more pressure. This is because the same number of rapidly moving molecules must occupy a smaller space. The pressure, volume, and temperature of a gas sample are related by the combined gas law.

Meteorologists track the variables that affect the gases in the atmosphere. For example, they record water vapor density, or humidity, and areas of differing air pressure. They study air masses and fronts where warm and cold air masses collide. They combine all the information and use it to make predictions about the weather.

REVIEW EXERCISES

General Review

Write a brief and clear answer to each question. Be sure to show your work.

1. Suppose that you have a 500 mL sample of liquid water and a 500 mL sample of ice. The ice has a density of 0.92 g/mL. Which has a greater mass? Explain your reasoning.

2. Convert 25 °F to kelvins. Show your work.

3. Use the kinetic theory of gases to explain why a gas expands when heated and contracts when cooled.

4. Suppose that you fill a balloon with 600 mL of air when the temperature is 27 °C. Later in the day, the temperature drops to 21 °C. Determine the new volume of the balloon.

5. A cylinder with a movable piston contains 2.5 L of gas at a pressure of 2 atm. If the volume is

348   Unit 3 | **Review**

INTRODUCE THE CLASSWORK

→ Let students know they will be working individually.

→ Pass out the student worksheet.

→ Students can use any worksheets, notes, or handouts they have gathered over the course of the unit to complete today's classwork.

EXPLAIN & ELABORATE (25 min)

REVIEW PHASE CHANGES

Ask: What happens on a particulate level when a substance is heated? Cooled?

Ask: Why does hot air rise?

Ask: What do phase changes have to do with the weather?

Key Points: Phase changes are a form of physical change. They are central to weather formation on our planet. Our planet is a watery one, and more than 70% of its surface covered with water in some form. Phase changes cause the transport of water from one part of the planet to another through evaporation and condensation. The Sun is the driving force in causing the water on the planet to change phase from a liquid to a gas. Once water vapor is in the atmosphere, it can easily move over the planet's surface.

decreased to 2.1 L and the temperature stays the same, what is the new pressure of the gas?

6. Which graph represents Boyle's law? Explain your thinking.

A.

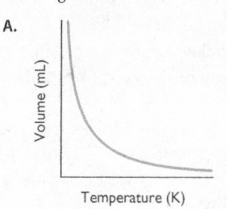

B.

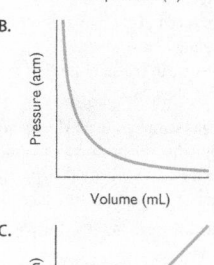

C.

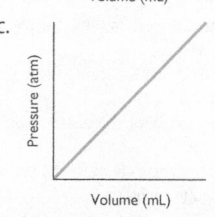

7. A propane gas tank contains propane gas at a pressure of 2.00 atm and a temperature of 30 °C. If the temperature is increased to 37 °C, what is the new pressure of the gas?

8. Suppose that you fill a balloon with 75 L of air at 28 °C and a pressure of 1 atm. You then take the balloon to the top of a mountain where the pressure is 0.8 atm and the temperature is 18 °C. What is the new volume of the balloon?

9. Suppose that you have 22.4 L of nitrogen gas, $N_2(g)$, at STP.
   a. What are the temperature and pressure of this sample?
   b. How many molecules of N_2 are in the sample?
   c. How many nitrogen atoms are in the sample?
   d. What is the number density, n/V, of nitrogen molecules in the sample?

10. Suppose that you have a 750 mL rigid container of hydrogen gas, $H_2(g)$, at 250 K and a 750 mL rigid container of nitrogen gas, $N_2(g)$, at 250 K. A pressure gauge on each container shows that they both have the same internal pressure.
    a. What do you know about the number of individual molecules of gas in each container? Explain your reasoning.
    b. Which container would have the greater mass? Explain your reasoning.

11. Suppose that you have 77.0 L of hydrogen, H_2, gas at a pressure of 3.5 atm and a temperature of 2 °C. How many moles of hydrogen are in this sample?

12. Would you feel cooler if the temperature were 28 °C at 100% relative humidity, or 28 °C at 45% relative humidity? Explain your reasoning.

13. Name the gas laws. Write the formula associated with each one, and list the factors that must stay constant for each law to apply.

STANDARDIZED TEST PREPARATION

Multiple Choice

Choose the best answer.

1. Which option has only units of pressure?
   (A) atm, lb/in², in Hg
   (B) lb/in², in Hg, mL/K
   (C) in Hg, mL/K, atm
   (D) mL/K, atm, lb/in²

Matter changes significantly when it is heated or cooled. When heated, the particles in a substance speed up, causing more collisions and increasing the average kinetic energy of the particles. If heated sufficiently, a substance can change phase, from a solid to a liquid, or from a liquid to a gas. The density of a substance also changes upon heating, usually decreasing because matter usually expands upon heating. When matter is cooled, its particles move more slowly, and the average kinetic energy decreases. Matter usually contracts when cooled, and its density becomes greater. Thermometers are designed to take advantage of the expansion and contraction of matter with temperature.

REVIEW THE KEY CHEMISTRY IDEAS

→ Create a list of topics and terms on the board as a study guide for the exam. Encourage note taking.

Ask: What four variables are tracked for gases?

Ask: What are the names and formulas of the gas laws? What must remain unchanged in the application of each gas law?

Ask: How do you use the proportionality constant, *k*, in solving a gas problem?

Ask: Describe a situation in which you would use each gas law.

Ask: What is the Kelvin scale? Why is it important to use kelvins with the gas laws?

Ask: How does the kinetic theory of gases explain temperature and pressure?

Ask: What is Avogadro's law?

Ask: What does the behavior of gases have to do with the weather?

Ask: What is water vapor density, and how is it measured?

Key Points: It is necessary to understand the behavior of gases if you want to understand the weather. Scientists generally track four variables when they deal with gases: temperature, volume, pressure, and number of gas particles. These variables are all mathematically related. Knowing the mathematical relationship can help you solve problems associated with gases. The temperature of a gas must be converted to the Kelvin scale to use the gas laws. For a specific gas sample, the proportionality constant, *k*, is always the same number. For a different gas sample, *k* is a different number.

Gas Laws

Charles's law $k = \dfrac{V}{T}$

Gay-Lussac's law $k = \dfrac{P}{T}$

Boyle's law $k = PV$

Combined gas law $k = \dfrac{PV}{T}$

Avogadro's law $k = \dfrac{n}{V}$

Ideal gas law $PV = nRT$ or $R = \dfrac{PV}{nT}$

The ideal gas law allows you to figure out problems that involve changes in the number of gas particles, *n*. The ideal gas law is represented by the equation $PV = nRT$, where *R* is a number that relates the different units to each other. For the units we have been using, $R = 0.082$ L · atm/mol · K.

EVALUATE

There is no Check-In for the review.

Homework

Assign Unit 3: Weather Review in the student text to help students prepare for the unit exam.

ANSWERS

GENERAL REVIEW

1. The ice has a lower mass because it has a lower density than liquid water. The density of water is 1.0 g/mL. Mass can be calculated by multiplying the density by the volume, and because the samples have the same volume, the ice has less mass.

2. The formula for converting Fahrenheit temperatures to Celsius temperatures is

$$C = \frac{9}{5}(F - 32)$$

$$K = C + 273$$
$$= \frac{5}{9}(F - 32) = 273$$
$$= \frac{5}{9}(25 - 32) + 273$$
$$= 269 \text{ K}$$

3. According to the kinetic theory of gases, gas particles move faster when they have more energy. When a gas is heated, its particles gain energy, move faster, and scatter farther apart, causing the gas to expand. When the gas is cooled, its particles lose energy, causing them to move more slowly and remain closer together.

4. 588 mL

5. 2.4 atm

6. The middle graph represents Boyle's law. According to Boyle's law, pressure and volume are inversely proportional while temperature stays the same.

7. 2.05 atm

8. 91 L

9. a. 273 K, 1.0 atm (STP) **b.** One mole of any gas occupies 22.4 L at standard temperature and pressure. **c.** Each nitrogen molecule has 2 atoms, so the sample has 2 moles of atoms. **d.** 0.0446 mol/L

10. a. Each container holds the same number of individual molecules of gas. According to Avogadro's law, the number of particles of gas under particular conditions of temperature, volume, and pressure is the same no matter what gas is being considered. **b.** The container of nitrogen would have the greater mass because each container holds the same number of particles and each nitrogen particle has more mass than each hydrogen particle.

11. 12 mol

2. A cylindrical rain gauge contains a volume of 6 mL for a height of 2 cm of rain. What is the volume of rain for a height of 9 cm in this rain gauge?
(A) 9 mL
(B) 18 mL
(C) 27 mL
(D) 54 mL

Use the graph below for Exercises 3 and 4.

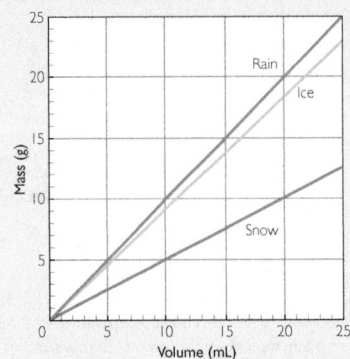

Mass Versus Volume of Water

3. For equal volume samples of snow, ice, and rain, which statement is correct in describing the mass of the samples?
(A) The masses of each will be the same because in each case, the substance is water, H_2O.
(B) The mass of snow will be smallest because the snow has the highest density.
(C) The mass of the ice will be smallest because the ice has the lowest density.
(D) The mass of the rain will be largest because the rain has the highest density.

4. What is the volume of 15.0 g of ice?
(A) 7.5 mL
(B) 13.8 mL
(C) 15.0 mL
(D) 16.3 mL

5. The density of solid nitrogen is 1.026 g/cm³ and the density of nitrogen gas at STP is 0.00125 g/cm³.

What is the volume of gas produced when 50 mL of solid nitrogen sublimes?
(A) 39 L
(B) 41 L
(C) 61 L
(D) 64 L

6. The average temperature at the top of Mt. Denali in Alaska can drop below −50 °F. What is this temperature in Kelvins?
(A) 125 K
(B) 263 K
(C) 227 K
(D) 283 K

7. Which of the following correctly describes the movement of gas particles according to the kinetic theory of gases?
(A) All gas particles move with the same speed at a given temperature.
(B) All gas particles move in curved lines in random directions.
(C) The average speed of gas particles increases with decreasing temperature.
(D) Gas particles are in constant motion at temperatures above 0 K.

8. Imagine that a gas sample is in a cylinder with a piston. Assuming that the pressure and number of molecules stays the same, what will happen to the volume of the gas if the temperature increases?
(A) The volume decreases because the particles are closer together.
(B) The volume decreases because the particles have collided less.
(C) The volume increases because the particles are moving faster.
(D) The volume increases because the particles are less attracted to each other.

9. A balloon is filled with air to a volume of 1.35 L at a temperature of 22 °C. The balloon is taken to a birthday picnic on a sunny day. After several hours in the Sun, the air inside of the balloon was warmed to 32 °C. What is the new volume of the balloon?
(A) 2.0 L
(B) 1.3 L
(C) 1.4 L
(D) 0.93 L

10. What phase change occurs when a cloud forms from water vapor?
(A) Evaporation
(B) Condensation
(C) Melting
(D) Freezing

12. *Possible answer:* People feel cooler at 45% humidity because more water evaporates from the skin, absorbing energy as it evaporates. At 100% humidity, the rate of evaporation is slower, equal to the rate of condensation, and heat energy is transferred away from the skin more slowly.

13. Charles's law: $V = kT$; P, n are constant

Gay-Lussac's law: $P = kT$; V, n are constant

Boyle's law: $P = \frac{k}{V}$; T, n are constant

STANDARDIZED TEST PREPARATION

1. A   8. C   15. A
2. C   9. C   16. B
3. D   10. B   17. C
4. D   11. A   18. A
5. B   12. D   19. B
6. C   13. D   20. C
7. D   14. C

11. A weather balloon expands as it travels to a higher altitude. Which of the following correctly describes the gas particles **inside** of the balloon?

(A) There is more space between the gas particles inside the balloon because the atmospheric pressure is less at higher altitudes.

(B) There is more space between the gas particles inside the balloon because the atmospheric pressure is greater at higher altitudes.

(C) There are more collisions between the gas particles inside of the balloon due to the lower atmospheric pressure.

(D) The average speed of the gas particles inside the balloon increases due to the decrease in temperature at higher altitudes.

12. A 0.75 L balloon has a pressure of 1.0 atm. The balloon is placed inside a vacuum chamber, in which air outside the balloon can be removed. After air has been removed from the chamber, the pressure inside the balloon is reduced to 0.50 atm. The temperature remains unchanged. What is the new volume of the balloon?

(A) 0.38 L   (B) 0.50 L
(C) 0.67 L   (D) 1.5 L

13. Which hypothesis best illustrates the particle view of what occurs when solid carbon dioxide sublimes?

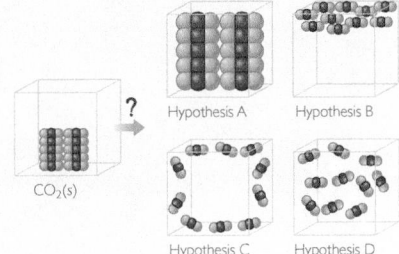

Hypothesis A   Hypothesis B

$CO_2(s)$

Hypothesis C   Hypothesis D

(A) Hypothesis A. The gas molecules individually expand when carbon dioxide sublimes.

(B) Hypothesis B. The gas molecules float to the top of the container when carbon dioxide sublimes.

(C) Hypothesis C. The gas molecules move to the surfaces of the container when carbon dioxide sublimes.

(D) Hypothesis D. The gas molecules move relatively far apart from one another when carbon dioxide sublimes.

14. An 8.5 L scuba tank is filled with air to a pressure of 150 atm at 26 °C. The scuba tank is cooled to a temperature of 15 °C. What is the pressure inside of the tank?

(A) 17 atm
(B) 86 atm
(C) 140 atm
(D) 160 atm

15. Which of the following statements correctly describes the relationship between gas measurements?

(A) When temperature and the amount of gas do not change, the pressure and volume of the gas are inversely proportional.

(B) When pressure and the amount of gas do not change, the volume and temperature are inversely proportional.

(C) When the amount of gas and volume do not change, the pressure and temperature are inversely proportional.

(D) When the temperature and pressure do not change, the number of particles and volume are inversely proportional.

16. A weather balloon is filled with helium gas to a volume of 7500 L at sea level where the pressure is 1 atm and the temperature is 24 °C. The balloon travels to the top of the stratosphere where the temperature is −15 °C and the pressure is 0.001 atm. What is the volume of the balloon while at this altitude?

(A) 4.7×10^6 L
(B) 6.5×10^6 L
(C) 8.6×10^6 L
(D) 1.2×10^7 L

ASSESSMENTS

Two Unit 3 Assessments (versions A and B) are available for download.

TRM   Unit 3 Assessment Answers

TRM   Unit 3 Assessments

A Lab Assessment for Unit 3 is available for download.

TRM   Unit 3 Lab Assessment Instructions and Answers

TRM   Unit 3 Lab Assessment

MIDTERM ASSESSMENTS

Two midterm assessments (versions A and B) covering topics from Units 1-3 are available for download.

TRM   Midterm Assessment Answers

TRM   Midterm Assessments

A midterm lab assessment covering topics from Units 1-3 is available for download.

TRM   Midterm Lab Assessment Instructions and Answers

TRM   Midterm Lab Assessment

Engineering Design Project: Hair-Dryer Hot-Air Balloon U3

As an extension to Unit 3: Weather, you can assign Engineering Design Project: Hair-Dryer Hot-Air Balloon. In this project, students use an engineering design cycle to build out of tissue paper an indoor hot-air balloon that uses a hair dryer as a heat source.

17. Suppose you have two balloons, one filled with helium, He, and one filled with argon, Ar. Each balloon has a volume of 22.4 L at STP. Which of the following correctly describes these samples?

(A) The helium balloon contains fewer gas molecules because the atomic number of helium is less than that of argon.

(B) The helium balloon contains fewer gas molecules because the mass number of helium is less than that of argon.

(C) Each balloon contains the same number of gas particles because the balloons are the same volume, at the same temperature and pressure.

(D) Each balloon contains a different number of gas particles because each balloon is filled with a different gas.

18. Complete the data set for a sample of gas.

Pressure	Volume	Temperature	Moles
0.75 atm	11.2 L	300 K	?

(A) 0.34 mol (B) 0.50 mol
(C) 2.93 mol (D) 244 mol

19. How many moles of N_2 gas are there in a container that holds 11.2 L at 0 °C and 1 atm?

(A) 0.25 mol (B) 0.50 mol
(C) 1.0 mol   (D) 2.0 mol

20. Find the pressure of 2.5 mol of gas if the gas temperature is 32.0 °C and the gas volume is 67.2 L.

(A) 10.2 atm (B) 1.07 atm
(C) 0.930 atm (D) 0.0976 atm

Unit 4 | Toxins

Stoichiometry, Solution Chemistry, and Acids and Bases

TOXINS AS CONTEXT

It is easy to see why the toxins context has such great appeal for students. Toxins can be mysterious and fascinating. What makes some substances so toxic? How do toxins work? Students might be curious about snakebites, poison oak, pollutants in our environment, and drugs and their side effects. Even more mysterious, some substances that are normally considered beneficial, such as iron and vitamin D, can be hazardous in large amounts. Many substances can be harmful or therapeutic, depending on the amount and the mode of exposure.

TOXICOLOGY AS A FIELD OF RESEARCH

Biologists define a toxin as a potentially lethal substance produced by living cells or organisms. For the purposes of this unit, however, we define a toxin as any substance that could potentially be harmful. We have purposefully avoided using the word *poison* because it implies an intention to harm. We want students to understand that every substance, be it water, sugar, or arsenic, is a potential toxin and that toxicity is connected to dosage.

Exposure to toxins can occur in a variety of ways. Toxic substances can be ingested, injected, inhaled, or absorbed. Some toxins, like ricin, a complex protein extracted from castor beans, require very little exposure to produce dramatic effects. Others, like mercury or lead, may accumulate slowly in the body, and their effects may not become apparent for a long time. Likewise, the net effects of toxins on living organisms are myriad, from rashes or difficulty breathing to unconsciousness or death.

It is natural to ask, "How much is too much?" Sometimes, lethal doses and ill effects are discovered accidentally. Sometimes, researchers test the lethal dose of a substance by delivering the compound to a group of test animals such as rats or mice. They use proportions to calculate the lethal dose for humans. However, a dose that kills one person or animal may not kill another. For this reason, toxicologists use the LD_{50}, or the amount that is lethal to 50% of the animals in a sample. The LD_{50} is expressed in units of mass per kilogram of body weight. Using this scale, the larger the LD_{50}, the safer the compound. As a measure of toxicity, the LD_{50} does have its limitations. Nevertheless, the information gleaned from these tests has undoubtedly saved countless lives.

CONTENT DRIVEN BY CONTEXT

Different classes of toxins interact with the body through different mechanisms. In this unit, we concentrate on changes in pH, blockages in the body due to the formation of solids, and exchange reactions that displace ions that the body needs.

These categories cover a surprising percentage of toxic responses. In the process of exploring these mechanisms, students investigate acids and bases, precipitation reactions, and reaction with ionic compounds. An investigation of toxins is intimately connected to amounts, a theme that lends itself well to a study of chemical reactions and stoichiometry.

The unit also includes comparisons of the nature of physical and chemical change. Just as there is no clear distinction between an ionic bond and a covalent bond, there is also some ambiguity in what constitutes a physical change versus a chemical change. Traditional definitions of chemical change cite such things as color changes, heat, and bubbles as evidence that one has occurred. However, it is possible to find physical changes that are accompanied by these same phenomena. A chemical change is traditionally defined as one that creates new substances with new properties. But many physical changes create substances that have strikingly different properties.

THE USE OF MODELS TO EXPLAIN CHANGE

Chemical reactions can be very abstract and puzzling for students without familiar context and a clear picture of what is happening on a particle level. In activities and in the student textbook, we frequently ask students to consider particle views and how the behavior of these particles relates to what is happening on an observable level.

BUILDING UNDERSTANDING

Students have seen a number of chemical and physical changes in earlier units. From Unit 2: Smells, they have a good understanding of molecules and bonding. In Unit 3: Weather, they were introduced to the concept of a mole. They now have the foundation they need to study chemical change and stoichiometry. Although some curricula introduce the study of stoichiometry much earlier in the course, we have found that balancing chemical equations is much more meaningful and understandable for students once they have the requisite background provided by the earlier units and can accurately interpret a chemical equation and what it represents.

Chapter Summary

Chapter	Description	Standard Schedule Days	Block Schedule Days
13	**Chapter 13** kicks off the study of toxic reactions. Students observe chemical and physical changes and come up with a working definition for each. They do a lab that reinforces conservation of mass, followed by an activity that makes the balancing of chemical equations concrete.	7	3.5
14	**Chapter 14** focuses on amount and different ways to measure substances. It is more practical to measure mass, but because of the way toxins work in our bodies, it is important to translate between grams and number.	7	4
15	**Chapter 15** focuses on solution chemistry. A strong connection is made between what we can observe and what is happening on a molecular or ionic level.	5	2
16	**Chapter 16** revisits acids and bases, focusing on categorizing and understanding these two important and complementary classes of substances. Once students have developed a good understanding of factors that affect pH, they carry out an acid-base titration.	7	4
17	**Chapter 17** focuses on solubility and precipitation as they relate to toxins. Precipitation reactions are used to develop the concept of mole ratio in a concrete way. Students practice gram-mole conversions and stoichiometry. By the end of the unit, they will have mastered balancing equations as well as have a solid foundation in interpreting chemical symbols and equations.	8	4.5

Pacing Guides

Standard Schedule

Day	Suggested Plan	Day	Suggested Plan
1	Chapter 13 Lesson 68	18	Chapter 15 Lesson 83, Chapter 15 Review
2	Chapter 13 Lesson 69	19	Chapter 15 Quiz, Project (optional)
3	Chapter 13 Lesson 70	20	Chapter 16 Lesson 84
4	Chapter 13 Lesson 71	21	Chapter 16 Lesson 85
5	Chapter 13 Lesson 72	22	Chapter 16 Lesson 86
6	Chapter 13 Lesson 73, Chapter 13 Review	23	Chapter 16 Lesson 87
7	Chapter 13 Quiz, Project (optional)	24	Chapter 16 Lesson 88
8	Chapter 14 Lesson 74	25	Chapter 16 Lesson 89, Chapter 16 Review
9	Chapter 14 Lesson 75	26	Chapter 16 Quiz, Project (optional)
10	Chapter 14 Lesson 76	27	Chapter 17 Lesson 90
11	Chapter 14 Lesson 77	28	Chapter 17 Lesson 91
12	Chapter 14 Lesson 78	29	Chapter 17 Lesson 92
13	Chapter 14 Lesson 79, Chapter 14 Review	30	Chapter 17 Lesson 93, Chapter 17 Review
14	Chapter 14 Quiz, Project (optional)	31	Chapter 17 Quiz, Project (optional)
15	Chapter 15 Lesson 80	32	Unit 4 Review
16	Chapter 15 Lesson 81	33	Unit Exam
17	Chapter 15 Lesson 82	34	Lab Exam (optional)

Block Schedule

Day	Suggested Plan	Day	Suggested Plan
1	Chapter 13 Lessons 68 and 69	10	Chapter 15 Quiz, Project (optional) Chapter 16 Lesson 84
2	Chapter 13 Lessons 70 and 71	11	Chapter 16 Lessons 85 and 86
3	Chapter 13 Lessons 72 and 73 Chapter 13 Review	12	Chapter 16 Lessons 87 and 88
4	Chapter 13 Quiz, Project (optional) Chapter 14 Lesson 74	13	Chapter 16 Lesson 89, Chapter 16 Review, Project (optional)
5	Chapter 14 Lessons 75 and 76	14	Chapter 16 Quiz Chapter 17 Lessons 90 and 91
6	Chapter 14 Lessons 77 and 78	15	Chapter 17 Lessons 92 and 93 Chapter 17 Review
7	Chapter 14 Lesson 79, Chapter 14 Review, Project (optional)	16	Chapter 17 Quiz, Project (optional)
8	Chapter 14 Quiz Chapter 15 Lessons 80 and 81	17	Unit 4 Review
9	Chapter 15 Lessons 82 and 83 Chapter 15 Review	18	Unit Exam, Lab Exam (optional)

Unit 4 | Toxins

Some animals produce toxins for self-defense, or to paralyze their prey. However, many toxins also have medicinal uses. For example, certain compounds in the venom of this African saw-scaled viper can help to treat a person having a heart attack.
© Tony Phelps/naturepl.com

© Charles D. Winters/Science Source

In this unit, you will learn:

- how toxins are defined
- how chemists determine toxicity
- the mechanisms by which toxic substances act in our bodies and what this has to do with chemical reactions

Why Toxins?

Chemical reactions help our bodies to process food and create new tissues. However, some chemical reactions have toxic and harmful outcomes. The toxicity of a substance is highly dependent on the dose size of that substance. Sometimes, a small amount of a compound, such as a vitamin, can be therapeutic, but a large amount can damage your health. This unit investigates chemical changes by exploring how toxic substances are measured and tracked through their transformations.

353

PD **CHAPTER 13 OVERVIEW**

Watch the video overview of Chapter 13 (for teachers) by clicking on the link in the TE-book, opening the TRFD, or logging onto the book's companion Web site bcs.whfreeman.com/livingbychemistry2e (teacher log-in required).

The first chapter of the Toxins unit more closely examines chemical equations. In Lesson 68, students interpret chemical equations and predict what they will observe when the reactions are carried out. Students use a set of Toxic Reactions cards, which they sort according to patterns they discover. As an ongoing assignment, students may each be assigned a toxic substance to explore over the course of the entire unit. In Lesson 69, students predict the outcome of a series of chemical reactions by examining the corresponding equations and then test their predictions in the lab. In Lesson 70, students compare chemical change and physical change and contemplate the gray area between the definitions. Students learn how the type of change is reflected in a chemical equation. Lesson 71, another lab, provides firsthand experience with conservation of mass and chemical change. In Lesson 72, students learn how to balance chemical equations using connecting blocks as manipulatives to represent atoms. Finally, in Lesson 73, students learn to classify reactions according to the type of change that takes place.

In this chapter, students will learn

- how chemical equations are related to observations
- to interpret chemical equations and make predictions
- to compare chemical and physical change
- that mass is conserved during chemical change
- how to balance chemical equations
- about combination, decomposition, and single and double displacement reactions

Ame v.d. Wolde/Shutterstock

CHAPTER 13

Toxic Changes

In this chapter, you will study

- how to write, balance, and interpret chemical equations
- definitions of chemical and physical change
- how to classify chemical reactions
- what happens to mass when a chemical change occurs

354

Biologists define toxins as harmful substances that come from living organisms, like these fly agaric mushrooms.

Chemical changes are necessary for life. However, some substances interact with the body in ways that have unhealthy effects. To monitor and understand toxic interactions, it is necessary to use chemical equations to track changes in matter.

Chapter 13 Lessons

Toxic Reactions

Chemical Equations

THINK ABOUT IT

In the world around you, matter undergoes many changes. Some of these changes have very little impact on living things on the planet. Other changes may be vital and necessary for life. Still other changes may threaten the health and well-being of plant and animal life. Chemists use chemical equations to keep track of all types of changes in matter, including those that are beneficial and those that are unsafe.

How do chemists keep track of changes in matter?

To answer this question, you will explore
- Chemical Equations
- Toxic Substances and Their Effects

EXPLORING THE TOPIC

Chemical Equations

A chemical equation is a chemical "sentence" that describes change, using numbers, symbols, and chemical formulas. Chemical equations describe what happens when a single substance is changed, or when two or more substances are combined and a change occurs. Once you understand how to decode chemical equations, you will be able to use them to predict what you might observe when substances are mixed.

INTERPRETING A CHEMICAL EQUATION

In some reactions that take place in your body, the element chromium is safe and even necessary. In other reactions, it is toxic. Consider what happens if you ingest chromium metal and it reacts with the hydrochloric acid in your stomach. A chemical equation can help you decode this reaction.

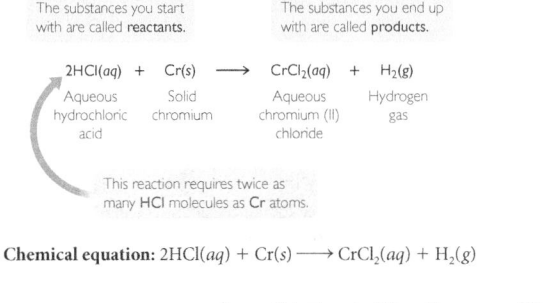

The substances you start with are called **reactants**.

The substances you end up with are called **products**.

$$2HCl(aq) + Cr(s) \longrightarrow CrCl_2(aq) + H_2(g)$$

Aqueous hydrochloric acid

Solid chromium

Aqueous chromium (II) chloride

Hydrogen gas

This reaction requires twice as many **HCl** molecules as **Cr** atoms.

Chemical equation: $2HCl(aq) + Cr(s) \longrightarrow CrCl_2(aq) + H_2(g)$

Lesson 68 | **Chemical Equations**   355

---

know about toxins. You then perform a demonstration to relate a chemical reaction to a chemical equation. During the activity, pairs of students are provided with sets of Toxic Reactions cards, and they are challenged to sort the cards in various ways, looking for patterns in the reactions. Over the course of this unit, five toxins projects are provided, one at the end of each Chapter Review. You can ask students to report their findings at the end of the unit. A complete materials list is available for download.

TRM Materials List 68

SETUP

Before class, prepare the substances needed and run the demo for your own benefit. For better viewing, you might want to run the reaction in a large test tube held in place with a test-tube holder.

> **SAFETY**
> - Hydrochloric acid is corrosive. Wear safety goggles and have baking soda on hand for spills.

CLEANUP

Dispose of demo products according to local guidelines.

Lesson Guide

ENGAGE (5 min)

TRM PowerPoint Presentation 68

> *ChemCatalyst*
> **1.** What toxins have you encountered in your life?
> **2.** How can toxins enter the body?
> **3.** How can toxins harm you?

Sample answers: **1.** Cleaning solutions, cigarette smoke, lead in paints, carbon monoxide, poison mushrooms, *E. coli*, viruses, and bacteria. **2.** Toxins can be inhaled, ingested, or absorbed through the skin and eyes. **3.** Some toxins act immediately and can cause death, make you sick, cause skin rashes, sting eyes, and so on. Some toxins, such as substances that cause cancer, act over a much longer period of time.

Ask: What experiences have you had with toxins?

Ask: What are some ways you come in contact with toxins?

Overview

DEMO: WHOLE CLASS

ACTIVITY: PAIRS

Key Question: How do chemists keep track of changes in matter?

KEY IDEAS

Some chemical reactions support life, yet others negatively affect the health and well-being of living things. Toxins are substances that react chemically with the human body, causing some sort of harm or pain. Chemical equations help us keep track of both the beneficial and the dangerous changes in matter that occur around us.

LEARNING OBJECTIVES

- Complete basic translations of chemical equations.
- Give a basic definition of a toxin.

FOCUS ON UNDERSTANDING

Students may have their own preconceived notions about toxins. Although the dictionary typically defines a toxin as a poisonous substance produced by living cells or organisms, we have adopted a broader definition of what constitutes a toxin. This definition will be developed over a series of lessons.

IN CLASS

During the ChemCatalyst, students are given a chance to discuss what they

COMPLETE THE DEMONSTRATION

→ Write the chemical equation on the board. Tell students that this chemical "sentence" describes a chemical reaction. Ask students to predict what they think the symbols and numbers mean.

$$HCl(aq) + NaHCO_3(aq) \rightarrow NaCl(aq) + H_2O(l) + CO_2(g)$$

Follow the procedure. Relate what you are doing to the chemical equation.

1. Put on goggles.
2. Use a graduated cylinder to measure 25 mL of 3.0 M sodium bicarbonate, $NaHCO_3$. Pour the $NaHCO_3$ into a 250 mL beaker.
3. Use a second cylinder to measure 25 mL of 3.0 M hydrochloric acid, HCl. Slowly add the HCl(aq) to the sodium bicarbonate in the beaker. Ask students to describe what they observe. (bubbling, a clear colorless solution when reaction is complete)

→ Ask students what they think is present in the beaker after the reaction. (dissolved salt)

→ Ask students how they could prove that NaCl is a product.

→ Place half the solution on a hot plate set on low heat to boil away the water while students complete the activity. Keep the other half of the solution as is for comparison. After the activity, show students the white, solid NaCl that is left in the container after the water has evaporated.

INTRODUCE THE ACTIVITY

→ Arrange students into pairs.

→ Pass out the student worksheet.

→ Leave the chemical equation from the demonstration on the board while students complete the activity. It is used again in the Explain & Elaborate section.

→ This is a two-part activity. In Part 1, students examine the chemical equation from the demo.

Interpretation: Hydrochloric acid reacts with solid chromium to produce a solution of chromium (II) chloride and bubbles of hydrogen gas.

A chemical equation can help you anticipate what you will observe when the reactants are combined. Examine the same equation, but focus on what you expect to observe.

$$2HCl(aq) + Cr(s) \longrightarrow CrCl_2(aq) + H_2(g)$$

According to the equation, when hydrochloric acid and chromium react, the solid chromium will disappear. You would expect to see the formation of a new aqueous solution as well as some evidence that a gas was produced.

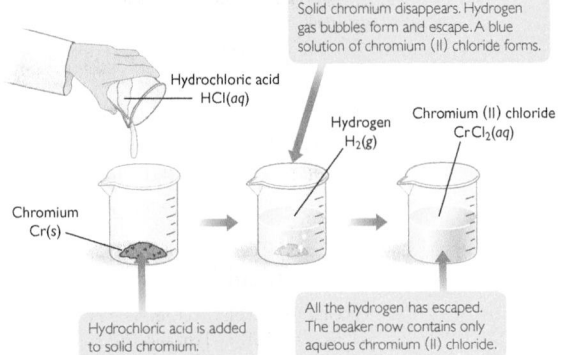

Sometimes, the changes that take place when substances are mixed are not visible to the eye. For example, death from poisoning occurs within minutes of swallowing a solution of sodium cyanide, NaCN. One successful treatment for this type of poisoning is injection of an *antidote*, a remedy that counteracts the poison. The antidote for sodium cyanide is a solution containing sodium thiosulfate, $Na_2S_2O_3$. But if you were to observe the reaction between sodium cyanide and sodium thiosulfate in a beaker, you would not be able to tell that a reaction had occurred because both reactants and both products are clear, colorless liquids.

Big Idea Chemical equations keep track of changes in matter.

↻ Toxic Substances and Their Effects

Toxic substances enter the body in limited ways. The most common methods are through ingestion (eating or swallowing), inhalation (breathing something in), or contact with the skin. Once toxic substances enter the body, they react in a variety of ways.

Toxic substances can react with water in mucous membranes, with oxygen carried through the blood, or with stomach acid. Some toxic substances have an immediate negative effect on the well-being of a person or living thing. Other

→ Students explore the Toxic Reactions cards in Part 2. Pass out one set of cards to each pair of students.

EXPLAIN & ELABORATE (15 min)

PROCESS THE REACTION FROM THE DEMONSTRATION

Refer students to the equation from the demonstration written on the board:

$$HCl(aq) + NaHCO_3(aq) \rightarrow NaCl(aq) + H_2O(l) + CO_2(g)$$

→ Show students the beaker containing the products of the reaction, which were evaporated to dryness. The solid salt, NaCl, should be visible.

Ask: What substances are formed by this reaction?

Ask: Can you tell what was made by observation alone? Explain.

Ask: How can you tell that sodium chloride is in the water if you cannot see it?

Ask: Would it be correct to put NaCl(s) on the right side of the equation? Why or why not?

Key Points: The chemical equation represents a change in matter using symbols and formulas. It shows the relative amounts of reactants and products and if the reactants and products are solid, liquid, or aqueous. In the demonstration, two liquids were mixed together, producing gas bubbles and a

toxic substances may stay a long time in the body, becoming part of the body's chemistry, perhaps damaging the body many years later.

Toxic substances may be molecular, ionic, or metallic substances. Many small molecular substances react with the body and create acid products, which can damage and irritate tissue or upset the acidity of the blood. Toxic metals often react to form ionic compounds, which move throughout the body and compete with "good" metals that are useful to the body. Some ionic compounds form solids that clog the body's filtering systems. Of course, these are just a few of the possible types of toxic reactions that occur. Chemical equations are the main tool you will use to track these changes.

LESSON SUMMARY

How do chemists keep track of changes in matter?

A chemical equation tracks changes in matter. The left side of the equation contains the chemical formulas for the reactants or the substances that are being combined. The right side of the equation contains the chemical formulas for the products or the substances that are produced. The equation also shows the phase of each reactant and product. Decoding an equation allows you to predict what substances may be made when the reactants are combined. Chemical equations often provide more information than what you can observe with your senses.

Exercises

Reading Questions

1. What is the difference between a reactant and a product?

2. Are chemicals and chemical reactions important for life? Why or why not?

3. Describe in your own words what a toxic substance is.

Reason and Apply

4. Both bleach and ammonia are used for cleaning. However, it is very dangerous to mix bleach with ammonia because they react to produce sodium hydroxide and the toxic gas chloramine.

$$NaOCl(aq) + NH_3(aq) \longrightarrow NaOH(aq) + NH_2Cl(g)$$

   a. Write an interpretation of the chemical equation.
   b. What do you expect to observe?

5. Poisoning with mercury chloride can be reversed by chelation therapy. The chelating agent called EDTA, $C_{10}H_{16}N_2O_8$, is injected into the bloodstream. EDTA forms a water-soluble compound with mercury ions, allowing removal from the body through the kidneys.

$$HgCl_2(s) + C_{10}H_{16}N_2O_8(aq) \longrightarrow HgC_{10}H_{12}N_2O_8(aq) + 4HCl(aq)$$

   a. Write an interpretation of the chemical equation.
   b. What do you expect to observe?

6. Describe at least three types of effects that a toxic substance can have on the body.

Toxins often react with water in the human body. Our bodies are about 60% water. Many toxins react with water found in the mucous membranes, such as in the eyes, nose, and lungs. The stomach contains very strong aqueous hydrochloric acid that can dissolve toxins. Toxins can be carried throughout the body in the bloodstream, which also contains a lot of water. Some toxins have an immediate, negative effect while others may persist for a long time in the body, becoming part of the body's chemistry.

Toxins may be molecular, ionic, or metallic substances. Many small molecular substances interact with the body by creating acids, which either damage and irritate tissue or alter the acidity of the blood. Toxic metals often react to form ionic compounds, which move throughout the body and compete with "good" metals that are useful to the body. Some ionic compounds will form solids that clog the body's filtering systems.

EVALUATE (5 min)

Check-In

Consider this reaction between sodium cyanide and a solution of hydrochloric acid:

$$NaCN(s) + HCl(aq) \rightarrow NaCl(aq) + HCN(g)$$

   a. Write an interpretation of the chemical equation.

   b. Sodium cyanide is highly toxic. What is the most likely way it will enter the body?

Answers: **a.** Interpretation: Solid sodium cyanide reacts with a solution of hydrochloric acid to produce a solution of sodium chloride and hydrogen cyanide gas. **b.** Sodium cyanide is a solid, so it most likely will enter the body by ingestion. It will react with the hydrochloric acid in the stomach.

Homework

Assign the reading and exercises for Toxins Lesson 13 in the student text.

Assign the Toxins Web Research Project in the Toxins Chapter 13 Summary in the student text (optional). You can assign the toxins by handing out one Toxic Reactions card to each student.

LESSON 68 ANSWERS

Answers continue in Answer Appendix (p. ANS-6).

liquid. The chemical equation indicates that carbon dioxide gas was produced along with a solution of sodium chloride. The sodium chloride is not visible to the eye when it is in solution. Thus, to label it as a solid, (s), would be incorrect. One way to provide evidence that the sodium chloride is present in the solution is to boil away the liquid, leaving behind solid sodium chloride.

DISCUSS PATTERNS FOUND IN THE TOXIC REACTIONS CARDS

→ Fill in the transparency Toxic Reactions Patterns as students tell you about the patterns they discovered. Download the Transparency—Toxic Reactions Patterns Completed 68 for a set of answers.

Ask: According to these patterns, what do metal toxins tend to do in the body?

Ask: What types of compounds affect the eyes, nose, and lungs?

Ask: Where in the body can substances react with oxygen, O_2? With water, H_2O? With hydrochloric acid, HCl?

Ask: What types of substances cause blockages in the body?

Key Points: Toxins can enter the body in a limited number of ways. The most common methods are through ingestion, inhalation, or contact with the skin. Once toxic substances enter the body, they react in a variety of ways.

Overview

LAB: GROUPS OF 4

Key Question: How can you predict what you will observe based on a chemical equation?

KEY IDEAS

Chemical equations describe changes in matter. To fully grasp the language of chemical equations, it is necessary to make real-world observations of processes described by these equations.

LEARNING OBJECTIVES

- Relate chemical equations to real-world observations.
- Make predictions based on chemical equations.

FOCUS ON UNDERSTANDING

- Note that when students are asked what they will observe, they often misinterpret this question to mean, "What products will be formed?" Remind students that observations are entirely about the information that comes in through the senses.
- Students often misinterpret a precipitate suspended in a solution as part of the solution, not as a solid.
- Students should understand that chemical equations are the result of observations of chemical changes and not the reverse. Just because two compounds are put together as reactants does not mean that they will react.

IN CLASS

Students are given chemical equations for nine different reactions or physical changes. They are asked to predict what they will observe, based solely on the information given by the equations. They work in groups to complete the nine reactions in the laboratory, recording their observations. Finally, they compare their observations with what they predicted. A complete materials list is available for download.

TRM Materials List 69

SAFETY

- Wear safety goggles at all times.
- Do not touch the dry ice with your fingers. It causes burns.
- $NaOH$, $Ca(OH)_2$, and NH_4OH can irritate or burn skin.
- In case of a spill or contact with skin, rinse with large amounts of water.
- Have citric acid available for spills.

Making Predictions

Observing Change

THINK ABOUT IT

Table salt, or sodium chloride, NaCl, can enhance the flavor of soup. Even though the salt dissolves in the soup, you can still taste it. This change is described by a chemical equation:

$$NaCl(s) \longrightarrow NaCl(aq)$$

It is also possible to change the salt in far more dramatic ways. For example, when an electric current is passed through a sodium chloride solution, a toxic, green gas bubbles out of the solution. The change is described by this chemical equation:

$$2NaCl(aq) + 2H_2O(l) \longrightarrow 2NaOH(aq) + Cl_2(g) + H_2(g)$$

> How can you predict what you will observe based on a chemical equation?

To answer this question, you will explore

↻ Predicting Change from Chemical Equations

↻ Information in Chemical Equations

EXPLORING THE TOPIC

↻ Predicting Change from Chemical Equations

While some chemical changes are difficult to see, many changes in matter are accompanied by observable evidence. For example, when a change takes place, you may hear fizzing, a pop, or an explosion. You might see changes in color or physical form. You may smell gases that escape, or you may feel heat. Perhaps there is even a fire.

But what if you don't have a laboratory and a lot of chemicals around? How can you predict what you might observe if you are provided only with a chemical equation?

SUGAR DISSOLVING

Consider these two chemical equations. What type of change does each describe?

Sugar dissolves in water.

$$C_{12}H_{22}O_{11}(s) \longrightarrow C_{12}H_{22}O_{11}(aq)$$

Low heat evaporates the water and leaves solid sugar.

$$C_{12}H_{22}O_{11}(aq) \longrightarrow C_{12}H_{22}O_{11}(s)$$

SETUP

Before class, set up the three stations: dry ice, calcium chloride, and copper (II) sulfate. Each station will have to accommodate roughly 12 students at a time. Three procedures will be performed at each station. Place station cards with instructions and safety information for these procedures at each station. Make sure all the appropriate beakers and dropper bottles are labeled as directed. *Note:* Under some circumstances, dry ice can generate enough carbonic acid to dissolve the calcium carbonate precipitate.

STATION I—DRY ICE

- dry ice (powdered)
- small cooler for dry ice, labeled $CO_2(s)$
- 2 labeled dropper bottles with saturated solution of $Ca(OH)_2$ (~50 mL in each bottle)
- 2 labeled dropper bottles with water (~50 mL in each bottle)
- 3 spatulas
- 12 small test tubes
- 3 test-tube racks (or beakers to hold the test tubes)
- 2 large beakers for disposing of contents of test tubes, labeled waste
- 3 wash bottles

The only difference between these two equations is the position of the (*aq*) and (*s*) symbols. Notice that the chemical formula does not change from one side of the equation to the other. The sugar molecule changes form but not identity. The first equation describes dissolving, in which a solid is mixed with water. The second equation describes removing water from an aqueous substance, leaving a solid.

Although the sugar changes appearance when it dissolves, the sugar itself has not changed into a different compound. The chemical equation indicates how the molecules have, or have not, changed.

SUGAR MELTING

Consider these two equations that involve changes to sugar.

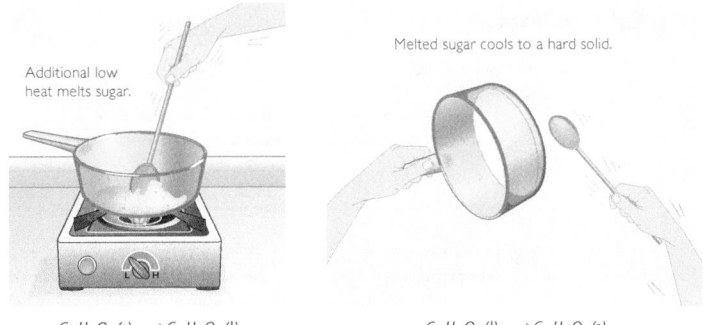

Additional low heat melts sugar.

Melted sugar cools to a hard solid.

$$C_{12}H_{22}O_{11}(s) \longrightarrow C_{12}H_{22}O_{11}(l)$$ $$C_{12}H_{22}O_{11}(l) \longrightarrow C_{12}H_{22}O_{11}(s)$$

Again the formulas are identical on both sides of the arrow, except for the symbols in parentheses, (*l*) and (*s*). These equations describe a phase change from a solid to a liquid and back to a solid. By heating sugar carefully, you can melt it to a clear liquid. When liquid sugar is cooled, it becomes a solid again.

The same compound shows up on both sides of the equation. Even though the tiny grains of sugar have been transformed into a single, large, chunk of solid sugar, the atoms within the sugar molecules have not rearranged into new compounds. Sugar is still sugar.

> **Big Idea** When a substance changes phase or dissolves, its chemical formula does not change.

SUGAR DECOMPOSING

Consider another transformation of sugar.

$$C_{12}H_{22}O_{11}(s) \longrightarrow 12C(s) + 11H_2O(g)$$

Notice that the two sides of the equation look very different. There is no sugar on the right side of the arrow. The sugar has been decomposed. It has been converted into two products: solid carbon and water vapor.

Lesson 69 | **Observing Change**   359

STATION 2—CALCIUM CHLORIDE

- 2 small labeled containers with $CaCl_2$ (anhydrous) (~2 g in each container)
- 2 small labeled containers with $CuSO_4$ (anhydrous or hydrated) (~2 g in each container)
- 2 labeled dropper bottles with 1.0 M $CaCl_2$ (~50 mL in each bottle)
- 2 labeled dropper bottles with 1.0 M NaOH (~50 mL in each bottle)
- 2 labeled dropper bottles with water (~50 mL in each bottle)
- 3 spatulas
- 12 small test tubes
- 3 test-tube racks (or beakers to hold the test tubes)
- 2 large beakers labeled waste for disposing of contents of test tubes
- 3 wash bottles

STATION 3—COPPER (II) SULFATE

- 2 small labeled containers with $CuSO_4$ (anhydrous or hydrated) (~5 g in each container)
- 2 small labeled containers with Zn (mossy or powdered) (~5 g in each container)
- 2 labeled dropper bottles with 1.0 M NH_4OH (~50 mL in each bottle)

- 2 labeled dropper bottles with 1.0 M $CuSO_4$ (~50 mL in each bottle)
- 2 labeled dropper bottles with water (~50 mL in each bottle)
- 3 spatulas
- 12 small test tubes
- 3 test-tube racks (or beakers to hold the test tubes)
- 2 large beakers labeled waste for disposing of contents of test tubes
- 3 wash bottles

CLEANUP

Evaporate the water from waste solutions containing copper compounds and dispose of solid according to local guidelines for metal wastes. The rest of the waste solutions can be diluted and poured down the drain.

Lesson Guide

ENGAGE (5 min)

TRM PowerPoint Presentation 69

> *ChemCatalyst*
>
> Consider this chemical equation:
>
> $$AgNO_3(aq) + KCl(aq) \rightarrow KNO_3(aq) + AgCl(s)$$
>
> **a.** What do you expect to *observe* if you carry out this reaction in a laboratory?
>
> **b.** Write an interpretation of the chemical equation, describing what is taking place.

Sample answers: **a.** Most students will say that they will see two liquids combine to form a liquid with a solid in it. **b.** Aqueous silver nitrate and aqueous potassium chloride are mixed, forming aqueous potassium nitrate and solid silver chloride.

Ask: What parts of the equation give you clues about what you will observe?

Ask: What is left in the test tube after you carry out the reaction?

Ask: How have the atoms of the reactants rearranged to form the products?

EXPLORE (20 min)

TRM Worksheet with Answers 69

TRM Worksheet 69

TRM Card Masters—Station Cards 69

INTRODUCE THE LAB

→ Arrange students into groups of four.

→ Pass out the student worksheet.

→ This is a two-part lab. Students should work on Part 1 individually to make predictions. They have five minutes to complete this portion of the worksheet.

→ Part 2 is a laboratory-station activity. Students should work in groups of four. Approximately three groups will be at each station at any one time. Each group has five minutes at each station.

→ Tell students that when they are finished at a station, they should dump the contents of the test tubes into the waste container provided and rinse the test tubes with water from the wash bottle.

→ Go over the safety considerations with students.

EXPLAIN & ELABORATE (15 min)

DISCUSS STUDENTS' PREDICTIONS AND OUTCOMES

→ Write the ChemCatalyst equation on the board to use as a concrete example:

$AgNO_3(aq) + KCl(aq) \rightarrow KNO_3(aq) + AgCl(s)$

Ask: How accurate were your predictions?

Ask: What clues did you look for in the chemical equations to help you predict what you might see?

Ask: What types of things did you observe? Did any of the observations surprise you?

Ask: Did some solids "disappear"? If so, where did they go?

Key Points: Chemical equations contain certain information that you can use to predict what you might observe if a procedure is performed. They identify the number of reactants and products. If two products are formed, look for them both. Equations also identify the phases of the reactants and products. You may see bubbles if a gas is produced, or some cloudiness in a solution if a solid is formed. The symbol (*aq*) indicates that a dissolved substance is present, which you may not be able to see. It also indicates that water is present.

Sometimes change is described by more than one chemical equation. This is possible when more than one change takes place during a reaction. When the

If you heat sugar sufficiently (beyond melting), it will turn black and it will release smoke. It will not taste sweet anymore, and there is no way to turn it back into sugar. These property changes, and the release of smoke, are signs that a more dramatic rearrangement of atoms has occurred.

Higher heat decomposes sugar, turning it into carbon and water vapor.

$C_{12}H_{22}O_{11}(s) \longrightarrow 12C(s) + 11H_2O(g)$

Adding water will not turn the carbon back into sugar.

$12C(s) + 11H_2O(l) \longrightarrow$ no change

This particular rearrangement of atoms is not easy to reverse. In other words, you cannot combine carbon with water and expect to form sugar.

SUGAR REACTING

Consider one more transformation of sugar, as described by this equation:

$$C_{12}H_{22}O_{11}(s) + 8KClO_3(s) \longrightarrow 12CO_2(g) + 11H_2O(g) + 8KCl(s)$$

Interpreting the chemical formulas tells you that sugar and potassium chlorate are added together. The result is the formation of two gases, carbon dioxide and water vapor, as well as the formation of some solid potassium chloride.

Sugar and potassium chlorate react explosively.

The sugar and potassium chlorate are converted to carbon dioxide gas, water vapor, and solid potassium chloride. This change happens very quickly and releases energy in the form of light and heat.

dry ice was placed into water, two things happened. Some of the dry ice sublimed to form a gas, while some of the dry ice reacted with water to form carbonic acid, $H_2CO_3(aq)$.

$CO_2(s) \rightarrow CO_2(g)$
$CO_2(s) + H_2O(l) \rightarrow H_2CO_3(aq)$

DISCUSS CHEMICAL EQUATIONS

Ask: What information is contained in a chemical equation?

Ask: What information is *not* contained in a chemical equation?

Ask: What information is best acquired through direct observation?

Ask: What information is best acquired from chemical equations?

Key Points: Chemical equations allow you to track changes in matter on an atomic level. The substances involved in a chemical change are tracked using their chemical formulas. With chemical equations, everything from dissolving and phase change to dramatic changes that create new compounds can be tracked.

There is some information that a chemical equation cannot provide. It

Example

Changing Salt

Predict what you would observe for the changes described by these equations:

a. $NaCl(s) \longrightarrow NaCl(aq)$ **b.** $NaCl(s) \longrightarrow NaCl(l)$

c. $2NaCl(l) \longrightarrow 2Na(l) + Cl_2(g)$

Solution

a. The solid sodium chloride dissolves in water.

b. The solid sodium chloride melts to become liquid sodium chloride.

c. The liquid sodium chloride decomposes into liquid sodium and chlorine gas.

◌ Information in Chemical Equations

Just as a chemical equation can help you predict what you might observe, it can also provide more information than you can get by observation alone. Consider the reaction described at the beginning of this lesson. The change that happens upon passing an electric current through aqueous sodium chloride is described by this equation:

$$2NaCl(aq) + 2H_2O(l) \longrightarrow 2NaOH(aq) + Cl_2(g) + H_2(g)$$

If you were to observe this reaction without knowing the equation, you would not be able to identify the compounds before and after the reaction. Many solutions are clear and colorless, and you might not realize that two gases are produced. A chemical equation identifies the reactants and products in a reaction. This chemical equation indicates that aqueous sodium chloride combines with water to form aqueous sodium hydroxide, chlorine gas, and hydrogen gas. An equation also identifies the form or phase a substance is in. This equation indicates that a liquid of some sort will be observed before and after the reaction, and gas bubbles will be seen as the reaction takes place.

Some information is not contained in a chemical equation. The equation above, for example, does not indicate that chlorine gas is green and toxic. The reaction of sugar with potassium chlorate, mentioned previously, is quite spectacular. The equation does not indicate that the reaction is explosive or that a purple flame is produced. An equation does not tell you whether a procedure will result in hot or cold sensations. It won't tell you how fast a change will occur or if a color change will be observed. It does not tell you if there will be an explosion, a loud noise, or an overflowing beaker. Only direct observation can provide you with all of the information about what you will see, hear, and feel.

LESSON SUMMARY

How can you predict what you will observe based on a chemical equation?

Chemical equations contain valuable information about what is happening during a change in matter. They track the identities of the substances involved in the

CONSUMER CONNECTION

When sugar molecules are allowed to crystallize, they form a hard candy, like lollipops. When lemon juice or fructose is added to the sugar, the molecules do not lock into place as in crystalline candy. Instead, they form soft candy like taffy or caramels, called amorphous candy.

does not contain information about energy, so it cannot tell you whether a procedure will result in hot or cold sensations. A chemical equation also will not tell you how fast a change will occur or if a color change will be observed. It does not tell you if there will be an explosion, a loud noise, or an overflowing beaker. Only direct observation can provide you with information about what you will see, hear, and feel.

Answer: Dissolved calcium chloride reacts with dissolved sodium hydroxide to form solid calcium hydroxide and dissolved sodium chloride. The calcium hydroxide solid should be visible in the solution.

Homework

Assign the reading and exercises for Toxins Lesson 69 in the student text.

EVALUATE (5 min)

Check-In

Examine this chemical equation. Write an interpretation of the chemical equation, describing what is taking place.

$CaCl_2(aq) + 2NaOH(aq) \rightarrow$
$Ca(OH)_2(s) + 2NaCl(aq)$

LESSON 69 ANSWERS

1. *Possible answer:* When sugar melts, it changes from the solid phase to the liquid phase. When sugar dissolves in water, the sugar molecules spread throughout the water.

2. sugar melting: $C_{12}H_{22}O_{11}(s) \rightarrow C_{12}H_{22}O_{11}(l)$

sugar decomposing:

$C_{12}H_{22}O_{11}(s) \rightarrow 12C(s) + 11H_2O(g)$

3. a. The solid magnesium will react and disappear into an aqueous compound and gas bubbles will appear. **b.** Gas bubbles will appear in the solution. **c.** A solid will form when the two solutions are mixed.

4. a. $NaCl(s) \rightarrow NaCl(aq)$

b. $MgS(s) \rightarrow Mg(s) + S(g)$

c. $Ti(s) + O_2(g) \rightarrow TiO_2(s)$

Students have not yet learned to balance equations, but if they remember that oxygen gas is $O_2(g)$, they should get the formulas correct.

5. The products are dilute solutions of bleach and ammonia. In both cases, an aqueous product and water form, which would be visually undetectable. The gas that is formed in each case would produce detectable bubbles.

6. Chloramine and hydrazine both are toxic materials, particularly in high concentration or as a gas. In lower concentrations, these materials are used in liquid form to disinfect water. Because the two compounds are in such wide use as disinfectants, students have to find information on the gaseous forms of chloramines and hydrazine to assess their toxicity. Otherwise, they may only give the inadequate answer that the two compounds are toxic in large quantities.

---

change, using chemical formulas. They also identify the phase of each substance. By carefully examining a chemical equation, you can predict what you might observe if the procedure were completed. A chemical equation cannot tell you everything that you will observe. It does not inform you about the speed of a reaction, if a color change will occur, or if energy is transferred.

Exercises

Reading Questions

1. In words, describe the difference between sugar melting and sugar dissolving in water. The formula for sugar is $C_{12}H_{22}O_{11}$.

2. Use chemical equations to describe the difference between sugar melting and sugar decomposing. The formula for sugar is $C_{12}H_{22}O_{11}$.

Reason and Apply

3. Describe what you think you would observe for these chemical equations.
    a. $Mg(s) + 2HCl(aq) \longrightarrow H_2(g) + MgCl_2(aq)$
    b. $2H_2O_2(aq) \longrightarrow 2H_2O(l) + O_2(g)$
    c. $2NaCl(aq) + Pb(NO_3)_2(aq) \longrightarrow 2NaNO_3(aq) + PbCl_2(s)$

4. Write a chemical equation for these reaction descriptions.
    a. Solid sodium chloride dissolves in water.
    b. Solid magnesium sulfide is heated to produce solid magnesium and sulfur gas.
    c. Solid titanium is heated in oxygen gas to produce titanium dioxide.

5. It's extremely dangerous to mix bleach and ammonia because highly toxic gas will form. So you should never mix these two chemicals. The two chemical reactions below describe what would happen if bleach and ammonia were mixed. If these reactions were carried out by a professional chemist using proper safety precautions in a controlled environment, what might that chemist observe in each case?

$$NaOCl(aq) + NH_4OH(aq) \longrightarrow NH_2Cl(g) + NaOH(aq) + H_2O(l)$$

$$NaOCl(aq) + 2NH_4OH(aq) \longrightarrow N_2H_4(g) + NaCl(aq) + 3H_2O(l)$$

6. **PROJECT**

 Look up chloramine, NH_2Cl, and hydrazine, N_2H_4, on the Internet. Describe the toxicity of each compound.

362   Chapter 13 | **Toxic Changes**

Spare Change

Physical versus Chemical Change

THINK ABOUT IT

You breathe in oxygen and it dissolves in your blood, where it binds to a molecule called hemoglobin. The hemoglobin transports the oxygen to where it is needed for reactions with carbohydrates. In the end, the oxygen and carbohydrates react to produce carbon dioxide and water. Some of these changes are physical changes and some are chemical changes.

How are changes in matter classified?

To answer this question, you will explore

- Defining Physical and Chemical Change
- Dissolving: A Special Case

EXPLORING THE TOPIC

Defining Physical and Chemical Change

Examine the chemical equations in the table.

Physical change	Chemical change
$H_2O(l) \longrightarrow H_2O(s)$	$2Na(s) + Cl_2(g) \longrightarrow 2NaCl(s)$
$Br_2(l) \longrightarrow Br_2(g)$	$CH_4(g) + 2O_2 \longrightarrow CO_2(g) + 2H_2O(l)$
$I_2(s) \longrightarrow I_2(g)$	$2KClO_3(s) \longrightarrow 2KCl(s) + 3O_2(g)$

PHYSICAL CHANGE

The changes in the left column are all **physical changes.** In each case, the substance on the left side of the equation changes phase (solid, liquid, or gas), but it does not change its identity. Also, there is only *one* substance involved in each of these equations, and that one substance is present on both sides of the arrow.

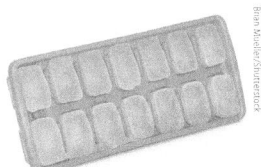

$H_2O(l) \longrightarrow H_2O(s)$
Liquid water freezes to become ice.

$Br_2(l) \longrightarrow Br_2(g)$
Liquid bromine evaporates to bromine gas.

$I_2(s) \longrightarrow I_2(g)$
Solid iodine sublimes to iodine gas.

Overview

DEMO: WHOLE CLASS

FOLLOW-UP: INDIVIDUAL

Key Question: How are changes in matter classified?

KEY IDEAS

Changes in matter can be divided into physical changes and chemical changes. Physical changes involve changes in form and do not involve changes in the identity of the substance. Chemical changes result in new substances (products) with new properties. Chemical equations can help identify which type of change is taking place. However, the dividing line between the two types of change is not clear-cut.

LEARNING OBJECTIVES

- Define physical and chemical change and explain the gray areas between them.
- Classify chemical equations as representing physical changes or chemical changes.

FOCUS ON UNDERSTANDING

Students will likely have discussed the concepts of physical and chemical change in earlier science classes, and they may have some ideas about how these terms are defined. In reality, the difference between physical change

---

and chemical change is not always clear-cut. All of the observable events usually considered a sign of chemical change can be found in many physical changes as well (color change, release of heat, change in properties, etc.).

KEY TERM

physical change

IN CLASS

During an initial demonstration, students see a compound that changes color when it dissolves. Students then work individually on a worksheet. In Part 1, students are exposed to equations that have been classified according to whether they represent physical change, chemical change, or some gray area of change. In Part 2, students use what they have learned to classify the chemical equations from the previous lesson as either physical change or chemical change. In the Explain & Elaborate discussion, students see the completion of the demonstration from the beginning of class. *Note:* Chemical change was introduced in Unit 1: Alchemy. A complete materials list is available for download.

TRM Materials List 70

SETUP

Gather materials for the demonstration. Make sure the solid cobalt (II) chloride is anhydrous (blue).

CLEANUP

Evaporate most of the liquid from the solutions containing cobalt compounds and dispose of according to local guidelines for metal wastes.

Lesson Guide

ENGAGE (5 min)

TRM PowerPoint Presentation 70

→ Complete the first half of the demonstration, placing anhydrous cobalt (II) chloride into a small amount of water in a beaker or petri dish. There should be an observable color change.

> *ChemCatalyst*
>
> A chunk of blue, solid cobalt (II) chloride, $CoCl_2$, is placed into some water. The water turns from clear to pink in color.
>
> **1.** What type of change has taken place? How do you know?
>
> **2.** Was something new made? Explain.

Ask: How do you think a physical change and a chemical change are different?

Ask: What are some examples of physical and chemical changes?

Ask: How would you classify the change that occurs when cobalt chloride is placed into water? Explain your thinking.

Ask: How might a chemical equation help you classify the change that occurred in the demonstration?

EXPLORE (15 min)

TRM **Worksheet with Answers 70**

TRM **Worksheet 70**

INTRODUCE THE FOLLOW-UP

→ Pass out the student worksheet.

→ Ask students to work individually on their worksheets. Tell them they will be classifying changes as physical or chemical.

EXPLAIN & ELABORATE (15 min)

DEFINE PHYSICAL AND CHEMICAL CHANGE

Ask: What kinds of equations and observations are associated with physical changes?

Ask: What kinds of equations and observations are associated with chemical changes?

Ask: How does a chemical equation help you determine what type of change takes place?

Key Points: Physical changes are changes in the appearance or form of a substance. Changes in phase (such as melting of ice or evaporation of water), shape (such as grinding or hammering a piece of metal), gas pressure, temperature, and so on are considered physical changes. Physical

When something changes in a physical way, the chemical formula does not change because you end up with the same substance you started with. Changing pressure or temperature, and grinding a substance into a powder, are other examples of physical change.

CHEMICAL CHANGE

The three equations in the right column of the table on the previous page are chemical changes. In each case, new substances are created. In the equations, the chemical formula(s) on the left side of each arrow are different from the formula(s) on the right side. New substances are formed during chemical changes, so we can expect the products to have properties significantly different from the reactants.

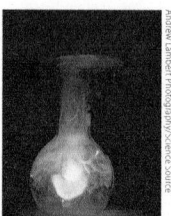

$2Na(s) + Cl_2(g) \longrightarrow 2NaCl(s)$
Solid sodium reacts with chlorine gas to produce solid sodium chloride.

$CH_4(g) + 2O_2(g) \longrightarrow CO_2(g) + 2H_2O(l)$
Methane gas and oxygen gas burn to produce carbon dioxide gas and water.

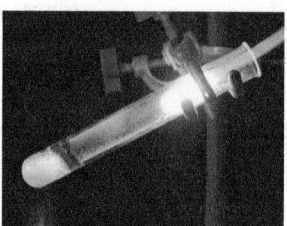

$2KClO_3(s) \longrightarrow 2KCl(s) + 3O_2(g)$
Solid potassium chlorate is heated to form solid potassium chloride and oxygen gas.

Heat, light, smoke, smells, bubbles, and changes in color often accompany chemical changes. When chemical change takes place, chemists say that a chemical reaction has occurred. In fact, the terms chemical *change* and chemical *reaction* mean the same thing.

⟳ Dissolving: A Special Case

Chemical equations can also represent the process of dissolving. Here are three equations that represent dissolving.

Liquid ethanol dissolves in water: $C_2H_6O(l) \longrightarrow C_2H_6O(aq)$

Gaseous hydrochloric acid dissolves in water: $HCl(g) \longrightarrow HCl(aq)$

Solid copper chloride dissolves in water: $CuCl_2(s) \longrightarrow CuCl_2(aq)$

These equations indicate that the identity of each dissolved substance has not changed. Each equation describes a physical change. The symbol (*aq*), for aqueous, on the product side of the equation means that each substance has dissolved in water.

Here is a different way to write the second and third equations listed above.

$$HCl(g) \longrightarrow H^+(aq) + Cl^-(aq)$$
$$CuCl_2(s) \longrightarrow Cu^{2+}(aq) + 2Cl^-(aq)$$

ART CONNECTION

Many ancient buildings and monuments such as the Great Sphinx in Egypt, shown here, are being destroyed by acid rain because they are made of materials like limestone that dissolve in sulfuric acid.

changes do not produce new substances. Chemists sometimes use chemical equations to describe phase changes. Only the phase information changes. Other types of physical changes typically are not represented by chemical equations. Unit 3: Weather was entirely about physical change.

Chemical changes produce new substances with new properties. Chemical changes are the result of chemical reactions. In a chemical equation describing a chemical change, the chemical formulas on either side of the equation do change. The products of the reaction differ from the reactants. Physical change

and chemical change are not always easy to distinguish visually.

Physical change: A change in matter in which a substance changes form but not identity.

Chemical change: A change in matter that results in the formation of a new substance or substances with new properties.

→ Complete the second half of the demonstration.

The dissolving of cobalt chloride in water results in a pink solution. The addition of cobalt chloride solution to

Important to Know You cannot always tell by observing a change whether it is a physical change or a chemical change. A chemical equation can provide more information than observation alone.

When hydrogen chloride gas and solid copper chloride dissolve in water, they break apart into ions. The second type of equation shows what happens during this change. These equations showing a substance breaking apart to form ions are necessary for the dissolving of ionic compounds and acids. Note that molecular substances like ethanol, C_2H_6O, do not break apart when they dissolve. So there is only one way to write the equation.

Ionic compounds dissolved in water have different properties from ionic solids. For example, ionic solutions conduct electricity, but the solid forms do not. This makes dissolving seem like a chemical change. However, the identity of the substances has not changed. You can recover solid compounds from solutions by evaporating the water they are dissolved in. So you can categorize dissolving as either physical or chemical change.

Example

Magnesium Chloride and Water

Write two equations that show what happens when magnesium chloride, $MgCl_2(s)$, dissolves.

Solution

$$MgCl_2(s) \longrightarrow MgCl_2(aq)$$
$$MgCl_2(s) \longrightarrow Mg^{2+}(aq) + 2Cl^-(aq)$$

LESSON SUMMARY

How are changes in matter classified?

KEY TERM

physical change

In general, changes in matter can be classified as either physical or chemical. Physical changes, including phase changes, involve a change in form without changing the identity of the substance. There is no change in chemical formula. Chemical changes, also called chemical reactions, involve the formation of new substances with new properties. The chemical formulas of these new products are different from the chemical formulas of the reactants. Dissolving can fall into both categories of change, though it is usually considered a physical change.

Exercises

Reading Questions

1. What is the difference between a physical change and a chemical change?
2. Explain why dissolving can be described as either a physical or chemical change.

Reason and Apply

3. Give five examples of physical changes. Explain why each example is a physical change.

Lesson 70 | **Physical versus Chemical Change**   365

---

calcium hydroxide results in the formation of a blue solid.

1. Start with two beakers, each containing what looks like a clear solution. One beaker contains water, and the other contains a saturated solution of calcium hydroxide.

2. Add solid cobalt chloride to the beaker containing the water.

3. Slowly add drops of 0.1 M cobalt chloride to the beaker containing the calcium hydroxide.

Write these equations on the board after initial discussion of the demonstration:

$$CoCl_2(s) \rightarrow CoCl_2(aq)$$
$$CoCl_2(aq) + Ca(OH)_2(aq) \rightarrow$$
$$Co(OH)_2(s) + CaCl_2(aq)$$

Ask: Are you observing chemical or physical changes? What is your reasoning?

Ask: What do you have to know to tell if a chemical change has taken place?

Ask: Which equation goes with which procedure? Explain your thinking.

Ask: Can you tell from the equations whether you are looking at physical changes or chemical changes?

Key Points: Many students learn in earlier grades that a chemical change is associated with a color change, a change in temperature, an explosion, or the production of a gas. However, similar observations sometimes can be associated with physical changes. For example, in the demonstration just completed, cobalt chloride dissolving in water resulted in a color change from a blue solid to a pink solution. And if liquid water is heated in a sealed container, the container will explode due to the higher pressure of the water vapor.

It is not always possible to distinguish between physical and chemical change based on observations alone. However, physical and chemical changes seem fairly easy to distinguish from each other using the chemical equations representing these changes. Simply check both sides of the equation to see if a new substance has formed (the chemical formula has changed). When new substances are formed, a chemical change clearly has taken place.

DISCUSS DISSOLVING AS A CHEMICAL OR PHYSICAL CHANGE

→ At the appropriate point in the discussion, write these chemical equations on the board:

$$CoCl_2(s) \rightarrow CoCl_2(aq)$$
$$CaCl_2(aq) \rightarrow Ca^{2+}(aq) + 2Cl^-(aq)$$

Ask: Do the properties of a substance change when it is dissolved in water?

Ask: What happens to ionic compounds when they are dissolved in water?

Ask: Why do you think some chemists might argue that dissolving is a type of chemical change?

Key Points: It is possible to argue that dissolving a substance in water changes the properties of that substance. For instance, when an ionic solid is dissolved in water, it conducts electricity. The ionic solid does not conduct electricity. And we just saw that aqueous cobalt chloride can be a different color from solid cobalt chloride.

Recall that ionic compounds do not dissolve in the same way as molecular solids. When ionic compounds are dissolved in water, they break apart into ions. Sometimes, interactions of these ions with water molecules, as in the case of the cobalt chloride, result in a colorful outcome. Thus, while dissolving is most often classified as a physical change, it clearly is a special case that could be considered a gray area between physical and chemical change.

The dissolving of ionic solids can be shown with a type of equation that stresses the formation of ions in solution.

$$CaCl_2(s) \rightarrow CaCl_2(aq)$$
$$CaCl_2(aq) \rightarrow Ca^{2+}(aq) + 2Cl^-(aq)$$

Both equations represent the same change. The first equation focuses on the change in form, from solid to aqueous. The second equation focuses more on the change in properties, from an ionic solid to ions in solution. In general, you can classify change as either physical or chemical, but some overlap does occur. It is best to rely on the information provided in the chemical equation to determine what type of change is being focused on.

EVALUATE (5 min)

Check-In

Does this chemical equation describe a physical change or a chemical change? Explain how you can tell.

$$C_{17}H_{17}O_3N(s) + 2C_4H_6O_3(l) \rightarrow$$
$$C_{21}H_{21}O_5N(s) + 2C_2H_4O_2(l)$$

Answer: The equation describes a chemical change. You can tell because the chemical formulas on the right side of the equation differ from the formulas on the left side. This means that new compounds with different properties have been created.

Homework

Assign the reading and exercises for Toxins Lesson 70 in the student text.

LESSON 70 ANSWERS

1. During a physical change, substances change form but new chemical substances are not made. During a chemical change, chemical substances are changed into other chemical substances.

2. Dissolving can be described as a physical change because the substance does not change identity during the change. However, for ionic compounds, dissolving can be described as a chemical change because the ions separate from one another as they form a solution.

3. *Possible answers (any 5):* melting ice • breaking glass • grinding pepper

---

Exercises

(continued)

4. Copy the equations 1–7 onto your paper. Match the chemical equations with their descriptions.

1. $2CH_4O(l) + O_2(g) \longrightarrow$
   $2CH_2O(l) + 2H_2O(l)$
2. $NH_2Cl(g) \longrightarrow NH_2Cl(aq)$
3. $2C_8H_{14}(l) + 23O_2(g) \longrightarrow$
   $16CO_2(g) + 14H_2O(l)$
4. $H_2SO_4(aq) + CaCO_3(s) \longrightarrow$
   $CaSO_4(aq) + CO_2(g) + H_2O(l)$
5. $Hg(l) \longrightarrow Hg(g)$
6. $C_{16}H_{30}O_2(s) + H_2 \longrightarrow$
   $C_{16}H_{32}O_2(s)$
7. $CaO(s) + H_2O(l) \longrightarrow$
   $Ca(OH)_2(s)$

A. The sulfuric acid dissolved in raindrops of acid rain reacts with the calcium carbonate in seashells and marble structures. This reaction produces calcium sulfate dissolved in water and carbon dioxide gas.

B. To balance the effect of acid rain, solid calcium oxide has been added to many lakes. It reacts with water to form solid calcium hydroxide, a strong base that is only slightly soluble in water.

C. Chloramine is added to our water supply in very small amounts to kill bacteria.

D. When octane and oxygen gas are burned in our cars, carbon dioxide and water come out in the exhaust.

E. Methanol, if ingested, reacts with oxygen to form formaldehyde, which is toxic. Water is also formed in this reaction.

F. Liquid mercury evaporates to produce mercury vapor.

G. Saturated fatty acids, like palmitic acid, tend to form long solids and clog people's arteries. These saturated fatty acids can be made from unsaturated fatty acid by adding hydrogen gas.

5. For equations 1–7 from Exercise 4, label each one as a physical or chemical change.

6. Classify the following two changes as physical or chemical. Explain your reasoning.
   a. $CaCO_3(s) + H_2SO_4(aq) \longrightarrow CaSO_4(aq) + CO_2(g) + H_2O(l)$
   b. $NaCl(s) \longrightarrow NaCl(l)$

• dissolving sugar • bending metal • boiling water • shaping molding clay. In each case, the material changes its appearance or breaks apart, but it does not change identity.

4. 1—E, 2—C, 3—D, 4—A, 5—F, 6—G, 7—B

5. Equation 1: chemical change

Equation 2: physical change

Equation 3: chemical change

Equation 4: chemical change

Equation 5: physical change

Equation 6: chemical change

Equation 7: chemical change

6. a. The equation shows calcium carbonate and sulfuric acid reacting to form calcium sulfate, carbon dioxide, and water. This reaction is a chemical change because the reactants are different in identity than the products. **b.** The equation shows sodium chloride, or table salt, dissolving in water. This reaction is a physical change because only the form of the substance changes, not its identity. This reaction could also be classified as a chemical change, because the sodium and chloride ions separate when they dissolve in the water.

Some Things Never Change

Conservation of Mass

LESSON 71

THINK ABOUT IT

During chemical and physical changes, the atoms in a substance rearrange. Sometimes, a solid forms or disappears. Sometimes, the substances change appearance or substances with new properties appear. But is matter really "appearing" or "disappearing"?

How does mass change during a chemical or physical change?

To answer this question, you will examine

○ Tracking Mass

○ Conservation of Mass

EXPLORING THE TOPIC

○ Tracking Mass

In Unit 1: Alchemy, you learned that matter is conserved during chemical reactions. If matter is not created or destroyed during physical and chemical changes, then there should be no measurable difference in mass before and after a change. Three chemical procedures are described here. The mass before and after each change is shown on the balances.

PHYSICAL CHANGE—DISSOLVING A SOLID

Consider what happens when water is added to solid sodium carbonate.

$$Na_2CO_3(s) \longrightarrow Na_2CO_3(aq)$$

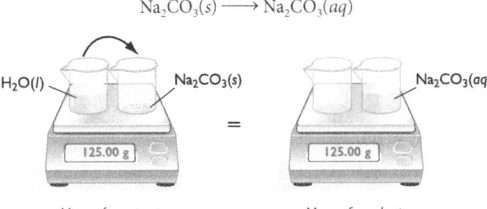

Mass of reactants   Mass of products

The solid seems to disappear into the liquid. However, notice that the total mass doesn't change. It is the same before and after the substances are mixed. The ions in the white solid have spread evenly throughout the water.

This change can also be represented by the chemical equation:

$$Na_2CO_3(s) \longrightarrow 2Na^+(aq) + CO_3^{2-}(aq)$$

Even though this compound breaks apart into ions, the mass is the same before and after the change.

Lesson 71 | **Conservation of Mass**   367

---

what happens to the mass when chemical and physical changes take place. A complete materials list is available for download.

TRM Materials List 71

SETUP

Gather materials for the lab. To save time during class, you can mark the plastic cups at the 20 mL level or measure out the 20 mL amounts. If you have enough equipment and materials, you can have students work in pairs.

CLEANUP

You might put out waste buckets for the three different procedures. The contents of each bucket can be disposed of according to your local guidelines.

Lesson Guide

ENGAGE (5 min)

TRM PowerPoint Presentation 71

ChemCatalyst

Consider this reaction:

$$Na_2CO_3(aq) + CaCl_2(aq) \rightarrow 2NaCl(aq) + CaCO_3(s)$$

1. Describe what you will observe when sodium carbonate, $Na_2CO_3(aq)$, and calcium chloride, $CaCl_2(aq)$, are mixed.

2. Will the mass increase, decrease, or stay the same after mixing? Explain.

Sample answers: **1.** Most students will suggest that this reaction will result in an aqueous salt solution, probably clear, and a solid within the solution **2.** Some students will say that the mass will stay the same because the same atoms are on both sides of the equation. Others will say that the mass is different because new compounds have been created or that the products weigh more because a solid has formed. (It will stay the same.)

→ Discuss the masses of the reactants and products.

Ask: Will the mass of the products be the same as the mass of the reactants? Explain your reasoning.

Ask: How will the mass change when a solid is formed? Explain your thinking.

Ask: Is there more, less, or the same amount of "stuff" after the reaction as before?

Overview

LAB: GROUPS OF 4

Key Question: How does mass change during a chemical or physical change?

KEY IDEAS

The law of conservation of mass applies to all chemical and physical changes. When substances are involved in chemical or physical changes, the atoms in those substances may be rearranged, but no atoms can be created or destroyed. Stated simply, during both chemical and physical change, mass is conserved. (The law of conservation of mass was introduced in Unit 1: Alchemy.)

LEARNING OBJECTIVE

Provide evidence that supports the law of conservation of mass.

IN CLASS

This is a two-part lab. First, students examine three chemical equations and predict what sort of changes in mass they might observe. Then, they carry out the three procedures and test their predictions. For the first two changes, the mass of the reactants and the products is identical. For the third change, the mass of the products is slightly less than that of the reactants because a gas is released. These firsthand experiences show students

Ask: Suppose that one of the products of a reaction is a gas. Would the mass change? Explain your thinking.

EXPLORE (20 min)

TRM Worksheet with Answers 71

TRM Worksheet 71

INTRODUCE THE LAB

→ Arrange students into groups of four.

→ Pass out the student worksheet.

→ This is a two-part lab. Part 1 involves prediction. Part 2 involves testing those predictions. Have students work in groups of four, or, if you have enough materials, in pairs.

→ You can have students find the mass of a clean empty cup and subtract it from their mass measurements. *Note:* Students may forget to measure the mass of both cups after the reaction. You can remind them.

SAFETY

• Safety goggles should be worn at all times.

EXPLAIN & ELABORATE (15 min)

REVIEW THE LAW OF CONSERVATION OF MASS

Ask: What happens to atoms when they go through a chemical or physical change?

Ask: Was anything added to or removed from the cups as a result of the reactions? From the universe?

Ask: If you were able to collect the gas that escaped during the third procedure and weigh it, what do you think you would find?

Ask: Is mass always conserved? Why or why not?

Key Points: In the first and second procedures, the mass of the reactants is equal to the mass of the products. In the first procedure, a solid dissolved. In the second procedure, a solid formed. The first change is a physical change, and the second is a chemical change.

In the third procedure, the measured mass decreases slightly because a gas escapes. In this case, one of the products of the reaction is carbon dioxide gas, which bubbles out into the air. If you were able to measure the mass of the carbon dioxide gas that escapes and add it to the mass of the other products, the result would equal the mass of the reactants.

© moodboard/Corbis

Important to Know It is difficult to measure the weight of a sample of gas, so it is tempting to conclude that gases do not have mass. However, as you saw in Unit 3: Weather, gases are made of molecules and they do have mass.

CHEMICAL CHANGE—PRODUCING A SOLID

Consider a chemical change in which a new solid forms. Adding an aqueous solution of sodium chloride to an aqueous solution of silver nitrate produces aqueous sodium nitrate and solid silver chloride.

$$NaCl(aq) + AgNO_3(aq) \longrightarrow NaNO_3(aq) + AgCl(s)$$
$$58\,g \quad + \quad 170\,g \quad = \quad 85\,g \quad + \quad 143\,g$$

Even though a solid is produced, the mass of the products is still equal to the mass of the reactants. No mass was gained or lost.

CHEMICAL CHANGE—PRODUCING A GAS

Now consider a chemical change that produces a gas as a product. Adding an aqueous solution of hydrochloric acid to solid magnesium metal produces an aqueous solution of magnesium chloride and bubbles of hydrogen gas.

$$2HCl(aq) + Mg(s) \longrightarrow H_2(g) + MgCl_2(aq)$$

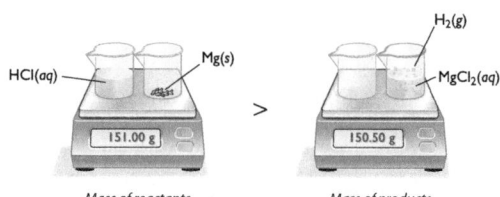

HCl(aq)   Mg(s)   151.00 g   >   150.50 g   H₂(g)   MgCl₂(aq)

Mass of reactants   *Mass of products*

According to the balance, some mass appears to have been lost. However, notice that some gas escaped from the beaker. If you were able to trap the gas that escaped and measure its mass, it would account for the difference in masses seen here.

Example

Chemical Change—Decomposing Calcium Carbonate

Examine this description of a chemical change along with its chemical equation.

Verbal description: Solid calcium carbonate is heated to produce solid calcium oxide and carbon dioxide gas.

Chemical equation: $CaCO_3(s) \longrightarrow CaO(s) + CO_2(g)$
$$50\,g \quad = \quad 28\,g \quad + \quad ?$$
mass of reactants = mass of products

What would you expect the mass of the carbon dioxide gas to be? Explain your reasoning.

Solution

The mass of the products on the right side must equal the mass of the reactants on the left side. So, the mass of the carbon dioxide produced is 22 grams.

Matter is neither created nor destroyed in physical and chemical changes. This means that atoms do not come in and out of existence in these processes. They are rearranged. Because atoms have mass, the mass does not change.

On the planet Earth, there is essentially an unchanging number of atoms. Everything on the planet is made up of these atoms. If atoms are not created or destroyed, they are recycled and rearranged to form new things. This includes atoms that you breathe in or eat and atoms we use in our daily lives or throw in the trash.

REVIEW THE DIFFERENCE BETWEEN MASS AND WEIGHT

Ask: How can you prove that gases have mass and take up space?

Ask: What happens to the mass of a glass of water when the water evaporates?

Ask: What is the difference between mass and weight?

Key Points: Mass and weight have different meanings. You typically determine the mass of an object or a substance by weighing it. However, the mass of an object does not change no matter where in the universe it is

Conservation of Mass

The fact that the mass of the reactants is equal to the mass of the products follows the law of conservation of mass. Mass is conserved during chemical and physical changes because no atoms are created or destroyed.

GETTING RID OF GARBAGE

People generate a lot of waste: cups, cans, bottles, candy wrappers, paper, and diapers. Most of this waste just keeps piling up in landfills. Some of the waste is quite toxic. In the United States alone, about 135 million tons of garbage are put into landfills each year. Will it be there forever?

Atoms are not created or destroyed; they are simply rearranged to form new substances. Therefore, everything on the planet is made up of a limited number of atoms. Even your body does not have the same atoms in it that it did several years ago. You are continually generating new cells of every kind and getting rid of the old ones.

The law of conservation of mass is important in considering waste disposal. You cannot get rid of the atoms in the waste that you throw away. Smelly, toxic garbage accumulates in large quantities in waste disposal sites. A small portion of it may biodegrade—that is, get broken down into harmless products by microorganisms—but the process can take decades. The waste will be there unless scientists find ways to convert more of our waste into useful products. Given this fact, it makes sense to reuse and recycle as much as possible.

LESSON SUMMARY

How does mass change during a chemical or physical change?

When chemical and physical changes take place, the atoms in substances rearrange. A rearrangement may be a physical change, such as the mixing of two substances, or a chemical change, the formation of new compounds. However, the atoms involved in these rearrangements cannot be created or destroyed. A chemical equation tracks the atoms involved in chemical and physical changes. All of the atoms are accounted for, so the mass of the products is identical to the mass of the reactants. This is known as the law of conservation of mass.

Exercises

Reading Questions

1. Explain the law of conservation of mass.
2. Explain how the law of conservation of mass applies to garbage.

Lesson 71 | **Conservation of Mass**   369

---

found, while the weight of an object depends on gravity.

Because a sample of gas does not appear to weigh much, it is tempting to conclude that gases do not have mass. However, as you learned in Unit 3: Weather, gases are made of molecules and do have mass. If gas molecules are compacted enough, they will become a liquid or a solid. This should convince you that they have mass.

Check-In

Consider this reaction:

$$CuCO_3(s) + H_2SO_4(aq) \rightarrow CO_2(g) + CuSO_4(aq) + H_2O(l)$$

1. Describe what you will observe when copper (II) carbonate, $CuCO_3(s)$, and sulfuric acid, $H_2SO_4(aq)$, are mixed.
2. Will the mass increase, decrease, or stay the same after mixing? Explain.

Answers: **1.** When solid $CuCO_3$ and aqueous H_2SO_4 are mixed, a solution and a gas form. The solution might be blue, because it is copper sulfate. **2.** The number of atoms stays the same. Mass is conserved. However, some of these atoms escape in the form of a gas. The mass of the products will appear to be less than the mass of the reactants unless the escaping gas also is trapped and weighed.

Homework

Assign the reading and exercises for Toxins Lesson 71 in the student text.

LESSON 71 ANSWERS

1. *Possible answer:* The law of conservation of mass states that matter is not created or destroyed during a chemical reaction. The atoms involved in the reaction are rearranged but are not created or destroyed.

2. *Possible answer:* The law of conservation of mass applies to garbage because material we discard does not go away. Everything we use must either be reused or it will pile up somewhere, and the more room we need for waste material, the less room we have to live in.

3. a. No matter was lost or gained during the reaction. You could prove this by measuring the mass of sulfur trioxide and water that reacted and the mass of sulfuric acid solution that was produced. These masses will be the same. **b.** The right side of the equation shows a total of one atom of sulfur, two atoms of hydrogen, and four atoms of oxygen. Because the same atoms are present in the reactants and the products, this provides evidence for the law of conservation of mass.

4. C

5. *Possible answer:* The number of atoms, the mass, and the weight of the water vapor are the same as those of the original liquid water, even though most or all of the water has left the glass. When water evaporates, the molecules of water do not change, but their arrangement with respect to one another changes. The density changes, but the mass and weight do not change. Even though the water vapor may drift out of the glass, the total mass of the water molecules alone is still the same.

6. A good paragraph will include: ● two or more examples of physical changes ● two or more examples of chemical changes ● an accounting of the mass before and after each change ● a reasonable, logical argument for how the evidence demonstrates a universal law.

7. To prove that matter is conserved when a piece of paper is burned, you would have to find a way to trap and measure the carbon dioxide, water vapor, and smoke produced by the reaction. If you can collect all of the products, it is possible to prove that no matter was created or destroyed by showing that the total mass of products is equal to the total mass of the reactants.

Exercises

(continued)

Reason and Apply

3. Below is a chemical equation along with a verbal description of the reaction.

Verbal description: Gaseous sulfur trioxide is added to liquid water to produce aqueous sulfuric acid.

$$\text{Chemical equation: } SO_3(g) + H_2O(l) \longrightarrow H_2SO_4(aq)$$

   a. Was matter lost or gained during this reaction? Explain how you could prove this by taking measurements.

   b. The left side of the equation shows one atom of S, two atoms of H, and four atoms of O. How many atoms of each element are on the right side of the equation? How does this provide evidence for the law of conservation of mass?

4. When an ice cube melts, which of these quantities will change?
   (A) The number of atoms it contains
   (B) Its mass
   (C) Its volume
   (D) All of the above
   (E) None of the above

5. Explain what happens to the number of atoms, the mass, and the weight of the water in a glass when it evaporates.

6. Write a paragraph to convince a friend that mass is conserved when chemical and physical changes take place. Give evidence.

7. What would you have to do to prove that matter is conserved when a piece of paper is burned?

Atom Inventory

Balancing Chemical Equations

THINK ABOUT IT

The law of conservation of mass states that mass is not lost or gained in a chemical reaction. When you write a chemical equation to describe change, it is important that the equation follows this law. So any equation describing a chemical change must account for every atom involved.

How do you balance atoms in a chemical equation?

To answer this question, you will explore
- Balancing Chemical Equations
- Coefficients Are Counting Units

EXPLORING THE TOPIC

Balancing Chemical Equations

Imagine you're in business to make ammonia, NH_3, for farmers to fertilize their crops. It would be helpful to know how much nitrogen, $N_2(g)$, and hydrogen, $H_2(g)$, to combine so that nothing is wasted. How do you make sure you have the correct amount of nitrogen and hydrogen so that you have enough of each with no extra?

A chemical equation represents the exact ratio in which reactants combine to form products. A chemical equation that accounts for all the atoms involved and shows them combining in the correct ratio is called a *balanced* chemical equation. Learning how to balance chemical equations is a necessary part of working with chemical reactions.

FORMATION OF AMMONIA

The unbalanced equation below describes how nitrogen gas and hydrogen gas react to make ammonia gas.

To balance this equation, first take an inventory of the atoms on each side.

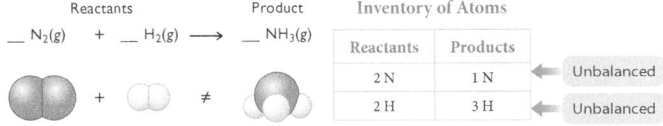

Reactants	+	Product		Inventory of Atoms		
__ $N_2(g)$	+ __ $H_2(g)$	$\longrightarrow$ __ $NH_3(g)$				

Reactants	Products	
2 N	1 N	← Unbalanced
2 H	3 H	← Unbalanced

Next, balance the atoms on each side by adding units of N_2, H_2, or NH_3. The product side needs one more N atom. The only way to accomplish this is to add a whole ammonia molecule, NH_3, to the product side.

Overview

ACTIVITY: PAIRS

Key Question: How do you balance atoms in a chemical equation?

KEY IDEAS

Balanced chemical equations reflect the law of conservation of mass. The same numbers of each kind of atom are present on both sides of a balanced equation. The coefficients in front of the chemical formulas in an equation indicate the correct ratio in which the reactants combine to form the products.

LEARNING OBJECTIVES

- Balance a simple chemical equation.
- Explain the role of coefficients in chemical equations.

FOCUS ON UNDERSTANDING

Students often try to change the number of atoms in chemical equations by changing subscripts rather than by changing coefficients. This activity may seem simple, but students benefit greatly from the concrete representations.

KEY TERMS

coefficient
formula unit

IN CLASS

Students work in pairs to balance four different chemical equations. Students use everyday items such as paper clips, beads, or interlocking cubes to represent the different atoms involved in each equation. Such models allow students to keep track of the atoms as they balance the equations. Several worked examples are completed after the discussion. A complete materials list is available for download.

TRM Materials List 72

SETUP

Obtain colored paper clips, pop beads, or other items that can be assembled and disassembled. Do a quick check of the bags to make sure there is enough of each item.

Differentiate

Some students may have an easier time using the blocks to balance equations because they are hands-on learners. You might encourage students also to use drawings of the blocks on whiteboards. On an assessment, provide students with space below an equation to draw the blocks if they wish. For those students quickly able to balance the chemical equations in this lesson, locate and provide more challenging equations.

Lesson Guide

ENGAGE (5 min)

TRM PowerPoint Presentation 72

> *ChemCatalyst*
>
> Does this equation obey the law of conservation of mass? Why or why not?
>
> $CuCl_2(aq) + Na_2S(aq) \rightarrow$
> $CuS(s) + NaCl(aq)$

Sample answer: Many students will not notice that the equation is not balanced and will say that it does obey the law of conservation of mass because matter cannot be created or destroyed during a chemical change. Other students will answer no, because the equation does not have the same number of each kind of atom on both sides. For example, there are two chlorine atoms and two sodium atoms on the left side of the equation and only one of each atom on the right side.

Ask: How many atoms of each element are there on each side of the equation?

Ask: Which side of the equation has more atoms? Can these extra atoms simply disappear?

Ask: What atoms do you have to add to the equation to make the number of atoms the same on both sides?

Ask: What is a balanced chemical equation?

EXPLORE (15 min)

TRM Worksheet with Answers 72

TRM Worksheet 72

INTRODUCE THE ACTIVITY

→ Arrange students into pairs.

→ Pass out the student worksheet.

→ Pass out sandwich bags containing the materials.

→ Inform students that the items in the bag represent atoms and are to be used in balancing the equations on the worksheet. Students should link the items to form a representation of each compound or element.

→ Explain to students that the goal is to have the same number of atoms on both sides of the equation. They can achieve this by adding more of a reactant or a product. They may not add individual atoms unless an atom is an element that is not diatomic.

GUIDE THE ACTIVITY

→ Go over Question 1 on the worksheet as a class so students understand how to use the items to balance the equations. Students can make their own tables for the equations in Question 2.

EXPLAIN & ELABORATE (20 min)

SHARE THE RESULTS OF THE WORKSHEET

Go over the worksheet, checking answers and inviting students to share their methods for balancing the equations. The balanced equations are repeated here, with the new coefficients in bold.

1: $Zn(s) + \textbf{2}HCl(aq) \rightarrow ZnCl_2(aq) + H_2(g)$

2: $O_2(g) + \textbf{2}H_2(g) \rightarrow \textbf{2}H_2O(l)$

3: $CH_4(g) + \textbf{2}O_2(g) \rightarrow CO_2(g) + \textbf{2}H_2O(l)$

4: $\textbf{3}NO_2(g) + H_2O(l) \rightarrow$
$\textbf{2}HNO_3(aq) + NO(g)$

Ask: What was the process you went through to balance the equations? Explain.

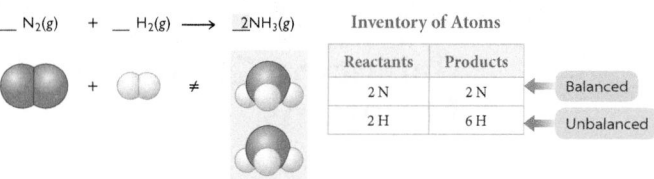

$\_\ N_2(g) + \_\ H_2(g) \longrightarrow \underline{2}NH_3(g)$

Inventory of Atoms		
Reactants	Products	
2 N	2 N	← Balanced
2 H	6 H	← Unbalanced

Take a new inventory. The N atoms are now balanced. Now the reactant side needs four more H atoms. The only way to accomplish this is to add two more molecules of H_2 to the reactant side.

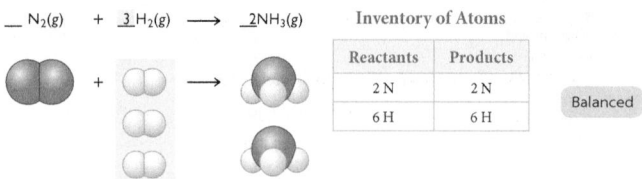

$\_\ N_2(g) + \underline{3}\ H_2(g) \longrightarrow \underline{2}NH_3(g)$

Inventory of Atoms		
Reactants	Products	
2 N	2 N	
6 H	6 H	Balanced

> **Important to Know** To balance a chemical equation, you can change only the coefficients, the numbers in front of each chemical formula. You cannot change the chemical formulas.

The equation is now balanced. The balanced equation shows that mass is conserved. There are the same numbers of N and H atoms on each side of the arrow.

In the balanced equation above, the 3 in front of the H_2 and the 2 in front of the NH_3 are called **coefficients.** Coefficients indicate how many units of each substance take part in the reaction. If there is no number in front of a chemical formula in an equation, the coefficient is understood to be a 1.

FORMATION OF RUST

Iron reacts with oxygen in the air to form iron (III) oxide, or rust. Iron (III) oxide is *not* a molecule. It is a highly organized collection of Fe and O atoms that are bonded to each other ionically. It is an ionic solid. There are a huge number of atoms in any piece of iron (III) oxide. But each piece will have two Fe atoms for every three O atoms. So Fe_2O_3 is the **formula unit** of iron oxide.

The unbalanced equation below describes how iron reacts with the oxygen in the air to form iron (III) oxide.

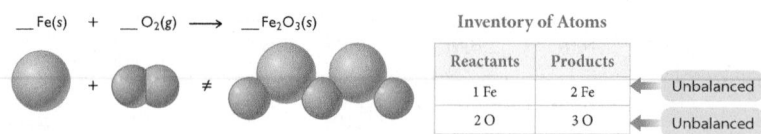

$\_\ Fe(s) + \_\ O_2(g) \longrightarrow \_\ Fe_2O_3(s)$

Inventory of Atoms		
Reactants	Products	
1 Fe	2 Fe	← Unbalanced
2 O	3 O	← Unbalanced

Balance the atoms on each side by adding molecules of O_2, atoms of Fe, or formula units of Fe_2O_3. If you add an Fe atom to the reactant side, the Fe atoms are balanced. However, the O atoms are still not equal.

Ask: How did you figure out how many multiples of each substance to include in the equations?

Key Points: A balanced chemical equation is one that shows the true mathematical relationship between the reactants and the products in a chemical reaction. The first step in balancing chemical equations is to take an inventory of the atoms on each side. When doing the atom inventory, you use the number in front of each unit (the coefficient) as well as the subscripts in the formula for each unit. If mass is conserved, numbers of each atom should be identical on both sides. If an imbalance is discovered, entire units of each

substance are added where necessary to achieve balance. The units can be atoms, molecules, or formula units, depending on the substances present in the reaction. Consider this example:

$$CH_4(g) + O_2(g) \rightarrow CO_2(g) + H_2O(g)$$

An atom inventory of this unbalanced equation is shown in the first table.

Inventory of Atoms	
Reactants	**Products**
1 C	1 C
4 H	2 H
2 O	3 O

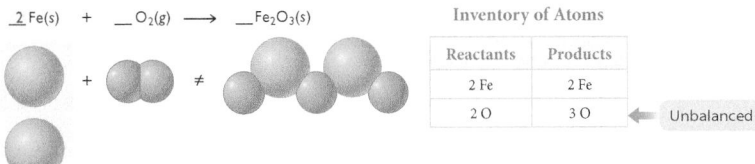

$\underline{2}$ Fe(s) + $\underline{}$ O$_2$(g) $\longrightarrow$ $\underline{}$Fe$_2$O$_3$(s)

Inventory of Atoms	
Reactants	Products
2 Fe	2 Fe
2 O	3 O	← Unbalanced

If you add two O$_2$ molecules to the reactant side and one Fe$_2$O$_3$ formula unit to the product side, the oxygen atoms are balanced.

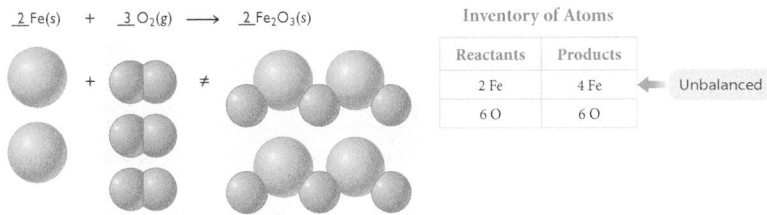

$\underline{2}$ Fe(s) + $\underline{3}$ O$_2$(g) $\longrightarrow$ $\underline{2}$ Fe$_2$O$_3$(s)

Inventory of Atoms	
Reactants	Products
2 Fe	4 Fe	← Unbalanced
6 O	6 O

But now the iron atoms are unbalanced again. Add two more iron atoms on the reactant side, increasing the total to four. The balanced chemical reaction is shown here.

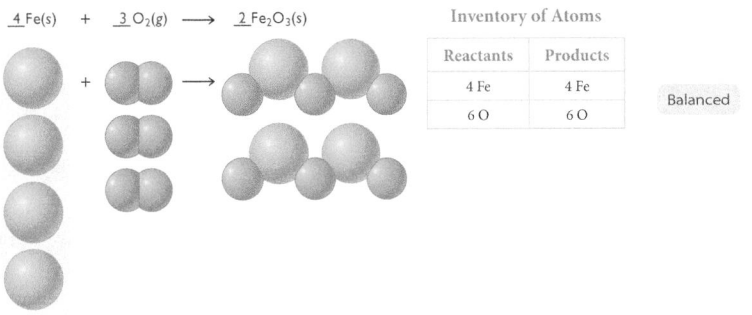

$\underline{4}$ Fe(s) + $\underline{3}$ O$_2$(g) $\longrightarrow$ $\underline{2}$ Fe$_2$O$_3$(s)

Inventory of Atoms	
Reactants	Products
4 Fe	4 Fe
6 O	6 O	Balanced

It took a few steps, but the equation is finally balanced. Four iron atoms combine with three oxygen molecules to form two formula units of iron oxide.

↻ Coefficients Are Counting Units

Once you have a balanced equation, multiplying all the coefficients by any counting unit will also give a balanced equation. You can multiply the coefficients by a dozen, or a thousand, or a million.

Lesson 72 | **Balancing Chemical Equations**   373

DISCUSS COEFFICIENTS AND SUBSCRIPTS

→ Write the equation for the combustion of propane on the board. Label the coefficients and the subscripts.

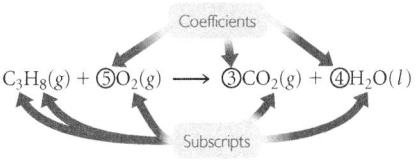

Coefficients

$C_3H_8(g) + \text{\textcircled{5}}O_2(g) \longrightarrow \text{\textcircled{3}}CO_2(g) + \text{\textcircled{4}}H_2O(l)$

Subscripts

Ask: In the equation shown, how many oxygen molecules react with one propane molecule, C$_3$H$_8$? (five) How many molecules of carbon dioxide and water are produced? (three CO$_2$ molecules, four H$_2$O molecules)

Ask: How many molecules of carbon dioxide will be produced if 1,000 molecules of propane react with 5,000 molecules of oxygen? (3,000 CO$_2$ molecules)

Ask: Can you substitute the number of moles for the coefficients in a chemical equation? Explain your reasoning. (Yes; any counting unit can be substituted.)

Key Points: There are two types of numbers in a chemical equation: coefficients and subscripts. A coefficient, like a number in a recipe, indicates how many of each ingredient can go together to make the products without having anything left over. (Not all reactions go to completion; this will be covered in Unit 6: Showtime.) Coefficients are the numbers you can change when you balance an equation. The subscript number is a part of the chemical formula and is not altered in balancing equations.

> **Coefficients:** The coefficients in a chemical equation are the numbers in front of the chemical formulas of the reactants and products. They show the correct ratio in which the reactants combine to form the products.

The coefficients in a chemical equation indicate how many "units" of an element or a compound you have. When there is no coefficient, it is understood that there is one unit. Note that the coefficients can represent any counting unit. For instance, the equation can show you that one propane molecule will react with five oxygen molecules to produce three carbon dioxide molecules and four water molecules. It can also show the number of moles of each reactant and product: One mole of propane reacts with five moles of oxygen to produce three moles of carbon dioxide and four moles of water.

To balance the equation, the product side needs more hydrogen atoms. This can be done only by adding H$_2$O molecules to the product side.

$$CH_4(g) + O_2(g) \rightarrow CO_2(g) + 2H_2O(g)$$

Now the hydrogen atoms are balanced, but the oxygen atoms still are not balanced, as shown in the second table.

Inventory of Atoms	
Reactants	**Products**
1 C	1 C
4 H	4 H
2 O	**4 O**

Adding an O$_2$ molecule to the reactant side balances the equation:

$$CH_4(g) + 2O_2(g) \rightarrow CO_2(g) + 2H_2O(g)$$

Balancing by inspection might have to proceed back and forth until the equation has the same number of each type of atom on both sides of the equation.

Inventory of Atoms	
Reactants	**Products**
1 C	1 C
4 H	4 H
4 O	4 O

COMPLETE SOME WORKED EXAMPLES

→ Ask students to come to the front of the class to work on each example and explain their process.

EXAMPLE 1

Balance this equation for the formation of nitrogen dioxide:

$$NO(g) + O_2(g) \rightarrow NO_2(g)$$

SOLUTION

First, take an inventory of the atoms on each side of the equation. The left side has more oxygen atoms than the right side. The right side needs more oxygen atoms. Add more oxygen atoms to the right side by changing the coefficient in front of NO_2 to 2. The number of oxygen atoms on the right has now increased to 4, and we have also increased the number of nitrogen atoms. Take a new inventory. Now, the left side needs one more nitrogen atom and one more oxygen atom to balance the equation. If we increase the number of NO molecules on the left side by adding the coefficient 2, the equation will be balanced.

$$2NO(g) + O_2(g) \rightarrow 2NO_2(g)$$

EXAMPLE 2

Balance this equation for rusting iron:

$$Fe(s) + O_2(g) \rightarrow Fe_2O_3(s)$$

SOLUTION

First, complete an inventory of each kind of atom on both sides of the equation. Next, proceed by balancing one type of atom. In this case, iron is a good atom to start with. There should be two iron atoms on each side of the equation. Thus, we place the coefficient 2 in front of Fe on the left. Balancing the oxygen atoms is a bit trickier. There are two oxygen atoms on one side of the equation and three oxygen atoms on the other side. One possibility is to put 3 in front of O_2 and 2 in front of Fe_2O_3. Now the iron atoms are no longer balanced. The coefficient in front of Fe has to be changed to 4.

$$4Fe(s) + 3O_2(g) \rightarrow 2Fe_2O_3(s)$$

EVALUATE (5 min)

Check-In

Balance this equation:

$$Ca(s) + O_2(g) \rightarrow CaO(s)$$

Original balanced equation	→	4 Fe atoms + 3 O_2 molecules ⟶ 2 Fe_2O_3 formula units
×2	→	8 Fe atoms + 6 O_2 molecules ⟶ 4 Fe_2O_3 formula units
×12	→	**4 dozen** Fe atoms + **3 dozen** O_2 molecules ⟶ **2 dozen** Fe_2O_3 formula units
×1000	→	4000 Fe atoms + 3000 O_2 molecules ⟶ 2000 Fe_2O_3 formula units

Recall that a mole is also a counting unit. So, the next equation is also balanced.

× 602,000,000,000,
000,000,000,000 → **4 mol** Fe atoms + **3 mol** O_2 molecules ⟶ **2 mol** Fe_2O_3 formula units

The coefficients in chemical equations can be multiples of any counting unit. The counting unit that chemists use most often is the mole. We can use moles to count the number of atoms, molecules, or formula units involved in a reaction.

LESSON SUMMARY

How do you balance atoms in a chemical equation?

To conserve mass, the number of atoms on both sides of a chemical equation must be equal. When an equation is balanced, it shows how many molecules, atoms, or formula units of an ionic compound take part in a reaction and how many are produced. You can balance a chemical equation by working with the coefficients in front of the chemical formulas. The coefficients can be any counting unit, including moles. Units of mass or volume *cannot* be used as coefficients in chemical equations.

KEY TERMS
coefficient
formula unit

Exercises

Reading Questions

1. Why do chemical equations need to be balanced?
2. How are the coefficients in chemical equations different from the subscripts in chemical formulas?

Reason and Apply

3. Your recipe for banana bread calls for two ripe bananas. However, you have six ripe bananas that you want to use before they go bad.
   a. How can you make banana bread with all six bananas?
   b. How is this related to balancing a chemical equation?
4. Copy and balance these chemical equations.
   a. $K(s) + I_2(s) \longrightarrow KI(s)$
   b. $Mg(s) + Br_2(l) \longrightarrow MgBr_2(s)$
   c. $KBr(aq) + AgNO_3(aq) \longrightarrow KNO_3(aq) + AgBr(s)$
   d. $KClO_3(s) \longrightarrow KCl(s) + O_2(g)$
   e. $C_2H_6(g) + O_2(g) \longrightarrow CO_2(g) + H_2O(l)$
   f. $Al(s) + O_2(g) \longrightarrow Al_2O_3(s)$
   g. $P_4(s) + H_2(g) \longrightarrow PH_3(g)$

Answer: $2Ca(s) + O_2(g) \rightarrow 2CaO(s)$

Homework

Assign the reading and exercises for Toxins Lesson 72 in the student text.

LESSON 72 ANSWERS

1. *Possible answer:* You have to balance chemical equations to show how many molecules, atoms, or formula units of each substance take part in the reaction or are produced by the reaction. A balanced chemical equation demonstrates that mass is conserved. An unbalanced equation violates the conservation of mass, so the equation does not accurately reflect the reaction.

2. The subscripts in a chemical formula show the number of each type of atom in a molecule or formula unit. These numbers cannot be changed without changing the identity of the substance. The coefficients in front of the chemical formula show how many units of the substance are involved in the chemical reaction. These numbers can be changed to balance an equation or to change the size of the reaction.

Answers continue in Answer Appendix (p. ANS-7).

What's Your Reaction?

Types of Reactions

THINK ABOUT IT

Toxins work by reacting with chemicals in the body. These toxic reactions can remove compounds that are important to body function, or they can create new compounds that interfere with normal body processes. Classifying toxic reactions according to the different ways they react can help us to understand how toxic substances work in the body. It can also help us to come up with suitable approaches for dealing with them.

How do atoms rearrange to form new products?

To answer this question, you will explore

↻ Combination and Decomposition Reactions

↻ Exchange Reactions

EXPLORING THE TOPIC

⟳ Combination and Decomposition Reactions

Imagine that you are an ocean researcher and you spend your days underwater in a submarine. As you breathe, you use up oxygen gas, $O_2(g)$, and release carbon dioxide gas, $CO_2(g)$. Eventually, the high level of carbon dioxide and low level of oxygen in your submarine will be dangerous to your health. How do you remove carbon dioxide and produce oxygen gas to sustain life in a submarine?

COMBINATION REACTIONS

Combination reactions are used in air scrubbers on submarines to remove $CO_2(g)$ from the air. In a **combination reaction,** two reactants combine to form a single product. In Reactions 1 and 2, carbon dioxide combines with another compound to form a new product.

Reaction 1: Carbon dioxide gas reacts with solid sodium oxide to produce sodium carbonate (washing soda):

$$CO_2(g) + Na_2O(s) \longrightarrow Na_2CO_3(s)$$

Reaction 2: Carbon dioxide gas reacts with aqueous sodium hydroxide to produce sodium bicarbonate (baking soda):

$$CO_2(g) + NaOH(aq) \longrightarrow NaHCO_3(aq)$$

A combination reaction is sometimes represented by the general equation

$$A + B \longrightarrow C$$

DECOMPOSITION REACTIONS

In Reactions 3 and 4, compounds containing oxygen atoms decompose to produce $O_2(g)$ and an ionic solid. These decomposition reactions are two ways to produce

Lesson 73 | **Types of Reactions**   375

FOCUS ON UNDERSTANDING

- Note that combination reactions sometimes are referred to as synthesis reactions and that exchange reactions sometimes are referred to as displacement reactions. We have chosen the words that we believe students will have the least trouble translating.
- Students sometimes are confused by generic equations, such as:

$$AB + CD \rightarrow AD + CB$$

Difficulties arise from the fact that letters can represent a single atom, a molecule, an ion, or a polyatomic ion.
- Students sometimes think that all reactions can be classified into these categories. There are other classification systems, subsystems, and overlapping systems for classifying reactions.

KEY TERMS

combination reaction
decomposition reaction
single exchange reaction (single
    replacement)
double exchange reaction (double
    replacement)

IN CLASS

Students are introduced to several general categories of chemical reactions through a worksheet. They work with the Toxic Reactions cards from Lesson 68, finding patterns in the way atoms rearrange to form new products. *Note:* Combustion reactions are covered in Unit 5: Fire. A complete materials list is available for download.

TRM Materials List 73

Overview

ACTIVITY: PAIRS

Key Question: How do atoms rearrange to form new products?

KEY IDEAS

Chemical reactions can be classified into general categories based on how the atoms in the reactants rearrange to form the products. Double exchange, single exchange, decomposition, and combination are four common categories of chemical reactions. Toxins can react in any of these ways in the body,

depending on the circumstances. Often, the result is that the toxins displace an atom or atoms from a compound that already exists in the body, creating a new compound with different properties.

LEARNING OBJECTIVES

- Identify patterns in chemical equations that reflect different types of reactions.
- Classify chemical equations as representing combination, decomposition, single exchange, or double exchange reactions.

Differentiate

Another way to assess student understanding about reaction types could be to ask students to develop their own method to explain the four types of reactions introduced in the lesson. For example, students might write a story using metaphors to characterize each reaction type, create a comic strip, or prepare a short skit. This allows students to showcase their learning and gives you the opportunity to gauge their progress.

Lesson Guide

TRM PowerPoint Presentation 73

ChemCatalyst

Consider these reactions:

$CaCO_3(aq) \rightarrow CaO(aq) + CO_2(g)$

$CO_2(g) + NaOH(aq) \rightarrow NaHCO_3(aq)$

1. How are these two reactions different?

2. How would you describe, in words, what happens to the reactants in each case?

Sample answers: **1.** The two reactions have different numbers of reactants and products; they involve different compounds; one forms a gas while the other starts with a gas. **2.** In the first reaction, aqueous calcium carbonate breaks up into two new substances, aqueous calcium oxide and carbon dioxide gas. In the second reaction, carbon dioxide gas and aqueous sodium hydroxide come together to form a single product, sodium bicarbonate.

Ask: Are these two chemical equations balanced? Are all the atoms accounted for?

Ask: What is each equation expressing?

Ask: What would you expect to see in each case?

Ask: How are these two reactions different in the way the atoms rearrange?

TRM Worksheet with Answers 73

TRM Worksheet 73

TRM Card Masters—Toxic Reactions 68

INTRODUCE THE ACTIVITY

→ Arrange students into pairs.

→ Pass out the student worksheet and Toxic Reactions cards.

INTRODUCE THE FOUR CLASSIFICATIONS OF REACTIONS

→ The reactions shown are samples you can use for the discussion.

Combination: $CO_2(g) + NaOH(aq) \rightarrow NaHCO_3(aq)$

Decomposition: $CaCO_3(aq) \rightarrow CaO(aq) + CO_2(g)$

Single exchange: $Cl_2(g) + 2NaBr(s) \rightarrow 2NaCl(s) + Br_2(l)$

Double exchange: $2AgCl(s) + BaBr_2(aq) \rightarrow 2AgBr(s) + BaCl_2(aq)$

→ You might want to show the four types of reactions as generic reactions.

Combination: $A + B \rightarrow AB$

Decomposition: $AB \rightarrow A + B$

Single exchange: $AB + C \rightarrow AC + B$

Double exchange: $AB + CD \rightarrow AD + CB$

oxygen. In a **decomposition reaction,** one reactant decomposes, or breaks down, to produce two or more products.

Reaction 3: Solid potassium chlorate decomposes to produce solid potassium chloride and oxygen gas:

$$2KClO_3(s) \longrightarrow 2KCl(s) + 3O_2(g)$$

Reaction 4: Solid barium peroxide decomposes to produce solid barium oxide and oxygen gas:

$$2BaO_2(s) \longrightarrow 2BaO(s) + O_2(g)$$

A decomposition reaction is sometimes represented by the general equation

$$A \longrightarrow B + C$$

On the International Space Station, oxygen is produced by a decomposition reaction that uses electrical energy to split water into hydrogen gas and oxygen gas:

$$2H_2O(l) \longrightarrow 2H_2(g) + O_2(g)$$

⟳ Exchange Reactions

Calcium is a necessary part of our diet, especially as we age and our bones become more fragile. An example of a calcium supplement is calcium carbonate, $CaCO_3(s)$. However, before the calcium in a vitamin tablet can become bone, it must be made more transportable in the bloodstream. It begins its chemical journey in the stomach.

Examine the reactions involving calcium below. What patterns do you notice in the ways the atoms in the reactants are rearranged to produce the products?

Reaction 5: Solid calcium reacts with hydrochloric acid in the stomach to produce aqueous calcium chloride and hydrogen gas:

$$Ca(s) + 2HCl(aq) \longrightarrow CaCl_2(aq) + H_2(g)$$

Reaction 6: Solid calcium carbonate reacts with hydrochloric acid (stomach acid) to produce aqueous calcium chloride and carbonic acid:

$$CaCO_3(s) + 2HCl(aq) \longrightarrow CaCl_2(aq) + H_2CO_3(aq)$$

In Reaction 5, calcium atoms replace the hydrogen atoms in the HCl molecules.

$$Ca(s) + 2\,H\,Cl(aq) \longrightarrow CaCl_2(aq) + H_2(g)$$

Ca replaces H.

In Reaction 6, the polyatomic ion, CO_3^{2-}, switches places with the chlorine ion, Cl^-. The second reaction is a common way in which calcium is made available to your body through chemical changes that take place in your stomach.

$$Ca\,CO_3(s) + 2H\,Cl\,(aq) \longrightarrow CaCl_2(aq) + H_2CO_3(aq)$$

CO_3^{2-} ions change places with Cl^- ions.

ENGINEERING CONNECTION

Inside submarines, air quality is a major concern. Not only must carbon dioxide be scrubbed out of the air but oxygen must be replenished and water vapor removed. Otherwise, the inside of the submarine becomes too damp from all the exhaled water vapor. For this reason, the air in a submarine is also pumped through dehumidifiers.

© Roger Ressmeyer/CORBIS

Ask: What is the difference between combination and decomposition reactions?

Ask: What is the difference between single exchange and double exchange reactions?

Ask: What is an example of a single exchange reaction and a double exchange reaction?

Key Points: Chemical reactions can be divided into categories based on how the atoms in the reactants rearrange to form the products. The four categories explored today are combination, decomposition, single exchange, and double exchange reactions.

In both cases, the two reactants exchange atoms. The first reaction is called a **single exchange reaction** or a single replacement reaction because atoms of an element are exchanged with atoms of another element in a compound. A single exchange reaction is sometimes represented by the general equation

$$A + BC \longrightarrow AC + B$$

The second reaction is called a **double exchange reaction** or a double replacement reaction because two compounds exchange atoms with each other. A double exchange reaction is sometimes represented by the general equation

$$AB + CD \longrightarrow AD + CB$$

Example

Classify Chemical Reactions

Classify each reaction as a combination reaction, a decomposition reaction, a single exchange reaction, or a double exchange reaction.

  a. $H_2(g) + Cl_2(g) \longrightarrow 2HCl(g)$
  b. $CaCO_3(s) \longrightarrow CaO(s) + CO_2(g)$
  c. $Sn(s) + O_2(g) \longrightarrow SnO_2(s)$
  d. $CaI_2(s) + Cl_2(g) \longrightarrow CaCl_2(s) + I_2(g)$
  e. $AgNO_3(aq) + NaOH(aq) \longrightarrow AgOH(s) + NaNO_3(aq)$

Solution

  a. Two reactants combine to produce one product. This is a combination reaction.
  b. One reactant decomposes to produce two products. This is a decomposition reaction.
  c. Two reactants combine to produce one product. This is a combination reaction.
  d. Chlorine, Cl, replaces iodine, I, in CaI_2. This is a single exchange reaction.
  e. Silver, Ag, and sodium, Na, exchange places. This is a double exchange reaction.

Toxins sometimes act in the body in an exchange reaction by replacing an atom or atoms from a compound that exists in the body. This can have the effect of destroying necessary compounds or creating a new compound with harmful properties, or both. One way to remove toxic substances from the environment is by using combination and decomposition reactions.

LESSON SUMMARY

How do atoms rearrange to form new products?

Chemical reactions can be classified into general categories based on how atoms in the reactants rearrange to form the products. Four general types of chemical change include combination reactions, decomposition reactions, single exchange reactions, and double exchange reactions. Toxins sometimes harm the body through chemical reactions. However, chemical reactions can also be used to remove toxins from the body or the environment.

KEY TERMS

combination reaction

decomposition reaction

single exchange reaction
  (single replacement)

double exchange reaction
  (double replacement)

Toxins can react in any of these ways in the body, depending on the toxin and the circumstances. The result is that toxins often displace an atom or atoms from a compound that exists in the body, creating a new compound with different properties that may be harmful.

EVALUATE (5 min)

Check-In

Examine this chemical equation, which describes a double exchange reaction between silver nitrate and sodium chloride. Predict the products. Make sure the equation is balanced.

$AgNO_3(aq) + NaCl(aq) \rightarrow$
$Ag\_\_\_\_(s) + Na\_\_\_\_(aq)$

Answer: This is a double exchange reaction. Because the silver and sodium atoms have a +1 charge, they combine with one atom of chlorine, Cl^-, and one unit of nitrate, NO_3^-, respectively.

$AgNO_3(aq) + NaCl(aq) \rightarrow$
$AgCl(s) + NaNO_3(aq)$

Homework

Assign the reading and exercises for Toxins Lesson 73 and the Chapter 13 Review.

Combination reaction: Several reactants combine to form a single product. Combination reactions are easy to spot because there is only one compound on the product side of the equation. The general reaction can be written as A + B → AB.

Decomposition reaction: A compound breaks down as a result of the chemical change. Decomposition reactions are easy to spot because there is only one reactant. The general reaction can be written as AB → A + B.

Single exchange reaction: A compound breaks apart, and one part combines with the other reactant— either an atom or a group of atoms such as OH^-, CO_3^2, or NO_3^-. Typically, one of the reactants is an element. The general reaction can be written as A + BC → AC + B, where A displaces B.

Double exchange reaction: Both reactants break apart. Their parts then recombine into two new products. Thus, the two reactants exchange parts. The general reaction can be written as AB + CD → AD + CB, where B and D exchange with each other (or A and C exchange with each other).

LESSON 73 ANSWERS

1. Combination reactions and decomposition reactions are opposites because combination reactions put two or more atoms or compounds together to make a new compound, while decomposition reactions break a compound into two or more atoms or compounds.

2. In a single exchange reaction, one reactant loses atoms and another reactant gains atoms. In a double exchange reaction, both reactants lose atoms and gain atoms.

3. a. double exchange reaction:

$NaOH(aq) + HNO_3(aq) \rightarrow NaNO_3(aq) + H_2O(l)$

b. combination reaction:

$C_2H_4(g) + Cl_2(g) \rightarrow C_2H_4Cl_2(g)$

c. single exchange reaction:

$Cl_2(g) + MgBr_2(s) \rightarrow Br_2(s) + MgCl_2(s)$

4. Molecules (any 4): HNO_3, H_2O, C_2H_4, $C_2H_4Cl_2$ ● Ionic compounds: NaOH, $NaNO_3$, $MgBr_2$, $MgCl_2$

5. $SO_3(g) + H_2O(l) \rightarrow H_2SO_4(aq)$. Because sulfur trioxide, SO_3, is a gas, it could enter the body through the nose and mouth and travel to the lungs. Sulfuric acid is an aqueous liquid that could enter the body by ingestion or through the skin.

6. $2Li(s) + 2HCl(aq) \rightarrow H_2(g) + 2LiCl(aq)$. Lithium, Li, could enter the body by ingestion. Then the reaction would take place in the hydrochloric acid in saliva and the stomach.

7. $AgNO_3(aq) + NaOH(aq) \rightarrow NaNO_3(aq) + AgOH(s)$. The reaction involves the removal of silver ions from the toxic substance silver nitrate, $AgNO_3$, dissolved in water. The result is a safer solution and a solid that can be easily removed.

Exercises

Reading Questions

1. How are combination reactions and decomposition reactions related?

2. What is the difference between a single exchange reaction and a double exchange reaction?

Reason and Apply

3. Classify these reactions as combination, decomposition, single exchange, or double exchange. Copy the equations, fill in any missing products, and write a balanced equation for each reaction.

  a. $NaOH(aq) + HNO_3(aq) \longrightarrow NaNO_3(aq) +$ _____ (l)
  b. $C_2H_4(g) + Cl_2(g) \longrightarrow$ _____ (g)
  c. $Cl_2(g) + MgBr_2(s) \longrightarrow Br_2(s) +$ _____ (s)

4. List four molecules and four ionic compounds in the reactions from Exercise 3.

5. Sulfur trioxide gas combines with liquid water to produce aqueous sulfuric acid:

$$\text{_____}(g) + H_2O(l) \longrightarrow H_2SO_4(aq)$$

Predict the missing reactant, balance the equation, and explain how these toxic substances could enter the body.

6. Solid lithium reacts with aqueous hydrochloric acid to produce hydrogen gas and aqueous lithium chloride:

$$Li(s) + HCl(aq) \longrightarrow H_2(g) + \text{_____}$$

Predict the missing product, balance the equation, and explain how these toxic substances could enter the body.

7. Aqueous silver nitrate and aqueous sodium hydroxide are mixed to produce aqueous sodium nitrate and solid silver hydroxide:

$$AgNO_3(aq) + NaOH(aq) \longrightarrow NaNO_3(aq) + \text{_____}$$

Predict the missing product, balance the equation, and explain how the reaction could help remove toxic substances from water.

CHAPTER 13

Toxic Changes

SUMMARY

KEY TERMS

physical change

coefficient

formula unit

combination reaction

decomposition reaction

single exchange reaction
  (single replacement)

double exchange reaction
  (double replacement)

Toxins Update

A toxic substance causes an undesirable chemical reaction, producing a harmful or unhealthy change in a living system. Chemical equations keep track of these changes and allow you to predict what you will observe when compounds combine. Becoming familiar with chemical equations is the first step in understanding these chemical changes.

REVIEW EXERCISES

1. Why is it difficult to identify a physical or chemical change through observations alone?

2. Based on the chemical equation for a reaction, can you tell if any of the substances are toxic?

3. Consider the equation for the formation of a kidney stone.

    $$Na_3PO_4(aq) + 3CaCl_2(aq) \longrightarrow Ca_3(PO_4)_2(s) + 6NaCl(aq)$$

    a. Is each reactant bonded ionically or covalently? How do you know?

    b. Is this is a combination reaction, decomposition reaction, single exchange reaction, or double exchange reaction?

    c. Is this a chemical change or a physical change?

    d. How does a balanced reaction show that matter is conserved?

    e. What is the chemical name of the solid that makes up a kidney stone?

4. Copy and balance these chemical equations.

    a. $N_2(g) + H_2(g) \longrightarrow N_2H_4(g)$
    b. $KNO_3(s) + K(s) \longrightarrow K_2O(s) + N_2(g)$
    c. $H_2SO_4(aq) + NaCN(aq) \longrightarrow HCN(g) + Na_2SO_4(aq)$
    d. $H_3PO_4(aq) + Ca(OH)_2(aq) \longrightarrow Ca_3(PO_4)_2(aq) + H_2O(l)$
    e. $C_3H_8(g) + O_2(g) \longrightarrow CO_2(g) + H_2O(l)$
    f. $H_2S(g) + O_2(g) \longrightarrow SO_2(g) + H_2O(l)$
    g. $H_2(g) + O_2(g) \longrightarrow H_2O(l)$

Toxins in the Environment

PROJECT
www

Research a potentially toxic substance (your teacher may assign you one). Find out where in your environment you might find this substance and describe its effects on the body. Prepare a short report.

e. $C_3H_8(g) + 5O_2(g) \rightarrow 3CO_2(g) + 4H_2O(l)$

f. $2H_2S(g) + 3O_2(g) \rightarrow 2SO_2(s) + 2H_2O(l)$

g. $2H_2(g) + O_2(g) \rightarrow 2H_2O(l)$

Project: Toxins in the Environment

A good report would include: what makes the substance toxic (its effect on the body) ● where the toxic substance is found in the environment ● the source of the toxic substance if it is not natural ● what form the toxic substance is in ● how the toxic substance typically enters the body ● steps that people and communities should take to avoid exposure to the toxic substance ● treatment for people or organisms exposed to the toxic substance ● a description of any relevant chemical reactions.

ASSESSMENTS

Two multiple-choice assessments (versions A and B) and two short-answer assessments (versions A and B) are available for download for Chapter 13.

TRM Chapter 13 Assessment Answers

TRM Chapter 13 Assessments

ANSWERS TO CHAPTER 13 REVIEW EXERCISES

1. *Possible answer:* Identifying physical or chemical changes through observations alone is difficult because often no change occurs that can be detected by sight or smell.

2. *Possible answer:* No, the chemical equation alone does not provide information about toxicity. Data on how each compound behaves in the body is necessary to determine if any of the substances are toxic.

3. a. Each reactant has ionic bonds because they both consist of metal atoms bonded to nonmetal atoms or polyatomic ions. **b.** Both reactants lose atoms and gain atoms, so this is a double exchange reaction. **c.** The reaction represents a chemical change because the products are different substances than the reactants. **d.** A balanced reaction shows the matter is conserved because all of the atoms in the reactants are accounted for in the products. **e.** calcium phosphate.

4. a. $N_2(g) + 2H_2(g) \rightarrow N_2H_4(g)$

b. $2KNO_3(s) + 10K(s) \rightarrow 6K_2O(s) + N_2(g)$

c. $H_2SO_4(aq) + 2NaCN(aq) \rightarrow 2HCN(g) + Na_2SO_4(aq)$

d. $2H_3PO_4(aq) + 3Ca(OH)_2(aq) \rightarrow Ca_3(PO_4)_2(aq) + 6H_2O(l)$

Watch the video overview of Chapter 14 (for teachers) by clicking on the link in the TE-book, opening the TRFD, or logging onto the book's companion Web site bcs.whfreeman.com/livingbychemistry2e (teacher log-in required).

The main focus of Chapter 14 is the measurement of matter in mass and moles. This chapter sets the groundwork for students to track amounts of matter through stoichiometric calculations. Lesson 74 is an introduction to lethal dose, or LD_{50}, the measurement system applied to toxic substances. In Lesson 75, students learn firsthand the value of counting large samples of small objects by weighing them. Avogadro's number is introduced in Lesson 76. Students are made aware of the magnitude of this number and how it translates into different representations, including scientific notation, and different units, including grams. In Lesson 77, students weigh out a mole of several different substances to get a tangible grasp of that unit. In Lesson 78, they use molar mass to complete basic mass-mole calculations. In Lesson 79, students use what they have learned to consider the pros and cons of two common food additives, fructose and aspartame. They complete calculations involving LD_{50}, molar mass, grams, and moles for these molecules.

In this chapter, students will learn

- the lethal dose measurement system, LD_{50}
- Avogadro's number and scientific notation
- counting by weighing
- percent error
- how to complete mass-mole conversions using molar mass

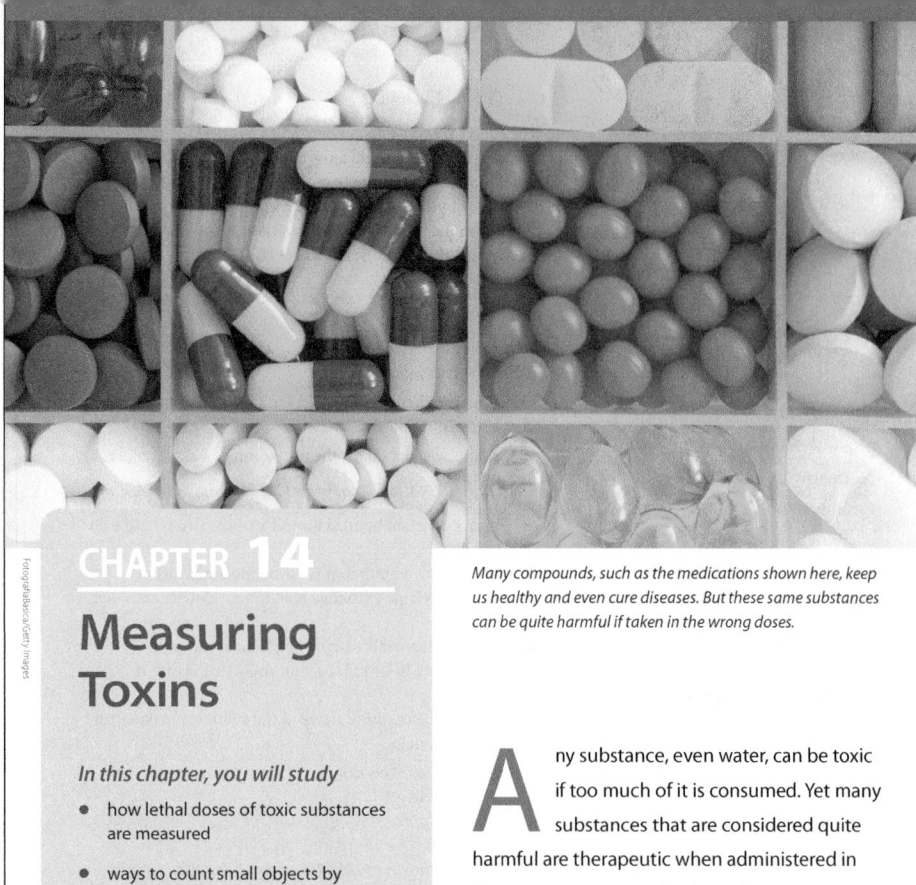

CHAPTER 14
Measuring Toxins

Many compounds, such as the medications shown here, keep us healthy and even cure diseases. But these same substances can be quite harmful if taken in the wrong doses.

In this chapter, you will study

- how lethal doses of toxic substances are measured
- ways to count small objects by weighing them
- converting between counting units (moles) and mass units
- how Avogadro's number is used

Any substance, even water, can be toxic if too much of it is consumed. Yet many substances that are considered quite harmful are therapeutic when administered in the correct amounts. The dose of a substance is key to determining the safety of that substance. This chapter concentrates on how amounts of matter are measured and counted.

Chapter 14 Lessons

Lethal Dose

Toxicity

THINK ABOUT IT

Poisonous snakebites are medical emergencies. The venom from the bite is very toxic and can be deadly, especially for small children or small animals such as rabbits and mice. However, tiny amounts of snake venom have been used therapeutically as a medicine to control high blood pressure. How much snake venom is toxic and how much is therapeutic?

How much is too much of a substance?

To answer this question, you will explore

- How Toxicity Is Measured
- Relationship to Body Weight

EXPLORING THE TOPIC

How Toxicity Is Measured

You have probably been warned about the dangers of dozens of different substances, from pesticides to poisonous wild mushrooms. You know better than to drink your shampoo or rub gasoline on your skin. But how is toxicity determined in the first place? And how is it that some things that are good for you, like vitamin tablets, can be toxic or even lethal in high doses?

MEASURING TOXICITY

For our safety, scientists try to determine the **toxicity** of the products we buy and the substances we come in contact with. Sometimes, they find the toxicity accidentally, when a person is exposed to a harmful substance. Other times, it takes a great deal of evidence before it is determined that a substance is toxic. This was the case with lead-based paints.

It is clearly not a good idea to determine how toxic something is by purposefully exposing humans. Yet it is vital to know how much of a particular substance is dangerous. So, toxicities of most substances are measured by exposing laboratory animals to toxic substances. Typically, rats or mice (sometimes rabbits, dogs, or monkeys) are used. The harm that might be done to the animals must be carefully weighed against potential harm to humans if the toxicities of substances, such as medicines, are unknown.

One way to measure toxicity is to measure the *lethal dose* of a substance, the amount of substance that can cause death. The LD_{50} of a substance is the amount of a substance that causes the death of half, or 50%, of the animals exposed to it.

Lesson 74 | **Toxicity** 381

Overview

CLASSWORK: INDIVIDUALS OR PAIRS

Key Question: How much is too much of a substance?

KEY IDEAS

In reality, every substance on the planet is potentially toxic. Similarly, many substances that are considered harmful can be beneficial when used properly. Toxicity is determined by dosage. Scientists use a variety of methods to measure toxicity. One measurement is called lethal dose, or LD_{50}. This is the amount of a substance that kills 50% of a test sample of animals. It is commonly expressed in milligrams of substance per kilogram of body weight.

LEARNING OBJECTIVES

- Calculate the toxic dose for a variety of substances, given the LD_{50}.
- Explain the role of dosage in toxicity.

KEY TERM

toxicity

IN CLASS

At the beginning of class, students discuss the relative toxicity of some common substances. They are introduced to a measurement used to quantify toxicity called LD_{50}. Students then complete a worksheet in which they calculate the toxicities of aspirin and acetaminophen for a child and for an adult. Finally, they examine and compare toxicity data for various substances from sugar to snake venom. The homework for this lesson includes an optional Web Research Project related to the toxin assigned in Lesson 68. A complete materials list is available for download.

TRM Materials List 74

Lesson Guide

ENGAGE (5 min)

TRM PowerPoint Presentation 74

ChemCatalyst

Which substance do you think is most toxic to you—alcohol (ethanol, C_2H_6O), aspirin (salicylic acid, $C_7H_6O_3$), or arsenic (III) oxide (As_2O_3)? Explain your thinking.

Sample answer: Some students will say that arsenic (III) oxide is the most toxic substance, simply because it has the word *arsenic* in it and it is usually identified as a "poison." Others will suggest that all three are toxic depending on how much is consumed. In reality, although a lethal dose of arsenic (III) oxide is smaller than a lethal dose of either aspirin or alcohol, all three substances can cause death. Some students may not realize that it is possible to die of alcohol poisoning. Some 50,000 cases of alcohol poisoning are reported in the United States each year, many of which are fatal.

→ Conduct an open-ended discussion of what constitutes toxicity.

→ An ethical debate about the pros and cons of using animals for laboratory testing has the potential to take a lot of class time. One possibility is to handle the topic as an assigned essay for homework.

Ask: How do you think toxicity is determined?

Ask: What toxic substances do you know about and avoid?

Ask: Can the same substance be both toxic and nontoxic—or even therapeutic—in different situations? Give examples.

Ask: If a substance harms lab animals, does that automatically mean it will also harm humans?

INTRODUCE THE CLASSWORK

→ Arrange students into pairs.

→ Introduce the term *lethal dose,* or LD$_{50}$ for short. A dose that kills one person or animal will not necessarily kill another, even if the subjects have the same mass. Lethal dose (LD$_{50}$) is the amount of an ingested substance that kills 50% of a test sample of animals. It is expressed in mg/kg, or milligrams of substance per kilogram of body weight. Provide a simple definition. Tell students that the word *lethal* means deadly. A lethal dose is a dose that kills.

→ Introduce the term *toxicity.* Toxicity is the measurement of how harmful or deadly a substance is to living things.

→ Pass out the student worksheet and the handout—Lethal Doses table.

EXPLAIN & ELABORATE (15 min)

DISCUSS TOXICITY

Ask: Why do lethal doses for children differ from those for adults?

Ask: Why can Vitamin A be good for you or be toxic?

Ask: Can you think of substances that are essential for life but toxic at higher doses?

Key Points: Toxicity depends on two quantities: amount of toxic substance and mass of the organism. This is why it is measured in milligrams per kilogram. The LD$_{50}$ for aspirin is 200 mg/kg, which means that if 200 mg of aspirin were given to a group of test animals, each of which have a mass of about 1 kg, half of them would die from this dosage. Because testing cannot be done on humans, the toxicity of substances is based on information from test animals (usually mice or rats, sometimes pigeons, dogs, or monkeys), even though the physiologies of humans and other animals differ in many ways. Other factors besides dosage play a role in toxicity. The method and site of delivery, the species of organism, and body mass

are all factors in determining if a dose will be lethal.

All substances are toxic in large enough doses, even sugar and water. Thus, it is very important to consider dosage when taking medication. From your calculations, it should be apparent that aspirin is more toxic to a child than acetaminophen. It takes approximately 48 of the 500 mg acetaminophen tablets to be lethal to a 22 lb child and only 4 of the 500 mg aspirin tablets. Because of their body mass, adults can tolerate a much larger dose of each substance.

The table provides the LD$_{50}$'s for several common substances in grams, milligrams, or micrograms (mcg) of substance per kilogram of body mass. A microgram is one millionth of a gram.

Lethal Doses

Common name	Chemical name and formula	Lethal dose (LD$_{50}$) per kg body mass	Toxic response
aspirin	acetylsalicylic acid, $C_9H_8O_4$	200 mg/kg (rat, oral)	gastric distress, confusion, psychosis, stupor, ringing in ears, drowsiness, hyperventilation
table salt	sodium chloride, NaCl	3 g/kg (rat, oral) 12,357 mg/kg (human, oral)	eye irritation, elevated blood pressure
castor beans	ricin—very large protein molecule, molecular mass 63,000 amu	30 mg/kg (human, oral) 3.0 mcg/kg (human, intravenous)	vomiting, diarrhea, internal bleeding, kidney and liver failure; death within minutes if injected
arsenic	arsenic (III) oxide As_2O_3	15 mg/kg (rat, oral)	*acute*: irritates eyes, skin, respiratory tract; nausea *chronic*: convulsions, tissue lesions, hemorrhage, kidney impairment
sugar	glucose, $C_6H_{12}O_6$	30 g/kg (rat, oral)	depressed activity, gastric disturbances; if diabetic: heart disease, blindness, nerve damage, kidney damage
snake venom	alpha-bungarotoxin, $C_{338}H_{529}N_{97}O_{105}S_{11}$	25.0 mcg/kg (rat, intramuscular)	paralysis, suffocation, loss of consciousness, seizures, hemorrhaging into tissues
coffee beans	caffeine, $C_8H_{10}N_4O_2$	140 mg/kg (dog, oral) 192 mg/kg (rat, oral)	acute kidney failure, nausea, psychosis, hemorrhage, increased pulse, convulsions

MEDICINE CONNECTION

Many toxins serve as medicines when used in moderation. Atropine, for example, extracted from the deadly nightshade plant, is used to treat cardiac arrest and poisoning by organophosphate insecticides and nerve gases. However, atropine is potentially addictive and an overdose can be fatal.

Notice the wide range of doses—from grams, to milligrams (thousandths of a gram), to micrograms (millionths of a gram). A small LD$_{50}$ indicates that it does not take very much of the substance to produce ill effects. Substances with small LD$_{50}$'s should definitely be avoided. Snake venom is one of the most dangerous toxins in the table, with an LD$_{50}$ of 25.0 mcg/kg, while sugar, with an LD$_{50}$ of 30 g/kg is one of the safer substances.

DOSAGE DETERMINES TOXICITY

You may be surprised to see lethal doses for substances like sugar and aspirin. These substances are usually considered beneficial. However, everything is toxic in a large enough dose, even life-sustaining substances like water and oxygen. And scientists have discovered that many highly toxic substances also have therapeutic value. Botulinum toxin (Botox) can help patients suffering from cerebral palsy. A very large protein isolated from the venom of the blue scorpion has been discovered to inhibit certain cancer tumors.

Castor oil is made from castor beans, which are the seeds of the castor plant.

The blue scorpion is found mainly in Cuba and other areas of the Caribbean.

DISCUSS THE LETHAL DOSES TABLE

Ask: What substances in the lethal doses table are fairly safe for consumption?

Ask: What substances in the lethal dose table are the most toxic? How can you tell? (ricin, snake venom, arsenic (III) oxide, caffeine, lorchel mushroom, aspirin)

Ask: Which is more lethal, a small LD$_{50}$ or a large LD$_{50}$? Explain.

Ask: Did the toxicity of any of the substances surprise you?

⟳ Relationship to Body Weight

You may be wondering how testing a substance on very small animals, like rats, will determine whether it is toxic to a larger mammal, such as a human. To take size into account, the LD_{50} is reported in milligrams of substance per kilogram of body weight, mg/kg. That way the lethal dose can be estimated for a mammal of any size.

Testing lab animals gives us only an approximate range for safety. Our bodies differ in many ways from the bodies of rats or rabbits. And it may seem particularly unkind to subject animals to testing. However, the alternative would be to risk human life with untested medications. Imagine if the safety of products on store shelves had never been confirmed in any way.

Example

> **Caffeine**
>
> The lethal dose for caffeine is approximately 150–200 mg/kg of body mass. How much caffeine would be lethal for a 120 lb person? How many cups of coffee is this?
>
>
>
> Structural formula for caffeine   I cup strong coffee   Contains ~150 mg caffeine
>
> *Solution*
>
> Lethal dose is reported in mg/kg, so first convert pounds to kilograms. There are 2.2 pounds in a kilogram.
>
> | | | |
> |---|---|---|
> | Convert body weight from pounds to kilograms. | $120 \text{ lb} \cdot \dfrac{1 \text{ kg}}{2.2 \text{ lb}} = 54.4 \text{ kg}$ | |
> | Multiply the person's weight by the lethal dose. | $54.4 \text{ kg} \cdot \dfrac{150 \text{ mg}}{1 \text{ kg}} = 8160 \text{ mg caffeine}$ | |
> | Convert number of milligrams to number of cups. | $8160 \text{ mg} \cdot \dfrac{1 \text{ cup}}{150 \text{ mg}} = 54.4 \text{ cups}$ | |
>
> So, it would be difficult to ingest a lethal dose of caffeine simply by drinking coffee. (It would take about 54 cups for the 120 lb person!)

[For a review of this math topic, see MATH Spotlight: Dimensional Analysis on page A-16.]

Lesson 74 | **Toxicity**   383

Ask: Some of the lethal doses are expressed in mg/kg, some in g/kg, and some in μg/kg. What is the difference between these units? Which is smallest?

Key Points: The smaller the lethal dose, or LD_{50}, the more potentially dangerous a substance is. The lethal dose values in the table range from 25.0 *micrograms* per kilogram of body mass to 30 *grams* per kilogram of body weight. This is a wide range. A microgram is one-millionth of a gram. A milligram is one-thousandth of a gram. Many toxic substances can be therapeutic at lower doses. For example, snake venom contains powerful neurotoxins that interfere with the cardiovascular and nervous systems, basically paralyzing a person and stopping the heart. However, some compounds in snake venom can have beneficial medicinal applications. Lower doses of certain compounds regulate blood pressure and lower heart rate. Thus, toxicity versus therapeutic effect really depends on dosage.

DISCUSS MATERIAL SAFETY DATA SHEETS (OPTIONAL)

→ Display an MSDS and introduce students to the concept of a Material Safety Data Sheet, or MSDS. This may be one of the best sources of information on the lethal dose of a substance. A related project appears in the student text on page 406.

EVALUATE (5 min)

> **Check-In**
>
> Methadone is a medication used as a painkiller and as a treatment for those recovering from heroin addiction. The LD_{50} for methadone is 95 mg/kg.
>
> **1.** Would you consider methadone to be more or less toxic than acetaminophen (LD_{50} = 2404 mg/kg)? Than aspirin (LD_{50} = 200 mg/kg)?
>
> **2.** Explain how you would calculate the amount of this substance that would be lethal to a 120 lb human.

Answers: **1.** From the LD_{50} presented here, methadone appears to be the most toxic substance of the three. **2.** To calculate the lethal dose for a 120 lb human, it is necessary to convert to kilograms: 120 lb = 54.5 kg. Thus, the lethal dose for methadone is 95 mg/kg · 54.5 kg, or 5178 mg.

Homework

Assign the reading and exercises for Toxins Lesson 74 in the student text. Assign the Web Research Project in the Chapter 14 Summary in the student text (optional). Have students continue research on the toxin assigned in Chapter 13.

Lesson 74 | **Toxicity**   383

LESSON 74 ANSWERS

1. *Possible answer:* Toxicities of most substances are measured by exposing laboratory animals to the substance in different dosages.

2. *Possible answer:* A person who has a larger body weight can generally tolerate a higher dose of a toxic substance than a person with a smaller body weight.

3. Answers will vary. Examples of products with warning labels include cleaning supplies, personal care products, detergents, medicines, bleach, paint, and fuels. A good answer for each product would include **a.** how to avoid exposure, **b.** recommended treatment if exposure occurs, **c.** what ingredient in the product is toxic, and what chemical properties the toxin has, **d.** the chemistry and the physical processes involved in the treatment.

4. a. 424,000 mg (to three significant figures) **b.** 5.4 glasses.

5. a. 90,600 mg **b.** 227,000 tablets **c.** *Possible answer:* Vitamin A is an important part of many of the body's functions. Vitamin A helps maintain good vision, a strong immune system, and strong bones and skin, and it is an antioxidant. It has also been linked to the prevention of heart disease and cancer.

How much is too much of a substance?

Every substance on the planet is a potential toxin. By the same token, many substances that are considered toxic have health benefits when taken in the correct dose. Thus, toxicity is determined by dose. Scientists use a variety of methods to measure toxicity. One measurement is called the LD_{50}. This is the amount of a substance that kills 50% of animals exposed to it. It is commonly expressed in milligrams of substance per kilogram of body weight. A low LD_{50} means a more toxic substance.

KEY TERM

toxicity

Exercises

Reading Questions

1. How can scientists determine the toxicity of a substance?
2. How is toxicity related to body weight?

Reason and Apply

3. Find at least five products at home with labels that warn of toxicity. Give the name of each product, and answer these questions:
   a. How does each label advise you to avoid harmful exposure to the product?
   b. What does each label tell you to do if a dangerous exposure does occur?
   c. Using all of this information, what can you hypothesize about the chemical properties of each product?
   d. Look at the recommended treatment for dangerous exposure to each product. What chemical and physical processes do you think might be involved in the treatment?

4. Ethanol is grain alcohol. The LD_{50} for ethanol is 7060 mg/kg (rat, oral).
   a. How many milligrams of ethanol would be lethal to a 132 lb adult?
   b. How many glasses containing 13,000 mg of ethanol would be lethal to a 22 lb child?

5. The LD_{50} for vitamin A is 1510 mg/kg (rat, oral).
   a. How many mg of vitamin A would be lethal to a 132 lb adult?
   b. How many vitamin tablets containing 0.40 mg of vitamin A would be lethal to an adult?
   c. PROJECT

   Use the Internet or other resource to research vitamin A. What are the benefits of vitamin A?

384   Chapter 14 | **Measuring Toxins**

always identical (because of the number of neutrons) to help them make the connection between this lab activity and counting atoms.

Counting by Weighing

Purpose

To count large numbers of small objects by weighing.

Procedure

1. Obtain a sandwich bag from your teacher. Your bag may contain items such as rice, beans, or paper clips.

2. Your challenge is to determine the number of objects in your bag without opening it.

3. Brainstorm how you will solve the challenge with the members of your group and decide what tools you might use.

Questions

1. Describe the method you are going to use to determine the number of objects you have in your bag. Be specific.

2. Find out what substances are in the sandwich bags of the other groups. Make a list of all the substances.

3. Which of the substances will be easiest to count? Explain your reasoning.

4. **Making Sense** What do you think this activity has to do with keeping track of chemical compounds?

5. **If You Finish Early** Are 100 molecules of snake venom equivalent in toxicity to 100 atoms of arsenic? Why or why not?

Featured ACTIVITY | **Counting by Weighing**   385

KEY TERM

percent error

IN CLASS

Students learned in Unit 3: Weather that a mole is a quantity used to count atoms or molecules. Here, they begin to make the connection between moles and mass. Sandwich bags containing different objects are distributed to groups of students. Each group works with a different substance, such as rice, rubber bands, or beads. The challenge is to determine exactly how many objects are in their sandwich bag without opening it. You determine the degree of open-endedness of this activity. You can provide students with the tools necessary to solve the challenge, or you can allow them to figure out what they need and then ask for the tools. In an optional extension of this lab activity, students can sort different substances into a "periodic table." A complete materials list is available for download.

TRM Materials List 75

SETUP

Before class, prepare 8–12 plastic sandwich bags with different objects in each so that each contains the same number of objects. We have chosen to use 237 objects in our examples. You can fill the sandwich bags quickly by weighing out samples that correspond to the masses in the table. Keep some of each object aside so that groups have an unbagged sample of 10–20 objects. Tell students that they may not open the sandwich bags. You may want to tape them shut.

Overview

ACTIVITY: GROUPS OF 4

Key Question: How can mass help you count large numbers of small objects?

KEY IDEAS

It is difficult to count very small objects like molecules. One way to get an accurate count of small objects is to find the mass of a large number of them and divide by the average mass of one object. This is how chemists determine the number of atoms in chemical compounds: by

finding the mass of a sample and dividing by the masses of the atoms.

LEARNING OBJECTIVES

- Explain how large numbers of small objects are determined.
- Calculate the percent error of a calculation.

FOCUS ON UNDERSTANDING

- Students might try to weigh single items. However, some of the masses will be below the detection limit of the balance.
- Students might have to be reminded that atoms of an element are not

Object	Mass per object	Mass per bag (237 objects)
seed bead	0.0056 g	1.33 g
sequin	0.0085 g	2.01 g
small pony bead	0.067 g	15.9 g
large pony bead	0.26 g	61.6 g
large plastic paper clip	0.45 g	107 g
grain of rice	0.022 g	5.21 g

Object	Mass per object	Mass per bag (237 objects)
split pea	0.090 g	21.3 g
elbow pasta	0.26–0.29 g	61.6–68.7 g
kidney bean	0.55–0.59 g	130–140 g
tiny rubber band	0.048 g	11.4 g
#10 small rubber band	0.126 g	29.9 g
#18 medium rubber band	0.25 g	59.3 g
#33 large rubber band	0.57 g	135 g
small sticky note	0.15 g	35.6 g
large sticky note	0.40 g	94.8 g
business card	0.80 g	19.0 g
#8 lock washer	0.136 g	32.2 g
small metal paper clip	0.94 g	223 g

Pre-AP® Course Tip

Students preparing to take an AP® chemistry course or college chemistry should be able to use models to design an experimental procedure that will generate useful data. In this lesson, students will be given a sandwich bag containing various items. Students are to work in groups to design a procedure for how to determine the number of objects in the bag without opening it.

Lesson Guide

ENGAGE (5 min)

TRM **PowerPoint Presentation 75**

ChemCatalyst

The LD_{50} of arsenic (III) oxide, As_2O_3, is 15 mg/kg.

1. Figure out the lethal dose for a 150 lb adult.

2. How many atoms do you think are in a lethal dose of arsenic (III) oxide, As_2O_3? What would you need to know in order to find out?

Sample answers: 1. A 150 lb adult weighs 68 kg. Multiplying the mass by the LD_{50} results in the lethal dose. The lethal dose will be about 1020 mg, or

---

Counting by Weighing

THINK ABOUT IT

When a toxin enters your body, it is important for a doctor to know exactly how much toxin there is. One molecule of snake venom has over 100 times as much mass as one atom of arsenic. So, 1 milligram of snake venom contains many fewer particles than 1 milligram of arsenic. To track the toxin, you need an inventory, or a count, of the atoms or molecules. However, atoms are so small that you cannot see them, let alone count them. Luckily, there are other ways of figuring out the number of atoms in a sample besides counting them one by one.

How can mass be used to count large numbers of small objects?

To answer this question, you will explore

⤴ Using Mass to Count

⤴ Weighing Atoms

EXPLORING THE TOPIC

⤴ Using Mass to Count

Let's explore how to count objects by weighing them. The masses of a number of different small objects are provided in the table.

Substance	Mass of 1000 pieces	Mass of 10 pieces	Mass of 1 piece
rice grains	22 g	0.22 g	0.022 g
lentils	56 g	0.56 g	0.056 g
rubber bands	260.5 g	2.60 g	0.260 g
paper clips	500 g	5.00 g	0.500 g
pennies	2500 g	25.00 g	2.500 g

It is difficult to accurately measure the mass of an object if the object has a mass that is below the detection limit of the balance. Many electronic balances have 0.001 g as a lower weight limit. So, weighing heavier objects tends to yield more accurate results.

One way around this difficulty is to count out a collection of identical (or nearly identical) objects and weigh the collection. You can then find the mass of a single object by dividing the mass of the collection by the number of objects. This gives you an average mass of one object that is more accurate than if you had weighed the object individually.

1.02 g. **2.** Students might guess that there are millions of atoms of arsenic in a lethal dose. This is a dramatic understatement. Some students will say that you need to know how many molecules are in 1 mg or 1 g of arsenic trioxide and then multiply or divide. Some students might mention the unit of the mole, which they learned about in Unit 3: Weather.

→ Write some of the students' predictions for 1 g of arsenic (III) oxide on the board.

Ask: How did you figure out the lethal dose for a 150 lb adult?

Ask: Make a prediction as to how many atoms of arsenic are in 1 g of arsenic (III) oxide. What did you base your prediction on? (Note that the answer is about 6.02×10^{21} atoms, but it is not necessary to share this information yet.)

Ask: Why would chemists even care how many molecules or atoms are in a sample?

Ask: How do you think chemists count the number of atoms in a sample? (Students will study mass-mole conversions in Lesson 78.)

Once you know the average mass of a single object, you can use it to determine exactly how many objects you have in a large sample. Consider a sample of pushpins.

10 pushpins have a mass of 4.7 g.	
So 1 pushpin has an average mass of 0.47 g.	
A large sample of pushpins has a mass of 84.6 g.	
Therefore this sample contains	$\dfrac{84.6\ g}{0.47\ g/pushpin}$ = 180 pushpins

In addition to helping you overcome the detection limit of the balance, there is a second reason that it is important to find the *average* mass of the objects in a sample: There is a slight variation in the masses of the objects. For example, not all grains of rice are identical in size. Some are slightly larger and others are slightly smaller (like isotopes of atoms). So, an average mass is a more accurate predictor of the mass of a *typical* object.

Example

A 5 lb Bag of Rice

How many grains of rice are there in a 5 lb bag of rice? The average mass of a rice grain is 0.0221 g, and the bag contains exactly 5.00 lb. (1 lb = 454 g)

Solution

First, determine how many grains of rice are in the bag.

Multiply by the conversion factor.
$$\text{mass of rice in grams} = 5.00\ lb \cdot \frac{454\ g}{1\ lb}$$
$$= 2270\ g\ rice$$

This is the mass of all the grains of rice. Next, determine how many grains of rice are in the bag.

Total mass of rice = average mass of 1 grain · number of grains

Divide the total mass of the rice by the average mass of 1 grain.
$$\text{number of grains} = \frac{\text{total mass of rice}}{\text{average mass of 1 grain}}$$

Calculate the answer.
$$= \frac{2270\ g}{0.0221\ g} = 103{,}181\ grains$$

So, rounded to three significant digits, there are about 103,000 grains.

[For a review of these math topics, see MATH Spotlight: Dimensional Analysis on page A-16 and MATH Spotlight: Accuracy, Precision, and Significant Digits on page A-1.]

Lesson 75 | **Counting by Weighing**   387

Substance	Grain of rice	#10 Small rubber band	Large sticky note
Mass of sandwich bag plus substance	8.11 g	32.8 g	97.7 g
Mass of empty sandwich bag	2.90 g	2.90 g	2.90 g
Mass of sample	5.21 g	29.9 g	94.8 g
Mass of 10 objects	0.23 g	1.25 g	4.00 g
Mass of 1 object	0.023 g	0.125 g	0.400 g
Count	227	239	237

Ask: What methods did you consider using to figure out the number of objects in your bag?

Ask: What do you think is the most difficult substance to count in today's activity?

Ask: Why is it more accurate to weigh ten objects and find the average mass of one of them than to simply weigh one object?

Ask: How are grains of rice similar to isotopes of one substance?

Key Points: The easiest way to determine the count of very small objects is to find their total mass and divide by the mass of one object. Of course, the mass of the container, in this case the plastic sandwich bag, must be subtracted from the total mass.

To get a more accurate average mass measurement for a tiny object, it is better to find the mass of 10 or 20 of the objects and divide by the number of objects to find the average mass, especially if the variation in size is slight. This is much like how the atomic mass of different elements is determined. Because each element usually has more than one isotope, the atomic mass is an average mass of the different isotopes.

Chemists use mass when measuring chemical compounds because it is not possible to count atoms directly. If they know the atomic mass, they can figure out how many atoms are in a compound. Recall that the atomic mass of each element is listed in the periodic table.

INTRODUCE THE ACTIVITY

→ Arrange students into groups of four.

→ Pass out the student worksheet.

→ Introduce the challenge. Pass out one sandwich bag of "stuff" to each team.

→ You might want to tell students what tools are available, or you could raise the level of challenge and wait for student groups to decide what tools to ask for.

DISCUSS HOW MASS CAN BE USED TO COUNT OBJECTS

→ Create a table on the board with three columns, headed Method, Mass, and How Many?

→ Have one member of each group come up to the board to fill in their information/data.

→ You might want to create a table similar to the one here to track how to determine the count.

Lesson 75 | **Counting by Weighing**   387

INTRODUCE PERCENT ERROR

→ Inform students that each bag contains the same number of objects. Share that number with the class. (237 for this particular set of data)

Ask: If all the bags contained the same number of objects, why didn't all the groups come up with the same number?

Ask: What are some sources of error in this procedure?

Ask: Calculate the percent error for your group's results.

Ask: What is the greatest percent error in the class? The least? Does percent error have anything to do with the specific objects being measured?

→ Introduce the equation for computing percent error.

Key Points: Chemists use percent error to express how close their measurements are to the accepted value. The lower the percent error, the more accurate your measurement. Note that you must know the actual value to calculate your percent error.

$$\text{Percent error} = \left| \frac{\text{observed value} - \text{actual value}}{\text{actual value}} \right| \cdot 100\%$$

For example, imagine that your sandwich bag contained exactly 340 beans. However, in calculating the number of beans using mass, you came up with 352. In this case, the percent error would be:

$$\text{Percent error} = \left| \frac{352 - 340}{340} \right| \cdot 100\%$$
$$= \left| \frac{12}{340} \right| \cdot 100\%$$
$$= 0.035 \cdot 100\%$$
$$= 3.5\%$$

CREATE A PERIODIC TABLE OF OBJECTS (OPTIONAL)

→ Display the transparency Periodic Table of Objects. Ask students to sort all the different objects into a table according to their properties, much as elements are sorted into a periodic table.

→ Have students share the organization of their periodic tables with the rest of the class.

→ As an extended activity, you might have groups create a card for each substance, similar to periodic table squares. These cards can then be sorted.

Ask: If you had to sort all the substances in the table, what would you base your sorting on?

BIOLOGY CONNECTION

Some snakebites can be harmless, others lethal. For example, the bites of garter snakes, found throughout North America, are generally harmless. However, it would take just 1/14,000 ounce of venom from an Australian brown snake, shown here, to kill a person.

William Weber/AP/Getty Images

KEY TERM

percent error

388 Chapter 14 | **Measuring Toxins**

PERCENT ERROR

Chemists use **percent error** to express how accurate their measurements are. For example, imagine a sandwich bag contains exactly 340 beans. Suppose you calculated the number of beans using mass and came up with 352. To figure out your percent error, you would use this formula:

$$\text{Percent error} = \left| \frac{(\text{observed value} - \text{actual value})}{\text{actual value}} \right| \cdot 100\%$$

In this case, the percent error is

$$\left| \frac{352 - 340}{340} \right| \cdot 100\% = \left| \frac{12}{340} \right| \cdot 100\% = 3.5\%$$

Notice that saying your answer was off by 3.5% is more meaningful than saying you were off by 12 beans, which could be a lot or a little, depending on how many beans you were counting. The smaller the percent error, the more accurate your answer.

Weighing Atoms

You can apply the same method you used to count tiny objects to counting atoms. To determine the number of atoms in a 1.0 g sample, you will need to know the average mass of the atoms in the sample. The average mass of the atoms of each element is given on the periodic table in atomic mass units, or amu.

Atomic mass units are special units used for atoms because the mass of an atom in grams is so tiny. Each atomic mass unit is equal to 0.0000000000000000000000166 g. There are 23 zeros in front of the 166, making this number inconvenient to use.

Consider hydrogen atoms. Hydrogen has an average atomic mass of about 1.0 amu, or 0.0000000000000000000000166 g. How many hydrogen atoms are in a 1.0 g sample?

$$\text{number of H atoms} = \frac{\text{total mass of the sample of H atoms}}{\text{average mass of 1 H atom}}$$
$$= \frac{1.0 \text{ g}}{0.0000000000000000000000166 \text{ g}}$$
$$= 602,000,000,000,000,000,000,000 \text{ H atoms}$$

So 1 gram of hydrogen atoms contains 602 sextillion, or 1 mole, of hydrogen atoms.

LESSON SUMMARY

How can mass be used to count large numbers of small objects?

It is difficult to count very small objects like atoms and molecules. One method for getting a count of small objects is to weigh a large number of the objects and divide by the average weight of one object. The average atomic mass on the periodic table is equivalent to the mass of one atom of each element. This number is an average because some atoms are isotopes with different masses. The mass of a single atom is expressed in atomic mass units rather than grams. To find the number of atoms in a sample of an element, you can divide the mass of the sample by the average mass of one atom.

Ask: How would you organize these substances into a table?

Key Points: The substances can be sorted by similar properties and by average mass per unit. This creates a periodic table of everyday objects. See page 389. Note that the average mass of a single unit of each substance is equivalent to the atomic mass of an element. The columns should be similar "families" of substances, such as food, beads, metal, and paper. The average mass of a single object should increase as you go across and down the table. There may be exceptions.

EVALUATE (5 min)

Check-In

You have a sandwich bag containing raisins. It weighs 24.6 g. A sample of ten raisins weighs 0.90 g. The empty bag has a mass of 2.90 g. How many raisins are in your sandwich bag?

Answer: Find the average mass of one raisin. Ten raisins have a mass of 0.90 g, so one raisin will weigh an average of 0.090 g. Next, subtract the mass of the sandwich bag: 24.6 g − 2.90 g = 21.7 g. Now, divide the total mass of the raisins by the mass of a single raisin to find out how many are in the sandwich bag: 21.7 g/0.09 g = 241 raisins.

Reading Questions

1. Explain how you can use mass to count large numbers of objects.

2. What does the percent error tell you?

Reason and Apply

3. Recall the method you used to count the objects in the sandwich bag you received in class and the results you obtained.
   a. Explain what you were trying to find out.
   b. Explain the method you used.
   c. Show your calculations and results.
   d. Calculate the percent error. Explain how you might modify your procedure to reduce the percent error.

4. Suppose that you have 50 grams of rice and 50 grams of beans. Which sample has more pieces? Explain your thinking.

5. Suppose that you have 740 tiny plastic beads and 740 marbles. Which sample has more mass? Explain your reasoning.

6. One bean weighs 0.074 g. How many beans are in a 50-pound bag?

7. Suppose you want to fill a 500 g bag with twice as many red jelly beans as yellow. What mass of red jelly beans do you need? A jelly bean weighs 0.65 g.

8. What is the mass in grams of one copper atom?

9. What is the mass in grams of one gold atom?

10. Suppose you have 50 grams of copper and 50 grams of gold. Which sample has more atoms? Explain your thinking.

11. Suppose you use mass to calculate the number of beans. Which of these experimental values has a smaller percent error?
    (A) A calculated value of 1342 when the actual value is 1327.
    (B) A calculated value of 1327 when the actual value is 1342.
    (C) They have the same percent error.
    (D) There is not enough information to answer the question.

Periodic Table of Objects (Sample)

	Plastic	Food	Rubber	Paper	Metal
Increasing mass/unit	seed bead 0.0056 g				
	sequin 0.0085 g	grain of rice 0.022 g	tiny rubber band 0.048 g		
	small pony bead 0.067 g	split pea 0.090 g	#10 small rubber band 0.126 g	small sticky note 0.15 g	#8 lock washer 0.136 g
	large pony bead 0.26 g	elbow pasta 0.287 g	#18 medium rubber band 0.27 g	large sticky note 0.40 g	
	large plastic paper clip 0.45 g	kidney bean 0.55 g	#33 large rubber band 0.57 g	business card 0.80 g	small metal paper clip 0.94 g

Increasing mass/unit

Homework

Assign the reading and exercises for Toxins Lesson 75 in the student text.

LESSON 75 ANSWERS

1. *Possible answer:* If you know the mass of an individual object in a large group of objects with the same mass, you can use it and the total mass of the group of objects to calculate the total number of objects.

2. *Possible answers:* The percent error tells you how accurate a measurement is. The percent error gives the amount of error relative to the actual value of the quantity being measured.

3. *Possible answers:* **a.** We were trying to find out the total number of objects in the bag. **b.** We measured the mass of a single object. Then we divided the total mass of all the objects in the bag by the mass of the single object to obtain the total number of objects in the bag. **c.** Use these equations to calculate the number of objects in a bag.

$$\frac{\text{Mass of}}{\text{one object}} = \frac{\text{total mass of sample objects}}{\text{number of objects in sample}}$$

$$\frac{\text{Number of}}{\text{objects in bag}} = \frac{\text{total mass of sample objects}}{\text{mass of one object}}$$

d. Use this equation to calculate percent error.

$$\% \text{ error} = \left| \frac{\text{calculated value} - \text{actual value}}{\text{actual value}} \right| \cdot 100\%.$$

Possible answer: We can reduce the percent error by making a more accurate measurement of the mass of a single object, either by using a more accurate scale or by using a larger sample of objects to calculate the individual mass.

4. The sample of rice has more pieces because each grain of rice has a smaller mass than a bean and the total mass of each sample is the same.

5. 740 marbles will have a greater mass than 740 plastic beads because the mass of a single marble is greater than the mass of a single bead.

6. 307,000 beans

7. The mass of the red jelly beans is two thirds of the total mass of 500 g, which is 333 g.

8. 0.00000000000000000000000106 g

9. 0.00000000000000000000000327 g

10. There are more atoms in 50 g of copper than in 50 g of gold because each copper atom has a smaller mass than a gold atom and the total mass of each sample is the same.

11. B

Overview

CLASSWORK: INDIVIDUALS OR PAIRS

Key Question: What is the relationship between mass and moles?

KEY IDEAS

Chemists use a unit called the *mole* to keep track of the number of molecules or atoms in a sample. A mole represents an extremely large number. Scientists use scientific notation to express extremely large or small numbers. The mass (in grams) of one mole of atoms of a substance can be found on the periodic table. Called the *molar mass,* it is the same number that represents the atomic mass but with different units.

LEARNING OBJECTIVES

- Translate numbers into scientific notation and vice versa.
- Explain the magnitude of a mole.
- Define molar mass for an element and find its value on the periodic table.

FOCUS ON UNDERSTANDING

Students may have difficulty grasping how the average atomic mass and the molar mass can be the same number. It is because 1 g is equivalent to 602 sextillion amu—but such a large number is difficult to grasp.

KEY TERMS

molar mass
scientific notation

IN CLASS

In this activity, students develop some proficiency in interpreting and translating very large and very small numbers and practice using scientific notation. They also are reintroduced to the concept of a mole in order to grasp its size in relation to mass amounts. During the discussion, molar mass is defined for a mole of atoms. A complete materials list is available for download.

TRM Materials List 76

Differentiate

Your students may display different levels of readiness with using scientific notation. Before the lesson, engage students in helping you create a table with the headings What I Know, What I Want to Know, and What I Learned about scientific notation, positive, zero, and negative exponents, place values using powers of ten, etc. Allow students to brainstorm in small groups about what they know and want to know before creating a class chart. After the lesson, students should be able to provide summary information for the learned portion of the chart. Display the table in your classroom and encourage students to refer to it as they continue with the Toxins unit.

Pre-AP® Course Tip

Students preparing to take an AP® chemistry or college chemistry course should know which data to collect to achieve a certain goal. In this lesson, students calculate the molar mass of common substances and use that knowledge to work in groups to create molar mass samples of various substances.

Lesson Guide

ENGAGE (5 min)

TRM PowerPoint Presentation 76

> ### ChemCatalyst
> Which do you think is more toxic, 1 mol of arsenic, As, or 10 g of arsenic? Explain your reasoning.

Sample answer: Students can estimate the answer by recalling how many

---

Billions and Billions

Avogadro's Number

THINK ABOUT IT

Mercury is a toxic substance that accumulates in the body and damages the central nervous system and other organs. Which do you think would be worse for you, 1 mole of mercury or 10 grams of mercury? To answer this question, it is necessary to understand the relationship between these two measures.

What is the relationship between mass and moles?

To answer this question, you will explore

⮑ Moles and Grams

⮑ Scientific Notation

EXPLORING THE TOPIC

⮑ Moles and Grams

In the world around you, many different measures are used to specify amounts. For example, at the store you might buy 10 apples or a 5 lb bag of apples. One measure is a counting unit and the other is a unit of weight. To compare the prices, you would probably want to know how many apples are in the 5 lb bag.

Likewise, chemists use both number and mass to specify the amount of a substance. For example, they might know the effect of either 1 mol of mercury atoms or of a 10 g sample of mercury. To compare these two quantities, they would need to know either how many moles of atoms are in each gram of mercury or the mass in grams of one mole of mercury.

The mole, abbreviated mol, is a counting unit used to count a large number of atoms. A mole of atoms is equal to 602 sextillion atoms. This number is also referred to as Avogadro's number.

One mole of O_2, Cu, NaCl, and H_2O

The table compares the average mass of one atom with the mass of 1 mol of atoms for five elements. Examine the table.

Comparing Atoms and Moles

Element	Average mass of one atom (amu)	Mass of one mole of atoms (g)
hydrogen, H	1.00 amu	1.00 g
carbon, C	12.01 amu	12.01 g
iron, Fe	55.85 amu	55.85 g
arsenic, As	74.92 amu	74.92 g
mercury, Hg	200.59 amu	200.59 g

Notice that the numerical value on the periodic table for the average atomic mass of an element is identical to the mass of one mole of atoms of that element. In other words, both values are represented by the same number. What is different about the two measurements is the units. The average mass of *one atom* of arsenic is 74.92 amu, and the mass of *one mole* of arsenic atoms is 74.92 grams. How convenient!

> **Big Idea** The mass of one mole of atoms of an element in grams is numerically the same as the average atomic mass of that element.

MOLAR MASS

80
Hg
Mercury
200.6

← Molar mass

The mass of one mole of atoms of an element is called the **molar mass.** You can find the molar mass of any element by looking up the average atomic mass on the periodic table. For example, the average atomic mass of mercury is 200.6 amu. The mass of 1 mol of mercury atoms is 200.6 g.

Example 1

Copper versus Arsenic

Which has more mass, 1 mol of copper atoms or 1 mol of arsenic atoms?

Solution

Consult the periodic table to find the molar mass of each substance. The molar mass of copper is 63.55 grams per mole, and the molar mass of arsenic is 74.92 grams per mole. A mole of arsenic atoms has more mass.

Example 2

Amount of Mercury, Hg

Which is more toxic, 1 mol or 10 g of mercury, Hg?

Solution

By looking on the periodic table, you can determine that the mass in grams of 1 mol of mercury is 200.6 g. So, 1 mol of Hg is more toxic than 10 g of Hg because 1 mol represents a much larger mass.

○ Scientific Notation

Consider the table on the next page. It shows the mass, in grams, of different numbers of atoms of mercury. Notice that the number of atoms and the mass are shown both in longhand and in **scientific notation.** Scientific notation is a shorthand method that uses an exponent to keep track of where the decimal point should be. Small numbers, between 0 and 1, have negative exponents. Numbers greater than 1 have positive exponents. The exponent indicates where the decimal point belongs in the longhand number.

Lesson 76 | **Avogadro's Number** 391

DISCUSS SCIENTIFIC NOTATION

Ask students to come to the board to complete some problems involving scientific notation. Ask students to translate longhand numbers into scientific notation and vice versa. If you wish, you can put numbers on a number line.

$$1.0 \times 10^4 = 10,000$$
$$1.0 \times 10^{-4} = 0.0001 = \% \text{ error}$$
$$= \frac{1}{10,000} = \frac{1}{1.0 \times 10^4}$$
$$2.65 \times 10^4 = 26,500$$

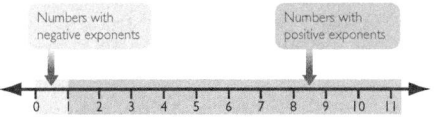

Ask: Why are numbers expressed in scientific notation?

Ask: What does the exponent tell you?

Ask: How are numbers in scientific notation added?

Ask: How are numbers in scientific notation multiplied? Divided?

Key Points: Scientific notation is a convenient way to write numbers that have lots of zeros, either because they are very large or because they are very small. Scientific notation keeps track of where the decimal point should be in any given number.

A number in standard notation can be converted to scientific notation by writing it as a decimal with one digit to the left of the decimal point times a power of ten. A number written in scientific notation has the form $a \times 10^n$ where $1 \leq a \leq 10$ or $-10 \leq a \leq -1$ and n is an integer (indicating what power of ten to multiply by). To convert a number (either positive or negative) in scientific notation back into standard notation, check the exponent, n, to find out how many places to move the decimal point. The sign of the exponent indicates which direction the decimal should be moved to obtain the correct value in standard notation.

For example, to convert 1.56×10^4 to standard notation, move the decimal four places to the right to get 15,600. To convert 1.56×10^{-4} to standard notation, move the decimal four places to the left to get 0.000156.

atoms are in a mole and how many atoms are in 1 mg of arsenic. One mole of arsenic weighs 74.9 g, so it is more toxic.

Ask: What do you have to know to answer this question?

Ask: Which do you think has more atoms, 1 mol of arsenic or 10 g of arsenic?

Ask: What do moles keep track of?

TRM Worksheet with Answers 76

TRM Worksheet 76

INTRODUCE THE CLASSWORK

➜ Arrange students into pairs.

➜ Pass out the student worksheet.

➜ Remind students that they can track the numbers of molecules and atoms with a unit called the *mole* (abbreviated mol). Write this on the board:

> **1 mole =** 602,000,000,000,000,000,000,000, **or 602 sextillion**

➜ Pass out worksheets. Students can work in pairs or individually on the worksheet.

You can remind students that if the absolute value of the number in standard notation is between 1 and 10, it can be written in scientific notation with an exponent equal to 0.

$$1.56 = 1.56 \times 10^0$$

DISCUSS THE MOLE AND AVOGADRO'S NUMBER

Ask: What is the mass of 602 sextillion atoms of iron?

Ask: What unit represents 602 sextillion objects?

Ask: What is Avogadro's number?

Ask: Why do chemists keep track of atoms and molecules by using the unit moles?

Key Points: Very small amounts of a substance contain enormously large numbers of atoms. For example, only 0.000000022 g of aluminum contains 500 trillion, or 500,000,000,000,000, atoms. Keeping track of all these atoms, even if you are using scientific notation, is cumbersome. So chemists use a unit called the *mole* (mol). You were introduced to the mole in Unit 3: Weather. One mole is equal to 602 sextillion objects. One mole is also referred to as Avogadro's number.

The mole is simply a counting unit. Just as one dozen is always equal to 12 objects, one mole is always equal to 6.02×10^{23} objects.

1 mole = Avogadro's number
= 602 sextillion
= 602,000,000,000,000,000,000,000
= 6.02×10^{23}

INTRODUCE MOLAR MASS

→ Quiz students on using the periodic table to find the mass of 1 mol of a variety of atoms.

→ Revisit the ChemCatalyst. Using the periodic table, you can determine that 1 mol of arsenic has a mass of 74.92 g. Thus, 1 mol of arsenic is more toxic than 10 g of arsenic.

Ask: Where can you find the mass of 602 sextillion atoms?

Ask: Where can you find the mass of 1 mol of a substance?

Ask: What is the mass of 1 mol of arsenic atoms? (74.92 g)

Ask: Would you rather be exposed to 1 billion arsenic atoms or 1 mol of arsenic atoms? Explain.

Number of Atoms and Mass of Mercury Atoms

Number of atoms	Number in scientific notation	Mass (g)	Mass in scientific notation (g)
1	1×10^0	0.00000000000000000000033 g	3.3×10^{-22} g
1000	1×10^3	0.00000000000000000033 g	3.3×10^{-19} g
1,000,000,000	1×10^9	0.00000000000033 g	3.0×10^{-13} g
602,000,000,000,000,000,000,000	6.02×10^{23} (1 mol)	200.59 g	2.0059×10^2 g

ASTRONOMY CONNECTION

Each galaxy in the universe is estimated to have 400 billion stars in it. There are about 130 billion galaxies in the known universe. This is about 5.2×10^{22} total stars in the universe, or 0.08 mol.

NASA/JPL-Caltech/ESA/Harvard-Smithsonian CfA

In scientific notation, ten thousand is written 1.0×10^4. One ten-thousandth is written 1.0×10^{-4}.

$$10,000 = 1.0 \times 10^4$$

$$1/10,000 = \frac{1}{1.0 \times 10^4} = 1.0 \times 10^{-4} = 0.0001$$

[For a review of this topic, see **MATH Spotlight**: Scientific Notation on page A-14.] The mass of 1000 atoms of arsenic is only 0.00000000000000000012 g or 1.2×10^{-19} g. This many arsenic atoms wouldn't even be visible to the naked eye. On the other hand, 74.92 g of arsenic contains 1 mol, or 6.02×10^{23} atoms.

When you are dealing with atoms, numbers tend to be either very tiny or very large. So it is more convenient to use scientific notation.

KEY TERMS

molar mass
scientific notation

LESSON SUMMARY

What is the relationship between mass and moles?

Chemists keep track of the number of molecules and atoms they are working with by using a unit called the mole. They also use scientific notation to express numbers that are very small or very large. One mole is equal to 602 sextillion. Using scientific notation, this number is written as 602,000,000,000,000,000,000,000, or 6.02×10^{23}. The mass (in grams) of one mole of atoms of an element can be found on the periodic table. This is called the molar mass and has the same numerical value as the average atomic mass.

Exercises

Reading Questions

1. How do you find the average atomic mass of atoms of an element? What unit of measure is this given in?
2. What is meant by the term *molar mass*? What unit of measure is this given in?
3. Why do chemists convert between moles and grams?

Key Points: The mass of 1 mol of a substance is called the *molar mass*. The molar mass of each element can be found on the periodic table. If you want to determine the mass of a mole of a particular type of atom, you can simply look up the average atomic mass of that element on the periodic table. For example, 1 mol of bromine atoms has a mass of 79.9 g, while 1 mol of potassium atoms has a mass of 39.1 g. Notice that the numbers for average atomic mass and molar mass are the same—only the units are different. In other words, the average mass of one atom of arsenic is 74.9 amu, and the mass of 1 mol of arsenic atoms is 74.9 g. (You might have students

round molar masses to the nearest tenth of a gram.)

EVALUATE (5 min)

Check-In

If you have 1 mol of aluminum and 1 mol of iron, which has more mass? How many atoms are in each sample?

Answer: By consulting the periodic table, you can see that 1 mol of iron has about twice as much mass as 1 mol of aluminum. A mole of aluminum atoms has a mass of 26.98 g. A mole of iron atoms has a mass of 55.85 g. Each sample contains 6.02×10^{23} atoms.

(continued)

Reason and Apply

4. Give the molar mass for the elements listed.
   a. nitrogen, N
   b. neon, Ne
   c. chlorine, Cl
   d. copper, Cu

5. Which has more mass?
   a. 1 mol of hydrogen, H, or 1 mol of carbon, C
   b. 1 mol of aluminum, Al, or 1 mol of iron, Fe
   c. 1 mol of copper, Cu, or 1 mol of gold, Au
   d. 5 mol of carbon, C, or 1 mol of gold, Au

6. Which contains more atoms?
   a. 12 g of hydrogen, H, or 12 g of carbon, C
   b. 27 g of aluminum, Al, or 27 g of iron, Fe
   c. 40 g of calcium, Ca, or 40 g of sodium, Na
   d. 40 g of calcium, Ca, or 60 g of zinc, Zn
   e. 10 g lithium, Li, or 100 g of lead, Pb

7. Which is more toxic? Explain your reasoning.
   a. 1 mol of beryllium, Be, or 10 g of beryllium
   b. 2 mol of arsenic, As, or 75 g of arsenic
   c. 2 mol of lead, Pb, or 500 g of lead

8. Copy and complete the table.

Amount	Number of moles	Number of atoms	Number of atoms in scientific notation
12 g carbon, C	1 mol	602,000,000,000,000,000,000,000 atoms	6.02×10^{23} atoms
24 g carbon, C			
40 g calcium, Ca			
20 g calcium, Ca			

has a higher number of atoms. **e.** The sample with 10 g of lithium has a higher number of atoms.

7. A higher amount of the substance is more toxic than a lower amount of the same substance. **a.** 1 mol of beryllium = 9.012 g. Therefore, 10 g of beryllium is more toxic than 1 mol. **b.** 2 mol of arsenic = 2(74.92 g/mol) = 149.8 g. Therefore, 2 mol of arsenic is more toxic than 75 g. **c.** 2 mol of lead = 2(207.2 g/mol) = 414.4 g. Therefore, 500 g of lead is more toxic than 2 mol.

8. 24 g carbon, C: 2 mol, 1,204,000,000,000,000,000,000,000 atoms, 1.204×10^{24} atoms ● 40 g calcium, Ca: 1 mol, 602,000,000,000,000,000,000,000 atoms, 6.02×10^{23} atoms ● 20 g calcium, Ca: 0.5 mol, 301,000,000,000,000,000,000,000 atoms, 3.01×10^{23} atoms

Homework

Assign the reading and exercises for Toxins Lesson 76 in the student text.

LESSON 76 ANSWERS

1. The average mass of atoms of an element is given on the periodic table in units of amu.

2. The molar mass of an element is the mass of one mole of atoms of that element. The units for molar mass are grams per mole.

3. *Possible answers:* Chemists use grams when they are comparing the mass of different samples, and they use moles when they are comparing the number density of different samples.

4. a. 14.01 g/mol **b.** 20.18 g/mol **c.** 35.45 g/mol **d.** 63.55 g/mol

5. a. 1 mol of carbon has more mass. **b.** 1 mol of iron has more mass. **c.** 1 mol of gold has more mass. **d.** 1 mol of gold has more mass.

6. a. The sample with 12 g of hydrogen has a higher number of atoms. **b.** The sample with 27 g of aluminum has a higher number of atoms. **c.** The sample with 40 g of sodium has a higher number of atoms. **d.** The sample with 40 g of calcium

Overview

LAB: GROUPS OF 4

Key Question: How can you convert between mass and moles?

KEY IDEAS

Chemists compare moles of substances rather than masses because moles are a way of counting atoms. When considering toxicity, or other properties of a substance, it is usually the number of particles that matters. The molar mass of a compound depends on the molar masses of the elements that make up the compound. The mass of a substance can be converted to moles using molar mass values from the periodic table.

LEARNING OBJECTIVES

- Calculate the molar mass of compounds.
- Describe the approximate magnitude of a mole of a substance.
- Complete simple conversions between mass and moles.

FOCUS ON UNDERSTANDING

The mole is a difficult unit for students to grasp. This lesson assists by having students create tangible molar samples.

IN CLASS

In this lab, students measure out a mole of several different substances, from water to aluminum, to gain a conceptual grasp of the quantity of a mole and to compare moles of different substances. In Part 1 of the worksheet, students figure out the molar mass of some common substances. In Part 2, they work in groups to create molar samples of several substances. A complete materials list is available for download.

TRM Materials List 77

SETUP

Before class, set up five weighing stations, one substance at each station: nickels, aluminum cans, water, table salt, and iron nails. You can set up more stations to avoid congestion. Also, you might set aside 1-mol samples of each substance for comparison. The amounts listed for the substances in the materials list are purposefully greater than 1 mol for each substance. This is to prevent giving away the amount needed for 1 mol. You can adjust according to your needs.

CLEANUP

Recycle materials, or save for reuse.

LESSON 77

What's in a Mole?

Molar Mass

THINK ABOUT IT

Lead is a highly toxic substance that can be accidentally ingested by humans and animals. Lead atoms interfere with normal processes in the body, causing disturbances in the nervous system. Suppose you have 100 g of lead (II) carbonate, $PbCO_3$, and 100 g of lead (II) chloride, $PbCl_2$. It is the lead atoms specifically that are toxic. If you are exposed to equal masses of lead carbonate and lead chloride, which substance exposes you to more lead atoms and is potentially more toxic? To find out, you need to determine the mass of 1 mol of each.

How can you convert between mass and moles?

To answer this question, you will explore

⤴ Molar Mass of Compounds

⤴ Comparing a Mole's Worth

EXPLORING THE TOPIC

⤴ Molar Mass of Compounds

COUNTING WITH MOLES

Any object can be counted with units such as a dozen or a million or a mole. Consider a dozen cheese sandwiches. Each sandwich has two slices of bread and one slice of cheese. Therefore, a dozen sandwiches would have a total of two dozen slices of bread and one dozen slices of cheese.

In a similar way, you can count atoms of lead in a lead compound. Because a compound consists of bonded atoms, you can simply add the molar masses of each atom in a molecule or formula unit of the compound to obtain the molar mass of the substance. For example, each formula unit of lead (II) chloride, $PbCl_2$, contains one atom of lead and two atoms of chlorine.

Molar mass of lead (II) chloride, $PbCl_2$

1 lead (II) chloride formula unit = 1 lead atom + 2 chlorine atoms

1 mol lead (II) chloride = 1 mol lead atoms + 2 mol chlorine atoms

The molar masses of lead and chlorine are 207.2 g/mol and 35.45 g/mol, respectively.

394   Chapter 14 | **Measuring Toxins**

Lesson Guide

ENGAGE (5 min)

TRM PowerPoint Presentation 77

ChemCatalyst

Consider 12 nickels, 2 empty aluminum cans, and a balloon full of carbon dioxide gas.

1. Which has the greatest mass?

2. Which has the greatest number of atoms?

3. Which has the greatest number of moles of atoms?

Explain the reasoning behind your answers.

Sample answers: Students can consult the periodic table to determine molar masses. Carbon dioxide has a molar mass of 44.01 g/mol; aluminum has a molar mass of 26.98 g/mol; nickel has a molar mass of 58.769 g/mol. However, if you do not know the mass of each sample, the masses cannot be converted to moles or atoms.

→ Stimulate a discussion on the relationship among mass, moles, and numbers of atoms.

Ask: What did you base your predictions on?

Ask: How could you find out which set of objects has the greatest mass? The greatest number of moles of atoms?

Adding the molar mass of lead and twice the molar mass of chlorine gives the molar mass of lead chloride, $PbCl_2$, which is 278.1 grams per mole, or 278.1 g/mol.

Example 1

The Mass of One Mole

Suppose you have 1 mol of lead, Pb, 1 mol of lead (II) chloride, $PbCl_2$, and 1 mol of lead (II) carbonate, $PbCO_3$. What is the mass of each sample?

Solution

You can find the molar mass of lead, Pb, on the periodic table, 207.2 g/mol.

A mole of $PbCl_2$ has 1 mol of lead atoms and 2 mol of chlorine atoms. You find the molar mass of the compound by adding these molar masses together.

$$\text{molar mass of } PbCl_2 = 207.2 + 2(35.45)$$
$$= 278.1 \text{ g/mol}$$

So the mass of 1 mol of lead (II) chloride is 278.1 g. The molar mass of $PbCO_3$ can be found similarly.

$$\text{molar mass of } PbCO_3 = 207.2 + 12.01 + 3(16.00)$$
$$= 267.2 \text{ g/mol}$$

So the mass of 1 mol of lead (II) carbonate is 267.2 g.

[For a review of this math topic, see **MATH Spotlight**: Solving Equations on page A-3.]

⤵ Comparing a Mole's Worth

A mole of atoms or molecules is usually an amount that you can hold in your hand if the substance is liquid or solid. A table with a "mole's worth" of a few substances is given below.

Molar Masses

Chemical formula	Molar mass (g/mol)	Moles of what?	Equivalent to
$O_2(g)$	32.00 g/mol	oxygen molecules	22.4 L oxygen gas
$Al(s)$	26.98 g/mol	Al atoms	2 aluminum cans
$H_2O(l)$	18.02 g/mol	H_2O molecules	18 mL water
$He(g)$	4.00 g/mol	He atoms	22.4 L helium gas
$NaCl(s)$	58.44 g/mol	sodium chloride units	1/4 cup salt
$C_{12}H_{22}O_{11}(s)$	342.3 g/mol	sugar molecules	0.75 lb sugar

Note that the molar mass of oxygen gas, $O_2(g)$, is double the molar mass of oxygen found on the periodic table because oxygen gas is diatomic. Also note that the volume of a mole of O_2 gas is the same as the volume of a mole of any other gas at standard temperature and pressure, or STP.

Lesson 77 | **Molar Mass** 395

Ask: How can you convert between mass and moles?

Ask: What makes predictions about carbon dioxide different from predictions about nickel or aluminum?

EXPLORE (15 min)

TRM Worksheet with Answers 77

TRM Worksheet 77

INTRODUCE THE LAB

⤵ Arrange students into groups of four.

⤵ Pass out the student worksheet.

EXPLAIN & ELABORATE (15 min)

DISCUSS THE MOLAR MASS OF COMPOUNDS

Ask: What do you have to know to figure out the mass of a mole of any compound on the planet?

Ask: How could you figure out the molar mass of carbon dioxide, CO_2?

Ask: Why is the molar mass listed for oxygen in the periodic table different from the molar mass of oxygen gas?

Key Points: You can figure out the mass of 1 mol of any element or compound

---

using a periodic table. For compounds, you must sum the molar masses of all the atoms in a molecule or a formula unit to find the molar mass of the compound. For example, each molecule of water contains two atoms of hydrogen and one atom of oxygen. The molar masses of hydrogen and oxygen are 1.008 g/mol and 16.00 g/mol, respectively. So, the molar mass of water, rounded to the correct number of significant digits, is 18.02 g/mol.

COMPARE A MOLE'S WORTH OF DIFFERENT SUBSTANCES

⤵ You might display a mole of each substance used for the lab.

Ask: What does 1 mol of water look like? What is its volume? (1 mol is 18 g, which has a volume of 18 mL when liquid—about the amount of an ice cube.)

Ask: What volume, in milliliters, would 10 mol of water occupy? (180 mL)

Ask: How many molecules of water are in your 1 mol sample? How many atoms of aluminum are in a 1 mol sample?

Ask: Which substance occupies the greatest volume of space per mole of substance? (any of the gases) The least? (the densest substance)

Ask: What is the volume of 1 mol of any gas at STP? (22.4 L)

Ask: Which has more mass, 1 mol of oxygen molecules, O_2, or 1 mol of aluminum atoms, Al? (oxygen molecules)

Key Points: A mole of atoms or molecules of a solid or a liquid is an amount you usually can hold in your hand. The mass per mole of the different elements in the periodic table ranges from 1.008 g/mol for hydrogen atoms to 238.0 g/mol for uranium atoms. Two aluminum cans contain about one mole of aluminum atoms. It is a convenient amount of substance to work with.

Recall Avogadro's law: A mole of any gas, if it is at standard temperature and pressure, always has a volume of 22.4 L. A mole of oxygen molecules occupies the same volume as a mole of helium atoms or a mole of chlorine gas molecules. A mole of a gas occupies a fairly large space—equivalent to ten 2 L soda bottles—because there is so much space between gas molecules. This is why it is difficult to compare the volume of a mole of a gas, like oxygen, and the volume of a mole of a solid, like aluminum.

Check-In

You have 1 mol of oxygen molecules, O_2, and 1 mol of carbon dioxide molecules, CO_2. Which has more mass? Which has a larger volume at room temperature?

Answer: CO_2 has a larger molar mass. Its molar mass is $12.02 + 2(16.00)$, or 42.02 g/mol. The molar mass of O_2 is $2(16.00)$, or 32.00 g/mol. They have equal volumes at room temperature.

Homework

Assign the reading and exercises for Toxins Lesson 77 in the student text.

LESSON 77 ANSWERS

1. To determine the molar mass of sodium chloride, add the atomic mass of sodium and chlorine to determine the mass of one unit of sodium chloride. The atomic mass in amu is the same as the molar mass in grams per mole.

2. *Possible answers:* For a gas, such as oxygen, 1 mol is always 22.4 L. ● For a liquid, such as water, 1 mol is a little more than a tablespoon. ● For a solid, such as salt 1 mol would be a small handful.

3. $Ne(g)$: 20.2 g/mol

$Ca(s)$: 40.1 g/mol

$CO_2(g)$: 44.0 g/mol

$CaCO_3(s)$: 100.1 g/mol

$CH_4O(l)$: 32.0 g/mol

$C_2H_6O(l)$: 46.1 g/mol

$Fe_2O_3(s)$: 159.7 g/mol

4. 1.0 g of methanol has more moles of molecules.

5. a. 10.0 g of calcium has more moles of metal atoms. **b.** 5.0 g of sodium fluoride has more moles of metal atoms. **c.** 2.0 g of iron oxide has more moles of metal atoms.

6. a. CH_4 has one carbon atom; 12.0 g. **b.** CH_4O has one carbon atom; 12.0 g. **c.** C_2H_6O has two carbon atoms; 24.0 g.

7. 798.5 g

8. C

Example 2

Toxicity of Lead Compounds

Which is potentially the most toxic: 1 g lead, Pb, 1 g lead (II) chloride, $PbCl_2$, or 1 g lead (II) carbonate, $PbCO_3$?

Solution

The molar mass of each compound was calculated in Example 1.

$$\text{molar mass of Pb} = 207.2 \text{ g/mol}$$
$$\text{molar mass of } PbCl_2 = 278.1 \text{ g/mol}$$
$$\text{molar mass of } PbCO_3 = 267.2 \text{ g/mol}$$

Because Pb has the lowest molar mass, 1 g lead, Pb, will have the largest number of moles of lead, so it will potentially be the most toxic, followed by lead (II) carbonate, $PbCO_3$.

LESSON SUMMARY

How can you convert between mass and moles?

Chemists compare moles of substances rather than masses of substances because moles are a way of counting atoms, molecules, or units in a compound. The molar mass of a substance is the mass, in grams, of one mole of the substance. The molar mass of a compound is the sum of the molar masses of the atoms in the compound.

Exercises

Reading Questions

1. Explain how to determine the molar mass of sodium chloride, NaCl.
2. Describe the approximate size of 1 mol of a solid, a liquid, and a gas. Give a specific example of each.

Reason and Apply

3. Copy this table and use a periodic table to complete the second column

Chemical formula	Molar mass (g/mol)	Moles of what?
$Ne(g)$		1 mol Ne atoms
$Ca(s)$		1 mol Ca atoms
$CO_2(g)$		1 mol carbon dioxide molecules
$CaCO_3(s)$		1 mol calcium carbonate units
$CH_4O(l)$		1 mol methanol molecules
$C_2H_6O(l)$		1 mol ethanol molecules
$Fe_2O_3(s)$		1 mol iron oxide units

(continued)

4. Which has more moles of molecules, 1.0 g methanol, CH_4O, or 1.0 g ethanol, C_2H_6O?

5. Which has more moles of metal atoms?
   a. 10.0 g calcium, Ca, or 10.0 g calcium chloride, $CaCl_2$
   b. 5.0 g sodium chloride, NaCl, or 5.0 g sodium fluoride, NaF
   c. 2.0 g iron oxide, FeO, or 2.0 g iron sulfide, FeS

6. How many grams of carbon molecules are in 1 mol of each substance?
   a. methane, CH_4
   b. methanol, CH_4O
   c. ethanol, C_2H_6O

7. What is the mass of 5 mol of iron (III) oxide, Fe_2O_3?

8. Which of these has the most chromium?
   (A) 1.0 g chromium (II) chloride, $CrCl_2$
   (B) 1.0 g chromium (III) chloride, $CrCl_3$
   (C) 1.0 g chromium (IV) oxide, CrO_2

Overview

CLASSWORK: PAIRS

Key Question: How are moles and mass related?

KEY IDEAS

To track amounts of compounds accurately, it is often necessary to convert mass to moles. Molar mass values allow us to convert back and forth between grams and moles.

LEARNING OBJECTIVES

- Convert the number of moles of a compound or an element to mass in grams.
- Convert the mass of a sample in grams to moles.

IN CLASS

Students complete a worksheet that gives them practice in relating mass to moles. They examine the amount of cyanide in a variety of cyanide compounds, comparing mass and moles. They also examine some over-the-counter pain medications, translating dosages into moles of each compound. Finally, they gain practice converting back and forth between mass and moles using common vitamin compounds. (*Note:* For convenience, in the remainder of the unit, we will work with molar mass values that have been rounded to one decimal place.) A complete materials list is available for download.

TRM Materials List 78

Differentiate

To help students develop scientific literacy and language skills, and to summarize their learning of mass-mole conversions, you can have them write about their understanding of mass-mole relationships. You can have students write an individual summary paragraph, or you might assign roles for students to write something as a group. You might also have students think of a particular audience they are addressing and how they might convey the information about mass-mole conversions creatively, such as in a poem, an advertisement, a letter, or a video.

Pre-AP® Course Tip

Students preparing to take an AP® chemistry or college chemistry course should be able to use mathematics to solve problems that describe the physical world. In this lesson, students use mathematics to convert between moles of a compound or an element to mass and vice versa.

Mass-Mole Conversions

THINK ABOUT IT

When you get a headache, you might reach for one of the many pain relievers on the market. A bottle of aspirin might direct you to take one 325 mg tablet. A bottle of acetaminophen might say to take one 500 mg tablet. Is one of these medications stronger than the other? One way to find out is to compare the number of moles of pain reliever in each dose.

How are moles and mass related?

To answer this question, you will explore

↻ Relating Mass and Moles

↻ Converting between Mass and Moles

EXPLORING THE TOPIC

↻ Relating Mass and Moles

You can find the mass of a substance by using a balance. However, you cannot directly measure the number of moles in a sample. You must convert the mass of a substance to moles to figure out how many atoms or molecules you have. If you want to compare amounts of two substances accurately, it is important to be able to convert between mass and moles.

Typically, you measure the mass of a substance in grams. Medications, like aspirin, are often measured in milligrams. Because there are 1000 milligrams in 1 gram, converting from milligrams to grams is simply a matter of dividing by 1000.

Consider three compounds commonly used in over-the-counter pain relievers.

EARTH SCIENCE CONNECTION

It is estimated that there are 2.0×10^{21} grains of sand on the entire Earth. This is only 0.003 mol of sand grains! In contrast, there is approximately 0.0002 mol of water molecules, or 1.2×10^{21} molecules, in a single drop of water.

Pain reliever	Molecular formula	Molar mass	Adult dose	Moles in a standard dose
ibuprofen	$C_{13}H_{18}O_2$	206.3 g/mol	400 mg	0.0019 mol
acetaminophen	$C_8H_9NO_2$	151.2 g/mol	500 mg	0.0033 mol
acetylsalicylic acid (aspirin)	$C_9H_8O_4$	180.2 g/mol	325 mg	0.0018 mol

Notice that each pain reliever has a different adult dose. It would appear that acetylsalicylic acid (aspirin), with a smaller, 325 mg dose, is stronger than either acetaminophen, 500 mg, or ibuprofen, 400 mg. But ibuprofen is a heavier molecule than acetylsalicylic acid. One gram of aspirin will not represent the same number of molecules as one gram of ibuprofen. To make

---

Lesson Guide

ENGAGE (5 min)

TRM PowerPoint Presentation 78

ChemCatalyst

Arsenic, As, arsenic (III) oxide, As_2O_3, and arsenic (III) sulfide, As_2S_3, are all toxic because they contain arsenic.

1. Which is more toxic, 1 mol of As or 1 mol of As_2O_3? Explain your thinking.
2. Which is more toxic, 1 g of As_2O_3 or 1 g of As_2S_3? Explain.

Sample answers: **1.** Students may say that 1 mol of arsenic is more toxic because it contains only pure arsenic atoms, or they may say that 1 mol of arsenic oxide is more toxic because it contains 2 mol of arsenic. **2.** Students may say that they have the same toxicity because each compound has two arsenic atoms. Others may say that the molar mass of arsenic sulfide is greater than the molar mass of arsenic oxide, so there will be fewer formula units of arsenic sulfide in a 1 g sample. This would make the arsenic oxide more toxic.

Ask: How do you compare moles of arsenic to moles of arsenic oxide?

an accurate comparison, you must convert the milligrams of each compound to moles.

Examine the table. Which medication contains the most moles of pain reliever? A 500 mg dose of acetaminophen contains almost twice as many moles of pain reliever as a 325 mg dose of acetylsalicylic acid. So aspirin does appear to be more powerful than acetaminophen. Acetylsalicylic acid has the lowest dose and uses the fewest molecules of pain reliever to get rid of your headache. Ibuprofen is almost identical in potency to acetylsalicylic acid. According to the data, acetaminophen is not as powerful as the other two compounds.

⟳ Converting between Mass and Moles

The relationship between the mass in grams of a substance and moles is proportional. The proportionality constant is the molar mass, in g/mol. The molar mass will be different for each substance.

Mass-Mole Conversions The molar mass is the mass per mole of a substance. You can use it to convert between the mass, m, and the number of moles, n.

Example 1

Converting from Moles to Mass

Imagine that a pharmaceutical company has 100 mol of acetaminophen available to be made into 500 mg tablets. The molar mass of acetaminophen is 151.2 g/mol. How many 500 mg tablets can be made?

Solution

First, figure out how many grams of acetaminophen there are in 100 mol.

Start with the mole to mass formula. $\qquad m = \text{molar mass} \cdot n$

Substitute values and solve for mass. $\qquad = (151.2 \text{ g/mol}) \, 100 \text{ mol}$

$\qquad\qquad\qquad\qquad\qquad\qquad\qquad = 15{,}120 \text{ g}$

Because each 500 mg tablet is 0.5 g, each gram of the compound will make two tablets.

$$(15{,}120 \text{ g}) \, (2 \text{ tablets/g}) = 30{,}240 \text{ tablets}$$

So 30,240 tablets can be made from 100 mol of acetaminophen.

Example 2

Converting from Mass to Moles

Restaurants usually provide small packets of sugar to sweeten coffee or tea. Each sugar packet contains approximately 1.0 g of sucrose, $C_{12}H_{22}O_{11}$. How many moles of sucrose does this represent?

Key Points: The relationship between mass (g) of a substance and number of moles is proportional.

$$\text{mass (g)} = k \cdot \text{moles}$$

The proportionality constant, k, is equal to the molar mass and is different for each substance. The molar mass of a substance is used to convert between moles and mass. Moles cannot be measured directly. Instead, we measure the mass of a substance and then convert the mass to moles using the molar mass.

To convert mass to moles, divide the mass in grams by the molar mass.

$$\text{number of moles} = \frac{\text{mass (g)}}{\text{molar mass}}$$

To convert moles to mass, multiply the number of moles by the molar mass.

$$\text{mass (g)} = \text{number of moles} \cdot \text{molar mass}$$

The relationship between the number of particles of a substance and the number of moles is also proportional.

$$\text{number of particles} = k \cdot \text{number of moles}$$

In this case, the proportionality constant, k, is Avogadro's number, 6.02×10^{23}. Note that "particles" can refer to atoms, ions, formula units, molecules, and so on. To convert from moles to particles, multiply the number of moles by Avogadro's number.

$$\text{number of particles} = 6.02 \times 10^{23} \cdot \text{number of moles}$$

To convert from particles to moles, divide the number of particles by Avogadro's number.

$$\text{number of moles} = \frac{\text{number of particles}}{6.02 \times 10^{23}}$$

COMPLETE SEVERAL WORKED EXAMPLES

→ Ask students to participate in working these examples on the board.

EXAMPLE 1

How many moles of arsenic, As, are there in a sample with a mass of 125 g?

Ask: Which substance has the most arsenic atoms in a 1 mol sample? Explain.

Ask: Which substance has the most arsenic atoms in a 1 g sample? Explain.

Ask: How can you convert from grams to moles?

EXPLORE (15 min)

TRM Worksheet with Answers 78

TRM Worksheet 78

INTRODUCE THE CLASSWORK

→ Arrange students into pairs.

→ Pass out the student worksheet.

EXPLAIN & ELABORATE (15 min)

DISCUSS HOW MOLES AND MASS ARE RELATED

Ask: If you have 1 mol of a substance, how can you determine the mass of the substance?

Ask: If you have the mass of a substance, how can you determine the number of moles of the substance?

Ask: Why do you think it is important to be able to convert back and forth between mass and number of moles?

SOLUTION

First, use the periodic table to determine the molar mass of arsenic.

> molar mass of As = 74.9 g/mol

Next, divide the mass of As by its molar mass.

$$\text{moles of As} = \frac{125 \text{ g}}{74.9 \text{ g/mol}} = 1.67 \text{ mol}$$

There are 1.67 moles of arsenic in a sample with a mass of 125 g.

EXAMPLE 2

How many moles of calcium chloride, $CaCl_2$, are there in a sample with a mass of 10 g?

SOLUTION

First, use the periodic table to determine the molar mass of calcium chloride.

> molar mass of $CaCl_2$ = 110.9 g/mol

Next, divide the mass of calcium chloride by its molar mass.

$$\text{moles of } CaCl_2 = \frac{10 \text{ g}}{110.9 \text{ g/mol}} = 0.090 \text{ mol}$$

There is 0.090 mol of $CaCl_2$ in a sample with a mass of 10 g.

EXAMPLE 3

What is the mass in grams of 3.0 moles of Mg?

SOLUTION

First, use the periodic table to determine the molar mass of magnesium, Mg.

> molar mass of Mg = 24.3 g/mol

Next, multiply the number of moles of magnesium by its molar mass.

> mass of Mg = 3.0 mol · 24.3 g/mol = 73 g

So, 3.0 moles of Mg is equal to 73 g of Mg.

EVALUATE (5 min)

Check-In

A sample of chlorine gas, $Cl_2(g)$, has a mass of 11 g. How many moles of $Cl_2(g)$ is this?

Answer: 0.16 mole of $Cl_2(g)$

(continued)

Solution

First, use the periodic table to determine the molar mass of sucrose. The molar mass of sucrose is 342.3 g/mol rounded to one decimal place. To convert from mass to moles, divide the mass in grams by the molar mass.

Rearrange the mole to mass formula to solve for *n*.

$$n = \frac{m}{\text{molar mass}}$$

Substitute values and solve for moles of sucrose.

$$= \frac{1.0 \text{ g}}{342.3 \text{ g/mol}} = 0.003 \text{ mol}$$

So one sugar packet contains 0.003 mol of sucrose.

LESSON SUMMARY

How are moles and mass related?

Molar mass values allow you to convert between mass in grams of a substance and moles. To convert mass to moles, divide the mass of a sample in grams by its molar mass. To convert moles to mass, multiply the number of moles of the substance by its molar mass.

Exercises

Reading Questions

1. How can you convert between moles of a substance and grams of a substance?
2. Why might a 200 mg tablet of aspirin not have the same effect as a 200 mg tablet of ibuprofen?

Reason and Apply

3. There are 8.0 mol of H atoms in 2.0 mol of CH_4O molecules. How many moles of H atoms are there in 2.0 mol of C_2H_6O molecules?
4. List these compounds in order of increasing moles of molecules: 2.0 g CH_4O, 2.0 g H_2O, 2.0 g C_8H_{18}. Show your work.
5. Which has more moles of oxygen atoms, 153 g of BaO, or 169 g of BaO_2? Show your work.
6. List these compounds in order of increasing mass in grams: 2.0 mol $SiCl_4$, 2.0 mol PbO, 2.0 mol Fe_2O_3. Show your calculations.
7. Copper is the third most used metal after iron and aluminum. Copper usage is rapidly expanding as more products are developed that contain electronic components. Suppose you run a company that buys copper compounds and then recycles the copper for resale. Your company wants to get the most pure copper for

---

Homework

Assign the reading and exercises for Toxins Lesson 78 in the student text.

LESSON 78 ANSWERS

1. To convert between moles of a substance and grams of the substance, use the formula:

mass (g) = molar mass (g/mol) · moles

2. A 200 mg tablet of aspirin would not have the same effect as a 200 mg tablet of ibuprofen because the aspirin molecules have a larger molecular mass, so there are fewer molecules of aspirin in a 200 mg tablet.

3. 12.0 mol H.

4. 1 mol CH_4O = 12.01 g + 4(1.008 g) + 16.00 g = 32.0 g

1 mol H_2O = 2(1.008 g) + 16.00 g = 18.0 g

1 mol C_8H_{18} = 8(12.01 g) + 18(1.008 g) = 114.1 g

$$\text{moles of } CH_4O = \frac{2.0 \text{ g}}{32.0 \text{ g/mol}} = 0.063 \text{ mol}$$

$$\text{moles of } H_2O = \frac{2.0 \text{ g}}{18.0 \text{ g/mol}} = 0.11 \text{ g/mol}$$

$$\text{moles of } C_8H_{18} = \frac{2.0 \text{ g}}{114.1 \text{ g/mol}} = 0.018 \text{ g/mol}.$$

(continued)

the lowest cost. Three different suppliers want to sell you 1 mol CuO(*s*), 1 mol CuCO$_3$(*s*), and 1 mol Cu$_2$O(*s*) for the same price.

a. Which compound has the greatest total mass? Show your work.

b. Which compound has the greatest mass of Cu? Show your work.

c. Assuming it costs the same to extract the copper from each compound, which compound represents the best deal for your company? Explain.

8. Suppose Container A contains 1 mol C(*s*) and 1 mol O$_2$(*g*), and Container B contains 1 mol CO$_2$(*g*). Both containers are closed and both are the same size. Compare the containers in each of these ways.

a. number of atoms

b. number of gas molecules

c. mass

d. gas pressure

1 mol Cu$_2$O has the most mass. **b.** One mole of CuO and one mole of CuCO$_3$ each contain one mole of copper atoms, so the mass of copper in each sample is 63.55 g. One mole of Cu$_2$O contains two moles of copper atoms, so the mass of copper in the sample is 2 · 63.55 g = 127.1 g. So, one mole of Cu$_2$O contains the most copper atoms. **c.** Cu$_2$O represents the best deal for the company because you get twice as much copper for the same price.

8. a. Both containers have the same number of atoms. **b.** Container A has 2 mol of gas molecules and Container B has 1 mol of gas molecules. **c.** Both containers have the same mass. **d.** Assuming both containers are at the same temperature, Container A has a greater pressure because it has twice as many gas molecules as Container B.

In order of increasing moles of molecules: C$_8$H$_{18}$, CH$_4$O, H$_2$O

5. 1 mol BaO = 137 g + 16.0 g = 153 g

1 mol BaO$_2$ = 137 g + 32.0 g = 169 g.

Each sample has 1 mole of molecules. However, the sample of 169 g of BaO$_2$ has more moles of oxygen atoms because it has 2 moles of oxygen atoms per molecule.

6. 1 mol SiCl$_4$ = 28.09 g + 4(35.45 g)
= 170.1 g

2 mol SiCl$_4$ = 2 mol(170.1 g/mol)
= 340.2 g

1 mol PbO = 207.2 g + 16.00 g
= 223.2 g

2 mol PbO = 2 mol(223.2 g/mol)
= 446.4 g

1 mol Fe$_2$O$_3$ = (55.85 g) + 3(16.00 g)
= 159.7 g

2 mol Fe$_2$O$_3$ = 2 mol(159.7 g/mol)
= 319.4 g

In order of increasing mass: Fe$_2$O$_3$, SiCl$_4$, PbO

7. a. 1 mol CuO = 63.55 g + 16.00 g
= 79.6 g

1 mol CuCO$_3$ = 63.55 g + 12.01 g
+ 3(16.00 g) = 123.6 g

1 mol Cu$_2$O = 2(63.55 g) + 16.00 g
= 143.0 g

Overview

CLASSWORK: INDIVIDUALS

DEMO: WHOLE CLASS

Key Question: How can you use moles to compare toxicity?

KEY IDEAS

Chemists often compare moles of substances rather than masses. Using moles allows scientists to compare numbers of particles or molecules. Substances with large molar masses will contain fewer particles per gram than substances with smaller molar masses.

LEARNING OBJECTIVES

- Use moles and molar mass to compare the amounts of different substances.
- Discuss the safety of sweeteners.

IN CLASS

This lesson further develops the concept of the mole and brings students' attention back to the topic of potentially toxic substances. It focuses on the issue of why counting particles can be important to chemists. Students find that because aspartame molecules (an artificial sweetener) are much sweeter than fructose molecules (sugar), very little aspartame is needed to sweeten soft drinks, compared with fructose. Students discuss the safety of sweeteners in soft drinks, both regular and diet. To do so, students take into account both the toxicity (LD_{50}) of fructose and aspartame and the amount of each sweetener in a can of soft drink. A complete materials list is available for download.

TRM Materials List 79

SETUP

It is important to test the cans of soft drinks you have chosen for the demo to make sure the one with artificial sweetener floats.

Lesson Guide

ENGAGE (5 min)

TRM PowerPoint Presentation 79

ChemCatalyst

Consider two cans of carbonated soft drink. One is regular, and the other is diet.

How Sweet It Is

Comparing Amounts

THINK ABOUT IT

The safety of artificial sweeteners has been the subject of debate for years. More recently, some schools have banned the sale of both diet and regular soft drinks. Regular soft drinks are sweetened with fructose, while diet soft drinks are usually sweetened with aspartame. If you want to make a healthy decision, you need to consider the toxicities of fructose and aspartame and the amount of each substance in a can of beverage.

How can you use moles to compare toxicity?

To answer this question, you will explore

⤵ Comparing Moles of Sweeteners

⤵ Toxicity of Sweeteners

EXPLORING THE TOPIC

⤵ Comparing Moles of Sweeteners

It is hard for many of us to imagine a day without sugar. From cereal to flavored drinks and candy, our intake of sugar adds up. So, when the first artificial sweeteners came onto the market in the 1950s, it seemed a good idea to use this low-calorie substitute for a favorite ingredient.

Two sweeteners, fructose, $C_6H_{12}O_6$, and aspartame, $C_{14}H_{18}N_2O_5$, are shown. Notice that they are both medium-sized molecules made up of carbon, hydrogen, and oxygen atoms. The artificial sweetener, aspartame, also has two nitrogen atoms in its molecules.

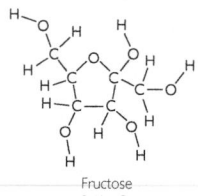

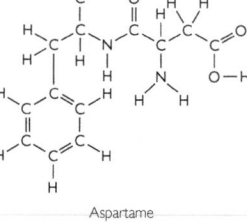

Fructose
$C_6H_{12}O_6$
Molar mass: 180.1 g/mol

Aspartame
$C_{14}H_{18}N_2O_5$
Molar mass: 294.3 g/mol

The molar masses of fructose and aspartame represent the mass per mole of molecules. Recall that you can obtain the molar mass of a molecule by adding the molar masses of each atom in the molecule.

> **CONSUMER CONNECTION**
>
> The compounds aspartame, saccharin, and sucralose are the most common artifical sweeteners in use today. Aspartame is the sweetener most commonly used in diet soft drinks. Saccharin is often used as a sweetener in toothpastes.

402   Chapter 14 | **Measuring Toxins**

---

1. Which do you think is more toxic, the regular soft drink or the diet soft drink? Explain your reasoning.
2. What information would help you answer this question?

Sample answers: **1.** Some students might think that the diet soft drink is more toxic because it contains a synthetic or artificial sweetener. Others might say that the sugar in the regular soft drink can lead to diabetes and obesity. **2.** It would be helpful to know the lethal doses of the ingredients in these soft drinks and the amount of each in a typical carbonated soft drink.

Ask: What did you base your answers on?

Ask: What do you have to know to answer this question?

Ask: How does the amount of each ingredient affect toxicity?

EXPLORE (15 min)

TRM Worksheet with Answers 79

TRM Worksheet 79

40 g
Fructose

0.225 g
Aspartame

The molar mass of fructose is less than that of aspartame. A 1 g sample of fructose will contain more molecules than a 1 g sample of aspartame because each fructose molecule has a smaller mass than an aspartame molecule.

A can of regular soft drink has about 40.0 g of fructose as a sweetener. A can of diet soft drink has about 0.225 g of aspartame as a sweetener. You can determine the number of molecules of sweetener in each can using the molar mass of each substance.

$$Fructose = (regular\ soft\ drink) \qquad Aspartame = (diet\ soft\ drink)$$
$$n = \frac{40\ g}{180.1\ g/mol} \qquad\qquad n = \frac{0.225\ g}{294.3\ g/mol}$$
$$= 0.22\ mol\ fructose \qquad\qquad = 0.00076\ mol\ aspartame$$

A diet soft drink needs only 0.00076 mol of aspartame molecules to replace the sweetness of 0.22 mol of fructose molecules in a regular soft drink. So you need significantly less aspartame to obtain the same sweetness provided by fructose.

○ Toxicity of Sweeteners

Artificial sweeteners allow people with diabetes, and others who must avoid sugar, to enjoy foods that would otherwise be dangerous for them to eat. Artificial sweeteners do not cause cavities, and because they are not digested, artificial sweeteners do not provide extra calories to the body. However, the safety of artificial sweeteners continues to be a subject of debate.

To make an informed decision about whether you want to consume artificial sweeteners, it is useful to compare the toxicities of fructose and aspartame. Take a moment to review this information about each sweetener.

Fructose (sugar)	Aspartame (artificial sweetener)
$C_6H_{12}O_6$	$C_{14}H_{18}N_2O_5$
$LD_{50} = 28.5\ g/kg = 0.158\ mol/kg$	$LD_{50} = 10\ g/kg = 0.034\ mol/kg$

The LD_{50} of fructose in grams is about three times larger than the LD_{50} for aspartame (30 g per kg versus 10 g per kg). The LD_{50} in moles is almost five times larger for fructose than for aspartame (0.158 mol per kg versus 0.034 mol per kg). So, the sweetener aspartame is more toxic than fructose if you consume similar amounts of each.

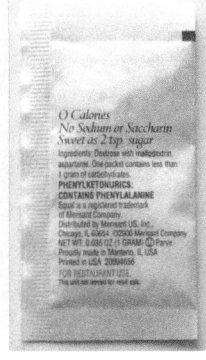

Example

How Many Cans?

How many cans of a regular soft drink does a 64 kg (141 lb) person have to drink in one sitting to exceed the lethal dose of fructose? How many cans of diet soft drink does the same person have to drink in a short time to exceed the lethal dose of aspartame?

Solution

First, determine how many grams of each sweetener are lethal for a 64 kg person. Then, take this number of grams and divide by the grams of each sweetener in a can of soft drink to get the number of cans of soft drink that would be lethal.

Lesson 79 | **Comparing Amounts**   403

RUN THE DEMONSTRATION

➔ Ask students to predict what will happen when two cans of carbonated soft drink are placed into a bucket of water. Will they float or sink? Ask for their reasoning.

➔ After students have discussed their predictions, place the cans of carbonated soft drink into a large container of water, one at a time. A translucent or transparent container works best for this demonstration so the class can see the outcome. Because of the difference in densities, the regular soft drink will sink and the diet soft drink will float.

➔ Allow students to share their explanations of what happened. You may want to let students feel the weight of each can.

INTRODUCE THE CLASSWORK

➔ Tell students they will work individually.

➔ Pass out the student worksheet.

EXPLAIN & ELABORATE (15 min)

PROCESS THE CLASSWORK

➔ You might want to share a restaurant sugar packet and an artificial sweetener packet with the

class for comparison. They have dramatically different masses.

Ask: Which molecule is sweeter, aspartame or fructose? Explain.

Ask: If you have 1 g of fructose and 1 g of aspartame, do you have the same amount of each substance? Explain.

Ask: Would you rather be exposed to a gram of a substance with a large molar mass or a small molar mass? Explain.

Ask: Why does a packet of sugar weigh more than a packet of aspartame?

Key Points: Aspartame molecules are much sweeter than fructose molecules. A diet soft drink uses only 0.00076 mol of aspartame to replace the sweetness of 0.22 mol of sugar in a regular soft drink. Thus, each molecule of aspartame is much sweeter than each sugar molecule.

DISCUSS THE SAFETY OF ARTIFICIAL SWEETENERS (OPTIONAL)

Ask: Do you think artificial sweeteners are safe for human consumption? How could scientists find out?

Ask: What would people with diabetes do if there were no artificial sweeteners?

Ask: How do artificial sweeteners contribute to the quality of life?

Ask: Do you think schools should ban carbonated soft drinks? Why or why not?

Key Points: The safety of artificial sweeteners has been a subject of debate for years. Many people enjoy having a substitute for sugar. Some people, such as those with diabetes, must avoid sugar. Artificial sweeteners offer many people an alternative to sugar. However, as a result of animal testing, several artificial sweeteners have been banned by the U.S. Food and Drug Administration. Aspartame is still considered safe for consumption, although it is a source of controversy.

It is interesting to compare the LD_{50} for fructose and for aspartame. The LD_{50} for aspartame is >10,000 mg per kilogram of body weight, and for fructose it is 29,700 mg per kilogram of body weight. Aspartame is more toxic because the LD_{50} is smaller. However, you use much less aspartame than fructose. For a 64 kg person, it would take about 48 cans of regular soft drink over a short period of time to reach a lethal dose, compared to about 280 cans of diet soft drink.

	Regular soft drink (fructose)	Diet soft drink (aspartame)
Lethal dose for a 64 kg person	29,700 mg/kg · 64 kg = 1,900,000 mg = 1,900 g	10,000 mg/kg · 64 kg = 64,000 mg = 64 g
Number of cans	1,900 g/40 g/can ~48 cans	64 g/0.225 g/can ~280 cans

EVALUATE (5 min)

Check-In

Imagine two substances of equal mass with molecules that are equally toxic. Would you rather be exposed to the substance with the larger molar mass or the smaller molar mass? Explain your reasoning.

Answer: It would be better to be exposed to 1 g of the substance with the larger molar mass than 1 g of the substance with the smaller molar mass. You would be exposed to fewer molecules.

Homework

Assign the reading and exercises for Toxins Lesson 79 and the Chapter 14 Summary in the student text.

LESSON 79 ANSWERS

1. *Possible answer:* A comparison of the amount of each compound needed to make a soft drink taste sweet provides evidence that aspartame is sweeter than fructose. It takes more molecules of fructose than aspartame to sweeten the drink.

2. *Possible answer:* The LD_{50} value of aspartame is much greater than the amount of aspartame in a soft drink. A person would have to drink nearly 3000 cans to consume the LD_{50} value.

3. a. 364 cans, or 360 cans (to two significant figures). **b.** 0.00072 mol/kg

4. aspartame LD_{50} = 0.034 mol/kg

1 mol $C_7H_5NO_3S$ = 7(12.01 g) + 5(1.008 g) + 14.01 g + 3(16.00 g) + 32.07 g = 183.2 g

saccharin LD_{50} = $\dfrac{14.2 \text{ g}}{\text{kg}} \cdot \dfrac{1 \text{ mol}}{183.2 \text{ g}}$

= 0.0775 mol/kg; It takes a smaller amount of aspartame to reach the LD_{50} value, so aspartame would be more toxic.

Example

(continued)

	Regular soft drink	Diet soft drink
Determine the lethal dose for each sweetener using LD_{50} values.	$64 \text{ kg}\left(25.8\dfrac{g}{kg}\right) = 1651 \text{ g}$	$(64 \text{ kg})\left(10\dfrac{g}{kg}\right) = 640 \text{ g}$
Divide by the mass of sweetener in one can to get number of cans.	$(1651 \text{ g})\left(\dfrac{1 \text{ can}}{40 \text{ g}}\right) = 41.3 \text{ cans}$	$640 \text{ g}\left(\dfrac{1 \text{ can}}{0.225 \text{ g}}\right) = 2884 \text{ cans}$

It would take about 41 cans of regular soft drink to reach the lethal dose of fructose and 2884 cans of diet soft drink to reach the lethal dose of aspartame.

HEALTH CONNECTION

In 1958, Congress passed the Delaney Clause of the Food, Drug, and Cosmetic Act. The clause prohibited the use of all food additives found to be carcinogenic. As a result, several artificial sweeteners have been banned by the Food and Drug Administration over the years.

It would be difficult to drink enough regular soda or diet soda to reach the lethal dose in a short period of time. So it is highly unlikely that exposure to either fructose or aspartame would be lethal in the *short term*. However, it is important to note that the LD_{50} is not a good measure of health effects from *long-term* exposure. It is probably best to limit intake of both sugar and aspartame.

LESSON SUMMARY

How can you use moles to compare toxicity?

Chemists often find it useful to compare moles of substances rather than masses of substances. Moles allow you to compare numbers of molecules or formula units in your sample. Substances with large molar masses contain fewer molecules per gram than substances with smaller molar masses. The health effect of exposure to a toxic substance depends on both the LD_{50} and the amount of the substance.

Exercises

Reading Questions

1. What evidence shows that aspartame is sweeter than fructose?
2. What evidence shows that it would be difficult to exceed the lethal dose of aspartame?

Reason and Apply

3. There are 25 mg of caffeine, $C_8H_{10}N_4O_2$, in a can of regular soft drink. The LD_{50} for caffeine, $C_8H_{10}N_4O_2$, is 140 mg/kg.
   a. How many cans of regular soft drink can a 65 kg person drink in a short period of time before exceeding the lethal dose?
   b. What is the toxicity of caffeine in moles per kilogram?
4. The LD_{50} for saccharin, $C_7H_5NO_3S$, is 14.2 g/kg. If you have 1 mol of aspartame and 1 mol of saccharin, which would be more toxic? Show your work.
5. PROJECT www Write an argument for or against the use of artificial sweeteners. Be sure to provide evidence to support your argument. Cite at least two references that you use.

5. A good argument will contain:
● A list of the risks and benefits of artificial sweeteners and a statement showing whichever side of the issue the student decides to take. For each item on the list that fails to support the student's argument, the student must show in what ways other factors outweigh the item. ● Two or more citations of authoritative sources, such as scientific journals, government studies, or interviews, should back up the argument.

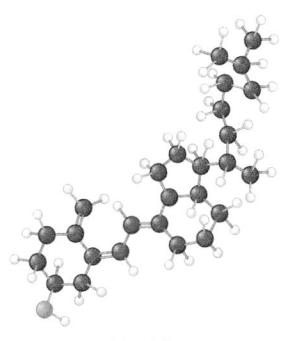

CHAPTER 14

Measuring Toxins

SUMMARY

KEY TERMS

toxicity

percent error

molar mass

scientific notation

Toxins Update

To keep the public safe, health experts measure the toxicity of substances that people might be exposed to in the course of their daily lives. The most common measurement of toxicity is called lethal dose, or LD_{50}, which is expressed in milligrams per kilogram of body weight. The higher the LD_{50} for a substance, the safer the compound is.

Chemical formulas track compounds in counting units called moles. To determine if a certain quantity of matter amounts to a lethal dose, it is necessary to convert between mass and moles using molar mass.

REVIEW EXERCISES

1. Mass and moles are proportionally related. Explain what this means.

2. Why is it necessary to take a person's weight into account when considering the correct dose for a medication?

3. Why is scientific notation a useful tool for chemists?

4. Consider vitamin D_3, $C_{27}H_{44}O$, cholecalciferol. Vitamin D_3 helps the body extract calcium and phosphorus from food, plays a role in mental health, and may prevent some cancers. Most of the vitamin D you get is produced in your skin when you are exposed to the Sun.

Vitamin D₃

Lethal dose: LD_{50} = 42.0 mg/kg (rat, oral)

Recommended Dietary Allowance: RDA = 0.010 mg/day

a. Determine the number of moles and molecules of vitamin D_3 in the Recommended Dietary Allowance.

b. Determine the lethal dose for a 140 lb person.

c. Determine the number of moles of vitamin D_3 in a lethal dose for a 140 lb person.

e. No, because 1×10^{20} molecules are less than the lethal doses of 4.2×10^{21} molecules for a 140 lb adult.

f. No, because 400 tablets contain $400 \cdot 10 \ \mu g = 4000 \ \mu g = 4$ mg of the Vitamin D_3. This value is less than the lethal dose of 2670 mg.

5. In each compound, one mole of the compound contains one mole of metal atoms.

1 mol NaCl = 22.99 g + 35.45 g = 58.4 g

1 mol AgCl = 107.9 g + 35.45 g = 143.4 g

1 mol LiCl = 6.941 g + 35.45 g = 42.4 g

moles of NaCl = $\dfrac{5.0 \text{ g}}{58.4 \text{ g/mol}}$ = 0.085 mol

moles of AgCl = $\dfrac{5.0 \text{ g}}{143.4 \text{ g/mol}}$

= 0.035 mol

moles of LiCl = $\dfrac{5.0 \text{ g}}{42.4 \text{ g/mol}}$ = 0.12 mol

In order of increasing moles of metal: AgCl, NaCl, LiCl.

6. a. $TiCl_4(s) + 2Mg(s) \rightarrow Ti(s) + 2MgCl_2(s)$

b. $Mg(s) + H_2O(l) \rightarrow MgO(s) + H_2(g)$

c. $Fe(s) + CuSO_4(s) \rightarrow FeSO_4(s) + Cu(s)$

d. $2Li(s) + 2H_2O(l) \rightarrow 2LiOH(aq) \pm H_2(g)$

Project: Lethal Dose of a Toxic Substance

A good report would include: ● a general description of the toxic substance, including its name, chemical formula, and its general uses ● the LD_{50} value for the toxic substance ● the type of lab study that was used to determine the LD_{50} value (the animal exposed and the method of exposure) ● a step-by-step calculation of the lethal dose for a 140 lb (63.6 kg) person ● two reputable sources for the information about the toxic substance, including at least one Material Safety Data Sheet.

ASSESSMENTS

Two multiple-choice assessments (versions A and B) and two short-answer assessments (versions A and B) are available for download for Chapter 14.

TRM Chapter 14 Assessment Answers

TRM Chapter 14 Assessments

ANSWERS TO CHAPTER 14 REVIEW EXERCISES

1. *Possible answer:* A proportional relationship means that as you increase the number of moles you will increase the mass by the same factor. If either the mass or the number of moles in a sample is known, it is always possible to determine the other value as long as the sample can be identified.

2. *Possible answer:* A person's weight must be taken into account when considering the correct dose because the amount of medication needed is proportional to the person's body weight. Medications have a stronger effect on small children than on large adults.

3. *Possible answer:* Scientific notation is a useful tool for chemists because it is often necessary to use very large or very small numbers in calculations. Scientific notation simplifies these calculations and also makes comparison of the values easier.

4. a. RDA of Vitamin D_3 = 1×10^{-5} g, so $1 \times 10^{-5} \text{ g}\left(\dfrac{1 \text{ mol}}{384.6 \text{ g}}\right) = 2.6 \times 10^{-8}$ mol

Number of molecules = 2.6×10^{-8} g $(6.02 \times 10^{23}) = 1.6 \times 10^{16}$

b. lethal dose = 2670 mg

c. 0.00695 mol

d. Number of molecules = 4.18×10^{21}

  d. Determine the number of molecules of vitamin D_3 in a lethal dose for a 140 lb person.

  e. Would 1.0×10^{20} molecules of vitamin D_3 represent a lethal dose for a 140 lb person? Explain your reasoning.

  f. Would 400 tablets (0.010 mg each) of vitamin D_3 be a lethal dose for a 140 lb person? Explain your reasoning.

5. List these compounds in order of increasing moles of metal in each compound: 5.0 g NaCl, 5.0 g AgCl, 5.0 g LiCl. Show your work.

6. Copy and balance these chemical equations.

  a. $TiCl_4(s) + Mg(s) \longrightarrow Ti(s) + MgCl_2(s)$

  b. $Mg(s) + H_2O(l) \longrightarrow MgO(s) + H_2(g)$

  c. $Fe(s) + CuSO_4(s) \longrightarrow FeSO_4(s) + Cu(s)$

  d. $Li(s) + H_2O(l) \longrightarrow LiOH(aq) + H_2(g)$

Lethal Dose of a Toxic Substance

PROJECT Research a toxic substance (your teacher may assign you one). Find some information about your substance. Your project should include

- the LD_{50} for the substance.

- the lethal dose for a 140 lb person expressed in grams and also in moles.

- two sources for the information you obtained. One source should be a Material Safety Data Sheet (MSDS).

CHAPTER 15

Toxins in Solution

In this chapter, you will study

- how to track dissolved substances in solutions
- a particle view of solutions
- how to create solutions

Water is one of the most valuable resources on the planet. It must be carefully monitored to make sure it is safe for public use.

Many compounds dissolve in water and can become part of our drinking water supply. An aqueous solution is one that contains water, with other ingredients uniformly mixed in. Understanding solutions allows you to monitor and track these substances.

407

PD CHAPTER 15 OVERVIEW

Watch the video overview of Chapter 15 (for teachers) by clicking on the link in the TE-book, opening the TRFD, or logging onto the book's companion Web site bcs.whfreeman.com/livingbychemistry2e (teacher log-in required).

Chapter 15 focuses on solution chemistry as students track toxic substances that become part of the water supply. Lesson 80 is a lab that challenges students to create saturated solutions of salt and sugar and compare their number density and mass density. Observations of gummy bears placed in the solutions provide information on differences in solution concentration. In Lesson 81, students examine solutions from a particulate point of view. They learn about molarity and are challenged to create solution samples of specific volume and concentration from a larger sample of known molarity. In Lesson 82, students work to create a water sample of a specific concentration for a saltwater aquarium. They use molarity calculations to determine the number of grams and moles of solute in their solutions and then check their results using a hydrometer. In Lesson 83, students use mass and moles to identify several solutions.

In this chapter, students will learn

- definitions of the terms *solute, solvent, saturated solution, concentration,* and *molarity*
- how to create a solution with a specific molarity
- how to calculate the number of moles or grams in a solution sample
- the difference between number density and mass density in a solution
- the proportional relationship between moles and liters

Chapter 15 Lessons

Overview

LAB: GROUPS OF 4

Key Question: How can you keep track of compounds when they are in solution?

KEY IDEAS

A solution is a mixture of two or more substances that is uniform throughout. The dissolved substance is generally referred to as the *solute,* and the medium the substance is dissolved in is called the *solvent.* The concentration of a liquid solution is a measure of the number of moles of a dissolved substance per liter of solution. The concentration of a solution is a measure of number density, *n/V,* called the *molarity* of the solution.

LEARNING OBJECTIVES

- Define the terms *solution, saturated solution, solute,* and *solvent.*
- Explain what the concentration and molarity of a solution represent.
- Complete calculations involving molarity.

KEY TERMS

solution
solvent
solute
homogeneous mixture
heterogeneous mixture
concentration
molarity
molar
saturated solution

IN CLASS

Students continue to explore moles in this lesson, focusing on the concentrations of substances in solutions. Students are challenged to create two saturated aqueous solutions, one using sugar, $C_{12}H_{22}O_{11}$, as the solute and the other using salt, NaCl. Students determine the molarity of each solution by measuring the mass of the water before and after adding the solute and calculating the number of moles of solute added. *Note:* Students are introduced to the terms *homogeneous mixture* and *heterogeneous mixture* in the reading in the student text. The homework in this lesson includes a Web Research Project related to the toxin assigned in Chapter 13. A complete materials list is available for download.

TRM Materials List 80

Featured LAB

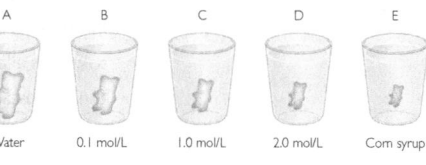

Solution Concentration

Purpose

To explore solutions and the concentration of dissolved solids in solution.

Part 1: Dissolving Solids

Materials
- sugar, $C_{12}H_{22}O_{11}$, 60 g
- salt, NaCl, 60 g
- water, 200 mL
- 250 mL beakers (2)
- balance
- stirring rod
- 2 spatulas or plastic spoons
- 2 green or red gummy bears

Procedure

1. Make predictions. How many grams of sugar do you think you can dissolve in 100 mL of water? (1 tsp = 4 g) How many grams of salt? (1 tsp = 6 g)
2. Determine how many grams of sugar will dissolve in 100 mL of water at room temperature. Keep adding sugar and stirring until you notice that the beaker contains solid sugar that will not dissolve.
3. Repeat this process using table salt, NaCl, instead of sugar.

Calculations

1. Compare the number of *grams* of dissolved solid per liter of solution.
2. Compare the number of *moles* of solid per liter of solution.
3. Explain why the sugar solution has the greater mass of dissolved solid per liter but the salt solution has the greater number of moles of solid per liter.

Part 2: Gummy Bears

Five gummy bears were placed overnight in aqueous sugar solutions as shown.

A	B	C	D	E	
Water	0.1 mol/L	1.0 mol/L	2.0 mol/L	Corn syrup	Dry gummy bear

1. Describe what happens to the bears in the solutions. Which solution had the greatest effect on the gummy bears? Why do you think this is so?
2. What evidence do you have that it is water and not sugar that is moving into and out of the bears?
3. Place a gummy bear in each saturated solution you created. Predict what they will look like after 24 hours. Explain your reasoning.

SETUP

Set up 2 (or more) gummy bear stations. The evening before, or several hours before class, prepare 50 mL of each of these solutions. Fill two sets of cups with 25 mL of each solution.

- cup A: water
- cup B: 0.1 M sugar solution (1.7 g sucrose diluted to 50 mL)
- cup C: 1.0 M sugar solution (17 g sucrose diluted to 50 mL)
- cup D: 2.0 M sugar solution (34 g sucrose diluted to 50 mL)
- cup E: corn syrup

After placing 25 mL of solution in each cup, place one gummy bear in each cup. A total of ten gummy bears

should be soaked in 25 mL of each solution for several hours (preferably overnight). Another two gummy bears should be glued to index cards so that students can compare the size and color of untreated gummy bears. You can use the same gummy bears/solutions for several classes on the same day. Green and red gummy bears work best.

Note: Certain brands of gummy bears dissolve more readily than others. You should test a few to be sure the demo will work. Be sure to keep the solutions with the gummy bears around room temperature—about 68 °F. On a very hot day, the gummy bears will dissolve.

Bearly Alive

Solution Concentration

THINK ABOUT IT

When a toxic substance is in solid form, keeping track of toxic amounts is fairly simple. But what happens if the toxic substance is dissolved in water? For example, the water that comes out of your tap is probably a solution containing small amounts of dissolved chlorine. Small amounts of chlorine benefit the water supply by killing bacteria, but large amounts would be toxic to people who drink the water. It is important to have methods for tracking substances dissolved in water.

> How can you keep track of compounds when they are in solution?

To answer this question, you will explore

- Solutions
- Solution Concentration

EXPLORING THE TOPIC

Solutions

A **solution** is a mixture of two or more substances that is uniform throughout. This means it is uniform at the molecular, ionic, or atomic level. The ocean, a soft drink, liquid detergent, window cleaner, and even blood plasma are all examples of solutions. A glass of powdered drink has many dissolved substances in it, but they are mixed so well that the individual substances in the solution are no longer distinguishable.

Each solution listed above consists of one or more substances dissolved in water. The water in these solutions is called the **solvent.** There are many other solvents besides water, such as alcohol, turpentine, and acetone. So the solvent is the substance that other substances are dissolved *in.* A solvent does not necessarily have to be a liquid. Solvents can be gases, liquids, or solids. Water is one of the most important solvents on Earth because it is so common in living systems.

A beverage made from a drink mix has many dissolved substances in it—not just sugar.

A substance that is dissolved in a solvent is called the **solute.** Solutes can be gases, liquids, or solids. In a sugar solution, the sugar is the solute and water is the solvent. In a carbonated soft drink, carbon dioxide gas is one of the solutes dissolved in the soft drink.

Lesson 80 | **Solution Concentration**   409

After soaking, gummy bears should look like the image on page 408 of the student edition.

CLEANUP

Save unused sugar and salt for future use. Put gummy bears in the trash. All the solutions can be disposed of through the drain.

Lesson Guide

ENGAGE (5 min)

TRM PowerPoint Presentation 80

Have two sets of the gummy bears in the five solutions available for students to examine. You might prompt students to walk up to a station to view the gummy bears more closely. Include the untreated gummy bear on the index card.

ChemCatalyst

Five gummy bears have been placed overnight in five different aqueous sugar solutions. Each solution contains a different amount of dissolved sugar.

1. Which solution do you think has the greatest amount of sugar in it? Explain your reasoning.

2. What do you think caused the bears to change size?

Sample answers: 1. Some students may say the corn syrup has the most sugar because it looks syrupy. Some may say the solution that caused the bear to expand the most has the most sugar, while others may say the solution that caused the bear to shrink has the most sugar. **2.** Students may hypothesize that the bears changed size because water moved into them or because sugar moved into them. It is the water that is moving in or out of the bears.

Ask: How would you describe the gummy bears?

Ask: What do you think happened to the bears?

Ask: How does the amount of sugar in each solution vary? Explain your thinking.

Ask: Which solution do you think has the greatest number of sugar molecules per volume? Explain.

EXPLORE (15 min)

TRM Worksheet with Answers 80

TRM Worksheet 80

INTRODUCE THE LAB

→ Arrange students into groups of four.

→ Pass out the student worksheets.

→ Define the term *solution* as a mixture of two substances that is uniform throughout.

→ Tell students that they will create two solutions in class. Their challenge will be to figure out the maximum number of moles of solid that will dissolve in 100 mL of water.

→ Note that students probably will stop dissolving sugar in the water long before it is completely saturated. Encourage them to keep stirring, but let them come up with their own answers. The water can take more than 12 tsp of sugar but only 1 tsp of salt.

EXPLAIN & ELABORATE (15 min)

INTRODUCE THE VOCABULARY OF SOLUTIONS

→ Write the words *solute, solvent, saturated, concentration,* and *molarity* on the board.

Ask: What is a solution?

Ask: What is a saturated solution? How can you tell when a solution is saturated?

Ask: You can add many more grams of sugar to 100 mL of water than you can salt. However, the number density of the salt in moles per liter is much higher than that of the sugar. Why? Explain.

Key Points: A solution is a mixture of two or more substances that is uniform throughout. This uniformity is found at the particle level. Solutions are formed when one substance is dissolved in another. The substance being dissolved, like the salt or sugar, is generally referred to as the *solute*. The substance that the solute dissolves in, such as water, is called the *solvent*.

> **Solute:** The substance dissolved in a solution.
>
> **Solvent:** The substance in which the solute dissolves in a solution.

A solution is saturated when no more solute will dissolve. At a given temperature and pressure, there is a maximum amount of solute that can dissolve in a particular amount of solvent. In today's lab, solid sugar and salt were added to water until no more would dissolve, making the two solutions saturated.

> **Saturated solution:** A solution that contains the maximum amount of solute for a given amount of solvent.

Concentration refers to the amount of solute that is dissolved in a solution. When the concentration of a solution is given in units of moles of solute per liters of solution, it is called the *molarity* (M). A 1.0 molar solution of salt in water, 1.0 M NaCl, has 1 mol of NaCl per liter of solution. Molarity is the number density of the solute.

> $$\text{Molarity (M)} = \frac{n \text{ (moles)}}{V \text{ (liters of solution)}}$$

> **Concentration:** A measure of the amount of solute dissolved in a specified volume of solution.
>
> **Molarity:** The concentration of a solution expressed in moles of solute per liter of solution.

When the concentration is given in units of moles of solute per kilogram of solution, it is called *molality* (m). A 1.0 molal solution of salt in water,

A tossed salad is a heterogeneous mixture.

Water-soluble dyes are used in preparing a lot of our foods. One form of blue dye that goes into our food is indigo, $C_{16}H_{10}N_2O_2$. Indigo is used to dye denim cloth for blue jeans. When this dye is put in food products, it is referred to as FD&C Blue No. 2.

A solution is considered a **homogeneous mixture** because of its uniformity. Mixtures that are not uniform throughout are called **heterogeneous.** Salad dressing or mixed nuts are examples of heterogeneous mixtures.

Solution Concentration

The **concentration** of a solution refers to the amount of a substance dissolved in a specified amount of solution. The concentration of a solution depends on both the quantity of solute and the quantity of solvent. There are several ways of measuring these quantities. A common way to measure concentration is to determine the number of moles of solute per total volume of solution in liters. This is called the **molarity** of a solution and is expressed in moles per liter, mol/L. Notice that solution concentration is the same as number density, n/V.

$$\text{Molarity} = \frac{\text{moles of solute}}{\text{liters of solution}}$$

Consider the two solutions in the illustration. Both contain a blue dye. Dyes are generally molecular substances that make a solution appear to be a certain color. The concentration of blue dye is greater in the first solution even though the volume of solution is smaller. You can tell this because the solution on the left is darker than the solution on the right.

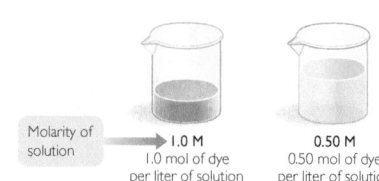

Molarity of solution → 1.0 M — 1.0 mol of dye per liter of solution — 0.50 M — 0.50 mol of dye per liter of solution

Dyes are often molecular solids. This is a sample of blue-colored dye.

The M is an abbreviation for **molar,** which means moles of solute per liter of solution. The dye solution on the left is 1.0 molar, 1.0 M, while the one on the right is 0.50 molar, or 0.50 M.

If you want to make a solution more concentrated, you need to add solute. If you want to make a solution more dilute, or *less* concentrated, you need to add water.

> **Example 1**
>
> **Molarity**
>
> Determine the molarity of each.
>
> a. 0.10 mol of NaCl in 1.0 L of solution
> b. 0.50 mol of NaCl in 1.0 L of solution
> c. 0.10 mol of NaCl in 0.50 L of solution
> d. 0.10 mol of NaCl in 250 mL of solution

1.0 *m* NaCl, has 1 mol of NaCl per kilogram of solution.

> $$\text{Molality } (m) = \frac{n \text{ (moles)}}{\text{kg of solution}}$$

PROCESS THE GUMMY BEARS LAB

Ask: What did you observe when you examined the gummy bears?

Ask: How do you know that sugar is not moving from the solution into the gummy bear?

Ask: Why do you think the gummy bears do not dissolve in the 2.0 M sugar solution or the corn syrup?

(The solutions are already saturated or near saturation.)

Ask: What happens to the sugar concentration in the gummy bear when the bear is placed in the corn syrup?

Ask: What do you think the gummy bear placed in your saturated sugar solution will look like tomorrow? What is your reasoning?

Key Points: The gummy bears changed size depending on the concentration of the solution they were soaked in. When the bears were placed into solutions that contained low amounts of dissolved sugar, they bloated up

(continued)

Solution

Molarity is the number of moles of solute divided by the liters of solution.

a. $\dfrac{0.10\ \text{mol}}{1.0\ \text{L}} = 0.10\ \text{mol/L} = 0.10\ \text{M}$

b. $\dfrac{0.50\ \text{mol}}{1.0\ \text{L}} = 0.50\ \text{mol/L} = 0.50\ \text{M}$

c. $\dfrac{0.10\ \text{mol}}{0.50\ \text{L}} = 0.020\ \text{mol/L} = 0.020\ \text{M}$

d. $\dfrac{0.10\ \text{mol}}{250\ \text{mL}} = \dfrac{0.10\ \text{mol}}{0.25\ \text{L}} = 0.40\ \text{mol/L} = 0.40\ \text{M}$

CALCULATING MOLARITY FROM MASS IN GRAMS

In the classroom, you were challenged to create a saturated sugar solution and a saturated salt solution. A **saturated solution** is one that contains the maximum amount of dissolved solute possible in a certain volume of liquid. To make a saturated solution, you must add solid and stir until no more solid dissolves. You can calculate how many grams of each solid you added if you find the mass of the solutions before and after you added the solute. The molar mass of a substance allows you to convert the mass of solute in grams to moles of solute. You can then determine the molarity of each solution.

Big Idea Molarity keeps track of the number of dissolved particles in a liquid.

Example 2

Finding Molarity

What is the concentration in moles per liter of a 250 mL solution that contains 16.0 g of dissolved sugar, $C_{12}H_{22}O_{11}$?

Solution

First, convert the mass of sugar to moles by dividing by the molar mass of sugar, $C_{12}H_{22}O_{11}$. Next, to determine molarity, divide the number of moles by the volume.

Convert mass to moles. $\dfrac{16.0\ \text{g}}{342.3\ \text{g/mol}} = 0.047\ \text{mol}$

Calculate the molarity. $\dfrac{0.047\ \text{mol}}{0.250\ \text{L}} = 0.188\ \text{M}$

Answer: Determine the number of moles of salt and divide by the volume of solution in liters: 10.0 g/58.45 g/mol = 0.17 mol; 0.171 mol/0.50 L = 0.34 mol/L, or 0.34 M.

Homework

Assign the reading and exercises for Toxins Lesson 80 in the student text. Assign the Web Research Project in the Chapter 15 Summary in the student text (optional). Have students continue research on the toxin assigned in Chapter 13.

more than when they were placed into solutions that contained high amounts of dissolved sugar. Because the bears placed into pure water were largest, it is reasonable to propose that water is moving into the bears to cause them to swell. This happens to a lesser extent in the more concentrated sugar because there is less difference between the concentrations inside and outside the bears.

Molecules move from areas of higher concentration to areas of lower concentration. In this case, the sugar molecules do not move through the gummy bear gel, probably because

they are too large. However, the water molecules do move. The gummy bears placed in the corn syrup actually lost water. Thus, you can conclude that the gummy bears must have a sugar concentration higher than 2.0 mol per liter but lower than that of corn syrup.

EVALUATE (5 min)

Check-In

Suppose 10.0 g of salt, NaCl, are dissolved in 0.50 L of water. What is the molarity of this solution?

LESSON 80 ANSWERS

1. *Possible answer:* "Uniform throughout" means that the components of the solution are so well mixed that two samples taken from anywhere in the solution will be identical.

2. *Possible answer:* The concentration of a solution refers to the relative amount of the solute to the solvent. The volume of a solution refers to the total amount of solution.

3. *Possible answer:* eyedrops: solvent is water; solute is salt ● bleach: solvent is water; solute is sodium hypochlorite ● vinegar: solvent is water; solute is acetic acid

4. *Possible answer:* salad dressing: vinegar, oil, spices ● potting soil: sand, clay, organic material ● garlic salt: garlic powder, salt

5. In order of increasing molarity, **c.** 1.0 mol per 10 L, **a.** 4.0 mol per 8.0 L, **b.** 6.0 mol per 6.0 L.

6. a. 1.0 M **b.** 2.0 M **c.** 4.9 M; in order of increasing molarity: **a.** 29.2 g per 0.50 L, **b.** 5.8 g per 50 mL, **c.** 2.9 g in 10.2 mL.

7. a. 0.56 M **b.** 1.1 M **c.** .58 M; in order of increasing molarity, **a.** 25 g of $C_6H_{12}O_6$ per 0.25 L, **c.** 50 g $C_{12}H_{22}O_{11}$ per 0.25 L, **b.** 50 g of $C_6H_{12}O_6$ per 0.25 L

8. B

9. A

10. C

KEY TERMS

solution
solvent
solute
homogeneous mixture
heterogeneous mixture
concentration
molarity
molar
saturated solution

LESSON SUMMARY

How can you keep track of compounds when they are in solution?

A solution is a mixture of two or more substances that is uniform throughout. To track substances that are in solution, chemists measure the number of particles per liter of liquid. This is called the concentration or the molarity of the solution. Molarity is a measure of number density and is expressed in units of moles per liter of solution.

Exercises

Reading Questions

1. What does it mean that a solution is "uniform throughout"?

2. What is the difference between the concentration of a solution and the volume of a solution?

Reason and Apply

3. Find at least three solutions at home. Identify the solute and the solvent for each.

4. Find at least three mixtures at home that are not solutions. Identify the substances in each mixture.

5. Place these salt solutions, NaCl(*aq*), in order of increasing molarity.
   a. 4.0 mol per 8.0 L
   b. 6.0 mol per 6.0 L
   c. 1.0 mol per 10 L

6. Determine the molarity for each of these salt solutions, NaCl(*aq*). Then list the solutions in order of increasing molarity.
   a. 29.2 g per 0.50 L
   b. 5.8 g per 50 mL
   c. 2.9 g in 10.2 mL

7. Determine the molarity for each of these solutions. Then list the solutions in order of increasing molarity.
   a. 25 g $C_6H_{12}O_6$ per 0.25 L
   b. 50 g $C_6H_{12}O_6$ per 0.25 L
   c. 50 g $C_{12}H_{22}O_{11}$ per 0.25 L

8. Which of the substances listed here represent a heterogeneous mixture? Choose all that apply.
   (A) the air in your classroom   **(B)** beef stew
   (C) mouthwash   **(D)** chocolate pudding

9. How can you increase the molarity of a solution?
   (A) Add solute.   **(B)** Add solvent.
   (C) Pour out some of the solution.   **(D)** All of the above.

10. When expressing molarity, M stands for
    (A) atoms per liter of water.   **(B)** moles per milliliter of solvent.
    (C) moles per liter of solution.   **(D)** atoms per milliliter of solute.
    (E) grams per liter of solution.

Drop In

Molecular Views

THINK ABOUT IT

A tiny drop of a concentrated solution of hydrochloric acid, HCl(*aq*), will cause a severe burn if it comes in contact with your skin. In contrast, a larger volume of a more dilute HCl(*aq*) solution will cause only a minor skin irritation. How can a single drop of a solution be so powerful?

> How can you use molarity to determine the moles of solute?

To answer this question, you will explore

- Particle Views of Solutions
- Molarity Calculations

EXPLORING THE TOPIC

Particle Views of Solutions

When a toxin is dissolved in water, the amount of toxin you are exposed to depends on both the concentration of the solution and the volume of the solution. To better understand the relationship between solution concentration and solution volume, it is useful to explore solutions with the same volume but different concentrations, as well as solutions with the same concentration but different volumes.

SAME VOLUME, DIFFERENT CONCENTRATIONS

This illustration shows four solutions of red dye, as well as particle views of each. Each red dot represents a red dye molecule dissolved in water. Take a moment to compare the solutions and molecular views.

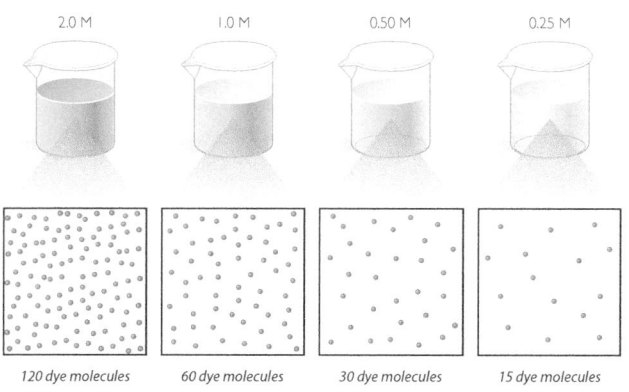

2.0 M	1.0 M	0.50 M	0.25 M
120 dye molecules	60 dye molecules	30 dye molecules	15 dye molecules

For example, a 1.0 M sodium chloride solution has 2 mol of ions per liter because it is made up of Na^+ and Cl^- ions in a 1:1 ratio.

IN CLASS

Students explore solution concentrations from a particulate point of view. Students are provided with illustrations depicting particle views of solutions with different molarities. They compute the number density per square inch of each illustration. They then use scissors to create samples of different sizes with similar numbers of moles of particles. Students apply their understanding to a real solution by creating samples containing specific numbers of moles. A complete materials list is available for download.

TRM Materials List 81

Differentiate

Students who need practice with terminology such as *molarity, solute, solvent,* and *concentration* and those who need support visualizing solutions on a molecular level may benefit from using the University of Colorado Boulder's PhET simulations: Concentration, Molarity, and Sugar and Salt Solutions. Students who are ready to go further can investigate solution equilibriums using the simulation: Salts & Solubility. A list of URLs can be found on the Chapter 15 Web Resources document.

TRM Chapter 15 Web Resources

Pre-AP® Course Tip

Students preparing to take an AP® chemistry or college chemistry course should be able to understand representations of common chemical systems on a particulate level. In this lesson, students explore solute concentrations from a particulate point of view and differentiate between concentrated and dilute.

SETUP

Prepare 8 L of red flavored drink-mix solution. The exact concentration is not critical. Adjust the concentration so that a 1:2 dilution produces an obvious color change. Copy two Handout—Particle Views per group. The handouts can be collected and reused for another class. Copy and cut out two Particle View (#2) cards per group. The Particle View (#2) cards have to be replenished for each class, because students cut them into smaller pieces.

Overview

ACTIVITY: GROUPS OF 4

Key Question: How can you use molarity to determine the moles of solute?

KEY IDEAS

The relationship between moles of particles in a solution and liters of solution is a proportional one described by the formula *number of moles = k · volume in liters*, where the proportionality constant, *k*, is the molarity of the solution, M. This formula can be used to calculate the exact number of moles in a sample of known molarity and volume. The concentration or molarity of a solution does not change with the size of the sample.

LEARNING OBJECTIVES

- Describe solution concentration on a particulate level.
- Calculate the number of moles of particles from the molarity and the volume of a solution.
- Differentiate between particles in ionic and molecular solutions.

FOCUS ON UNDERSTANDING

Students might be confused about the molarity of ionic solutions and the number of moles of ions in solution.

Lesson Guide

ENGAGE (5 min)

TRM PowerPoint Presentation 81

Ask students to check on their gummy bears from the previous lesson. The first questions of the ChemCatalyst relate to their bears.

ChemCatalyst

Examine the gummy bear in the sugar solution you prepared in class yesterday.

1. What does the gummy bear's appearance suggest about the solution?

2. Is the solution saturated? Why or why not?

Imagine that you have 1 L of a 2.0 M sugar solution in a large container. You pour out 100 mL into a beaker.

3. Did the concentration of sugar in the large container change?

4. Did the number of moles of sugar in the large container change?

Sample answers: **1.** The plumpness of the gummy bear indicates the concentration of sugar in the solution. **2.** The solution is not saturated, because the bear is larger than the bear in a saturated solution. **3.** The concentration does not change. **4.** The number of moles of sugar decreases.

→ Discuss the relationship among molarity, number of moles, and sample size.

→ You might demonstrate the ChemCatalyst using a large container of solution and a glass or small beaker.

Ask: What happens to the concentration of a solution when you pour a sample of it from a large container into a beaker?

Ask: What happens to the number of moles of solute in the original container?

Ask: Do the large container and the beaker have the same amount of sugar?

Ask: How do the molarities compare of the solution in the large container and of that in the beaker?

EXPLORE (20 min)

TRM Worksheet with Answers 81

TRM Worksheet 81

TRM Handout—Particle Views 81

TRM Card Masters—Particle View (#2) 81

SPORTS CONNECTION

Chlorine, found in hypochlorous acid, HOCl, is toxic to humans in high concentration. However, a tiny amount can be added to a swimming pool to make the water safe. The low concentration is deadly to bacteria and other microorganisms but not to humans—though it can irritate the eyes and skin.

Notice that each time the concentration is halved, there are half as many particles per unit of volume. So, as the solution concentration decreases, equal volumes of solution will contain fewer molecules.

The color of the red dye solution provides evidence to support this conclusion. When there are more dye molecules per unit of volume, the red color of the solution is more intense.

SAME CONCENTRATION, DIFFERENT VOLUMES

The particle views of the two solutions in the next illustration show what happens when you pour a portion of a solution into another container. Take a moment to compare the solutions and molecular views.

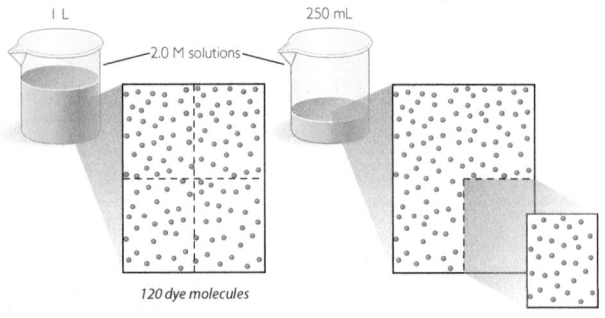

2.0 M solutions

120 dye molecules

30 dye molecules

There are fewer total dye molecules when you pour a portion of the solution into the smaller container. However, the concentration of the solution has not changed. If you remove 250 mL from a 1 L solution, you have one quarter as many dye molecules. But because the concentration has not changed, the color of the red dye solution does not change.

↻ Molarity Calculations

Consider the solutions below. Suppose the first two contain a concentrated solution of a toxin. The third and fourth contain a more dilute, or watered down, solution of the same toxin. Take a moment to try to put the solutions in order of most total moles of toxin to least total moles of toxin.

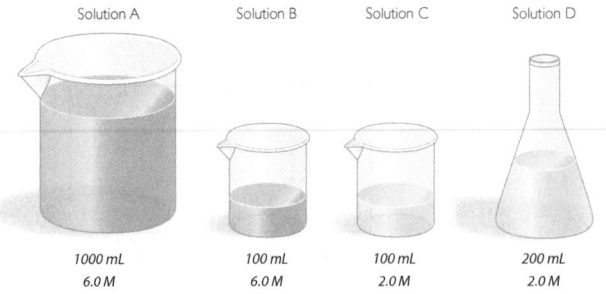

| Solution A | Solution B | Solution C | Solution D |

| 1000 mL 6.0 M | 100 mL 6.0 M | 100 mL 2.0 M | 200 mL 2.0 M |

INTRODUCE THE ACTIVITY

→ Arrange students into groups of four.

→ Pass out the student worksheets, handouts, and cards.

→ If students have difficulty determining the density of the particles, suggest that they divide each view into quarters using the ruler and count the particles in each quarter view.

→ Tell students that you will provide 1000 mL of a 3.0 M red dye solution to groups of four students. They will use this solution to make three solutions with different total numbers of moles.

Be sure to leave enough time for this portion of the class.

→ Students might recognize that the solution is a beverage. Remind them of the dangers of tasting anything in a laboratory setting.

EXPLAIN & ELABORATE (15 min)

TRM Transparency—Particle Views 81

DISCUSS SOLUTION CONCENTRATION ON THE PARTICULATE LEVEL

→ Have the transparency Particle Views and the Particle Views handout available to use as visuals.

The relationship between moles of solute and volume of solution in liters is proportional. Recall that the molarity of a solution in moles/liter is the proportionality constant, *k,* for that solution. So, if you multiply the molarity by the volume in liters, you can determine the number of moles of solute.

moles of solute = *k* · volume of solution in liters

= 6.0 mol/L · 1.0 L = **6 mol**   Solution A

= 6.0 mol/L · 0.10 L = **0.60 mol**   Solution B

= 2.0 mol/L · 0.10 L = **0.40 mol**   Solution C

= 2.0 mol/L · 0.20 L = **0.20 mol**   Solution D

You can see that even though the volume of Solution B is smaller than the volume of Solution D, there are more moles of toxin dissolved in Solution B. This is because Solution B is more concentrated than Solution D.

A small drop of highly concentrated toxin may be worse for you than a large quantity of dilute toxin. For example, a 2 mL drop of 12 M HCl solution is much more toxic than 50 mL of 0.10 M HCl solution.

Big Idea The concentration of a solution does not change with the size of the sample.

Example

Comparing Moles of Solute

Two solutions of a toxic substance are shown in the illustration.

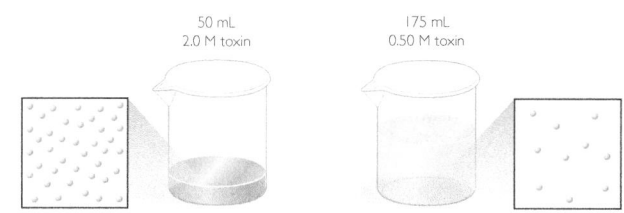

50 mL
2.0 M toxin

175 mL
0.50 M toxin

Which beaker contains the larger dose of the toxic substance?

Solution

Use the concentration of each solution to determine the moles of solute.

2.0 M solution: 2.0 mol/L · 0.050 L = 0.10 mol

0.50 M solution: 0.50 mol/L · 0.175 L = 0.0875 mol

So, the beaker on the left, the more concentrated solution, contains more moles of toxic substance.

Lesson 81 | **Molecular Views**   415

Ask: How would you describe the concentrations of the four solutions in number of particles per square inch?

Ask: What happened to the concentration of the particles when you cut up the paper? Explain.

Ask: How can two samples have different numbers of particles but the same concentration?

Ask: How many pages of Particle View 4 would be needed to have the same number of particles as in Particle View 1? (eight pages)

Key Points: The particle views represent molecules dissolved in water. Each dot represents a red dye molecule dissolved in water. You can compare the dot concentrations, or number density, of each "solution."

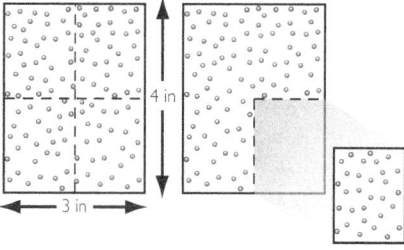

Any sample cut out of the original Particle View 1 has the same concentration of dots per square inch as the large sample. However, if you count the total number of particles in the small sample, it is less than that in the original sample. The sample cut from Particle View 1 contains 30 particles. Thus, one-quarter of solution 1 contains the same number of particles as *all* of solution 3.

RELATE SOLUTION CONCENTRATION TO REAL LIFE

→ Ask students to instruct you on how to make samples identical to solutions B, C, and D using solution A.

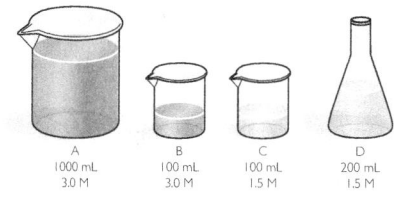

A
1000 mL
3.0 M

B
100 mL
3.0 M

C
100 mL
1.5 M

D
200 mL
1.5 M

Ask: How can two samples have different numbers of molecules but the same concentration?

Ask: Is the total amount of red dye in beaker B different from the total amount of red dye in container A? Explain.

Ask: How can you make a 1.5 M red dye solution from a 3.0 M red dye solution?

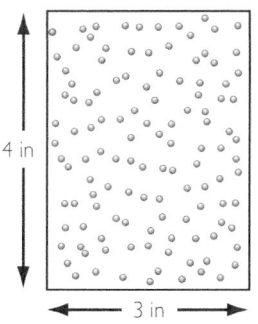

4 in

← 3 in →

Particle view 1
120 particles
Number density:
10 particles/in^2

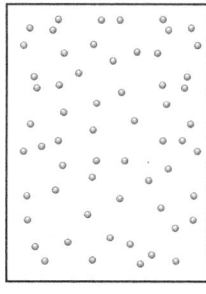

Particle view 2
60 particles
Number density:
5 particles/in^2

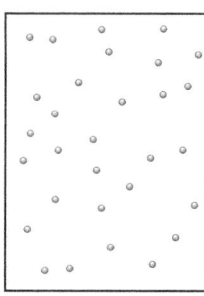

Particle view 3
30 particles
Number density:
2.5 particles/in^2

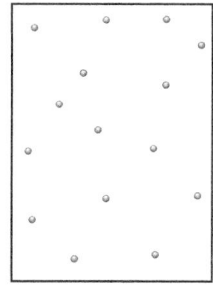

Particle view 4
15 particles
Number density:
1.25 particles/in^2

Key Points: Concentration does not depend on the size of the sample. If you take a sample of an existing solution and simply put it in a different container, you have not changed the concentration of the solution. On a particulate level, there is the same number of red dye molecules per unit of volume. However, the volume has decreased, and thus, the total number of moles has decreased.

To create solutions C and D, you must dilute the original solution to half its molarity. Adding 50 mL of water to 50 mL of solution A results in solution C. Adding 100 mL of water to 100 mL of solution A results in solution D.

COMPLETE SOME WORKED EXAMPLES

→ Complete several sample calculations on the board. Be sure to include an example that uses ionic compounds in solution.

Ask: Describe how you can calculate the number of moles you have in a sample of a solution.

Ask: How many moles of sugar molecules are in 1 L of a 2.5 M sugar solution?

Ask: How many moles of ions are in 1 L of a 1.0 M solution of sodium chloride, NaCl? (2 mol of ions)

Ask: Why must you take extra care when figuring out the number of particles of dissolved ionic solids?

Key Points: The relationship between the number of moles of particles in a solution and the volume of the solution is proportional.

> Number of moles = $k \cdot$ volume of solution (in liters)

The proportionality constant, k, is equal to M, the molarity of the solution. If you know the molarity of a solution, you can figure out exactly how many moles of particles are in the solution. The equation for molarity lets you solve many problems involving solutions.

EXAMPLE

Imagine that you have 100 mL of a 2.5 M $CaCl_2$ solution. How many moles of $CaCl_2$ formula units are in the solution? How many moles of Cl^- ions are in the solution? How many total moles of ions are in the solution?

HEALTH CONNECTION

Fructose is a natural sugar found in fruits and vegetables. However, when it is highly concentrated, as it is in high-fructose corn syrup, it is difficult for your body to process. Too much of it can cause health problems.

Dean Zider/Shutterstock

Scientists also measure the concentration of substances in solution using parts per million (ppm) or parts per billion (ppb). When water is tested, the results are often reported in ppm. A level of 1 ppm means that there is 1 mg of a dissolved solute per liter of solution.

LESSON SUMMARY

How can you use molarity to determine the moles of solute?

The concentration of a solution does not change with the size of the sample. This is because concentration is a measure of number density, or moles of particles per unit of volume, a property that does not change with sample size. However, a large sample of a solution will contain more total particles than a small sample of the same solution. The relationship between moles of particles in a solution and liters of solution is a proportional one. It is described by the formula, moles of solute = $k \cdot$ volume of solution, where k is the molarity of the solution. This equation can be used to calculate the exact number of moles in a sample of known concentration and volume.

Exercises

Reading Questions

1. How can two solutions with different volumes have the same concentration?

2. How can you figure out how many moles of solute you have in a solution with a specific concentration?

Reason and Apply

3. Draw molecular views for blue dye solutions that are 0.50 M, 0.25 M, and 0.10 M.

4. What portion of 1.0 L of 0.50 M blue dye solution has the same number of moles as 1.0 L of 0.25 M blue dye solution?

5. Glucose and sucrose are two different types of sugar. Consider these aqueous solutions:

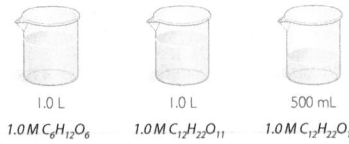

1.0 L	1.0 L	500 mL
1.0 M $C_6H_{12}O_6$ (glucose)	1.0 M $C_{12}H_{22}O_{11}$ (sucrose)	1.0 M $C_{12}H_{22}O_{11}$ (sucrose)

a. Which solution has the most molecules? Explain.
b. Which solution has the greatest concentration? Explain.
c. Which solution has the most mass? Explain.

SOLUTION

Moles $CaCl_2$ = (2.5 M)(0.10 L)
$\quad$ = 0.25 mol

Moles Cl^- = (2)(moles $CaCl_2$)
$\quad$ = (2)(0.25 mol) = 0.50 mol

Moles Ca^+ = (1)(moles $CaCl_2$)
$\quad$ = (1)(0.25 mol) = 0.25 mol

Total moles of ions = moles of Cl^- + moles of Ca^+ = 0.50 mol + 0.25 mol = 0.75 mol

---

EVALUATE (5 min)

Check-In

1. How many moles of sugar, $C_{12}H_{22}O_{11}$, are in 52 mL of a 0.50 M solution?

2. How many moles of sugar, $C_{12}H_{22}O_{11}$, are in 26 mL of a 0.50 M solution?

Answers: Multiply the volume in liters by the molarity to determine the number of moles in the sample. **1.** (0.052 L)(0.50 mol/L) = 0.026 mol of sugar. **2.** (0.026 L)(0.50 mol/L) = 0.013 mol of sugar.

Exercises

(continued)

6. How many moles of sodium cations, Na$^+$, are dissolved in a 1.0 L sample of each solution listed below?
   a. 0.10 M NaCl
   b. 3.0 M Na$_2$SO$_4$
   c. 0.30 M Na$_3$PO$_4$

7. Draw particle views for 1.0 M NaCl, 2.0 M NaCl, and 1.0 M Na$_2$S. Use different symbols for each type of ion. Circle the solution(s) with the least total number of ions.

8. Determine the number of moles of solute in each of these aqueous solutions.
   a. 50 L of 0.10 M NaCl
   b. 0.25 L of 3.0 M C$_6$H$_{12}$O$_6$
   c. 35 mL of 12.0 M HCl
   d. 300 mL of 0.025 M NaOH

9. How many liters of each solution do you need to get 3.0 mol HCl?
   a. 12.0 M HCl
   b. 2 M HCl
   c. 0.5 M HCl
   d. 0.010 M HCl

product of the molarity and the volume. **b.** All three solutions have the same concentration, 1.0 M. Molarity is a measure of concentration. **c.** The 1.0 L sucrose solution has the greatest mass because each sucrose molecule has more mass than each glucose molecule. Therefore, one mole of sucrose has more mass than one mole of glucose. Also, 1 L of an aqueous solution has more mass than 0.5 L of an aqueous solution.

6. a. 0.10 mol **b.** 6.0 mol **c.** 0.90 mol

7.

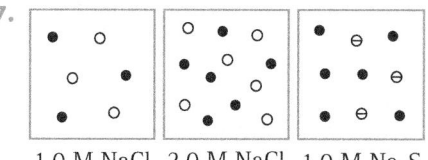

1.0 M NaCl   2.0 M NaCl   1.0 M Na$_2$S

● Na$^+$ ○ Cl$^-$ ⊖ S^{2-}

The ratio of particles in each drawing should meet the requirements listed below. ● The first drawing should show an equal number of Na$^+$ ions and Cl$^-$ ions (this should be circled) ● The second drawing should show double the number of Na$^+$ ions and Cl$^-$ ions as in the first drawing. ● The third drawing should show double the number of Na$^+$ ions as the first drawing and twice as many Na$^+$ ions as S$^-$ ions.

8. a. Moles of NaCl = 5.0 mol
b. Moles of C$_6$H$_{12}$O$_6$ = 0.75 mol
c. Moles of HCl = 0.42 mol
d. Moles of NaOH = 0.0075 mol

9. a. 0.25 L **b.** 1.5 L **c.** 6.0 L **d.** 300 L

Homework

Assign the reading and exercises for Toxins Lesson 81 in the student text.

LESSON 81 ANSWERS

1. Two solutions of different volumes can have the same concentration if the proportion of the substance to the volume is the same.

2. Determine the number of moles of solute in a solution by multiplying the concentration by the volume.

3. *Possible answers:* Drawings should show the least number of particles in the 0.10 M drawing, 2.5 times as many particles in the 0.25 M drawing, and 5 times as many particles in the 0.50 M drawing.

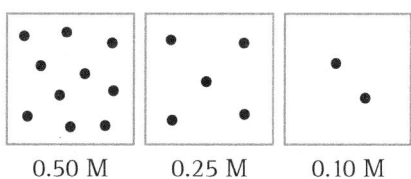

0.50 M   0.25 M   0.10 M

4. 0.5 L of the 0.50 M solution has the same number of moles of dye as 1 L of the 25 M solution.

5. a. The first two solutions have the same number of molecules. Each of these solutions has more molecules than the third solution. The number of molecules in each sample is the

Overview

LAB: GROUPS OF 4

Key Question: How can you make a solution with a specific molarity?

KEY IDEAS

To create solutions of specific molarities, you can calculate the amount of solid needed by using the molar mass of the solid. In addition, you can use the equation *molarity = moles/liters* and the molar mass to figure out how many grams of a substance are contained in a sample of a solution with a specific molarity and volume.

LEARNING OBJECTIVES

- Calculate the amounts necessary to create a solution with a desired volume and molarity.
- Determine the number of grams of solute in a sample with a specific volume and molarity.

IN CLASS

This activity is organized around a challenge to find the appropriate (nontoxic) solution for a saltwater fish tank. Students work in groups of four to create salt solutions. Each group of students is responsible for making one solution using a specific mass of solid and then determining the molarity of that solution. After all eight solutions have been made, a sample of each is placed into a test tube and tested for accuracy with a hydrometer (see figure) or a conductivity meter. The solutions are placed in order of concentration, and the one deemed nontoxic for a marine aquarium is identified. A complete materials list is available for download.

TRM Materials List 82

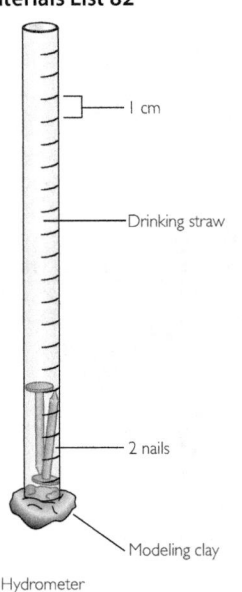

- 1 cm
- Drinking straw
- 2 nails
- Modeling clay

Hydrometer

Preparing Solutions

THINK ABOUT IT

CONSUMER CONNECTION

Water that has a high mineral content is called hard water, while water with a low mineral content is called soft water. Soap and shampoo do not lather up well in very hard water and don't rinse out easily in very soft water. Minerals commonly found in tap water are calcium and magnesium carbonates, and they sometimes build up as deposits in pipes and on faucets.

© Tony Freeman/PhotoEdit

Human blood plasma contains a certain molarity of dissolved salt, NaCl. If you are taken to the hospital in an ambulance and given an intravenous, or IV, solution, it is critical that the salt concentration match that of your blood, about 0.15 M NaCl. Otherwise, the IV solution could be toxic to you.

How can you create a solution with a specific molarity?

To answer this question, you will explore

⮕ Creating Solutions

⮕ Relating Mass, Moles, and Volume

EXPLORING THE TOPIC

⮕ **Creating Solutions**

Suppose a lab technician wants to make a 0.15 M sodium chloride solution. How should this solution be prepared?

PREPARING A 0.15 M SALT SOLUTION

A concentration of 0.15 M means that 0.15 mol of salt is dissolved in 1 L of solution. Therefore, if you want to make 1 L of solution, you will need to measure out 0.15 mol of solid sodium chloride, NaCl.

To know how much salt to weigh out, you must convert moles to mass using the molar mass of the compound. In this case, simply add the molar masses for Na and Cl from the periodic table. The molar mass of NaCl is 58.4 g/mol to one decimal place. Now multiply the moles of NaCl you want by the molar mass to get the number of grams of NaCl you need for your solution.

$$\text{Mass} = \text{molar mass} \cdot n = (58.4 \text{ g/mol})(0.15 \text{ mol}) = 8.76 \text{ g NaCl}$$

If you dissolve 8.76 g NaCl in water so that the total volume of solution is 1 L, you will have a solution that is 0.15 M, or 0.15 mol/L NaCl.

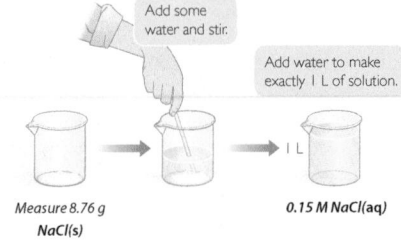

Add some water and stir.

Add water to make exactly 1 L of solution.

1 L

Measure 8.76 g NaCl(s)

0.15 M NaCl(aq)

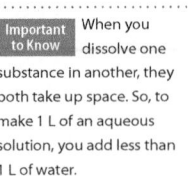
Important to Know When you dissolve one substance in another, they both take up space. So, to make 1 L of an aqueous solution, you add less than 1 L of water.

Differentiate

To support students in building mathematical fluency in the context of chemistry, you might have students use a graphic organizer to summarize their understanding of solution chemistry and molarity calculations. The graphic organizer can be designed as a diagram, flow-chart, or concept map that helps students show the relationship between new and existing concepts. You can provide students with a ready-made template they can fill in or allow them to develop their own. Encourage students to be explicit in showing how molarity calculations connect to their prior learning about mass-to-moles

calculations. Guide advanced learners to make connections between their learning from Chapter 12, which involved a form of number density of gases, and molarity, which involves a form of number density for solutions.

Pre-AP® Course Tip

Students preparing to take an AP® chemistry or college chemistry course should be able to use mathematics to solve problems that describe the physical world. In this lesson, students use mathematics to calculate the amount (moles) of solute in a solution, and they determine the number of grams of solute necessary to make a solution of known molarity.

Example 1

Dextrose Solution

Sometimes doctors administer an IV solution containing a solution of dextrose to a patient who has low sugar or high sodium in the bloodstream. Dextrose is a type of sugar. The structural formula for dextrose is shown in the illustration.

The IV solution contains 150 g of dextrose in 3 L of solution. What is the molarity of this solution?

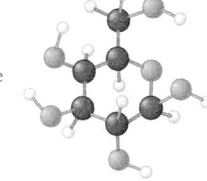

Dextrose

Solution

You need to convert grams of dextrose to moles of dextrose to determine the molarity. You know there are 150 g of dextrose in 3 L, or 50 g/L.

First, determine the molar mass of dextrose. As you can see from the ball-and-stick model, there are 6 C atoms, 12 H atoms, and 6 O atoms, so the formula is $C_6H_{12}O_6$.

Find the molar mass of dextrose, $C_6H_{12}O_6$.	180.2 g/mol
Divide grams of solute by the molar mass to obtain moles.	$n = \dfrac{50 \text{ g}}{180.2 \text{ g/mol}} = 0.28 \text{ mol}$

There is 0.28 mol of dextrose in 1 L of solution, so the molarity is 0.28 M.

⟳ Relating Mass, Moles, and Volume

Just as it is important to convert mass to moles, it is also important to be able to convert moles to mass. The table below shows data for four samples of sucrose, $C_{12}H_{22}O_{11}$, each with a different volume and molarity. How many grams of sucrose are dissolved in each sample?

To find the moles of solute in each sample, multiply the values in the second and third columns. To find the mass of solute in each sample, multiply the fourth and fifth columns.

		Molarity • Volume			Moles • Molar mass
Sample	Molarity (moles/liter)	Volume (liters)	Number (moles of solute)	Molar mass of solute (grams/mole)	Mass of solute (grams)
1	1.0 M	1.0 L	1.0 mol	342.3 g/mol	342.3 g
2	0.10 M	1.0 L	0.10 mol	342.3 g/mol	34.23 g
3	1 M	0.1 L	0.10 mol	342.3 g/mol	34.23 g
4	0.10 M	0.1 L	0.010 mol	342.3 g/mol	03.423 g

Notice that there is more than one way to make a 0.10 M solution. For example, you can dissolve 0.10 mol to make 1.0 L or you can dissolve 0.010 mol to make 100 mL (0.10 L).

Lesson 82 | **Preparing Solutions**   419

Lesson Guide

ENGAGE (5 min)

TRM PowerPoint Presentation 82

ChemCatalyst

Imagine that you want to make 5 L of 0.60 M NaCl solution for a saltwater aquarium. You have to determine the correct mass of NaCl needed to obtain the correct concentration, or the fish will die.

1. What information do you have?

2. What information do you need?

Sample answers: **1.** You know the number of liters and the molarity of the final solution. **2.** You need to know the number of moles and the molar mass of NaCl.

→ Focus the discussion on the process of converting among mass, number, and volume.

Ask: If you know the number of moles of a substance, what do you do to find the number of grams?

Ask: If you know the volume of a solution, what do you have to know to find the number of moles?

Ask: How might you go about solving the problem?

EXPLORE (15 min)

TRM Worksheet with Answers 82

TRM Worksheet 82

INTRODUCE THE LAB

→ Arrange students into groups of four.

→ Pass out the student worksheets.

→ Tell students that they are searching for a nontoxic environment for saltwater fish. Their goal is to create one of eight solutions and determine its molarity.

→ Below are the eight different gram amounts of NaCl. Assign one amount to each group. Keep track of the amount you assign to each group so you can check the accuracy of the group's results.

Group	Mass of NaCl (g)
1	1.75 g
2	2.34 g
3	5.85 g
4	7.01 g
5	3.50 g
6	4.68 g
7	0.58 g
8	8.77 g

GUIDE THE LAB

→ After groups have made their solution and calculated its molarity, they should turn in a 50 mL sample of their solution in a test tube, clearly labeled with the molarity they have calculated. This sample will be tested with a hydrometer.

→ Create a table on the board for students to enter their results.

EXPLAIN & ELABORATE (15 min)

PROCESS THE EIGHT SALTWATER SOLUTIONS

→ Have each group place a 50 mL sample of its solution in a test-tube rack at the front of the room. Make sure each group has clearly labeled its sample with the molarity.

→ Place the test tubes in order of concentration according to the students' calculations. Use the hydrometer to test each solution and check the accuracy of the students' results.

→ You might have a sample of marine aquarium water at the front of the room. The hydrometer can be placed in this sample for comparison.

Ask: Which solution do you think is the one that is safe for a marine aquarium? What are you basing your answer on?

Ask: The hydrometer floats in each solution. Does it float higher or lower in a solution with a higher concentration of salt?

Ask: Were any of the solutions out of order according to the hydrometer?

Ask: What are some possible sources of error in this experiment?

Key Points: You can use a hydrometer to measure the density of a solution. The hydrometer floats higher in solutions with higher density. Some hydrometers are calibrated so that you can read the grams of salt per cubic centimeter from a scale. The density in grams per cubic centimeter can be converted to moles per liter. Thus, a solution that is denser is also more concentrated. Salt solutions that are more concentrated will cause the hydrometer to float higher.

The 0.60 M solution comes closest to resembling ocean water. The ocean has 0.60 mol per liter of salt, assuming all the salt is sodium chloride, NaCl. Note that in addition to NaCl, there are traces of other dissolved salts in the ocean. Ocean water also has variations in salt concentration depending on water temperature, depth, and location.

There are several possible sources of error in this experiment. One is inaccurate weighing. Another is using too much water to create the solution (using 100 mL). A third is the molarity calculations. The hydrometer cannot tell you where you made an error, only that an error was made.

DISCUSS MOLARITY CALCULATIONS

→ Complete the sample problems on the board.

Ask: How did you determine the molarity of your group's solution? Explain.

Ask: How might you determine the number of grams of solute in a sample with a certain molarity and volume? Explain.

Key Points: Molarity relates moles and liters. To figure out the molarity of a solution, it is necessary to know the number of moles of dissolved

substance and divide that number by the volume of the solution in liters. The mass in grams first must be converted to moles using the molar mass of the substance.

$$\text{molarity} = \text{moles/liters}$$

If you know the molarity of a solution and the size of the sample, you can calculate the number of grams of substance dissolved in that solution.

$$\text{grams} = \text{moles} \cdot \text{molar mass}$$

Remember that molarity tells you the concentration of a solution in units of moles per liter. Milliliters must be

converted to liters using the conversion formula 1 mL = 0.001 L.

COMPLETE SOME WORKED EXAMPLES

→ Ask students to come to the front of the class to work on each example and explain their process.

EXAMPLE 1

You want to make 0.10 L of a 0.50 M solution of sodium bromide, NaBr. What mass of solid should you use?

SOLUTION

First, figure out how many moles of solute are in 0.10 L of a 0.50 M solution.

$$(0.10 \text{ L})(0.50 \text{ mol/L}) = 0.050 \text{ mol solute}$$

Example 2

Guidelines for Safe Drinking Water

Arsenic, As, is an element that is found naturally in the soil. As a result, tiny levels of arsenic are often found in some water supplies. Like that of all substances, the toxic effect of arsenic is dependent on the dosage you receive. For this reason, the Environmental Protection Agency, or EPA, sets limits on the amounts of dissolved substances that can be present in our drinking water. The EPA has set the upper limit for arsenic in drinking water at 0.00010 g/L.

Is it safe to drink water that has a concentration of 0.000020 M As?

Solution

The EPA limit is given in grams/liter, and the concentration of the solution is in moles/liter. So you need to convert moles to grams (or vice versa) to make a comparison.

The proportionality constant between grams of arsenic and moles of arsenic is the molar mass, which you can find on the periodic table. To find the mass of arsenic, multiply the molar mass by the number of moles.

Look up the molar mass of As.   74.9 g/mol

Mass = molar mass · n (74.9 g/mol)(0.000020 mol) = 0.0015 g As

The 0.000020 M arsenic solution has 0.0015 g of As per liter of solution. This amount is 15 times the concentration allowed by the EPA. So, the solution is not safe for drinking.

LESSON SUMMARY

How can you create a solution with a specific molarity?

Molar mass values on the periodic table allow you to convert between grams and moles. The amount in grams of solid needed to create solutions of specific molarities can be calculated using the molar mass of the solid. Likewise, the mass of solid dissolved in a solution can be determined if the concentration and volume of the solution are known.

Exercises

Reading Questions

1. Explain how you would prepare a solution of sucrose with a molarity of 0.25.
2. How are mass of solute, moles of solute, and volume of solution related?

Reason and Apply

3. How many grams of solute do you need to make 1 L of each of the solutions listed?
   a. 0.50 M NaCl
   b. 2.0 M $CaCl_2$
   c. 1.5 M NaOH

Exercises *(continued)*

4. Copy this table and complete it for solutions of glucose, $C_6H_{12}O_6$. Remember to convert milliliters to liters.

Molarity (moles/liter)	Volume (liters)	Number (moles of solute)	Molar mass of solute (grams/mole)	Mass of solute (grams)
	2.0 L			180 g
0.40 M		0.10 mol		
0.10 M		0.03 mol		
	100 mL			0.180 g

5. What volume of each of these solutions would contain 58.44 g of NaCl?
   a. 0.10 M NaCl
   b. 3.0 M NaCl
   c. 5.5 M NaCl

6. How many grams of fructose, $C_6H_{12}O_6$, are in 1 L of soft drink if the molarity of fructose in the soft drink is 0.75 M?

7. Which is more concentrated: a 1.0 L solution with 20 g of sucrose, $C_{12}H_{22}O_{11}$, or a 1.0 L solution with 20 g of glucose, $C_6H_{12}O_6$?

8. The Environmental Protection Agency has set the upper limit for fluorine, F, in drinking water at 0.0040 g/L of water. Is it safe to drink a glass of water in which the concentration of fluorine is 0.00010 M?

Next, figure out the molar mass of sodium bromide, using the periodic table.

$$\text{molar mass} = 22.99 \text{ g/mol} + 79.90 \text{ g/mol} = 102.9 \text{ g/mol}$$

Multiply the number of moles of solute by the molar mass of the solute to get the number of grams of solid.

$$(102.9 \text{ g/mol})(0.050 \text{ mol}) = 5.15 \text{ g}$$

You should use 5.15 g of solid sodium bromide.

EXAMPLE 2

You have a 25 mL sample of a 0.01 M sodium bromide solution. What mass of NaBr is in the 25 mL sample?

SOLUTION

Figure out how many moles of sodium bromide are in your 25 mL sample.

$$(0.025 \text{ L})(0.01 \text{ mol/L}) = 0.00025 \text{ mol of solute}$$

Convert the number of moles of solute to grams using the molar mass of NaBr.

$$(0.00025 \text{ mol})(102.8 \text{ g/mol}) = 0.026 \text{ g}$$

There is 0.026 g of sodium bromide in the sample.

EVALUATE (5 min)

Check-In

How many grams of salt are in 0.100 L of 2.50 M NaCl?

Answer: Molarity = moles/liters, and $2.50 = n/0.100$ L and $n = 0.250$ mol. Now convert moles to grams using the molar mass of salt: $(0.250 \text{ mol})(58.5 \text{ g/mol}) = 14.6$ g of NaCl.

Homework

Assign the reading and exercises for Toxins Lesson 82 in the student text.

LESSON 82 ANSWERS

1. *Possible answer:* To prepare a 0.25 M solution of sucrose, you would calculate the mass of 0.25 mol of sucrose, then measure the amount of sucrose and dissolve it in enough water to make 1 L of solution.

2. The mass of solute and the moles of solute are proportional. The proportionality constant is the molecular mass of the compound. The moles of solute and the volume of the solutions are also proportional. The proportionality constant is the molarity, or concentration, of the solution.

3. a. 29 g (two significant figures)
b. 220.0 g (two significant figures)
c. 60.0 g (two significant figures)

4.

Molarity	Volume	Number	Molar mass	Mass of solute
0.50 M	2.0 L	1.0 mol	180 g/mol	180 g
0.40 M	250 mL	0.10 mol	180 g/mol	18 g
0.10 M	300 mL	0.03 mol	180 g/mol	5.4 g
0.01 M	100 mL	0.001 mol	180 g/mol	0.180 g

5. a. 10 L **b.** 0.33 L **c.** 0.18 L

6. 135.2 g

7. For each sample, the molarity of the sample is the same as the moles of solute because the volume of the solution is 1 L. Therefore, the glucose sample, $C_6H_{12}O_6$, is more concentrated.

8. The molarity and the upper limit do not depend on the volume of the glass, so assume the glass of water is 1 L in volume. If the glass contains 0.00010 M of fluorine, it also contains 0.00010 mol of fluorine. The molar mass of fluorine is about 19 g/mol, so the mass of the fluorine is 0.00010 mol · 19 g/mol = 0.0019 g. This is a little less than half the safe limit, 0.0040 g for a 1 L glass, set by the Environmental Protection Agency (EPA).

Overview

CLASSWORK: PAIRS

Key Question: Can molarity calculations be used to identify a toxin?

KEY IDEAS

The density (and therefore the mass) of an aqueous solution depends on the molar mass of the solute. Generally, for solutions with identical molarities and volumes, the solute with the greater molar mass will have a greater mass density.

LEARNING OBJECTIVE

Explain the relationship between the mass of a solution sample and the molar mass of the solute.

IN CLASS

Students are shown three solutions labeled 1, 2, and 3. Each is a 1.0 M ionic solution, and all three have the same volume but different masses. Students are asked to match the solutions with the compound they contain: KCl, NaBr, or NaOH. Because the molar masses of the solutes are different, the solutions can be identified by their total masses. After completing their calculations, students determine which of the solutions would be least harmful to drink by reviewing toxicity information. A complete materials list is available for download.

TRM Materials List 83

SETUP

Prepare 100 mL of each of the three solutions:

Solution 1: Dissolve 7.46 g KCl in a small amount of water and bring to 100 mL.

Solution 2: Dissolve 10.3 g NaBr in a small amount of water and bring to 100 mL.

Solution 3: Dissolve 4.00 g NaOH in a small amount of water and bring to 100 mL.

Do not label the KCl, NaBr, and NaOH solutions with their contents. Label them only as solutions 1, 2, and 3. Place the three beakers of solution on three balances at the front of the room, where students can see them.

CLEANUP

The solutions can be diluted and disposed of according to local guidelines.

Is It Toxic?

Mystery Solutions

THINK ABOUT IT

Selling bottled water is a big business. Some companies claim to get their water from mountain streams. Some say that their water is quite pure. Still others claim their water contains beneficial minerals. How could you determine how pure a water sample is?

Can molarity calculations be used to identify a toxic substance?

To answer this question, you will explore

↪ Accounting for the Mass of a Solution

↪ Drinking Water Standards

EXPLORING THE TOPIC

↪ Accounting for the Mass of a Solution

Suppose that you want to know how pure a sample of water is. Examine the illustration and consider how you might use mass to make a determination of purity.

Three 100 mL samples

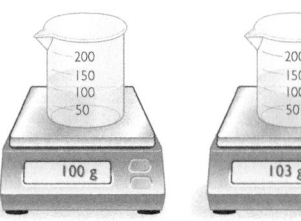 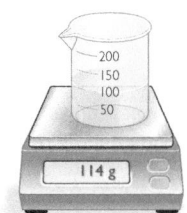

What masses do you expect if the samples are all pure water? Recall that mass per volume is equal to density. The density of pure water is 1 g/mL. If a 100 mL sample is pure water, then you expect a mass of 100 g.

The sample in the first beaker may be pure water. The other two samples must contain solutes that add more mass to the same volume, making them denser (and therefore heavier) than pure water.

THE DENSITY OF A SOLUTION

Suppose you have three samples of salt water with the concentrations shown in the illustration at the top of the next page. Take a moment to consider how the densities of the three solutions compare. Why does the ball float higher in the solution on the right?

Lesson Guide

ENGAGE (5 min)

TRM PowerPoint Presentation 83

ChemCatalyst

Suppose you wanted to determine if your tap water contains lead (II) nitrate, $Pb(NO_3)_2$, which is toxic.

1. Do you expect the mass of 100 mL of pure water to be the same as that of a solution containing $Pb(NO_3)_2$? Explain your reasoning.

2. What is the molar mass of lead nitrate? What would be the mass of 0.100 mol of lead (II) nitrate?

Sample answers: **1.** Most students will say that a solution containing lead (II) nitrate will have more mass than a solution of pure water because the solution contains both water molecules and heavy lead and nitrate ions. Some students may reason that the solutions will have the same mass because the lead (II) nitrate takes up space and may displace less water in the sample. **2.** The molar mass of lead (II) nitrate is 331.2 g/mol. A 0.100 mol sample of lead (II) nitrate would have a mass of 33.1 g.

Ask: What are some ways you might be able to tell if a substance is dissolved in water?

The ball floats in each solution.

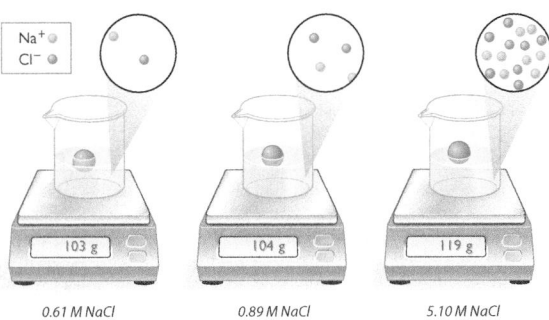

0.61 M NaCl   0.89 M NaCl   5.10 M NaCl

It's easy to float in the Dead Sea, which is a salt lake and one of the world's saltiest bodies of water.

As the density of the solution increases, an object floats higher. That is why when you swim in very salty water, you float with more of your body out of the water. As the concentration increases, there are more units of NaCl per volume of solution. The higher the salt concentration, the denser the solution.

SAME MASS, DIFFERENT SOLUTES

Imagine that you add 10 g of three different solids to three different beakers and then add enough water to make 100 mL of solution. The solutions all have approximately the same mass—the mass of about 100 mL of water plus 10 g of solute. But do they have the same concentration?

Compound	Molar mass
$CuSO_4$	249.7 g/mol
NaOH	40.0 g/mol
KCl	74.6 g/mol

Ions in 100 mL of solution

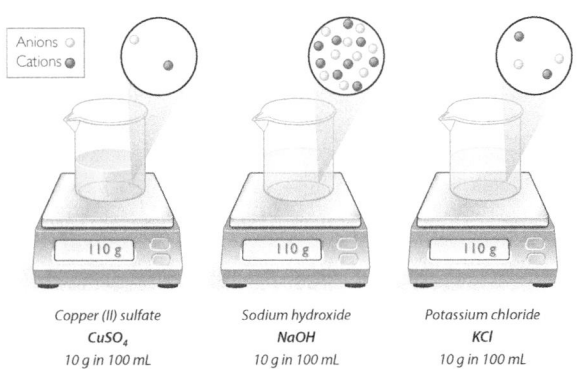

Copper (II) sulfate
$CuSO_4$
10 g in 100 mL

Sodium hydroxide
NaOH
10 g in 100 mL

Potassium chloride
KCl
10 g in 100 mL

Copper (II) sulfate, $CuSO_4$ has the largest molar mass. That means there are fewer formula units of $CuSO_4$ in 10 grams compared with NaOH. While you've added

Lesson 83 | **Mystery Solutions**   423

Ask: What would happen if you evaporated the water from a solution containing a dissolved substance?

Ask: Can you determine what compound is in a solution based on its molarity and its mass? Explain.

Ask: Does 10 mL of ocean water have the same mass as 10 mL of tap water? Explain.

EXPLORE (20 min)

TRM Worksheet with Answers 83

TRM Worksheet 83

INTRODUCE THE CLASSWORK

→ Arrange students into pairs.

→ Pass out the student worksheets.

→ Remind students the importance of never tasting anything in a laboratory.

DISCUSS THE IMPORTANCE OF MOLAR MASS

Ask: Which 100 mL sample of solution contained the greatest number of moles of solute? (all the same)

Ask: How do you determine the mass of 1 mol of a solute?

Ask: How can you determine the identities of the solutions?

Ask: Which do you think is more important in determining toxicity, the mass or the number of moles? Explain your thinking.

Ask: Why do you think LD_{50} values are expressed in milligrams per kilogram of body weight and not in moles?

Key Points: The molarities of the 100 mL samples of solute solution were all 1.00 M. This means that 1 L of each solution contains 1.00 mol of dissolved solute. Each beaker held 100 mL samples of solution, so each beaker contained 0.10 mol of solute.

To figure out the identities of the solutions, we determined the molar mass of each from the periodic table.

> Molar masses: KCl = 74.6 g/mol
>   NaBr = 102.9 g/mol
>   NaOH = 40.0 g/mol

If we assume that the water in each sample has approximately the same mass, the remaining mass is due to the dissolved solute. Although this is an approximation, it allows us to identify the solutions based on which solutes have the greatest and least molar mass.

It is important to know the toxicity of a substance in moles per kilogram. Because of the way toxic substances work, the number of moles in a sample matters more than the mass, because it indicates the number of atoms or molecules that interact with your body. However, for consumers, it is more practical to know the LD_{50} in grams per kilogram, because it is easier to determine the mass of a sample than the number of moles.

EVALUATE (5 min)

Check-In

You have three aqueous solutions, 200 mL of each: 1.5 M NaCl, 1.0 M KCl, and 1.0 M $CaCl_2$. Which has the greatest mass? Which has the least?

Answer: The molar masses are 58.5 g/mol for NaCl, 74.6 g/mol for KCl, and 111.0 g/mol for $CaCl_2$. The NaCl solution is 1.5 times as concentrated as the other two, so for the same volume it will have the greatest mass. The KCl solution will have the least mass.

Homework

Assign the reading and exercises for Toxins Lesson 83 and the Chapter 15 Summary in the student text.

the same mass to each sample, you have not added the same number of moles. Dissolving 10 grams of the compound with the higher molar mass will result in a lower concentration.

Example

Different Solutions, Same Concentration

Suppose you create 0.10 M solutions of copper (II) sulfate, sodium hydroxide, and potassium chloride. For equal volumes, which solution will have the most mass?

Solution

The mass of each solution is the mass of the water plus the mass of solute. The mass of the solute added depends on its molar mass. To determine the mass corresponding to 0.10 mol of each solute, multiply the number of moles by the molar mass of each compound.

$$\text{Mass of } CuSO_4 = (249.7 \text{ g/mol})(0.10 \text{ mol}) = 25.0 \text{ g } CuSO_4$$

$$\text{Mass of NaOH} = (40.0 \text{ g/mol})(0.10 \text{ mol}) = 4.0 \text{ g NaOH}$$

$$\text{Mass of KCl} = (74.6 \text{ g/mol})(0.10 \text{ mol}) = 7.5 \text{ g KCl}$$

So, for equal volumes of solution, the $CuSO_4(aq)$ will have the most mass.

⟳ Drinking Water Standards

Environmental Protection Agency standards protect public health by limiting contaminants in drinking water. The maximum contaminant levels for fluorine, iron, and barium in drinking water are given in mg/L in the table.

Element	Maximum contaminant level
fluorine, F	4.0 mg/L
iron, Fe	0.30 mg/L
barium, Ba	2.0 mg/L

Based on the numbers in the table, you might conclude that iron, Fe, is the most toxic because the allowable level is the smallest by mass. However, these contaminants interact with your body atom by atom. So you need to compare the contaminants by number of atoms.

Find the concentration of the potential toxins in moles per liter. To do so, convert the maximum contaminant level milligrams to grams, and then divide by the molar mass.

fluorine, F $\quad \dfrac{0.0040 \text{ g/L}}{19.0 \text{ g/mol}} = 0.00021 \text{ mol/L} = 0.00021 \text{ M} = 2.1 \times 10^{-4} \text{ M}$

iron, Fe $\quad \dfrac{0.00030 \text{ g/L}}{55.8 \text{ g/mol}} = 0.000054 \text{ mol/L} = 0.000054 \text{ M} = 5.4 \times 10^{-5} \text{ M}$

barium, Ba $\quad \dfrac{0.0020 \text{ g/L}}{137.3 \text{ g/mol}} = 0.000014 \text{ mol/L} = 0.000014 \text{ M} = 1.4 \times 10^{-5} \text{ M}$

In moles per liter, the maximum contaminant level for barium, Ba, is the smallest numerical value. This means less of it is tolerated in our drinking water, making barium the most toxic of the three elements.

ENVIRONMENTAL CONNECTION

The Cuyahoga River in northeastern Ohio used to be so polluted that it caught fire several times between 1936 and 1969. In 1969, an article about the fires in *Time* magazine led to the Clean Water Act, the Great Lakes Water Quality Agreement, and the creation of the Environmental Protection Agency, or EPA.

© Bettmann/CORBIS

LESSON SUMMARY

Can molarity calculations be used to identify a toxic substance?

When a substance is dissolved in water, it adds to the mass of the solution. One mole of one substance may have a very different molar mass than one mole of another substance. So different solutions of the same concentration can be differentiated by weighing them. Comparing contaminants by number of atoms may give a different picture of toxicity than comparing them by mass.

Exercises

Reading Questions

1. Explain how you might use mass to determine if a sample of water is contaminated.
2. Explain why the density of a solution increases as the concentration increases.
3. Explain why 0.10 M $CuCl_2$ has a greater density, m/V, than 0.10 M KCl.

Reason and Apply

4. Find the molarity of each solution. Which solution has the highest molarity? The highest density? Give your reasoning.
   a. 1.0 g KCl in 1.0 L water
   b. 5.0 g KCl in 0.5 L water
   c. 10.0 g KCl in 0.3 L water

5. Find the molarity of each solution. Which solution has the highest molarity? The highest density? Give your reasoning.
   a. 10 g $NaNO_3$ in 1.0 L water
   b. 10 g $Cd(NO_3)_2$ in 1.0 L water
   c. 10 g $Pb(NO_3)_2$ in 1.0 L water

6. Find the mass of solute in each solution. Which solution has the highest density? Is it possible to calculate the density of each solution? Why or why not?
   a. 1.0 L of 0.10 M $NaNO_3$
   b. 1.0 L of M 0.10 M $Cd(NO_3)_2$
   c. 1.0 L of 0.10 M $Pb(NO_3)_2$

7. Which would weigh the most? Explain your thinking.
   (A) 1.0 L of 1.0 M NaBr
   (B) 500 mL of 1.0 M KCl
   (C) 1.0 L of 0.5 M NaOH

8. PROJECT Look up the MSDS or Materials Safety Data Sheet, for sodium chloride, NaCl, and sodium sulfate, Na_2SO_4.
   www
   a. Describe the toxic effects of each salt.
   b. What are the LD_{50}'s (mouse, oral)?
   c. Compare the toxicities based on weight in grams.
   d. Compare the toxicities based on moles.
   e. Compare the toxicities based on volume of 0.10 M solution.

5. a. 0.118 M, **b.** 0.0424 M, **c.** 0.0302 M. The solution of $NaNO_3$ has the highest molarity. All three solutions have the same density because the same mass of solute was added to the same volume of water.

6. a. 8.5 g **b.** 24 g **c.** 33 g. The 0.10 M solution of $Pb(NO_3)_2$ has the greatest mass. The 0.10 M solution of $Pb(NO_3)_2$ also has the greatest density because a greater mass of solute was added to the 1 L of solution. It is possible to calculate the density of each solution because the mass of the solute and the mass of the water are known and the volume of the solution is also known.

7. (A) The 1.0 L of 1.0 M NaBr solution would weigh the most because it has the more solute added than the 1.0 L of 0.5 M NaOH solution.

8. a. Toxic effects of sodium chloride include vomiting, diarrhea, and dehydration. Toxic effects of sodium sulfate include temporary asthma and eye irritation from dust containing the salt. **b.** LD_{50} of NaCl = 3000 mg/kg, LD_{50} of Na_2SO_4 = 5989 mg/kg. **c.** Sodium chloride is more toxic based on weight, because its LD_{50} value is lower. **d.** LD_{50} of NaCl = 0.0513 mol; LD_{50} of Na_2SO_4 = 0.0422 mol. Sodium sulfate is more toxic based on moles. **e.** If the two substances are in the same concentration in the same volume of fluid, the substance that is more toxic by mole will be more toxic by volume of fluid. Sodium sulfate is more toxic based on volume.

LESSON 83 ANSWERS

1. A substance dissolved in the water adds mass to the solution. A sample of contaminated water should have a greater density than a sample of pure water. If this difference can be measured, the greater density indicates that the water is not pure.

2. Concentration is a measure of mass per unit volume. If the volume remains the same, the mass of the solute increases when the concentration increases. Because the mass of the whole solution also increases, the density of the solution must increase as well.

3. The 0.10 M copper chloride, CuCl, solution has a greater density than the 0.10 M potassium chloride, KCl, solution. Copper chloride has a molar mass of about 134 g and potassium chloride has a molar mass of about 75 g. So, if you have equal volumes of each solution, there would be more mass in the copper chloride solution. If the mass is greater, the density is greater.

4. a. 0.013 M, **b.** 0.13 M, **c.** 0.45 M. The 10 g solution of KCl in 0.3 L water has the greatest molarity and the greatest density.

ANSWERS TO CHAPTER 15 REVIEW EXERCISES

1. *Possible answer:* Knowing exactly how many moles or grams of contaminant are in a water sample determines if the sample is toxic because toxicity is based on dosage.

2. *Possible answer:* Lead has a zero tolerance level because the toxicity of lead is very high. Because lead easily interferes with body functions, every effort should be made to ensure that lead is not ingested. Lead poisoning is especially dangerous for infants and small children. Keeping lead out of the environment has been an emphasis of governments at all levels since the discovery in the 1960s and 1970s of lead's danger.

3. a. Each unit of lead (II) nitrate has one lead cation, so the number of moles of lead is also 0.000020 mol. **b.** 0.0066 g. **c.** 0.0041 g.

4. a. The 1.0 M NaOH and 1.0 M KCl solutions have the greatest concentration. **b.** Because each solution has the same mass of water, the solution with the most solute added will have the greatest mass. The $PbCl_2$ solution has the greatest mass. **c.** Because each solution has the same volume, the solutions with the highest molarity will have the most moles of solute. The 1.0 M NaOH and 1.0 M KCl solutions have the most moles of solute. **d.** The first two solutions separate into two ions in solution, so each solution has 0.20 mol of ions. The $PbCl_2$ solution separates into three ions and has 0.15 mol of ions. The 1.0 M NaOH and 1.0 M KCl solutions have the most ions.

5. a. To make a 0.5 M solution of NaOH, 0.5 mol of NaOH is needed for every liter of solution. The molar mass of NaOH is 40 g/mol, and you could make a 0.5 M solution of NaOH by adding 20 g of NaOH to a container and adding enough water to make 1 L of solution. **b.** No. The amount of solution can vary as long as the number of moles of solute divided by the volume of solution is 0.5 M.

6. 0.96 L

7. a. 0.250 M **b.** 0.750 M **c.** 0.428 M **d.** 0.214 M

8. 1.2 g

CHAPTER 15

Toxins in Solution

SUMMARY

KEY TERMS

solution

solvent

solute

homogeneous mixture

heterogeneous mixture

concentration

molarity

molar

saturated solution

Toxins Update

It is easy for toxins to dissolve in and enter our water supply. In fact, it would be challenging to find or create a sample of "pure water"—one that had nothing else in it but H_2O molecules. To track the concentrations of different substances in solutions, scientists use moles of solute per liter of solvent. This is also called the molarity of the solution. If you know the molarity of a solution and its volume, you can determine the exact number of grams, moles, or even molecules or ions of a substance in your water sample. Sometimes, toxic substances in solution are tracked in parts per million, or ppm. A solution with a toxic concentration of 1 ppm would have 1 mL of toxic substance per 1000 L of solution.

REVIEW EXERCISES

1. Why is it useful to know exactly how many moles or grams of contaminant are in a water sample?

2. Lead is one of the few substances with a zero tolerance level in our drinking water, meaning that there should be no lead at all in our water. Why do you think this is?

3. Consider a 10 mL water sample from a nearby creek that has a 0.002 M lead (II) nitrate, $Pb(NO_3)_2$, concentration.
   a. How many lead ions are in this sample?
   b. How many grams of lead nitrate are in this sample?
   c. How many grams of lead ions are in this sample?

4. Suppose you measure 100 mL each of 1.0 M NaOH, 1.0 M KCl, and 0.5 M $PbCl_2$ solutions.
   a. Which has the greatest concentration?
   b. Which has the most mass?
   c. Which has the most moles?
   d. Which has the most ions?

5. **a.** Suppose you need to make a 0.5 M solution of sodium hydroxide, NaOH. Explain step by step how you would go about making this solution.
   b. Do you have to use 1 L of water when you make this solution? Explain your reasoning.

6. A mass of 47 g of sulfuric acid, H_2SO_4, is dissolved in water to prepare a 0.50 M solution. What is the volume of the solution?

7. Determine the molarity of the solutions created by following each set of directions.
   a. Dissolve 7.30 g NaCl to make 500 mL of solution.
   b. Dissolve 43.84 g NaCl to make 1 L of solution.
   c. Dissolve 25.00 g NaCl to make 1 L of solution.
   d. Dissolve 25.00 g NaCl to make 2 L of solution.

8. How many grams of magnesium sulfate, $MgSO_4$, do you need to make 100 mL of a 0.1 M solution?

Project: Types of Bonding

A good report would include:
● the chemical formula for the toxic substance, ● the type of bonding in the toxic substance, ● an appropriate sketch of the toxic substance, depending on its method of bonding, ● a list of basic properties of the toxic substance, including conductivity.

ASSESSMENTS

Two multiple-choice assessments (versions A and B) and two short-answer assessments (versions A and B) are available for download for Chapter 15.

TRM Chapter 15 Assessment Answers

TRM Chapter 15 Assessments

Types of Bonding

PROJECT Research a toxic substance (your teacher may assign you one). Your report should cover these topics.

- Is the bonding in your substance ionic, covalent, or metallic?

- If your substance is ionic, determine the cation and the anion. Write and label the ions and their charges. Include a particle view drawing of your substance.

- If your substance is covalent, determine if it is polar or nonpolar. Explain your reasoning. Include the structural formula and show the dipole(s).

- If your substance is metallic, find out its density. Include a particle view drawing of your substance.

- Does your substance conduct electricity? Explain your reasoning.

Watch the video overview of Chapter 16 (for teachers) by clicking on the link in the TE-book, opening the TRFD, or logging onto the book's companion Web site bcs.whfreeman.com/ livingbychemistry2e (teacher log-in required).

Acid-base chemistry is the focus of this chapter. In Lesson 84, students use a cabbage juice indicator to sort and classify common household solutions according to their color and behavior. A second test with a universal indicator allows students to position these solutions on the pH scale. Lesson 85 introduces the Arrhenius and Brønsted-Lowry definitions of acids and bases. Students examine particle models of acids and bases and begin to create definitions. Lesson 86 focuses on the pH scale and connects it to both H^+ and OH^- concentrations. Lesson 87 explores the effect of dilution on solution pH. In this lab, students complete a serial dilution of an acid and a base. Neutralization reactions are introduced in Lesson 88, in which students complete a lab that requires them to mix acids and bases and test the resulting pH. In Lesson 89, students complete three titrations and calculate the molarities of three unknown acid solutions.

In this chapter, students will learn

- Arrhenius and Brønsted-Lowry definitions of acids and bases
- the difference between strong and weak acids and bases
- the effects of dilution and neutralization on pH
- how to complete a titration procedure
- stoichiometric calculations associated with acids and bases

A vinegaroon is similar to a scorpion, but instead of stinging, it squirts acetic acid, or vinegar, on its prey.

CHAPTER 16

Acidic Toxins

In this chapter, you will study

- how the acidity of solutions is tracked
- theories of acids and bases
- differences between acids and bases
- the effect of dilution and neutralization on acid-base solutions
- a lab procedure for determining acid concentration

Some solutions can be classified as either *acids* or *bases*. They have a unique set of properties and are an extremely important part of solution chemistry. Toxic levels of acids and bases can burn the skin and damage living tissue. However, without acids and bases, the body would not function properly. This chapter introduces you to the properties of acids and bases and shows how chemists monitor them.

428

Chapter 16 Lessons

Heartburn

Acids and Bases

THINK ABOUT IT

Solutions have a wide variety of properties. Acetic acid, or vinegar, is used to flavor salad dressing, while citric acid gives a sour taste to lemon juice and orange juice. Ammonium hydroxide is used to clean windows, and sodium hydroxide is used to open clogged drains. All of these solutions are either acids or bases and can be classified into these categories based on their general properties.

What are the properties of acids and bases?

To answer this question, you will explore

↻ Acids and Bases

↻ Indicators

↻ The pH Scale

EXPLORING THE TOPIC

↻ Acids and Bases

GENERAL PROPERTIES

Acids and bases are special categories of solutions. They are extremely useful to us in our everyday lives precisely because of their unique properties. The term *acid* comes from the Latin word *acidus,* which means sour. Many of the sour tastes in our food come from the acids found in those foods.

For much of history, the term *alkaline* was used instead of *base.* Bases are found in many household cleaners, from soaps to drain openers to oven cleaners. Bases have a bitter taste, as you may have noticed if you have ever accidentally tasted soap. Bases usually feel slippery to the touch. The slipperiness of bases arises from the fact that they are reacting with the fats in your skin and turning them into soap.

In general, acids and bases are toxic, especially large quantities of concentrated solutions. It is important that acids and bases not splash on your skin or in your eyes. Acids and bases are both corrosive and can cause a *chemical burn.* A chemical burn is one in which living tissue is damaged.

↻ Indicators

Many acidic and basic solutions are colorless and odorless, which can make them difficult to detect by their appearance alone. Because these solutions can be toxic, it is useful to be able to monitor them. Luckily, there are molecular substances called **indicators** that change color when they come into contact with acids and bases. If you add a drop or two of an indicator to an unknown solution, you can tell if you have an acid or a base by the color that results.

Lesson 84 | **Acids and Bases**   429

Overview

LAB: GROUPS OF 4

Key Question: What are the properties of acids and bases?

KEY IDEAS

Acids and bases are classified according to their unique set of properties and reactivity. Chemical compounds called *indicators* help chemists identify the presence of acids and bases in solution and also indicate the relative acidity or basicity of a substance. The pH scale is a number scale that assigns values to acids or bases. Substances with a pH below 7 are considered acidic, substances with a pH at or near 7 are considered neutral, and substances with a pH greater than 7 are considered basic at 25 °C.

LEARNING OBJECTIVES

- Identify acids and bases based on general observable properties.
- Explain how an indicator is used to determine whether a solution is acidic, basic, or neutral.

KEY TERMS

indicator
pH scale
acid
base

IN CLASS

Students test a group of common substances with cabbage juice indicator and with universal indicator. They classify the substances based on the results of these tests. Students also test a smaller sample of the substances with calcium carbonate and observe differences in reactivity. *Note:* pH is introduced in the student text for this lesson and is covered in detail in Lesson 86: pHooey! A complete materials list is available for download.

TRM Materials List 84

Pre-AP® Course Tip

Students preparing to take an AP® chemistry or college chemistry course should be able to collect data and analyze it for patterns or relationships. In this lesson, students conduct a lab with various household chemicals and classify them as acidic, basic, or neutral based on the color of the cabbage juice indicator.

SETUP

SAFETY

Acids and bases are corrosive. In case of a spill, rinse with large amounts of water.

Wear safety goggles.

Have baking soda on hand for acid spills.

Have citric acid on hand for base spills.

Before class, fill eight sets of labeled dropper bottles with each solution listed. You will need about 200 mL of each solution to divide among eight bottles. In advance, prepare red cabbage juice for use as an indicator. Cut red cabbage into wedges and immerse them in several quarts of water. Simmer the cabbage in water at a low boil for approximately 30 minutes. As an alternative, you can grind the uncooked cabbage in a blender or food processor and drain off the liquid. Place the cabbage juice in eight dropper bottles. *Note:* It is best to prepare the cabbage juice within 24 hours of using it. Keep it refrigerated if possible. It smells bad after a few days.

CLEANUP

Dilute wastes and pour them down the drain.

Lesson Guide

ChemCatalyst

Countless products are advertised on TV with the promise of reducing acid indigestion.

1. What is acid indigestion?

2. What does acid have to do with your stomach?

3. How do you think antacids work?

Sample answers: **1.** Acid indigestion is when your stomach hurts because of what you ate or because there is too much acid in it. **2.** The stomach contains acid to help break down food. **3.** Students might speculate that antacids coat the stomach or neutralize the acid. Antacids are bases, which neutralize acids.

Ask: How can you tell if a substance is an acid?

Ask: What does it mean to "neutralize" an acid?

Ask: How is an acidic solution different from pure water?

Ask: Is acid bad for your stomach? Explain.

TRM Worksheet with Answers 84

TRM Worksheet 84

INTRODUCE THE LAB

→ Arrange students into groups of four.

→ Pass out the student worksheet.

→ Each pair within the group will share some materials with the other pair.

→ Briefly introduce the concept of indicators. Tell students that indicators are a set of compounds that respond to certain solutions with vivid color changes. In this activity, students will use two different indicators and record the various colors they observe.

> **Indicator:** An indicator is a molecular substance that changes color when it comes into contact with an acid or a base.

→ You might demonstrate an indicator for the entire class, either in a beaker or on an overhead for good visibility. Use a sample of the hydrochloric acid solution and add two drops of universal indicator.

→ Briefly outline the procedure students will follow. In Part 1, students will test one set of each solution with the cabbage juice indicator and one set of each solution with the universal indicator. In Part 2, students will test several of the solutions with solid calcium carbonate.

→ Remind students about safety.

INTRODUCE ACIDS AND BASES

→ You can draw the number line from the worksheet on the board for easy reference.

Ask: Which solutions did you group together? What do they have in common?

Ask: Which solutions are on the left side of the number line? On the right side? In the middle?

Ask: Which solutions do you think are acids? Explain your reasoning.

Ask: How did the substances react with calcium carbonate? What conclusions did you draw from these observations?

Key Points: Note: Students are introduced to the concept of the pH scale on page 430 of the student text. For now, simply refer to it as a number line. The substances on the left side of

Cabbage juice is a natural acid-base indicator obtained by boiling or grinding up red cabbage. The cabbage juice changes color as it is added to various solutions, as shown in the illustration. Take a moment to look for patterns.

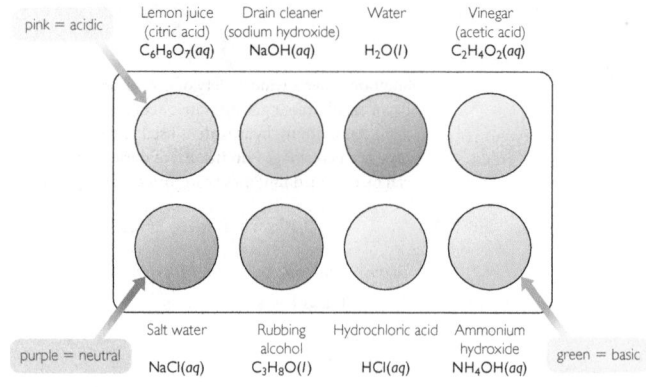

pink = acidic

Lemon juice (citric acid) $C_6H_8O_7(aq)$ Drain cleaner (sodium hydroxide) $NaOH(aq)$ Water $H_2O(l)$ Vinegar (acetic acid) $C_2H_4O_2(aq)$

purple = neutral

Salt water $NaCl(aq)$ Rubbing alcohol $C_3H_8O(l)$ Hydrochloric acid $HCl(aq)$ Ammonium hydroxide $NH_4OH(aq)$ green = basic

Cabbage juice turns pink when it is in contact with acids, such as citric acid, acetic acid, and hydrochloric acid. It turns green or yellow in contact with bases, such as sodium hydroxide and ammonium hydroxide. And it remains purple or blue in contact with *neutral* substances such as pure water, salt water, and rubbing alcohol. Neutral solutions are neither acidic nor basic. So cabbage juice can be used to indicate if a solution is acidic or basic or neither.

↻ The pH Scale

There are dozens of different types of acid-base indicators, each with a unique color scheme. The colors associated with "universal indicator" are shown in the illustration. Notice that there is a number associated with each color. This number is referred to as the *pH number*, or just as the *pH*.

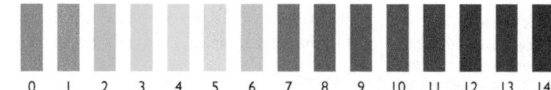

0 1 2 3 4 5 6 7 8 9 10 11 12 13 14

Another way to measure the pH for a solution is with paper coated with indicator, referred to as pH paper.

One common type of pH paper is called *litmus paper*. Litmus paper turns red in acidic solutions and blue in basic solutions.

The **pH scale** is a number line that assigns number values from 0 to 14 to acids and bases.

Substances with a pH below 7 at 25 °C are **acids.** Substances with a pH above 7 at 25 °C are **bases.** Substances with a pH at or near 7 at 25 °C are considered neutral. Stomach acid is extremely acidic, with a pH around 1. Lemon juice is the

Litmus Paper

Acid   Base

> **HEALTH CONNECTION**
>
> Too much acid in the stomach may result in indigestion, heartburn, or ulcers. Antacids are bases that work in the stomach to neutralize the excess acid there. Other remedies work on receptor sites in your body, preventing the secretion of excess acid in the first place.

The pH scale

Stomach acid | Lemon juice | Vinegar | Salt solution, water, rubbing alcohol | Washing soda | Ammonium hydroxide | Drain cleaner

0 — Increasing [H$^+$] — Acids — 7 — Neutral — Increasing [OH$^-$] — Bases — 14

next most acidic substance shown here, with a pH near 2.4. The most basic substance shown here is drain cleaner. The substances on either end of the pH scale are potentially more dangerous and more toxic than substances found in the middle of the scale.

LESSON SUMMARY

What are the properties of acids and bases?

KEY TERMS

indicator
pH scale
acid
base

Acids and bases are special categories of solutions. The sour tastes in our food come from the acids found in those foods. Bases cause the slippery feel of soaps and detergents. Chemical compounds called indicators help chemists to identify the presence of acids and bases in solution. Indicators change color in response to acids and bases. The pH scale is a number scale that assigns values to acids and bases, between 0 and 14. Substances with a pH below 7 at 25 °C are acids. Substances with a pH above 7 at 25 °C are bases. Substances with a pH at or near 7 at 25 °C are neutral.

Exercises

Reading Questions

1. What are some of the observable properties of acids and bases?
2. What is the pH scale?
3. What does it mean to say that a substance has a neutral pH?

Reason and Apply

4. **Lab Report** Write a lab report for the Lab: Acids and Bases. In your report, give the title of the experiment, purpose, procedure, observations, analysis, and conclusion.

5. Classify each of the following solutions at 25 °C as acidic or basic based on the information provided.
   a. lemon juice tastes sour
   b. a solution of washing soda turns cabbage juice green
   c. a dilute solution of potassium hydroxide feels slippery
   d. a sugar solution has a pH of 7
   e. drain cleaner has a pH of 12

(Title)

Purpose: (Explain what you were trying to find out.)

Procedure: (List the steps you followed.)

Observations: (Describe your observations.)

Analysis: (Explain what you observed during the experiment.)

Conclusion: (What can you conclude about what you were trying to find out? Provide evidence for your conclusions.)

Lesson 84 | **Acids and Bases**   431

---

the number line (from 0 to 7) are all acids. The acids tested today were vinegar (acetic acid), hydrochloric acid (found in stomach acid), and lemon juice (citric acid). Cabbage juice turns pink when in contact with acids. When tested with a universal indicator, the colors range from red to yellow. Also, these substances react vigorously with calcium carbonate, releasing carbon dioxide gas.

The substances on the right side of the number line (from 7 to 14) are called bases. The bases tested today were washing soda (sodium carbonate), drain cleaner (sodium hydroxide), and window cleaner (ammonium hydroxide). Cabbage juice turns green when in contact with bases (yellow when a substance is highly basic). When tested with a universal indicator, the colors range from pale green to blue. When calcium carbonate is added to a base, nothing is observed—it does not react.

The substances located in the middle of the number line (around 7) are called neutral substances. Cabbage juice does not change color when in contact with a neutral substance. So a neutral substance in cabbage juice is purple (and sometimes blue). The

neutral substances examined today were distilled water, rubbing alcohol, and salt solution. The universal indicator turns greenish yellow when in contact with neutral substances.

EVALUATE (5 min)

Check-In

An unknown substance is mixed with cabbage juice, and the solution turns purple. The substance does not react with calcium carbonate. Is it an acid, a base, or a neutral substance? Explain.

Answer: This substance is probably a neutral substance, based on its behavior. The neutral substances that were tested did not alter the color of the purple cabbage juice.

Homework

Assign the reading and exercises for Toxins Lesson 84 in the student text. Assign the Web Research Project in the Chapter 16 Summary in the student text (optional). Have students continue research on the toxin assigned in Chapter 13.

LESSON 84 ANSWERS

1. *Possible answer:* Acids have a sour taste and can burn the skin in concentrated form. Bases have a bitter taste and a slippery feel. Bases can also cause skin burns in concentrated form. Most acidic and basic solutions are colorless and odorless.

2. *Possible answer:* The pH scale is a numeric scale used to classify acids and bases.

3. A substance with a neutral pH is neither acidic nor basic.

4. A good lab report will contain ● a title (Lab: Acids and Bases) ● a statement of purpose (*Possible answer:* To explore the properties of acids and bases) ● a procedure (a summary of the steps followed in the experiment) ● results (a list of observations made during each step in the procedure) ● a conclusion (*Possible answer:* Different solutions have different acidities, which can be measured by an indicator. Acids can be neutralized by calcium carbonate.)

5. a. Acids taste sour. **b.** Cabbage juice turns green in a base. **c.** Bases feel slippery. **d.** A solution with a pH of 7 is neutral. **e.** A solution with a pH of 12 is basic.

6. Possible answers (any 3): • All of the names are two words and end with *acid*. • The first word always ends in "-ic." • All of the acids have at least one hydrogen atom. • All of the elements in the chemical formulas are on the right side of the periodic table.

7. Possible answers (any 3): • All of the base names are two words. • The second word is always *hydroxide*. • All of the bases have at least one hydroxide ion. • The first word in the name is usually an element in Group 1A or Group 2A.

8. Answers will vary depending on the products chosen. Check the chemical names and formulas for each compound chosen. • Examples of acids include vinegar, citrus juices, and metal polish. Uses include the removal of corrosion and as an additive to food for taste. • Examples of bases include baking soda, drain cleaner, and antacid tablets. Uses of bases include neutralizing acids and cleaning.

Exercises

(continued)

6. Examine the list of acids below. List three patterns that you notice in the names and chemical formulas.

hydrochloric acid, HCl   nitric acid, HNO_3

carbonic acid, H_2CO_3 acetic acid, $C_2H_4O_2$

sulfuric acid, H_2SO_4 citric acid, $C_6H_8O_7$

7. Examine the list of bases below. List three patterns that you notice in the names and chemical formulas.

sodium hydroxide, NaOH   magnesium hydroxide, $Mg(OH)_2$

potassium hydroxide, KOH   barium hydroxide, $Ba(OH)_2$

lithium hydroxide, LiOH   ammonium hydroxide, NH_4OH

8. PROJECT Find three acids and three bases used in your home.

www **a.** What are these acids and bases used for?

b. Look up the chemical name and chemical formula of each acid and base.

Pass the Proton

Acid and Base Theories

THINK ABOUT IT

People have known for thousands of years that vinegar, lemon juice, green apples, and many other foods taste sour. However, it is only recently that scientists have presented theories to explain what all these substances have in common. In what ways are these sour-tasting substances similar, and why are they defined as acids?

How are acids and bases defined?

To answer this question, you will explore

- Chemical Makeup of Acids and Bases
- Acid and Base Theories
- Strong and Weak Acids and Bases

EXPLORING THE TOPIC

Chemical Makeup of Acids and Bases

Many substances dissolve in water to produce solutions that are acidic. Because the behavior of all acidic solutions is similar, you might expect some similarity in the chemical makeup of these substances.

Take a moment to look for patterns in the table of common acids.

BIOLOGY CONNECTION

The human stomach secretes hydrochloric acid, HCl, which begins the digestion process for large protein molecules and also provides a barrier against bacteria, fungi, and other microorganisms naturally found in food and water. Stomach acid is also necessary for the proper absorption of certain minerals such as iron and zinc as well as some of the B vitamins.

Common Acids

Name	Found in	Chemical formula	Ions in solution
hydrochloric acid	stomach acid	HCl	$H^+ \ Cl^-$
nitric acid	acid rain	HNO_3	$H^+ \ NO_3^-$
sulfuric acid	acid rain	H_2SO_4	$H^+ \ H^+ \ SO_4^{2-}$
phosphoric acid	cola	H_3PO_4	$H^+ \ H^+ \ H^+ \ PO_4^{3-}$
acetic acid	vinegar	CH_3COOH	$H^+ \ CH_3COO^-$

Here are some patterns you might notice.

- The acids are made of main group nonmetal atoms, such as carbon, oxygen, and chlorine, bonded covalently.
- The only element that they all have in common is hydrogen.
- They all break apart, or **dissociate,** in solution to form at least one hydrogen cation, H^+, and an anion.

Lesson 85 | **Acid and Base Theories**   433

Overview

ACTIVITY: GROUPS OF 4

DEMO: WHOLE CLASS

Key Question: How are acids and bases defined?

KEY IDEAS

According to the Arrhenius definition, an acid is a substance that adds hydrogen ions, H^+, to an aqueous solution, and a base is a substance that adds hydroxide ions, OH^-, to an aqueous solution. However, not all bases contain hydroxide ions. Some substances add OH^- to solution by

removing H^+ from water. According to the Brønsted-Lowry definition, acids are proton donors and bases are proton acceptors. In this case, *proton* refers to H^+ ions.

LEARNING OBJECTIVES

- Define Arrhenius and Brønsted-Lowry acids and bases.
- Explain the behavior of acids and bases on a particulate level.
- Explain the difference between strong and weak acids and bases.

FOCUS ON UNDERSTANDING

Where appropriate, we use extended chemical formulas, such as CH_3OH for

methanol and $CHOOH$ for formic acid. Students might see the OH and assume that the substance is a base.

KEY TERMS

dissociate
strong acid
weak acid
strong base
weak base

IN CLASS

Students view a demonstration of the conductivity of four different solutions. Three of the solutions conduct electricity, even though only one is an ionic compound. Students speculate on what these observations might mean. Students then work in groups to sort 12 cards with particulate views of various acid, basic, and neutral solutions. Students work to explain the difference between acids and bases using the solution cards. (*Note:* Dissociation of water into H^+ and OH^- ions is covered in Lesson 86: pHooey!) A complete materials list is available for download.

TRM Materials List 85

Differentiate

Use a 3-2-1 strategy to assess students' understanding of acid-base theories. Ask students to identify 3 differences between the Arrhenius and Brønsted-Lowry acid-base theories, 2 ways these acid-base theories connect to previous concepts in chemistry, and 1 question students have about what they learned.

Pre-AP® Course Tip

Students preparing to take an AP® chemistry or college chemistry course should be able to use representations or models to make predictions about chemical phenomena. In this lesson, students use particle diagrams to explain how acids and bases differ on a particulate level and differentiate between strong and weak acids and bases.

SETUP

Prepare the materials for the demo that takes place during the ChemCatalyst. Before class, test the conductivity apparatus to make sure it is functioning properly. You can dim the lights during the demo to allow students to see the light bulb better.

CLEANUP

Dilute the waste solutions and pour them down the drain.

Lesson Guide

ChemCatalyst

Which of these four solutions conduct electricity? 0.10 M HCl (hydrochloric acid), 0.10 M CH_3COOH (acetic acid), 0.10 M NaCl (sodium chloride), or 0.10 M $C_{12}H_{22}O_n$ (sugar). Explain.

Sample answer: The first three solutions conduct, and the fourth does not. Students may say that the solutions that conduct electricity must make ions in solution. Some may speculate that the conducting solutions are all ionic, even though evidence suggests that only one solution is. They probably will state that sugar does conduct electricity because it is molecular.

➜ Use the conductivity apparatus to test the four solutions in the ChemCatalyst.

Ask: What causes the bulb to light up?

Ask: What types of solutions conduct electricity?

Ask: Do you think there are ions in the HCl solution? Explain your thinking.

Ask: Do you think there are ions in all the solutions that conduct electricity? Explain your thinking.

🔲 **TRM** Worksheet with Answers 85

🔲 **TRM** Worksheet 85

🔲 **TRM** Card Masters—Acid-Base Solution Cards 85

INTRODUCE THE ACTIVITY

➜ Arrange students into groups of four.

➜ Pass out the student worksheet.

➜ Provide groups of four students with a set of 12 Acid-Base Solution cards from the Teacher's Resource Materials. Explain that the square on each card represents a tiny volume of solution, approximately 4.2×10^{-21} L.

➜ Explain to students that sometimes an extended chemical formula is used to represent a compound with more than one isomer. Thus, CH_3OH identifies methanol, CH_3COOH identifies acetic acid, and CHOOH identifies formic acid. Students should not confuse these with hydroxides.

Lye, or sodium hydroxide, is a corrosive substance found in many household items including oven cleaner and drain opener. Lye is also used for curing foods such as green olives and century eggs and for glazing pretzels. It is also used in soap making.

© Rachel Epstein/PhotoEdit

You might also notice that some of the molecular formulas are written in a way that gives some information about the molecular structure. For example, acetic acid, $C_2H_4O_2$, is written as CH_3COOH.

Many other substances dissolve in water to produce solutions that are basic. If acidic solutions have H^+ ions in common, what is similar in the chemical makeup of basic solutions?

Take a moment to look for patterns in the names and chemical formulas of the common bases listed in this table.

Common Bases

Name	Found in	Chemical formula	Ions in solution
sodium hydroxide	drain cleaner	NaOH	Na^+ $(OH)^-$
magnesium hydroxide	antacid tablets	$Mg(OH)_2$	Mg^{2+} $(OH)^-$ $(OH)^-$
ammonia	window cleaner	NH_3	NH_4^+ $(OH)^-$
sodium carbonate	washing soda	Na_2CO_3	Na^+ Na^+ $(HCO_3)^-$ $(OH)^-$

Here are some things you might notice.

• Many of the bases contain a metal atom and a hydroxide ion, OH^-.

• The metal atoms are in the first two columns of the periodic table.

• Some compounds, such as ammonia and sodium carbonate, produce OH^- when dissolved in water, even though there is no OH^- in the chemical formula.

• Except for the two hydroxides, their names do not reveal that they are bases.

⟳ Acid and Base Theories

ARRHENIUS THEORY OF ACIDS AND BASES

One of the early theories of acids and bases was first proposed in 1884 by a Swedish chemist named Svante Arrhenius. He defined an acid as a substance that adds hydrogen ions, H^+, to an aqueous solution. He defined a base as a substance that adds hydroxide ions, OH^-, to an aqueous solution.

BRØNSTED-LOWRY THEORY OF ACIDS AND BASES

The Arrhenius definitions of acids and bases do not explain how substances without hydroxide ions, OH^-, in their formula, like ammonia, NH_3, can be bases. In the early 1920s, a Danish chemist, Johannes Brønsted, and an English chemist, Thomas Lowry, proposed a slightly different definition for acids and bases to account for this chemical behavior. They defined an acid as a substance that can donate a proton to another substance. They defined a base as a substance that can accept a proton from another substance. A proton in this case is actually a hydrogen ion, H^+. Because a hydrogen atom typically has no neutrons, if you remove the electron from a hydrogen atom to form an ion, all that's left is a proton.

When ammonia, NH_3, is added to water, it removes a hydrogen ion, H^+, from some of the water molecules, leaving behind some hydroxide ions, OH^-. An NH_3

➜ Remind students that all of these are aqueous solutions. The water molecules are not shown because there are so many of them—approximately 140,000 per square.

DISCUSS THE PATTERNS FOUND IN THE CARDS

➜ Allow students to assist you in sorting the Acid-Base Solution cards and identifying the common features of the 12 solutions.

➜ Be sure to acknowledge different sorting schemes, because they are all useful. The cards can be sorted by the

type of ions (H^+, OH^-, and neither), by concentration, or by whether or not molecules are present.

Ask: What were some discoveries you made as a result of sorting the cards?

Ask: Did you discover similar ions in some of the solutions?

Ask: What do all the solutions with *acid* in their name have in common?

Ask: How are the two sodium hydroxide solutions different from each other?

Key Points: Acids are substances that add H^+ to solution. When an acid molecule dissolves in water, a hydrogen ion, H^+, splits off, leaving the rest of

molecule accepts a hydrogen ion and becomes NH_4^+. This makes NH_3 a Brønsted-Lowry base.

$$NH_3(g) + H_2O(l) \longrightarrow NH_4^+(aq) + OH^-(aq)$$

So dissolved ammonia really consists of ammonium and hydroxide ions. For this reason, it is often referred to as ammonium hydroxide solution. Many substances that do not contain OH^- can act as bases.

Strong and Weak Acids and Bases

These two illustrations show particle views of a 0.010 M hydrochloric acid, HCl, solution and a 0.010 M formic acid, HCOOH, solution. The water molecules are not shown. Take a moment to examine them.

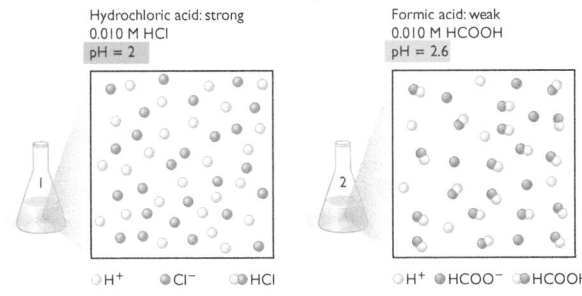

Particle Views of a Strong and Weak Acid

Hydrochloric acid: strong
0.010 M HCl
pH = 2

Formic acid: weak
0.010 M HCOOH
pH = 2.6

○ H^+ ● Cl^- ◐ HCl   ○ H^+ ● $HCOO^-$ ◐ HCOOH

Notice that there are no molecules of HCl in the solution on the left. The HCl has dissociated completely into H^+ and Cl^- ions. However, the solution on the right contains formic acid molecules. Only some of the HCOOH molecules have dissociated into ions. This means that the concentration of H^+ ions is smaller in the formic acid solution than in the hydrochloric acid solution, even though the solutions have the same molarities.

Acids that dissociate completely in solution are called **strong acids.** Strong acids include hydrochloric acid, HCl, nitric acid, HNO_3, sulfuric acid, H_2SO_4, and hydrobromic acid, HBr. Strong acids are good conductors of electricity.

Acids that dissociate only partially in solution are called **weak acids.** These solutions are only moderate conductors of electricity. Some common weak acids are formic acid, HCOOH, acetic acid (vinegar), CH_3COOH, citric acid, $C_3H_5O(COOH)_3$, and phosphoric acid, H_3PO_4. Weak acids tend to be less corrosive because they do not dissociate completely into ions.

Bases can also be classified as strong or weak. A **strong base** dissociates completely into ions in solution and **weak bases** dissociate only partially. Some examples of strong bases include sodium hydroxide, NaOH, and barium hydroxide, $Ba(OH)_2$. Examples of weak bases include ammonia, NH_3, and aniline, $C_6H_5NH_2$.

Big Idea Strong acids and strong bases dissociate completely into ions. Weak acids and weak bases dissociate only partially into ions.

Lesson 85 | **Acid and Base Theories**   435

---

the molecule with a negative charge. For example, HCl dissociates into H^+ and Cl^-. Formic acid splits apart into H^+ and $CHOO^-$. Because acids add ions to water, acidic solutions are electrolytes and will conduct electricity.

Bases are substances that add OH^- to solution. Many bases are ionic solids called metal hydroxides. When these bases dissolve in water, they split apart into metal cations and hydroxide anions, OH^-. For example, sodium hydroxide, NaOH, breaks apart in water into Na^+ cations and OH^- anions. Notice that some acids such as vinegar, CH_3COOH, have OH in their structures

but are not bases. The OH does not break off. In fact, whenever you see two oxygen atoms together in a chemical formula, you can assume you are dealing with an acid. When you see an OH in a chemical formula, you may be dealing with an alcohol or a base. Because bases add ions to water in solution, basic solutions are electrolytes and conduct electricity.

Neutral substances do not add H^+ or OH^- to solution. Ionic compounds like sodium chloride produce Na^+ and Cl^- ions in solution. This does not change the amount of either H^+ or OH^-. Most molecules dissolve in

water without dissociating into ions. For example, methanol, CH_3OH, being an alcohol, does not add H^+ or OH^- ions to solution. Only those molecules that contain a carbon atom with two oxygen atoms attached will release H^+.

INTRODUCE ACID-BASE DEFINITIONS

→ Write the Arrhenius and Brønsted-Lowry definitions of acids and bases on the board at the appropriate point in the discussion.

→ Make sure students understand what the word *dissociate* means.

Ask: Based on the Acid-Base Solution cards, how are acids and bases different from each other?

Ask: Why do you think acids are often referred to as proton donors? (*Hint:* What is the structure of a hydrogen atom?)

Ask: What is an example of a base that does not have OH^- in the chemical formula?

Ask: How is it possible that ammonia, NH_3, dissolves in water to produce a basic solution? (*Hint:* What else is present that might release OH^-?)

Key Points: The definitions of acids and bases have changed over time. One early definition was described by the Swedish chemist Svante Arrhenius around 1890. It is still known as the Arrhenius definition.

> **Arrhenius Definition of Acids and Bases:** An acid is any substance that adds hydrogen ion (H^+) to the solution. A base is any substance that adds hydroxide ion (OH^-) to the solution.

Two chemists, Johannes Brønsted (Danish, 1879–1947) and Thomas Lowry (English, 1874–1936), proposed a slightly different definition for acids and bases around 1923. They referred to H^+ as a proton, because a hydrogen atom has a proton and an electron. If you remove the electron to form H^+, all that is left is a proton. (Stress to students that this is *not* a nuclear change.)

> **Brønsted-Lowry Definition of Acids and Bases:** An acid is a proton donor. A base is a proton acceptor.

According to the Brønsted-Lowry definition, bases are not simply those substances that release OH^-. They are substances that accept a proton, H^+, from another substance. This explains why NH_3 is a base even though it does not have OH^-. NH_3 accepts H^+ from water, leaving NH_4^+ and OH^-. When the water molecule donates a proton, H^+, to NH_3, the water is acting as an acid.

$$NH_3 + H_2O \rightarrow NH_4^+ + OH^-$$

INTRODUCE STRONG AND WEAK ACIDS AND BASES

Ask: How are 0.010 M solutions of hydrochloric acid, HCl, formic acid, CHOOH, and acetic acid, CH_3COOH, similar to one another?

Ask: How do they differ from one another?

Ask: HCl is a strong acid. What do you think this means?

Ask: Formic acid and acetic acid are weak acids. What do you think this means?

Ask: NaOH is a strong base, but NH_3 is a weak base. Explain why.

Key Points: Acids and bases that break apart (*dissociate*) completely in solution are called *strong acids* and *strong bases*. Strong acids include hydrochloric acid, HCl; nitric acid, HNO_3; sulfuric acid, H_2SO_4; and hydrobromic acid, HBr. Strong bases include sodium hydroxide, NaOH; potassium hydroxide, KOH; and magnesium hydroxide, $Mg(OH)_2$.

Acids and bases that do not dissociate completely in solution are called *weak acids* and *weak bases*. Solutions of weak acids and bases contain undissociated molecules in addition to ions. Because weak acids and weak bases do not dissociate completely, they put fewer H^+ and OH^- ions into solution. Some weak acids include formic acid, CHOOH; acetic acid, CH_3COOH; citric acid, $C_3H_5O(COOH)_3$; and phosphoric acid, H_3PO_4. Some weak bases include ammonia, NH_3, and aniline, $C_6H_5NH_2$.

EVALUATE (5 min)

Check-In

Which substances do you expect will conduct electricity? Hydrocyanic acid, HCN; magnesium hydroxide, $Mg(OH)_2$; or methanol, CH_3OH. Explain your thinking.

Sample answer: HCN will dissociate into H^+ and CN^- ions and conduct

---

How are acids and bases defined?

The Arrhenius definition of an acid is a compound that dissociates in solution to form a hydrogen ion, H^+, and an anion. The Arrhenius definition of a base is a substance that dissociates in solution to form a hydroxide ion, OH^-, and a cation. However, not all bases contain OH^-. Brønsted and Lowry defined acids as proton donors and bases as proton acceptors. The word "proton" in this case refers to an H^+ ion. A strong acid or base dissociates completely into ions in solution. A weak acid or base dissociates only partially into ions in solution.

KEY TERMS
dissociate
strong acid
weak acid
strong base
weak base

Exercises

Reading Questions

1. According to the Arrhenius theory, what is an acid and what is a base?
2. How is the Brønsted-Lowry theory of acids similar to the Arrhenius theory, and how is it different?
3. What is the difference between a strong acid and a weak acid? Draw a picture showing the particle view of each.

Reason and Apply

4. Label the substances listed below as acids or bases. In each case, list the ions you would expect to find in solution.
   a. hydroiodic acid, HI
   b. formic acid, HCOOH
   c. rubidium hydroxide, RbOH
   d. hypochlorous acid, HOCl
   e. selenous acid, H_2SeO_4
   f. phosphine, PH_3
   g. perchloric acid, $HClO_4$
   h. calcium hydroxide, $Ca(OH)_2$
5. Consider a solution of hydrobromic acid, HBr.
   a. If you draw a particle view of this acid with 10 H^+ ions, how many Br^- ions would you need? Explain your thinking.
   b. Sketch a particle view of an HBr solution with 10 H^+ ions.
6. Consider a solution of magnesium hydroxide, $Mg(OH)_2$.
   a. If you draw a particle view of this base with 10 Mg^{2+} ions, how many OH^- ions would you need? Explain your thinking.
   b. Sketch a particle view of a $Mg(OH)_2$ solution with 10 Mg^{2+} ions.
7. Explain why aqueous washing soda, Na_2CO_3, is a basic solution.
8. A solution of 0.10 M hydrochloric acid, HCl, is a better conductor of electricity than 0.10 M acetic acid, CH_3COOH. Sketch the ions and molecules in both solutions to explain this observation.

---

electricity. $Mg(OH)_2$ will dissociate into Mg^{2+} and OH^- ions and conduct electricity. CH_3OH will not dissociate into ions, so it will not conduct electricity.

Homework

Assign the reading and exercises for Toxins Lesson 85 in the student text.

LESSON 85 ANSWERS

1. According to the Arrhenius theory, an acid is a molecule that dissociates to form hydrogen ions in solution, and a base is a substance that dissociates to form hydroxide ions in solution.

2. According to the Brønsted-Lowry theory, an acid is a substance that donates a proton in solution, while a base is a substance that accepts protons. This is similar to the Arrhenius theory because a proton is a hydrogen ion, but it is different because it also accounts for bases that do not dissociate to form hydroxide ions.

Answers continue in Answer Appendix (p. ANS-7).

pHooey!

[H$^+$] and pH

THINK ABOUT IT

During delivery, a baby can sometimes go into distress. So doctors monitor the vital signs to make sure the baby is okay. One test is called a fetal blood test. This test involves taking the pH of the baby's blood. If the baby's blood has a pH of 7.2 or lower, the doctors know that the baby is not getting enough oxygen and is in danger. A pH of around 7.3 indicates that the baby is safe.

> How is pH related to the acid or base concentration of a solution?

To answer this question, you will explore

⟳ Acid Concentration

⟳ Relationship between [H$^+$] and pH

⟳ pH of Water and Basic Solutions

EXPLORING THE TOPIC

⟳ Acid Concentration

If you are testing the acidity of a solution by determining pH, you are determining the hydrogen ion concentration in moles per liter. When a solution has a greater hydrogen ion concentration than water does, it is an acid. Chemists use a standard notation with brackets to symbolize concentration in moles per liter. For example, [H$^+$] symbolizes the concentration of hydrogen ions, H$^+$, in moles per liter.

PARTICLE VIEWS OF ACIDIC SOLUTIONS

To get an idea of how pH numbers relate to hydrogen ion concentration, it is useful to compare particle views of solutions. The illustration shows the ions in two hydrochloric acid solutions, one that is 0.010 M and the other 0.002 M. How is the hydrogen ion concentration, [H$^+$], related to the pH?

How Does H$^+$ Concentration Relate to Molarity?

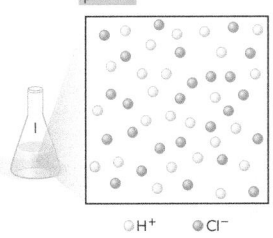

0.010 M HCl
pH = 2

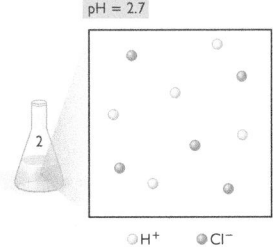

0.002 M HCl
pH = 2.7

○ H$^+$ ● Cl$^-$ ○ H$^+$ ● Cl$^-$

FOCUS ON UNDERSTANDING

● The pH scale can be confusing to students, for a number of reasons. For one thing, it is a logarithmic scale, so tiny changes in pH value reflect large changes in solution concentration. For another, it focuses on H$^+$ ions, even for basic solutions.

● Students often have difficulty with the idea that a lower pH value indicates that a solution is more acidic.

IN CLASS

Students begin by considering why pure water has a pH of 7. They examine the dissociation of H$_2$O into H$^+$ and OH$^-$. They then examine a data table that relates the molarity of acid and base solutions to the H$^+$ concentration, the OH$^-$ concentration, and the pH. They explore the relationships among pH, [H$^+$], and [OH$^-$]. Students are then challenged to place the 12 Acid-Base Solution cards from Lesson 85: Pass the Proton in order of decreasing H$^+$ concentration. A complete materials list is available for download.

TRM Materials List 86

Differentiate

Many students may find the pH scale challenging because they do not have a solid understanding of logarithms. Consult a mathematics teacher at your school to learn how logarithms were introduced to students. Consider using a diagnostic assessment to gauge students' initial understanding of logarithms. Depending on students' readiness with logarithms, you might display a wall chart that explains them or give individual students the option of using "help cards" with examples and explanations of base 10 logarithms and inverse logs.

SETUP

Set up two stations for students to test the pH of 11 solutions. Prepare 100 mL of the 11 solutions listed on the Acid-Base Solution cards (excluding formic acid). Be sure to use water with a pH close to 7. Add 50 mL of each solution to the beakers at each station. You may adjust concentrations where necessary to reflect the pH specified. Solution preparation directions are given below. Alternatively, you can test the solutions as a demonstration for your class. *Note:* If using pH meters to test each solution, have a rinsing container to rinse the probe and a container of distilled water to place it in between tests.

Overview

LAB: GROUPS OF 4

Key Question: How is pH related to the acid or base concentration of a solution?

KEY IDEAS

The pH of a solution is a reflection of its acidity or basicity. The pH changes by 1 unit for every tenfold change in the H$^+$ concentration. This relationship is logarithmic: pH = $-\log$[H$^+$]. As the H$^+$ concentration decreases, the pH number increases. Both H$^+$ and OH$^-$ are always present in aqueous solutions,

because water dissociates a tiny bit. The H$^+$ and OH$^-$ concentrations in any solution are related to each other: [H$^+$][OH$^-$] = 1.0×10^{-14}. In pure water, [H$^+$] = [OH$^-$] = 1.0×10^{-7} M.

LEARNING OBJECTIVES

● Explain the mathematical relationship between the H$^+$ and OH$^-$ concentrations in a solution.

● Define pH and explain the relationship between H$^+$ concentration and pH.

● Determine the H$^+$ concentration of a solution given the [OH$^-$], and vice versa.

SAFETY

When making up solutions, always put acid or base into water rather than the other way around.

Wear safety goggles at all times.

Acids and bases are corrosive. Do not get any on skin or near eyes.

In case of a spill, rinse with large amounts of water.

Have baking soda and citric acid on hand for spills.

- 100 mL of 0.010 M HCL: Dilute 1.0 mL of concentrated (12 M) HCl to 100 mL to prepare a 0.12 M solution. Dilute 8.3 mL of the 0.12 M HCL to 100 mL. Adjust the pH to 2.
- 100 mL of 0.005 M HCl: Dilute 50.0 mL of 0.010 M HCl to 100 mL. Adjust the pH to 2.3.
- 100 mL of 0.001 M HCl: Dilute 10.0 mL of 0.010 M HCl to 100 mL. Adjust the pH to 3.
- 100 mL of 0.010 M CH_3COOH: Dilute 1.0 mL vinegar (about 1 M) to 100 mL. Adjust the pH to 3.4. *Or:* Dilute 1.0 mL concentrated acetic acid (17 M) to 100 mL to prepare a 0.17 M solution. Dilute 5.9 mL of the 0.17 M CH_3COOH to 100 mL. Adjust the pH to 3.4.
- 100 mL of 0.010 M NaCl: Dissolve 0.584 g of NaCl in water to a total volume of 100 mL. Adjust the pH to 7.
- 100 mL of 0.010 M CH_3OH: Dissolve 0.4 mL of methanol to a total volume of 100 mL. Adjust the pH to 7.
- 100 mL of 0.005 M CH_3OH: Dilute 50 mL of 0.010 M CH_4O to 100 mL. Adjust the pH to 7.
- 100 mL of 0.010 M NaOH: Dissolve 0.40 g of NaOH to a total volume of 100 mL. Adjust the pH to 12.
- 100 mL of 0.001 M NaOH: Dilute 10 mL of 0.010 M NaOH to 100 mL. Adjust the pH to 11.
- 100 mL of 0.010 M NH_3: Concentrated ammonia is about 18 M. Dilute 1.0 mL to 100 mL to prepare a 0.18 M solution. Dilute 5.6 mL of 0.18 M NH_3 to 100 mL. Adjust the pH to 10.6.

CLEANUP

Dilute the waste solutions and dispose according to local guidelines.

ENGAGE (5 min)

ChemCatalyst

Pure water has an H+ concentration of 1.0×10^{-7} M and an OH− of 1×10^{-7} M. What does this mean?

In the illustration, solution 1 has five times as many H+ cations and five times as many Cl− anions as solution 2. However, the difference in the pH is only 0.7. Small changes in pH signal large differences in hydrogen ion concentrations.

Big Idea The greater the concentration of H+ ions in a solution, the lower its pH, and the more acidic it is.

Relationship between [H+] and pH

The hydrogen ion concentration of a solution is directly related to the pH of that solution. The table shows the H+ concentration and the pH number for a few different concentrations of hydrochloric acid.

The Relationship between [H+] and pH Number

Molarity	[H+]	pH
1.0 M HCl	1.0×10^{0} M	0
0.1 M HCl	1.0×10^{-1} M	1
0.01 M HCl	1.0×10^{-2} M	2
0.001 M HCl	1.0×10^{-3} M	3
0.0001 M HCl	1.0×10^{-4} M	4
0.00001 M HCl	1.0×10^{-5} M	5

Notice that each time the H+ concentration changes by a factor of 10, the pH number changes by 1 unit. You can also see that the pH number is equal to the exponent, without the negative sign. For example, 1.0×10^{-4} M HCl has a pH of 4. This is true whenever the H+ concentration is expressed in scientific notation and the coefficient is 1.0.

Each unit of change in pH represents a tenfold difference in H+ concentration. This is called a *logarithmic* relationship. The mathematical formula connecting pH to [H+] is

$$pH = -\log [H^+]$$

You can use this formula and a calculator to convert between pH and [H+].

[For review of this math topic, see **MATH Spotlight**: Logarithms on page A-17.]

Example 1

Calculating pH for an HCl Solution

a. What is the pH of a 0.000001 M HCl solution?

b. What is the pH of a 4.2×10^{-3} M HCl solution?

Sample answer: Some students will say that water must dissociate into ions. Others may speculate that something else is present in the water besides H_2O molecules. The number itself probably will not have much meaning for students.

→ Discuss the source of H+ and OH− in pure water.

Ask: Where would H+ and OH− come from in a sample of water?

Ask: Are the concentrations of H+ and OH− large or small in a sample of pure water?

EXPLORE (15 min)

TRM Worksheet with Answers 86

TRM Worksheet 86

TRM Handout—H+ Concentration, OH− Concentration, and pH 86

TRM Handout—pHooey! Lab Procedure 86

TRM Card Masters—Acid-Base Solution Cards 85

INTRODUCE THE LAB

→ Arrange students into groups of four.

Example 1

(continued)

Solution

a. Each HCl formula unit dissociates into an H^+ and a Cl^- ion. So the concentration of H^+ ions is 0.000001 M, which can also be written 1.0×10^{-6} M. Because the coefficient is 1.0, the pH is the exponent without the minus sign. The pH is 6.

b. Use the relationship $pH = -\log[H^+]$.

$$pH = -\log[4.2 \times 10^{-3}] = 2.4$$

The pH of the 4.2×10^{-3} molar HCl solution is between 2 and 3. This makes sense because the H^+ concentration is between 1.0×10^{-2} M and 1.0×10^{-3} M.

⟳ pH of Water and Basic Solutions

Basic solutions also have pH numbers associated with them. This is because they also have an H^+ concentration and pH value.

The number line below shows the relationship between pH and H^+ concentration for acidic, basic, and neutral solutions at 25 °C. The H^+ concentration in basic solutions is quite small. For example, a solution with a pH of 14 has a hydrogen ion concentration of 1.0×10^{-14} M, or 0.00000000000001 M.

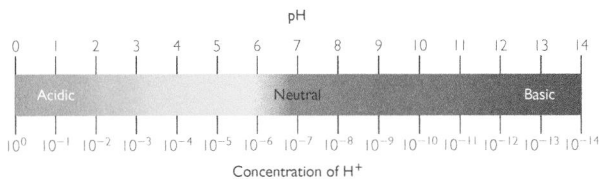

DISSOCIATION OF WATER

It may seem odd that a basic solution has hydrogen ions in it. To understand this, it is necessary to consider water on a molecular level.

Notice that the hydrogen ion concentration in a neutral solution, such as water, is listed as 1.0×10^{-7} M H^+ at 25 °C. Water molecules  dissociate slightly, forming H^+ and OH^- ions. The dissociation is so slight that there are only 2 H^+ ions and 2 OH^- ions for every 1 billion water molecules. This translates to an H^+ concentration of 1×10^{-7} M, or a pH of 7. Water is considered neutral because it contains equal amounts of H^+ and OH^-.

When a substance dissolves in water and the H^+ concentration, $[H^+]$, remains equal to the OH^- concentration, $[OH^-]$, we say that substance is neutral. For example, NaCl dissolves in water to form Na^+ cations and Cl^- anions. This does not change the balance of H^+ and OH^- ions already present in the water. NaCl(*aq*) is neutral.

Lesson 86 | **[H⁺] and pH**   439

→ Pass out the student worksheet, handouts, and cards.

→ This is a two-part lab. In Part 1, students explore the relationship between H^+ and OH^- concentrations in solutions. In Part 2, groups of four students work with the Acid-Base Solution cards from the previous lesson. They sequence the cards in order of decreasing H^+ concentration and predict the pH for each solution.

→ Introduce students to the standard notation for concentration in moles per liter. For example, $[H^+]$ means hydrogen ion concentration in moles per liter.

→ After students have made their pH predictions for the 12 Acid-Base Solution cards, they will test these solutions with pH paper.

→ Remind students about safety with acids and bases.

TRM Transparency—pH and H⁺ Concentration 86

TRM Transparency—H⁺ Concentration and OH⁻ Concentration 86

RELATE THE pH SCALE TO H⁺ CONCENTRATION

→ Display the transparency showing the relationship between pH value and H^+ concentration.

→ Show students how to calculate the pH using the formula $pH = -\log[H^+]$ for the solutions on the cards so they can check their predictions.

Ask: What information did you use to place the Acid-Base Solution cards in order of decreasing H^+ concentration?

Ask: When the pH increases by 1 unit, how much does the H^+ concentration change?

Ask: If you know the molarity of a hydrochloric acid, HCl, solution, how do you determine its pH? What is the pH of 0.000001 M HCl?

Ask: Estimate the pH of a 0.0005 M HCl solution.

Key Points: The pH scale is a logarithmic scale that describes the concentration of H^+ ions in solution. Each unit of change in pH represents a tenfold difference in H^+ concentration. A solution with a pH of 3 has an H^+ concentration ten times greater than a solution with a pH of 4. When the concentration of H^+ expressed in scientific notation has a coefficient of 1.0, the pH is equal to the exponent without the minus sign. For example, if $[H^+]$ is 1.0×10^{-4} M, the pH is 4.

> pH is related to $[H^+]$ by the formula
> $$pH = -\log[H^+]$$

By entering the formula, you can use a calculator to determine the pH of a 0.056 M HCl solution as $pH = -\log[5.6 \times 10^{-2}] = 1.25$

RELATE THE H⁺ CONCENTRATION TO THE OH⁻ CONCENTRATION

→ Display the transparency showing the relationship between H^+ concentration and OH^- concentration.

Ask: How does the H^+ concentration change as the OH^- concentration increases?

Ask: How does the pH change as the OH^- concentration increases?

Ask: If you know the OH^- concentration, how can you determine the H^+ concentration, and vice versa?

Ask: What does the pH value tell you about the H^+ concentration? The OH^- concentration?

Ask: There is a pH associated with water. What does this indicate about water?

Key Points: In any solution, the product of the hydrogen ion, H^+, concentration and hydroxide ion, OH^-, concentration is a constant.

$$[H^+][OH^-] = 1.0 \times 10^{-14} \text{ at } 25 \text{ °C}$$

When the hydrogen ion concentration in any solution is multiplied by the hydroxide ion concentration, the result is always 1.0×10^{-14} at 25 °C.

For example:

$$0.01 \text{ M HCl, pH: 2}$$
$$[H^+][OH^-] = (1.0 \times 10^{-2} \text{ M})(1 \times 10^{-12} \text{ M})$$
$$= 1.0 \times 10^{-14}$$

$$0.00001 \text{ M NaOH, pH: 9}$$
$$[H^+][OH^-] = (1 \times 10^{-9} \text{ M})(1 \times 10^{-5} \text{ M})$$
$$= 1.0 \times 10^{-14}$$

Note that as the H^+ concentration of a solution decreases, the OH^- concentration increases, and vice versa.

Water dissociates into H^+ and OH^- ions. A small percentage of water molecules breaks apart to form H^+ and OH^- ions. See the image on page 439 in the student text. Water has 1.0×10^{-7} moles of H^+ ions per liter and 1.0×10^{-7} moles of OH^- per liter at 25 °C. Water is considered neutral because the number of H^+ ions is equal to the number of OH^- ions. This is why water is in the middle of the pH scale, with a pH of 7.

Note: What we call H^+ is really H_3O^+, the hydronium ion. The equation that describes most accurately how water dissociates is

$$H_2O(l) + H_2O(l) \rightleftharpoons H_3O^+(aq) + OH^-(aq)$$

where $[H_3O^+][OH^-] = 1.0 \times 10^{-14}$.

When an acid such as HCl dissolves in water, it breaks apart into H^+ and Cl^- ions, increasing the ratio of H^+ ions to OH^- ions. Likewise, when a substance like NaOH dissolves in water, it breaks apart into Na^+ and OH^- ions. This also alters the balance of H^+ and OH^- ions in the water. In the case of NaOH(*aq*), the hydrogen ion concentration, $[H^+]$, is less than the hydroxide ion concentration, $[OH^-]$, so NaOH is considered a base.

The table shows the concentration of H^+ and OH^- ions in a variety of sodium hydroxide solutions.

The Relationship between $[OH^-]$ and pH Number

Molarity	$[OH^-]$	$[H^+]$	pH
water	1.0×10^{-7} M	1.0×10^{-7} M	7
0.00001 M NaOH	1.0×10^{-5} M	1.0×10^{-9} M	9
0.0001 M NaOH	1.0×10^{-4} M	1.0×10^{-10} M	10
0.001 M NaOH	1.0×10^{-3} M	1.0×10^{-11} M	11
0.01 M NaOH	1.0×10^{-2} M	1.0×10^{-12} M	12
0.1 M NaOH	1.0×10^{-1} M	1.0×10^{-13} M	13
1.0 M NaOH	1.0×10^{0} M	1.0×10^{-14} M	14

Notice that the H^+ concentration decreases as the OH^- concentration increases and vice versa. There is a mathematical relationship between $[H^+]$ and $[OH^-]$:

$$[H^+][OH^-] = 1 \times 10^{-14}$$

In *any* aqueous solution at 25°C, the hydrogen ion concentration multiplied by the hydroxide ion concentration is equal to 1.0×10^{-14}. For example, when the OH^- concentration is 1×10^{-3} M, the H^+ concentration is 1×10^{-11} M.

$$(1 \times 10^{-11})(1 \times 10^{-3}) = 1 \times 10^{-14}$$

You can also see that adding the values of the exponents always results in a total of -14.

Example 2 | Calculating pH for a Solution of Sodium Hydroxide, NaOH

What is the pH of a 0.000001 M NaOH solution?

Solution

A solution with a molarity of 0.000001 M NaOH has an OH^- concentration of 1.0×10^{-6} moles per liter. The H^+ concentration must be 1.0×10^{-8} M because the exponents must sum to -14. The pH is just the exponent of the H^+ concentration without the negative sign. So the pH is 8.

EVALUATE (5 min)

Check-In

If you know the pH of a solution, what else do you know?

Sample answer: If the solution has a pH between 0 and 7, it is an acid solution. If the pH is between 7 and 14, it is a basic solution. If you know the pH of a solution, you can figure out the H^+ and OH^- concentrations.

Homework

Assign the reading and exercises for Toxins Lesson 86 in the student text.

LESSON SUMMARY

How is pH related to the acid or base concentration of a solution?

The pH number and the hydrogen ion concentration of a solution are related logarithmically: $pH = -\log[H^+]$. This means that when the hydrogen ion concentration changes by a factor of 10, the pH changes by 1. Water and basic solutions also have pH numbers because they contain small concentrations of H^+ ions. Water dissociates slightly into H^+ and OH^- ions. In pure water, the H^+ concentration is 1×10^{-7} M and the pH is 7. The hydroxide ion concentration and the hydrogen ion concentration in a solution are related mathematically: $[H^+][OH^-] = 1.0 \times 10^{-14}$.

Exercises

Reading Questions

1. What is the relationship between pH and H^+ concentration?
2. How is the H^+ concentration related to the OH^- concentration in a solution?

Reason and Apply

3. What pH would you expect for solutions with the concentrations given here?
   a. $[H^+] = 1.0 \times 10^{-4}$ M
   b. $[H^+] = 1.0 \times 10^{-12}$ M
   c. $[OH^-] = 1.0 \times 10^{-8}$ M

4. Determine the pH for solutions with the H^+ concentrations given here.
   a. $[H^+] = 0.0014$ b. $[H^+] = 6.0 \times 10^{-8}$ M
   c. $[H^+] = 4.2 \times 10^{-11}$ M   d. $[H^+] = 1.5 \times 10^{-1}$ M

5. Which of the solutions in Exercise 4 are acids? Which are bases?

6. What H^+ concentration would you expect for the solutions below?
   a. $pH = 9$ b. $pH = 7.3$
   c. $pH = 2.9$ d. $pH = 10.2$

7. What is the pH of a 2.5 M HCl solution?

8. What is the pH of a 0.256 M NaOH solution?

9. **PROJECT** www The Richter scale is a logarithmic scale for measuring the magnitude of earthquakes. How much more intense is a Richter 7.0 earthquake than a Richter 5.0 earthquake?

9. An increase of 1.0 on the Richter scale is equivalent to an increase by a factor of 10 for the magnitude of an earthquake. An increase of $1.0 + 1.0 = 2.0$ is equivalent to an increase by a factor of $10 \cdot 10 = 100$. A magnitude 7.0 earthquake is 100 times more intense than a magnitude 5.0 earthquake.

LESSON 86 ANSWERS

1. The pH of a solution is equal to the negative logarithm of the H^+ concentration. Therefore, the pH increases by 1 as the hydrogen ion concentration decreases by a factor of 10.

2. The product of the H^+ concentration and the OH^- concentration in an aqueous solution is always equal to 1.0×10^{-14}.

3. a. 4 b. 12 c. 6

4. a. 2.9 b. 7.2 c. 10.4 d. 0.8

5. The acid solutions in Exercise 4 are **a** and **d**. The basic solutions are **b** and **c**.

6. a. 10^{-9} M $= 1.0 \times 10^{-9}$ M

   b. $10^{-7.3}$ M $= 5.0 \times 10^{-8}$ M

   c. $10^{-2.9}$ M $= 1.3 \times 10^{-3}$ M

   d. $10^{-10.2}$ M $= 6.3 \times 10^{-11}$ M

7. -0.40. Although a negative pH is theoretically possible, in general a strong acid at this concentration does not dissociate completely, so the measured pH would be higher.

8. 13.4

Overview

LAB: GROUPS OF 4

Key Question: How does dilution affect acids and bases?

KEY IDEAS

Adding water to an acid or base solution will change the pH of the solution. As an acid solution is diluted with water, its pH increases toward 7. As a basic solution is diluted with water, its pH decreases toward 7. The pH of a neutral solution is not affected by dilution.

LEARNING OBJECTIVES

- Complete a serial dilution of a solution.
- Explain the effect of dilution on the acidity or basicity of a solution.

FOCUS ON UNDERSTANDING

It may be hard for students to grasp that adding pure water to a base increases the number of H^+ ions in solution. You might remind them of the dissociation in water.

KEY TERM

dilution

IN CLASS

Students examine the effect of dilution on the acidity or basicity of solutions of HCl and NaOH. They work in groups of four to complete a serial dilution of 0.01 M HCl. Students test each dilution with an indicator and record the pH value. In the second part of the lab, they continue diluting to determine if an acid can "become" a base and vice versa. A complete materials list is available for download.

TRM Materials List 87

SETUP

Prepare 100 mL of 0.010 M HCl: Dilute 1.0 mL of concentrated (12 M) HCl to 100 mL to get a 0.12 M solution. Then dilute 8.3 mL of the 0.12 M HCl to 100 mL. To prepare 100 mL of 0.00001 M NaOH, dissolve 0.40 g of NaOH to a total volume of 1.0 L. Then take 1.0 mL of this solution and dilute to a total volume of 1.0 L. You may prepare more than 100 mL so you have some to spare for groups that make mistakes.

Watered Down

Dilution

THINK ABOUT IT

In Unit 1: Alchemy, a "golden" penny was made with the help of a strong base called sodium hydroxide. In Unit 2: Smells, you added concentrated sulfuric acid to a mixture of chemicals to make a sweet-smelling ester. Whenever you work with very concentrated acids and bases, you are instructed to rinse with large amounts of water if you get any on your skin or clothing. How does this help?

> How does dilution affect acids and bases?

To answer this question, you will explore

- Diluting Acidic Solutions
- Approaching pH 7

EXPLORING THE TOPIC

Diluting Acidic Solutions

If you spill concentrated acid on your skin, it will begin to sting and burn right away. If you spill it on your clothing, you will eventually see holes in the cloth. The best thing you can do for this situation is to put a lot of water on the part of your skin or clothing where the acid spilled.

The process of adding more and more water to a solution is called **dilution.** You could say that the solution is becoming more dilute or "watered down." This changes the number density of the solute particles in solution.

The effect of diluting a concentrated solution can be demonstrated with a glass of grape juice. The color of the grape juice becomes paler as more water is added. And the more dilute the grape juice becomes, the weaker it tastes.

Ken Karp Photography

Acidic solutions are not necessarily colorful like grape juice, but the concept of dilution is the same: The solution becomes less concentrated. When you dilute an acid, you can track what is happening by testing the pH.

The illustration at the top of the next page shows a particle view of what happens when you dilute an acidic solution. A few drops of red dye have been added so that you can visually keep track of the different solutions. The solution is diluted with water in the beaker. The starting solution in the test tube has a pH of 2. The particle-view squares show how the H^+ ions and Cl^- ions end up farther apart as more water is added.

SAFETY

When making up solutions, always put acid and base into water rather than the other way around.

Wear safety goggles at all times.

Acids and bases are corrosive. Do not get any on skin or near eyes.

In case of a spill, rinse with large amounts of water.

Have baking soda and citric acid on hand for spills.

CLEANUP

Dilute waste solutions and pour them down the drain.

ChemCatalyst

Imagine that a sample of drinking water has been contaminated with an acid. The pH is now 4.0.

1. What do you think would happen to the pH if you added more water to the sample?
2. Do you think you could ever get the pH back to 7? Explain why or why not.

Sample answers: **1.** Some students will say that adding a neutral substance to an acid will not change its pH.

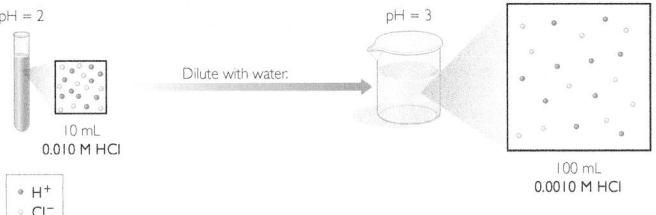

pH = 2

10 mL
0.010 M HCl

Dilute with water.

pH = 3

100 mL
0.0010 M HCl

• H⁺
• Cl⁻

Mixing the 10 mL of 0.010 M HCl with 90 mL of water dilutes the original solution by a factor of 10.

Dilution of a strong acid or base can make it safer. This is why you can swim in a swimming pool even though the water contains both acids and bases.

Example 1

Diluting HCl

Imagine you have a 1.0 mL water sample containing 0.10 M hydrochloric acid, HCl. You want to get it to a safe pH of 6. How much water do you need to add?

Solution

The pH of 0.10 M HCl is 1. The H^+ concentration needs to change from 1.0×10^{-1} M to 1.0×10^{-6}, a factor of 10^5, or 100,000. So you need to add 99,999 mL of water to 1 mL 0.10 M HCl to get it to a safe pH of 6.

RECREATION CONNECTION

Swimming pools require a slightly basic pH, between 7.2 and 7.8. If the pH is too low, eye irritation occurs and metal pumps may corrode. If the pH is too high, the water may become cloudy and the chlorine may not work to combat algae. Sodium carbonate is added to pools to keep the pH in the right range. Indicators are used to test the pH.

↻ **Approaching pH 7**

Suppose you continue to dilute the 0.0010 M HCl solution in the large beaker in the previous illustration to a concentration of 0.00010 M HCl and pH of 4. If you continue to dilute, what is the lowest H^+ concentration you can obtain? What is the lowest pH you can reach?

If you take 1 mL of 0.00010 M HCl and add 9 mL of water, the new solution will have a concentration of 0.000010 M HCl and a pH of 5. If you take 1 mL of 0.000010 M HCl and add 9 mL of water, the new solution will have a concentration of 0.0000010 M HCl and a pH of 6. You can continue this process, diluting the acid until the pH becomes 7. At this point, the liquid is no longer acidic. It is neutral.

As you dilute an acidic solution, its H^+ concentration gets closer and closer to the H^+ concentration of water, 1×10^{-7} moles per liter. However, the pH does not increase beyond 7, no matter how much you dilute the solution. In other words, the solution does not become basic. You cannot start with an acidic solution and dilute it to make a basic solution.

Likewise, you cannot dilute a basic solution to make an acidic solution. When you dilute a basic solution, the OH^- concentration decreases, the H^+ concentration increases, and the pH decreases toward 7.

Lesson 87 | **Dilution** 443

Others will say that adding water will dilute the H⁺ concentration in the solution and lower the acidity (making the pH go up). **2.** If you add enough water, you should be able to get a pH very close to but not past 7.

→ Sample students' ideas about acidity and dilution.

Ask: Can you change the pH of an acid solution simply by adding water? Explain your thinking.

Ask: What does the word *dilution* mean?

Ask: As you add water to an acid, will the pH go up, go down, or stay the same? Explain.

Ask: What would happen to the pH if you added water to a basic solution?

EXPLORE (15 min)

TRM Worksheet with Answers 87

TRM Worksheet 87

INTRODUCE THE LAB

→ Arrange students into groups of four.

→ Pass out the student worksheet.

→ Briefly go over the procedure with the class. Tell students that this is a two-part lab. In Part 1, they will repeatedly dilute a solution of HCl and measure its pH. In Part 2, they will repeatedly dilute HCl and NaOH until the pH no longer changes.

→ Review safety considerations.

EXPLAIN & ELABORATE (15 min)

DISCUSS DILUTION OF ACIDS AND BASES

Ask: What effect does adding water have on a hydrochloric acid solution?

Ask: Does the acid solution become neutral when water is added? Explain.

Ask: What happens to the pH if you dilute a sodium hydroxide solution with water?

Ask: During dilutions, can the pH ever go past 7? Explain.

Ask: Can dilution of a basic solution result in an acidic solution, or vice versa? Explain.

Key Points: Adding water to an acid or a base dilutes the solution, making it less acidic or less basic. As an acidic solution is diluted, there are fewer hydrogen ions per liter of solution. Thus, the H^+ concentration decreases, and the pH increases toward 7. As a basic solution is diluted, the H^+ concentration actually increases, and the pH decreases toward 7. More water is required to dilute a highly concentrated solution. An acid can never be turned into a base by diluting it with water, and a base can never be turned into an acid by diluting it with water. Both the dilution of an acid and the dilution of a base result in the pH approaching a neutral value of 7. In actuality, the number density of the ions becomes so low that eventually you simply are testing the pH of the water.

EXPLORE THE MATH BEHIND DILUTIONS

→ Create a table like the one below on the board.

pH	H⁺ concentration (decimal number)	H⁺ concentration (scientific notation)
2	0.010 mol/L	1.0×10^{-2} mol/L
3	0.0010 mol/L	1.0×10^{-3} mol/L
4	0.00010 mol/L	1.0×10^{-4} mol/L
5	0.000010 mol/L	1.0×10^{-5} mol/L
6	0.0000010 mol/L	1.0×10^{-6} mol/L
7	0.00000010 mol/L	1.0×10^{-7} mol/L

Ask: If you dilute an acid solution by a factor of 10, what happens to the pH? (It increases by 1 unit.)

Ask: If you dilute an acid or a base solution by a factor of 100, what happens to the pH? (It changes by 2.)

Ask: When the pH changes from 2 to 4, what is the H^+ concentration changing by?

Key Points: Each time the H^+ concentration is diluted tenfold, the pH number goes up 1 unit. So, if you want to dilute a solution down by 1 pH, you have to add water to increase the volume by ten times. If you want to dilute a solution down by 2 pH, you have to add water to increase its volume by 100 times. A change in pH from 3 to 7 means a 10,000-fold change in volume. This means that 1 mL of a solution with a pH of 3 would have to be diluted by 9,999 mL of water to get to a pH of 7.

EVALUATE (5 min)

Check-In

If you get lemon juice in your eye, you should flush the eye with lots of water. Why?

Sample answer: Adding water decreases the concentration of H^+ ions per unit of volume. Flushing with lots of water decreases the acidity and the solution becomes safer.

Homework

Assign the reading and exercises for Toxins Lesson 87 in the student text.

Big Idea An acid can never be made into a base by diluting with water. A base can never be made into an acid by diluting with water.

Example 2

Diluting NaOH

Imagine that you have 10 mL of a 0.010 M solution of sodium hydroxide, NaOH.

a. What is the H^+ concentration of this solution?

b. What is the pH?

c. What is the NaOH concentration if you dilute the 0.010 M solution by adding 90 mL of water?

d. Does the pH increase or decrease after diluting the 0.010 M NaOH solution?

e. What is the pH of the solution after adding 90 mL of water?

Solution

a. The OH^- concentration is 0.010 mole per liter, which is equivalent to 1×10^{-2} M. Because $[H^+][OH^-] = 1 \times 10^{-14}$, the H^+ concentration is 1×10^{-12} M. The exponents sum to -14.

b. pH $= -\log[H^+] = -\log 10^{-12} = 12$

c. If you add 90 mL water to 10 mL of 0.010 M NaOH, then you have 100 mL. The volume has increased tenfold. So the concentration after diluting is 0.0010 M NaOH.

d. As the OH^- concentration decreases, the H^+ concentration increases. So the pH decreases when you dilute 0.010 M NaOH to 0.0010 M.

e. The OH^- concentration is now 1.0×10^{-3} M. Because $[H^+][OH^-] = 1.0 \times 10^{-14}$, $[H^+] = 1 \times 10^{-11}$ M.

$$pH = -\log[1 \times 10^{-11}] = 11$$

So the pH has decreased, from 12 to 11, as you might expect. Diluting with water makes a solution more neutral.

LESSON SUMMARY

How does dilution affect acids and bases?

When water is added to an acidic solution, the result is a "watering down" of the solution. The H^+ concentration of the solution decreases, and the pH increases toward 7. When water is added to a concentrated basic solution, the OH^- concentration decreases. This increases the H^+ concentration and the pH decreases toward 7. Acids cannot be made into bases and bases cannot be made into acids by adding water.

KEY TERM
dilution

Exercises

Reading Questions

1. When you add water to an acidic solution, describe what happens to the pH of that solution.
2. Explain why you cannot turn an acid into a base by diluting it with water.

Reason and Apply

3. Copy this table and fill in the missing information.

Solution	$[H^+]$ in scientific notation	pH number	$[OH^-]$ in scientific notation
1.0 M HCl	1×10^0	0	1×10^{-14}
0.10 M HCl	1×10^{-1}	1	1×10^{-13}
0.010 M HCl			
0.0010 M HCl			
0.00000010 M HCl			
0.000000010 M HCl	1×10^{-7} M		
0.0000000010 M HCl	1×10^{-7} M		
0.0010 M NaCl			
0.00000010 M NaOH			1×10^{-7} M
0.0000010 M NaOH	1×10^{-8} M		1×10^{-6} M
0.000010 M NaOH			

4. How much water do you need to add to 10 mL of a solution of HCl with a pH of 3 to change the pH to 6?
5. Explain why the pH of a 1×10^{-9} M HCl solution is 7 even though the exponent is −9.
6. Explain why the pH of a NaCl solution does not change if you dilute the solution.
7. Imagine that you have 0.75 L of a 0.10 M HCl solution.
   a. How many moles of H^+ are in the 0.75 L?
   b. If you add 0.35 L of water, what is the new concentration of the solution?
   c. What is the pH after adding 0.35 L?
8. Imagine that you have 15 mL of a 0.025 M HCl solution.
   a. How many moles of H^+ are in the 15 mL?
   b. If you add 25 mL of water, what is the new concentration of the solution?
   c. What is the final pH after adding 25 mL?

LESSON 87 ANSWERS

1. When you add water to an acid solution, the pH of the solution increases to a maximum of pH = 7.

2. *Possible answer:* Because the pH of pure water is 7, you can increase the hydroxide ion concentration and decrease the hydrogen ion concentration of the solution until you reach a pH of 7. At that point, adding more water does not change the ratio of hydroxide ions to hydrogen ions, so the pH cannot change.

4. 9,990 mL

5. Because the concentration of hydrogen ions contributed by the acid is less than the concentration of hydrogen ions in pure water, the addition of 1×10^{-9} moles of HCl does not affect the pH of pure water.

6. The pH of a NaCl solution does not change when you dilute the solution because the NaCl solution is neutral. It has the same concentration of hydrogen and hydroxide ions as pure water.

7. a. 0.075 mol **b.** 0.068 M **c.** 1.2

8. a. 0.000375 mol **b.** 0.0094 M **c.** 2.0

3.

Acid concentration	$[H^+]$	pH	$[OH^-]$
1.0 M HCl	1×10^0	0	1×10^{-14}
0.10 M HCl	1×10^{-1}	1	1×10^{-13}
0.010 M HCl	1×10^{-2}	2	1×10^{-12}
0.0010 M HCl	1×10^{-3}	3	1×10^{-11}
0.00000010 M HCl	1×10^{-7}	7	1×10^{-7}
0.000000010 M HCl	1×10^{-8}	8	1×10^{-6}
0.0000000010 M HCl	1×10^{-9}	9	1×10^{-5}
0.0010 M NaCl	1×10^{-7}	7	1×10^{-7}
0.00000010 M NaOH	1×10^{-7}	7	1×10^{-7}
0.0000010 M NaOH	1×10^{-8}	8	1×10^{-6}
0.000010 M NaOH	1×10^{-9}	9	1×10^{-5}

Overview

LAB: PAIRS

Key Question: What happens when acids and bases are mixed?

KEY IDEAS

When an acid and a base are mixed, an ionic salt and water are produced. The hydrogen ions, H^+, from the acid and the hydroxide ions, OH^-, from the base combine to form water. As a result, the pH of the product solution approaches 7. Reactions between acids and bases are referred to as neutralization reactions.

LEARNING OBJECTIVES

● Write a chemical equation for an acid-base neutralization reaction.
● Describe how the pH changes when acids and bases are mixed.

KEY TERM

neutralization reaction

IN CLASS

Within groups of four, students work in pairs to examine reactions between $HCl(aq)$, $HNO_3(aq)$, $NaOH(aq)$, and $NH_4OH(aq)$. They mix various combinations of these four substances. By measuring the pH before and after mixing, students can determine whether or not a reaction has occurred. They write balanced chemical equations for the reactions that do occur and discover that when an acid reacts with a base, a salt and water are produced. The concept of neutralization is introduced. A complete materials list is available for download.

TRM Materials List 88

SETUP

Before class, prepare 200 mL of each solution listed. Each solution is 0.10 M. Once you have prepared the HCl, HNO_3, and NaOH solutions, you have to test them by mixing equal volumes of acid and base. Add an indicator to measure the pH. If the mixtures are not at pH 7, adjust the concentration of one of the solutions by adding drops of concentrated acid or base. When 0.10 M NH_4OH is mixed with an equal volume of HCl or HNO_3, the mixture should be pH 5. Divide each solution into eight labeled dropper bottles.

0.10 M HCl: 1.7 mL of 12 M HCl (conc.) diluted to 200 mL

0.10 M HNO_3: 1.2 mL of 16 M HNO_3 (conc.) diluted to 200 mL

0.10 M NaOH: 8.0 g dissolved in total volume of 200 mL

0.10 M NH_4OH: 1.1 mL of 18 M NH_4OH (conc.) diluted to 200 mL

Also prepare eight dropper bottles with bromothymol blue indicator. *Note:* You can use universal indicator if bromothymol blue is unavailable.

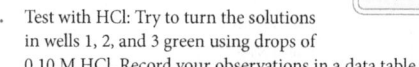

Featured LAB

Neutralization Reactions

Purpose

To examine reactions between acids and bases.

Procedure

1. Add 20 drops of each solution to the well plate as specified in the illustration.

2. Add 1 drop of bromothymol blue indicator to each well. Bromothymol blue is yellow in acid, blue in base, and green in neutral solution.

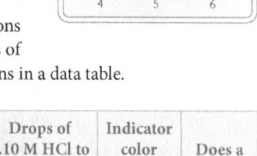

3. Test with HCl: Try to turn the solutions in wells 1, 2, and 3 green using drops of 0.10 M HCl. Record your observations in a data table.

Solution in the well plate	Indicator color	Acid, base, or neutral?	Drops of 0.10 M HCl to turn solution green	Indicator color after mixing	Does a reaction occur?
20 drops 0.10 M HCl					

4. Test with NaOH: Try to turn the solution in wells 4, 5, and 6 green using drops of 0.10 M NaOH.

Analysis

1. What did you observe when you mixed an acid with a base?

2. Acids react with bases to form an ionic salt and water. Copy this chemical equation and label the acid, base, and ionic salt.

$$HCl(aq) + NaOH(aq) \longrightarrow NaCl(aq) + H_2O(l)$$

3. Complete and balance the equation for the reaction in each well. If no reaction occurred, simply write "no reaction" on the products side of the equation.

4. What combinations of reactants do not yield new products?

5. Name the three salts produced in the reactions.

6. **Making Sense** List three things you learned as a result of performing this lab procedure.

SAFETY Instructions

Safety goggles must be worn at all times.

Acids and bases are corrosive. Do not get any on skin or near eyes.

In case of a spill, rinse with large amounts of water.

Ken Karp Photography

Materials

● well plate
● set of five labeled dropper bottles: 0.10 M HCl (hydrochloric acid), 0.10 M HNO_3 (nitric acid), 0.10 M NaOH (sodium hydroxide), 0.10 M NH_4OH (ammonium hydroxide), and bromothymol blue indicator

---

SAFETY

When making up solutions, always put acid and base into water rather than the other way around.

Wear safety goggles at all times.

Acids and bases are corrosive. Do not get any on skin or near eyes.

In case of a spill, rinse with large amounts of water.

Have baking soda and citric acid on hand for spills.

CLEANUP

Dilute waste solutions and dispose according to local guidelines.

Neutral Territory

Neutralization Reactions

THINK ABOUT IT

There are many advertisements on television about products that are designed to treat "acid indigestion." Some of these substances claim to neutralize extra acid that may have built up in your stomach.

What happens when acids and bases are mixed?

To answer this question, you will explore

- Neutralization Reactions
- Predicting the Products of Neutralization Reactions

EXPLORING THE TOPIC

Neutralization Reactions

Pure water is neutral because it has equal numbers of H^+ and OH^- ions.

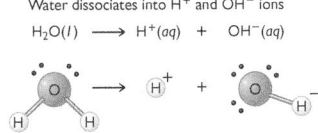

Water dissociates into H^+ and OH^- ions

$$H_2O(l) \longrightarrow H^+(aq) + OH^-(aq)$$

In pure water at 25 °C, $[H^+]$ and $[OH^-]$ are both 1.0×10^{-7} mol/L. But if an acid or a base is added to water, it upsets the balance of ions in solution. Acidic solutions have an excess of H^+ and basic solutions have an excess of OH^-.

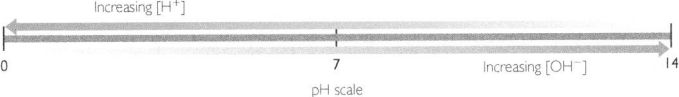

Increasing $[H^+]$

0   7   Increasing $[OH^-]$ 14

pH scale

One way to reduce the toxicity of an acidic or basic solution is by dilution. If you add enough water, you can reduce the concentration of the acid or base. Extremely dilute acids and bases have a pH of 7 at 25 °C because this is the pH of pure water.

However, there is another way you can reduce the toxicity of an acidic or a basic solution. To counteract the effects of an acid in solution, you can add a base. Likewise, to counteract the effects of a base in solution, you can add an acid.

Lesson 88 | **Neutralization Reactions**   447

Lesson Guide

ENGAGE (5 min)

ChemCatalyst

Excess stomach acid, HCl, can cause extreme discomfort and pain. Milk of magnesia, $Mg(OH)_2$, is often taken to reduce stomach acid.

1. What products do you think are produced when $Mg(OH)_2$ and HCl are mixed?

2. What products do you think are produced when HNO_3 and HCl are mixed?

Sample answers: **1.** Students might speculate that magnesium chloride and water are created. The products are $MgCl_2$ and H_2O. **2.** No reaction takes place. Both substances are acids.

- Discuss student ideas about the mixing of acids and bases.

Ask: How do you think milk of magnesia works?

Ask: What do you think is produced when milk of magnesia mixes with stomach acid?

Ask: What happens to the pH of an acidic solution when you add a base? Explain your thinking.

Ask: What happens when two acids are mixed?

EXPLORE (15 min)

TRM   Worksheet with Answers 88

TRM   Worksheet 88

TRM   Transparency—Well Plate Setup 88

INTRODUCE THE LAB

→ Arrange students into pairs.

→ Pass out the student worksheet.

→ Let students know that they will work in pairs and share one set of solutions with another pair.

→ Briefly explain the procedure. Use the transparency Well Plate Setup or draw a representation of the well plate on the board, showing where the different solutions will be placed.

→ Go over safety considerations with students.

SAFETY

Wear safety goggles at all times.

Acids and bases are corrosive. Do not get any on skin or near eyes.

In case of a spill, rinse with large amounts of water.

EXPLAIN & ELABORATE (15 min)

INTRODUCE NEUTRALIZATION REACTIONS

→ Assist students in writing the four equations for the reactions they observed in the activity on the board. Label the acids, bases, and ionic salts. Show the double exchange reaction that is occurring.

Ask: What happens when an acid and a base are mixed?

Ask: Which combinations of reactants did not result in new products?

Ask: Why does the pH change when an acid is added to a base? Explain.

Ask: Why do you think acid-base reactions are called neutralization reactions?

Ask: What is the balanced equation for the reaction of milk of magnesia with stomach acid? ($Mg(OH)_2(aq)$ + $2HCl(aq) \rightarrow 2H_2O(l) + MgCl_2(aq)$)

Lesson 88 | **Neutralization Reactions**   447

Ask: How might a neutralization reaction help you deal with a toxic substance?

Key Points: A neutralization reaction between a strong acid and a strong base in aqueous solution produces an ionic compound (salt) and water. When an acid and an acid or a base and a base are mixed, no chemical reaction occurs. However, when a strong acid and a strong base are mixed, there is a dramatic change in pH. The acid produces H^+ ions, while the base produces OH^- ions in aqueous solution. Together, these ions form water, H_2O.

$$H^+ + OH^- \rightarrow H_2O$$

A neutralization reaction can be described as a double exchange reaction in which the two compounds exchange cations. The formation of water during this reaction removes both H^+ ions and OH^- ions from solution, causing the pH to move toward 7 at 25 °C.

$$HNO_3(aq) + NaOH(aq) \rightarrow NaNO_3(aq) + H_2O(l)$$

EVALUATE (5 min)

Check-In

Sulfuric acid, $H_2SO_4(aq)$, reacts with magnesium hydroxide, $Mg(OH)_2(aq)$. Write a balanced equation for the reaction that occurs.

Answer: $H_2SO_4(aq) + Mg(OH)_2(aq) \rightarrow MgSO_4(aq) + 2H_2O(l)$

Homework

Assign the reading and exercises for Toxins Lesson 88 in the student text.

The acid and base react in solution and neutralize each other. The chemical equation for the reaction between an acid and base is shown below.

A Neutralization Reaction

Acid	Base		Salt	Water

$$HCl(aq) + NaOH(aq) \longrightarrow NaCl(aq) + H_2O(l)$$

Cl^- changes place with OH^-

Notice that this reaction is a double exchange reaction. The H^+ and Na^+ cations exchange anions. The result is the production of a salt, NaCl, and water. The excess H^+ from the acid combines with OH^- from the added base to form H_2O. The reaction is referred to as a **neutralization reaction**.

Big Idea Acids and bases neutralize each other, producing a salt and water.

So, one way to make an acidic solution safe is by adding a base. Likewise, you can make a basic solution safe by adding an acid.

Not all neutralization reactions produce neutral solutions. For example, if you use an acidic solution that is very concentrated and a basic solution that is not very concentrated, there will be leftover H^+ ions after mixing. There will not be enough OH^- ions to neutralize all of the H^+ ions. But some neutralization will have taken place. As a result of mixing, the solution will be closer to neutral than either of the starting solutions.

↪ Predicting the Products of Neutralization Reactions

Several neutralization reactions are shown. Notice the patterns in the products of the reactions.

$$HCl(aq) + KOH(aq) \longrightarrow KCl(aq) + H_2O(l)$$
$$HNO_3(aq) + NaOH(aq) \longrightarrow NaNO_3(aq) + H_2O(l)$$
$$H_2SO_4(aq) + 2NaOH(aq) \longrightarrow Na_2SO_4(aq) + 2H_2O(l)$$
$$2HCl + Mg(OH)_2(aq) \longrightarrow MgCl_2(aq) + 2H_2O(l)$$
$$H_2SO_4(aq) + Ca(OH)_2(aq) \longrightarrow CaSO_4(aq) + 2H_2O(l)$$

Each reaction results in the production of an ionic compound (a salt) plus water. The salt consists of the cation of the base and the anion of the acid. For example, the cation in potassium hydroxide, KOH, is potassium, K^+. The anion in nitric acid, HNO, is nitrate, NO_3^-. The K^+ and NO_3^- produce the ionic compound, potassium nitrate, KNO_3, which is a salt.

Some acids transfer more than one H^+ ion. For example, each mole of sulfuric acid, H_2SO_4, transfers two moles of H^+ ions. This is because there are two

HEALTH CONNECTION

Some antacids are made of calcium carbonate. Calcium carbonate is the main component of eggshells and seashells and can also be used as an inexpensive calcium supplement.

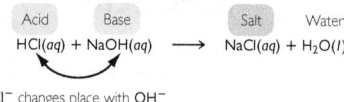

Christian Delbert/Shutterstock

hydrogen ions in solution for every one sulfate anion, SO_4^{2-}. Likewise, each mole of calcium hydroxide, $Ca(OH)_2$, transfers two moles of OH^- ions. There are two hydroxide ions in each formula unit of calcium hydroxide to balance the charge on the calcium cations, Ca^{2+}.

Example

Producing Potassium Sulfate, K_2SO_4

Which of the acids and bases in this table could you mix to make potassium sulfate, K_2SO_4? Write a balanced chemical equation for the reaction.

Acids	Bases
HNO_3	KOH
H_2SO_4	NaOH
HBr	$Mg(OH)_2$

Solution

One way to use a neutralization reaction to make potassium sulfate is

$$H_2SO_4(aq) + 2KOH(aq) \longrightarrow K_2SO_4(aq) + 2H_2O(l)$$

Notice that you need 2 mol of KOH to balance the equation properly.

The appearance of the white solid in suspension gives this remedy the name "milk."

STOMACH ACID RELIEF

A common over-the-counter remedy for excess stomach acid uses a neutralization reaction. Milk of magnesia is a white mixture containing magnesium hydroxide, $Mg(OH)_2$. The mixture looks like milk because magnesium hydroxide, a white solid, does not dissolve completely in water. Instead, the white solid is suspended in the liquid.

Magnesium hydroxide is a base. It neutralizes stomach acid, HCl, to produce a salt and water as shown by this reaction.

$$2HCl(aq) + Mg(OH)_2(aq) \longrightarrow MgCl_2(aq) + 2H_2O(l)$$

stomach acid + milk of magnesia $\longrightarrow$ magnesium chloride + water

LESSON SUMMARY

What happens when acids and bases are mixed?

A neutralization reaction is a reaction in which an acid and a base react in aqueous solution to produce an ionic compound (a salt) and water. The pH approaches 7 because the H^+ from the acid and the OH^- from the base combine to form H_2O. It is relatively easy to predict the products of a neutralization reaction. The salt that forms is made from the cation of the base and the anion of the acid.

KEY TERM

neutralization reaction

Exercises

Reading Questions

1. Describe two ways to make a strong acidic solution safer.
2. What is a neutralization reaction?

Lesson 88 | **Neutralization Reactions** 449

LESSON 88 ANSWERS

1. Two ways to make a strong acid solution safer are dilution with water and neutralization with a base.

2. A neutralization reaction is the combination of an acid and a base to form a salt and water.

3. A good lab report will contain ● a title (Lab: Neutralization Reactions) ● a statement of purpose (*Possible answer:* To explore reactions between acids and bases) ● a procedure (a summary of the steps followed in the experiment) ● results (a list of observations made during each step in the procedure) ● a conclusion (Student analysis should include how they can use their results to predict the products of other reactions between acids and bases. *Possible answer:* The results of a reaction between an acid and a base can be predicted from their chemical formulas and concentrations.)

4. a. $HF(aq) + NaOH(aq) \rightarrow$ $NaF(aq) + H_2O(l)$

b. $2HCl(aq) + Mg(OH)_2(aq) \rightarrow$ $MgCl_2(aq) + 2H_2O(l)$

c. $HF(aq) + NH_4OH(aq) \rightarrow$ $NH_4F(aq) + H_2O(l)$

d. $CH_3COOH(aq) + NaOH(aq) \rightarrow$ $NaCH_3COO(aq) + H_2O(l)$

e. $2HNO_3(aq) + Mg(OH)_2(aq) \rightarrow$ $Mg(NO_3)_2(aq) + 2H_2O(l)$

f. $HNO_3(aq) + NH_4OH(aq) \rightarrow$ $NH_4NO_3(aq) + H_2O(l)$

g. $2CH_3COOH(aq) + Mg(OH)_2(aq) \rightarrow$ $Mg(CH_3COO)_2(aq) + 2H_2O(l)$

h. $CH_3COOH(aq) + NH_4OH(aq) \rightarrow$ $NH_4CH_3COO(aq) + H_2O(l)$

i. $H_2SO_4(aq) + HNO_3(aq) \rightarrow$ $3H^+(aq) + SO_4^{2-}(aq) + NO_3^-(aq)$

j. $H_2SO_4(aq) + Mg(OH)_2(aq) \rightarrow$ $MgSO_4(aq) + 2H_2O(l)$

5. Each mole of sulfuric acid forms two moles of hydrogen ions in solution, while each mole of sodium hydroxide forms one mole of hydroxide ions. Therefore, the final solution has more hydrogen ions than hydroxide ions, so it is acidic and the pH is less than 7.

6. Both of the solutions are basic, so none of the hydroxide ions is neutralized. The final solution is basic and the pH is greater than 7.

7. D

8. B

Exercises

(continued)

Reason and Apply

3. Lab Report Complete a lab report for the Lab: Neutralization Reactions. In your report, give the title of the experiment, purpose, procedure, observations, and conclusion. In your conclusion, explain how you can use chemical equations to predict the products of the reactions you carried out.

(Title)

Purpose: (Explain what you were trying to find out.)

Procedure: (List the steps you followed.)

Observations: (Describe your observations.)

Conclusion: (What can you conclude about what you were trying to find out?)

4. Predict the products for the reactions given below. Be sure to balance each equation.
  a. $HF(aq) + NaOH(aq) \longrightarrow$
  b. $HCl(aq) + Mg(OH)_2(aq) \longrightarrow$
  c. $HF(aq) + NH_4OH(aq) \longrightarrow$
  d. $CH_3COOH(aq) + NaOH(aq) \longrightarrow$
  e. $HNO_3(aq) + Mg(OH)_2(aq) \longrightarrow$
  f. $HNO_3(aq) + NH_4OH(aq) \longrightarrow$
  g. $CH_3COOH(aq) + Mg(OH)_2(aq) \longrightarrow$
  h. $CH_3COOH(aq) + NH_4OH(aq) \longrightarrow$
  i. $H_2SO_4(aq) + HNO_3(aq) \longrightarrow$
  j. $H_2SO_4(aq) + Mg(OH)_2(aq) \longrightarrow$

5. Suppose you mix 1 mol of sulfuric acid, H_2SO_4, with 1 mol of sodium hydroxide, NaOH. Why does the pH of the solution remain below 7?

6. Suppose you mix 1 mol of sodium hydroxide, NaOH, with 1 mol of magnesium hydroxide, $Mg(OH)_2$. Why does the pH of the solution remain above 7?

7. Which of these substances might be useful in neutralizing a lake damaged by acid rain?
  (A) H_2SO_4
  (B) CH_3COOH
  (C) $CaCl_2$
  (D) $Ca(OH)_2$

8. Which combination of reactants would result in a neutralization reaction with sodium nitrate, $NaNO_3$, as one of the products?
  (A) $Mg(NO_3)_2 + NaOH$
  (B) $HNO_3 + NaOH$
  (C) $CH_3OH + NaOH$
  (D) $HNO_3 + NaCl$

Drip Drop

Titration

THINK ABOUT IT

When sulfur dioxide, SO_2, and nitrogen oxide, NO, two components of air pollution, come in contact with water in the atmosphere, they are converted to sulfuric acid, H_2SO_4, and nitric acid, HNO_3. The acids then fall as acid rain or snow. Acidic precipitation is highly destructive, sometimes causing the complete elimination of fish and insect species from lakes or streams. Acid rain also makes the soil unsuitable for plant life, killing off whole sections of forest. Water scientists, called hydrologists, regularly study the acidity of lakes and streams to monitor the extent of the acid rain problem.

> How can you use a neutralization reaction to figure out acid or base concentration?

To answer this question, you will explore

◌ Titrations

◌ Particle Views of Titrations

◌ Titration Calculations

EXPLORING THE TOPIC

◌ **Titrations**

To track the spread of acid rain, scientists take water samples from lakes and test them for H^+ concentration. One way to determine the concentration of a strong acid in a water sample is to use a method called a **titration.** A titration is a neutralization reaction that is monitored with an acid-base indicator. For example, a strong base with a known concentration is added to a strong acid sample with an unknown concentration. The base is added until the indicator changes color.

The indicator provides a visual signal that the solution has reached the **equivalence point,** the point at which the moles of base added have neutralized the moles of acid. By keeping track of the exact volume of base that is added to a known volume of acid, you can figure out the unknown concentration of acid in the sample.

The illustration shows a titration setup. The long thin tube is a *burette*. A valve on the burette allows you to regulate how much solution goes into the beaker. During a titration, solution is added from the burette until the indicator changes color. You determine the volume of solution added by noting the change in the height of the solution in the burette.

A Titration Setup

Base

Acid + indicator

Phenolphthalein is an indicator that changes from colorless to bright pink in a base.

Lesson 89 | **Titration**   451

Overview

LAB: PAIRS

Key Question: How can a neutralization reaction help you figure out acid or base concentration?

KEY IDEAS

A titration is a type of chemical procedure carried out between an acid and a base to determine the concentration of either the acid or the base. A titration uses a neutralization reaction that is monitored with an indicator. When the molarity and volume of a known solution are used in a titration, the molarity of a known volume of another solution can be determined.

LEARNING OBJECTIVES

● Explain and complete a titration procedure.

● Use titration data to determine the molarity of a solution of unknown concentration.

FOCUS ON UNDERSTANDING

● The equivalence point of a titration does not always occur at a pH of 7 for titrations of weak acids and bases. We only consider titrations of strong acids and bases.

● Different indicators change color at different pH values. For simplicity, we use a strong acid-strong base neutralization reaction and an indicator that changes color at a pH of 7.

KEY TERMS

titration
equivalence point

IN CLASS

Three hydrochloric acid, HCl, solutions are labeled A, B, and C. Within groups of four, students work in pairs to conduct a titration and determine which is the most concentrated (or toxic) sample. Students test a 20-drop sample of all three solutions, measuring the number of drops of 0.10 M NaOH required to neutralize each solution to an equivalence point indicated by a color change. They complete molarity calculations using their titration data. A complete materials list is available for download.

TRM Materials List 89

Differentiate

Students may need appropriate scaffolding for titration problems depending on their readiness level. For English-language learners or students who struggle with literacy, write straightforward example word problems to assess a student's ability to determine an unknown concentration rather than interpret a complicated sentence. Assess vocabulary such as *titration* or *end point* separately from mathematical calculations. For advanced students, provide problems that involve bases such as calcium hydroxide or diprotic acids such as sulfuric acid.

Pre-AP® Course Tip

Students preparing to take an AP® chemistry or college chemistry course should be able to tell what data to collect to achieve a certain goal, such as monitoring the titration of an acid with a base. In this lesson, students conduct a simplified titration of a strong acid with a strong base to determine the molarity of a solution of unknown concentration.

SETUP

Before class, prepare 200 mL of each solution listed. Divide each solution into eight labeled dropper bottles.

HCl solution A (0.050 M HCl)

HCl solution B (0.10 M HCl)

HCl solution C (0.020 M HCl)

0.10 M NaOH (8.0 g in 200 mL)

bromothymol blue indicator

Label the HCl solutions A, B, and C. Do not tell students the concentrations.

SAFETY

When making up solutions, always put acid or base into water rather than the other way around.

Wear safety goggles at all times.

Acids and bases are corrosive. Do not get any on skin or near eyes.

In case of a spill, rinse with large amounts of water.

Have baking soda and citric acid on hand for spills.

CLEANUP

Neutralize and dilute wastes, then dispose according to local guidelines.

Lesson Guide

ENGAGE (5 min)

ChemCatalyst

A student mixes 100 mL of 0.10 M HCl with different volumes of 0.10 M NaOH.

1. 100 mL of 0.10 M HCl + 50 mL of 0.10 M NaOH

2. 100 mL of 0.10 M HCl + 100 mL of 0.10 M NaOH

3. 100 mL of 0.10 M HCl + 150 mL of 0.10 M NaOH

a. Which solution is the least toxic?

b. Are the final solutions acidic, basic, or neutral?

Sample answers: **a.** Solution B is the least toxic, because the H⁺ and OH⁻ ions neutralize each other. Equal concentrations of ions come together to form a salt, NaCl, and water. **b.** Solution A is acidic, solution B is neutral, and solution C is basic.

Ask: What is the equation for the neutralization reaction between NaOH and HCl?

Ask: Will adding HCl to a solution containing a toxic amount of NaOH make the solution safer? Explain.

Ask: How can you figure out how much HCl to add to make the solution neutral?

Ask: How many moles of HCl must be added to a water sample containing 0.010 mol of NaOH to neutralize it? (0.010 mol)

Ask: How can you determine the number of moles of hydroxide ions, OH⁻, in a sample if you know the concentration?

➲ Particle Views of Titrations

This illustration shows what happens as equal volumes of 1.0 M HCl, hydrochloric acid, and 1.0 M NaOH, sodium hydroxide, are mixed together.

$$HCl(aq) + NaOH(aq) \longrightarrow NaCl(aq) + H_2O(l)$$

● H⁺   ○ Na⁺
□ Cl⁻   ■ OH⁻

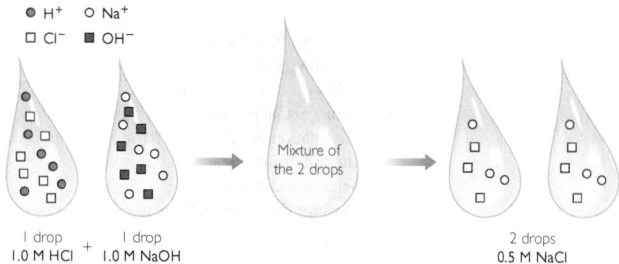

| 1 drop 1.0 M HCl | + | 1 drop 1.0 M NaOH | → | Mixture of the 2 drops | → | 2 drops 0.5 M NaCl |

Neutralization of 1.0 M NaOH by 1.0 M HCl

When you add one drop of 1.0 M HCl to one drop of 1.0 M NaOH, H⁺ ions from the acid and OH⁻ ions from the base combine to form H₂O. The Na⁺ ions and Cl⁻ ions from the original drops are dissolved in the water. Notice that the concentration of Na⁺ ions and Cl⁻ ions has decreased because each ion has spread out over twice the volume, equal to two drops of solution.

The next illustration shows what happens when equal volumes of 1.0 M HCl and 2.0 M NaOH are mixed together.

● H⁺   ○ Na⁺
□ Cl⁻   ■ OH⁻

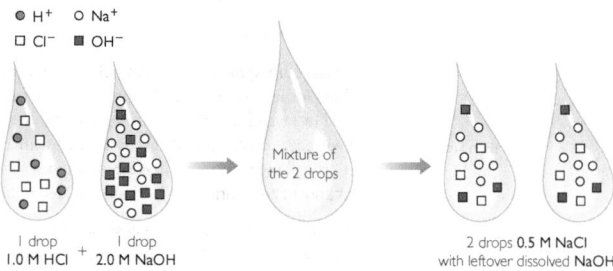

| 1 drop 1.0 M HCl | + | 1 drop 2.0 M NaOH | → | Mixture of the 2 drops | → | 2 drops 0.5 M NaCl with leftover dissolved NaOH |

Neutralization of 1.0 M HCl by 2.0 M NaOH

In this example, the base is more concentrated than the acid so the two-drop mixture will have leftover dissolved NaOH in addition to dissolved 0.5 M NaCl.

Because the 2.0 M NaOH is twice as concentrated as the 1.0 M HCl, you only need half as much of it to neutralize the HCl.

Moles of H⁺ in 1.0 M HCl (10 mL)	Moles of OH⁻ in 2.0 M NaOH (5 mL)
moles H⁺ = molarity · volume = 1.0 M · 0.010 L = 0.005 mol	moles OH⁻ = molarity · volume = 2.0 M · 0.005 L = 0.005 mol

---

EXPLORE (15 min)

TRM Worksheet with Answers 89

TRM Worksheet 89

INTRODUCE THE LAB

→ Arrange students into pairs. Have one pair share a set of solutions with another pair.

→ Pass out the student worksheet.

→ Tell students that their goal in today's lab is to figure out the concentrations of three different solutions of hydrochloric acid using a procedure called a titration.

→ Go over safety considerations with students.

EXPLAIN & ELABORATE (15 min)

PROCESS THE RESULTS OF THE LAB

Create a table like this one on the board. Share students' results, and help students to show how they arrived at the concentration of each solution.

HCl solution	Drops of 0.10 M NaOH	Concentration of HCl
solution A—20 drops	~10 drops	0.05 M
solution B—20 drops	~20 drops	0.10 M
solution C—20 drops	~4 drops	0.02 M

Titration Calculations

Suppose you take a 100 mL water sample from a lake contaminated with sulfuric acid, H_2SO_4. You want to determine the concentration of the sulfuric acid in the lake. So you titrate the sample with 2.0 molar sodium hydroxide, 2.0 M NaOH, until the solution is neutral. At the equivalence point, you know that the moles of base you added neutralized moles of acid that were in the unknown solution.

Bromothymol blue indicator is yellow in acidic solution, green in neutral solution, and blue in basic solution.

You need an indicator that changes color when the solution is neutral, so you add a drop of bromothymol blue to the lake water sample. You add NaOH slowly until the indicator turns green. Because you know the volume and molarity of the NaOH added, you can calculate the moles of acid per liter of solution in the lake water sample.

Example

Titration of H_2SO_4

A 100 mL sample of aqueous sulfuric acid, H_2SO_4, is titrated with 2.0 M NaOH. After adding 50 mL of NaOH, the indicator changes color at pH 7. What was the starting concentration of the H_2SO_4?

Solution

Begin by writing a balanced chemical equation.

$$H_2SO_4(aq) + 2NaOH(aq) \longrightarrow Na_2SO_4(aq) + 2H_2O(l)$$
$$\text{sulfuric acid} + \text{sodium hydroxide} \longrightarrow \text{sodium sulfate} + \text{water}$$

Find the total number of moles of NaOH that were used to neutralize the H_2SO_4.

$$\text{moles NaOH} = (\text{molarity of NaOH})(\text{volume of NaOH added})$$
$$= (2.0 \text{ mol/L})(0.050 \text{ L}) = 0.10 \text{ mol NaOH}$$

Since 2 mol of NaOH are needed to neutralize every 1 mol of H_2SO_4, the moles of H_2SO_4 in the sample must be half the number of moles of the added NaOH.

$$\text{moles of } H_2SO_4 = \frac{1}{2}(\text{moles of NaOH}) = \frac{1}{2}(0.10 \text{ mol}) = 0.05 \text{ mol}$$

Finally, find the molarity of the sulfuric acid solution.

$$\text{molarity of sulfuric acid} = \frac{n}{V}$$
$$= 0.050 \text{ mol}/0.100 \text{ L}$$
$$= 0.50 \text{ M } H_2SO_4$$

LESSON SUMMARY

How can you use a neutralization reaction to figure out acid or base concentration?

A titration is a chemical procedure carried out between an acid and a base to determine the concentration of either the acid or the base. A titration is a

formal titration. See the illustration on page 451 of the student textbook.

Ask: What would make this experiment more precise?

Ask: Why is it important to keep track carefully of the amount of base you add to the solution you are titrating?

Key Points: A titration is a procedure in which a neutralization reaction is monitored with an indicator allowing you to calculate the unknown concentration of an acid or base. You do this by reacting a known volume of a solution with an unknown concentration with a known volume of a solution with known concentration. When the number of moles of acid is equal to the number of moles of base in the solution, the reaction has reached the equivalence point of the titration. An appropriate indicator is used to provide a visual signal that the equivalence point has been reached. The pH at which an indicator changes color is called the *end point*. Typically, titrations are carried out using special glassware that allows you to measure volumes precisely. A pH meter also can be used to determine the equivalence point. Because the molarity and volume of one solution are known and the volume of the second solution is known, the molarity of the second solution can be calculated.

DISCUSS THE CALCULATIONS ASSOCIATED WITH TITRATIONS

→ Complete a sample titration calculation on the board.

Ask: When the titration is at its equivalence point, what do you know about the number of moles of acid and base?

Key Points: When the equivalence point is reached in a titration between a strong acid and a strong base, the number of moles of H^+ ions equals the number of moles of OH^- ions. Remember that the number of moles in a solution is related mathematically to its volume by the proportionality constant, M, the molarity in moles per liter.

EXAMPLE

If 25.0 mL of 0.500 M nitric acid, HNO_3, solution is required to neutralize 62.0 mL of sodium hydroxide, NaOH, what is the molarity of the NaOH?

First, determine the number of moles of HNO_3 that were added to the base.

$$\text{Moles } HNO_3 = (0.500 \text{ mol/L})(0.0250 \text{ L})$$
$$= 0.0125 \text{ mol}$$

Ask: What does the color change indicate in the activity?

Ask: How many drops of base did it take to neutralize each of the three solutions?

Ask: Which solution needed the most base to be neutralized? What does this tell you about the molarity of acid in that solution?

Ask: Why didn't everyone get the same answer? What are some possible sources of error in this procedure?

Key Points: The chemical equation for the neutralization reaction shows the ratio in which the substances combine.

$$HCl(aq) + NaOH(aq) \rightarrow$$
$$NaCl(aq) + H_2O(l)$$

In this case, the reactants are in a 1:1 ratio. This means that 1 mol of NaOH is required to react with 1 mol of HCl. Thus, if 20 drops of 0.10 M NaOH are required to neutralize 20 drops of HCl, you can assume that the concentrations of the two solutions are identical. If 10 drops of 0.10 M NaOH are required to neutralize 20 drops of HCl, then the HCl concentration is half as much as the NaOH concentration, or 0.050 mol per liter.

EXPAND ON THE CONCEPT OF TITRATION

→ You might have traditional titration equipment, such as a burette and a pH meter, on hand to demonstrate a more

Second, determine the number of moles of NaOH that were neutralized.

At the equivalence point, moles of H^+ = moles of OH^-

Moles NaOH = moles HNO_3
= 0.0125 mol

Third, determine the concentration of the original NaOH solution.

Concentration of NaOH =
0.0125 mol/0.0620 L = 0.201 M NaOH

EVALUATE (5 min)

Check-In

A beaker has 50 drops of HCl, along with a drop of phenolphthalein indicator. After 100 drops of 0.10 M NaOH are added, the color changes from clear to bright pink. What is the concentration of the original HCl solution?

Answer: 0.20 M

Homework

Assign the reading and exercises for Toxins Lesson 89 and the Chapter 16 Summary in the student text.

LESSON 89 ANSWERS

1. *Possible answer:* You can add an indicator to a known volume of the water sample. Then add a measured amount of a solution of a strong acid until the indicator changes color. Calculate the number of moles of hydrogen ions added, which is equal to the number of moles of hydroxide ions in the original sample. Use this value to calculate the concentration of potassium hydroxide in the original sample.

2. The indicator is used to show when the titration has reached the equivalence point.

3. 1000 mL

4.

Initial volume of HCl	Volume 0.10 NaOH added	Moles of NaOH	Mole of HCl	Initial [HCl]
1.0 L	1.0 L	0.10 mol	0.10 mol	0.10 M
100 mL	200 mL	0.20 mol	0.020 mol	0.20 M
50 mL	200 mL	0.20 mol	0.020 mol	0.40 M
50 mL	25 mL	0.0025 mol	0.0025 mol	0.050 M
100 mL	73 mL	0.0073 mol	0.0073 mol	0.073 M

neutralization reaction that is monitored with an indicator. The volume of acid and the volume of base used in the procedure are carefully recorded. If the molarity of either the acid or the base is known, the molarity of the other can be determined.

Exercises

Reading Questions

1. Describe how you might use a titration to figure out the concentration of potassium hydroxide in a water sample.
2. What is the role of an indicator in titration?

Reason and Apply

3. How many mL of 0.1 M NaOH would be required to neutralize 2.0 L of 0.050 M HCl?
4. The table represents a series of titrations using 0.10 M NaOH to determine the concentrations of five different samples of HCl. Copy the table and fill in the missing values.

Initial volume of HCl	Volume of 0.10 M NaOH added	Total moles of NaOH	Total moles of HCl	Initial HCl concentration
1.0 L	1.0 L	0.10 mol	0.10 mol	0.10 M
100 mL	200 mL	0.020 mol	0.020 mol	
50 mL	200 mL			
50 mL		0.0025 mol	0.0025 mol	0.050 M
100 mL	73 mL			

5. A student mixes 100 mL of 0.20 M HCl with different volumes of 0.50 M NaOH. Are the final solutions acidic, basic, or neutral? Explain your thinking.
   a. 100 mL of 0.20 M HCl + 20 mL of 0.50 M NaOH
   b. 100 mL of 0.20 M HCl + 40 mL of 0.50 M NaOH
   c. 100 mL of 0.20 M HCl + 60 mL of 0.50 M NaOH
6. Imagine that you use 0.95 M NaOH to titrate several water samples. The volume of base needed to neutralize a specified amount of acid is given. Determine the acid concentration and the pH for each.
   a. 25 mL acid, 46 mL NaOH
   b. 10 mL acid, 17 mL NaOH
   c. 25 mL acid, 12 mL NaOH

5. a. The final solution is acidic because the solution has more H^+ ions than OH^- ions. **b.** The final solution is neutral because the solution has equal numbers of H^+ ions and OH^- ions.

c. The final solution is basic because the solution has fewer H^+ ions than OH^- ions.

6. a. Molarity = 1.7 M; pH = -0.23
b. Molarity = 1.6 M; pH = -0.20
c. Molarity = 0.46 M; pH = 0.34

CHAPTER 16

Acidic Toxins

SUMMARY

KEY TERMS

indicator

pH scale

acid

base

dissociate

strong acid

weak acid

strong base

weak base

dilution

neutralization reaction

titration

equivalence point

Toxins Update

Acids and bases are present in living systems and are a valuable part of the chemistry of life. They are toxic under certain conditions, such as when they are too concentrated or when they upset the pH balance that must be maintained for proper health. The pH number is a measure of the H^+ concentration of a solution. The lower the pH, the more acidic a solution is.

There are two main approaches to dealing with toxic acids and bases. One is dilution. This moves the pH toward neutral. Another approach is neutralization. Acids and bases neutralize one another.

REVIEW EXERCISES

1. What are the main differences between the Arrhenius and Brønsted-Lowry definitions of acids and bases?

2. Name three substances that you might use to neutralize a hydrochloric acid solution. Write the balanced chemical equation for each reaction.

3. Explain how a titration procedure works to help you identify the H^+ concentration of a sample solution.

4. Lemon juice has a pH around 2. This is quite acidic and can damage the tissue of the eye.
   a. What is the concentration of hydrogen ions, $[H^+]$, in lemon juice?
   b. What is the concentration of hydroxide ions, $[OH^-]$?

5. Sodium bicarbonate, $NaHCO_3$, is a base.
   a. Explain how sodium bicarbonate can be classified as a base, even though it has no OH^- ion.
   b. Explain why you had sodium bicarbonate on hand during the Lab: The Copper Cycle from Unit 1: Alchemy.

Dissolving Toxins

PROJECT

www

Choose a toxic substance (your teacher may assign you one). Find some information about your toxin. Your report should cover these topics.

• Does your substance dissolve in water? If it does not dissolve, would it sink or float?

• Make a particle view drawing of your substance in water, even if it does not dissolve.

• Is your substance acidic, basic, or neutral? If your substance cannot be described in this way, explain why.

with a measured amount of base to form a neutral solution. The number of H^+ ions in the acid must be equal to the measured number of OH^- ions in the base. This number can be used with the volume to determine the concentration of H^+ in the original solution.

4. a. $[H^+] = 1 \times 10^{-2}$ M
b. $[OH^-] = 1 \times 10^{-12}$ M

5. a. Sodium bicarbonate can be classified as a base because it accepts a proton from water, forming hydroxide ions. Sodium bicarbonate forms Na^+ cations. The bicarbonate anions react with water to form H_2CO_3 and OH^- ions. b. One of the steps in the Lab: The Copper Cycle was to add sulfuric acid to the copper solution. Sodium bicarbonate was available in case sulfuric acid spilled on a person. The sodium bicarbonate could have been used to neutralize any acid that was spilled.

Project: Dissolving Toxins

A good report would include ● the solubility and density of the toxic substance ● a description of what happens to the toxic substance when it is added to water, including a sketch of the particle view ● an appropriate sketch for the toxic substance, depending on its method of bonding ● a description of whether the toxic substance is acidic, basic, neutral, or none of the above ● any chemical equations that may be relevant to the topics above.

ASSESSMENTS

Two multiple-choice and two short-answer assessments for Chapter 16 are available for download.

TRM Chapter 16 Assessment Answers

TRM Chapter 16 Assessments

ANSWERS TO CHAPTER 16 REVIEW EXERCISES

1. According to the Arrhenius theory, an acid is a substance that dissociates to form hydrogen ions in solution, and a base is a substance that dissociates to form hydroxide ions in solution. According to the Brønsted-Lowry theory, an acid is a substance that donates protons in solution, while a base is a substance that accepts protons. This is similar to the Arrhenius theory, but it accounts for bases that do not dissociate to form hydroxide ions.

2. *Possible answer:*
potassium hydroxide, KOH:
$HCl(aq) + KOH(aq) \rightarrow KCl(aq) + H_2O(l)$
ammonium hydroxide, NH_4OH:
$HCl(aq) + NH_4OH(aq) \rightarrow NH_4Cl(aq) + H_2O(l)$
calcium hydroxide, $Ca(OH)_2$:
$2HCl(aq) + Ca(OH)_2(aq) \rightarrow CaCl_2(aq) + 2H_2O(l)$

3. *Possible answer:* A titration of an acid is a neutralization reaction in which a known volume of acid reacts

Watch the video overview of Chapter 17 (for teachers) by clicking on the link in the TE-book, opening the TRFD, or logging onto the book's companion Web site bcs.whfreeman.com/livingbychemistry2e (teacher log-in required).

The final chapter of Unit 4: Toxins deals with precipitation reactions and stoichiometric problems. Lesson 90 is a lab activity that introduces students to precipitates and solubility. In Lesson 91, students use varying amounts of reactants to explore mole ratios and their effect on product formation. In Lesson 92, students practice completing gram-mole-gram conversions. Students are presented in Lesson 93 with problems in which reactants are not mixed in correct mole ratios. They determine the limiting reactant, the amount of product formed, and the exact amount of reactant needed to remove a toxic substance from a water sample. In the unit review, students explore ways of using chemical reactions to remove toxic substances from living systems and water supplies.

In this chapter, students will learn

- about precipitation reactions
- about soluble and insoluble compounds
- the role of equation coefficients in stoichiometry
- how to complete limiting reactant problems
- gram-mole-gram conversions

Chapter D. Winters/Science Source

CHAPTER **17**

Toxic Cleanup

Chemical reactions are often used to remove toxic substances from a solution. Here, a solid forms when an ionic solution is added.

In this chapter, you will study

- solids in solutions
- soluble and insoluble ionic compounds
- how to determine the product yield of a reaction
- how to use the "mole tunnel" to perform stoichiometric calculations

One way to remove specific ions from a solution is to use chemical reactions to transform these ions into solid products. Solid compounds can be easily filtered out of a solution. The human body uses this same approach with many toxic substances, by forming solids that are then passed out of the body as waste. Mass-mole conversions allow you to determine how to create the maximum amount of product from a chemical reaction.

456

Chapter 17 Lessons

Solid Evidence

Precipitation Reactions

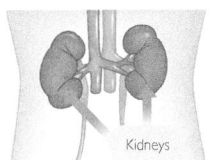
Kidneys

THINK ABOUT IT

Sometimes, reactions between dissolved substances result in the formation of a solid. In your kidneys, for example, dissolved substances can react to form solid calcium oxalate. If there is enough of this solid, a painful blockage called a kidney stone can form.

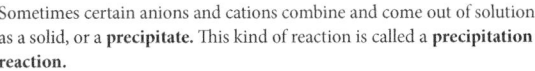
Which substances precipitate from aqueous solutions?

To answer this question, you will explore

- Precipitation of Ionic Solids
- Solubility
- Toxicity of Precipitates

EXPLORING THE TOPIC

↻ Precipitation of Ionic Solids

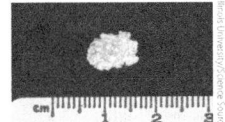
Kidney stone crystal

Sometimes certain anions and cations combine and come out of solution as a solid, or a **precipitate**. This kind of reaction is called a **precipitation reaction.**

The formation of kidney stones is an example of a precipitation reaction. If the concentrations of calcium ions, Ca^{2+}, and polyatomic oxalate ions, $C_2O_4^{2-}$, in urine become too great, calcium oxalate, CaC_2O_4, precipitates as kidney stones.

$$CaCl_2(aq) + Na_2C_2O_4(aq) \longrightarrow 2NaCl(aq) + CaC_2O_4(s)$$

| Dissolved calcium chloride | Dissolved sodium oxalate | Dissolved sodium chloride | Solid calcium oxalate |

The equation for the formation of kidney stones shows that it is a double exchange reaction. Many other similar chemical combinations result in precipitation reactions.

Example 1

Precipitation of Lead Iodide

An aqueous solution of lead nitrate, $Pb(NO_3)_2(aq)$, is mixed with an aqueous solution of potassium iodide, $KI(aq)$. The result is a bright yellow solid, lead iodide, $PbI_2(s)$, in a clear solution. Write a balanced chemical equation for this precipitation reaction.

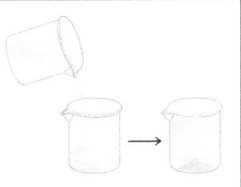

Overview

LAB: PAIRS

Key Question: Which substances precipitate from aqueous solutions?

KEY IDEAS

A precipitation reaction is a reaction in which a solid forms when two solutions are mixed. When two aqueous salt solutions are mixed, the product can be an aqueous solution with the salts dissolved, or one of the salts can be insoluble and precipitate out. The metallic elements in Periods 4 and 5 on the periodic table tend to form insoluble compounds.

LEARNING OBJECTIVES

- Explain what a precipitation reaction is.
- Write net ionic equations for precipitation reactions.

FOCUS ON UNDERSTANDING

- Precipitation was mentioned in the context of weather in Unit 3. Here, it is used in the context of solids coming out of an aqueous solution.
- Solubility was defined loosely in Unit 1: Alchemy.

KEY TERMS

precipitate
precipitation reaction

complete ionic equation
spectator ion
net ionic equation

IN CLASS

Students balance chemical equations for reactions in which two aqueous salt solutions are mixed. Students then work in pairs to perform these same reactions in a well plate to verify whether or not a precipitate forms. A complete materials list is available for download.

TRM Materials List 90

Differentiate

While discussing the solubility patterns as a class, sketch a sample 4 × 4 well plate on the board and label columns with one of the four sodium salts and the rows with one of the four nitrate salts. This will help students who are visual learners discern overall trends and will translate to the solubility chart in the text.

SETUP

Before class, prepare 200 mL of each solution listed. Each solution is 1.0 M. Divide the solutions into 50 mL samples in labeled dropper bottles.

KNO_3 (20.2 g in 200 mL)
$Mg(NO_3)_2 \cdot 6H_2O$ (36.9 g in 200 mL)
$Cu(NO_3)_2 \cdot 3H_2O$ (48.2 g in 200 mL)
$AgNO_3$ (34.0 g in 200 mL)
$NaCl$ (11.7 g in 200 mL)
Na_2CO_3 (21.2 g in 200 mL)
$NaOH$ (8.0 g in 200 mL)

CLEANUP

Solutions in dropper bottles can be reused in subsequent classes. Each class should consume less than 5 mL of each solution. Evaporate the water from any remaining solutions and dispose of solids according to local guidelines for metal waste.

ENGAGE (5 min)

TRM PowerPoint Presentation 90

ChemCatalyst

Kidney stones are solid blockages that sometimes occur in the kidney. Below is the chemical equation for a double exchange reaction between calcium chloride and sodium oxalate that results in the formation of one type of kidney stone.

$$CaCl_2(aq) + Na_2C_2O_4(aq) \rightarrow$$
$$2NaCl(aq) + CaC_2O_4(s)$$

1. What do you expect to see in the beaker if you complete this reaction?

2. Which compound is the kidney stone? What is your reasoning?

Sample answers: **1.** You will be able to see an aqueous solution and a solid. **2.** The word *stone* implies something solid, so the solid calcium oxalate is the kidney stone.

→ Assist students in interpreting the chemical equation.

Ask: Is sodium chloride, NaCl, soluble in water? How do you know? What about sodium oxalate, $Na_2C_2O_4$? Calcium oxalate, CaC_2O_4?

Ask: Why do you think the formation of calcium oxalate causes problems in the kidneys?

Lesson Guide

EXPLORE (15 min)

TRM Worksheet with Answers 90

TRM Worksheet 90

INTRODUCE THE LAB

→ Tell students to work in pairs.

→ Pass out the worksheet to each group.

→ Introduce the word *precipitate,* a term commonly used to describe a solid substance that forms when two solutions are mixed. If you like, you can make the connection to precipitation in a weather context. In all cases, precipitation involves a substance coming out of solution.

> **Precipitate:** A solid produced in a chemical reaction between two solutions.

→ Tell students that they will work with aqueous salts, which are ionic solutions. Students should complete Part 1 on their own, then work in pairs on the lab procedure. Students can work more efficiently if they use the well plate as a grid, with the sodium compounds in rows and the nitrate compounds in columns.

Example 1 *(continued)*

Solution

This is a double exchange reaction. The lead and potassium cations exchange anions with one another. Begin by writing an equation for what you know.

$$Pb(NO_3)_2(aq) + KI(aq) \longrightarrow PbI_2(s) + ?$$

| lead nitrate | potassium iodide | lead iodide | clear solution |

The potassium ions, K^+, and the nitrate ions, NO_3^-, combine in a 1:1 ratio to form KNO_3.

$$Pb(NO_3)_2(aq) + KI(aq) \longrightarrow PbI_2(s) + \mathbf{KNO_3(aq)}$$

| lead nitrate | potassium iodide | lead iodide | potassium nitrate |

Balance the equation.

$$Pb(NO_3)_2(aq) + 2KI(aq) \longrightarrow PbI_2(s) + 2KNO_3(aq)$$

| lead nitrate | potassium iodide | lead iodide | potassium nitrate |

↻ Solubility

Ionic substances vary significantly as to how much they will dissolve in water. For some ionic substances, large quantities dissolve in water. These substances have a high solubility. For other ionic substances, only very small quantities dissolve. These substances have a low solubility. Substances with a low solubility tend to form precipitates in aqueous solutions.

> **Big Idea** Some ionic solids are more soluble than others. When a compound reaches the limits of its solubility, undissolved solid is visible.

You can use a solubility table such as the one shown here to determine the solubility of various ionic compounds. To use the table, combine a cation from the rows on the left with an anion from the columns on the right to determine if the compound formed is either very soluble (S), or insoluble/not very soluble (N).

> **WEATHER CONNECTION**
> Precipitation can refer to either water coming out of the atmosphere as rain or snow or as compounds coming out of an aqueous solution as solids.

EXPLAIN & ELABORATE (15 min)

DISCUSS PRECIPITATION REACTIONS

Write the equation from the ChemCatalyst on the board as an example of a precipitation reaction.

> $$CaCl_2(aq) + Na_2C_2O_4(aq) \rightarrow$$
> $$2NaCl(aq) + CaC_2O_4(s)$$

Ask: What types of products form when two ionic solutions are mixed?

Ask: What type of reaction does this represent? (double exchange)

Key Points: The mixing of two ionic solutions sometimes results in the formation of a solid precipitate. A solid precipitate is a solid substance that separates out of a solution because the solid is not very soluble in water. Chemical reactions that result in the formation of a precipitate are called *precipitation reactions*. Note that all of our examples are double exchange reactions, in which cations and anions are switched. In the ChemCatalyst, two aqueous solutions containing dissolved ionic solids—calcium chloride, $CaCl_2$, and sodium oxalate, $Na_2C_2O_4$—were combined. The result was an aqueous salt, NaCl, and a

Solubility Trends

Cations		Anion						
		NO_3^-	Cl^-	OH^-	SO_4^{2-}	CO_3^{2-}	$C_2O_4^{2-}$	PO_4^{3-}
	Most alkali metals, such as Li^+, Na^+, K^+, NH_4^+	S	S	S	S	S	S	S
	Most alkaline earth metals, such as Mg^{2+}, Ca^{2+}, Sr^{2+}	S	S	N	S	N	N	N
	Some Period 4 transition metals, such as Fe^{3+}, Co^{3+}, Ni^{2+}, Cu^{2+}, Zn^{2+}	S	S	N	S	N	N	N
	Other transition metals, such as Ag^+, Pb^{2+}, Hg^{2+}	S	N	N	N	N	N	N

Recall that in Unit 1: Alchemy, you characterized ionic solids as soluble in water. This is true for many ionic solids. But as you can see from the solubility table, some ionic solids are not very soluble.

Example 2

Predicting Solid Products

Suppose that you combine aqueous calcium nitrate, $Ca(NO_3)_2$, with aqueous sodium phosphate, Na_3PO_4, and there is a double exchange reaction. Do you expect a precipitate to form?

Solution

First, write a balanced chemical equation for this reaction. You know that the two aqueous cations exchange anions:

$$3Ca(NO_3)_2(aq) + 2Na_3PO_4(aq) \longrightarrow Ca_3(PO_4)_2(?) + 6NaNO_3(?)$$

Next, use the solubility table to determine if there is a precipitate. According to the table, $NaNO_3$ is soluble in water, so it remains dissolved. $Ca_3(PO_4)_2$ is not very soluble, so it forms a precipitate.

$$3Ca(NO_3)_2(aq) + 2Na_3PO_4(aq) \longrightarrow Ca_3(PO_4)_2(s) + 6NaNO_3(aq)$$

EARTH SCIENCE CONNECTION

Stalactites and stalagmites are caused by precipitation of solids from water that drips from cave walls. The word *stalagmite* comes from the Greek word for "drip." These beautiful cave structures are often formed from calcium carbonate.

Reactions in aqueous solutions usually involve ions in solution. To describe what is happening, you can write a **complete ionic equation,** which shows all of the dissolved ions involved in the reaction. Examine the complete ionic equation for the reaction in Example 2.

$$3Ca^{2+}(aq) + 6NO_3^-(aq) + 6Na^+(aq) + 2PO_4^{3-}(aq) \longrightarrow$$
$$Ca_3(PO_4)_2(s) + 6Na^+(aq) + 6NO_3^-(aq)$$

An ion that appears on both sides of a complete ionic equation and does not directly participate in the reaction is called a **spectator ion.** For example, Na^+ and NO_3^- are spectator ions in this reaction. To write a more efficient equation for this reaction, you can cancel the spectator ions on each side of the equation:

$$3Ca^{2+}(aq) + \cancel{6NO_3^-(aq)} + \cancel{6Na^+(aq)} + 2PO_4^{3-}(aq) \longrightarrow$$
$$Ca_3(PO_4)_2(s) + \cancel{6Na^+(aq)} + \cancel{6NO_3^-(aq)}$$

Lesson 90 | **Precipitation Reactions**   459

white solid, calcium oxalate, CaC_2O_4. In this reaction, the precipitate is the calcium oxalate.

Precipitation is not limited to solids. Any substance of a contrasting phase can come out of a solution. Rain is called precipitation because it is liquid coming out of air, a gaseous solution. In this unit, the focus is on precipitation reactions that involve ionic solids in solution.

DISCUSS THE SOLUBILITY OF IONIC COMPOUNDS

→ *Note:* The solubility table on page 459 of the student text is for

convenience. It is *not* recommended that you require students to learn these rules.

Ask: What happens when you mix aqueous salts?

Ask: Do you see any patterns in the compounds that are soluble? Insoluble?

Ask: What appears to be true about the solubility of compounds that contain carbonate or hydroxide? (They are insoluble if the metal is not an alkali metal.)

Key Points: The degree to which a compound dissolves in water is called its *solubility.*

Some ionic salts dissolve to a greater extent in water than others. For example, sodium chloride, NaCl, is highly soluble in water. When a compound does not dissolve in water or dissolves only slightly, it is considered insoluble. Silver chloride, AgCl, is a compound that is considered insoluble. The formation of a precipitate from the mixing of two ionic solutions depends on the solubility of the products that form.

Solubility varies from compound to compound. However, you can make some generalizations. You might have noticed that all the nitrate compounds are aqueous. So one general solubility rule is that compounds containing nitrate, NO_3^-, are soluble in water. Notice that some hydroxides, such as NaOH and KOH, are soluble, while others are not, such as $Mg(OH)_2$, AgOH, and $Cu(OH)_2$. Another general rule is that hydroxides generally are insoluble, except for hydroxides with metal cations from Group 1A.

The rules for predicting which compounds will form solids in ionic solutions are not simple. Most chemists either memorize which compounds are insoluble (and thus form precipitates) or use solubility tables to help them make predictions. The solubility trends in the table on page 459 of the student text are for reactions at 25 °C.

DISCUSS THE RELATIONSHIP BETWEEN SOLUBILITY AND TOXICITY

Ask: How do you think the human body filters out solid toxins? Aqueous toxins?

Ask: What might be the danger of a precipitate or solid forming within the body?

Ask: What might be the danger of a substance being soluble in the human body?

Ask: Alkali metal compounds are not very toxic. In contrast, metal compounds of Ag, Hg, and Pb are much more toxic. How can you explain this, based on what you know about solubility?

Key Points: There are positive and negative aspects to the solubility of toxic substances. If a substance is not very soluble, it may pass through the body without reacting with anything. For example, some metals can be ingested and pass through our bodies without causing harm. However, many metal compounds are soluble, and once our bodies absorb them, they can do enormous long-term damage.

Lesson 90 | **Precipitation Reactions**   459

Once a substance is absorbed, it may be difficult for the body to get rid of it through its natural filtration systems, the kidneys and the liver. Substances that are not very soluble can cause problems. Solid substances can build up in the fine network of tubes and filters in the body, causing blockages.

EVALUATE (5 min)

Check-In

A solution of K_2SO_4 is mixed with a solution of $Pb(NO_3)_2$, and a precipitate forms.

1. Write the chemical equation for this reaction.

2. What is the precipitate that forms? How do you know?

Sample answers: 1. $K_2SO_4(aq)$ + $Pb(NO_3)_2 \rightarrow 2KNO(aq) + PbSO_4(s)$.
2. $PbSO_4$, lead sulfate. Potassium nitrate is soluble, because nitrates in general are soluble, and the lead sulfate is insoluble. Thus, the precipitate is the lead sulfate.

Homework

Assign the reading and exercises for Toxins Lesson 90 in the student text.

LESSON 90 ANSWERS

1. A precipitate is an insoluble compound that forms as a product of a chemical reaction in solution.

2. Possible answer: A precipitate forms when two soluble compounds react with one another in solution to produce a compound that is not soluble. The product then comes out of solution as a precipitate.

3. Possible answer: A spectator ion is an ion that is present in both the reactants and the products of a chemical reaction. Spectator ions do not participate in the reaction.

4. soluble: $LiNO_3$, KCl, $MgCl_2$, $RbOH$, Li_2CO_3

5. Possible answers: a. $CaCO_3$: $Ca(NO_3)_2$ and Li_2CO_3; **b.** $FePO_4$: $FeCl_3$ and Na_3PO_4; **c.** $HgCl_2$: $Hg(NO_3)_2$ and $NaCl$; **d.** $AgCl$: $AgNO_3$ and KCl

6. a. $2NaCl(aq) + Hg(NO_3)_2(aq) \rightarrow 2NaNO_3(aq) + HgCl_2(s)$

Doctors use a type of treatment called *chelation therapy* to treat patients who have had long-term exposure to metals like arsenic, mercury, or lead, usually in the workplace. In chelation therapy, a compound is introduced into the bloodstream of the patient. This compound bonds with heavy metals, forming water-soluble products that are then passed out of the body.

KEY TERMS

precipitate

precipitation reaction

complete ionic equation

spectator ion

net ionic equation

Exercises

The result is a **net ionic equation** that describes the reaction in terms of only the ions that are involved in the reaction:

$$3Ca^{2+}(aq) + 2PO_4^{3-}(aq) \longrightarrow Ca_3(PO_4)_2(s)$$

Toxicity of Precipitates

There are positive and negative aspects to the solubility of toxic substances. On the one hand, if a substance is not very soluble, it might not react with anything, and it might pass through the body relatively unnoticed. For example, some metals that can be ingested go through our bodies without causing harm. (Imagine a child swallowing a nickel.) However, many metal compounds are soluble, and once our bodies absorb them, some can do long-term damage.

LESSON SUMMARY

Which substances precipitate from aqueous solutions?

A precipitate is a solid that forms when a reaction in an aqueous solution produces a compound that is not very soluble. Most precipitation reactions are double exchange reactions involving ionic compounds. The cations exchange anions in solution, and one of the substances precipitates as a solid. Precipitation depends on the solubility of a compound. Some ionic compounds are not very soluble, so they precipitate from aqueous solutions. Precipitates in your body include bones, teeth, and kidney stones.

Reading Questions

1. Describe what a precipitate is and how it forms.
2. What does solubility have to do with precipitation from solution?
3. Explain what a spectator ion is.

Reason and Apply

4. Circle the ionic solids that are soluble in water.

$LiNO_3$	KCl	$MgCl_2$	$Ca(OH)_2$	$RbOH$
$CaCO_3$	Li_2CO_3	$PbCl_2$	$AgCl$	

5. Name solutions you can combine that will precipitate each of the compounds listed.
   a. $CaCO_3$ **b.** $FePO_4$ **c.** $HgCl_2$ **d.** $AgCl$

6. Write balanced chemical equations for these reactions. Be sure to indicate which substances are aqueous and which are solid.
   a. $NaCl(aq) + Hg(NO_3)_2(aq)$ **b.** $NaOH(aq) + CaCl_2(aq)$
   c. $K_2CO_3(aq) + LiNO_3(aq)$ **d.** $Zn(NO_3)_2(aq) + Na_3PO_4(aq)$
   e. $NaCl(aq) + Ca(NO_3)_2(aq)$

7. Refer to your answers for Exercise 6. For each equation that represents a precipitation reaction, write both complete and net ionic equations.

b. $2NaOH_3(aq) + CaCl_2(aq) \rightarrow 2NaCl(aq) + Ca(OH)_2(s)$

c. $K_2CO_3(aq) + 2LiNO_3(aq) \rightarrow 2KNO_3(aq) + Li_2CO_3(aq)$

d. $3Zn(NO_3)_2(aq) + 2Na_3PO_4(aq) \rightarrow Zn_3(PO_4)_2(s) + 6NaNO_3(aq)$

e. $2NaCl(aq) + 2Ca(NO_3)_2(aq) \rightarrow 2NaNO_3(aq) + CaCl_2(aq)$

7. a: $2Na^+(aq) + 2Cl^-(aq) + Hg^{2+}(aq) + 2NO_3^-(aq) \rightarrow 2Na^+(aq) + 2NO_3^-(aq) + HgCl_2(s)$; $2Cl^-(aq) + Hg^{2+}(aq) \rightarrow HgCl_2(s)$

b: $2Na^+(aq) + 2OH^-(aq) + Ca^{2+}(aq) + 2Cl^-(aq) \rightarrow 2Na^+(aq) + 2Cl^-(aq) + Ca(OH)_2(s)$; $Ca^{2+}(aq) + 2OH^-(aq) \rightarrow Ca(OH)_2(s)$

d: $3Zn^{2+}(aq) + 6NO_3^-(aq) + 6Na^+(aq) + 2PO_4^{3-}(aq) \rightarrow 6Na^+(aq) + 6NO_3^-(aq) + Zn_3(PO_4)_2(s)$; $3Zn^{2+}(aq) + 2PO_4^{3-}(aq) \rightarrow Zn_3(PO_4)_2(s)$

**SAFETY
Instructions**

Safety goggles must be
worn at all times.

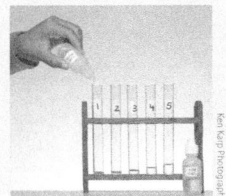

Mole Ratios

Purpose

To examine how the ratio of reactants affects the amount of products.

Materials

- 12 × 75 mm test tubes (10)
- test-tube rack
- marker (to label test tubes)
- 2 dropper bottles with 0.10 M solutions of calcium chloride, $CaCl_2$, and sodium carbonate, Na_2CO_3

Procedure

1. Number five small test tubes from 1 to 5.
2. Add drops of 0.10 M $CaCl_2(aq)$ and 0.10 M Na_2CO_3 to the test tubes as indicated in the table below.
3. Swirl each test tube gently to mix the reactants. Allow the solids to settle for about ten minutes while you complete the table and continue the analysis.

	Tube 1	Tube 2	Tube 3	Tube 4	Tube 5
Drops of 0.10 M $CaCl_2$	4	6	12	18	20
Drops of 0.10 M Na_2CO_3	20	18	12	6	4
Ratio of $CaCl_2$ to Na_2CO_3	1:5				

Observations and Analysis

1. Write the balanced chemical equation for the exchange reaction of aqueous calcium chloride with aqueous sodium carbonate. Predict the ratio of moles of $CaCl_2(aq)$ to $Na_2CO_3(aq)$ that will produce the maximum amount of $CaCO_3(s)$.
2. After the solids have settled, determine which tube has the most $CaCO_3(s)$. Identify the ratio in the table that resulted in the largest volume of solid.
3. **Making Sense** How do the balanced chemical equations compare to your observations?
4. **If You Finish Early** Suppose that you mix calcium chloride, $CaCl_2(aq)$, with sodium oleate, $NaC_{18}H_{33}O_2(aq)$. What products do you expect? What ratio of drops would give you the largest amount of precipitate? (*Note:* One product is calcium oleate, commonly referred to as soap scum.)

Overview

LAB: GROUPS OF 4

Key Question: How can you convert all the reactants to products?

KEY IDEAS

The coefficients in chemical equations represent the number ratio in which reactants combine and products form. To produce the maximum amount of product, reactants should be combined in the correct ratios.

LEARNING OBJECTIVES

- Define a mole ratio.
- Explain how to combine reactants to make the most product from a reaction.
- Identify a limiting reactant.

FOCUS ON UNDERSTANDING

Students might have the faulty impression that a chemical reaction will not proceed unless the reactants are mixed in the appropriate proportions. In fact, any amount of two reactants will result in some of the products. The mole ratio represents the amounts that ideally will convert all the reactants to products.

KEY TERMS

mole ratio
limiting reactant (limiting reagent)

IN CLASS

In this lesson, students complete a short lab and answer questions related to their observations. During the lab, students focus on two precipitation reactions, one forming solid calcium carbonate and the other forming solid copper hydroxide. They use varying ratios of reactants to arrive at the optimal combination—the ratio that produces the maximum amount of product. *Note:* Limiting reactants are covered in greater detail in Lesson 93: Get the Lead Out. A complete materials list is available for download.

TRM Materials List 91

Pre-AP® Course Tip

Students preparing to take an AP® chemistry course or college chemistry should be able to collect data and analyze it for patterns or relationships. In this lesson, students conduct a lab with various chemicals to determine the maximum yield that can be obtained from a precipitation reaction. Based on their data, students should be able to determine that the ratio of reactants combined is the same as in the balanced equation, which yielded the maximum amount of product.

SETUP

Before class, prepare 200 mL of each of the solutions listed. Each solution is 0.10 M. Divide each solution into eight labeled dropper bottles, approximately 25 mL per bottle. Note that copper (II) sulfate often comes in a hydrated form, as $CuSO_4 \cdot 5H_2O$, in which case 200 mL of a 0.10 M solution will require 5.00 g of solute.

$CaCl_2$ (2.22 g in 200 mL)
Na_2CO_3 (2.12 g in 200 mL)
$CuSO_4$ (3.20 g in 200 mL)
$NaOH$ (0.80 g in 200 mL)

SAFETY

Students must wear safety goggles at all times.

Sodium hydroxide is corrosive. Do not get any on skin or near eyes.

In case of spills, rinse with large amounts of water.

Have citric acid on hand for spills.

Lesson Guide

ENGAGE (5 min)

TRM PowerPoint Presentation 91

ChemCatalyst

One way to remove potentially toxic substances from a water source is to precipitate out the harmful ions. Consider the removal of copper ions by precipitation.

$2NaOH(aq) + Cu(NO_3)_2(aq) \rightarrow$
$2NaNO_3(aq) + Cu(OH)_2(s)$

Suppose you add 100 mL of 0.10 M NaOH(aq) to 100 mL of 0.10 M $Cu(NO_3)_2(aq)$.

1. Have you added equal numbers of moles of the two substances? Explain your thinking.

2. Have you added equal numbers of grams of the two substances? Explain your thinking.

3. Have you added enough NaOH(aq) to remove all the copper from the solution? Explain your thinking.

Sample answers: **1.** Yes. Equal numbers of moles have been added because the volume and concentrations are the same. **2.** No. An equal number of moles does not provide an equal number of grams. Each substance has a different molar mass, which means that a mole of each substance weighs a different amount. **3.** No. The coefficients of the balanced chemical equation indicate that you have to have 2 mol of NaOH for every mole of $Cu(NO_3)$.

→ Assist students in considering how amounts are represented in chemical equations.

Ask: What information is contained in the coefficients in a chemical equation?

Ask: What do you expect to observe as you add NaOH(aq) to a solution of $Cu(NO_3)_2(aq)$?

Ask: If you keep adding NaOH, is there a maximum amount of $Cu(OH)_2(s)$ that you can obtain? Explain your thinking.

EXPLORE (20 min)

TRM Worksheet with Answers 91

TRM Worksheet 91

Mole Ratios

Water treatment plants in our communities process millions of gallons of water a day. These plants remove toxic substances from the water through a series of chemical and physical treatments. For example, impure water often contains dissolved metal cations. These can be removed from water through precipitation of insoluble metal salts. Once the metals are in solid form, they can be safely separated by filtration. Of course, for maximum effectiveness, the treatment plant should remove as much of each metal from the water as possible.

A worker takes a sample to determine water quality at a treatment plant.

How can you convert all the reactants to products?

To answer this question, you will explore

⟳ Moles of Product

⟳ Excess and Limiting Reactant

EXPLORING THE TOPIC

⟳ Moles of Product

The list of metals that can show up in our water supplies is quite long, and it includes lead, silver, mercury, chromium, copper, zinc, cadmium, and tin. According to the solubility table in the previous lesson, many metals form hydroxides and carbonates that are not soluble. So a good way to remove the more toxic metal cations from our water supplies is to add hydroxides or alkali metal carbonates to precipitate hazardous metal cations.

Imagine a water sample that contains unwanted copper ions. The goal is to figure out how to remove as many of those ions per liter of solution as possible through precipitation. Sodium hydroxide, NaOH, is chosen as a reactant because it is soluble and contains a hydroxide ion. When hydroxide ions enter the solution,

INTRODUCE THE LAB

→ Tell students to work in groups of 4.

→ Pass out the worksheet to each group.

→ Go over the general procedure and safety for the lab.

→ Tell students to complete the procedures fairly quickly, because the solids that form will require 10–15 minutes to settle.

→ Tell students to complete the tables and answer the worksheet questions while waiting for the precipitates to settle.

EXPLAIN & ELABORATE (15 min)

RELATE CHEMICAL EQUATIONS TO THE AMOUNT OF PRODUCT

→ Ask student volunteers to write the balanced chemical equation for each procedure on the board.

$CaCl_2(aq) + Na_2CO_3(aq) \rightarrow$
$CaCO_3(s) + 2NaCl(aq)$

$CuSO_4(aq) + 2NaOH(aq) \rightarrow$
$Cu(OH)_2(s) + Na_2SO_4(aq)$

Ask: What ratios of reactants produce the greatest amounts of solid products for the two reactions you studied?

solid copper (II) hydroxide, $Cu(OH)_2$, will precipitate. The question remains: How much NaOH should be added, and in what concentration?

The next illustration shows what happens when a solution of 1.0 M $CuSO_4$ is mixed with a solution of 1.0 M NaOH in different ratios. Take a moment to examine the outcomes.

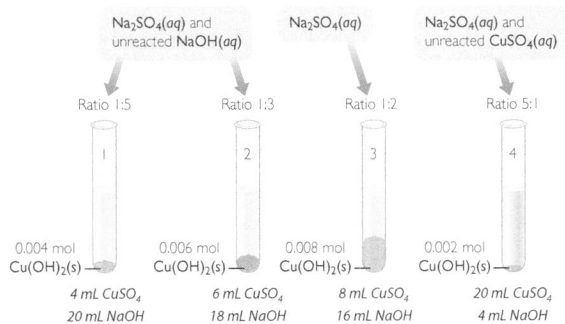

In each case, a total of 24 mL of solution are combined. However, the amounts of 1.0 M $CuSO_4$ and 1.0 M NaOH that are combined vary from test tube to test tube. The combination that produces the maximum amount of solid precipitate is in test tube 3, where 8 mL of 1.0 M $CuSO_4$ combined with 16 mL of 1.0 M NaOH. This results in 0.008 mol of solid copper hydroxide.

After the reaction, all four test tubes contain aqueous sodium sulfate, $Na_2SO_4(aq)$, and a blue solid, copper (II) hydroxide, $Cu(OH)_2(s)$. Therefore, sodium sulfate and copper (II) hydroxide must be products of this reaction. The solutions in three of the test tubes are almost colorless, while the solution in test tube 4 is pale blue. This means there is unreacted $CuSO_4(aq)$ left over in the last test tube.

Chemical equation: $CuSO_4(aq) + 2NaOH(aq) \longrightarrow Cu(OH)_2(s) + Na_2SO_4(aq)$

MOLE RATIO

The coefficients in a chemical equation indicate the proportions in which substances react, in moles or other counting units. The coefficients in this particular equation indicate that for every 1 mol of $CuSO_4$, 2 mol of NaOH are required to make the maximum amount of products. The test tube containing the most amount of solid (test tube 3) corresponds to this 1:2 ratio of reactants. This ratio is often referred to as the **mole ratio**. So, using the ratio given by the balanced chemical equation produces the maximum amount of product.

> **Big Idea** To get the maximum amount of product from a reaction, reactants must be mixed in the correct proportions.

Ask: What do the coefficients in a chemical equation indicate? What don't the coefficients indicate?

Key Points: The coefficients in chemical equations indicate how many counting units (such as moles) of each reactant combine and how many counting units of each product formed in a single reaction.

In the first chemical equation above, units of calcium chloride combine in a 1:1 ratio with units of sodium carbonate. The products, calcium carbonate and sodium chloride, form in a 1:2 ratio. So, 1 unit of calcium chloride reacts with 1 unit of sodium carbonate to form 1 unit of calcium carbonate and 2 units of sodium chloride. Coefficients can be replaced with any counting unit. So, 1 mol of calcium chloride combines with 1 mol of sodium carbonate to produce 1 mol of calcium carbonate and 2 mol of sodium chloride. In the second equation above, the reactants combine in a 1:2 ratio and form 1 unit of copper hydroxide and 2 units of sodium sulfate.

Coefficients in chemical equations always refer to the number of units (molecules, ions, moles) that are

combining. They never refer to the volume or mass of a substance. It is vital to keep this fact in mind when completing calculations.

INTRODUCE MOLE RATIOS

➔ Write the equation for the formation of ammonia from nitrogen and hydrogen on the board. Ask students for help in balancing the equation.

$$N_2(g) + 3H_2(g) \rightarrow 2NH_3(g)$$

Ask: Why is it important to balance chemical equations?

Ask: When ammonia is formed, how many molecules of hydrogen are required to make one molecule of ammonia?

Ask: Imagine that you have 1.5 mol of hydrogen gas. How many moles of nitrogen gas do you need to react all of the hydrogen? (0.5 mol)

Ask: What happens if you have 3 mol of hydrogen gas and 2 mol of nitrogen gas? Which reactant would run out? Which would be left over?

Key Points: The mole ratios are the proportions in which two substances (reactants, products, or both) combine or form. Consider the equation for the formation of ammonia. The mole ratio of nitrogen to hydrogen for this reaction is 1:3. In other words, 1 unit of nitrogen gas combines with 3 units of hydrogen gas in this particular reaction. The mole ratio of hydrogen to ammonia for this reaction is 3:2. For every 3 units of hydrogen used, 2 units of ammonia can be formed.

> Mole ratio: The ratio represented by the coefficients in a chemical equation showing how many units of each substance must combine to make the maximum amount of product.

When reactants are not combined in their exact mole ratios, one of the reactants runs out and the other is left over. The reactant that runs out is referred to as the *limiting reactant*, because it limits the amount of product that can be made. Once one reactant runs out, the other reactant becomes excess reactant. For example, in the ammonia example, if 2 mol of nitrogen and 3 mol of hydrogen are combined, there will be excess nitrogen. The hydrogen gas is the limiting reactant because there is not enough of it for the number of moles of nitrogen available. To use up all the nitrogen, 6 mol of hydrogen would be required.

EVALUATE (5 min)

Check-In

The reaction to form silver phosphate, $Ag_3PO_4(s)$, is given by this chemical equation:

$AgNO_3(aq) + Na_3PO_4(aq) \rightarrow Ag_3PO_4(s) + NaNO_3(aq)$

1. Balance the equation.

2. Which combination of reactants results in the maximum amount of product?

   a. 1.0 g $AgNO_3$ to 1.0 g Na_3PO_4

   b. 3.0 g $AgNO_3$ to 1.0 g Na_3PO_4

   c. 1.0 mol $AgNO_3$ to 1.0 mol Na_3PO_4

   d. 3.0 mol $AgNO_3$ to 1.0 mol Na_3PO_4

Sample answers: **1.** $3AgNO_3(aq) + Na_3PO_4(aq) \rightarrow Ag_3PO_4(s) + 3NaNO_3(aq)$. **2.** The ratio 3 $AgNO_3$ to 1 Na_3PO_4 is the mole ratio for this reaction, so the correct answer is **d**.

Homework

Assign the reading and exercises for Toxins Lesson 91 in the student text.

Example 1

Precipitation of Silver Hydroxide

An aqueous solution contains 0.020 mol of silver nitrate, $AgNO_3(aq)$. Predict the number of moles of $NaOH(aq)$ you will need to precipitate all of the silver, Ag, as silver hydroxide, AgOH.

Solution

First, write a balanced chemical equation for the reaction.

$$AgNO_3(aq) + NaOH(aq) \longrightarrow AgOH(s) + NaNO_3(aq)$$

The mole ratio of $AgNO_3$ to NaOH is 1:1.

The ratio given by the balanced equation is 1:1, and you have 0.020 mol of $AgNO_3$. So you will need 0.020 mol of NaOH to precipitate all of the silver.

⮎ Excess and Limiting Reactant

Look again at the four test tubes. Using the molarity and volume data for each reactant, you can determine the number of moles of each reactant. The table shows the moles of each reactant and the ratios in which they are combined.

Test tube	Moles of $CuSO_4$	Moles of NaOH	Ratio	Left over	Runs out
1	0.004 mol	0.020 mol	1:5	NaOH	$CuSO_4$
2	0.006 mol	0.018 mol	1:3	NaOH	$CuSO_4$
3	0.008 mol	0.016 mol	**1:2**	neither	neither
4	0.020 mol	0.004 mol	5:1	$CuSO_4$	NaOH

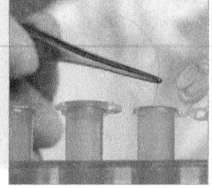

Notice that in test tubes 1, 2, and 4, one of the reactants runs out before the other. Only in test tube 3 have both reactants been used up completely. Test tubes 1 and 2 no longer have copper ions in solution, but the solution is now toxic due to the excess NaOH. Test tube 4 still has copper ions in solution.

If you mix the reactants in a ratio other than the mole ratio, a reaction will still occur, but one of the reactants will run out and the other one will have some left over. The reactant that runs out is called the **limiting reactant**, or **limiting reagent**. This is because it limits how much product you can make.

To remove all of the copper ions from the solution, it is best to mix the reactants in the mole ratio determined by the balanced chemical equation.

If you want to remove copper ions (or any other metal ions) from water in a water treatment plant, it would be best to add sodium hydroxide in the mole ratio specified by the chemical equation. To do so, you will first have to determine the concentration of the copper ions in your water in moles per liter so you can match that concentration with the appropriate concentration of reactant.

Example 2

Precipitation of Calcium Phosphate

Aqueous calcium chloride, $CaCl_2(aq)$, reacts with aqueous sodium phosphate, $Na_3PO_4(aq)$, forming a precipitate of calcium phosphate, $Ca_3(PO_4)_2(s)$.

a. If 12 mol of $CaCl_2(aq)$ react with 8 mol of $Na_3PO_4(aq)$, will there be any reactant left over? If so, which reactant?

b. If 12 mol of $CaCl_2(aq)$ react with 16 mol of $Na_3PO_4(aq)$, will there be any reactant left over? If so, which reactant?

c. If 12 mol of $CaCl_2(aq)$ react with 4 mol of $Na_3PO_4(aq)$, will there be any reactant left over? If so, which reactant?

Solution

First, write a balanced chemical equation for the reaction. Because atoms are conserved, $NaCl(aq)$ must be the other product.

$$3CaCl_2(aq) + 2Na_3PO_4(aq) \longrightarrow Ca_3(PO_4)_2(s) + 6NaCl(aq)$$

The mole ratio for $CaCl_2$ and Na_3PO_4 is 3:2. For any other ratio, there will be leftover reactant.

a. 12 mol $CaCl_2$ to 8 mol Na_3PO_4 is a ratio of 3:2. So all the reactants will be converted to products.

b. 12 mol $CaCl_2$ to 16 mol Na_3PO_4 is a ratio of 3:4. There will be some Na_3PO_4 left over.

c. 12 mol $CaCl_2$ to 4 mol Na_3PO_4 is a ratio of 4:1. There will be some $CaCl_2$ left over.

LESSON SUMMARY

How can you convert all the reactants to products?

KEY TERMS

mole ratio

limiting reactant

(limiting reagent)

Coefficients in chemical equations represent the proportions in which moles of reactants combine and products form. The ratio of two coefficients is also called the mole ratio. To produce the maximum amount of product, moles of reactants should be combined in the exact proportions specified by the coefficients. Mass and volume amounts cannot be substituted for coefficients. If there is not enough of a reactant, resulting in other reactants being left over, that reactant is called a limiting reactant.

Reading Questions

1. Why are coefficients important in chemical equations?

2. Explain how to create the maximum amount of product from a reaction.

LESSON 91 ANSWERS

1. *Possible answer:* Coefficients are important in chemical equations because they represent the proportion in which reactants combine to form products.

2. *Possible answer:* To create the maximum amount of product from a reaction, you must mix the reactants in the correct mole ratio as indicated by the chemical equation.

3. By balancing the equations, you can find the mole ratio by comparing the coefficients of the reactants.
a. $AgNO_3(aq) + NaCl(aq) \rightarrow AgCl(s) + NaNO_3(aq)$. The mole ratio of reactants is 1:1. **b.** $Cu(NO_3)_2(aq) + K_2CO_3(aq) \rightarrow CuCO_3(s) + 2KNO_3(aq)$. The mole ratio of reactants is 1:1. **c.** $ZnCl_2(aq) + 2NaOH(aq) \rightarrow Zn(OH)_2(s) + 2NaCl(aq)$. The mole ratio of reactants is 1:2. **d.** $CaCl_2(aq) + Na_2C_2O_4(aq) \rightarrow CaC_2O_4(s) + 2NaCl(aq)$. The mole ratio is 1:1.

4. A good lab report will contain ● a title (Lab: Mole Ratios) ● a statement of purpose (***Possible answer:*** To examine how the ratio of reactants affects the amount of products) ● a procedure (a summary of the steps followed in the experiment) ● results (a list of observations made during each step in the procedure) ● a conclusion (Student analysis should include a correlation between the chemical equation for each reaction and their numerical data from the experiment. ***Possible answer:*** A chemical formula can be used to determine the amounts of the products that will be formed from the reactants.)

5. a. 0.10 mol of NaCl is needed. **b.** 5.8 g

6. 200 mL

Exercises

(continued)

Reason and Apply

3. For each of the reactions listed, determine the mole ratio of reactants that produces the maximum amount of precipitate. Be sure to balance the equations.
   a. $AgNO_3(aq) + NaCl(aq) \longrightarrow AgCl(s) + NaNO_3(aq)$
   b. $Cu(NO_3)_2(aq) + K_2CO_3(aq) \longrightarrow CuCO_3(s) + KNO_3(aq)$
   c. $ZnCl_2(aq) + NaOH(aq) \longrightarrow Zn(OH)_2(s) + NaCl(aq)$
   d. $CaCl_2(aq) + Na_2C_2O_4(aq) \longrightarrow CaC_2O_4(s) + NaCl(aq)$

4. Lab Report Write a lab report for the experiment you did in which you found the mole ratio by precipitating solids.

(Title)

Purpose: (Explain what you were trying to find out.)

Procedure: (Describe the steps you followed.)

Observations: (Describe what you observed. Include a data table of your results.)

Analysis: (Write a balanced chemical equation for each reaction.)

Conclusion: (Explain what happened in the different test tubes.)

5. Aqueous silver nitrate reacts with aqueous sodium chloride producing a precipitate of silver chloride.

$$AgNO_3(aq) + NaCl(aq) \longrightarrow AgCl(s) + NaNO_3(aq)$$

   a. How many moles of NaCl do you need to react with 0.10 mol of $AgNO_3$?
   b. How many grams of NaCl does this represent?

6. Imagine that you have 500 mL of 0.20 M silver nitrate solution. How many milliliters of 0.50 M sodium chloride solution should you add to remove all the silver ions from the solution?

BIOLOGY CONNECTION

Calcium phosphate is similar to the solid that forms bones and teeth. The calcium cations and phosphate anions dissolved in your blood deposit as solids to form your skeleton.

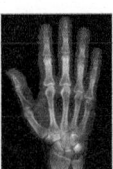

Ted Kinsman/Science Source

Mole Tunnel

Stoichiometry

THINK ABOUT IT

Sales of medications bring in billions of dollars per year. Most of these medications are produced through chemical reactions that are performed in a laboratory. In manufacturing the medications, it is important to know exactly how much reactant you need to prepare a specified mass of product. The balanced chemical equation for the reaction serves as a guide. However, the calculation requires conversion from mass to moles and back to mass.

> How do you convert between grams and moles to determine the mass of product?

To answer this question, you will explore
- Gram-Mole Conversions
- Solving Stoichiometry Problems

EXPLORING THE TOPIC

➔ Gram-Mole Conversions

One type of medication available is an antacid made of aluminum hydroxide, $Al(OH)_3$. It neutralizes excess acid in the stomach. To make aluminum hydroxide, many manufacturers start with aluminum chloride, which is commercially available in large quantities. When solutions of aluminum chloride are mixed with sodium hydroxide, a precipitate of aluminum hydroxide is formed. The balanced chemical equation for the reaction is

$$AlCl_3(aq) + 3NaOH(aq) \longrightarrow Al(OH)_3(s) + 3NaCl(aq)$$

The coefficients in the equation indicate how many moles to mix together. But in the laboratory, reactants are measured in grams, kilograms, or pounds.

GOING THROUGH THE MOLE TUNNEL

A chemical equation is like a recipe for making a product. It indicates how much of the reactants to mix. If you want to make aluminum hydroxide, $Al(OH)_3$, you need one formula unit of aluminum chloride, $AlCl_3$, for every three formula units of sodium hydroxide, NaOH. Before the recipe can be put into practice, it must be converted to mass. This is because in the laboratory, substances are weighed, not counted.

1 mol $AlCl_3(aq)$ + 3 mol $NaOH(aq)$ ⟶ 1 mol $Al(OH)_3(s)$ + 3 mol $NaCl(aq)$

__?__ g $AlCl_3(aq)$ + __?__ g $NaOH(aq)$ ⟶ __?__ g $Al(OH)_3(s)$ + __?__ g $NaCl(aq)$

CONSUMER CONNECTION

Aluminum chloride, $AlCl_3$, and other aluminum compounds are used in antiperspirants to control underarm wetness. When the aluminum ions enter the skin cells, they cause more water to pass into the cells. The cells swell and squeeze the sweat ducts closed, trapping the sweat inside the sweat glands.

Overview

CLASSWORK: PAIRS

Key Question: How do you convert between grams and moles to determine the mass of product?

KEY IDEAS

Stoichiometric calculations involve quantitative relationships of reactants and products in chemical reactions. To determine the mass of product made by a certain mass of reactant (and vice versa), it is necessary to convert mass to moles and then back again to mass. It is also necessary to account for the proportions in which substances combine and are formed.

LEARNING OBJECTIVE

Complete stoichiometric calculations for a variety of chemical reactions.

FOCUS ON UNDERSTANDING

Despite being told otherwise, students will continue to try to convert grams of reactant directly into grams of product. The *mole tunnel* helps by reinforcing that they must first go through moles to cross the arrow of a chemical equation.

IN CLASS

In this lesson, students complete stoichiometric problems involving gram-mole conversions. They work in pairs to complete a worksheet. They are introduced to strategies that will help them tackle these problems. A complete materials list is available for download.

TRM Materials List 92

Differentiate

You might use an ICE (Initial, Change, Eventual/Equilibrium) table to support student mathematical reasoning skills. ICE tables help students focus on how to use mole ratios to predict what will happen during a chemical reaction. Also, this scaffold can help students as they learn about limiting/excess reactants in Lesson 93 or chemical equilibrium in later chapters. There are many Web-based tutorials on how to use ICE tables. Try adding an ICE template to homework or assessment questions to prompt students to convert from mass to moles.

Pre-AP® Course Tip

Students preparing to take an AP® chemistry course or college chemistry should be able to use mathematics to solve problems that describe the physical world, such as stoichiometry. In this lesson, students use mathematics to complete stoichiometric calculations for various chemical reactions.

Lesson Guide

ENGAGE (5 min)

TRM PowerPoint Presentation 92

ChemCatalyst

This reaction produces the main substance found in human bones (calcium phosphate):

$3CaCl_2(aq)$ + $2Na_3PO_4(aq) \rightarrow$
calcium chloride   sodium phosphate

$Ca_3(PO_4)_2(s)$ + $6NaCl(aq)$
calcium phosphate   sodium chloride

1. How many moles of calcium phosphate can you make using 6 mol of calcium chloride, $CaCl_2$?

2. How many moles of calcium phosphate can you make using 111 g of $CaCl_2$?

Sample answers: **1.** 2 mol of calcium phosphate. **2.** One hundred eleven grams of calcium chloride is equivalent to 1 mol and will make 1/3 mol of calcium phosphate. Assuming that every 3 grams of calcium chloride makes 1 gram of calcium phosphate is incorrect reasoning that this lesson should remedy.

➜ Discuss how to complete the calculations in the ChemCatalyst.

Ask: How many moles of calcium phosphate can you make from 6 mol of calcium chloride? How do you know?

Ask: How did you determine how many moles of calcium phosphate you could make from 111 g of calcium chloride?

Ask: How are moles of $CaCl_2$ related to moles of calcium phosphate, $Ca_3(PO_4)_2$?

Ask: Are grams of $CaCl_2$ related to grams of calcium phosphate, $Ca_3(PO_4)_2$, in the same way?

EXPLORE (15 min)

TRM Worksheet with Answers 92

TRM Worksheet 92

INTRODUCE THE CLASSWORK
➜ Tell students to work in pairs.
➜ Pass out the worksheet to each group.
➜ Tell students they will be working with gram-mole conversions.
➜ Introduce the term *stoichiometry,* which refers to the quantitative relationship between reactants and products in a chemical reaction. The types of problems students will work on today are called stoichiometric calculations.

GUIDE THE CLASSWORK
➜ As students are working, check that they first convert gram amounts to moles. If they are not, ask them if their answers agree with the mole ratio for the reaction.
➜ You might stop students if they are having trouble and introduce the mole tunnel on page 468 of the student text before they go on to Part 2.

EXPLAIN & ELABORATE (15 min)

TRM Transparency—The Mole Tunnel 92

DISCUSS GRAM-MOLE CONVERSIONS
➜ Display the transparency The Mole Tunnel while students work on these

If you know the mass of reactants, you can determine the mass of products that can be made.

This requires a trip through the "mole tunnel." Starting with grams of reactant and the balanced chemical equation, there are three steps you must take to convert from grams of reactants to grams of products.

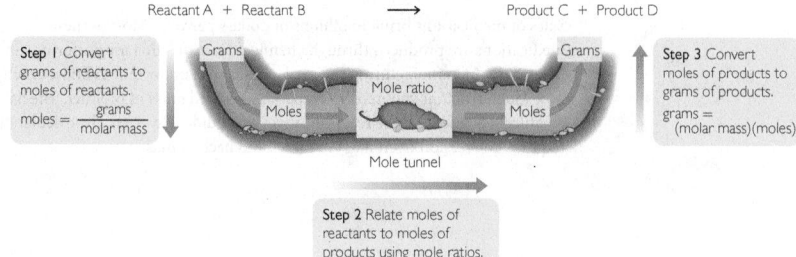

To go from one side of the equation to the other, you must go through moles, represented by the tunnel under the equation. The "mole tunnel" in the illustration is one way to remember these steps. These types of calculations, involving quantities of reactants and products, are referred to as gram-mole conversions, or **stoichiometry** calculations. The word *stoichiometry* comes from the Greek words *stoikheion,* meaning "element," and *metrein,* meaning "to measure."

The mole tunnel also works in reverse. If you want a certain mass of a product, you can determine the mass of reactants required.

↻ Solving Stoichiometry Problems

The chemical equation for the formation of aluminum hydroxide, $Al(OH)_3$, shows the reactants combining in a mole ratio of 1:3. The chemical equation also indicates that the mole ratio of the $AlCl_3$ reactant to the $Al(OH)_3$ product is 1:1. So, for every mole of $AlCl_3$ that reacts, one mole of $Al(OH)_3$ is produced. If you want to make 100 mol of $Al(OH)_3$, you will need 100 mol of $AlCl_3$ and 300 mol of NaOH.

The molar masses on the periodic table allow you to convert between grams and moles.

Molar Masses

Reactants: $AlCl_3$ = 133.5 g/mol   NaOH = 40.0 g/mol

Products: $Al(OH)_3$ = 78.0 g/mol   NaCl = 58.5 g/mol

Next, the mole amounts shown in the chemical equation can be converted to gram amounts. The table summarizes these calculations.

	Reactants		Products	
	$AlCl_3$	NaOH	$Al(OH)_3$	NaCl
Moles	1 mol	3 mol	1 mol	3 mol
Grams	133.5 g	120.0 g	78.0 g	175.5 g
Total mass	253.5 g		253.5 g	

HISTORY CONNECTION
Early black-and-white photographs, called *calotypes,* were made on paper prepared with silver nitrate. The paper was loaded into a camera under near-total darkness, then exposed to a scene for 10–60 seconds to make a photo negative. Prints could then be produced from the negative. An early photograph of President Abraham Lincoln is shown.

problems. This image is a duplicate of what is on page 468 of the student text.

Ask: What does the molar mass help you do in these calculations?

Ask: Why can't you simply convert grams of reactant into grams of product?

Ask: When is it useful to consider mole ratios?

Key Points: When working stoichiometric problems, grams must be converted to moles and then moles must be converted back to grams. Grams of reactant cannot be directly converted to grams of product or vice versa.

COMPLETE A WORKED EXAMPLE
➜ Work through this gram-mole conversion example on the board.

Ask: Imagine that you want to know how much product you can make with a certain mass of reactant. What is the first thing you should do?

Ask: Describe the steps you went through in calculating how many grams of calcium chloride were required to create 10 g of calcium phosphate.

EXAMPLE
How many grams of solid calcium phosphate can be made when 25 g of

Example 1

From Moles to Grams

Suppose you have 5.00 mol of $AlCl_3$ in solution and an unlimited supply of NaOH. How many grams of $Al(OH)_3$ and NaCl can you make?

$$AlCl_3(aq) + 3NaOH(aq) \longrightarrow Al(OH)_3(s) + 3NaCl(aq)$$

Solution

Use mole ratios to determine the moles of product you can make.

$$AlCl_3(aq) + 3NaOH(aq) \longrightarrow Al(OH)_3(s) + 3NaCl(aq)$$

1:1   1:3

So, with 5.00 mol of $AlCl_3$, you can make 5.00 mol of $Al(OH)_3$ and 15.0 mol of NaCl. Use molar mass to convert these quantities to grams.

Multiply the number of moles by the molar mass.   5.0 mol $Al(OH)_3 \cdot$ 75.0 g/mol = 375 g

15.0 mol NaCl $\cdot$ 58.5 g/mol = 878 g

If you start with 5.0 mol $AlCl_3$, you can react it with 15.0 mol NaOH to prepare 375 g $Al(OH)_3$ and 878 g NaCl.

Example 2

From Grams to Grams

Suppose you have 500 g of $AlCl_3$ in solution and an unlimited supply of NaOH. How many grams of each product can you make?

Solution

You cannot convert directly from grams of reactant to grams of product. You must use the "mole tunnel."

Use molar mass to convert from grams to moles of reactants.

Divide by the molar mass. $\dfrac{500 \text{ g AlCl}_3}{133.5 \text{ g/mol}}$ = 3.75 mol $AlCl_3$

Use mole ratios to determine the moles of each product.

So 3.75 mol $AlCl_3$ will make 3.75 mol $Al(OH)_3$ and 11.25 mol NaCl.

Use molar mass to convert between moles and grams.

Multiply the number of moles by the molar mass.   3.75 mol $Al(OH)_3 \cdot$ 75.0 g/mol = 281 g $Al(OH)_3$
11.25 mol NaCl $\cdot$ 58.5 g/mol = 658 g NaCl

So, 500 g $AlCl_3$ produces 281 g $Al(OH)_3$ and 658 g NaCl.

Lesson 92 | **Stoichiometry**   469

$$0.23 \text{ mol CaCl}_2 \cdot \dfrac{1 \text{ mol Ca}_3(PO_4)_2}{3 \text{ mol CaCl}_2}$$
$$= 0.077 \text{ mol Ca}_3(PO_4)_2$$

Step 3: Moles to grams. Convert moles of calcium phosphate to grams of calcium phosphate by multiplying the number of moles of calcium phosphate by its molar mass.

$$0.077 \text{ mol Ca}_3(PO_4)_2 \cdot \dfrac{310 \text{ g Ca}_3(PO_4)_2}{1 \text{ mol Ca}_3(PO_4)_2}$$
$$= 23.9 \text{ g Ca}_3(PO_4)_2$$

So, 24 g of calcium phosphate can be made with 25 g of calcium chloride.

EVALUATE (5 min)

Check-In

Consider this reaction:

$$Mg(s) + 2HCl(aq) \rightarrow MgCl_2(aq) + H_2(g)$$

How many grams of magnesium, Mg, are required to produce 190 g of magnesium chloride, $MgCl_2$?

Sample answer: 190 g $MgCl_2$ = 2 mol $MgCl_2$; the mole ratio is 1:1, so 2 mol of magnesium metal are needed. This is equal to 2 times the molar mass of magnesium or

(2 mol)(24.3 g/mol) = 48.6 g of magnesium.

Homework

Assign the reading and exercises for Toxins Lesson 92 in the student text.

---

dissolved calcium chloride are combined with excess aqueous sodium phosphate?

SOLUTION

Consult the chemical equation to find the mole ratio.

$$3CaCl_2(aq) + 2Na_3PO_4(aq) \rightarrow$$
$$Ca_3(PO_4)(s) + 6NaCl(aq)$$

The mole ratio of $CaCl_2$ to $Ca_3(PO_4)_2$ is 3:1.

Step 1: Grams to moles. Convert grams of calcium chloride to moles of

calcium chloride by dividing grams of calcium chloride by its molar mass.

$$\dfrac{25 \text{ g}}{111 \text{ g/mol}} = 0.23 \text{ mol CaCl}_2$$

Step 2: Moles to moles. Convert moles of calcium chloride to moles of calcium phosphate. Go through the mole tunnel. Using mole ratios, 3 mol of calcium chloride are used for every mole of calcium phosphate made. So one-third as many moles of calcium phosphate are made.

LESSON 92 ANSWERS

1. *Possible answer:* You have to convert between grams and moles to use mole ratios to find the right mass proportions of reactants.

2. a. 2:1 **b.** 1:1 **c.** The mole ratio of $AgNO_3$ to Cu is 2:1, so 12.0 mol of $AgNO_3$ is needed. The mole ratio of $Cu(NO_3)$ to Cu is 1:1, so 6.0 mol of $Cu(NO_3)$ are produced. **d.** 0.278 mol Ag, 0.139 mol of Cu is needed.

3. a. 12.6 g NaCl **b.** 19.8 g NaCl react to produce 47.0 g $PbCl_2$.

4. a. Each reactant gains an ion and loses an ion, so the reaction is a double-exchange reaction. **b.** mass of Na_2CO_3 = 9540 g, mass of $CaSO_4$ = 6130 g

How do you convert between grams and moles to determine the mass of product?

To determine the mass of product produced by a certain mass of reactant (and vice versa), it is necessary to convert mass to moles and then back to mass. This is because the balanced chemical equation is written in moles. Typically, when you're working with chemical reactions, you know the grams of reactant you need to produce a certain number of grams of product. Calculations involving mole ratios and masses of reactants and products are referred to as gram-mole conversions, or stoichiometry calculations.

KEY TERM

stoichiometry

Exercises

Reading Questions

1. Explain why you need to do gram-mole conversions when carrying out chemical reactions.

Reason and Apply

2. Consider this balanced chemical equation.

$$Cu(s) + 2AgNO_3(aq) \longrightarrow Cu(NO_3)_2(aq) + 2Ag(s)$$

copper   silver nitrate   copper (II) nitrate   silver

 a. Find the mole ratio of $AgNO_3$ to $Cu(NO_3)_2$.
 b. Find the mole ratio of $AgNO_3$ to Ag.
 c. Suppose you have 6.0 mol of Cu. How many moles of $AgNO_3$ are needed to react completely with 6.0 mol of Cu? How many mol of $Cu(NO_3)_2$ are produced?
 d. Suppose you want to make 30.0 g Ag. How many moles of Ag is that? How many moles of Cu do you need?

3. Consider this balanced chemical equation.

$$2NaCl(aq) + Pb(NO_3)_2(aq) \longrightarrow 2NaNO_3(aq) + PbCl_2(s)$$

sodium chloride   lead (II) nitrate   sodium nitrate   lead (II) chloride

 a. Suppose you want to make 30.0 g $PbCl_2$. How many grams of NaCl do you need?
 b. Suppose you have 56 g $Pb(NO_3)_2$. How many grams of NaCl are needed to react completely with 56 g $Pb(NO_3)_2$? How many grams of $PbCl_2$ are produced?

4. An aqueous solution of sodium carbonate reacts with an aqueous solution of calcium sulfate to produce solid calcium carbonate and an aqueous solution of sodium sulfate.

$$2Na_2CO_3(aq) + CaSO_4(aq) \longrightarrow Na_2SO_4(aq) + CaCO_3(s)$$

sodium carbonate   calcium sulfate   sodium sulfate   calcium carbonate

 a. What type of reaction is this?
 b. How many grams of each reactant are needed to produce 4500 g of $CaCO_3$?

Get the Lead Out

Limiting Reactant and Percent Yield

THINK ABOUT IT

Imagine you want to make as many cheese sandwiches as possible from a loaf of bread and a large package of sliced cheese. The package of cheese contains 15 slices. The loaf of bread contains 24 slices. Which ingredient will you run out of first, and how many sandwiches can you make?

> Which reactant determines how much product you can make?

To answer this question, you will explore

- Limiting Reactants
- Solving Limiting Reactant Problems
- Percent Yield

EXPLORING THE TOPIC

⟳ Limiting Reactants

Mixing reactants to form products is a little like making cheese sandwiches from specific quantities of cheese and bread. In the example above, you could make 12 sandwiches before running out of bread. You would have 3 slices of cheese left over. In the world of chemistry, substances rarely come together in the exact mole ratios specified by a chemical equation. One of the reactants usually runs out first. As you learned in Lesson 91: Mole to Mole, the reactant that runs out is called the limiting reactant.

Imagine that you have a beaker with an aqueous solution of sulfuric acid containing a total of 4 mol of H_2SO_4. You add 112 g of solid potassium hydroxide, KOH. Will the addition of this amount of KOH(s) completely neutralize the $H_2SO_4(aq)$? Will either reactant be left over?

The balanced chemical equation for the reaction you are examining is

$$H_2SO_4(aq) + 2KOH(s) \longrightarrow K_2SO_4(aq) + 2H_2O(aq)$$

You must first find out how many moles are represented by 112 g of KOH. The molar mass of KOH is 56 g/mol, so 112 g KOH is equal to 2 mol of KOH.

The chemical equation shows that you will need 2 mol of KOH to neutralize 1 mol of H_2SO_4. There are 4 mol of H_2SO_4 in the aqueous solution, and you have only 2 mol of KOH. You will be able to neutralize only 1 mol of H_2SO_4. There will be 3 mol of H_2SO_4 left over.

Lesson 93 | **Limiting Reactant and Percent Yield** 471

IN CLASS

Students consider how to determine if one reactant will run out before the other when the reactants are combined to make products. They observe a reaction between sodium bicarbonate and acetic acid. They then complete a worksheet, determining the identity of the limiting reactant for various reactions that involve the removal of toxic metals from solutions. Percent yield is covered in the student text. A complete materials list is available for download.

TRM Materials List 93

Differentiate

Limiting reactants are not intuitive to all students. You might direct students to the University of Colorado Boulder's PhET simulation titled Reactants, Products, and Leftovers. You can also engage students in scenarios related to an everyday experience such as making sandwiches or salads or building a structure out of building blocks. For advanced learners or to help students build connections to other chemistry content, develop question sets that involve stoichiometry and solutions, titrations, or gases. You can find a list of URLs for this lesson on the Chapter 17 Web Resources document.

TRM Chapter 17 Web Resources

Pre-AP® Course Tip

Students preparing to take an AP® chemistry course or college chemistry should be able to use mathematics to solve problems that describe the physical world, such as stoichiometry. In this lesson, students use mathematics to complete stoichiometric calculations involving limiting reactants and calculate percent yield.

SETUP

Label three Erlenmeyer flasks with the numbers 1, 2, and 3. Pour 50 mL of 1.0 M acetic acid into Flask 1 and Flask 2. Pour 100 mL of 1.0 M acetic acid into Flask 3. Fill one balloon with 4.2 g sodium bicarbonate (0.050 mol). Fit this balloon on the mouth of Flask 1. Be careful not to spill the sodium bicarbonate into the acetic acid. Fill two more balloons with 8.4 g sodium bicarbonate (0.10 mol) each and fit these balloons on Flasks 2 and 3. It is important that your measurements of the sodium bicarbonate and the acetic acid be accurate. You will have to refill the balloons and flasks for each class.

Overview

DEMO: WHOLE CLASS

CLASSWORK: PAIRS

Key Question: Which reactant determines how much product is made?

KEY IDEAS

When reactants are combined to complete a chemical reaction, they are not always combined in the exact ratios specified by the chemical equation. In such cases, one reactant runs out first, leaving excess of the other reactant(s). The limiting reactant determines the theoretical yield, how much product

can be made. In reality, the yield of a reaction usually is less than the theoretical yield. Chemists measure actual yield and calculate percent yield to evaluate the efficiency of processes.

LEARNING OBJECTIVES

- Complete stoichiometric calculations involving limiting reactant.
- Calculate percent yield when the actual yield is known.

KEY TERMS

theoretical yield
actual yield
percent yield

CLEANUP

Dilute the solutions and dispose of according to local guidelines.

Lesson Guide

ENGAGE (5 min)

TRM Transparency—ChemCatalyst 93

TRM PowerPoint Presentation 93

ChemCatalyst

Sodium bicarbonate, $NaHCO_3$, and acetic acid, CH_3COOH, react to generate CO_2 gas:

$CH_3COOH(aq) + NaHCO_3(s) \rightarrow$
$NaCH_3COO(aq) + H_2O(l) + CO_2(g)$

You set up the reactions as shown. The gas generated by the reaction will inflate the balloon when the sodium bicarbonate is poured into the acetic acid.

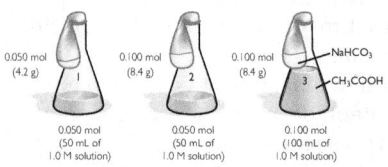

0.050 mol (4.2 g)	0.100 mol (8.4 g)	0.100 mol (8.4 g) — NaHCO₃ ... CH₃COOH
0.050 mol (50 mL of 1.0 M solution)	0.050 mol (50 mL of 1.0 M solution)	0.100 mol (100 mL of 1.0 M solution)

Predict the relative order of the balloon sizes. Justify your choice.

a. 1 = 2 = 3 **b.** 1 < 2 < 3 **c.** 1 < 2 = 3
d. 1 = 2 < 3

Sample answer: **d.** Reaction 1 produces 0.050 mol CO_2. Reaction 2 produces 0.050 mol CO_2. Reaction 3 produces 0.100 mol CO_2.

→ Assist students in figuring out what happens when different numbers of moles of reactants are combined. You can point out that sodium bicarbonate is baking soda and acetic acid is the acid in vinegar. This reaction resembles the one that makes bubbles in the mix for bread and baked goods.

Ask: How can you figure out how many moles of CO_2 are produced?

Ask: Is there a limiting reactant in any of the reactions? Explain your thinking.

Ask: What sizes do you expect for the balloons? (Remind students that the volume of any gas at STP is 22.4 L/mol, so the volume of each balloon is proportional to the number of moles of gas it contains.)

The illustration represents what happens in your beaker. Each unit represents 1 mol.

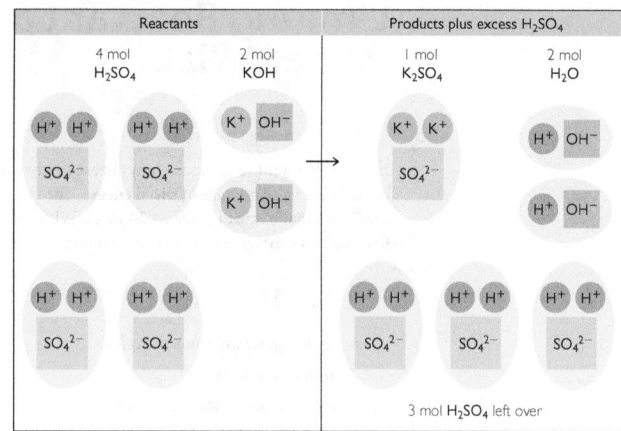

3 mol H_2SO_4 left over

There is not enough KOH to react with all of the H_2SO_4. KOH is the limiting reactant, and at the end of the reaction, there will be excess H_2SO_4.

⟳ Solving Limiting Reactant Problems

Knowing the identity of the limiting reactant allows you to figure out the maximum amount of product you can make with the substances you have. Mole ratios from the balanced chemical equation are the key to solving limiting reactant problems.

> **Big Idea** The limiting reactant determines how much product you can make from a chemical reaction.

The steps to complete a limiting reactant problem are shown here.

Solving a Limiting Reactant Problem

Step 1: Write the balanced chemical equation.

Step 2: Determine the molar mass of each compound.

Step 3: Determine the number of moles of each reactant that you have.

Step 4: Use the mole ratio to identify the limiting reactant.

Step 5: Use the limiting reactant to determine the maximum amount of product.

EXPLORE (15 min)

TRM Worksheet with Answers 93

TRM Worksheet 93

PERFORM THE DEMONSTRATION

→ Demonstrate the ChemCatalyst. Pour the sodium bicarbonate into the acetic acid, one flask at a time. The balloons will inflate. The balloons on Flask 1 and Flask 2 will be similar in size. The balloon on Flask 3 will be larger.

INTRODUCE THE CLASSWORK

→ Tell students to work in pairs.

→ Pass out the student worksheet to each group.

→ Tell students that today's activity involves heavy-metal compounds that can get into our water supply. These toxic substances often are precipitated out of solution to make our water safe to drink.

EXPLAIN & ELABORATE (15 min)

REVIEW THE CONCEPT OF LIMITING REACTANTS

Ask: What determines how much product will be made when substances react?

SOLUTION

Example 1

Preparing Magnesium Chloride

Adding hydrochloric acid, HCl, to magnesium hydroxide, $Mg(OH)_2$, produces magnesium chloride, $MgCl_2$, and water. Imagine you start with 10.0 g of $Mg(OH)_2$ and 100 mL of 4.0 M HCl. How much $MgCl_2$ will be produced?

Solution

Step 1: Write the balanced chemical equation.

$$Mg(OH)_2(s) + 2HCl(aq) \longrightarrow MgCl_2(aq) + 2H_2O(l)$$

Step 2: Determine the molar mass of each compound.

Reactants: $Mg(OH)_2$ = 58.3 g/mol   HCl = 36.5 g/mol
Products: $MgCl_2$ = 95.3 g/mol H_2O = 18.0 g/mol

Step 3: Determine the number of moles of each reactant that you have.

$$\text{Moles of } Mg(OH)_2 = \frac{10.0 \text{ g}}{58.3 \text{ g/mol}} = 0.17 \text{ mol}$$

$$\text{Moles of HCl} = (0.100 \text{ L})(4.0 \text{ mol/L}) = 0.40 \text{ mol}$$

Step 4: Use the mole ratio to identify the limiting reactant.

The reactants combine in a 1:2 ratio. So you need 0.17 mol · 2 = 0.34 mol of hydrochloric acid, HCl, to react with the 0.17 mol of magnesium hydroxide, $Mg(OH)_2$.

You have 0.40 mol of HCl, so the limiting reactant is $Mg(OH)_2$. When the reaction is complete, there will be 0.06 mol of HCl left over.

Step 5: Use the limiting reactant to determine the maximum amount of product.

For each mole of $Mg(OH)_2$, 1 mol of $MgCl_2$ is produced. So, 0.17 mol of $MgCl_2$ is produced. Use the molar mass to determine the mass of $MgCl_2$.

$$(0.17 \text{ mol}) (95.3 \text{ g/mol}) = 16.2 \text{ g } MgCl_2$$

So starting with 10.0 g $Mg(OH)_2$ and 100 mL of 4.0 M HCl, you can make 16.2 g $MgCl_2$.

> **Important to Know** The reactant present in the smallest amount is not necessarily the limiting reactant. The mole ratio specified by the coefficients of the balanced equation must be taken into consideration.

⟳ Percent Yield

When you use the limiting reactant to calculate the amount of a product produced in a chemical reaction, you are calculating the **theoretical yield** for that product or what you should be able to produce in theory. For a variety of reasons, including experimental error, reactions rarely produce the predicted theoretical yield when run in the laboratory. The amount of a product produced when the reaction is run is called the **actual yield** for that product and is usually a bit less than the theoretical yield. When you compare the actual yield with the theoretical yield and express it as a percentage, it is called the **percent yield.**

$$\text{Percent yield} = \frac{\text{actual yield}}{\text{theoretical yield}} \cdot 100\%$$

SOLUTION

Step 1: Write the balanced chemical equation.

$$CaCl_2(aq) + Na_2CO_3(aq) \rightarrow$$
$$CaCO_3(s) + 2NaCl(aq)$$

Step 2: Find the molar mass of the reactants.

$CaCl_2$: 111.0 g/mol
Na_2CO_3: 106.0 g/mol

Step 3: Determine the number of moles of each reactant.

$$\text{Moles } CaCl_2 = \frac{13.3 \text{ g}}{111.0 \text{ g/mol}} = 0.12 \text{ mol}$$

$$\text{Moles } Na_2CO_3 = \frac{9.84 \text{ g}}{106.0 \text{ g/mol}}$$
$$= 0.093 \text{ mol}$$

Step 4: Check the mole ratio to see how the two reactants combine. They combine in a 1:1 ratio.

Step 5: Determine the limiting reactant. At 0.093 mol, the sodium carbonate is the limiting reactant.

Step 6: Determine the maximum amount of product in grams.

With these starting products, you should be able to make 0.093 mol of calcium carbonate and 0.186 mol of sodium chloride. (These are the theoretical yields. Actual yield and percent yield are covered in the student text.)

$CaCO_3$: 100.1 g/mol   NaCl: 58.5 g/mol
grams $CaCO_3$ = 0.093 mol · 100.1 g/mol
      = 9.3 g
grams NaCl = 0.186 mol · 58.5 g/mol
      = 10.88 g

EVALUATE (5 min)

Check-In

Consider this reaction:

$$N_2(g) + 3H_2(g) \rightarrow 2NH_3(g)$$

1. If you mix 28.0 g of nitrogen gas, N_2, and 12.0 g of hydrogen gas, H_2, which reactant is the limiting reactant? Show your work.

2. How much ammonia can be made from these amounts of reactants?

Ask: How can you determine which reactant is a limiting reactant?

Key Points: In the real world, substances are rarely mixed in the exact mole ratios specified by a chemical equation. If reactants are not present in the exact mole ratio, one reactant will get used up, limiting the amount of product that can be formed. There will be an excess of the other reactant.

To identify the limiting reactant, the number of moles of the reactants on hand must be compared with the mole ratio of the reactants. If enough moles of one reactant are available to react with all the moles of the other reactant, there is no limiting reactant. However, if there is less of one reactant than required, it is the limiting reactant. The amount of product is determined by the limiting reactant.

COMPLETE A WORKED EXAMPLE

→ Ask students to assist with each step of the example problem.

EXAMPLE

How many grams of $CaCO_3$ can be made with 13.3 g of $CaCl_2$ and 9.84 g of Na_2CO_3 in solution?

Sample answers: 1. The molar mass of nitrogen gas is 28.0 g/mol. The molar mass of hydrogen gas is 2.0 g/mol. So you have 1 mol of N_2 and 6 mol of H_2. The mole ratio of the reactants is 1:1. Therefore, nitrogen is the limiting reactant. **2.** Based on the mole ratio, 1 mol of nitrogen will make 2 mol of ammonia, NH_3.

Homework

Assign the reading and exercises for Toxins Lesson 93 and the Chapter 17 Summary in the student text.

The percent yield tells you how successful or efficient your procedure was in the laboratory. The closer your percent yield is to 100%, the more efficient your procedure was.

Example 2

Percent Yield of $MgCl_2$

Suppose that you ran the reaction from Example 1 and 15.4 g of $MgCl_2$ are produced. What is the percent yield of your reaction?

Solution

Calculate the percent yield for the reaction you ran. The theoretical yield was 16.2 g $MgCl_2$.

$$Percent\ yield = \frac{actual\ yield}{theoretical\ yield} \cdot 100\%$$
$$= \frac{15.4\ g}{16.2\ g} \cdot 100\%$$
$$= 95.1\%$$

Your percent yield was very close to the theoretical yield.

LESSON SUMMARY

Which reactant determines how much product you can make?

KEY TERMS

theoretical yield

actual yield

percent yield

In the real world, substances rarely come together in the exact mole ratios specified by a chemical equation. Usually, there is more of one reactant than is required and less of the other. If reactants are not present in the exact mole ratio, one reactant, the limiting reactant, will get used up, therefore limiting the amount of product that can be produced. There will be excess of the other reactant. The percent yield of product is the actual amount of product produced in a chemical reaction as compared with the theoretical yield expressed as a percent. The theoretical yield of a chemical reaction is determined by the limiting reagent and the molar ratios in the chemical equation.

Exercises

Reading Questions

1. How can you determine how much product you can make from two compounds?
2. What is percent yield?

Exercises

(continued)

Reason and Apply

3. Consider the reaction to remove mercury from a water source through precipitation:

$$Hg(NO_3)_2(aq) + 2NaCl(aq) \longrightarrow HgCl_2(s) + 2NaNO_3(aq)$$

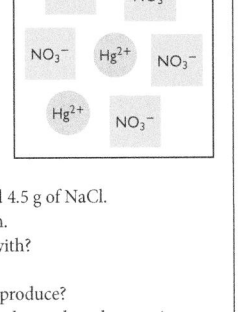

a. On your paper, draw a diagram that shows the correct amount of sodium chloride needed to react with the amount of mercury (II) nitrate represented in the box on the right.

b. What else is present in the beaker at the end of the reaction, besides the solid $HgCl_2$?

c. Why aren't there any mercury ions left over at the end?

4. Suppose you were trying to remove 50.3 g of mercury (II) nitrate, $Hg(NO_3)_2$, from a water supply using the reaction described in Exercise 3. How many grams of sodium chloride would you need to add?

5. Silver nitrate, $AgNO_3$ reacts with sodium chloride, NaCl, in aqueous solution to form solid silver chloride, $AgCl(s)$, and aqueous sodium nitrate, $NaNO_3(aq)$. Suppose you start with 6.3 g of $AgNO_3$ and 4.5 g of NaCl.

a. Write a balanced chemical equation for this reaction.

b. How many moles of each reactant are you starting with?

c. What is the limiting reactant?

d. How many grams of each product do you expect to produce?

e. How many grams of excess reactant do you expect to have when the reaction is complete?

6. In the laboratory, you run a procedure for the reaction described in Exercise 5 and produce 2.9 g of silver chloride. What is the percent yield for your procedure?

INDUSTRY CONNECTION

Pharmaceutical companies can calculate percent yield for each chemical reaction involved in creating a medicine. A low percent yield can be an indication of possible problems with the manufacturing process. Correcting these problems helps the manufacturer to make a safer, better product and to make the most of it at the lowest cost of raw materials.

LESSON 93 ANSWERS

1. *Possible answer:* Determine the number of moles of each reactant. Compare the number of moles of each compound to the chemical equation to determine the limiting reactant. Use the number of moles of the limiting reactant to determine the amount of product that can be made, assuming all of the limiting reactant is used up in the reaction.

2. The percent yield is the actual amount of product produced divided by the theoretical amount expected from the chemical equation.

3. a.

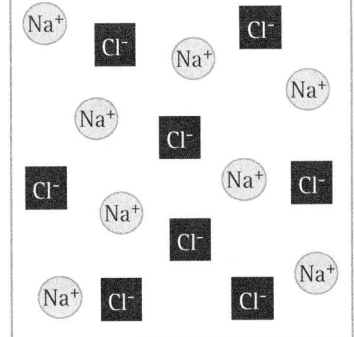

b. A solution of sodium nitrate, $NaNO_3$
c. All the mercury ions react with chloride ions to form solid mercury chloride, which precipitates out of solution.

4. 18.1 g

5. a. $AgNO_3(aq) + NaCl(aq) \rightarrow AgCl(s) + NaNO_3(aq)$ **b.** moles of $AgNO_3$ = 0.037 mol, moles of NaCl = 0.077 mol **c.** Silver nitrate, $AgNO_3$, is the limiting reactant. **d.** mass of AgCl = 5.3 g, mass of $NaNO_3$ = 3.1 g **e.** 2.3 g

6. 55%

ANSWERS TO CHAPTER 17 REVIEW EXERCISES

1. *Possible answer:*

$2NaOH(aq) + NiCl_2(aq) \rightarrow$
$2NaCl(aq) + Ni(OH)_2(s)$

$2Na^+(aq) + 2OH^-(aq) + Ni^{2+}(aq) +$
$2Cl^-(aq) \rightarrow 2Na^+(aq) + 2Cl^-(aq) +$
$Ni(OH)_2(s)$

$2OH^-(aq) + Ni^{2+}(aq) \rightarrow Ni(OH)_2(s)$

$K_2CO_3(aq) + NiSO_4(aq) \rightarrow$
$K_2SO_4(aq) + NiCO_3(s)$

$2K^+(aq) + CO_3^{2+}(aq) + Ni^{2+}(aq) +$
$SO_4^{2+}(aq) \rightarrow 2K^+(aq) + SO_4^{2+}(aq) +$
$NiCO_3(s)$

$CO_3^{2+}(aq) + Ni^{2+}(aq) \rightarrow NiCO_3(s)$

$2Li_3PO_4(aq) + 3NiCl_2(aq) \rightarrow$
$6LiCl(aq) + Ni_3(PO_4)_2(s)$

$6Li^+(aq) + 2PO_4^{3+}(aq) + 3Ni^{2+}(aq) +$
$6Cl^-(aq) \rightarrow 6Li^+(aq) + 6Cl^-(aq) +$
$Ni_3(PO_4)_2(s)$

$2PO_4^{3+}(aq) + 3Ni^{2+}(aq) \rightarrow Ni_3(PO_4)_2(s)$

2. a. $3CaCl_2(aq) + 2Na_3PO_4(aq) \rightarrow$
$Ca_3(PO_4)_2(s) + 6NaCl(aq)$ **b.** 9.31 g
c. 3.55 g **d.** 88%

3. a.

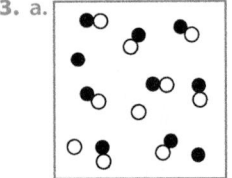

slightly soluble
compound

b. The resulting mixture would be heterogeneous because the undissolved units of the compound would be visible.

Project: Removing Toxins

A good report would include ● a general description of the toxic substance, including its name, chemical formula, and general uses ● information from the EPA on the toxic substance, including a Maximum Contaminant Level, if applicable ● a procedure to remove the toxic substance from a solution in water, including relevant chemical equations ● a procedure to remove the toxic substance from soil, including relevant chemical equations.

CHAPTER 17

Toxic Cleanup

SUMMARY

KEY TERMS

precipitate

precipitation reaction

complete ionic equation

spectator ion

net ionic equation

mole ratio

limiting reactant (limiting reagent)

stoichiometry

theoretical yield

actual yield

percent yield

Toxins Update

One way to remove toxic substances from water supplies is to add the appropriate ions to create a solid precipitate that can be filtered out. In other situations, the appropriate ions can be added to dissolve a solid into solution. A solubility table can help identify which ionic compounds can help remove a toxic ion from solution.

How do you know how much reactant you need? Sometimes, it is necessary to translate between moles and mass. The relationship between mass and moles is a proportional one, connected by molar mass. When comparing quantities of reactants and products, it is necessary to consider the mole ratios.

REVIEW EXERCISES

1. Write a complete ionic equation and a net ionic equation for three precipitation reactions that involve the formation of solid nickel compounds.

2. Calcium chloride reacts with sodium phosphate to produce calcium phosphate and sodium chloride. The unbalanced chemical equation is

   $\underline{\quad} CaCl_2(aq) + \underline{\quad} Na_3PO_4(aq) \longrightarrow$
   $\underline{\quad} Ca_3(PO_4)_2(s) + \underline{\quad} NaCl(aq)$

   a. Balance the chemical equation for this reaction.
   b. If you start with 13.5 g of $CaCl_2$ and 9.80 g of Na_3PO_4, how many grams of calcium phosphate can you expect to produce?
   c. How many grams of excess reactant will be left over?
   d. In the laboratory, you run a procedure for this reaction, and you produce 8.20 g of calcium phosphate. What is the percent yield for your procedure?

3. Suppose that you mix a slightly soluble compound with water.
   a. Draw a particle view diagram of the resulting mixture.
   b. Would the resulting mixture be homogeneous or heterogeneous? Explain.

Removing Toxins

PROJECT Choose a toxin (your teacher may assign you one). Find some information about your toxin. Your report should cover these topics.

- According to the Environmental Protection Agency, what is the maximum contaminant level for your toxin in the drinking water supply? How is it reported?
- If your toxin dissolves in water, how would you remove it from water?
- If it does not dissolve in water, how would you remove it from soil?

ASSESSMENTS

Two multiple-choice assessments (versions A and B) and two short-answer assessments (versions A and B) are available for download for Chapter 17.

TRM Chapter 17 Assessment Answers

TRM Chapter 17 Assessments

Toxins | REVIEW

UNIT 4

© Tony Phelps/naturepl.com

Chemical changes make life possible on this planet. Through chemical reactions, our bodies process the food we eat, make new cells and tissue, and eliminate waste products. Chemical equations allow us to track these changes on an atomic level.

However, some substances cause chemical changes that affect your body in negative ways, causing damage or harm. Acids and bases can react with your skin, causing burns and destroying tissue. Metal salts can interfere with ionic substances in the bloodstream, upsetting the natural balance. Precipitates can form in your body and block passages—examples are cholesterol in the bloodstream and kidney stones in the urinary tract.

But chemical reactions can be used to address toxic reactions. Acids and bases can be dealt with through neutralization reactions or by dilution with water. Toxic substances can often be safely removed from your body through precipitation reactions. Other times, the best approach is to dissolve solid toxic substances and then remove them from the body in liquid form. These same methods also have other applications, such as purifying the drinking water that comes to your home.

Every substance on the planet has the potential to be either toxic or therapeutic. The outcome depends on the dosage or amount and the way it is administered. Health officials test substances and set limits on exposure to all the different compounds that we might encounter in our tap water, medications, atmosphere, food, or the household products we buy.

REVIEW EXERCISES

General Review

Write a brief and clear answer to each question. Be sure to show your work.

1. Balance the following chemical equations:
   a. $Zn(s) + HCl(aq) \longrightarrow ZnCl_2(aq) + H_2(g)$
   b. $KClO_3(s) \longrightarrow KCl(s) + O_2(g)$
   c. $S_8(s) + F_2(g) \longrightarrow SF_6(g)$
   d. $Fe(s) + O_2(g) \longrightarrow Fe_2O_3(s)$

2. For each of the four equations in Exercise 1, name the type of reaction it represents.

3. Using the same equations you balanced in Exercise 1, suppose that you start with 1 g of each reactant. Identify the limiting reactant and determine the mass of each product that can be made.

4. Write a balanced chemical equation for these reactions:
   a. Calcium metal reacts with water to form calcium hydroxide and hydrogen gas.
   b. A solution of hydrochloric acid reacts with solid calcium bicarbonate to produce water, carbon dioxide, and calcium chloride.

5. What is a neutralization reaction?

6. Write out the complete and net ionic equations for reactions between these compounds in solution:
   a. magnesium nitrate and calcium chloride
   b. silver nitrate and potassium iodide
   c. sodium sulfate and barium chloride

7. Does a precipitate form in any of the reactions from Exercise 6? Circle them.

Unit 4 | **Review**   477

Overview

CLASSWORK: PAIRS

Key Question: What is the best way to deal with a toxic substance?

KEY IDEAS

Students have learned about many types of reactions. Depending on the toxic substance and the specific situation, different approaches may work best in dealing with a toxic substance.

LEARNING OBJECTIVE

Create a comprehensive list of topics to use as a study aid for a unit exam.

FOCUS ON UNDERSTANDING

You might inform students that Poison Control is not a place to get research information on individual toxic substances. Its job is to save lives and help those in need.

IN CLASS

Students complete a worksheet to review the unit in preparation for the unit exam. The topics covered are summarized and the sample problems completed during the discussion portion of the class. If you assigned the Toxins projects, you might have students report on their assigned toxic substances. Ask them to describe the

approach that will help clean up their toxic substances. A complete materials list is available for download.

TRM Materials List R4

GUIDE TO UNIT REVIEW

ENGAGE (5 min)

ChemCatalyst

Imagine that your little brother has spilled a toxic substance on his skin. You decide to call Poison Control right away.

1. What are some things you want to know about the substance?

2. What kinds of questions do you think Poison Control will ask you? What might you be told to do? Explain your thinking.

Sample answers: 1. Students may want to know the danger level and how to treat the victim. They may say they would like to know the LD_{50}. 2. Poison Control will want to know things such as the age of the victim, the extent of the spill, the identity of the substance, and any symptoms. If the substance came from a bottle, the label will identify key ingredients and provide instructions on what to do in case of contact. Dilution with lots of water is one standard approach.

→ Discuss what to do if you encounter a toxic substance.

Ask: Where could you get the information Poison Control might want?

Ask: How would it help to know if the substance was an acid or a base?

Ask: What might Poison Control ask you to do for your brother?

EXPLORE (15 min)

TRM Worksheet with Answers R4

TRM Worksheet R4

INTRODUCE THE REVIEW

→ Tell students to work in pairs.
→ Pass out the worksheet to each group.

EXPLAIN & ELABORATE (15 min)

DISCUSS THE WORKSHEET (OPTIONAL)

→ Go over molarity and gram-mole conversions as a class.

REVIEW THE KEY CHEMISTRY IDEAS

Make a list of the key chemistry content areas on the board.

Ask: What chemistry concepts are covered in this unit?

Ask: What does a chemical equation keep track of?

Ask: Explain how matter is conserved in a chemical equation.

Ask: What is Avogadro's number? How many atoms are in 2 mol of helium?

Ask: What is an example of a combination reaction?

Ask: What type of reaction are most precipitation reactions?

Ask: What does molarity describe?

Ask: How can you determine the concentration of a solution? What information do you have to have?

Ask: What is pH?

Ask: What is scientific notation? How does it work?

Ask: What is a limiting reactant?

Ask: What makes a substance harmful to living things? What is a lethal dose?

EVALUATE (5 min)

Check-In

1. Balance this equation and fill in the phases for the products:

   $Ba(NO_3)_2(aq) + Na_2SO_4(aq) \rightarrow BaSO_4 + NaNO_3$

2. What solid is formed?

3. How many grams of barium nitrate must be in the solution to make 3 mol of barium sulfate?

Sample answers: **1.** $Ba(NO_3)_2(aq) + Na_2SO_4(aq) \rightarrow BaSO_4(s) + 2NaNO_3(aq)$. **2.** The solid is barium sulfate. Sodium nitrate is soluble. **3.** To make 3 mol of barium sulfate, you will need 3 mol of barium nitrate; 3 mol · 261.3 g/mol = 784 g.

Homework

Assign the Unit 4: Toxins Review in the student text in preparation for the unit exam.

ANSWERS

GENERAL REVIEW

1. a. $Zn(s) + 2HCl(aq) \rightarrow ZnCl_2(aq) + H_2(g)$

---

8. Describe the steps you would follow to make a 1.0 M solution of silver nitrate.

9. How many grams of solute are dissolved in a 500.0 mL sample of 0.50 M sodium sulfate?

10. What is the pH of a solution with $[H^+] = 2.0 \times 10^{-8}$? What is the $[OH^-]$ of the solution?

11. This reaction is completed in a factory to create potassium phosphate:

   ___ $H_3PO_4(aq) + $ ___ $KOH(aq) \longrightarrow$
   ___ $K_3PO_4(aq) + $ ___ $H_2O(l)$

   a. What type of reaction is represented by the equation?
   b. Copy the equation and balance it.
   c. If you want to make 5.0 mol of potassium phosphate, how many moles of phosphoric acid, H_3PO_4, do you need?
   d. If 628.0 g of phosphoric acid are reacted with 1122.0 g of potassium hydroxide, what is the maximum amount of potassium phosphate that can be produced?
   e. Which is the limiting reactant in part d? How much (in moles and in grams) of the excess reactant is left over?

12. The reaction shown here is completed by a chemical company to create phosphoric acid, H_3PO_4.

   ___ $K_3PO_4(aq) + $ ___ $H_2SO_4(aq) \longrightarrow$
   ___ $H_3PO_4(aq) + $ ___ $K_2SO_3(aq)$

   a. What type of reaction is represented by the equation?
   b. Copy the equation and balance it.
   c. How many moles of potassium phosphate, K_3PO_4, are required to produce 6 mol of phosphoric acid?
   d. How many moles of sulfuric acid, H_2SO_4, are required to produce 6 mol of phosphoric acid?
   e. If 750.0 g of potassium phosphate are reacted with 0.75 L of 18.4 M sulfuric acid, what is the maximum number of moles of phosphoric acid that can be produced?
   f. Which is the limiting reactant in part e? How much (in moles and in grams) of the excess reactant is left over?

STANDARDIZED TEST PREPARATION

Multiple Choice

Choose the best answer.

1. What is a correct interpretation of the following chemical equation?

   $K_2CO_3(aq) + Mg(NO_3)_2(aq) \longrightarrow$
   $2KNO_3(aq) + MgCO_3(s)$

   (A) A solution of potassium nitrate is added to solid magnesium carbonate to produce a solution of potassium carbonate and magnesium nitrate.
   (B) Solutions of potassium carbonate, magnesium nitrate, and potassium nitrate are mixed to produce solid magnesium carbonate.
   (C) A solution of potassium carbonate is added to a solution of magnesium nitrate to produce a solution of potassium nitrate and solid magnesium carbonate.
   (D) Solid magnesium carbonate dissolves in potassium nitrate to form a solution of potassium carbonate and magnesium nitrate.

2. What information can be obtained from a written chemical equation?

   (A) the color change of the reactant
   (B) the speed of the overall reaction
   (C) the states of the products
   (D) the probability of an explosion

3. For the following equation, what would you expect to observe when this reaction is performed in the laboratory?

   $Na_2CO_3(aq) + 2HCl(aq) \longrightarrow$
   $2NaCl(aq) + CO_2(g) + H_2O(l)$

   (A) The resulting solution contains bubbles.
   (B) The resulting solution freezes.
   (C) The resulting solution changes color.
   (D) The resulting solution contains white crystals.

4. Which of the following balanced chemical equations represents a physical change?

   (A) $2H_2(g) + O_2(g) \longrightarrow 2H_2O(g)$
   (B) $N_2(l) \longrightarrow N_2(g)$
   (C) $2KClO_3(s) \longrightarrow 2KCl(s) + 3O_2(g)$
   (D) $Zn(s) + 2HCl(aq) \longrightarrow ZnCl_2(aq) + H_2(g)$

---

b. $2KClO_3(s) \rightarrow 2KCl(s) + 3O_2(g)$

c. $S_8(s) + 24F_2(g) \rightarrow 8SF_6(g)$

d. $4Fe(s) + 3O_2(g) \rightarrow 2Fe_2O_3(s)$

2. a. single exchange **b.** decomposition **c.** combination **d.** combination

3. a. HCl is the limiting reactant. Mass of $ZnCl_2 = 1.87$ g. Mass of $H_2 = 0.0552$ g. **b.** There is only one reactant, so it must be the limiting reactant, $KClO_3$. Mass of KCl = 0.608 g. Mass of $O_2 = 0.390$ g. **c.** F_2 is the limiting reactant. Mass of $SF_6 = 1.28$ g. **d.** Fe is the limiting reactant. Mass of $Fe_2O_3 = 1.43$ g.

4. a. $Ca(s) + 2H_2O(l) \rightarrow Ca(OH)_2(s) + H_2(g)$

b. $2HCl(aq) + Ca(HCO_3)_2(s) \rightarrow 2H_2O(l) + 2CO_2(g) + CaCl_2(aq)$

5. A neutralization reaction is a chemical reaction in which an acid and a base combine to form a salt and water.

6. a. $Mg^{2+}(aq) + 2NO_3^-(aq) + Ca^{2+}(aq) + 2Cl^-(aq) \rightarrow Mg^{2+}(aq) + 2NO_3^-(aq) + Ca^{2+}(aq) + 2Cl^-(aq)$. The net ionic equation does not exist because all of the ions are spectator ions.

b. $Ag^+(aq) + NO_3^-(aq) + K^+(aq) + I^-(aq) \rightarrow AgI(s) + NO_3^-(aq) + K^+(aq)$
$Ag^+(aq) + I^-(aq) \rightarrow AgI(s)$

c. $2Na^+(aq) + SO_4^{2-}(aq) + Ba^{2+}(aq) + 2Cl^-(aq) \rightarrow BaSO_4(s) + 2Na^+(aq) + 2Cl(aq)$
$Ba^{2+}(aq) + SO_4^{2-}(aq) \rightarrow BaSO_4(s)$

5. Which of the following correctly matches a balanced chemical reaction to its type of reaction?

(A) single exchange:
$$NaCl(aq) + AgNO_3(aq) \longrightarrow NaNO_3(aq) + AgCl(aq)$$

(B) double exchange:
$$2AlBr_3(aq) + 3Cl_2(g) \longrightarrow 2AlCl_3(aq) + 3Br_2(g)$$

(C) decomposition:
$$Na_2CO_3(s) \longrightarrow Na_2O(s) + CO_2(g)$$

(D) combination:
$$2PbSe(s) + O_2(g) \longrightarrow 2PbO(s) + 2Se(s)$$

6. The LD_{50} for lead toxicity is 450 mg/kg. How many milligrams would be a lethal dose for a 132 lb adult?

(A) 0.13 mg
(B) 50 mg
(C) 27 g
(D) 59 g

7. Which of the following correctly describes the comparison of one mole of copper, Cu, magnesium, Mg, and neon, Ne?

(A) Each mole sample has the same mass.
(B) Each mole sample has the same number of atoms.
(C) Each mole sample has the same molar mass.
(D) Each mole sample has a different number of atoms based on the atomic mass.

8. Which correctly describes a 500 mL bottle filled completely with water?

(A) There are 9000 mol of water in this bottle.
(B) There are 1.67×10^{25} molecules of water in this bottle.
(C) The mass of the water in the bottle is 500 mL.
(D) The molarity of the water in this bottle is 1.00 M.

9. How many moles of $Ca(NO_3)_2$ are there in a sample with a mass of 50.0 g?

(A) 0.305 mol
(B) 0.333 mol
(C) 0.490 mol
(D) 8200 mol

10. What is the molarity of a sugar solution that has 45.0 g of sugar, $C_{12}H_{22}O_{11}$, dissolved in 250 mL of water?

(A) 0.53 M
(B) 0.016 M
(C) 0.18 M
(D) 0.0010 M

11. You want to make 100.0 mL of a 0.75 M solution of potassium sulfate. What mass of solid potassium sulfate should you use?

(A) 430 mg
(B) 1.3 g
(C) 13 g
(D) 13 kg

12. What is the pH of a 1.0×10^{-5} M HCl solution?

(A) −5
(B) 5
(C) 9
(D) −9

13. What is the pH of a 0.001 M NaOH solution?

(A) −3
(B) 2
(C) 3
(D) 11

14. Which of the following is a balanced neutralization reaction?

(A) $2HNO_3(aq) + CaOH_2(aq) \longrightarrow 2H_2O(l) + Ca(NO_3)_2(aq)$

(B) $H_2SO_4(aq) + KOH(aq) \longrightarrow H_2O(l) + K_2SO_4(aq)$

(C) $HC_2H_3O_2(aq) + NH_4OH(aq) \longrightarrow 2H_2O(l) + NH_2C_2H_3O(aq)$

(D) $2H_3PO_4(aq) + 3Mg(OH)_2(aq) \longrightarrow 6H_2O(l) + Mg_3(PO_4)_2(aq)$

15. If 15.0 mL of 0.50 M HCl solution is required to neutralize 34.1 mL of NaOH, what is the molarity of the base?

(A) 0.22 M NaOH
(B) 0.26 M NaOH
(C) 1.14 M NaOH
(D) 4.6 M NaOH

Engineering Project: Toxic Clean Up

As an extension to Unit 4: Toxins, you might want to assign the Engineering Project: Toxic Clean Up. In this project, students expand on the project from Unit 1: Alchemy to include ways of purifying water with chemical reactions.

TRM Engineering Project: Toxic Clean Up U4

ASSESSMENTS

Two Unit 4 Assessments (versions A and B) are available for download.

TRM Unit 4 Assessment Answers

TRM Unit 4 Assessments

A Lab Assessment for Unit 4 is available for download.

TRM Unit 4 Lab Assessment Instructions and Answers

TRM Unit 4 Lab Assessment

---

7. Yes, AgI and $BaSO_4$ are precipitates.

8. *Possible answer:* ● Determine the molecular mass of $AgNO_3$. ● Measure exactly one mole of $AgNO_3$ into a 1.0 L calibrated flask. ● Add water to the 1.0 L mark. ● Shake to dissolve and mix.

9. Mass of Na_2SO_4 = 36 g

10. pH = 7.7, $[OH^-] = 5.0 \times 10^{-7}$

11. a. double-exchange reaction b. $2H_3PO_4(aq) + 6KOH(aq) \rightarrow 2K_3PO_4(aq) + 6H_2O(l)$ c. 5 mol of phosphoric acid is needed to make 5 mol of potassium phosphate. d. Mass of K_3PO_4 = 1360 g e. H_3PO_4 is the limiting reactant. Moles of excess KOH = 0.77 mol. Mass of excess KOH = 43.2 g.

12. a. double exchange b. $2K_3PO_4(aq) + 3H_2SO_4(aq) \rightarrow 2H_3PO_4(aq) + 3K_2SO_4(aq)$ c. 6 mol of potassium phosphate is needed to make 6 mol of phosphoric acid. d. 9 mol of sulfuric acid is needed to make 6 mol of phosphoric acid. e. 5.5 mol is the maximum amount of phosphoric acid that can be made. f. K_3PO_4 is the limiting reactant. Moles of excess H_2SO_4 = 5.53 mol. Mass of excess H_2SO_4 = 542 g.

16. Which solid would be best to add to a solution of $Cu(NO_3)_2$ in order to remove the Cu^{2+} from the solution?

(A) $NaCl(s)$
(B) $KOH(s)$
(C) $Mg(NO_3)_2(s)$
(D) $PbCO_3(s)$

17. The reaction to form ammonia gas, NH_3, is shown in the chemical equation below. Which statement correctly describes this chemical reaction?

$$N_2(g) + 3H_2(g) \longrightarrow 2NH_3(g)$$

(A) The mole ratio cannot be determined because the chemical equation is not balanced.
(B) More molecules of N_2 are needed than H_2 molecules to maximize the amount of NH_3 produced.
(C) To produce 2 mol of NH_3, three times as many moles of H_2 are needed as moles of N_2.
(D) If 6 mol of H_2 and 2 mol of N_2 are used, then 2 mol of NH_3 will be produced.

18. Consider the reaction below. What mass of sodium chloride will be produced if 5.0 g of calcium oxalate is made?

$$CaCl_2(aq) + Na_2C_2O_4(aq) \longrightarrow$$
$$2NaCl(aq) + CaC_2O_4(s)$$

(A) 2.3 g NaCl
(B) 4.6 g NaCl
(C) 11 g NaCl
(D) 22 g NaCl

19. Consider the reaction below. How many grams of $Zn_3(PO_4)_2$ can be made with 26.5 g of $Zn(NO_3)_2$ and 19.7 g of Na_3PO_4?

$$3Zn(NO_3)_2(aq) + 2Na_3PO_4(aq) \longrightarrow$$
$$Zn_3(PO_4)_2(s) + 6NaNO_3(aq)$$

(A) 12.0 g $Zn_3(PO_4)_2$ because $Zn(NO_3)_2$ is the limiting reagent
(B) 18.0 g $Zn_3(PO_4)_2$ because $Zn(NO_3)_2$ is the limiting reagent
(C) 23.2 g $Zn_3(PO_4)_2$ because Na_3PO_4 is the limiting reagent
(D) 34.8 g $Zn_3(PO_4)_2$ because Na_3PO_4 is the limiting reagent

20. Consider the balanced chemical equation below. Which substance is the limiting reactant, and how many moles of Fe_3O_4 can be obtained by reacting 16.8 g Fe with 10.0 g H_2O?

$$3Fe(s) + 4H_2O(g) \longrightarrow Fe_3O_4(s) + 4H_2(g)$$

(A) Iron is the limiting reactant and 0.100 mol Fe_3O_4 can be formed.
(B) Water is the limiting reactant and 0.100 mol Fe_3O_4 can be formed.
(C) Iron is the limiting reactant and 0.902 mol Fe_3O_4 can be formed.
(D) Water is the limiting reactant and 0.902 mol Fe_3O_4 can be formed.

Unit 5 | Fire

Energy, Thermodynamics, and Oxidation-Reduction

FIRE AS CONTEXT

Any time matter moves or changes in any way, energy is involved. Fire is an impressive expression of energy in our daily lives. Who doesn't enjoy a good fireworks show or a crackling campfire? The fire theme drives the investigation of energy. The unit begins by asking: What is fire? Students observe a set of demonstrations and consider commonalities and differences. Then students explore the question: Is one fuel better than another? Next, we ask: Where does the energy from a fire come from, and how can we best make use of it? Students explore the question: Is it possible to harness the energy of a combustion reaction and make it controllable, safe, and portable? This leads to an investigation of oxidation and reduction and electrochemistry. The unit concludes with an investigation of the question: How does light interact with matter? This final chapter focuses on light energy, more broadly referred to as electromagnetic radiation.

THE IMPORTANCE OF UNDERSTANDING ENERGY

Fire is central to our energy consumption, from the internal combustion engine and kitchen stoves to coal-fired power plants. Because of concerns about global warming, an exploration of combustion takes on special significance. Projects in this unit that explore alternative energy sources such as nuclear power engage students in considering the big picture.

CONTENT DRIVEN BY CONTEXT

In our daily language, *heat* is commonly used as a verb, as when people talk about heating something up. However, there is also a tendency to speak of heat as if it is a substance, as in "heat rises." In reality, heat is a process of energy transfer. It is important to help students understand that although heat is not a substance, it cannot exist independent of matter. Whenever possible in this unit, we talk about heat being *transferred* rather than released or absorbed, because the latter words imply that energy is a substance. While the term *heat transfer* is technically redundant since heat *is* transfer, we use it on occasion to help students focus on the process.

It is also common for people to talk about energy as a thing that can be stored and used up, much like fuel. It is more accurate to say that energy is an attribute of an object or system that can be quantified. In physics, energy is simply defined as the capacity to do work. In this unit, we consider both aspects of energy, making connections between chemical energy and mechanical energy.

To further clarify conceptual confusion, the curriculum provides students with opportunities to experience heat transfer directly and acquaints them with the concepts of system and surroundings. Students discover firsthand that heat transfer is a result of differences in temperature and that cold is a sensation that is only experienced when heat is transferred out of the body.

THE USE OF MODELS TO EXPLAIN ENERGY

Throughout this unit, we use models that show the motions of atoms and molecules to explain various phenomena. Considering the motions of atoms and molecules is particularly important while explaining what happens to heat during a phase change. Several computer simulations help students to grasp the intricacies of heating, temperature change, phase change, and potential and kinetic energy. Students also learn various models to describe electromagnetic radiation and how it interacts with matter. As students grapple with energy concepts, it is worthwhile to continually take them back to basic explanations of what is going on at the molecular or atomic level.

BUILDING UNDERSTANDING

In the first four units, students have built an understanding of atomic and molecular properties, the behavior of gases, and the chemistry behind physical and chemical change. These earlier units set the conceptual foundation for studying the kinetic aspects of energy in Unit 5: Fire.

Chapter Summary

Chapter	Description	Standard Schedule Days	Block Schedule Days
18	In **Chapter 18,** students explore fire as a dramatic chemical transformation accompanied by enormous transfers of energy. Students are introduced to exothermic and endothermic changes, both physical and chemical. The inquiry expands to include other types of energy transfers and the first and second laws of thermodynamics.	6.5	3.5
19	In **Chapter 19,** students are introduced to units of measurement for energy and the concept of heat of reaction. They have an opportunity to design and build a simple calorimeter to measure and quantify heat transfer, allowing them to compare the efficiency of different fuels.	5	3
20	In **Chapter 20,** students explore the source of energy from chemical change. This requires them to compare the composition of different fuels and to investigate average bond energies. Students also investigate factors that affect the rate of a reaction as well as ways that chemical energy can be harnessed for work.	4	2
21	In **Chapter 21,** students move beyond combustion to investigate redox reactions, focusing on energy from reactions that involve metal compounds. Students learn to write net ionic equations to track the transfer of electrons in chemical reactions. They also explore the relative reactivity of metals, which in turn allows them to build an electrochemical cell.	5	3
22	In **Chapter 22,** students explore how fire and extremely hot objects transfer energy as electromagnetic radiation. Students explore emission, absorption, transmission, and reflection of light. This final chapter also focuses on models to describe electromagnetic radiation and its interaction with matter.	7.5	3.5

Pacing Guides
Standard Schedule

Day	Suggested Plan	Day	Suggested Plan
1	Chapter 18 Lesson 94	15	Chapter 20 Lesson 107, Chapter 20 Review
2	Chapter 18 Lesson 95	16	Chapter 20 Quiz, Chapter 21 Lesson 108
3	Chapter 18 Lesson 96	17	Chapter 21 Lesson 109
4	Chapter 18 Lesson 97	18	Chapter 21 Lesson 110
5	Chapter 18 Lesson 98	19	Chapter 21 Lesson 111
6	Chapter 18 Lesson 99, Chapter 18 Review	20	Chapter 21 Lesson 112, Chapter 21 Review
7	Chapter 18 Quiz, Chapter 19 Lesson 100	21	Chapter 21 Quiz, Project (optional), Chapter 22 Lesson 113
8	Chapter 19 Lesson 101	22	Chapter 22 Lesson 114
9	Chapter 19 Lesson 101 (continued)	23	Chapter 22 Lesson 115
10	Chapter 19 Lesson 102	24	Chapter 22 Lesson 116, Chapter 22 Review
11	Chapter 19 Lesson 103, Chapter 19 Review	25	Chapter 22 Quiz, Project (optional)
12	Chapter 19 Quiz, Chapter 20 Lesson 104	26	Unit 5 Review
13	Chapter 20 Lesson 105	27	Unit Exam
14	Chapter 20 Lesson 106	28	Lab Exam (optional)

Block Schedule

Day	Suggested Plan	Day	Suggested Plan
1	Chapter 18 Lessons 94 and 95	9	Chapter 20 Quiz, Chapter 21 Lessons 108 and 109
2	Chapter 18 Lessons 96 and 97	10	Chapter 21 Lessons 110 and 111
3	Chapter 18 Lessons 98 and 99, Chapter 18 Review	11	Chapter 21 Lesson 112, Chapter 21 Review
4	Chapter 18 Quiz, Chapter 19 Lesson 100	12	Chapter 21 Quiz, Project (optional), Chapter 22 Lessons 113 and 114
5	Chapter 19 Lesson 101	13	Chapter 22 Lessons 115 and 116, Chapter 22 Review
6	Chapter 19 Lessons 102 and 103, Chapter 19 Review	14	Chapter 22 Quiz, Project (optional), Unit 5 Review
7	Chapter 19 Quiz, Chapter 20 Lessons 104 and 105	15	Unit Exam, Lab Exam (optional)
8	Chapter 20 Lessons 106 and 107, Chapter 20 Review		

Unit 5 | Fire

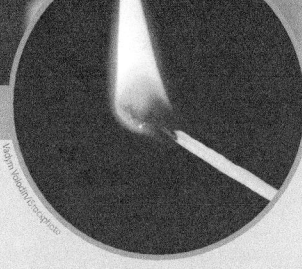

Fire is a fascinating phenomenon and a vital source of energy on our planet.

In this unit, you will learn

- the nature of heat, energy, and fire
- how to keep track of and measure changes in energy
- about energy exchanges associated with chemical changes
- how chemical energy is transformed into work
- about energy exchanges during reactions with metals and ionic compounds
- how light energy interacts with matter

Why Fire?

Every change that happens to matter is accompanied by a change in energy. Fire is visible evidence of the energy associated with one particular type of chemical change. When a compound burns, it is broken down into smaller, less complex substances, and heat and light are released. When fire is uncontrolled, it can be destructive. However, this same chemical reaction can also provide heat, light, and mechanical or electrical energy. This unit explores how the energy from chemical and physical change can be observed, measured, understood, and controlled.

481

PD CHAPTER 18 OVERVIEW

Watch the video overview of Chapter 18 (for teachers) by clicking on the link in the TE-book, opening the TRFD, or logging onto the book's companion Web site bcs.whfreeman.com/ livingbychemistry2e (teacher log-in required).

The Fire unit opens with observing expressions of energy and exploring its effects on different substances. In Lesson 94, students observe demonstrations featuring a wide variety of exothermic reactions. Lesson 95 introduces the concept of heat as a process. In a lab procedure, students use thermometers to measure energy transfers into and out of a system, while a computer simulation connects energy changes to the motions of particles. In Lesson 96, students explore sensory experiences of heat and cold and provide explanations for their observations. A simple lab procedure in Lesson 97 allows students to investigate the difference between heat and temperature. Students are introduced to the calorie and to the equation describing heat transfer in water. In Lesson 98, students observe heat transfer in different materials and are introduced to the concept of specific heat capacity. Lesson 99 wraps up the chapter with a look at the energy involved in phase changes.

In this chapter, students will learn

- exothermic and endothermic reactions
- the definitions of heat, temperature, and thermal energy
- the first and second laws of thermodynamics
- thermal equilibrium
- simple heat transfer calculations
- specific heat capacity and latent heat

CHAPTER **18**

Observing Energy

In this chapter, you will study

- energy exchange in chemical reactions
- heat, temperature, and thermal energy
- the effect of heat on different substances
- the energy associated with phase changes

482

The snow on the branch is changing from the solid to the liquid phase. Energy is involved in phase changes.

All around you is evidence of energy exchanges. Any time matter moves or changes, energy is involved. Some expressions of energy are familiar, like the heat from a campfire or the radiant energy from the Sun. But energy can be a tricky concept to define. Chapter 18 of Unit 5: Fire introduces you to energy through observations of its effects on matter.

Chapter 18 Lessons

Fired Up!

Energy Changes

THINK ABOUT IT

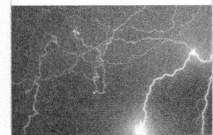

Fire is central to our lives. The most familiar fires are wood, oil, and gas fires. When these substances are ignited, brilliant flames leap into the air and there is intense heat. The energy from fire is used to warm homes, to cook food, and to move cars. For millennia, fire has been a vital energy source for humankind.

> **What reactions are sources of heat?**

To answer this question, you will explore

↻ Energy

↻ Exothermic Processes

EXPLORING THE TOPIC

↻ Energy

The universe consists of matter and energy. As discussed in Unit 1: Alchemy, matter is made up of atoms of different elements. So, what is energy?

Because energy is not matter, it is not a substance. Energy does not have mass and it does not take up space. Energy is difficult to describe and define. Take a moment to examine how the word *energy* is used in the following sentences. What do these situations have in common?

- A plant needs energy from the Sun to grow.
- An athlete eats a snack bar for energy to continue running.
- A campfire provides energy to heat water.
- Energy from flowing water makes the waterwheel rotate.
- Electrical energy causes the filament in the light bulb to glow.
- Pressure from steam provides the energy to move a locomotive.

You may notice that all of the statements are about energy causing something to happen to matter. Matter moves, falls, glows, melts, breaks apart, or burns. In each case, matter changes in some way and energy causes the change to happen. So, one definition of energy is that it is about change. **Energy** is a measure of the ability to cause change to occur. Even though you cannot see energy or hold it in your hand, you can measure amounts of energy transferred to a substance or from a substance.

Overview

DEMOS: WHOLE CLASS

Key Question: What reactions are sources of heat?

KEY IDEAS

Changes in matter involve changes in energy. One way to define energy is as a measure of the work involved when change occurs. A chemical reaction that produces products that are hotter than the reactants is called *exothermic*. Flames, sparks, a glow, or some other release of light accompanies many exothermic chemical reactions. Fire is a specific type of exothermic reaction that produces flames, heat, and light.

LEARNING OBJECTIVES

- Define energy.
- Define exothermic reactions.
- Describe fire.

FOCUS ON UNDERSTANDING

- Energy is a difficult concept to define. Energy is not a substance. It is a concept that describes motion or the potential for motion.
- Energy is often defined as the ability to do work, but this definition does not apply very well to a chemistry context. In this unit, we focus mainly on thermal energy.

IN CLASS

Students observe a series of demonstrations, all of which are exothermic reactions that result in products that are hotter than the reactants. Based on their observations and on the chemical equations of the reactions involved, students identify patterns in the changes that take place. During the discussion, students consider how to define the word *energy*. The instructions and list of materials to run six demos is available for download.

TRM Materials List 94

TRM Demo Instructions 94

Pre-AP® Course Tip

Students preparing to take an AP® chemistry course or college chemistry should be able to collect data and analyze it for patterns or relationships. In this lesson, students observe various exothermic reactions. Based on the reactions, students identify patterns in the changes that take place and consider the role that energy plays in these reactions.

SETUP

Before class, collect and organize the materials and review the instructions for each demo. Make sure you have a fireproof surface of some sort for the sugar sparks demo. Have a fire extinguisher available. Select the demos you want to perform and practice them.

Jet Engine: If you try to repeat this demo, you must wait until enough oxygen has diffused back into the bottle.

Sparklers: You can purchase sparklers if they are permitted in your state and county.

CLEANUP

Dispose of any remaining solid and liquid waste according to local guidelines.

ENGAGE (5 min)

TRM PowerPoint Presentation 94

ChemCatalyst

Several people are stranded on a cold, remote, deserted island with only the clothes on their backs. They must stay warm and purify some water to drink. One of the first tasks they all agree on is to try to build a fire.

1. How could the survivors go about starting a fire?
2. What is fire? Describe it.
3. What makes a fire hot?

Sample answers: **1.** Fires can be started using friction or, if there is sunlight, a lens from some reading glasses. The survivors must find fuel of some sort to burn. **2.** A fire is a chemical change that produces flames, light, smoke, and ash, and it is a source of heat. The flames from fires look yellow-orange. **3.** Heat is a result of the chemical reaction of the fuel burning. Students may have a variety of explanations for the heat that accompanies a fire, including the mistaken idea that it is due to the breaking of chemical bonds.

Ask: Why is fire so useful?

Ask: Where does fire come from?

Ask: What is happening to matter in a fire?

Ask: Is fire necessary to our survival on the planet? Explain.

EXPLORE (20 min)

TRM Worksheet with Answers 94

TRM Worksheet 94

INTRODUCE THE DEMOS

→ Pass out the student worksheets.

→ Brief students on fire safety guidelines and behavior expectations for this unit.

→ Tell students that this unit focuses on fire and will sometimes require the burning of some substances and the use of open flames. Caution is always a must when dealing with fire. Any careless or irresponsible behavior will result in nonparticipation.

→ Pass out worksheets to individual students. Inform students that they will record their observations of a number of demonstrations.

HISTORY CONNECTION

The Great Fire of London occurred in 1666. The fire lasted for four days, destroying 13,200 homes. Historians credit this fire with eradicating the rat population and helping to end the Great Bubonic Plague that had killed almost 20 percent of the city's populace.

Exothermic Processes

The chemical reactions listed here all release energy. This is evident in the flames, the sparks, the light you see, the sounds you hear, and the heat you feel when you observe these reactions. A reaction that creates products that are hotter than the reactants is called **exothermic**. Take a moment to examine these exothermic reactions.

HEAT, LIGHT, FLAMES

The first two equations involve molecular covalent compounds. They react with oxygen, O_2. The products include carbon dioxide and water. Heat, light, and flames accompany these reactions.

Methane: $CH_4(g) + 2O_2(g) \longrightarrow CO_2(g) + 2H_2O(g)$

Propanol: $2C_3H_8O(l) + 9O_2(g) \longrightarrow 6CO_2(g) + 8H_2O(g)$

HEAT, LIGHT, SPARKS

The second set of equations involves elemental substances. Elemental metals and many nonmetals also react with oxygen to produce heat and light. There are no flames associated with these reactions, but sometimes sparks are emitted.

Phosphorus: $P_4(s) + 5O_2(g) \longrightarrow 2P_2O_5(s)$

Iron: $4Fe(s) + 3O_2(g) \longrightarrow 2Fe_2O_3(s)$

In all of these exothermic reactions, the reactants change into products that are very hot. These hot products cool down over time as they transfer heat to the surroundings. Exothermic changes are associated with big changes to matter.

FIRE

When you observe fire, matter is changing drastically. As a result of a fire, an entire forest can be reduced to smoke and ashes. During this change, the trees and brush are converted to carbon dioxide and water, which spread out in the atmosphere in gaseous form. Some ash and charred fuel remain after a fire. These are the result of reactants that did not react completely.

→ You can complete all or some of the demonstrations depending on local guidelines, the availability of materials, and time constraints. Allow students time between demos to record what they have seen.

LIST OF DEMOS

1. Safety Match
2. Candle Flame
3. Jet Engine
4. Sugar Sparks
5. Flaming Bubbles
6. Sparklers

EXPLAIN & ELABORATE (15 min)

ANALYZE OBSERVATIONS

Ask: What are some of the changes you observed during today's demos? Describe them.

Ask: What common features did you notice?

Ask: What types of reactions resulted in flames?

Ask: Is energy associated with these reactions? Explain.

Key Points: Changes in matter involve changes in energy. The reactions in the demonstrations all release energy. This is evident in the dancing flames,

Flames are generally what distinguish fires from other exothermic reactions. Flames consist mainly of hot gases. The glowing yellow color of a wood fire is caused by small particles of carbon carried into the air by the gaseous products of this exothermic reaction.

> **Big Idea** Changes in matter are accompanied by changes in energy.

LESSON SUMMARY

What reactions are sources of heat?

Energy is not a substance, so it does not have mass and it does not take up space. However, any change in matter is accompanied by a change in energy. Many chemical reactions are exothermic, which means they result in products that are hotter than the reactants. Fire is an important example of an exothermic reaction that transfers energy in the form of light and heat. In a fire, reactants combine with oxygen to form carbon dioxide and water. The energy associated with fires has many uses such as heating your home, cooking your food, and moving cars.

KEY TERMS
energy
exothermic

Exercises

Reading Questions

1. In your own words, define the word *energy*.

2. Use your own words to write a short paragraph describing fire to someone who has never seen it before.

Reason and Apply

3. Write five sentences that use the word *energy*.

4. Describe three ways in which fire is central to life on our planet.

5. Overnight, the ashes of a fire cool down. What has happened to the products and the heat?

6. PROJECT
www Research uses of fire on the Web. Describe how fire is used in gas heaters to heat homes.

7. PROJECT
www Research uses of fire on the Web. Describe how fire is used to generate electricity in a coal-burning power plant.

Lesson 94 | **Energy Changes**   485

the sparks that fly, the light you see, the sound you hear, and the heat you feel. The products are "energetic." A reaction that creates products hotter than the reactants is *exothermic*.

> **Exothermic process:** A process that releases energy in the form of heat.

Reactions of elements and compounds with oxygen typically are exothermic. Hydrocarbon molecules can combine with oxygen to produce carbon dioxide and water. The products are hot gases that emit light, transfer heat, and sometimes produce explosions. The flames are composed mostly of gaseous

reactants and products. When metals react with oxygen, as in the sparklers, solid products are formed. The sparks are small chunks of the hot solid products that fly into the air and emit light.

Combustion requires oxygen. The combustion demos required air, which is mostly nitrogen, $N_2(g)$, and not very reactive. The oxygen, $O_2(g)$, component of air is involved in all combustion reactions.

DISCUSS ENERGY

→ Write the definition of energy (given below) on the board.

Ask: How do you use the word *energy* in your daily life?

Ask: In your experience, how is energy related to change and motion?

Ask: How would you define energy?

Key Points: Energy is a measure of how much you can move or change something. Moving or changing anything involves energy. This includes both the motions of the objects in the world around us and the motions of atoms and molecules associated with physical and chemical changes. Energy is one of those words that most people know but is hard to define.

> **Energy:** A measure of the work involved in moving or changing something.

Energy is not a substance. When things change, energy is transferred. Having a stack of firewood is like having a large energy bank account. Putting the wood on the fire is like spending the assets in the energy account. Scientists keep track of energy by monitoring how much energy is transferred into or out of the energy account when something moves or changes.

EVALUATE (5 min)

There is no Check-In for this lesson.

Homework

Assign the reading and exercises for Fire Lesson 94 in the student text.

LESSON 94 ANSWERS

1. *Possible answer:* Energy is a measure of the ability to cause a change in matter. The given answer reflects the definition in the text. Energy is a difficult concept to define precisely and is sometimes defined differently in different subject areas.

2. *Possible answer:* Fire is a flame or a series of flames that is caused by a substance reacting with oxygen in the air and producing a large amount of heat. If the substance gets too hot, it bursts into flames, and much of the substance is converted into smoke, light, gas, and water.

Answers continue in Answer Appendix (p. ANS-7).

Lesson 94 | **Energy Changes**   485

Overview

LAB: GROUPS OF 4

COMPUTER ACTIVITY: GROUPS OF 4

Key Question: In what direction is heat transferred during a chemical process?

KEY IDEAS

During changes to matter, heat is often transferred between a chemical system and its surroundings. Exothermic processes result in the transfer of heat from the system to the surroundings and feel hot to the observer. Endothermic processes result in the transfer of heat from the surroundings to the system and feel cold to the observer. The heat transfer that is experienced is related to the motions of the product molecules.

LEARNING OBJECTIVES

- Distinguish between exothermic and endothermic processes.
- Explain sensory experiences of hot and cold.
- Discuss energy changes from a molecular viewpoint.

FOCUS ON UNDERSTANDING

- Students often think the direction of heat transfer for endothermic processes is in the opposite direction, believing that they "throw off" heat, causing the temperature of the system to decrease.
- Solutions are created at stations 2 and 3 of the laboratory. The process of forming solutions is in the gray area between chemical and physical changes.

KEY TERMS

heat
system
surroundings
endothermic
kinetic energy

IN CLASS

Students conduct procedures involving exothermic or endothermic processes at three stations and record temperature changes. During the discussion, students view a computer simulation of hydrogen and oxygen molecules reacting to form water, and they examine the meaning of hot and cold. A complete materials list is available for download.

TRM Materials List 95

Differentiate

To help students distinguish *exothermic* and *endothermic,* you can build connections to other words with exo- and endo- prefixes. Ask students to come up with a list of other examples such as exoskeleton or endocrine and help students relate new vocabulary to old vocabulary.

Pre-AP® Course Tip

Students preparing to take an AP® chemistry course or college chemistry should be to provide evidence based reasoning for claims. In this lesson, students carry out simple chemical and physical reactions involving heat transfer. Students classify the reactions as either endothermic or exothermic based on evidence of temperature changes.

SETUP

Place the materials for the three procedures at stations around the room. Two groups can work at each station at one time. Photocopy two station cards for each station, one for each group.

Simulation: Prepare to show a computer simulation during the discussion. Use a simulation that demonstrates the

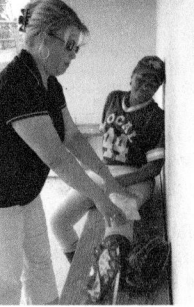

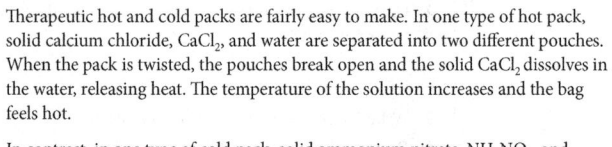

LESSON 95

Not So Hot

Exothermic and Endothermic

THINK ABOUT IT

You just received a bad bruise on your leg playing sports. The bruise is swelling and is quite painful. Your coach pulls out a disposable cold pack from the first-aid kit. A quick twist of the package activates it, and the temperature of the cold pack suddenly decreases. It feels nice and cold on your injury. What is the source of this cold sensation?

In what direction is heat transferred during a chemical process?

To answer this question, you will explore

- Heat
- Exothermic and Endothermic Processes
- Kinetic Energy

EXPLORING THE TOPIC

Heat

Therapeutic hot and cold packs are fairly easy to make. In one type of hot pack, solid calcium chloride, $CaCl_2$, and water are separated into two different pouches. When the pack is twisted, the pouches break open and the solid $CaCl_2$ dissolves in the water, releasing heat. The temperature of the solution increases and the bag feels hot.

In contrast, in one type of cold pack, solid ammonium nitrate, NH_4NO_3, and water are separated into two different pouches. When you twist the pack and the NH_4NO_3 dissolves in the water, the temperature of the solution decreases and the bag feels cold. If you perform these two reactions in beakers, you can measure the temperature changes with a thermometer.

What the thermometer records and your hand experiences is energy transferred into or out of the products of the chemical change. This process of energy transfer is called heat. **Heat** is a transfer of energy between two objects due to temperature differences. Heat always transfers from a higher temperature to a lower temperature. So, for instance, when the beaker feels cold to your hand, it is because heat transfers from your hand to the beaker.

Temperature drops

When you dissolve ammonium nitrate in water, the temperature decreases. The beaker feels cold.

SYSTEM AND SURROUNDINGS

To communicate clearly about heat transfer, it is necessary to specify where the heat is transferring to or from. The matter that you are focusing on is referred to as the **system**. Once you have defined the system, everything else is referred to as

486   Chapter 18 | **Observing Energy**

the **surroundings,** or environment. If the solution of ammonium nitrate and water is the system, then the beaker, the thermometer, the air around the beaker, your hand, and everything else in the universe constitute the surroundings. Heat transfers between a system and its surroundings.

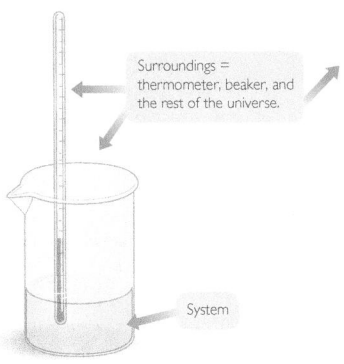

Surroundings = thermometer, beaker, and the rest of the universe.

System

Big Idea Heat is a transfer of energy due to temperature differences.

CONSUMER CONNECTION

Most household refrigerators contain a liquid called a *refrigerant* that circulates in a series of coiled tubes. As it evaporates, the refrigerant absorbs heat from inside of the refrigerator. The refrigerant is then condensed back into a liquid in tubes located in back of the refrigerator, releasing heat, and the process starts all over again.

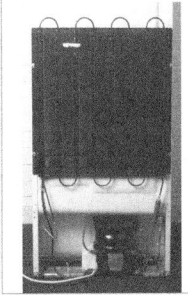

Exothermic and Endothermic Processes

Chemists categorize chemical changes according to the direction of heat transfer. As you learned in the previous lesson, when heat is transferred out of the system to the surroundings, the process is exothermic. When heat is transferred from the surroundings to the system, the process is **endothermic.** (In Greek, *exo* means outside and *endo* means inside.) Exothermic processes are experienced as warm by an observer. Endothermic processes are experienced as cold by an observer.

Some examples of exothermic and endothermic changes are listed below. The word *heat* is included in the chemical equation to highlight the direction of heat transfer.

EXOTHERMIC PROCESSES

Burning methane gas

$$CH_4(g) + 2O_2(g) \longrightarrow CO_2(g) + 2H_2O(g) + heat$$

Dissolving calcium chloride in water

$$CaCl_2(s) \longrightarrow Ca^{2+}(aq) + 2Cl^-(aq) + heat$$

Neutralization of sodium hydroxide with hydrochloric acid

$$NaOH(aq) + HCl(aq) \longrightarrow H_2O(l) + NaCl(aq) + heat$$

Lesson 95 | **Exothermic and Endothermic**   487

kinetics of chemical reactions, such as the reaction of hydrogen gas and oxygen to form water, preferably one that allows you to simulate the kinetics of an explosion and the effect of a spark. You can find a list of URLs for this lesson on the Chapter 18 Web Resources document.

TRM **Chapter 18 Web Resources**

CLEANUP

At the end of class, the beakers for waste at each station can be diluted and disposed of according to local regulations.

ENGAGE (5 min)

TRM **PowerPoint Presentation 95**

Demonstrate the hand warmers before, during, or after the ChemCatalyst. Pass one or more around the classroom. You might use a rubber band to secure one hand warmer to the end of a thermometer and note the temperature change.

ChemCatalyst

1. Name at least three ways you can warm yourself on a cold day.

2. What do you think is the source of the heat in a hand warmer?

Sample answers: **1.** You can sit near a fire or a heater, drink a warm liquid, cover yourself with a blanket, run, or rub your hands together. **2.** Two substances are mixed together in a hand warmer, causing an exothermic chemical reaction.

→ Assist students in exploring the source of the heat in the hand warmers.

Ask: Why does rubbing your hands together make you warmer?

Ask: What does twisting the hot pack do?

Ask: A toe warmer will work for up to six hours if it is in your shoe, but it will not last as long out of your shoe. How can you explain this?

EXPLORE (15 min)

TRM **Worksheet with Answers 95**

TRM **Worksheet 95**

TRM **Card Masters—Station Cards 96**

INTRODUCE THE LAB

→ Arrange students into groups of four.

→ Pass out the student worksheets.

→ Tell students that they will observe three different chemical processes. They will focus on the heat transfer associated with each one.

→ Groups will follow the directions on the station cards.

SAFETY

Wear safety goggles at all times.

HCl can irritate skin. If you get it on your skin, rinse the affected area immediately and thoroughly with water.

Dispose of chemicals in the waste containers provided for each station.

Have baking soda on hand for acid spills.

EXPLAIN & ELABORATE (15 min)

DISCUSS HEAT TRANSFER ASSOCIATED WITH THE CHEMICAL PROCESSES

→ You might create drawings to help you define system and surroundings.

Ask: What did you observe at the three stations? What did the thermometer tell you?

Ask: How would you describe the system and surroundings for the reaction between hydrochloric acid, HCl, and sodium bicarbonate, $NaHCO_3$?

Key Points: A chemical process involves energy transfer. At each station, the products were at a different temperature from the reactants. You can tell because the liquid in the thermometer moved and your hand sensed hot or cold. What the thermometer records and your hand experiences is energy transferred into or out of the products of the chemical change. This process of energy transfer is called *heat*. Using the term *heat transfer* helps students remember that heat involves a transfer of energy.

> **Heat:** A transfer of energy between two substances due to temperature differences.

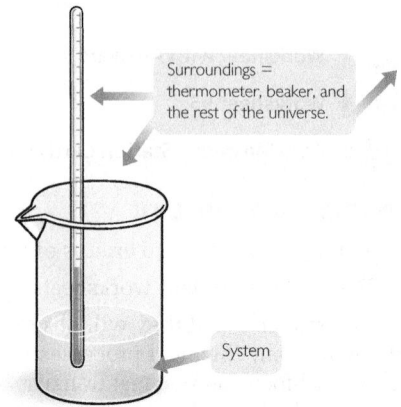

Surroundings = thermometer, beaker, and the rest of the universe.

System

The chemical process you are focusing on is referred to as the *system*. In today's lab, the system is the solution in each of the three beakers. Once the system is defined, everything else is referred to as the *surroundings,* or environment. The beaker, the thermometer, the air around the beaker, your hand, and everything else constitute the surroundings. Heat transfer occurs between the system and the surroundings. The liquid in a thermometer expands when heat is transferred to it from an exothermic reaction.

> **System:** The process you are focusing on (for example, a water molecule, a beaker of solution, a campfire, air in a balloon).
>
> **Surroundings:** Everything outside the system (for example, containers, your body, a thermometer, the atmosphere).

DEFINE EXOTHERMIC AND ENDOTHERMIC

→ Write the words *exothermic* and *endothermic* on the board. Help students to classify each of the three procedures under one of these headings.

ENDOTHERMIC PROCESSES

Dissolving ammonium nitrate in water

$$NH_4NO_3(s) + heat \longrightarrow NH_4^+(aq) + NO_3^-(aq)$$

Decomposition of mercury oxide

$$2HgO(l) + heat \longrightarrow 2Hg(l) + O_2(g)$$

Notice that heat transfer is not limited to chemical reactions. The process of dissolving is also exothermic or endothermic depending on the substance. This is because energy is involved every time atoms are rearranged.

⟳ Kinetic Energy

Changes in temperature due to a chemical reaction are associated with changes in motion. **Kinetic energy** is the energy of motion, and temperature is a reflection of the average kinetic energy of a sample. If the products of a reaction are hotter than the reactants, they must be moving faster.

The reaction between H_2 and O_2 to produce H_2O is an extremely exothermic reaction. It is the reaction used to launch the space shuttle. The water molecules produced in the reaction are so hot and they expand so rapidly that the water vapor thrusts the space shuttle up into the air. The photo here provides evidence that the product molecules are moving with explosive speed.

A great deal of the heat energy of this reaction is converted to kinetic energy in the form of rapidly moving water molecules. While the water molecules push the space shuttle up, they transfer some of their kinetic energy to the space shuttle. They also transfer some of their energy to the surroundings as heat. With less kinetic energy, the water molecules gradually become cooler and move more slowly.

© CORBIS

The shuttle launches due to the reaction of H_2 and O_2. The water vapor produced expands rapidly to thrust the shuttle upward.

KEY TERMS
heat
system
surroundings
endothermic
kinetic energy

LESSON SUMMARY

> In what direction is heat transferred during a chemical process?

Heat is a transfer of energy between two objects due to temperature differences. Heat transfer accompanies all chemical changes. The heat is transferred either into the system or out of the system. Exothermic reactions are chemical processes that result in the transfer of heat *from* the products of the reaction (the system) *to* the surroundings. These reactions feel hot to the observer. Endothermic reactions are chemical processes that result in the transfer of heat *from* the surroundings *to* the products of the reaction. These reactions feel cold to the observer. Because temperature is directly related to the motion of molecules, hotter products mean faster moving molecules. Colder products mean slower moving molecules.

Ask: Did your hand sense the same temperature change that the thermometer recorded?

Ask: How can you explain the sensation of hot and cold in terms of heat transfer?

Key Points: Some net transfer of energy occurs during all chemical processes. When heat is transferred from the system to the surroundings, the process is called *exothermic*. When heat is transferred from the surroundings to the system, the process is called *endothermic*. An observer experiences exothermic processes as hot and endothermic processes as cold.

> **Endothermic process:** A change that results in the transfer of heat to the system from the surroundings.

SHOW A COMPUTER SIMULATION OF AN EXOTHERMIC REACTION

→ Show students a computer simulation of an exothermic reaction, such as the reaction between hydrogen and oxygen to produce water. Some things to point out:
- A spark breaks an H—H bond.
- The H atoms collide with oxygen and hydrogen molecules and break O=O bonds and other H—H bonds.

Reading Questions

1. In your own words, define exothermic and endothermic chemical changes.
2. Why does an endothermic reaction that takes place in a beaker cause the beaker to feel cold?

Reason and Apply

3. Methane, $CH_4(g)$, reacts with oxygen, $O_2(g)$, to produce carbon dioxide, $CO_2(g)$, and water, $H_2O(g)$. You observe flames.
   a. How does the average kinetic energy of the reactants differ from the average kinetic energy of the products?
   b. Describe what you would expect to see if you had a molecular view of the products and reactants.

4. You mix solid copper (II) sulfate, $CuSO_4(s)$, with a small amount of water, $H_2O(l)$ in a beaker. Hydrated copper (II) sulfate, $CuSO_4 \cdot 5H_2O(s)$ is produced. A thermometer inside the beaker indicates that the temperature has increased from 19 °C to 48 °C.

$$CuSO_4(s) + 5H_2O(l) \longrightarrow CuSO_4 \cdot 5H_2O(s)$$

   a. List all the substances that are part of the system.
   b. List at least four things that are part of the surroundings.
   c. Which is at a lower temperature: $CuSO_4(s)$ or $CuSO_4 \cdot 5H_2O(s)$? Explain your thinking.
   d. What will you feel if you touch the beaker?
   e. Is the process endothermic or exothermic? Explain your thinking.
   f. What evidence do you have that heat is transferred *to* the surroundings *from* the products of the reaction?

5. You mix solid hydrated barium hydroxide, $Ba(OH)_2 \cdot 8H_2O(s)$, and solid ammonium nitrate, $2NH_4NO_3(s)$, in a beaker. The reaction is shown here. A small pool of water in contact with the outside of the beaker freezes.

$$Ba(OH)_2 \cdot 8H_2O(s) + 2NH_4NO_3(s) \longrightarrow 2NH_3(g) + 10H_2O(l) + Ba(NO_3)_2(aq)$$

   a. List all the substances that are part of the system.
   b. List at least four objects that are part of the surroundings.
   c. Which is at a lower temperature: $NH_4NO_3(s)$ or $Ba(NO_3)_2(aq)$? Explain your thinking.
   d. What will you feel if you touch the beaker?
   e. Is the reaction endothermic or exothermic? Explain your thinking.
   f. What evidence do you have that heat is transferred *from* the surroundings *to* the products of the reaction?

- Eventually, fast-moving molecules of H_2O form.
- Slow-moving H_2 and O_2 molecules form fast-moving H_2O molecules.

Ask: What reaction is taking place?

Ask: How would you describe the reaction from a molecular point of view?

Ask: How do you know that the temperature increases as the reaction proceeds?

Ask: What would a simulation of an endothermic reaction look like?

Key Points: Changes in temperature are associated with changes in the motions of particles. Kinetic energy is the energy of motion, and temperature is a reflection of the average kinetic energy of a sample. If the products of a reaction are hotter than the reactants, they are moving faster. In the simulation, the H_2 and O_2 reactants rearrange to form H_2O. The water molecules move much more rapidly than the reactant molecules. This indicates that the product water molecules are at a higher temperature and thus the reaction is exothermic. In an endothermic process, the products are colder than the reactants and would move more slowly.

Check-In

You have water at 25 °C. You dissolve ammonium acetate, $NH_4C_2H_3O_2$, in the water and find that the temperature decreases to 15 °C.

1. Is heat transferring from the solution to the surroundings or from the surroundings to the solution?
2. Is the process exothermic or endothermic?

Answers: 1. Heat is transferring from the surroundings to the solution. 2. The process is endothermic.

Homework

Assign the reading and exercises for Fire Lesson 95 in the student text.

LESSON 95 ANSWERS

1. *Possible answer:* Exothermic chemical changes are changes that transfer heat from the system into the environment, and endothermic chemical changes are changes in which heat is transferred from the environment into the system.

2. *Possible answer:* As the endothermic reaction takes place, heat is transferred to the reaction mixture from its surroundings. Because the surroundings include the beaker itself, the material of the beaker transfers some heat to the reaction. This causes the beaker to feel cold. The key point is that energy transfers from the beaker into the solution as heat because the heat is necessary for the reaction to take place.

3. a. The production of flames by the reaction shows that the reaction is exothermic, because energy is being transferred into the surrounding environment. Temperature is a measure of the average kinetic energy of the material, so in an exothermic reaction the average kinetic energy of the products must be greater than the average kinetic energy of the reactants. b. In a molecular view, the gas particles in the reactants would be moving more slowly than the particles in the products. Because the average kinetic energy is faster in the products, the particles are moving faster.

Answers continue in Answer Appendix (p. ANS-7).

Overview

ACTIVITY: PAIRS

Key Question: What do temperature differences indicate about heat transfer?

KEY IDEAS

The first law of thermodynamics states that when energy is transferred, it is conserved. The second law of thermodynamics states that thermal energy is always transferred from a hotter object to a cooler object, dispersing the energy. This energy transfer continues until the objects in contact are at the same temperature—called thermal equilibrium.

LEARNING OBJECTIVES

- Explain situations involving energy transfer.
- Define thermal equilibrium.
- State the first and second laws of thermodynamics.

FOCUS ON UNDERSTANDING

The difference between energy and heat is subtle. Energy is a measure of how much you can move or change something. Heat is the energy that is transferred due to a temperature difference.

KEY TERMS

thermal equilibrium
first law of thermodynamics
second law of thermodynamics
entropy

IN CLASS

Students complete a worksheet that focuses on heat transfer and thermal equilibrium. While half of the pairs work on the worksheets, the other half work at stations where they explore sensory experiences of temperature. Students realize that what they experience as hot and cold is an exchange of energy in both cases, but in opposite directions. During the discussion, students are introduced to the first and second laws of thermodynamics. A complete materials list is available for download.

TRM Materials List 96

SETUP

Create three water stations and three alcohol stations around the room. Each water station should have three large containers with hot water (safe to touch), lukewarm water, and ice water. Each alcohol station should have a small beaker of alcohol with a dropper, along with a thermometer with cheesecloth attached at the tip with a rubber band.

LESSON 96

Point of View

First and Second Laws

THINK ABOUT IT

Suppose that you have just come inside on a very cold day. There is a fire in the fireplace. If you go near the fire, you will feel the warmth. Heat is transferred to your body by the warm gases of the flames and the energy radiating from the burning wood.

> What do temperature differences indicate about heat transfer?

To answer this question, you will explore

- Heat Transfer
- The First and Second Laws of Thermodynamics

EXPLORING THE TOPIC

Heat Transfer

CONSUMER CONNECTION

Materials that do not conduct heat well are called *thermal insulators*. We use them in the walls and ceilings of our homes to slow down the transfer of heat from our homes in the wintertime and into our homes in the summertime. A thermos bottle is composed of several layers of insulating material, often with a vacuum in the middle, to limit heat transfer and to keep cold liquids cold and hot liquids hot.

© Ole Graf/Corbis

There are three main ways heat is transferred: conduction, convection, and radiation.

- Conduction takes place when a substance transfers heat to another substance or object that it is in contact with. For example, the handle on a frying pan gets hot because the frying pan is hot.
- Convection takes place when a warm substance changes location, such as when warm air rises.
- Radiation takes place when electromagnetic waves carry energy from an energy source, such as the Sun.

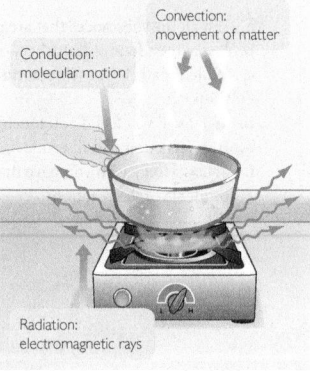

Convection: movement of matter

Conduction: molecular motion

Radiation: electromagnetic rays

TRM PowerPoint Presentation 96

ChemCatalyst

1. Imagine that you are sitting near a campfire. You feel warm. Explain how you think the warmth gets from the burning wood to your body.

2. The fire goes out. The next morning, you find ashes that are the same temperature as the air. Explain why the ashes no longer are hot.

Sample answers: **1.** Students might say that fire heats up the air molecules and the air molecules heat up the body. (This description of conduction is only part of the answer. Heat transfer also occurs through radiation and convection.) **2.** Students might say the ashes are no longer hot because the fire is not burning anymore; the chemical reaction ran out of reactants; the heat from the fire dispersed to the surrounding air; or the air cooled down the ashes.

Ask: Do you think the campfire is at the same temperature as your body? Why or why not?

SYSTEM AND SURROUNDINGS

Consider each of the three illustrations. Assume water is the system. Try to identify the direction of heat transfer, the type of heat transfer, and the surroundings.

①

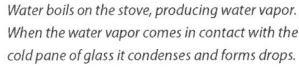

②

Water is placed in the freezer to make ice cubes.

③

An ice cube is on a table and melts.

Water boils on the stove, producing water vapor. When the water vapor comes in contact with the cold pane of glass it condenses and forms drops.

① In the first illustration, water molecules are in the form of both water vapor and liquid water. For the water vapor, the surroundings include the stove, the pot, the window, and the air in the room. Heat transfers from the stove to the pot and then to the water by conduction, then to the air in the room by convection. Heat also transfers from the water vapor to the window by conduction.

② In the second illustration, water takes the form of both liquid and solid. For the liquid water in the ice cube tray, the surroundings include the ice cube tray, the freezer, and the air in the freezer. Heat transfers from the water to the surroundings by conduction. When enough heat energy transfers from the liquid water to the interior of the freezer, the water freezes and becomes solid.

③ For the ice cube on the table, the surroundings include the table and the air in the room. Heat transfers to the ice cube by conduction. When enough energy is transferred into the ice cube on the table, it melts.

HOT AND COLD

The sensation of hot or cold depends on the direction of heat transfer. For example, if you hold an ice cube, your hand will feel cold as heat transfers from your hand to the ice cube. If you hold a cup of hot tea, your hand will feel warm as heat transfers from the hot tea to the cup and then to your hand. However, "hot" and "cold" are relative.

Lesson 96 | **First and Second Laws**   491

Ask: What happens to the temperatures of two objects when they are in contact for a long time?

Ask: Describe in detail all the energy transfers that occur when water is boiled on a gas stove. What happens to the surrounding air? What happens to the water vapor?

Ask: If you know what phase change is occurring, do you know the direction of heat transfer? Explain.

Key Points: Heat is not a substance. It is a process of energy transfer. The direction of energy transfer is always from an object with a higher temperature to an object with a lower temperature. You experience cold when you touch an object that is at a lower temperature than your skin and energy is transferred from your skin to the colder substance. You experience hot when you touch an object that is at a higher temperature than your skin and energy is transferred to your skin from the hotter object. In both cases, there is heat transfer but in opposite directions. There is no such thing as "cold transfer." Cold is the experience of heat transfer away from your body.

Objects in contact with each other tend to undergo heat transfer until thermal equilibrium is reached. Energy is always transferred when two objects with different temperatures are in contact with each other. Over time, the two objects move toward the same temperature. When there is no more difference in temperature between the two objects, the heat transfer stops. This end result of heat transfer is called *thermal equilibrium*.

> **Thermal equilibrium:** When two objects in contact with each other reach the same temperature, they are at thermal equilibrium.

INTRODUCE THE FIRST AND SECOND LAWS OF THERMODYNAMICS

Ask: What happens if you leave a cup of hot tea on the counter overnight? What happens to the energy?

Ask: How can you provide evidence to support the statement that energy is conserved?

Ask: How can you provide evidence to support the statement that energy tends to disperse?

Ask: How do you think an exothermic reaction can cause a physical change?

Ask: Why do you think the campfire eventually burns out?

EXPLORE (15 min)

TRM Worksheet with Answers 96

TRM Worksheet 96

TRM Transparency—Energy Transfer 96

INTRODUCE THE ACTIVITY

→ Arrange students into pairs.

→ Pass out the student worksheets.

→ Tell students that some pairs will work on the worksheet while other pairs are at the activity stations.

→ To avoid congestion at the stations, designate when certain pairs of students can go to the stations.

EXPLAIN & ELABORATE (15 min)

DISCUSS THE DIRECTION OF HEAT TRANSFER

→ At the appropriate time, display the Transparency—Energy Transfer showing the illustrations from the worksheet.

Ask: How can you explain your experiences at the water station in terms of heat transfer?

Key Points: The study of energy and its transformations is generally referred to as *thermodynamics*. The first and second laws of thermodynamics summarize the observations from today's activities. These two laws cannot be proved directly, but they are consistent with observations of a large number of systems.

First law of thermodynamics: Energy is conserved. The heat transferred *from* a hotter object is equal to the heat transferred *to* a colder object. There is no loss of energy. Likewise, you cannot create energy. It is only transferred from somewhere to somewhere else. In the case of fire, the energy released as heat is a consequence of the rearrangements of the atoms from reactants to products. This will be covered in later lessons.

Second law of thermodynamics: Energy tends to disperse. Thermal energy cannot be transferred from a cold object to a hot object, because this would concentrate the energy in one place rather than spread it out. This concept helps explain why the energy from a fire disperses and spreads from its source and explains why you cannot put ashes and gases back together to make a log. This concept is also related to *entropy,* which is covered in the reading for this lesson.

EVALUATE (5 min)

Check-In

Imagine that a thermometer is placed in a beaker of water and the temperature is noted. An ice cube is dropped into the water, and after ten minutes, the temperature is noted again.

1. What will you observe?

2. What is the direction of energy transfer?

Answers: **1.** The ice will begin to melt as energy is transferred from the water to the ice cube, causing the water temperature to decrease. **2.** The liquid in the thermometer will transfer energy to the water, causing the thermometer to register a lower temperature.

Homework

Assign the reading and exercises for Fire Lesson 96 in the student text.

Important to Know There is no such thing as "cold transfer." Cold is the experience of heat transfer away from our bodies.

RECREATION CONNECTION

Hypothermia occurs when the body loses heat faster than it can generate heat. A body temperature below 95 °F (35 °C) is considered dangerous. People who are outdoors in winter without enough protective clothing, or who accidentally fall into very cold water, may develop hypothermia. Hot liquids and external heat sources can help a person recover from hypothermia.

Melissa McManus/Getty Images

Just because something is at a low temperature does not mean it will feel cold. For example, if your hands are cold because you have been out in the snow all day, even holding them under cold water can feel painfully hot because the water is at a higher temperature than your hands. The experience of hot or cold depends on whether energy is being transferred to or from the observer.

THERMAL EQUILIBRIUM

Heat transfer occurs between two substances that are in contact with each other until there is no longer a temperature difference between them. If you take a pot of hot soup off the stove and place it on the counter, it will transfer heat to the air and the countertop. The soup (and the metal pot) will continue to transfer heat until they are at the same temperature as the room.

If two objects at different temperatures are in contact with one another long enough, the two substances will reach the same temperature. When they reach the same temperature, they are at **thermal equilibrium.**

Big Idea The direction of heat transfer is always from a hotter substance or object to a colder one.

⟳ The First and Second Laws of Thermodynamics

The study of heat transfer is called *thermodynamics.* The principles behind thermodynamics have been summarized into scientific laws that are consistent with observations of a large number of systems.

The **first law of thermodynamics** states that energy is always conserved. The heat transferred *from* a hotter object is equal to the heat transferred *to* the colder objects surrounding it. There is no loss of energy. Likewise, you cannot create energy. It can only be transferred from one place to another place, and it can change form, such as when solar energy is converted to electrical energy.

Big Idea Energy is conserved. It cannot be created or destroyed.

The **second law of thermodynamics** states that energy tends to disperse, or spread out. Energy transfers from hotter objects to colder objects until the temperature evens out. It does not transfer from a cold object to a hot object because this would concentrate the energy in one place rather than spread it out. This concept helps to explain why you can sit far away from a fire and become warmed by it. The energy from the fire disperses, or spreads out, from its source.

Another way to explain this concept is to say that the **entropy** of a system tends to increase over time. This means that energy and matter have a natural tendency to become more dispersed and disordered rather than more collected and ordered. For example, when you open a bottle of perfume, the molecules disperse throughout the room. The molecules will not naturally collect back into the bottle.

LESSON SUMMARY

What do temperature differences indicate about heat transfer?

The sensory experience of hot or cold depends on the direction of heat transfer. If your hand is at a higher temperature than what you are touching, the object will feel cold. If your hand is at a lower temperature, the object will feel warm. Heat is always transferred from a hotter object or to a colder one. Heat transfer is associated with phase changes. An ice cube melts because heat is transferred into it from the surroundings. If objects remain in contact with each other, heat is transferred until objects are at the same temperature. This balanced state is referred to as thermal equilibrium. Two laws of thermodynamics are related to these observations: The first law states that energy is conserved, and the second law states that energy tends to disperse, or spread out.

KEY TERMS

thermal equilibrium

first law of thermodynamics

second law of thermodynamics

entropy

Exercises

Reading Questions

1. Why does an ice cube feel cold to your hand?
2. What is thermal equilibrium?

Reason and Apply

3. Is energy transferred to a liquid or from a liquid when it evaporates? When it solidifies?
4. Is energy transferred to a gas or from a gas when it condenses?
5. What is evaporation and why is it a cooling process? What cools during evaporation?
6. What determines the direction of heat transfer?
7. Do humans ever reach thermal equilibrium with the surrounding air? Explain.
8. Provide evidence to support the first law of thermodynamics.
9. Provide evidence to support the second law of thermodynamics.
10. Explain why a campfire burns out. Explain why the ashes eventually cool to the temperature of the surrounding air.
11. Could there ever be a situation in which an ice cube actually heats up something else? Explain.

LESSON 96 ANSWERS

1. The ice cube feels cold to your hand because heat is being transferred from your hand to the ice cube, which is at a lower temperature.

2. Thermal equilibrium is the state that is reached when two objects in contact with each other exchange heat until they are at the same temperature.

3. When a liquid evaporates, energy is transferred to the liquid from the surroundings. When a liquid solidifies, energy is transferred from the liquid to the surroundings.

4. When a gas condenses, energy is transferred from the gas to the surroundings.

5. Evaporation is the process of a liquid changing into a gas. It is a cooling process because the substance gains energy during the phase change, so that energy must be removed from the surrounding environment. Objects and substances in the surroundings of the evaporating liquid are cooled.

6. The relative temperatures of two objects or substances determine the direction of heat transfer. Heat always flows from the warmer object to the cooler object.

7. *Possible answer:* Yes, humans reach thermal equilibrium with the surrounding air when they do not feel too hot or too cold. When humans are not in equilibrium, their brains tell them they are too hot or too cold so that they adjust the amount of clothing they are wearing or they move to a different location. Another possible answer is that a person reaches thermal equilibrium when his or her body temperature is the same as the outside environment. Note that body temperature is about 98 °F, so this equilibrium would feel hot to most people. The body has to maintain a higher temperature than the environment to carry out the processes necessary to survive.

8. *Possible answer:* When an object is warmer than its environment, the object cools and the environment warms until they are at the same temperature. For example, when a metal pot is taken off a hot stove and placed on a cool countertop, the metal pot gets cooler and the countertop gets warmer. This observation supports the first law of thermodynamics, which states that energy is never lost but is conserved.

9. *Possible answer:* When an object is warmer than its environment, the heat spreads into the environment. For example, a fire will eventually cause air in all parts of a room to feel warmer. The observation that heat energy always flows out of an object into its surroundings or into a cooler object from its surroundings supports the second law of thermodynamics, which states that energy tends to disperse.

10. A campfire burns out when there is not enough fuel available to keep it going. Either all of the substance is used up or no oxygen is available. The ashes eventually cool to the temperature of the surrounding air because energy tends to disperse as stated in the second law of thermodynamics.

11. Yes, if the ice cube is warmer than the air around it or an object that is touching it, heat will flow from the ice cube to its surroundings or into the other object. The first law of thermodynamics implies that temperature is relative, so that only a comparison of the temperatures between two objects is important, not the absolute temperature of an object.

Overview

COMPUTER ACTIVITY: WHOLE CLASS

LAB: PAIRS

Key Question: What is the difference between temperature and heat?

KEY IDEAS

Temperature is a measure of the average motion (kinetic energy) of the atoms and molecules in a sample. Thermal energy is the total kinetic energy of the atoms and molecules in a sample. Thermal energy depends on the amount of a sample, while temperature does not. During heat transfer, thermal energy is measured as heat in units of energy called calories. When water is heated, the amount of heat transfer (q) in calories is equal to the mass of the water times the change in temperature, or $q = (1 \text{ cal/g °C}) \, m\Delta T$.

LEARNING OBJECTIVES

- Define a calorie.
- Complete simple calculations involving heat transfer for water samples.
- Describe basic differences between thermal energy, heat, and temperature.

FOCUS ON UNDERSTANDING

- Vocabulary probably will remain an issue for students, because the word *heat* is used inconsistently and, sometimes, inaccurately in science literature as well as in everyday language.
- Heat is a process that describes energy transfer, whereas thermal energy refers to the energy content of a sample relative to its mass and temperature.

KEY TERMS

thermal energy
calorie

IN CLASS

Students view a computer simulation of the mixing of two gas samples at different temperatures from a molecular viewpoint. Students then work in pairs to measure what happens when they mix two water samples of different temperatures. They design experiments to determine the effects of mixing water of different masses and different temperatures. The unit *calorie* is introduced, along with the mathematical relationship used to calculate the heat transfer to and from water. Specific heat capacity is introduced. A complete materials list is available for download.

TRM Materials List 97

SETUP

Simulation: Prepare to show a computer simulation during the discussion. Use a simulation that shows two gas samples coming to thermal equilibrium, preferably one that allows you to vary the starting amounts of gas. You can find a list of URLs for this lesson on the Chapter 18 Web Resources document.

TRM Chapter 18 Web Resources

ENGAGE (5 min)

TRM PowerPoint Presentation 97

Featured LAB

Heat Transfer

Purpose

To explore the factors that affect heat transfer between two samples of water.

Materials

- large foam cup
- 250 mL beakers (2)
- 100 mL graduated cylinder
- thermometer
- hot plate
- beaker tongs
- ice in a large beaker or plastic tub

Procedure

1. Measure and record a volume of water. Pour it into the foam cup and record the temperature.
2. Measure and record a second volume of water. Pour it into a beaker and either heat or cool the water using a hot plate or ice. Record the temperature.
3. Add the second sample to the first. Stir and watch the thermometer closely. Record the temperature after the temperature has stabilized.
4. Repeat the experiment using different volumes and temperatures. Create a data table with results from at least three trials.
5. Convert your volume measurements to mass. The density of water is 1 g/mL.

Analysis

1. Look for relationships in your data table. What factors affect the temperature change?
2. Chemists measure the amount of heat transfer from one sample to another in units of calories. A *calorie* is the amount of energy needed to raise the temperature of 1 g of water by 1 °C. How does this relate to your data?
3. Write an equation relating the masses of the two water samples, m_1 and m_2, and the temperature changes, ΔT_1 and ΔT_2, of these two masses of water.
4. **Making Sense** What happens to the motions of the water molecules if you mix hot and cold water? Use changes in molecular motion to explain how energy from the hot water transfers to the cold water.
5. **If You Finish Early** Predict the final temperature if you mix 15 g of water at 20 °C with 45 g of water at 80 °C. Show your work.

494 Chapter 18 | **Observing Energy**

ChemCatalyst

Which will melt more of a snowman, a cup of water at 90 °C or a large barrel of water at 20 °C? Explain your thinking.

Sample answer: Although the barrel of water is at a lower temperature, it contains a very large volume of water that is warmer than the snow and will affect a larger amount of snow.

→ Discuss heat and temperature.

Ask: What factors are important in heat transfer?

Heat versus Temperature

Heat Transfer

THINK ABOUT IT

Imagine that you want two samples of really hot water. You want a large sample to make soup and a smaller sample for a cup of hot cocoa. Will it take the same amount of heat to get both samples to the same temperature?

What is the difference between temperature and heat?

To answer this question, you will explore

- Particle View of Energy
- Measuring Heat Transfer

EXPLORING THE TOPIC

○ Particle View of Energy

Consider a sample of water molecules that is heated. What happens to the molecules as more energy is transferred to the sample?

Same volume, different amounts of thermal energy

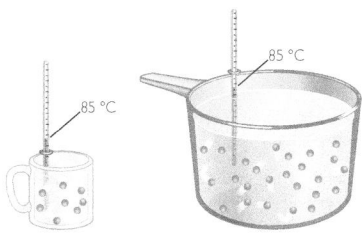

Same temperature, different amounts of thermal energy

As shown in the illustration (left), atoms and molecules move faster as they are heated. As the movement of the molecules increases, the temperature of the sample goes up.

Now consider two samples of water that contain different numbers of molecules. While the temperatures might be identical, the amount of energy represented by the two samples is not the same. In one sample, you have many quickly moving particles. In the other sample, you have fewer particles, even though they are moving just as quickly. There will be more total energy in the larger sample.

If the water samples in the large pot and the small cup shown in the illustration (bottom left) are both at 85 °C, you could say that they both have the same "degree of hotness," or temperature. However, the water in the large pot has a greater total "quantity of hotness" because it is a larger sample. This second concept is referred to as **thermal energy.** Thermal energy describes the total amount of energy in the particles of a sample. Because the temperature is the same, the gas molecules in both samples are moving with the same average kinetic energy. Because their volumes are different, the two samples have different amounts of thermal energy. So thermal energy is dependent on both the temperature of the sample *and* on the number of particles in the sample.

Lesson 97 | **Heat Transfer**   495

Ask: What do you think happens from a molecular viewpoint when energy is transferred to a water sample?

Ask: Can you give an example of how heat transfer and temperature are different from each other? (Eating a large bowl of soup makes you warmer than taking a sip of hot tea; a tiny piece of ice does not cool the ground as much as 3 ft of snow.)

TRM **Worksheet with Answers 97**

TRM **Worksheet 97**

SHOW THE COMPUTER SIMULATIONS

→ Show the first experiment. There is a temperature difference between two gas samples separated by a removable barrier. The number of gas particles in each sample is the same. Ask students to predict what will happen if the barrier is removed. Then remove the barrier between the two samples.

→ Show another experiment in which the temperature difference is the same as in the first experiment but the number of gas particles on the colder side is less than the number of gas particles on the hotter side. Ask students to predict the outcome before you remove the barrier between the two samples.

→ Discuss what happens when heat transfer occurs from a molecular point of view.

INTRODUCE THE LAB

→ Arrange students into pairs.

→ Pass out the student worksheets.

→ Explain to students that they should mix water samples of different sizes and temperatures to discover what affects the final temperature of the mixture.

→ Tell students to keep careful notes on their procedures and to create a data table showing their results.

TRM **Transparency—Temp Work Class Data 97**

DISCUSS STUDENTS' OBSERVATIONS

→ Use the Transparency—Temp Work Class Data to fill in the data students have collected. Make sure students understand why the numbers for volume of water and mass of water are identical.

Sample 1 mass (g)/ temp. (°C)	Sample 2 mass (g)/ temp. (°C)	Final temp. (°C)
100 g/20 °C	100 g/80 °C	50 °C
100 g/20 °C	100 g/70 °C	45 °C
100 g/20 °C	100 g/60 °C	40 °C
100 g/20 °C	300 g/80 °C	65 °C
200 g/20 °C	100 g/80 °C	40 °C
300 g/20 °C	100 g/80 °C	35 °C

Ask: How is the temperature of the final mixture related to the temperatures of the two samples?

Ask: How is the temperature of the final mixture related to the mass of the two samples?

Ask: What is a molecular explanation of why heat transfer depends on both the temperature and the mass of a sample?

Ask: What is the difference between heat transfer and temperature?

Key Points: The final temperature of a mixture depends on both the mass and the temperature of the samples. When two samples of equal volume are mixed, the final temperature is halfway between the two initial temperatures. If one sample is larger, the final temperature will be closer to the initial temperature of the larger sample.

Heat and temperature are different concepts. Temperature is a measure of the average kinetic energy of the atoms and molecules in a sample. Heat is related to the total kinetic energy of all the atoms and molecules in the sample. *Thermal energy* refers to the amount of energy in a sample. Thermal energy is dependent on both the mass and the temperature of the sample and can only be measured as energy transfer between two samples of known temperature and mass. Because heat is a *process* of energy transfer, it is incorrect to refer to the "heat energy of a sample." The correct term is "thermal energy of a sample." A larger sample of cool water may have more thermal energy than a smaller sample of very hot water. The thermal energy depends on both the motions and the number of water molecules. A sample with greater thermal energy has the potential to transfer more energy as heat.

> Thermal energy: The total kinetic energy associated with the mass and motions of the particles in a sample of matter.

INTRODUCE THE CALORIE AS A UNIT OF ENERGY

Ask: What is a calorie?

Ask: How does the unit of a calorie help to quantify the energy required to raise the temperature of a certain quantity of water?

Key Points: The *calorie* is a unit of energy used to measure heat transfer and to express thermal energy. It takes 2 calories of energy to raise the temperature of 2 g of water by 1 °C or to raise the temperature of 1 g of water 2 °C. The equation for heat transfer in water is

> $q = m(1 \text{ cal/g °C})\Delta T$
>
> heat transferred = (mass)(1 cal/g °C)(change in temperature)

The value 1 cal/g °C applies only to water.

> calorie: The amount of energy it takes to raise the temperature of 1 gram of water by 1 Celsius degree.

Big Idea Temperature depends on the average kinetic energy of matter. Thermal energy depends on the average kinetic energy and the mass of the sample.

↻ Measuring Heat Transfer

Unfortunately, you cannot measure the thermal energy of a sample of matter directly. A thermometer gives you only an indication of the average kinetic energy. However, changes in thermal energy can be determined using heat transfer.

Imagine that you have three samples of water at different temperatures: a pitcher of water at 80 °C, a beaker of water at 80 °C, and a pitcher of water at 50 °C. Your common sense tells you that the largest 80 °C sample is the one with the greatest thermal energy. But can that energy be measured or quantified?

You can calculate the amount of thermal energy in each sample by determining how much energy can be transferred from each one. One way to do this is to use the energy in the three samples to warm identical samples of cold water.

Warming 100 mL samples of water

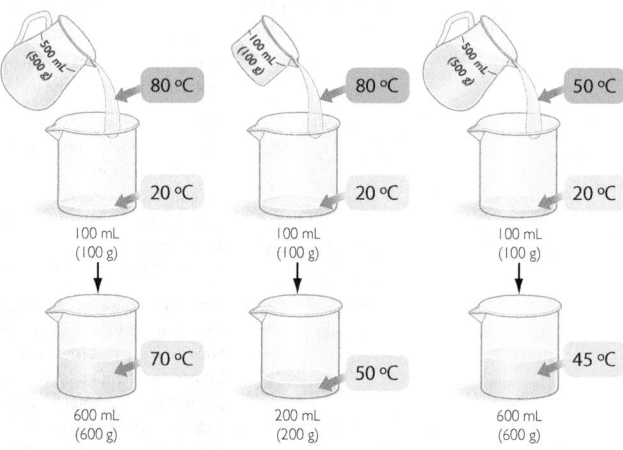

Notice that the large sample of water at 50 °C warms the cold water sample nearly as much as the small sample of water at 80 °C.

In each case, the mixture reaches thermal equilibrium. That is, the faster moving molecules in the hot sample slow down and the slower moving molecules in the cold sample speed up until they are all moving with the same average kinetic energy. The transfer of energy happens through collisions of the molecules as the

EXAMPLE

How much heat transfer is required to raise the temperature of 5.0 g of water from 25 °C to 75 °C?

SOLUTION

> Mass = 5.0 g
>
> $\Delta T = 50$ °C
>
> $q = (5.0 \text{ g})(1 \text{ cal/g °C})(50 \text{ °C})$
>
> $q = 250$ calories

EVALUATE (5 min)

Check-In

You have one beaker containing 500 g of water at 75 °C. You also have a beaker with 2000 g of water at the same temperature. Which will melt more ice? Explain your answer in terms of thermal energy, heat transfer, and temperature.

samples mix. The temperature of the final mixture depends on the temperatures and the masses of the samples that were mixed.

A UNIT OF ENERGY: THE CALORIE

Scientists define a unit called a **calorie** to measure heat transfer. A calorie is the amount of energy needed to raise the temperature of 1 gram of water by 1 Celsius degree. So, this unit of energy is based on heat transfer to or from water. To determine the number of calories needed, you multiply the mass of the water by the temperature increase in Celsius degrees. This can be summarized by the mathematical equation for the heat transfer, q, in terms of the mass of water, m, and the temperature change, ΔT. The uppercase Greek letter delta, Δ, is used to represent a change in quantity.

$$q = m(1 \text{ cal/g} \cdot {}^{\circ}\text{C})\Delta T$$

(Remember, 1 g of water is equivalent to 1 mL of water at room temperature.)

So, for example, to raise the temperature of 500 grams of water by 10 °C, from a temperature of 20 °C to a temperature of 30 °C, it would take 5000 calories of energy.

[For a review of this math topic, see **MATH Spotlight**: Averages on page A-7.]

Example 1

Calories

Suppose that you need to heat some water.

a. How many calories of energy does it take to heat a 75 mL sample of water for cocoa from 25 °C to 38 °C?

b. How many calories of energy does it take to heat a 2 L sample of water for a footbath from 25 °C to 38 °C?

c. Which sample of water contains more thermal energy: the water for cocoa or the water for the footbath?

Solution

Convert volume to mass in grams. Determine the change in temperature ΔT for both situations. Then use the equation for heat transfer to solve for q.

a. 75 mL water = 75 g water

$\Delta T = 38 - 25 = 13\,{}^{\circ}\text{C}$

$q = m(1 \text{ cal/g} \cdot {}^{\circ}\text{C})\Delta T$

$\quad = 75 \text{ g}(1 \text{ cal/g} \cdot {}^{\circ}\text{C})13\,{}^{\circ}\text{C}$

$\quad = 975 \text{ calories}$

b. 2 L water = 2000 g water

$\Delta T = 38 - 25 = 13\,{}^{\circ}\text{C}$

$q = m(1 \text{ cal/g} \cdot {}^{\circ}\text{C})\Delta T$

$\quad = 2000 \text{ g}(1 \text{ cal/g} \cdot {}^{\circ}\text{C})13\,{}^{\circ}\text{C}$

$\quad = 26{,}000 \text{ calories}$

c. The water in the footbath represents a greater amount of thermal energy.

Lesson 97 | **Heat Transfer**   497

Answer: The beaker with the greater number of water molecules has more thermal energy and will transfer more energy. Even though the temperatures of the two samples are equal, the larger sample will melt more ice.

Homework

Assign the reading and exercises for Fire Lesson 97 in the student text.

Example 2

Conservation of Energy

You mix 300 mL of water at 60 °C with 100 mL of water at 20 °C.

a. Find the final temperature of the mixture.

b. Show that the heat transferred *from* the hot water is equal to the heat transferred *to* the cold water.

Solution

a. The final temperature will be a weighted average of the two initial temperatures. (Each temperature is "weighted" by multiplying it by the mass of each sample that is mixed.) The masses of the samples are 300 g and 100 g.

$$\frac{300 \text{ g } (60 \text{ °C}) + 100 \text{ g}(20 \text{ °C})}{400 \text{ g}} = 50 \text{ °C}$$

The final temperature is 50 °C. It makes sense that the final temperature is closer to 60 °C, since there was more of the 60 °C water.

b. The calories of energy transferred is given by this equation:

$$q = m(1 \text{ cal/g} \cdot \text{°C})\Delta T$$

Remember, 300 mL of water = 300 g of water.

Hot water: $q = (300 \text{ g})(1 \text{ cal/g} \cdot \text{°C})(50 \text{ °C} - 60 \text{ °C}) = -3000$ calories

Cold water: $q = (100 \text{ g})(1 \text{ cal/g} \cdot \text{°C})(50 \text{ °C} - 20 \text{ °C}) = 3000$ calories

The negative sign means that energy is transferred *from* the hot water. The positive sign for the cold water indicates that energy is transferred *to* the cold water. Based on the calculated answers, energy is conserved.

LESSON SUMMARY

What is the difference between temperature and heat?

KEY TERMS

thermal energy

calorie

Temperature and heat are different, although they are both related to the motion of atoms and molecules. The temperature of a substance is a measure of the average kinetic energy of its particles. Heat is a process of energy transfer to or from a sample of matter. Thermal energy is a term often used to describe the "energy content" of a sample. The thermal energy of a sample depends on the number of particles in a sample, whereas temperature is the same regardless of the size of the sample. A calorie is a unit of energy. It is defined as the energy required to raise the temperature of 1 g of water by 1 °C. The energy transferred is referred to as heat.

Reading Questions

1. If you heat up a large and a small sample of cold tap water to the same temperature, explain why it takes more energy to heat up the larger sample.

2. What is the difference between thermal energy and temperature?

Exercises *(continued)*

Reason and Apply

3. How many calories of energy do you need to transfer for each of the following changes?
   a. Raise the temperature of 1 g of water by 5 °C.
   b. Raise the temperature of 2 g of water by 5 °C.
   c. Raise the temperature of 9 g of water by 35 °C.

4. Which will warm a child's inflatable pool more: adding 500 g of water at 50 °C or 100 g at 95 °C? The temperature of the water in the pool is 20 °C. Explain your reasoning.

5. Suppose that two water samples are mixed. Copy the data table and fill in the amounts of heat transferred.
   a. How do the values in the last two columns compare?
   b. How does the amount of heat transferred depend on the change in temperature?
   c. How does the amount of heat transferred depend on the mass of each sample?

	Sample 1 mass (g)	Sample 1 temperature (°C)	Sample 2 mass (g)	Sample 2 temperature (°C)	Final temperature (°C)	Heat transfer to sample 1 (cal)	Heat transfer from sample 2 (cal)
Trial 1	100 g	20 °C	100 g	80 °C	50 °C		
Trial 2	100 g	20 °C	100 g	70 °C	45 °C		
Trial 3	100 g	20 °C	300 g	80 °C	65 °C		
Trial 4	200 g	20 °C	100 g	80 °C	40 °C		

6. Suppose that you mix two water samples: 300 g of water at 20 °C and 200 g of water at 50 °C. What do you expect the final temperature of the water to be?

7. From a molecular viewpoint, explain how thermal equilibrium is reached when hot and cold water mix.

Lesson 97 | **Heat Transfer** 499

requires 7500 calories. Therefore, the 500 g of water would have to transfer more heat into the pool for the temperature of the water in the pool to reach equilibrium.

5.

	Heat transfer to sample 1 (cal)	Heat transfer from sample 2 (cal)
Trial 1	3000 cal	−3000 cal
Trial 2	2500 cal	−2500 cal
Trial 3	4500 cal	−4500 cal
Trial 4	4000 cal	−4000 cal

a. The values in the last two columns are equal but with opposite signs. The values for sample 1 are always positive, and the values for sample 2 are always negative. **b.** The amount of heat transferred is proportional to the change in temperature. **c.** The amount of heat transferred is proportional to the mass of water in each sample.

6. The final temperature is a weighted average of the two initial temperatures, 32 °C.

7. *Possible answer:* Thermal equilibrium is reached when the particles in the hot water lose thermal energy and move more slowly, while the particles of cold water gain thermal energy and move more rapidly. The particles that originally move more slowly will collide more frequently with other particles moving more rapidly and will eventually speed up. The particles that are originally moving more rapidly will collide with particles moving more slowly and eventually will slow down.

LESSON 97 ANSWERS

1. Because the larger sample has more particles, it takes more energy to increase the average kinetic energy of that sample by the same amount as the smaller sample.

2. Thermal energy is a measure of the total kinetic energy of all the particles of a sample. Temperature is a measure of the average thermal energy of those particles.

3. a. 5 cal **b.** 10 cal **c.** 315 cal

4. $q = m(1 \text{ cal/g} \cdot {}^\circ\text{C})\Delta T$

500 g of water at 50 °C:
$q = (500 \text{ g})(1 \text{ cal/g} \cdot {}^\circ\text{C})(50 \text{ °C} - 20 \text{ °C})$
$= 15{,}000 \text{ cal}$

100 g of water at 95 °C:
$q = (100 \text{ g})(1 \text{ cal/g} \cdot {}^\circ\text{C})(95 \text{ °C} - 20 \text{ °C})$
$= 7500 \text{ cal}$

The 500 g of water at 50 °C will warm the water in the pool more. Cooling down 500 g of water at 50 °C to 20 °C requires 15,000 calories. Cooling down 100 g of water at 95 °C to 20 °C

LESSON 97

Lesson 97 | **Heat Transfer** 499

Overview

LAB: GROUPS OF 4

Key Question: How do different substances respond to heat?

KEY IDEAS

Substances differ in their response to heat. This property is called the *specific heat capacity* of a substance. Specific heat capacity is the amount of energy required to raise the temperature of 1 g of a substance 1 °C and is measured in calories. The specific heat capacity of a substance depends on its atomic composition and structure.

LEARNING OBJECTIVES

- Define specific heat capacity.
- Complete simple specific heat capacity calculations.

KEY TERMS

specific heat capacity
hydrogen bond

IN CLASS

Students place a metal sample at room temperature into a foam cup and add a measured quantity of hot water. They record the temperature of the water when it has reached the same temperature as the metal. Students compare the amount of heat transferred out of the hot water and into the metal sample and are introduced to the concept of specific heat capacity. A complete materials list is available for download.

TRM Materials List 98

Differentiate

Some students will understand that energy transferred from one object is equivalent to the energy transferred to another object. However, students may struggle to set up the mathematical equations to predict temperature changes or determine the amount of energy transferred. Create "math help" index cards with the procedure written out step by step. Punch holes in the index cards and connect them using a metal ring, tied loop of string, or pipe-cleaner loop. Generate several sets of these cards and have them available to students to use as needed.

Pre-AP® Course Tip

Students preparing to take an AP® chemistry course or college chemistry should know which data to collect to achieve a certain goal. In this lesson, students perform a coffee-cup calorimetry experiment to determine how different objects respond to heat. From the data collected, students complete simple calculations of specific heat capacity.

SETUP

Place large beakers of water on hot plates so that the water is at about 70 °C when students begin the lab.

SAFETY

Wear safety goggles. Have oven mitts available at each station, and instruct students to use them when they are handling hot water. Remind students to pour the hot water slowly, and warn group members to stand back.

The Heat Is On

Specific Heat Capacity

THINK ABOUT IT

Imagine that you put a metal pan and a pizza in a hot oven. Both the pizza and the pan start out at room temperature. You leave them in the oven for the same amount of time. Will they heat up to the same temperature? And once they are taken out of the oven, will they cool off differently?

How do different substances respond to heat?

To answer this question, you will explore

- Specific Heat Capacity
- Bonding, Numbers, and Heat

EXPLORING THE TOPIC

Specific Heat Capacity

Every substance has a unique response to heat. Suppose that you place an aluminum pot of water on the stove to heat. After a minute, the metal pot will be too hot to touch, but the water will still be cool. This is because different amounts of energy are needed to raise the temperature of each type of substance.

To compare heat transfer to different substances, you can measure the energy required to raise the temperature of the same mass of each substance by the same number of degrees. The amounts of energy needed to raise the temperature of 1 g of several sample substances by 1 °C are given in the table.

Sample	Mass (g)	ΔT (°C)	Energy required (cal)
water, $H_2O(l)$	1.0 g	1 °C	1.0 cal
methanol, $CH_3OH(l)$	1.0 g	1 °C	0.58 cal
aluminum, $Al(s)$	1.0 g	1 °C	0.21 cal
copper, $Cu(s)$	1.0 g	1 °C	0.09 cal

Notice that fewer calories of energy are required to raise the temperature of methanol, aluminum, and copper from 20 °C to 21 °C, compared with water. This is consistent with the observation that the aluminum pot heats up and cools off faster than the water in the pot. Less energy is involved in changing the temperature of the aluminum.

ENGAGE (5 min)

TRM PowerPoint Presentation 98

ChemCatalyst

Put a pot of water on the stove. Before long, you cannot touch the metal pot, but you still can put your finger comfortably into the water. Propose at least two different hypotheses to explain why.

Sample answers: More energy is needed to raise the temperature of water than to raise that of metal. The mass of the water is larger, so more

The heat required to raise the temperature of 1 g of a substance by 1 °C is called the **specific heat capacity**. Every substance has a specific heat capacity. Some values are shown in the table.

Substance	Specific heat capacity (cal/g · °C)
aluminum, Al(s)	0.21
water, $H_2O(l)$	1.00
copper, Cu(s)	0.09
iron, Fe(s)	0.11
wood (cellulose), $(C_6H_{10}O_5)_n(s)$	0.41
glass, $SiO_2(s)$	0.16
nitrogen, $N_2(g)$	0.24
ethanol, $CH_3CH_2OH(l)$	0.57
hydrogen, $H_2(g)$	3.34

The greater the specific heat capacity of a substance, the less its temperature will rise when it absorbs a given amount of energy. If a substance has a low specific heat capacity, it will change temperature easily, with a small transfer of energy. This applies whether the energy is being transferred into or out of the substance. This is why the aluminum pot heats up faster, and also cools down faster, than water. Water does not change temperature as easily as most substances.

Big Idea Substances with low specific heat capacities can heat up and cool down easily with only a small transfer of energy.

The specific heat capacity of a substance can be used to determine the amount of energy needed to raise the temperature of any mass by any number of degrees. The formula for heat transfer to any substance is a product of the mass, m, its specific heat capacity, C_p, and the temperature change, ΔT.

$$q = mC_p \Delta T$$

Example 1

Cooling

Suppose that you have 15 g of methanol, CH_3OH, and 15 g of water, H_2O, both at 75 °C. You want to cool both samples to 20 °C. How much energy (in calories) do you need to remove from each sample?

Solution

The energy transferred is a product of the mass, the specific heat capacity, and the change in temperature.

$$q = mC_p \Delta T$$

Lesson 98 | **Specific Heat Capacity**   501

---

energy is required to raise its temperature. The amount of energy needed to raise the temperature of a substance depends on the type of matter in the object. Water is a liquid and evaporates, a process that requires more energy. Heating liquids requires more energy than heating solids. Students also might say that heat transfers to metals more rapidly than to other substances.

Ask: Do you expect the burner on the stove to transfer heat to all of the surroundings at an equal rate? Why or why not?

Ask: Would a metal jacket keep you warm on a cold day? Explain your thinking.

Ask: If you left a set of keys and a glass of water out in the Sun for a while, would you expect them to be at the same temperature? Explain.

EXPLORE (15 min)

TRM Worksheet with Answers 98

TRM Worksheet 98

INTRODUCE THE LAB

→ Arrange students into groups of four.

→ Pass out the student worksheets.

→ Tell students that they will determine how much the temperature of a sample of brass is changed by placing the sample in hot water.

EXPLAIN & ELABORATE (15 min)

INTRODUCE SPECIFIC HEAT CAPACITY

→ Put a chart on the board with the specific heat capacity of several substances.

Ask: What did you observe when you poured hot water over a cold metal sample?

Ask: Why do you think the final temperature of the water was above the midpoint temperature of the two substances?

Ask: How can you determine the amount of heat transfer from the hot water to the brass sample?

Key Points: Different substances respond to heat transfer differently. When you add hot water to a metal sample, the temperature of the water decreases by only a small amount, but the temperature of the metal changes a great deal. This indicates that less thermal energy is required to change the temperature of a metal than to change the temperature of water. Brass has a lower specific heat capacity than water, which means that it takes fewer calories of energy per gram of brass to change its temperature.

Specific heat capacity: The heat required to raise the temperature of 1 g of a substance 1 °C.

Specific heat capacity is used to measure heat transfer. In today's lab, the heat transferred from the water is equal to the heat transferred to the metal rod, assuming no loss of heat to the surroundings. You know the mass of water and its temperature change, so you can use the equation for heat transfer, in which the variable C_p stands for the specific heat capacity.

$$q = mC_p \Delta T$$

EXAMPLE

You pour 50.0 g of hot water at 70.0 °C on 59.0 g of brass at 20.0 °C. The final temperature is 65.2 °C. Show that energy is conserved.

SOLUTION

First, determine the number of calories transferred from the water. The temperature change is 65.2 °C minus 70.0 °C, or −4.8 °C. The negative sign indicates that the heat is transferred from the hot water.

$$q = mC_p \Delta T$$
$$= (50.0 \text{ g})(1 \text{ cal/g °C})(-4.8 \text{ °C})$$
$$= -240 \text{ calories}$$

Next, determine the number of calories transferred to the brass. The temperature change is 65.2 °C minus 20.0 °C, or 45.2 °C.

$$q = mC_p \Delta T$$
$$= (59.0 \text{ g})(0.090 \text{ cal/g °C})(45.2 \text{ °C})$$
$$= 240 \text{ calories}$$

So, the amount of heat transferred from the hot water is equal to the amount of heat transferred to the brass. Because the specific heat capacity of brass, 0.090 cal/g °C, is lower than that of water, 1.0 cal/g °C, fewer calories are needed to raise the temperature of the brass. More calories of energy would need to be transferred from the water to change its temperature by the same amount.

EVALUATE (5 min)

Check-In

1. The specific heat capacity of copper is 0.09 cal/g °C. How many calories of energy are needed to heat 10 g of copper from 25 °C to 35 °C?

2. Do you need more or fewer calories to heat 10 g of water from 25 °C to 35 °C? Explain.

Answers: 1. $q = mC_p \Delta T = (10 \text{ g})$ (0.09 cal/g °C)(10 °C) = 9 calories.
2. More calories are needed to heat the same amount of water, because the specific heat capacity of water is greater:

$$q = mC_p \Delta T$$
$$= (10 \text{ g})(1.0 \text{ cal/g °C})(10 \text{ °C})$$
$$= 100 \text{ calories}$$

Homework

Assign the reading and exercises for Fire Lesson 98 in the student text.

Example 1

(continued)

The mass of each sample is 15 g, and the change in temperature is 20 °C minus 75 °C, or −55 °C. However, the specific heat capacities of water and methanol are different. The specific heat capacity of water is 1 cal/g · °C. The specific heat capacity of methanol is 0.58 cal/g · °C.

Water:
$$q = (15 \text{ g})(1 \text{ cal/g · °C})(-55 \text{ °C})$$
$$= -830 \text{ cal}$$

Methanol:
$$q = (15 \text{ g})(0.58 \text{ cal/g · °C})(-55 \text{ °C})$$
$$= -480 \text{ cal}$$

It takes a greater transfer of energy to cool water than to cool methanol. Because of this, water is a very good insulator and retains its temperature. Note that the negative signs indicate that energy is transferred from the system to the surroundings.

Example 2

Final Temperature

Imagine that you have 3.5 g of copper and 3.5 g of water. Both are at 25 °C. Which sample will be at a higher temperature if you transfer 150 cal to each? Show your work.

Solution

The energy transferred is a product of the mass, the specific heat capacity, and the change in temperature.

$$q = mC_p \Delta T$$

You know the value of q and the masses of the water and copper samples. You want to find the temperature change. Solve for ΔT. You can rearrange the equation:

$$\Delta T = \frac{q}{mC_p}$$

You can look up the specific heat capacities of water and copper in a table like the one on page 501. The specific heat capacity of water is 1 cal/g · °C. The specific heat capacity of copper is 0.09 cal/g · °C.

Water:
$$\Delta T = \frac{q}{mC_p} = \frac{150 \text{ cal}}{(3.5 \text{ g})(1 \text{ cal/g · °C})}$$
$$= 43 \text{ °C}$$

Copper:
$$\Delta T = \frac{q}{mC_p} = \frac{150 \text{ cal}}{(3.5 \text{ g})(0.09 \text{ cal/g · °C})}$$
$$= 480 \text{ °C}$$

The temperature of copper increases about ten times as much as the temperature of the same quantity of water when 150 calories of energy are transferred.

To find the final temperature, add the temperature change to the initial temperature.

Water: Final temperature = 25 °C + 43 °C = 68 °C

Copper: Final temperature = 25 °C + 480 °C = 505 °C

For the same amount of heat transfer, the copper is much hotter.

↻ Bonding, Numbers, and Heat

It may seem strange that different amounts of energy are needed to raise the temperature of different substances. Why would you need more energy for water, compared with the same mass of methanol, aluminum, or copper to heat them all to the same temperature?

The answer is partially related to molar mass. Look back at the table of specific heat capacities on page 501. While the molar mass of aluminum is 27.0 g/mol, the molar mass of copper is 63.6 g/mol. So, there are more atoms in 1 g of aluminum than in 1 g of copper. It makes sense that more energy is required to cause a larger number of atoms to move faster.

Bonding also has an effect on the specific heat capacity of a substance. Substances that are polar tend to have high specific heat capacities. This means that molecules will have higher specific heat capacities than metals.

Finally, complicated molecules with large numbers of atoms have a variety of internal vibrational motions. The atoms within the molecule can vibrate back and forth like balls on a spring. And the entire molecule can rotate. All of these motions need to be increased to raise the temperature. So, substantially more heat is required to increase the average kinetic energy of molecules compared with individual atoms.

HIGH SPECIFIC HEAT CAPACITY OF WATER

Water has a particularly high specific heat capacity for several reasons. First, the molar mass of water, 18.0 g/mol, is quite small. So there are a lot of water molecules to get moving in a 1 g sample. Second, because water is molecular, it has complex internal movements. Third, water consists of H_2O molecules with strong intermolecular attractions called **hydrogen bonds.** A hydrogen bond is the attraction between the positive and negative dipoles of different water molecules. The hydrogen bonding restricts the motions of the molecules. Extra heat needs to be added to overcome these attractions and raise the kinetic energy. This is why the water in the metal pot heats up so much more slowly than the metal pot itself.

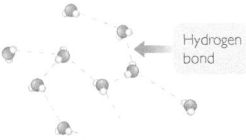

Hydrogen bond

LESSON SUMMARY

How do different substances respond to heat?

To increase the temperature of a sample of matter, you must supply energy to it. But the addition of a given amount of energy does not always result in the same rise in temperature. Substances differ in their responses to heat. Some substances, like metals, change temperature more than others in response to the transfer of the same amount of energy. This is because they have a lower specific heat capacity. Specific heat capacity is the amount of energy needed to raise the temperature of 1 g of a substance by 1 °C. The specific heat capacity of a substance depends on a number of factors, including the number of atoms or molecules in a sample and the type of bonding.

CONSUMER CONNECTION

To make ice cream, the ingredients must be mixed at a temperature below 0 °C. This is why you add salt to the ice in the outermost chamber of an ice-cream maker. Salt water has a lower freezing point than water without salt. The salt water with ice cubes is at a lower temperature than pure water with ice cubes. Heat is transferred from the ice cream to the salt-ice mixture, lowering the temperature to below 0 °C.

KEY TERMS

specific heat capacity

hydrogen bond

LESSON 98 ANSWERS

1. *Possible answer:* The specific heat capacity of a substance is the amount of thermal energy needed to raise the temperature of 1 g of the substance by 1 °C. For example, the specific heat capacity of aluminum is 0.21 cal/g · °C, which means that 0.21 cal of thermal energy is needed to increase the temperature of 1 g of water by 1 °C.

2. More thermal energy is needed to raise the temperature of 1.0 g of water compared with 1.0 g of aluminum because water has a higher specific heat capacity. It requires more energy to cause polar water molecules held together by hydrogen bonds to move faster than is required to cause the metallic bonded atoms in aluminum to move faster.

3. a.

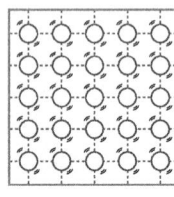 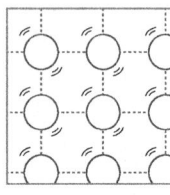

Al   Pb

Aluminum, Al, has a higher specific heat capacity because more atoms are in 1.0 g of aluminum than in 1.0 g of lead. Lead atoms have a greater mass than aluminum atoms.

b.

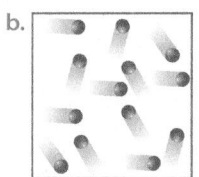 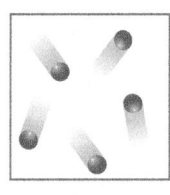

H₂   Ar

Hydrogen, H₂, has a higher specific heat capacity than argon, Ar, because more particles are in 1.0 g of hydrogen gas than are in 1.0 g of argon gas. Argon atoms have a greater mass than hydrogen gas molecules. Also, molecules require more heat for their temperature to increase than do atoms.

c.

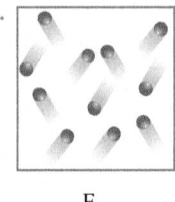 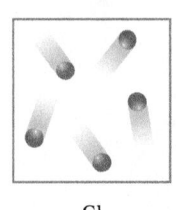

F₂   Cl₂

Fluorine, F₂, has a higher specific heat capacity than chlorine, Cl₂ because more particles are in 1.0 g of fluorine gas than

in 1.0 g of chlorine gas. Chlorine gas molecules have a greater mass than fluorine gas molecules.

4. Copper would be hottest to the touch because the specific heat capacity of copper is the lowest among them. Because less energy is required to raise 1 g of copper by 1 °C than for all the other substances in the table, its temperature would increase the most. Hydrogen gas would be the coolest to the touch because the specific heat capacity of hydrogen is the highest. Because more energy is required to raise 1 g of hydrogen by 1 °C than for all the other substances in the table, its temperature would increase the least.

5. *Possible answer:* Because metals have a lower specific heat capacity than plastic, they experience a greater change in temperature when they absorb a specific amount of energy from sunlight. Plastics are composed of large molecules that require more energy to rise in temperature than a metal does.

6. 1500 cal (to two significant figures)

7. The copper will cool faster because it has a lower specific heat capacity. Copper does not have to lose as much thermal energy as aluminum does to cool to the temperature of the ice water.

Answers continue in Answer Appendix (p. ANS-8).

Exercises

Reading Questions

1. What is specific heat capacity? Give the specific heat capacity of a substance, and explain what that means in terms of that substance.
2. Why is more energy required to raise the temperature of 1.0 g of water compared with 1.0 g of aluminum?

Reason and Apply

3. Which substance in each pair has the higher specific heat capacity? Justify your choice with an atomic view.
   a. Al or Pb   **b.** H₂ or Ar   **c.** F₂ or Cl₂
4. Imagine that you transfer the same amount of energy to all of the substances in the table of specific heat capacities on page 501. Which one would be hottest to the touch? Coolest? Explain your reasoning.
5. Use specific heat capacity to explain why metal keys left in the sun are hotter than a plastic pair of sunglasses left in the sun.
6. How much energy is required to raise the temperature of 50 g of methanol from 20 °C to 70 °C?
7. If you place 20 g of aluminum and 20 g of copper, both at 20 °C, into ice water, which will cool faster? Explain.
8. Imagine that you have 50 g of water and 50 g of methanol, each in a 100 mL beaker. You place both on a hot plate on low heat.
   a. Which sample will be at a higher temperature after 5 min? Explain your thinking.
   b. If the initial temperature of each liquid is 23 °C, what is the temperature of each after 25 cal of energy are transferred from the hot plate to each sample?
9. Would water make a good thermal insulator for a home? Explain your reasoning.
10. Why do farmers spray water on oranges to protect them from frost? Give two reasons why the water on the orange might help.

504   Chapter 18 | **Observing Energy**

Where's the Heat?

Heat and Phase Changes

THINK ABOUT IT

Water boils at 100 °C, at 1 atmosphere pressure. Once it reaches that temperature, liquid water does not get any hotter, no matter how big the flames are under its container or how much you heat it. If the temperature of the boiling water is not changing, where is all that heat going?

What happens to the heat during a phase change?

To answer this question, you will explore

- The Heating Curve of Water
- Heat and Phase Changes

EXPLORING THE TOPIC

The Heating Curve of Water

Imagine that you have an ice cube that you have just removed from the freezer. The temperature of the freezer is below zero. The graph shows the changes in temperature as you heat the ice cube from below 0 °C to above 100 °C. The *x*-axis represents time. Because the water is being heated steadily, as time goes by, more and more heat is transferred to the water. This graph is the heating curve of water.

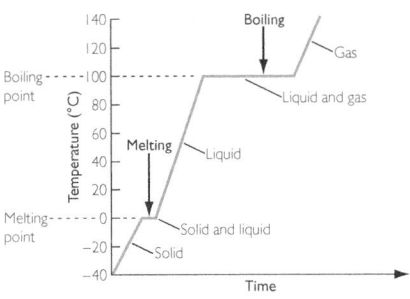

Heating Curve of Water

As you can see, the heating curve of water is a series of steep inclines with flat horizontal parts in between. Along the horizontal parts, energy continues to be transferred into the water, but the temperature does not change. The flat lines occur when melting or boiling is taking place, at 0 °C or 100 °C.

Lesson 99 | **Heat and Phase Changes** 505

- The energy associated with phase change is sometimes called latent heat, though this term is not much in use anymore.

IN CLASS

Students view a demonstration of the mixing of equal masses of ice and hot water. They explore a computer simulation of kinetic energy, potential energy, and particle behavior during phase changes. Students then carry out two experiments: They monitor the temperature of a beaker filled with ice water until the water has warmed to room temperature, and they determine how much ice is required to cool hot water to a specified temperature. The concept of latent heat is introduced, and students determine the latent heat of fusion for water. During the discussion, students learn about the heating curve of water. A complete materials list is available for download.

TRM Materials List 99

Differentiate

If students might struggle with the concept that temperature remains the same during phase change, have them use a temperature probe and software to gather a large set of data with short time intervals rather than use alcohol thermometers. This will allow students to focus on interpreting the data without having to take frequent measurements. A large data set can help make the plateau and sloped regions of the heating curve more obvious. If equipment is limited, have some groups use alcohol-thermometers and others use the digital setup. Students can then compare results obtained with methods of data collection.

SETUP

Demonstration: Heat 100 mL of water to about 85 °C. Have ice handy to put into the water.

Simulation: Prepare to show a computer simulation during the discussion. Use a simulation that shows latent heat of a nanoparticle as well as changes in kinetic and potential energy. You can find a list of URLs for this lesson on the Chapter 18 Web Resources document.

TRM Chapter 18 Web Resources

Lesson Guide

ENGAGE (5 min)

TRM PowerPoint Presentation 99

Overview

DEMO: WHOLE CLASS

COMPUTER ACTIVITY: WHOLE CLASS

LAB: GROUPS OF 4

Key Question: What happens to the heat during a phase change?

KEY IDEAS

During a phase change, the transfer of heat into a substance does not cause its temperature to rise. Instead of increasing molecular motion, the transferred energy alters the attractions between atoms and molecules.

LEARNING OBJECTIVES

- Define latent heat.
- Describe the energy exchanges that take place during a phase change.
- Explain the heating curve of water.

KEY TERMS

heat of vaporization
heat of fusion

FOCUS ON UNDERSTANDING

- Students often believe that temperature increases or decreases during phase change. They often revert to this belief even after witnessing firsthand that it is not the case.

ChemCatalyst

Imagine that you have a sample of 100 g of water at 85 °C. You place 100 g of ice at 0 °C into the water.

1. Predict the final temperature from the choices given. Explain your choice.

(A) 80 °C

(B) 60 °C

(C) 42.5 °C

(D) 20 °C

(E) 5 °C

2. Describe how you might measure the amount of heat required to melt the ice.

Sample answers: **1. (E)** Heat is required to melt the ice and to raise the temperature of the mixture to the final temperature. (It is not necessary to share the correct answer with students who may choose **(C)** because equal quantities of water and ice are mixed.) **2.** To measure the amount of heat required to melt the ice, measure the final temperature of the mixture, then calculate the heat transferred from the water: $q = m(1 \text{ cal/g °C})\Delta T$.

→ Discuss heat transfer and phase change.

Ask: Will you need to transfer more heat, less heat, or the same amount of heat to raise the temperature of ice at 0 °C to 1 °C compared to raising the temperature of liquid water at 0 °C to 1 °C? Explain your thinking.

Ask: Provide evidence that a transfer of heat is needed to melt ice and to boil water.

Ask: Why do you think liquid water does not get hotter than 100 °C or colder than 0 °C?

EXPLORE (20 min)

TRM Worksheet with Answers 99

TRM Worksheet 99

DO THE DEMONSTRATION

→ Mix equal masses of ice and hot water. Measure the final temperature. It will be close to 0 °C. Ask students to explain why the final temperature is so low. Why was so much of the heat transferred to the ice?

SHOW THE COMPUTER SIMULATION

→ Run the simulation. Heat up the nanoparticles gradually. Wait.

Note that whenever the temperature is not changing, two phases are present. As soon as only one phase—solid, liquid, or gas—is present, the temperature is able to change again. Something is happening to the energy going into the water during phase change.

↺ Heat and Phase Changes

During melting and boiling, there is no doubt that energy is going into the substance. However, this energy is doing something besides making the molecules move faster. Remember, temperature is a measure of the average kinetic energy of the molecules. If the temperature is not going up, then the molecules are not moving any faster on average. Where is the energy going?

The answer lies in the intermolecular forces. These attractive forces hold particles in a solid or a liquid close to each other. When heat is transferred into a substance, like ice, the molecules in the ice move faster and the temperature rises. Once the temperature rises to 0 °C, the ice begins melting. When heat is transferred into melting ice, the average kinetic energy does not change, and the temperature does not rise. Instead, the energy transferred weakens the attractive forces between the molecules so that they can wiggle and rotate. Once the ice has melted completely, the energy transferred causes the molecules in the liquid to move faster, and the temperature rises once again.

During the boiling process, all of the energy transferred is used to completely break the intermolecular attractions so that molecules can move freely as a gas. As a result, no temperature change is observed during boiling.

Big Idea Heat transfer does not always result in a temperature change.

When a substance changes phase from solid to liquid or from liquid to gas, a certain amount of energy must be supplied to overcome the molecular or ionic attractions between the particles. This energy is changing the solid into the liquid or the liquid into a gas, without changing the temperature. For example, the heat required to change 1 gram of a substance from a liquid into a gas is called the **heat of vaporization.** The heat required to change 1 gram of a substance from a solid to a liquid is called the **heat of fusion.**

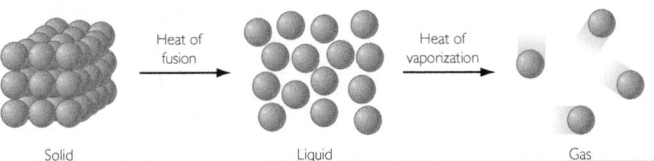

Solid   Heat of fusion →   Liquid   Heat of vaporization →   Gas

When a substance changes phase from a solid to a liquid, or from a liquid to a gas, energy is transferred *to* the substance from the surroundings. When a substance changes phase from a gas to a liquid, or from a liquid to a solid, energy is transferred *from* the substance to the surroundings.

→ Ask students what is happening to the kinetic and potential energy. Add more heat until the phase change begins. You will notice that the potential energy is increasing but the kinetic energy is not. Ask students what this indicates about the temperature. (If the average kinetic energy does not change, the temperature does not change.)

→ You will also notice more and more particles escaping into the gas phase. Ask students what this represents. (The substance is boiling.)

→ Keep adding heat until the kinetic energy begins to increase. Ask students what is happening to the particles. (All the particles are moving freely. Their speed is increasing.)

→ Ask students why they think the potential energy does not change for the gas. (The potential energy increases as the atoms in the nanoparticle break apart.)

Optional: Cool the particles to observe the reverse trends.

INTRODUCE THE LAB

→ Arrange students into groups of four.

→ Pass out the student worksheets.

→ Briefly go over the procedure: Part 1: Tell students that they will be monitoring the temperature of ice water until the water reaches 5–10 °C. This should take approximately 10 minutes, during which time students should complete

The table below shows some phase-change data for five common substances. Notice that it takes very little energy per gram to melt mercury (2.7 calories per gram), whereas it takes a great deal of energy to melt aluminum (94.5 calories per gram). Substances also require significantly more energy to change into gases than to melt.

Phase-Change Data

	Water	Ethanol	Mercury	Aluminum	Gold
Heat of fusion	79.9 cal/g	24.9 cal/g	2.7 cal/g	94.5 cal/g	15.4 cal/g
Melting point	0 °C	−117 °C	−39 °C	659 °C	1063 °C
Heat of vaporization	539 cal/g	205 cal/g	70 cal/g	2500 cal/g	377 cal/g
Boiling point	100 °C	78 °C	357 °C	2467 °C	2660 °C

Example

Heat of Fusion

You place 11 g of ice at 0 °C into 100 g of liquid water at 20 °C in an insulated water bottle. After a few minutes, the ice melts completely. Calculate the final temperature of the water in the bottle.

Solution

The heat of fusion of water given in the table is 79.9 cal/g. This means that, to melt 11 g of ice, 79.9 cal/g · 11 g, or 879 calories, are transferred. This energy is transferred from the 100 g of liquid water in the bottle. Use the equation for heat transfer to determine how much this amount of energy changes the temperature.

$$q = mC_p \Delta T$$
$$-879 \text{ cal} = (100 \text{ g})(1 \text{ cal/g} \cdot °C)\Delta T$$
$$\Delta T = -8.8 °C$$

The final temperature of the 100 g of water is 20 °C − 8.8 °C, or 11.2 °C. But there are also 11 g melted ice at 0 °C, so you need to find the weighted average of all the water in the bottle.

$$\frac{100 \text{ g } (11.2 °C) + 11 \text{ g } (0 °C)}{111 \text{ g}} = 10 °C$$

So, the final temperature of the water in your insulated water bottle will be around 10 °C.

LESSON SUMMARY

What happens to the heat during a phase change?

KEY TERMS
heat of vaporization
heat of fusion

Heat transfer does not always result in a temperature increase. During a phase change, the transfer of energy into a substance does not cause any rise in temperature. Instead, this energy is used to overcome the attractions between atoms and molecules.

Lesson 99 | **Heat and Phase Changes**   507

Part 2. Part 2: Ask students to determine how much ice is needed to cool water from 60 °C to 10 °C. They will have to design their own experiment.

EXPLAIN & ELABORATE (15 min)

TRM Transparency—Heating Curve of Water 99

TRM Transparency—Phase Changes 99

DISCUSS THE HEATING CURVE OF WATER

→ Display the transparency Heating Curve of Water. This image also appears on page 505 of the student text.

Ask: What did you discover when you heated the ice?

Ask: Why does the graph have places where heat is added but there is no rise in temperature?

Ask: What phases are present when the temperature is not changing?

Ask: From a molecular point of view, what does each part of the heating curve mean?

Key Points: The graph showing changes in temperature as you heat water from solid to liquid to gas is called the *heating curve of water*. The temperature of the ice rises until the ice begins to melt. When ice and water are present together, the temperature does not

change, and the graph is horizontal. Once the ice has melted completely, the temperature of the liquid rises. Once the water begins to boil, the temperature levels off again even though the water is still being heated.

During a phase change, the heat transferred to or from a substance does not result in a change in temperature. Instead of increasing the average kinetic energy of the water molecules, heat is causing a phase change. During melting and boiling, the heat transferred to the water changes the intermolecular interactions so that the water molecules are more free to move. The horizontal portions of the graph, where the temperature is not changing, are associated with the melting and boiling temperatures of water, 0 °C and 100 °C.

INTRODUCE HEAT OF FUSION AND HEAT OF VAPORIZATION

→ Display the transparency Phase Changes

Ask: What happens to heat that is transferred if the temperature of the substance does not change?

Ask: What happens to heat that is transferred when you reverse the process and go from a gas phase to a liquid phase?

Ask: How can you determine the heat of fusion for water?

Key Points: When a substance changes phase during heating, energy must be supplied to overcome the attractions between molecules. Heat is transferred, but there is no change in temperature. For example, when water boils, the heat transferred breaks water molecules apart so they can move freely as a gas. Heat that is required to change a substance from a liquid to a gas is called the *heat of vaporization*. The heat that is required to change a substance from a solid to a liquid is called the *heat of fusion*. This is illustrated on page 506 of the student text.

Heat involved in phase changes can be exothermic or endothermic, depending on the phase change. When a substance changes phase from a gas to a liquid or from a liquid to a solid, heat is transferred out of the substance and is negative. The cooling curve for water is the reverse of the heating curve.

EXAMPLE

You place 11 g of ice at 0 °C into 100 g of liquid water at 20 °C. The final temperature is 10 °C. Calculate the heat of fusion for water.

Lesson 99 | **Heat and Phase Changes**   507

SOLUTION

First, calculate the heat transferred from the liquid water to the ice:

Mass = 100 g

$\Delta T = 20\ °C - 10\ °C = 10\ °C$

$q = mC_p\ \Delta T$
$\quad = (100\ g)(1\ cal/g\ °C)(10\ °C)$
$\quad = 1000\ cal$

Next, calculate the heat transferred to the liquid water from the melted ice:

Mass = 11 g

$\Delta T = 10\ °C - 0\ °C = 10\ °C$

$q = mC_p\ \Delta T$
$\quad = (11\ g)(1\ cal/g\ °C)(10\ °C)$
$\quad = 110\ cal$

The difference in the two values is the heat of fusion for 11 g of ice.

Heat of fusion for 11 g of ice =
1000 cal − 110 cal = 890 cal

Heat of fusion for 1 g of ice =
890 cal/11 g = 81 cal/g

The actual value of the heat of fusion of ice is 79.9 cal/g, which means that adding 79.9 cal will melt 1 g of ice and removing 79.9 cal will freeze 1 g of liquid water at 0 °C.

EVALUATE (5 min)

Check-In

If you boil water on a stove, the temperature of the water will remain close to 100 °C. As long as some liquid water remains in the pot, the temperature of the water will stay the same regardless of the outside temperature. Explain why this is true.

Answer: As long as some liquid water remains in the pot, the phase change is not complete. Any energy entering the system will go into continuing to boil the water.

Homework

Assign the reading and exercises for Fire Lesson 99.

LESSON 99 ANSWERS

1. Water temperature does not change during boiling because the energy added to the water weakens the intermolecular interactions instead of adding to the kinetic energy of molecules.

2. Possible answer: Heat of vaporization is the amount of energy required to change 1 g of a material from the liquid

The heat of fusion is the energy required to change a substance from solid to liquid. The heat of vaporization is the energy required to change a substance from liquid to gas.

Exercises

Reading Questions

1. Why doesn't the temperature of water change during boiling?

2. What is heat of vaporization? Use a substance from the phase-change data table on page 507 as an example.

Reason and Apply

3. Use the table of phase-change data on page 507 to sketch a heating curve for mercury.

4. A jeweler pours liquid gold into a mold to make a 6 g wedding ring. How much energy is transferred from the ring when it solidifies? What temperature is the gold at the instant all of its atoms become solid?

5. Which of the following processes are exothermic? Explain your thinking.
   (A) boiling ethanol
   (B) freezing liquid mercury
   (C) subliming carbon dioxide

6. How much energy is required to melt these quantities of ice?
   a. 56 g
   b. 56 mol

7. If you transfer 5000 cal, how many grams of ethanol can you vaporize?

8. a. How much energy do you need to transfer to raise the temperature of 150 g of aluminum from 20 °C to its melting point?
   b. How much energy do you need to transfer to melt 150 g of aluminum?

9. You place 25.0 g of ice at 0 °C into 100 g of liquid water at 45 °C. The final temperature is 20 °C. Show that energy is conserved.

phase to the gas phase once the temperature of the material is at the boiling point. For example, the heat of vaporization of ethanol is 205 cal/g, requiring a heat input of 205 cal to change 1 g of ethanol from a liquid to a gas.

3.

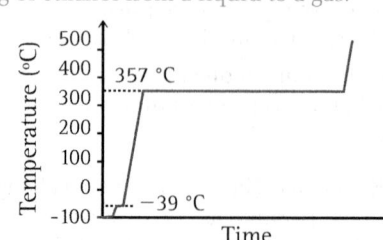

4. 92.4 cal. During the entire process of melting or freezing, the temperature of the gold remains at its melting point 1063 °C.

5. (B) Only freezing liquid mercury is exothermic. Boiling ethanol is endothermic. Energy must be added to break the intermolecular bonds that hold a liquid together and turn it into a gas. Freezing liquid mercury is exothermic. Energy is released when a material changes from a liquid to a solid because solid particles do not wiggle and rotate as much as liquid particles. Subliming carbon dioxide is endothermic. Before a solid can turn into a gas, energy must be added to break the intermolecular bonds that hold the particles of a solid in place.

Answers continue in Answer Appendix (p. ANS-8).

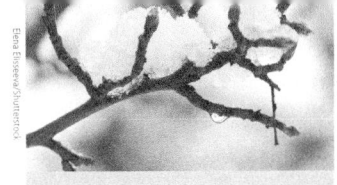

CHAPTER 18

Observing Energy

SUMMARY

KEY TERMS

energy
exothermic
heat
system
surroundings
endothermic
kinetic energy
thermal equilibrium
first law of thermodynamics
second law of thermodynamics
entropy
thermal energy
calorie
specific heat capacity
hydrogen bond
heat of vaporization
heat of fusion

Fire Update

Energy and heat can be difficult to define. However, one thing is certain: Whenever matter changes in any way, energy is involved. Here are some generalizations about heat:

- Fire is an exothermic chemical change. Fire releases energy in the form of heat and light.

- Heat is a transfer of thermal energy. Heat always transfers from a system at a high temperature to a system at a lower temperature until both systems are at thermal equilibrium.

- In an endothermic reaction, energy is transferred from the surroundings to the system. An ice pack feels cold because your body transfers heat to the pack.

- The energy transferred by heating does not always result in temperature change. Some of that energy goes into breaking intermolecular attractions during phase change.

- A metal pot will heat up and cool off more quickly than the water inside the pot. This is because different substances have different specific heat capacities.

REVIEW EXERCISES

1. Provide evidence to support the first and second laws of thermodynamics.

2. Why is it possible to touch a piece of aluminum foil shortly after it has been removed from a hot oven, but a piece of hot cheese pizza can burn you long after it has been removed?

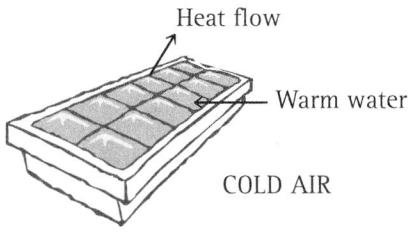

3. Thermal energy is transferred to raise the temperature of several substances as specified here. List the substances in order of increasing amount of energy that needs to be transferred. Show your work by determining the number of calories of energy transferred.
   a. Raise the temperature of 20 g of water by 25 °C.
   b. Raise the temperature of 200 g of copper by 50 °C.
   c. Raise the temperature of 20 g of methanol by 40 °C.

4. An ice cube tray can make 16 ice cubes. Each mold is filled with 25 mL of water, and the ice cube tray is placed in the freezer.
   a. As the water is freezing, is energy transferred from the freezer to the water, or from the water to the freezer? Draw a diagram of this scenario with an arrow showing the direction of heat transfer.
   b. Would the freezing of the water in the tray be considered endothermic or exothermic for the ice cube?
   c. Based upon your answer to part a, calculate the amount of heat transfer needed to freeze one ice cube.
   d. Based upon your answer to part a, calculate the amount of heat transfer needed to freeze all the ice cubes in the tray.

In order of increasing amount of energy transferred to the liquid: methanol, water, copper.

4. a. Heat is transferred from the water into the freezer. Heat is always transferred from warmer objects to colder objects.

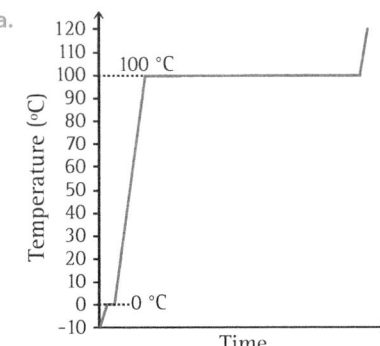

b. Freezing is exothermic because the process of a liquid becoming a solid releases energy.

c. The density of water is 1.0 g/mL, so 25 mL of water is equal to 25 g of water. q = (heat of fusion) · m = (79.9 cal/g)(25 g) = 2000 cal

d. 16 cubes · 2000 cal per cube = 32,000 cal

5. a. 69,000 cal (to two significant figures) **b. *Possible answer:*** After three minutes, the pot will be hotter than the water. The pot has a much lower specific heat capacity than the water, so it can change temperature much more rapidly. The water will take longer to reach a temperature of 150 °C than the iron pot. Also, the water would turn into steam at 100 °C, so the water could not possibly be at a temperature of 150 °C.

6. a. 35 °C **b.** The final temperature occurs when the water and the coin are at thermal equilibrium. Thermal equilibrium means that the two substances are at the same temperature, so the final temperature of the coin will also be 35 °C. **c.** 0.088 cal/g · °C

7. 57 °C

8. a.

```
Temperature (°C)
120
110
100  ···· 100 °C
 90
 80
 70
 60
 50
 40
 30
 20
 10
  0  ···· 0 °C
-10
              Time
```

b. ~37,000 cal (to two significant figures).

c. The part requiring the most heat is vaporizing water because much heat is required to overcome the forces that hold water molecules together in the

ANSWERS TO CHAPTER 18 EXERCISES

1. *Possible answer:* In accordance with the first law of thermodynamics, when an object is warmer than its environment, the object cools and the environment warms until they are at the same temperature. For example, when a metal pot is taken off a hot stove and placed on a cool countertop, the metal pot gets cooler and the countertop gets warmer. In accordance with the second law of thermodynamics, when an object is warmer than its environment, the heat spreads into the environment. For example, a fire will eventually cause air in all parts of a room to feel warmer.

2. The aluminum has a lower specific heat capacity compared to the cheese and sauce on a piece of pizza. Because the aluminum stores less heat energy, it loses energy more quickly and becomes cool enough to touch sooner.

3. a. Heat 20 g of water by 25 °C:

q = (20 g)(1.00 cal/g · °C)(25 °C) = 500 cal

b. Heat 200 g of copper by 50 °C:

q = (200 g)(0.09 cal/g · °C)(50 °C) = 900 cal

c. Heat 20 g of methanol by 40 °C:

q = (20 g)(0.58 cal/g · °C)(40 °C) = 460 cal

liquid state. These intermolecular bonds must break completely for the water molecules to move freely as they do in a gas.

Project: Uses of Fire

A good report would include: ● a physical description of the product, including images and/or diagrams when applicable ● uses of the product ● the sources of material for the product ● how the product is manufactured, harvested, or otherwise produced ● the typical path a product takes from its place of manufacture to the place at which it is used ● an analysis of how fire is involved in any of the above processes.

ASSESSMENTS

Two multiple-choice assessments (versions A and B) and two short-answer assessments (versions A and B) are available for download for Chapter 18.

TRM **Chapter 18 Assessment Answers**

TRM **Chapter 18 Assessments**

5. A 5 kg iron pot is filled with 500 mL of water, placed on the stove, and the unit is turned on. After three minutes have passed, the temperature of the iron pot has increased from 25 °C to 150 °C.
   a. How much heat was transferred to the pot?
   b. After three minutes, would you expect the water in the pot to be the same temperature as the iron pot? Or hotter or colder than the iron pot? Why?

6. A 30.4 g copper coin is heated to 400 °C with a Bunsen burner and then dropped into a beaker containing 100 mL of water at 25 °C, room temperature. The water absorbs 1000 cal of heat when this happens.
   a. What is the final temperature of the water in the beaker?
   b. Based upon your answer to part a, what will be the final temperature of the coin?
   c. What is the specific heat of the copper coin?

7. If 70 g of hot water at 85 °C are poured into 50 g of cold water at 18 °C, what is the temperature of the water mixture after it has come to thermal equilibrium?

8. A 50.0 g sample of pure water is heated from 10 °C to 120 °C.
   a. Draw and label a heating curve that represents the different stages of this process.
   b. If the specific heat capacity is 0.49 cal/g · °C for ice, 1.0 cal/g · °C for liquid water, and 0.50 cal/g · °C for steam, how much energy is added to a 50.0 g sample of water to raise its temperature from 10 °C to 120 °C?
   c. Use the heating curve and your calculations to determine which part of the process requires the most heat transfer. Explain your reasoning.

Uses of Fire

PROJECT

Pick an object that you might purchase, and consider different ways that fire might have been used to make it, harvest it, transport it, and so on. Describe step by step how fire may be involved in getting that object from its natural beginnings to you.

510   Chapter 18 | **Observing Energy**

CHAPTER 19

Measuring Energy

In this chapter, you will study

- substances that will and will not burn
- experimental methods for measuring heat
- how to compare different fuels
- quantifying energy

When charcoal burns in a barbecue grill, energy is released, allowing us to cook a tasty meal.

O
ne way to make use of the energy from chemical change is to burn a fuel. In homes, methane gas is burned in water heaters and kitchen stoves. In most automobiles, gasoline or some other fuel is burned. However, not all substances burn, and not all fuels are equally productive from an energy standpoint. Chapter 19 of Unit 5: Fire introduces you to ways that energy from chemical change can be observed and measured.

511

PD CHAPTER 19 OVERVIEW

Watch the video overview of Chapter 19 (for teachers) by clicking on the link in the TE-book, opening the TRFD, or logging onto the book's companion Web site bcs.whfreeman.com/livingbychemistry2e (teacher log-in required).

This chapter revolves around combustion reactions and calorimetry. Students learn how energy from chemical reactions is measured. In Lesson 100, students investigate the types of substances that burn and explore the chemistry of combustion. Lesson 101 is a double-period, open-ended lab that allows student groups to design their own calorimetry experiment. Lesson 102 is a follow-up lesson that debriefs, summarizes, and explains the calorimetry lab. In Lesson 103, students use calorimetry to compare two different alcohols as fuels. They are introduced to heat of reaction and consider different units of energy and different ways to compare fuels.

In this chapter, students will study

- combustion reactions
- heat of combustion values
- how calories are measured
- units of energy in calories and joules
- how to use heat of reaction to compare fuels

Chapter 19 Lessons

Overview

LAB: GROUPS OF 4

DEMO: WHOLE CLASS

Key Question: What types of substances burn?

KEY IDEAS

Combustion is informally called "burning." Combustion is defined as the combining of a compound with oxygen through a chemical reaction that releases energy as heat and light. Most ionic substances do not combust. Most molecular covalent compounds containing carbon are combustible. Carbon dioxide and water often are the products of combustion.

LEARNING OBJECTIVES

- Define and describe combustion.
- Describe the chemical composition of substances that burn.

FOCUS ON UNDERSTANDING

Students may find it surprising that metals combust, because in everyday use they do not. The reason is that it is very difficult to ignite metals unless they are finely divided and any surface oil or oxidized layer is cleaned off.

KEY TERM

combustion

IN CLASS

This lesson focuses on the fuel portion of fire. Students work in groups of four to discover what types of substances will combust. In Part 2, students make predictions about four more substances based on the patterns they have found and then confirm or refute their predictions based on teacher demonstrations. The term *combustion* is introduced. A complete materials list is available for download.

TRM Materials List 100

Differentiate

This lesson is an opportunity to reflect on previously learned material. Ask students to label their worksheets each time they find a concept, formula, or term that is a connection to prior units or lessons. More than one link may exist for a particular concept. Help students recognize, for example, that combustion is another reaction type, like those covered in Unit 4: Toxins. After students make connections, have them identify areas of strength and areas to improve.

You're Fired!

Combustion

THINK ABOUT IT

An automobile engine burns gasoline or ethanol. In a fireplace, you burn paper and wood. In a barbecue grill, you burn charcoal, lighter fluid, and an occasional marshmallow. While many substances burn, others—like sodium chloride, table salt, and water—do not.

What types of substances burn?

To answer this question, you will explore

⤷ Substances That Burn

⤷ Combustion Reactions

EXPLORING THE TOPIC

⤷ Substances That Burn

A wide variety of substances can burn, or combust. Solids, such as wood and coal, liquids, such as methanol or kerosene, and gases, such as methane or hydrogen, all combust. Something besides phase must determine whether or not a material can burn.

Use the table to compare substances that combust with those that do not combust.

HISTORY CONNECTION

From the 1880s until the 1930s, photographers would ignite a sample of magnesium to produce the flash needed to take a picture. For safety and simplicity, flash bulbs, which contained magnesium wire or thin sheets of magnesium foil, were then invented. In the 1980s, these bulbs were replaced with electronic flashes.

© Steve Hamblin/Alamy

Combustion and Bond Type

Substance	Chemical formula	Type of bond	Does it combust?
sodium chloride	$NaCl(s)$	ionic	no
iron (III) oxide	$Fe_2O_3(s)$	ionic	no
nickel	$Ni(s)$	metallic	yes
magnesium	$Mg(s)$	metallic	yes
water	$H_2O(l)$	molecular covalent	no
cellulose (paper)	$C_6H_{12}O_6(s)$	molecular covalent	yes
octane	$CH_3(CH_2)_6CH_3(l)$	molecular covalent	yes
carbon dioxide	$CO_2(g)$	molecular covalent	no
hydrogen	$H_2(g)$	molecular covalent	yes
silicon dioxide (sand)	$SiO_2(s)$	network covalent	no

SETUP

Set up distribution stations around the room so students can get the materials quickly and safely. There are five demos. At the start of class, you will demonstrate the combustion of a marshmallow. At the end of class, you will test paraffin oil, sodium chloride, calcium carbonate, and steel wool so students can compare their predictions to their observations. Clean the steel wool with vinegar before class so it has time to dry.

CLEANUP

Cooled products of combustion may be put in the trash can. Dispose of any remaining chemicals according to local regulations.

Lesson Guide

ENGAGE (5 min)

TRM PowerPoint Presentation 100

ChemCatalyst

You are on a camping trip with your family. While you are toasting marshmallows, one catches on fire.

1. List three more items you might have that will burn.

2. List three items that will not burn.

Sample answers: **1.** food, matches, dry leaves. **2.** Metal pot, rocks, wet matches.

Using the information in the table, it is possible to make some generalizations about whether or not a substance will combust.

- Most ionic compounds do not combust.
- Most molecular covalent compounds do combust, especially those that contain carbon and hydrogen.
- Most elemental metals combust.
- Water and carbon dioxide do not combust.

Combustion Reactions

Combustion is a chemical reaction in which an element or a compound reacts with oxygen and releases energy in the form of heat and light. When molecular covalent substances containing mainly carbon and hydrogen burn, they produce carbon dioxide and water. When metallic substances combust, they produce a metal oxide.

These are the chemical equations for the combustion of methane and of magnesium.

$$CH_4(g) + 2O_2(g) \longrightarrow CO_2(g) + 2H_2O(g) + \text{heat} + \text{light}$$
$$2Mg(s) + O_2(g) \longrightarrow 2MgO(s) + \text{heat} + \text{light}$$

Oxygen is always a reactant in a combustion reaction. The other reactant in a combustion reaction is usually called a *fuel*. In these reactions, methane and magnesium are considered the fuels.

Notice that both reactions have heat and light on the product side of the equation. Remember, heat is not a substance, but chemists sometimes include it in the equation to keep track of energy changes.

If a compound is already combined with oxygen, it is not likely to combust and combine with *more* oxygen. This is why some molecular covalent compounds like water, H_2O, and carbon dioxide, CO_2, do not burn. Other examples include sulfur trioxide, $SO_3(g)$, and nitrogen dioxide, $NO_2(g)$.

FIRE—A SPECIAL TYPE OF COMBUSTION

The photo shows the flame produced when methane burns. When we call something a fire, it is because a flame is present during the combustion reaction. The combustion of methane produces a fire.

Not all combustion reactions produce fire. The burning of magnesium produces a bright light but no flame. The light produced is so bright that it can damage your eyes. Most elemental metals are combustible but do not produce fire.

Ash is often left over after a fire. For molecular compounds, ash is composed of unburned or partially burned reactants. After combustion of metallic substances, the product may look like ash, but it is actually the metal oxide product of the reaction.

HISTORY CONNECTION

Around 1680, Irish physicist Robert Boyle experimented with creating a ready source of fire by coating a small piece of wood with sulfur to be struck against phosphorus-coated paper. Yet it was not till the mid-1800s that efficient manufacturing of the safety match was developed by Johan Lundstrom of Sweden.

DISCUSS THE CHEMCATALYST

➜ Write students' ideas about what will or will not burn on the board under the headings "combust" and "will not combust."

Ask: What do you think is similar about the substances that burn or combust?

Ask: What is similar about the substances that do not burn or combust?

Ask: Are wet things combustible? What about liquids? Solids? Metals?

Ask: What differences do you notice in the substances that burn and those that do not burn?

Combustion: An exothermic reaction between a fuel and oxygen, often producing flames.

EXPLORE (15 min)

TRM Worksheet with Answers 100

TRM Worksheet 100

INTRODUCE THE LAB

➜ Arrange students into groups of four.

➜ Pass out the student worksheets.

➜ Tell students that combustion is a reaction between oxygen and fuel. Fuel is the substance that burns.

➜ Briefly go over the procedure and safety with the class.

SAFETY

Students must wear safety goggles at all times. Remind them about the proper use of fire extinguishers and fire blankets. All liquids must be tested by lighting the wick of the burners, not in the watch glass. Tongs should be used when testing the copper wire.

GUIDE THE DEMONSTRATION

➜ Ignite a marshmallow (optional). Have an aluminum pan available to catch any drips.

➜ Ask students about their predictions before you test each of the four substances in Part 2. Students should record their observations in the data table on the worksheet.

➜ To test the paraffin oil, light the wick in the burner.

➜ To test the sodium chloride and calcium carbonate, scoop up a tiny amount with the spatula and hold it in the flame of the Bunsen burner.

➜ To test the steel wool, use the tongs to hold it in the flame of the Bunsen burner to ignite it. Place it in the crucible to complete the reaction and to cool it.

EXPLAIN & ELABORATE (15 min)

DISCUSS COMBUSTION

➜ You may return to the lists you started on the board during the ChemCatalyst discussion. What burns? What does not burn?

Ask: What types of substances combust?

Ask: What types of substances do not combust?

Ask: On what did you base your predictions in Part 2 of the lab?

Ask: How does combustion seem to be related to bonding?

Ask: According to what you learned in today's class, what types of substances would make good fuels?

Key Points:

Substances that combust
wood, $C_6H_{12}O_6(s)$
ethanol, $CH_3CH_2OH(l)$
hexane, $CH_3CH_2CH_2CH_2CH_2CH_3(l)$
oil, $C_{21}H_{39}O_6(l)$
hydrogen, $H_2(g)$
copper, $Cu(s)$
iron, $Fe(s)$
magnesium, $Mg(s)$

Substances that do not combust
water, $H_2O(l)$
carbon dioxide, $CO_2(g)$
zinc oxide, $ZnO(s)$
sodium bicarbonate, $NaHCO_3(s)$
sodium chloride, $NaCl(s)$
calcium carbonate, $CaCO_3(s)$

Ionic salts rarely combine with oxygen in a combustion reaction. Ionic compounds do not make good fuels. In general, molecular covalent substances that contain mainly carbon and hydrogen will combust. Exceptions are water and carbon dioxide, which are molecular covalent molecules that do not combust. Carbon dioxide and water are the products of combustion. Metals combust, but the combustion of most metals does not produce a flame. While the reaction of metals will produce heat and usually light, sometimes the light released is not detectable to the human eye. At other times, the light released is almost blinding, as when magnesium metal reacts with oxygen. Remember that not everything that combusts results in what we call fire. We will explore this later in the unit.

→ Write the chemical equations given below on the board at the appropriate point in the discussion.

Ask: Did everything that burned produce a flame? Explain.

Ask: What is the ash or charred material that is left over when things burn?

Ask: If you react magnesium with oxygen, what product or products do you think will be created? Try to write a chemical equation describing this.

Ask: What generalizations can you make about the products of combustion reactions?

Key Points: Combustion is a type of chemical reaction in which an element or a compound reacts with oxygen and releases energy in the form of heat and often light.

Everything that burned today reacted with oxygen. Oxygen is always one of the reactants in combustion. The other reactant usually is referred to as a fuel. When molecular covalent substances burn, they produce carbon dioxide and water. When metallic substances burn, they produce a metal oxide. Ash often is composed of a carbon-based material that did not burn completely. Substances that already contain a high percentage of

oxygen atoms are less likely to be combustible. Here are the equations for the combustion of methane and of magnesium.

$$CH_4(g) + 2O_2(g) \rightarrow CO_2(g) + 2H_2O(g) + heat$$

$$2Mg(s) + O_2(g) \rightarrow 2MgO(s) + heat$$

EVALUATE (5 min)

Check-In

Which of these substances are likely to combust? What is your reasoning?

methane, CH_4
calcium bromide, $CaBr_2$
sodium, Na

What types of substances burn?

The process of burning is also referred to as combustion. Combustion is the combining of compounds or elements with oxygen through a chemical reaction that produces energy in the form of heat and light. Most ionic compounds do not combust. Most molecular covalent compounds are combustible. When molecular substances that are made mostly of hydrogen and carbon combust, they produce carbon dioxide and water. When metals combust, they form metal oxides, producing heat and light but no flame.

KEY TERM
combustion

Exercises

Reading Questions

1. What should you look for in a chemical formula to decide if a compound or element will not combust?

2. Name three things that are true of every combustion reaction.

Reason and Apply

3. Which of the substances listed below will combust? Explain your reasoning.
   a. $Na_2SO_4(s)$ b. $Cu(s)$ c. $MgO(s)$
   d. $Na(s)$ e. $C_2H_6O(l)$ f. $Ar(g)$

4. Balance the following equations for combustion reactions. Circle the fuel for each reaction.
   a. $C_2H_6(g) + O_2(g) \longrightarrow CO_2(g) + H_2O(l)$
   b. $C_6H_{12}O_6(s) + O_2(g) \longrightarrow CO_2(g) + H_2O(l)$
   c. $C_2H_5OH(l) + O_2(g) \longrightarrow CO_2(g) + H_2O(l)$
   d. $C_{21}H_{24}N_2O_4(s) + O_2(g) \longrightarrow CO_2(g) + H_2O(l) + NO_2(g)$
   e. $C_2H_5SH(l) + O_2(g) \longrightarrow CO_2(g) + H_2O(l) + SO_2(g)$

5. What are the products of these combustion reactions? Write balanced chemical equations for each reaction.
   a. $C_7H_6O(l) + O_2(g) \longrightarrow$ b. $CH_3COCH_3(l) + O_2(g) \longrightarrow$
   c. $H_2C_2O_4(s) + O_2(g) \longrightarrow$ d. $Ca(s) + O_2(g) \longrightarrow$
   e. $Li(s) + O_2(g) \longrightarrow$ f. $Si(s) + O_2(g) \longrightarrow$

Answer: The methane molecular covalent compound is composed entirely of carbon and hydrogen, so it will combust. Sodium is a metal, so it probably will combust. Calcium bromide will not combust, because it is an ionic compound.

Homework

Assign the reading and exercises for Fire Lesson 100 in the student text.

LESSON 100 ANSWERS

Answers continue in Answer Appendix (p. ANS-8).

Ken Karp Photography

Calorimetry

Purpose

To determine which snack food, cheese puffs or toasted corn snacks, transfers the most heat in a combustion reaction.

Materials

- cheese puffs, toasted corn snacks
- ring stand with ring and clamp
- Bunsen burner
- wire mesh, paper clips, safety pins, corks, modeling clay, rubber bands, tape
- matches
- crucible
- thermometer
- 150 mL beaker
- balance
- water
- empty tuna fish can, empty soft drink cans with tabs, aluminum foil

SAFETY Instructions

Safety goggles must be worn at all times. Be sure to know the location of the fire blanket and fire extinguisher.

Procedure

1. Design an experiment using any of the materials above to determine which snack food will transfer more energy when burned. You must provide measurements and calculations to support your conclusion.

2. Write a detailed materials list of what you will use and a step-by-step procedure. You must get teacher approval before beginning your test.

3. When your teacher approves your procedure, conduct your experiment and record your observations and data.

Analysis

1. List all the types of data your team collected.

2. How did you decide what data were important for your experiment?

3. Why do you need to know the mass of the snack food burned?

4. **Making Sense** Which snack food supplies more energy to your body, cheese puffs or toasted corn snacks? What evidence do you have to support your answer?

5. **If You Finish Early** Repeat the experiment to determine whether the volume of water or the mass of fuel used makes a difference in the outcome.

Lesson 101 | **Calorimetry**   515

IN CLASS

Students begin a quantitative investigation of heat transfer in this lesson. The challenge for students is to develop a method for determining which is a better fuel, cheese puffs or toasted corn snacks. Most student groups ultimately will arrive at a calorimetry procedure to measure the rise in temperature of a known quantity of water due to heat transfer.

Note: This is a two-day, open-ended lab experiment. In the spirit of open inquiry, this lesson is not about immediately doing things correctly as much as it is about allowing students to explore their own ideas. Students will have a chance to do a standard calorimetry lab in Lesson 103: Fuelish Choices. You might read Lesson 102: Counting Calories to familiarize yourself with the debriefing that follows this lab experiment, and photocopy the worksheets for Lessons 101 and 102 together. The Explain & Elaborate section in this lesson is for day 2 of this experiment. A complete materials list is available for download.

TRM Materials List 101

Differentiate

Students may have varying levels of readiness for this open-inquiry experiment. To scaffold for students who might not be fully ready, you can include pre-made soda can calorimeters. This would allow students to focus only on designing a procedure for how they will burn the food and what to measure. For advanced students, you can challenge them to run more experiments to refine their apparatus and technique.

Pre-AP® Course Tip

Students preparing to take an AP® chemistry course or college chemistry should be able to design a plan that will generate useful data. In this lesson, students must design a plan for measuring the heat transferred from two different fuels during combustion.

SETUP

Place the materials so that students can see what is available. Groups must design their own setup, requesting anything they might need from the instructor. Some of the items listed are superfluous, intended to get groups to come up with their own approach.

Overview

LAB: GROUPS OF 4

Key Question: How are food calories measured?

KEY IDEAS

The products of a combustion reaction are at a high temperature. These products transfer energy to the surroundings to reach thermal equilibrium. Different fuels transfer different amounts of heat during a combustion reaction. Calorimetry is a lab procedure that measures the amount of heat transferred to the surroundings by a reaction. During a calorimetry procedure, the heat released during a chemical or physical change is transferred to another substance, such as water, which undergoes a temperature change.

LEARNING OBJECTIVES

- Design a procedure to compare the amount of heat transferred from two different fuels during combustion.
- Explain how calorimetry can be used to determine the amount of heat transferred during a combustion reaction.

KEY TERM

calorimetry

Lesson Guide

ENGAGE (5 min)

TRM PowerPoint Presentation 101

ChemCatalyst

Imagine that you heat water over a campfire.

1. How can you tell that energy is transferred to the water?
2. How can you you measure the amount of energy transferred to the water?
3. How can you measure the amount of energy one stick transfers to the water when it burns?

Sample answers: 1. The water getting hot, or boiling, is proof that energy is transferred to the water. **2.** Using a thermometer can help you measure the amount of energy transferred. **3.** You can measure the increase in the temperature of the water caused by the burning of one stick.

→ Assist students in sharing their ideas about how to measure energy transfer.

Ask: How can you measure the amount of energy transferred when you burn something?

Ask: What types of fuels do you use in your daily life?

Ask: Is food a form of fuel? Explain.

Ask: How could you test a variety of foods to find out which would give you more energy?

EXPLORE (25–30 min)

TRM Worksheet with Answers 101

TRM Worksheet 101

INTRODUCE THE INQUIRY LAB

→ Arrange students into groups of four.

→ Pass out the student worksheets.

→ Let students know that this is a two-day lab.

→ Tell students that they have to get your approval of their procedure before conducting any experiment.

Now We're Cooking

Calorimetry

THINK ABOUT IT

Foods are sources of energy, and certain foods are better sources of energy than others. What are the energy differences between a granola bar, a handful of nuts, and a few potato chips? Burning dry food items is one way to discover the possible energy potential of each type of food.

How are food Calories measured?

To answer this question, you will explore

➲ Foods as Fuel

➲ Calorimetry

EXPLORING THE TOPIC

➲ Foods as Fuel

The foods you eat are sources of energy. There are no fires in your stomach, but your body processes the food you eat to generate energy. Burning two foods and comparing the energy transfer that results is a way to measure the energy of each food.

FOOD CALORIES

Food labels contain a wide variety of nutritional data, including Calorie content, which is a measure of energy. Food Calories, or Calories with a capital C, are kilocalories of energy, abbreviated kcal or Cal. This product label tells you that 1 ounce of potato chips has 160 Cal. In comparison, 1 ounce of walnuts has 190 Cal.

HEALTH CONNECTION

A healthy diet for the average adult consists of an intake of about 2000 food Calories per day. One fast-food double cheeseburger contains about 750 Calories, a medium soda is about 210 Calories, and a medium order of French fries is 450 Calories. If you ate this for lunch, you would have only 600 Calories left for the rest of the day.

Nutrition Facts
Serving Size 1 oz.

Amount Per Serving

Calories 160 Calories from fat 80

	% Daily Value*
Total Fat 10g	16%
Saturated Fat 1g	5%
Polyunsaturated Fat 3g	
Monounsaturated Fat 6g	
Trans Fat 0g	
Cholesterol 0mg	0%
Sodium 160mg	7%
Potassium 340mg	10%
Total Carbohydrate 14g	5%
Dietary Fiber 1g	4%
Sugars 0g	
Protein 2g	

Vitamin A 0%	•	Vitamin C 10%	
Calcium 0%	•	Iron 0%	
Vitamin E 8%	•	Thiamin 2%	
Niacin 4%	•	Vitamin B₆ 6%	

Phosphorus 4%

*Percent Daily Values are based on a 2,000 calorie diet. Your daily values may be higher or lower depending on your calorie needs:

		Calories	2,000	2,500
Total Fat	Less than	65g	80g	
Sat Fat	Less than	20g	25g	
Cholesterol	Less than	300mg	300mg	
Sodium	Less than	2,400mg	2,400mg	
Potassium		3,500mg	3,500mg	
Total Carbohydrate		300g	375g	
Dietary Fiber		25g	30g	

Calories per gram:
Fat 9 • Carbohydrate 4 • Protein 4

Example

Comparing Calories per Gram

Are there more Calories in 1.0 g of potato chips or 1.0 g of walnuts? (28.3 g = 1.0 oz)

Solution

There are 160 Cal in 1.0 oz of potato chips. There are 190 Cal in 1.0 oz of walnuts. An ounce is the same as 28.3 g.

→ Tell students to complete at least one preliminary experiment on day 1 to work out bugs. On the next day, they should do a second run-through to finish gathering data.

→ Remind students about lab safety when using fire.

GUIDE THE LAB

→ Move around the lab, monitoring the progress of each group. If a group appears stuck, ask questions to stimulate thinking. The goal is for them to problem-solve and design their own experiment, even if it is somewhat flawed.

→ Allow groups to go in their own directions on day 1. On day 2, you can ask them questions that will guide them more directly toward a successful calorimetry experiment.

Ask: What data do you want to collect in this experiment? Why?

Ask: How can you measure the amount of energy in a fuel?

Ask: What is a calorie?

Ask: What do you know about water that makes water good for heat-transfer experiments?

Ask: Does one cheese puff weigh the same as one toasted corn snack? Do you think mass matters in this experiment?

Example

(continued)

> Potato chips: 160 Cal/28.3 g = 5.6 Cal/g
>
> Walnuts: 190 Cal/28.3 g = 6.7 Cal/g
>
> So, there are more Calories available in 1.0 g of walnuts than 1.0 g of potato chips.

According to these calculations, an ounce of walnuts contains more energy than an ounce of potato chips. However, the number of Calories per gram is fairly close. In contrast, raisins contain 3.0 Cal per gram, and raw carrots contain 0.4 Cal per gram.

○ Calorimetry

To measure the Calories in food, you have to find out how much energy is released when the food reacts with oxygen. One way to determine the amount of heat transfer during a combustion reaction is to use the reaction between the fuel and oxygen to heat a measured amount of some other substance, such as water. You can then use a thermometer to measure the change in temperature of the water. The science of measuring the energy released or absorbed in a chemical reaction or physical change is called **calorimetry.** (In Latin, *calor* means heat, and *meter* comes from the Greek word for measure.) Chemists and nutritionists alike use a bomb calorimeter to figure out the calorie content of combustible substances.

The illustration shows a bomb calorimeter. The combustion reaction takes place in an inner reaction chamber called a bomb, which is completely surrounded by water. The sample is ignited by electrical energy, and the energy, due to the burning of the fuel, is transferred to the water. A stirring rod makes sure the heat is uniform in the water. If you determine the change in temperature of the water and you know the quantity of water in the calorimeter, you can determine the heat transferred during the combustion reaction.

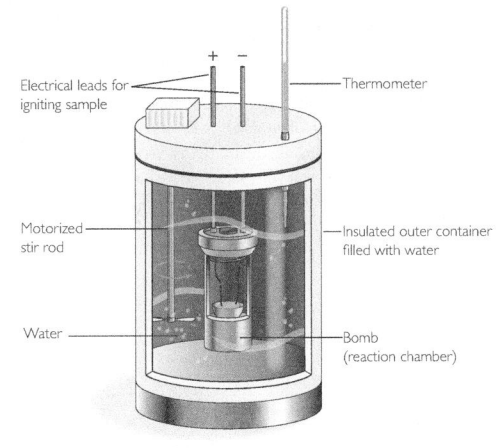

Electrical leads for igniting sample

Thermometer

Motorized stir rod

Insulated outer container filled with water

Water

Bomb (reaction chamber)

Ask: How can you use the temperature data to calculate heat energy?

→ Here are some directions in which students might go (see Lesson 102 for for more details).

- Many groups eventually design some variation of a setup with the food placed under a container of water.
- Groups might try to measure the heat directly from the flame instead of measuring the heat transfer to water.
- Some groups will believe that the food that burns longest has the most energy, while others will think that the food that produces the greatest change in temperature provides the most energy.

- Groups might not weigh their fuels at all, or they might weigh the fuel at the beginning but not at the end of the combustion. Some will track the number of pieces of food.
- Groups will try different ways of suspending the fuel below a container of water, from stabbing the pieces with a paper clip to creating a pedestal with clay.
- Groups might want to crush the cheese puff and the toasted corn snack and place them in a crucible to burn. Usually, they do this because they feel compelled to burn the exact same mass of each fuel. Crushing can make the fuel difficult to ignite.

Ask: What was the goal of your experiment?

Ask: What data did you collect during your experiment?

Ask: What types of things did you modify or revise in your procedure?

Ask: What things are causing difficulty in your experiment?

Key Points: Accurately measuring the energy from a combustion reaction has its challenges. Some challenges associated with this lab procedure are how much of each substance to burn, how to minimize heat loss to the air and surroundings, and how to account for incomplete combustion. Allow groups to try to come up with solutions to these challenges on their own.

EVALUATE (5 min)

There is no Check-In for this lesson.

Homework

Assign the reading and exercises for Fire Lesson 101 in the student text. (You can assign the lab report on the second day.)

LESSON 101 ANSWERS

1. *Possible answer:* Calories are a measure of heat. Because a combustion reaction is exothermic, the heat that is produced by the reaction of the material with oxygen during combustion can be measured in Calories.

2. *Possible answer:* The calorimeter has two chambers so that all of the heat produced by combustion inside the inner chamber is captured in the outer chamber and not lost to the environment.

3. No, placing a thermometer in the flames measures the temperature of the combustion reaction, but it cannot measure the total heat energy given off by the reaction.

4. Calorimetry is the science of measuring the amount of heat released or absorbed during a chemical reaction or physical change.

5. *Possible answers:* **a.** Build a calorimeter using water, a small open metallic container, and a beaker. ● Place the open metal container in the beaker. ● Fill the beaker with a measured amount of water that almost covers the container. ● Use a match to burn a measured amount of the cheese puffs in the metal container and measure the temperature change of the water in the beaker. ● Use the mass and the temperature change of the water to determine the number of calories of heat absorbed by the water. ● Measure the mass of the unburned puffs. ● Divide the energy absorbed by the water by the difference in the measured mass of the puffs to determine the energy in calories per gram within the food. ● Repeat the experiment on the roasted corn snack. **b.** No, because the calculation gives a result in calories per gram, making the mass of the fuel unnecessary for the comparison. **c.** Only the amount of food that is burned should be included in the calculations, so the unburned amount should be subtracted from the original amount to determine what portion of the food was burned. **d.** Sources of error include: ● heat lost to the environment ● incomplete combustion of fuel ● measurement errors in the mass of the foods ● measurement errors in the temperature ● change of

How are food Calories measured?

KEY TERM
calorimetry

There are a number of different ways to compare the energy transferred when food reacts with oxygen, as it does in the overall digestion process. Some foods can be burned and then their energies can be measured and compared. Calorimetry is the science of measuring the energy released or absorbed by a chemical or physical change. You can use a bomb calorimeter to measure the amount of heat transferred to the surroundings in a combustion reaction.

Exercises

Reading Questions

1. What do Calories have to do with combustion?
2. Why does a bomb calorimeter have an inner chamber and an outer chamber?

Reason and Apply

3. Can you measure the heat energy of a combustion reaction by placing a thermometer directly into the flames? Explain.
4. What is calorimetry?
5. Most school laboratories do not have bomb calorimeters.
   a. Design an experiment that would allow you to compare whether a cheese puff snack or a roasted corn snack transfers more energy when burned.
   b. In your experiment, does it matter if you burn the same mass of cheese puff and roasted corn snack? Why or why not?
   c. Why is it important to measure the mass of each snack food at the end of your experiment?
   d. What are some possible sources of error that could occur in your experiment?
6. **Lab Report** Write a lab report for the Lab: Calorimetry. In the procedure section, describe your procedure in detail and include a sketch of your setup. In your conclusion, include a discussion of how well your setup worked and whether or not you would change anything.

(Title)

Purpose: (Explain what you were trying to find out.)

Procedure: (List the steps you followed.)

Observations: (Describe your observations.)

Analysis: (Explain what you observed during the experiment.)

Conclusion: (What can you conclude about what you were trying to find out? Provide evidence for your conclusions.)

the water ● measurement error in the amount of water in the beaker.

6. A good lab report will contain: ● a title (Lab: Calorimetry) ● a statement of purpose (*Possible answer:* to determine which snack food releases the most energy when burned in a combustion reaction) ● a procedure (a summary of the steps followed in the experiment; student written procedures should be detailed enough to allow duplication of the experiment and include a sketch of the setup) ● results (check student observations to make sure that all of the necessary data were collected during the experiment and that observations of how the food changes during the experiment were recorded.) ● a conclusion (should include a statement of which food released the most energy and an analysis of the experimental procedure and setup).

Calorimetry Calculations

THINK ABOUT IT

Heating a beaker of water with a burning potato chip seems like an unusual way to determine the number of Calories in this snack food. However, this is very close to the actual procedure used by nutritionists.

How does a calorimetry experiment translate into Calories?

To answer this question, you will explore

- Calorimetry in the Lab
- Calorimetry Calculations

EXPLORING THE TOPIC

Calorimetry in the Lab

Suppose that you wanted to measure the amount of energy transfer from combustion of a cashew nut. Because the digestive processes of the human body cannot be duplicated in your classroom, the next best approach is to determine the energy that can be transferred when food reacts directly with oxygen. To do this, you burn the food item and transfer the energy released due to combustion to another substance. Many substances can be used, but the most common one to use is water.

One feature of this type of experiment that is difficult to control in the lab is heat transfer to the surrounding air and equipment. Not all of the energy from burning is transferred directly to the water. Some thermal energy is transferred to the container, to the air, and even to you. One way to control this loss of energy is to burn the food sample very close to the bottom of the water container. Another way is to create some sort of shield to keep the energy transfer focused on the water. All of these details must be taken into account when designing a procedure.

Building a shield of some sort helps to direct the heat from burning.

Calorimetry Calculations

There are three measurements and one property that help you to determine the energy transferred during a calorimetry procedure.

Data needed:

1. Mass of water heated in grams; 1 mL = 1 g
2. Temperature change of the water in degrees Celsius, ΔT (delta T)
3. Mass of fuel burned in grams
4. Specific heat capacity of water = 1.00 cal/g °C at 25 °C and 1 atm pressure

---

Overview

FOLLOW-UP: GROUPS OF 4

Key Question: How does a calorimetry experiment translate into calories?

KEY IDEAS

The temperature change, fuel mass, and water volume data from a calorimetry procedure can be used to determine how much heat is transferred during a combustion reaction. The amount of energy transferred from a substance during combustion depends on the identity and mass of the substance.

LEARNING OBJECTIVES

- Calculate the amount of heat transferred in a combustion reaction.
- Use calculated values of heat transfer to determine which of two foods is a better fuel.

FOCUS ON UNDERSTANDING

Some students will mistakenly put the mass of the fuel burned into the equation for heat transfer, rather than the mass of the water heated.

IN CLASS

This is the debriefing of the two-day snack-food comparison lab in

---

Lesson 101: Now We're Cooking. Students complete a worksheet to summarize their results. Student groups present evidence to support their conclusions regarding the two snack foods. The instructor should then guide the class through a correct calorimetry calculation. A complete materials list is available for download.

TRM Materials List 102

Pre-AP® Course Tip

Students preparing to take an AP® chemistry course or college chemistry should be able to justify the use of particular mathematical formulas and procedures. In this lesson, students are continuing their calculations from the previous lesson. Students present evidence to support their conclusions regarding the snack foods and then use mathematics to solve calorimetry problems.

Lesson Guide

ENGAGE (5 min)

TRM PowerPoint Presentation 102

> *ChemCatalyst*
>
> You read the wrapper of a candy bar and find out that the bar has 300 Calories.
>
> **1.** What do you think this means?
>
> **2.** Is a candy bar a good source of energy? Why or why not?

Sample answers: **1.** Students might correctly say that 300 Calories is equal to 300,000 calories. Some students will say that 300 Calories describes how fattening the food is. **2.** Students might say that a candy bar is a good source of energy because it contains so many Calories. Others might say that a candy bar contains too much sugar and fat to be a good source of energy.

→ Remind students that 1 food Calorie = 1000 calories. (This was covered in the student text for Lesson 102: Now We're Cooking.)

→ Discuss the meaning behind food Calories. Help students understand the connection between food Calories and the calories released from burning fuels.

Ask: Is energy transferred when food is digested? Explain.

Ask: Why are Calories listed on food wrappers?

Ask: How do you think the number of Calories is determined?

Ask: What is a good source of food energy in your diet?

TRM Worksheet with Answers 102

TRM Worksheet 102

INTRODUCE THE FOLLOW-UP

→ Arrange students into groups of four.

→ Pass out the student worksheets.

→ Tell students that they will use their group's data but will work individually on the worksheet.

EXPLAIN & ELABORATE (25 min)

ASSIST STUDENTS IN SHARING THEIR EXPERIMENTAL APPROACHES

→ You might construct your own experimental setup at the front of the room, as shown here, or reconstruct a setup used by a student group.

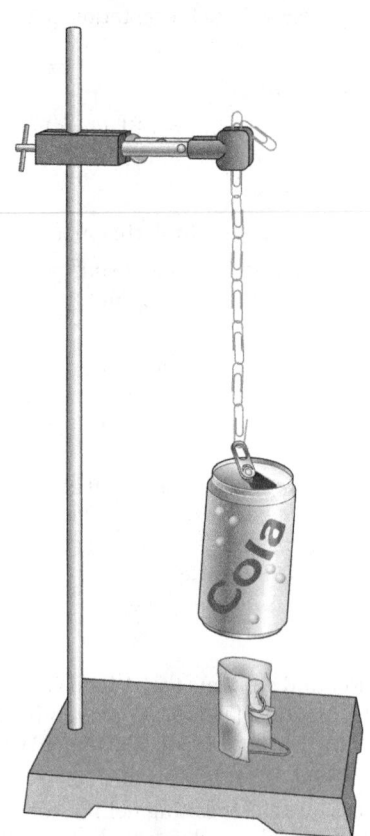

Ask: What was the purpose of your experiment?

Ask: How would you describe your experimental setup?

Ask: What possible sources of error did you have to consider?

With the first two pieces of data, you can calculate how many calories of thermal energy are transferred to the water. Recall that the equation for heat transfer is

$$q = mC_p \Delta T$$

where C_p is the specific heat capacity of the substance being heated.

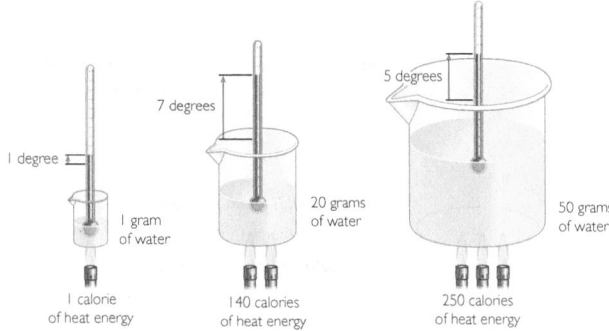

The number of calories of energy transferred is dependent on the mass of the water and the temperature change.

The illustration shows that if you multiply the change in temperature by the mass of the water, you can calculate the number of calories of heat transferred to the water. This is because the specific heat capacity of water is 1.00 cal/g · °C.

Example 1

Calories from a Cashew

These data were collected from the burning of one cashew. How much thermal energy was transferred?

Initial T of water = 19.0 °C

Final T of water = 34.5 °C

Volume of water = 30.0 mL

Solution

First, figure out the temperature change.

Subtract the initial temperature from the final temperature.

$$\Delta T = 34.5\,°C - 19.0\,°C = 15.5\,°C$$

Since 1 mL of water is equal to 1.0 g of water, the mass of the water heated is 30.0 g.

Use the equation for heat transfer.

Substitute values and solve.

$$q = mC_p \Delta T = 30.0\ \text{g}(1.00\ \text{cal/g} \cdot °C)15.5\,°C = 465\ \text{cal}$$

Ask: What issues arose in trying to design your experiment?

Ask: How did you modify your experiment to improve it?

Key Points: Experimental setups may vary from group to group. The illustration shows a soda can hanging by its pop-tab on linked paper clips hanging from a clamp on a ring stand. The can holds a measured amount of water. To allow for easy lighting, the snack food is suspended in the air using an unfolded paper clip. To minimize heat loss, the food is surrounded by foil.

There are two main sources of experimental error with this setup. First,

some of the energy from the burning food does not warm the water. Besides the water, the heat is transferred to the surroundings, so the food will contain more calories than were measured. Second, some of the mass of the food does not burn. There is a residue or ash left over after the flame extinguishes. The mass of the residue must be subtracted from the original mass to determine the amount of food that actually burned.

Typical experimental results indicate that toasted corn snacks contain more energy per gram than cheese puffs. The product labels indicate that cheese puffs contain more food Calories per

One cashew transferred 435 cal of energy to the water. To compare different foods, it is necessary to figure out the calories *per gram* of food burned.

Example 2

Calories per Gram of Cashew

How much energy per gram was released by the combustion of the cashew?

$$\text{Initial mass of cashew} = 0.66 \text{ g}$$

$$\text{Final mass of cashew} = 0.06 \text{ g}$$

Solution

Figure out the mass of the cashew burned.

Subtract final mass from initial mass.

$$\text{Mass of cashew burned} = 0.66 \text{ g} - 0.06 \text{ g} = 0.60 \text{ g}$$

Determine the number of calories per gram.

$$465 \text{ cal}/0.60 \text{ g} = 775 \text{ cal/g}$$

The number of calories per gram is also called the food energy density. Cashews have an energy density of 775 cal/g of cashew burned. Remember, these are chemist's calories. To convert them to food Calories, you must divide by 1000 to get 0.775 Cal/g.

The actual amount of energy a human obtains after the digestive processes are completed is about 85% of the Calorie content listed on nutrition labels. Fats have high energy densities, around 9 Cal/g. Sugars and proteins are around 4 Cal/g.

COMPARING FOODS

In the real world, nutritionists use calorimeters to figure out the Calorie content of foods. The word *content* is misleading. The food does not actually contain the Calories. The Calorie content represents the energy transferred specifically by *combustion* of these compounds.

Bomb calorimeters minimize the heat transfer to the surroundings, and the results are much more precise than our lab experiments. The method of calculation is very similar to the one used here. If a bomb calorimeter is used, a cashew would be found to have closer to 5500 chemists' calories per gram, or 5.5 Cal/g. Our experimental results came up with 775 cal/g. There is a lot of error in the classroom procedure.

LESSON SUMMARY

How does a calorimetry experiment translate into Calories?

Different fuels transfer different amounts of energy during a combustion reaction. Data about the temperature change, fuel mass, and water volume from a calorimetry procedure can be converted into Calories. Using the specific heat

Lesson 102 | **Calorimetry Calculations**   521

gram than toasted corn snacks. The measurement error for the cheese puffs typically is greater than that for the toasted corn snacks. This could be because the heat transfer to the water is more efficient for the slower-burning toasted corn snacks.

PROCESS STUDENTS' DATA, CALCULATIONS, AND CONCLUSIONS

➡ Create a data table on the board listing all the types of data collected by the groups. Some of the data may not be necessary for making the final calculations. Because this was an open-inquiry lab, this is all right as

long as students use valid reasoning in their approach.

SAMPLE DATA

	Cheese puffs	**Toasted corn snacks**
Mass burned (g)	0.90 g	0.34 g
ΔT (°C)	14.5 °C	12 °C
Volume of water (mL)	30 mL	30 mL
Calories	435 cal	360 cal
Calories per gram	483 cal/g	1059 cal/g

Ask: How did you calculate the amount of heat transferred from the data you gathered?

Ask: What conclusions did you reach about the two snack foods?

Ask: Why is it important to measure the mass of the fuel that did not burn?

Ask: Will timing the burning of the two snack foods allow you to compare the amount of energy in them? Explain.

Key Points: One common approach is to time the burning of the two fuels. In general, the toasted corn snacks burn quite a bit longer than the cheese puffs. While it is valid to compare the amount of time each fuel burned, this does not provide a measure of the amount of energy "contained" in each fuel.

LEAD STUDENTS THROUGH A STANDARD CALORIMETRY CALCULATION

➡ Review the definition of a calorie. You might want to write the definition on the board.

➡ Complete sample calculations on the board for both snack foods.

Ask: What is the definition of a calorie?

Ask: Why is it important to measure the amount of water used?

Ask: Why is it important to know the mass of the fuel burned?

Ask: How should you use the sample data to calculate the calorie content of a cheese puff?

Key Points: Calorimetry is the science of measuring the amount of heat transferred in a chemical or physical change. To determine the heat exchanged in the burning of a food, in calories per gram, three measurements are required: the mass of fuel burned, the volume of water heated, and the temperature change of the water. Use the energy transfer equation $q = mC_p \Delta T$, where C_p is the specific heat capacity of water.

EXAMPLE

Initial mass of cheese puff = 1.21 g

Initial temperature of water = 19.5 °C

Volume of water = 100 mL

Final mass of cheese puff = 0.15 g

Final temperature of water = 28.5 °C

C_p of water = 1 cal/g °C

SOLUTION

Mass of cheese puff burned = 1.06 g

$\Delta T = 9\ °C$

Mass of water heated = 100 g

$q = mC_p\,\Delta T = (100\ g)(1\ cal/g\ °C)(9\ °C)$

$q = 900$ calories

$\dfrac{900}{1.06\ g} = 849\ cal/g$

$849\ cal/g \sim 0.8\ Cal/g$

EVALUATE (5 min)

> **Check-In**
>
> How many calories are in each gram of peanut if the following data are collected?
>
> Mass of peanut burned = 0.5 g
>
> Volume of water heated = 50 mL
>
> Temperature change = 10 °C

Answer: 500 calories of heat are transferred when you burn 0.5 g of peanut. This is equivalent to 1000 cal/g, or 1 food Calorie per gram of peanut.

Homework

Assign the reading and exercises for Fire Lesson 102 in the student text.

LESSON 102 ANSWERS

1. *Possible answers (any 3):* ● loss of heat to the air ● heat lost by evaporation of water ● incomplete combustion of fuel ● measurement error in the mass of fuel, mass of water, or the temperature change.

2. The specific heat of the substance being heated by combustion of the fuel is necessary to calculate the total amount of energy transferred into the calorimeter. The energy transferred depends on the mass of the substance and the specific heat capacity of the substance.

3. *Possible answer:* If you know the specific heat of ethanol, you can measure the temperature difference in the ethanol and calculate the total calories absorbed. However, ethanol would be a poor choice because it is combustible and can catch fire during the experiment.

4. % error = 85.9%

5. 150 cal

6. 13,400,000 cal

7. Temperature is a measure of the average kinetic energy of the particles of water in the beakers. It takes more energy to increase the temperature

---

capacity of the substance that was heated by the reaction allows you to calculate the exact number of calories transferred by the combustion of the fuel. This provides you with a measure of the amount of thermal energy "contained" in a fuel. The number of calories transferred from a substance that burns depends on the identity of the substance and its mass.

Exercises

Reading Questions

1. Name three possible sources of error in a calorimetry experiment.
2. Why is it important to know the specific heat of the substance being heated by the combustion of a fuel?

Reason and Apply

3. Could you use ethanol instead of water in a calorimetry experiment? Explain.
4. The experimental value for the Calorie content of a cashew is 0.775 Cal/g. The actual value provided on the package label is 5.5 Cal/g. Determine the percent error.
5. A cereal flake is burned under a beaker containing 25 mL of water. If the water temperature goes up 6 °C, how many calories of energy were transferred to the water?
6. Fuel pellets are used in modern energy-saving wood stoves. If the pellets used for these stoves release 742 cal/g, how many calories of energy will be released by combustion of an entire 40 lb sack of pellets?
7. Examine the illustration on page 520. Use a kinetic molecular view to explain why it takes more calories of heat to raise the water temperature in the third beaker by 5 °C than it does to raise the temperature of the water in the second beaker by 7 °C.
8. The calorie content of a peanut is measured by burning it beneath a can of water and measuring the temperature change of the water. Which of these is a possible source of error?
   (A) The initial mass of the peanut is measured incorrectly.
   (B) Some of the heat of combustion is transferred to the air.
   (C) Some unburned remnants of the peanut are lost before finding the final mass.
   (D) All of the above.
   (E) None of the above.

---

of the third beaker because there are many more particles of water, making the amount of thermal energy required to increase the temperature greater even though there is less change in temperature.

  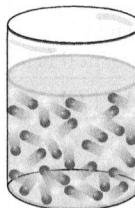

8. D

Fuelish Choices

Heat of Combustion

THINK ABOUT IT

Combustion reactions are used to power vehicles, heat water and homes, and
create electricity. But not all combustion reactions provide the same amount of
energy. Some fuels heat water or power furnaces better than others.

How do different fuels compare?

To answer this question, you will explore

- Burning Fuels
- Heat of Reaction
- Comparing Fuels

EXPLORING THE TOPIC

Burning Fuels

There are a number of different choices of fuels to power race cars. Ideally, you
want to choose a fuel that transfers the most energy per gram. That way, the fuel
does not unnecessarily add to the weight of the car. You can compare the energy
from combustion of a variety of fuels using calorimetry to see which one would
be best for a race car.

ETHANOL VERSUS METHANOL

Both ethanol, C_2H_6O, and methanol, CH_4O, have been used to power race cars.
Suppose that you put the two fuels into alcohol burners and use them to heat two
identical 50 mL samples of water by 30 °C.

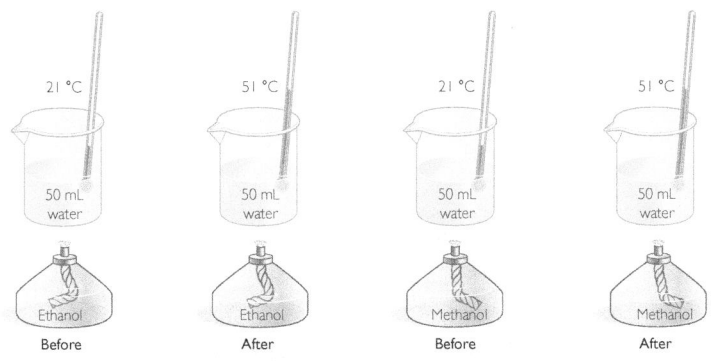

21 °C	51 °C	21 °C	51 °C
50 mL water	50 mL water	50 mL water	50 mL water
Ethanol	Ethanol	Methanol	Methanol
Before	After	Before	After
Burner plus ethanol = 128.0 g	Burner plus ethanol = 126.8 g	Burner plus methanol = 128.0 g	Burner plus methanol = 119.4 g

---

Overview

LAB: GROUPS OF 4

Key Question: How do different
fuels compare?

KEY IDEAS

Different fuels can be compared by
measuring the amount of thermal
energy they transfer during combustion.
Calorimetry is the most common
method of measuring this transferred
energy. The energy transferred as heat
can be expressed as kilocalories per
mole, kilocalories per gram, kilojoules
per mole, or kilojoules per gram of
substance. One calorie is equal to
approximately 4.184 joules. The
amount of heat transferred by burning
is commonly called the *heat of
combustion,* or ΔH.

LEARNING OBJECTIVES

- Define heat of reaction and heat
of combustion.
- Perform a calorimetry procedure to
determine the heat of combustion
of two different alcohols.
- Compare the heats of combustion
of different carbon-based fuels.

FOCUS ON UNDERSTANDING

The positive and negative numbers
associated with heat transfer and heat
of combustion may confuse some
students. The sign of the energy
value reflects whether the reaction
is exothermic or endothermic.

KEY TERMS

heat of reaction
heat of combustion
joule

IN CLASS

Students complete a standard calorimetry
lab comparing methanol and isopropanol
as fuels. A worksheet allows students
to compare the energy released by the
combustion of a variety of carbon-based
fuels. Students are introduced to the
joule as another unit of energy, as
well as to heat of reaction and heat
of combustion. A complete materials
list is available for download.

TRM Materials List 103

Differentiate

To help struggling readers, provide
students a printout of Transparency—
Heat of Combustion Lab Setup to help
students get a visual sense of what to
do. Consider creating "fuel cards"
for the 10 fuels listed in the student
worksheet. On each fuel card, show the
name, chemical formula, energy (kcal/
mol) and a picture of the fuel where
students might find it used (e.g., a
gas pump for octane or a candle for
paraffin).

SETUP

Before class, prepare alcohol burners for
each group. Fill the burners with 25 mL
of each alcohol.

SAFETY

Have available the fire blanket and
proper fire extinguisher for alcohol
fires. Ensure students wear safety
goggles at all times.

CLEANUP

Save the remaining methanol and
isopropanol for use in another class.

Lesson Guide

ENGAGE (5 min)

TRM Transparency—ChemCatalyst 103

TRM PowerPoint Presentation 103

LESSON 103

ChemCatalyst

An engineer wants to know which of these two alcohols is a better rocket fuel.

Ethanol

Isopropanol

1. What do you think "better" means in this case?

2. What measurements should you make to determine the better choice?

Sample answers: **1.** Students may say that "better" means more thermal energy per gram or per molecule. They may say that it is important for the fuel to burn completely or that you do not want the fuel to produce a lot of carbon dioxide. **2.** You have to measure the thermal energy per gram to compare the two fuels.

Ask: What properties does a good fuel have?

Ask: Which do you expect to react with more oxygen, ethanol or isopropanol? Explain your thinking.

Ask: Why might it be important to measure the mass of the fuels?

Ask: Why do you think octane is a major component of gasoline?

EXPLORE (20 min)

TRM Worksheet with Answers 103

TRM Worksheet 103

TRM Transparency—Heat of Combustion Lab Setup 103

INTRODUCE THE LAB

→ Arrange students into groups of four.

→ Pass out the student worksheets.

→ Review the general lab procedure.

→ Show the Transparency—Heat of Combustion Lab Setup. Make sure

ENVIRONMENTAL CONNECTION

A biofuel is any plant substance that can be used as a fuel. For example, some people have customized their cars to burn vegetable oils. These people can collect the oil that would have been discarded after deep frying by restaurants for use in their fuel tanks.

© Gary Yeowell/CORBIS

Important to Know The heat of combustion is reported with a negative sign to indicate that energy is released by the reaction.

The chemical equations for the combustion of each type of fuel are shown here.

Combustion of ethanol: $C_2H_6O(l) + 3O_2(g) \longrightarrow 2CO_2(g) + 3H_2O(l)$

Combustion of methanol: $2CH_4O(l) + 3O_2(g) \longrightarrow 2CO_2(g) + 4H_2O(l)$

	Mass of fuel burned (g)	Mass of water heated (g)	ΔT for the water (°C)	Energy transferred to water (cal)	Energy per gram of fuel (cal/g)
Ethanol	1.2 g	50 g	30 °C	1500 cal	1250 cal/g
Methanol	1.6 g	50 g	30 °C	1500 cal	938 cal/g

The experimental results show that the combustion of ethanol delivers more energy per gram of fuel consumed, compared with the combustion of methanol. So you would want to use ethanol, not methanol, in a race car if weight were a concern.

⟳ Heat of Reaction

The amount of energy transferred during a chemical reaction is called the **heat of reaction**. It is often designated with the symbol ΔH. The heat of reaction for a combustion reaction is usually referred to as the **heat of combustion**. Heat of reaction, ΔH, is expressed as a negative number when the reaction is exothermic. This is because energy is transferred from the system. When a reaction is endothermic, the system gains energy. Heat of reaction, ΔH, is expressed as a positive number when the reaction is endothermic.

Scientists have done carefully controlled experiments to measure the heat of combustion for many different fuels. Accurate measurements for the combustion of methanol result in a heat of reaction that is much larger in magnitude than the experimental value reported earlier.

Combustion of methanol: $2CH_4O + 3O_2 \longrightarrow 2CO_2 + 4H_2O$
$$\Delta H = -5400 \text{ cal/g}$$

Combustion of ethanol: $C_2H_6O(l) + 3O_2(g) \longrightarrow 2CO_2(g) + 3H_2O(l)$
$$\Delta H = -7100 \text{ cal/g}$$

UNITS

The metric unit that scientists use to express energy is the joule, J. A **joule** is a unit of energy that is 4.184 times larger than a calorie. The heat of reaction can be expressed in kilocalories per gram, kilocalories per mole, kilojoules per gram, or kilojoules per mole.

$$1 \text{ joule} = 0.239 \text{ calorie}$$
$$1 \text{ calorie} = 4.184 \text{ joules}$$

students know that the beaker of water should be located approximately 2–3 cm above the top of the alcohol burner.

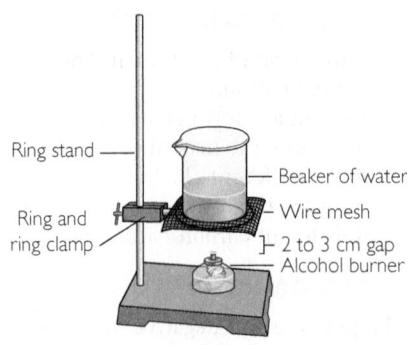

Ring stand —
Ring and ring clamp —
— Beaker of water
— Wire mesh
— 2 to 3 cm gap
— Alcohol burner

→ Review safety instructions for working with flame in the laboratory.

GUIDING THE LAB

→ Students' calculated values for methanol and isopropanol will be quite far off from those given in the table of energy data on the worksheet. This is addressed in Question 10 in the Analysis section.

EXPLAIN & ELABORATE (15 min)

TRM Transparency—Heat of Combustion of Common Fuels 103

⊙ Comparing Fuels

One way to compare fuels is to calculate the number of kilojoules of energy per mole (kJ/mol) of fuel. Simply multiply the kilojoules per gram (kJ/g) by the molar mass (g/mol) of each fuel.

This table provides the heats of combustion for several fuels in both kilojoules per gram and kilojoules per mole.

Heats of Combustion

	Fuel	Chemical formula	ΔH (kJ/g)	ΔH (kJ/mol)
elements	hydrogen	$H_2(g)$	−121.3	−242
	carbon (coal)	$C(s)$	−32.6	−393
alkanes	methane	$CH_4(g)$	−55.6	−891
	ethane	$C_2H_6(g)$	−51.9	−1560
	butane	$C_4H_{10}(l)$	−49.6	−2882
	hexane	$C_6H_{14}(l)$	−48.1	−4163
	octane	$C_8H_{18}(l)$	−47.7	−5508
alcohols	methanol	$CH_4O(l)$	−22.6	−724
	ethanol	$C_2H_6O(l)$	−29.7	−1368
	butanol	$C_4H_{10}O(l)$	−36.0	−2669
sugar	glucose	$C_6H_{12}O_6(s)$	−15.9	−2828

Notice that when the number of carbon atoms increases, the energy per mole increases. Also, the energy transferred per mole is smaller for the alcohols compared with the alkanes. This makes sense because combustion is a reaction in which a substance combines with oxygen, and the alcohols already contain an oxygen atom. Also notice that elemental hydrogen releases the highest energy per gram. This is why hydrogen is used as the fuel to launch a space shuttle. You get a lot of energy without adding much weight.

LESSON SUMMARY

How do different fuels compare?

One way to compare different fuels is to measure the amount of energy they transfer as a result of combustion. Calorimetry is the most common method of measuring this energy. The energy transferred can be expressed in kcal/mol, kcal/g, kJ/mol, or kJ/g of substance. The amount of energy transferred per gram or per mole during a chemical change is commonly called the heat of reaction, ΔH. For combustion reactions, the heat of reaction is negative and is also referred to as the heat of combustion.

KEY TERMS
heat of reaction
heat of combustion
joule

INTRODUCE HEAT OF COMBUSTION, ΔH

→ Display the Transparency—Heat of Combustion of Common Fuels.

Ask: What results did you obtain for the amount of energy transferred when you burned methanol and isopropyl alcohol?

Ask: How did your experimental results compare to the values given in the table? How do you explain this difference?

Ask: What would you do to improve on the experiment?

Ask: Why are the heats of combustion negative numbers?

Key Points: The amount of energy transferred during any reaction is referred to as the *heat of reaction*. The energy transferred during burning is called the *heat of combustion*. Both are designated by the symbol ΔH. Different fuels have different heats of combustion. Notice that the heat of combustion for each fuel is expressed as a negative number. This is because the reaction is exothermic, and the system—the fuel—is "losing" thermal energy to the surroundings.

> **Heat of reaction, ΔH:** The amount of heat transferred during a chemical reaction.

> **Heat of combustion, ΔH:** The amount of heat transferred when a compound undergoes complete combustion with oxygen.

Chemists use a bomb calorimeter to measure the amount of heat released by fuels. In a bomb calorimeter, fuel is burned in a chamber and the heat is transferred to water. No heat is lost to the surrounding air, allowing an accurate energy calculation. Calorimetry calculations measure energy transferred into a volume of water. Chemists also talk about fuels in terms of the amount of energy they *release* per gram or mole.

COMPARE DIFFERENT FUELS

→ Display the Transparency—Heat of Combustion of Common Fuels

Ask: Why do you think the magnitude of the value of heat of combustion per mole of octane is greater ("more negative") than the value of the heat of combustion per mole of methane?

Ask: Why do you think the magnitude of the value of the heat of combustion per gram of methane is greater ("more negative") than the value of the heat of combustion per gram of methanol?

Ask: Which substance transfers the most heat per gram? (hydrogen)

Ask: What is the advantage of choosing a fuel that has less mass?

Key Points: Hydrocarbons make good fuels. They combine readily with oxygen to produce carbon dioxide and water. As the number of carbon atoms in a compound increases, the energy output per mole from combustion generally increases. This makes sense, because more bonds have to be broken and more new bonds have to be made. Fuels containing oxygen atoms have a lower energy output from combustion per mole.

The energy output from combustion of each substance is just one consideration in rating fuels. The results of the lab show that some fuels may produce more energy per mole but less energy per gram. Hydrogen gas is an extremely lightweight fuel compared to carbon compounds. It produces much more energy per gram than any other fuel listed in the table. This is why it is used in rocket engines.

INTRODUCE THE JOULE AND THE KILOJOULE

→ Write the conversion factors for calories to joules and joules to calories on the board.

1 joule ≈ 0.239 calorie
1 calorie ≈ 4.184 joules

Ask: How many kilojoules are in 100 kilocalories? (418.4 kJ)

Ask: How many kilocalories are in 100 kilojoules? (23.9 kcal)

Key Points: Scientists use *joules* as a unit of energy. The *calorie* is a unit related to calorimetry and the heating of water. The metric unit used by scientists is the kilojoule, or kJ.

Joule: The metric unit of energy commonly used by scientists. One calorie is equal to 4.184 joules.

EVALUATE (5 min)

Check-In

1. Which fuel provides more energy per mole, paraffin or propane?

2. How much energy would you expect to be released by burning 1 g of each of these substances?

3. Which substance is a better fuel by weight?

Answers: **1.** Paraffin. Paraffin transfers 3493 kcal/mol. Propane transfers only 526 kcal/mol. **2.** Paraffin transfers 11.3 kcal/g. Propane transfers 11.9 kcal/g. **3.** Comparing the energy transferred per gram, propane is a slightly better fuel by weight.

Homework

Assign the reading and exercises for Fire Lesson 103 in the student text.

LESSON 103 ANSWERS

1. *Possible answer:* You can compare two fuels by burning a measured amount of each fuel in a calorimeter. By measuring the temperature change of the water in the calorimeter you can determine the energy per gram that is released when each fuel is burned.

Reading Questions

1. How can you use water to compare two fuels?

2. What is heat of combustion?

3. Explain why heat of combustion is expressed as a negative number.

Reason and Apply

4. The combustion of 4.0 g of propanol transfers energy to 150 mL of water. The temperature of the water rises from 20 °C to 45 °C.
   a. How many calories of energy are transferred to the water?
   b. How many kilocalories of energy are transferred to the water?
   c. How many kilojoules of energy are transferred to the water?
   d. What temperature increase do you expect for 300 mL of water?
   e. What temperature increase do you expect for 2.0 g of propanol and 150 mL of water?

5. Write balanced chemical equations for the combustion of the fuels in the heats of combustion table on page 525. What generalization can you make about the amount of oxygen required for combustion of hydrocarbons?

6. Explain why the combustion of glucose provides a large energy per mole but a rather small energy per gram.

7. PROJECT In the heats of combustion table, hydrogen has the highest energy per gram of fuel that is combusted. Research to find three reasons why it is challenging to use hydrogen as a fuel for your automobile.

---

2. Heat of combustion is the amount of heat transferred to the environment when one mole or one gram of a substance burns.

3. Heat of combustion is expressed as a negative number because energy transfers out of the system and into the environment.

4. a. 3800 cal (to two significant figures) **b.** 3.8 kcal **c.** 16 kJ **d.** Because the equation for the heat absorbed is $q = mC_p \Delta T$, temperature increase is inversely proportional to the mass of water. So, if the amount of water is doubled, the temperature increase will be halved, or 12.5 °C. **e.** Temperature increase is directly proportional to the mass of fuel. So, if the amount of propanol is halved, the temperature increase will also be halved, or 12.5 °C.

Answers continue in Answer Appendix (p. ANS-8).

CHAPTER 19

Measuring Energy

SUMMARY

KEY TERMS

combustion

calorimetry

heat of reaction

heat of combustion

joule

Fire Update

Combustion can be defined as a reaction in which a substance combines with oxygen, releasing energy. The process of combustion is commonly referred to as burning, and the substance that burns is called a fuel. Some fuels are better sources of energy than others. To measure the energy from a fire, or any chemical reaction, you must measure it indirectly by measuring its effect on matter. Energy is usually measured in units of calories or joules.

Calorimetry is the procedure used to measure the transfer of energy. The amount of energy transferred during a chemical reaction is called the heat of reaction, ΔH. The energy released by the combustion of a fuel is called the heat of combustion.

REVIEW EXERCISES

1. Describe an experimental setup that would help you compare two fuels. What would you measure?

2. A chemist wants to find out how many Calories are in a bag of peanuts. He sets up a calorimeter and burns a peanut underneath it. This is his data table.

Initial mass of peanut (g)	Mass of water (g)	ΔT of water (°C)	Final mass of peanut (g)
3.75 g	100 g	10 °C	1.20 g

a. Calculate the amount of energy, in calories, that can be transferred from the combustion of one peanut.
b. Calculate the amount of calories per gram of peanut.
c. If a bag of peanuts is 500 g, how many food Calories does it contain?

3. If iron nails are left outside, they will rust according to the reaction: $4Fe(s) + 3O_2(g) \longrightarrow 2Fe_2O_3(s)$. Why do you think this reaction does not appear to release light and heat?

4. Propane, $C_3H_8(g)$, is used as fuel in small gas stoves and barbecue grills.
a. Write the balanced equation for the combustion of propane.
b. The heat of combustion of propane is -2220 kJ/mol. How many kJ/g would combusting propane release?
c. According to the table on page 525, is the heat of combustion per gram of propane greater than the heat of combustion per gram of methane? Per mole? Can you explain why?

b. -50.3 kJ/g **c.** No, the heat of combustion for methane is -50.4 kJ/g from part b, which is less negative than the heat of combustion of methane per gram, -55.6 kJ/g. The heat of combustion per gram of propane releases less energy per gram. However, the table gives the heat of combustion of methane as -891 kJ/mol, which is less negative than -2220 kJ/mol. So, combustion of propane does release more energy per mole. *Possible answer:* Propane is a larger molecule than methane, so a mole of propane has more atoms available to react with oxygen than a mole of methane. However, if we assume the energy released comes from the molecular bonds of each molecule, methane has 4 carbon-hydrogen bonds and a mass of 16 grams and propane has 8 carbon-hydrogen bonds and a mass of 44 grams. Methane has more bonds per gram than propane, so it has more chemical potential energy.

ASSESSMENTS

Two multiple-choice and two short-answer assessments for Chapter 19 are available for download.

TRM Chapter 19 Assessment Answers

TRM Chapter 19 Assessments

ANSWERS TO CHAPTER 19 EXERCISES

1. *Possible answer:* ● Build a calorimeter using water, a small open metallic container, and a beaker. ● Place the open metal container in the beaker. ● Fill the beaker with a measured amount of water that almost covers the container. ● Use a match to burn a measured amount of a fuel in the metal container and measure the temperature change of the water in the beaker. ● Use the mass and the temperature change of the water to determine the number of calories of heat absorbed by the water. ● Measure the mass of the unburned fuel.

● Divide the energy absorbed by the water by the difference in mass of the fuel to determine the energy in calories per gram within the fuel. ● Repeat the procedure above on the second fuel. ● Compare the results.

2. a. 1000 cal **b.** 392 cal/g **c.** 196 Cal

3. *Possible answer:* The reaction does not appear to release light and heat because it occurs slowly, so the energy is transferred into the environment without being noticed.

4. a. $C_3H_8(l) + 5O_2(g) \rightarrow 3CO_2(g) + 4H_2O(l)$

CHAPTER 20

Watch the video overview of Chapter 20 (for teachers) by clicking on the link in the TE-book, opening the TRFD, or logging onto the book's companion Web site bcs.whfreeman.com/livingbychemistry2e (teacher log-in required).

This chapter seeks to explain the chemistry behind the energy transfers that result from chemical change. Lesson 104 starts off the chapter by introducing bond energies and the energy transfers associated with breaking and making bonds. Students compare the net energy transfers of exothermic and endothermic reactions. In Lesson 105, students investigate the energetics of trying to reverse reactions. This lesson provides evidence to support the first law of thermodynamics and connects changes in kinetic and potential energy to chemical reactions. Demos in Lesson 106 allow students to observe the effects of certain changes on the rate of a reaction. Lesson 107 defines work and shows how the energetic and gaseous products of combustion can be harnessed to do work.

In this chapter, students will study

- average bond energies
- the energy of bond breaking and bond making
- tracking energy exchanges with diagrams
- kinetic and potential energy changes in reactions
- factors affecting the kinetics of reactions
- the process of converting chemical energy to work

CHAPTER **20**

Understanding Energy

In this chapter, you will study
- bond energies and heat of reaction
- reversing chemical reactions
- rates of chemical reactions
- how energy is converted to work

528

Early trains were powered by steam engines. A steam engine burns coal or wood to boil water, then uses the pressure from the water vapor to drive the engine. Modern trains run on diesel or electricity instead.

When chemical change occurs, new substances with new properties are produced. This means some chemical bonds must break and new chemical bonds must form. The energy transferred during chemical change is related to bond making and bond breaking. People have figured out how to make productive use of the energy that accompanies chemical changes. Chapter 20 of Unit 5: Fire focuses on the source of energy in chemical change. It also explores how easily or how quickly a reaction might happen.

Chapter 20 Lessons

Featured ACTIVITY

Make It or Break It

Purpose
To explore bond energies and calculate the net energy exchange for several reactions.

Materials
- ball-and-stick model kit

Questions

1. List at least two patterns you see in the table of average bond energies.

Average Bond Energies (per mole of bonds)

Bond	H—H	C—H	C—C	C≡C	O—H	C—O	C=O	O=O	O—O
Bond energy (kJ/mol)	432	413	347	614	467	358	799	495	146

2. Use ball-and-stick models to create the reactants for this reaction.

   Burning methane: $CH_4(l) + 2O_2(g) \longrightarrow CO_2(g) + 2H_2O(g)$

3. Rearrange the reactant models into carbon dioxide and water. Count how many of each type of bond you must break when this reaction takes place. Record the information in a table. Use the average bond energies to determine the amount of energy transferred to break all the bonds.

4. Count how many of each type of bond you must make to form the products. Record this information in your table. Use the average bond energies to determine the amount of energy transferred out when the new bonds are formed.

5. What is the net energy you expect to be transferred to the surroundings by this reaction?

6. The energy of reaction for exothermic reactions is normally expressed as a negative number. Why do you think it is negative?

7. **Making Sense** When a substance combusts, energy is transferred to the surroundings as heat and light. Where does that energy come from?

8. **If You Finish Early** What is the energy of reaction for 1 g of methane burned, in kilocalories per gram?

Lesson 104 | **Bond Energy**   529

Overview

DEMO: WHOLE CLASS

ACTIVITY: PAIRS

Key Question: Where does the energy from an exothermic reaction come from?

KEY IDEAS

When substances react, bonds between atoms are broken and new bonds are formed. To break a bond, energy must be transferred from the surroundings to a reactant. The average energy required to break a specific bond is called the *bond energy*. Conversely, when new bonds are made, the bond energy is transferred from the products to the surroundings. The heat transferred by a reaction is approximately equal to the difference in energy between the average bond energies of the reactants and the average bond energies of the products.

LEARNING OBJECTIVES

- Calculate the net energy exchange due to a chemical reaction.
- Draw and interpret energy exchange diagrams.
- Explain the source of heat in a combustion reaction.

FOCUS ON UNDERSTANDING

- Strictly speaking, the energy required to break a certain bond in a certain compound is called the bond *dissociation* energy. The bond energies used here are averages.
- Average bond energies allow you to predict only approximate heats of reaction. For more accurate predictions, you have to use bond enthalpies for specific bonds, which take pressure-volume work into account.

KEY TERM

bond energy

IN CLASS

The class starts with a quick demo by the instructor using bar magnets to simulate bonds between atoms. Students complete a worksheet that focuses on bond energies and on how average bond energies are related to the overall energy transfer of a reaction. Pairs of students then use molecular models to simulate a combustion reaction, tracking the breaking of bonds and the formation of new bonds. They calculate the total energy required to break all the reactant bonds and the total energy required to make all the product bonds. These data let them compute the energy exchanges for the reaction. A complete materials list is available for download.

TRM Materials List 104

Differentiate

Students can benefit from drawing structural formulas to help determine the net energy exchange for a reaction. Direct students to draw, in the margins of the student worksheet, the structural formulas for all of the reactant and product molecules according to the balanced chemical equation. This way, students can visually count how many of each bond is broken or made during the reaction. Doing so helps students to avoid overlooking C—C bonds in larger molecules such as ethanol.

Pre-AP® Course Tip

Students preparing to take an AP® or college chemistry course should be able to use representations and models to make predictions and solve problems. In this lesson, students use models to draw and interpret energy diagrams to get an overall picture of the energy changes that occur from a chemical reaction. They use these diagrams to calculate the net energy exchanged.

SETUP

The molecular models must have at least 4 carbon atoms, 12 hydrogen atoms, and 14 oxygen atoms. Each pair should have 14 single bonds and 8 double bonds.

Lesson Guide

ENGAGE (5 min)

TRM PowerPoint Presentation 104

ChemCatalyst

1. When a fuel is burned, energy is released. Where do you think that energy comes from?

2. Consider this reaction:

$CH_4(g) + 2O_2(g) \rightarrow CO_2(g) + 2H_2O(l)$.

What is happening to the methane molecules as they combust?

Sample answers: **1.** Students may think that the breaking apart of the bonds in the molecules of the reactants is the source of the energy. This is a common misconception. Some students will come up with alternative explanations having to do with electrons or with the motions of the particles (kinetic energy). Some will say that bond making is the source of the energy. **2.** The methane molecules are breaking apart into atoms and rearranging into new substances by forming new bonds between atoms.

→ Briefly discuss the ChemCatalyst. Students will explore this reactant in depth in today's activity.

Ask: What is happening to the reactants during a combustion reaction?

EXPLORE (15 min)

TRM Worksheet with Answers 104

TRM Worksheet 104

COMPLETE THE DEMONSTRATION

→ Use the two bar magnets to simulate the breaking and making of bonds. The idea is that breaking a bond requires an input of energy. Conversely, the two bar magnets are attracted to each other. They come together with a snap. Similarly, making bonds releases energy.

INTRODUCE THE ACTIVITY

→ Arrange students into pairs.

→ Pass out the student worksheets.

Make It or Break It

Bond Energy

THINK ABOUT IT

In a chemical reaction, bonds in the reactants are broken and the atoms rearrange to form new bonds in the products. Flames associated with a combustion reaction indicate that energy is released in the form of heat and light. All combustion reactions are exothermic.

Where does the energy from an exothermic reaction come from?

To answer this question, you will explore

↻ Bond Breaking and Bond Making

↻ Energy of Change

↻ Energy Exchange Diagrams

EXPLORING THE TOPIC

↻ Bond Breaking and Bond Making

In chemical compounds, atoms are bonded together. When the compound methane burns, it combines with oxygen to make new products. To make carbon dioxide and water from molecules of methane and oxygen, the atoms must rearrange as shown. This means that covalent bonds in the methane and oxygen molecules must be broken and new covalent bonds must be formed.

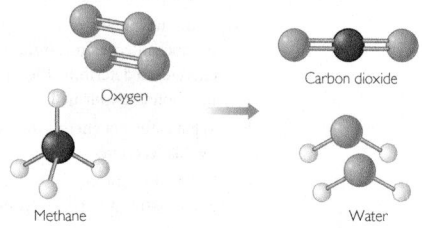

Energy is required to break bonds. Atoms that are bonded together are attracted to each other, so it takes effort to get these atoms apart. So bond breaking is an endothermic process.

In contrast, when new bonds are formed, energy is released. In terms of energy, bond making is the opposite of bond breaking. Because energy is conserved, the amount of energy released when a bond is made is assumed to be equal to the amount of energy required to break that same bond.

EXPLAIN & ELABORATE (15 min)

TRM Transparency—Energy Exchange Diagrams 104

TRM Transparency—Energy Exchange Diagram: Formation of H_2O 104

DISCUSS BOND ENERGIES

→ Write these statements on the board.

Bond breaking requires energy.

Bond making releases energy.

Ask: What patterns did you notice in the average bond energies?

Ask: How would you explain the order of increasing bond energies for C—C, H—H, and C≡C?

Ask: Which bonds are easiest to break? Hardest?

Ask: Do you expect product molecules in a combustion reaction to have strong or weak bonds compared to reactant molecules? Explain your thinking.

Key Points: It takes energy to break bonds between atoms. An easy way to remember this is that it takes effort to break something apart, just as it takes effort to pull apart magnets attracted to each other. The amount of energy

During a combustion reaction, energy is required to break the bonds in the reactant molecules and energy is released when the bonds form in the product molecules. The net energy released is observed as heat and light.

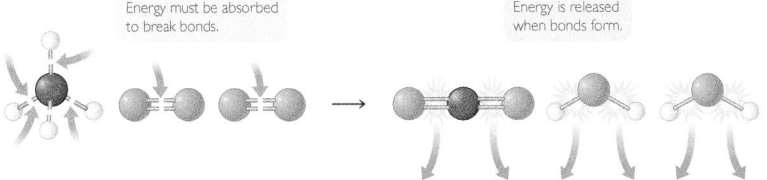

Energy must be absorbed to break bonds.

Energy is released when bonds form.

Big Idea Bond breaking requires energy. Bond making releases energy.

BOND ENERGY

The amount of energy it takes to break a specific bond, such as a carbon-hydrogen bond, is called the **bond energy**. Bond energy is considered a measure of the strength of that bond. However, not all C—H bonds have the same bond energy. For example, it takes 435 kJ/mol to break the first C—H bond in a methane molecule, as compared to 339 kJ/mol to break the last C—H bond.

For this reason, bond energies are reported as averages. The bond energy of an average carbon-hydrogen bond is approximately 413 kJ/mol (or 99 kcal/mol). This means that it takes 413 kilojoules of energy to break one mole of carbon-hydrogen bonds.

This table shows the bond energies of some common bonds.

Average Bond Energies (per mole of bonds)

Bond	H—H	C—H	C—C	C=C	C≡C	O—H	C—O	C=O	O=O	O—O
Bond energy (kJ/mol)	432	413	347	614	811	467	358	799*	495	146

*C=O in CO_2: 799, C=O in organic molecules: 745

Example

Methane Molecule

How much energy would it take to break all the bonds in 1 mol of methane molecules?

Solution

Each molecule of methane has four C—H bonds. Therefore, each mole of methane molecules has 4 mol of carbon-hydrogen bonds.

$$413 \text{ kJ/mol} \cdot 4 \text{ mol C—H bonds} = 1652 \text{ kJ}$$

In reality, all of the bonds of the reactants do not necessarily break to form the products in a chemical reaction. However, this is a useful model to explain energy exchanges related to a chemical reaction.

Lesson 104 | **Bond Energy**   531

to break a specific bond in a specific compound is called the *bond energy*. The bond energy of a particular bond may vary somewhat depending on the compound it is in. A C=O bond in a carbon dioxide molecule has a slightly different bond energy from a C=O bond in a larger molecule. The bond energies you have been working with all are averages. The bond energy of an average carbon-carbon bond is 347 kJ/mol. This means that 347 kJ of energy are needed to break 1 mol of C—C bonds.

Making bonds between atoms releases energy. When C—C bonds form, 347kJ of energy are released per mole of bonds formed. An analogy would be two magnets coming together with a loud click.

This table shows the average bond energies of some common bonds. Note that double bonds generally have higher bond energies than single bonds. Double bonds are stronger than single bonds, and triple bonds are stronger than double bonds. Bonds between smaller atoms also tend to be stronger.

Average Bond Energies (per mole of bonds)

Bond	Bond Energy (kJ/mol)
H—H	432
C—H	413
C—C	347
C=C	614
O—H	467
C—O	358
C=O	799*
O=O	495
O—O	146

*C=O in CO_2: 799; C=O in organic molecules: 745.

Bond energy: The amount of energy required to break a specific chemical bond.

DISCUSS THE ENERGY EXCHANGES OF CHEMICAL CHANGE

→ Write this statement on the board.

The net energy exchange of a reaction is the difference between the energies of the bond-breaking and bond-making processes.

→ Show the Transparency—Energy Exchange Diagram: Formation of H_2O that tracks the energy changes in the formation of water.

Ask: Why is the net energy of reaction for an exothermic process is always negative? Explain.

Ask: What does it mean if the net energy of reaction is a positive number?

Key Points: Each chemical reaction has a net energy exchange, which determines whether a reaction is exothermic or endothermic. Chemical change is always accompanied by energy exchanges. Energy transfer in the process of bond breaking is given a positive sign because energy is added to the system. Energy transfer in the process of bond making is given a negative sign because energy is released from the system. What an observer experiences following completion of a reaction (hot or cold) depends on whether there is more energy in or more energy out.

An energy exchange diagram is a helpful way to get an overall picture of the energy changes that occur during a

Lesson 104 | **Bond Energy**   531

chemical reaction. Consider the formation of water from hydrogen and oxygen.

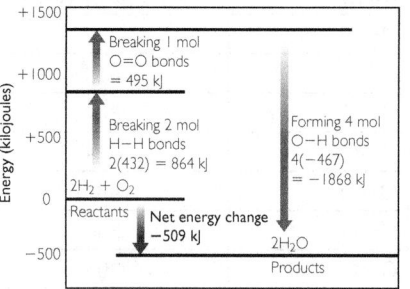

Formation of water from hydrogen and oxygen

Energy goes into the system to break apart 2 mol of H—H bonds and 1 mol of O=O bonds. Energy comes out of the system with the formation of 4 mol of O—H bonds. The net result is more energy out. This reaction is exothermic, releasing −509 kilojoules of energy.

DISCUSS EXOTHERMIC AND ENDOTHERMIC ENERGY EXCHANGES

→ Display the Transparency—Energy Exchange Diagrams, which shows general diagrams for an exothermic reaction and an endothermic reaction.

Ask: For the reaction of methane and oxygen, how much energy does it take to break all the bonds in the reactants? How did you figure this out?

Ask: For the reaction of methane with oxygen, which involves more energy: bond breaking or bond making?

Key Points: See the diagrams on page 533 of the student text. For an exothermic reaction, the total energy out is greater than the total energy in. Therefore, the observer experiences a release of energy, usually in the form of heat or light. For an endothermic reaction, the total energy in is greater than the total energy out. The observer experiences a sensation of cold. Note that in using bond energies to determine heat of reaction, ΔH, it is assumed that all bonds break and all bonds are newly made. However, this is not the actual mechanism by which most products are made. (Recall the formation of esters in Unit 2: Smells, in which only a couple of bonds break in the molecules.) These bond energy values also do not reflect any energy exchanges associated with phase changes. Nevertheless, this method provides a very close estimate of the actual energy transfer.

Energy of Change

To estimate the energy of an entire chemical reaction, you can consider the reaction as if it takes place in two parts—*energy in* for bond breaking and *energy out* for bond making.

COMBUSTION OF METHANE

Consider the reaction for the combustion of 1 mol of methane molecules with 2 mol of oxygen molecules. What is the net energy exchange?

$$CH_4 + 2O_2 \longrightarrow CO_2 + 2H_2O$$

First, figure out how much energy is required to break all the bonds in the reactants.

Bond Breaking—Energy In

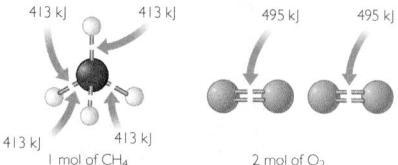

413 kJ 413 kJ 495 kJ 495 kJ

413 kJ 413 kJ
1 mol of CH₄ 2 mol of O₂

Total energy input = 4(413 kJ) + 2(495 kJ) = 2642 kJ

Notice that the energy required to break the bonds of the reactants is positive because energy is added to the system from the surroundings. Now figure out how much energy is released by the formation of new bonds.

Bond Making—Energy Out

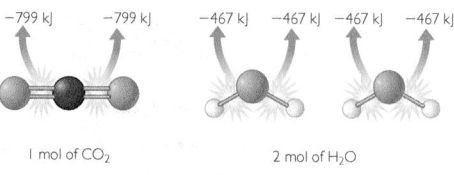

−799 kJ −799 kJ −467 kJ −467 kJ −467 kJ −467 kJ

1 mol of CO₂ 2 mol of H₂O

Total energy output = 2(−799 kJ) + 4(−467 kJ) = −3466 kJ

Notice that the energy required to make the bonds of the reactants is negative because energy leaves the system and goes to the surroundings. The net energy exchange is equal to the sum of the energy input and the energy output.

$$\begin{aligned} \text{Net energy exchange} &= (2642 \text{ kJ}) + (-3466 \text{ kJ}) \\ &= -824 \text{ kJ} \end{aligned}$$

The magnitude of energy out is greater than the magnitude of energy in, so this reaction is exothermic. Net energy is expressed as a negative number, because the system loses energy to the surroundings. This value is very close to the actual value for the heat of combustion, ΔH, for this reaction.

EVALUATE (5 min)

Check-In

How do you think the energy exchange diagram for the combustion of ethane would differ from the diagram for methane?

Answer: An energy exchange diagram for the combustion of ethane would look much like the energy exchange diagram for methane. Both are exothermic, so both will have net energy out.

However, ethane is a larger molecule than methane. There are more bonds to break and make when ethane is burned because more atoms are involved. So the top line will be higher up, and the energy out will be greater.

Homework

Assign the reading and exercises for Fire Lesson 104 in the student text.

Energy Exchange Diagrams

An energy exchange diagram is a way to keep track of the energy changes during a chemical reaction. The arrows pointing up represent the energy going into the system to break the bonds. The arrows pointing down represent the energy released when new bonds are made. The net energy change determines if a reaction is exothermic or endothermic.

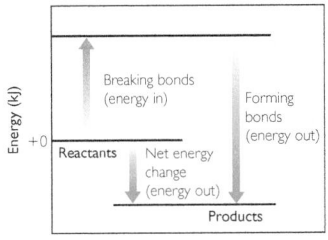

Energy exchange for an exothermic reaction

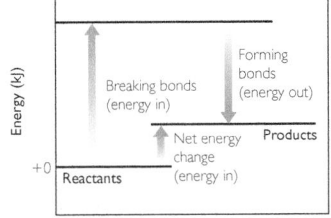

Energy exchange for an endothermic reaction

LESSON SUMMARY

Where does the energy from an exothermic reaction come from?

KEY TERM

bond energy

When reactions take place, bonds between atoms are broken and new bonds are formed. Energy must be transferred *to* a reactant to break a bond. The average energy required to break a certain type of bond is called its bond energy. Conversely, when new bonds are made, energy is released. The heat transferred during a chemical reaction is equal to the difference in energy between the bond-breaking process and the bond-making process. An energy diagram can be used to keep track of energy exchanges in a chemical reaction.

Exercises

Reading Questions

1. How can you determine the net energy exchange of a chemical reaction using average bond energies?

2. Use the energies involved in bond breaking and bond making to explain why combustion reactions are exothermic.

Reason and Apply

3. Place these bonds in order of increasing energy required to break the bond: H—H, C—C, C=C, O—O, O=O.

4. Place these bonds in order of increasing energy released when the bond is formed: H—H, C—H, O—H, C=O.

5. What is the total energy required to break all the bonds in 1 mol of ethanol, C_2H_6O?

6. **a.** 1359 kJ/mol **b.** 1868 kJ/mol **c.** Net energy exchange is −509 kJ/mol. The reaction is exothermic.

7. **a.** methanol combustion: −660 kJ/mol; butanol combustion: −2509 **b.** From the heats of combustion table: methanol: −724 kJ/mol; butanol: −2669 kJ/mol. The calculated values are slightly higher (slightly less negative) than the values given in the table.

8. 127 kJ/mol

9. A

LESSON 104 ANSWERS

1. *Possible answer:* You can determine the net energy exchange of a chemical reaction by adding the energy input into the reaction and subtracting the energy output from the reaction. Breaking the bonds in the reactants requires an energy input. Forming the bonds in the products causes an energy output. The average bond energies can be used to determine the amount of energy input or output as each bond is broken or formed.

2. Combustion reactions are exothermic because it takes less energy to break the bonds in the reactants than is released by the formation of the chemical bonds in the products. For example, in the combustion of a hydrocarbon, the C—C bonds and C—H bonds have bond energies around 400 kJ/mol. The C=O double bond in carbon dioxide has a bond energy of 799 kJ/mol.

3. In order of increasing energy required to break the bond: O—O, C—C, H—H, O=O, C=C.

4. In order of increasing energy released when the bond is formed: C—H, H—H, O—H, C=O.

5. 3237 kJ/mol

Exercises *(continued)*

6. Consider the combustion of hydrogen. Use the table of average bond energies to answer the questions.

$$2H_2(g) + O_2(g) \longrightarrow 2H_2O(l)$$

   a. What is the total energy required to break the bonds in the reactant molecules?
   b. What is the total energy released by the formation of the bonds in the product molecules?
   c. Use the values from (a) and (b) to determine if this reaction is exothermic or endothermic.

7. Consider the combustion of methanol and butanol. Use the table of average bond energies to answer these questions.

$$\text{Methanol: } 2CH_4O(l) + 3O_2(g) \longrightarrow 2CO_2(g) + 4H_2O(l)$$
$$\text{Butanol: } C_4H_{10}O(l) + 6O_2(g) \longrightarrow 4CO_2(g) + 5H_2O(l)$$

   a. Calculate the net energy exchange for each reaction.
   b. Compare your answers to the heat of combustion values for methanol and butanol in Lesson 103: Fuelish Choices. What did you discover?

8. Calculate the net energy change for this reaction:

$$C_2H_6(g) \longrightarrow C_2H_4(g) + H_2(g)$$

9. In exothermic reactions,
   (A) It takes more energy to make all the new bonds in the products than to break all the bonds in the reactants.
   (B) There are more bonds made than broken.
   (C) There are more bonds broken than made.
   (D) It takes more energy to break all the bonds in the reactants than to make all the new bonds in the products.

Over the Hill

Reversing Reactions

THINK ABOUT IT

So far, you have been shown chemical reactions that proceed in one direction. This means that reactants are converted to products until the reaction is complete. Some reactions can be reversed. If a reaction is run in the reverse direction, the products of the reaction are converted back into reactants.

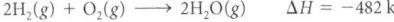

What is the energy associated with reversing reactions?

To answer this question, you will explore

↻ Reversing Reactions

↻ Kinetic and Potential Energy

↻ Activation Energy

EXPLORING THE TOPIC

↻ Reversing Reactions

If you place a lit candle under a balloon filled with hydrogen, you can observe the reaction between hydrogen and oxygen. When hydrogen combines with oxygen to form water there is a large "bang."

$$2H_2(g) + O_2(g) \longrightarrow 2H_2O(g) \qquad \Delta H = -482 \text{ kJ}$$

The energy released by this reaction is the heat of reaction, −482 kJ. This energy corresponds to 2 mol of $H_2(g)$ reacting with 1 mol of $O_2(g)$ to form 2 mol of $H_2O(g)$. This converts to −241 kJ per mole of $H_2(g)$, the value given in the heat of combustion table in Lesson 103: Fuelish Choices.

Important to Know Heat of reaction values are determined experimentally by calorimetry. You can use average bond energies to estimate the heat of reaction.

Notice that the decomposition of water into molecules of hydrogen and oxygen is the reverse of the combustion of hydrogen to form water. What is the heat of reaction for this change?

$$2H_2O(g) \longrightarrow 2H_2(g) + O_2(g) \qquad \Delta H = +482 \text{ kJ}$$

The first law of thermodynamics states that energy is conserved. So, the energy exchange of the reaction in the forward direction is equal and opposite to the energy exchange required for the reverse of the reaction. Notice that when the forward reaction is exothermic, the reverse reaction must be endothermic.

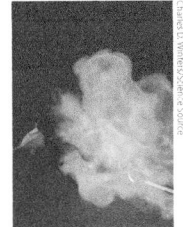

Formation of water in a balloon filled with hydrogen

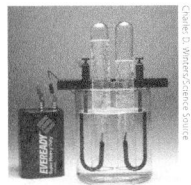

Hydrogen combustion

Decomposition as the reverse of hydrogen combustion

Lesson 105 | **Reversing Reactions** 535

Overview

DEMOS: WHOLE CLASS

CLASSWORK: INDIVIDUALS

Key Question: What is the energy associated with reversing reactions?

KEY IDEAS

Energy is conserved in chemical processes. This means that the net energy exchange of the reverse reaction is equal and opposite to the net energy exchange of the forward reaction. Some reactions are easier to reverse than others. The energy in a system at any given time is in the form of

either kinetic energy (energy of heat and motion) or potential energy (energy of position or composition). When exothermic reactions occur, the potential energy of the system decreases. When endothermic reactions occur, the potential energy of the system increases. Activation energy is a measure of the minimum amount of energy required to start a reaction.

LEARNING OBJECTIVES

● Use energy diagrams to compare forward and reverse reactions.
● Explain changes in kinetic and potential energy in chemical reactions.
● Define activation energy.

FOCUS ON UNDERSTANDING

Many factors contribute to the likelihood or favorability of a reaction. The heat of reaction is only one predictor.

KEY TERMS

potential energy
activation energy

IN CLASS

At the beginning of the lesson, students observe two demonstrations: a video showing the formation of water from the combination of hydrogen and oxygen in a hydrogen balloon, the other showing the formation of hydrogen and oxygen from the decomposition of water. These two reactions are the reverse of each other. However, the formation of water from hydrogen and oxygen is energetically favored over the reverse reaction. Students complete a worksheet that focuses on the energetics of reversing reactions. The discussion relates energy diagrams to changes in potential energy and introduces activation energy. A complete materials list is available for download.

TRM Materials List 105

Pre-AP® Course Tip

Students preparing to take an AP® or college chemistry course should be able to label representations of common chemical systems. In this lesson, students observe two demonstrations and use energy diagrams to explain the energetics of the reactions. They also learn how to label the activation energy in an energy diagram and its role in chemical reactions.

SETUP

Hydrogen Balloon Video: Search online for a demonstration of the igniting of a hydrogen balloon. Use the search term Hydrogen Balloon Demo. There are many available. Students will compare this to the reverse reaction in the demo below.

Decomposition of water: Set up a standard electrolysis apparatus using the materials from the downloadable materials list. You can place the 9-volt battery directly in the salt solution and try to collect the two gases from the two terminals, or you can set up the apparatus shown here for ease of viewing and gas collection:

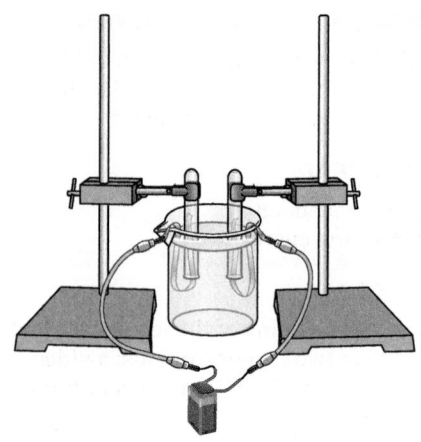

Hang the two strips of aluminum foil on the beaker so that a sufficient length of aluminum is in the salt solution. The two strips should not touch each other. With the alligator clips, connect the battery to each aluminum strip. Bubbles will form around both pieces of aluminum. To collect the gases, you can submerge two test tubes in the beaker so that they are filled completely with salt water. Turn the test tubes upside down and clamp them to the ring stands. Position the electrodes inside the tubes to capture the bubbles. The gas volume will be in a 2:1 ratio, forming 2 mol of H_2 for every mole of O_2. To make sure all the gas goes into the test tubes, you can coat most of the aluminum foil with nail polish, leaving exposed only the parts inside the test tubes and in contact with the alligator clips.

SAFETY

Wear safety goggles during preparation and during the demonstration. Have students wear goggles if they will be viewing from close by.

Lesson Guide

ENGAGE (5 min)

TRM **PowerPoint Presentation 105**

ChemCatalyst

Hydrogen is considered the clean fuel of the future for automobiles. The simple chemistry behind the technology is the combustion of hydrogen to produce water and 286 kJ of heat (per mole of hydrogen used).

1. Write the chemical equation for this reaction. Balance it.

2. What is the net energy input or output of the reaction as written?

Diagrams showing the energy exchange for the formation of water and the decomposition of water are provided here. These energy diagrams are similar to the ones in Lesson 104: Make It or Break It.

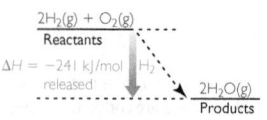

Combustion of hydrogen is exothermic.   *Decomposition of water is endothermic.*

↻ Kinetic and Potential Energy

The energy in a system is a combination of kinetic energy and potential energy. Kinetic energy is the energy of motion. **Potential energy** is energy that is stored within a physical system and can be converted into another type of energy, such as kinetic energy.

Potential energy can also be defined as the energy associated with the composition of a substance or the position or location of an object in space. For example, a ball at the top of a hill has high potential energy. This potential energy is converted into kinetic energy if the ball rolls down the hill. Likewise, some molecules have a high potential energy. For an exothermic reaction, this potential energy is converted into kinetic energy by converting reactants into products.

CONSERVING ENERGY

These diagrams show the changes in potential energy for an exothermic reaction and an endothermic reaction.

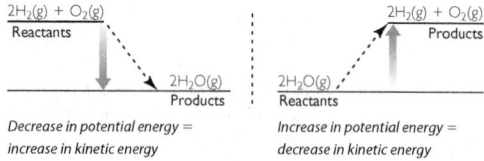

Decrease in potential energy =   *Increase in potential energy =*
increase in kinetic energy   *decrease in kinetic energy*

For an exothermic reaction, the potential energy of the system decreases in converting from reactants to products. Because energy is conserved, the sum of the kinetic energy and the potential energy must remain the same, so the kinetic energy increases. The products are hotter than the reactants, because they have a higher average kinetic energy. To reach thermal equilibrium with the surroundings, the hot products transfer energy to the surroundings.

For an endothermic reaction, the potential energy of the system increases. This means that kinetic energy must be converted into potential energy to conserve the total energy. The products are colder than the reactants because they have a lower kinetic energy. To reach thermal equilibrium with the surroundings, energy is transferred from the surroundings to the cold products.

---

3. What kind of energy exchange would you expect if you completed the reverse reaction and formed hydrogen and oxygen from water? Explain your thinking.

Sample answers: **1.** The equation for this reaction is $2H_2(g) + O_2(g) \rightarrow 2H_2O(l)$. **2.** The amount of energy released by this reaction is approximately -572 kJ. This is double the amount of energy released per mole of hydrogen (2 times -286 kJ/mol H_2). **3.** Some students will say that the reverse reaction will be endothermic. They may surmise that the energy exchanges for the reverse reaction are

just the opposite of those for the original reaction. Students may go through the process of working it out mathematically using bond energies, in which case the value will differ slightly.

Ask: What is the chemical equation for the reverse reaction? $(2H_2O(l) \rightarrow 2H_2(g) + O_2(g))$

Ask: How can you determine the net energy input or output from this reaction?

Ask: What type of reaction is this? (decomposition)

Ask: Do you think one reaction is energetically more likely to occur than the other? Explain your thinking.

Activation Energy

A fire can be started with a match. The combustion reaction in an automobile engine is started with a spark plug. All chemical reactions, not just combustion reactions, require some energy input to get started. This is called the **activation energy.**

Examine the energy diagram for an exothermic reaction. The potential energy of the products is lower than that of the reactants. However, notice that the potential energy first increases on the pathway from reactants to products before it is converted to kinetic energy.

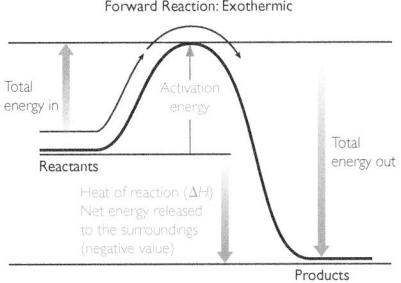

The potential energy between the reactants and products for this reaction is called the activation energy. Energy must be supplied to the reactants to get over this energy barrier. This is why reactions often require a spark to get started. Once an exothermic reaction is started, the reaction itself can provide the necessary energy for more molecules to react. This is why a single spark can cause an entire forest to burn down.

Big Idea Some reactions are energetically favored over other reactions.

In contrast, energy must be supplied continuously to cause the reverse of an exothermic reaction. You can see from the diagram that the reverse of the exothermic reaction is an endothermic reaction. Notice also that the activation energy, or energy barrier, is much greater in the reverse direction.

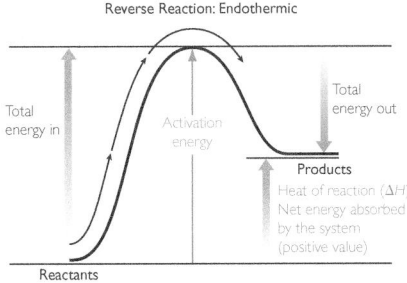

Lesson 105 | **Reversing Reactions**   537

TRM Worksheet with Answers 105

TRM Worksheet 105

COMPLETE THE DEMONSTRATIONS

→ Pass out the student worksheets. Students will work individually.

→ Write the chemical equations for the forward and reverse reactions on the board.

→ Tell students that you will show them this reaction going both forward and backward. The first demo is the video of combustion of hydrogen:

$$2H_2(g) + O_2(g) \rightarrow 2H_2O(l)$$

→ Set up a video of a hydrogen balloon being ignited. Ask students to predict what they will observe when a flame is applied to the balloon. Run the video. (A loud explosive pop and a significant burst of flame is observed.)

→ The second demo is the decomposition of water:

$$2H_2O(l) \rightarrow 2H_2(g) + O_2(g)$$

• Tell students that you will use electrical energy to split molecules of water into hydrogen and oxygen.

Tell them you are performing the reverse reaction of the one they just observed in the video.

• Use a standard setup for the electrolysis of water. You will see bubbles forming around both pieces of aluminum after a minute or so. One set of bubbles is hydrogen gas, the other oxygen. Unplug the battery to show that bubbles appear only when energy is supplied (in this case in the form of electrical energy).

• If you have collected hydrogen and oxygen in the test tubes, you can demonstrate their presence with a glowing splint. The hydrogen will emit a little pop. The oxygen will cause the splint to glow brighter.

DISCUSS THE REVERSING REACTIONS

→ Draw energy diagrams on the board as needed to illustrate what you are talking about. Draw a forward reaction and ask students to draw the reverse reaction, and vice versa.

Ask: Why are the energy diagrams for a reaction and its reverse reaction mirror images of each other? (conservation of energy)

Ask: Do you think the reaction forming nitrogen tetroxide from nitrogen dioxide is easily reversed? Explain.

Ask: Why is it that once you spark an exothermic reaction, the reaction continues without additional energy input? Explain.

Ask: Why, for many endothermic reactions, do you have to provide energy continuously? Explain.

Key Points: The energy diagrams for a reaction and its reverse reaction are mirror images of each other because energy is conserved. The net energy exchange involved in rearranging reactants to form products in an exothermic reaction is the same as that of the reverse reaction to convert the products back to reactants. Consider the forward and reverse reactions of nitrogen dioxide, the combustion of methane, and the combustion of hydrogen.

Nitrogen dioxide. The reaction of nitrogen dioxide to produce dinitrogen tetroxide takes place in our atmosphere and is part of the chemistry of smog and air pollution. This reaction is fairly easy to reverse. The conversion of two molecules of NO_2 to one molecule of N_2O_4 requires the formation of a single N—N bond. When smog is brown, it is a sign that NO_2 is present.

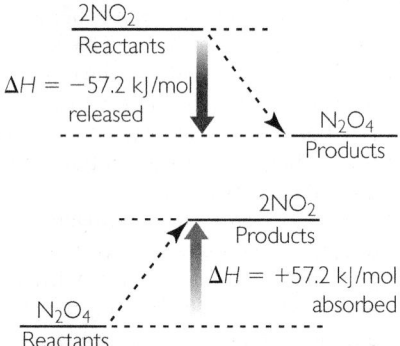

Combustion of methane. The large heat of reaction, ΔH, for the combustion of fuels indicates that reversing these reactions is not energetically easy. It takes much energy input to reverse combustion reactions. The formation of many hydrocarbon fuels is more complicated than a simple reversal of a reaction. On Earth, it has taken millions of years to create the planet's natural-gas and petroleum reserves. That is why these fuels are considered nonrenewable resources.

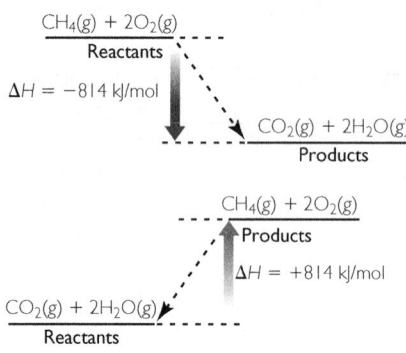

Combustion of hydrogen. In the demo video, a balloon filled with hydrogen exploded once it was sparked. Electricity was supplied continuously to the water to provide the energy to split apart the water molecules. See the diagram on p. 536 of the student text.

At any point in time, all the energy in a system can be considered either kinetic energy (energy of heat and motion) or *potential energy* (energy of position or composition). When substances burn, potential energy is converted to kinetic energy. Thus, whenever there is an exothermic reaction, the potential energy of the system decreases. Likewise, when there is an endothermic reaction, the potential energy of the system increases. These simplified energy

diagrams also can be used to reflect changes in the potential energy of a system. See the diagram on p. 536 of the student text.

When the potential energy changes, the kinetic energy also changes because energy is conserved. This means that when the potential energy decreases from reactants to products, the kinetic energy increases and the products are hotter than the reactants. Likewise, when the potential energy increases from reactants to products, the kinetic energy decreases and the products are colder than the reactants. In all cases, the products eventually reach thermal equilibrium with the surroundings.

The net energy change from reactants to products of this reaction does not provide the necessary energy to activate more molecules to react. This is why you need to continuously supply energy to decompose water to hydrogen and oxygen. Endothermic reactions *generally* are less likely to occur than exothermic reactions.

LESSON SUMMARY

What is the energy associated with reversing reactions?

Energy is conserved in chemical processes. This means that the net energy exchange in a forward process is equal and opposite to the net energy exchange in the reverse process. If a forward reaction is exothermic, the reverse reaction is endothermic. Energy in a chemical system is in the form of either kinetic energy or potential energy. When substances burn, a great deal of potential energy is converted into kinetic energy. Reactions with lower activation energies are easier to get started.

KEY TERMS

potential energy

activation energy

Exercises

Reading Questions

1. What happens to the potential energy and the kinetic energy of a chemical system during an exothermic reaction?

2. Explain why you do not have to keep sparking a fire once it is started.

Reason and Apply

3. Consider the combustion of ethane, $C_2H_6(g)$.
   a. Write a balanced equation for the reaction.
   b. Draw an energy diagram for the forward reaction. Label the reactants and the products. Use the table of heats of combustion from Lesson 103: Fuelish Choices to label the heat of reaction.
   c. What is the net energy exchange of the reverse reaction? Explain.

4. Consider the energy diagrams below for the combustion of one mole of methane and the reverse reaction.

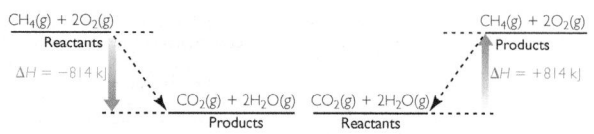

   a. Which diagram represents an endothermic reaction?
   b. Which substances have the lowest potential energy?
   c. When methane reacts with oxygen to form carbon dioxide and water, the potential energy decreases. What happens to the kinetic energy?
   d. Explain why the reverse reaction requires a constant input of energy.
   e. Use the table of average bond energies from Lesson 104: Make It or Break It to estimate the heat of reaction. How does your value compare to the heat of reaction given in the diagram?

DISCUSS ACTIVATION ENERGY

Ask: What started the combustion of hydrogen?

Ask: What started the electrolysis of water?

Ask: Does this energy show up on our net energy diagrams?

Key Points: Every chemical reaction (not just combustion reactions) requires some sort of energy input to start it. This energy input is called the *activation energy.* The match that lights a fire is responsible for delivering the activation energy that gets the combustion reaction going. In the hydrogen-balloon demo, the candle provided the spark

Exercises

(continued)

5. Copy the three energy diagrams shown here.

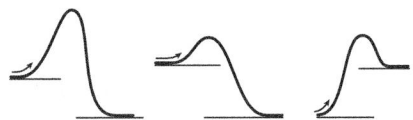

a. Label the heat of reaction on each.
b. Label the activation energy on each.
c. Which diagram represents the most exothermic reaction?
d. Which reactions require energy to get started?

to overcome the activation-energy barrier for this reaction. Once an exothermic reaction is started, it continues on its own without further energy input. This is because the heat produced by the reaction itself supplies any additional energy needed. In contrast, energy has to be supplied continuously to drive an endothermic reaction.

Activation energy: The minimum amount of energy required to initiate a chemical process or reaction.

The activation energy of a reaction depends on factors other than the value of the heat of reaction, ΔH. The formation of nitrogen tetroxide from nitrogen dioxide has a low activation energy. This reaction proceeds easily, with a small input of energy required to overcome the activation barrier, and the heating of the atmosphere in the afternoon is enough to reverse the reaction. In contrast, the electrolysis of water has a relatively high activation energy. Merely knowing if a reaction is exothermic or endothermic does not help us to predict if its activation energy is high or low. See the diagrams on page 537 of the student text.

EVALUATE (5 min)

Check-In

1. Draw an energy diagram for a reaction that releases heat. Label the energy involved in the reaction.

2. Explain how your diagram shows that energy is conserved.

Answers: 1.

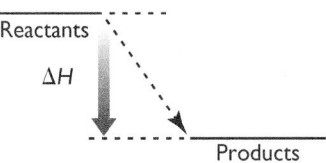

2. The potential energy of the products is lower than the potential energy of the reactants. The difference in potential energy is converted into kinetic energy, so energy is conserved.

Homework

Assign the reading and exercises for Fire Lesson 105 in the student text.

LESSON 105 ANSWERS

1. During an exothermic reaction, the potential energy of a chemical system decreases and the kinetic energy of the system increases. Some of the potential energy in the system is converted to kinetic energy.

2. A fire is a combustion reaction, so it is exothermic. It will keep going as the energy released by the reaction provides the energy to cause other molecules to react.

3. a. $2C_2H_6(l) + 7O_2(g) \rightarrow 4CO_2(g) + 6H_2O(l)$

b.

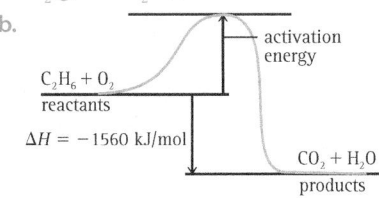

c. The net energy exchange of the reverse reaction is 1560 kJ/mol. The amount of energy used to make ethane from carbon dioxide and water is equal to the amount of energy released when ethane is converted to carbon dioxide and water.

Answers continue in Answer Appendix (p. ANS-8).

Overview

DEMOS: WHOLE CLASS

CLASSWORK: INDIVIDUALS

Key Question: How can you control the speed of a reaction?

KEY IDEAS

For molecules to react with one another, they must collide with a sufficient energy to stimulate bond breaking or bond making. The rate of a reaction depends on how much these collisions can be accommodated. There are a number of ways to increase the rate of a reaction, such as increasing the temperature or changing the concentrations of the reactants. When such methods increase favorable collisions, they increase the rate of reaction. The rate can also be increased with a catalyst, a substance added to a reaction mixture to facilitate the reaction.

LEARNING OBJECTIVES

- Describe the conditions that speed up reactions.
- Explain how a catalyst works.

IN CLASS

There are five teacher-led demonstrations at the beginning of class. Three of the demonstrations show the same chemical reaction twice—once under normal conditions, and again with the rate increased. Students then complete a worksheet that focuses on the factors that affect reaction rates. Where appropriate, reaction rates are related to activation energy. The instructions and list of materials to run five demos is available for download.

TRM Materials List 106

TRM Demo Instructions 106

Differentiate

To develop students' inquiry and engineering skills, consider asking students to design, execute, and analyze experiments to optimize the rate of reaction between baking soda (sodium carbonate, $NaHCO_3$) and vinegar (acetic acid, $C_2H_4O_2$), the rate of dissolving of sugar (use sugar cubes), the decomposition of hydrogen peroxide, or some other easy-to-observe chemical process. Let students brainstorm which independent variables they want to test (such as temperature or mixing), choose an appropriate dependent variable, and set up experimental controls. Depending on students' readiness, provide more or less guidance in designing the experiment.

LESSON 106

Speed Things Up

Rate of Reaction

THINK ABOUT IT

© Roger Ressmeyer/CORBIS

Looking north from high up on the Golden Gate Bridge

The Golden Gate Bridge in San Francisco is a national landmark. However, it is slowly eroding because of a chemical reaction in which iron in the steel beams combines with oxygen in the air to create iron oxide. This iron oxide falls away from the metal underneath. Is there a way to slow down the rate of this reaction?

How can you control the speed of a reaction?

To answer this question, you will explore

⤷ Reaction Rates

⤷ Catalysts

EXPLORING THE TOPIC

⤷ Reaction Rates

Some reactions proceed quickly and others proceed slowly. When a balloon full of hydrogen is ignited, it reacts rapidly, exploding into flame. The formation of iron oxide on steel bridges is a slower process called rusting.

Combustion of hydrogen occurs rapidly. $2H_2(g) + O_2(g) \longrightarrow 2H_2O(g)$

Rusting of iron occurs slowly. $4Fe(s) + 3O_2(g) \longrightarrow 2Fe_2O_3(s)$

Both of these reactions are exothermic, yet the rates of each reaction are very different.

You might notice that the reactants are both gases for the quicker reaction. Gases mix on a molecular scale, so there is a lot of contact between the reactants. In contrast, the slow reaction involves a solid. The reactants are in contact with one another only at the surface of the solid. There are many more iron atoms on the interior of the solid that are not exposed to oxygen and therefore do not react. When a layer of the iron oxide, rust, flakes off, the underlying iron is then exposed to oxygen and will react.

COLLISION THEORY

For a reaction to occur, the reactant particles must collide with enough energy to cause reactant bonds to break and product bonds to form. This energy is the activation energy of the reaction. Also, the reactant particles must collide in the correct orientation with one another. When this happens, the reactant particles are able to form a temporary arrangement called an *activated complex,* which facilitates bond breaking and bond making.

540 Chapter 20 | **Understanding Energy**

Pre-AP® Course Tip

Students preparing to take an AP® or college chemistry course should be able to justify claims with evidence. In this lesson, students observe various demonstration reactions. Based on their observations, students determine which factors affect the rate of a reaction.

SETUP

Gather materials for the demonstrations and find a suitable video for the "rocket engine" demo. We recommend that you check that you are following local safety regulations and that you prepare and practice the demos ahead of time. Take care not to inhale any of the lycopodium powder. As an alternative, videos of most of these demos can easily be found online. Optional: Cut out and assemble the fire tetrahedron. Or you can hand out the tetrahedron template to students to construct in class or later on their own.

SAFETY

Keep the area around the demonstrations free of flammable substances. You and the students should wear safety goggles for the demos involving flame. Wear rubber gloves during Demo 4, Elephant's Toothpaste.

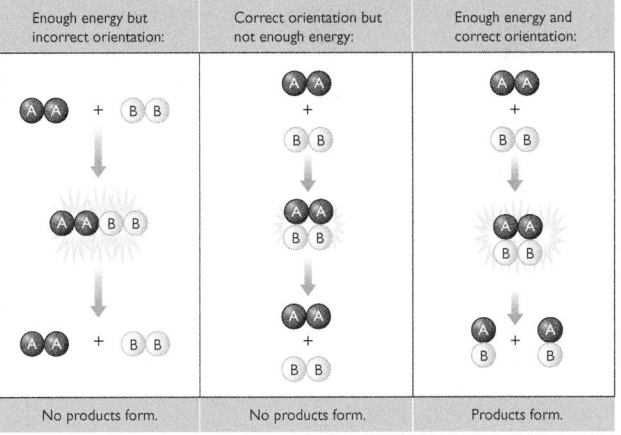

Enough energy but incorrect orientation:	Correct orientation but not enough energy:	Enough energy and correct orientation:
No products form.	No products form.	Products form.

So, for a reaction to occur, the reactant particles must collide in the correct orientation and with enough energy. If they do not, they will bounce apart and products will not form. Various factors can increase the frequency of collisions between particles and the energy of the collisions. These factors will speed up a reaction.

- **Surface area:** The exposed surface area of the reactants can affect the rate of a reaction. For example, in combustion reactions, fuels with larger exposed surface area, like sawdust or twigs, burn more rapidly than fuels with smaller exposed surface area, like an entire tree or a wooden table.

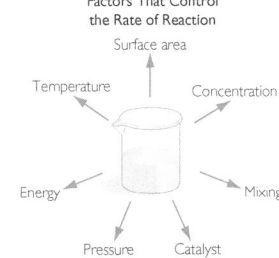

Factors That Control the Rate of Reaction

Surface area
Temperature
Concentration
Energy
Mixing
Pressure
Catalyst

- **Mixing:** When reactants are mixed well, the reaction rate increases. This is related to the increased probability that the appropriate reactants will collide.

- **Temperature:** When the temperature is raised, the average kinetic energy of the reactant particles is also raised. The particles move more rapidly and more collisions between particles are possible.

- **Concentration or pressure:** Reactions usually proceed faster as the concentration of the reactants in solution increases or as the pressure of gases increases. Again, more reactants in a small volume result in more collisions.

- **Energy:** Adding energy, such as microwave radiation or light, can increase the rate of a reaction.

Catalysts

Most of the methods of increasing the rate of reaction involve an increase in the number of collisions of the molecules or an increase in the speeds of the

---

CLEANUP

Dispose of any remaining chemicals according to local regulations.

Lesson Guide

TRM PowerPoint Presentation 106

ChemCatalyst

Paper and a log are made from the same compounds. Why don't they burn at the same speed?

Sample answer: More surface area of the paper is in contact with oxygen.

Students might say that a log is denser or that it is easier for oxygen to reach more molecules of paper.

→ Assist students in discussing what factors affect the speed of a reaction.

Ask: What could you do to speed up a reaction?

Ask: What makes an explosion happen so quickly?

Ask: Why won't a wet fuel burn?

EXPLORE (15 min)

TRM Worksheet with Answers 106

TRM Worksheet 106

INTRODUCE THE DEMONSTRATION

→ Pass out the student worksheets. Students will work individually on the worksheet.

→ Tell students that they will observe a series of reactions and that you will perform two versions of some reactions.

COMPLETE THE DEMONSTRATIONS

1. Breath of Fire
2. Glow Sticks
3. Rocket Engine
4. Elephants toothpaste
5. Blazing Paper

EXPLAIN & ELABORATE (20 min)

TRM Handout—Fire Tetrahedron Pattern 106

DISCUSS RATES OF REACTION

→ Create a list on the board of factors that affect the rate of a reaction. You can use models to demonstrate collision theory as described below.

Ask: Why didn't the pile of lycopodium powder burn as well as the powder blown through the plastic tubing?

Ask: Which changes increased the mixing of the reactants?

Ask: Which changes increased the average kinetic energy of the reactants?

Key Points: For a reaction to occur, the reactants must collide with one another with enough energy and in the correct orientation to break bonds in the reactants and form new bonds. This idea is called *collision theory*. Some reactions proceed very quickly, and others proceed at a slower rate. Even reactions that involve burning do not all happen quickly. Alcohol will burn faster than a piece of paper, and scraps of paper will burn faster than a small log of equal mass. See the diagram on page 541 of the student text.

A variety of factors affect the rate at which a reaction will proceed. These demonstrations focused on identical reactions proceeding at two different rates. In each pair of demos, the reactants were the same but the conditions of the reaction were changed.

- **Surface area:** The exposed surface area of the reactants can affect the rate of a reaction. For example, in combustion reactions, smaller-size fuels with larger surface areas, such as sawdust or twigs, burn more rapidly than large fuels, such as an entire tree or a wooden table. The size of the fuel affects the mixing of the reactants.

- **Mixing:** When reactants are mixed well, the reaction rate increases. This result is related to the increased probability that the appropriate reactants will collide.

- **Temperature:** When the temperature is raised, the average kinetic energy of the reactant particles is also raised. The particles move more rapidly, and more collisions between particles are possible.

- **Concentration or pressure:** Reactions usually proceed faster as the concentration of the reactants in solution increases or as the pressure of gases increases. Again, having more reactants in a smaller volume results in more collisions.

- **Energy:** Adding energy, such as microwave radiation or light, can increase the rate of a reaction.

In general, these factors affect the number of collisions of the molecules, the speeds at which the molecules collide, or the probability of a collision between the appropriate particles.

DISCUSS THE ROLE OF CATALYSTS

→ Draw an energy diagram showing an exothermic reaction with and without a catalyst. See the image on page 542 of the student text.

Ask: What happened when a catalyst was added to the hydrogen peroxide in Demo 4?

Ask: What would an energy diagram look like for an endothermic reaction with and without a catalyst?

Key Points: A catalyst is a substance that is added to a reaction mixture to speed up the reaction. A catalyst takes part in the reaction without being permanently changed or consumed by the reaction. Catalysts have the effect of lowering the activation energy of a reaction. Thus, there is less of an energetic barrier to get past for the reaction to proceed, and the reaction proceeds more rapidly. Catalysts are widely used in manufacturing to speed up desired reactions. Platinum is a particularly good catalyst for many reactions.

Catalysts work in two main ways. Most reactions occur in a sequence of intermediate steps. (Bonds do not break and

CONSUMER CONNECTION

The opposite of a catalyst is an *inhibitor*, a substance added to slow down a reaction. Sodium nitrite, ascorbic acid, and cinnamaldehyde are used to slow down rusting reactions. Inhibitors such as lemon juice, which contains ascorbic acid or 1-methylcyclopropene, can be added to fruit to slow down the reactions that cause browning.

Ken Karp Photography

collisions. Another way to increase the rate of reaction is to add a catalyst to the reactants.

A catalyst is a substance that assists a reaction but is neither consumed nor permanently changed by the reaction. A catalyst lowers the activation energy for the reaction. In Unit 2: Smells, sulfuric acid was used to catalyze the reaction that produced sweet-smelling ester molecules.

Chemical reactions may go through a series of intermediate steps to get from reactants to products. Some catalysts work by changing the sequence of steps to shorten the reaction path. Catalysts may undergo chemical transformations during a reaction, but they always return to their original form at the end of the reaction.

Another way a catalyst works is to provide a surface on which the reaction takes place. A catalytic converter in a car provides a platinum surface where pollutants can react to form harmless gases.

The energy diagram shows the same reaction with and without a catalyst. The catalyst lowers the energy barrier for the reaction.

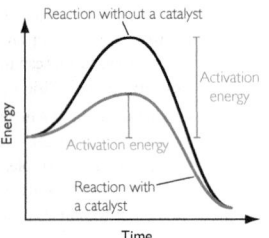

LESSON SUMMARY

How can you control the speed of a reaction?

For molecules to react with each other, they must collide with a sufficient amount of energy for bond breaking or bond making. There are a number of ways to increase the rate of a reaction, such as increasing the temperature of the system or changing the concentration of the reactants. These methods increase collisions between reactants. One way catalysts increase the rate of a reaction is by lowering the activation energy to get the reaction started. Catalysts are substances that speed up a reaction but are not permanently changed or consumed by the reaction.

Exercises

Reading Questions

1. Name three things you could do to increase the rate of a reaction between two liquids.
2. Name three things you could do to increase the rate of a reaction between two solids.
3. Name two ways you could decrease the rate of a reaction.

542   Chapter 20 | **Understanding Energy**

form all at once.) A catalyst can change the sequence of steps to make the reaction more efficient. The other way a catalyst works is by providing a surface on which the reaction can take place. A catalytic converter in the exhaust pipe of a car provides a platinum surface where conversion of pollutants into harmless gases takes place.

RELATE RATES OF REACTION SPECIFICALLY TO FIRE

→ Optional: Show students either a three-dimensional paper model or an illustration of the fire tetrahedron. Handout—Fire Tetrahedron Pattern 106.

Ask: Are fires easy to start? Explain your thinking.

Ask: If you provide a spark to only one corner of a piece of paper, why does the whole paper burn?

Key Points: A spark or other input of energy is needed to start a fire. Even exothermic reactions need some input of energy to start the reaction. Once the reaction begins, the energy released is enough to spark further reaction. This is why the whole paper will burn when you light a piece of paper in only one corner. Reactions that cause fires provide their own energy to keep them going and sometimes are referred to as chemical chain reactions.

The fire tetrahedron represents the four factors necessary for fire to occur.

Exercises

(continued)

Reason and Apply

4. What could you do to speed up the reactions listed here?
   a. Lighting charcoal for a barbecue
   b. Baking cupcakes
   c. Dissolving sugar in water
   d. Removing a stain from clothing

5. Which will burn faster, wood chips or a tree? Explain the difference in the rates of reaction.

6. Which will dissolve faster, a lollipop or powdered sugar? Explain the difference in the rates of reaction.

7. Which will react faster, a steel beam or powdered iron? Explain the difference in the rates of reaction.

8. Why do you think you can extinguish a flame with a carbon dioxide fire extinguisher?

9. What makes an explosion happen so quickly?

10. PROJECT
    Research a particular enzyme in your body. Explain how the enzyme catalyzes a reaction in your body.

11. The rate of a reaction can be affected by
    (A) The phase of the reactants.
    (B) The temperature of the reactants.
    (C) The presence of a catalyst.
    (D) All of the above.
    (E) None of the above.

Lesson 106 | **Rate of Reaction**   543

A fire must have more than just a spark, fuel, and oxygen to be sustained. An uninhibited chemical chain reaction also is necessary. The heat from the match starts the paper burning, and then the burning of the paper provides the energy to get more of the paper burning, and so on.

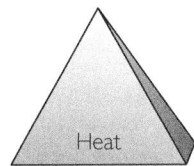

EVALUATE (5 min)

Check-In

You want to burn a piece of binder paper. Name three things you can do to increase the rate of combustion of the paper.

Sample answer: The paper can be placed in an oxygen-rich environment, coated with a catalyst, heated, or torn into smaller pieces.

Homework

Assign the reading and exercises for Fire Lesson 106 in the student text.

LESSON 106 ANSWERS

1. *Possible answers (any 3):* ● Mix the liquids well. ● Add heat. ● Increase the concentration of the reactants. ● Add a catalyst.

2. *Possible answers (any 3):* ● Increase the surface area. ● Add heat. ● Add energy such as microwave radiation or light. ● Add a catalyst.

3. *Possible answers (any 2):*
● Decrease the temperature. ● Decrease the pressure. ● Dilute the solution formed by two liquid reactants.

4. *Possible answers:* a. Increase the surface area by crushing the charcoal. b. Increase the oven temperature. c. Stir the mixture of sugar and water. d. Increase the concentration of the detergent.

5. Wood chips will burn faster because they have a much greater surface area than a tree, allowing the molecules in the wood chips more contact with the oxygen in the air.

6. Powdered sugar will dissolve faster than a lollipop because the sugar has a larger surface area, allowing it more contact with the water molecules.

7. The powdered iron oxidizes faster than the steel beam because the small particles have much more total surface area than the beam. Because a greater percentage of the iron is exposed to the oxygen in the air, the reaction is faster.

8. The large amounts of carbon dioxide foam that the fire extinguisher sprays on the flames reduce the concentration of oxygen around the fire, slowing the combustion reaction.

9. During an explosion, a large amount of energy is released very quickly from chemical bonds in molecules. This energy acts as activation energy to other molecules and causes the reaction to occur rapidly throughout the original fuel for the explosion.

10. A complete answer should contain: ● the name of the enzyme ● how the body obtains or produces the enzyme ● reactions that occur involving the enzyme, including a list of the reactants and products ● how the enzyme lowers the activation energy of these reactions.

11. D

Overview

DEMOS: WHOLE CLASS

COMPUTER ACTIVITY: WHOLE CLASS

CLASSWORK: INDIVIDUALS

Key Question: How can a chemical reaction be used to do work?

KEY IDEAS

Work is the transfer of energy that results from a force acting through a distance ($W = Fd$). Work is expressed in units of energy (such as joules). The combustion of fuels is one of the most common ways humans transform chemical energy into work. The heat of the combustion reaction produces expanding gaseous products ($W = P \Delta V$). Sometimes the energy from chemical change is used to cause a physical change to do work.

LEARNING OBJECTIVES

- Define work.
- Explain how chemical reactions are used to do work.

FOCUS ON UNDERSTANDING

- We have chosen to introduce the concept of work by beginning with the mechanical definition and then building to include chemical systems.
- The concept of work resembles the concept of heat in that both describe energy transfer.

KEY TERM

work

IN CLASS

The class opens with a discussion of mechanical work and what it takes to move objects. Then you demonstrate an aluminum-can steam engine and a baking-soda rocket for the class. Students speculate on how the energy from chemical changes can be harnessed to do work. Finally, the class views a couple of online computer simulations showing how chemical change is used to do mechanical work (optional) and students complete the worksheet. A complete materials list is available for download.

TRM Materials List 107

SETUP

Set up your computer and projector. Find Web sites that demonstrate how steam engines and rocket engines work, and familiarize yourself with these sites. There are many available.

Work

THINK ABOUT IT

All around us are machines doing work for us. They transport us from place to place, they help us to manufacture products, and they allow us to listen to music, wash our clothes, and watch television. The source of energy for most of these machines is chemical change.

How can a chemical reaction be used to do work?

To answer this question, you will explore

- Work
- Converting Chemical Energy into Work

EXPLORING THE TOPIC

Work

To move anything, whether it is a truck or an atom, from one place to another, requires work. **Work,** like heat, is a transfer of energy. Specifically, work is the result of a force acting through a distance. A force is a push or a pull upon an object as a result of an interaction with another object. Force is measured in newtons, N. Scientists define mechanical work as the product of force and distance.

$$\text{Work} = \text{force} \cdot \text{distance}$$
$$W = Fd$$

LIFTING AN OBJECT

Imagine that you pick up your baby sister, who weighs 22 pounds, or 10 kilograms, and lift her 1.2 meters. To lift her, you exert 100 newtons of force. This force is equal to the force of gravity acting on your sister, or her weight if it were expressed in newtons.

The work done picking up your 22-pound sister is equal to the force of her weight in newtons multiplied by the distance she was lifted.

$$\text{Work} = 100 \text{ N} \cdot 1.2 \text{ m} = 120 \text{ N} \cdot \text{m}$$

Work is expressed in units of newton-meters, N · m. Work and energy are related. One newton-meter is equal to one joule.

Prepare the aluminum can for the steam-engine demo. Holding the can over a bucket or sink, use the nail to make a small hole in the side of the can, in the middle. Push in the nail partway, and then press it all the way to the right so that it creates a hole facing to the right. Pull out the nail and make a second hole on the opposite side of the can, again using the nail to create a hole that points to the right. Drain all the soft drink out of the can. Submerge the can in water until you have 5–15 mL of water inside it.

5. Bend tab up and tie a string to it.

1. Pierce side of can with nail.

2. Repeat on opposite side.

3. Drain soda from can.

4. Use nail to pry both openings to the right.

LESSON 107

Machines (or tools) help us to do work. Levers, ramps, pulleys, wheels, and wedges are all examples of simple machines. They give you a mechanical advantage to perform work. Consider how easy it is to lift a great deal of weight using a seesaw. A claw hammer acts as both a wedge and a lever, allowing you to pull a nail out of a board, something that would be nearly impossible with your bare hands.

A hammer acts as a lever when you use it to pull out a nail.

ENERGY FROM CHEMICAL REACTIONS

Chemical reactions are sources of energy. Matter has the potential to transfer energy when it is transformed into new compounds. You have seen that burning different fuels can transfer large quantities of energy in the form of heat and light. But heat is not the only useful source of energy from chemical reactions. Consider the products of a combustion reaction. Both the gases formed and the heat produced are potential sources of work.

$$CH_4(g) + 2O_2(g) \longrightarrow \underbrace{CO_2(g) + 2H_2O(g)} + \underbrace{heat}$$

Two potential sources of work from a chemical reaction

Recall from Unit 3: Weather that the pressure from expanding gases can be a powerful force. In the case of a combustion reaction, the products are in the form of heated gases that can expand and push on objects. The work done by expanding gases is often referred to as PV work, where P and V stand for pressure and volume. PV work is equal to pressure times the change in volume.

$$W = P\Delta V$$

If pressure is expressed in atmospheres and the volume is expressed in liters, PV work is expressed in liter-atmospheres. Liter-atmospheres can also be converted to units of energy in joules.

1 liter-atmosphere = 101 joules

⟳ Converting Chemical Energy into Work

Chemical systems are often paired up with machines to transform chemical change into useful work. The heat transferred, q, from a reaction or process can be utilized in some way, or the chemical products themselves can be used.

STEAM ENGINE

A steam engine uses a combustion reaction to heat water in a cylinder with a piston. When the heat is transferred to the water, the liquid changes phase to a gas that expands and moves the piston.

Important to Know A system cannot contain or store either heat or work. So, heat and work both represent energy in transition.

Lesson 107 | **Work** 545

Lesson Guide

ENGAGE (5 min)

TRM PowerPoint Presentation 107

ChemCatalyst

Describe how you might use a chemical reaction to move something.

Sample answer: A chemical reaction can release gases or heat that causes a physical change (expansion of gases or phase change), which in turn can move an object.

Ask: How could you use heat transfer to move an object?

Ask: How could you use the products of a chemical reaction to move an object?

Ask: How do you think a steam engine works? An automobile engine?

EXPLORE (20 min)

TRM Worksheet with Answers 107

TRM Worksheet 107

INTRODUCE THE DEMONSTRATIONS

→ Pass out the student worksheets. Students will work individually on the worksheet.

→ Hold up a Ping-Pong ball or tennis ball. Optional: Ask students to brainstorm as many ways as they can think of to move the ball. Accept all reasonable answers. Ask students if each method requires an expenditure of energy.

PERFORM THE DEMONSTRATIONS

→ **Baking-Soda Rocket.** Tell students that you will use a chemical change to make something move.

1. Place about 1 tsp of baking soda onto a piece of paper towel about 4 in square. Fold the paper around the baking soda, making a packet to hold it.

2. Put 1/4 cup of vinegar into a 2 L plastic bottle. Place the bottle on the counter and make sure it is not pointing at anyone.

3. Drop the paper-towel package into the bottle and quickly insert a cork into the opening of the bottle. The gases that are produced by the reaction should pop the cork out of the bottle.

SAFETY

Wear goggles during the demo. Have your students wear goggles if they will be near the demonstration.

→ **Steam Engine.** Tell students that this is a second way to use a chemical change to do work. Set up the aluminum-can steam engine and demonstrate it.

1. Show students the aluminum can you have prepared. Tell them that there is a small amount of water inside.

2. Without opening the can, tie the string to the tab at the top of the can so the can will hang vertically when suspended by the string.

3. Turn on the gas burner and suspend the can over the flames. Soon, the can will start to spin.

SAFETY

Objects will become very hot in this experiment. Water vapor can cause burns. Use tongs to handle the hot can. You and your students should wear goggles for demos with flame.

GUIDE THE COMPUTER SIMULATION (OPTIONAL)

→ Inform students that they will view some computer simulations (or direct students to diagrams in the student text on page 546) and answer related questions.

→ Run a simulation that models a simple steam engine.

● Point out the piston and cylinder to the class. Allow students time to make visual sense of the simulation and to answer questions on the worksheet.

EXPLAIN & ELABORATE (15 min)

DEFINE WORK

Ask: If you push a pencil across a desk, are you doing work? Explain. What if you are holding a book? Explain.

Ask: What does it take to get a Ping-Pong ball to move?

Ask: What is one tool that helps us to do work?

Key Points: To move the Ping-Pong ball, it is necessary to expend some energy.

Energy is defined as the ability to do work. Thus, work is energy transferred. When you pick up an object and raise it off the floor, you are doing work. If you hit the Ping-Pong ball with a paddle, you do work by swinging the paddle. You must transfer energy to the paddle to get it to move. When the paddle hits the Ping-Pong ball, it transfers energy to the ball.

> **Work:** The transfer of energy that causes an object to move. When a force acts to move an object, work is done on that object.

If no motion happens, work is not done. Imagine that you push against a wall with all your strength. The wall does not move. According to our definition, no work has been done.

Mechanical work is defined as force times distance: $W = F \cdot d$. Force is measured in newtons and distance is expressed in meters, so work can be expressed in newton-meters, or joules.

$$1\ N \cdot m = 1\ J$$

One joule is approximately the work required to raise a 2.2 lb (1 kg) book 4 in (10 cm) off the floor.

Machines can be used to help convert energy to work. Machines give humans an advantage by changing the amount, speed, or direction of a force. Some machines (which we often call tools)—such as levers, pulleys, or wedges—depend on the shape or arrangement

The earliest recorded steam engine was built by an ancient Greek mathematician named Hero of Alexandria. The steam came out through two pipes, making a metal arm spin around. There is little evidence that this engine was used for anything more than amusement.

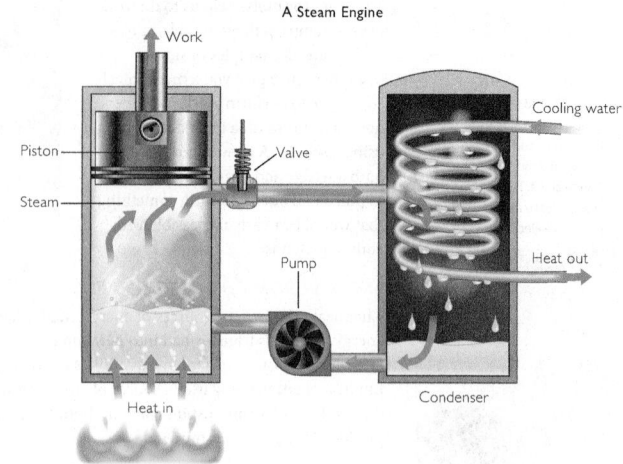

A Steam Engine

Notice that this is a two-step process as the heat from a chemical change is used to cause a physical change. The expanding gas (water vapor) does the work in this type of engine. Early steam engines were used to pump water and to power steamboats, trains, farm machinery, factories, and the first automobiles.

INTERNAL COMBUSTION ENGINE

Internal combustion engines like those in cars also make use of expanding gases. However, this type of engine uses the gases produced in the combustion reaction, carbon dioxide and water vapor, to do the work.

The illustration shows an internal combustion engine going through a four-step process.

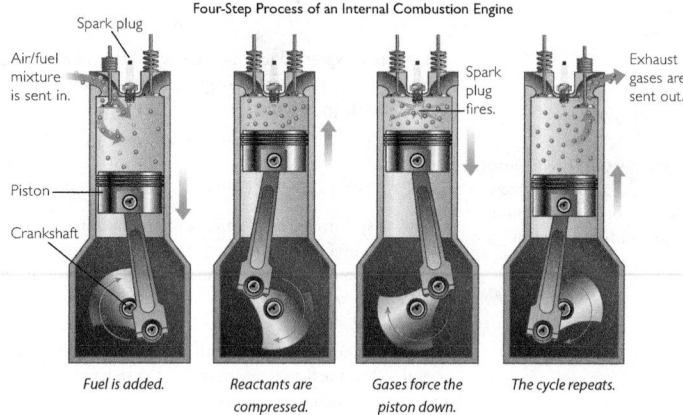

Four-Step Process of an Internal Combustion Engine

Fuel is added.   *Reactants are compressed.*   *Gases force the piston down.*   *The cycle repeats.*

of physical objects to provide a work advantage. Consider how easy it is to lift another person using a teeter-totter, which is a type of lever.

DISCUSS HOW CHEMICAL ENERGY IS CONVERTED TO WORK

Ask: How can chemical energy be used to do work?

Ask: What is the original source of the energy in the demos you watched today?

Ask: What is a steam engine? How does it do work?

Key Points: There are two main ways to convert chemical energy into work. In one demo, the gaseous products of the reaction caused the cork to fly out of the bottle. In the other demo, the heat released by the combustion of gases caused a physical change, and water vapor escaped out the small holes in the can, causing the can to spin. In this case, the source of work from chemical systems is the expansion or contraction of matter.

The energy that is transferred when gases change volume is called *PV* work. Recall that gas pressure is defined as the force per unit area. When a gas is

This four-step process is repeated at an extremely rapid rate to run the engine. The expanding carbon dioxide and water vapor from the combustion reaction move the pistons up and down, turning a large crank that ultimately makes tires, wheels, propellers, or turbines go around.

ENTROPY

Internal combustion and steam engines convert thermal energy into work. As required by the first law of thermodynamics, the total energy is conserved in the process. However, it is impossible to completely convert all that energy into work. Some of the energy is lost to the surroundings as heat.

This loss of heat to the surroundings is related to entropy and the second law of thermodynamics. As you recall, the second law states that energy and matter tend to disperse or become more disordered. Entropy is the measure of the unavailability of a system's energy to do work due to this dispersal. So, when designing an efficient machine, scientists and engineers must account for entropy.

Big Idea Energy disperses; it does not collect.

LESSON SUMMARY

How can a chemical reaction be used to do work?

KEY TERM
work

Work is force acting upon an object causing it to move in the direction of the force. So work is energy in motion. The heat transferred, q, from chemical reactions can be used to do work. The work done by expanding gases is called PV work. The combustion of fuels is one of the most common ways we transform chemical energy into work. However, some energy is always lost, due to entropy.

Exercises

Reading Questions

1. What is the scientific meaning of the word *work*?
2. Give an example of how a chemical reaction can be used to do work.

Reason and Apply

3. The expansion of gases pushes a piston. The volume of the gases increases from 0.5 L to 12.5 L. If the piston is pushing against a pressure of 1.2 atmospheres, how much work is done by the gases?
4. If it takes 45 N of force to lift a box 3.8 m, how much work is done to the box?
5. PROJECT Research and describe three ways that a rocket engine is different from an internal combustion engine. What extra item must a rocket engine carry with it into space?

---

allowed to change volume and expand, it exerts pressure. The work done by a gas is the product of the pressure and the change in volume.

$$W = P\,\Delta V$$

This work has units of liter-atmospheres, which can be converted to joules.

$$1\ \text{L}\cdot\text{atm} = 101\ \text{J}$$

EXAMPLE

Consider a piston that is free to move up and down. The pressure inside and outside the piston is 1.2 atm. An explosion inside the piston causes the gases to expand from a volume of 4.0 L to a volume of 6.0 L. Calculate the work done by the gas inside the piston.

SOLUTION

The change in volume, ΔV, is 2.0 L. The pressure outside the piston is still 1.2 atm. The work done by the expanding gases is

$$W = P\,\Delta V = 1.2\ \text{atm}\cdot 2.0\ \text{L}$$
$$= 2.4\ \text{L}\cdot\text{atm}$$

This can be converted to joules. One L · atm = 101 joules.

$$2.4\ \text{L}\cdot\text{atm}\cdot\frac{101\ \text{joules}}{1.0\ \text{L}\cdot\text{atm}} = 240\ \text{joules}$$

Note that the value of P is the pressure the system is pushing against, in other words, the pressure outside the system.

EVALUATE (5 min)

Check-In

What is the source of energy in a natural-gas power plant?

Answer: Natural gas is burned in a natural-gas power plant. Thus, the energy released by the combustion of natural gas is the main energy source.

Homework

Assign the reading and exercises for Fire Lesson 107 and the Chapter 20 Summary in the student text.

LESSON 107 ANSWERS

1. Work is the result of a force acting through a distance. Work is a transfer of energy, like heat.

2. *Possible answer:* The internal combustion engine in a car is an example of using a chemical reaction to do work. The combustion of the fuel creates hot gases that push on a piston. The chemical energy in the fuel is converted to mechanical energy.

3. 14 L · atm

4. 171 J

5. Possible ways that a rocket engine is different from an internal combustion engine (any 3): Rocket engines use a different combination of fuels from those internal combustion engines use. ● Ejecting gases from the engine away from the direction of flight is what propels rockets. Internal combustion engines convert the chemical energy of the fuel into mechanical energy. ● Rocket engines require a large proportion of the mass of the vehicle to be fuel. Internal combustion engines do not require as much fuel. ● Combustion is continuous in a rocket engine, but not continuous in an internal combustion engine. ● Rocket engines often do not emit pollutants in their exhaust, while internal combustion engines that run on hydrocarbon fuels emit pollutants. Possible answer for the extra item a rocket engine must carry: The rocket engine must carry oxygen gas or some other oxidizing agent because there is no air available in space for use in the combustion reaction.

ANSWERS TO CHAPTER 20 EXERCISES

1. Possible answer: ● Mix the reactants. Mixing increases the probability that the reactants will collide. ● Add heat. Raising the temperature of the reactants increases the average kinetic energy of the particles, causing more collisions to occur and increasing the probability that the kinetic energy in the collision will yield the activation energy for the reaction. ● Add energy in some form other than heat. Energy such as microwave radiation or light increases the kinetic energy of particles in the reactants. ● Increase the surface area. A greater surface area on the reactants increases the number of collisions between reactants. ● Increase the concentration. A higher concentration of the reactants in a solution or in a gas increases the number of collisions between the reactants. ● Increase the pressure. Increasing the pressure of reactants that are gases increases the concentration and therefore the number of collisions between the reactants. ● Add a catalyst. A catalyst lowers the activation energy of a reaction, increasing the probability that a reaction will occur with each collision.

2. Possible answers (any 3): ● Mix the coffee and the sugar by stirring the coffee rapidly. Mixing increases the probability that the sugar molecules will collide with liquid molecules. ● Increase the surface area of the sugar by crushing the sugar into smaller pieces. A greater surface area on the sugar particles increases the number of collisions between the sugar molecules and the liquid molecules. ● Increase the concentration. Adding more sugar cubes increases the number of collisions between sugar molecules and liquid molecules. ● Add heat. Raising the temperature of coffee increases the average kinetic energy of its particles, causing more collisions to occur and increasing the probability that the sugar will dissolve. If the objective is to have iced coffee, however, this may not be a plausible idea.

3. Possible answer: Paige's answer is correct. The latent heat of vaporization of water is 539 cal/g. The amount of energy required to break two H—O bonds is 934 kJ/mol. The heat of vaporization can be converted into the same units. 539 cal/g · 18.0 g/mol · 4.19 J/cal = 40,700 J/mol = 40.7 kJ/mol. The water will evaporate before any of the hydrogen-oxygen bonds are broken.

CHAPTER 20

Understanding Energy

SUMMARY

KEY TERMS

bond energy

potential energy

activation energy

work

Fire Update

All chemical reactions involve the breaking of bonds and the making of bonds. The net energy difference between these two processes determines if a reaction is exothermic or endothermic. This is because bond breaking requires energy and bond making releases energy. By using average bond energies, you can closely approximate the heat of reaction, ΔH.

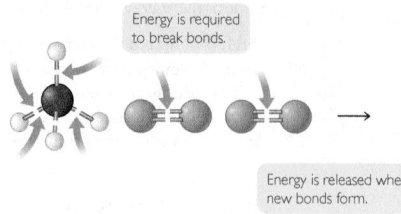

Energy is required to break bonds.

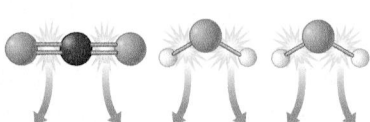

Energy is released when new bonds form.

Each chemical system expresses energy in the motions of its molecules or particles. This is known as kinetic energy. In addition, each chemical system has potential, or the potential for further energy in rearranging its chemical bonds. When an exothermic reaction occurs, the system's potential energy decreases and its kinetic energy increases. The reverse is true in an endothermic reaction. An increase in kinetic energy is accompanied by an increase in temperature.

It generally takes an input of energy, called the activation energy, to get a reaction started. This is why a fire requires a match or a spark. The overall conditions and the form that the reactants are in can affect the rate of a reaction. For example, raising the temperature, reducing the surfaces of reactants, or adding a catalyst can increase the rate of a reaction.

Chemical reactions can be used to do work. For example, the internal combustion engine uses the gaseous products formed in a combustion reaction to do work on pistons.

REVIEW EXERCISES

1. List seven ways you could speed up a chemical reaction. Explain how each method can increase the rate of reaction.

2. You have a cup of iced coffee that you are going to sweeten with a sugar cube. Name three ways that you could speed up the rate of dissolving sugar in the coffee. Explain why each of your methods works according to collision theory.

4. a.

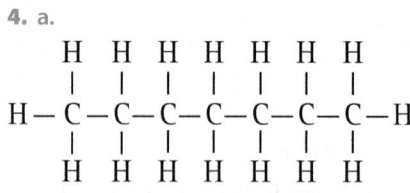

b. $C_7H_{16}(l) + 11O_2(g) \rightarrow 7CO_2(g) + 8H_2O(l)$

c. net energy exchange: $= -4523$ kJ/mol

5. a.

H
|
H — C — H   C = O = C
|
H

Methane   Carbon dioxide

b. methane: 4 C—H bonds: 4 · 413 kJ/mol = 1652 kJ/mol; carbon dioxide: 2 C=O bonds: 2 · 799 kJ/mol = 1598 kJ/mol **c.** Even though methane and carbon dioxide require a similar amount of energy to break all of their bonds, the individual bonds of the methane molecule are much easier to break than the double bonds of the carbon dioxide molecule. Therefore, methane is more likely to combust than carbon dioxide.

6. a. 1.7 L · atm **b.** Nothing else has to be taken into account to calculate the work done by the expanding gas. Work is defined as the product of the pressure and the change in volume of

3. You are walking by a classroom, and you overhear two students having a discussion about boiling water. Whom do you agree with and why?

   Trey: When you're boiling water, you're breaking the bonds between the hydrogen and oxygen atoms in water. The steam you see above the boiling water is hydrogen and oxygen gas. Heating the water provides the energy to break these bonds.

   Paige: Boiling water is not providing nearly enough energy to decompose water. The steam you see above the boiling water is still water. You have not broken any bonds to make it. Heating the water provides the energy to vaporize the water.

4. The chemical formula for heptane is C_7H_{16}.
   a. Draw the molecular structure for heptane.
   b. Write the balanced equation for the combustion of heptane.
   c. Using the table of average bond energies (located on page 531) and your answers to parts a and b, calculate the heat of reaction for heptane.

5. Methane, CH_4, is a combustible gas used in high school and college labs to burn in Bunsen burners. Carbon dioxide, CO_2, is an incombustible gas used in fire extinguishers.
   a. Draw the molecular structures for methane and carbon dioxide.
   b. Using the table of average bond energies on page 531, calculate the amount of energy that would be needed to break the bonds in a mole of methane and a mole of carbon dioxide.
   c. Why do you think that methane combusts, but carbon dioxide doesn't?

6. A balloon is placed over a flask with water in it. As the flask is heated on a hot plate, the balloon expands from a volume of 0.1 L to 1.6 L at a constant 1.1 atm.
   a. Calculate the amount of work that the water molecules did on the balloon.
   b. Is there anything else that needs to be taken into account when calculating the amount of work that the water vapor did?

Uses of Fire

PROJECT

Write a report describing where coal comes from and how the energy from coal is converted into electricity in a power plant. Include a diagram of the process in your report.

the gas. If these quantities can be measured accurately, nothing else has to be taken into account. It is not relevant that some energy is transferred to the environment, due to the second law of thermodynamics.

Project: Uses of Fire

A good report would include: ● a definition of coal and the process by which coal forms ● how coal is extracted from the Earth and brought to a power plant ● the relevant chemical equations that describe the combustion of coal ● the process by which the heat from the combustion of coal is converted into the mechanical energy of a turbine, including reference to *PV* work, and then to electrical energy ● a diagram of the process that converts the heat of combustion to electrical energy.

ASSESSMENTS

Two multiple-choice assessments (versions A and B) and two short-answer assessments (versions A and B) are available for download for Chapter 20.

TRM **Chapter 20 Assessment Answers**

TRM **Chapter 20 Assessments**

Watch the video overview of Chapter 21 (for teachers) by clicking on the link in the TE-book, opening the TRFD, or logging onto the book's companion Web site bcs.whfreeman.com/livingbychemistry2e (teacher log-in required).

Chapter 21 focuses on reactions that involve metals and metal compounds. These are the reactions typically found in batteries. In Lesson 108, students observe several demos showing the oxidation of metals and explore the equations associated with this type of reaction. Heat of formation is introduced in Lesson 109, and students compare the oxidation of different metals in terms of the amount of energy transferred. Lesson 110 introduces reduction, the counterpart to oxidation. Students write net ionic equations to track the transfer of electrons. The relative activity of metals is explored in Lesson 111 with a lab. In Lesson 112, students set up an electrochemical cell that converts chemical energy to electrical energy.

In this chapter, students will study
- oxidation-reduction reactions
- heat of formation
- activity series
- how to track electron transfer
- electrochemical cells

CHAPTER 21

Controlling Energy

In this chapter, you will study
- exothermic reactions of metals and metal compounds
- the energy it takes to extract pure metals from compounds
- how chemical change is converted into electrical energy
- combinations of reactants that make useful batteries

The battery in your car converts chemical reactions into electricity.

Fires are not the only source of energy resulting from chemical change. Energy from all types of reactions can be made useful. Reactions involving metals and metal compounds are particularly valuable sources of energy. Using certain techniques, these reactions can be made portable, controllable, and safe. This chapter of Unit 5: Fire focuses on ways in which the energy from chemical change is contained and made easily available.

550

Chapter 21 Lessons

Metal Magic

Oxidation

THINK ABOUT IT

If you examine a handful of pennies, chances are some of them will be a bright coppery color and others will be discolored and dull. These tarnished pennies have reacted with the oxygen in the air, just as a fuel does when it burns. But is this a form of combustion?

What happens when metals react with oxygen?

To answer this question, you will explore

- Oxidation
- Electron Transfer

EXPLORING THE TOPIC

Oxidation

No penny stays bright and shiny forever. Eventually, they all become dark and dull when the copper metal on the surface forms copper (II) oxide and other copper compounds. This reaction happens very slowly at room temperature. The reaction is also exothermic, but it happens so slowly the heat is difficult to detect.

The reaction of copper with oxygen is described by this equation.

$$2Cu(s) + O_2(g) \longrightarrow 2CuO(s)$$

On the other hand, if magnesium is held to a flame, it reacts rapidly and violently with oxygen, producing an almost blinding light. This metal is also combining with oxygen, but it is a much more exothermic reaction and happens quite rapidly once it gets started.

$$2Mg(s) + O_2(g) \longrightarrow 2MgO(s)$$

Copper and magnesium are just two examples of metals reacting with oxygen. Under the right conditions, *all* metals will react with oxygen. This process is called oxidation. Both combustion of molecules and tarnishing of metals are forms of oxidation. Compare the oxidation of two metals with a common combustion reaction:

Oxidation of copper: $2Cu(s) + O_2(g) \longrightarrow 2CuO(s)$

Oxidation of magnesium: $2Mg(s) + O_2(g) \longrightarrow 2MgO(s)$

Combustion of methane: $CH_4(g) + 2O_2(g) \longrightarrow CO_2(g) + 2H_2O(g)$

The reaction of oxygen with a metal produces a solid, whereas the reaction of oxygen with a molecular substance produces two gases. But the reactions have

HEALTH CONNECTION

Antioxidants are molecules that can slow or prevent the oxidation of other molecules. Antioxidants found in fruits and vegetables can cancel out the cell-damaging effects of free radicals. This is one of the reasons why eating fruits and vegetables can lower a person's risk for heart disease, neurological disease, and some forms of cancer. Berries are high in antioxidants.

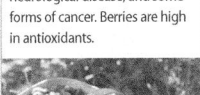

Students complete a worksheet and are introduced to the chemical equations that accompany the oxidation of metals. The instructions and list of materials to run three demos is available for download.

TRM Materials List 108

TRM Demo Instructions 108

SETUP

Right before class or during the ChemCatalyst, set up materials for the demos.

Flaming Steel: Prepare the potato and hydrogen peroxide to produce oxygen gas. Peel and cut up a potato ahead of time. Place it in a large Erlenmeyer flask. Pour in 1/2 cup of hydrogen peroxide, H_2O_2. Cover the flask with a watch glass. Allow this to sit about 15 minutes before conducting the demonstration.

Iron Lung: Prepare the steel wool. Fluff up a piece of steel wool to the size of a small apple. Place it in a 250 mL beaker and cover it with vinegar. Soak for 10 minutes. This will remove any oxide coating.

SAFETY

The potato and hydrogen peroxide reaction produces O_2 gas. Do not leave it sealed up for more than 40 minutes.

Have a type D (dry chemical) fire extinguisher on hand. A regular CO_2 extinguisher will not work on a magnesium fire. Have a container of sand and a fire blanket on hand.

Avoid looking directly at the light given off when the magnesium burns and be sure your students do the same. Protect your eyes with polarized sunglasses and set up the demo so that students view it through colored acrylic glass.

CLEANUP

Products of combustion may be cooled and thrown in the trash.

Lesson Guide

ENGAGE (5 min)

TRM PowerPoint Presentation 108

Overview

DEMOS: WHOLE CLASS

Key Question: What happens when metals react with oxygen?

KEY IDEAS

Most metals react with oxygen to form metal oxides, releasing heat. This process is called *oxidation*. For some metals, intense light is also emitted. Oxidation occurs when any substance combines chemically with oxygen, transferring electrons in the process. Combustion is a form of oxidation.

LEARNING OBJECTIVES

- Define and generally explain oxidation.
- Write chemical equations for the oxidation of metals.

KEY TERM

oxidation

IN CLASS

In this lesson, students observe the oxidation of metals. Students witness three different demonstrations showing the oxidation of metals. Two of the demonstrations involve steel wool, and the third involves magnesium.

ChemCatalyst

Fireworks contain small metal particles that react with oxygen. For example, when copper, Cu, reacts with oxygen there is a green glow.

$$2Cu(s) + O_2(g) \rightarrow 2CuO(s)$$

1. Explain why copper powder reacts vigorously with oxygen, while a copper saucepan does not.

2. Why do fireworks produce a glow rather than flames?

3. If copper powder reacted with oxygen in a sealed container, would the container be in danger of exploding? Explain your reasoning.

Sample answers: **1.** The copper powder has a higher surface area than the copper saucepan, so it mixes well with oxygen. **2.** The product is a solid that glows because it is hot. When you see flames, the products are gases. **3.** The air pressure would decrease as oxygen gas is removed.

Ask: What evidence is there that the reaction of a metal with oxygen is exothermic?

Ask: What is the difference between the glow of a hot object and a flame?

Ask: What factors might contribute to the pressure increasing or decreasing in the reaction container? (Temperature is increasing, while the number of oxygen gas particles is decreasing.)

EXPLORE (15 min)

TRM Worksheet with Answers 108

TRM Worksheet 108

COMPLETE THE DEMONSTRATIONS
→ Arrange students into pairs.
→ Pass out the student worksheet.

LIST OF DEMOS
1. Iron Lung
2. Flaming Steel
3. Magnesium No-Peekie

EXPLAIN & ELABORATE (15 min)

DISCUSS COMBUSTION OF METALS

Ask: How does the reaction of metals with oxygen differ from the combustion of carbon-based compounds?

Over time, silver oxidizes and becomes dull. One way to make it shiny again is to place it in a solution of sodium bicarbonate (baking soda) with aluminum foil. Within a few minutes, the silver becomes shiny again and the aluminum oxidizes instead.

Ken Karp Photography

Ask: Are gases produced when metals react with oxygen, O_2? (generally, no)

Ask: What types of compounds are created when metals react with oxygen? (oxides)

Ask: Was a flame produced by any of the reactions in the demonstrations? Explain.

Key Points: Under the right conditions, metals react with oxygen. These reactions release energy in the form of heat. Sometimes, they are accompanied by bright light. This chemical change can be considered a type of combustion. Flames typically are not observed when metals combust. In contrast to the combustion of

---

several characteristics in common. This table shows a comparison of the two types of reactions:

	Combustion of molecules	Oxidation of metals
combines with	oxygen	oxygen
releases heat?	yes	yes
releases light?	yes	sometimes
produces flames?	yes	no
product	gaseous molecules	solid metal oxides

Oxidation of metals transfers energy to the surroundings, just like combustion of molecular covalent compounds. This energy is in the form of heat and sometimes a bright light. However, you can see that the products of oxidation of metals and combustion of molecular compounds are different.

FLAME VERSUS GLOW

Flames are composed largely of superheated gases. When metals oxidize, they might glow with a very bright light, but there is no flame. The bright light is due to the excitation of electrons. Recall that the colored flame tests in Unit 1: Alchemy Lesson 17 were also the result of excited electrons changing energy levels. The bright light from the oxidation of metals may resemble a flame, but it is not a flame.

Electron Transfer

All of the oxidation reactions discussed so far involve oxygen. However, **oxidation** is defined more broadly as any reaction in which an atom transfers electrons to oxygen or another atom. When magnesium oxide is formed, the magnesium atoms transfer electrons to the oxygen atom. When this happens, we say that the magnesium has been oxidized.

Big Idea When an atom or ion is oxidized, it loses electrons.

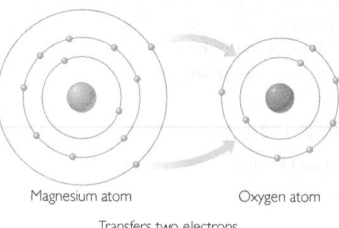

Magnesium atom   Oxygen atom   Magnesium ion, Mg^{2+} Oxygen ion, O^{2-}

Transfers two electrons   Ionic bond   Forms magnesium oxide

Oxidation of magnesium

---

molecules, the reactions involving metals and oxygen all produce a solid metal oxide. No gases are part of the products. Thus, the presence of flames is not likely unless some sort of impurity is present. The bright light that is visible when some metals combust is due to the excitation of electrons.

Two variables in particular have the greatest effect on the rate of the reaction of metals with oxygen. First, finer particles provide more overall surface area for better mixing with oxygen. That is why the tongs do not burn but the steel wool does. Second, many metals that ordinarily would react vigorously with oxygen do not

The shell models show a magnesium atom being oxidized. When magnesium is oxidized, it forms Mg^{2+} ions.

Recall that metal elements generally have one, two, or three electrons in their valence electron shells. Having such a small number of electrons in the valence shell makes it more likely that metal atoms will lose electrons, forming cations that have a charge of +1, +2, or +3. Through electron transfer, metal atoms form ionic bonds with nonmetal atoms.

Molecular covalent compounds also transfer electrons to oxygen when they react. Therefore, combustion reactions are considered oxidation reactions.

LESSON SUMMARY

What happens when metals react with oxygen?

Metal atoms react with oxygen to form metal oxides. When metal oxides form, heat is released, and sometimes, an intense bright light is emitted. This process is called oxidation. Molecular substances can also be oxidized, as is the case in combustion. However, no flames are present when metals oxidize. This is partly what differentiates the oxidation of metals from the oxidation of molecular substances. Oxidation occurs whenever an atom, ion, or molecule transfers electrons to another atom, ion, or molecule.

KEY TERM
oxidation

Exercises

Reading Questions

1. How does oxidation of a metal compare to combustion of a covalent compound?

2. Explain what happens when magnesium metal is oxidized.

Reason and Apply

3. **a.** Write balanced equations for the reactions of calcium, chromium, and lithium with oxygen. Identify the charge on the cation and the charge on the anion for each product.

$Ca(s) + O_2(g) \longrightarrow CaO(s)$

$Cr(s) + O_2(g) \longrightarrow Cr_2O_3(s)$

$Li(s) + O_2(g) \longrightarrow Li_2O(s)$

   b. Circle the elements that are oxidized in the chemical equations in part a.

4. Draw shell models for lithium and oxygen atoms to show the transfer of electrons during oxidation.

5. Draw shell models for silicon and oxygen atoms to show the transfer of electrons during oxidation.

Lesson 108 | **Oxidation**   553

---

react because they are coated with a protective layer. For example, aluminum is highly reactive with oxygen. However, we use it all the time in cooking and can even put aluminum foil directly into a campfire to cook a potato or an ear of corn. Aluminum does not react with the oxygen in the air around it because a microscopically thin layer of aluminum oxide on the surface protects the aluminum atoms below the surface. Magnesium is also very reactive. However, it did not react with oxygen until it was put into the flame because it was coated with magnesium oxide.

DEFINE OXIDATION

→ Write the chemical equation for the oxidation of magnesium on the board.

→ Draw the shell models for Mg and O on the board—something similar to the diagram below.

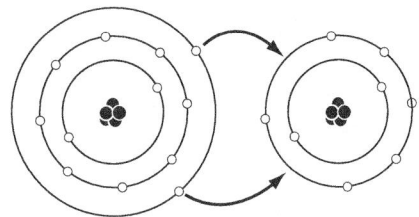

Electrons transferred during formation of magnesium oxide, MgO.

Ask: Why does the word *oxidation* make sense for these reactions?

Ask: What other oxidation reactions can you think of besides the combination of metals with oxygen?

Ask: Is the burning of paper an oxidation reaction? Explain.

Ask: What kind of bond is formed when magnesium reacts with oxygen? (ionic)

Key Points: The reaction of metals with oxygen is called *oxidation*. However, the modern definition of oxidation includes any reaction in which electrons are transferred.

For example, magnesium transfers two electrons to oxygen to form Mg^{2+} and O^{2-}. When this happens, we say that the magnesium has been oxidized.

$$2Mg(s) + O_2(g) \rightarrow 2MgO(s)$$

Oxidation: The reaction of a substance with oxygen. Also, the loss of electrons by an atom or ion during a reaction.

Metals are especially susceptible to oxidation because they have one, two, or three electrons in their outermost (valence) electron shell. Having such a small number of electrons in the valence shell makes it more likely that metal atoms will transfer electrons to a nonmetal atom, forming positive metal ions. When metals transfer electrons to oxygen, an ionic bond forms between the metal and the oxygen. All combustion reactions are oxidation reactions. Any time an element or a compound reacts with and combines with oxygen, it is considered oxidation. Oxidation is not limited to combustion.

EVALUATE (5 min)

Check-In

Write the chemical equation for the oxidation of zinc. Make sure it is balanced.

Answer: $2Zn(s) + O_2(g) \rightarrow 2ZnO(s)$

Homework

Assign the reading and exercises for Fire Lesson 108 in the student text.

LESSON 108 ANSWERS

Answers continue in Answer Appendix (p. ANS-9).

Overview

CLASSWORK: PAIRS

Key Question: How much energy is transferred during oxidation of metals?

KEY IDEAS

Almost all oxidation reactions release energy. The amount of energy transfer that occurs during the oxidation of a metal to a metal oxide is called the *heat of formation* of the metal oxide. This energy is expressed in kilocalories or kilojoules per mole (or gram) of the metal oxide product. The energy required to complete the reverse reaction, in which pure metals are extracted from metal oxide ores, is equal and opposite to the heat of formation required to create the metal oxide.

LEARNING OBJECTIVES

- Define heat of formation.
- Use heats of formation to compare the formation of metal oxides.
- Use chemical equations to compute heat of formation values.

FOCUS ON UNDERSTANDING

Students might mistakenly think that heat of formation values apply to all reactions. It is important to specify that these values apply only to those reactions in which a single product is formed from its constituent elements.

KEY TERM

heat of formation

IN CLASS

Students investigate the energy associated with the oxidation of metals. They compare metals by examining the energy transfer associated with their oxidation. Part 2 of the worksheet focuses on the energies associated with reversing these reactions. During the discussion, students are introduced to heat of formation. Note that oxidation-reduction is introduced later, in Lesson 110: Electron Cravings. A complete materials list is available for download.

TRM Materials List 109

Lesson Guide

ENGAGE (5 min)

TRM PowerPoint Presentation 109

Pumping Iron

Heat of Formation

THINK ABOUT IT

With all the heat and light it gives off, fire is an obvious source of energy. Less obvious is the possible use of other exothermic chemical reactions as energy sources. Because the oxidation of metals is exothermic, it may be possible to use oxidation reactions to generate power.

> How much energy is transferred during oxidation of metals?

To answer this question, you will explore

↻ Heat of Formation

↻ Reversing Oxidation

EXPLORING THE TOPIC

↻ Heat of Formation

The oxidation of most metals is exothermic, just like the combustion of fuels. So, oxidation reactions are a potential source of energy for our daily lives. To explore the usefulness of metal oxidation as an energy source, we must examine and compare the amount of thermal energy transferred during these reactions.

COPPER VERSUS ZINC

Copper metal and zinc metal react with oxygen to form copper (II) oxide, CuO, and zinc oxide, ZnO.

$$2Cu(s) + O_2(g) \longrightarrow 2CuO(s) \qquad \Delta H = -314 \text{ kJ}$$
$$2Zn(s) + O_2(g) \longrightarrow 2ZnO(s) \qquad \Delta H = -696 \text{ kJ}$$

Notice that ΔH is negative for both reactions. As you recall, this indicates that the reactions are exothermic and heat is transferred to the surroundings. Also notice that in both cases the heat of reaction is associated with *two* moles of metal oxide being produced. So, the formation of *one* mole of each oxide results in the release of half as much energy.

The energy released when one mole of a compound is formed from elements is called the **heat of formation,** or ΔH_f. Compare the heats of formation of copper (II) oxide, CuO, and zinc oxide, ZnO:

Heat of formation of copper (II) oxide: $\Delta H_f(\text{CuO}) = -157 \text{ kJ/mol}$

Heat of formation of zinc oxide: $\Delta H_f(\text{ZnO}) = -348 \text{ kJ/mol}$

These values indicate how much heat transfer to expect when one mole of each compound is formed from its elements. The formation of 1 mol of zinc oxide

---

ChemCatalyst

Explain the differences in the energy values given.

$4Al(s) + 3O_2(g) \rightarrow 2Al_2O_3(s) + 3352$ kJ thermal energy

$4Al(s) + 3O_2(g) \rightarrow 2Al_2O_3(s) + \Delta H = -1676$ kJ/mol Al_2O_3

Sample answer: The energy value in the first equation shows energy as one of the products of this exothermic reaction. The value represents the amount of energy produced when 4 mol of aluminum react with 3 mol

of oxygen. The value listed with the second equation is negative to represent that the system is losing energy. The value is half that of the first value because it is expressed per mole of aluminum oxide formed.

Ask: Is this an exothermic reaction or an endothermic reaction? How do you know?

Ask: Why do you think the amount of energy is positive in one equation and negative in the other?

Ask: Why aren't the numerical amounts the same in both cases?

releases over twice as much energy as the formation of 1 mol of copper (II) oxide. You can find heats of formation values of some metal oxides in the table below. These values are determined under standard conditions, at 1 atm and 25 °C. Note that standard conditions are different from STP, standard temperature and pressure, which is 1 atm and 0 °C.

Heats of Formation

Compound	ΔH_f(kJ/mol)
$Fe_2O_3(s)$	−826
$Al_2O_3(s)$	−1676
$PbO(s)$	−218
$MgO(s)$	−602
$CuO(s)$	−157
$Au_2O_3(s)$	+81

Example 1

Oxidation of Mg and Fe

Which reaction will release more energy?

$$2Mg(s) + O_2(g) \longrightarrow 2MgO(s)$$
$$4Fe(s) + 3O_2(g) \longrightarrow 2Fe_2O_3(s)$$

Solution

The heat of formation per mole of product can be found in the table above. Multiply ΔH_f by the number of moles of each product to get the heat of the reaction, ΔH:

$$(-602 \text{ kJ/mol})(2 \text{ mol MgO}) = -1204 \text{ kJ}$$
$$(-826 \text{ kJ/mol})(2 \text{ mol Fe}_2O_3) = -1652 \text{ kJ}$$

The oxidation of iron, or the formation of iron (III) oxide, Fe_2O_3, produces more energy than the oxidation of magnesium. Heats of formation are not limited to metal oxides. For example, the heat of formation of magnesium chloride, $MgCl_2$, is −2686 kJ/mol.

⟳ Reversing Oxidation

When an oxidation reaction for a metal is reversed, the resulting products are a pure metal and oxygen gas. This reaction is a decomposition reaction. The metal oxide decomposes and one of the products is a metal element. The decomposition of a metal oxide is one way to extract pure metals from metal compounds. But how difficult is it to accomplish energetically? Consider the decomposition of aluminum oxide, Al_2O_3:

$$2Al_2O_3(s) \longrightarrow 4Al(s) + 3O_2(g)$$

Key Points: Like all chemical reactions, the formation of metal oxides involves an energy transfer. With very few exceptions, the formation of metal oxides is exothermic. Unlike carbon compounds, the metals that are being oxidized are not usually called fuels, even though these reactions can release a great deal of energy. Oxidation of metals often involves the combining of two elemental substances. The formation of 1 mol of aluminum oxide results in the release of 1676 kJ of energy. Likewise, the formation of 2 mol of aluminum oxide results in the release of 3352 kJ of energy, and so on.

$$4Al(s) + 3O_2(g) \rightarrow 2Al_2O_3(s)$$
$$\Delta H_f = -1676 \text{ kJ/mol Al}_2O_3$$

The heat of formation can be determined for any compound. These values are determined under standard conditions of 1 atm pressure and 25 °C, using only pure elements as reactants. You can find values for the heat of formation in data tables.

The heat of formation of 1 mol of liquid water is −286 kJ. This means that 286 kJ of energy are released for each mole of water formed from elemental hydrogen and oxygen. Note that H_2 and O_2 are both gases at 25 °C and 1 atm and that H_2O is a liquid.

$$2H_2(g) + O_2(g) \rightarrow 2H_2O(l)$$
$$\Delta H_f = -286 \text{ kJ/mol H}_2O$$
$$\Delta H = -572 \text{ kJ}$$

You can point out to students that the reaction can also be written with a fractional coefficient. Then ΔH_f and ΔH will have the same value.

$$H_2(g) + \tfrac{1}{2}O_2(g) \rightarrow H_2O(l)$$
$$\Delta H_f = -286 \text{ kJ/mol H}_2O$$
$$\Delta H = -286 \text{ kJ}$$

Note: Standard heat of formation values cannot be determined experimentally for most compounds. Many are determined theoretically using Hess's law of heat summation, which is not covered formally in the Fire unit.

REVERSING OXIDATION REACTIONS

→ Draw on the board a simple energy diagram for the formation of a metal oxide and its reverse reaction.

EXPLORE (15 min)

TRM Worksheet with Answers 109

TRM Worksheet 109

INTRODUCE THE CLASSWORK

→ Arrange students into pairs.
→ Pass out the student worksheet.

EXPLAIN & ELABORATE (15 min)

INTRODUCE HEAT OF FORMATION

→ Write on the board one or more reactions from the worksheet showing the oxidation of metals.

Ask: Is the oxidation of metals exothermic or endothermic? How do you know?

Ask: Which metals are most easily oxidized? (Al, Fe, Mg, Sn)

Ask: Which metals are most difficult to oxidize? (Au, Ag, Hg)

Ask: How much energy is released by the formation of 1 mol of magnesium oxide? (602 kJ) One mole of silver oxide? (31 kJ)

Ask: Which oxidation reaction appears to be the best source of energy per mole of oxide formed? (oxidation of aluminum)

Ask: How would you draw the energy diagram for the oxidation of aluminum?

Ask: How would you draw the energy diagram for the reverse reaction?

Ask: How difficult is it to reverse oxidation reactions?

Ask: Which elemental metals are most difficult to extract from metal oxides? How do you know?

Key Points: A simple energy diagram can be used to show the progress of an oxidation reaction. For the most part, the oxidation of metals is an exothermic reaction, and many metals are easily oxidized. This means that the reverse reaction, in which the metal elements are extracted from the metal oxides, is endothermic and requires an input of energy.

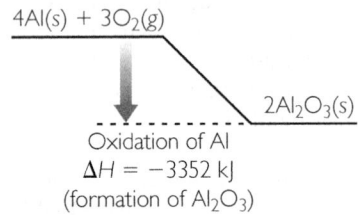

$4Al(s) + 3O_2(g)$

$2Al_2O_3(s)$

Oxidation of Al
$\Delta H = -3352$ kJ
(formation of Al_2O_3)

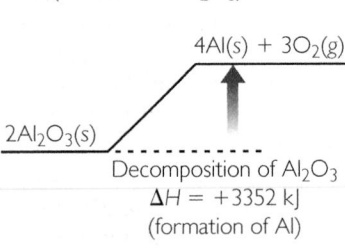

$4Al(s) + 3O_2(g)$

$2Al_2O_3(s)$

Decomposition of Al_2O_3
$\Delta H = +3352$ kJ
(formation of Al)

Metals that are easily oxidized, such as magnesium, aluminum, and tin, are never found in their pure elemental forms in nature. Metals that are easily oxidized have the most negative heats of formation. Much energy is required to reverse these reactions and extract the pure metal. On the other hand, metals that are less easily oxidized, such as silver, copper, and gold, are sometimes found in nature in their pure, uncombined state. In fact, gold has a positive heat of formation, ΔH_f, and requires an input of energy to combine with oxygen.

EVALUATE (5 min)

Check-In

1. Write the chemical equation for the oxidation of nickel. Assume that nickel ions have a charge of +2 in the oxide.

Energy diagrams for the oxidation of aluminum and for the decomposition of aluminum oxide are shown here. Take a moment to examine the energy changes.

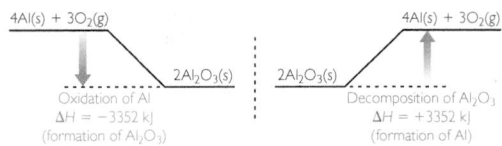

$4Al(s) + 3O_2(g)$

$2Al_2O_3(s)$

Oxidation of Al
$\Delta H = -3352$ kJ
(formation of Al_2O_3)

$2Al_2O_3(s)$

$4Al(s) + 3O_2(g)$

Decomposition of Al_2O_3
$\Delta H = +3352$ kJ
(formation of Al)

The oxidation of aluminum to form aluminum oxide is very exothermic. The energy for the reverse reaction is equal and opposite. So the decomposition of aluminum oxide is highly endothermic. This means that a large amount of energy is required to extract pure aluminum from aluminum oxide.

Indeed, aluminum was not discovered and brought into use until about 200 years ago, partly because so much energy is required to extract the pure aluminum metal from aluminum ores. The simple heating of aluminum ores does not provide enough energy to decompose the aluminum compounds and extract aluminum metal.

Example 2

Extracting Al

How much energy is required to extract a mole of aluminum metal from aluminum oxide, Al_2O_3?

Solution

The equation for the decomposition of aluminum oxide is

$$2Al_2O_3 \longrightarrow 4Al(s) + 3O_2(g)$$

This process is the reverse of the formation of $Al_2O_3(s)$, for which $\Delta H_f = -1676$ kJ/mol. (See the table on page 555.)

Because 2 mol of $Al_2O_3(s)$ are decomposed in the equation, the heat of reaction, ΔH, is 2(1676 kJ/mol), or 3352 kJ/mol. Note that the sign is positive because the reaction is reversed.

Because 4 mol of aluminum are produced, the energy required to extract 1 mol of aluminum metal is one-fourth of the heat of reaction.

3352 kJ/mol/4 = 838 kJ/mol Al

THE HISTORY OF METAL USE

It is apparent from archeological sites that gold was in common use thousands of years before iron. This makes sense from the energy data for these metals. The metals that have been in use the longest are the ones that are easiest to extract from their compounds.

A Roman gilded-silver cup showing a scene from the Iliad.

---

2. Assume that the ΔH_f for this reaction is -239.7 kJ/mol of nickel oxide. What is the value of ΔH for the reverse reaction, per mole of nickel metal?

Answers: **1.** $2Ni(s) + O_2(g) \rightarrow 2NiO(s)$.
2. Because 2 mol of nickel would be extracted from 2 mol of nickel oxide, the value of ΔH per mole of nickel for the reverse reaction is simply $+239.7$ kJ/mol Ni.

Homework

Assign the reading and exercises for Fire Lesson 109 in the student text.

This table shows the energy data for the decomposition of metal oxides to metals. To extract the metals, people had to construct furnaces that transferred a sufficient amount of heat.

Extraction of Metals from Metal Oxides

Metal	ΔH kJ/mol	Date
gold, Au	−41 kJ/mol	6000 B.C.E.
silver, Ag	+15 kJ/mol	4000 B.C.E.
copper, Cu	+156 kJ/mol	4000 B.C.E.
lead, Pb	+218 kJ/mol	3500 B.C.E.
iron, Fe	+413 kJ/mol	2500 B.C.E.
tin, Sn	+581 kJ/mol	1800 B.C.E.
magnesium, Mg	+602 kJ/mol	1755 C.E.
aluminum, Al	+838 kJ/mol	1825 C.E.

Easier to extract

Harder to extract

The difficulty of extraction of various metals from metal compounds is reflected in the chronological "discovery" and use of these metals through history. The more energy it takes to extract a metal, the later the metal was put into practical use.

Metals that are easily oxidized are hard to extract. In other words, the oxidation reaction is harder to reverse. These are also the metals that tarnish and rust easily. Gold is often found in its pure form in nature. This is because gold is not easily oxidized and is fairly unreactive.

LESSON SUMMARY

How much energy is transferred during oxidation of metals?

The formation of metal oxides is usually an exothermic reaction. This means oxidation reactions may be valuable sources of energy. The heat of formation of a compound is the thermal energy change associated with the formation of *one mole* of that compound from its constituent elements. The decomposition of metal oxides into pure metals and oxygen is the reverse of an oxidation reaction. These reactions are generally endothermic. This means it requires an input of energy to extract a metal from a metal compound. Metals that are easily oxidized (with high heats of formation) are more difficult to extract from their metal compounds than metals that are less easily oxidized (with low heats of formation).

KEY TERM

heat of formation

Exercises

Reading Questions

1. Of all the possible reactions in this lesson, which ones have the potential of being the best sources of energy? Explain your answer.

2. What does the heat of formation indicate about a metal in terms of its practical use as a metal?

LESSON 109 ANSWERS

1. Possible answer: The oxidation reactions are better sources of energy than the reversed reactions because they are almost always exothermic. The metals that are the hardest to extract have the lowest (most negative) heats of formation, which means that they release the most energy into the environment. The oxidation of aluminum is the best source of energy listed in the lesson.

2. Possible answer: The heat of formation is an indication of the difficulty of isolating the metal from the compounds in which it naturally occurs. The lower (more negative) the heat of formation for the compound, the more energy must be available for extracting the metal from the compound. To justify the expense of extracting these metals, the properties of the metals must be very desirable. Metals with less extreme heats of formation require less justification for their extraction.

3. 2180 kJ

4. Metals that are easily oxidized tend to react with the oxygen in the air to form an oxide. Because this reaction releases energy, it occurs spontaneously. The same amount of energy must be added to the oxide to form the pure metal, making it hard to form pure metal by natural processes. These metals include aluminum and magnesium.

5.

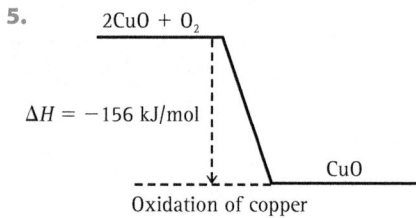

Oxidation of copper

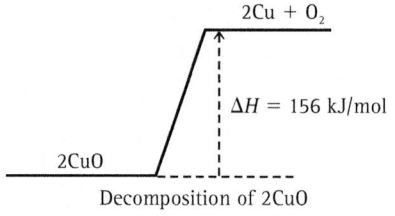

Decomposition of 2CuO

6. 3486 kJ

7. 24.6 kJ

8. Formation of 6 mol of magnesium oxide releases 3612 kJ of energy. Formation of 2 mol of aluminum oxide releases 3352 kJ of energy. So, the formation of 6 mol of magnesium oxide is a source of more energy than the formation of 2 mol of aluminum oxide.

(continued)

Reason and Apply

3. How much energy is released when 10 mol of lead (II) oxide, PbO, are formed?

4. Explain why metals that are easily oxidized are not found in nature in their pure forms. Name two of these metals.

5. Draw the energy diagrams for the oxidation of copper and its reverse reaction. Include the energy data.

6. How many kilojoules of energy does it take to extract 6 mol of tin from tin (II) oxide, SnO?

7. How many kilojoules of energy does it take to extract 10 g of copper from copper (II) oxide, CuO?

8. Which would be a source of more energy, the formation of 6 mol of magnesium oxide, MgO, or the formation of 2 mol of aluminum oxide, Al_2O_3? Explain.

9. PROJECT   Research how one of the metals of antiquity was extracted from metal compounds dug out of the Earth. Describe the process and draw a sketch.

9. A good report will include: ● the name of the metal ● the names of some of the compounds, or ores, in which the metal was found ● some of the cultures that used the metal ● where the metal was found and how it was transported ● a description of the process the ancients used to extract the metal from its ore, including a diagram of the structures and any additional material that was necessary for the extraction.

Oxidation-Reduction

THINK ABOUT IT

A silvery or reddish mineral called hematite, Fe_2O_3, is the main source of iron metal on Earth. When this iron (III) oxide is decomposed to iron and oxygen, the oxygen atoms lose electrons. This means that the iron atoms gain electrons.

What happens to electrons during oxidation?

To answer this question, you will explore

↪ Oxidation-Reduction

↪ Redox Reactions

EXPLORING THE TOPIC

↪ Oxidation-Reduction

When iron oxides are decomposed to iron metal, the iron cations gain electrons and the oxygen anions lose electrons.

$$2Fe_2O_3(s) \longrightarrow 4Fe(s) + 3O_2(g)$$

The oxygen anions in $Fe_2O_3(s)$ have a charge of -2. In the decomposition of $Fe_2O_3(s)$, each O^{2-} anion loses two electrons to form neutral oxygen atoms. The oxygen atoms bond in pairs to form elemental oxygen, O_2 (to satisfy the octet rule). In this process, the oxygen anions, O^{2-}, are oxidized to elemental oxygen, O_2.

The iron cations in $Fe_2O_3(s)$ have a charge of $+3$. The iron atoms in $Fe(s)$ have a charge of 0. In the decomposition of $Fe_2O_3(s)$, each Fe^{3+} cation gains three electrons to form neutral iron atoms in the elemental metal. Just as there is a term for losing electrons (oxidation), there is a term for gaining electrons: **reduction.** In this process, the iron atoms, Fe^{3+}, are reduced to elemental iron, Fe.

Whenever metals are extracted from their compounds, they are reduced. While the term *reduction* refers to the process of gaining electrons, its origins have to do with the concept of reducing metal ores to their simplest elements. Whenever you see a metal element as the product of a reaction, you can be fairly certain that it has been reduced.

Sample of hematite, Fe_2O_3

Reduction—gains electrons

$$2\overset{2+}{Fe_2}\overset{2-}{O_3}(s) \longrightarrow 4\overset{0}{Fe}(s) + 3\overset{0}{O_2}$$

Oxidation—loses electrons

Lesson 110 | **Oxidation-Reduction**   559

Overview

CLASSWORK: PAIRS

Key Question: What happens to electrons during oxidation?

KEY IDEAS

Whenever an atom loses electrons, another atom gains electrons. The losing of electrons is called *oxidation,* and the gaining of electrons is called *reduction*. Oxidation and reduction always occur together. This process is known as *oxidation-reduction,* or *redox*. When metals react with oxygen to form metal oxides, the metal atoms are oxidized and the oxygen is reduced.

LEARNING OBJECTIVES

● Define reduction.
● Explain oxidation and reduction in terms of electron exchange.
● Write net ionic equations.

KEY TERMS

reduction
oxidation-reduction reaction (redox)

IN CLASS

On the worksheet, students explore what happens during oxidation and reduction. The definition of oxidation is expanded on, and reduction is introduced. Students identify which

---

substances in a chemical equation are oxidized and which are reduced. A complete materials list is available for download.

TRM Materials List 110

Lesson Guide

EXPLORING THE TOPIC (5 min)

TRM PowerPoint Presentation 110

> *ChemCatalyst*
>
> These reactions are both considered oxidation reactions.
>
> $$2Ca(s) + O_2(g) \rightarrow 2CaO(s)$$
> $$2Fe(s) + 3CuSO_4(aq) \rightarrow$$
> $$3Cu(s) + Fe_2(SO_4)_3(aq)$$
>
> **1.** What makes both reactions oxidation reactions?
>
> **2.** Examine each reaction and determine where electrons are being lost and where electrons are being gained. How many electrons are transferred?

Sample answers: **1.** They are both oxidation reactions because in each case an element (calcium or iron) is losing electrons. **2.** In the first equation, calcium loses two electrons and oxygen gains two electrons. In the second equation, each iron atom loses three electrons and each copper ion gains two electrons.

Ask: Where do the electrons that are "lost" end up? How do you know?

Ask: What is the charge on the calcium atom in CaO? $(2+)$ On the oxygen atom? $(2-)$

Ask: When a substance is oxidized, another substance is reduced. What do you think this means?

EXPLORE (15 min)

TRM Worksheet with Answers 110

TRM Worksheet 110

INTRODUCE THE CLASSWORK

➔ Arrange students into pairs.

➔ Pass out the student worksheet.

➔ Show students how to track the movement of electrons with net ionic equations. Write the two ChemCatalyst equations on the board. Write the charges and explain how to write net ionic equations for aqueous solutions.

$$2\overset{0}{Ca}(s) + \overset{0}{O_2}(g) \rightarrow 2\overset{2+\ 2-}{CaO}(s)$$

The charge on calcium changed from 0 to +2, so it must have lost two electrons. Calcium was oxidized. The charge on oxygen changed from 0 to −2, so it must have gained two electrons. The charge on oxygen was reduced.

The full ionic equation for the reaction between iron and copper (II) sulfate is

$$2\overset{0}{Fe}(s) + 3\overset{2+}{Cu}(aq) + 3\overset{2-}{SO_4}(aq) \rightarrow$$
$$3\overset{0}{Cu}(s) + 2\overset{3+}{Fe}(aq) + 3\overset{2-}{SO_4}(aq)$$

→ Point out to students that the sulfate ion is a spectator ion. You can ignore it when you are figuring out where the electrons went.

Net ionic equation: $2\overset{0}{Fe}(s) + 3\overset{2+}{Cu}(aq) \rightarrow$
$3\overset{0}{Cu}(s) + 2\overset{3+}{Fe}(aq)$

Iron was oxidized; copper was reduced.

→ Work an example problem on the board.

EXAMPLE: NET IONIC EQUATION

Write the net ionic equation for this reaction:

$$Cu(s) + 2AgNO_3(aq) \rightarrow$$
$$2Ag(s) + Cu(NO_3)_2(aq)$$

SOLUTION

Step 1: Write out the equation, expanding ionic substances into their appropriate ions.

$$Cu(s) + 2Ag^+(aq) + 2NO_3^-(aq) \rightarrow$$
$$2Ag(s) + Cu^{2+}(aq) + 2NO_3^-(aq)$$

Step 2: Cross out any spectator ions that are unchanged by the reaction.

$$Cu(s) + 2Ag^+(aq) + 2\cancel{NO_3^-}(aq) \rightarrow$$
$$2Ag(s) + Cu^{2+}(aq) + 2\cancel{NO_3^-}(aq)$$

Step 3: Write the net ionic equation.

$$Cu(s) + 2Ag^+(aq) \rightarrow 2Ag(s) + Cu^{2+}(aq)$$

Charles D. Winters/Science Source

....
Important to Know The atom that is being reduced gains electrons. The atom that is oxidized loses electrons.
....

Every time a substance loses electrons, another substance gains electrons and vice versa. These two processes always take place together: oxidation and reduction. **Oxidation-reduction reactions** are often referred to as **redox** reactions.

REACTION OF ZINC AND COPPER SULFATE

A strip of zinc metal is placed in a light blue solution of copper (II) sulfate, $CuSO_4$, as shown in the illustration. Soon, a reddish coating begins to form on the zinc strip. The blue color of the solution begins to fade. What is going on?

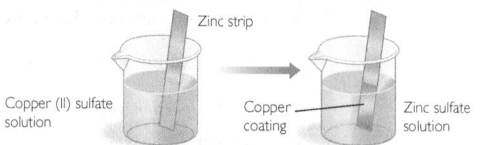

Start with the balanced chemical equation for this reaction:

$$Zn(s) + CuSO_4(aq) \longrightarrow ZnSO_4(aq) + Cu(s)$$

You know from Unit 1: Alchemy that when salts dissolve in water, the compound dissociates into ions. Often, chemical equations are written in ionic form to specify the charges of the ions in solution as shown here.

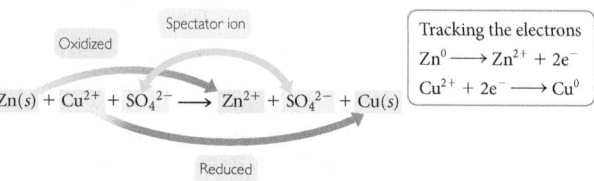

Notice that the sulfate ion is a spectator ion. The net ionic equation is therefore

$$Zn(s) + Cu^{2+} \longrightarrow Zn^{2+} + Cu(s)$$

The next illustration shows this reaction from a particle view. Take a moment to follow the transfer of the electrons.

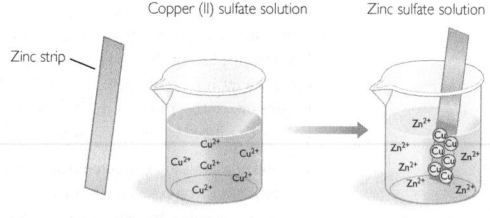

The copper ions, Cu^{2+}, in solution are reduced to copper metal, Cu^0. The zinc metal atoms on the strip, Zn^0, are oxidized to zinc ions, Zn^{2+}, and enter the solution.

→ Pass out worksheets. Students can work in pairs.

INTRODUCE OXIDATION-REDUCTION

→ You may write a few equations from the worksheet on the board. Have students help you transform these into net ionic equations. Track the electron transfer.

Ask: How do you know that a substance has lost electrons? Gained electrons?

Ask: Do gaining electrons and losing electrons always occur together? Explain.

Ask: How would you show that the charges are balanced on both sides of an ionic equation?

Key Points: Every time a substance loses electrons, another substance gains electrons. Losing electrons is called *oxidation*. Gaining electrons is called *reduction*. These two processes—oxidation and reduction—always occur together and often are referred to as *redox reactions* for short.

Redox Reactions

Many chemical reactions are oxidation-reduction reactions. In fact, any reaction in which electrons are transferred from one reactant to another is a redox reaction. This includes most reactions that have an element as a reactant or product, which are often single exchange reactions.

Typically, double exchange reactions are *not* redox reactions. For example, in acid-base reactions, H^+ trades places with another cation. In precipitation reactions, an aqueous ion from one compound combines with an aqueous ion from another. So, double exchange reactions often involve transfer of ions, while single exchange reactions involve transfer of electrons.

SINGLE EXCHANGE: REDOX REACTION

$$Mg(s) + 2HCl(aq) \longrightarrow H_2(g) + MgCl_2(aq)$$

DOUBLE EXCHANGE: ACID-BASE REACTION

$$NaOH(aq) + HCl(aq) \longrightarrow H_2O(l) + NaCl(aq)$$

DOUBLE EXCHANGE: PRECIPITATION REACTION

$$AgNO_3(aq) + KCl(aq) \longrightarrow AgCl(s) + KNO_3(aq)$$

Oxidation-reduction reactions are not limited to metals and ionic compounds. Molecular covalent compounds also take part in oxidation-reduction reactions.

COMBUSTION: REDOX REACTION

$$CH_4 + 2O_2 \longrightarrow CO_2 + 2H_2O$$

In this reaction, carbon is oxidized. Electron transfer is trickier to identify in reactions with covalent compounds. But we already know that combustion is a form of oxidation. For molecules, oxidation often involves a loss of hydrogen and a gain of oxygen, and reduction often involves a gain of hydrogen and the loss of oxygen.

Example

Oxidation-Reduction

Solid magnesium, Mg, is placed in a solution of iron (III) nitrate, $Fe(NO_3)_3$. Magnesium nitrate and solid iron are formed. Show the ionic equation and determine what is oxidized and what is reduced.

Solution

Write and balance the chemical equation for this reaction:

$$3Mg(s) + 2Fe(NO_3)_3(aq) \longrightarrow 3Mg(NO_3)_2(aq) + 2Fe(s)$$

Translate this equation into an ionic equation. Ignore the spectator ions, in this case, the NO_3^- ions.

$$3Mg(s) + 2Fe^{3+}(aq) \longrightarrow 3Mg^{2+}(aq) + 2Fe(s)$$

Lesson 110 | **Oxidation-Reduction** 561

Reduction: The process of gaining electrons during a chemical reaction. Reduction is always accompanied by oxidation.

Redox: A short name given to oxidation-reduction reactions.

If you write the net ionic equation, it is easier to see the electron transfers.

$$Mg(s) + Cu(NO_3)_2(aq) \rightarrow Mg(NO_3)_2(aq) + Cu(s)$$

becomes

$$Mg(s) + Cu^{2+}(aq) + 2NO_3^-(aq) \rightarrow Mg^{2+}(aq) + 2NO_3^-(aq) + Cu(s)$$

Mg oxidized

$$Mg(s) + Cu^{2+}(aq) \longrightarrow Mg^{2+}(aq) + Cu(s)$$

Cu reduced

Notice that the magnesium atoms lose electrons and become ionic. The magnesium atoms have been oxidized. At the same time, the copper cations gain electrons, becoming neutral. The copper cations have been reduced. Nothing has happened to the nitrate ions.

Oxidation is not limited to metals and ionic compounds. Most of the equations in this lesson involve reactants that contain metal atoms.

However, in these two equations, magnesium and methane are both oxidized:

$$2Mg(s) + O_2(g) \rightarrow 2MgO(s)$$
$$CH_4(g) + 2O_2(g) \rightarrow CO_2(g) + 2H_2O(l)$$

Electron transfer is relatively easy to identify in reactions that involve ionic substances, but covalent compounds also are oxidized.

CLARIFY THE LANGUAGE ASSOCIATED WITH REDOX

→ Write the mnemonics OIL RIG and LEO GER on the board (optional).

Ask: When a metal is being extracted from an ionic compound, is it being oxidized or reduced?

Ask: If the atom is gaining electrons, why is it called reduction?

Key Points: Sometimes the terms used with oxidation and reduction can seem a bit confusing. The thing that is being reduced is gaining electrons. The thing that is being oxidized is losing electrons. This language originated because a metal ore is "reduced" to its simplest elements by this process.

A couple of mnemonics can help students recall which is which. OIL RIG stands for Oxidation Is Losing—Reduction Is Gaining. LEO GER stands for Losing Electrons is Oxidation—Gaining Electrons is Reduction.

Check-In

Liquid bromine, Br_2, is added to a solution of potassium iodide, KI, forming potassium bromide, KBr, and iodine, I_2. What is oxidized and what is reduced? Show your work or give your reasoning.

Answer: Start with the chemical equation and balance it.

$$Br_2(l) + 2KI(aq) \rightarrow 2KBr(aq) + I_2(aq)$$

Write the ionic equation (aqueous salts ionize).

$$Br_2(l) + 2K(aq) + 2I^-(aq) \rightarrow 2K(aq) + 2Br^-(aq) + I_2(aq)$$

The potassium ions do not change, so we can ignore them. The bromine atoms gain electrons and thus are reduced. The iodide ions lose electrons and thus are oxidized.

Homework

Assign the reading and exercises for Fire Lesson 110 in the student text.

LESSON 110 ANSWERS

1. In this reaction, copper is oxidized because the neutral copper atom loses the electrons and becomes a positively charged copper ion. The oxygen atom is reduced because it gains electrons and becomes a negatively charged oxygen ion.

2. In this reaction, copper is reduced because the copper ion gains electrons to become a neutral atom. The oxygen is oxidized because the oxygen ion loses electrons to become a neutral atom, which then combines with another oxygen atom to form a molecule.

3. *Possible answer:* A metal can be extracted from a metal salt solution by a single exchange reaction in which the electrons are transferred from a different metal. The electrons are added to a positively charged metal ion, forming a neutral atom that precipitates out of the solution. The other metal replaces the extracted metal in the solution.

4. a. Hydrogen is oxidized and carbon is reduced. **b.** Sulfur is oxidized and fluorine is reduced. **c.** Silver is oxidized and oxygen is reduced. **d.** Carbon is oxidized and oxygen is reduced.

5. a. $Zn(s) + Cu^{2+}(aq) \rightarrow Cu(s) + Zn^{2+}(aq)$

b. $Ba^{2+}(aq) + SO_4^{2-}(aq) \rightarrow BaSO_4(s)$

6. The reaction between barium nitrate and copper sulfate is not a redox reaction because no electrons are transferred between atoms during the reaction. The net ionic equation shows that the charge on the copper and zinc changes during the first reaction, but the charges on the barium and the sulfate ions do not change during the second reaction.

7. a. Because this is a double exchange reaction, the charges on the ions do not change. This is not a redox reaction. **b.** Because this is a double exchange reaction, the charges on the ions do not change. This is not a redox reaction. **c.** Because this is a double exchange reaction, the charges on the ions do not change. This is not a redox reaction. **d.** Because this is a double exchange reaction, the charges on the ions do not change. This is not a redox reaction.

Example

(continued)

The metal losing electrons is oxidized. The metal gaining electrons is reduced.

Each magnesium atom loses two electrons, so magnesium is oxidized. Two iron ions accept three electrons each, so iron is reduced.

> **Tracking the electrons**
> $3Mg^0 \longrightarrow 3Mg^{2+} + 6e^-$
> $2Fe^{3+} + 6e^- \longrightarrow 2Fe^0$

LESSON SUMMARY

What happens to electrons during oxidation?

KEY TERMS
reduction
oxidation-reduction
  reaction (redox)

Whenever an atom or ion loses electrons, another atom or ion gains electrons. The loss of electrons is called oxidation, and the gain of electrons is called reduction. Oxidation and reduction always occur together. This process is known as oxidation-reduction, or redox. When metals react with oxygen to form metal oxides, the metal atoms are oxidized and the oxygen is reduced. To extract elemental metals, the reverse reaction must occur and metal ores must be reduced.

Exercises

Reading Questions

1. Explain what is oxidized and what is reduced when copper reacts with oxygen to form copper (II) oxide, CuO.

2. Explain what is oxidized and what is reduced when copper (II) oxide is decomposed to copper and oxygen.

Reason and Apply

3. Describe how you might extract a metal from a metal salt solution.

4. Determine what is oxidized and what is reduced in these reactions:
   a. $CO_2(g) + H_2(g) \longrightarrow CO(g) + H_2O(g)$
   b. $SF_4(g) + F_2(g) \longrightarrow SF_6(g)$
   c. $4Ag(s) + 2H_2S(g) + O_2(g) \longrightarrow 2Ag_2S(s) + 2H_2O(g)$
   d. $C_6H_{12}O_6(aq) + 6O_2(g) \longrightarrow 6CO_2(g) + 6H_2O(l)$

5. Write the net ionic equations for these reactions:
   a. $Zn(s) + CuCl_2(aq) \longrightarrow Cu(s) + ZnCl_2(aq)$
   b. $Ba(NO_3)_2(aq) + CuSO_4(aq) \longrightarrow BaSO_4(s) + Cu(NO_3)_2(aq)$

6. Determine which of the reactions in Exercise 5 is not a redox reaction. Explain your reasoning.

7. Which of these reactions from Unit 4: Toxins can be classified as redox reactions? For each redox reaction, show what is oxidized, what is reduced, and the total number of electrons transferred.
   a. $Tl_2O(s) + 2HCl(l) \longrightarrow 2TlCl(aq) + H_2O(l)$
   b. $HgS(s) + 2HCl(aq) \longrightarrow HgCl_2(s) + H_2S(aq)$
   c. $PbCO_3(s) + 2HCl(aq) \longrightarrow PbCl_2(aq) + H_2CO_3(aq)$
   d. $Na_2C_2O_4(aq) + CaCl_2(aq) \longrightarrow CaC_2O_4(s) + 2NaCl(aq)$
   e. $Pb(s) + 2HCl(aq) \longrightarrow PbCl_2(aq) + H_2(g)$
   f. $2As(s) + 6HCl(aq) \longrightarrow 2AsCl_3(aq) + 3H_2(g)$

562   Chapter 21 | **Controlling Energy**

e. This is a redox reaction. Lead loses electrons to become a cation, so it is oxidized. Hydrogen gains electrons to become hydrogen gas, so it is reduced. Two electrons are transferred. **f.** This is a redox reaction. Arsenic loses electrons to become a cation, so it is oxidized. Hydrogen gains electrons to become hydrogen gas so it is reduced. Six electrons are transferred.

The Active Life

Activity of Metals

THINK ABOUT IT

Imagine that you set up three experiments. In one, a strip of copper is placed in a solution of silver nitrate. In the second, a strip of zinc is placed in a solution of copper (II) nitrate. In the third, a strip of silver is placed in a solution of zinc nitrate. Only two of these setups will result in reactions. How can you predict which reactions will occur?

Which metal atoms are most easily oxidized?

To answer this question, you will explore

- Comparing Metals
- Activity Series

EXPLORING THE TOPIC

Comparing Metals

When metal atoms are combined with metal ions in solution, you may or may not end up with a reaction. Some combinations of metals and metal salts result in redox reactions and others don't. It all depends on which combinations of metals are used.

Consider the three solutions shown on the next page. Each solution contains metal cations and a strip of elemental metal. If the metal atoms on the strip transfer electrons to the metal cations in solution, a reaction will occur.

INDUSTRY CONNECTION

Redox reactions have been used for decades to coat metals with chromium. Chrome plating improves appearance and durability. Chromium sulfate, or chromium chloride, is often used in industrial plating solutions.

Lesson 111 | **Activity of Metals**   563

Overview

LAB: GROUPS OF 4

Key Question: Which metal atoms are most easily oxidized?

KEY IDEAS

Oxidation-reduction reactions are best understood in terms of comparisons between metals. The relative activity of a metal is demonstrated when it is combined with another metal cation. The metal that gives up electrons is considered more active. The ranking of metals in order of their activity is referred to as an activity series.

LEARNING OBJECTIVES

- Compare the relative activity of various metals and explain why some metals displace others in reactions.
- Predict the outcome of reactions between metals and metal salts by using an activity series.

FOCUS ON UNDERSTANDING

The term *activity* refers to the tendency to react, given favorable thermodynamics. Although the concepts are related, we have chosen not to use the term reactivity because it depends on both kinetics and thermodynamics.

KEY TERM

activity series

IN CLASS

Students combine metals and metal compounds to determine relative activities. The lab is divided into two parts. Part 1 focuses on the favorability of a reaction or its reverse. Students use chemical equations and their knowledge of redox reactions to predict what they will observe. In Part 2, they test their predictions by combining the appropriate reactants. In the well-plate activity, students compare how copper, magnesium, and zinc react in relation to one another. This allows students to create a mini activity series and to make predictions about other reactions. A complete materials list is available for download.

TRM Materials List 111

Differentiate

To deepen students' understanding of the activity series, you can have them explore the simulations called "Metals in Aqueous Solutions" from the Chemical Education Research Group at Iowa State University. Students can view a simulation of how the redox reaction occurs at the molecular level. This can help students understand that oxidation and reduction occur simultaneously. Students can also test additional metals, which sets up later understanding of how standard reduction potentials use hydrogen as a reference when they test how metals react in hydrogen chloride solution. A list of URLs for this lesson is in the Chapter 21 Web Resources document.

TRM Chapter 21 Web Resources

SETUP

Prepare the solutions before class. Label the pipettes and place them in the appropriate solutions.

CLEANUP

Place leftover metal solutions in a metal waste container. Dispose of the evaporated solids according to state or district guidelines.

Lesson Guide

TRM PowerPoint Presentation 111

ENGAGE (5 min)

ChemCatalyst

Under standard conditions, one of these reactions will occur and the other will not.

$$Cu(s) + FeSO_4(aq) \rightarrow CuSO_4(aq) + Fe(s)$$

$$CuSO_4(aq) + Fe(s) \rightarrow Cu(s) + FeSO_4(aq)$$

1. Write the net ionic equations.

2. Which substance is oxidized and which is reduced in each reaction?

3. How can you find out which reaction will take place? Explain your reasoning.

Sample answers:
1. $Cu(s) + Fe^{2+}(aq) \rightarrow Cu^{2+}(aq) + Fe(s)$; $Cu^{2+}(aq) + Fe(s) \rightarrow Cu(s) + Fe^{2+}(aq)$.
2. In the first reaction, the copper is oxidized and the iron is reduced. In the second reaction, copper is reduced and iron is oxidized. **3.** Answers will vary. One could try to carry out the reactions. Students may recall that iron is more easily oxidized than copper.

Ask: What would you observe for each reaction if it occurred?

Ask: How did you determine which substance was oxidized?

Ask: Why won't both reactions occur under normal circumstances?

Ask: How could you determine which metal is more easily oxidized?

EXPLORE (15 min)

TRM Worksheet with Answers 111

TRM Worksheet 111

INTRODUCE THE ACTIVITY

→ Arrange students into groups of four.
→ Pass out the student worksheet.

EXPLAIN & ELABORATE (15 min)

COMPARE THE SOLUTIONS FROM PART I: COPPER VERSUS IRON

→ Write the appropriate equations on the board. Each type of equation provides different information.

Ask: What did you observe?

Ask: How can you tell when a metal is being reduced? Oxidized?

Ask: Which metal is more active, copper or iron?

Beaker 1: Copper in silver nitrate solution

Beaker 2: Zinc in copper (II) nitrate solution

Beaker 3: Silver in zinc nitrate solution

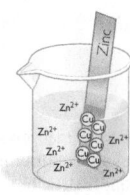
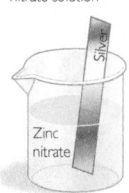

Silver: reduced
Copper: oxidized

Copper: reduced
Zinc: oxidized

No reaction

Beaker 1: $Cu(s) + 2AgNO_3(aq) \longrightarrow Cu(NO_3)_2(aq) + 2Ag(s)$

$Cu(s) + 2Ag^+(aq) \longrightarrow Cu^{2+}(aq) + 2Ag(s)$

Beaker 2: $Zn(s) + Cu(NO_3)_2(aq) \longrightarrow Zn(NO_3)_2(aq) + Cu(s)$

$Zn(s) + Cu^{2+}(aq) \longrightarrow Zn^{2+}(aq) + Cu(s)$

Beaker 3: $Ag(s) + Zn(NO_3)_2(aq) \longrightarrow$ no reaction

In the first beaker, the copper atoms transfer electrons to the silver ions. Copper atoms are oxidized and silver cations are reduced. As a result, silver metal can be seen as a coating on the copper strip. In the second beaker, zinc atoms transfer electrons to the copper ions in solution. Zinc atoms are oxidized and copper is reduced. Copper metal coats the zinc strip. However, nothing happens in the third beaker.

In the third beaker, there is no reaction. Silver does not give up its electrons in the presence of zinc ions. Chemists say that silver is not as active as zinc.

⟳ Activity Series

By combining different metals and metal ions, you can determine experimentally which metals are more active than other metals. The result is called an **activity series.** Working with the data we have so far, zinc, copper, and silver can be placed in order of their activity.

Zinc

gives up electrons to

Copper

which gives up electrons to

Silver

More active

Less active

Where would magnesium fit on this list? In the lab, you combined magnesium with zinc nitrate. The magnesium was oxidized and formed Mg^{2+} ions.

$$Mg(s) + Zn(NO_3)_2(aq) \longrightarrow Mg(NO_3)_2(aq) + Zn(s)$$

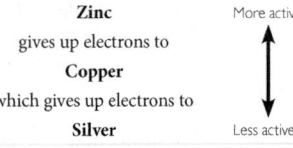

Ask: What ions were spectators in the reaction?

Key Points: In a single exchange reaction, one metal replaces another in a compound. All the redox reactions focused on in this lesson were single replacement reactions. In each case, an elemental metal replaced a metal cation in a compound. When you place an iron nail in a solution of copper (II) sulfate, the nail becomes coated with copper metal, and the solution begins to change color from dark blue to a lighter blue-green. The iron atoms are oxidized. They give up their electrons and become ions. The

copper ions are reduced. They gain the electrons given up by iron and become neutral, elemental atoms. The equations for this reaction are these:

$$Fe(s) + CuSO_4(aq) \rightarrow FeSO_4(aq) + Cu(s)$$

$$Fe(s) + Cu^{2+}(aq) + \cancel{SO_4^{2-}}(aq) \rightarrow Fe^{2+}(aq) + \cancel{SO_4^{2-}}(aq) + Cu(s)$$

$$Fe(s) + Cu^{2+}(aq) \rightarrow Fe^{2+}(aq) + Cu(s)$$
(copper reduced, iron oxidized)

When you put copper metal in a solution of iron (II) sulfate, nothing happens.

Because silver is easily reduced, it is relatively easy to plate metals with silver for ornamentation.

So, magnesium belongs above zinc on the list.

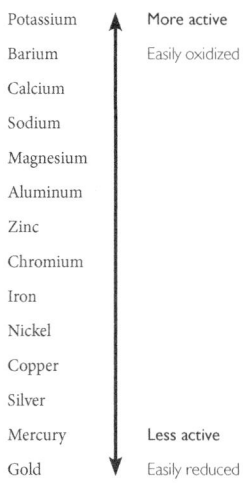

Magnesium
gives up electrons to
Zinc
which gives up electrons to
Copper
which gives up electrons to
Silver

More active

Less active

You may have noticed that all the reactions in this section have been single exchange reactions in which the more easily oxidized metal displaces the other metal. In this way, the more active metal forms cations while the less active metal is reduced to a solid.

A series of experiments can reveal where other metals belong on the list. The more active atoms at the top of the list will always displace the less active ones below.

Activity Series

Potassium	More active
Barium	Easily oxidized
Calcium	
Sodium	
Magnesium	
Aluminum	
Zinc	
Chromium	
Iron	
Nickel	
Copper	
Silver	
Mercury	Less active
Gold	Easily reduced

On this list, gold is at the bottom. Gold atoms are very difficult to oxidize, and gold cations are very easy to reduce. This is one reason why gold is rarely found in compounds combined with other atoms.

A metal high in the activity series

• reacts vigorously with compounds
• readily gives up electrons in reactions to form positive ions
• is corroded easily

Lesson 111 | **Activity of Metals**   565

Magnesium	More easily oxidized
Aluminum	**More active**
Zinc	
Iron	Less easily oxidized
Copper	**Less active**

Key Points: Combining aqueous metal salts with metals enables us to determine which metals are most easily oxidized and therefore most active. In Part 2 of the experiment, nitrates of magnesium, copper, and zinc were combined with magnesium, copper, and zinc metals. Magnesium oxidized, replacing both zinc ions and copper ions to form Mg^{2+} ions. Zinc replaced copper ions but not magnesium. Finally, copper did not replace either magnesium ions or zinc ions. Copper was the least active metal of the three. Magnesium was the most active metal of the three.

You can place metals in order of the ease of oxidation, creating an activity series. The metal at the top is considered more active than those below it. This metal is most easily oxidized. It gives up electrons most easily and replaces other metal ions. Metals toward the bottom are considered the least active. These metals do not give up electrons easily and thus do not react as easily. Metals that are more active will replace metals that are less active in reactions. You can look up the activities of metals in a table.

Activity Series

Potassium	**More active**
Barium	Easily oxidized
Calcium	
Sodium	
Magnesium	
Aluminum	
Zinc	
Chromium	
Iron	
Nickel	
Copper	
Silver	
Mercury	**Less active**
Gold	Easily reduced

$$Cu(s) + Fe^{2+}(aq) + SO_4^{2-}(aq) \rightarrow \text{no reaction}$$

Copper is not as easily oxidized as iron. It does not give up its electrons to iron ions. By pairing different metals and metal ions, you can determine experimentally which metals are more active than others in giving up electrons. More active metals are more easily oxidized.

INTRODUCE ACTIVITY SERIES

→ Create a mini activity series on the board as you discuss the outcome of the lab procedure. Place the most easily oxidized metal at the top of the list.

Ask: What happened when you tried different combinations of metals and metal nitrates?

Ask: Which metal is most easily oxidized? Least easily oxidized? What do the results tell you about the activity of these metals in relation to one another?

Ask: How could you determine where other metals would fit on the list?

Ask: If aluminum is oxidized by nitrates of zinc, iron, and copper but not by magnesium nitrate, where would it belong in the activity series? (above Zn and below Mg)

> Activity series: A table listing metals in order of their ease of oxidation.

EVALUATE (5 min)

Check-In

1. If silver does not react with magnesium, zinc, or copper nitrate, where does it belong in the activity series you created in class? Explain your thinking.

2. Name three things you know about silver based on its placement in the list.

Answers: **1.** Silver belongs at the bottom of the list, below copper, because it is not very active.

2. *Sample answers:* Silver is not easily oxidized. Silver probably will be found in nature in its elemental form. Silver will not react with copper, iron, zinc, aluminum, or magnesium salts. Silver is more easily reduced than the other metals listed.

Homework

Assign the reading and exercises for Fire Lesson 111 in the student text.

A metal low in the activity series
- does not react vigorously with chemicals
- does not readily give up electrons in reactions to form positive ions
- is not corroded easily

Example

Activity of Lead

Lead, Pb, will give up electrons to copper cations, Cu^{2+}, but not to zinc cations, Zn^{2+}.

a. Write the net ionic equation for the reaction between Pb and Cu^{2+}.

b. Do you expect that Zn will transfer electrons to lead cations, Pb^{2+}? Explain your reasoning.

Solution

a. $Pb(s) + Cu^{2+}(aq) \longrightarrow Pb^{2+}(aq) + Cu(s)$

b. Pb does not give up electrons to Zn^{2+}. This means that the reverse reaction *will* happen. Zn will give up electrons to Pb^{2+}. Zn is more active than Pb, so it displaces Pb in the ionic compound.

$$Zn(s) + Pb^{2+}(aq) \longrightarrow Zn^{2+}(aq) + Pb(s)$$

All of the metals on the periodic table can be placed in an activity series table. Chemists are interested in the activity of metals, because this allows them to control the outcome of different reactions. Metal compounds can be combined in different ways to form other compounds. Also, the most easily oxidized metals turn out to be a good source of electrons. You'll learn more about this in Lesson 112: Current Events.

LESSON SUMMARY

Which metal atoms are most easily oxidized?

Metals that are easily oxidized give up their electrons easily. They are considered more active than other metals. Comparing metals experimentally allows you to rank the metals in terms of their ease of oxidation. The activity of a metal is demonstrated when it is combined with another metal cation. The more active metal will lose electrons, displacing the other metal in the compound. The displaced metal is reduced. The ranking of metals in order of their activity is referred to as an activity series.

KEY TERM

activity series

Exercises

Reading Questions

1. Explain how you would figure out where tin should be listed in the activity series.

2. Why are metals that are more active also considered less stable?

(continued)

Reason and Apply

3. If zinc nitrate, $Zn(NO_3)_2$, is paired with solid iron, what would you expect to observe? Explain your reasoning.

4. If iron (III) nitrate, $Fe(NO_3)_3$, were combined with solid zinc, what would you expect to observe? Explain your reasoning

5. Consider any reactions that occur in Exercises 3 and 4.
   a. Write chemical equations.
   b. Write the net ionic equation.
   c. Identify the more active metal.
   d. Which metal is oxidized? Which metal is reduced?

6. Platinum is more active than gold but less active than silver. Where does it belong in the activity series?

7. If cobalt ions are replaced by iron to form iron ions and solid cobalt, is cobalt above or below iron in the activity series? Explain.

8. Write net ionic equations for two reactions between magnesium and less active metals.

9. Name three combinations of ionic compounds and metals that will not react.

is less active than zinc. Therefore, electrons will be transferred from zinc atoms to the iron ions, forming a solid iron precipitate and zinc ions in solution.

5. No reaction occurs in Exercise 3.
a. The chemical equation for the reaction in Exercise 4 is:

$2Fe^{3+}(aq) + 6NO_3^{-}(aq) + 3Zn(s) \rightarrow 2Fe(s) + 3Zn^{2+}(aq) + 6NO_3^{-}(aq)$

b. The net ionic equation for the reaction in Exercise 4 is:

$2Fe^{3+}(aq) + 3Zn(s) \rightarrow 2Fe(s) + 3Zn^{2+}(aq)$

c. Zinc is more active than iron.
d. Zinc is oxidized, and iron is reduced. Zinc nitrate forms in a cation to anion ratio of 1:2.

6. Platinum belongs near the bottom of the activity series, above gold and below silver.

7. Cobalt is below iron on the activity series because the iron is oxidized by the cobalt ions, making iron the more active metal.

8. *Possible answers (any 2):*

$Ni^{2+}(aq) + Mg(s) \rightarrow Ni(s) + Mg^{2+}(aq)$
$Ag^{2+}(aq) + Mg(s) \rightarrow Ag(s) + Mg^{2+}(aq)$
$Cu^{2+}(aq) + Mg(s) \rightarrow Cu(s) + Mg^{2+}(aq)$
$2Fe^{3+}(aq) + 3Mg(s) \rightarrow 2Fe(s) + 3Mg^{2+}(aq)$

9. *Possible answers (any 3):*
- $Hg^{2+}(aq)$ and $Au(s)$
- $Ni^{2+}(aq)$ and $Ag(s)$
- $Al^{3+}(aq)$ and $Fe(s)$
- $Ni^{2+}(aq)$ and $Au(s)$
- $Al^{3+}(aq)$ and $Cr(s)$

LESSON 111 ANSWERS

1. *Possible answer:* Figure out where tin should be listed in the activity series by placing strips of tin in beakers containing solutions of salts of various metals. If the metal ion is less active than tin, electrons will be transferred from the tin atoms to the metal ions and the metal will coat the tin strip. If the metal ion is more active than tin, there is no change. Tin can be placed on the activity series above the highest metal that coats the strip and below the lowest metal that does not coat the strip.

2. *Possible answer:* Metals that are more active are also considered less stable because they oxidize easily, causing them to react readily with oxygen in the air. The metals do not remain in elemental form as long as more stable metals do.

3. No change would be observed because iron is less active than zinc. Therefore, iron cannot be oxidized by the zinc nitrate solution, and the iron strip would not be affected.

4. The solid zinc would become coated with iron, and some of the zinc would dissolve into the solution. Iron

Overview

LAB: GROUPS OF 4

Key Question: How can you use a redox reaction as an energy source?

KEY IDEAS

Electrons are transferred during redox reactions. Chemists can turn this transfer of electrons into electrical energy. By separating the reactants, chemists can control and, in a sense, "store" electrical energy. An activity series allows us to predict which chemical combinations will result in the most useful reactions for electricity production. The specific identity of the reactants and their concentrations affect the amount of energy produced.

LEARNING OBJECTIVES

- Describe an electrochemical cell and how it works.
- Calculate the potential of an electro-chemical cell.
- Use an activity series to help design electrochemical cells.

KEY TERMS

electrolyte
electrochemical cell
half-cell
half-reaction
voltage

IN CLASS

Students complete an experimental procedure that instructs them in the creation of a copper-zinc electrochemical cell. Students observe the cell before and after the introduction of a salt bridge connecting the two beakers. Then they insert a tiny light bulb in the circuit to demonstrate the formation of electricity by the cell. Some of the key terms are introduced more fully in the student text. A complete materials list is available for download.

TRM Materials List 112

Pre-AP® Course Tip

Students preparing to take an AP® chemistry course or college chemistry should know which data to collect to achieve a certain goal. In this lesson, students use their knowledge of the activity series to build a voltaic cell using copper and zinc. Based on their design, they can calculate the cell potential and describe how it generates electricity.

SETUP

Prepare the copper (II) sulfate, zinc sulfate, and potassium nitrate solutions before class. Set up one electrochemical cell, following the procedure on the worksheet. You can set up one or two alternate cells using different metals or concentrations to demonstrate the differences in voltage. (See the voltage discussion in Explain & Elaborate.)

SAFETY

Students should wear safety goggles. Students should use surgical gloves to handle the salt bridge.

CLEANUP

Place leftover metal solutions in a metal waste container. Salt bridges should be disposed of according to local regulations for metal waste.

Featured LAB

Electrochemical Cell

SAFETY Instructions
Wear safety goggles. Wear protective gloves for handling the salt bridge.

Salt bridge

Copper (II) sulfate   Zinc sulfate

Purpose

To explore how to get electrical energy from a redox reaction.

Materials

- zinc strip
- copper strip
- 250 mL beakers (2)
- 500 mL beaker
- 2 connecting wires with alligator clips
- tiny LED light bulb
- 1.0 M $CuSO_4$, 100 mL
- 1.0 M zinc sulfate solution, $ZnSO_4$, 100 mL
- saturated potassium nitrate (KNO_3) solution, 20 mL—for salt bridge
- filter paper approximately 1 in wide and 6 in long—for salt bridge
- gloves for handling salt bridge

Procedure

1. Make a salt bridge. Place a piece of folded-up filter paper in the bottom of an empty 500 mL beaker. Soak it thoroughly with KNO_3. Set aside for later.

2. Carefully pour 100 mL of $CuSO_4$ into one beaker and 100 mL of $ZnSO_4$ into another beaker.

3. Set up the zinc strip in the $ZnSO_4$ solution and the copper strip in the $CuSO_4$ solution. Use a wire with alligator clips to connect the two metal strips. The clips should not touch the solutions.

4. Use gloves to place the salt bridge between the two beakers as shown.

Observation and Analysis

1. What do you observe?

2. Why is it necessary to have an ionic solution in the salt bridge?

3. Which substance is being oxidized? Which substance is being reduced? How do you know?

4. Connect the tiny LED light into your circuit using both sets of alligator clips. What do you observe? What does that prove?

5. Explain how you might reverse this reaction.

6. **Making Sense** Explain where the electricity in the electrochemical cell is coming from.

568   Chapter 21 | **Controlling Energy**

Lesson Guide

ENGAGE (5 min)

TRM Transparency—ChemCatalyst 112

TRM PowerPoint Presentation 112

ChemCatalyst

The table shows the oxidation half of some oxidation-reduction reactions. Suppose you want to set up an oxidation-reduction reaction. Which pairing of metal and metal ion will result in the greatest energy release? Explain your reasoning.

Electrochemical Cell

THINK ABOUT IT

Batteries are used extensively in our daily lives. They are in our cell phones, automobiles, electronic toys, mobile music players, and wristwatches. A battery contains a controlled redox reaction.

How can you use a redox reaction as an energy source?

To answer this question, you will explore

↻ Electrochemical Cells

↻ Maximizing the Potential

↻ Batteries

EXPLORING THE TOPIC

↻ Electrochemical Cells

The combustion of gasoline powers cars. The combustion of methane is used to cook food and heat homes. But it would be hard to imagine a cell phone or a toy running off a combustion reaction.

There is another way, besides burning, to make use of the energy of a reaction. In the case of many oxidation-reduction reactions, the transfer of electrons can be converted into electrical energy using an apparatus called an *electrochemical cell.*

CONTROLLING REDOX REACTIONS

The first step to controlling a redox reaction is to separate the oxidation and reduction reactions. One possible way to do this is shown. Two strips of metal called *electrodes* are placed in two beakers. Both beakers are filled with aqueous salt solutions called **electrolytes.** Electrons move through a wire connecting the two metal strips, and ions move through a salt bridge connecting the two electrolyte solutions.

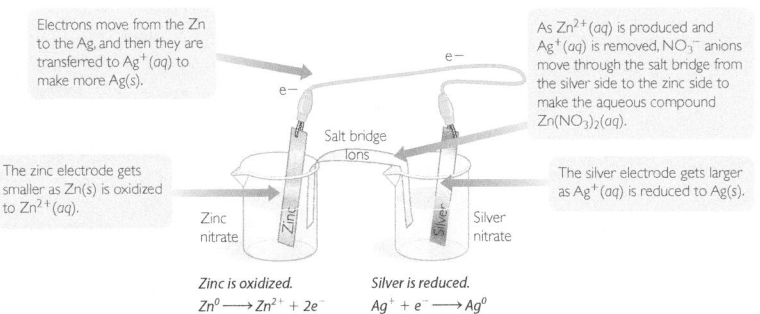

Electrons move from the Zn to the Ag, and then they are transferred to $Ag^+(aq)$ to make more Ag(s).

As $Zn^{2+}(aq)$ is produced and $Ag^+(aq)$ is removed, NO_3^- anions move through the salt bridge from the silver side to the zinc side to make the aqueous compound $Zn(NO_3)_2(aq)$.

The zinc electrode gets smaller as Zn(s) is oxidized to $Zn^{2+}(aq)$.

The silver electrode gets larger as $Ag^+(aq)$ is reduced to Ag(s).

Salt bridge Ions

Zinc nitrate

Silver nitrate

Zinc is oxidized.
$Zn^0 \longrightarrow Zn^{2+} + 2e^-$

Silver is reduced.
$Ag^+ + e^- \longrightarrow Ag^0$

Lesson 112 | **Electrochemical Cell**   569

Ask: What will happen when aluminum metal is mixed with zinc ions in solution? (Al is oxidized) When aluminum metal is mixed with magnesium ions? (no reaction)

TRM **Worksheet with Answers 112**

TRM **Worksheet 112**

INTRODUCE THE LAB

➡ Arrange students into groups of four.

➡ Pass out the student worksheet.

➡ You might show students how to fold a filter paper to make a salt bridge. An accordion fold works well. Address safety issues.

> **SAFETY**
> Remind students to wear safety goggles and use surgical gloves to handle the salt bridge.

TRM **Transparency—Electrochemical Cell 112**

DISCUSS ELECTROCHEMICAL CELLS

➡ Display the transparency Electrochemical Cell

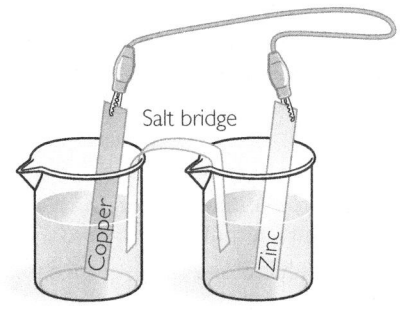

Salt bridge

Copper (II) sulfate   Zinc sulfate

➡ Use the activity series tables on the ChemCatalyst transparency when needed.

Ask: How does an electrochemical cell work?

Ask: How did you prove that the electrical energy was being created by your electrochemical cell? (The light bulb lit up.)

Ask: How can you start and stop this reaction? (Hook it together to start it, unhook it to stop it.)

Activity Series

Oxidation half-reactions	
sodium	$Na \rightarrow Na^+ + e^-$
magnesium	$Mg \rightarrow Mg^{2+} + 2e^-$
aluminum	$Al \rightarrow Al^{3+} + 3e^-$
zinc	$Zn \rightarrow Zn^{2+} + 2e^-$
iron	$Fe \rightarrow Fe^{2+} + 2e^-$
lead	$Pb \rightarrow Pb^{2+} + 2e^-$
copper	$Cu \rightarrow Cu^{2+} + 2e^-$
silver	$Ag \rightarrow Ag^+ + e^-$
gold	$Au \rightarrow Au^+ + e^-$

Sample answers: Students might correctly speculate that the farther apart the metals are on the list, the more energy will be generated by their reaction. You will have to pair a metal above with a cation below. This, combining sodium metal, Na, and gold ions, Au^+, should produce the most energy. However, because gold and silver ions are hard to find and sodium is reactive, it can be easier to pair magnesium and copper.

Ask: What is an oxidation half-reaction? (the electron transfer that occurs when a metal is oxidized)

Ask: What would a reduction half-reaction look like? (e.g., $Na^+ + e^- \rightarrow Na$)

Ask: What would you discover if you weigh the metal strips before and after the procedure? (Cu strip gained mass, Zn strip lost mass.)

Ask: What would happen if you left the reaction hooked up for a long time?

Ask: What was the role of the salt bridge?

Key Points: Whenever there is chemical change, there is also some sort of energy transfer. When there is a transfer of energy, there is always the potential to make use of that energy. Oxidation reactions are no exception. In the case of many oxidation-reduction reactions, the transfer of electrons can be converted to electrical energy. Using chemical reactions to create electrical energy is called electrochemistry.

An electrochemical cell creates electrical energy from chemical energy. In an electrochemical cell, the oxidation reaction is separated from the reduction reaction. This separation allows control over the reaction—the reaction will proceed only when the two halves of the reaction are connected somehow. Separating the two reactions forces the electrons to travel through the connecting wire, producing an electric current. Each half of an electrochemical cell is referred to as a *half-cell*.

DISCUSS DIFFERENT COMBINATIONS OF METALS

Ask: What are some possible metals and metal ions to pair in an electro-chemical cell?

Ask: Which metal should be in ionic form?

Ask: Do you think all the pairings result in the same output of energy?

Ask: Which atoms give up electrons? Which gain electrons?

Key Points: Different pairs of metals will produce different amounts of energy. The activity series can help you figure out which metals to use. The farther apart two metals are on the list, the greater the amount of energy that will be released when they are paired in an electrochemical cell. You can pick any two metals, but the metal that is higher on the list will be oxidized. The amount of energy produced by an electro-chemical cell is also dependent on the concentrations of the ions and the temperature at which the reaction occurs.

The forward reaction of a car battery is used to start the engine, but once the engine is running, energy from the combustion of gasoline is transferred to mechanical energy and then converted to electrical energy by the alternator to reverse the reaction, recharging the battery. When you jump-start a car, you leave the engine running for a while to recharge the battery.

© Tony Freeman/PhotoEdit

You can use the activity series table below to help you predict what will happen in the **electrochemical cell.** Because zinc is higher on the chart than silver, zinc is the more active metal and you can expect it to be oxidized. Likewise, because silver is not a very active metal, you can expect it to be reduced.

$$Zn(s) + 2Ag^+(aq) \longrightarrow Zn^{2+}(aq) + 2Ag(s)$$

If you place $Zn(s)$ and $Ag^+(aq)$ in the same beaker, the reaction will occur with direct transfer of electrons from the $Zn(s)$ to the $Ag^+(aq)$. The reaction just goes until one or both reactants are used up.

Separating the reactants allows you to control the reaction and to create an electrical current. Each half of an electrochemical cell is called a **half-cell.** The reaction in the electrochemical cell can be stopped and started when desired. It will proceed only when the circuit is closed and the two halves of the reaction are connected. The reaction that takes place in each half-cell is called a **half-reaction.**

⟳ Maximizing the Potential

Not every combination of metals and metal salts results in the same output of energy. Generally, the farther away two metals are from each other on the activity series list, the more energy that will be produced by a reaction between them. For example, the energy released when calcium and silver are paired is greater than the energy released when iron and lead are paired.

Activity Series

Metal	Oxidation half-reactions
potassium	$K \longrightarrow K^+(aq) + 1e^-$
calcium	$Ca \longrightarrow Ca^{2+}(aq) + 2e^-$
sodium	$Na \longrightarrow Na^{2+}(aq) + 1e^-$
magnesium	$Mg \longrightarrow Mg^{2+}(aq) + 2e^-$
aluminum	$Al \longrightarrow Al^{3+}(aq) + 3e^-$
zinc	$Zn \longrightarrow Zn^{2+}(aq) + 2e^-$
iron	$Fe \longrightarrow Fe^{2+}(aq) + 2e^-$
tin	$Sn \longrightarrow Sn^{2+}(aq) + 2e^-$
lead	$Pb \longrightarrow Pb^{2+}(aq) + 2e^-$
hydrogen	$H_2 \longrightarrow 2H^+(aq) + 2e^-$
copper	$Cu \longrightarrow Cu^{2+}(aq) + 2e^-$
silver	$Ag \longrightarrow Ag^+(aq) + 1e^-$
mercury	$Hg \longrightarrow Hg^{2+}(aq) + 2e^-$
gold	$Au \longrightarrow Au^+(aq) + 1e^-$

Ca and Ag$^+$: Large energy release

Fe and Pb^{2+}: Small energy release

Activity series	Oxidation half-reactions
Sodium	$Na \rightarrow Na^+ + e^-$
Magnesium	$Mg \rightarrow Mg^{2+} + 2e^-$
Aluminum	$Al \rightarrow Al^{3+} + 3e^-$
Zinc	$Zn \rightarrow Zn^{2+} + 2e^-$
Iron	$Fe \rightarrow Fe^{2+} + 2e^-$
Lead	$Pb \rightarrow Pb^{2+} + 2e^-$
Copper	$Cu \rightarrow Cu^{2+} + 2e^-$
Silver	$Ag \rightarrow Ag^+ + e^-$
Gold	$Au \rightarrow Au^+ + e^-$

Mg and Ag$^+$ Large energy release

Fe and Pb^{2+} Small energy release

Example 1

Magnesium-Tin Cell

Imagine you want to create an electrochemical cell using magnesium, Mg, and tin, Sn.

  a. Which metal would be oxidized and which would be reduced?
  b. What ion will form and what metal will be plated out of solution as a solid?
  c. Write the half-reactions and the net ionic equation.

Solution

Magnesium is higher in the activity series.

  a. Magnesium will be oxidized. Tin will be reduced.
  b. Tin will be plated out onto the tin strip. Magnesium ions, $Mg^{2+}(aq)$, will form.
  c. The half-reactions are $Mg^0 \longrightarrow Mg^{2+} + 2e^-$ and $Sn^{2+} + 2e^- \longrightarrow Sn^0$.
     Adding these, the net reaction is $Mg(s) + Sn^{2+}(aq) \longrightarrow Mg^{2+}(aq) + Sn(s)$.

The concentrations of the reactants, as well as the identities of the metals and metal ions used, can affect the amount of energy that is produced.

VOLTAGE

The energy that comes out of an electrochemical cell is expressed in volts. The **voltage** of a battery lets you know how much energy is "stored" in it, that is, how much potential it has to produce electricity.

Chemists can predict the voltage of an electrochemical cell by examining the two half-reactions involved. The measurement of the voltage of half-cells takes place under uniform conditions. These voltages are normally expressed as *reduction* half-reactions and are listed in a standard reduction potentials table.

Standard Reduction Potentials at 25 °C, 1 atm, and 1 M Ion Concentration

Reduction Reaction	Half-cell Potential
$Li^+ + e^- \longrightarrow Li$	−3.05 volts
$Mg^{2+} + 2e^- \longrightarrow Mg$	−2.37 volts
$Zn^{2+} + 2e^- \longrightarrow Zn$	−0.76 volt
$Ni^{2+} + 2e^- \longrightarrow Ni$	−0.26 volt
$Pb^{2+} + 2e^- \longrightarrow Pb$	−0.13 volt
$Sn^{2+} + 2e^- \longrightarrow Sn$	0.15 volt
$Cu^{2+} + 2e^- \longrightarrow Cu$	0.34 volt
$Ag^+ + e^- \longrightarrow Ag$	0.80 volt
$Au^+ + e^- \longrightarrow Au$	1.69 volts

The overall voltage of an electrochemical cell can be calculated by finding the difference between the potentials of the half-reactions. This means subtracting the voltage of the oxidation process from the voltage of the reduction process.

In the activity series, zinc is higher than copper. This means that it gives up electrons—it is oxidized whenever it is paired with anything below it on the list. The half-reaction equation for the zinc half-cell is

> $Zn \rightarrow Zn^{2+} + 2e^-$
> (oxidation half-reaction)

The copper ions receive electrons when they are paired with anything above copper on the list. Thus, copper cations are reduced, and copper metal forms on the copper side of the cell. The half-reaction equation for the copper half-cell is

> $Cu^{2+} + 2e^- \rightarrow Cu$
> (reduction half-reaction)

DISCUSS VOLTAGE AND BATTERIES

→ You can demonstrate voltage by placing a voltmeter in the copper-zinc circuit. Having a second cell set up will allow you to demonstrate differences in voltage and relate them to the activity series (optional).

Ask: What household items can you name that require a battery?

Ask: Do batteries last forever? Why or why not?

Ask: Are all batteries the same? Explain some differences.

Key Points: Electrochemical cells are used in batteries. The batteries in our automobiles, watches, TV remote controls, flashlights, and portable music players all use the electrochemistry of oxidation-reduction reactions. Batteries provide a way to store chemical energy until it is needed. The chemical reaction within a battery is separated into half-cells and does not take place until the two reactants are connected with a wire to allow electrons to move. If a battery remains connected, the chemical reaction will continue until the battery is dead. There are many types of batteries, but most use metals and metal ions to transfer electrons.

The potential energy of an electro-chemical cell is expressed in volts. The *voltage* on a battery tells you its potential to produce electricity. The potential voltage of half-cells has been measured by chemists under uniform conditions. Normally, these voltages are listed in a table of standard reduction potentials and are shown as reduction half-reactions.

Standard Reduction Potentials at 25 °C, 1 atm, and 1 M Ion Concentration

Reduction Reaction	Half-Cell Potential
$Li^+ + e^- \rightarrow Li$	−3.05 volts
$Mg^{2+} + 2e^- \rightarrow Mg$	−2.37 volts
$Zn^{2+} + 2e^- \rightarrow Zn$	−0.76 volt
$Cu^{2+} + 2e^- \rightarrow Cu$	0.34 volt
$Ag^+ + e^- \rightarrow Ag$	0.80 volt

When an electrochemical cell is generating electricity, the total overall voltage can be calculated by finding the difference between the potentials of the two half-reactions. This means subtracting the voltage of the oxidation process from the voltage of the reduction process. For a copper-magnesium electro-chemical cell, the voltage would be calculated as follows:

> $Cu^{2+} + 2e^- \rightarrow Cu$ 0.34 volt
> (reduction)
> $Mg^{2+} + 2e^- \rightarrow Mg$ −2.37 volts
> (occurs in reverse direction, as oxidation)

Subtract the voltage of the oxidation process from the voltage of the reduction process:

$$0.34 - (-2.37) = 0.34 + 2.37$$
$$= 2.71 \text{ volts}$$

EVALUATE (5 min)

Check-In

Draw an electrochemical cell that uses magnesium and copper. Label the parts. What is being oxidized? What is being reduced? Label each beaker with its half-reaction.

Answer: Magnesium is oxidized. Copper is reduced.

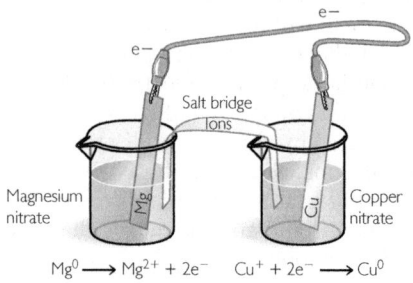

Salt bridge
Ions

e−

Magnesium nitrate

Copper nitrate

$$Mg^0 \longrightarrow Mg^{2+} + 2e^- \qquad Cu^+ + 2e^- \longrightarrow Cu^0$$

Homework

Assign the reading and exercises for Fire Lesson 112 and the Chapter 21 Summary in the student text.

KEY TERMS
electrolyte
electrochemical cell
half-cell
half-reaction
voltage

Example 2

Silver-Zinc Electrochemical Cell

Determine the voltage of a silver-zinc electrochemical cell.

Solution

Write the half-reactions for silver and zinc from the standard reduction potentials table on page 571. The half-reaction that is more positive will proceed as a reduction, and the half-reaction that is more negative will proceed as an oxidation (in the reverse direction).

This half-reaction proceeds as reduction. $Ag^+ + e^- \longrightarrow Ag$ 0.80 volt

This half-reaction proceeds in the opposite direction as oxidation. $Zn^{2+} + 2e^- \longrightarrow Zn$ −0.76 volt

Next, subtract the reduction potential of the oxidation half-reaction from the reduction potential of the reduction half-reaction.

$$0.80 - (-0.76) = 0.80 + 0.76 = +1.56 \text{ volts}$$

This cell has a voltage of +1.56 volts.

⊃ Batteries

Electrochemical cells have been developed into batteries, which are basically containers of chemicals just waiting to be converted to electrical energy. The chemical reaction within a battery does not take place until the two reactions are connected to one another. When you flip the switch on a flashlight, you are connecting the halves of the battery inside. If you forget and leave the flashlight on, the chemical reaction inside will continue until the battery is dead.

Some batteries are rechargeable. This means that once the battery has died, an electric current can be run through it in the opposite direction to push the reverse reaction to occur. This restores the original reactants, and the battery is good once again.

A battery that is not connected to anything still has a voltage, even though it is not yet doing any work. Most of the batteries in your household can produce voltages from about 1.25 volts to 9 volts.

LESSON SUMMARY

How can you use a redox reaction as an energy source?

You can convert the transfer of electrons that occurs during oxidation-reduction into electrical energy. By separating the reactants into half-cells, you can control and "store" electrical energy. The activity series allows you to choose electrodes and electrolytes that will produce electricity. The voltage of an electrochemical cell can be calculated from the half-cell potentials.

Exercises

Reading Questions

1. Explain how to determine the direction of electron transfer in an electrochemical cell.

2. Explain how you might choose which reactants to put together in an electrochemical cell.

Reason and Apply

3. Suppose that you make an electrochemical cell. One beaker has a calcium metal electrode and a solution of aqueous calcium ions, $Ca^{2+}(aq)$. The other beaker has an iron metal electrode and a solution of aqueous iron ions, $Fe^{3+}(aq)$.
   a. Write the half-reactions for each beaker.
   b. Use the activity series table on page 570 to decide which metal is oxidized and which metal cation is reduced.
   c. Make a sketch of the electrochemical cell. Show the reaction that occurs in each beaker.
   d. Write a net ionic equation for the reaction that takes place to generate electricity.

4. Repeat Exercise 3 with the pairs of metals here.
   a. copper and tin
   b. silver and aluminum
   c. copper and lead

5. Predict which combination of metals in Exercises 3 and 4 will result in the greatest voltage. Explain your prediction. Then calculate the voltages of the redox reactions. Was your prediction correct?

6. What are the pros and cons of using combustion reactions as a source of energy?

7. What are the pros and cons of using electrochemistry as a source of energy?

8. PROJECT Research common batteries, such as AA or D cell batteries, to find out the chemistry behind them. List the metals and compounds commonly used in their construction.

9. PROJECT Car batteries are usually 12-volt. Research car batteries and describe how such a high voltage is achieved.

10. PROJECT When car lights are left on for a long time, this can cause a "dead" battery, which will require a jump-start from another car battery. In chemical terms, what does it mean to have a "dead" car battery? What does "jump-starting" a car battery mean in terms of the overall redox reaction? Research and report on your findings.

Lesson 112 | **Electrochemical Cell**   573

LESSON 112 ANSWERS

1. In an electrochemical cell, electrons are transferred from the more active metal to the less active metal.

2. *Possible answer:* Choose reactants to put together in an electrochemical cell by finding metals whose reduction potentials differ by the desired voltage. Place each metal in a solution containing ions of that metal to make a half-cell. When the half-cells are connected, electrons will flow from the half-cell containing the more active metal to the half-cell containing the less active metal.

3. a. $Ca \rightarrow Ca^{2+} + 2e^-$
    $Fe \rightarrow Fe^{3+} + 3e^-$

b. Calcium loses electrons, so it is oxidized, and iron gains electrons, so it is reduced.

c.

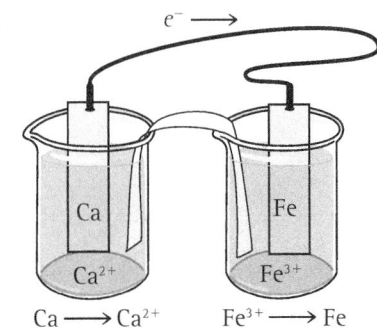

$Ca \longrightarrow Ca^{2+}$ $Fe^{3+} \longrightarrow Fe$

Diagrams should show the separate half-cells, a salt bridge to connect the solutions in the half-cells, and a wire to connect the two electrodes.

d. $3Ca(s) + 2Fe^{3+}(aq) \rightarrow 3Ca^{2+}(aq) + 2Fe(s)$

4. copper and tin:

a. $Cu \rightarrow Cu^{2+} + 2e^-$
    $Sn \rightarrow Sn^{2+} + 2e^-$

b. Tin is oxidized and copper is reduced.

c.

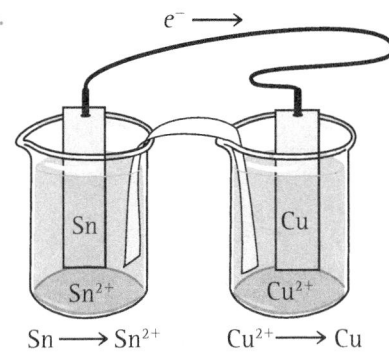

$Sn \longrightarrow Sn^{2+}$ $Cu^{2+} \longrightarrow Cu$

d. $Sn(s) + Cu^{2+}(aq) \rightarrow Sn^{2+}(aq) + Cu(s)$

silver and aluminum:

a. $Ag \rightarrow Ag^+ + e^-$
    $Al \rightarrow Al^{3+} + 3e^-$

b. Aluminum is oxidized and silver is reduced.

c.

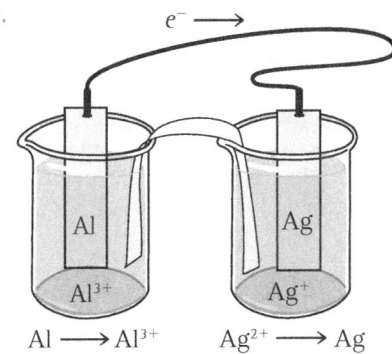

$Al \longrightarrow Al^{3+}$ $Ag^{2+} \longrightarrow Ag$

d. $Al(s) + 3Ag^+(aq) \rightarrow Al^{3+}(aq) + 3Ag(s)$

Copper and lead:

a. $Cu \rightarrow Cu^{2+} + 2e^-$
    $Pb \rightarrow Pb^{2+} + 2e^-$

b. Lead is oxidized and copper is reduced.

c.

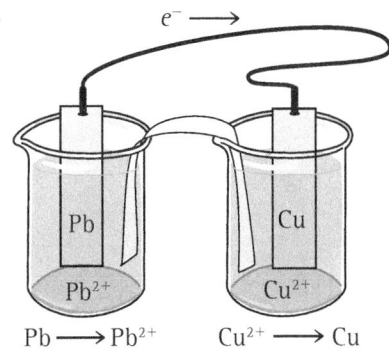

$Pb \longrightarrow Pb^{2+}$ $Cu^{2+} \longrightarrow Cu$

d. $Pb(s) + Cu^{2+}(aq) \rightarrow Pb^{2+}(aq) + Cu(s)$

Answers continue in Answer Appendix (p. ANS-9).

ANSWERS TO CHAPTER 21 EXERCISES

1. *Possible answer:* An electrochemical cell works by separating the two parts of a redox reaction. As electrons are transferred from one metal to the other, they travel through a wire connecting the two half-reactions. The electrical energy is the stream of moving electrons. The chemical energy is the voltage potential of the redox reaction.

2. *Possible answer:* Some metal elements are easier to extract than others because it takes less energy to reduce the metal ore to the pure metal. In other words, the heat of reaction for the reduction reaction is larger for some metals than for other metals.

3. The extraction of metals from the metal oxides table lists the heat of reaction of iron (III) oxide decomposing into pure iron and oxygen gas as +413 kJ/mol. Because this value is positive, the reaction is endothermic. Energy must be continuously added into the system for the reaction to occur.

4. Aluminum is a more active metal than silver, according to the activity series table. If a piece of aluminum foil is placed in a silver nitrate solution, aluminum will be oxidized and silver reduced in a single replacement reaction. If a piece of silver foil is placed in an aluminum nitrate solution, no reaction will occur because the more active metal is already ionized.

5. a. Lead will be oxidized because it is higher on the activities series table than silver. The chemical formulas for lead nitrate and silver nitrate yield the charges on each metal cation. The net ionic equations for each half-reaction are

$Pb \rightarrow Pb^{2+} + 2e^-$

$Ag^+ + e^- \rightarrow Ag$

b. $Pb(s) + 2Ag^+(aq) \rightarrow$
$Pb^{2+}(aq) + 3Ag(s)$

c. Two moles of electrons will flow from the anode to the cathode for each mole of iron that is oxidized.

d. $Ag^+ + e^- \rightarrow Ag$ 0.80 volt (reduction)
$Pb \rightarrow Pb^{2+} + 2e^-$ −0.13 volt (oxidation)
0.80 volt − (−0.13 volt) = 0.93 volt

Project: Alternative Energy

A good project includes: ● a clear description of the energy source and what the origin of the energy within the

source is ● how the energy would be harnessed or extracted ● a comparison of the energy source to traditional energy sources, in terms of cost, efficiency, and environmental impact ● the current state of development for the energy source, including avenues of potential future research and issues that must be overcome for the development of the energy source to occur ● a poster that promotes the energy source in a creative, appealing, and accurate way and includes diagrams and photographs that

describe the energy source and how it works.

CHAPTER 21

Controlling Energy

SUMMARY

KEY TERMS

oxidation

heat of formation

reduction

oxidation-reduction
  reaction (redox)

activity series

electrolyte

electrochemical cell

half-cell

half-reaction

voltage

Fire Update

Every chemical change is accompanied by an energy transfer. The chemical energy from reactions between metals and metal compounds can be converted to electrical energy with the appropriate apparatus.

Oxidation-reduction reactions are a particularly useful source of energy. Oxidation is the loss of electrons by an atom, ion, or molecule during a chemical change. Its counterpart, reduction, is the gaining of electrons by an atom, ion, or molecule during a chemical change. Combustion is one type of oxidation-reduction reaction in which the fuel is oxidized and the oxygen is reduced.

Metals that are easily oxidized give up their electrons easily. These metals are considered more active.

REVIEW EXERCISES

1. Explain how an electrochemical cell works to convert energy from chemical reactions into electrical energy.

2. Why are some metal elements easier to extract from metal ores than others?

3. Based upon the value in the heats of formation table on page 555, determine the heat of reaction for the decomposition of rust into iron and oxygen gas. Is this reaction endothermic or exothermic? Explain.

$$2Fe_2O_3 \longrightarrow 4Fe(s) + 3O_2(g)$$

4. If you place a piece of aluminum foil in a beaker of 1 M $AgNO_3$, would a single exchange reaction occur? What if you place a piece of silver foil in a beaker of 1 M $Al(NO_3)_3$? How can you explain these two different results?

5. An electrochemical cell is set up with 1 M $Pb(NO_3)_2$ on one side, and 1 M $AgNO_3$ on the other.
   a. Which metal will be oxidized? Which will be reduced? Write out balanced half-reactions for these reactions.
   b. Write out the balanced overall reaction for this electrochemical cell.
   c. How many moles of electrons will flow from the anode to the cathode?
   d. Calculate the voltage for this electrochemical cell.

Alternative Energy

 PROJECT Investigate an energy source that is an alternative to fossil fuels. Create a poster promoting the benefits of the energy source and explaining how it works. Write a short paper on the pros and cons of your energy source.

ASSESSMENTS

Two multiple-choice assessments (versions A and B) and two short-answer assessments (versions A and B) are available for download for Chapter 21.

TRM Chapter 21 Assessment Answers

TRM Chapter 21 Assessments

CHAPTER 22

Radiating Energy

In this chapter, you will study

- the emission, reflection, transmission, and absorption of light
- how color arises
- the path of light as a ray, as a wave, and as a particle
- how to detect electromagnetic radiation

Leonard Lessin/Photo Researchers/Getty Images

Bees and humans do not see the patterns on flowers in the same way. The photograph of this flower has been altered to show the way a bee sees light. The tan areas help to lead the bee to the nectar at the flower's center.

F ire produces bright light and transfers energy. When hot gases produced in a fire come into direct contact with objects at lower temperatures, some energy is transferred. In addition, fire and extremely hot objects transfer energy as electromagnetic radiation. Fire transfers energy efficiently as infrared radiation that makes you feel warm. Chapter 22 of Unit 5: Fire focuses on models to describe light energy and its interaction with matter.

575

PD CHAPTER 22 OVERVIEW

Watch the video overview of Chapter 22 (for teachers) by clicking on the link in the TE-book, opening the TRFD, or logging onto the book's companion Web site bcs.whfreeman.com/livingbychemistry2e (teacher log-in required).

The final chapter of Unit 5: Fire focuses on light, more broadly referred to as *electromagnetic radiation*. The dancing colors in a fire are caused by the *emission* of light from gases in the flames, while the warmth of a fire is due to the emission of infrared radiation caused by the combustion reaction that is taking place. In Lesson 113, students use a ray model to describe the path of light and find that white light is a mixture of colors. A wave model of light is introduced in Lesson 114, and students compare the wavelength, frequency, and speed of various colors of light. Lesson 115 introduces electromagnetic *radiation*. Students learn about different types of electromagnetic radiation and relate frequency to energy. A particle model of light is introduced. Lesson 116 provides students with an opportunity to use UV-sensitive paper to detect UV radiation. Students design their own project to measure the amount of UV radiation from various light sources or to test the efficacy of substances in protecting against UV radiation.

In this chapter, students will study

- absorption, transmission, and reflection of light
- ray, wave, and particle models of light
- electromagnetic radiation
- light energy

Chapter 22 Lessons

Overview

ACTIVITY: GROUPS OF 4

Key Question: How can you describe light shining on matter?

KEY IDEAS

Light is a form of energy emitted from sources such as the Sun and light bulbs. The light travels through space undetected until it strikes an object. The object can reflect, transmit, or absorb the light. Your eye intercepts emitted, reflected, and transmitted light that travels toward you. Colored objects reflect or transmit the color you perceive, while the other colors are absorbed. Black objects absorb all colors. The energy from light that is absorbed is transferred to the object, typically as heat. In your eye, the energy is transferred as a signal to your brain.

LEARNING OBJECTIVES

- Use light rays to indicate the path light travels from a source to an object to your eye.
- Describe emission, reflection, transmission, and absorption of light.

FOCUS ON UNDERSTANDING

- While it is possible to see a glow at the light source (e.g., the Sun, a light bulb), the path of traveling light is not visible. In this lesson, students will use a ray model to trace the invisible path of the light.
- Students often confuse emitted light and reflected light. Light sources emit light. We see objects that do not emit light by reflected light. Once the light source is extinguished, these objects that only reflect light are no longer visible.
- Students think of their reflection as being "in the mirror." The perception of a reflection is due to light that "bounces off" their bodies, to the mirror, and back to their eyes.
- Matter does not have color. It reflects colored light to our eyes, and our eyes detect it. We see the colors that are reflected.

KEY TERMS

emission
radiation
light ray
white light
reflection
transmission
transparent
opaque
absorption

Light Energy

THINK ABOUT IT

Light from the Sun travels a great distance through nearly empty space to reach Earth. When the Sun's light shines on objects, those objects can reflect light into your eyes, and you are able to see the objects. Although the light from the Sun is white, the light that is reflected from objects is often colorful. What is the path of the light such that you can see white, clear, and colored objects?

> How can you describe light shining on matter?

To answer this question, you will explore

- Light Emission and Light Rays
- Reflection, Transmission, and Absorption

EXPLORING THE TOPIC

↻ Light Emission and Light Rays

At a very young age, children learn to draw the Sun as a circle with lines radiating outward to indicate that the Sun is glowing. These lines of light represent *light rays,* and they show that light is traveling away from the Sun. Light rays provide a useful way of indicating the path that light travels and how light interacts with matter.

LIGHT SOURCES

Light is emitted from the Sun because the Sun is extremely hot. Nuclear reactions transfer heat to raise the temperature of the surface of the Sun to nearly 6000 K. The Sun transfers energy to Earth by the **emission** of light.

Other hot objects, such as the thin wire inside a light bulb, an electric burner on a stove, and burning wood, transfer energy as light. Creating a light source requires transfer of energy to the source. Thermal energy, electricity, exothermic chemical reactions, and striking a substance to make sparks can all generate light.

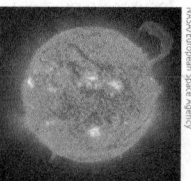  

Extremely hot objects emit light.

IN CLASS

Students consider the path of traveling light from a flashlight that creates a bright white circle on a sheet of paper. The teacher makes the path of the light visible with a white powder and then shows that white light is a mixture of colors. Students examine *reflection* of light from smooth and rough opaque surfaces, and then they explore transmission and absorption of light with a flashlight and colored plastic that is transparent. They use a ray model to sketch the path of light from the Sun to an object to their eyes. A complete materials list is available for download.

TRM Materials List 113

Differentiate

There is much vocabulary in this lesson, and students may benefit from using graphic organizers targeted at their literacy level. Design the graphic organizer to help students to connect ideas and link concepts to prior learning. Depending on the needs of your students, the graphic organizer could be used during the student activity, during whole-class discussion, or with the student textbook for Lesson 113.

SETUP

For the student activity, prepare materials to show light shining through a prism, as well as a path of light you

As you learned in previous lessons in this unit, hot substances transfer energy to reach thermal equilibrium. So far, you have considered conduction, which relies on a hot object being in direct contact with a colder object. **Radiation** refers to energy that travels outward from a hot substance even when the substance is isolated. Visible light is one form of radiation.

> **Big Idea** One way that hot substances transfer energy is by the emission of light, called radiation.

RAY MODEL OF LIGHT

Generally, it is not possible to see the path of light from a source such as the Sun. The path of light becomes visible as a glowing sunbeam only if there are dust particles or water droplets in the air. The path of light is commonly referred to as a **light ray.** Light rays do not describe the complexities of light, but they do provide a simple model with which to track the path of light as it shines from the source onto an object and then to your eye.

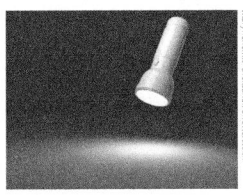
Light from this flashlight produces a bright spot on a surface. Notice that while the light is visible at the flashlight, the path of light is invisible between the flashlight and the surface. The light becomes visible when it strikes the surface.

WHITE LIGHT IS A MIXTURE OF COLORS

In the late 1600s, Sir Isaac Newton showed that white light produces a rainbow of colors after passing through a piece of glass in the shape of a triangular prism. At the time, other hypotheses suggested that the prism might somehow distort the light or make it less pure. However, when Newton placed a second prism in the path of the rainbow of light created by the first prism, he observed a single beam of white light. This experiment convinced Newton that the colors are already present in the light.

Sunbeams are visible through the trees when there is a mist of water in the air. Small particles show the path of the light.

> **GEOGRAPHY CONNECTION**
>
> Normally, we think of lightning as accompanying a thunderstorm. But volcanoes spew lots of dusty matter, which forms a plume of highly charged particles much like the droplets in a storm cloud. The clouds of particles from erupting volcanoes can unleash spectacular lightning bolts, like this one in Indonesia.
>
>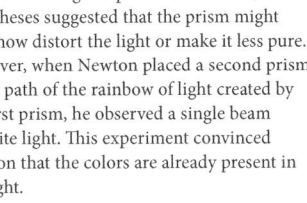

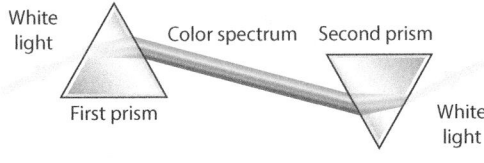

When white light moves through a prism, it spreads out into colors in a sequence: red, orange, yellow, green, blue, and purple. A second prism can recombine the colors back into white light.

---

make visible by using chalk dust or flour. You can find thin sheets of colored plastic in art supply stores. You can also use clear plastic and color the plastic with markers. Squares that are 2 in × 2 in are a sufficient size. It is preferable to have 3 to 4 different colors for each group to examine. A small inexpensive flashlight will suffice.

LESSON GUIDE

TRM PowerPoint Presentation 113

ChemCatalyst

How does light travel to your eye? Draw a picture that shows the path of light from a light bulb to a book on a table and then to your eye.

Sample answers: Student drawings will vary. Students might draw straight lines coming out of the light bulb in all directions. One or more lines will point toward the book. Students might then show lines reflecting off the book and into their eyes.

Ask: How did you draw the path of the light from the light bulb to the book and to your eye?

Ask: In what direction is the light emitted from the light bulb? (all directions) How do you know?

Ask: Why does turning on a light allow you to see an object? (Light strikes the object to make it visible.)

Ask: How do you know that the object is not generating its own visible light? (It is not visible in a dark room.)

EXPLORE (15 min)

TRM Worksheet with Answers 113

TRM Worksheet 113

INTRODUCE THE ACTIVITY

➡ Tell students to work in groups of four.

➡ Pass out the worksheet to each group.

GUIDE THE ACTIVITY

➡ **Make light visible demo:** Turn on a flashlight. Turn out the room lights. Make the path of the light from the flashlight visible with chalk dust or a small amount of flour tossed into the light path.

➡ **Rainbow Demo:** Shine a flashlight through a prism to make a rainbow of colors on a piece of white paper. If you do not have a prism, fill a shallow pan with water. Place a mirror at an angle in the water. Shine a flashlight onto a part of the mirror that is underwater. Hold the white paper above the mirror. Move the paper around until you see a rainbow appear.

EXPLAIN & ELABORATE (15 min)

DISCUSS LIGHT EMISSION AND LIGHT RAYS

Ask: What are some sources of light?

Ask: What do you have to do to make light?

Ask: What evidence do you have that light is a form of energy?

Ask: What is the ray model of light?

Ask: How can you show that white light is a mixture of colors?

Key Points: Light is emitted from a light source, such as the Sun or an electric light bulb. Nuclear reactions on the Sun transfer an enormous amount of energy to raise the temperature of the Sun to nearly 6000 K. The thermal energy is transferred through space to Earth as light. Other hot objects transfer energy as light, such as the thin piece of tungsten wire in a light bulb. The tungsten wire gets very hot when electricity is passed through the wire. The hot wire transfers energy as light.

It is not possible to see light as it travels through space. It is only possible to detect light when it strikes an object. Even sunbeams are visible only when the light hits dust particles in the air. The path of light is commonly called a *light ray*. It is typical to use straight lines to indicate the path of the light. Even young children are taught to draw lines coming from the Sun to show that it is glowing.

White light is a mixture of colors. When white light passes through a prism or a water droplet in the sky, the white light spreads out into a rainbow of colors. The colors are always in the same sequence: red, orange, yellow, green, blue, and violet. The acronym ROYGBV (pronounced "roy-g-biv") is useful for remembering the sequence. In a rainbow, red is on the top. Before Sir Isaac Newton first passed light through a prism, in 1670–1672, people believed that light was colorless. Newton's experiments demonstrated that the colors are already present in the light.

DISCUSS REFLECTION, TRANSMISSION, AND ABSORPTION OF LIGHT

➜ You can have students draw sketches of light interacting with matter to assist with the discussion.

Ask: What is the difference between emission and reflection of light?

Ask: Can you use light rays to show why sand is white and glass is clear?

Ask: What happens to white light when it shines on a black road?

Ask: How is it possible that you see a green color when white light shines on grass?

Key Points: Most objects reflect some light. When light shines on an object, some or all of the light is reflected. When light reflects from a smooth surface, all the light rays are parallel to one another. When light reflects from a rough surface, the light rays go off in many different directions. See the illustration on page 578 of the student text. You can distinguish between a light source (emission of light) and an object that reflects light in a room that is pitch black. Objects that reflect light are not visible in the pitch-black room.

Some objects transmit light. Light passes through certain types of objects such as glass, thin fabric, and thin sheets of plastic. See the illustration on page 578 of the student text. These objects are *transparent*. Very little light is reflected during

Reflection, Transmission, and Absorption

In a room that is dark and sealed off so that no light leaks in, you see nothing. Turn on the lights, and objects in the room become visible. The once invisible objects are now illuminated. Moreover, many of them are colorful even under light that appears white at the source, referred to as **white light**. What are the ways that white light interacts with matter to cause these perceptions of object's visibility and color?

OPAQUE OBJECTS REFLECT LIGHT

When light shines on an object, some or all of the light is reflected. **Reflection** of light refers to light that "bounces" off a surface. When light from one source reflects from a smooth surface, all the light rays are parallel to one another. When light reflects from a rough surface, the light rays go off in many different directions. It is typical that you can see a reflection on a smooth surface such as a mirror. You can also see a faint image of a reflection in a window. Rough surfaces do not create as clear an image because the light goes off in multiple directions.

Important to Know Only light sources emit, or give off, their own light. Most objects only reflect light and, therefore, are not visible in a pitch-black room.

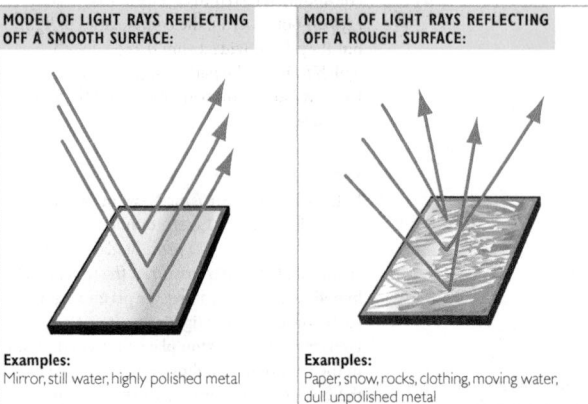

MODEL OF LIGHT RAYS REFLECTING OFF A SMOOTH SURFACE:

Examples:
Mirror, still water, highly polished metal

MODEL OF LIGHT RAYS REFLECTING OFF A ROUGH SURFACE:

Examples:
Paper, snow, rocks, clothing, moving water, dull unpolished metal

CLEAR OBJECTS TRANSMIT LIGHT

Light passes through certain types of objects, such as glass, thin fabric, and thin sheets of plastic, by **transmission**. Objects that transmit light are **transparent**. Very little light is reflected during transmission. However, most objects do not transmit light, and are referred to as **opaque**. It is not possible to see through opaque objects.

Some light moves through objects that are transparent, such as sunglasses.

matteeock/ Getty Images

transmission. However, most objects do not transmit light, and they are referred to as *opaque*. It is not possible to see through opaque objects.

Colored objects partially absorb light. When white light shines on an object that appears green, it is because green light is reflected and the other colors are absorbed. See the illustration on page 579 of the student text. When white light shines on a black object, all of the light is absorbed. Light that is absorbed no longer exists as light. The light energy is transferred to the object, which can result in an increase in the temperature of the object. This

is especially evident if you walk on a black road on a summer day. The road surface has become hot.

EVALUATE (5 min)

Check-In

Consider a pair of sunglasses that are tinted green.

1. Draw a picture to show what happens when sunlight shines on the sunglasses.

2. What color of light is transmitted?

3. What color of light is absorbed?

COLORED OBJECTS PARTIALLY ABSORB LIGHT

Consider green leaves. You cannot see through leaves, so light is not transmitted. If our eyes detect a green color, then it must mean that rays of green light reflect from the leaves and enter our eyes. How is it possible that you detect green when the leaves are illuminated with white light?

For green leaves, colors other than green are taken in, or absorbed, by the leaves while green light rays are reflected. This process of taking in certain colors of light is called light **absorption.** The light that is absorbed transfers energy to the object.

White light, which is a mixture of colors, shines on a leaf. The leaf reflects green light rays. The other colors are absorbed by the leaf.

Big Idea An object is colored because it reflects or transmits certain colors of light (these are the colors you see), while it absorbs other colors of light (these light rays transfer energy back to the object).

Example

Yellow Streetlamps and Car Colors

Imagine that you are walking down a street at night. The streetlamps emit yellow light.
   a. Explain why a blue car appears black.
   b. Explain why a yellow car appears yellow.

Solution

   a. A blue car absorbs all colors of light except for blue light. The streetlamps emit yellow light, which is absorbed by the blue car. So, no light is reflected into your eyes, and the car appears black.
   b. A yellow car absorbs all colors of light except for yellow light. The streetlamps emit yellow light, which is reflected by the yellow car. So, the car appears yellow as it would during the day in sunlight.

KEY TERMS

emission
radiation
light ray
white light
reflection
transmission
transparent
opaque
absorption

LESSON SUMMARY

How can you describe light shining on matter?

Light is a form of energy emitted from sources such as the Sun and light bulbs. The light travels through space undetected until it strikes an object. The object can reflect, transmit, or absorb the light. Your eye detects reflected and

Answers: **1.** The picture should show that light is transmitted and absorbed. **2.** Green light is transmitted. **3.** Colors other than green are absorbed.

Homework

Assign the reading and exercises for Fire Lesson 113 in the student text.

LESSON 113 ANSWERS

1. Light shining on matter raises its temperature.

2. The color we perceive is the color of light that is reflected off an object, or transmitted through an object. For example, when white light shines on a green, opaque object, only green light is reflected. When white light shines through a piece of green colored glass, only green light is transmitted.

3. Blue

4. Blue

5. Green

6. No color or black

7. a. Orange. **b.** Black

8. The dust particles reflect light in all directions. The ray-model drawing should show light from a light source, such as the Sun reflecting off the dust particles and into the eye.

transmitted light that travels toward you. A colored object reflects or transmits the color or colors you perceive, while it absorbs the other colors. Black objects absorb all colors. The energy from light that is absorbed by an object is transferred to the object.

Exercises

Reading Questions

1. How could you provide evidence that light is a form of energy?

2. Provide an explanation as to why we see colors.

Reason and Apply

3. What color will you see if you shine white light through a piece of thin blue plastic?

4. What color will you see if you shine blue light through a piece of thin blue plastic?

5. What color will you see if you shine white light from a flashlight onto a green car at night?

6. What color will you see if yellow light from a streetlamp shines on a red car at night?

7. Imagine that you are wearing a pair of orange-tinted sunglasses.
   a. What color will you see when you look at an orange object? Explain your thinking.
   b. What color will you see when you look at a blue object? Explain your thinking.

8. Explain how it is possible to see a sunbeam in a room that has dust in the air. Draw a ray model to show how the light path travels from its source, to the dust, and then to your eye.

Now You See

Light Waves

THINK ABOUT IT

Light from the Sun travels a great distance through nearly empty space to reach Earth. It is hard to imagine how the light gets from one place to another. How does it travel and with what speed?

How is a wave model used to describe light?

To answer this question, you will explore

⟳ The Wave Model of Light

⟳ Wavelength, Frequency, and Speed

EXPLORING THE TOPIC

⟳ **The Wave Model of Light**

The ray model of light indicates the light path, but it does not reveal much about what light is and how it travels through space. The Greek philosopher Aristotle, born in 384 B.C.E., thought of light as wavelike in nature, traveling through space similar to the way ripples spread across a pond's surface. Even though this view has undergone modification and elaboration through the centuries, it remains a useful model.

WAVES

Waves are created by a disturbance. For example, throw a pebble in a pond, and water waves spread out in circles of up-and-down motion. Or pluck a guitar string to make the entire string move back and forth.

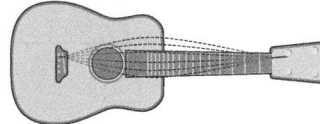

Throw a pebble in a pond.   *Pluck a guitar string.*

You can also place a piece of rope on the floor and move one end back and forth while another person holds the other end fixed to create a single pulse of back-and-forth motion. This pulse travels along the rope. If you move one end

---

FOCUS ON UNDERSTANDING

Students may wonder what makes "waves" when light travels. It is challenging even for scientists to imagine a physical reality for waves of light. Let students know that this model provides a way to distinguish among different colors of light.

KEY TERMS

light wave
wavelength
frequency
speed of light

IN CLASS

Students consider how to apply a wave model of light. First, they explore creating waves along a rope and observe how the speed of the back-and-forth motion on one end of the rope changes the dimensions of the wave. Next, students examine the wavelength and frequency of waves of colored light and find out that frequency and wavelength vary inversely to one another. Finally, students learn that all light waves travel with the same speed and that wavelength $\times$ frequency = speed of light. A complete materials list is available for download.

TRM Materials List 114

Differentiate

Spectroscopy is a tool chemists and other scientists use to explore properties of matter. To help teach students about potential careers in STEM fields, assign groups to research various spectrometers (e.g., infrared, UV/Vis, NMR), as well as the scientists who use them, becoming "class experts" for their assigned instrument. Students can create posters, Wiki-pages, PowerPoint presentations, or other visual aids and then jigsaw with other expert groups to expand their learning.

Pre-AP® Course Tip

Students preparing to take an AP® chemistry course or college chemistry should be able to use representations and models to make predictions. In this lesson, students learn how light behaves as a wave using a rope as a way of modeling the wave nature of light. From their model, they learn the interrelationship between wavelength and frequency and how to calculate one or the other mathematically.

Overview

ACTIVITY: GROUPS OF 4

Key Question: How is a wave model used to describe light?

KEY IDEAS

It is useful to characterize different colors of light with a wave model. This model treats light as a wave, or oscillation, that travels through space. Light waves have a wavelength, λ (lambda), which is the distance between the peaks of the wave. The number of waves that pass by per second is called the *frequency, f*. Wavelength is typically reported in meters, m, and frequency is reported in hertz, Hz = 1/s. While different colors of light have different wavelengths and frequencies, all colors of light travel at the same speed in a vacuum, called the *speed of light, c*. A relationship exists between the wavelength, frequency, and speed of light waves: $\lambda \times f = c$, where $c = 3 \times 10^8$ m/s. Assigning a wavelength to different colors of light provides a precise continuous scale for all the different colors of the rainbow.

LEARNING OBJECTIVES

● Use a wave model of light to distinguish light of different colors.
● Relate wavelength, frequency, and speed of light waves.

SETUP

Provide a piece of rope to each group of four students.

Lesson Guide

TRM PowerPoint Presentation 114

ChemCatalyst

Imagine that the colors of the rainbow are given numbers with red = 1, orange = 2, and green = 4.

● What number would you give to blue?
● What number would you give to yellow?
● What number would you give to red-orange?

Sample answers: **1.** Blue = 5.
2. Yellow = 3. **3.** Red-orange = 1.5.

Ask: What numbers did you assign to blue, yellow, and red-orange? Explain your choices.

Ask: Point out objects in the room that you would call red. Do they all have the identical color?

Ask: Why might it be useful to have a number scale to distinguish colors between red and orange?

TRM Worksheet with Answers 114

TRM Worksheet 114

INTRODUCE THE LAB

➜ Tell students to work in groups of four.

➜ Pass out the worksheet to each group.

DISCUSS THE WAVE MODEL OF LIGHT

➜ You might have students draw sketches of what they observed with the rope demo.

Ask: What did you observe with the rope activity?

Ask: Describe what is meant by the wavelength and frequency of a wave.

Ask: What happens to the wavelength and frequency of the wave moving along the rope if you move one end of the rope back and forth faster?

back and forth in a continuous motion, the entire rope will move with a wave motion. Certain parts of the rope move left, while other parts move right.

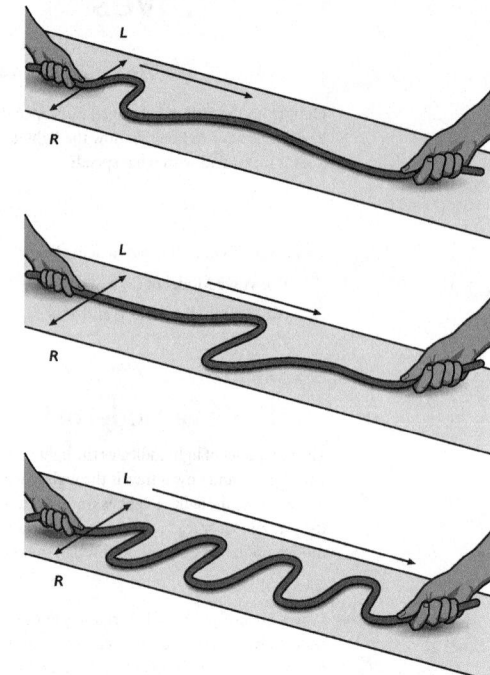

The pulse that you create along the rope models the movement of a wave. As seen here from above, the rope moves first to the left and then to the right.

Important to Know Light waves differ from other waves in one important way: they can carry energy through empty space. All other waves can only travel through matter.

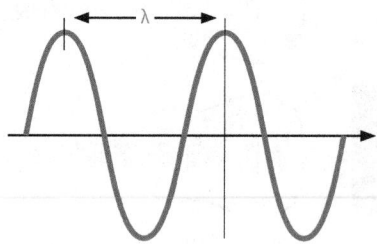

The wave shown is traveling from left to right. The wavelength, λ, is the distance between two peaks. The frequency, f, is the number of full waves (from peak to peak) that pass by the dashed line in a given amount of time.

Notice that when a disturbance is created on one end of the rope, a person watching from the other end can detect the disturbance once the wave moves along the rope. Even from a large distance, it is possible to detect a disturbance as a wave reaches you and transfers its energy.

Light waves are thought to be ripples in space created by forces that oscillate up and down at a source, such as the Sun. The oscillations move as a wave over extremely long distances to Earth.

WAVE MEASUREMENTS

Two important measurements of waves are wavelength, λ, and frequency, *f*. The **wavelength** is the distance between two peaks or between two troughs of the wave measured in

Ask: How do the wavelength and frequency vary from red to violet light?

Ask: What does the wave model of light indicate about the differences between red and yellow light?

Key Points: It is useful to characterize different colors of light with a wave model. This model treats light as a wave, or oscillation, that travels through space. The rope provides an example of a wave. If you move one end of the rope back and forth one time, a pulse of motion moves along the rope. If you move one end of the rope back and forth continuously,

there is a back-and-forth motion all along the rope. This motion is called a *wave*. One advantage of the wave model over the ray model of light is that the wave model allows you to distinguish light of different colors: The dimensions of the wave are distinct for each color of light.

Light waves are characterized by a wavelength and frequency. Light waves are characterized by a wavelength, λ (lambda), which is the distance between the peaks of the wave. The number of waves that pass by per second is called the *frequency, f*. Wavelength is typically reported in meters, m. The **frequency** is the measure of the number of waves that pass by a certain point in space per second, measured in hertz, Hz.

The wave below has different dimensions. Notice that the wavelength is shorter compared with the wave above. It also has more up-and-down motion. If you pick a place along the line and count waves that pass by, you will notice more waves per second for the wave with the shorter wavelength.

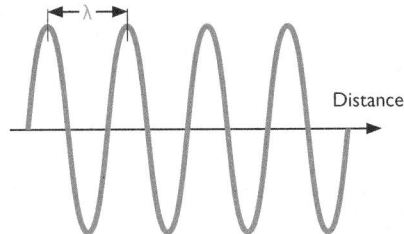

Distance

This wave has a shorter wavelength, λ, and higher frequency, f.

Notice that as the wavelength decreases, the frequency increases. And as the wavelength increases, the frequency decreases.

Wavelength, Frequency, and Speed

The bar provides the sequence of colors of the rainbow from red to violet. There are names for many of the colors, but not all of them. The six main categories are red, orange, yellow, green, blue, and violet. We tend to use labels such as red-orange and blue-green to refer to the colors in between the six main categories.

Lowest frequency	Highest frequency

Longest wavelength	Shortest wavelength

The wave model of light is useful in considering the differences and similarities of light of different colors. What are the dimensions of the light waves of various colors?

WAVES OF COLORED LIGHT

The light waves of different colors have characteristic wavelengths and frequencies. The table below gives the wavelength, frequency, and **speed of light** in a vacuum for the waves of five different colors. Take a moment to notice the trends.

Wavelengths and Frequencies of Colored Light

Color of light	Wavelength, λ	Frequency, f	Speed of light, c
deep red	7.5×10^{-7} m	4.0×10^{14} Hz	3×10^8 m/s
red-orange	6.5×10^{-7} m	4.6×10^{14} Hz	3×10^8 m/s
yellow-green	5.5×10^{-7} m	5.5×10^{14} Hz	3×10^8 m/s
blue-violet	4.5×10^{-7} m	6.7×10^{14} Hz	3×10^8 m/s
deep violet	3.9×10^{-7} m	7.7×10^{14} Hz	3×10^8 m/s

Key Points: The wavelengths of colored light are tiny. The wavelengths of visible light measured in meters, m, range from about 7.5×10^{-7} m for red light to 3.9×10^{-7} m for violet light. This dimension on the order of 10^{-7} m, or 0.0000001 m, is only hundreds of nanometers, which is about the size of a virus, one of the smallest things that can be seen in an optical microscope. Notice that the wavelength decreases from red to violet.

The frequencies of colored light are very large. The frequencies are measured in inverse seconds, 1/s or s⁻¹, called *hertz*, Hz, and range from 4.0×10^{14} Hz for red light to 7.7×10^{14} Hz for violet light. This dimension on the order of 10^{14} Hz is huge. This is 100 trillion or 100,000,000,000,000 oscillations per second. Notice that the frequency increases from red to violet.

All colors of light travel with the same speed in a vacuum. A relationship exists between the wavelength, frequency, and speed of light waves: $\lambda \times f = c$, where $c = 3 \times 10^8$ m/s. This speed is extremely fast. It is not possible to cause visible objects to move this fast.

EVALUATE (5 min)

Check-In

Imagine a wave of light with the wavelength of 5.0×10^{-7} m.

1. What is the frequency of this light wave?

2. Use the table from the worksheet to determine the approximate color of the light.

Answers: **1.** 6.0×10^{14} Hz. **2.** Blue-green.

Homework

Assign the reading and exercises for Fire Lesson 114 in the student text.

meters, m, and frequency is reported in hertz, Hz, or inverse seconds, 1/s. See the diagram on page 582 of the student edition.

Different colors of light have different wavelengths and frequencies. Thus, the wave model creates a numerical scale that identifies different colors. It is no longer necessary to group colors into only six categories (red, orange, yellow, green, blue, and violet). By assigning a number, it is possible to uniquely identify all colors in these six groups on a continuous scale.

DISCUSS THE RELATIONSHIP BETWEEN WAVELENGTH, FREQUENCY, AND SPEED OF LIGHT WAVES

Ask: What are typical values with units for the wavelength and frequency of colored light? Are these big or small numbers?

Ask: What is meant by the speed of light?

Ask: How do wavelength, frequency, and speed of colored light vary?

Ask: How are wavelength, frequency, and speed related to one another?

Unlike light, sound does not travel through empty space—only through gases, liquids, and solids. This is because sound causes wavelike motion of the atoms and molecules in these substances. The speed of sound waves is much slower than light waves, and vehicles, like this aircraft, can reach the speed of sound.

The wavelengths of visible light measured in meters, m, range from about 7.5×10^{-7} m for deep red light to 3.9×10^{-7} m for deep violet light. This dimension on the order of 10^{-7} m, or 0.0000001 m, is about the size of a large virus, one of the smallest things that can be seen using an optical microscope. Notice that the wavelength decreases from red to violet.

The frequencies, measured in inverse seconds, 1/s or s^{-1} or hertz, Hz, range from 4.0×10^{14} Hz for deep red light to 7.7×10^{14} Hz for deep violet light. This dimension on the order of 10^{14} Hz is huge. It is 100 trillion or 100,000,000,000,000 up-and-down motions, or oscillations, per second. Notice that the frequency increases from red to violet.

In empty space, all colors of light travel with the same speed, called the speed of light, $c = 3 \times 10^8$ m/s. The distance of 10^8 m is 100 million meters, or 100,000,000 meters. If you could travel at the speed of light, you would be able to make seven-and-a-half trips around Earth's equator in 1 second. This speed is extremely fast. It is not possible to cause visible objects to move at this speed!

Big Idea   **All colors of light travel with the same speed in a vacuum, called the speed of light.**

A relationship exists among the wavelength, frequency, and speed of light waves:

$$\lambda \times f = c = 3 \times 10^8 \text{ m/s}$$

If you know the wavelength, you can calculate the frequency of colored light. And if you know the frequency, you can calculate the wavelength of colored light. The speed of light is the proportionality constant: frequency and wavelength are inversely proportional.

$$\text{Wavelength} = \lambda = \frac{c}{f} \qquad \text{Frequency} = f = \frac{c}{\lambda}$$

Example

Wavelength of Colored Light

Imagine a light wave with the frequency of 7.5×10^{14} Hz.

  a. What is the wavelength of this light wave?

  b. Use the table on page 583 to determine the approximate color of the light.

Solution

  a. The wavelength and frequency of colored light are related by the equation:

$$\lambda \times f = c$$

Rearrange the equation to solve for wavelength:

$$\text{Wavelength} = \lambda = \frac{c}{f}$$

Substitute values for the frequency and speed of light:

$$\text{Wavelength} = \lambda = \frac{3 \times 10^8 \text{ m}}{7.5 \times 10^{14} \text{ Hz}} = 4.0 \times 10^{-7} \text{ m}$$

  b. The wavelength and frequency indicate that the color is at the violet end of the rainbow.

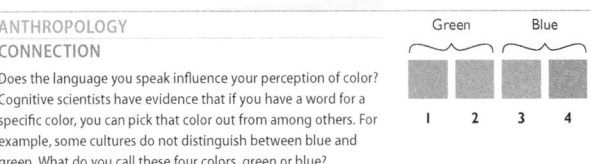

ANTHROPOLOGY
CONNECTION

Does the language you speak influence your perception of color? Cognitive scientists have evidence that if you have a word for a specific color, you can pick that color out from among others. For example, some cultures do not distinguish between blue and green. What do you call these four colors, green or blue?

LESSON SUMMARY

How is a wave model used to describe light?

KEY TERMS

light wave

wavelength

frequency

speed of light

It is useful to characterize different colors of light with a wave model. This model treats light as a wave that travels through space. Light waves have a wavelength, λ (pronounced *lambda*), which is the distance between the peaks of the waves. The number of waves that pass by per second is called the frequency, *f*. Wavelength is typically reported in meters, m, and frequency is reported in hertz, Hz = 1/s. Different colors of light have different wavelengths and frequencies. However, all colors of light travel at the same speed, called the speed of light, *c*. A relationship exists among the wavelength, frequency, and speed of light waves: $\lambda \times f = c$, where $c = 3 \times 10^8$ m/s. Assigning a wavelength to different colors of light provides a precise continuous scale for all the different colors of the rainbow.

Exercises

Reading Questions

1. Explain how to determine the wavelength, λ, of a wave. What would you need to know?
2. Explain how to determine the frequency, *f*, of a wave. What would you need to know?

Reason and Apply

3. Draw sketches of two waves to show how green light and blue light differ from one another.
   a. What are the approximate wavelengths of these two waves?
   b. What are the approximate frequencies of these two waves?
   c. How fast are these waves traveling?
4. Imagine a wave of light with the frequency of 4.1×10^{14} Hz.
   a. What is the wavelength of this light wave?
   b. Use the table on page 583 to determine the approximate color of the light.
5. Imagine a wave of light with the wavelength of 5.1×10^{-7} m.
   a. What is the frequency of this light wave?
   b. Use the table on page 583 to determine the approximate color of the light.
6. Why might it be useful to describe colors in terms of wavelength rather than simply using words?

Lesson 114 | **Light Waves** 585

LESSON 114 ANSWERS

1. $\lambda = c/f$; You have to know the frequency of the light, *f*, and the speed of light, *c*

2. $f = c/\lambda$; You have to know the wavelength of the light, λ, and the speed of light, *c*

3. The sketches should show that green light has a longer wavelength compared with blue light.
a. *Sample answers:* λ(green) = 5.1×10^{-7} m and λ(blue) = 4.7×10^{-7} m.
b. *Sample answers:* *f*(green) = 5.9×10^{14} Hz and *f*(blue) = 6.4×10^{14} Hz.
c. Both waves are traveling at the speed of light, $c = 3 \times 10^8$ m/s.

4. a. 7.3×10^{-7} m b. Red
5. a. 5.9×10^{14} Hz b. Green
6. Instead of saying simply red-orange, which might be more on the red side or more on the orange side, each color between red and orange can be described precisely with a number (the wavelength).

Lesson 114 | **Light Waves** 585

Overview

CLASSWORK: GROUPS OF 4

Key Question: What is the evidence for light waves that are invisible to your eyes?

KEY IDEAS

The human eye is capable of detecting visible light. But it turns out that visible light is only a small portion of an extensive collection of types of light referred to as *electromagnetic radiation.* The wavelength regions include radio waves, microwave radiation, infrared radiation, visible light, ultraviolet radiation, x-rays, and gamma rays. The entire collection is related by the speed at which the waves travel in a vacuum, the speed of light, $c = 3 \times 10^8$ m/s. However, the frequencies and wavelengths vary greatly, by many orders of magnitude. The energy transferred by electromagnetic radiation is proportional to the frequency of the wave. Electromagnetic waves of high frequency transfer more energy than waves of low frequency. High frequency electromagnetic waves such as gamma rays and x-rays are quite dangerous because of the large amount of energy transferred by even a dim beam.

LEARNING OBJECTIVES

- Describe electromagnetic radiation.
- Relate wavelength and frequency of light waves to energy transfer.

FOCUS ON UNDERSTANDING

- Students may not be familiar with the word *radiation,* which derives from the verb *to radiate,* or to give off.
- Point out that the names of various wavelength regions use a variety of words, including *ray, radiation,* and *wave* (e.g., x-ray, UV radiation, and microwave).
- Electromagnetic waves of higher frequency transfer more energy per a single up-and-down motion of the wave, called a *photon,* than those of lower frequency.
- Having more photons in a beam of bright light transfers more total energy than having fewer photons in a beam of dim light.

KEY TERMS

electromagnetic radiation
radio wave
microwave radiation
infrared radiation
visible light
ultraviolet radiation
x-ray

gamma ray
Planck's constant
photon model

IN CLASS

Students consider if there might be light waves outside the range of wavelengths detected by the human eye. They examine evidence such as sunburn and the difference between what bee eyes and human eyes can detect. On the worksheet, students consider a chart showing a collection of waves with a wide range of wavelengths, called *electromagnetic radiation.* These waves have an important feature in common: They all travel at the speed of light, $c = 3 \times 10^8$ m/s, in a vacuum. Next, students learn that the energy of electromagnetic waves is proportional to the frequency of the wave and inversely proportional to wavelength. Thus, high-frequency waves, such as gamma rays and x-rays, are much more harmful compared with low-frequency waves, such as radio waves and microwaves, because high-frequency waves transfer more energy. Finally, in the discussion, a photon model of light is introduced. A complete materials list is available for download.

TRM Materials List 115

Beyond What You See

Electromagnetic Radiation

THINK ABOUT IT

On a dark night, it is very difficult to find your way without bumping into objects that you cannot see. Put on a pair of night vision goggles and you can detect objects fairly easily. What do the goggles detect so that you can see at night?

> What is the evidence for light waves that are invisible to your eyes?

To answer this question, you will explore

- Electromagnetic Radiation
- Light Energy

Night vision goggles detect light that you cannot see. They convert it into light that you can see so that you can locate objects in the dark.

BIOLOGY CONNECTION

Unlike human beings, bees can see ultraviolet light. Bee vision is shifted toward the blue range, and they do not see red. Flowers have adapted to guide bees toward pollen and nectar. The center of this creeping zinnia flower is an unmistakable target for a bee, which sees two-toned petals that humans do not see. As part of its reproduction process, the zinnia advertises that it has pollen for bees to collect for the hive and deposit in other flowers.

EXPLORING THE TOPIC

➔ Electromagnetic Radiation

The human eye is able to detect a range of light waves with wavelengths from about 7.5×10^{-7} m for red light to 3.9×10^{-7} m for violet light. You might wonder if these are the only light waves that exist or if there are light waves with longer and shorter wavelengths outside your range of vision. The example of night vision goggles indicates that there is light even in the dark. However, human eyes cannot detect this type of light without the assistance of technology.

A LARGE COLLECTION OF WAVES

The light that we see is a small part of a large collection of waves called **electromagnetic radiation.** As indicated in the figure, electromagnetic radiation exists with a huge range of wavelengths and frequencies. Its frequency ranges from a low frequency of less than 10^5 Hz to a high frequency of greater than 10^{20} Hz. Its wavelength ranges from longer than 10^3 m to shorter than 10^{-12} m. These wavelengths have dimensions that are roughly equivalent to mountains for the longest wavelengths and to smaller than atoms for the shortest wavelengths.

All of these electromagnetic waves have one important thing in common: They all travel through space with the same speed, the speed of light, $c = 3 \times 10^8$ m/s. In all cases, the frequency times the wavelength is equal to the speed of light.

$$f \times \lambda = c = 3 \times 10^8 \text{ m/s}$$

Only waves that are classified as electromagnetic radiation travel at the speed of light.

Big Idea Electromagnetic waves all travel with the same speed in a vacuum, the speed of light.

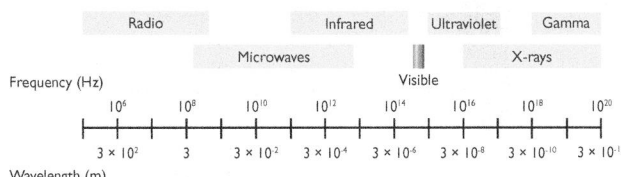

Frequency (Hz)

10^6	10^8	10^{10}	10^{12}	10^{14}	10^{16}	10^{18}	10^{20}
3×10^2	3	3×10^{-2}	3×10^{-4}	3×10^{-6}	3×10^{-8}	3×10^{-10}	3×10^{-12}

Wavelength (m)

The figure shows the range of wavelengths and frequencies that are observed for electromagnetic waves.

NAMES FOR ELECTROMAGNETIC RADIATION

The various wavelength ranges are classified with names, some of which are familiar because they have entered into everyday language. Here is a description of the names for the various wavelength ranges from the longest wavelengths to the shortest wavelengths. Notice that the words *ray, wave,* and *radiation* are used in the names.

Radio waves: Radio waves are emitted at a radio or television station. Radio waves are all around you all the time, but your body is unable to detect them. A device such as a radio or television converts radio waves into sound and picture. Wireless routers and wireless pointers also use radio waves.

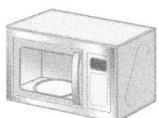

Microwave radiation: Microwaves are emitted by the power source in a microwave oven. Water molecules absorb microwaves. This causes the water molecules to move. The motion raises the temperature of the water contained in food, which in turn causes the temperature of the food to increase.

Infrared radiation: Your body senses infrared radiation as heat. For example, if you stand in direct sunlight, you will feel hotter than if you stand in the shade. In both cases, the air temperature is the same. The difference is that in direct sunlight, you are exposed to infrared radiation in addition to visible light. Infrared, or IR, radiation causes molecules to vibrate. This vibration motion also causes the temperature to increase. You may be familiar with heat lamps, such as those used to keep pet reptiles warm. These lamps use IR radiation.

Lesson 115 | **Electromagnetic Radiation**   587

Lesson Guide

ENGAGE (5 min)

TRM PowerPoint Presentation 115

ChemCatalyst

Each situation described below involves light.

1. Both the sun and an ordinary indoor lamp emit light. When you are outside in the Sun, you can get a sunburn. But when you are inside in a room lit by an ordinary lamp, you will not get a sunburn. Propose a hypothesis to explain how this is possible.

2. Bees detect different colors on the petals of a flower. A human looking at the same flower sees the petals as being one uniform color. Propose a hypothesis to explain how this occurs.

Sample answers: **1.** In addition to all the colors of the rainbow, sunlight must have light of wavelengths that we do not see that cause sunburn. These wavelengths must be missing from indoor lighting, even when the light looks the same. **2.** The bees must also see reflected light that human eyes do not detect.

Ask: What is the evidence that light that you cannot see causes sunburn?

Ask: What is the evidence that the structure of the eyes of bees allows them to detect light that humans cannot see?

Ask: What do you think allows nocturnal animals to find prey at night?

EXPLORE (15 min)

TRM Worksheet with Answers 115

TRM Worksheet 115

INTRODUCE THE CLASSWORK

→ Tell students to work in groups of four.

→ Pass out the student worksheet to each group.

EXPLAIN & ELABORATE (15 min)

EVIDENCE FOR ELECTROMAGNETIC RADIATION

Ask: Name some types of electromagnetic radiation. Where have you heard these names before? (Examples would be radio, microwave ovens, heat lamp or infrared lamp, sunscreens that provide UV protection, x-rays of bone fracture.)

Ask: What evidence can you provide that there are light waves that we cannot see?

Ask: How do the wavelength and frequency vary from gamma rays to radio waves?

Ask: What do all electromagnetic waves have in common?

Key Points: The light that we see is a tiny part of a large collection of waves called *electromagnetic radiation.* Electromagnetic radiation exists with a huge range of wavelengths and frequencies. The various wavelength regions are classified with names such as *gamma rays, x-rays, ultraviolet rays, visible light, infrared rays, microwaves,* and *radio waves.* Notice that both the words *ray* and *wave* are used. These words are consistent with the two models we have discussed, the ray model and the wave model. The wavelength of electromagnetic radiation increases for the sequence as listed, and the frequency decreases. All of these electromagnetic waves have one important property in common: They all travel through space with the same speed, the speed of light. Only waves that travel at the speed of light are classified as electromagnetic radiation.

Lesson 115 | **Electromagnetic Radiation**   587

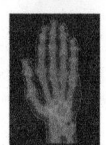

There are views of the world that are invisible to our eyes. Our eyes are limited detectors of electromagnetic radiation. We see only a narrow band of electromagnetic radiation as visible light. However, the eyes of owls and butterflies allow them to detect ultraviolet light. Sunburn is evidence that the Sun emits more than just visible light. The ultraviolet light from the Sun that we cannot see causes sunburn, and long-term exposure to ultraviolet light can cause more damaging effects to skin, such as skin cancer. It is possible to build instruments that detect wavelengths of electromagnetic radiation that are outside the visible range. There are devices similar to digital cameras that can detect infrared light. These devices detect the infrared light and generate an image of what is invisible to our eyes. These images greatly expand our understanding of the world.

LIGHT ENERGY

Ask: What evidence can you provide that light transfers energy?

Ask: How is the energy transferred related to the frequency and wavelength of light?

Ask: Which transfers more energy, red or blue light? Explain your thinking.

Ask: Why is it important to limit exposure to x-rays?

Key Points: Electromagnetic radiation transfers energy when it interacts with matter. One way that energy is transferred is by direct contact between two objects at different temperatures. Energy is transferred by conduction until the two objects are at the same temperature. Energy is also transferred when light interacts with matter. This is referred to as transfer of energy by radiation. The energy of light increases as the frequency increases and the wavelength decreases. One way to remember this is by considering what you did to make a high-frequency wave with the rope—a rapid hand motion back and forth. It required a greater input of energy to generate a higher frequency wave.

Planck's constant relates the frequency of the electromagnetic wave to its energy. Planck's constant is given the symbol h, where $h = 6.63 \times 10^{-34}$ J · s. The energy of the electromagnetic wave is calculated with the formula given below. For a frequency in hertz (inverse seconds, s^{-1}), the energy transferred by each wave is given in joules.

Visible light: Your eyes detect visible light as various colors. The colors of the rainbow are in the sequence of the shortest wavelength closest to Earth (violet) to the longest wavelength at the top of the arc (red).

Ultraviolet radiation: If you are out in the sun too long, your skin will turn red with sunburn. This does not happen with room lights, indicating that sunlight has additional waves with wavelengths outside the visible region. The waves that cause sunburn are called ultraviolet, or UV, radiation. Sunburn is damage to your skin. The UV rays cause molecules in your skin to break apart. Sunscreens provide protection by blocking the UV rays from getting to your skin.

X-rays: If you have ever broken a bone, then you have probably had an x-ray photograph taken. X-rays are dangerous, so the x-ray technician keeps the exposure time short. The x-rays fly through most substances, but they are blocked by bones and teeth. By detecting the x-rays that come through your body, it is possible to capture a photograph of your bones to locate a break or a photo of your teeth to locate cavities. X-ray technicians and dentists place a lead apron on you to block x-rays from reaching parts of the body that are not under examination.

Gamma rays: Gamma rays have the highest frequencies and shortest wavelengths, and they are extremely dangerous. Gamma rays are emitted during a nuclear explosion, in which the nuclei of atoms are falling apart.

⟲ Light Energy

Consider a fire caused by a chemical reaction between wood and oxygen to produce carbon dioxide and water. The reaction is exothermic, and it transfers a lot of heat to the surroundings. How does this transfer of heat occur?

So far in this unit, you have considered energy transferred by direct contact between two objects at different temperatures. Energy is transferred by conduction until the two objects are at the same temperature.

Energy is also transferred when light interacts with matter. This is referred to as transfer of energy by radiation. IR radiation is especially efficient at transferring energy to make you feel warm. Fire emits a lot of IR radiation in addition to emitting visible light. Again, while you cannot see IR radiation, your body does detect it as heat.

Kevin Schafer/Getty Images

Big Idea Electromagnetic radiation is one way that energy is transferred from a source to an object.

RELATIONSHIP BETWEEN FREQUENCY AND ENERGY

The table shows examples of values for the wavelength, frequency, and energy of various types of electromagnetic radiation. The energy is measured in joules, J. One joule is roughly equivalent to 0.24 calories. This is enough energy to raise the temperature of 1 mL of water by 0.24 degrees.

$$E = h \times f$$

Frequency and wavelength are related by the equation $c = f \times \lambda$.

Rearranging the equation gives $f = \dfrac{c}{\lambda}$.

Substituting $\dfrac{c}{\lambda}$ for the frequency in the equation $E = h \times f$ gives the relationship between energy and wavelength.

$$E = hf = \frac{hc}{\lambda}$$

A photon model is a third model to describe light. It is useful to think about one up-and-down motion of the wave as a "particle" or photon of light. Photons associated with waves of higher frequency transfer a greater amount of energy per photon. If the light of a specific frequency is brighter, there is a greater number of photons. Thus, more total energy is transferred. For example, bright red light has a greater number of photons compared to dim red light. Each photon of a specific color of light transfers the same amount of energy, but there are more photons in the

Take a moment to notice the relationships among wavelength, frequency, and energy.

Sample Values for Types of Radiation

Type of EM radiation	Wavelength, λ	Frequency, f	Energy, E	Size of Wavelength (approx.)
gamma rays	3×10^{-12} m	10^{20} Hz	6.63×10^{-14} J	
x-rays	3×10^{-10} m	10^{18} Hz	6.63×10^{-16} J	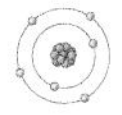
ultraviolet	3×10^{-8} m	10^{16} Hz	6.63×10^{-18} J	
visible light	6×10^{-7} m	5×10^{14} Hz	3.32×10^{-19} J	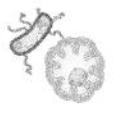
infrared	3×10^{-5} m	10^{13} Hz	6.63×10^{-21} J	
microwaves	3×10^{-2} m	10^{10} Hz	6.63×10^{-24} J	
radio waves	3×10^{3} m	10^{5} Hz	6.63×10^{-29} J	

The energy of light increases as the frequency increases and as the wavelength decreases. One way to remember this is by considering that to make a high-frequency wave with the rope, you made a rapid back-and-forth hand motion. It definitely required a greater input of energy to generate a higher frequency.

PLANCK'S CONSTANT

Planck's constant relates the frequency of the electromagnetic wave with its energy. It is given the symbol, h, defined as $h = 6.63 \times 10^{-34}$ J · s. The energy of the electromagnetic wave is calculated with the formula given below. For a frequency in hertz (s^{-1}), the energy transferred by each wave (one up-and-down motion), is given in joules.

$$E = h \times f$$

Frequency and wavelength are related by the equation $c = f \times \lambda$. Rearranging the equation gives

$$f = c/\lambda.$$

Substituting c/λ for the frequency in the equation $E = h \times f$ gives the relationship between energy and wavelength.

$$E = \frac{hc}{\lambda}$$

Lesson 115 | **Electromagnetic Radiation** 589

Answers: **1.** The first beam is called *ultraviolet radiation* and the second is called *microwaves.* **2.** Ultraviolet radiation transfers more energy because the wavelength is shorter. Energy is inversely proportional to wavelength.

Homework

Assign the reading and exercises for Fire Lesson 115 in the student text.

bright light. In contrast, even dim x-rays transfer a lot of energy because each photon transfers a much larger amount of energy than a photon of red light. This is because the frequency of an x-ray is 10^{18} Hz, whereas the frequency of red light is only 10^{14} Hz. Each x-ray photon transfers about 10,000 times more energy compared with a photon of red light. This is why x-rays are dangerous. The wave model and the particle model are both useful. Scientists refer to the wave-particle duality of light, which means that they and you can consider either model in describing the behavior of light.

EVALUATE (5 min)

Check-In

Imagine two beams of electromagnetic radiation. The wavelength of one beam is 10^{-8} m. The wavelength of the second beam is 10^{-2} m.

1. Use the information on the worksheet to classify these two waves.

2. Which wave transfers more energy? Explain your thinking.

While each wave (one up-and-down motion) only transfers a tiny amount of energy, keep in mind that the waves are coming at a high frequency and with a high speed. So, the total energy transferred can be quite large.

PHOTON MODEL OF LIGHT

So far, you have considered a ray model and a wave model to describe light. A **photon model** is a third model to describe light. It is useful to think about one up-and-down motion of the wave as a "particle," or photon of light. Instead of imagining light as a continuous light ray, the photon model suggests that a beam of light is composed of small particles.

According to the photon model, photons associated with waves of higher frequency transfer a greater amount of energy per photon. An x-ray photon transfers more energy than an infrared photon because the x-ray has a higher frequency compared with IR radiation.

Further, the photon model helps to explain the brightness of a beam of light. If the light is brighter, there should be a greater number of photons in a given area in one second. Light that is dimmer has fewer photons in the same area in one second.

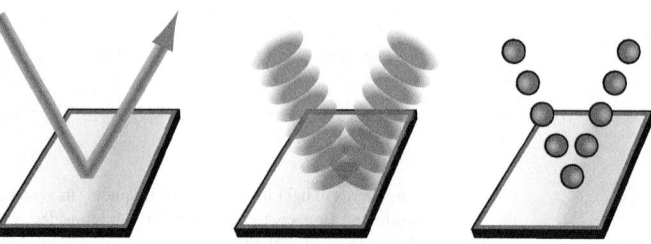

Light reflected as a strong beam.

Light as a weak beam.

There are two ways of transferring more total energy in a given amount of time: (1) make the light brighter so that there are more photons, or (2) choose light of a higher frequency such that each photon transfers more energy. For example, bright red light has a greater number of photons compared with dim red light. Each photon transfers the same amount of energy, but there are more photons in the bright light. In comparison, even dim x-rays transfer a lot of energy because each photon transfers a much larger amount of energy than a photon of red light. The frequency of an x-ray is 10^{18} Hz, whereas the frequency of red light is only 10^{14} Hz. Each x-ray photon transfers about 10,000 times more energy compared with a photon of red light. This is why x-rays are dangerous.

The wave model and the photon model are both useful. Scientists refer to the *wave-particle duality* of light, but this simply means that you can choose either model. Observations of radio waves indicate that they behave most like waves. In contrast, observations of the behavior of x-rays indicate that they are more particle-like.

> **Important to Know** Wave-particle duality simply indicates that both a wave model and a particle model describe electromagnetic radiation.

Example

Energy of Colored Light

Imagine a light wave with the frequency of 10^{17} Hz.

   a. What is the energy of this electromagnetic wave?

   b. Use the figure on page 587 and the table on page 589 to determine the type of electromagnetic radiation.

Solution

   a. The energy and frequency of electromagnetic radiation are related by the equation:

$$E = h \times f, \text{ where } h = 6.63 \times 10^{-34} \text{ J} \cdot \text{s}$$

   Substitute values for the Planck's constant, h, and the frequency of light:

$$\text{Energy} = E = (6.63 \times 10^{34} \text{ J} \cdot \text{s})(10^{17} \text{ Hz}) = 6.63 \times 10^{-17} \text{ J}$$

   Notice that seconds cancel because Hz = 1/s, giving units of joules.

   b. The energy and frequency indicate that the electromagnetic radiation is x-rays.

BIOLOGY CONNECTION

The human eye detects colored light by using three receptors that have maximum frequencies for red, green, and blue. When light strikes these receptors at the back of the eye, a signal is sent along nerves to the brain. The brain then takes the information from the three receptors to create the perception of various colors. (And unlike bees, human beings can see red.)

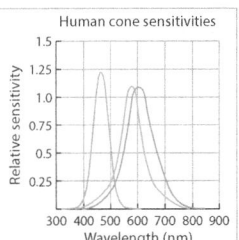

Human cone sensitivities

Relative sensitivity vs *Wavelength (nm)*

LESSON 115 ANSWERS

1. Electromagnetic radiation exists with a huge range of wavelengths and frequencies for which $\lambda \times f = c$, the speed of light. As the frequency increases and the wavelength decreases, the energy transferred by electromagnetic radiation increases.

2. The ray model indicates the path of electromagnetic radiation using straight lines. The wave model indicates the wave nature of light rays. The photon model suggests that each up-and-down motion of an electromagnetic wave can be described as a particle, or photon.

3. $E = hf$; energy and frequency are proportional to one another.

4. Radio waves

5. *Sample answer:* An insect

6. 3.8×10^{-1} m

7. 2×10^{-13} J

KEY TERMS

electromagnetic radiation
radio wave
microwave radiation
infrared radiation
visible light
ultraviolet radiation
x-ray
gamma ray
Planck's constant
photon model

LESSON SUMMARY

What is the evidence for light waves that are invisible to your eyes?

The light that we see is a tiny part of a large collection of waves called *electromagnetic radiation*. Electromagnetic radiation exists with a huge range of wavelengths and frequencies. The wavelength regions include radio waves, microwave radiation, infrared radiation, visible light, ultraviolet radiation, x-rays, and gamma rays. The entire collection of waves travels at the same speed, the speed of light, $c = 3 \times 10^8$ m/s. The energy transferred by electromagnetic radiation is proportional to the frequency of the wave. Electromagnetic waves of high frequency transfer more energy than waves of low frequency. High-frequency electromagnetic waves such as gamma rays and x-rays are quite dangerous because of the large amount of energy they transfer, even with a dim beam.

Exercises

Reading Questions

1. Summarize what you learned about electromagnetic radiation.
2. Describe the ray model, the wave model, and the photon model of light.
3. Describe the relationship between energy and frequency.

Reason and Apply

4. What type of radiation has a wavelength of 6.12×10^2 m?
5. Identify an object that is similar in size to the wavelength of a microwave.
6. A radio station broadcasts at a frequency of 790 MHz (7.9×10^8 s^{-1}). What is the wavelength of the electromagnetic radiation generated by this radio station?
7. The gamma rays emitted by ^{60}Co are used in radiation treatment for cancer. The frequency of these gamma rays is 3×10^{20} s^1. What is the energy of this radiation?

How Absorbing
Spectroscopy

THINK ABOUT IT

While enjoying a day at the beach, you notice a person whose skin has turned quite red. In contrast, your skin looks the same as it did in the morning when you left your house. This is because you have applied sunscreen several times during the day. The claim on the package is that sunscreen protects against harmful UV radiation. Although you cannot see UV radiation, how is it possible to determine which substances block UV radiation effectively?

How is it possible to measure electromagnetic radiation that you cannot see?

To answer this question, you will explore

- Light Detectors
- Spectroscopy: Studying Matter with Light

EXPLORING THE TOPIC

Light Detectors

A **light detector** is a device that can sense electromagnetic radiation. The detector provides some type of signal when it comes into contact with electromagnetic radiation. Some detectors, such as solar cells, produce an electrical signal that converts visible light into electricity. Radio antennas detect radio waves. The radio then converts the signal detected by the antenna to sound. Your skin signals exposure to UV radiation by turning red. In this lesson, you will explore what happens when paper coated with certain chemical mixtures responds to electromagnetic radiation by changing color.

Solar cells detect visible light.

The radio detects radio waves.

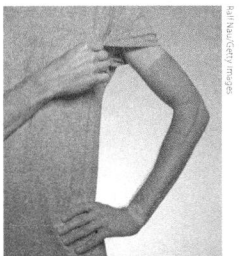
Your skin detects UV radiation.

- Describe different ways that scientists can detect radiation that is not in the visible region.

FOCUS ON UNDERSTANDING

- Students may be surprised to find that direct sunlight has much more UV radiation than indoor lights.
- It is necessary to make a special detector of UV radiation. One possibility is to choose a mixture of chemicals that changes color under UV radiation.
- Students may have difficulty relating the ability of a substance to block UV radiation with absorption, transmission, and reflection. Substances that block UV radiation absorb the radiation.

KEY TERMS

light detector
spectrometer
spectroscopy

IN CLASS

This is a lab that will take two 45-minute class periods. Day 1: Students begin by considering how sunscreen protects their skin against sunburn. Students then experiment with UV-sensitive paper to measure the amount of UV radiation in direct sunlight as opposed to indoor lighting. The UV-sensitive paper turns a different darkness of blue, depending on the amount of UV exposure. At the end of this class, students begin their experimental design for day 2 and complete it for homework. Day 2: Students run their own experiment to test which substances are best at blocking UV radiation. They place the substance between the light source and the UV-sensitive paper. A complete materials list is available for download.

TRM Materials List 116

Pre-AP® Course Tip

Students preparing to take an AP® chemistry course or college chemistry should be able to design an experimental procedure that will generate useful data. In this lesson, students design a procedure for determining how well different substances can transmit or absorb UV radiation.

SETUP

Day 1: Cut small pieces of UV-sensitive paper (roughly 2 in × 2 in) in a dimly lit area away from direct sunlight. Place the pieces of paper into small envelopes to keep them from light exposure until needed. Set out shallow dishes with

Overview

LABORATORY: GROUPS OF 4

Key Question: How is it possible to measure electromagnetic radiation that you cannot see?

KEY IDEAS

Various sensors are used to detect electromagnetic radiation. Human eyes detect visible light, as does the detector in a digital camera. It is also possible to use certain chemicals that change color on illumination as detectors to make visible electromagnetic radiation outside the range detected by the human eye. For example, UV-sensitive paper is coated with a mixture of two chemicals

that react on exposure to UV radiation. The products of the reaction are a different color from the reactants, and thus, the UV radiation is detected by a color change. For longer exposures, the color change is darker. This allows examination of the amount of UV radiation emitted from various light sources and the determination of the ability of various substances placed on top of the paper to block UV radiation.

LEARNING OBJECTIVES

- Identify whether sunlight or indoor light emits more UV radiation.
- Design an experiment to explore the ability of various substances to transmit/absorb UV radiation.

water. Day 2: Students design an experiment to test the ability of a substance to block UV radiation. Place whatever materials you can provide where students can see what is available. Groups must design their own setup, requesting what they might need from the instructor. Before students conduct their experiment, students should get your approval to make sure their design is feasible and safe.

SAFETY

Students must wear safety goggles at all times. Remind students that, in the laboratory, the household products should be treated as carefully as any other lab supplies. Do not apply any materials to skin, and keep out of eyes. Students must obtain permission from the teacher before conducting any procedures they design.

CLEANUP

The paper and the household materials can be placed in a trash can.

Lesson Guide

ENGAGE (5 min)

TRM PowerPoint Presentation 116

ChemCatalyst

1. List three factors to consider in choosing a sunscreen.

2. How do you think sunscreen works?

Sample answers: **1.** Sunscreens protect your skin from sunburn, have to be easy to spread, and should not be too greasy. **2.** Sunscreens might block the light so it does not get to your skin. Or perhaps sunscreens "toughen" your skin so that the light does not cause sunburn.

Ask: What types of sunscreens do you buy?

Ask: Why is it important to use sunscreen if you are outside all day long?

Ask: Which sunscreens provide the best protection?

Ask: How do you think sunscreen protects your skin?

EXPLORE (15 min)

TRM Worksheet with Answers 116

TRM Worksheet 116

UV-SENSITIVE PAPER

For more than 150 years, people have been creating images and photos by using light-sensitive paper and film to capture light. One way to create an image this way is to place an object that blocks the incoming electromagnetic radiation onto the light-sensitive paper. The electromagnetic radiation only strikes the paper where it is not blocked by the object. This creates a silhouette of the object.

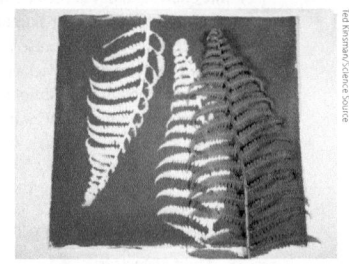

The leaves block the UV radiation from the Sun, resulting in a white image. The background is blue due to a chemical reaction caused by the UV radiation.

Some UV-sensitive paper is coated with a mixture of compounds. Before exposure to UV radiation, the mixture of compounds is pale green. This color indicates that one of the compounds in the mixture absorbs all colors except for green. Green light is reflected.

The green reactant compounds are soluble in water. If the paper is submerged in water, the green compounds dissolve and wash away. With the green reactants removed, the paper is no longer UV-sensitive. If you expose it to UV radiation, it will not turn blue. Instead, it will remain white.

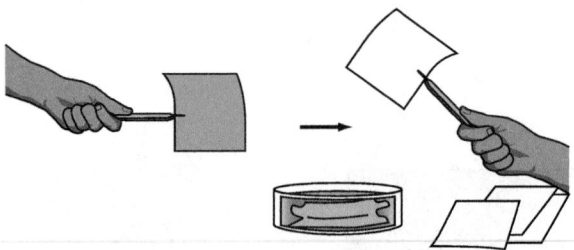

The green coating on the UV-sensitive paper can be washed away in water to make the paper white.

If you shine UV radiation on the green UV-sensitive paper, the energy that is transferred provides the energy needed to cause a chemical reaction to occur. The

INTRODUCE THE LAB

→ Tell students to work in groups of four.

→ Pass out the worksheet to each group.

→ Go over the lab procedure so that students understand how to handle the UV-sensitive paper. You can demonstrate a 1-minute exposure along with rinsing so students have a clear idea of what to do.

EXPLAIN & ELABORATE (15 min)

OBSERVATIONS BASED ON UV-SENSITIVE PAPER

→ You can show some sample pieces of UV-sensitive paper after exposure to UV radiation and rinsing in water.

Ask: How does the UV-sensitive paper provide evidence for UV radiation?

Ask: What do you observe in this experiment if a substance blocks UV radiation effectively?

Ask: Why do you think it was necessary to rinse the paper in water?

reactant that causes the green color changes into a new compound that is blue. The paper is blue because the new compound reflects blue light.

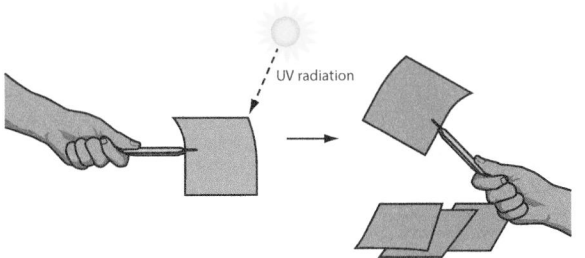

mixture of reactant compounds ⟶ mixture of product compounds

Shining UV radiation on the UV-sensitive paper causes a chemical reaction that turns the paper from green to blue.

The longer the paper is exposed to UV radiation, the deeper the shade of blue it turns. The reaction will continue until all the reactants are used up and the paper has the deepest blue color. The amount of the green compound is a limiting reactant. So one way to stop the reaction is to expose the paper to UV radiation until all limiting reactant is used up in the reaction.

Longer exposure to UV radiation causes the blue color to be a deeper blue.

Another way you could stop the reaction is related to the difference in solubility between the green reactant compound and blue product compound. The green reactant compound is soluble in water, but the blue product compound is insoluble in water. If you expose the paper to UV radiation for a very short time, only some of the reactant compound will turn blue. The paper will appear to be a pale blue color and is a mixture of green reactant compound that has not yet reacted and blue product compound. If you then submerge the partially reacted paper in water, the green reactant compound will dissolve in the water and can be washed away. The blue product compound remains attached to the paper because it does not dissolve in the water. Now, even if you were to expose the partially reacted and washed paper to UV radiation again, it would no longer change to a

Lesson 116 | **Spectroscopy**   595

Ask: Is the coating soluble in water before exposure to UV radiation? Is the coating soluble in water after exposure to UV radiation? Explain what happens to the coating on the paper from start to finish.

Key Points: The UV radiation causes the compounds in the coating to react and change color. The initial green color fades because the absorbed UV radiation provides the energy to cause a reaction to create a new compound that reflects blue light. For longer exposure times to UV radiation, more of the green compound reacts to form the blue compound. The amount of

green compound is a limiting reactant. Once there is no green compound left to react, the paper is as dark blue as it can get.

The green coating on the paper is water-soluble, but the blue compound is not. After rinsing in water, the green color from unexposed areas (like the area underneath an object) dissolves and washes away, leaving those areas white. The blue areas remain after rinsing because the compound that gives rise to the blue color is not soluble in water. The intensity of the blue color depends on how many of the initial molecules of the green compound react

to become molecules of the blue compound. This depends on the length of the exposure to UV radiation. If all the molecules in the exposed area react, the paper will change to be the darkest possible blue color. If only some of the molecules convert, those that do not convert are washed away and the shade of blue that remains is lighter.

The shade of blue makes it possible to compare the UV-blocking ability of various substances. For identical exposure times and conditions, the paper will turn a darker shade of blue if the light source emits a greater amount of UV radiation. The paperclip did not transmit any UV radiation. This created a white image of the paperclip after unexposed green compound was washed away. A substance that blocks UV radiation well, such as a sunscreen with a high SPF number, causes the paper to turn white or very light blue after washing away the unexposed green compound. Conversely, a substance that blocks UV radiation poorly, such as a sunscreen with a low SPF number, causes the paper to turn dark blue because the UV light is not being blocked effectively. The paper will be various shades of blue for substances in between.

Scientists make use of light detectors to create instruments called *spectrometers* to study properties of matter. A spectrometer has four main components: a source of electromagnetic radiation, a device, such as a prism, to separate the wavelengths of light, a sample holder, and a light detector. See the diagram on page 597 of the student text. By using electromagnetic radiation of different wavelengths, it is possible to gain a lot of information about atoms, molecules, and compounds. These methods for analyzing compounds are referred to as *spectroscopy*.

Point out to students that the experiments with UV have similarities to spectroscopy. For example, various samples are placed between a light source (the Sun or room lights) and a detector (the UV-sensitive paper). The results of the experiment help to determine how well certain substances transmit UV light.

ASSIST STUDENTS IN SHARING THEIR RESULTS AND CONCLUSIONS (OPTIONAL)

Ask: What was the purpose of your experiment?

Ask: How would you describe your experimental setup?

Ask: What issues came up while trying to design your experiment?

Ask: What variables did you control?

Ask: What did you conclude based on your experiment?

Encourage students to state their conclusions. Challenge students to provide convincing evidence to support their conclusions. Engage other students in asking questions if they are not convinced, as well as to raise issues with the experimental design. Emphasize that the goal is to understand experimental design more deeply through constructive criticism and through refinement of experimental design.

EVALUATE (5 min)

Check-In

Draw a sketch to describe how you arranged the light source, the sample, and the UV-sensitive paper to measure how much UV radiation the sample absorbs. Show the path of the light.

Answer: The light travels from the light source to the sample. Light that is transmitted by the sample goes to the UV-sensitive paper.

Homework

Assign the reading and exercises for Fire Lesson 116 in the student text.

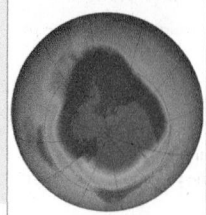

Photographers are able to make beautiful, complex images with UV-sensitive paper.

deeper blue. It would remain the same pale blue color because there would no longer be any green reactant compound available to change to blue product compound.

CREATING AN IMAGE

Now imagine that you place a small object on top of the UV-sensitive paper and expose the paper and object to UV radiation. The paper turns blue in the areas surrounding the object. But underneath the object, the paper remains green. This is because the object blocks UV radiation, preventing it from shining on the paper directly beneath the object. There is a green image of the object on a blue background after exposure to UV radiation.

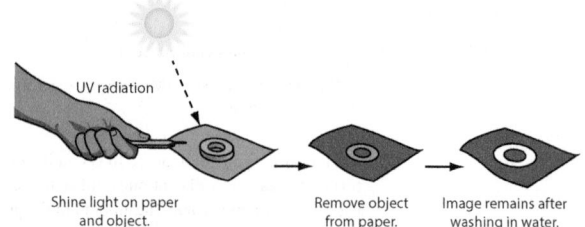

Shine light on paper and object. — Remove object from paper. — Image remains after washing in water.

If the paper with the green image is exposed to light, the green compounds will react and the paper will be completely blue. However, if you protect the paper with the green image from UV radiation and then wash away the green compounds, a white image of the object will remain permanently.

➲ Spectroscopy: Studying Matter with Light

Scientists make use of light detectors to create instruments called **spectrometers** to study properties of matter. A spectrometer has four main components: a source of electromagnetic radiation, a device, such as a prism, to separate the wavelengths of light, a sample holder, and a light detector. By using electromagnetic radiation of different wavelengths, it is possible to gain a lot of information about atoms, molecules, and compounds. It is also possible to learn about the concentration of a substance and its identity depending on the choice of electromagnetic radiation. These methods for analyzing compounds are referred to as **spectroscopy.**

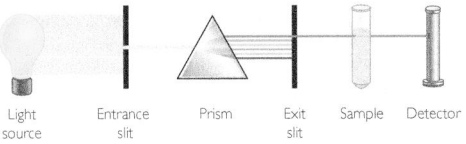

Light source | Entrance slit | Prism | Exit slit | Sample | Detector

Schematic of a basic spectrometer: light, sample, prism and detector

STUDYING WHAT YOU CANNOT SEE

Some substances, such as sunscreen, clothing, glass, and plastic, absorb UV radiation. Other substances, such as the atmosphere, transmit some UV radiation. You can use UV-sensitive paper as a simple UV detector to see how a substance behaves when exposed to UV light. Although it's not a spectrometer, this tool is similar to one in some ways. It allows you to study what you can't see. The light source for this tool is UV radiation, the sample is the substance you test, and the detector is the UV-sensitive paper.

First, you have to place your sample substance on a piece of UV-sensitive paper. It is best to spread the sunscreen on a thin piece of glass or plastic. Next, expose the sample on the paper to a source of UV radiation. If the substance absorbs UV radiation, the paper under the substance will remain green when exposed to the UV radiation source. If the substance transmits UV radiation, the paper under the substance will turn blue when exposed to the UV light source.

You can also use this setup to determine how different amounts of a substance block or transmit UV radiation. To do this, you need to examine samples with different concentrations of the substance being tested. For example, you might test sunscreens with the same active ingredient but different sun protection factor (SPF) ratings to see which ones block the most UV radiation. The SPF rating indicates the ability of the sunscreen to block UV radiation. Sunscreens with higher SPF ratings are more concentrated in the substance providing protection.

You could also use this tool to examine various light sources to compare the amount of UV radiation they emit. For example, the Sun emits a lot of UV radiation. In contrast, indoor lighting emits much less UV radiation. This is why you can get a sunburn when you are outside and exposed to the Sun but do not get a sunburn from indoor light sources.

> **Big Idea** Spectroscopy is a method of using electromagnetic radiation to study properties of atoms, molecules, and compounds.

LESSON SUMMARY

How is it possible to measure electromagnetic radiation that you cannot see?

Various sensors are used to detect electromagnetic radiation. Human eyes detect visible light as does the detector in a digital camera. It is also possible to use certain chemicals as light detectors that respond to different wavelengths of

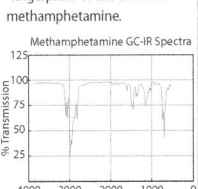

LESSON 116 ANSWERS

1. The green coating on the paper absorbs UV radiation causing a chemical reaction, which changes the green compound into a blue product compound.

2. The sketch should include a light source, a prism to separate the wavelengths of light, a slit, to select a single wavelength, a sample, and a detector.

3. It will turn dark blue.

4. *Sample answers:* Sunscreen, metal, clothing.

5. It remains green. No reaction occurs.

6. a. Red **b.** The sample is between the light source and the detector. The light follows a path from the light source, through the sample, and to the detector. **c.** If the solutions are different, a different amount of light will strike the detector in each case. **d.** UV-sensitive paper will not work because it does not detect red light.

7. The sunglasses with UV protection will block UV light such that the UV-sensitive paper remains green. If washed away, the paper will turn white. The regular reading glasses will transmit more UV light, thereby exposing the UV-sensitive paper and turning it blue.

8. A good lab report will contain:
● a title (*Possible answer:* UV Light Detection) ● a statement of purpose (*Possible answer:* To determine which substances best block UV radiation.)
● a procedure (a summary of the steps followed in the experiment)
● observations (descriptions of what was observed) ● conclusions (that include statements that support what students were trying to determine in the experiment.

KEY TERMS
light detector
spectrometer
spectroscopy

electromagnetic radiation. For example, UV-sensitive paper is coated with a chemical that changes color upon exposure to UV light. Longer exposures result in a color that is a deeper shade. This allows you to examine the amount of UV radiation emitted from various light sources, and to determine the ability of various substances to block UV radiation.

Exercises

Reading Questions

1. Why does UV radiation change the color of UV-sensitive paper?

2. Make a sketch of a spectrometer and label the main parts.

Homework

3. What happens if you leave a piece of UV-sensitive paper out in the Sun?

4. List three materials that block UV radiation.

5. What happens if you leave a piece of UV-sensitive paper out in a room with a dim red light?

6. Imagine that you want to measure two red solutions to determine if they have the same concentration of red dye or not.
   a. What type of light source do you need for the spectrometer?
   b. Sketch the placement of the light source, the detector, and the sample. Show the path of the light.
   c. What do you predict you will observe if the two solutions are different?
   d. Can you use UV-sensitive paper for the detector? Why or why not?

7. What differences do you predict if you use UV-sensitive paper to test sunglasses with UV protection compared with regular reading glass without UV protection?

8. Write a lab report for the experiment that you designed in class. In your report, include the title of the experiment, purpose, procedure, observations and conclusions.

(Title)

Purpose: (Explain what you were trying to find out.)

Procedure: (List the steps you followed.)

Observations: (Describe what you observed during the experiment.)

Conclusions: (What can you conclude about what you were trying to find out? Provide evidence for your conclusions.)

Leonard Lessin/Photo Researchers/Getty Images

CHAPTER 22
Radiating Energy

SUMMARY

KEY TERMS

emission
radiation
light ray
white light
reflection
transmission
transparent
opaque
absorption
light wave
wavelength
frequency
speed of light
electromagnetic radiation
radio wave
microwave radiation
infrared radiation
visible light
ultraviolet radiation
x-ray
gamma ray
Planck's constant
photon model
light detector
spectrometer
spectroscopy

Fire Update

The hot gases emitted from the reaction of wood and oxygen in a fireplace transfer thermal energy by conduction when they collide with your body. Upon collision, the faster-moving gas molecules slow down, the atoms moving more slowly in your body speed up, and you sense warmth. Conduction is one way that a fire transfers thermal energy, but this is not the only way that energy is transferred.

You feel the warmth of the Sun even though its light travels great distances through nearly empty space to reach Earth. This occurs by the emission of electromagnetic radiation. We perceive energy transfer from the Sun by radiation as visible light with our eyes, as the warmth due to infrared radiation, and, sometimes, as the sunburn caused by ultraviolet radiation. The color of a fire's flames and much of its warmth are due to energy transfer by the radiation emitted by the hot gas molecules produced by the combustion reaction.

There are a number of models that lend understanding to the nature of light. The ray model suggests that light travels in straight lines until it is reflected, transmitted, or absorbed when it strikes an object. If white light is partially absorbed, the reflected or transmitted light has a color. The wave model is used to explain that different colors are waves all traveling at the speed of light but of varying wavelength and frequency. The photon model suggests that energy is transferred in small units, called *photons*. Photon energy increases with frequency.

Electromagnetic radiation exists with a huge range of wavelengths and frequencies. The wavelength regions include radio waves, microwave radiation, infrared radiation, visible light, ultraviolet radiation, x-rays, and gamma rays. Instruments, called spectrometers, use various wavelengths of electromagnetic radiation to probe matter. In essence, spectroscopy extends our ability to see light that we would otherwise not be able to detect with only our eyes.

REVIEW EXERCISES

1. Why are some objects colored when illuminated with white light?
2. Explain why the dentist covers you with a lead apron before taking an x-ray image of your teeth.
3. What is the wavelength of a light wave with the frequency 1.2×10^6 Hz?
4. Imagine that you want to measure two green solutions to determine if they have the same concentration of green dye or not.
   a. What type of light source do you need for the spectrometer?
   b. What do you predict you will observe if the two solutions are different?
5. What is the frequency of a light wave with energy 2.7×10^{-19} J?

ANSWERS TO CHAPTER 22 REVIEW EXERCISES

1. Colored objects reflect the color we see and absorb other colors.

2. The lead apron blocks x-rays from entering other parts of your body.

3. 2.5×10^3 m

4. a. Green **b.** The two solutions will transmit different amounts of light.

5. 4.1×10^{14} Hz

ASSESSMENTS

Two multiple-choice assessments (versions A and B) and two short-answer assessments (versions A and B) are available for download for Chapter 22.

TRM **Chapter 22 Assessment Answers**

TRM **Chapter 22 Assessments**

Project: Wavelength and Energy

A good project would include ● answers to debate questions, including evidence to support answers (**Possible answer:** If you ask a scientist, he or she might say that when referring to the reflection of light, white light is the combination of all colors and black is the absence of color. However, if you ask an artist, he or she might say that a white pigment is the absence of all color and a black pigment is made by mixing all colors. Black pigments absorb all colors of white light.) ● a description of the experiment including what images were captured and what conclusions were drawn.

Wavelength and Energy

 PROJECT

• How would you answer the question: "Are black and white colors?" Debate the two questions posed below.

1. A light bulb manufacturer might claim: "Black *is not* a color, but white *is* a color." Is this correct or not? Explain your thinking.

2. An artist with paints might claim: "Black *is* a color, but white *is not* a color." Is this correct or not? Explain your thinking.

• Research the type of light emitted by three different light sources. Take digital images of white objects and two colored objects illuminated by these three light sources. Write a report explaining what you learned about light sources. Include a description of the experiment you carried out and an explanation of the results.

Fire | REVIEW

UNIT 5

Phil McDonald/Shutterstock

Combustion reactions are extremely useful in our daily lives. Most cars, planes, and trains move by the energy transferred from a combustion reaction. Even the digestion of our food can be considered a very controlled combustion reaction.

Calorimetry is one technique used to figure out how much energy is transferred from combustion reactions. It is possible to measure the energy transferred by measuring the temperature change of a known quantity of water. The heat required to change the temperature of 1 g of water by 1 degree, called the heat capacity of water, is needed to convert between the measured temperature change of the water and energy.

Different substances respond differently to heating. Substances that change temperature easily, with very little energy transfer, have low specific heat capacities. It takes more energy to change the temperature of substances with higher specific heat capacities.

Once the data from calorimetry experiments are collected, the data can be used to compare and classify reactions. Exothermic reactions transfer energy *to* the surroundings. Exothermic reactions are often sensed as "hot" by the observer. Endothermic reactions require energy transfer *from* the surroundings. Endothermic reactions are often sensed as "cold" by the observer.

The heat of a reaction comes from bond breaking and bond making. It takes energy to break the bonds in a compound. In contrast, bond making is a process that releases energy. An energy diagram can be used to show the energetic process of a reaction.

When the energy of making the new bonds in the products is greater than the energy that it takes to break the bonds in the reactants, energy is transferred *to* the surroundings, and the reaction is exothermic. When it takes more energy to break the bonds, energy is transferred *from* the surroundings, and the reaction is endothermic.

The net energy of a reaction is referred to as the heat of the reaction, or ΔH. In the case of an exothermic reaction, ΔH is negative, because energy is transferred *from* the system. In the case of an endothermic reaction, ΔH is positive, because energy is transferred *to* the system.

The energy associated with the formation of any compound from its elements is referred to as the heat of formation, or ΔH_f. Heat of formation is expressed in kilojoules per mole of product formed. There is also an energy associated with getting a reaction started, called the activation energy. For example, it takes a match or a spark to get a fire started. Once a few bonds are broken by the spark, combustion reactions proceed on their own because of the heat released by the formation of the products.

Unit 5 | **Review**   601

ChemCatalyst

1. Name three specific changes to matter that are useful sources of energy. Describe a situation in which each change is applied to do work.

2. What makes these reactions good sources of energy?

Sample answers: **1.** Burning methane in your stove at home to cook dinner; dissolving calcium chloride in water for a hot pack; a lead–sulfuric acid battery in your car; a lithium battery in your laptop computer; burning gasoline in your car engine. (Most combustion reactions, redox reactions involving metal compounds, and some dissolving changes can be useful sources of energy transfer.) **2.** These reactions are exothermic. The redox reactions can be converted to electrical energy. The combustion reactions transfer heat and light, and produce expanding gases.

DISCUSS THE CHEMCATALYST

Ask: How are the chemical changes that you listed similar? How are they different?

Ask: What types of reactions are useful for energy production?

Ask: Are exothermic reactions the only types of chemical change useful for harnessing energy? Explain.

TRM Worksheet with Answers R5

TRM Worksheet R5

INTRODUCE THE REVIEW

→ Tell students to work in pairs.

→ Pass out the worksheet to each pair.

SUMMARIZE THE KEY CHEMISTRY IDEAS

Make a list of the key chemistry content areas on the board.

thermal equilibrium

specific heat capacity

heat versus temperature

heating curve of water

system and surroundings

Overview

CLASSWORK: INDIVIDUALS OR PAIRS

Key Question: What is the relationship between energy and chemical change?

KEY IDEAS

Students have learned about how energy is related to changes in matter.

LEARNING OBJECTIVE

Review the material from the Fire unit to create a comprehensive list of topics to study for a unit exam.

IN CLASS

This class gives students a chance to summarize and review the ideas about energy learned over the course of the unit. The emphasis is on five main areas: heat transfer, combustion, energy exchanges during chemical change, redox reactions, and electromagnetic radiation. Students work by themselves or in pairs on a worksheet. A complete materials list is available for download.

TRM Materials List R5

first and second laws of
    thermodynamics

combustion reactions

exothermic and endothermic

calorimetry calculations

heat of combustion

bond energies

reversing reactions

heat of reaction

work

oxidation-reduction

heat of formation

activity series

electrochemical cells

half-reactions

heat of fusion

light energy

light waves

electromagnetic radiation

spectroscopy

Ask: Do energy exchanges take place during physical changes? Explain.

Ask: What conditions are necessary for combustion to occur?

Ask: What is a calorie?

Ask: In a calorimetry experiment, what constitutes "the system"?

Ask: What are the main ways chemical energy is converted to work?

Ask: What is PV work?

Ask: Why is combustion considered oxidation? What is reduced in a combustion reaction?

Ask: Why are redox reactions a useful source of energy?

Ask: What is an electrochemical cell? Describe how to set one up.

Ask: What is an activity series?

Ask: What are the models that describe the behavior of light?

Ask: How can you show that white light is a mixture of colors?

Ask: How are wavelength, frequency, and speed of light related?

Ask: When wavelength increases, what happens to frequency?

Ask: How does light transfer energy? What evidence can you provide to show this?

Ask: What is spectroscopy?

Ask: How can UV paper be used as a type of light detector? How is this similar to spectroscopy?

Combustion is an oxidation-reduction reaction. Oxidation is defined as the loss of electrons by an atom, an ion, or a molecule during a chemical change. Reduction is when an atom, an ion, or a molecule gains electrons in a chemical change.

Reactions are controlled efficiently in batteries. A battery keeps the reactants separated from one another, preventing the reactants from mixing together and starting a fire. When a wire connects to the battery, the reaction occurs and energy is transferred. The reaction stops when the wire is disconnected, or when the reactants are used up.

Aside from transferring energy by direct contact between a hot substance and a cold substance, called conduction, energy is transferred by radiation. Radiation refers to the generation of electromagnetic radiation. When described by the wave model, electromagnetic waves have a large range of frequency and wavelength. However, all of these waves travel at the speed of light. The energy transferred depends on the frequency of the electromagnetic waves. A fire transfers energy both by conduction and radiation.

REVIEW EXERCISES

General Review

Write a brief and clear answer to each question. Be sure to show your work.

1. What have you learned about fire in this unit?

2. Generally describe how our energy needs are satisfied by chemical changes.

3. Where does the thermal energy from an exothermic reaction come from?

4. What do the first and second laws of thermodynamics indicate about energy?

5. How many calories of energy are transferred to
   a. raise the temperature of 50 grams of water from 25 °C to boiling?
   b. lower the temperature of 100 grams of water from boiling to 25 °C?

6. The specific heat capacity of aluminum is 0.21 cal/g °C. How many calories of thermal energy must be transferred to heat up a 100 g aluminum pot from 20 °C to 80 °C?

7. For the same number of grams, does it take more or less energy to raise the temperature of aluminum metal or liquid water? Explain.

8. The heat of fusion for water is 79.9 cal/g. The heat of fusion for ethanol is 24.9 cal/g. What does this tell you about these two substances?

9. Name three ways you can increase the rate of a reaction.

10. Define work. How many joules of work are done if a 500 lb piano is lifted by a pulley to a height of 20 m?

11. Write and balance the equation for the oxidation of calcium, Ca.

12. Write and balance a chemical equation for a redox reaction.

Key Points: Fire is a special category of chemical change that produces large amounts of observable energy in the form of heat and light. The energy from fire composes a large portion of the energy used on our planet. Fire is the result of combustion, which is a type of oxidation-reduction reaction. The hot gases emitted from the reaction of wood and oxygen in a fire transfer thermal energy by conduction. Upon collision, the faster-moving gas molecules slow down, the atoms moving more slowly in your body speed up, and you sense warmth. Conduction is not the only way that energy is transferred. The color of a fire's flames and much of its warmth are due to energy transfer by the radiation emitted by the hot gas molecules produced by the combustion reaction.

Heat is a process of energy transfer that depends on differences in temperature. This energy transfer occurs only from hotter objects or substances to cooler ones. Different substances have different responses to heat transfer, with some changing temperature more readily than others. This property is called *specific heat capacity*.

The heat associated with the combustion of a certain amount of fuel is

13. Use net ionic equations to show which reactant atoms are oxidized and which atoms are reduced in the following chemical equations.

$$Zn(s) + 2MnO_2(s) \longrightarrow ZnO(s) + Mn_2O_3(s)$$

$$3CaO(s) + 2Al(s) \longrightarrow Al_2O_3(s) + 3Ca(s)$$

14. Explain how you know which atoms are oxidized and which are reduced in a redox reaction.

15. Explain the ray model, the wave model, and the photon model of light. Draw a sketch of each model showing transmission through a thin red plastic film.

16. What types of electromagnetic radiation are least harmful? Explain why in terms of wavelength, frequency, and energy.

17. Describe the main components of a spectrometer.

STANDARDIZED TEST PREPARATION

Multiple Choice

Choose the best answer.

1. Which of these is best classified as energy as opposed to matter?
   - (A) safety match
   - (B) potassium chlorate, $KClO_3(s)$
   - (C) fireworks
   - (D) light

Use the following to answer Exercises 2 and 3.

A solution of sodium bicarbonate is mixed with a solution of hydrochloric acid in a beaker. A solution of sodium chloride and carbonic acid then forms. A thermometer inside the beaker indicates that the temperature decreased from 24 °C to 17 °C.

$$NaHCO_3(aq) + HCl(aq) \longrightarrow$$
$$NaCl(aq) + H_2CO_3(aq)$$

2. What evidence would indicate that energy was transferred from the surroundings to the products of the reaction in the beaker?
   - (A) The products are aqueous. No precipitate formed.
   - (B) Both the products and reactants include acids.
   - (C) The beaker would feel cold to the touch.
   - (D) The beaker would feel hot to the touch.

3. Which correctly names this process?
   - (A) endothermic process
   - (B) exothermic process
   - (C) equilibrium process
   - (D) combustion process

4. Water condenses on the outside of a cold glass of lemonade on a warm day. Which of the following correctly describes this process?
   - (A) When water vapor condenses, energy is transferred *from* the surroundings to the gaseous water molecules, thereby lowering the temperature.
   - (B) When water vapor condenses, energy is transferred *to* the surroundings, thereby lowering the temperature of the water molecules.
   - (C) Condensation is a cooling process because hydrogen and oxygen react to form water molecules.
   - (D) Condensation is a cooling process because gaseous water molecules collide with one another during the phase change.

5. Choose the best description of the final temperature if 100 mL of water at 20 °C is mixed with 50 mL of water at 80 °C.
   - (A) The final temperature will be 50 °C because this is the average temperature.
   - (B) The final temperature will less than 50 °C because the 100 mL sample has the greater volume.
   - (C) The final temperature will greater than 50 °C because the 50 mL sample has the higher temperature.
   - (D) The final temperature cannot be determined with the information given.

---

known as the *heat of combustion*. Heat of combustion is often expressed as kilocalories per mole or kilojoules per mole. However, to compare one fuel to another, it is useful to compare kilojoules *per gram*.

The energy transferred during a chemical change comes from bond breaking and bond making. It takes energy to break bonds. When new bonds are formed, energy is released. The net energy exchanged is the heat of the reaction, ΔH. When the reaction is exothermic, ΔH is negative. When the reaction is endothermic, ΔH is positive.

Reactions involving metals and metal compounds are good sources of energy. Oxidation-reduction reactions can be used to produce energy in electro-chemical cells. Large outputs of energy (measured as voltage) can be produced by pairing metals that are easily oxidized with metals that are not easily oxidized.

Energy can be transferred by the generation of electromagnetic radiation or light. Several models give insight into the nature of light. The ray model suggests that light travels in straight lines until it is reflected, transmitted, or absorbed when it strikes an object. If

white light is partially absorbed, the reflected or transmitted light has a color. The wave model is used to explain that different colors are waves all traveling at the speed of light but of varying wavelength and frequency. The photon model suggests that energy is transferred in small units, called *photons*. Photon energy increases with frequency. Electromagnetic radiation exists with a huge range of wavelengths and frequencies. Instruments, called *spectrometers*, use various wavelengths of electromagnetic radiation to probe matter. Spectroscopy extends our ability to see light that we would otherwise not be able to detect with only our eyes.

EVALUATE (5 min)

No check-in for this review.

Homework

Assign the Unit 5: Fire Review in the student text.

ANSWERS

GENERAL REVIEW

1. *Possible answers:* ● Fire is a combustion reaction that occurs when molecular covalent compounds combine with oxygen. ● Fire is composed of super-heated gases that emit light. ● A fire can be put out by removing its access to fuel or preventing oxygen from reaching the fuel. ● Fire is a redox reaction. Redox reactions are also important to internal combustion engines and extracting metals from their ores. ● Not all substances can burn. Some do not combust at all, and others only emit a flash of heat and light but do not burst into flames.

2. *Possible answer:* Chemical changes in which chemical bonds break and re-form during exothermic reactions release useful energy. Redox reactions such as combustion supply energy. These reactions release the potential chemical energy in fuels to heat our homes and power our technology. Our bodies also use redox reactions to release the potential chemical energy in the food we eat.

3. The thermal energy of an exothermic reaction is the result of bond making releasing more energy than is required for bond breaking for the reaction.

4. *Possible answer:* The first law of thermodynamics states that energy is conserved during chemical or physical changes. The second law of thermodynamics says that energy tends to disperse and only flows from a hotter object to a colder object.

5. a. 3,800 cal **b.** −7,500 cal

6. 1,300 cal

7. More energy is required to raise the temperature of water than to raise the temperature of aluminum by the same amount because water has a higher specific heat capacity than aluminum.

8. *Possible answer:* The difference in the latent heat of fusion tells you that it takes more than three times as much energy to melt a given mass of ice than to melt the same mass of solid ethanol. Also, three times as much energy is given off when a given mass of water freezes than when the same mass of ethanol freezes.

9. *Possible answers (any 3):* ● Mix the reactants. ● Add heat. ● Add energy in some form other than heat. ● Increase the surface area. ● Increase the concentration. ● Increase the pressure. ● Add a catalyst.

10. Work is a transfer of energy resulting from a force acting through a distance. 44,000 J

11. *Possible answer:* $Ca \rightarrow Ca^{2+} + 2e^-$

12. *Possible answer:*
$2Al(s) + 3Cu^{2+}(aq) \rightarrow 2Al^3(aq) + 3Cu(s)$

13. $Zn(s) + 2Mn^{4+}(aq) \rightarrow Zn^{2+}(aq) + 2Mn^{3+}(aq)$

Zinc loses electrons, so it is oxidized. Magnesium gains electrons, so magnesium ions are reduced.

$2Al(s) + 3Ca^{2+}(aq) \rightarrow 2Al^{3+}(aq) + 3Ca(s)$

Aluminum loses electrons, so it is oxidized. Calcium gains electrons, so calcium ions are reduced.

14. Atoms or ions that lose electrons are oxidized. Atoms or ions that gain electrons are reduced.

15. The ray model shows the light path using straight lines. The wave model shows light as waves and indicates that different types of light have different wavelengths and frequencies. The photon model shows light as particles. The drawing should show red light rays, waves, and photons moving through the thin red plastic film.

16. Radio waves are least harmful. This is because they have a long wavelength, a low frequency, and low energy.

17. A spectrometer consists of a light source, a prism to spread out the light, a sample, and a detector.

---

6. Which cup of water gets the hottest, if the initial temperature is the same?
(A) 160 calories of energy are transferred to 32 mL of water.
(B) 525 calories of energy are transferred to 105 mL of water.
(C) 2375 calories of energy are transferred to 475 mL of water.
(D) Each cup of water will reach the same final temperature.

7. The specific heat capacities for several substances are listed in the following table.

Substance	Specific heat capacity
water, H_2O	1.00 cal/g °C
methanol, CH_3OH	0.58 cal/g °C
brass, CuZn	0.090 cal/g °C
aluminum, Al	0.21 cal/g °C

Which substance would require the most energy to be tranferred to raise the temperature from 20 °C to 30 °C?
(A) water
(B) methanol
(C) brass
(D) aluminum

8. If the specific heat capacity is 0.49 cal/g °C for ice and 1.0 cal/g °C for liquid water, how much energy must be transferred to a 32 g sample of ice to raise its temperature from −4.0 °C to 24 °C? The heat of fusion of water is 79.9 cal/g °C.
(A) 770 cal
(B) 820 cal
(C) 3300 cal
(D) 3400 cal

9. The heating curve shows what happens as heat is added to water at a constant rate, starting with a piece of ice. What is present during time D?

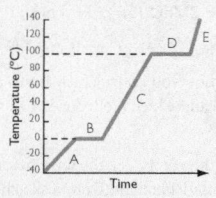

(A) both liquid water and ice
(B) both liquid water and water vapor
(C) only water vapor
(D) only liquid water

10. Which of the following describes a plausible combustion reaction with a balanced chemical equation?
(A) $C_6H_{14}(l) + 6 O_2(g) \longrightarrow 6 CO_2(g) + 7 H_2O(l)$
(B) $C_2H_4O_2(l) + 2 O_2(g) \longrightarrow 2 CO_2(g) + 2 H_2O(l)$
(C) $2 Li(s) + O_2(g) \longrightarrow Li_2O(s)$
(D) $2 MgO(s) + O_2(g) \longrightarrow 2 MgO_2(s)$

11. A student performed a calorimetry experiment where 1.3 g of crushed potato chips were burned to warm 25.0 mL of water. If the temperature of the water increased by 6.00 °C, how much energy per gram was released by the combustion of the crushed chips?
(A) 0.416 cal/g
(B) 115 cal/g
(C) 150 cal/g
(D) 195 cal/g

12. The combustion of 2.6 g of octane transferred energy to 926 mL of water. If the temperature of the water increased from 22 °C to 54 °C, how many kilojoules of energy were transferred to the water?
(A) 0.083 kJ
(B) 0.35 kJ
(C) 29 kJ
(D) 120 kJ

STANDARDIZED TEST PREPARATION

1. D	9. B	16. B
2. C	10. B	17. D
3. A	11. B	18. C
4. B	12. D	19. D
5. B	13. B	20. A
6. D	14. A	21. C
7. A	15. C	22. C
8. D		

ASSESSMENTS

Two Unit 5 Assessments (versions A and B) are available for download.

TRM Unit 5 Assessment Answers

TRM Unit 5 Assessments

A Lab Assessment for Unit 5 is available for download.

TRM Unit 5 Lab Assessment Answers

TRM Unit 5 Lab Assessment

Use the following table to answer Exercises 13 and 14.

Average Bond Energies (per mole of bonds)

Bond	H–H	C–H	C–C	C=C	C≡C	O–H	C–O	C=O	O=O	O–O
Bond energy (kJ/mol)	432	413	347	614	811	467	358	799*	495	146

*C=O in CO_2: 799, C=O in organic molecules: 745

13. What is the total energy required to break all the bonds in 1 mole of ethanol, C_2H_6O?

(A) 880 kJ/mol   (B) 1120 kJ/mol
(C) 3240 kJ/mol   (D) 4150 kJ/mol

14. Based on bond energies, what is the energy change for the combustion of propane?

$$C_3H_8(g) + 5\,O_2(g) \longrightarrow 3\,CO_2(g) + 4\,H_2O(g)$$

(A) −2060 kJ/mol
(B) 2060 kJ/mol
(C) −15,000 kJ/mol
(D) 15,000 kJ/mol

15. Which process could be represented correctly by the energy diagram?

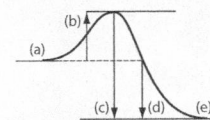

(A) An ice cube is melting at 0 °C.
(B) $H_2O(g) \longrightarrow H_2(g) + O_2(g)$ $\Delta H = +482$ kJ
(C) $2\,C_4H_{10}(g) + 13\,O_2(g) \longrightarrow$
$$8\,CO_2(g) + 10\,H_2O(g)$$
(D) Ammonium chloride is dissolved in water. The temperature cools by 16 °C.

16. Which of these processes will result in an increase of the speed of the process?

(A) Use cool water to dissolve sugar crystals.
(B) Crush charcoal into smaller pieces before lighting.
(C) Decrease the concentration of detergent to remove a stain.
(D) Put a rusty nail in a small container sealed with a lid.

17. What would you expect to observe if you added aluminum metal to a solution of copper (II) chloride? Use the activity series diagram below to answer.

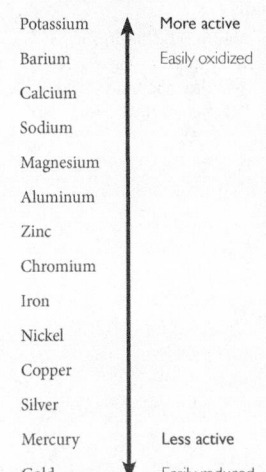

Activity Series

Potassium	More active
Barium	Easily oxidized
Calcium	
Sodium	
Magnesium	
Aluminum	
Zinc	
Chromium	
Iron	
Nickel	
Copper	
Silver	
Mercury	Less active
Gold	Easily reduced

(A) Aluminum would dissolve as copper oxidized.
(B) Copper would dissolve as aluminum oxidized.
(C) Copper would precipitate as copper oxidized.
(D) Copper would precipitate as aluminum oxidized.

Engineering Project: U5 Build a Better Calorimeter

As an extension to Unit 5: Fire, you can assign the Engineering Design Project: Build A Better Calorimeter. In this project, students extend what they have learned so far to refine their calorimeters using an engineering design cycle.

18. What occurs in an electrochemical cell using the two half-reactions shown below?

$$Li^+ + e^- \longrightarrow Li \qquad -3.05 \text{ V}$$
$$Zn^{2+} + 2e^- \longrightarrow Zn \qquad -0.76 \text{ V}$$

(A) Li is reduced; Zn is oxidized; cell is 2.29 V.

(B) Li is reduced; Zn is oxidized; cell is 3.81 V.

(C) Zn is reduced; Li is oxidized; cell is 2.29 V.

(D) Zn is reduced; Li is oxidized; cell is 3.81 V.

19. When visible light strikes a mirror, the light is mostly

(A) Absorbed by the mirror.

(B) Transmitted by the mirror.

(C) Absorbed and transmitted by the mirror.

(D) Reflected by the mirror.

20. Red light has a wavelength of 7.5×10^{-7} m and a frequency of 4.0×10^{14} Hz. Blue light has a shorter wavelength. Which statement is true about blue light?

(A) Blue light has a higher frequency compared with red light.

(B) Blue light has a greater speed compared with red light.

(C) Blue light has a lower energy compared with red light.

(D) Blue light has the same speed as red light but a lower frequency.

21. The wavelength of a radio wave is 300 m. What is the frequency of this wave?

(A) 10 Hz   (B) 1,000 Hz

(C) 1,000,000 Hz   (D) 100,000,000 Hz

22. Spectroscopy is a method that can be used to determine

(A) The speed of light.

(B) The density of a substance.

(C) The concentration of a solution.

(D) The mass of a substance.

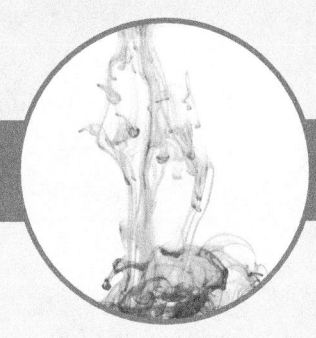

Unit 6 | Showtime

Reversible Reactions and Chemical Equilibrium

SHOWTIME AS CONTEXT

As the *Living by Chemistry* course reaches the culminating Unit 6: Showtime, it is time for students to reflect on what they have learned throughout the year. This unit serves as a capstone for the whole course. The unit reflects on topics from earlier units, such as precipitation reactions and acid-base chemistry. Students go further with these topics to explore the concept of reversible processes that have a dynamic balance between starting substances and products. The unit introduces chemical equilibrium and Le Châtelier's principle.

IT'S SHOWTIME!

Throughout the unit, students will consider the behavior of molecules in the human body, the reversible processes of dissolving and precipitation, weak acids in aqueous solutions, and colorful molecules called indicators. At the end of the unit, students have the opportunity to complete a research project that involves designing a lab procedure for extracting colorful molecules (pigments) from flowers, leaves, fruits, berries, vegetables, spices, bark, or roots to examine the acid-base properties of a naturally occurring pigment. Designing this lab experiment with these colorful molecules ties together concepts from the entire course and allows students to *show what they know* in a fun and colorful way.

CONTENT DRIVEN BY CONTEXT

The context of this unit allows students an opportunity to investigate the abstract and often daunting concept of equilibrium in a way that is enjoyable and meaningful. To help keep students focused, they are first given observable changes to explain, rather than complex calculations.

Once a conceptual understanding of equilibrium is established, students move on to the quantitative aspect of equilibrium as they write equilibrium-constant equations and solve for K. In Unit 5: Fire, we discussed the possibility of *reversing* reactions, from an energetic point of view. In Unit 6: Showtime, the focus is on reversible processes. A reversible process is one in which the forward and reverse processes are both occurring simultaneously, and products and starting substances are both present in the system. The color changes of acid-base indicator molecules in aqueous solutions as the solution pH varies provide vivid examples of reversible processes. Many naturally occurring pigments are indicator molecules.

THE USE OF MODELS TO DESCRIBE EQUILIBRIUM

The subject of equilibrium can be quite challenging for students as the dynamic nature of equilibrium is not always apparent. Often, students think of balance as a static situation. However, systems at equilibrium are continuously changing. A tightrope walker can be a good analogy for dynamic equilibrium. The performer may move an arm or may waver but continues to stay balanced. One of the practical issues in teaching about systems at equilibrium is that changing conditions are usually unobservable to the eye. This is why color-changing processes are so powerful in demonstrating the effects of changing conditions. Students can make observations, then try to speculate about what must be happening on a particle level. To further assist students with the idea of dynamic equilibrium, students model reversible systems with manipulatives to highlight what is occurring on the particle level. This unit also continues to make use of particle-level diagrams to help students visualize what is occurring.

BUILDING UNDERSTANDING

The study of ions in Unit 1: Alchemy, of molecular structure and intermolecular attractions in Unit 2: Smells, of phase changes in Unit 3: Weather, of reactions, solutions, and acid-base chemistry in Unit 4: Toxins, and of thermal equilibrium, and color and light in Unit 5: Fire have all laid the foundation for the study of equilibrium in Unit 6: Showtime.

Chapter Summary

Chapter	Description	Standard Schedule Days	Block Schedule Days
23	In **Chapter 23** students explore reactions that are reversible and what goes on in an equilibrium mixture. The mathematical relationship between concentrations of starting substances and products in an equilibrium mixture is introduced.	5.5	3.5
24	In **Chapter 24**, students consider what happens when conditions are changed, for example, by adding an acid or base to an equilibrium mixture. Le Châtelier's principle is introduced to assist students in predicting what changes might occur. Students use the concepts that they have learned to design a lab experiment to extract a naturally occurring pigment and examine its properties in order to show what they have learned in the course.	6.5	4.5

Pacing Guides

Standard Schedule

Day	Suggested Plan	Day	Suggested Plan
1	Chapter 23 Lesson 117	7	Chapter 24 Lesson 123
2	Chapter 23 Lesson 118	8	Chapter 24 Review
3	Chapter 23 Lesson 119	9	Chapter 24 Quiz
4	Chapter 23 Lesson 120	10	Unit Review
5	Chapter 23 Lesson 121, Chapter 23 Review	11	Chapter 24 Presentations
6	Chapter 23 Quiz, Chapter 24 Lesson 122	12	Unit Exam

Block Schedule

Day	Suggested Plan	Day	Suggested Plan
1	Chapter 23 Lessons 117 and 118	5	Chapter 24 Review
2	Chapter 23 Lessons 119 and 120	6	Chapter 24 Quiz, Unit Review
3	Chapter 23 Lesson 121, Chapter 23 Review	7	Chapter 24 Presentations
4	Chapter 23 Quiz Chapter 24 Lessons 122 and 123	8	Unit Exam

Unit 6 | Showtime

Coral reefs support a wondrous diversity of life in a rich habitat. Yet coral reefs are highly sensitive to the acidity of ocean water. As oceans become more acidic because of carbon dioxide released into the atmosphere by human activity, the balance favored by corals is thrown off, affecting their well-being.

In this unit, you will learn

- what happens in a chemical system at equilibrium

- how the balance in a reversible process is maintained

- the mathematical relationship between starting substances and products in a system at equilibrium

- about variables that affect reversible processes

Why Showtime?

As the *Living by Chemistry* course comes to the culminating Unit 6: Showtime, it is time for you to reflect on what you have learned and show what you know. You will explore reversible processes that have a dynamic balance between starting substances and products. This balance is known as chemical equilibrium. You will consider the behavior of molecules in the human body, the reversible processes of dissolving and precipitation, weak acids in aqueous solutions, and colorful molecules called indicators. You'll complete a research project that involves extracting colorful molecules from flowers or berries and then manipulating the solution color in a showy ending to the course.

607

Watch the video overview of Chapter 23 (for teachers) by clicking on the link in the TE-book, opening the TRFD, or logging onto the book's companion Web site bcs.whfreeman.com/livingbychemistry2e (teacher log-in required).

Chapter 23 begins by introducing students to the types of reactions that are reversible and what goes on in an equilibrium mixture. In Lesson 117, students explore the color changes that occur with indicators in solution. In Lesson 118, students apply the concept of reversibility to substances that cause a sweet taste when the molecules in these substances bind to taste receptors on the tongue. Lesson 119 is a paper-clip activity that allows students to simulate and compare reversible and irreversible processes. In Lesson 120, students learn about the equilibrium constant, *K*, and how it relates to pain-reducing medicine and the degree of dissociation of strong versus weak acids. In Lesson 121, students begin to solve problems using the equilibrium constant.

In this chapter, students will learn

- to distinguish between reversible and irreversible processes
- to write equations for reversible processes
- to define chemical equilibrium and describe it on a particle level
- to characterize the mathematical relationship between reactants and products in an equilibrium mixture
- to write, interpret, and use equilibrium expressions

CHAPTER 23
Chemical Equilibrium

In this chapter, you will study

- reversible chemical and physical processes
- the dynamic nature of chemical equilibrium
- the mathematical relationship between products and starting substances at equilibrium

Hot water running through the limestone (calcium carbonate) under parts of Yellowstone National Park dissolves the limestone. Hot springs leave enormous deposits above ground as the water cools and limestone precipitates as a result of this reversible process.

Many chemical processes are reversible. Unlike most combustion reactions, reversible processes do not proceed all the way to the products. In reversible chemical systems, both the forward and reverse processes occur at the same time, so starting substances and products exist in the same mixture. When two opposite processes are in balance, the result is an equilibrium mixture. To the eye, the equilibrium mixture may appear unchanging, but this is not the case. Changes are continuously happening on the atomic scale in systems at equilibrium.

608

Chapter 23 Lessons

How Awesome

Acid-Base Indicators

THINK ABOUT IT

The same "red" cabbage plant is known by different colors in various regions, depending on the pH of the soil in which it is grown and how people describe the color in their native language. The leaves can be reddish, purple, or blue. If you cook red cabbage, it normally turns blue. But you can get the red color back by adding vinegar to the pot. What is the chemistry behind these color changes?

The color of red cabbage depends on compounds in the soil where it is grown.

What do color changes indicate?

To answer this question, you will explore

⟳ Molecular Views of Indicator Color Changes

⟳ Chemical Symbols to Represent Indicator Color Changes

Red cabbage turns blue when you cook it in water. It remains red if you add vinegar.

EXPLORING THE TOPIC

Colors of cabbage juice solutions at different pH values

⟳ **Molecular Views of Indicator Color Changes**

In Unit 4: Toxins, you explored cabbage juice as an **indicator** of the pH of a solution. As shown in the figure, the color of the cabbage juice indicator changes from red at pH ~3, to purple at pH ~6, and to blue at pH ~9. At higher pH values, the yellow color of a second indicator molecule is evident.

THE PIGMENT ANTHOCYANIN

A **pigment** is a substance that is colored because of the light it reflects. Instead of reflecting all the colors in white light from the light source, a pigment only reflects certain colors. For example, the leaves of red cabbage contain a pigment that reflects red light.

Lesson 117 | **Acid-Base Indicators**   609

Overview

LAB: GROUPS OF 4

Key Question: What do color changes indicate?

KEY IDEAS

Color changes can provide information about chemical reactions on a molecular level, while fascinating observers at the same time with unexpected outcomes. Certain molecules change color in solution depending on the environment. One class of color-changing molecules is called an *acid-base indicator,*

HIn, where In⁻ represents a cluster of atoms. The molecule HIn is a weak acid that dissociates (breaks apart) into H⁺ and In⁻ ions. In an acidic solution, there is plenty of H⁺ such that the indicator molecule is present as HIn. In a basic solution, the base reacts with H⁺ to produce In⁻ ions. Because the HIn molecules and In⁻ ions have different colors, the color of the solution changes when base is added. Thus, the color of the solution indicates the acid-base properties of the solution. A goal of this unit is to develop an understanding of these color changes.

LEARNING OBJECTIVES

● Describe color changes of an indicator as the pH changes.

● Explain what happens on a molecular level when an indicator changes color.

FOCUS ON UNDERSTANDING

● There are many different indicator molecules. Because they have large molecular formulas, we will abbreviate the formulas as H*In*. Universal indicator is a mixture of several different indicator molecules.

● Students have difficulty understanding that the color of an indicator solution is due to a mixture of H*In* molecules and *In*⁻ ions, which have two different colors.

● The indicator that changes from yellow to blue in universal indicator is bromothymol blue. For this indicator, *In*⁻ represents $C_{27}H_{28}Br_2O_5S^-$.

● Effervescent tablets typically consist of sodium bicarbonate and citric acid; the tablets release carbon dioxide when dissolved in water.

● Effervescent tablets are often called *antacids* because sodium bicarbonate, $NaHCO_3$, neutralizes excess stomach acid. However, sodium bicarbonate *also neutralizes base.*

Neutralizing acid: $HCO_3^-(aq) + H^+(aq) \rightarrow H_2CO_3(aq)$

Neutralizing base: $HCO_3^-(aq) + OH^-(aq) \rightarrow CO_3^{2-}(aq) + H_2O(l)$

KEY TERMS

indicator

pigment

IN CLASS

Students first consider why it is possible to make green food coloring by mixing yellow and blue. They describe the green solution as a mixture of yellow and blue, which our brain perceives as green. Students then explore the color changes that occur when an effervescent tablet is added to a basic solution mixed with universal indicator. The solution color changes from blue to green to yellow as the antacid reacts with the solution. In this case, the initial blue color of the solution is due to the presence of *In*⁻ ions. The effervescent tablet adds H⁺ ions, which react with the *In*⁻ ions to form H*In* molecules, thereby turning the solution yellow. The solution color changes as the amounts of *In*⁻ ions (blue) and H*In* molecules (yellow) in the mixture changes. A complete materials list is available for download.

TRM Materials List 117

SETUP

Prepare 16 100 mL beakers with 25 mL of NaOH solution that has a pH ~ 9–10. You will need about 500 mL of NaOH solution to divide among 16 beakers. Prepare 8 sets of dropper bottles containing 10 mL of universal indicator solution. You will need about 100 mL of universal indicator solution to divide among eight bottles.

SAFETY

Students must wear safety goggles at all times. The base used is dilute (pH ~ 9–10). If students spill base on their skin, rinse thoroughly. Use citric acid to neutralize any spills.

CLEANUP

The solutions can be diluted disposed of according to local guidelines.

Lesson Guide

ENGAGE (5 min)

ChemCatalyst

Imagine that you have yellow and blue solutions of food coloring. The yellow solution contains dye molecules that reflect yellow light. The blue solution contains dye molecules that reflect blue light. To make frosting for cupcakes, you mix the yellow and blue solutions to make a green solution and add it to white frosting. The result is a green-colored frosting.

● What do you think happens on a molecular level to make the solution appear green when yellow and blue are mixed? Provide two possible explanations.

Sample answers: **1.** The yellow and blue dye molecules might react to form green dye molecules. **2.** The yellow and blue dye molecules simply mix, and your eyes mix yellow and blue so that your brain perceives green.

Ask: Food coloring is a collection of molecules that impart color. Why does your skin appear blue if you spill blue food coloring on your hand?

Ask: Over time, the blue color on your skin fades. Why do you think this happens?

Ask: If you spill yellow food coloring on top of the blue stain on your hand, do you think you will observe both yellow and blue? Why or why not?

Ask: How would you explain why yellow and blue dye solutions mix to give a green solution?

Optional demonstration: Place a drop of a green solution of food coloring on a piece of filter paper. As the liquid spreads, you will see separation into regions of blue and yellow, showing

Anthocyanins give these berries their colors.

The pigments in red cabbage are called *anthocyanins*. Anthocyanins are found in blueberries, cranberries, raspberries, cherries, grapes, eggplant, and the flower petals of violets.

The anthocyanins that give red cabbage its color are large, complex molecules. Instead of writing out their molecular formulas, it is common to abbreviate the anthocyanin molecule as H*In*. H*In* is a weak acid, which means that it partially dissociates (breaks apart) into H$^+$ cations and *In*$^-$ anions when dissolved in water. The H*In* anthocyanin molecule is a red color. The *In*$^-$ anions that form when the anthocyanin molecules dissociate are a blue color.

Anthocyanins are called indicators because they change color when the pH of the solution they are dissolved in changes. For simplicity, all indicators in this unit will be abbreviated as H*In*. And for the purpose of this lesson, we will refer to the H*In* molecule as one form of the indicator, and the *In*$^-$ ion as a second form. The color of a solution that the indicator is added to depends on how much of each form of the indicator is present in that solution.

The anthocyanin indicator is abbreviated as H*In* when in acidic solution (structure on the left) and as *In*$^-$ when in basic solution (structure on the right). Notice that the *In*$^-$ anion is missing an H$^+$.

CHANGES IN COLOR

When an acid-base indicator is placed in an acidic solution, mostly H*In* molecules will be present in the solution. When the indicator is placed in a basic solution, mostly *In*$^-$ anions will be present in the solution. The anthocyanin indicator is red in acid (H*In*) and blue in base (*In*$^-$). When the indicator is placed in a neutral solution, both molecules and anions, H*In* and *In*$^-$, are present. Because both the red and blue forms are present, the mixture appears purple. If the indicator is placed in a solution that is slightly acidic, the solution contains slightly more red H*In* molecules, and its color is red-purple. If it is placed in a solution that is slightly basic, there are more blue *In*$^-$ anions, and its color is blue-purple.

Molecular views of a tiny volume of solution are shown in the figure on the next page. There are 20 anthocyanin molecules in this tiny volume, either as H*In* molecules (red circles) or as *In*$^-$ anions (blue circles). Notice that there are no purple circles. Solutions that are varying shades of purple do not contain molecules that reflect purple light. Instead, these solutions reflect red and blue light. Our brains perceive the combination of red and blue as purple.

ENVIRONMENTAL CONNECTION

Besides its use as an indicator, anthocyanin is responsible for spectacular fall foliage. At the end of the growing season, when the chlorophyll in plants and trees decays, the red and purple colors of anthocyanins become visible, contributing to colorful foliage displays. Every autumn, "leaf peeping" draws many people to Vermont.

Big Idea The color of an indicator solution varies with pH because the solution contains varying mixtures of H*In* molecules and *In*$^-$ anions.

that the blue and yellow dye molecules are still present in the green solution.

EXPLORE (15 min)

TRM Worksheet with Answers 117

TRM Worksheet 117

INTRODUCE THE LAB

➜ Tell students to work in groups of four.

➜ Pass out the student worksheet to each group.

➜ Remind students about safety.

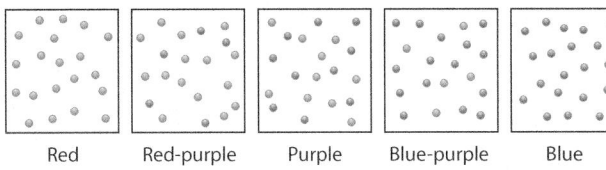

| Red | Red-purple | Purple | Blue-purple | Blue |

Notice that the square on the left has only red circles, representing HIn. The square on the right has only blue circles, representing In⁻. The three in the middle are mixtures of HIn and In⁻. The mixtures are varying shades of purple.

Chemical Symbols to Represent Indicator Color Changes

Molecular views are a good way to illustrate how the mixture of molecules in solution relates to the observation of solution color. Chemical equations are also useful for expressing the color changes that occur with indicator molecules in solution.

ANTHOCYANIN IN A BASIC SOLUTION

When anthocyanin is added to a solution that is basic, with pH ~9, the solution turns blue because the indicator is present mostly as In^- anions. If H^+ ions are added to the basic solution with In^- ions, H^+ combines with In^- anions to form HIn molecules. The HIn molecule is red. As blue In^- anions are converted to red HIn molecules, the solution appears varying shades of purple. Once all the In^- anions have combined with H^+, the solution turns red. The following balanced chemical equation describes the process that occurs as H^+ ions are added.

$$H^+(aq) + In^-(aq) \longrightarrow HIn(aq)$$
$$\text{blue} \qquad\qquad \text{red}$$

ANTHOCYANIN IN AN ACIDIC SOLUTION

When anthocyanin is added to a solution that is acidic, the solution turns red because the indicator is present mostly as HIn molecules. If OH^- anions are then added to the acidic solution, the solution changes color from red to purple to blue. This is evidence that the number of In^- anions in the solution increases as the pH increases. A reasonable conclusion to draw from these observations is that adding base to an acidic solution reverses the process described above of adding an acid to a basic solution. The following balanced chemical equation describes the process that occurs as OH^- is added.

$$HIn(aq) + OH^-(aq) \longrightarrow H_2O(l) + In^-(aq)$$
$$\text{red} \qquad\qquad\qquad\qquad \text{blue}$$

Although this equation does not appear to be an exact reversal of the first equation that describes adding H^+, it can be written as two equations: one that shows that OH^- combines with H^+ to make H_2O, and another that shows the HIn molecules dissociate into H^+ and In^- ions.

$$HIn(aq) \longrightarrow H^+(aq) + In^-(aq)$$
$$\text{red} \qquad\qquad\qquad \text{blue}$$

$$H^+(aq) + OH^-(aq) \longrightarrow H_2O(l)$$

When an acid-base indicator is placed in an acidic solution, mostly HIn molecules will be present in the solution. When the indicator is placed in a basic solution, mostly In^- anions will be present in the solution. When the indicator is placed in a neutral solution, both molecules and anions, HIn and In^-, are present. The indicator bromothymol blue is yellow in solutions that are acidic and blue in solutions that are basic. In acid, HIn molecules are mostly present; these molecules reflect yellow. In base, In^- ions are mostly present; this is the cause of the blue color. In acid solutions, there is plenty of H^+, so the HIn molecules (yellow) stay intact. The OH^- in base solutions reacts with H^+ from HIn to form water, producing In^- ions (blue). Neutral solutions consist of a mixture of both yellow HIn molecules and blue In^- ions. The mixture in neutral solution appears green.

DISCUSS ACID-BASE REACTIONS OF INDICATORS

→ Write the reactions on the board:

$$H^+(aq) + In^-(aq) \to HIn(aq)$$
$$HIn(aq) + OH^-(aq) \to H_2O(l) + In^-(aq)$$
$$HIn(aq) \to H^+(aq) + In^-(aq)$$
$$H^+(aq) + OH^-(aq) \to H_2O(l)$$

Ask: What do you observe when you add the effervescent tablet to the basic solution with universal indicator?

Ask: Does the effervescent tablet add acid or base to the solution? Explain how you know.

Ask: How can you represent the reaction when H^+ is added to In^-?

Ask: How can you represent the reaction when OH^- is added to HIn?

Ask: How would you explain why the reaction with OH^- is the reverse of the reaction with H^+?

Key Points: When the effervescent tablet is added to the basic solution, the pH decreases, indicating that the tablet is adding H^+. As H^+ combines with In^- in the blue basic solution, the number of HIn molecules increases. As blue In^- ions are converted to yellow HIn molecules, the solution appears varying shades of green. Once all the In^- ions have combined with H^+, the solution is yellow. The reaction that occurs is shown below.

$$H^+(aq) + In^-(aq) \to HIn(aq)$$
$$\text{blue} \qquad\qquad\qquad \text{yellow}$$

DISCUSS MOLECULAR VIEWS OF THE COLOR CHANGES OF INDICATORS

→ Ask students to draw molecular views of solutions that are yellow, blue, and various shades of green.

Ask: Which is present in larger quantity in basic solution: HIn or In^-? Explain your thinking.

Ask: Draw a molecular view of the indicator in basic solution.

Ask: Which is present in larger quantity in acid solution: HIn or In^-? Explain your thinking.

Ask: Draw a molecular view of the indicator in acid solution.

Ask: Bromothymol blue is blue in base and yellow in acid. Draw a molecular view of a green solution.

Key Points: Acid-base indicators change color depending on solution pH. Because acid-base indicators are large, complex molecules, we will use the symbol HIn to represent various indicators such as methyl orange, bromothymol blue, phenolphthalein, and cabbage juice. The universal indicator used in the experiment for this lesson is a mixture of several different indicators. The specific indicator that changes color from blue to green to yellow is called bromothymol blue.

When OH⁻ is added to an acidic solution with H*In*, the reverse reaction occurs. The OH⁻ added reacts with H*In* to form H_2O and *In*⁻. As the reaction proceeds, the yellow H*In* molecules are converted to the blue *In*⁻ ions. Once all the H*In* molecules have reacted with OH⁻, the solution is blue. The reaction that occurs is shown below.

$$HIn(aq) + OH^-(aq) \rightarrow H_2O(l) + In^-(aq)$$
yellow   blue

These two reactions are the reverse of one another. At first glance, the second reaction might appear different from the first. However, it can be written as two equations: one that shows that OH⁻ combines with H⁺ to make H_2O and another that shows the H*In* molecules dissociate into H⁺ and *In*⁻ ions.

$$HIn(aq) \rightarrow H^+(aq) + In^-(aq)$$
$$H^+(aq) + OH^-(aq) \rightarrow H_2O(l)$$

These two chemical equations describe the same process as the single chemical equation above for adding OH⁻. It is possible to show this by adding the two chemical equations and canceling identical substances on both sides of the arrow.

$$HIn(aq) + OH^-(aq) + \cancel{H^+(aq)} \rightarrow$$
$$H_2O(l) + \cancel{H^+(aq)} + In^-(aq)$$

Bromothymol blue when base is added to an acidic solution:

$$HIn(aq) \rightarrow H^+(aq) + In^-(aq)$$
yellow   blue

Bromothymol blue when acid is added to a basic solution:

$$H^+(aq) + In^-(aq) \rightarrow HIn(aq)$$
blue   yellow

EVALUATE (5 min)

Check-In

Cabbage juice can be used as an indicator. The H*In* molecules are red and the *In*⁻ ions are blue.

1. Complete the equation below to show what happens when H⁺ is added slowly to a blue solution of the cabbage juice indicator, *In*⁻.

$$H^+(aq) + In^-(aq) \rightarrow \underline{\hspace{1cm}}$$

2. Predict the color changes that you will observe as the reaction proceeds.

In a few species of plants, the color of the flower depends on chemistry. The flowers of hydrangeas, especially the species *Hydrangea macrophylla*, tend to be blue in acidic soil (pH below about 6) and pink in neutral and alkaline soil (pH above about 7). What does the color of these blossoms indicate?

Tohoku Color Agency/Getty Images

KEY TERMS

indicator

pigment

These two chemical equations describe the same process as the single chemical equation above for adding OH⁻. It is possible to show this by adding the two chemical equations and canceling identical substances on both sides of the arrow.

$$HIn(aq) + \cancel{H^+(aq)} + OH^-(aq) \longrightarrow H_2O(l) + \cancel{H^+(aq)} + In^-(aq)$$

It is easy to see now that adding acid to a basic solution of anthocyanin and adding base to an acidic solution of anthocyanin are processes that are the reverse of one another.

Anthocyanin in a Basic Solution

$$H^+(aq) + In^-(aq) \longrightarrow HIn(aq)$$
            blue   red

Anthocyanin in an Acidic Solution

$$HIn(aq) \longrightarrow H^+(aq) + In^-(aq)$$
   red   blue

LESSON SUMMARY

What do color changes indicate?

Color changes can provide information about chemical processes on a molecular level. One class of color-changing molecules is called an indicator. An indicator is a large, complex molecule that is abbreviated with the symbol H*In*. When placed in solution, indicator molecules can exist as H*In* molecules or as *In*⁻ anions. This is because the H*In* molecules can dissociate into H⁺ and *In*⁻ ions. The reverse process can also occur in which the H⁺ and *In*⁻ ions combine to form H*In* molecules. In an acidic solution, there is plenty of H⁺ such that the indicator is present as H*In* molecules. Adding base removes H⁺ to produce *In*⁻ anions. Because the H*In* molecules and *In*⁻ anions are different colors, the color of the solution indicates the acid-base properties of the solution.

Exercises

Reading Questions

1. Explain how molecules account for the purple cabbage leaves.

2. Explain how the two forms of an indicator are related to one another.

Reason and Apply

3. Methyl red is an indicator that is red in acidic solution and yellow in basic solution. Draw three molecular views to show how the color changes from red to yellow. What color do you observe in between?

4. Write a balanced chemical equation to describe what happens as the color changes from red to yellow for methyl red indicator using the general symbols for an indicator molecule and its anion.

5. Write a balanced chemical equation to describe what happens as the color changes from yellow to red for methyl red indicator.

6. **PROJECT** Research an acid-base indicator and explain how it works. (Do not use the universal indicators bromothymol blue, methyl red, or anthocyanin.)

Answers: **1.** $H^+(aq) + In^-(aq) \rightarrow$ H*In*(aq). **2.** The color changes from blue to purple to red.

Homework

Assign the reading and exercises for Showtime Lesson 117 in the student text.

LESSON 117 ANSWERS

1. Both H*In* (red) and *In*⁻ (blue) are present.

2. *In*⁻ ions are present in solution when H*In* molecules dissociate.

3. The view for a red acidic solution should show mainly H*In* molecules. The view for a yellow basic solution should show mainly *In*⁻ ions. The color for the solution in between is orange. The view for the solution in between should show both H*In* molecules and *In*⁻ ions.

4. $HIn \rightarrow H^+ + In^-$

5. $H^+ + In^- \rightarrow HIn$

6. A good report could include:
● a description of the indicator molecule and how it changes in solutions of different pH ● correct molecular formula for the indicator molecule ● balanced equation to show what happens to the molecule in acidic, basic, and neutral solutions ● a molecular view drawing of what occurs in solutions of differing pH.

How Backward

Reversible Processes

THINK ABOUT IT

Imagine that you open a bottle of perfume. You detect a sweet smell that remains quite intense as long as the bottle is open. This indicates that molecules are attaching to receptors in your nose and causing a sensation by stimulating nerves that send signals to your brain. Once you close the bottle, the sweet smell grows fainter over time. Why don't you continue to detect the sweet smell?

What is a reversible process?

To answer this question, you will explore

⟳ Reversible Processes: The Double Arrow

⟳ Examples of Reversible Processes

⟳ Degree of Sweetness

EXPLORING THE TOPIC

⟳ Reversible Processes: The Double Arrow

Based on what you have learned so far in this course, it might seem that chemical processes proceed only in the direction in which they are written as a chemical equation. For example, here is the chemical equation that describes gasoline burning.

$$2C_8H_{18}(l) + 25O_2(g) \longrightarrow 16CO_2(g) + 18H_2O(g)$$

When gasoline burns, the reaction goes in the forward direction, from reactants to products. This reaction is said to *go to completion,* which means that all reactants are completely converted to products. Although it *is* possible for gasoline to be created through chemical reactions, it takes many chemical steps and requires a great deal of energy. So this reaction, as it is written, is not reversible. Once you burn the gasoline, there is no going back.

In contrast, there are many **reversible processes** that proceed in both the forward and reverse directions. The term *process* can describe either a physical process or a chemical process (chemical reaction). In Lesson 70, you learned that chemical equations are used to represent both chemical and physical processes.

Physical process	Chemical process
$H_2O(l) \longrightarrow H_2O(s)$	$2Na(s) + Cl_2(g) \longrightarrow 2NaCl(s)$
$Br_2(l) \longrightarrow Br_2(g)$	$CH_4(g) + 2O_2(g) \longrightarrow CO_2(g) + 2H_2O(l)$
$I_2(s) \longrightarrow I_2(g)$	$N_2O_4(g) \longrightarrow 2NO_2(g)$

Lesson 118 | **Reversible Processes**   613

Overview

WORKSHEET: GROUPS OF 4

Key Question: What is a reversible process?

KEY IDEAS

A reversible process is one that proceeds freely in both the forward and the reverse directions. Many chemical processes are reversible given the appropriate conditions. For reversible processes, it is possible to obtain mainly products or to drive the process in the reverse direction to obtain mainly starting substances. For

example, it is possible to evaporate water by heating and then to condense the water vapor by cooling. Under certain conditions, the forward and reverse processes can occur simultaneously, such that both starting substances and products are present in the same mixture (for instance, bubbles of gaseous water in liquid water when water boils). In a chemical equation, a double arrow (⇌) is used to symbolize reversible processes. Several types of reversible processes will be considered in this unit, including phase changes, dissolving and precipitation, acid-base reactions, and binding to protein receptors.

LEARNING OBJECTIVES

● Explain and give examples of reversible processes.
● Define and describe reversible processes.

FOCUS ON UNDERSTANDING

● Some processes can be reversed if the conditions are manipulated.
● In reversible processes, the meaning of reactant (or starting substances) and product becomes blurred. Both can be mixed together.
● Reversible processes are very important in the human body.

KEY TERM

reversible process

IN CLASS

Reversible processes are introduced, including phase changes, dissolving and precipitation, and further analysis of the indicator experiment in the previous lesson. Students consider how to represent these reversible processes by a chemical equation with double arrows, and they describe what happens in the forward and reverse directions. Students apply the concept of reversibility to substances that cause a sweet taste when the molecules in these substances bind to taste receptors on the tongue. The sweet taste "wears off" with time as the process is reversed and the sweet molecules are removed from the taste receptors and swallowed. Students then consider substances with different levels of sweetness for the same number of molecules. One hypothesis is that the molecules of the sweeter substances remain bound to the taste receptors for a longer period of time. The more that are bound for the same number, the sweeter the taste sensation. A complete materials list is available for download.

TRM Materials List 118

Pre-AP® Course Tip

Students preparing to take an AP® chemistry course or college chemistry should be able to relate atomic level models to macroscopic phenomena. In this lesson, students consider how certain molecules create different degrees of sweetness when the molecules bind to the receptors on their tongue. The idea of reversible processes is explored through this context.

Lesson Guide

TRM PowerPoint Presentation 118

ChemCatalyst

If you were to place a sugar cube in your mouth, the sugar would dissolve in your saliva. Some of the molecules of sucrose would attach to specific taste receptors on your tongue. This produces a signal to the brain of a sweet taste. Eventually, the sweet taste will fade.

● Why do you think that the perception of sweetness wears off after a certain span of time?

Sample answers: **1.** The molecules come back off the taste receptor, and when you swallow them, they are no longer present to be detected. **2.** The molecules break apart into smaller molecules that do not taste sweet. **3.** The molecules modify the taste receptor so that it no longer works, and your body has to make more receptors.

Ask: Why do only certain molecules produce a sweet taste?

Ask: Once you perceive sweetness, why do you think it wears off?

Ask: How would you explain why it is an advantage to the body that molecules can bind and unbind to receptor sites that send signals to the brain?

TRM Worksheet with Answers 118

TRM Worksheet 118

INTRODUCE THE LAB

→ Tell students to work in groups of four.

→ Pass out the student worksheet to each group.

DISCUSS REVERSIBLE PROCESSES

→ Write these processes on the board:

$$H_2O(l) \rightleftharpoons H_2O(g)$$
$$NaCl(s) \rightleftharpoons Na^+(aq) + Cl^-(aq)$$
$$HIn(aq) \rightleftharpoons H^+(aq) + In^-(aq)$$
$$\text{geraniol} + \text{receptor} \rightleftharpoons [\text{geraniol}:\text{receptor}]$$

As written, the processes in the table appear to move in one direction only. However, some of these processes are reversible.

Physical process	Chemical process
$H_2O(l) \rightleftharpoons H_2O(s)$	$2Na(s) + Cl_2(g) \longrightarrow 2NaCl(s)$
$Br_2(l) \rightleftharpoons Br_2(g)$	$CH_4(g) + 2O_2(g) \longrightarrow CO_2(g) + 2H_2O(l)$
$I_2(s) \rightleftharpoons I_2(g)$	$N_2O_4(g) \rightleftharpoons 2NO_2(g)$

Notice that for a reversible process, a double arrow is used in the equation. The double arrow indicates that both a forward and a reverse process can occur. A reversible process can be written in either the forward or reverse direction. All phase changes are reversible, while many chemical processes are not.

Big Idea All physical processes and some chemical processes are reversible.

↻ Examples of Reversible Processes

As is evident from the second table above, phase changes are reversible. Other common reversible processes involve processes in aqueous solution, such as dissolving and precipitation and acid-base reactions, and processes in the human body involving binding to a receptor. Reversible processes can occur in both the forward and reverse directions. Depending on the conditions, it is possible to cause these processes to proceed exclusively in one direction, or to create a mixture of starting substances and products. This is explained further with the examples below.

PHASE CHANGES

Water going from liquid to gaseous state.

Consider water boiling in a pot on the stove. The liquid water evaporates and spreads out in the air. If the gaseous water is cooled, the reverse process occurs and water droplets form. Intermolecular forces, including hydrogen bonds and dipole-dipole attractions, account for the attraction between water molecules. Heating easily breaks these forces of attraction, and they re-form by cooling. One way to represent this reversible process of evaporating and condensing is with a double arrow as shown in the equation below.

$$H_2O(l) \rightleftharpoons H_2O(g)$$

DISSOLVING AND PRECIPITATION

Suppose you dissolve a small amount of salt in water to create an aqueous salt solution. If you evaporate the water, the reverse process occurs and solid salt precipitates. The dominant force that accounts for the formation of the solid is cation-anion attractions. Upon dissolving, the ions are attracted to the water molecules by ion-dipole attractions. The difference in the strength of these forces is small enough such that the reaction is easy to reverse. One way to represent this

Ask: How would you describe what you would observe if you carried out these processes?

Ask: Why is a double arrow used?

Ask: What evidence do you have that both starting substances and products are present in a reversible process?

Ask: How were you able to reverse the acid-base indicator reaction in the previous lesson?

Ask: Are all processes reversible? Explain your thinking.

Key Points: A reversible process is one that proceeds freely in both the

forward and the reverse directions. For such processes, it is possible to obtain mainly products or to drive the process in the reverse direction to obtain mainly starting substances. For example, it is possible to evaporate water and then condense the water vapor. Under certain conditions, the forward and reverse processes can occur simultaneously, such that both starting substances and products are present in the same mixture (for instance, the bubbles of gaseous water in liquid water when water boils). For a salt dissolving, you can add water to dissolve the salt and then evaporate the water to reverse the process. For acid-base indicators,

In a flask of salt solution, some of the salt remains on the bottom of the flask.

reversible process of dissolving and precipitating is with a double arrow as shown in the chemical equation below.

$$NaCl(s) \rightleftharpoons Na^+(aq) + Cl^-(aq)$$

ACID-BASE REACTIONS

When an acid is dissolved in water, the acid molecules dissociate into H^+ and an anion. This process can also go in the reverse direction, in which the ions are attracted to one another to form the acid molecule, depending on the pH of the solution.

In the previous lesson, you examined a solution of an acid-base indicator, HIn. Adding base to an aqueous solution causes the HIn molecules to dissociate into H^+ and In^- ions. Adding acid causes the reverse process in which the ions combine to form HIn molecules.

A solution of the indicator bromothymol blue is yellow in acidic solutions due to the presence of mostly HIn molecules, blue in basic solutions due to the presence of mostly In^- anions, and green at pH = 7 because both HIn molecules and In^- anions are present in roughly equal amounts. A double arrow, as shown in the chemical equation below, is one way to represent the reversible process of acid molecules dissociating into ions and ions attracting one another to re-form molecules.

Bromothymol blue displays as a different color for acidic (yellow), neutral (green), and basic (blue) solutions.

$$HIn(aq) \rightleftharpoons H^+(aq) + In^-(aq)$$

BINDING TO A RECEPTOR IN THE BODY

Processes involving binding to receptor sites in the human body are reversible. Prime examples include taste and smell, in which molecules attach to and reversibly release from receptors on the tongue and in the nose. The receptors are made of large protein molecules with a specific shape and sections that are polar, charged, or nonpolar.

Small molecules are attracted selectively to the protein by intermolecular attractions and how well the molecule fits. When molecules attach to the receptor, this initiates a signal sent along the nervous system to the brain. The brain receives the signal and it is perceived as a certain taste or smell. The signal is no longer detected when the molecules leave the receptor.

The general process for reversible binding between a molecule and a receptor is shown below. The blue shape represents a small molecule that is soluble in blood, saliva, or fluids in the cell. The red shape is a piece of a large protein molecule that forms the receptor. In the forward process, the molecule attaches to the receptor. In the reverse process, the molecule separates from the receptor.

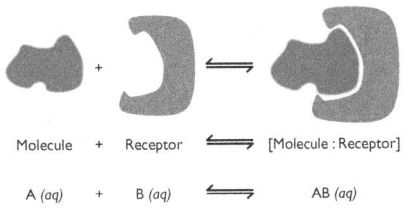

| Molecule | + | Receptor | $\rightleftharpoons$ | [Molecule : Receptor] |

| A (aq) | + | B (aq) | $\rightleftharpoons$ | AB (aq) |

Lesson 118 | **Reversible Processes**   615

Ask: What is a simple explanation of what causes a sweet sensation?

Ask: Why does the sensation of sweetness go away over time?

Ask: For the same number of molecules, why do certain compounds taste sweeter than others?

Ask: How would you explain why packets of artificial sweetener have less mass compared with packets of sugar (sucrose)?

Key Points: Many processes in the body involve binding of a molecule to a receptor. In Unit 2: Smells, students explored molecules binding to smell receptor sites in the nose. Receptors are typically large protein molecules that selectively attract a certain class of molecules. For example, the taste receptors on the tongue attract sugar molecules. When a sugar molecule is attached to a taste receptor, this initiates a series of steps that send a signal along nerves to the brain. Over time, the process reverses, and the molecule leaves the receptor. When this occurs, the sweet sensation stops.

The general process for the binding is shown below, where A is a small molecule and B is the receptor. AB indicates that the small molecule is attached to the receptor.

$$A + B \rightleftharpoons AB$$

Some molecules stay in the receptor longer, thereby creating a stronger sensation. This is because some molecules are attracted more strongly by intermolecular attractions to the receptor. Sucrose (table sugar) is attracted more strongly than maltose (the sugar in malt) and more strongly than lactose (the sugar in milk). For the same number of molecules, more sucrose molecules are detected because more are attached to the sweet receptors. For lactose, more molecules remain in the saliva. While both molecules bind reversibly, the process involving the sucrose favors attached molecules, while the process with lactose favors free molecules. Artificial sweeteners are 200 times sweeter than sugar. This is why packets of artificial sweetener have very little mass.

---

adding base causes more In^- ions to form, while adding acid causes more HIn molecules to form. Smell molecules bind to receptors when you breathe in and are removed when you breathe out.

Reactions that cannot be reversed are said to proceed completely to products. In Unit 5: Fire, students learned about many chemical reactions that go from reactants to products but are not easily reversed. For example, when methane burns, the reaction proceeds only in the forward direction with reactants being turned to products. It is not possible to react carbon dioxide and water to

reverse the reaction and produce methane and oxygen.

Burning methane goes to completion: $CH_4(g) + 2O_2(g) \rightarrow CO_2(g) + 2H_2O(l)$

DISCUSS MOLECULES BINDING TO RECEPTORS

→ Write the general process for a molecule attaching to a receptor on the board. During the discussion, ask students to identify A and B.

$$A + B \rightleftharpoons AB$$

Check-In

Consider the reversible process for the binding of a smell molecule:

menthol + receptor ⇌ [menthol : receptor]

1. What will happen to the perception of a minty smell if you inhale after opening a bottle containing menthol?

2. What will happen to the perception of a minty smell if you exhale?

Answers: **1.** Inhaling menthol will cause the forward process to occur. The perception of a minty smell will increase. **2.** Exhaling will cause the reverse process to occur. The perception of a minty smell will diminish.

Homework

Assign the reading and exercises for Showtime Lesson 118 in the student text.

There are many reversible receptor processes in the body. The smell of menthol is detected in the nose when menthol molecules are attracted to smell receptor sites. A protein in the blood called *hemoglobin* reversibly binds oxygen, O_2. The pain reliever ibuprofen binds reversibly to an enzyme to block it from producing molecules that cause the pain signal. The pain signal returns as the ibuprofen molecules leave the receptor. In the chemical equations below, the symbol ":" represents an attraction between the molecule and receptor.

$$menthol + receptor \rightleftharpoons [menthol : receptor]$$
$$O_2 + hemoglobin \rightleftharpoons [O_2 : hemoglobin]$$
$$ibuprofen + receptor \rightleftharpoons [ibuprofen : receptor]$$

Some of the reversible processes described above can be represented as two components, A and B, forming one component, AB, and then breaking apart as A and B. Here is a general form of a chemical equation describing this type of process:

$$A + B \rightleftharpoons AB$$

Important to Know If a process is reversible, the equation describing it can be written in either the forward or reverse direction.

Because this describes a reversible process, the equation can also be written in the other direction:

$$AB \rightleftharpoons A + B$$

This model can be used to describe salt dissolving and precipitating, acid-base reactions, and molecules binding to and releasing from receptors. For example, for salt dissolving, AB represents the salt, and A and B are the ions that it separates into when dissolved. For sugar molecules binding to a taste receptor, A represents a sugar molecule and B represents the receptor.

⟲ Degree of Sweetness

According to the receptor site theory, the attraction of sugar molecules to sweet receptors in the taste buds on the tongue involves the formation of two hydrogen bonds on two adjacent carbon atoms in the sugar molecule. Recall from Unit 2: Smells that a hydrogen bond forms when an O−H or N−H on one molecule is attracted to an O or N atom on another molecule.

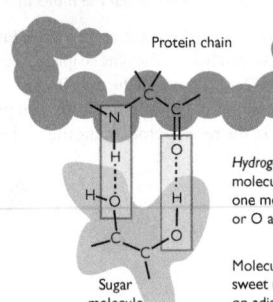

Protein chain

The receptor site is a large protein molecule consisting of a chain of C and N atoms.

Hydrogen bonds are between two molecules. An −OH or −NH on one molecule is attracted to an N or O atom on another molecule.

Molecules that attach to the sweet receptor have an −OH on adjacent C atoms.

Sugar molecule

The molecules in substances that taste sweet, such as cake with strawberries and whipped cream, are attracted to the sweet receptor on the tip of the tongue by two hydrogen bonds formed on adjacent carbon atoms.

616 Chapter 23 | **Chemical Equilibrium**

The table below shows the relative sweetness for four substances. The molecules in these substances all attach to the same type of sweet receptors. Why do they vary in relative sweetness?

Sweetener	Source	Relative sweetness
lactose	milk and other dairy products	2
maltose	beer, cereal, pasta, potatoes	3
sucrose	table sugar	10
aspartame	artificial sweetener	1800

One explanation for the data is that some molecules stay in the sweet receptor longer, thereby creating a stronger sensation. This can be explained by differences in the strength of the intermolecular attractions between the molecule and the receptor, as well as a better fit. Sucrose (table sugar) is attracted more strongly to a sweet receptor than maltose (the sugar in malt) and more strongly still than lactose (the sugar in milk). The result is that, for the same number of molecules and the same number of sweet receptors, more sucrose molecules are detected because more are attached to the sweet receptors. For lactose, more molecules remain in the saliva.

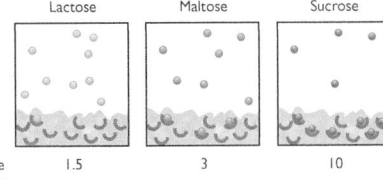

Lactose	Maltose	Sucrose

Relative sweetness: 1.5   3   10

The squares show that as more molecules bind to the sweet receptors, the relative sweetness of the substance increases. The small circles represent the sweet molecules, and the dark area at the bottom represents the area in the tongue with the sweet receptors. The white space represents saliva.

Artificial sweeteners are about 180 times sweeter than table sugar (sucrose) per mole. This is why packets of artificial sweeteners are much smaller than packets of table sugar.

Big Idea Some reversible processes favor the forward direction, while others favor the reverse direction.

LESSON SUMMARY

What is a reversible process?

A reversible process is one that can proceed in both the forward and the reverse directions. Many physical and chemical processes are reversible given the appropriate conditions. In a chemical equation, a double arrow is used to symbolize reversible processes. Examples of reversible processes include phase

LESSON 118 ANSWERS

1. *Possible answer:* evaporation and condensation of water.

2. The process is reversible.

3. *Possible answer:* freezing and melting, evaporation and condensing.

4. *Possible answer:* Boiling water involves the breaking of hydrogen bonds, which re-form when water condenses.

5. Burning octane (gasoline) in oxygen involves breaking many bonds and then making new ones to produce carbon dioxide and water. This reaction is not reversible.

6. a. $H_2O(s) \rightarrow H_2O(l)$ **b.** Put the liquid water in the freezer.

7. a. $C_{12}H_{22}O_{11}(s) \rightleftharpoons C_{12}H_{22}O_{11}(aq)$ **b.** forward. **c.** reverse.

8. a. $CH_4(g) + 2O_2(g) \rightarrow CO_2(g) + 2H_2O(l)$ **b.** Carbon dioxide and water do not react to form methane and oxygen.

9. a. $O_2(g)$ + hemoglobin $\rightarrow$ [O_2 : hemoglobin] **b.** The reverse reaction occurs in which O_2 is released from hemoglobin. **c.** You have to keep breathing in O_2 to bring more O_2 to the cells.

---

changes, dissolving and precipitation, acid-base reactions, and binding to protein receptors. In each case, it is possible to find conditions to cause these processes to proceed almost exclusively in one direction or the reverse, or to create a mixture of starting substances and products.

KEY TERM

reversible process

Exercises

Reading Questions

1. What evidence is there that some processes are reversible?

2. What does a double arrow in a chemical equation tell you?

Reason and Apply

3. Provide examples to support the claim that phase changes are reversible.

4. Provide examples to support the claim that processes involving the breaking and forming of intermolecular attractions are reversible.

5. Provide examples to support the claim that processes involving breaking and making of several covalent bonds are sometimes not reversible.

6. Imagine that you take an ice cube out of the freezer and place it on a counter at room temperature.
   a. Write a balanced equation for the process that occurs.
   b. Describe what you would do to reverse the process.

7. Imagine that you dissolve sucrose, $C_{12}H_{22}O_{11}(s)$, in water.
   a. Write a balanced equation for this reversible process.
   b. In which direction does the process proceed if you place a tiny amount of sugar in a large pot filled with water?
   c. In which direction does the process proceed if you allow the water to evaporate from a sugar solution?

8. Methane, $CH_4(g)$, burns in oxygen, $O_2(g)$, to produce carbon dioxide, $CO_2(g)$, and water, $H_2O(l)$.
   a. Write a balanced equation for this process.
   b. This process is described as irreversible. What does this mean?

9. Imagine that you breathe in oxygen and it binds to hemoglobin in your blood to form [O_2 : hemoglobin].
   a. Write a balanced equation for the process.
   b. The [O_2 : hemoglobin] delivers oxygen to your cells. Describe this process.
   c. Provide evidence that this process is reversible.

How Dynamic
Dynamic Equilibrium

THINK ABOUT IT

A tightrope walker moves carefully and purposefully, keeping his balance as he goes. He tips back and forth while he walks, sometimes using an outstretched arm or a leg to counterbalance any movement one way or the other. He is maintaining his balance but not keeping still. Chemical systems that are at equilibrium maintain a similar type of dynamic balance.

What is chemical equilibrium?

To answer this question, you will explore

- Models of Equilibrium
- Chemical Equilibrium

EXPLORING THE TOPIC

◑ Models of Equilibrium

If you put a small amount of sodium chloride, $NaCl(s)$, in water, the process goes toward forming $Na^+(aq)$ and $Cl^-(aq)$ ion products. If you keep adding more and more $NaCl(s)$, the solution will become saturated. Once this happens, no more $NaCl(s)$ will dissolve. The excess $NaCl(s)$ remains on the bottom of the container.

THE VIEW AT AN ATOMIC SCALE

To gain a better understanding of why there is a limit to the amount of $NaCl(s)$ that dissolves in a given quantity of water, it is useful to consider what is happening at the atomic scale. The figure on the next page shows a model of how water molecules bump into the surface of the solid and carry the ions into the solution. This is due to ion-dipole intermolecular attractions between the $Na^+(aq)$ and $Cl^-(aq)$ ions and the water dipoles.

It is also possible to carry out the reverse process by evaporating the water. As the water evaporates, there is less space for the $Na^+(aq)$ and $Cl^-(aq)$ ions to spread out, and they bump into each other more often. The forces of attraction between ions cause the salt to re-form.

Lesson 119 | **Dynamic Equilibrium**   619

- Processes such as dissolving/precipitating and binding/detaching are typically reversible. These processes involve changes in intermolecular attractions.
- The words *reactant* and *product* are associated with a chemical reaction. Yet not every process is a chemical reaction, so we use *starting substance* converted to *product* to be more general.
- Students who are progressing slowly do not have to complete a hundred tries in the activities to get the tactile experience of what happens in an irreversible and reversible process.
- Some students may believe incorrectly that when a process reaches equilibrium, it means there are equal amounts of reactants and products present.

KEY TERM

chemical equilibrium

IN CLASS

In Part 1 of the activity, students use a paper-clip model to simulate an irreversible process. Pairs of paper clips (AB) in a bag are taken apart (A + B). Although the rate of the process of taking apart the pairs slows down over time, there are eventually only A and B single paper clips in the bag. In Part 2, students simulate a reversible process. AB pairs are taken apart, and single A and B clips are also put back together. The two students continue taking turns at the tasks until a balance is reached between the forward and reverse processes. This is seen when the numbers of AB pairs and A + B pairs no longer change over time. Although the numbers of AB, A, and B no longer change, the forward and reverse processes continue to occur. A complete materials list is available for download.

TRM Materials List 119

SETUP

Prepare 16 paper bags with 50 pairs of linked paper clips (AB). Make sure the pair consists of one each of the two different colors (A and B). This setup will act as a model for studying a reversible and an irreversible process.

CLEANUP

Ask students to recreate the original setup by making 50 pairs of paper clips of two different colors.

Overview

ACTIVITY: PAIRS

Key Question: What is chemical equilibrium?

KEY IDEAS

A reversible process is in a state of dynamic equilibrium. In this state, a balance is reached between the forward and the reverse processes. The rate of the forward process slows down as the concentrations of starting substances decrease. The rate of the reverse process speeds up as the concentrations of the products increase. At equilibrium, the rate of the forward process is equal to the rate of the reverse process. No macroscopic changes are visible at equilibrium, but changes continue to occur at the atomic level.

LEARNING OBJECTIVES

- Describe chemical equilibrium from a macroscopic and a particle viewpoint.
- Explain the dynamic nature of chemical equilibrium.

FOCUS ON UNDERSTANDING

- Many chemical reactions are not reversible, especially when there are changes in covalent bonds.

Differentiate

To get a sense of what students take away from this lesson, use an index-card summary strategy. Give each of the students an index card and have them use their own words to describe what new learning they gained and how it relates to prior learning. You can provide sentence starters to English language learners and others with literacy issues, such as, "I learned that …" or "Equilibrium is reached when …". You might also offer an option to complete a graphic organizer that shows how concepts are related.

Pre-AP® Course Tip

Students preparing to take an AP® chemistry course or college chemistry should be able to relate atomic-level models to macroscopic phenomena. In this lesson, use paper-clip models to describe chemical equilibrium from a macroscopic and microscopic point of view and how this relates to the dynamic nature of chemical equilibrium.

Lesson Guide

ENGAGE (5 min)

TRM PowerPoint Presentation 119

ChemCatalyst

Consider the reversible process represented by the chemical equation:

$$AB \rightleftharpoons A + B$$

You mix 1 mol of A and 1 mol of B. After the process described by the equation occurs, the amount of AB will be:

a. exactly 0.

b. between 0 and 1 mol.

c. exactly 1 mol.

d. exactly 2 mol.

Justify your choice.

Sample answers: Possible justifications for each choice.

a. Because A and B are products, no AB will be produced. (*Comment:* This response would only be true if the process is not reversible.)

b. Because the process is reversible, some AB will form. (*Comment:* This is

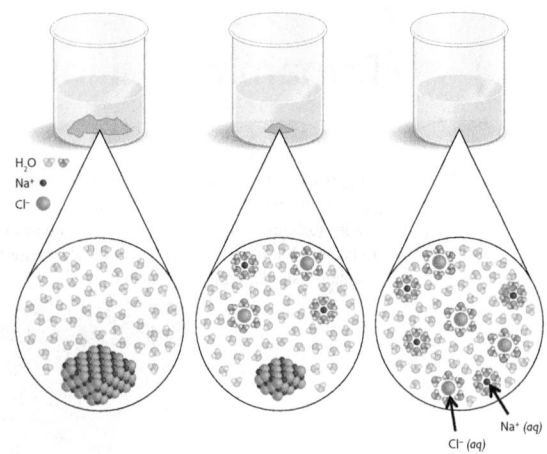

H_2O
Na^+ ●
Cl^- ◐

Na^+ (aq)
Cl^- (aq)

The water molecules are attracted to Na^+ (purple spheres) and Cl^- (green spheres) ions in the solid NaCl. Once surrounded by water, the ions move away from the solid and into the solution.

A DYNAMIC BALANCE

In a dilute aqueous solution of NaCl, the process of dissolving has gone entirely to the Na^+(aq) and Cl^-(aq) ion products. The ions do bump into one another, but enough of them do not cluster together to form a solid. For dilute solutions, you can represent the process of dissolving NaCl(s) by using a chemical equation with a single arrow in the direction of forming Na^+(aq) and Cl^-(aq) ions.

When the concentration of Na^+(aq) and Cl^-(aq) ions is large enough, solid salt is visible at the bottom of the container. The solution is referred to as saturated. For an observer, it appears that nothing is happening. However, on an atomic scale, the dissolving and precipitation processes are still happening. When some Na^+(aq) and Cl^-(aq) ions precipitate, there is enough room for more NaCl(s) to dissolve, and so on, back and forth. This explains why all the salt does not dissolve: some keeps precipitating. The saturated solution with excess solid NaCl is represented by a chemical equation with a double arrow to show that both the forward and reverse processes are occurring.

The saturated NaCl solution with excess solid on the bottom is at **chemical equilibrium**. Equilibrium is a dynamic state in which opposing processes occur at the same time. At equilibrium, the rate of the forward process is equal to the rate of the reverse process. Because these rates are equal, no change is visible.

In contrast, in the case of a dilute solution, the rate of the reverse process (precipitation) is never equal to the rate of the forward process (dissolution). In this case, all of the solid dissolves.

The flask on the left shows a solution of NaCl. Because the entire solid has dissolved, the process has gone in the forward direction. The flask on the right shows a saturated NaCl solution with solid NaCl at the bottom. The solid has not dissolved completely because the process has reached equilibrium.

Ken Karp Photography

620   Chapter 23 | **Chemical Equilibrium**

---

the best response, given the double arrows in the chemical equation.)

c. Because all the coefficients are 1, 1 mol of A and 1 mol of B will form 1 mol of AB. (*Comment:* This response would only be true if the reverse process converted all of A and B to AB.)

d. 1 mol of A and 1 mol of B will form 2 mol of AB. (*Comment:* This is not correct.)

Ask: How would you explain what it means that the process is reversible?

Ask: What do you expect will happen if you combine A and B and the process is reversible?

Ask: What would you observe if you combine A and B and the process is not reversible?

Ask: What would you observe if the process reverses completely?

EXPLORE (15 min)

TRM Worksheet with Answers 119

TRM Worksheet 119

INTRODUCE THE ACTIVITY

→ Tell students to work in pairs.

→ Pass out the student worksheet and bags filled with paired paper clips to each group.

Big Idea At equilibrium, the rate of the forward process is equal to the rate of the reverse process.

Example

Evaporation of Water

Consider the evaporation of water at 25° C.

a. Right after a rainfall, there are water puddles on the sidewalk. Once the rain stops and the sun is shining, the water puddles evaporate completely. Explain why. Provide a chemical equation to describe the process.

b. The water in the bottle shown in the picture does not evaporate completely. Explain why. Provide a chemical equation to describe the process.

c. How does chemical equilibrium explain the observation of droplets of water on the sides of the bottle above the water level?

Solution

a. The puddles evaporate and the gaseous water molecules spread out in the atmosphere. The reverse process (condensation) does not happen because the rate at which molecules attract one another is too slow to form water droplets. A chemical equation with a single arrow describes the process.

$$H_2O(l) \longrightarrow H_2O(g)$$

b. The water in the bottle does not evaporate completely because the closed container limits the space in which the water molecules can travel. When there are enough gaseous water molecules in the small space above the liquid water, the molecules bump into each other frequently enough to condense. This means that while water continues to evaporate, the gaseous water molecules are also condensing. So, the forward and reverse processes of evaporation and condensation are occurring simultaneously. A chemical equation with a double arrow describes the process.

$$H_2O(l) \Longleftrightarrow H_2O(g)$$

c. The water in the bottle is at equilibrium. This means that the rate of evaporation is equal to the rate of condensation. When the gaseous water condenses, it can form droplets on the side of the bottle above the water level, or simply fall onto the surface of the liquid water.

Chemical Equilibrium

You can use a paper-clip model to describe an irreversible process and a reversible process. The paper-clip model is a general example of any process involving an AB pair that separates into A and B.

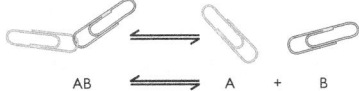

AB   A   +   B

Lesson 119 | **Dynamic Equilibrium**   621

→ Explain to students that the bags contain paper-clip pairs made of clips of different colors called AB pairs. Review the procedures for Parts 1 and 2.

→ Discuss ways of keeping track of the number of AB pairs. After 10 tries, students can dump out the bag and count. Or they can keep track of the number of AB pairs taken apart in 10 tries and calculate the number of AB pairs remaining by difference. The number of AB pairs taken apart is equal to the number of A and B single clips added.

→ For students who finish quickly, you might suggest that they repeat the reversible process in Part 2 starting with only single paper clips of each color in the bag (A + B).

EXPLAIN & ELABORATE (15 min)

DISCUSS THE IRREVERSIBLE PROCESS

→ Show students a general chemical equation for an irreversible process.

$$AB \rightarrow A + B$$

Ask: How would you describe your experience in completing the irreversible process?

Ask: If you continue the activity for long enough, how many AB would remain?

Ask: At any given time, how are the numbers of AB, A, and B related? (That is, if you start with 50 AB and 20 break apart, how many AB, A, and B are in the bag?)

Ask: Why did the process slow down over time?

Ask: Imagine you have just eaten a piece of candy. You drink some water, and the sweet taste slowly goes away. How does this relate to the paper-clip model?

Key Points: An irreversible process converts all the starting substance(s) to products. In the activity, all of the pairs of paper clips (AB) are converted into single paper clips (A and B). This is a model for both chemical and physical changes. For example, it could represent a salt, such as NaCl dissociating into Na^+ and Cl^- ions if the solution is not saturated. It could also represent sugar molecules that are bound to the receptors on your tongue. The sugar molecules will dissolve away if you drink a lot of water. The water molecules bump into the sugar molecules on the receptors. There is an attraction between the sugar molecules and the water, and it causes the sugar molecules to break away from the receptor and dissolve. In the end, the molecules are removed by swallowing.

The chemical equation relates amounts at different time points. Because the reaction stoichiometry is 1:1:1 for AB:A:B, the 50 AB pairs are converted to 50 A and 50 B. At any point as the process proceeds, the number of AB plus the number of A must equal 50. Likewise, the number of AB plus the number of B must equal 50. For example, if there are 15 AB, there must be 35 A and 35 B. The number of A is equal to the number of B.

An irreversible process slows down over time. In the activity, it is easy at first to find pairs of paper clips to take apart. Over time, though, the process slows down because you often grab an individual paper clip rather than a pair. This resembles what happens in a chemical reaction. Why? Imagine that AB only falls apart if it collides with other molecules or the walls of the container. If there are fewer AB in the container, there will be fewer collisions and fewer will break apart.

Lesson 119 | **Dynamic Equilibrium**   621

DISCUSS THE REVERSIBLE PROCESS

➜ Show students a general chemical equation for a reversible process.

$$AB \rightleftharpoons A + B$$

Ask: How would you describe your experience in completing the reversible process?

Ask: If you continue the activity for long enough, are all AB pairs converted to products? Why or why not?

Ask: What do you predict will happen if you do the same experiment starting with 50 A and 50 B?

Ask: Imagine that you have a piece of candy in your mouth and you do not swallow. The sweet taste remains. Explain how this relates to the reversible process.

Key Points: A reversible process does not convert all of the starting substance(s) to products. At first, the forward process proceeds rapidly, while the reverse process hardly occurs. This is because there are lots of AB and very few A and B at the beginning. Over time, the forward process slows down and the reverse process speeds up. At some point, the rate of the forward process is equal to the rate of the reverse process. Even though paper-clip pairs are still being pulled apart, and single paper clips are being put together, the number of AB pairs and the numbers of A and B no longer change.

A reversible process results in a mixture of starting substance(s) and products, called an *equilibrium mixture*. A reversible process reaches equilibrium when the amounts of the starting substance(s) and products no longer change. However, this does not mean that the process has stopped. To the contrary, both the forward and reverse processes continue, but they do not produce a net change. *Chemical equilibrium* is dynamic, involving continuous change, but no net change.

Equilibrium is characteristic of the resulting mixture, not the starting point. If you start the reversible process with only products, the result will be the same as if you started with only starting substance(s). For example, 50 A and 50 B will form the identical mixture as the one you obtain if you start with 50 AB.

The paper-clip pair, AB, might represent:

- a salt such as NaCl, which dissociates to $Na^+(aq) + Cl^-(aq)$
- a strong acid such as HCl, which dissociates to $H^+(aq) + Cl^-(aq)$
- an acid-base indicator, H*In*, which dissociates to $H^+(aq) + In^-(aq)$
- a molecule bound to a receptor, which separates into a molecule and the empty receptor
- a molecule such as N_2O_4, which breaks apart into two molecules of NO_2

It is useful to examine the features of the model for both irreversible and reversible processes.

IRREVERSIBLE VERSUS REVERSIBLE PROCESSES

The two graphs below show what happens for two different starting materials, one labeled XY and the second labeled AB. Imagine that the first graph represents a collection of chocolate pieces, XY, with each piece broken into two smaller pieces, X and Y. The second graph might represent a collection of paper-clip pairs, AB, of two different colors that are pulled apart into single paper clips, A and B, and put back together again.

Take a moment to examine the graphs and describe the similarities and differences.

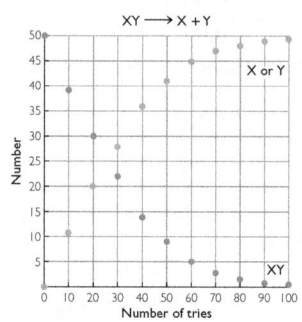
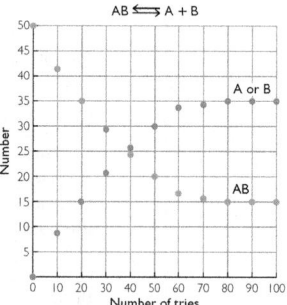

Here are some similarities that you might notice:

- In both cases, the starting substance is breaking apart into two pieces.
- At the start of the process, there are 50 XY pairs and 50 AB pairs.
- The number of XY pairs decreases as the number of X or Y pieces increases. The number of AB pairs decreases as the number of A or B pieces increases.

Here are some differences that you might notice:

- The number of XY pairs goes to 0 over time. Once you break the chocolate, you cannot put it back together again. The process is irreversible.
- The number of AB pairs does not go to 0. There are 15 AB pairs remaining, even after a long time. This is because the reverse reaction makes new AB

BIOLOGY CONNECTION

The word *equilibrium* also refers to the sense of balance that keeps people and animals from falling over. This sense of equilibrium is a result of combined input from the eyes, inner ears, pressure sensors on the skin, and the cerebellum in the brain. When your sense of equilibrium is impaired, you can experience motion sickness.

Maka photo/Getty Images

EVALUATE (5 min)

TRM Transparency—Check-In 119

Check-In

Consider this chemical reaction:
$X + Y \rightleftharpoons XY$

You mix 30 mol of X with 30 mol of Y in a closed rigid container. The amount of XY is measured over time, and the values are recorded in a data table.

Time (s)	Amount of XY (mol)
30	7
60	12
90	15
120	17
150	19
180	20
210	20
240	20
270	20

pairs. At equilibrium, there is a mixture of 15 AB pairs, 35 individual A pieces, and 35 individual B pieces.

If you examine the graph on the right, you might notice that the amounts of AB, A, and B in the mixture no longer change after about 80 tries. This indicates that equilibrium is reached: the amounts remain fixed.

Example

Strong versus Weak Acid

Consider the observations reported here for two aqueous solutions: one with hydrochloric acid, HCl, and a second with formic acid, HCOOH.

a. The process of dissolving hydrochloric acid, HCl, in water results in a solution of $H^+(aq)$ and $Cl^-(aq)$. If there are any HCl molecules remaining, the numbers are so tiny that they are not detected. Explain why. Base your explanation on the rates of the forward and reverse processes.

b. The process of dissolving formic acid, HCOOH, in water results in a solution of HCOOH, $H^+(aq)$, and $HCOO^-(aq)$. There are many HCOOH molecules in the solution compared with ion products. Explain why. Base your explanation on the rates of the forward and reverse processes.

Solution

a. Only the forward process occurs at a high rate. Once $H^+(aq)$ and $Cl^-(aq)$ form, they do not stay together long enough to produce a measurable amount of HCl(aq). (See the graph for XY.) So nearly all of the HCl is converted to products. This process is often represented with a single arrow. Because very nearly all the HCl molecules dissociate, HCl is called a strong acid.

$$HCl(aq) \longrightarrow H^+(aq) + Cl^-(aq)$$

b. The process does not go to completion to become products because the reverse process also occurs at a high rate. Once $H^+(aq)$ and $HCOO^-(aq)$ form, they do go back together again to make HCOOH. (See the graph for AB.) At first, the rate of the forward reaction is fast and the rate of the reverse reaction is slow. The rate of the forward reaction slows down as more HCOOH molecules dissociate. The rate of the reverse reactions speeds up as the number of H^+ and $HCOO^-$ ions increases. Over time, the rates become equal and the process reaches equilibrium. At equilibrium, the amounts of HCOOH(aq), $H^+(aq)$, and $HCOO^-(aq)$ no longer change. This reversible process is represented with a double arrow. Because very few of the HCOOH molecules dissociate, HCOOH is called a weak acid.

$$HCOOH(aq) \Longleftrightarrow H^+(aq) + HCOO^-(aq)$$

STARTING WITH ALL PRODUCTS

The two graphs below compare the same reversible processes but for different starting amounts. In the graph on the left, there are 50 AB pairs at the start but no A or B pieces. In the graph on the right, there are 50 A and 50 B pieces at the start

After 270 seconds, which statement is true?

a. The concentrations of X, Y, and Z are equal.

b. The concentration of XY is greater than the concentrations of X and Y.

c. The system is not at equilibrium.

Answer: **b.** At 270 s, the concentration of XY is no longer changing. This indicates that the reaction is at equilibrium. The concentration of XY is greater than that of X and Y because if 20 mol of XY are produced, then only 10 mol of X and 10 mol of Y remain.

Homework

Assign the reading and exercises for Showtime Lesson 119 in the student text.

LESSON 119 ANSWERS

1. Dynamic equilibrium means that the rate of a forward process and the rate of its reverse process are equal.

2. Salt dissolves until the rate of the forward process of dissolving is equal to the rate of the reverse process of precipitation.

3. a. Many more artificial sweetener molecules bind to the sweet receptors than do the same number of sucrose molecules. So, AB is more favored for the artificial sweetener compared with sucrose. **b.** More products form for NaCl, whereas AgCl remains mostly as the starting substance. So, AB is more favored for AgCl compared with NaCl.

4. a. $[PCl_5] = 0.29$ mol/L, $[PCl_3] = 0.71$ mol/L, $[Cl_2] = 0.71$ mol/L. **b.** 80 sec, when the concentrations no longer change. **c.** At 20 sec, the forward process is faster. At 100 sec, the rates of the forward and reverse processes are equal. **d.** At equilibrium, the rates of the forward and reverse processes are equal, but greater concentrations of the products are needed for the molecules to collide and change back to the starting material. The table shows that the concentrations are not equal. **e.** Each time that the forward process occurs, one PCl_5 molecule breaks apart to form one PCl_3 molecule and one Cl_2 molecule.

5. a. AB $\rightleftharpoons$ A + B. **b.** green curve = starting substance (AB), red curve = products (A or B). **c.** $[AB] = 20$ millimole/L, $[A] = [B] = 30$ millimole/L. **d.** about 75 milliseconds. **e.** the rates of the forward and reverse processes are equal.

but with no AB pairs. Take a moment to examine the graphs. What do you notice about the amounts at equilibrium?

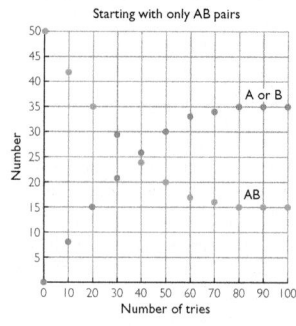

Starting with only AB pairs

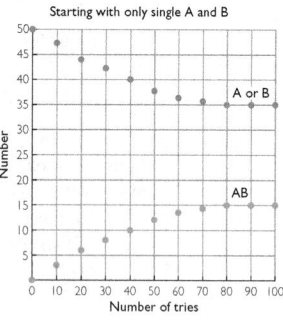

Starting with only single A and B

After 100 tries, the amounts of AB, A, and B are the same in both cases. This indicates that equilibrium is a property of the mixture. It does not matter if you start with the substances on the left or right of the double arrow; the resulting mixture is the same.

Important to Know Chemical equilibrium means equal rates of forward and reverse processes, not equal amounts of reactants and products. The amounts remain fixed at equilibrium, but they are not necessarily equal.

LESSON SUMMARY

What is chemical equilibrium?

Equilibrium is a dynamic process. The word *dynamic* means that things are constantly changing. Equilibrium is a balance that is reached between the forward and reverse processes in a reversible change. To an observer, it seems that things are no longer changing. In reality, however, changes are continuously occurring at an atomic scale, but the changes balance each other out. When a system is at chemical equilibrium, the rate of the forward reaction is equal to the rate of the reverse reaction. Equilibrium is a property of the mixture, meaning that it does not matter how you start the process.

KEY TERM

chemical equilibrium

Exercises

Reading Questions

1. What does it mean when a system is in a state of dynamic equilibrium?
2. Why is there a limit as to how much salt you can dissolve?

Reason and Apply

3. Explain these observations in terms of the amounts of starting substances (AB) and products (A + B) in an equilibrium mixture.
   a. Artificial sweetener is sweeter than the same amount of sucrose.
   b. Much more sodium chloride, NaCl, dissolves in 1 L of water compared with silver chloride, AgCl.

624   Chapter 23 | **Chemical Equilibrium**

Exercises

(continued)

4. When phosphorus pentachloride, $PCl_5(g)$, is placed in a sealed container, it breaks apart reversibly. The concentrations of the starting substance and products are monitored. Use the data in the table to create a graph showing the concentrations of the starting substance and products over time.

$$PCl_5(g) \rightleftharpoons PCl_3(g) + Cl_2(g)$$

Time (s)	[PCl₅] (mol/L)	[PCl₃] (mol/L)	[Cl₂] (mol/L)
0	0.00	1.00	1.00
20	0.10	0.90	0.90
40	0.20	0.80	0.80
60	0.25	0.75	0.75
80	0.29	0.71	0.71
100	0.29	0.71	0.71
120	0.29	0.71	0.71

   a. What are the equilibrium concentrations of the starting substance and the products?
   b. At what time did the mixture reach equilibrium? How do you know?
   c. At 20 seconds, which process is faster, the forward or the reverse process? At 100 seconds?
   d. Why is the amount of starting substance, PCl_5, not equal to the amounts of each of the products, PCl_3 and Cl_2, at equilibrium? Support your answer with evidence.
   e. Why are the amounts of PCl_3 and Cl_2 always equal to one another?

5. The graph below shows the reversible separation of a molecule bound to a receptor.

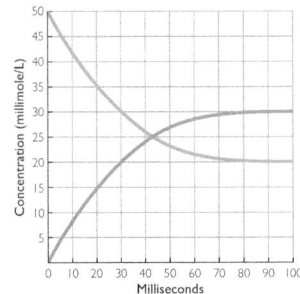

   a. The process begins with all the molecules attached to receptors. Write a chemical equation to represent the process.
   b. Label the two curves as either starting substance (with the molecule attached to the receptor) or products (molecule or empty receptor).
   c. What are the equilibrium concentrations of the starting substance and the two products?
   d. How long did it take for this process to reach equilibrium?
   e. What is "equal" at equilibrium?

Overview

CLASSWORK: GROUPS OF 4

Key Question: How can you predict if products are favored in a reversible process?

KEY IDEAS

Chemical processes are characterized by an equilibrium constant, *K*. The value of *K* is a measure of the degree to which the starting substances are converted to products. A large value for *K* indicates a large amount of product in the equilibrium mixture. A small value for *K* indicates very little product in the equilibrium mixture. For strong acids, the acid molecules break apart (dissociate) into ions completely, and the value of *K* is very large. For weak acids, the acid molecules only break apart (dissociate) to a small degree and the value of *K* is small. The equilibrium constant, *K*, provides information about what is present in the solution as well as whether starting substances or products are favored in reversible processes.

LEARNING OBJECTIVES

- Explain how the equilibrium constant, *K*, is a measure of whether starting substances or products are favored in an equilibrium mixture.
- Make connections between the value of the equilibrium constant, *K*, and the H^+ concentration in strong and weak acid solutions.

FOCUS ON UNDERSTANDING

- The word *dissociation* refers to aqueous molecules breaking into ions. Students often confuse this with the term *dissolving*, which refers only to a molecule or particles of solid that become surrounded by water molecules. Some substances will both *dissolve* and *dissociate* (e.g., acids), while others will only *dissolve* (e.g., sugar).
- *Solution molarity* is a measure in moles per liter of acid molecules dissolved initially in water to make the solution. This measure is independent of the degree of dissociation of acid molecules in the solution.
- The word *acid* refers both to undissociated acid molecules and to H^+ dissolved in water. However, pH is only a measure of the H^+ concentration in solution.
- One type of acid will have a different value of *K* for the acid dissociation

---

How Favorable

Equilibrium Constant *K*

THINK ABOUT IT

HEALTH CONNECTION

Sneezing? When the body releases histamine into the nasal passages, the well-known "runny" nose occurs. Histamine is important as the immune system fights allergens. So histamine isn't just a nuisance. It is a neurotransmitter involved in regulating many bodily responses. We can take antihistamine medicine to block the binding of histamines to shut down the "runny" nose response to the flood of histamines released.

In a reversible chemical process at equilibrium, the starting substances and the products are both present as a mixture. In some processes at equilibrium, there are more products than starting substances present in the mixture. In other processes, there are more starting substances than products in the equilibrium mixture. Is there some way to predict whether more starting substances or products will be present when a chemical process reaches equilibrium?

How can you predict if products are favored in a reversible reaction?

To answer this question, you will explore

- Products versus Starting Substances
- Equilibrium Constant *K*
- Dissociation of Strong and Weak Acids into Ions

EXPLORING THE TOPIC

Products versus Starting Substances

Certain pain relievers act by attaching reversibly to receptors to block molecules that cause the pain signals from being sent to the brain. The strength of the binding between the pain reliever and its receptor site influences how effective the pain reliever will be. The stronger the binding, the longer the pain reliever stays in the receptor site, thereby blocking the molecules that cause the pain signal.

Consider the particle views of the binding of three different pain relievers. What do you think the *equilibrium constant K* indicates about the pain reliever?

Pain reliever X	Pain reliever Y	Pain reliever Z
K = 0.024	*K* = 0.031	*K* = 9

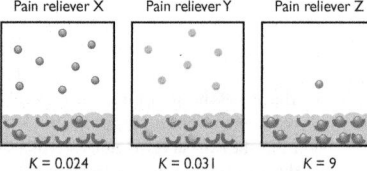

pain reliever + receptor $\rightleftharpoons$ [pain reliever : receptor]

A + B $\rightleftharpoons$ AB

The figures show particle views of the binding for three different pain relievers. The pain reliever molecules are represented as spheres, and the receptors are shown as half circles. The square represents a very tiny volume of solution.

---

than another type of acid. However, the value of *K* is the same for one acid at the same temperature, even for solutions of differing concentrations of that one acid.

- There are two common ways to write a chemical equation to describe the dissociation of an acid in water as shown below.

$$HA(aq) \rightleftharpoons H^+(aq) + A^-(aq)$$
$$HA(aq) + H_2O(aq) \rightleftharpoons H_3O^+(aq) + A^-(aq)$$

- To emphasize acid dissociation, we have chosen to leave water out of the equations to describe dissociation of acid in water.

KEY TERM

equilibrium constant *K*

IN CLASS

Students begin by examining the efficacy of pain relievers, then learn how an equilibrium constant, *K*, is a measure of the extent to which the pain reliever is attached to a receptor to block the pain signal to the brain. Students reexamine five of the solution cards, all of which are acids, from Unit 4: Toxins, Lesson 85. They arrange the cards in order, based on H^+ concentration, and review that strong acids dissociate completely (the process is nearly irreversible),

The figure shows that the equilibrium constant K is smallest for pain reliever X and greatest for pain reliever Z. For pain reliever X, 8 out of 10 pain reliever molecules are unbound. For pain reliever Z, only 1 molecule out of 10 pain reliever molecules is unbound. Although other factors might be important, it is likely that pain reliever Z is more effective at relieving pain per mole because more molecules are bound to receptors. It is reasonable to conclude that a small value of K suggests that fewer pain reliever molecules are bound, and therefore, the pain reliever is less effective.

Equilibrium Constant K

As you have learned, many processes in the human body involve the attachment of molecules to receptors. For example, molecules bind reversibly to receptors to regulate sleep, memory, mood, muscle contraction, and pain. In studying these processes in the human body, the concepts of equilibrium and reversible processes are relevant.

The **equilibrium constant K** is a measure of whether starting substances or products are favored in an equilibrium mixture. A large value of K indicates a large amount of product in the equilibrium mixture. A small value of K indicates very little product in the equilibrium mixture. A value of K can be determined experimentally for each reversible process. Because K describes the concentrations of starting substances and products in the equilibrium mixture, it is a valuable concept in making predictions about equilibrium mixtures.

> **Big Idea** The equilibrium constant K is a measure of whether starting substances or products are favored in an equilibrium mixture.

Example 1

Pain Relievers and K

Explain how the value of K relates to the particle views for pain relievers X, Y, and Z shown on page 626.

Solution

The smaller the value of K is, the more the starting substance is favored in the equilibrium mixture. In this case, the starting substances are individual molecules and empty receptor sites. For the larger value of K, the products are favored in the equilibrium mixture. In this case, the product is the pain reliever attached to the receptor.

Dissociation of Strong and Weak Acids into Ions

The molecules in substances called acids both *dissolve* in water and *dissociate* into H^+ and an anion. The word *dissolve* indicates that water molecules surround the acid molecules, and the acid molecules spread out in the solution. While salts dissolve and dissociate into ions, most molecular substances *dissolve*, without dissociating, like sugar. Nothing further happens.

(circles) to receptors (half circles) are shown in the figure. A number called K is given for each. Below the figure is an equation that represents the process.

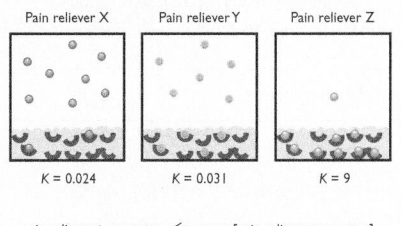

Pain reliever X	Pain reliever Y	Pain reliever Z
$K = 0.024$	$K = 0.031$	$K = 9$

pain reliever + receptor ⇌ [pain reliever : receptor]
A + B ⇌ AB

1. How many pain-reliever molecules and receptors are there in each case?

2. How is the value for K related to the amount of pain reliever that is bound to a receptor?

3. Based on the information provided, generate a hypothesis as to which pain reliever is the most effective for the smallest dose. Explain your thinking.

Sample answers: **1.** There are 10 pain-reliever molecules and 10 receptors in each case. **2.** The value for K gets larger as more molecules are attached. **3.** Pain reliever Z is the most effective for the smallest dose because more of the molecules are attached to the receptors.

Ask: What is the same in each case? What is different?

Ask: For pain reliever X, are the starting substances or products favored for the process as written? Explain your thinking.

Ask: How does the value for K relate to the amount of bound pain reliever?

---

while weak acids dissociate reversibly, resulting in an equilibrium mixture of the acid molecule, H^+, and an anion. Students relate the extent of dissociation of the acid molecule and the pH with the equilibrium constant K. A complete materials list is available for download.

TRM Materials List 120

SETUP

Prepare 8 sets of 5 of the Card Masters—Acid-Base Solutions 85 from Unit 4: Toxins, Lesson 85. The cards needed are Solution 1, Solution 2, Solution 3, Solution 4, and Solution 5.

Lesson Guide

ENGAGE (5 min)

TRM Transparency—ChemCatalyst 120

TRM PowerPoint Presentation 120

➡ Display the ChemCatalyst transparency.

ChemCatalyst

Some pain relievers, such as ibuprofen, bind reversibly to receptors to block the pain signal from being sent to the brain. Particle views of the binding of three different pain relievers

EXPLORE (15 min)

TRM Worksheet with Answers 120

TRM Worksheet 120

TRM Card Masters—Acid-Base Solutions 85

INTRODUCE THE CLASSWORK

➡ Tell students to work in groups of four.

➡ Pass out the worksheet to each group.

➡ Pass out the sets of solution cards to each group. Remind them that all 5 cards represent particle views of acids.

DISSOCIATION OF STRONG AND WEAK ACIDS INTO IONS

→ Put the general equation for acid dissociation on the board.

$$HA(aq) \rightleftharpoons H^+(aq) + A^-(aq)$$

Ask: How would you describe a particle view of what happens when an acid dissolves in water?

Ask: Why is the concentration of H^+ equal to the solution molarity in an aqueous solution of HCl?

Ask: Why is the concentration of acetic acid molecules, CH_3COOH, in an aqueous solution much greater than the concentration of H^+?

Ask: What does the value of pH indicate about the degree of acid dissociation?

Key Points: Acid molecules dissociate into ions in aqueous solution. When an acid molecule is dissolved in water, the molecule dissociates into H^+ and an anion. The word *dissolve* means that the molecules go into solution where they are surrounded by water molecules. Molecules labeled as acids *dissolve,* and they also *dissociate.* This is what happens for hydrochloric acid, HCl. The solution is entirely H^+ and Cl^- ions. Some molecules, like sugar, *dissolve* but do not *dissociate* into ions. This is why sugar is not an acid.

Strong acids dissociate irreversibly to the ion products. Weak acids dissociate reversibly. The *solution molarity* is the number of moles of acid molecules that dissolve to make the solution. When strong acids dissolve in water, all of the molecules dissociate such that the solution molarity is equal to the H^+ concentration. For example, in a 0.10 M HCl solution, there are no HCl molecules, only 0.10 M H^+ and 0.10 M Cl^-. When weak acids dissolve in water, the concentrations of the products, H^+ and A^- ions, are much less than the solution molarity. This is because the reverse process in which the ions recombine to form the molecule occurs at an appreciable rate.

The pH of an acid solution is a measure of the degree of *dissociation.* For a strong acid, the concentration of H^+ is equal to the solution molarity, so the pH is equal to $-\log$ (solution molarity). For weak acids, the concentration of H^+ is much less than the total acid concentration, so the pH is greater than $-\log$ (solution

Acids are special because even though they are molecular substances, they *dissociate* into cations and anions when they dissolve. This is shown in the particle views for tiny volumes of hydrochloric acid, HCl, solution. The solution is mostly H^+ and Cl^- ions, with no measurable number of HCl molecules.

Because acid molecules *dissolve* in an aqueous solution, and they also *dissociate,* you might wonder what solution molarity refers to in these cases. For example, the particle view of 0.010 M HCl does not have any molecules of HCl. Instead, the solution consists of 0.010 M H^+ and 0.010 M Cl^- because all of the molecules are *dissociated* into ions.

Formic acid, HCOOH, also *dissolves* and *dissociates.* However, as shown in the particle view, the solution is a mixture of HCOOH molecules, H^+ ions, and $HCOO^-$ ions. This is because the reaction is reversible. For this solution, the concentration of acid molecules, HCOOH, remains close to the solution molarity of 0.010 M because so few molecules *dissociate.* The H^+ concentration is only 0.0013 M. Therefore, for the same solution molarity, the pH of the HCl solution is lower than the pH of the HCOOH solution.

> **Important to Know** Hydrochloric acid is referred to as a strong acid, which means a strongly dissociated acid. Formic acid is called a weak acid, which means a weakly dissociated acid.

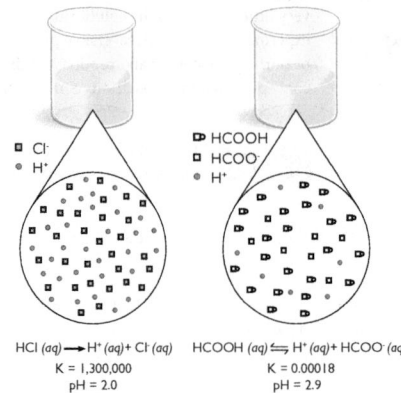

HCl (aq) ⟶ H^+ (aq) + Cl^- (aq)
K = 1,300,000
pH = 2.0

HCOOH (aq) ⇌ H^+ (aq) + $HCOO^-$ (aq)
K = 0.00018
pH = 2.9

Example 2

The Value of K

A student prepares two aqueous solutions with a total acid concentration of 0.10 M. One solution is prepared with chlorous acid, HOClO, and the other with hypochlorous acid, HOCl. The equilibrium constant for HOClO is $K = 0.011$. The equilibrium constant for HOCl is $K = 0.000000029$. Which solution will have a lower pH value? Explain your thinking.

Solution

A larger value of K indicates more products and a greater degree of dissociation. This means that there will be more H^+ in the HOClO solution, and the pH will be lower.

molarity). If you prepare a strong acid solution and a weak acid solution that have equivalent solution molarity, the strong acid will have more H^+ in solution than the weak acid. The pH will be lower for the strong acid. To obtain the same pH, a weak acid solution must be much more concentrated compared with a strong acid solution.

INTRODUCE THE EQUILIBRIUM CONSTANT, K

Ask: What does the value of K indicate about the equilibrium mixture?

Ask: Which has a larger value of K, a strong acid solution or a weak acid solution? Explain your thinking.

Ask: How does the value of K compare for a 0.10 M acetic acid solution and a 0.010 M acetic acid solution? Explain your thinking.

Key Points: The equilibrium constant, K, is a measure of whether starting substances or products are favored in an equilibrium mixture. A large value for K indicates a large amount of product in the equilibrium mixture. A small value for K indicates very little product in the equilibrium mixture. For strong acids, the acid molecules dissociate into ions completely, and the value of K is very large. For weak acids, the acid molecules only dissociate to a small degree, and the

KEY TERM

equilibrium constant K

LESSON SUMMARY

How can you predict if products are favored in a reversible process?

Chemical processes are characterized by an equilibrium constant K. The value of K is a measure of the degree to which the starting substances are converted to products. A large value of K indicates a large amount of product in the equilibrium mixture. A small value of K indicates very little product in the equilibrium mixture. Strong acids have a value of K that is very large, so products are favored. This means that very nearly all of the molecules are dissociated into ions. Weak acids have a value of K that is very small, which means that reactants are favored. This means that at any given time, only a small number of molecules are dissociated. The equilibrium constant K provides information about what is present in the solution and whether starting substances or products are favored in reversible processes.

Reading Questions

1. Why is it useful to know the value of the equilibrium constant K for a reversible process?

2. How does the equilibrium constant K distinguish between strong and weak acids?

Reason and Apply

3. Hemoglobin is a large molecule in red blood cells that transports O_2 from the lungs to cells in the human body. Consider the two reversible processes shown below involving oxygen, O_2, and carbon monoxide, CO, attaching to hemoglobin.

$$\text{hemoglobin} + O_2 \rightleftharpoons [\text{hemoglobin} : O_2]$$
$$\text{hemoglobin} + CO \rightleftharpoons [\text{hemoglobin} : CO]$$

The binding of CO to hemoglobin is more than 200 times greater than the binding of O_2 to hemoglobin.
   a. Which reversible process has the larger equilibrium constant? Explain your thinking.
   b. Sketch a particle view for both processes.
   c. Explain why CO is extremely toxic to humans.

value of K is small. The equilibrium constant, K, provides information about what is present in the solution and whether starting substances or products are favored in reversible processes.

EXAMPLE

Suppose pain reliever A is being tested. The value of the equilibrium constant, K, is 0.11. Use the figure in the ChemCatalyst to estimate how many molecules of pain reliever A are attached to a receptor. Explain your thinking.

Answer: Based on the value for K, there are fewer molecules of pain

reliever A that are attached to a receptor compared with Y and Z. For every 10 molecules of pain reliever A, probably only 4 are attached to a receptor.

EVALUATE (5 min)

Check-In

For hypochlorous acid, HOCl, $K = 0.000000030$.

1. Write a chemical equation to show the acid dissociation.

2. Do you expect mainly HOCl in the solution or the ions H^+ and OCl^-? Explain your thinking.

Answers: 1. $HOCl(aq) \rightleftharpoons H^+(aq) + OCl^-(aq)$. 2. Mainly HOCl because K is small.

Homework

Assign the reading and exercises for Showtime Lesson 120 in the student text.

LESSON 120 ANSWERS

1. A large value of K indicates that there are more products in the equilibrium mixture, while a small value of K indicates that more starting substance is present.

2. The value of K is large for a strong acid and small for a weak acid.

3. a. The binding of CO to hemoglobin has the larger equilibrium constant because [hemoglobin:CO] is greater than [hemoglobin:O_2]. **b.** Each particle view should show hemoglobin binding to a particle. The view with oxygen should have more free oxygen and hemoglobin particles than the view for carbon monoxide. **c.** CO is toxic because it takes up binding sites in hemoglobin so that they are not available to oxygen. Because the [hemoglobin:CO] product is favored, fewer free hemoglobin molecules are available to bind oxygen.

4. a. Each H^+ concentration is equal to their respective solution molarity. **b.** Each acid has a specific value of K. The value of K for a specific acid does not depend on the solution molarity. **c.** When an acid molecule dissociates, the products are one H^+ ion and one A^- ion. **d.** A particle view would show that the solution is mainly acid molecules with very tiny amounts of H^+ and anions from the dissociation of the acid. **e.** Solution 1: pH = 0, Solution 2: pH = 0.30, Solution 3: pH = 1.0, Solution 4: pH = 1.7, Solution 5: pH = 2.1. **f.** Although Solution 3 is less concentrated, there are more H^+ ions, and the pH is lower.

4. Use the table to answer the questions below. The brackets, [ ], indicate concentrations in the equilibrium mixture.

#	Solution	Formula	Solution molarity	[HA]	[H⁺]	[A⁻]	K
1	nitric acid	HNO_3	1.0 M	~0 M	1.0 M	1.0 M	40
2	nitric acid	HNO_3	0.50 M	~0 M	0.50 M	0.50 M	40
3	nitric acid	HNO_3	0.10 M	~0 M	0.10 M	0.10 M	40
4	nitrous acid	HNO_2	1.0 M	0.98 M	0.022 M	0.022 M	0.00056
5	benzoic acid	C_6H_5COOH	1.0 M	0.99 M	0.0079 M	0.0079 M	0.000063

a. Describe how the H^+ concentration for Solutions 1, 2, and 3 relates to the solution molarity.

b. Explain how the value of K depends on the specific acid and the solution molarity.

c. Explain why $[H^+] = [A^-]$.

d. For formic acid and acetic acid, describe a particle view of the equilibrium mixture.

e. Determine the pH for all five solutions.

f. If spilled, Solution 3 will cause more harm than Solution 5. Use this information to explain why pH, as opposed to molarity, is important in determining how dangerous an acidic solution is.

How Balanced

Equilibrium Calculations

THINK ABOUT IT

One reason that aqueous acidic solutions can be dangerous is because of the sensitivity of the human body to high concentrations of H^+, or low pH. Because weak acids only dissociate to a small degree, the molarity of a weak acid solution alone does not provide information about the concentration of H^+. So far, you have considered that the equilibrium constant K is a measure of the degree of dissociation. But is there a more specific relationship between K and the solution molarity?

> How does K describe the concentrations in an equilibrium mixture?

To answer this question, you will explore

⟳ The Equilibrium-Constant Equation for a Weak Acid

⟳ The General Form of the Equilibrium-Constant Equation

EXPLORING THE TOPIC

⟳ **The Equilibrium-Constant Equation for a Weak Acid**

Many parts of the human body can only tolerate a specific range of pH values. For example, the normal pH range of your skin is 4 to 6, due to the presence of lactic acid from perspiration and of the acids that make up the outer skin layer. When you wash your hands with soaps with pH values that are much higher than that of your skin, this can cause skin irritation and harm beneficial microorganisms that reside in the skin. For this reason, small amounts of weak acids such as citric acid are added to hand soaps to adjust their pH.

Because weak acids do not dissociate completely, you might wonder how to determine the pH of a solution for a certain concentration of a weak acid. One way is to prepare solutions of various molarities and measure the pH. Another way is to calculate the pH by using the equilibrium constant K for a specific acid and the **equilibrium-constant equation**.

RELATIONSHIP BETWEEN K AND EQUILIBRIUM CONCENTRATIONS

In the previous lesson, K was introduced as a number that describes what is present in the equilibrium mixture. The equation below defines K more specifically for an acidic solution. The equation relates the equilibrium constant K, the concentrations of an acid, HA, and the ion products, H^+ and A^-, in an

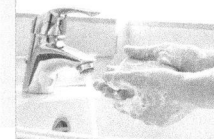

LEARNING OBJECTIVES

● Solve problems that relate K to the total acid concentration and the solution pH.

● Understand that K is a property of a specific process and remains the same for various starting concentrations.

FOCUS ON UNDERSTANDING

● Students may have trouble grasping the idea that many different starting substances and product concentrations will satisfy the equilibrium constant equation for a specific reversible process.

● Students may think that the same value for K applies to different reversible processes. However, different processes have different equilibrium constants.

● The equilibrium constant equation given for weak acids in this lesson is only an approximation. Because so little of HA dissociates, the solution molarity is used for [HA] in the expression instead of the equilibrium concentration of HA.

● The lesson provides numbers both in scientific notation and written out in decimal form. It is best to allow students to choose which they prefer to use.

KEY TERM

equilibrium-constant equation

IN CLASS

Students begin by considering atomic-scale views of three acid solutions to foster a discussion of the significance of K. They learn that K is the same for two solutions of the same acid at equilibrium, even though the concentrations are different. Students then complete a worksheet that introduces the relationship between K for dissociation of a monoprotic weak acid in water, the total acid concentration, and the H^+ concentration, called the *equilibrium-constant equation*. Students use this equation to determine the value of K given [HA] and [H^+] and to determine the pH of a solution given K and [HA]. There is an optional discussion regarding writing *equilibrium-constant equations* for reversible processes of all types. A complete materials list is available for download.

TRM **Materials List 121**

Overview

CLASSWORK: INDIVIDUALS OR PAIRS

Key Question: How does K describe the concentrations in an equilibrium mixture?

KEY IDEAS

A system at equilibrium reaches a balance between starting substances and products characterized by the equilibrium constant, K. The concentrations adjust until the rates of the forward and reverse processes are equal. At equilibrium, there is a specific relationship between K and the concentrations in the equilibrium mixture, called the *equilibrium constant equation*. For example, for a dissociation reaction, the relationship is:

$$AB \rightleftharpoons A + B$$
$$K = \frac{[A][B]}{[AB]}$$

Once K is known for a reversible process at a given temperature, it can be used in calculations for various equilibrium concentrations for this same system.

Differentiate

You may want to spend more than one day on this lesson, as the calculations are complex. To help students work with equilibrium calculations, have them use whiteboards and verbalize their problem-solving process. Go through the student worksheet or other equilibrium-constant calculation problems, and have students work in groups of 3–4 to prepare whiteboards to present to the class. Tell students that the emphasis is on being able to explain how they arrived at their answer and that their whiteboard is meant to support that.

Lesson Guide

ENGAGE (5 min)

TRM Transparency—ChemCatalyst 121

TRM PowerPoint Presentation 121

→ Display the ChemCatalyst transparency.

ChemCatalyst

The three squares below show particle views of a tiny volume of three acid solutions.

$$HA(aq) \rightleftharpoons H^+(aq) + A^-(aq)$$

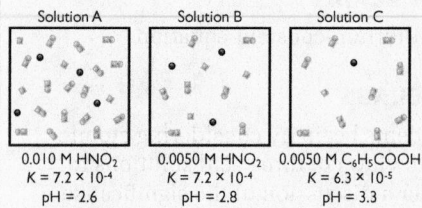

Solution A	Solution B	Solution C
0.010 M HNO_2	0.0050 M HNO_2	0.0050 M C_6H_5COOH
$K = 7.2 \times 10^{-4}$	$K = 7.2 \times 10^{-4}$	$K = 6.3 \times 10^{-5}$
pH = 2.6	pH = 2.8	pH = 3.3

1. Solutions A and B contain the same acid at different concentrations. Explain why Solution B has a higher pH than Solution A.

2. Solutions B and C contain different acids at the same concentration. Why do you think the solutions have different values for the pH in this case?

Sample answers: **1.** Because Solution B is less concentrated for the same acid, the concentration of H^+ is less and the pH is higher. **2.** Although the total acid concentrations are the same for Solutions B and C, the equilibrium constant for C_6H_5COOH is smaller than for HNO_2. This means that C_6H_5COOH does not dissociate as much, so that the concentration of H^+ is less and the pH is higher.

Ask: What are the similarities and differences among the three solutions?

Ask: How would you explain why Solution A has the smallest value of the pH?

Ask: How would you explain why Solution C has the largest value of the pH?

Ask: What do you notice about the value of K for the three solutions?

EXPLORE (15 min)

TRM Worksheet with Answers 121

TRM Worksheet 121

---

equilibrium mixture of an acid dissolved in water. The brackets, [ ], indicate the concentrations at equilibrium in moles per liter.

$$HA(aq) \rightleftharpoons H^+(aq) + A^-(aq)$$

$$K = \frac{[H^+][A^-]}{[HA]}$$

For a weak acid solution, K is a very small number. There are very few ions in the solution because very few acid molecules dissociate. It is possible to determine the value of K for a specific weak acid by making a solution of known molarity and measuring the solution pH.

- $[H^+]$ at equilibrium is equal to 10^{-pH}. This relationship is determined by rearranging $pH = -\log[H^+]$. For the dissociation of a weak acid, $[H^+]$ is a tiny number.
- $[A^-] = [H^+]$ for a weak acid dissolved in water. For every H^+ ion, there is one A^- ion.
- $[HA] \approx$ solution molarity. The equilibrium concentration of HA is approximately equal ("$\approx$") to the solution molarity because such a small amount dissociates.

Example 1

Determination of the Value of K

A student prepares a 0.10 M solution of acetic acid. The pH of the solution is 2.9. Determine the equilibrium constant K for lactic acid.

Solution

$$[H^+] = 10^{-pH} = 10^{-2.9} = 1.3 \times 10^{-3} = 0.0013 \text{ M}$$

$$[A^-] = [H^+] = 0.0013 \text{ M}$$

$$[HA] \approx [HA]_{total} \approx 0.10 \text{ M}$$

Substituting the values in the equilibrium-constant equation gives:

$$K = \frac{[H^+][A^-]}{[HA]} \approx \frac{(0.0013)(0.0013)}{0.10} \approx 0.000017 \approx 1.7 \times 10^{-5}$$

Note that all these calculations are approximate because they use the solution molarity to approximate the concentration of HA at equilibrium. Since very little HA dissociates, the equilibrium concentration of HA does not differ significantly from the solution molarity. In this case, the concentration of HA is ~75 times the concentration of H^+.

USING K TO CALCULATE THE pH OF A WEAK ACID SOLUTION

Values of K for various weak acids are listed in the table. If you look up the value of K for a specific weak acid, you can use it to determine the solution pH. (The temperature is assumed to be 25° C.)

632   Chapter 23 | **Chemical Equilibrium**

INTRODUCE THE CLASSWORK

→ Tell students to work in pairs.

→ Pass out the student worksheet to each pair.

EXPLAIN & ELABORATE (15 min)

EQUILIBRIUM CONSTANT EQUATION FOR A WEAK ACID, HA

→ Write the general form of the equilibrium constant equation for a monoprotic weak acid, HA, on the board.

concentration of H⁺ is so small that you might wonder why it matters at all. It turns out that the human body easily detects even small changes in [H⁺], for example, as a sour taste, a skin irritation, or as a stinging sensation in your eyes.

Acid	Formula	K
hydrochloric acid	HCl	1,300,000
nitric acid	HNO₃	40
nitrous acid	HNO₂	0.00056
formic acid	HCOOH	0.00018
benzoic acid	C₆H₅COOH	0.000063
acetic acid	CH₃COOH	0.000018
hydrogen sulfide	H₂S	0.00000010
hypochlorous acid	HOCl	0.000000030

First, it is necessary to rearrange the equilibrium-constant equation to solve for [H⁺].

$$K = \frac{[H^+][A^-]}{[HA]}$$

For an acid dissolved in water: $[H^+] = [A^-]$

Substituting for $[A^-]$ gives

$$K = \frac{[H^+]^2}{[HA]}$$

Solving for $[H^+]$ gives

$$[H^+]^2 = K[HA]$$

$$[H^+] = \sqrt{K[HA]}$$

$$pH = -\log[H^+]$$

The next step is to find the value of K in a reference table and decide on a solution molarity.

Big Idea The equilibrium constant K is useful in predicting the pH of a weak acidic solution.

Example 2

Determination of the pH of Artificial Lemon Juice

Citric acid is often added to juice to give it a sour taste. However, if the pH is below 3, many people will find that the juice is too sour to drink. Given that $K = 7.2 \times 10^{-4} = 0.00072$ for the dissociation of citric acid, determine if the solutions listed below will have a pH that is below 3.

a. 1.0 M

b. 0.50 M

c. 0.10 M

Lesson 121 | **Equilibrium Calculations**   633

---

concentration of undissociated weak acid molecules. For a weak acid solution, K is a very small number. There are very few ions in the solution because very few acid molecules dissociate. The value of K for a weak acid solution can be determined if the concentrations of HA, H⁺, and A⁻ at equilibrium are known.

- The value for [H⁺] at equilibrium is a tiny number. The concentration of H⁺ in the equilibrium mixture is so small that you might wonder why it matters at all. It turns out that the human body easily detects even small changes in [H⁺] as a sour taste or as a stinging sensation on skin or eyes. The value for [H⁺] can be determined by measuring the pH of the solution, where $[H^+] = 10^{-pH}$.

- The value for [HA] at equilibrium is approximately equal to the solution molarity. Because only a tiny amount of HA dissociates, the concentration of HA at equilibrium is just slightly smaller than the solution molarity. Thus, it is reasonable to approximate [HA] at equilibrium with the solution molarity. This approximation is good to two significant figures for most weak acid calculations.

- The value for [A⁻] at equilibrium is equal to [H⁺] for a weak acid dissolved in pure water. Each time a molecule of HA dissociates, the products are one H⁺ and one A⁻. For a weak acid dissolved in pure water, this means that $[H^+] = [A^-]$. This is not always true. In the next lesson, you will encounter situations in which the water is not pure but instead contains a salt, such as NaA, thereby adding to the concentration of A⁻.

These values can all be put together to calculate the value of K.

EXAMPLE 1

A solution of hydrofluoric acid, HF, is prepared by dissolving 0.10 mol in 1 L of solution. The pH is measured and this provides a value for the H⁺ concentration equal to 0.0082 M. Write the chemical equation and determine the equilibrium constant, K, for the dissociation of HF.

ANSWER:

Chemical equation for the dissociation process: $HF(aq) \rightleftharpoons H^+(aq) + F^-(aq)$

[HF] at equilibrium is approximately the solution molarity = 0.10 M

$[H^+] = 0.0082$

$[F^-] = [H^+] = 0.0082$

$$HA(aq) \rightleftharpoons H^+(aq) + A^-(aq)$$

$$K = \frac{[H^+][A^-]}{[HA]}$$

$$[H^+] = \sqrt{K[HA]}$$

Ask: What measurement can you make to determine [H⁺]?

Ask: For a weak acid solution, which is larger at equilibrium, [HA] or [H⁺]? Explain your thinking.

Ask: How does the solution molarity compare with [HA] at equilibrium if HA is a weak acid?

Ask: For a weak acid dissolved in water, how are [H⁺] and [A⁻] related? Explain.

Ask: How would you explain what numbers you need to determine K for the dissociation of a weak acid?

Ask: How would you explain what numbers you need to determine the pH of a weak acid solution?

Key Points: The equilibrium constant equation relates the value of K to the concentrations in an equilibrium mixture. In the previous lesson, K was defined only as a number that describes what is present in the equilibrium mixture. In this lesson, K is defined more specifically for a weak acid solution as the product of the concentrations of the ions divided by the

Substituting into the equilibrium constant expression:

$$K = \frac{[H^+][A^-]}{[HA]}$$

$$K = \frac{[H^+][F^-]}{[HF]}$$

$$K \approx \frac{(0.0082)(0.0082)}{(0.10)}$$

$$K \approx 0.00067 = 6.7 \times 10^{-4}$$

The value of K and the solution molarity can be used to predict the pH of a weak acid solution. The values of K are found in tables in a variety of sources. This is helpful in making predictions about the pH of a solution. Imagine that you want to prepare a solution of a specific pH with a certain weak acid. Rather than making up solutions of lots of different total acid concentrations, and then measuring pH, you can look K up in a reference table and use the equilibrium-constant equation to determine the pH value. What makes it particularly useful is that if you know K, you can predict the pH for any concentration of that same weak acid at the same temperature. Knowing the pH is important for substances applied to the body (e.g., contact lens solution, shampoo, lotion, medicine), as the human body is quite sensitive to variations in pH.

EXAMPLE 2

What is the pH of a 0.010 M solution of acetylsalicylic acid, commonly called aspirin? ($K = 0.00030 = 3.0 \times 10^{-4}$)

ANSWER:

$$[H^+] = \sqrt{K[HA]}$$
$$[H^+] \approx \sqrt{(0.00030)(0.10)}$$
$$[H^+] \approx 0.0055 \text{ M} = 5.5 \times 10^{-3} \text{ M}$$
$$pH = -\log[H^+]$$
$$pH = -\log(0.0055)$$
$$pH \approx 2.3$$

WRITING EQUILIBRIUM CONSTANT EQUATIONS FOR OTHER PROCESSES (OPTIONAL)

→ Write the general form for the equilibrium constant equation for any reversible process on the board.

$$aA + bB \rightleftharpoons cC + dD$$
$$K = \frac{[C]^c[D]^d}{[A]^a[B]^b}$$

Ask: How is the equilibrium constant equation related to the chemical equation?

Solution

a. $[H^+] = \sqrt{K[HA]} \approx \sqrt{(0.00072)(1.0)} \approx 0.027$
   $pH = -\log[H^+] \approx -\log(0.027) \approx 1.6$

b. $[H^+] = \sqrt{K[HA]} \approx \sqrt{(0.00072)(0.50)} \approx 0.019$
   $pH = -\log[H^+] \approx -\log(0.019) \approx 1.7$

c. $[H^+] = \sqrt{K[HA]} \approx \sqrt{(0.00072)(0.10)} \approx 0.0084$
   $pH = -\log[H^+] \approx -\log(0.0084) \approx 2.0$

All three solutions have a pH less than 3, so they may all be too sour to drink.

⟳ The General Form of the Equilibrium-Constant Equation

So far, the discussion has centered on reversible processes involving dissociation of a substance into two pieces. However, the form of the equilibrium-constant equation changes depending on the balanced chemical equation for the reversible process. The table shows how the equilibrium-constant equation changes with different types of chemical processes. Examine the table to look for patterns.

Reaction	Equilibrium-constant equation
$A \rightleftharpoons B$	$K = \frac{[B]}{[A]}$
$2A \rightleftharpoons B$	$K = \frac{[B]}{[A]^2}$
$A + B \rightleftharpoons C$	$K = \frac{[C]}{[A][B]}$
$C \rightleftharpoons A + B$	$K = \frac{[A][B]}{[C]}$
$A_2 + B_2 \rightleftharpoons 2AB$	$K = \frac{[AB]^2}{[A_2][B_2]}$

You might notice two important patterns.

- The products are always in the numerator, and the starting substances are always in the denominator.
- The coefficients in the chemical equation show up in the equilibrium-constant equation as exponents.

These patterns can be generalized as shown below.

$$aA + bB \rightleftharpoons cC + dD$$

$$K = \frac{[C]^c[D]^d}{[A]^a[B]^b}$$

HEALTH CONNECTION

Is a super-sour taste a sign that your food has too much acid? Sour candy has always been popular, but daredevils now savor super-sour candies. Dentists say that foods with a pH below 4.0 have a bad effect on tooth enamel. Some super-sour candy, including super-sour gummy candies, have a pH between 2 and 3. A pH below 2 is approaching the level of sulfuric acid used in car batteries.

Ask: Which concentrations are in the numerator, the starting substances or the products?

Ask: If the equilibrium favors products, do you expect a large or small value for K? Explain.

Ask: What is the equilibrium constant expression for the decomposition of HCl to H_2 and Cl_2? (*Hint:* First write a balanced chemical equation.)

Key Points: The *equilibrium-constant equation* varies for different types of processes. There are some simple guidelines you can follow in writing equilibrium-constant equations. Products are always above reactants.

The square brackets represent concentration in moles per liter. The coefficients in the chemical equation (the number that tells how many moles are involved) show up in the equilibrium constant equation as exponents. For example, a coefficient of 2 in front of a compound means that its concentration is squared in the equilibrium constant equation. Pure liquids and solids do not appear in the equation because they do not have a concentration that varies. At first glance, it appears that the units for K should be moles per liter to some power or other, depending on the chemical equation. However, K is reported as a dimensionless number, so there are no units.

PURE SOLIDS AND LIQUIDS

The equilibrium-constant equation keeps track of concentrations that vary. This includes concentrations of substances in solution and concentrations of gases, which vary as gases expand and contract. However, pure solids and pure liquids do not appear in the equation. They are assigned a value of 1 because they don't have a "concentration" that changes.

Here are two examples.

$$MgCO_3(s) \Longleftrightarrow MgO(s) + CO_2(g) \qquad K = [CO_2]$$

$$PbCl_2(s) \Longleftrightarrow Pb^{2+}(aq) + 2Cl^-(aq) \qquad K = [Pb^{2+}][Cl^-]^2$$

LESSON SUMMARY

How does K describe the concentrations in an equilibrium mixture?

A system at equilibrium reaches a balance between starting substances and products characterized by the equilibrium constant K. The concentrations adjust until the rates of the forward and reverse processes are equal. At equilibrium, there is a specific relationship between K and the concentrations of starting substances and products in an equilibrium mixture, called the *equilibrium-constant equation*. For a dissociation reaction, the relationship is:

$$HA(aq) \Longleftrightarrow H^+(aq) + A^-(aq) \qquad K = \frac{[A][B]}{[AB]}$$

KEY TERM

equilibrium-constant equation

If K is known for a reversible process at a given temperature, it can be used in calculations for various equilibrium concentrations for this same system.

Exercises

Reading Questions

1. How can you use the equilibrium constant K to determine the pH of a weak acid solution?

2. Explain how to write the equilibrium-constant equation for a reversible process.

Reason and Apply

3. Use the information provided in the table to determine the equilibrium constant K for these weak acid solutions.

Weak acid solution	$[H^+]$	K
0.10 M chloroacetic acid	0.0118 M	
0.10 M formic acid	0.00424 M	
0.10 M acetic acid	0.00134 M	
0.050 M acetic acid	0.000949 M	

Reaction	Equilibrium-constant equation
$A \rightleftharpoons B$	$K = \dfrac{[B]}{[A]}$
$2A \rightleftharpoons B$	$K = \dfrac{[B]}{[A]^2}$
$A + B \rightleftharpoons C$	$K = \dfrac{[C]}{[A][B]}$
$C \rightleftharpoons A + B$	$K = \dfrac{[A][B]}{[C]}$
$A_2 + B_2 \rightleftharpoons 2AB$	$K = \dfrac{[AB]^2}{[A_2][B_2]}$

EXAMPLE 3

Write the equilibrium constant expression for the decomposition of $HCl(g)$ to $H_2(g)$ and $Cl_2(g)$.

ANSWER

$$2HCl(g) \rightleftharpoons H_2(g) + Cl_2(g)$$

$$K = \frac{[H_2][Cl_2]}{[HCl]^2}$$

EVALUATE (5 min)

Check-In

You prepare a solution of 1.0 M HOCl. What is the pH of the solution? ($K = 4 \times 10^{-8}$)

Answer:

$[H^+] = \sqrt{K[HA]}$

$[H^+] \approx \sqrt{(0.00000004)(1.0)}$

$[H^+] \approx 0.0002 \text{ M} = 2 \times 10^{-4} \text{ M}$

$pH = -\log[H^+]$

$pH = -\log(0.0002)$

$pH \approx 3.7$

Homework

Assign the reading and exercises for Showtime Lesson 121 in the student text.

LESSON 121 ANSWERS

1. You can rearrange the equilibrium constant equation to solve for the hydrogen ion concentration. Once you have the hydrogen ion concentration, you can solve for pH. $[H^+] = \sqrt{K[HA]}$, $pH = -\log[H^+]$

2. K equals the concentrations of the products multiplied together, divided by the concentrations of the reactants multiplied together, but with each individual concentration raised to the power of that substance's coefficient in the chemical equation.

3. Solution 1: 1.4×10^{-3}
Solution 2: 1.8×10^{-4}
Solution 3: 1.8×10^{-5}
Solution 4: 1.8×10^{-5}

4. Solution 1: $[H^+] = 2.5 \times 10^{-3}$ M, $pH = 2.6$
Solution 2: $[H^+] = 3.2 \times 10^{-3}$ M, $pH = 2.5$
Solution 3: $[H^+] = 6.1 \times 10^{-3}$ M, $pH = 2.2$
Solution 4: $[H^+] = 4.3 \times 10^{-3}$ M, $pH = 2.4$

5. a. $K = \dfrac{[NH_4^+][OH^-]}{[NH_3]}$
b. $K = [Na^+]^2[SO_4^{2-}]$
c. $K = \dfrac{[C_2H_6]}{[C_2H_4][H_2]}$

(continued)

4. Use the information provided in the table to determine the H^+ concentration and the pH for these weak acid solutions.

Weak acid solution	K	$[H^+]$	pH
0.10 M benzoic acid	6.3×10^{-5}		
0.10 M p-chlorobenzoic acid	1.0×10^{-4}		
0.10 M p-nitrobenzoic acid	3.7×10^{-4}		
0.050 M p-nitrobenzoic acid	3.7×10^{-4}		

5. Write equilibrium-constant equations for each reversible process.
   a. $NH_3(aq) + H_2O(l) \rightleftharpoons NH_4^+(aq) + OH^{2-}(aq)$
   b. $Na_2SO_4(s) \rightleftharpoons 2Na^+(aq) + SO_4^{2-}(aq)$
   c. $C_2H_4(g) + H_2(g) \rightleftharpoons C_2H_6(g)$

CHAPTER 23

Chemical Equilibrium

SUMMARY

KEY TERMS

indicator
pigment
reversible process
chemical equilibrium
equilibrium constant K
equilibrium-constant equation

Showtime Update

Many chemical processes proceed in the forward and reverse directions. Colorful indicator molecules dissolved in solutions of varying acidity can provide evidence of what is happening on the atomic scale. The color of the solution shows conditions that favor the forward process and other conditions that favor the reverse process.

Unlike combustion reactions, which go to completion to become products, in reversible processes both the forward and reverse processes occur simultaneously. Reversible processes are written with a double arrow to show that both the forward and reverse processes occur at the same time.

When both the forward and reverse processes occur at equal rates, the mixture is at chemical equilibrium. Examples of equilibrium mixtures include the back-and-forth evaporation of water in a closed container or the dissociation of an acid such as hydrofluoric acid, HF into ions in solution:

$$HF(aq) \rightleftharpoons H^+(aq) + F^-(aq)$$

At equilibrium, the rate of the forward process is equal to the rate of the reverse process, but this does not mean that the amounts of starting substances and products are necessarily equal. While there is change on an atomic scale, the amounts of starting substances and products do not change once equilibrium is reached.

There is a mathematical relationship between the amounts of starting substances and products in an equilibrium mixture. This relationship can be described by the equilibrium-constant equation. The equilibrium constant K is the same for a specific reversible process at a specified temperature.

REVIEW EXERCISES

1. Suppose that you have a 0.10 M HCl solution and a 0.10 M HF solution. The HCl solution has an H^+ concentration that is 0.10 M. The HF solution has an H^+ concentration that is 0.010 M.
   a. Hydrochloric acid is a strong acid. Write a chemical equation to describe what happens in solution.
   b. Draw a particle view showing what happens to 10 HCl molecules when they are dissolved in a tiny volume of water.
   c. For the same concentration of acid, the H^+ concentration is smaller. Write a chemical equation to describe what happens in solution.
   d. Draw a particle view showing what happens to 10 HF molecules when they are dissolved in a tiny volume of water.

Chapter 23 | **Summary**   637

ASSESSMENTS

Two multiple-choice assessments (versions A and B) and two short-answer assessments (versions A and B) are available for download for Chapter 23.

TRM   Chapter 23 Assessment Answers

TRM   Chapter 23 Assessments

ANSWERS TO CHAPTER 23 REVIEW EXERCISES

1. a. $HCl(aq) \rightarrow H^+(aq) + Cl^-(aq)$.
b. The particle view should show 10 H^+ ions, 10 Cl^- ions, and no HCl.
c. $HF(aq) \rightleftharpoons H^+(aq) + F^-(aq)$. **d.** The particle view should show mostly HF molecules as well as a few H^+ and F^- ions.

2. a. The ammonium ion, NH_4^+, is a weak acid. The pH is below 7, but it is greater than 1 (pH = 1 is expected for a 0.10 M solution of a strong acid).
b. $NH_4^+(aq)$, $H^+(aq)$, $NH_3(aq)$, H_2O.
c. $K = [H^+][NH_3]/[NH_4^+] = 1.8 \times 10^{-5}$.
d. 2.5 **e.** Because, in solution, NH_3

combines with H^+ to make more NH_4^+. When you add ammonia, NH_3, the hydrogen ion concentration decreases because NH_3 combines with H^+ to make NH_4^+.

3. a. $K = [C][D]/[A][B]$
b. $K = [CO][Cl_2]/[COCl_2]$
c. $K = [Pb^{2+}][Cl^-]^2$

4. a. $K = [NH_3]^2/[N_2][H_2]^3$
b. $K = 12.5$ **c.** both

Project: Neurotransmitters

● A good project should explain the function of a neurotransmitter in the body and address all points in the bulleted list.

2. Suppose that you are studying an aqueous solution with the ammonium ion, NH_4^+. The chemical equation describing the equilibrium mixture is

$$NH_4^+(aq) \rightleftharpoons H^+(aq) + NH_3(aq)$$

   a. A 0.10 M solution has a pH of 2.9. Is the ammonium ion, NH_4^+, a strong acid, a strong base, a weak acid, or a weak base? Explain how you know.

   b. List all the molecules and ions that would be present in a 0.1 M solution of the ammonium ion.

   c. Write the equilibrium-constant equation for the ammonium ion ($K = 1.8 \times 10^{-5}$).

   d. If $[NH_4^+] = 0.50$ M at equilibrium, what is the pH of the solution?

   e. If you add ammonia, NH_3, to an acidic solution, the pH increases. Explain why.

3. Write the equilibrium-constant equations for these reversible processes.

   a. $A(aq) + B(aq) \rightleftharpoons C(aq) + D(aq)$

   b. $COCl_2(g) \rightleftharpoons CO(g) + Cl_2(g)$

   c. $PbCl_2(s) \rightleftharpoons Pb^{2+}(aq) + 2Cl^-(aq)$

4. For the reaction $N_2(g) + 3H_2(g) \rightleftharpoons 2NH_3(g)$, complete these exercises:

   a. Write the equilibrium-constant equation.

   b. Solve for K if the equilibrium concentrations are $[N_2] = 0.1$ M, $[H_2] = 0.2$ M, $[NH_3] = 0.1$ M.

   c. Based on your answer to part b, do you think that this reaction produces an equilibrium mixture that favors starting substances, products, or both about equally?

Neurotransmitters

 PROJECT   Do Web research on one neurotransmitter. What is the function of the neurotransmitter in the body? What type of receptors are involved and what happens when the neurotransmitter binds to those receptors? Write a report.

Your report should include

- a description of the structure of the neurotransmitter molecule.

- a description of the receptors for the neurotransmitter and where they are found in the body.

- an explanation of what happens when neurotransmitter molecules bind to the receptor.

- a description of how the binding process is reversed.

- a discussion of what happens if the body does not produce enough of the neurotransmitter.

Watch the video overview of Chapter 24 (for teachers) by clicking on the link in the TE-book, opening the TRFD, or logging onto the book's companion Web site bcs.whfreeman.com/livingbychemistry2e (teacher log-in required).

In Chapter 24, students explore how changing the conditions of a system at equilibrium causes changes in the balance between reactants and products in an equilibrium mixture. Lesson 122 is a lab that introduces Le Châtelier's principle by showing that reactions at equilibrium shift when stressors are applied. These concepts and skills are put to a practical application in Lesson 123 as students investigate acid-base indicators. Finally, as a review, students have the chance to design their own lab procedure that ties together concepts from the entire unit and course and allows students to *show what they know* in a fun and colorful way.

In this chapter, students will learn

- about changing conditions at equilibrium
- to explain and apply Le Châtelier's principle
- about the chemical equilibrium of acid-base indicators

CHAPTER 24

Changing Conditions

In this chapter, you will study

- changing conditions for an equilibrium mixture
- how to use Le Châtelier's principle to explain how processes respond to stresses
- the chemical equilibrium of acid-base indicators
- equilibrium constants for different types of reactions

If starting substances or products are added to an equilibrium mixture, the composition adjusts to maintain equilibrium. Sometimes, this change is visible as a color change.

W hen a system is at equilibrium, there is a dynamic balance between the forward process and the reverse process. At equilibrium, the rate of the forward process is equal to the rate of the reverse process. However, conditions do not always stay the same for systems at equilibrium. When conditions change, an equilibrium mixture will adjust in a predictable way to those changes so as to reduce the effect of the change and maintain equal rates.

639

Chapter 24 Lessons

Overview

DEMO (VIDEO): WHOLE CLASS
CLASSWORK: PAIRS

Key Question: What happens to a system at equilibrium when conditions change?

KEY IDEAS

When conditions are changed for a system at equilibrium, the system will respond by reducing the effect of the change. This is known as Le Châtelier's principle. Changing conditions (or stresses) may be changes to concentration or temperature. Equilibrium is regained by rebalancing the concentrations of starting substances and products. The nature of the change is predictable and depends on the type of change encountered by the system. A temperature change affects the value of K. A change in concentration or pressure stresses the system, but it does not change the value of K.

LEARNING OBJECTIVES

- Explain how a reversible process will shift in response to certain stresses.
- Articulate Le Châtelier's principle.

FOCUS ON UNDERSTANDING

- To say that the "equilibrium shifts" may be confusing to students, especially when the value of K stays the same and the system remains at equilibrium. It is helpful to explain that the composition of the equilibrium mixture changes due to an external influence, such as adding more of one of the substances in the mixture.
- Students may be confused by the fact that the value of K stays the same when a process is carried out with different concentrations of starting substances and products, but the value of K changes when the process is carried out at different temperatures. The value of K for a reversible process is dependent on temperature but not on concentration.

KEY TERM

Le Châtelier's principle

IN CLASS

Students observe a demonstration that shows a solution of iron (III) thiocyanate ions, $FeSCN^{2+}$, under different conditions. The solution is an equilibrium mixture in which $FeSCN^{2+}$ dissociates to iron (III) ions, Fe^{3+}, and thiocyanate ions, SCN^-. The equilibrium mixture changes color between red, orange, and yellow,

depending on what is added to the mixture. In the demonstration, "stresses" are applied to the equilibrium mixture by adding $Fe^{3+}(aq)$ and $SCN^-(aq)$, and by changing the temperature. Students use evidence of color changes to analyze how the composition of the equilibrium mixture changes in each case. Le Châtelier's principle is introduced. A complete materials list is available for download.

TRM Materials List 122

Differentiate

To help students make sense of the different effects of changing temperature, pressure, or concentration on a system

at equilibrium, encourage students to make a concept map showing what occurs when different stresses are applied. To support students in science literacy and learners of the English language, offer help in breaking down what it means to use the term *dynamic equilibrium*. Help students to articulate what is meant by *dynamic* at the particle level, even if on a macroscopic level, it appears that a system is no longer changing.

SETUP

Search for a video online that shows changes in the equilibrium system for a solution of iron (III) thiocyanate ions, $FeSCN^{2+}$, under different conditions.

LESSON 122

How Pushy

Le Châtelier's Principle

THINK ABOUT IT

Human blood has a certain pH. When people exercise, they produce a lot of carbon dioxide, which in turn builds up carbonic acid in their bodies. A build-up of carbonic acid can alter the blood's pH, which can put stress on the body. How does the body maintain a healthy equilibrium?

> **What happens to a system at equilibrium when conditions change?**

To answer this question, you will explore
- Changing Conditions
- Le Châtelier's Principle

EXPLORING THE TOPIC

Changing Conditions

When a mixture is at equilibrium, the forward and reverse processes are dynamically balanced, going backward and forward at equal rates. For example, if an indicator is dissolved in water, the indicator, H*In*, dissociates in the forward process and the ions, H^+ and In^-, attract one another in the reverse process. Although equilibrium is dynamic, once equilibrium is established, the concentrations of H*In*, H^+, and In^- in the aqueous solution do not change over time, as long as the conditions stay the same.

However, conditions do not always stay the same for an equilibrium mixture. Something new might be added to the mixture or the mixture might be heated or cooled. For example, HCl might be added to the indicator solution described above. Such changes disturb the equilibrium, which changes in a predictable way by adjusting to new concentrations to counteract the disturbance.

Here are three types of changes that can be made to an equilibrium mixture:

- A change in the concentrations—this applies when more of the starting substances or products are added to an equilibrium mixture.
- A change in external pressure—this applies to processes involving gases when the container volume is increased or decreased.
- A change in temperature—heating or cooling an equilibrium mixture changes the value of K.

Changing the conditions for an equilibrium mixture will disturb the system. The examples here demonstrate how each of the three types of changes (or disturbances) affects a reversible process at equilibrium.

COBALT CHLORIDE EQUILIBRIA

The dissociation of tetrachlorocobalt (II) ions, $CoCl_4^{2-}(aq)$, to aqueous cobalt (II) ions, $Co^{2+}(aq)$ and chloride ions, $Cl^-(aq)$, is especially useful in demonstrating changes in equilibrium because it involves a striking color change.

$$CoCl_4^{2-}(aq) \rightleftharpoons Co^{2+}(aq) + 4Cl^-(aq) \qquad K = 4000$$
blue   pink

Cobalt chloride solutions can be pink, blue, or various shades of purple. This figure shows molecular views of tiny volumes of a pink and blue cobalt solution. Notice that there are more Co^{2+} ions (pink circles) than $CoCl_4^{2-}$ ions (blue circles) in the pink solution, but more $CoCl_4^{2-}$ ions in the blue solution. Notice also that the blue solution has a larger number of Cl^- ions (gray circles).

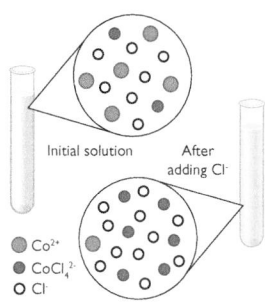

Initial solution   After adding Cl^-

● Co^{2+}
● $CoCl_4^{2-}$
○ Cl^-

There are more Co^{2+} ions in the pink solution, but more $CoCl_4^{2-}$ ions in the blue solution.

It is possible to predict how a disturbance, such as the addition of Cl^- ions to the pink cobalt solution, will change the equilibrium mixture.

CHANGES IN CONCENTRATION

Imagine that you have a pink sample of a cobalt chloride solution. The pink color indicates that the sample has more Co^{2+} ions (one of the products) than $CoCl_4^{2-}$ ions (starting substance). If you change the conditions by adding more Cl^- ions to the mixture, the sample turns blue. So adding more products results in the formation of more starting substance.

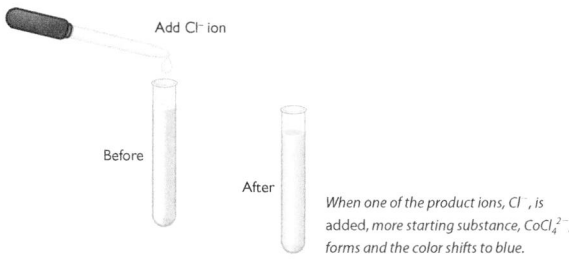

Add Cl^- ion

Before

After

When one of the product ions, Cl^-, is added, more starting substance, $CoCl_4^{2-}$, forms and the color shifts to blue.

If you take this blue sample and add silver ions, $AgNO_3(aq)$, to remove chloride ions, Cl^-, by precipitating $AgCl(s)$, the mixture changes back to pink. So removing

Important to Know Equilibrium mixtures can have various concentrations of starting substance and products. For example, equilibrium mixtures of cobalt chloride can be pink, blue, or various shades of purple.

There are many available. Use the search terms "iron (III) thiocyanate" and "Le Châtelier's principle." If you decide to do the demonstration live rather than show a video, you can find the materials, procedure, and safety instructions in the Lesson 122 Materials document.

Lesson Guide

ENGAGE (5 min)

TRM PowerPoint Presentation 122

ChemCatalyst

Imagine that you have a test tube with a solution of iron (III) thiocyanate ions, $FeSCN^{2+}$. The solution is an equilibrium mixture in which $FeSCN^{2+}$ dissociates to iron (III) ions, Fe^{3+}, and thiocyanate ions, SCN^-.

$$FeSCN^{2+}(aq) \rightleftharpoons Fe^{3+}(aq) + SCN^-(aq)$$
red   yellow $K = 0.025$

1. Write the equilibrium constant equation for the equilibrium mixture.

---

2. If $[Fe^{3+}] = 0.0060$ M and $[SCN^-] = 0.0060$ M, determine the concentration of $[FeSCN^{2+}]$.
3. Explain why the equilibrium mixture is orange.

Sample answers: 1. $K = \dfrac{[Fe^{3+}][SCN^-]}{[FeSCN^{2+}]}$

2. $0.025 = \dfrac{(0.0060)(0.0060)}{[FeSCN^{2+}]}$

$[FeSCN^{2+}] = \dfrac{(0.0060)(0.0060)}{(0.025)}$

$= 0.0014$

3. The solution is orange because there is a mixture of 0.0014 M $FeSCN^{2+}(aq)$, which is red, and 0.0060 M $Fe^{3+}(aq)$, which is yellow.

Ask: How did you determine the concentration of $FeSCN^{2+}(aq)$?

Ask: Why is the solution orange and not red?

Ask: What would have to happen for the solution to turn red?

EXPLORE (15 min)

TRM Worksheet with Answers 122

TRM Worksheet 122

INTRODUCE THE DEMO

➡ Tell students to work in pairs.

➡ Pass out the worksheet to each group.

➡ Pause the demo video after each step to allow students to record their observations.

EXPLAIN & ELABORATE (15 min)

DEBRIEF THE EFFECT OF CHANGING CONDITIONS IN THE DEMO REACTION

➡ Write the chemical equation for the equilibrium mixture on the board.

$$heat + FeSCN^{2+}(aq) \rightleftharpoons Fe^{3+}(aq) + SCN^-(aq)$$
red   yellow

$\Delta H > 0 \quad K = 0.025$

Ask: What does the color of a solution of $FeSCN^{2+}$ indicate?

Ask: What does it mean that the solution of $FeSCN^{2+}$ is an equilibrium mixture?

Ask: How can you change the equilibrium mixture to produce more $FeSCN^{2+}$?

Ask: How does temperature change the amount of $FeSCN^{2+}$ in the equilibrium mixture?

Ask: How would you prepare a solution to maximize the concentration of FeSCN²⁺ and minimize the concentration of Fe³⁺?

Key Points: Adding more product increases the concentration of starting material. When SCN⁻ is added to the equilibrium mixture, the rate of the reverse process increases initially. This produces more FeSCN²⁺. As the SCN⁻ concentration decreases because it reacts to form FeSCN²⁺, the rate of the reverse process decreases over time. As more FeSCN²⁺ is produced, the rate of the forward process increases. At some point, the rates become equal and there is a new equilibrium mixture with more FeSCN²⁺ and less Fe³⁺. The solution turns from orange to red. The same color change from orange to red is observed if Fe²⁺ is added to the original mixture. Adding a lot of one of the products will maximize the concentration of FeSCN²⁺. It is important to note that the equilibrium constant, K, does not change as concentrations in the mixture change.

For an endothermic process, raising the temperature increases the concentrations of products. Raising the temperature for an endothermic process increases the rate of the forward process, thereby producing more products. The increase in concentrations of the products causes the rate of the reverse process to increase until the rates are equal and there is a new equilibrium mixture. The solution turns from orange to yellow because there is less FeSCN²⁺ in the mixture and more Fe²⁺. The reverse happens if the temperature is lowered. Keep in mind that for an endothermic process, heat is a reactant. Heat is a product in an exothermic reaction, so the effect of temperature is the reverse. It is important to note that when the temperature changes, the equilibrium constant takes on a new value.

INTRODUCE LE CHÂTELIER'S PRINCIPLE

→ Write a general chemical equation for an exothermic process on the board.

$$A + B \rightleftharpoons C + D + \text{heat} \qquad \Delta H < 0$$

Ask: What do you predict will happen if you add more A to the equilibrium mixture?

Ask: What do you predict will happen if you remove B from the equilibrium mixture?

Ask: What do you predict will happen if you add C to the equilibrium mixture?

Cl⁻ ions (one of the products) results in the formation of more products (Co²⁺ and Cl⁻ ions). The illustration shows the results of these two changes.

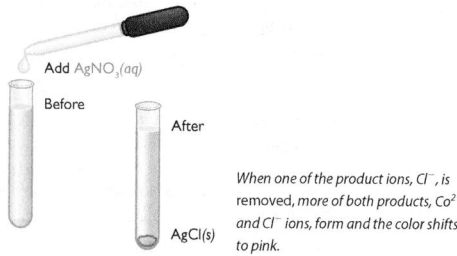

Add AgNO₃(aq)

Before

After

AgCl(s)

When one of the product ions, Cl⁻, is removed, more of both products, Co²⁺ and Cl⁻ ions, form and the color shifts to pink.

The color changes indicate that adding or removing one of the products in an equilibrium mixture has the effect of changing the final concentrations of the new equilibrium mixture. Likewise, the equilibrium mixture also changes if the concentration of starting substance is changed. The change depends on which side of the reversible process gets the stress.

CHANGES IN TEMPERATURE

This same reversible process can be used to test the effects of changing temperatures on an equilibrium mixture. If a pink starting sample of cobalt chloride solution is heated, it turns blue. If a blue sample of this mixture is cooled, it turns back to pink or pinkish-purple. In this particular reversible process, heating favors the starting substance.

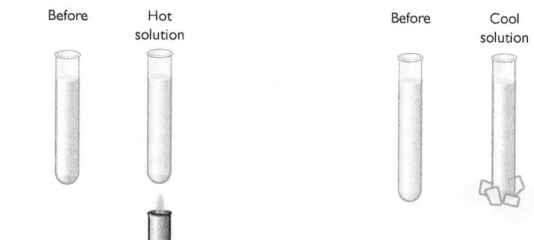

Before   Hot solution

For this process, heating produces more starting substance, CoCl₄²⁻, and the color shifts to blue.

Before   Cool solution

For this process, cooling produces more product ions, Co²⁺ and Cl⁻, and the color shifts to pink.

> **Important to Know** Changes in temperature change the value of K. Other changes, such as changes in pressure or concentration, stress the system, but they do not change the value of K.

Heating does not always cause more starting substance to form. The direction of the change depends on whether the process is exothermic or endothermic. For an endothermic process, heating produces more products. For an exothermic process, cooling produces more products. The reversible cobalt process is exothermic in the directions the chemical equation is written, so heating causes the formation of more starting substance.

CHANGES IN PRESSURE

Pressure changes are particularly important when considering equilibrium mixtures that are in the gas phase. Recall from Unit 3: Weather that it is common to use a cylinder and piston as a flexible container for a gas. Pushing down on the

Ask: What do you predict will happen if you raise the temperature?

Key Points: When a system at equilibrium is stressed, it responds by reducing the effect of the change. A reversible process operates somewhat like a seesaw with a movable base, always seeking a balance. When one side of the seesaw is overloaded, the base of the seesaw can be moved to compensate and find a new balance. If you add more starting substance, more product will form. If you add more product, more starting substance will form. You can also change the concentration of gases by altering the volume to change the pressure. If you transfer

heat to an *endothermic* reaction, more product will form to absorb the added energy. If you transfer heat to an *exothermic* reaction, more starting substance will form to remove the added energy. This is referred to as *Le Châtelier's principle.*

> **Le Châtelier's principle:** When conditions change for a system at equilibrium, the system will respond by reducing the effect of the change.

Changing the concentration of one substance causes a change in the composition of the equilibrium mixture. However, the equilibrium constant, K,

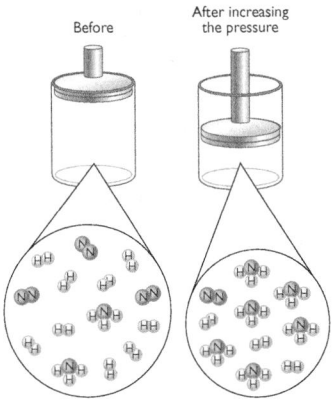

Before   After increasing the pressure

When the volume is made smaller, more ammonia forms, decreasing the number of moles of gas.

piston to decrease the gas volume increases the external pressure on a gaseous equilibrium mixture. A change will occur in the direction that decreases the number of moles of gas and relieves some of the pressure. Recall that the pressure decreases as the number of moles decreases for the same temperature and volume.

For example, consider the process for the formation of ammonia gas from nitrogen and hydrogen gases.

$$N_2(g) + 3H_2(g) \rightleftharpoons 2NH_3(g)$$

Decreasing the volume of the container increases the pressure on the gas mixture. The system will change so that more ammonia forms. This is because there are four moles of gas on the starting substance side of the equation and only two moles of gas on the product side.

The creation of more ammonia relieves the stress of the added pressure by reducing the total number of moles of gas in the mixture.

↻ Le Châtelier's Principle

These very predictable changes to equilibrium mixtures were first noted around 1885 by a French chemist named Henry Le Châtelier. He observed that when the conditions of an equilibrium mixture are modified, the system responds in such a way as to reduce the effect of the change. This idea is known as **Le Châtelier's principle.**

Le Châtelier's Principle
When conditions change for a system at equilibrium, the system will respond by reducing the effect of the change.

Le Châtelier's principle allows you to predict qualitatively how the system responds to changes in concentration, pressure, or temperature.

AN EXPLANATION OF LE CHÂTELIER'S PRINCIPLE

If more of one of the products is added to the equilibrium mixture of starting substances and products, this momentarily increases the rate of the reverse process. The concentrations in the mixture adjust until the rates of the forward and reverse processes are equal. This adjustment results in a new equilibrium mixture that has more starting substance compared with that of the initial solution. The value of the equilibrium constant K is the same for both the adjusted mixture and the initial solution.

A second change is to raise or lower the solution's temperature, which also changes the composition of the equilibrium mixture of starting substances and products. In this case, the changes in composition result in a new value of K depending on if the forward process is exothermic or endothermic. Because there is a new value of K, the concentrations of starting substances and products change to satisfy the new equilibrium constant equation. If the forward process is exothermic (heat is released),

Check-In

Imagine you have an orange solution of 0.001 M $FeSCN^{2+}(aq)$. You add $Ag^+(aq)$ to the solution. This causes the precipitation of $AgSCN(s)$, thereby removing $SCN^-(aq)$ from the solution. Use Le Châtelier's principle to predict the color change.

Answer: If a product is removed from the equilibrium mixture, more of the starting substance will break apart to $Fe^{3+}(aq)$ and $SCN^-(aq)$. This decreases the concentration of the $FeSCN^{2+}(aq)$, which is red, and increases the concentration of Fe^{3+}, which is yellow. The solution will turn from orange to yellow.

Homework

Assign the reading and exercises for Showtime Lesson 122 in the student text.

does not change. The composition of the equilibrium mixture changes to reduce the effect of adding more of one component. For example, suppose you prepare a solution of a weak acid, HA. The weak acid dissociates to a small extent, producing H^+ and A^-. Now imagine that you add H^+ to the solution (such as adding HCl). This will cause more HA to form to reduce the effect of the stress. This is because the rate of the reverse process will be momentarily larger than the rate of the forward process. The concentrations change until the rates are once again equal. This will also happen if you add A^- (for instance, by adding a soluble salt such as NaA). The equilibrium

concentrations in the new mixture satisfy the equilibrium constant equation and K remains the same.

Changing the temperature of an equilibrium mixture causes a change in the composition of the equilibrium mixture. But in this case, the value of the equilibrium constant, K, changes as well. For endothermic processes, heat is considered a reactant. Raising the temperature increases the value of K. More products form to reduce the effect of the change. For exothermic processes, heat is considered to be a product. Raising the temperature decreases the value of K. More starting substance forms to reduce the effect of the change.

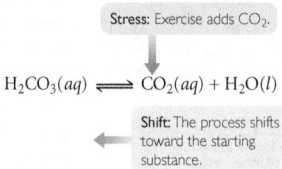

the value of K gets smaller as the temperature is raised. Likewise, if the forward process is endothermic (heat is required), the value of K gets larger as the temperature is raised.

BLOOD pH AND EXERCISE

The human body is a large collection of reversible processes. For processes to proceed at the appropriate rates, and for substances to stay in the appropriate concentrations within our bodies, a balance must be maintained. For example, the pH of the body's blood is normally around 7.4, slightly basic. When you exercise, more carbon dioxide is produced, and it dissolves in the bloodstream. This is a natural waste product that you normally exhale from the lungs. However, carbon dioxide in the bloodstream also leads to the formation of carbonic acid, H_2CO_3. A large build up of carbonic acid would increase the acidity of the blood and lower the pH.

Stress: Exercise adds CO_2.

$$H_2CO_3(aq) \rightleftharpoons CO_2(aq) + H_2O(l)$$

Shift: The process shifts toward the starting substance.

One of the ways your body deals with a change in the blood's pH is to exhale more rapidly, removing CO_2 from the bloodstream and shifting the process back toward the right. This is one reason why we breathe more heavily when exercising—to rid the body of excess carbon dioxide and increase the blood's pH.

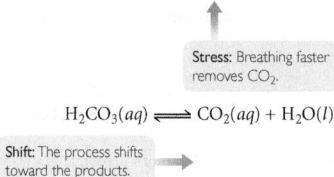

Stress: Breathing faster removes CO_2.

$$H_2CO_3(aq) \rightleftharpoons CO_2(aq) + H_2O(l)$$

Shift: The process shifts toward the products.

Example

Decomposition of SO_3

Gaseous sulfur trioxide, SO_3, decomposes to form sulfur dioxide, SO_2, and oxygen, O_2. The process is endothermic. Predict how the changes listed here will affect the process.

$$2SO_3(g) \rightleftharpoons 2SO_2(g) + O_2(g)$$

a. Increasing the pressure on the system   **b.** Adding more O_2

c. Removing O_2 from the system   **d.** Adding heat

Solution

a. In the forward process, two moles of gas decompose to form three moles of gas. The system can minimize the effect of a pressure increase by converting some of the products back to SO_3 to decrease the number of moles of gas molecules present. This will push the process more toward the starting substances.

(continued)

b. Adding more O_2 to the system will cause more starting substance to form. By making more SO_3, some of the excess O_2 is removed, relieving the stress on the system.

c. Removing O_2 from the system has the opposite effect; it causes the formation of more products. This change has the effect of restoring some of the O_2 that was removed.

d. Because energy is absorbed in the forward process, heating will favor products.

LESSON SUMMARY

What happens to a system at equilibrium when conditions change?

When conditions are changed for a system at equilibrium, the system will respond by reducing the effect of the change. This is known as Le Châtelier's principle. Changing conditions (or stresses) include changes to temperature, concentration, or pressure. A system regains equilibrium by rebalancing the concentrations of starting substances and products. The exact nature of the effects on a system are predictable because they depend on the type of change in condition that the system encounters. A temperature change affects the value of K. A pressure change or a change in concentration stresses the system, but does not change the value of K.

KEY TERM

Le Châtelier's principle

Reading Questions

1. Name two things that you can do to form more starting substances in an equilibrium mixture.

2. Based on the evidence provided in this lesson, explain how you can tell if a process is exothermic as written.

Reason and Apply

3. If you add more ammonia, $NH_3(aq)$, to the equilibrium mixture described by the chemical equation given below, what will happen? Explain your thinking.

$$Cu(NH_3)_6^{2+}(aq) \rightleftharpoons Cu^{2+}(aq) + 6NH_3(aq)$$

4. If you remove ammonia, $NH_3(g)$, from the equilibrium mixture described by the equation here, what will happen? Explain your thinking.

$$N_2(g) + 3H_2(g) \rightleftharpoons 2NH_3(g)$$

5. If you heat the equilibrium mixture for this endothermic process, what will happen? Explain your thinking.

$$2NiO(s) \rightleftharpoons 2Ni(s) + O_2(g)$$

6. Suppose that you want to dissolve more copper (II) hydroxide, $Cu(OH)_2(s)$, in an aqueous solution. Would you add acid or base? Explain your thinking.

$$Cu(OH)_2(s) \rightleftharpoons Cu^{2+}(aq) + 2OH^-(aq)$$

7. For each reversible process listed, describe *two* things you can do to change the concentration of one of the products.
a. $2NO_2(g) \rightleftharpoons N_2O_4(g)$
b. $HF(aq) \rightleftharpoons H^+(aq) + F(aq)$
c. $PbCl_2(s) \rightleftharpoons Pb^{2+}(aq) + 2Cl^-(aq)$
d. $CaCO_3(s) \rightleftharpoons CaO(s) + CO_2(g)$

LESSON 122 ANSWERS

1. Add more of one of the products. Remove some of the starting substance.

2. If you add heat to the system while it's at equilibrium, more starting substance(s) will form.

3. More $Cu(NH_3)_6^{2+}(aq)$ forms to reduce the concentration of NH_3.

4. If you remove NH_3, the starting substances will react to form more NH_3. The lower concentration of ammonia makes the reverse reaction slower than the forward reaction.

5. Because this is an endothermic process, heating favors products. Therefore, more products will form to relieve the stress.

6. Adding acid will decrease the concentration of OH^-. This will cause more $Cu(OH)_2$ to dissolve.

7. *Possible answers:* **a.** Raise the temperature, increase the pressure. **b.** Add a base, add water. **c.** Add NaCl, add water. **d.** Remove the $CO_2(g)$, change the temperature.

Overview

CLASSWORK: PAIRS

Key Question: How do acid-base indicators work?

KEY IDEAS

Acid-base indicators, H*In,* are themselves weak acids. The molecule, H*In,* has one color in solution, and the anion, *In*⁻, that forms has a different color. The amounts of the molecule and the anion in any given solution depend on the concentration of H⁺ in that solution. Thus, the color of the indicator is a measure of the acidity of the solution. There is a pH range over which each indicator changes color.

LEARNING OBJECTIVES

- Explain generally how acid-base indicators work.
- Apply Le Châtelier's principle to predict the color of an indicator solution as the pH changes.
- Complete equilibrium problems involving equilibria with acid-base indicators.

FOCUS ON UNDERSTANDING

When the weak acid, HA, is dissolved in water, $[H^+] = [A^-]$. In this lesson, indicators, H*In,* are dissolved in solutions in which the $[H^+]$ is fixed independently such that $[H^+] \neq [In^-]$.

IN CLASS

Students consider the colors when several different indicators are dissolved in solutions with different values of the pH. In this case, other dissolved substances control the pH independently. Students use Le Châtelier's principle to explain how the color of an indicator changes in basic and acidic solutions. They also relate the color to the magnitude of the equilibrium constant, *K,* for the dissociation of the indicator molecule into H⁺ and *In*⁻. A complete materials list is available for download.

TRM Materials List 123

Pre-AP® Course Tip

Students preparing to take an AP® chemistry course or college chemistry should be able to use models and theories to generate predictions or claims. In this lesson, students observe what happens when different indicators are dissolved in solutions with varying pH values. They apply Le Châtelier's principle to predict the response of

the system when an external stress is applied.

Lesson Guide

ENGAGE (5 min)

TRM PowerPoint Presentation 123

ChemCatalyst

The chemical equation for the equilibrium mixture of an acid-base indicator is given below. A solution at pH = 7 is green.

$$HIn(aq) \rightleftharpoons H^+(aq) + In^-(aq)$$
yellow   blue

How Colorful

Applying Le Châtelier's Principle

THINK ABOUT IT

Acid-base indicators, such as red cabbage juice, change color as the H⁺ concentration in a solution is changed. The pH at which the color changes is different for different indictors. For example, cabbage juice changes color from red to blue at pH ~7, whereas phenolphthalein changes color from colorless to pink at pH ~9. How do equilibrium considerations predict the pH for the color change?

How do acid-base indicators work?

To answer this question, you will explore

- Applying Le Châtelier's Principle to Acid-Base Indicator Solutions
- Relating the Value of *K* and the pH to the Indicator Color
- Types of Equilibrium Constants

EXPLORING THE TOPIC

Applying Le Châtelier's Principle to Acid-Base Indicator Solutions

Acid-base indicators are weak acids, chosen for the bright colors of the H*In* molecule and *In*⁻ ion. Like other weak acids, the dissociation process of the indicator molecules to H⁺ and an anion in aqueous solutions is reversible.

Indicator molecule (one color)   Anion when the indicator molecule dissociates (a different color)

$$HIn(aq) \rightleftharpoons H^+(aq) + In^-(aq)$$

COLORS AT DIFFERENT pH VALUES

Each indicator has a specific value for the equilibrium constant *K* and a certain color that depends on the pH of the solution. Examine the information in the table to look for a relationship between *K*, pH, and color.

Indicator	H*In* color	*In*⁻ color	*K*	Color at different pH values		
				pH = 4	pH = 7	pH = 10
methyl orange	red	yellow	3.4×10^{-4}	orange	yellow	yellow
litmus	red	blue	3.2×10^{-7}	red	purple	blue
phenolphthalein	colorless	pink	3.2×10^{-10}	colorless	colorless	pink

Here are some things you might notice:

- **The value of *K*:** The equilibrium constants are all much less than 1, indicating that the concentrations of the products are quite small. This is because these

1. Use Le Châtelier's principle to explain what happens if you add concentrated acid to the solution.
2. Use Le Châtelier's principle to explain what happens if you add concentrated base to the solution.

Sample answers: **1.** If you add concentrated acid, there will be extra H⁺ ions, which, according to Le Châtelier's principle, will favor the starting substance in the equilibrium mixture. This is why acidic solutions with the indicator are yellow. **2.** If you add base, the base will react with the H⁺ ions,

indicators are very weak acids that do not dissociate very much. The value of K for methyl orange is the largest, so, of these three indicators, it dissociates the most. Phenolphthalein, with the smallest K value, dissociates the least. The value of K for litmus is in between the values for methyl orange and phenolphthalein.

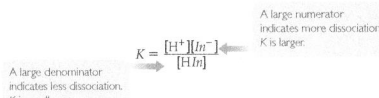

$$K = \frac{[H^+][In^-]}{[HIn]}$$

A large numerator indicates more dissociation. K is larger.

A large denominator indicates less dissociation. K is smaller.

• **The color at pH = 7**: A methyl orange solution is yellow at pH = 7, showing that the yellow In^- ion is present in relatively large concentration. A phenolphthalein solution is colorless, showing that the colorless HIn molecules is present in relatively large concentration. A litmus solution, being purple, is a mixture of red HIn molecules and blue In^- ions in roughly equal concentrations. Notice that the methyl orange solution, with the largest value of K, has the most product ions. The phenolphthalein solution has the smallest value of K and the least number of product ions.

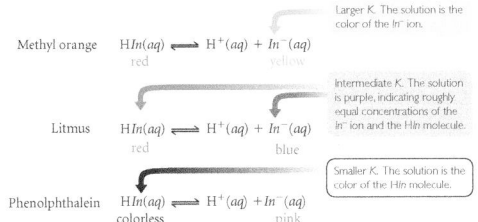

Methyl orange $HIn(aq) \rightleftharpoons H^+(aq) + In^-(aq)$
    red   yellow

Larger K. The solution is the color of the In^- ion.

Litmus $HIn(aq) \rightleftharpoons H^+(aq) + In^-(aq)$
   red   blue

Intermediate K. The solution is purple, indicating roughly equal concentrations of the In^- ion and the HIn molecule.

Phenolphthalein $HIn(aq) \rightleftharpoons H^+(aq) + In^-(aq)$
    colorless   pink

Smaller K. The solution is the color of the HIn molecule.

• **The color at different pH values**: As the pH value increases, the solution color changes from the color of HIn in acidic solutions to the color of In^- in basic solutions. The illustration below shows the color changes for the three indicators in the table. Notice that the color change occurs at higher pH values as the equilibrium constant K gets smaller.

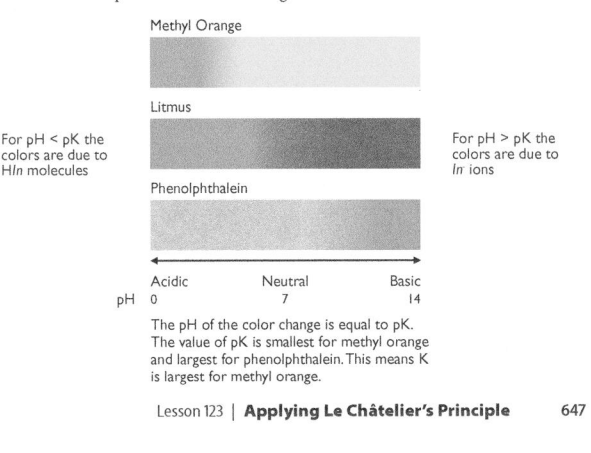

Methyl Orange

Litmus

Phenolphthalein

For pH < pK the colors are due to HIn molecules

For pH > pK the colors are due to In^- ions

Acidic   Neutral   Basic
pH   0   7   14

The pH of the color change is equal to pK. The value of pK is smallest for methyl orange and largest for phenolphthalein. This means K is largest for methyl orange.

Ask: How can you use Le Châtelier's principle to explain the color changes for bromothymol blue in solutions with different pH values?

Ask: Why don't all three indicators change color at pH = 7?

Ask: How can you make a red solution of methyl orange?

Key Points: Acid-base indicators are weak acids that dissociate into H^+ and In^-. The indicator molecule, HIn, and the anion, In^-, are different colors. You can tell the degree to which the indicator molecule dissociates by the color of the solution. For example, the methyl orange solution is yellow at pH = 7, indicating the dominance of In^- ions. The bromothymol blue solution is green at pH = 7 because both the blue HIn molecule and the yellow In^- ion are present in roughly equal concentrations.

The value of the equilibrium constant, K, is also a measure of the degree to which an acid-base indicator dissociates. When the equilibrium constant is small, the indicator does not dissociate very much at pH = 7. When the equilibrium constant is larger, the indicator dissociates to a greater degree. In solution with pH = 7, indicators with larger equilibrium constants will be present as In^- ions. Indicators with smaller equilibrium constants will mostly be undissociated HIn molecules.

When indicators are mixed with acids and bases, the degree of dissociation depends on the amount of H^+ in the solution. All indicators dissociate to a greater degree in very basic solutions, because the indicator reacts with the OH^- ions. Because product is removed from the equilibrium mixture, more product forms, according to Le Châtelier's principle. Likewise, indicators do not dissociate in concentrated acid because the In^- form reacts with the excess H^+ to produce more HIn.

RELATE THE VALUE OF K AND THE pH TO THE INDICATOR COLOR

Ask: How does the H^+ concentration at which the indicator changes color depend on the value of K?

Ask: What do you notice about the pH value for the color change and the exponent for the equilibrium constant, K?

Ask: If the equilibrium constant for a certain indicator is equal to 1×10^{-4}, would you use it to detect an acid or a base? Explain your thinking.

---

which, according to Le Châtelier's principle, will favor products in the equilibrium mixture. This is why basic solutions with the indicator are blue.

Ask: What is the equilibrium-constant equation for the dissociation of an indicator, HIn?

Ask: How can you explain the color changes using the equilibrium-constant equation?

Ask: Why is bromothymol blue called an acid-base indicator?

EXPLORE (15 min)

TRM   **Worksheet with Answers 123**

TRM   **Worksheet 123**

INTRODUCE THE CLASSWORK

→ Tell students to work in pairs.

→ Pass out the worksheet to each group.

EXPLAIN & ELABORATE (15 min)

APPLY LE CHÂTELIER'S PRINCIPLE TO ACID-BASE INDICATOR SOLUTIONS

→ Refer students to the first table from the worksheet.

Ask: What can you add to a solution to lower the pH value?

Ask: How can you raise the pH value of a solution?

Key Points: The equilibrium constant equation is useful in explaining the color changes of an indicator solution as the pH varies:

$$K = \frac{[H^+][In^-]}{[HIn]}$$

If the numerator gets smaller because H^+ is removed, the system responds in such a way that the denominator is decreased and the numerator is increased to keep the ratio equal to K. Thus, more HIn molecules dissociate to reduce the denominator, resulting in more In^- ions and thereby increasing the numerator. The ratio of the numerator and the denominator remains equal to K. Likewise, if more H^+ is added to the solution, the numerator becomes larger. Thus, more HIn is produced, thereby increasing the denominator until the ratio is equal to K.

The color change for an indicator occurs when the H^+ ion concentration is equal to the value of K. When the concentration of HIn molecules is equal to the concentration of In^- ions, the solution color is a mixture of the colors of the two substances.

When $[HIn] = [In^-]$, $K = [H^+]$.

$$K = \frac{[H^+][In^-]}{[HIn]}$$

When $[H^+] > K$, $[HIn] > [In^-]$. The solution has the color of the HIn molecules.

When $[H^+] < K$, $[HIn] < [In^-]$. The solution has the color of the In^- ions.

EVALUATE (5 min)

Check-In

For the indicator phenol red, HIn is yellow and In^- is red. The equilibrium constant, K, is equal to 1.3×10^{-8}.

1. What color do you expect phenol red to be when it is dissolved in a solution of pH = 10?

2. How could you make a solution of phenol red turn yellow?

MINIMIZING CHANGES AT EQUILIBRIUM

One way to understand the color differences for indicators in acidic and basic solutions is to apply Le Châtelier's principle. When an acid is added to a solution with an indicator, $[H^+]$ increases. Because H^+ is a product of the dissociation of the indicator, the system responds to minimize the change by forming HIn molecules. Alternatively, if a base is added to a solution with an indicator, $[H^+]$ decreases. The system responds by making more product ions, H^+ and In^-.

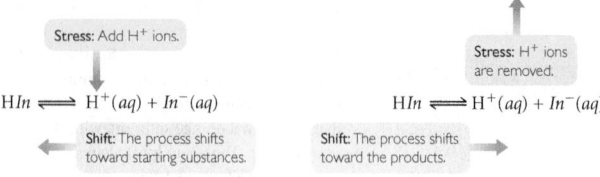

Relating the Value of K and the pH to the Indicator Color

The color change of the indicator signals the shift that has occurred in the mixture to minimize the changes that occur. This can also be seen mathematically if you examine the equilibrium-constant equation.

$$K = \frac{[H^+][In]}{[HIn]}$$

Since K is a constant, it does not change. If any variable in the equation changes, the other variables must adjust. For example, if $[H^+]$ increases, $[In^-]$ must decrease and $[HIn]$ must increase to keep K the same.

pH RANGE FOR COLOR CHANGE

Take a moment to examine the information for the color change for several indicators in the table. Notice that the color change occurs at different pH values for different indicators.

Indicator	HIn color	In⁻ color	K	[H⁺] when [HIn] = [In⁻]	pH range for color change
methyl orange	red	yellow	3.4×10^{-4}	3.4×10^{-4}	3.2–4.4
bromocresol green	yellow	blue	1.3×10^{-5}	1.3×10^{-5}	3.8–5.4
methyl red	red	yellow	1.0×10^{-5}	1.0×10^{-5}	4.8–6.0
bromothymol blue	yellow	blue	5.0×10^{-8}	5.0×10^{-8}	6.0–7.6
phenol red	yellow	red	1.0×10^{-8}	1.0×10^{-8}	6.6–8.0
phenolphthalein	colorless	pink	3.2×10^{-10}	3.2×10^{-10}	8.2–10.0
alizarin yellow	yellow	red	1.0×10^{-11}	1.0×10^{-11}	10.0–12.0

648   Chapter 24 | **Changing Conditions**

Answers: **1.** The equilibrium constant suggests that there will be mostly red In^- ions in a basic solution with pH = 10. **2.** To make the phenol red turn yellow, there must be more HIn in solution. According to Le Châtelier's principle, you have to add concentrated acid.

Homework

Assign the reading and exercises for Showtime Lesson 123 in the student text.

The color change for an indicator occurs when the H^+ ion concentration is equal to K. This happens when the concentration of HIn molecules is equal to the concentration of the In^- ions. Rearranging the equilibrium-constant equation proves this to be true.

$$K = \frac{[H^+][In^-]}{[HIn]}$$

$$= [H^+]\frac{[In^-]}{[HIn]}$$

When $[In^-] = [HIn]$, then $K = [H^+]$

When $[H^+] > K$, then $[HIn] > [In^-]$. The solution has the color of HIn molecules. When $[H^+] < K$, then $[HIn] < [In^-]$. The solution has the color of In^- ions. When $[H^+] = K$, then $[HIn] = [In^-]$. The solution color is a mixture of the color of HIn molecules and In^- ions.

Example 1

The Color of Congo Red

Determine the pH value at which the indicator Congo red changes color. The equilibrium constant for the dissociation of the indicator is $K = 10^{-4}$.

Solution

The color change happens when the concentration of HIn equals the concentration of In^-. If $[HIn] = [In^-]$, then $K = [H^+] = 10^{-4}$.

$$pH = -\log[H^+]$$
$$= -\log[10^{-4}]$$
$$= 4$$

So the indicator changes color at pH = 4.

Types of Equilibrium Constants

The equilibrium constant K can be used to solve a variety of equilibrium problems. In every case, the equilibrium-constant equation can be used to figure out the concentrations of one or more of the compounds in the equilibrium mixture. The table shows some specific types of chemical equilibrium constants. Each of these constants is designated by a unique subscript.

Equilibrium constant	Chemical equation	Equilibrium-constant equation
Solubility product constant K_{sp}	$MX(s) \rightleftharpoons M^+(aq) + X^-(aq)$	$K_{sp} = [M^+][X^-]$
Acid dissociation constant K_a	$HA(aq) \rightleftharpoons H^+(aq) + A^-(aq)$	$K_a = \frac{[H^+][A^-]}{[HA]}$
Base dissociation constant K_b	$B(aq) + H_2O(l) \rightleftharpoons BH^+(aq) + OH^-(aq)$	$K_b = \frac{[BH^+][OH^-]}{[B]}$
Ionization of water constant K_w	$H_2O(l) \rightleftharpoons H^+(aq) + OH^-(aq)$	$K_w = [H^+][OH^-]$

Notice that the equilibrium-constant equations are the same as those that you might have written, except that they have a subscript attached to the symbol K. As discussed previously, liquids and solids are not included in the equilibrium-constant equation because their concentrations do not change. So, equilibrium-constant equations can be used to determine the concentrations of molecules and ions in solution and the concentrations of gaseous molecules in systems at equilibrium.

The equilibrium constant K_{sp} can be used to determine the solubility of an ionic compound.

Example 2

Comparing Solubility

The solubility product constants K_{sp} for several alkaline earth metal sulfates at 25 °C are given in the table.

Metal sulfate	Solubility product constant K_{sp}
magnesium sulfate, $MgSO_4(s)$	4.7
calcium sulfate, $CaSO_4(s)$	4.9×10^{-5}
barium sulfate, $BaSO_4(s)$	1.1×10^{-10}

a. Put the three salts in order of increasing solubility from least soluble to most soluble. Explain your thinking.

b. Use the periodic table to predict whether strontium sulfate, $SrSO_4(s)$, is more soluble than calcium sulfate, $CaSO_4(s)$. Explain your thinking.

c. Suppose you add sodium sulfate, $Na_2SO_4(s)$, to a saturated solution of calcium sulfate, $CaSO_4(aq)$. Use Le Châtelier's principle to explain why a small amount of $CaSO_4(s)$ precipitates from the solution.

Solution

a. The chemical equation for the solubility of an alkaline earth metal sulfate, $MSO_4(s)$, is

$$MSO_4(s) \rightleftharpoons M^{2+}(aq) + SO_4^{2-}(aq) \qquad K_{sp} = [M^{2+}][SO_4^{2-}]$$

If the solubility product constant K_{sp} is large, products are favored (the salt is more soluble). Therefore, the solubility from least soluble to most soluble is $BaSO_4 < CaSO_4 < MgSO_4$.

b. All of the metals are in Group 2A, the alkaline earth metals. The solubility decreases as you go down the group. So, you expect strontium sulfate, $SrSO_4(s)$, to be less soluble than calcium sulfate, $CaSO_4(s)$, but more soluble than barium sulfate, $BaSO_4(s)$.

c. Adding sodium sulfate, $Na_2SO_4(s)$, adds more sulfate ions. Because $SO_4^{2-}(aq)$ is a product of dissolving calcium sulfate, $CaSO_4(aq)$, the system will reduce the stress by forming more reactants. This is why some $CaSO_4(s)$ precipitates.

The equilibrium constant K_a can be used to determine the strength of an acid or a base.

Example 3

Strength of an Acid

Imagine that you prepare a 0.10 M solution of HCl and a 0.10 M solution of HOCl. The concentrations of the ions in equilibrium are given in this table.

The general chemical equation for acid dissociation is:

$$HA(aq) \rightleftharpoons H^+(aq) + A^-(aq)$$

Acid	[HA]	[H$^+$]	[A$^-$]
hydrochloric acid, HCl(aq)	~0	0.10 M	0.10 M
hypochlorous acid, HOCl(aq)	0.099994 M	6.0×10^{-5} M	6.0×10^{-5} M

a. Determine K_a for hypochlorous acid, HOCl(aq).

b. Explain why K_a for HCl(aq) is a very large number.

c. Which acid is stronger? Explain your thinking.

Solution

a. $HOCl(aq) \rightleftharpoons H^+(aq) + OCl^-(aq)$

$$K_a = \frac{[H^+][OCl^-]}{[HOCl]}$$

$$= \frac{(6.0 \times 10^{-5})(6.0 \times 10^{-5})}{0.099994}$$

$$= 3.6 \times 10^{-8}$$

b. The value [HCl] is extremely small since it is approximately zero. Dividing by such a small number gives a very large number, so K for HCl must be extremely large.

So K_a for this acid is very large.

c. The very large K_a for HCl indicates that HCl fully dissociates in water. The tiny K_a for HOCl indicates that HOCl does not dissociate as much as HCl. So HCl is a stronger acid than HOCl.

LESSON 123 ANSWERS

1. Indicator molecules have a bright color that changes to a different color when the H*In* molecules dissociate into H$^+$ and *In*$^-$. When the solution is more acidic, the color of H*In* molecules is observed. In basic solution, the solution displays the color of *In*$^-$.

2. The value of K does not change, but the concentrations adjust such that the equilibrium constant equation is satisfied.

3. a. In order of increasing solubility, AgCl, CuCl, TlCl, the most soluble is TlCl. A small K_{sp} indicates mostly undissolved solid. **b.** CuCl (*s*) will precipitate.

4. a. $K = 6.9 \times 10^{-5}$. **b.** K is a very large number for HNO$_3$ because all of it dissociates. **c.** HNO$_3$ is stronger because nearly all of it dissociates. **d.** The 0.10 M HNO$_3$ solution has the higher concentration of H$^+$ because nearly all of it dissociates.

5. a. yellow, due to the large amount of H*In*. **b.** blue, due to the large amount of *In*$^-$. **c.** yellow, due to large amount of *In*$^-$. **d.** orange, due to presence of both H*In* and *In*$^-$.

6. Hydroxide ions in the base combine with H$^+$ ions from the indicator molecule to form H$_2$O. This causes more H*In* to dissociate, increasing the concentration of *In*$^-$, which makes the solution turn blue.

7. pH ~ 10

8. $K = 10^{-1}$

9. A good description will include the source, molecular structure and properties of a particular indicator. The student might also draw a molecular level diagram.

LESSON SUMMARY

How do acid-base indicators work?

Acid-base indicators are weak acids or bases that change color depending on pH. The concentrations of an indicator molecule and its anion depend on the concentration of H$^+$ ions in solution. Therefore, the color of an indicator is a measure of the acidity of the solution. The color of an indicator changes when the concentrations of the H*In* molecules and *In*$^-$ anions are identical. This occurs when $K = [H^+]$. The lessons learned from applying equilibrium ideas to indicators are general, and they can be applied to other types of equilibria.

Exercises

Reading Questions

1. Explain how an acid-base indicator works to show the pH of a solution.
2. Does the value of K for an indicator change when the indicator is added to an acid or a base solution? Explain.

Reason and Apply

3. The solubility product constants K_{sp} at 25 °C for several chlorides are provided in the table.

Metal chloride	Solubility product constant K_{sp}
thallium (I) chloride, TlCl(*s*)	1.86×10^{-4}
copper (I) chloride, CuCl(*s*)	1.72×10^{-7}
silver (I) chloride, AgCl(*s*)	1.77×10^{-10}

   a. Put the three salts in order of increasing solubility—from least soluble to most soluble. Explain your thinking.
   b. Suppose that you add sodium chloride, NaCl(*s*), to a saturated solution of copper chloride, CuCl(*aq*). Use Le Châtelier's principle to explain what happens. Assume that NaCl is very soluble.

4. Imagine you prepare a 0.10 M solution of nitric acid, HNO$_3$, and a 0.10 M solution of hydrofluoric acid, HF. The concentrations of the ions in equilibrium are given in the table.
   General equation for acid dissociation: HA(*aq*) $\rightleftharpoons$ H$^+$(*aq*) + A$^-$(*aq*)

Acid	[HA]	[H$^+$]	[A$^-$]
nitric acid, HNO$_3$(*aq*)	~0	0.10 M	0.10 M
hydrofluoric acid, HF(*aq*)	0.0974 M	2.6×10^{-3} M	2.6×10^{-3} M

   a. Determine K for hydrofluoric acid, HF.
   b. Explain why K for nitric acid, HNO$_3$, is a very large number.
   c. Which acid is stronger? Explain your thinking.
   d. Which solution has the higher concentration of H$^+$(*aq*), 0.10 M HNO$_3$ or 0.10 M HF? Explain your thinking.

Exercises

(continued)

5. Use the table on page 648 to predict the color of these indicators if you dissolve them in water. Explain your thinking.
   a. alizarin yellow
   b. bromocresol green
   c. methyl red
   d. phenol red

6. Use Le Châtelier's principle to explain why bromocresol green is blue if you dissolve it in a base solution, such as 0.10 M sodium hydroxide, NaOH.

7. Determine the pH value at which thymolphthalein changes color. The equilibrium constant for the dissociation of the indicator is $K = 10^{-10}$.

8. Methyl violet changes color from the yellow H*In* molecule to the blue-violet *In*⁻ ion at a pH of 1. What is the equilibrium constant for the dissociation of methyl violet?

9. PROJECT

   Look up the source, structure, and properties of a particular indicator. Write a short description of the molecule.

ANSWERS TO CHAPTER 24 REVIEW EXERCISES

1. a. $HIn(aq) \leftrightharpoons H^+(aq) + In^-(aq)$
b. blue, basic solutions are blue. **c.** no color change, the pH decreases. **d.** the solution turns blue, the pH increases. **e.** The particle view should show the same number of red HIn, blue In^-, and H^+ particles.

2. a. $K = [H^+][F^-]/[HF]$ **b.** $[H^+] = [F^-]$. Each HF molecule dissociates to one H^+ and one F^-. **c.** $[H^+] < [HF]$. Because $K < 1$, very little HF dissociates. **d.** [HF] increases.

3. a. No more solid will dissolve. Excess solid remains. In a saturated solution, no more solid can be dissolved. Excess solid will remain at the bottom of the container. **b.** More $CaCl_2$ precipitates. **c.** More $CaCl_2$ precipitates. **d.** More $CaCl_2$ precipitates. **e.** More $CaCl_2$ dissolves.

4. a. $K = 1.0 \times 10^{-4} = [NO]^2/[N_2][O_2]$ **b.** starting substances, because K is very small. **c.** less NO because $[O_2]$ is lower. **d.** $K = 1.0 \times 10^{-31} = [NO]^2/[N_2][O_2]$ **e.** There will not be much NO due to this reaction at 25 °C because K is very low. **f.** endothermic. **g.** No change is expected because the number of moles of starting substances is the same as the number of moles of product.

5. a. $K = \dfrac{[CO][Cl_2]}{[COCl_2]}$

b. 8.6×10^{-2} M **c.** More $COCl_2$ forms to minimize the effect of adding one of the products. **d.** 0.14 M

ASSESSMENTS

Two multiple-choice assessments (versions A and B) and two short-answer assessments (versions A and B) are available for download for Chapter 24.

TRM Chapter 24 Assessment Answers

TRM Chapter 24 Assessments

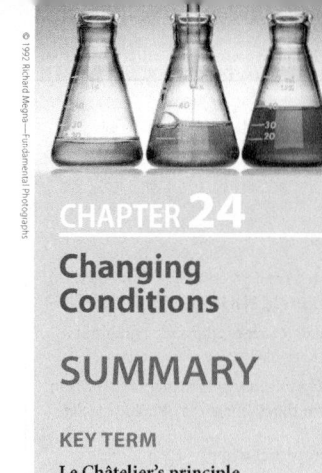

© 1992 Richard Megna—Fundamental Photographs

CHAPTER 24

Changing Conditions

SUMMARY

KEY TERM

Le Châtelier's principle

Showtime Update

Le Châtelier's principle states that when conditions change for an equilibrium mixture, the amounts of starting substances and products will adjust to reduce the effect of the change.

If you add more of a starting substance to an equilibrium mixture, the amounts will shift in the direction of forming products. If you remove a starting substance from an equilibrium mixture, the amounts will shift in the direction of forming more starting substances. The same is true for adding or removing products.

Changing temperature is another way to place a stress on an equilibrium mixture. If the process is exothermic, heating will shift the amounts in the direction of forming starting substances. For an endothermic process, heating will shift the amounts in the direction of forming products.

For equilibrium mixtures with starting substances and products that are gases, changes in volume that cause changes in pressure place a stress on the system at equilibrium. If the pressure increases, a change will occur in the direction that decreases the total number of moles of gas to reduce the pressure.

REVIEW EXERCISES

1. The indicator called litmus, H*In*, is red in acidic solution and blue in basic solution.
   a. Write a chemical equation to describe the equilibrium mixture.
   b. What color is *In*⁻ ? Explain how you know.
   c. What happens if you add acid to a red solution of litmus?
   d. What happens if you add base to a red solution of litmus?
   e. Draw a particle view of a purple litmus solution.

2. Hydrofluoric acid, HF, is a weak acid with $K = 6.3 \times 10^{-4}$. The chemical equation below describes the dissociation of HF in water:

$$HF(aq) \leftrightharpoons H^+(aq) + F^-(aq)$$

   a. Write the equilibrium-constant equation for the dissociation of HF.
   b. At equilibrium, is $[H^+]$ greater than, less than, or equal to $[F^-]$? Explain your answer.
   c. At equilibrium, is $[H^+]$ greater than, less than, or equal to [HF]? Explain your answer.
   d. How does [HF] change if H^+ is added to the equilibrium mixture?

3. The dissolution of calcium chloride, $CaCl_2$, in water is used to create hot packs for foot and hand warmers. The chemical equation below describes a saturated solution.

$$CaCl_2(s) \leftrightharpoons Ca^{2+}(aq) + 2Cl^-(aq)$$

   a. Describe what is meant by a saturated solution. How do you know the solution is saturated?

   b. Describe what happens if you add more Ca^{2+} to a saturated solution.
   c. Describe what happens if you add more Cl^- to a saturated solution.
   d. Describe what happens if you raise the temperature.
   e. Describe what happens if you lower the temperature.

4. Nitrogen, N_2, and oxygen, O_2, react in a car engine to produce nitric oxide, NO, as described by the chemical equation given below:

$$N_2(g) + O_2(g) \rightleftharpoons 2NO(g)$$

   a. In an engine at 1800 °C, $K = 1.0 \times 10^{-4}$. Write the equilibrium-constant equation for this reaction at 1800 °C.
   b. Are the starting substances or products favored in an equilibrium mixture at 1800 °C? Explain your thinking.
   c. If you drive the car to a mountaintop where the concentration of oxygen is much lower, will there be more or less NO in the equilibrium mixture? Explain your thinking.
   d. At 25 °C, $K = 1.0 \times 10^{-31}$. Write the equilibrium constant equation for this reaction at 25 °C.
   e. Do you expect to find much NO in the atmosphere due to the reaction of N_2 and O_2 at 25 °C? Why or why not?
   f. Given that the value of K is much smaller at lower temperatures, is the reaction exothermic or endothermic?
   g. If the volume of the container with the equilibrium mixture is expanded, do you expect a change in the equilibrium mixture? Why or why not?

5. Suppose this chemical reaction is in a state of dynamic equilibrium in a sealed container:

$$COCl_2(g) \rightleftharpoons CO(g) + Cl_2(g) \qquad K = 4.63 \times 10^{-3}$$

   a. Write the equilibrium-constant equation for this reaction.
   b. At equilibrium, what would $[COCl_2]$ be if $[Cl_2] = [CO] = 2.0 \times 10^{-2}$ M?
   c. In which direction would the equilibrium shift if some additional CO were added to the system? Explain why.
   d. Suppose that the CO added in part c brought $[COCl_2]$ to 4.0 M at equilibrium. What would $[Cl_2]$ and $[CO]$ be?

Project: It's Showtime!

● A good project should include student research and cited references for their pigment ● a description of how the pigment behaves in different solutions ● a detailed procedure to safely extract the pigment for testing ● a detailed procedure outlining how to test acid-base properties of substances with the pigment including all materials and safety considerations ● results from testing of the pigment ● appropriate demonstration or presentation designed to show what was learned about the pigment

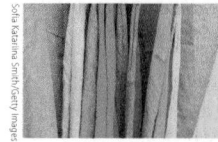

It's Showtime!

PROJECT   It is time for you to show what you have learned both in Unit 6: Showtime and in the entire *Living by Chemistry* course. For this project, you will do research and design your own lab. The goal of this project is to examine the acid-base indicator properties of a naturally occurring pigment.

Research

- Begin by doing literature research to identify a pigment molecule that gives color to a common substance and behaves as an indicator in solution. Choose a substance that is safe to handle and that you can obtain easily, such as flowers, leaves, fruits, berries, vegetables, spices, bark, or roots.

- Find out how it might be possible to safely extract this indicator molecule, so you can prepare a solution. For example, you might crush rose petals in rubbing alcohol or boil beets in water.

Design a Lab Experiment

- Based on your research, write a detailed lab procedure that explains how to prepare a solution of the pigment.

- Next, write a lab procedure for how to test the acid-base properties of the pigment with common substances. You might want to refer to Unit 4: Toxins, Lesson 84, when preparing this procedure.

- Finally, write a lab procedure for testing a small piece of fabric, such as wool, cotton, or nylon, to determine if the pigment will stain the fabric and if the stain is permanent.

- Be sure to include a materials list and any safety considerations.

Teacher and Peer Review

- Bring your proposed lab experiment to class and have your classmates review and critique your project.

- Submit your research and lab experiment to your teacher for review and approval.

- With teacher supervision, conduct your experiment in your school lab.

- Make changes based on feedback and the results of your testing. Once your experiment is a success, your teacher may ask you to lead a small group of students in carrying out the lab, or ask you to do a presentation to show what you learned about pigments, extraction, acid-base properties, and staining fabric.

Showtime | REVIEW

UNIT 6

Carlos Villoch/Robert Harding

David Ash/Getty Images

Many chemical processes are reversible. These processes do not convert all of the starting substances to products. Instead, the result is an equilibrium mixture of starting substances and products. In reversible chemical processes, the forward and reverse processes continue. When the forward and reverse processes occur at equal rates, these systems are said to be at equilibrium. Reversible processes are written with a double arrow to show that the forward and reverse processes occur at the same time.

Common equilibrium mixtures involve dissociation processes in which a starting substance breaks apart into two or more pieces. The back-and-forth evaporation and condensation of water on the planet is a form of phase equilibrium. Salts dissolve in water and dissociate into cations and anions, and the cations and anions combine to re-form the salt. Weak acids in water dissociate into H^+ and an anion, and they recombine to form the acid molecule. Molecules in the body attach and detach from receptor sites. In all of these cases, there is a dynamic balance between the starting substances and products.

There is a mathematical relationship between the products and starting substances in an equilibrium mixture. The equilibrium constant K and the equilibrium-constant equation can be used to solve problems involving systems at equilibrium.

$$aA + bB \rightleftharpoons cC + dD$$

$$K = \frac{[C]^c[D]^d}{[A]^a[B]^b}$$

In general, a large value of K indicates that products are favored in the equilibrium mixture, and a small value of K indicates that starting substances are favored.

Le Châtelier's principle states that when conditions are changed for an equilibrium mixture, the system will respond by reducing the effect of the change. When conditions such as temperature, pressure, and concentration are changed in a system, the system regains equilibrium by rebalancing the concentrations of starting substances and products.

REVIEW EXERCISES

General Review

Write a brief and clear answer to each question. Be sure to show your work.

1. Write the reversible reactions for these processes.
   a. Chlorine gas becomes liquid chlorine.
   b. Nitrogen dioxide becomes dinitrogen tetroxide.
   c. Silver chloride precipitates from aqueous solution. Water evaporates.

2. Write the equilibrium-constant equations for the processes in Exercise 1.

3. Write two statements that are always true of a system at equilibrium.

Unit 6 | **Review**   657

Overview

LAB: GROUPS OF 2

Key Question: Which natural pigments are acid-base indicators?

KEY IDEAS

Many pigments found in nature are acid-base indicators. The colorful pigments can be extracted and their colors can be changed by adding acids or bases to the solution. Le Châtelier's principle predicts the outcome of these changes.

IN CLASS

For this project, students will show what they have learned throughout the year by researching and designing a lab experiment. The goal of this project is to examine the acid-base indicator properties of a naturally occurring pigment. Students should be asked to research how to extract pigments from plant matter and submit a procedure to extract a specific pigment before coming to class.

You may want to spread the in-class part of the lesson over a few days, allowing at least one or two class periods for students to extract the pigment and test its acid-base properties, and at least one class period for groups to present their experiments.

TRM Materials List Unit 6 Review

SAFETY

Students should wear safety glasses at all times. Provide baking soda to neutralize acid spills and citric acid to neutralize base spills.

SETUP

Provide students with a selection of plant matter such as flowers, leaves, fruits, berries, vegetables, spices, bark, or roots. To extract the pigments, water and/or alcohol is required. Cheesecloth or nylon stockings are useful for pressing the juice from certain berries to get the pigment. Provide acid (pH ~ 3) and base (pH ~ 10) solutions for testing the acid-base properties of the pigments. Gather a selection of white fabrics to test if the pigments create stains and if the stains are permanent.

CLEANUP

The solutions can be diluted and poured down the drain. Unused plant matter can be placed in the trash or compost.

GUIDE TO UNIT REVIEW

ENGAGE (5 min)

ChemCatalyst

In the research you did before coming to class, what procedures did you find for extracting a pigment from plant matter?

Sample answers: You can boil berries in water to make a colored berry juice. You can grind leaves in alcohol with a mortar and pestle to extract the green pigment, chlorophyll. You can press red grapes through cheesecloth to a get a red liquid.

Ask: What are some ways of extracting pigments from plants?

Ask: How can you test the pH of the extracted pigment solution?

Ask: How can you test the acid-base properties?

Ask: How does Le Châtelier's principle apply?

EXPLORE (60 min)

INTRODUCE THE LAB

→ Let students know they will be working in groups of two.

→ Tell students how much time they have to work on developing their experiment. (At least 1 hour of class time is recommended.)

→ Hand out the worksheets to assist students in organizing their experiments. Make sure each student leaves class knowing what part of the project he or she is responsible for.

→ Each pair of students will:

● Conduct a literature research to identify a pigment molecule that gives color to plant matter and find out how to safely extract the pigment molecules.

● Develop and write an experimental procedure, including safety instructions, for extracting a pigment from plant matter, testing the acid-base properties of the pigment, testing if the pigment stains fabric, and testing if the stain is permanent.

● Show the procedure to peers in class for review and critique.

● Make changes based on feedback.

● Obtain instructor approval of the procedure.

● Test the procedure in class.

● Make changes based on results of testing.

● Obtain instructor approval of revised procedure and test again.

● Present findings to the class. The presentation should include ideas involving equilibrium to explain the extraction procedure, the acid-base properties, and the permanence of stains.

EXPLAIN & ELABORATE (40 min)

→ Have students present their experiments. Check that students have shared the work equally and that both group members can explain the experiments.

Ask: What can you explain about the structure and properties of the pigment you extracted?

Ask: Did the color of the pigment change with pH?

---

4. When a dissolving and precipitation process is at equilibrium, the solution is saturated. Explain why this is true.

5. What is the concentration of copper ion, Cu^{2+}, in an aqueous solution of copper (II) carbonate, $CuCO_3$, if $K_{sp} = 1.4 \times 10^{-10}$ at $25\,°C$?

6. What does an acid-base indicator have to do with equilibrium?

7. The chemical equation for the dissolution of lead chloride, $PbCl_2(s)$, in water is given here.

$$PbCl_2(s) \longleftrightarrow Pb^{2+}(aq) + 2Cl^-(aq)$$

Determine the concentration of Pb^{2+} ions in a saturated $PbCl_2$ solution if $K_{sp} = 1.7 \times 10^{-5}$.

8. Solid lead (II) sulfate, $PbSO_4$, is dissolved in water to form a solution with excess undissolved solute at the bottom of the container. The value of K_{sp} for lead sulfate is 2.5×10^{-8}.

   a. How can you tell that this solution is saturated?

   b. Is the solution at equilibrium? How do you know?

   c. Write the chemical equation for the reversible change that is taking place.

   d. What is the equilibrium-constant equation for this equilibrium?

   e. What is the equilibrium concentration of each ion?

   f. Dissolving lead (II) sulfate is an endothermic change. Predict what will happen to the value of $[Pb^{2+}]$ and $[SO_4^{2-}]$ if the solution is heated.

   g. What would happen if you added more sulfate ions to the solution?

9. The value of K_{sp} for cadmium sulfide, CdS, is 8.0×10^{-27}. Compare this value with the K_{sp} for lead (II) sulfate given in Exercise 8.

   a. Which compound will have a higher concentration of ions in solution at equilibrium? Explain your reasoning.

   b. Calculate the concentration of cadmium ions in solution at equilibrium.

---

10. For the acid-base indicator litmus, the molecule, $HIn(aq)$, is red and the anion, $In^-(aq)$, is blue.

    a. Write a reaction for the dissociation of the HIn molecule in solution.

    b. Why is litmus purple when you dissolve it in water? Explain your reasoning.

    c. What color do you think litmus is in a basic solution? Use Le Châtelier's principle to explain your thinking.

STANDARDIZED TEST PREPARATION

Multiple Choice

Choose the best answer.

Use the following to answer Exercises 1–3.

$$H^+(aq) + In^-(aq) \rightleftharpoons HIn(aq)$$
$$\qquad\qquad yellow \qquad\quad red$$

1. Which correctly identifies the acid and base forms of the indicator?

   (A) H^+ is the acid form of the indicator; HIn^- is the base form of the indicator.

   (B) In^- is the acid form of the indicator; HIn is the base form of the indicator.

   (C) HIn is the acid form of the indicator; In^- is the base form of the indicator.

   (D) HIn is the acid form of the indicator; H^+ is the base form of the indicator.

2. Which statement correctly describes the color of the solution containing the indicator when it is orange?

   (A) Only the H^+ ions are present in the solution.

   (B) Only In^- anions are present in the solution.

   (C) Only the HIn molecules are present in the solution.

   (D) There is a mixture of both the In^- anions and HIn molecules in solution.

3. What will happen to the reaction as H^+ ions are added to the solution?

   (A) The color of the solution will change from red to yellow.

   (B) The color of the solution will change from yellow to red.

   (C) The color of the solution will remain yellow.

   (D) The color of the solution will remain red.

Ask: How can you estimate the K for acid-dissociation of the pigment based on your observations of the color changes?

Ask: Explain how Le Châtelier's principle applies to your procedure.

Ask: Is the process of the pigment binding to the fabric reversible? Explain how you know.

Ask: Is the equilibrium constant for the pigment binding to the cloth large or small? Explain.

Key Points: The primary chemistry behind these experiments is reversible processes and equilibrium. At first, the pigment molecules are bound to

plant matter. If the pigments are soluble in water or alcohol, the reverse reaction occurs, and the pigments detach from the plant matter and dissolve in the solvent. By adding acid or base, it is possible to reversibly change the color of the pigment if it is a weak acid. When the color is a mixture of the colors in acid or base, pH = pK. Thus, it is possible to determine K for acid dissociation of the pigment by measuring the pH at which the color changes. The pigment in solution is attracted to cloth. If the attraction is strong, it will be difficult to reverse the process, and the stain is permanent. In this case, the equilibrium

4. For the cabbage juice indicator, what might be the pH of the solution when the color is red?

(A) 2
(B) 7
(C) 9
(D) 12

5. Which of these is a reversible process?

(A) lighting a safety match
(B) burning a log on a fire
(C) water freezing to become ice
(D) baking a cake

6. Carbon dioxide binds to hemoglobin, Hb, in your blood to form [CO_2 : hemoglobin] in a reversible process. Which of the following correctly represents a balanced chemical equation for this process?

I. [CO_2 : hemoglobin] $\rightleftharpoons CO_2$ + hemoglobin
II. CO_2 + hemoglobin $\rightleftharpoons$ [CO_2 : hemoglobin]
III. CO_2 + hemoglobin $\longrightarrow$ [CO_2 : hemoglobin]

(A) I only
(B) II only
(C) III only
(D) Both I and II

7. The particle views below show 10 molecules of each of three sweeteners with their relative sweetness. Which statement correctly describes sucrose?

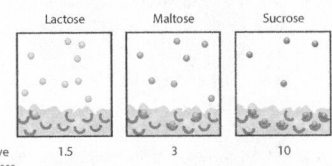

| Lactose | Maltose | Sucrose |

Relative sweetness   1.5   3   10

(A) Sucrose molecules are detected as sweeter because more are attached to the receptors.
(B) Sucrose molecules have weaker intermolecular attractions with receptors than maltose molecules.
(C) More sucrose molecules remain in the saliva than either maltose or lactose molecules.
(D) Sucrose molecules are the sweetest because they bind to all available receptors.

8. An octane molecule is reversibly held weakly to a smell receptor in the nose. What might be responsible for reversibly holding octane to smell receptors?

(A) hydrogen bonds
(B) covalent bonds
(C) dipole-dipole interactions
(D) intermolecular forces

9. Which correctly describes a reversible process?

(A) The forward and reverse reactions can occur at the same time.
(B) A reversible process proceeds in both the forward and reverse directions.
(C) In a reversible chemical reaction system, the products and the starting substances are both present.
(D) All of the above are true of a reversible process.

10. Which chemical process might this graph represent?

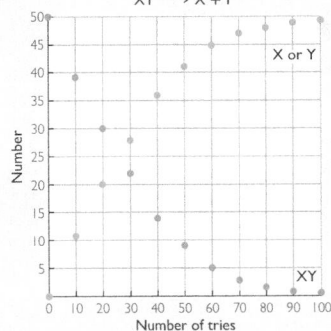

$XY \longrightarrow X + Y$

(A) a salt such as NaCl, which dissolves partially to Na$^+$(aq) + Cl$^-$(aq) to form a saturated solution
(B) a strong acid such as HBr, which separates to H$^+$(aq) + Br$^-$(aq)
(C) an acid-base indicator, HIn, which separates to H$^+$(aq) + In^-(aq)
(D) a molecule bound to a receptor, which separates to a molecule and the empty receptor

4. If the solution is not saturated, more salt will dissolve. Once the solution is saturated more salt will not dissolve because the rate of precipitation is equal to the rate of dissolution.

5. 1.2×10^{-5} M

6. An acid–base indicator in solution is a mixture of HIn and In^- depending on the H$^+$ concentration. The concentrations change depending on the pH of the solution to satisfy the equilibrium constant equation: $K = [\text{H}^+][In^-]/[\text{H}In]$

7. [Pb^{2+}] = 1.6×10^{-2} M

8. a. The solution is saturated when there is undissolved solid. **b.** Yes, the solution is at equilibrium if there is undissolved solid.
c. PbSO$_4$(s) $\rightleftharpoons$ Pb^{2+}(aq) + SO$_4^{2-}$(aq)
d. $K = $ [Pb^{2+}(aq)][SO$_4^{2-}$(aq)]
e. [Pb^{2+}(aq)] = [SO$_4^{2-}$(aq)] = 1.6×10^{-4} M.
f. The concentrations will increase because more PbSO$_4$ will dissolve.
g. PbSO$_4$(s) will precipitate.

9. a. PbSO$_4$ will have more ions in solution because K_{sp} is larger.
b. [Cd^{2+}] = 8.9×10^{-14} M.

10. a. HIn(aq) $\rightleftharpoons$ H$^+$(aq) + In^-(aq)
b. There is a mixture of red HIn molecules and blue In^- ions. **c.** blue, because OH$^-$ reacts with H$^+$ from HIn molecules.

STANDARDIZED TEST PREPARATION

1. C

2. D

3. B or D (the question does not specify the starting solution)

4. A	10. B	16. D
5. C	11. B	17. A
6. D	12. B	18. C
7. A	13. A	19. B
8. D	14. C	20. B
9. D	15. A	

---

constant K for binding is large. If the equilibrium constant for binding is small, the stain is readily washed away. In all cases, Le Châtelier's principle can be used to make predictions and offer explanations of the outcomes of these experiments.

There is no Check-In for this review.

Homework

Assign the Unit 6 Review in the student text to help students prepare for the unit exam.

ANSWERS

1. a. Cl$_2$(g) $\rightleftharpoons$ Cl$_2$(l)

b. 2NO$_2$(g) $\rightleftharpoons$ N$_2$O$_4$(g)

c. AgCl(s) $\rightleftharpoons$ Ag$^+$(aq) + Cl$^-$(aq)

2. a. $K = \dfrac{[1]}{[\text{Cl}_2]}$

b. $K = \dfrac{[\text{N}_2\text{O}_4]}{[\text{NO}_2]^2}$

c. $K = [\text{Ag}^+][\text{Cl}^-]$

3. *Possible answers:* The reaction is reversible. The equilibrium constant does not change as the concentrations of products and reactants change.

ASSESSMENTS

Two Unit 6 Assessments (versions A and B) are available for download.

TRM Unit 6 Assessment Answers

TRM Unit 6 Assessments

The Unit 6 Project: Showtime can be used as the lab exam to cover topics explored in this unit.

Two final assessments covering all units are available for download.

TRM Final Assessment Answers

TRM Final Assessments

A final lab assessment covering topics from all units is available for download.

TRM Final Lab Assessment Instructions and Answers

TRM Final Lab Assessment

Engineering Project U6: It's Showtime!

The Unit 6 Project: It's Showtime allows students a chance to design their own experiment. To give this project an engineering design focus, have students test their modified design and write a reflection on how these engineering practices helped to improve their experiment.

---

11. Consider this chemical reaction: $AB \rightleftharpoons A + B$. In a rigid container, 25 mol of AB decompose to reach equilibrium. The amounts of A and B are measured over time, and the values are recorded in the table below. At 240 s, which statement is correct?

Time (s)	30	60	90	120	150	180	210	240	270
Amount of A (mol)	4	6	8	12	14	15	15	15	15
Amount of B (mol)	4	6	8	12	14	15	15	15	15

(A) The concentrations of AB, A, and B are equal.
(B) The concentration of A is greater than the concentration of AB.
(C) The concentration of AB is greater than the concentration of B.
(D) The system is not at equilibrium.

Use this graph to answer Exercises 12–14. The graph represents the chemical reaction:

$$A_2(g) + 3B_2(g) \rightleftharpoons 2AB_3(g)$$

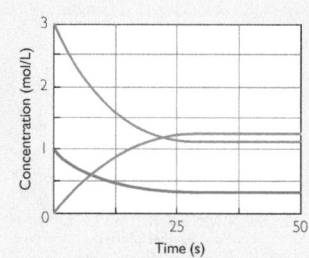

12. Given that the initial concentration of A_2 is 1.0 M and the initial concentration of B_2 is 3.0 M, what are the equilibrium concentrations of the starting substances and the products?

(A) $[A_2] = 0.37$ M $[B_2] = 1.1$ M $[AB_3] = 1.2$ M
(B) $[A_2] = 0.40$ M $[B_2] = 1.5$ M $[AB_3] = 0.50$ M
(C) $[A_2] = 1.0$ M $[B_2] = 1.5$ M $[AB_3] = 0.50$ M
(D) $[A_2] = 1.2$ M $[B_2] = 1.1$ M $[AB_3] = 0.37$ M

13. Which correctly describes the reaction process at equilibrium?

(A) The rate of the forward reaction is equal to the rate of the reverse reaction.
(B) The concentration of the starting substances equals the concentration of the products.
(C) The rate of the forward reaction is faster than the rate of the reverse reaction.
(D) The concentrations of the starting substances are greater than the concentration of the products.

14. How long did it take for this reaction to reach equilibrium?

(A) 6 s
(B) 12 s
(C) 25 s
(D) 50 s

Use this table to answer Exercises 15–18.

Solution	Formula	Solution molarity	[HA]	[H⁺]	[A⁻]	K
nitric acid	HNO_3	1.0 M	~0 M	1.0 M	1.0 M	40
nitrous acid	HNO_2	1.0 M	0.97 M	0.023 M	0.023 M	0.00056
benzoic acid	C_6H_5COOH	1.0 M	0.99 M	0.0079 M	0.0079 M	0.000063

15. Which is the correct chemical equation for the dissociation of nitrous acid, HNO_2?

(A) $HNO_2(aq) \rightleftharpoons H^+(aq) + NO_2^-(aq)$
(B) $H^+(aq) + NO_2^-(aq) \rightleftharpoons HNO_2(aq)$
(C) $HNO_2(aq) \longrightarrow H^+(aq) + NO_2^-(aq)$
(D) $H^+(aq) + NO_2^-(aq) \longrightarrow HNO_2(aq)$

16. Which correctly describes the dissociation of benzoic acid at equilibrium?

(A) There are equal concentrations of $C_6H_5COO^-$ and H^+ ions.
(B) There are more C_6H_5COOH molecules than $C_6H_5COO^-$ and H^+ ions.
(C) The pH of the solution will be between 2 and 3.
(D) All of the above are true statements.

17. Which correctly describes the three reactions based on the value of their equilibrium constants?

(A) Both nitrous acid and benzoic acid dissolve and partially dissociate in water.

(B) Both nitrous acid and benzoic acid dissolve and completely dissociate in water.

(C) Both nitrous acid and nitric acid dissolve and partially dissociate in water

(D) Both nitrous acid and nitric acid dissolve and completely dissociate in water.

18. Which of the following is the correct equilibrium-constant expression for the dissociation of benzoic acid?

(A) $K = \dfrac{[C_6H_5COO^-]}{[C_6H_5COOH]}$

(B) $K = \dfrac{[C_6H_5COOH][H^+]}{[C_6H_5COO^-]}$

(C) $K = \dfrac{[C_6H_5COO^-][H^+]}{[C_6H_5COOH]}$

(D) $K = \dfrac{[C_6H_5COO^-][H^+]^2}{[C_6H_5COOH]}$

19. The dissociation of alizarin yellow indicator, a weak acid, is given below.

$$HIn(aq) \rightleftharpoons H^+(aq) + In^-(aq) \quad K = 3.4 \times 10^{-11}$$

What is the pH value of a 1.0 M solution of alizarin yellow, HIn?

(A) pH = 3.3

(B) pH = 5.2

(C) pH = 10.5

(D) pH = 10.7

20. Which is the correct equilibrium-constant expression for the following reaction?

$$(NH_4)_3PO_4(s) \rightleftharpoons 3NH_4^+(aq) + PO_4^{3-}(aq)$$

(A) $K = [(NH_4)_3PO_4]$

(B) $K = [NH_4^+]^3\,[PO_4^{3-}]$

(C) $K = \dfrac{[NH_4^+]^3[PO_4^{3-}]}{[(NH_4)_3PO_4]}$

(D) $K = \dfrac{[(NH_4)_3PO_4]}{[NH_4^+]^3[PO_4^{3-}]}$

Appendix A: Math Spotlights

Quantity	Unit (abbreviation)
length	meter (m)
mass	kilogram (kg)
time	second (s)
temperature	Kelvin (K)
amount of substance	mole (mol)

SI Units of Measure

Scientists rely on repeatable measurements as they study the physical world. It is important that they use consistent units of measure worldwide. In 1960 an international council standardized the metric system, creating the *Système International d'Unités* (International System of Units), abbreviated as SI.

In the table at left, there are the basic SI units used in chemistry. Other units such as density and volume are combinations of these.

SI units are based on powers of 10. Larger and smaller units get their names by combining standard prefixes with these basic units. For example, the word *centimeter* is a combination of *centi-* and *meter*. A centimeter is one one-hundredth of a meter. These are the prefixes used in the SI, along with their abbreviations and their meanings. (A kilogram is the only basic SI unit that has a prefix as part of its name.)

Prefix	Multiple	Scientific Notation	Prefix	Multiple	Scientific Notation
tera- (T-)	1,000,000,000,000	10^{12}	pico- (p-)	0.000 000 000 001	10^{-12}
giga- (G-)	1,000,000,000	10^{9}	nano- (n-)	0.000 000 001	10^{-9}
mega- (M-)	1,000,000	10^{6}	micro- (μ-)	0.000 001	10^{-6}
kilo- (k-)	1,000	10^{3}	milli- (m-)	0.001	10^{-3}
hecto- (h-)	100	10^{2}	centi- (c-)	0.01	10^{-2}
deka- (da-)	10	10^{1}	deci- (d-)	0.1	10^{-1}

Example 1

Length Conversions

How many meters does each of these lengths represent?

a. 562 centimeters   **b.** 2.5 kilometers

Solution

a. $562 \text{ cm} \cdot \dfrac{1 \text{ m}}{100 \text{ cm}} = 5.62 \text{ m}$ **b.** $2.5 \text{ km} \cdot \dfrac{1000 \text{ m}}{1 \text{ km}} = 2500 \text{ m}$

Example 2

The Mass of One Liter

One milliliter, or cubic centimeter, of water at 4 °C weighs 1 g. How much does 1 L of water at 4 °C weigh?

(continued)

Solution

One milliliter is one one-thousandth of a liter; multiply the mass of 1 mL by 1000 to get the mass of 1 L.

$$1000 \cdot 1 \text{ g} = 1000 \text{ g} = 1 \text{ kg}$$

Practice Exercises

Convert these measurements to the indicated units.

1. 7 m = _____ cm   **2.** 3200 mL = _____ L

3. 20,012 cm = _____ km   **4.** 0.003 kg = _____ g

5. 16 m² = _____ cm²   **6.** 2 m³ = _____ dm³

Answers

1. 700 cm **2.** 3.2 L **3.** 0.20012 km **4.** 3 g **5.** 160,000 cm² **6.** 8000 dm³

Accuracy, Precision, and Significant Digits

There are two kinds of numbers in the world—exact and inexact. For instance, counting is exact because you can safely say that there are exactly 12 eggs in a dozen. However, no measurement with a ruler, a balance, or a graduated cylinder is ever exact. So, when a measurement or the average of several measurements comes out extremely close to the actual true value, we say that the measurement or average is *accurate.*

If you measure something several times and get very similar answers each time, your measurements are *precise.*

The ability to make precise measurements depends partly on the equipment used. For example, a graduated cylinder is more precise than a beaker. Precision also depends on how carefully a measurement was made. For example, a measurement of 23.76 mL is more precise than a measurement of 24 mL.

To understand the difference between precision and accuracy, imagine a lab experiment to measure the boiling point of water at sea level. Several readings are taken: 97.2 °C, 97.0 °C, and 97.1 °C. The measurements are close to each other; repeating the experiment would likely give similar results, so they are precise. However, they are not accurate; the boiling point of water at sea level is known to be 100 °C. Perhaps the thermometer was faulty, or the person taking the measurements consistently read the thermometer incorrectly.

Using significant digits in a measurement allows you to indicate the degree of certainty in the measurements. In general, the last digit of any measurement is uncertain. For example, suppose you use a meterstick, marked in millimeters, to measure the length of an object. If you record a measurement of 24.33 cm, the last digit is an estimate based on the closest millimeter markings and is not certain. Another person might measure the length as 24.34 cm or 24.32 cm. However, you will both agree on the 24.3 because you can read the meterstick accurately to the millimeter, or 0.1 cm. The measurement 24.33 has four significant digits.

The rules for determining the number of significant digits are complicated. Nonzero digits always count. Zeros sometimes count depending on where they are. Zeros might be leading, trapped, or trailing.

- *Leading zeros,* such as those in 0.004728, never count as significant digits.
- *Trapped zeros,* as in 1.08, always count as significant digits.
- *Trailing zeros,* or those at the end of a number, count only when there is a decimal point. In the numbers 20.0, 300.00, and even 50., the zeros count as significant digits. If there is no decimal point, as in 500, then it isn't possible to tell whether the zeros are significant. Sometimes you can deduce that zeros are significant based on the instrumentation. For example, a thermometer generally measures to the nearest degree. You can use scientific notation to avoid ambiguity. The measurement 5.00×10^2 has three significant digits, whereas the measurement 5×10^2 has only one significant digit.

As you do calculations involving numbers with different numbers of significant digits, follow these two rules.

- Adding or subtracting: Your final answer will have only as many *decimal places* as the measurement with the fewest decimal places.
- Multiplying or dividing: The result can have only as many significant digits as the number with the fewest significant digits.

As you add, multiply, or combine measurements in other ways, you will often have to round the result of your calculation to obtain the correct number of significant digits.

Example 1

Rounding the Sum or Difference

Calculate each sum or difference and round the answer to the appropriate number of significant digits.

    a. $2.24 + 3.4 + 5.231$ **b.** $10.5 \text{ cm} - 3.36 \text{ cm}$

Solution

    a. First add the numbers to get the sum 10.871. Then consider the decimal places to arrive at the final answer.

2 decimal places	2.24
1 decimal place	3.4
3 decimal places	+ 5.231
	10.871

Because 3.4 has only one decimal place, the sum should be rounded to 10.9. Always complete the calculation before rounding.

    b. First subtract the numbers.

1 decimal place	10.5 cm
2 decimal places	− 3.36 cm
	7.14 cm

Because 10.5 has only one decimal place, round the final answer to one decimal place: 7.1 cm.

Example 2

Rounding a Product

Calculate each product and round to the appropriate number of significant digits.

    a. $12.34 \cdot 1.6$ **b.** $4.71 \text{ m} \cdot 5.28 \text{ m}$

Solution

    a. First multiply the two decimals.

Four significant digits	Two significant digits

$$12.34 \cdot 1.6 = 19.744$$

Because there are only two significant digits in 1.6, you can have only two significant digits in your answer. You must round the product to 20. You might write this answer as $2.0 \cdot 10$ to make it clear that both digits are significant.

    b. First multiply: $4.71 \text{ m} \cdot 5.28 \text{ m} = 24.8688 \text{ m}^2$.

Both factors have three significant digits, so the answer must be rounded to three significant digits: 24.9 m^2.

Practice Exercises

  1. How many significant digits are in each of these numbers?
    a. 20.1   **b.** 300.0   **c.** 0.0031   **d.** 0.03010

  2. Complete these calculations.
    a. $25.14 + 3.4 + 15.031$ **b.** $100.04 \text{ cm} - 7.362 \text{ cm}$ **c.** $3005 \cdot 45.20$

  3. What is the volume of a box 14.5 cm by 15.9 cm by 21.1 cm?

  4. The density of copper is 8.92 g/cm^3. What is the mass of 24 cm^3 of copper?

Answers

1a. 3 **1b.** 4 **1c.** 2 **1d.** 4 **2a.** 43.5 **2b.** 92.68 **2c.** 135,800
3. 4860 cm^3 **4.** 210 g

Solving Equations

Solving chemistry problems sometimes involves solving a math equation. When the quantity you are looking for is isolated, or alone on one side of the equation, all you need to do is complete the calculations. A simple example is finding the Fahrenheit equivalent of 28 °C using the equation

$$F = \frac{9}{5}C + 32°$$

Sometimes the quantity that answers your question is not alone on one side of the equation. To solve these problems, isolate the variable you are looking for on one side of the equation, so that calculations that will lead to your answer are on the other side of the equation. For example, to find a temperature in degrees Celsius, rearrange the equation in terms of C.

Start with the known relationship	$F = \dfrac{9}{5}C + 32°$
Subtract 32° from both sides	$F - 32° = \dfrac{9}{5}C$
Multiply both sides by $\dfrac{5}{9}$.	$\dfrac{5}{9}(F - 32°) = C$

Now that C is isolated, you can substitute any Fahrenheit temperature and carry out the calculations to find the corresponding Celsius temperature.

Example 1

Solving for x

Solve these equations for x.

a. $0.1x + 12 = 2.2$ **b.** $\dfrac{12 + 3.12x}{3} = -100$

Solution

a. Original equation. $\qquad 0.1x + 12 = 2.2$

Subtract 12 from both sides. $\quad 0.1x + 12 - 12 = 2.2 - 12$

$\qquad\qquad\qquad\qquad\qquad\quad 0.1x = -9.8$

Divide both sides by 0.1. $\qquad\qquad x = -98$

b. Original equation. $\qquad\qquad \dfrac{12 + 3.12x}{3} = -100$

Multiply both sides by 3. $\qquad 12 + 3.12x = -300$

Subtract 12 from both sides. $\quad -12 + 12 + 3.12x = -300 - 12$

$\qquad\qquad\qquad\qquad\qquad\quad 3.12x = -312$

Divide both sides by 3.12. $\qquad\qquad x = -100$

Example 2

Solving for V_2

Solve this equation for V_2.

$$\frac{V_1 P_1}{T_1} = \frac{V_2 P_2}{T_2}, \text{ for } V_2$$

Solution

Original equation.

$$\frac{V_1 P_1}{T_1} = \frac{V_2 P_2}{T_2}$$

Multiply both sides by T_2; remove the factor of 1.

$$\frac{T_2 V_1 P_1}{T_1} = \frac{T_2 V_2 P_2}{T_2}$$

Divide both sides by P_2; remove the factor of 1.

$$\frac{T_2 V_1 P_1}{T_1 P_2} = \frac{V_2 P_2}{P_2}$$

$$V_2 = \frac{T_2 V_1 P_1}{T_1 P_2}$$

The equation is now expressed in terms of V_2. Substituting in values for the other five variables will give a value for V_2.

Practice Exercises

1. Solve these equations. Indicate the action you take at each stage.

    a. $144x + 33 = 45$ **b.** $\frac{1}{6}x + 2 = 8$ **c.** $5(x - 7) = 15 + 5^2$

2. Solve these equations for the variable indicated.

    a. $d = rt$, for t **b.** $P = 2(l + w)$, for w **c.** $A = \frac{1}{2}h(a + b)$, for h

---

Answers

1a. $x = \frac{1}{12}$ (On both sides, subtract 33 and divide by 144.)

1b. $x = 36$ (On both sides, subtract 2 then multiply by 6.)

1c. $x = 15$ (Combine 15 and 25 on the right side, divide both sides by 5, and add 7 to both.)

2a. $t = \dfrac{d}{r}$ **2b.** $l = \dfrac{P}{2} - w$ **2c.** $h = \dfrac{2A}{a} + b$

Order of Operations

Math expressions and equations often involve several operations. For example, to convert from Celsius to Fahrenheit, you first multiply the number of Celsius degrees by the fraction $\frac{9}{5}$ and then add 32 degrees. To convert from Fahrenheit to Celsius, you subtract 32 degrees from the number of Fahrenheit degrees and then multiply the result by $\frac{5}{9}$. A rule called the *order of operations* is used to write math expressions clearly so that anyone seeing the formula or equation would know whether multiplication was the first step or the second step.

Order of Operations

1. Evaluate all expressions within parentheses.
2. Evaluate all terms with exponents.
3. Multiply and divide from left to right.
4. Add and subtract from left to right.

Example 1

Temperature Conversions

Convert these temperatures.

   a. 37 °C to degrees Fahrenheit   **b.** 48 °F to degrees Celsius

Solution

Substitute the known value into each equation and then solve using the order of operations.

   a. Substitute 37° into the equation. $F = \frac{9}{5} \cdot 37° + 32°$

   Multiply. $F = 66° + 32°$

   Add. $F = 98\ °\text{F}$

(continued)

b. Substitute 48° into the equation. $\quad\quad C = \dfrac{5}{9}(48° - 32°)$

Subtract. $\quad\quad\quad\quad\quad\quad\quad\quad\quad C = \dfrac{5}{9}(16°)$

Multiply. $\quad\quad\quad\quad\quad\quad\quad\quad\quad C = 9\,°C$

Example 2

Parentheses, Exponents, and Fractions

Evaluate these expressions.

a. $\dfrac{5}{9}(96 - 15)$ $\quad$ **b.** $\dfrac{(70 - 64)^2}{2 \cdot 5} + 12$

Solution

Evaluate the expression in the parentheses first.

a. Original expression $\quad\quad\quad\quad\quad\quad\quad \dfrac{5}{9}(96 - 15)$

Subtract the numbers within the parentheses. $\quad = \dfrac{5}{9}(81)$

Multiply $\frac{5}{9}$ by 81. $\quad\quad\quad\quad\quad\quad\quad\quad\quad = 45$

b. The fraction line acts like parentheses. In fact, when the expression is entered into a calculator, parentheses are required around the $2 \cdot 5$.

Original expression. $\quad\quad\quad\quad\quad\quad\quad \dfrac{(70 - 64)^2}{2 \cdot 5} + 12$

Evaluate the expressions above and below the fraction line. $\quad = \dfrac{6^2}{10} + 12$

Divide 36 by 10 and then add 12. $\quad\quad = 15.6$

Practice Exercises

1. Evaluate these expressions.
   a. $3 \cdot 24 \div 8$ $\quad\quad\quad$ **b.** $3 + 24 \cdot 8$ $\quad\quad\quad$ **c.** $3 - 24 + 8$
   d. $(3 + 24) \cdot 8$ $\quad\quad$ **e.** $(3 + 21) \div 8$ $\quad\quad$ **f.** $3 - (24 + 8)$

2. Calculate the value of each expression.
   a. $-2 + 5 - (-8)$ $\quad$ **b.** $(-5^2) - (-3)^2$ $\quad$ **c.** $-0.3 \cdot 20 + 15$

3. Insert parentheses as needed to make each equation true.
   a. $15 \div 3 + 7 - 4 = -48$ $\quad$ **b.** $15 \div 3 + 7 - 4 = 8$ $\quad$ **c.** $-4^2 + -3^2 = -7$

Answers

1a. 9 $\quad$ **1b.** 195 $\quad$ **1c.** -13 $\quad$ **1d.** 216 $\quad$ **1e.** 3 $\quad$ **1f.** -29

2a. 11 $\quad$ **2b.** -34 $\quad$ **2c.** 9

3a. $(15 \div 3 + 7)(-4) = -48$ $\quad$ **3b.** $15 \div 3 + 7 - 4 = 8$

3c. $-4^2 + (-3)^2 = -7$

Averages

Scientists are often interested in the typical result of a repeated experiment. The average, also called the *mean,* is one way to determine a typical value. The average is calculated by totaling the data values and dividing by the number of values.

Example 1

The Mean

Find the average of these numbers: 14, 23, 10, 21, 7, 80, 32, 30, 92, 14, 26, 21, 38, 20, 35, 21.

Solution

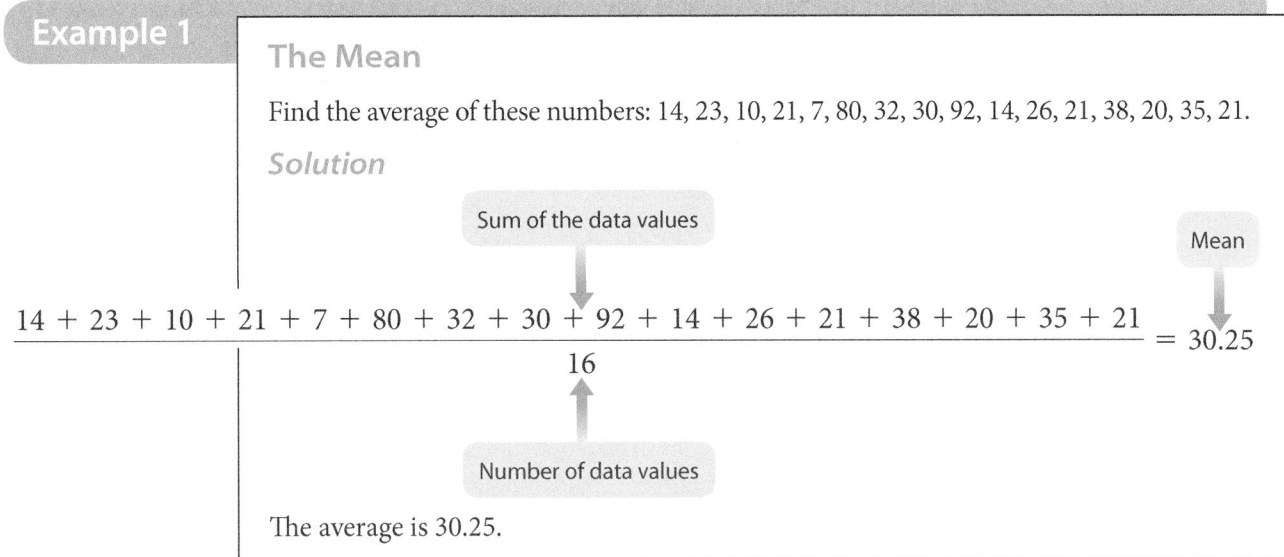

The average is 30.25.

Sometimes the data values are not of equal importance in contributing to the average. You need to weight the values differently.

Example 2

The Weighted Average

On a chemistry quiz, 1 student got 100%, 7 students got 95%, 12 students got 90%, 1 student got 85%, 5 students got 80%, 3 students got 75%, and 1 student got 70%. What was their average (mean) score?

Solution

One solution would be to change the list so that 95% appeared 7 times, 90% appeared 12 times, and so on. However, it is more efficient to use a weighted average as shown here.

$$(1)(100) + (7)(95) + (12)(90) + (1)(85) + (5)(80) + (3)(75) + (1)(70) = 2625$$

Thirty students took the quiz, so divide 2625 by 30.

The average score was 87.5%.

Example 3

Average Atomic Mass

The element silver, Ag, has two naturally occurring isotopes. Approximately 52% of all silver consists of atoms with 60 neutrons, and 48% consists of atoms with 62 neutrons. Calculate the average atomic mass of silver atoms.

(continued)

Solution

Silver atoms have 47 protons. The atoms with 60 neutrons have masses of 47 + 60 = 107 amu. The atoms with 62 neutrons have masses of 47 + 62 = 109 amu. If you have a sample of 100 silver atoms, 52 atoms will have a mass of 109 amu and 48 atoms will have a mass of 109 amu. Use a weighted average. The total mass is 52(107 amu) + 48(109 amu) = 10,796 amu. The average mass is 10,796 amu/100 = 107.96. This is close to the atomic mass of 107.9 amu shown on the periodic table.

Practice Exercises

1. Find the average of these numbers: 52.3, 18.91, 35.66, 4.35.

2. A student had these scores on chapter tests: 87%, 90%, 95%, 92%. He got 92% on one unit test and 86% on the other. His final exam score was 91%. Unit tests count twice as much as chapter tests, and the final exam counts four times as much as a chapter test. What is his average score?

3. About 76% of all chlorine atoms have an atomic mass of 35.00 amu; about 24% have an atomic mass of 37.00 amu. Use this information to calculate the average atomic mass of 100 chlorine atoms.

Answers

1. 27.8   **2.** 90%   **3.** 35.48 amu

Graphing

Many chemical experiments involve changing one variable and seeing how another variable changes as a result. The results of experimental procedures can be listed in a table, but it is often helpful to look on a graph for any trends in the data.

Data consisting of two variables, such as temperature and volume measurements, can be graphed using a coordinate plane. Each point on the coordinate plane can be identified by a pair of numbers (x, y) called *coordinates*. The first number, the x-coordinate, describes how far the point is to the right or left of the origin; the second number, the y-coordinate, describes how far up or down the point is.

In the example below, a coordinate pair represents one (temperature, volume) data point. The quantity that the experimenter is changing, in this case temperature, is called the *independent variable*. It is graphed on the horizontal axis, the x-axis. The resulting measurement, or the *dependent variable*, is graphed on the vertical axis, or y-axis. The dependent variable in this example is the volume. If the experiment had been set up to measure changes in temperature as a result of changes in volume, the independent variable would have been volume, and temperature would have been the dependent variable.

Normally, you graph the dependent variable versus the independent variable. (*Versus* means "compared with" and is abbreviated "vs.")

Example 1

Graphing Coordinate Pairs

Graph these results for the heating of a gas:

Temperature (°C)	Volume (mL)	Temperature (°C)	Volume (mL)
0	465	15	491
5	475	20	498
10	481	25	509

Solution

The temperature is being changed, so temperature is the independent variable and should be graphed on the x-axis. The volume changes as a result, so it is the dependent variable and should be graphed on the y-axis. To graph the data, you need to decide on the scale for each axis. The x-axis can start at 0 °C and go up to 30 °C. To save space, you don't need to show the y-axis starting at 0; it can go from 450 to 520. Label the axes "Temperature (°C)" and "Volume (mL)." Graph the coordinate pairs. Title the graph. In the title, the dependent variable is usually named first.

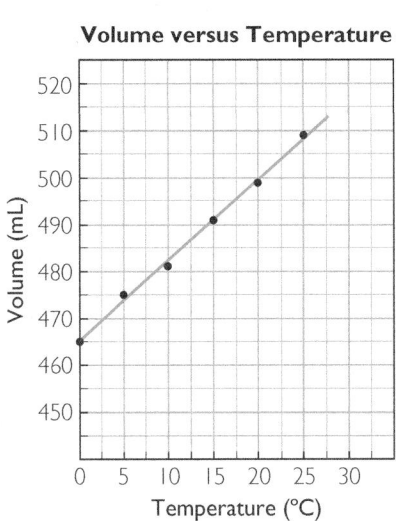

Notice that the points in the graph of volume versus temperature lie nearly in a straight line. Small errors in measurement account for the variation. The points are close enough to a line to indicate that the relationship between temperature and volume is linear. As one quantity increases, the other increases at a constant rate. So the quantities vary directly, or are *directly proportional*.

Example 2

Volume versus Pressure

Graph the results of an experiment in which a gas's volume is measured as pressure is applied. The temperature of the gas is kept from changing. This table shows the data collected.

Pressure (atm)	Volume (L)	Pressure (atm)	Volume (L)
1.0 atm	10.0 L	3.0 atm	3.3 L
1.5 atm	6.7 L	3.5 atm	2.9 L
2.0 atm	5.0 L	4.0 atm	2.5 L
2.5 atm	4.0 L		

(continued)

Solution

An appropriate scale for the independent variable on the *x*-axis is 0 atm to 4.5 atm. The scale for the dependent variable on the *y*-axis can go from 0 L to 12 L. Label the axes and indicate the units: "Pressure (atm)" and "Volume (L)." The coordinate pairs to graph are (1.0, 10), (1.5, 6.7), (2.0, 5.0), and so on. Title the graph *Volume versus Pressure*.

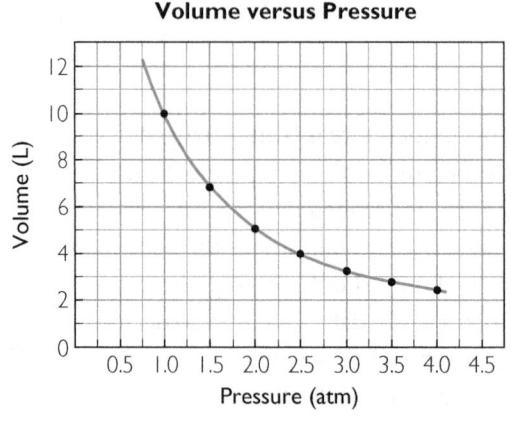

The points in the graph of volume versus pressure do not lie in a line, but you can draw a smooth curve through the points to show the trend. The shape of the graph is typical of relationships that are inversely proportional.

Practice Exercises

1. The examples all used coordinate pairs with positive values. But points can also have negative coordinates. The coordinates of point *A* are (−7, −4). Name the (*x*, *y*) coordinates of each point pictured.

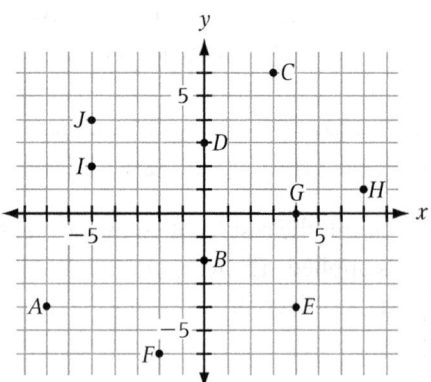

Temperature (°C)	Pressure (atm)
22.0 °C	3.20 atm
35.0 °C	3.34 atm
45.0 °C	3.45 atm
60.0 °C	3.61 atm

2. A sealed rigid container is heated and the pressure in the container is measured. Describe the graph of the data (your scales, axes titles, graph title, plotted points, and shape of the curve through the points).

3. These coordinate pairs represent measurements of volume and mass of copper: (1.1, 9.8), (2.3, 20.5), (2.7, 24.1), (4.5, 40.1). If these are graphed, what will be the shape of the graph? Are mass and volume directly proportional or inversely proportional?

Answers

1. $B\,(0, -2)$, $C\,(3, 6)$, $D\,(0, 3)$, $E\,(4, -4)$, $F\,(-2, -6)$, $G\,(4, 0)$, $H\,(7, 1)$, $I\,(-5, 2)$, $J\,(-5, 4)$
2. Sample answer: The x-axis is "Temperature (°C)" with a scale from 20 to 65. The y-axis is "Pressure (atm)" with a scale of 3.10 to 3.70. Plotted points are (22, 3.20), (35, 3.34), (45, 3.45), (60, 3.61). The graph title might be *Pressure versus Temperature.* The points lie in a straight line; the quantities are directly proportional.
3. A straight line through the origin; directly proportional.

Ratios and Proportions

As chemists study matter, they are often working with ratios. The four statements here all express some sort of ratio.

Density is the ratio of mass to volume: $D = m/V$.

The units of molar mass are g/mol.

The mole ratio of H_2 to O_2 is 2:1.

The reaction gave a 78% yield.

Ratios can be written as fractions. For instance, 78% is $\frac{78}{100}$. The ratio 2:1 can be written $\frac{2}{1}$. Ratios can be reduced like fractions, or they can be added once they have a common denominator. Unit fractions, which equal 1 because the numerator is equivalent to the denominator, can be used to convert between units or to create equivalent fractions.

Whenever you are working with ratios, use the rules for working with fractions.

- *To create equivalent fractions,* multiply a fraction by a unit fraction. In a unit fraction, the numerator is equivalent to the denominator. A unit fraction is equal to 1 and can be used to convert between units or create equivalent fractions.

Unit fraction

$$\frac{1}{2} = \frac{1}{2} \cdot \frac{5}{5} = \frac{5}{10}$$

Unit fraction

$$300\text{ m} = 300\text{ m} \cdot \frac{1\text{ km}}{1000\text{ m}} = 0.3\text{ km}$$

- *To add or subtract fractions,* find equivalent fractions with a common denominator, then add or subtract the numerators and put that number over the common denominator.

Multiply by unit fractions.

Fractions equivalent to $\frac{1}{3}$ and $\frac{2}{5}$

$$\frac{1}{3} + \frac{2}{5} = \left(\frac{1}{3} \cdot \frac{5}{5}\right) + \left(\frac{2}{5} \cdot \frac{3}{3}\right) = \frac{5}{15} + \frac{6}{15} = \frac{11}{15}$$

Sum

- To *multiply fractions*, multiply the two numerators, and multiply the two denominators. You may want to reduce the fraction so that there are no common factors between the numerator and the denominator.

$$\frac{1}{3} \cdot \frac{3}{5} = \frac{1 \cdot 3}{3 \cdot 5} = \frac{3}{15} = \frac{1}{5}$$

- To *divide fractions*, multiply by the reciprocal of the divisor.

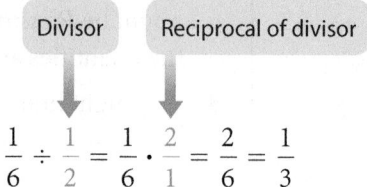

$$\frac{1}{6} \div \frac{1}{2} = \frac{1}{6} \cdot \frac{2}{1} = \frac{2}{6} = \frac{1}{3}$$

- To *reduce a fraction*, divide the numerator and denominator by any factor they have in common; this is often referred to as *canceling*.

$$\frac{6}{15} = \frac{2 \cdot 3}{3 \cdot 5} = \frac{2}{5} \qquad \frac{300 \, \cancel{m} \cdot 1 \, km}{1000 \, \cancel{m}} = 0.3 \, km$$

Proportions are equations that state two ratios are equal. Each of the equations shown here is a proportion.

$$\frac{P_1 V_1}{T_1} = \frac{P_2 V_2}{T_2} \qquad \frac{3.20 \, atm}{22 \, °C} = \frac{P_2}{50 \, °C}$$

$$\frac{1 \, mol \, H_2O}{18.0 \, g \, H_2O} = \frac{x \, mol \, H_2O}{51.2 \, g \, H_2O}$$

Proportions can be solved by multiplying (or dividing) both sides of the equation by the same value. As you solve chemistry problems, you will often use ratios and proportions.

Example 1

Limiting Reactant

Imagine you start with 10.0 g of $Mg(OH)_2$ and 100.0 mL of 4.0 M HCl. How much $MgCl_2$ will be produced?

Solution

Write the balanced chemical equation.

$$Mg(OH)_2(s) + 2HCl(aq) \longrightarrow MgCl_2(aq) + 2H_2O(l)$$

Determine the molar masses of each compound.

Reactants: $Mg(OH)_2$ = 58.3 g/mol   HCl = 36.5 g/mol

Products: $MgCl_2$ = 95.2 g/mol H_2O = 18.0 g/mol

Determine the number of moles of each reactant that you have.

$$\text{Moles of } Mg(OH)_2 = \frac{10.0 \, g}{58.3 \, g/mol} = 0.17 \, mol$$

Moles of HCl = (0.100 L)(4.0 mol/L) = 0.40 mol

Use the mole ratio to identify the limiting reactant.

Example

(continued)

The reactants combine in a 1:2 ratio. So you need $0.17 \text{ mol} \cdot 2 = 0.34 \text{ mol}$ of hydrochloric acid, HCl, to react with the 0.17 mol of magnesium hydroxide, $Mg(OH)_2$.

You have 0.40 mol of HCl, which is plenty, so the limiting reactant is $Mg(OH)_2$. When the reaction is complete, there will be 0.06 mol of HCl left over.

Use the limiting reactant to determine the maximum amount of product.

For every 1 mol of $Mg(OH)_2$, 1 mol of $MgCl_2$ is produced.

So 0.17 mol of $MgCl_2$ is produced. Use the molar mass of $MgCl_2$ to determine the mass of $MgCl_2$.

$$\frac{95.2 \text{ g } MgCl_2}{1 \text{ mol}} = \frac{x \text{ g } MgCl_2}{0.17 \text{ mol}}$$

$$0.17 \text{ mol} \cdot \frac{95.2 \text{ g}}{1 \text{ mol}} = x$$

$$x = 16 \text{ g } MgCl_2$$

Example 2

Percent Yield

Suppose you ran the reaction from Example 1 again, this time starting with 35.0 g $Mg(OH)_2$ and the same amount of HCl as before. In the laboratory, 15.4 g of $MgCl_2$ are produced. What is the percent yield of your reaction?

Solution

First, determine the number of moles of each reactant you have. This will make it possible to identify the limiting reactant.

$$\frac{58.3 \text{ g}}{1 \text{ mol } Mg(OH)_2} = \frac{35.0 \text{ g}}{x \text{ mol } Mg(OH)_2}$$

$$x = 0.60 \text{ mol } Mg(OH)_2$$

$$\frac{y \text{ mol HCl}}{0.100 \text{ L}} = \frac{4.0 \text{ mol}}{1 \text{ L}}$$

$$y = 0.40 \text{ mol HCl}$$

The mole ratio of reactants is 1:2, so you need 1.2 mol of HCl to react with 0.60 mol of $Mg(OH)_2$. You have only 0.40 mol of HCl available, so this time HCl is the limiting reagent.

It takes 2 mol of HCl to produce each mole of $MgCl_2$, so you can expect to produce 0.20 mol of $MgCl_2$: $\left(\frac{2}{1} = \frac{0.40}{0.20} \right)$.

Next, use the molar mass of $MgCl_2$ to calculate the theoretical yield.

$$\frac{x \text{ g } MgCl_2}{0.20 \text{ mol}} = \frac{95.2 \text{ g}}{1 \text{ mol}}$$

$$x = \frac{95.2 \text{ g}}{1 \text{ mol}} \cdot 0.20 \text{ mol}$$

$$x = 19 \text{ g}$$

(continued)

Now you can calculate the percent yield for the reaction you ran. Percent yield is yield per 100, or yield/100.

$$\frac{\text{yield}}{100} = \frac{\text{actual yield}}{\text{theoretical yield}}$$

$$\text{yield} = \frac{15.4 \text{ g}}{19 \text{ g}} \cdot 100$$

Percent yield = 81%

Your percent yield was significantly lower than 100%.

Practice Exercises

1. What ratios and proportions appear in Examples 1 and 2?

2. Write each expression as a fraction.
   a. 68%
   b. 3:4
   c. the ratio of volume to temperature
   d. grams per cubic centimeter

3. Complete these calculations.
   a. $\dfrac{2}{3} + \dfrac{1}{5}$ **b.** $\dfrac{4}{5} - \dfrac{1}{6}$ **c.** $\dfrac{1}{3} \cdot \dfrac{9}{11}$ **d.** $\dfrac{3}{8} \div \dfrac{1}{4}$

Answers

1. $\dfrac{10 \text{ g}}{58.3 \text{ g/mol}}$, 1:2, $\dfrac{95.3 \text{ g MgCl}_2}{1 \text{ mol}} = \dfrac{x \text{ g MgCl}_2}{0.17 \text{ mol}}$,

$\dfrac{58.3 \text{ g}}{1 \text{ mol Mg(OH)}_2} = \dfrac{35.0 \text{ g}}{x \text{ mol Mg(OH)}_2}$ $x = 0.60 \text{ mol Mg(OH)}_2$, $\dfrac{15.4 \text{ g}}{19.4 \text{ g}}$

2a. $\dfrac{68}{100}$ **2b.** $\dfrac{3}{4}$ **2c.** volume/temperature, or $\dfrac{V}{T}$ **2d.** $\dfrac{\text{g}}{\text{cm}^3}$

3a. $\dfrac{13}{15}$ **3b.** $\dfrac{19}{30}$ **3c.** $\dfrac{3}{11}$ **3d.** $\dfrac{3}{2}$, or $1\dfrac{1}{2}$

Scientific Notation

The mass of a hydrogen atom is 0.0000000000000000000000014 g. The number of atoms in 1 mol of carbon is 602,000,000,000,000,000,000,000. To make these numbers easier to read, compare, and use in calculations, scientists use scientific notation. The long numbers above can be written as 1.4×10^{-24} and 6.02×10^{23}.

Each number is written as a decimal with one digit to the left of the decimal point times a power of 10. A number written in scientific notation has the form $a \times 10^n$, where $1 \leq a < 10$ or $-10 < a \leq -1$, and n is an integer. In other words, if a is a positive number, it is greater than or equal to 1 and less than 10. If a is a negative number, it is less than or equal to -1 and greater than -10.

Use the properties of exponents as you combine numbers written in scientific notation.

- For addition and subtraction, convert all the numbers so that the powers of 10 you are combining are the same. (The converted numbers might not be in scientific notation.)

- For multiplication, add the exponents. For any values of b, m, and n, $b^m \cdot b^n = b^{m+n}$.

- For division, subtract the exponents. For any nonzero value of b, and any values of m and n, $\frac{b^m}{b^n} = b^{m-n}$.

Example 1

Billions

Write these numbers in scientific notation.

a. 2,110,000,000   **b.** 0.0074

Solution

a. Move the decimal point so that there is only one digit to the left of it, in this case 2. The decimal point moved nine places. The number in scientific notation is 2.11×10^9. This means that to get the number in standard notation you start with 2.11 and move the decimal nine places to the right.

b. Move the decimal after the 7 to get 7.4. From here, the decimal will need to move four places to the left, so the number in scientific notation is 7.4×10^{-4}.

Example 2

Operations with Scientific Notation

Calculate these values.

a. $(3.2 \times 10^5)(4.0 \times 10^{-8})$ **b.** $\dfrac{1.4 \times 10^3}{2.8 \times 10^5}$ **c.** $(3.2 \times 10^5) + (4.0 \times 10^4)$

Solutions

a. 1.28×10^{-2}. Multiplying the decimals gives 12.8. The product of the powers of 10 is 10^{-3}. To put this into scientific notation, divide 12.8 by 10; to keep things balanced, increase the power of 10 by a factor of 10 to -2.

b. 5.0×10^7. The division of the numbers gives 0.50. To divide the powers of 10, subtract the exponents: $3 - (-5)$, or 8. To rewrite in scientific notation, multiply 0.50 by 10, so subtract 1 from the power of 10.

c. 3.6×10^5. Change the second number to 0.4×10^5 before you add.

Practice Exercises

1. Use scientific notation to write the speed of light, 299,792,458 meters per second, accurate to five digits.

2. How many zeros would follow the final 2 if 6.022×10^{23} were written without scientific notation?

3. Calculate these values.

    a. $(3.0 \times 10^{14})(4.0 \times 10^{-4})$ **b.** $\dfrac{2.8 \times 10^{-31}}{7 \times 10^{-28}}$

    c. $(1.21 \times 10^{-4})(4.18 \times 10^4)$ **d.** $(3.61 \times 10^7) - (2.5 \times 10^6)$

Answers

1. 2.9979×10^8 **2.** 20

3a. 1.2×10^{11} **3b.** (4×10^{-4}) **3c.** 5.06×10^0, or 5.06 **3d.** (3.36×10^7)

Dimensional Analysis

As you work on problems that involve numbers with units of measurement, it is convenient to consider the units (or dimensions) as factors. For example, you might want to change 0.002 kilometer to millimeters. You can always multiply by 1 without changing a value, so to convert between units you can multiply by a unit fraction. A unit fraction's value is 1 because its numerator and denominator are equivalent. Some unit fractions are shown here.

$$\frac{1000\text{ m}}{1\text{ km}} = 1 \qquad \frac{1000\text{ mm}}{1\text{ m}} = 1 \qquad \frac{1\text{ m}}{1000\text{ mm}} = 1$$

To convert from kilometers to millimeters, use the unit fraction with meters in the numerator and the equivalent kilometers in the denominator and a unit fraction with millimeters in the numerator and equivalent meters in the denominator.

$$0.002\text{ km} \cdot \frac{1000\text{ m}}{1\text{ km}} \cdot \frac{1000\text{ mm}}{1\text{ m}} = 2000\text{ mm}$$

You can see that the unit fractions have been chosen so that most of the units cancel out. The answer is then in millimeters as they are the only remaining units.

Dimensional analysis is also helpful when you need to convert between two different systems of measurement. For example, 1 inch is equivalent to 2.54 centimeters. To convert from centimeters to inches, use the unit fraction $\frac{1\text{ in}}{2.54\text{ cm}}$. To convert from inches to centimeters, use the unit fraction $\frac{2.54\text{ cm}}{1\text{ in}}$.

Example 1

Calculating Volume

What is the volume of a box 15 cm by 10 cm by 5.12 in.?

Solution

The volume is the product of the three dimensions, but you need to include a factor to convert the inches to centimeters.

$$15\text{ cm} \cdot 10\text{ cm} \cdot 5.12\text{ }\cancel{\text{in.}} \cdot \frac{2.54\text{ cm}}{1\text{ }\cancel{\text{in.}}} \approx 1950\text{ cm}^3$$

If the problem specified the answer in cubic inches, you would have used this product.

$$15\text{ }\cancel{\text{cm}}\frac{1\text{ in.}}{2.54\text{ }\cancel{\text{cm}}} \cdot 10\text{ }\cancel{\text{cm}} \cdot \frac{1\text{ in.}}{2.54\text{ }\cancel{\text{cm}}} \cdot 5.12\text{ in} \approx 119\text{ in}^3$$

Example 2

Speed

A radio-controlled car travels 30 feet across the room in 1.6 seconds. How fast is it traveling in miles per hour?

Solution

From the information, the rate of the car is $\frac{30\ ft}{1.6\ s}$. Multiply by unit fractions such as $\frac{60\ s}{1\ min}$ to go from seconds to minutes to hours and from feet to miles. These fractions are chosen so that the final units are miles per hour.

$$\frac{30\ ft}{1.6\ s} \cdot \frac{60\ s}{1\ min} \cdot \frac{60\ min}{1\ h} \cdot \frac{1\ mi}{5{,}280\ ft} = \frac{108{,}000\ mi}{8{,}448\ mi} \approx \frac{12.8\ mi}{1\ h}$$

or 12.8 miles per hour

Practice Exercises

1. Show how you would use dimensional analysis to convert 6 cm/s to km/h.

2. Use dimensional analysis to change
   a. 50 meters per second to kilometers per hour
   b. 0.025 day to seconds
   c. the speed 60 miles per hour to kilometers per hour. (1609 meters = 1 mile)

3. The equation for the universal gas law is $PV = nRT$. Suppose the pressure P is in pascals (Pa), the volume V is in cubic meters (m^3), the amount of substance n is in moles (mol), and the temperature T is in kelvins (K). What must be the units of the universal gas law constant R?

Answers

1. $\frac{6\ cm}{s} \cdot \frac{1\ m}{100\ cm} \cdot \frac{1\ km}{1000\ m} \cdot \frac{60\ s}{1\ min} \cdot \frac{60\ min}{1\ hr} = 0.216$ km/hr

2a. 180 km/h   2b. 2160 s   2c. 97 km/h   3. $\frac{m^3 \cdot Pa}{mol \cdot K}$

Logarithms

Logarithms can be used to solve problems involving exponential functions, such as the half-life of a radioactive substance or the pH of a solution.

A *logarithm* is an exponent and is abbreviated *log*. For example, $\log_{10} x$ is the exponent you put on 10 to get x. So, if $x = 100$, then $\log_{10} x = 2$ because $10^2 = 100$. In the same way, $\log_{10} 1000 = 3$, $\log_{10} 10 = 1$, $\log_{10} 1 = 0$, $\log_{10} \frac{1}{10} = -1$. Ten is a common base for logarithms, so $\log x$ is called a *common logarithm* and is shorthand for $\log_{10}x$. Check out these log equivalents on your calculator, and use your calculator to determine the logs of numbers that are not multiples of 10.

Because logarithms are exponents, they follow the properties of exponents.

Product property $\quad a^m \cdot a^n = a^{m+n}$ $\quad \log xy = \log x + \log y$
Quotient property $\quad a^m / a^n = a^{m-n}$ $\quad \log x/y = \log x - \log y$
Power property $\quad\quad\quad\quad\quad\quad\quad\quad\quad \log x^n = n \log x$

In some problems you know what the logarithm is, but you want to find the number that has that logarithm. What you are looking for is called the *antilogarithm,* abbreviated antilog. Use the $10x$ key on your calculator to find an antilog.

Example 1

Comparing Logarithms

Find the values *a–f.* Which values are equal?

$a = \log 18 \quad b = \log 71 \quad c = a + b \quad d = \text{antilog } c \quad e = 18 \cdot 71 \quad f = \log(8 \cdot 71)$

Solution

$a \approx 1.255; \quad b \approx 1.851; \quad c \approx 3.106; \quad d = 1278; \quad e = 1278; \quad f \approx 3.106$
$d = e$, or antilog $(\log 18 + \log 71) = 18 \cdot 71$
$c = f$, or $\log 18 + \log 71 = \log(8 \cdot 71)$

Example 2

pH

What is the pH of a solution in which $[\text{H}^+]$ is 5.3×10^{-4}?

Solution

Start with the definition of pH. $\quad\quad\quad\quad \text{pH} = -\log[\text{H}^+]$
Substitute the value of $[\text{H}^+]$. $\quad\quad\quad\quad\quad\quad = -\log[5.3 \times 10^{-4}]$
Find the log, then change the sign. $\quad\quad\quad\quad = 3.35$

Practice Exercises

1. Find the log.

   a. 456 $\quad$ **b.** $\dfrac{x}{15}$ $\quad$ **c.** 3.4×10^{-6}

2. Find the antilog
   a. 2 $\quad$ **b.** -3 $\quad$ **c.** 3.4

3. The pH of a solution is 8.25. What is the concentration of hydrogen ions in the solution?

Answers

1a. 2.659 $\quad$ **1b.** $\log x - 1.176$ $\quad$ **1c.** 5.469

2a. 100 $\quad$ **2b.** $\dfrac{1}{1000}$ $\quad$ **2c.** 2512 $\quad$ **3.** 5.62×10^{-9}

Appendix B: Connecting Chemical Concepts

Chemical Names and Formulas

Recall from Unit 1: Alchemy that all matter is made up of elements that can combine with one another to form compounds. Chemists have created an organized system of chemical names and formulas that specify what elements are present in a given compound. In Unit 2: Smells, chemical names and formulas were used to predict the smell of different substances. So knowing the chemical name and formula of a compound can often be useful in predicting properties of substances. Understanding some basic patterns in naming and writing formulas is also useful because it allows chemists from different parts of the world to communicate chemical information easily. Examine the table that contains binary ionic compounds and binary covalent compounds.

Binary Compounds

Chemical name	Chemical formula	Type of compound
carbon dioxide	CO_2	molecular covalent
calcium chloride	$CaCl_2$	ionic
carbon monoxide	CO	molecular covalent
dinitrogen tetroxide	N_2O_4	molecular covalent
sodium iodide	NaI	ionic

Here are some patterns you might notice.

- Each compound is made up of two different elements.
- The name of each compound ends in the suffix *-ide*.
- The names of the molecular covalent compounds include prefixes: *di-*, *mono-*, *tri-*, and *tetr-* correspond to the number of atoms of each element in the formula.
- The names of the ionic compounds do not include prefixes.

The compounds in the table are called *binary compounds* because they are made up of only two different elements. Binary ionic compounds are made up of metal and nonmetal atoms and binary covalent compounds are made up of nonmetals only. Notice that with binary ionic compounds, the name alone does not tell you how many atoms of each element are in the compound. However, with binary covalent compounds, the prefixes in the name indicate how many of each atom are present in the formula.

Be careful not to confuse binary compounds with elements that exist as diatomic molecules such as hydrogen, H_2, nitrogen, N_2, and oxygen, O_2. These diatomic molecules are made up of only one type of element. The elements that exist in nature as diatomic molecules include hydrogen, nitrogen, oxygen, and the halogens.

NAMES OF BINARY COVALENT COMPOUNDS

If you know the formula for a binary covalent compound, some simple rules can help you to name it correctly.

1. Write the name of the first element in the formula. Use a prefix in front of the element name to indicate if there are two or more atoms of that element in the formula.

2. Write the name of the second nonmetal in the formula and change the ending of the name to the suffix *-ide*. Use a prefix to indicate how many atoms of that element are in the formula, even if there is only one.

The table lists common prefixes and the number of atoms the prefix represents in the formula for a binary covalent compound.

Prefixes for Binary Covalent Compounds

Prefix	Number
mono-	one
di-	two
tri-	three
tetra-	four
penta-	five
hexa-	six
hepta-	seven
octa-	eight
nona-	nine
deca-	ten

The first molecular covalent compound in the binary compounds table on page B-1 is CO_2. The compound is named *carbon dioxide* following the rules for naming binary covalent compounds. Notice that this name helps to distinguish CO_2 from carbon monoxide, CO, which has one oxygen atom in the formula. Notice also that the prefix *mono-* is used in carbon monoxide to indicate that there is only one oxygen atom in that compound but that the prefix is not used in front of carbon in the name of either compound.

FORMULAS OF BINARY COVALENT COMPOUNDS

If you know the name of a binary molecular compound, you can determine the correct chemical formula.

1. Write the symbol for the first element in the name. Use a subscript to indicate the number of atoms of that element in the formula, based on the prefix in the name.

2. Write the symbol for the second element. Use a subscript to indicate the number of atoms of the second element, based on the prefix in the name.

3. Remember that, in formulas, the numeral 1 is not written as a subscript.

For example, the compound dinitrogen tetroxide contains nitrogen, N, and oxygen, O. The *di-* prefix indicates that there are two nitrogen atoms in the formula and the *tetra-* prefix indicates that there are four oxygen atoms in the formula. (Note that in this case, the "a" is dropped.) So the correct formula is N_2O_4.

Practice Exercises

1. Write the name for these binary covalent compounds.
   a. CO_2 **b.** S_2O **c.** PCl_3 **d.** N_2O **e.** SO_2 **f.** NF_3
2. Write the formula for these binary covalent compounds.
   a. dinitrogen tetrafluoride **b.** dinitrogen pentoxide
   c. carbon monoxide **d.** diphosphorus trisulfide
   e. carbon tetrabromide **f.** sulfur trioxide

Answers

1a. carbon dioxide **1b.** disulfur monoxide **1c.** phosphorus trichloride
1d. dinitrogen monoxide **1e.** sulfur dioxide **1f.** nitrogen trifluoride
2a. N_2F_4 **2b.** N_2O_5 **2c.** CO **2d.** P_2S_3 **2e.** CBr_4 **2f.** SO_3

NAMES OF BINARY IONIC COMPOUNDS

If you know the formula for a binary ionic compound, some simple rules can help you to name it correctly. Recall that a binary ionic compound is a combination of a positively charged metal cation and a negatively charged nonmetal anion.

1. Write the element name of the cation followed by the element name of the anion.

2. Change the ending of the element name of the anion to the suffix *-ide*.

For example, NaCl is the formula for sodium chloride, and $CaCl_2$ is the formula for calcium chloride. Notice that no prefixes are used in naming binary ionic compounds.

FORMULAS OF BINARY IONIC COMPOUNDS

If you know the name of a binary ionic compound, you can use it to determine the correct chemical formula.

1. Write the symbol for the cation, including the charge.

2. Write the symbol for the anion, including the charge.

3. Use subscripts in the formula so the net charge of the compound is zero.

4. Remember not to use a subscript to indicate the numeral 1, and do not include charges in the formula.

For example, the compound strontium chloride is made up of a combination of the cation strontium, Sr^{2+}, and the anion chloride, Cl^-. For the net charge to be zero, the formula for strontium chloride requires one strontium cation and two chloride anions:

$$1(+2) + 2(-1) = (+2) + (-2) = 0$$

So the formula for strontium chloride is $SrCl_2$.

Practice Exercises

1. Write the name for the following binary ionic compounds.
   a. KBr **b.** MgO **c.** Na_3P **d.** NaCl **e.** Ca_3N_2 **f.** LiS_2

2. Write the formula for the following binary ionic compounds.
   a. calcium sulfide **b.** beryllium nitride **c.** aluminum oxide
   d. potassium oxide **e.** strontium sulfide **f.** barium bromide

---

Answers

1a. potassium bromide **1b.** magnesium oxide **1c.** sodium phosphide
1d. sodium chloride **1e.** calcium nitride **1f.** lithium sulfide

2a. CaS **2b.** Be_3N_2 **2c.** Al_2O_3 **2d.** K_2O **2e.** SrS **2f.** $BaBr_2$

NAMES OF IONIC COMPOUNDS WITH TRANSITION METALS

Unlike main group atoms, most transition metals can have more than one ion charge. When naming ionic compounds that contain transition metals, Roman numerals are used to indicate the charge on the cation. If you know the formula for an ionic compound that contains a transition metal, you can determine its name.

1. Follow the same basic rules for naming binary ionic compounds. Name the cation first, followed by the anion, changing the ending of the element name of the cation to the suffix *-ide*.

2. To indicate the charge on the cation, use a Roman numeral in parentheses. Use the rule of zero charge and the charge on the anion to determine the charge on the cation.

3. Note that if the compound contains silver, Ag, or zinc, Zn, there is no need to use a Roman numeral in the name. Silver typically has a charge of $+1$ and zinc has only one cation with a charge of $+2$.

For example, the formula for Fe_2O_3 contains three oxide anions, each with a charge of -2. The formula shows that there are two Fe ions. For the net charge on the compound to be zero, each iron cation must have a charge of $+3$:

$$2(+3) + 3(-2) = (+6) + (-6) = 0$$

So, the name of this compound is *iron (III) oxide*.

FORMULAS OF IONIC COMPOUNDS WITH TRANSITION METALS

If you know the name of an ionic compound that contains a transition metal, you can use the name to determine the correct formula.

1. Write the symbol for the metal cation. The Roman numeral in parentheses indicates the charge on the cation. Remember that silver has a charge of $+1$ and zinc has a charge of $+2$.

2. Write the symbol for the anion, including the charge.

3. Use subscripts in the formula to indicate that the net charge of the compound is zero.

4. Remember not to use a subscript to indicate the numeral 1, and do not include charges in the formula.

For example, the compound chromium (III) oxide contains a chromium cation, Cr^{3+} as indicated by the Roman numeral III in parentheses. The oxide anion has a charge of (−2). Using the rule of zero charge, you can determine how many of each ion is in one formula unit of chromium oxide.

$$2(+3) + 3(-2) = (+6) + (-6) = 0$$

So the formula contains two chromium cations and three oxygen anions: Cr_2O_3.

Practice Exercises

1. Write the name for the following binary ionic compounds.
   a. CuF_2 **b.** $AgCl$ **c.** $FeCl_3$ **d.** PbS **e.** SnO_2 **f.** Mn_3P_2

2. Write the formula for the following binary ionic compounds.
   a. cobalt (III) chloride   **b.** chromium (IV) sulfide   **c.** copper (I) nitride
   d. zinc oxide   **e.** cobalt (II) iodide   **f.** lead (IV) sulfide

---

Answers

1a. copper (II) fluoride **1b.** silver chloride **1c.** iron (III) chloride
1d. lead (II) sulfide **1e.** tin (IV) oxide **1f.** manganese (II) phosphide
2a. $CoCl_3$ **2b.** CrS_2 **2c.** Cu_3N **2d.** ZnO **2e.** CoI_2 **2f.** PbS_2

NAMES OF IONIC COMPOUNDS WITH POLYATOMIC IONS

Some ionic compounds contain ions made up of two or more elements called polyatomic ions.

1. Follow the basic rules for naming binary ionic compounds.

2. Insert the name of any polyatomic ions in the formula. There is a list of common polyatomic ions in the reference tables at the end of your textbook.

For example, $Ca(NO_3)_2$ contains more than two elements, so it's not a binary ionic compound. It's a compound that contains the polyatomic ion nitrate, $(NO_3)^-$. The metal cation is calcium, so the name of this compound is *calcium nitrate*.

FORMULAS OF IONIC COMPOUNDS WITH POLYATOMIC IONS

If you know the name of an ionic compound containing a polyatomic ion, you can use it to determine the correct chemical formula. Remember that, in the formula, the charge on the polyatomic ion is for the entire group of atoms as a unit. When there is more than one of the same polyatomic ion in a formula, the ion is enclosed in parentheses and a subscript number indicates how many ions are in the compound. Follow the same rules for writing formulas for other types of compounds, but make sure polyatomic ions are treated as a unit.

1. Write the symbol for the cation, including the charge.

2. Write the symbol for the anion, including the charge.

3. Use subscripts in the formula so the net charge of the compound is zero.

For example, in the compound magnesium hydroxide, the cation is magnesium, Mg^{2+}, and the anion is hydroxide, $(OH)^-$. For the charge on the compound to be zero, the formula must contain one magnesium ion and two hydroxide polyatomic ions:

$$1(+2) + 2(-1) = (+2) + (-2) = 0$$

So the formula for magnesium hydroxide is $Mg(OH)_2$. Notice that parentheses and a subscript are needed to show that there are two hydroxide units in the formula. An ionic compound could contain both a transition metal ion and a polyatomic ion. If that's the case, follow the rules for both transition metal ions and polyatomic ions.

Practice Exercises

1. Write the name for the following ionic compounds with polyatomic ions.
    a. $Ca(NO_3)_2$ b. Ag_2SO_4 c. $Ba(OH)_2$
    d. $CuCO_3$ e. $Ca_3(PO_4)_2$ f. $Pb(HCO_3)_4$

2. Write the formula for the following ionic compounds with polyatomic ions.
    a. calcium bromate   b. ammonium hydroxide   c. potassium nitrate
    d. nickel (II) cyanide   e. iron (III) sulfate   f. copper (II) phosphate

---

Answers

1a. calcium nitrate   **1b.** silver sulfate   **1c.** barium hydroxide
1d. copper (II) carbonate   **1e.** calcium phosphate
1f. lead (IV) hydrogen carbonate

2a. $Ca(BrO_3)_2$ **2b.** NH_4OH **2c.** KNO_3 **2d.** $Ni(CN)_2$
2e. $Fe_2(SO_4)_3$ **2f.** $Cu_3(PO_4)_2$

NAMES OF ACIDS

In Unit 4: Toxins, you learned that acids could be defined as molecules that break apart, or dissociate, in solution to form at least one hydrogen cation, H^+, and an anion. Examine the table of acids. Try to find patterns in the names, formulas, and ions in solution.

Acids

Acid name	Chemical formula	Cations in solution	Anions in solution
sulfuric acid	H_2SO_4	2 hydrogen ions, H^+	1 sulfate ion, SO_4^{2-}
sulfurous acid	H_2SO_3	2 hydrogen ions, H^+	1 sulfite ion SO_3^{2-}
hydroiodic acid	HI	1 hydrogen ion, H^+	1 iodide ion, I^-
hydrochloric acid	HCl	1 hydrogen ion, H^+	1 chloride ion, Cl^-
chloric acid	$HClO_3$	1 hydrogen ion, H^+	1 chlorate ion, ClO_3^-
perchloric acid	$HClO_4$	1 hydrogen ion, H^+	1 perchlorate ion, ClO_4^-
hypochlorous acid	$HClO$	1 hydrogen ion, H^+	1 hypochlorite ion, ClO^-
chlorous acid	$HClO_2$	1 hydrogen ion, H^+	1 chlorite ion, ClO_2^-

Here are some patterns you might notice.

- The formula for each acid contains the element hydrogen, and each acid dissociates to form H^+ ions in solution.
- When dissociated in solution, some of the anions are polyatomic ions, while others are monatomic ions.
- The names of acids containing only two elements begin with "hydro."
- The names of the acids with polyatomic ions start with the root of the name of the polyatomic ion.
- Some acids have a name that ends *-ic acid*; others have names that end in *-ous acid*.
- There is no net charge on the compound before it dissociates, and the net charge of the ions in solution is zero.

NAMES OF BINARY ACIDS

In the table, some of the acids are made up of covalent molecules that contain only two elements. These compounds dissociate into hydrogen ions and nonmetal anions that are not polyatomic ions. To name these binary acids, you need to know the root of the name of the anion. For example, the root of the chloride ion is *chlor-*, and the root of the iodide ion is *iod-*.

1. Add the prefix *hydro-* to the root name of the anion.
2. Add the suffix *-ic* to the root of the anion name.
3. Add the word *acid* to the name.

For example, the molecule HBr is an acid that dissociates into H^+ and Br^- ions in solution. The root of the anion name is *brom-*, so the name of this acid is *hydrobromic acid*. Likewise, HF is hydrofluoric acid.

FORMULAS OF BINARY ACIDS

Acids have formulas that begin with hydrogen, H. Binary acids are named using the root of the anion that forms in solution, so you can work backward from the root to determine the name.

1. Identify the root of the anion name by removing the prefix *hydro-* and the suffix *-ic* from the root of the anion name. Rename the anion using the suffix *-ide*.
2. Write the symbol for the anion, including the charge.
3. Write the symbol for the hydrogen ion, including the charge.
4. Use subscripts to write a chemical formula for a molecule that has no charge and when dissociated in solution has no net charge.

For example, if you remove the prefix *hydro-* and the suffix *-ic* from hydrobromic acid, you are left with the root, *brom-*. Adding back the *-ide* ending indicates that the anion is bromide, Br^-. This acid is a molecule with no charge that dissociates into hydrogen ions, H^+, and Br^- anions so the formula can be determined using the charges of the ions in solution, such that they add to zero charge.

$$1(+1) + 1(-1) = 0$$

This means that when one molecule of this acid dissociates, one hydrogen ion and one bromide ion form in solution, so the compound is the molecule, HBr. Consider hydroselenic acid. The root of the anion name is *selen-*, which corresponds to the selenide ion, Se^{2-}. For the solution to have no net charge, the molecule must dissociate into 2 hydrogen ions, H^+ and 1 selenide ion, Se^{2-}.

$$2(+1) + 1(-2) = (+2) + (-2) = 0$$

So the formula for this acid is H_2Se.

NAMES OF ACIDS THAT CONTAIN POLYATOMIC ANIONS WITH OXYGEN

In the table, some of the acids are made up of covalent molecules that dissociate into hydrogen ions and polyatomic anions that contain oxygen. To name these acids, you use the root of the name of the polyatomic anion. For example, the root of the sulfate ion is *sulf-* and the root of the chlorite ion is *chlor-*.

1. Identify the root of the name of the polyatomic anion.
2. If the polyatomic anion ends in *-ate*, add the suffix *-ic* to the root of the polyatomic anion name.
3. If the polyatomic anion ends in *-ite*, add the suffix *-ous* to the root of the polyatomic anion name.
4. Add the word *acid* to the name.

For example, HClO dissociates into one hydrogen ion, H^+, and the polyatomic ion hypochlorite, ClO^-, which has the root, *hypochlor-*. Since the polyatomic anion ends in *-ite*, the suffix *-ous* is added to the root, and the word *acid* is added to form hypochlorous acid. As another example, $HClO_2$ dissociates into one hydrogen H^+ ion and the polyatomic ion chlorite, with the root, *chlor-*. Because the anion name ends in *-ite*, this acid is called *chlorous acid*. Likewise, $HClO_4$ with the perchlorate ion is perchloric acid, and $HClO_3$ with the chlorate ion is chloric acid.

FORMULAS OF ACIDS THAT CONTAIN POLYATOMIC ANIONS WITH OXYGEN

Remember that these acids are molecules with no charge that dissociate into hydrogen ions and polyatomic anions that contain oxygen in solution. They are named using the root of the polyatomic anion that forms in solution. So work backward from the root of the anion to determine the name.

1. Identify the name and charge of the polyatomic ion from the root of the anion name.
2. Write the symbol for the polyatomic anion, including the charge.
3. Write the symbol for the hydrogen ion, including the charge.
4. Use subscripts to write a chemical formula for a molecule that has no charge and when dissociated in solution has no net charge.

For example, sulfuric acid does not have the prefix *hydro-*, so it must contain a polyatomic ion. Both sulfite, $(SO_3)^{2-}$ and sulfate $(SO_4)^{2-}$ have the root *sulf-*, but the name of this acid ends in *-ic*. The correct anion is sulfate, because its name ends in *-ate*. The sulfate ion has a charge of $+2$, so when it dissociates, the original molecule must break apart into two hydrogen ions, H^+, and one sulfate ion, $(SO_4)^{2-}$.

$$2(+1) + 1(-2) = (+2) + (-2) = 0$$

So the formula for this acid contains two hydrogen atoms and one sulfate ion, H_2SO_4.

Practice Exercises

1. Write the name for these acids.
   a. HBr **b.** H_2SO_4 **c.** HNO_3 **d.** HCl **e.** HBrO **f.** H_3PO_3

2. Write the formula for these acids.
   a. carbonic acid **b.** hydrofluoric acid **c.** phosphoric acid
   d. sulfurous acid **e.** perbromic acid **f.** nitrous acid

---

Answers

1a. hydrobromic acid **1b.** sulfuric acid **1c.** nitric acid
1d. hydrochloric acid **1e.** hypobromous acid **1f.** phosphorous acid
2a. H_2CO_3 **2b.** HF **2c.** H_3PO_4 **2d.** H_2SO_3 **2e.** $HBrO_4$ **2f.** HNO_2

NAMES AND FORMULAS FOR ARRHENIUS BASES

According to the Arrhenius theory of acids and bases, a base is defined as a substance that adds OH^- to an aqueous solution. Examine the table of Arrhenius bases.

Arrhenius Bases

Acid name	Chemical formula	Cations in solution	Anions in solution
potassium hydroxide	KOH	1 potassium ion, K^+	1 hydroxide ion, OH^-
calcium hydroxide	$Ca(OH)_2$	1 calcium ion, Ca^{2+}	2 hydroxide ions, OH^-
aluminum hydroxide	$Al(OH)_3$	1 aluminum ion, Al^{3+}	3 hydroxide ions, OH^-

Here are some patterns you might notice.

- They each contain the hydroxide polyatomic ion.
- They each dissociate into metal cations and hydroxide polyatomic anions.
- There are as many hydroxide ions as the charge on the metal.
- The names all contain the word *hydroxide*.

When naming and writing formulas for Arrhenius bases, you simply follow the rules for naming ionic compounds that contain a polyatomic ion. For these bases, the polyatomic ion is the hydroxide ion, OH^-.

Practice Exercises

1. Write the name for these bases.
   a. $Mg(OH)_2$ **b.** NaOH **c.** $Sr(OH)_2$

2. Write the formula for these bases.
   a. potassium hydroxide
   b. barium hydroxide
   c. lithium hydroxide

Answers

1a. magnesium hydroxide **1b.** sodium hydroxide **1c.** strontium hydroxide
2a. KOH **2b.** $Ba(OH)_2$ **2c.** LiOH

Appendix B: Connecting Chemical Concepts

Percent Composition

Elements combine in specific ratios to form compounds, so the ratio of one element to another in a compound is always the same. For example, water, H_2O, is made up of the elements hydrogen and oxygen in a 2:1 ratio. If the ratio of hydrogen to oxygen were different, the compound would not be water. For example, hydrogen peroxide, H_2O_2, is also made up of hydrogen and oxygen atoms, but the ratio of hydrogen to oxygen in this compound is 1:1. In one mole of water, there are always two moles of hydrogen atoms and one mole of oxygen atoms, as indicated by the ratios of elements in the chemical formula.

Examine the table that shows the composition of water in terms of moles, mass, and percent mass of each element in the compound.

Composition of Different Samples of Water, H_2O

Sample size	Moles of H_2O	Moles of hydrogen atoms	Moles of oxygen atoms	Mass of sample	Mass of hydrogen in sample	Mass of oxygen in sample	Percent by mass of hydrogen	Percent by mass of oxygen
1 mol H_2O	1 mol	2 mol	1 mol	18.02 g	2.02 g	16.00 g	11.2%	88.8%
5 mol H_2O	5 mol	10 mol	5 mol	90.10 g	10.1 g	80.00 g	11.2%	88.8%
18.02 g H_2O	1 mol	2 mol	1 mol	18.02 g	2.02 g	16.00 g	11.2%	88.8%
200 g H_2O	11.1 mol	22.2 mol	11.1 mol	200 g	22.4 g	177.6 g	11.2%	88.8%

Notice that for each sample size of water, the percent of hydrogen in the sample and the percent of oxygen in the sample are always the same. This is because these elements always combine in the same molar ratio to form water. The percent composition of elements in a compound is the percent by mass of each element in the compound. A percent is an amount per 100. You can calculate a percent by taking part of a whole amount, dividing it by the whole amount, and multiplying by 100%.

$$\text{percent} = \left(\frac{\text{part}}{\text{whole}}\right) \cdot 100\%$$

In terms of percent composition of elements in a compound, the "part" is the mass of one of the elements in a sample of a compound and the "whole" is the mass of the entire sample of the compound. If you add the masses of each element present in the sample of the compound, the result is the mass of the compound. If you add the percent by mass of each element in a compound the total should be 100%.

DETERMINING PERCENT COMPOSITION USING THE CHEMICAL FORMULA

If you know the chemical formula for a compound, you can determine the percent composition of each element in it by using the mass of one mole of the compound and the mass of each element that would be found in one mole of the compound.

Example 1

Percent Composition of CO

What is the percent composition of carbon monoxide, CO?

Solution

Assume that you have one mole of CO.

Step 1: Use the periodic table to calculate the molar mass of CO.

$$\text{molar mass of CO} = 12.01 \text{ g/mol} + 16.00 \text{ g/mol} = 28.01 \text{ g/mol}$$

One mole of CO is 28.01 g.

Step 2: Determine the percent mass of each element in the compound.

Divide the mass of each element in one mole of the compound by the mass of one mole of the compound and multiply by 100%. The molar ratio of C:O in the compound is 1:1, so there is one mole of each element in one mole of CO.

The mass of one mole of carbon, C, is 12.01 g. The mass of one mole of oxygen, O, is 16.00 g. Use these values to determine the percent mass of each compound in one mole of the compound.

$$\text{percent by mass of C} = \left(\frac{12.01 \text{ g}}{28.01 \text{ g}}\right) \cdot 100\% = 42.88\%$$

$$\text{percent by mass of O} = \left(\frac{16.00 \text{ g}}{28.01 \text{ g}}\right) \cdot 100\% = 57.12\%$$

Step 3: Check your answer by adding the percent by mass of each element. The total should be 100%.

42.88% + 57.12% = 100.00% (or 100% when rounded to the nearest whole number)

Example 2

Percent Mass of Oxygen in $KClO_3$

What is the mass of oxygen that can be produced from the decomposition of 150.0 g of potassium chlorate, $KClO_3$? (*Hint:* Find the percent composition of oxygen in the compound to figure out the answer.)

Solution

When potassium chlorate, $KClO_3$, decomposes, the oxygen produced will be the amount present in the 150.0 g sample of the compound.

Step 1: Calculate the mass of one mole of potassium chlorate, $KClO_3$.

$$\begin{aligned} \text{molar mass of KClO}_3 &= 1(39.10 \text{ g/mol}) + 1(35.45 \text{ g/mol}) + 3(16.00 \text{ g/mol}) \\ &= 39.10 \text{ g/mol} + 35.45 \text{ g/mol} + 48.00 \text{ g/mol} \\ &= 122.55 \text{ g/mol} \end{aligned}$$

One mole of potassium chlorate, $KClO_3$, is 122.55 g.

(continued)

Step 2: Determine the percent mass of oxygen, O, in one mole of the compound.

The mass of one mole of oxygen is 16.00 g. In one mole of potassium chlorate, $KClO_3$, there are three moles of oxygen, O.

$$\text{mass of O in one mole of compound} = 3(16.00 \text{ g})$$
$$= 48.00 \text{ g}$$

Divide the mass of oxygen, O, in one mole of potassium chlorate, $KClO_3$, by the molar mass of the potassium chlorate, $KClO_3$, and multiply by 100%.

$$\text{percent by mass O} = \left(\frac{48.00 \text{ g}}{122.55 \text{ g}}\right) \cdot 100\% = 39.17\%$$

Step 3: Use the percent composition of oxygen, O, to determine the amount of oxygen produced in the decomposition reaction.

The mass of oxygen produced when 150.0 g of $KClO_3$ decomposes is 39.17% of 150.0 g. Convert the percentage to a decimal and multiply by 150.0 g.

$$0.3917 \times 150.0 \text{ g} = 58.75 \text{ g}$$

The decomposition of 150.0 g of potassium chlorate, $KClO_3$, produces 58.75 g of oxygen, O.

Practice Exercises

1. What is the percent composition of HCl?
2. What is the percent composition of ammonium phosphate $(NH_4)_3PO_4$?
3. What is the mass of aluminum needed to produce 298 g of $Al_2(SO_4)_3$?

Answers

1. H = 2.7%, Cl = 97.3%
2. N = 28.19%, H = 8.113%, P = 20.77%, O = 42.93%. The percent by mass values add to 100.10% or 100% if rounded to the nearest whole number.
3. 47.0 g Al

Molecular and Empirical Formulas

In Unit 2: Smells, you learned that the chemical formula for a molecular covalent compound is also called the *molecular formula*. The molecular formula indicates the number of atoms of each element present in one molecule of a compound. Examine the table of formulas. Notice that for each compound, there is a molecular formula and what is called the *empirical formula*.

Molecular and Empirical Formulas

Name	Molecular formula	Empirical formula
ethyne	C_2H_2	CH
benzene	C_6H_6	CH
carbon monoxide	CO	CO
carbon dioxide	CO_2	CO_2
tetraphosphorus decaoxide	P_4O_{10}	P_2O_5

Here are some patterns you might notice.

- For each compound, the molecular formula and the empirical formula both have the same type of atoms.

- Some compounds have a molecular formula that is different from the empirical formula.

- Some compounds have a molecular formula that is the same as the empirical formula.

- The molecular formula and the empirical formula are related mathematically.

- If the empirical formula is different from the molecular formula, you can multiply the subscripts in the empirical formula by a single number to get the values for subscripts in the molecular formula. For example, if you multiply the subscripts for CH by 6, you get the molecular formula for benzene, C_6H_6. You can multiply the subscripts in P_2O_5 by 2 to get the molecular formula P_4O_{10}.

The empirical formula is the simplest formula for a compound and shows the smallest whole-number ratio of the atoms present in that compound. There is a relationship between molecular formula and empirical formula that can be seen in the patterns of the subscripts. If you multiply the subscripts in the empirical formula by the number of empirical units, you get the molecular formula. Remember if there is no subscript, the value is 1.

$$(\text{empirical formula})_{(\text{number of empirical units})} = \text{molecular formula}$$

ethyne: $(CH)_2 = C_2H_2$

benzene: $(CH)_6 = C_6H_6$

carbon monoxide: $(CO)_1 = CO$

carbon dioxide: $(CO_2)_1 = CO_2$

If the molecular formula is different from the empirical formula, you can determine the empirical formula by dividing each subscript by the largest whole-number factor they have in common. For example, you can divide subscripts 4 and 10 in P_4O_{10} by the number 2 to get the empirical formula P_2O_5.

Example 1

Determining Empirical Formula from Molecular Formula

Write the empirical formula for these compounds.

    a. $C_6H_{12}O_2$

    b. $C_4H_8O_2$

    c. H_2SO_4

Solution

    a. The largest whole-number factor the subscripts have in common is 2. You can divide each subscript by 2 to determine that the empirical formula is C_3H_6O.

    b. The largest whole-number factor the subscripts have in common is 2. You can divide each subscript by 2 to determine that the empirical formula is C_2H_4O.

    c. The largest whole-number factor the subscripts have in common is 1. You can divide each subscript by 1 to determine that the empirical formula is H_2SO_4, which is the same as the molecular formula.

If you know the amount in grams of each element present in a sample of a compound, or you know the percent composition for a compound, you can determine its empirical formula. If you also know the molar mass of the compound, you can use it along with the empirical formula to determine the molecular formula of the compound.

Example 2

Determining the Molecular Formula Using Mass Data

A 500.0 g sample of hexanoic acid contains 310.2 g of carbon, 52.1 g of hydrogen, and 137.8 g of oxygen. The molar mass of this compound is 116.16 g/mol. Determine the empirical and molecular formulas for this compound.

Solution

Step 1: The empirical and molecular formulas show the relative numbers of atoms in an element, not the relative amount of mass in each element. So the first step is to convert the amount of each element in grams to the amount in moles. You can do this using the molar mass of the element from the periodic table.

$$\text{moles of C} = \left(\frac{310.2 \text{ g}}{12.01 \text{ g/mol}} \right) = 25.83 \text{ mol}$$

$$\text{moles of H} = \left(\frac{52.1 \text{ g}}{1.008 \text{ g/mol}} \right) = 51.69 \text{ mol}$$

$$\text{moles of O} = \left(\frac{137.8 \text{ g}}{16.00 \text{ g/mol}} \right) = 8.62 \text{ mol}$$

Example

(continued)

Step 2: The molecular formulas show the relative amounts of moles in whole-number ratios. To convert the amounts of moles to whole numbers, divide each value by the lowest number of moles.

$$\text{moles of C} = \frac{25.83 \text{ mol}}{8.62} = 2.997 \text{ mol, which rounds to 3 mol}$$

$$\text{moles of H} = \frac{51.69 \text{ mol}}{8.62} = 5.997 \text{ mol, which rounds to 6 mol}$$

$$\text{moles of O} = \frac{8.62 \text{ mol}}{8.62} = 1 \text{ mol}$$

Use the ratio of moles, 3:6:1, to write the empirical formula, C_3H_6O.

Step 3: You can calculate the number of empirical units by dividing the molar mass of the compound by the empirical molar mass. The empirical molar mass is the sum of the molar masses of the elements in the empirical formula.

$$\begin{aligned}\text{empirical molar mass} &= 3(12.01 \text{ g/mol}) + 6(1.008 \text{ g/mol}) + (16.00 \text{ g/mol}) \\ &= 58.08 \text{ g/mol}\end{aligned}$$

$$\begin{aligned}\text{number of empirical units} &= \frac{\text{molar mass}}{\text{empirical molar mass}} \\ &= \frac{116.16 \text{ g/mol}}{58.08 \text{ g/mol}} \\ &= 2 \text{ units}\end{aligned}$$

$$\begin{aligned}(\text{empirical formula})_{(\text{number of empirical units})} &= \text{molecular formula} \\ (C_3H_6O)_2 &= C_6H_{12}O_2\end{aligned}$$

The molecular formula for the compound is $C_6H_{12}O_2$.

Example 3

Determining the Molecular Formula Using Percent Composition Data

A compound contains 71.65% Cl, 24.27% C, and 4.07% H. The molar mass of the compound is 98.96 g/mol. Determine the empirical formula and molecular formula for this compound.

Solution

Step 1: When amounts in mass are not provided, assume that you have 100 g of the compound. Find the mass of each element in grams by converting the percent values to decimals and multiplying by the total mass.

$$\begin{aligned}\text{mass of Cl} &= 0.7165 \cdot (100 \text{ g}) = 71.65 \text{ g of Cl} \\ \text{mass of C} &= 0.2427 \cdot (100 \text{ g}) = 24.27 \text{ g of C} \\ \text{mass of H} &= 0.0407 \cdot (100 \text{ g}) = 4.07 \text{ g of H}\end{aligned}$$

Example

(continued)

Step 2: Convert the amounts in grams to moles.

$$\text{moles of Cl} = \left(\frac{71.65 \text{ g}}{35.45 \text{ g/mol}}\right) = 2.02 \text{ mol}$$

$$\text{moles of C} = \left(\frac{24.27 \text{ g}}{12.01 \text{ g/mol}}\right) = 2.02 \text{ mol}$$

$$\text{moles of H} = \left(\frac{4.07 \text{ g}}{1.008 \text{ g/mol}}\right) = 4.04 \text{ mol}$$

Step 3: Convert the amounts in moles to whole numbers by dividing by the lowest value.

$$\text{moles of Cl} = \left(\frac{2.02 \text{ mol}}{2.02}\right) = 1 \text{ mol}$$

$$\text{moles of C} = \left(\frac{2.02 \text{ mol}}{2.02}\right) = 1 \text{ mol}$$

$$\text{moles of H} = \left(\frac{4.04 \text{ mol}}{2.02}\right) = 2 \text{ mol}$$

The empirical formula is CH_2Cl.

Step 4: Determine the empirical molar mass.

$$\text{empirical molar mass} = (12.01 \text{ g/mol}) + 2(1.008 \text{ g/mol}) + (35.45 \text{ g/mol})$$
$$= 49.48 \text{ g/mol}$$

Step 5: Calculate the number of empirical units.

$$\text{number of empirical units} = \frac{\text{molar mass}}{\text{empirical molar mass}}$$
$$= \frac{98.96 \text{ g/mol}}{49.48 \text{ g/mol}}$$
$$= 2 \text{ units}$$

Step 6: Multiply the subscripts in the empirical formula by the number of empirical units to determine the molecular formula.

$$(\text{empirical formula})_{(\text{number of empirical units})} = \text{molecular formula}$$
$$(CH_2Cl)_2 = C_2H_4Cl_2$$

The molecular formula for the compound is $C_2H_4Cl_2$.

Practice Exercises

1. What is the molecular formula for a compound with an empirical formula of C_4H_9 and a molar mass of 114.22 g/mol?

2. A compound has a mass percent composition of 5.927% hydrogen and 94.073% oxygen. The molecular mass of this compound is 34.016 g/mol. Find both the empirical and molecular formulas.

3. A compound has a mass percent composition of 40.00% carbon, 6.72% hydrogen, and 53.28% oxygen. The molecular mass of this compound is 150.13 g/mol. Determine both the empirical and molecular formulas.

4. A compound contains 0.384 g carbon, 0.048 g hydrogen, and 0.5674 g chlorine. The molecular mass of this compound is 125 g/mol. Calculate the empirical and molecular formulas.

Answers

1. C_8H_{18}
2. empirical formula: HO, molecular formula: H_2O_2
3. empirical formula: CH_2O, molecular formula: $C_5H_{10}O_5$
4. empirical formula: C_2H_3Cl, molecular formula: $C_4H_6Cl_2$

More Stoichiometry Practice

Recall from Unit 4: Toxins that stoichiometry involves calculations based on the quantitative relationship between amounts of reactants and products in a chemical reaction. Using mole ratios from a balanced chemical equation, it's possible to determine amounts of reactants and products that will be used or formed during a reaction.

When the calculations involve amounts in mass, you need to convert given amounts in mass to moles. To do this, you use the molar mass for each substance, because the molar mass is the amount in grams of exactly one mole of that substance.

$$\text{number of moles} = \frac{\text{mass in g}}{\text{molar mass in g/mol}}$$

$$\text{mass in g} = (\text{molar mass in g/mol}) \cdot (\text{number of moles})$$

Once you convert amounts to moles, you can use mole ratios from the balanced chemical equation to relate reactants to products. The "mole tunnel" is a way to help you remember these steps.

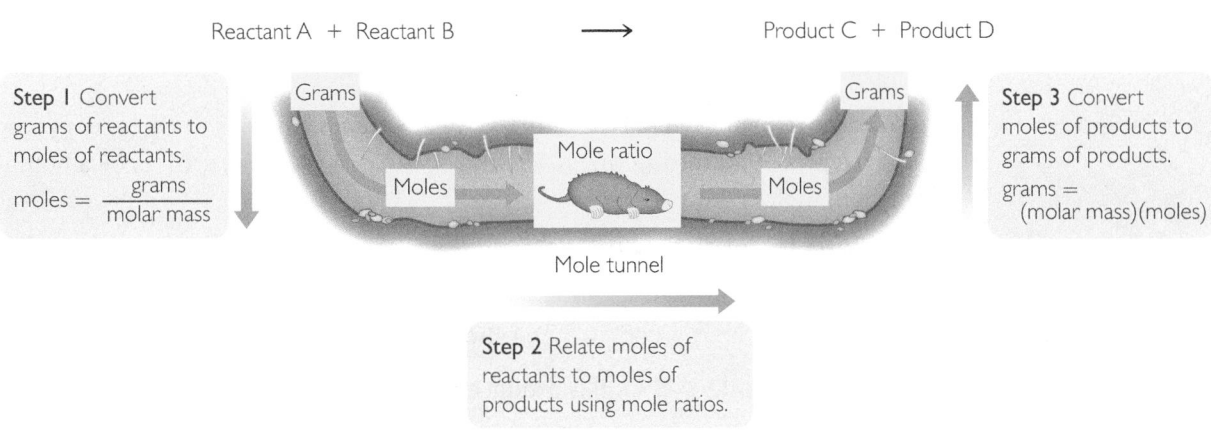

Reactant A + Reactant B $\longrightarrow$ Product C + Product D

Step 1 Convert grams of reactants to moles of reactants.

$$\text{moles} = \frac{\text{grams}}{\text{molar mass}}$$

Grams → Moles → Mole ratio → Moles → Grams

Mole tunnel

Step 2 Relate moles of reactants to moles of products using mole ratios.

Step 3 Convert moles of products to grams of products.

$$\text{grams} = (\text{molar mass})(\text{moles})$$

If you are solving a problem that involves a particular volume of gas, you also need to convert to moles. Recall from Unit 3: Weather that at standard temperature and pressure, STP, one mole of gas occupies a volume of 22.4 liters.

$$\text{number of moles} = \frac{\text{volume at STP in L}}{22.4 \text{ L/mol}}$$

$$\text{volume at STP in L} = \text{number of moles} \cdot (22.4 \text{ L/mol})$$

If the reaction is not carried out under STP conditions, you can use the ideal gas law, $PV = nRT$, to solve for the number of moles, n, as long as you are given the temperature, volume, and pressure.

$$n = \frac{PV}{RT}$$

$$V = \frac{nRT}{P}$$

In the above equations, the number of moles is represented by n, pressure by P, volume by V, temperature by T, and R is the ideal gas constant.

It is also possible to solve stoichiometry problems in terms of the numbers of particles of each substance because one mole of a substance = 6.02×10^{23} particles (such as atoms, molecules, ions, or formula units).

$$\text{number of particles} = 6.02 \times 10^{23} \text{ particles/mol} \cdot (\text{number of moles})$$

$$\text{number of moles} = \frac{\text{number of particles}}{6.02 \times 10^{23} \text{ particles/mol}}$$

Example 1

Volume of Gas Produced in a Reaction at STP

What volume of H_2O vapor is produced when 0.35 mol of O_2 gas reacts with an excess of H_2 gas at STP? How many molecules of H_2O is this?

Solution

Step 1: Write the balanced equation for this reaction.

$$2H_2(g) + O_2(g) \longrightarrow 2H_2O(g)$$

Step 2: The oxygen gas, O_2, is already reported in moles. There is an excess of hydrogen, H_2, so oxygen, O_2, is the limiting reactant. Use mole ratios to determine the amount of product in moles.

The mole ratio of O_2 to H_2O is 1:2.

$$0.35 \text{ mol } O_2 \cdot \left(\frac{2 \text{ mol } H_2O}{1 \text{ mol } O_2} \right) = 0.70 \text{ mol } H_2O$$

So 0.70 mole of $H_2O(g)$ is produced in this reaction.

(continued)

Step 3: Convert moles of $H_2O(g)$ to volume of $H_2O(g)$. This reaction takes place at standard temperature and pressure, STP.

> volume of H_2O at STP = number of moles · (22.4 L/mol)
> volume of H_2O at STP = 0.70 mol H_2O · (22.4 L/mol)
> volume of H_2O at STP = 15.68 L, rounded to two significant figures, 16 L

Step 4: Convert moles of H_2O (g) to number of molecules.

> number of molecules H_2O = 6.02×10^{23} particles/mol · (number of moles)
> number of molecules H_2O = 6.02×10^{23} particles/mol · (0.70 mol H_2O)
> = 4.21×10^{23} molecules H_2O, rounded to two significant figures, 4.2×10^{23} molecules H_2O

Recall from Unit 3: Weather that if there is a mixture of nonreacting gases, the pressure exerted by one gas in the mixture is called the *partial pressure*. Dalton's law of partial pressures states that the partial pressures of all the gases add up to the total pressure exerted by that mixture of gases. This information can also be helpful in solving stoichiometry problems involving gases.

Dalton's Law of Partial Pressures

$$P_{total} = P_1 + P_2 + P_3 \ldots$$

In the above equation, the numbered subscripts refer to each individual gas in the mixture.

Example 2

Stoichiometry with Partial Pressures and the Ideal Gas Law

A student carried out the experiment from the illustration with an excess of hydrochloric acid. The student added a step to bubble the hydrogen gas formed through a closed container of water to collect it in the space above the water. When finished, a 75 mL mixture of water vapor and hydrogen gas was above the water. If the temperature in the container was 30 °C, the total pressure of the gas mixture was 1.015 atm, and the pressure of water vapor was 0.042 atm, what mass of chromium must have reacted?

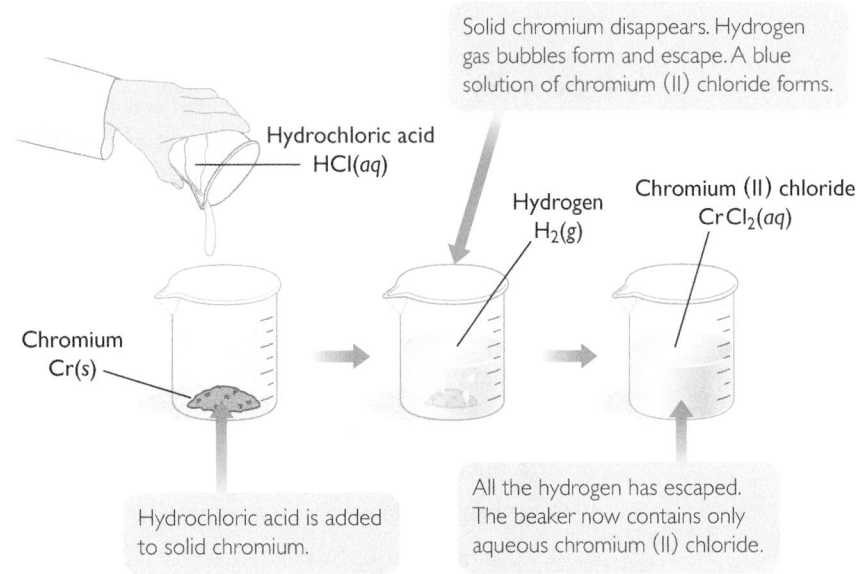

Solid chromium disappears. Hydrogen gas bubbles form and escape. A blue solution of chromium (II) chloride forms.

Hydrochloric acid
HCl(*aq*)

Hydrogen
$H_2(g)$

Chromium (II) chloride
$CrCl_2(aq)$

Chromium
Cr(s)

Hydrochloric acid is added to solid chromium.

All the hydrogen has escaped. The beaker now contains only aqueous chromium (II) chloride.

Example

(continued)

Solution

Step 1: Start by writing the balanced chemical equation for the reaction.

$$Cr(s) + 2HCl(aq) \longrightarrow H_2(g) + CrCl_2(aq)$$

Step 2: To determine the mass of chromium reacted, you need to determine how much hydrogen gas was formed. The problem states the total pressure of the final mixture of gas, the total volume of the mixture of gas, and the partial pressure of water vapor under these conditions. You can use Dalton's law of partial pressures to determine the pressure of the hydrogen gas in the mixture and then the ideal gas law to determine the amount of hydrogen in moles.

$$P_{total} = P_{H_2O} + P_{H_2}$$
$$P_{H_2} = P_{total} - P_{H_2O}$$
$$P_{H_2} = 1.015 \text{ atm} - 0.042 \text{ atm}$$
$$P_{H_2} = 0.973 \text{ atm}$$

Step 3: Use the ideal gas law to calculate the number of moles of hydrogen gas produced.

$$T = 30\,^{\circ}C + 273 = 303 \text{ K}$$
$$75 \text{ mL} = 0.075 \text{ L}$$
$$n = \frac{PV}{RT}$$
$$n = \frac{0.973 \text{ atm} \cdot 0.075 \text{ L}}{\left(0.082\dfrac{L \cdot atm}{mol \cdot K}\right) \cdot 303K}$$
$$= 2.9 \times 10^{-3} \text{ mol H}_2$$

Step 4: Use mole ratios to work backward to determine the number of moles of chromium that reacted.

The mole ratio of Cr to H_2 is 1:1

$$2.9 \times 10^{-3} \text{ mol H}_2 \cdot \left(\frac{1 \text{ mol Cr}}{1 \text{ mol H}_2}\right) = 2.9 \times 10^{-3} \text{ mol Cr}$$

Step 5: Convert moles of Cr to grams.

$$\text{mass of Cr} = (52.00 \text{ g/mol}) \cdot (2.9 \times 10^{-3} \text{ mol})$$
$$= 0.15 \text{ g Cr}$$

BASIC STEPS FOR SOLVING STOICHIOMETRY PROBLEMS

In general, you can follow some basic steps to solve stoichiometry problems.

1. If not given, write a balanced chemical equation.

2. Convert amounts given in mass, volume, or number of particles to moles.

3. If needed, determine the limiting reactant using mole ratios.

4. Use mole ratios to relate reactants to products.

5. Convert amounts in moles back to mass, volume, or number of particles.

Sometimes, you have to incorporate other chemistry concepts, as in Example 2, before you can solve the problem. So before you begin each problem, determine:

- What information is given in the problem?
- What exactly are you being asked to calculate or determine?
- Does the problem contain all the necessary information or should you think about other chemistry concepts that are related to the question?
- What unit conversions will you have to perform?
- What mathematical relationships or formulas are relevant?

The problems that follow give you an opportunity to practice these problem-solving skills.

Practice Exercises

1. Using the balanced equation from Example 1, determine the volume of O_2 gas required to produce 5.50 moles of H_2O vapor at STP.

2. Consider this balanced chemical equation.

$$2C_2H_2(g) + 5O_2(g) \longrightarrow 4CO_2(g) + 2H_2O(g)$$

   a. Find the mole ratio of C_2H_2 to O_2.
   b. Find the mole ratio of C_2H_2 to CO_2.
   c. Suppose you have 2.0 L of ethyne, C_2H_2. What volume of oxygen gas would be needed to react completely with 2.0 L of ethyne at STP?
   d. What volume of carbon dioxide would be produced from 2.0 L of ethyne at STP?

3. Consider this balanced chemical equation.

$$Mg(s) + 2HCl(aq) \longrightarrow MgCl_2(aq) + H_2(g)$$

   A 120 mL sample of hydrogen gas was collected over water at 24 °C. The vapor pressure of water at 24 °C is 0.030 atm. The total pressure of gas collected was 1.010 atm. What mass of magnesium must have reacted to produce this amount of hydrogen gas with excess hydrochloric acid?

4. Consider this balanced chemical equation.

$$2KClO_3(s) \longrightarrow 2KCl(s) + 3O_2(g)$$

   a. How many moles of $KClO_3$ are needed to produce 0.60 mol of O_2?
   b. What mass of $KClO_3$ is needed to produce 0.20 mol of KCl?
   c. How many molecules of O_2 are produced from 0.50 mol $KClO_3$?

5. Consider this balanced chemical equation.

$$2Mg(s) + O_2(g) \longrightarrow 2MgO(s)$$

   a. What volume of oxygen gas is needed to burn 0.50 g of magnesium ribbon at STP?
   b. What mass of MgO will be produced from the 0.50 g magnesium ribbon?

6. Consider this balanced chemical equation.

$$4Al(s) + 3O_2(g) \longrightarrow 2Al_2O_3(s)$$

   What mass of aluminum would be used in the production of 10.2 g of aluminum oxide if aluminum reacted directly with unlimited oxygen?

7. Consider this balanced chemical equation.

$$C(s) + O_2(g) \longrightarrow CO_2(g)$$

a. If 6.00 g of carbon were burned to form CO_2, what mass of CO_2 would be expected?

b. If the actual yield of CO_2 was measured to be 19.0 g, what is the percent yield?

8. Consider this balanced chemical equation.

$$CH_4(g) + 2O_2(g) \longrightarrow CO_2(g) + 2H_2O(g)$$

a. If 50.0 g of methane, CH_4, were burned, what mass of CO_2 would be expected?

b. If the actual yield of CO_2 was measured to be 98.0 g, what is the percent yield?

9. Consider this balanced chemical equation.

$$3Fe(s) + 4H_2O(g) \longrightarrow Fe_3O_4(s) + 4H_2(g)$$

a. How many moles of Fe_3O_4 can be obtained by reacting 16.8 g Fe with 10.0 g H_2O?

b. Which substance is the limiting reactant?

c. Which substance is in excess?

10. The following reaction was performed with 36.0 g Mg and 0.25 L of a 2.0 M HCl.

$$Mg(s) + 2HCl(aq) \longrightarrow MgCl_2(aq) + H_2(g)$$

a. Which compound is the limiting reactant?

b. What mass of hydrogen gas was produced?

c. What mass of magnesium chloride was produced?

Answers

1. 61.6 L O_2

2a. $\dfrac{2 \text{ mol } C_2H_2}{5 \text{ mol } O_2}$ 2b. $\dfrac{1 \text{ mol } C_2H_2}{2 \text{ mol } CO_2}$ 2c. 5.0 L O_2 2d. 4.0 L CO_2

3. 0.12 g Mg

4a. 0.40 mol $KClO_3$ 4b. 24.5 g $KClO_3$ 4c. 4.52×10^{23} molecules O_2

5a. 0.23 L O_2 5b. 0.83 g MgO

6. 5.40 g Al

7a. 22.0 g CO_2 7b. 86.4%

8a. 137 g CO_2 8b. 71.5%

9a. 0.100 mol Fe_3O_4 9b. iron, Fe   9c. water, H_2O, is in excess

10a. HCl   10b. 0.50 g H_2

10c. 23.8 g $MgCl_2$, rounded to two significant figures, 24 g $MgCl_2$

Appendix C: Answers to Selected Exercises

Unit 1: Alchemy

Lesson 1

1. Possible answer: We use glass because substances in a glass container are visible and glass containers are relatively easy to clean and reuse. Tempered glass containers can be heated over flames without shattering.

3. Possible answers: Know the location of safety equipment. Read lab instructions carefully. Check to be sure that you are using the right chemicals and equipment.

5. Possible answers: Put all equipment in its proper place. Clean your work area. Make sure all bottles and containers holding chemicals are closed and stored properly.

Lesson 2

1. Alchemists developed some of the first laboratory tools and chemistry techniques. They classified substances into categories and experimented with mixing and heating different substances to create something new.

5. Observable changes that involve chemistry often involve an alteration in the appearance of matter. Examples include metal rusting, cookies baking, and ice cubes forming. Changes that do not involve chemistry only involve matter moving to a different location. Examples include the Sun going down, objects falling, and hands moving on a clock.

Lesson 3

1. Mass is the amount of material in an object. Volume is the amount of space the object takes up.

5. Possible answer: Examples of things that are matter: a car, a tree, a person, water. Each of these objects has mass and volume. Examples of things that are not matter: sound, movement, feelings, energy. These objects do not have mass or volume.

Lesson 4

1. Measure the dimensions of the object and calculate its volume using a geometric formula, or if it does not float or dissolve, measure the amount of liquid that it displaces when it is submerged.

3. Yes, the volume of an object is the amount of space it fills. You can usually see how much space an object fills and estimate its volume based on its dimensions. However, for objects that have an irregular shape, are very thin, or have a surface with lots of holes or pits, determining the volume of the object by sight may be difficult.

5. Yes, the mass of the rubber band is the same because only the shape of the rubber band changes, not the amount of matter in it.

11. Possible answer: The balance or scale that was used to measure the two different cubes may not be able to measure accurately to a hundredth of a gram. The two cubes could have exactly the same mass because the measurement error is greater than 0.03 g.

Lesson 5

1. Possible answer: Density is the mass of an object divided by its volume.

3. The density of aluminum is less than the density of gold. More matter is present in a given volume of gold than in the same volume of aluminum.

5. Possible answer: The object that has a density of 2.7 g/cm^3 has the larger volume. The two objects have the same mass, but the mass is packed into a smaller space in the denser object.

7. **a.** 2.87 g/cm^3 **b.** 111 g, 38.7 cm^3, 2.87 g/cm^3
   c. The density is the same.

Chapter 1 Review Exercises

1. Possible answer: Determine the volume of a powdered solid or of a liquid by pouring the substance into a graduated cylinder or beaker and reading the markings on the side. Determine the volume of a rock by submerging the rock in a graduated cylinder partially filled with water and then reading how much the water level changes.

3. Density is no help in determining which object will displace more water. A large object will displace more water than a small object no matter how dense the two objects are.

Lesson 6

1. Elements are the building materials of all matter. A compound is matter that is made up of two or more elements combined in a specific ratio. An element cannot be broken down into simpler substances by chemical means, but a compound can be broken down into elements.

3. Sodium nitrate contains three elements: sodium, nitrogen, and oxygen.

5. The chemical formula for cubic zirconia is ZiO_2. The chemical formula for a diamond is C. The stone cannot be a diamond because it has a different chemical formula than a diamond.

Lesson 7

1. Possible answer: A chemical reaction is a change leading to the final substance or substances being different from the original substance or substances. Some of the signs that a new substance has formed include color changes, formation of a new solid or gas, and the release of energy as heat or light.

3. Possible answer: A chemical reaction combined the zinc with part of the dissolved copper compound. The resulting zinc compound was also dissolved in water. The zinc was still present, but not as a pure element.

5. a. The baking soda is a solid, the vinegar is a liquid, the clear colorless liquid is a liquid, and the CO_2 is a gas.
   b. Yes, the production of carbon dioxide gas is evidence that a chemical change has occurred after the original liquid and solid substances were mixed.
   c. Before the change, the sodium is in the solid baking soda. After the change, the sodium must be in the clear colorless liquid.

Lesson 8

1. Possible answer: The chemical names and symbols indicate what compounds were combined in each step. Because matter is conserved, the products must contain the same elements as the original compounds, which enables you to make a reasonable guess about the products. For example, when sulfuric acid combines with copper oxide, it is likely that copper sulfate and water are the products.

5. The solution would be yellow because the combination of nickel, Ni, and hydrochloric acid, HCl, can only produce a solution containing compounds with nickel, hydrogen, and chlorine. Nickel chloride, $NiCl_2$, is a possible product. Nickel sulfate, $NiSO_4$ is not.

Lesson 9

1. Three useful properties for sorting elements are reactivity, formulas of their compounds, and atomic mass.

5. a. CaS
   b. The compound with sulfur will have more mass for a given amount of calcium. The two compounds have the same number of atoms, but the atomic mass of sulfur is 32, and the atomic mass of oxygen is only 16.

Lesson 10

1. Within Group 1A, the elements tend to get more reactive as you move from the top of the column to the bottom.

5. (B) titanium   (C) lead   (E) potassium   (F) silicon

7. Elements copper and mercury are the least reactive. On the periodic table the least reactive elements (aside from the noble gases) are the transition metals that are located in the center of the table. The other elements listed are from more reactive groups near the edge of the table: alkali metals (potassium and rubidium), alkaline earth metals (barium), and halogens (chlorine).

Chapter 2 Review Exercises

1. Gold, represented by the symbol Au, is a transition metal that is a solid at room temperature. It has an atomic number equal to 79 and an average atomic mass of 196.97. Gold is nonreactive, a good conductor of heat, and a good conductor of electricity. It has properties similar to those of copper and silver.

3. Possible answer: A chemical formula is a symbol that represents a compound. The chemical formula shows what elements are in the compound and the ratio in which the elements combined. It can also show what physical form the compound is in.

Lesson 11

1. When Thomson zapped atoms with electricity, he found that a negatively charged particle was removed. Because the solid sphere model does not allow for particles splitting off atoms, he created the plum pudding model.

3. Bohr revised the nuclear model of the atom when he noticed different atoms giving off different colors of light when exposed to flame or electric fields. Because the nuclear model fails to account for this process, he created the solar system model.

5. Possible answer: The two types of atoms could have different amounts of positive fluid and different numbers of electrons. Each atom would still have a net charge of zero.

11. No, the Greeks were not correct. It is now known that atoms are made up of smaller particles such as electrons, protons, and neutrons. Each of these particles has a mass and a volume, so they are matter.

Lesson 12

1. The atomic number indicates the number of protons in the nucleus of an atom.

3. magnesium

5. The atomic mass is the sum of the number of protons and the number of neutrons in the nucleus of an atom. Although boron and carbon each have six neutrons, carbon has six protons while boron has only five protons.

Lesson 13

1. Atomic number refers to the number of protons in the nucleus of an atom. Atomic mass, when expressed in amu, is the sum of the number of protons and the number of neutrons.

3. Possible answer: The isotopes differ from one another in the number of neutrons in their nuclei. The isotopes are $^{39}_{19}K, ^{40}_{19}K, ^{41}_{19}K$.

5. a. 58 amu   **b.** $^{58}_{26}Fe$

7. 35.48 amu

9. **(B)** Because nitrogen has an atomic number of 7 on the periodic table, any possible isotopes must have an atomic number of 7.

Lesson 14

1. Possible answer: An element is a fundamental building block of matter. An atom is the smallest possible unit of an element. An atom is the smallest unit of an element that still has the same characteristics as the element.

3. Oxygen has three stable isotopes. Neodymium has five stable isotopes. Copper has two stable isotopes. Tin has ten stable isotopes.

5. The diagonal line on the graph represents isotopes that have equal numbers of protons and neutrons, because the line passes through points that have the same x- and y-coordinates.

7. No, because no isotope is indicated on the graph of isotopes of the first 95 elements that has 31 protons and 31 neutrons.

11. no

Lesson 15

1. A nuclear reaction is a change in the nucleus of an atom.

3. Possible answer: Gamma radiation is the most harmful because it has the most power to penetrate living tissues and cause damage.

5. The mass number of the atom does not change during beta decay because the nucleus loses one electron, which has a mass that is only a tiny fraction of the total mass of the nucleus.

7. a. calcium-42, $^{42}_{20}Ca$✓   **b.** xenon-131, $^{131}_{54}Xe$✓
   c. cobalt-52, $^{52}_{27}Co$ **d.** magnesium-24, $^{24}_{12}Mg$✓

Lesson 16

1. Possible answers: In alpha decay, a nucleus emits a particle consisting of two protons and two neutrons. The atomic number decreases by 2 and the atomic mass decreases by 4. In beta decay, a nucleus emits an electron. The atomic number increases by 1 as one of the neutrons becomes a proton. The atomic mass does not change. In nuclear fission, a nucleus splits apart to form the nuclei of two or more lighter elements. In nuclear fusion, two nuclei combine to form the nucleus of a heavier element.

3. $^{141}_{58}Ce \longrightarrow ^{0}_{-1}e + ^{141}_{59}Pr$

5. Possible answer: $^{50}_{24}Cr + ^{4}_{2}He \longrightarrow ^{54}_{26}Fe$

Chapter 3 Review Exercises

1. Possible answer: An atom changes identity when the number of protons in its nucleus changes. Processes in which this can occur are radioactive decay, fission, and fusion. In radioactive decay, emission of an alpha particle decreases the atomic number by 2, while emission of a beta particle increases the atomic number by 1. Nuclear fission is the process in which a nucleus breaks apart, forming two or more smaller nuclei. In nuclear fusion, two nuclei join to form one larger nucleus.

Lesson 17

1. The color of the flame produced during a flame test is a characteristic of particular metallic elements. When a compound containing one of the metallic elements is heated, its atoms emit light of a specific color.

5. Possible answer: Red fireworks could contain lithium chloride, and purple fireworks could contain a mixture of lithium chloride and copper chloride. Red fireworks could contain lithium sulfate, and purple fireworks could contain a mixture of lithium sulfate and copper sulfate.

7. No, the flame color of each nitrate compound is different and matches the flame color of the metal in the compound. This indicates that the nitrate is not responsible for the color of the flame.

9. **a.** yellow-orange **b.** green **c.** pink-lilac
   d. pink-lilac **e.** green

Lesson 18

1. For main group elements, the number of shells containing electrons is equal to the period number. All of the shells, except the highest, are completely filled. The group number indicates the number of electrons in the outermost shell.

3. Possible answer: Beryllium, magnesium, and calcium are all alkaline earth metals. They are located in Group 2A, so their atoms all have two valence electrons.

5. The number of core electrons does not change across a period.

7. **a.** Element number 17 is chlorine. It has the chemical symbol Cl and is located in Group 7A. This information comes directly from square 17 on the periodic table.
   b. The nucleus contains 17 protons. The number of protons is equal to the atomic number.
   c. Possible answer: The nucleus can contain either 18 or 20 neutrons. This information is given in the graph of isotopes of the first 95 elements.
   d. The number of electrons in a neutral atom of chlorine is 17. In a neutral atom, the number of electrons equals the number of protons.
   e. Chlorine has 7 valence electrons. The number of valence electrons is equal to the group number.
   f. Chlorine has 10 core electrons. The number of core electrons equals the difference between the total number of electrons and the number of valence electrons, or $17 - 7$.
   g. Possible answer (any three elements): Fluorine, bromine, iodine, and astatine all have the same number of valence electrons. All of these elements are in the same Group, 7A, as chlorine.

Lesson 19

1. A cation is an ion that has a positive charge. An anion is an ion that has a negative charge.

3. 2 electrons, 3 protons, and either 3 or 4 neutrons

7. The noble gas closest to sulfur is argon. A sulfur atom gains two electrons to have an electron arrangement similar to that of argon.

9. Possible answers: S^{2-}, Cl^-, K^+, Ca^{2+}

11. Ti^{4+}

13. Elements on the right side of the table gain electrons to have a noble gas arrangement. They do not tend to lose electrons because the charge would be too large.

Lesson 20

1. The number of valence electrons can predict whether an atom will form a cation or an anion, as well as the size of the charge on the ion. Ionic compounds form between cations and anions in a ratio so that the charges are balanced.

3. **a.** $+1$ **b.** -3
   c. Lithium nitride has three lithium ions with charge $+1$ and one nitride ion with charge -3.
   $3(+1) + (-3) = 3 + (-3) = 0$
   d. 8 valence electrons

5. **a.** KBr has one potassium ion with charge $+1$ and one bromide ion with charge -1.
   $+1 + (-1) = 0$
   b. CaO has one calcium ion with charge $+2$ and one oxide ion with charge -2.
   $+2 + (-2) = 0$
   c. Li_2O has two lithium ions with charge $+1$ and one oxide ion with charge -2.
   $2(+1) + (-2) = 2 + (-2) = 0$
   d. $CaCl_2$ has one calcium ion with charge $+2$ and two chloride ions with charge -1.
   $+2 + 2(-1) = 2 + (-2) = 0$
   e. $AlCl_3$ has one aluminum ion with charge $+3$ and three chloride ions with charge -1.
   $+3 + 3(-1) = 3 + (-3) = 0$

7. **a.** $NaCl_2$ does not form because it has a net charge of -1. The sodium ion has a charge of $+1$ and each chloride ion has a charge of -1.
   b. CaCl does not form because it has a net charge of $+1$. The calcium ion has a charge of $+2$ and the chloride ion has a charge of -1.
   c. AlO does not form because it has a net charge of $+1$. The aluminum ion has a charge of $+3$ and the oxide ion has a charge of -2.

Lesson 21

1. For main group elements, the group number shows the number of valence electrons. Metal atoms lose all of their valence electrons when they form an ion, adding a positive charge for each electron lost. Nonmetal atoms gain enough electrons to have eight valence electrons, adding one negative charge for each electron gained.

3. **a.** LiCl is possible because the total of the charges on the ions equals zero. The lithium ion has a charge of $+1$ and the chloride ion has a charge of -1.

b. $LiCl_2$ is not possible because the total of the charges on the ions equals -1. **c.** MgCl is not possible because the total of the charges on the ions equals $+1$. Magnesium ions have a charge of $+2$. **d.** $MgCl_2$ is possible because the total of the charges on the ions equals zero. **e.** $AlCl_3$ is possible because the total of the charges on the ions equals zero. Aluminum ions have a charge of $+3$.

7. **a.** $AlBr_3$, aluminum bromide
   b. Al_2S_3, aluminum sulfide
   c. AlAs, aluminum arsenide
   d. Na_2S, sodium sulfide
   e. CaS, calcium sulfide
   f. Ga_2S_3, gallium sulfide

Lesson 22

1. A polyatomic ion is an ion that consists of two or more elements.

3. **a.** ammonium chloride **b.** potassium sulfate
   c. aluminum hydroxide **d.** magnesium carbonate

5. -1

Lesson 23

1. The Roman numeral indicates the charge on the transition metal cation in the compound.

3. **a.** $+2$, mercury (II) sulfide **b.** $+2$, copper (II) carbonate
   c. $+2$, nickel (II) chloride **d.** $+3$, cobalt (III) nitrate
   e. $+2$, copper (II) hydroxide **f.** $+2$ iron (II) sulfate

5. $Co_3(PO_4)_2$

Lesson 24

1. Electron subshells are divisions within a specific electron shell of an atom.

3. As the number of electrons in an element increases, they are added in a specific sequence that is illustrated by the position of the element on the periodic table. Each section of the table corresponds to a particular subshell of electrons.

5. 15 subshells

7. $4p$

9. **a.** $1s^2 2s^2 2p^6 3s^2 3p^1$ **b.** Element 13 has 3 valence electrons because it has three electrons in the outer shell, $n = 3$. **c.** Element 13 has 10 core electrons because it has two shells filled completely: $n = 1$ with 2 electrons and $n = 2$ with 8 electrons.

11. **a.** $1s^2 2s^2 2p^4$, [He] $2s^2 2p^4$
    b. $1s^2 2s^2 2p^6 3s^2 3p^5$, [Ne] $3s^2 3p^5$
    c. $1s^2 2s^2 2p^6 3s^2 3p^6 4s^2 3d^6$, [Ar] $4s^2 3d^6$
    d. $1s^2 2s^2 2p^6 3s^2 3p^6 4s^2$, [Ar] $4s^2$

    e. $1s^2 2s^2 2p^6 3s^2$, [Ne] $3s^2$
    f. $1s^2 2s^2 2p^6 3s^2 3p^6 4s^2 3d^{10} 4p^6 5s^2 4d^9$, [Kr] $5s^2 4d^9$
    g. $1s^2 2s^2 2p^6 3s^2 3p^2$, [Ne] $3s^2 3p^2$
    h. $1s^2 2s^2 2p^6 3s^2 3p^6 4s^2 3d^{10} 4p^6 5s^2 4d^{10} 5p^6 6s^2 4f^{14} 5d^{10}$, [Xe] $6s^2 4f^{14} 5d^{10}$

13. **a.** chromium **b.** silicon **c.** nitrogen **d.** cesium
    e. lead **f.** silver

Chapter 4 Review Exercises

1. Valence electrons are important because they are in the outermost electron shell of an atom and will interact with other atoms. This interaction is what determines the properties of an element.

3. As the number of electrons in an element increases, they are added in a specific sequence that is illustrated by the position of the element on the periodic table. Each section of the table corresponds to a particular subshell of electrons.

5. **a.** The anion is Mg^{2+} and has a charge of $+2$. The cation is Cl^- and has a charge of -1.
   b. The anion is Ca^{2+} and has a charge of $+2$. The cation is NO_2^- and has a charge of -1.

Lesson 25

1. A substance is insoluble if it fails to dissolve in a particular solvent.

3. Possible answer: The substance is most likely an ionic compound. Many compounds dissolve in water, but electrical conductivity is a characteristic of compounds that separate into ions, such as ionic compounds.

5. No, though ionic compounds generally do not conduct electricity as solids, they do conduct electricity as aqueous solutions.

Lesson 26

1. The atoms that make up substances are held together by chemical bonds. The bond is an attraction between the positively charged nuclei of atoms and the valence electrons of other atoms.

3. **a.** metallic **b.** molecular covalent **c.** ionic

5. NO_2 is a gas that is made up of nonmetal atoms, so it has molecular covalent bonds.

7. Possible answer: Drop the mixture into a beaker of water. The sodium chloride will dissolve in the water, but the carbon will not.

9. Carbon is a solid because many carbon atoms are held together in a large array by network covalent bonds.

Lesson 27

1. Possible answers: finding pure metals in nature, heating ionic compounds to separate the metal, extracting metals with electricity.

3. Attach the coated object to the positive terminal of a battery in an electroplating circuit.

5. The copper sulfate solution is composed of cations (Cu^{2+}) and anions (SO_4^-) dissolved in water. The only thing that is added to the solution during the experiment is a stream of electrons. When the electrons are added to the copper ions, copper atoms (Cu) are formed on the metal strip. This indicates that the Cu^{2+} ions are simply copper ions that are missing electrons.

7. Possible answer: Nickel cannot change into copper unless the number of protons in the nucleus changes. The plating apparatus only adds electrons to the nickel strip, causing the plating to occur. Adding electrons does not change the nucleus, so it cannot change the identity of the atoms.

Chapter 5 Review Exercises

1. Possible answer: While it is not practical to try to make gold, many substances that are quite valuable can be made through chemistry.

Unit 1 Review Exercises

General Review

1. Possible answer: An element is the basic building block of compounds. An element has only one type of atom, while a compound has at least two types of atoms held together by chemical bonds.

3. **a.** Lithium, Li, has atomic number 2 and is in Group 2A. It has 2 protons and 2 electrons. **b.** Bromine, Br, has atomic number 35 and is in Group 7A. It has 35 protons and 35 electrons. **c.** Zinc, Zn, has atomic number 30 and is in Group 2B. It has 30 protons and 30 electrons. **d.** Sulfur, S, has atomic number 16 and is in Group 6A. It has 16 protons and 16 electrons. **e.** Barium, Ba, has atomic number 56 and is in Group 2A. It has 56 protons and 56 electrons. **f.** Carbon, C, has atomic number 6 and is in Group 4A. It has 6 protons and 6 electrons.

5. An isotope is an atom of an element with a specific number of neutrons in its nucleus. Predict the most common isotope by rounding the average atomic mass of the element to the nearest whole number.

7. Cations are ions that have lost electrons, causing them to have a positive charge. Anions are ions that have gained electrons, causing them to have a negative charge.

9. **a.** Magnesium chloride has ionic bonding and conducts electricity in solution only. **b.** Oxygen has molecular covalent bonding and does not conduct electricity. **c.** Silver (I) hydroxide has ionic bonding and conducts electricity in solution only. **d.** Platinum has metallic bonding and conducts electricity.

11. A material that does not dissolve in water and does not conduct electricity is held together by network covalent bonds or molecular covalent bonds.

Standardized Test Preparation

1. C	3. C	5. B
7. A	9. D	11. C
13. D	15. A	17. B
19. B, C	21. B	

Unit 2: Smells

Lesson 28

1. Possible answer: Scientists classify smells by placing similar types of smells in a category. This allows scientists to talk about smells in a consistent way.

7. **a.** Methyl octenoate probably smells sweet because it has two oxygen atoms and twice as many hydrogen atoms as carbon atoms in each molecule. The chemical name also ends in "-*ate*." **b.** Monoethylamine probably smells fishy because it has one nitrogen atom and no oxygen atoms in each molecule. **c.** Ethyl acetate probably smells sweet because it has two oxygen atoms and twice as many hydrogen atoms as carbon atoms in each molecule. The chemical name also ends in "-*ate*."

Lesson 29

1. Structural formulas show what atoms are present in a molecule and how the atoms are bonded to one another.

3. Yes, the structural formula shows all of the atoms in a molecule, enabling use of the structural formula to determine the molecular formula.

5. **a.** $C_3H_6O_2$ **b.** $C_5H_{10}O$ **c.** $C_4H_{10}O$ **d.** $C_4H_8O_2$ **e.** $C_4H_{11}N$ **f.** $C_2H_4O_2$

Lesson 30

1. The HONC 1234 rule describes the bonding patterns in molecules. Hydrogen forms one bond with other atoms. Oxygen forms two bonds with other atoms.

Nitrogen forms three bonds with other atoms. Carbon forms four bonds with other atoms.

3. Possible answers:

a.
$$H-O-\overset{\overset{\displaystyle H}{|}}{C}-\overset{\overset{\displaystyle H}{|}}{C}-\overset{\overset{\displaystyle H}{|}}{C}-O-H$$

b.

c.

d.

5. Possible answer: Atoms on the molecules might react with smell receptors inside the nose. Different receptors may detect different types of molecular structures, leading to different sensations of smell.

Lesson 31

1. Possible answer: In an ionic bond, a valence electron is transferred from one metal atom to a nonmetal atom. In a covalent bond, two nonmetal atoms share a pair of electrons. Ionic and covalent bonds are similar because both involve the valence electrons of the atoms that are bonded together. The result of the bond is that the two atoms have an outer shell that is filled. In both types of bonds, electrical forces hold the atoms together.

3. a.

K·	In·	·Pb·	·Bi·	·Te:	:I:
Group I	3	4	5	6	7

b. two covalent bonds: Te, one covalent bond: I, does not form covalent bonds: K, In, Pb, Bi

5. a. :Te:Cl: b. :I:H c. :As:Br: (with :Br: above and :Br: below)

d. :F:Si:F: (with :F: above and :F: below) e. :F:F:

Lesson 32 Exercises

1. Possible answer: Nitrogen has three unpaired electrons, as shown in the Lewis dot structure. Hydrogen atoms have one unpaired electron, so three hydrogen atoms can form bonds with one nitrogen, giving the nitrogen an octet of electrons and each hydrogen two electrons. In NH_2, nitrogen would have only seven electrons, and in NH_4, an extra hydrogen atom is left over after all the unpaired electrons in the nitrogen atom have formed bonds.

·N·   ·N:H   H:N:H   H:N:H· (with H above and H below)

3. Possible answer: hydrogen, chlorine, or another fluorine

5. a. :F:C:F: (with :F: above and :F: below) b. H:C:Cl: (with H above and H below) c. :Cl:Si:Cl: (with H above and H below)

d. H:C:O:H (with H above and H below) e. H:O:Cl: f. H:C:N: (with H H above and H H below)

9. CH_4 would form a stable compound because all of the atoms in the molecule are surrounded by the most stable number of valence electrons—eight for carbon, two for hydrogen. In CH_3, the carbon atom would have only seven electrons in its outer shell.

Lesson 33

1. A functional group is a portion of a molecular structure that is the same in all molecules of a certain type.

3. C_2H_4 has fewer hydrogen atoms than C_2H_6 because the two carbon atoms are held together by a double bond. That means there are only two unpaired electrons in each carbon atom available to form bonds with hydrogen.

5. a. $C_6H_{14}O$

b.

c. The structural formula is more useful because it shows the −OH functional group that is characteristic of alcohols, and alcohols have a characteristic smell.

11. a. $C_3H_6O_2$ b. $C_5H_{10}O$ c. $C_5H_{10}O$
    d. $C_4H_8O_2$ e. $C_4H_{11}N$ f. $C_2H_4O_2$

Lesson 34

1. Combine the alcohol and acid with a strong acid and heat the mixture.

3. Possible answer: When butyric acid is heated with methanol, the two compounds react to form a new compound, which is a sweet-smelling ester. The butyric acid, which has a foul smell, is no longer present.

5. Possible answer: Combine the acetic acid with an alcohol and sulfuric acid and then heat the mixture. The hydrogen atom of the acid is replaced by the part of the alcohol molecule that is attached to the $-OH$ functional group.

Lesson 35

1. Possible answer: Converting a carboxylic acid into an ester is a way to make a new compound with more desirable properties. For example, eliminating a putrid odor is one possible goal of converting a carboxylic acid into an ester.

3. A catalyst is a chemical that is added to a reaction mixture to help get the reaction started, but is not consumed by the reaction.

5. **a.** $C_8H_{16}O_2$ **b.** $C_7H_{14}O_2$ **c.** $C_9H_{18}O_2$ **d.** $C_6H_{12}O_2$

Chapter 6 Review Exercises

1. **a.** ester **b.** ketone **c.** carboxyl **d.** amine

3. **a.** 4 lone pairs **b.** 12 lone pairs **c.** no lone pairs

5. Possible answer: The smell of a compound is strongly related to the functional group of the molecule.

Lesson 36

1. Possible answer: A structural formula shows the arrangement of atoms and the types of bonds within a molecule in a two-dimensional form. A ball-and-stick model adds information about the arrangement of the atoms in three dimensions.

3. 10 black balls (carbon), 18 white balls (hydrogen), 1 red ball (oxygen), 30 connectors (chemical bonds)

7. Possible answer: The three molecules geraniol, menthol, and fenchol each have ester functional groups in their structural formulas. However, each of the three molecules has a distinctive smell.

Lesson 37

1. A tetrahedral shape has four single bonds spaced equally around one central atom.

5. Possible answer: CH_4, $SiCl_4$, CF_4

7. AsH_3 will have a pyramidal shape because the three bonded pairs and the one lone pair on the arsenic atom

form a tetrahedron. Therefore, the three single bonds form a pyramid shape with the arsenic atom at its top.

9.

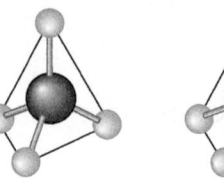 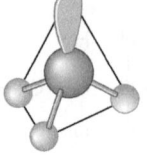 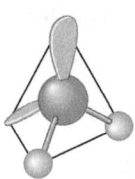

Methane   Ammonia   Water

Lesson 38

1. If a molecule has three electron domains, its shape will be trigonal planar.

3. Cl_2 and CO_2 are linear and H_2O is bent.

5. A molecule with two atoms will be linear. A molecule with three atoms can be either linear or bent. A molecule with four atoms can be linear, bent, trigonal planar, or pyramid-shaped. A molecule with five atoms can be linear, bent, tetrahedral, or trigonal planar or pyramidal with the fifth molecule attached to one of the triangle's or pyramid's vertices.

7. **a.**
$$\begin{array}{c} \text{H} \quad \text{H} \quad \text{H} \quad \text{H} \\ \text{H}:\ddot{\text{C}}:\ddot{\text{C}}:\ddot{\text{C}}:\ddot{\text{C}}:\text{H} \\ \text{H} \quad \text{H} \quad \text{H} \quad \text{H} \end{array}$$
b. 13

   c. Possible answer: a crooked, zigzag shape

   d. Each carbon atom is surrounded by four electron domains. Therefore, the shape of the molecule is a series of linked tetrahedra.

Lesson 39

1. A space-filling model is a more accurate model of how the atoms of the molecule are arranged in space. Space-filling models give a better picture of the shape of the molecule than ball-and-stick models.

3. Possible answer:

$$\begin{array}{c} \quad\text{H} \ \ \text{H} \ \ \text{H} \ \ \text{H} \ \ \text{H} \ \ \text{H} \ \ \text{H} \ \ \text{H} \ \ \text{H} \\ \quad | \ \ \ | \ \ \ | \ \ \ | \ \ \ | \ \ \ | \ \ \ | \ \ \ | \ \ \ | \\ \text{H}-\text{C}-\text{C}-\text{C}-\text{C}-\text{C}-\text{C}-\text{C}-\text{C}-\text{C}-\text{O}-\text{H} \\ \quad | \ \ \ | \ \ \ | \ \ \ | \ \ \ | \ \ \ | \ \ \ | \ \ \ | \ \ \ | \\ \quad\text{H} \ \ \text{H} \ \ \text{H} \ \ \text{H} \ \ \text{H} \ \ \text{H} \ \ \text{H} \ \ \text{H} \ \ \text{H} \end{array}$$

   $C_9H_{20}O$ is long and stringy, with a zigzag shape.

5. Possible answer:

$C_{10}H_{16}O$ has a ring-shaped structure of carbon atoms with a "handle." The overall shape of the molecule resembles a frying pan.

Lesson 40

1. The molecular formula of a compound can provide some information about its smell but is not useful in most cases because it fails to give information about functional groups. For alkanes and amines, the chemical formula is a good indicator of smell.

3. Possible answer: The minimum information required to determine that a molecule smells sweet is the shape and the functional group or, if it is an ester, just the functional group.

Lesson 41

1. According to the receptor site theory, the nose is lined with sites that match the shapes of molecules that have smells. When a molecule fits into this site, it stimulates nerves to send a message to the brain. The brain interprets the message as a smell.

3. Possible answer: The compound would most likely have a minty smell. The nose detects the minty smell when the molecule shown fits into a receptor site in the nose that detects minty smells. In order for this to happen, some of the rub must change phase and become a gas.

Chapter 7 Review Exercises

1. **a.** $C_5H_{10}O_2$

   b.

   c. ester
   d. The molecule probably has a sweet smell.

3. The shape of a molecule is determined by the number of electron domains around its atoms. The electron domains are all negatively charged, so they are positioned as far apart as possible.

5. Possible answer: Although the functional group appears to be a main factor that determines the smell of a molecule, its shape also affects the smell because it helps determine which receptors will be stimulated to send a signal to the brain.

Lesson 42

1. Possible answer: A polar molecule is a molecule in which the charge is not evenly distributed around the molecule. This means that different portions of the molecule will have partial electric charges.

5. Hexane would not be expected to dissolve in water because the information given indicates that it is a nonpolar compound. Compounds with nonpolar molecules do not tend to dissolve in water.

Lesson 43

1. Possible answer: In a polar covalent bond, the electron is attracted more by one atom than by the other, so one of the atoms has a partial negative charge and the other has a partial positive charge. In a nonpolar covalent bond, in which the two atoms share the electrons equally, there are no partial charges.

3. Although the two carbon-oxygen bonds of carbon dioxide are polar, the molecule itself has a linear shape, so the partial negative charges are on opposite sides of the molecule. These charges balance one another, so there is no dipole.

Lesson 44

1. Electronegativity values help determine the polarity of the bond between two atoms because they can be used to determine the tendency of an electron to be attracted to one atom rather than to another atom. The greater the difference in the electronegativities of the two atoms, the more polar the bond that forms between the atoms.

3. **a.** Li—F, Na—F, K—F, and Rb—F (same polarity), Cs—F; For an alkali metal bonding with fluorine, the polarity increases from the top of the group to the bottom.

   b. P—S, N—F, Al—N, Mg—O, K—Cl; The polarity of the molecules increases as the distance between the atoms on the periodic table increases.

5. No, hydrogen will have a partial positive charge only when it is bonded to an atom that has a greater electronegativity. If the hydrogen atom is bonded to an atom with a smaller electronegativity, such as boron, then the hydrogen atom will have a partial negative charge. If a hydrogen atom is bonded with another hydrogen atom, the bond will be nonpolar.

7. Possible answer: O—F, 0.54; C—H, 0.45; S—F, 1.40

9. To say that bonding is on a continuum means that the type of bonding changes gradually as the difference in electronegativity between atoms increases. There is no sharp distinction between polar covalent and ionic bonds.

Lesson 45

1. To determine whether a molecule is polar, use a Lewis dot structure or structural diagram to figure out the shape of the molecule. Then look at whether the

individual bonds are polar. If the molecule is asymmetrical and has polar bonds, then it will be a polar molecule.

3. a.

$$: \overset{\cdot\cdot}{\underset{\cdot\cdot}{Se}} : H$$
(H above Se)

H_2Se is a bent molecule. Because its bonds are polar and it is asymmetrical, it is likely to have a smell.

b. $H : H$

H_2 is a linear molecule. Its bond is nonpolar. Because it is symmetrical and has a nonpolar bond, it is not likely to have a smell.

c. $: \overset{\cdot\cdot}{\underset{\cdot\cdot}{Ar}} :$

Because Ar consists of a single atom, it is not polar and is not likely to have a smell.

d.

$$: \overset{\cdot\cdot}{\underset{\cdot\cdot}{F}} :$$
$$: \overset{\cdot\cdot}{O} : H$$

HOF is a bent molecule. Its bonds are polar and it is asymmetrical, so it is likely to have a smell.

e.

$$: \overset{\cdot\cdot}{\underset{\cdot\cdot}{F}} : \overset{H}{\underset{\cdot\cdot}{C}} : \overset{\cdot\cdot}{\underset{\cdot\cdot}{Cl}} :$$
$$: \overset{\cdot\cdot}{\underset{\cdot\cdot}{F}} :$$

$CHClF_2$ is a tetrahedral molecule and it has polar bonds. Its shape is symmetrical, but because the atoms at each point in the tetrahedron are different, the bonds are not symmetrical. It is likely to have a smell.

f.

$$\overset{H}{\underset{H}{\,}} \overset{\cdot\cdot}{C} :: \overset{\cdot\cdot}{\underset{\cdot\cdot}{O}} :$$

CH_2O is a trigonal planar molecule. Its bonds are polar and it is asymmetrical, so it is likely to have a smell.

5. Possible answer: Water does not have a smell because there are no receptors in the nose that are sensitive to water molecules. If the nose had receptor sites that were sensitive to water, they would always be filled by the water in the mucous membrane.

Lesson 46

3. Decanol has a smell because it is a medium-sized molecular compound and because it is polar. Lead does not have a smell because it is a metal. Iron oxide does not have a smell because it is an ionic compound. Potassium chloride does not have a smell because it is an ionic compound.

5. Possible answer: If a substance can become a gas under normal conditions, you should be able to smell it as long as receptor sites in the nose can detect it. In general, small nonpolar molecules are odorless.

7. Possible answer: When the T-shirt comes out of the clothes dryer, it is so warm that molecules from the detergent and fabric softener are likely to have changed phase and become gases. When they reach your nose, you are able to smell them.

Chapter 8 Review Exercises

1. a. Br   b. Li   c. Au

3. a. $\overset{\delta+}{H} - Cl \,^{\delta-}$

b.

$$\overset{\delta+}{H}$$
$$\overset{\delta+}{H} - \overset{\delta-}{C} \overset{\delta+}{-H}$$
$$\overset{\delta+}{H}$$

c.

$$\overset{\delta-}{O} \overset{\overset{\delta+}{H}}{\underset{\overset{\delta+}{H}}{\,}}$$

HCl and H_2O are polar molecules because they are asymmetrical and have polar bonds. CH_4 is not a polar molecule because its polar bonds are symmetrically arranged around the carbon atom.

Lesson 47

1. The mirror image of the letter "D" can be superimposed on the original as long as the mirror is placed horizontally with respect to the letter. Then the mirror image and the original image are identical.

3. Examples of objects that look different in a mirror include a glove, a written sentence, a pair of scissors. Examples of objects that look the same in a mirror include a spoon, an empty glass, and a pencil with no writing on it.

9. a. citronellol: $C_{10}H_{20}O$; geraniol: $C_{10}H_{18}O$
   b. Citronellol has two distinct smells because the structural formulas have mirror-image isomers that interact with different receptors in the nose. Citronellol has mirror-image isomers because the third carbon atom from the right in the structural formula is bonded to four different groups.
   c. The geraniol molecule has only one smell because the molecule and its mirror image are identical. None of the carbon atoms in the structure are bonded to four different groups.

Lesson 48

1. Amino acids are the building materials for many of the structures in the body, especially those related to the functioning of cells. An amino acid has a carbon backbone and two functional groups, an amine group and a carboxylic acid group.

Chapter 9 Review Exercises

1. Possible answer: Two molecular models represent mirror-image isomers if they are mirror images of each other and one cannot be rotated in space to be superimposed on top of the other. When the models have a carbon atom that is attached to four different groups, then they will be mirror-image isomers.

3. Possible answer: A molecule fits into its receptor site only when the receptor has the same "handedness" as the molecule. This is similar to the way that each of your feet will fit into only one shoe of a pair of shoes, even though the shoes are mirror images of one another.

Unit 2 Review Exercises

General Review

1. Isomers are molecules that have the same molecular formula but different structural formulas.

3. A molecular formula provides the elements and number of atoms per element in a compound. A structural formula shows the way in which these atoms are bound to each other. A ball-and-stick model provides a three-dimensional image of the bonds. A space-filling model shows the actual amount of space each atom takes up, with atoms that are sharing electrons overlapped in space.

5. **a.** yes **b.** $C_8H_8O_3$
   c. The molecule has a flat section with the ester functional group attached to its side, similar to the shape of a frying pan.
   d. The bond is bent because there are four electron domains. The oxygen atom has two bonded pairs and two lone pairs. The shape that separates the four domains as much as possible is tetrahedral, which causes the two bonds to be bent.

Standardized Test Preparation

1. D	3. B	5. D
7. C	9. C	11. C
13. C	15. D	17. B
19. B		

Unit 3: Weather

Lesson 49

1. Possible answer: To have weather, a planet must have a layer of gases surrounding its surface.

3. A physical change is a change in the form of a substance that does not change the identity of the substance.

Lesson 50

1. Possible answer: One way to determine the volume of 3 cm of water in the rain gauge is to use the graph. Find 3 cm on the x-axis, move up to the line on the graph, and then read the corresponding y-value. A second way is to use a proportionality constant. Examining the data in the table shows that the value of the volume is always twice the value of the height.

3. Possible answer: Both the volume and height of water in the rain gauge will keep track of the increase because they both increase by the same proportion as additional rain falls into the gauge.

5. Possible answer: The three containers would have different volumes of water in them after the storm because their bases have different areas. The greatest volume of water will be in the washtub and the least will be in the graduated cylinder. However, the height of water in each container will be the same.

Lesson 51

1. You can convert the volume of snow in a snowpack to a volume of water by using the mathematical equation for density, $D = \frac{m}{V}$. If you know the volume and density of snow, you can determine the mass of the snow in the snowpack. If the snow melts, the mass of the water will be the same as the mass of the snow. Then you can calculate the volume of water by dividing the mass by the density of water, which is equal to 1 g/mL.

7. **a.** 16.1 g **b.** 17.5 g

9. **a.** 4.3 g of iron occupies a greater volume because volume increases as density decreases, and iron is less dense than lead.
   b. 2.6 mL of lead has a larger mass because mass increases as density increases, and lead is more dense than iron.

Lesson 52

1. Possible answer: Liquids usually expand when they are heated and contract when they are cooled. At any given temperature the volume of a fixed amount of liquid will always be the same. If the liquid is in a tube, the height of the column of liquid will change in a predictable way with changes in temperature.

5. $-40\,°F$

7. Normal body temperature is about 98.6 °F. A body temperature of 40 °C is the same as 104 °F, which indicates fever and probable illness.

9. a.

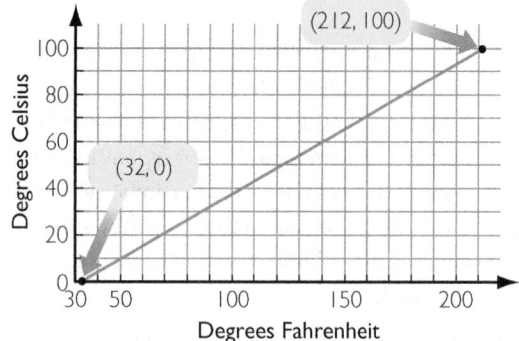

b. 50 °F   **c.** 13 °C

Lesson 53

1. Possible answer: Absolute zero is the temperature at which the volume of a gas would be equal to zero. It is considered a theoretical temperature because a real gas would condense into a liquid and then a solid before reaching absolute zero, making a volume of zero impossible.

3. Possible answer: The kinetic theory of gases defines the temperature of a gas as the average kinetic energy of the particles of the gas.

5. The smallest unit is 1 °F because there are 180 Fahrenheit degrees between the freezing temperature of water and the boiling temperature of water, while there are only 100 Celsius degrees or kelvins between the freezing temperature of water and the boiling temperature of water.

7. **a.** −173 °C   **b.** 333 K   **c.** −23 °C   **d.** 298 K
   e. 27 °C   **f.** 173 K   **g.** 127 °C

9. **a.** 308 K   **b.** 450 K   **c.** 258 K

11. B

Lesson 54

1. Determine the proportionality constant, k, for a gas by dividing the volume of the gas by its temperature measured on the Kelvin temperature scale.

3. 689 mL

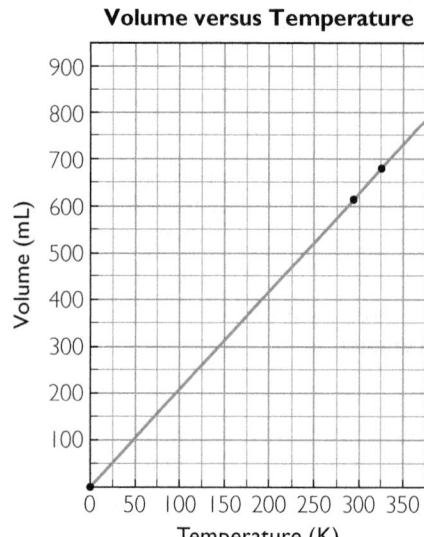

The answer is reasonable because 325 K corresponds to about 700 mL.

5. 186 K

Lesson 55

1. When air gets warmer, its volume increases and its density decreases. Air that is less dense than the surrounding air will rise.

3. A

Chapter 10 Review Exercises

1. 283 K

3. 760 mL

5. **a.** 3.3 mL **b.** 2700 mL. The volume of gas can be calculated using the density equation. The mass of the carbon dioxide does not change when it changes phase and becomes a gas.

Lesson 56

1. The density of a gas is much lower than the density of a solid.

3. Possible answer: Sublimation occurs when a substance changes from the solid phase directly to the gas phase without becoming a liquid. Evaporation occurs when a substance changes from the liquid phase to the gas phase.

5. Possible answer: Ice floats in water so it must have a slightly lower density than water.

7. 4.1 mL

9. 12 g

Lesson 57

3. Because the air inside the balloon exerts pressure on the inside wall but there is no pressure balancing it on the outside wall, the balloon will expand and most likely pop.

Lesson 58

1. As you push down on the plunger of the syringe, the gas pressure inside the syringe increases and the pressure on the scale increases. The pressure inside the syringe is equal to the pressure you are exerting on the plunger, which can be read from the scale. To calculate the pressure of the gas, divide the weight on the scale by the surface area of the plunger. Then add atmospheric pressure.

5. Possible answer: The relationship between gas pressure and gas volume is different from the relationship

between gas volume and gas temperature because an increase in pressure causes the volume to decrease, while an increase in temperature causes the volume to increase.

Lesson 59

1. Possible answer: You can change the pressure, the volume, or the temperature of a gas sample.

3. 720 L

5. **a.** The volume of air inside the bottle stays the same because the bottle is a rigid container.
   b. Yes, the air inside the bottle will lose energy until it has the same temperature as its surroundings because glass does not insulate against temperature changes.
   c. Pressure and temperature are directly proportional. The pressure of the air inside the bottle will decrease as its temperature decreases.
   d. 0.93 atm

Lesson 60

1. According to the kinetic theory of gases, increasing the gas volume decreases the gas pressure because the molecules of gas have more room in which to travel and therefore they strike the walls of the container less often.

3. Possible answer: An increase in temperature probably caused the change in pressure because pressure and temperature are proportional and the volume of a gas cylinder doesn't change.

5. The volume of a gas cannot decrease to zero because the particles of the gas themselves occupy space.

7. **a.** The volume of your lungs at 3.5 atm is much lower than the volume at sea level.
   b. Holding your breath when you ascend quickly is dangerous because the pressure decreases rapidly as you rise. If you hold your breath, your lungs will act as sealed containers, making the pressure inside the lungs much greater than the pressure outside the lungs, possibly causing your lungs to rupture.

Lesson 61

1. The combined gas law is a mathematical equation that relates gas pressure, temperature, and volume. This law applies when all three variables can change at the same time.

3. **a.** The volume of the gas will decrease because pressure and volume are inversely proportional when the temperature and the amount of gas remain unchanged.
   b. $PV = k$ **c.** 0.49 L

5. **a.** The volume of the balloon will increase because, based on the values given, the change in pressure has a greater effect on the final volume than the change in temperature.
   b. $\dfrac{PV}{T} = k$ **c.** 1.6 L

Lesson 62

1. Possible answer: High-pressure areas are generally associated with clear skies. Low-pressure areas are generally associated with cloudy weather and precipitation.

Chapter 11 Review Exercises

1. The pressure of a gas is proportional to the temperature of the gas. As a gas becomes warmer, its molecules move faster and collide with each other and other objects more frequently.

3. 0.80 atm

5. 2.0 atm

Lesson 63

1. Possible answer: Air pressure decreases with increasing altitude because the number of molecules of air in a given volume decreases. This means that there are fewer collisions of air molecules with one another and with other objects, and therefore the air exerts less pressure.

5. Possible answers: Increase the amount of air in the tire. Increase the temperature of the air in the tire. Press on the outside of the tire to decrease its volume.

7. When a high-pressure system moves into a region, the air pressure increases. When the air pressure on the open end of the barometer increases, it pushes on the mercury harder than it did before. Mercury is pushed higher into the closed end of the barometer until the downward pressure caused by the weight of the column of mercury is equal to the atmospheric pressure.

Lesson 64

1. Possible answer: Chemists invented the mole so as to have a sufficiently large unit with which to count the enormous number of particles in a sample of gas.

3. **a.** Each gas sample contains 1 mol of atoms.
   b. 0.0446 mol/L **c.** The xenon sample has the largest mass because the mass of each xenon atom is greater than the mass of each neon atom or argon atom.
   d. the xenon sample

5. 8.0 g of helium

7. The two balloons hold the same volume. The volumes of the balloons are the same because the number of moles of He is equal to the number of moles of O_2.

Lesson 65

1. The ideal gas law is an equation that relates the variables pressure, volume, number of moles, and temperature for any sample of any gas. The equation is $PV = nRT$.

3. 0.13 mol

5. 3.90 atm

7. Possible answers: Reduce the number of particles to 0.5 mol. Reduce the temperature to 137 K. Increase the volume to 44.8 L.

Lesson 66

1. Humidity is a measure of the number density of molecules of water vapor in the air.

3. about 1.7 mol per 1000 L

5. 2.2 mol per 1000 L

7. 0.43 mol per 1000 L

11. a. No, if the amount of water vapor in the air remains constant, it will still be less than the maximum possible vapor.
  b. Yes, the amount of water vapor in the air cannot remain constant, because the maximum possible vapor at 20 °C is less than the original amount of water vapor. Therefore, some water vapor will condense and form fog.

Lesson 67

1. Possible answer: For a hurricane to form, the temperature at the ocean's surface must be at least 80 °F and the air must contain a lot of moisture.

Chapter 12 Review Exercises

3. D

5. 68%

Unit 3 Review Exercises

General Review

1. The ice has a lower mass because it has a lower density than liquid water. The density of water is 1.0 g/mL. Mass can be calculated by multiplying the density by the volume, and because the samples have the same volume, the ice has less mass.

3. According to the kinetic theory of gases, gas particles move faster when they have more energy. When a gas is heated, its particles gain energy, move faster, and

scatter farther apart, causing the gas to expand. When the gas is cooled, its particles lose energy, causing them to move more slowly and remain closer together, so the gas contracts.

7. 2.05 atm

9. a. 273 K, 1.0 atm   **b.** 1 mol
  c. 2 mol   **d.** 0.0446 mol/L

11. 12 mol

Standardized Test Preparation

1.	A	**3.**	D	**5.**	B
7.	D	**9.**	C	**11.**	A
13.	D	**15.**	A	**17.**	C
19.	B				

Unit 4: Toxins

Lesson 68

1. Reactants are the substances that are present before a chemical reaction, and products are the substances that are formed during the reaction.

3. Possible answer: Toxic substances cause harm to living organisms.

5. a. Solid mercury chloride is added to a solution of EDTA to produce a solution containing a compound of mercury with EDTA and hydrochloric acid.
  b. If the two reactants are mixed, the solid will disappear and a mixture of two liquids remains.

Lesson 69

1. Possible answer: When sugar melts, it changes from the solid phase to the liquid phase. When sugar dissolves in water, the sugar molecules spread throughout the water.

3. a. The solid magnesium will react and disappear into an aqueous compound, and gas bubbles will appear.
  b. Gas bubbles will appear in the solution.
  c. A solid will form when the two solutions are mixed.

5. In both cases, the mixture of two clear solutions produces gas bubbling out of the resulting solution.

Lesson 70

1. During a physical change, substances change form but new chemical substances are not made. During a chemical change, chemical substances are changed into other chemical substances.

3. In each case, the material changes its appearance or breaks apart, but it does not change identity. Examples include melting ice, breaking glass, grinding pepper, dissolving sugar, and bending metal.

5. 1: chemical change; 2: physical change; 3: chemical change; 4: chemical change; 5: physical change; 6: chemical change; 7: chemical change

Lesson 71

1. Possible answer: The law of conservation of mass states that matter is not created or destroyed during a chemical reaction. The atoms involved in the reaction are rearranged but are not created or destroyed.

Lesson 72

1. Possible answer: You need to balance chemical equations to show how many molecules, atoms, or formula units of each substance take part in the reaction or are produced by the reaction.

Lesson 73

1. Combination reactions and decomposition reactions are opposites because combination reactions put two or more atoms or compounds together to make a new compound, while decomposition reactions break a compound into two or more atoms or compounds.

3. a. double exchange reaction
$NaOH(aq) + HNO_3(aq) \longrightarrow$
$\qquad\qquad NaNO_3(aq) + H_2O(l)$
b. combination reaction
$C_2H_4(g) + Cl_2(g) \longrightarrow C_2H_4Cl_2(g)$
c. single exchange reaction
$Cl_2(g) + MgBr_2(s) \longrightarrow Br_2(s) + MgCl_2(s)$

5. $SO_3(g) + H_2O(l) \longrightarrow H_2SO_4(aq)$
Because sulfur trioxide is a gas, it could enter the body through the nose and mouth and travel to the lungs. Sulfuric acid is an aqueous liquid that could enter the body by ingestion or through the skin.

Chapter 13 Review Exercises

1. Possible answer: Identifying physical or chemical changes through observations alone is difficult because often no change occurs that can be detected by sight or smell.

3. a. Each reactant has ionic bonds because they both consist of metal atoms bonded to nonmetal atoms or polyatomic ions. **b.** double exchange reaction **c.** chemical change **d.** A balanced reaction shows that matter is conserved because all of the atoms in the reactants are accounted for in the products. **e.** calcium phosphate

Lesson 74

1. Possible answer: Toxicities of most substances are measured by exposing laboratory animals to the substance in different dosages.

5. a. 90,600 mg   **b.** 227,000 tablets

Lesson 75

1. Possible answer: If you know the mass of an individual object in a large group of objects with the same mass, you can use it and the total mass to calculate the total number of objects.

5. 740 marbles will have a greater mass than 740 tiny plastic beads because the mass of a single marble is greater than the mass of a single bead.

7. 333 g of red jelly beans

9. 0.000000000000000000000327 g

Lesson 76

1. The average atomic mass of atoms of an element is given on the periodic table in units of amu.

5. a. 1 mol carbon   **b.** 1 mol iron
   c. 1 mol gold   **d.** 1 mol gold

Lesson 77

1. To determine the molar mass of sodium chloride, add the atomic mass of sodium and the atomic mass of chlorine to determine the mass of one unit of sodium chloride. The atomic mass in amu is the same as the molar mass in moles.

3. Ne(g): 20.2 g/mol, Ca(s): 40.1 g/mol, $CO_2(g)$: 44.0 g/mol, $CaCO_3(s)$: 100.1 g/mol, $CH_4O(l)$: 32.0 g/mol, $C_2H_6O(l)$: 46.1 g/mol, $Fe_2O_3(s)$: 159.7 g/mol

5. a. 10.0 g calcium, Ca
   b. 5.0 g sodium fluoride, NaF
   c. 2.0 g iron oxide, FeO

Lesson 78

1. To convert between moles of a substance and grams of the substance, use the formula: mass (g) = molar mass · moles.

3. 12.0 mol

5. 169 g BaO_2

7. a. 1.0 mol $Cu_2O(s)$
   b. 1.0 mol $Cu_2O(s)$
   c. $Cu_2O(s)$ represents the best deal for the company because you get twice as much copper for the same price.

Lesson 79

1. Possible answer: A comparison of the amount of each compound needed to make a soft drink taste sweet provides evidence that aspartame is sweeter than fructose. It takes many more molecules of fructose than aspartame to sweeten the drink.

3. **a.** about 360 cans   **b.** 0.00072 mol/kg

Chapter 14 Review Exercises

3. Possible answer: Scientific notation is a useful tool for chemists because it is often necessary to use very large or very small numbers in calculations. Scientific notation simplifies these calculations and also makes comparison of the values easier.

5. AgCl, NaCl, LiCl

Lesson 80

1. Possible answer: "Uniform throughout" means that the components of the solution are so well mixed that two samples taken from anywhere in the solution will be identical.

5. In order of increasing molarity, the solutions are c, a, b.

7. In order of increasing molarity, the solutions are a, 0.56 M, c, 0.58 M, and b, 1.1 M.

Lesson 81

1. Concentration refers to the amount of a substance divided by the volume in which the substance is dissolved. Two solutions of different volumes can have the same concentration if the proportion of the substance to the volume is the same.

5. **a.** The first two solutions have the same number of molecules. Each of these solutions has more molecules than the third solution. The number of molecules in each sample is the product of the molarity and the volume. The first two solutions have 1 mol of molecules, and the third solution has 0.5 mol of molecules. **b.** All three solutions have the same concentration, 1.0 M. Molarity is a measure of concentration. **c.** The 1.0 L sucrose solution has the greatest mass because each sucrose molecule has more mass than each glucose molecule. Therefore, one mole of sucrose has more mass than one mole of glucose. Also, one mole of sucrose has more mass than half a mole of sucrose.

9. **a.** 0.25 L   **b.** 1.5 L   **c.** 6.0 L   **d.** 300 L

Lesson 82

1. Possible answer: To prepare a 0.25 molar solution of sugar, you would calculate the mass of 0.25 mol of sugar, then measure the amount of sugar and dissolve it in enough water to make 1 L of solution.

3. **a.** 29 g   **b.** 220 g   **c.** 60 g

5. **a.** 10 L   **b.** 0.33 L   **c.** 0.18 L

7. 1.0-liter solution with 20 g of glucose

Lesson 83

1. A substance dissolved in the water adds mass to the solution. A sample of contaminated water should have a greater density than a sample of pure water. If this difference can be measured, the greater density indicates that the water is not pure.

3. The 0.10 M copper chloride ($CuCl_2$) solution has a greater density than the 0.10 M potassium chloride (KCl) solution. Copper chloride has a molar mass of 134.5 g and potassium chloride has a molar mass of 75.6 g. So if you have equal volumes of each solution there would be more mass in the copper chloride solution. And if the mass is greater, the density is greater.

7. The sodium bromide (NaBr) solution would weigh the most because it has more solute added than the sodium hydroxide (NaOH) solution. The KCl weighs about half as much as the other two samples because a 500 mL aqueous solution will always weigh less than a 1 L aqueous solution.

Chapter 15 Review Exercises

1. Possible answer: Knowing exactly how many moles or grams of contaminant are in a water sample determines whether the sample is toxic because toxicity is based on dosage.

3. **a.** 0.00002 mol Pb^{2+} **b.** 0.0066 g $Pb(NO_3)_2$
   c. 0.0041 g Pb^{2+}

Lesson 84

1. Possible answer: Acids have a sour taste and they can burn the skin in concentrated form. Bases have a bitter taste and a slippery feel. Bases can also cause skin burns in concentrated form. Most acidic and basic solutions are colorless and odorless.

Lesson 85

1. According to the Arrhenius theory, an acid is a molecule that dissociates to form hydrogen ions in solution and a base is a substance that dissociates to form hydroxide ions in solution.

7. Washing soda, Na_2CO_3, forms a basic solution because it accepts a proton from water, thereby forming a hydroxide ion in solution.

Lesson 86

1. The pH of a solution is equal to the negative logarithm of the H^+ concentration. Therefore, the pH increases by 1 as the hydrogen ion concentration decreases by a factor of 10.

7. $pH = -0.40$

Lesson 87

1. When you add water to an acidic solution, the pH of the solution increases to a maximum of 7.

5. Because the concentration of hydrogen ions contributed by the acid is less than the concentration of hydrogen ions in pure water, the addition of 1×10^{-9} moles of HCl does not affect the pH of pure water.

7. **a.** 0.075 mol   **b.** 0.068 M   **c.** $pH = 1.2$

Lesson 88

1. Two ways to make a strong acid solution safer are dilution with water and neutralization with a base.

5. Each mole of sulfuric acid forms two moles of hydrogen ions in solution, while each mole of sodium hydroxide forms one mole of hydroxide ions. Therefore, the final solution has more hydrogen ions than hydroxide ions, so it is acidic and the pH is less than 7.

Lesson 89

3. 1000 mL

5. **a.** The final solution is acidic because the solution has more H^+ ions than OH^- ions. **b.** The final solution is neutral because the solution has equal numbers of H^+ ions and OH^- ions. **c.** The final solution is basic because the solution has fewer H^+ ions than OH^- ions.

Chapter 16 Review Exercises

1. According to the Arrhenius theory, an acid is a substance that dissociates to form hydrogen ions in solution and a base is a substance that dissociates to form hydroxide ions in solution. According to the Brønsted-Lowry theory, an acid is a substance that donates protons in solution while a base is a substance that accepts protons. This is similar to the Arrhenius theory, but it accounts for bases that do not dissociate to form hydroxide ions.

5. **a.** Sodium bicarbonate can be classified as a base because it accepts a proton from water, forming hydroxide ions. **b.** Sodium bicarbonate was available in case sulfuric acid spilled. The sodium bicarbonate could have been used to neutralize any acid that was spilled.

Lesson 90

1. A precipitate is an insoluble compound that forms as a product of a chemical reaction in solution.

3. Possible answer: A spectator ion is an ion that is present in both the reactants and the products of a chemical reaction. Spectator ions do not participate in the reaction.

Lesson 91

1. Possible answer: Coefficients are important in chemical equations because they represent the proportion in which reactants combine to form products.

5. **a.** 0.10 mol   **b.** 5.8 g

Lesson 92

1. Possible answer: You need to convert between grams and moles to use mole ratios to find the right mass proportions of reactants.

3. **a.** 12.6 g   **b.** 19.8 g NaCl, 47.0 g $PbCl_2$

Lesson 93

5. **a.** $AgNO_3(aq) + NaCl(aq) \longrightarrow$
$AgCl(s) + NaNO_3(aq)$
   b. 0.037 mol $AgNO_3$; 0.077 mol NaCl
   c. silver nitrate, $AgNO_3$
   d. 5.3 g AgCl; 3.1 g $NaNO_3$
   e. 2.3 g NaCl

Chapter 17 Review Exercises

1. Possible answer:

$2NaOH(aq) + NiCl_2(aq) \longrightarrow$
$$2NaCl(aq) + Ni(OH)_2(s)$$
$2Na^+(aq) + 2OH^-(aq) + Ni^{2+}(aq) + 2Cl^-(aq) \longrightarrow$
$$2Na^+(aq) + 2Cl^-(aq) + Ni(OH)_2(s)$$
$2OH^-(aq) + Ni^{2+}(aq) \longrightarrow Ni(OH)_2(s)$
$K_2CO_3(aq) + NiSO_4(aq) \longrightarrow$
$$K_2SO_4(aq) + NiCO_2(s)$$
$2K^+(aq) + CO_3^{2}(aq) + Ni^{2+}(aq) + SO_4^{2+}(aq) \longrightarrow$
$$2K^+(aq) + SO_4^{2+}(aq) + NiCO_3(s)$$
$CO_3^{2+}(aq) + Ni^{2+}(aq) \longrightarrow NiCO_3(s)$
$2Li_3PO_4(aq) + 3NiCl_2(aq) \longrightarrow$
$$6LiCl(aq) + Ni_3(PO_4)_2(s)$$
$6Li^+(aq) + PO_4^{3+}(aq) + 3Ni^{2+}(aq) + 6Cl^-(aq) \longrightarrow$
$$6Li^+(aq) + 6Cl^-(aq) + Ni_3(PO_4)_2(s)$$
$2PO_4^{3+}(aq) + 3Ni^{2+}(aq) \longrightarrow Ni_3(PO_4)_2(s)$

General Review

1. **a.** $Zn(s) + 2HCl(aq) \longrightarrow ZnCl_2(aq) + H_2(g)$
   b. $2KClO_3(s) \longrightarrow 2KCl(s) + 3O_2(g)$
   c. $S_8(s) + 24F_2(g) \longrightarrow 8SF_6(g)$
   d. $4Fe(s) + 3O_2(g) \longrightarrow 2Fe_2O_3(s)$

3. **a.** limiting reactant: HCl; 1.87 g $ZnCl_2$ made
   b. limiting reactant: $KClO_3$; 0.608 g KCl made
   c. limiting reactant: F_2; 1.28 g SF_6
   d. limiting reactant: Fe; 1.43 g Fe_2O_3

5. A neutralization reaction is a chemical reaction in which an acid and a base combine to form a salt and water.

9. 36 g

Standardized Test Preparation

1. C	3. A	5. C
7. B	9. A	11. C
13. D	15. A	17. C
19. B		

Unit 5: Fire

Lesson 94

1. Possible answer: Energy is a measure of the ability to cause a change in matter.

5. Possible answer: Most of a fire is hot gases that become part of the atmosphere. Ashes are leftover material that was not converted to light and heat by a fire. The heat from the ashes transfers into the air and other materials around the site of the fire as it cools.

Lesson 95

1. Possible answer: Exothermic chemical changes are changes that transfer heat into the environment, and endothermic chemical changes are changes that absorb heat from the environment.

3. **a.** The average kinetic energy of the products is greater than the average kinetic energy of the reactants because the reaction is exothermic.
   b. In a molecular view, the gas particles in the reactant would be moving more slowly than the particles in the products.

5. **a.** barium hydroxide, $Ba(OH)_2 \cdot 8H_2O$; ammonium nitrate, NH_4NO_3; ammonia, NH_3; water, H_2O; barium nitrate, $Ba(NO_3)_2$
   b. Possible answer: the beaker, the pool of water outside the beaker, air, my hand

c. The $Ba(NO_3)_2$ is at a lower temperature than the NH_4NO_3 because the water freezing outside the beaker shows the temperature decreased after the reactants were mixed.
d. The beaker will feel cold.
e. The process is endothermic because the temperature decreases during the reaction. Energy was transferred from the surrounding environment into the reaction.
f. A chemical reaction must have occurred for a change in energy to occur. The reaction must have been the cause of the water outside the beaker turning into ice.

Lesson 96

1. The ice cube feels cold to your hand because heat is being transferred from your hand to the ice cube, which is at a lower temperature.

3. When a liquid evaporates energy is transferred to it, and when it solidifies it transfers energy to the surroundings.

5. Evaporation is the process of a liquid changing into a gas. It is a cooling process because the substance absorbs energy during the phase change, so that energy must be transferred from the surrounding environment. Objects and substances in the surroundings of the evaporating liquid are cooled.

9. Possible answer: When an object is warmer than its environment, the heat spreads into the environment. For example, a fire will eventually cause air in all parts of a room to feel warmer.

11. Yes, if the ice cube is warmer than the air around it or an object that is touching it, heat will transfer from the ice cube to its surroundings or into the other object.

Lesson 97

1. Because the larger sample has more particles, it takes more energy to increase the average kinetic energy of that sample by the same amount as the smaller sample.

3. **a.** 5 cal   **b.** 10 cal   **c.** 315 cal

Lesson 98

1. Possible answer: The specific heat capacity of a substance is the amount of thermal energy needed to raise the temperature of 1 g of the substance by 1 °C. For example, the specific heat capacity of aluminum is 0.21 cal/g °C, which means that 0.21 cal of thermal energy is needed to increase the temperature of 1 g of aluminum by 1 °C.

5. Possible answer: Because metals have a lower specific heat capacity than plastic, they experience a greater

change in temperature when they absorb a specific amount of energy from sunlight. Plastics are composed of large molecules that require more energy to cause a rise in temperature than a metal does.

7. The copper will cool faster because it has a lower specific heat capacity. Copper does not need to transfer as much thermal energy as aluminum does to cool to the temperature of the water.

Lesson 99

1. Water temperature doesn't change during boiling because the energy added to the water doesn't increase the kinetic energy of molecules.

3.
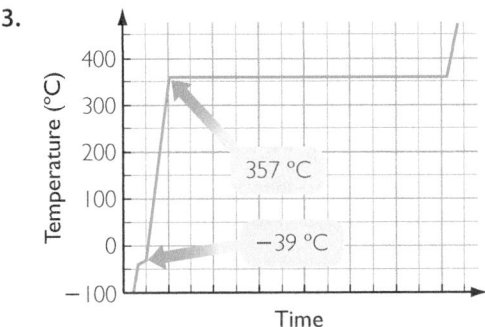

5. B.

7. 24.4 g

Chapter 18 Review Exercises

3. methanol (460 cal), water (500 cal), copper (900 cal)

7. 57 °C

Lesson 100

1. If a compound is ionic, or a substance contains water or carbon dioxide, it is unlikely to combust.

3. a. Na_2SO_4 will not combust because it is an ionic compound.
   b. Cu will combust because it is a metallic element.
   c. MgO will not combust because it is an ionic compound.
   d. Na will combust because it is a metallic element.
   e. C_2H_6O will combust because it is a molecular substance containing carbon and hydrogen.
   f. Ar will not combust because it is a noble gas.

Lesson 101

1. Possible answer: Calories are a measure of heat. The heat that is produced by the reaction of the substance with oxygen during combustion can be measured in calories.

3. No, placing a thermometer in the flames measures the temperature of the combustion reaction, but it

cannot measure the total heat energy given off by the reaction.

Lesson 102

5. 150 cal

7. Temperature is a measure of the average kinetic energy of the particles of water in the beakers. It takes more energy to increase the temperature of the third beaker because there are many more particles of water, making the amount of thermal energy needed to increase the temperature greater even though there is less temperature change.

Lesson 103

3. Heat of combustion is expressed as a negative number because energy transfers out of the system and into the environment.

Chapter 19 Review Exercises

3. Possible answer: The reaction does not appear to release light and heat because it occurs slowly, so the energy is lost to the environment without being noticed.

Lesson 104

3. O—O, C—C, H—H, O=O, C=C

5. 3237 kJ

7. a. methanol: -660 kJ/mol; butanol: -2509 kJ/mol
   b. The calculated values are slightly higher (slightly less negative) than the values given in the table.

Lesson 105

1. During an exothermic reaction, the potential energy of the system decreases and the kinetic energy of the system increases. Some of the potential energy in the system is converted into kinetic energy.

3. a. $2C_2H_6(g) + 7O_2(g) \longrightarrow 4CO_2(g) + 6H_2O(g)$
   b.
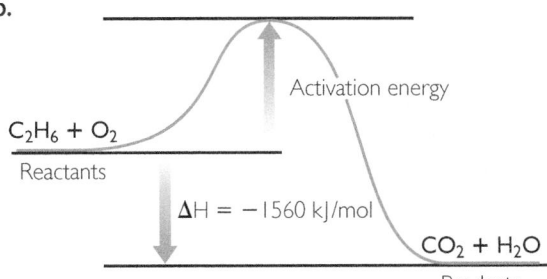

   c. The net energy exchange of the reverse reaction is 1560 kJ/mol. Conservation of energy requires that the amount of energy used to make ethane and oxygen from carbon dioxide and water is equal to the amount of energy released when ethane and oxygen are converted into carbon dioxide and water.

5. **a.** and **b.**

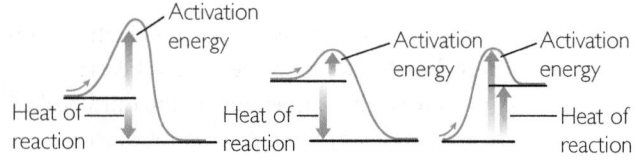

   c. the second diagram

   d. All of the reactions require energy to get started.

Lesson 106

5. Wood chips will burn faster because they have a much greater surface area than a tree, allowing the molecules in the wood chips more contact with the oxygen in the air.

7. The powdered iron oxidizes faster than the steel beam because the small particles have much more total surface area than the beam. Because a greater percentage of the iron is exposed to the oxygen in the air, the reaction is faster.

9. During an explosion, a large amount of energy is released very quickly from chemical bonds in molecules. This energy acts as activation energy to other molecules and causes the explosion to occur rapidly.

Lesson 107

1. Work is the result of a force acting through a distance. It is a transfer of energy, like heat.

3. $14 \text{ atm} \cdot \text{L}$

Chapter 20 Review Exercises

3. Paige's answer is correct. The water will evaporate before any hydrogen-oxygen bonds are broken.

Lesson 108

3. **a.** $2Ca(s) + O_2(g) \longrightarrow 2CaO(s); Ca^{2+}, O^{2-}$
$4Cr(s) + 3O_2(g) \longrightarrow 2Cr_2O_3(s); Cr^{3+}, O^{2-}$
$4Li(s) + O_2(g) \longrightarrow 2Li_2O(s); Li^+, O^{2-}$

   b. Calcium, Ca, is oxidized. Chromium, Cr, is oxidized. Lithium, Li, is oxidized.

Lesson 109

1. Possible answer: The oxidation reactions are better sources of energy than the reverse reactions because they

are almost always exothermic. The metals that are the hardest to extract have the lowest (most negative) heats of formation, which means that they release the most energy into the environment. The oxidation of aluminum is the best source of energy listed in the lesson.

3. 2180 kJ

5.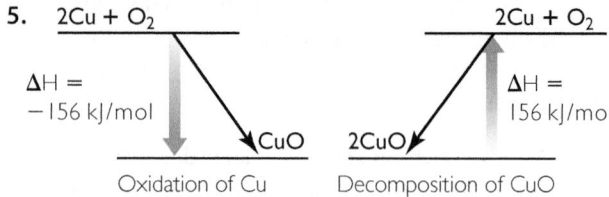

7. 24.5 kJ

Lesson 110

1. In this reaction, copper is oxidized because the neutral copper atom loses the electrons and becomes a positively charged copper ion. The oxygen atom is reduced because it gains electrons and becomes a negatively charged oxygen ion.

5. **a.** $Zn(s) + Cu^{2+}(aq) \longrightarrow Cu(s) + Zn^{2+}(aq)$
   b. $Ba^{2+}(aq) + SO_4^{2-}(aq) \longrightarrow BaSO_4(s)$

7. **a.** not a redox reaction **b.** not a redox reaction **c.** not a redox reaction **d.** not a redox reaction **e.** redox reaction, lead is oxidized and hydrogen is reduced, 2 electrons are transferred **f.** redox reaction, arsenic is oxidized and hydrogen is reduced, 6 electrons are transferred

Lesson 111

3. No change would be observed because iron is less active than zinc. Therefore, iron cannot be oxidized by the zinc nitrate solution and the iron strip would not be affected.

5. **a.** No reaction occurs in Exercise 3. The chemical equation for the reaction in Exercise 4 is
$2Fe(NO_3)_2(aq) + 3Zn(s) \longrightarrow$
$$2Fe(s) + 3Zn(NO_3)_2(aq)$$
   b. $2Fe^{3+}(aq) + 3Zn(s) \longrightarrow 2Fe(s) + 3Zn^{2+}(aq)$
   c. zinc
   d. Zinc is oxidized, and iron is reduced.

7. Cobalt is below iron in the activity series because the iron is oxidized by the cobalt ions, making iron the more active metal.

Lesson 112

1. In an electrochemical cell, electrons are transferred from the more active metal to the less active metal.

3. **a.** $Ca \longrightarrow Ca^{2+} + 2e^-$
   $Fe \longrightarrow Fe^{3+} + 3e^-$
   b. Calcium is oxidized and iron is reduced.

c.

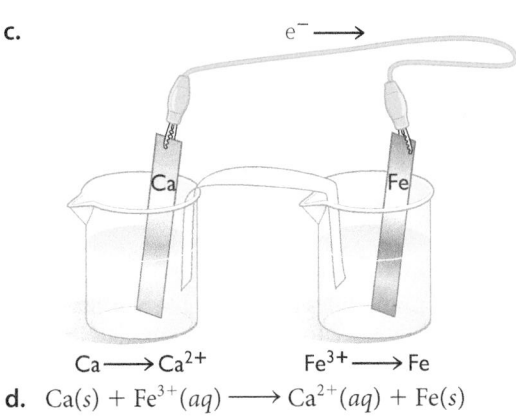

Ca $\longrightarrow$ Ca^{2+} Fe^{3+} $\longrightarrow$ Fe

d. Ca(s) + Fe^{3+}(aq) $\longrightarrow$ Ca^{2+}(aq) + Fe(s)

Chapter 21 Review Exercises

1. Possible answer: An electrochemical cell works by separating the two parts of a redox reaction. As electrons are transferred from one metal to the other, they travel through a wire connecting the two half-reactions. The electrical energy is the stream of moving electrons. The chemical energy is the voltage potential of the redox reaction.

3. In the extraction of metals from metal oxides table, the heat of reaction of iron (III) oxide decomposing into pure iron and oxygen gas is listed as -413 kJ/mol. Because this value is positive, the reaction is endothermic. Energy must be continuously added into the system for the reaction to occur.

Lesson 113

1. Light shining on matter raises its temperature.

3. Blue

5. Green

7. **a.** Orange, glasses transmit orange light, so an orange object would look orange.
   b. Black, because blue light would not be transmitted.

Lesson 114

1. $\lambda = c/f$; you need to know the frequency of the light, f, and the speed of light, c

5. **a.** 5.9×10^{14} Hz   **b.** Green

Lesson 115

3. $E = hf$; energy and frequency are proportional to one another

5. Possible answer: an insect

7. 2×10^{-13} J

Lesson 116

1. A chemical reaction happens.

5. It remains green. No reaction occurs.

7. The sunglasses with UV protection will block UV light such that the UV-sensitive paper remains green. The regular reading glasses will transmit more UV light, thereby exposing the UV-sensitive paper and turning it blue.

Chapter 22 Review Exercises

1. Colored objects reflect the color we see and absorb other colors.

3. 2.5×10^2 m

5. 4.1×10^{14} Hz

Unit 5 Review Exercises

General Review

3. The thermal energy of an exothermic reaction is the result of bond making releasing more energy than is required for bond breaking for the reaction.

5. **a.** 3800 cal
   b. -7500 cal

7. More energy is required to raise the temperature of water than to raise the temperature of aluminum by the same amount because water has a higher specific heat capacity than aluminum.

13. Zn(s) + 2Mn^{4+}(aq) $\longrightarrow$ Zn^{2+}(aq) + 2Mn^{3+}(aq)
    Zinc loses electrons, so it is oxidized. Magnesium gains electrons, so magnesium ions are reduced.
    2Al(s) + 3Ca^{2+}(aq) $\longrightarrow$ 2Al^{3+}(aq) + 3Ca(s)
    Aluminum loses electrons, so it is oxidized. Calcium gains electrons, so calcium ions are reduced.

Standardized Test Preparation

1. D	3. A	5. B
7. A	9. B	11. B
13. C	15. C	17. D
19. D	21. C	

Unit 6: Showtime

Lesson 117

1. Both H*In* (red) and *In*$^-$ (blue) are present.

5. H$^+$ + *In*$^-$ $\longrightarrow$ H*In*

Lesson 118

1. Possible answer: evaporation and condensation of water

3. freezing and melting, evaporation and condensing

7. **a.** C$_{12}$H$_{22}$O$_{11}$(s) $\rightleftharpoons$ C$_{12}$H$_{22}$O$_{11}$(aq)
   b. forward   **c.** reverse

Lesson 119

1. Dynamic equilibrium means that the rate of a forward process and the rate of its reverse process are equal.

3. **a.** Artificial sweetener is sweeter than the same amount of sucrose. Answer: Many more artificial sweetener molecules bind to the sweet receptors than do the same number of sucrose molecules. So AB is more favored for the artificial sweetener compared with sucrose. **b.** Much more sodium chloride, NaCl, dissolves in 1 L of water compared with silver chloride, AgCl. Answer: More products form for NaCl, whereas AgCl remains mostly as the starting substance. So AB is more favored for AgCl compared with NaCl.

Lesson 120

1. A large value of K indicates that there are more products in the equilibrium mixture, while a small value of K indicates that more starting substance is present.

3. **a.** The binding of CO to hemoglobin has the larger equilibrium constant because [hemoglobin:CO] is greater than [hemoglobin:O_2]. **b.** Each particle view should show a hemoglobin binding to a particle. The view with oxygen should have more free oxygen and hemoglobin particles than the view for carbon monoxide. **c.** CO is toxic because it takes up binding sites in hemoglobin so that they are not available to oxygen. Because the [hemoglobin:CO] product is favored, fewer free hemoglobin molecules are available to bind oxygen.

Lesson 121

1. You can rearrange the equilibrium constant equation to solve for hydrogen ion concentration. Once you have the hydrogen ion concentration, you can solve for pH. $[H^+] = \sqrt{K[HA]}$, $pH = -\log[H^+]$

3. Solution 1: 1.4×10^{-3}, Solution 2: 1.8×10^{-4}, Solution 3: 1.8×10^{-5}, Solution 4: 1.8×10^{-5}

Chapter 23 Review Exercises

1. **a.** $HCl(aq) \longrightarrow H^+(aq) + Cl^-(aq)$ **b.** The particle view should show 10 H^+ and 10 Cl^- ions and no HCl molecules. **c.** $HF(aq) \rightleftharpoons H^+(aq) + F^-(aq)$ **d.** The particle view should show mostly HF molecules as well as a few H^+ and F^- ions.

3. **a.** $K = [C][D]/[A][B]$ **b.** $K = [CO][Cl_2]/[COCl_2]$ **c.** $K = [Pb^{2+}][Cl^-]^2$

Lesson 122

1. Add more of one of the products. Remove some of the starting substance.

3. More $Cu(NH_3)_6^{2+}(aq)$ forms to reduce the concentration of NH_3.

5. Since this is an endothermic process, heating favors products. Therefore more products will form to relieve the stress.

Lesson 123

1. Indicator molecules have a bright color that changes to a different color when the H*In* molecules dissociate into H^+ and *In*$^-$. When the solution is more acidic, the color of H*In* molecules is observed. In basic solution, the solution displays the color of *In*$^-$.

3. **a.** In order of increasing solubility, AgCl, CuCl, TlCl, the most soluble is TlCl. A small K_{sp} indicates mostly undissolved solid. **b.** CuCl(*s*) will precipitate.

5. **a.** yellow, due to the large amount of H*In* **b.** blue, due to the large amount of *In*$^-$ **c.** yellow, due to large amount of *In*$^-$ **d.** orange, due to presence of both H*In* and *In*$^-$

7. pH $\sim$ 10

Chapter 24 Review Exercises

3. **a.** No more solid will dissolve. Excess solid remains. In a saturated solution, no more solid can be dissolved. Excess solid will remain at the bottom of the container. **b.** More $CaCl_2$ precipitates. **c.** More $CaCl_2$ precipitates. **d.** More $CaCl_2$ precipitates. **e.** More $CaCl_2$ dissolves.

5. **a.** $K = \dfrac{[CO][Cl_2]}{[COCl_2]}$ **b.** $8.6 \times 10^{-2}\,M$ **c.** More $COCl_2$ forms to minimize the effect of adding one of the products. **d.** 0.14 M

Unit 6 Review Exercises

General Review

1. **a.** $Cl_2(g) \rightleftharpoons Cl_2(l)$
   b. $2NO_2(g) \rightleftharpoons N_2O_4(g)$
   c. $Ag^+(aq) + Cl^-(aq) \rightleftharpoons AgCl(s)$
   $H_2O(l) \rightleftharpoons H_2O(g)$

3. Possible answers: The reaction is reversible. The equilibrium constant does not change as the concentrations of products and reactants change.

5. $1.2 \times 10^{-5}\,M$

Standardized Test Preparation

1. C	3. B	5. C
7. A	9. D	11. B
13. A	15. A	17. A
19. B		

Appendix D: Reference Tables

Table of Average Atomic Masses

Element and symbol	Atomic number	Average atomic mass (amu)	Element and symbol	Atomic number	Average atomic mass (amu)
Actinium, Ac	89	(227)	Dysprosium, Dy	66	162.5
Aluminum, Al	13	26.98	Einsteinium, Es	99	(252)
Americium, Am	95	(243)	Erbium, Er	68	167.3
Antimony, Sb	51	121.8	Europium, Eu	63	152.0
Argon, Ar	18	39.95	Fermium, Fm	100	(257)
Arsenic, As	33	74.92	Flerovium, Fl	114	(289)
Astatine, At	85	(210)	Fluorine, F	9	19.00
Barium, Ba	56	137.3	Francium, Fr	87	(223)
Berkelium, Bk	97	(247)	Gadolinium, Gd	64	157.3
Beryllium, Be	4	9.012	Gallium, Ga	31	69.72
Bismuth, Bi	83	209.0	Germanium, Ge	32	72.64
Bohrium, Bh	107	(272)	Gold, Au	79	197.0
Boron, B	5	10.81	Hafnium, Hf	72	178.5
Bromine, Br	35	79.90	Hassium, Hs	108	(277)
Cadmium, Cd	48	112.4	Helium, He	2	4.003
Calcium, Ca	20	40.08	Holmium, Ho	67	164.9
Californium, Cf	98	(251)	Hydrogen, H	1	1.008
Carbon, C	6	12.01	Indium, In	49	114.8
Cerium, Ce	58	140.1	Iodine, I	53	126.9
Cesium, Cs	55	132.9	Iridium, Ir	77	192.2
Chlorine, Cl	17	35.45	Iron, Fe	26	55.85
Chromium, Cr	24	52.00	Krypton, Kr	36	83.80
Cobalt, Co	27	58.93	Lanthanum, La	57	138.9
Copernicium, Cn	112	(285)	Lawrencium, Lr	103	(262)
Copper, Cu	29	63.55	Lead, Pb	82	207.2
Curium, Cm	96	(247)	Lithium, Li	3	6.941
Darmstadtium, Ds	110	(281)	Livermorium, Lv	116	(292)
Dubnium, Db	105	(268)	Lutetium, Lu	71	175.0

The values in parentheses are the average atomic mass of the longest-lasting isotope of the element at the time of writing.

Element and symbol	Atomic number	Average atomic mass (amu)	Element and symbol	Atomic number	Average atomic mass (amu)
Magnesium, Mg	12	24.31	Rubidium, Rb	37	85.47
Manganese, Mn	25	54.94	Ruthenium, Ru	44	101.1
Meitnerium, Mt	109	(276)	Rutherfordium, Rf	104	(267)
Mendelevium, Md	101	(258)	Samarium, Sm	62	150.4
Mercury, Hg	80	200.6	Scandium, Sc	21	44.96
Molybdenum, Mo	42	95.94	Seaborgium, Sg	106	(271)
Moscovium, Mc	115	(289)	Selenium, Se	34	78.96
Neodymium, Nd	60	144.2	Silicon, Si	14	28.09
Neon, Ne	10	20.18	Silver, Ag	47	107.9
Neptunium, Np	93	(237)	Sodium, Na	11	22.99
Nickel, Ni	28	58.69	Strontium, Sr	38	87.62
Nihonium, Nh	113	(286)	Sulfur, S	16	32.07
Niobium, Nb	41	92.91	Tantalum, Ta	73	180.9
Nitrogen, N	7	14.01	Technetium, Tc	43	(98)
Nobelium, No	102	(259)	Tellurium, Te	52	127.6
Oganesson, Og	118	(294)	Tennessine , Ts	117	(293)
Osmium, Os	76	190.2	Terbium, Tb	65	158.9
Oxygen, O	8	16.00	Thallium, Tl	81	204.4
Palladium, Pd	46	106.4	Thorium, Th	90	232.0
Phosphorus, P	15	30.97	Thulium, Tm	69	168.9
Platinum, Pt	78	195.1	Tin, Sn	50	118.7
Plutonium, Pu	94	(244)	Titanium, Ti	22	47.87
Polonium, Po	84	(209)	Tungsten, W	74	183.8
Potassium, K	19	39.10	Uranium, U	92	238.0
Praseodymium, Pr	59	140.9	Vanadium, V	23	50.94
Promethium, Pm	61	(145)	Xenon, Xe	54	131.3
Protactinium, Pa	91	231.0	Ytterbium, Yb	70	173.0
Radium, Ra	88	(226)	Yttrium, Y	39	88.91
Radon, Rn	86	(222)	Zinc, Zn	30	65.39
Rhenium, Re	75	186.2	Zirconium, Zr	40	91.22
Rhodium, Rh	45	102.9			
Roentgenium, Rg	111	(280)			

Table of Electronegativities

H 2.10																	He
Li 0.98	Be 1.57											B 2.04	C 2.55	N 3.04	O 3.44	F 3.98	Ne
Na 0.93	Mg 1.31											Al 1.61	Si 1.90	P 2.19	S 2.58	Cl 3.16	Ar
K 0.82	Ca 1.00	Sc 1.36	Ti 1.54	V 1.63	Cr 1.66	Mn 1.55	Fe 1.83	Co 1.88	Ni 1.91	Cu 1.90	Zn 1.65	Ga 1.81	Ge 2.01	As 2.18	Se 2.55	Br 2.96	Kr
Rb 0.82	Sr 0.95	Y 1.22	Zr 1.33	Nb 1.60	Mo 2.16	Tc 1.90	Ru 2.2	Rh 2.28	Pd 2.20	Ag 1.93	Cd 1.69	In 1.78	Sn 1.96	Sb 2.05	Te 2.1	I 2.66	Xe
Cs 0.79	Ba 0.89	La* 1.10	Hf 1.30	Ta 1.50	W 2.36	Re 1.90	Os 2.20	Ir 2.20	Pt 2.28	Au 2.54	Hg 2.00	Tl 1.62	Pb 2.33	Bi 2.02	Po 2.00	At 2.20	Rn
Fr 0.70	Ra 0.89	Ac* 1.10															

* Electronegativity values for the lanthanides and actinides range from about 1.10 to 1.50.

Bonding Continuum

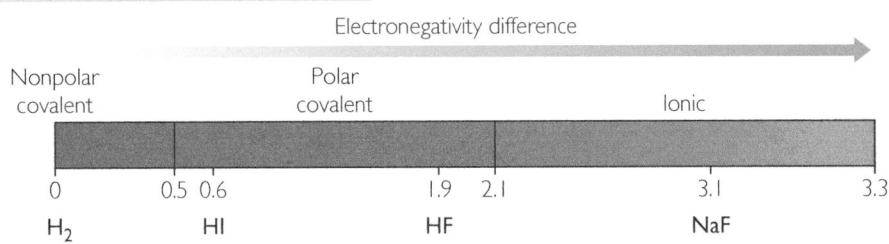

Electronegativity difference

Nonpolar covalent · Polar covalent · Ionic

0	0.5 0.6	1.9 2.1	3.1 3.3
H₂	HI	HF	NaF

Average Bond Energies (per mole of bonds)

Bond	H—H	C—H	C—C	C=C	C≡C	O—H	C—O	C=O	O=O	O—O
Bond energy (kJ/mol)	432	413	347	614	811	467	358	799*	495	146

*C=O in CO_2: 799, C=O in organic molecules: 745

Heats of Combustion

	Fuel	Chemical formula	ΔH (kJ/mol)	ΔH (kJ/g)
Elements	hydrogen	$H_2(g)$	−243	−121.3
	carbon (coal)	$C(s)$	−393	−32.6
Alkanes	methane	$CH_4(g)$	−891	−55.6
	ethane	$C_2H_6(g)$	−1560	−51.9
	butane	$C_4H_{10}(l)$	−2882	−49.6
	hexane	$C_6H_{14}(l)$	−4163	−48.1
	octane	$C_8H_{18}(l)$	−5508	−47.7
Alcohols	methanol	$CH_4O(l)$	−724	−22.6
	ethanol	$C_2H_6O(l)$	−1368	−29.7
	butanol	$C_4H_{10}O(l)$	−2669	−36.0
Sugar	glucose	$C_6H_{12}O_6(s)$	−2828	−15.9

Solubility Trends

			Anions						
			NO_3^-	Cl^-	OH^-	SO_4^{2-}	CO_3^{2-}	$C_2O_4^{2-}$	PO_4^{3-}
Cations		Most alkali metals, such as Li^+, Na^+, K^+, NH_4^+	S	S	S	S	S	S	S
		Most alkaline earth metals, such as Mg^{2+}, Ca^{2+}, Sr^{2+}	S	S	N	S	N	N	N
		Some Period 4 transition metals, such as Fe^{3+}, Co^{3+}, Ni^{2+}, Cu^{2+}, Zn^{2+}	S	S	N	S	N	N	N
		Other transition metals, such as Ag^+, Pb^{2+}, Hg^{2+}	S	N	N	N	N	N	N

S = very soluble, N = not very soluble or insoluble

Polyatomic Ions

Anions

Charge: 1^-

Name	Formula
acetate	$CH_3CO_2^-$
chlorate	ClO_3^-
chlorite	ClO_2^-
hypochlorite	ClO^-
perchlorate	ClO_4^-
cyanide	CN^-
formate	HCO_2^-
hydroxide	OH^-
nitrate	NO_3^-
nitrite	NO_2^-
permanganate	MnO_4^-
thiocyanate	SCN^-

Charge: 2^-

Name	Formula
oxalate	$C_2O_4^{2-}$
carbonate	CO_3^{2-}
chromate	CrO_4^{2-}
sulfate	SO_4^{2-}
sulfite	SO_3^{2-}

Charge: 3^-

Name	Formula
phosphite	PO_3^{3-}
phosphate	PO_4^{3-}

Cations

Charge: 1^+

Name	Formula
ammonium	NH_4^+

Charge: 2^+

Name	Formula
hexaaquocobalt	$Co(H_2O)_6^{2+}$

Specific Heat Capacities at 20 °C

Substance	Specific heat capacity (cal/g °C)	Specific heat capacity (J/g °C)
aluminum	0.21	0.88
water	1.00	4.18
brass	0.09	0.38
iron	0.11	0.46
rubber	0.41	1.72
glass	0.16	0.67
sand	0.19	0.95
mercury	0.03	0.13
hydrogen	3.34	13.97
copper	0.09	0.38
wood	0.41	1.72
silver	0.05	0.21
concrete	0.20	0.84
air (dry, sea level)	0.24	1.00
zinc	0.09	0.38
ethanol	0.57	2.38
lead	0.03	0.13
helium	1.24	5.19
silver	0.05	0.23
tin	0.05	0.21
gold	0.03	0.13

SI Units and Conversion Factors

Length

SI unit: meter (m)

1 meter = 100 cm = 1.0936 yards
1 inch = 2.54 centimeters (exactly)
1 kilometer = 1000 m = 0.62137 mile

Mass

SI unit: kilogram (kg)

1 kilogram = 1000 grams = 2.2046 pounds
1 pound = 453.59 grams = 0.45359 kilogram = 16 ounces
1 atomic mass unit = 1.66054×10^{-27} kilogram

Volume

SI unit: cubic meter (m^3)

1 liter = 1000 milliliters = 0.001 cubic meter
1 cubic centimeter = 1 milliliter

Pressure

SI unit: pascal (Pa)

1 Pa = 1 newton/m^2 = 0.01 millibar
1 atmosphere = 101.325 kilopascals
= 760 torr (mm Hg)
= 14.70 pounds per square inch
= 1.01 bar

Energy

SI unit: joule (J)

1 joule = 0.23901 calorie
1 calorie = 4.184 joules

Physical Constants

Constant	Symbol	Value
atomic mass unit	amu	1.66054×10^{-24} g
Avogadro's number	N_A	6.02214×10^{23} particles/mole
charge of an electron	e	-1.60218×10^{-19} coulomb
universal gas constant	R	8.31451 J/K $\cdot$ mol $0.08206 \dfrac{\text{L} \cdot \text{atm}}{\text{mol} \cdot \text{K}}$
mass of an electron	m_e	9.10938×10^{-28} g 5.48580×10^{-4} amu
mass of a neutron	m_n	1.67493×10^{-24} g 1.00866 amu
mass of a proton	m_p	1.67262×10^{-24} g 1.00728 amu
density of water (at 4 °C)	—	1.0000 g/cm^3
mass of carbon-12 atom	—	12 amu (exactly)
ideal gas molar volume at STP	V_m	$22.4 \dfrac{\text{L}}{\text{mol}}$

Symbols and Abbreviations

m	mass		F	force
V	volume		A	area
D	density		h	height
α	alpha particle		LD_{50}	lethal dose 50%
β	beta particle		[ ]	concentration in moles per liter
γ	gamma ray		q	heat
$\delta^+ \, \delta^-$	partial ionic charge		E	energy
n	number of moles		C_p	specific heat capacity
T	temperature		ΔH	heat of reaction (change in enthalpy)
ΔT	change in temperature		w	work
P	pressure		K_{eq}	equilibrium constant
STP	standard temperature and pressure		K_{sp}	solubility product constant
R	universal gas constant		K_a, K_b	acid/base dissociation constant

Proportional Relationships

Mathematical relationship	Variables	Proportionality constant
$m = D \cdot V$ $D = \dfrac{m}{V}$	mass, m volume, V	density, D
$n = M \cdot V$ $M = \dfrac{n}{V}$	moles, n volume, V	molarity, M
$m = \text{molar mass} \cdot n$ $\text{molar mass} = \dfrac{m}{n}$	mass, m moles, n	molar mass (in g/mol)
$F = P \cdot A$ $P = \dfrac{F}{A}$	force, F area, A	pressure, P
$V = A \cdot h$ $A = \dfrac{V}{h}$	volume, V height, h	area, A

Proportional Relationships in the Gas Laws

Law	Variables	Mathematical relationships	Variables held constant
Charles's law	volume, V temperature, T	$V = k \cdot T$ $k = \dfrac{V}{T}$	pressure, P number of moles, n
Gay-Lussac's law	pressure, P temperature, T	$P = k \cdot T$ $k = \dfrac{P}{T}$	volume, V number of moles, n
Boyle's law	pressure, P volume, V	$k = P \cdot V$ $p = \dfrac{k}{V}$	temperature, T number of moles, n
combined gas law	pressure, P volume, V temperature, T	$PV = kT$ $k = \dfrac{PV}{T}$	number of moles, n
ideal gas law	pressure, P volume, V temperature, T moles, n	$PV = nRT$ $R = \dfrac{PV}{nT}$	none

Other Important References

Table	Location
Isotopes of the Elements	Unit 1: Alchemy, Chapter 3, Lesson 14
Four Models of Bonding	Unit 1: Alchemy, Chapter 5, Lesson 26
Water Vapor Density Versus Temperature	Unit 3: Weather, Chapter 12, Lesson 66
Lethal Doses	Unit 4: Toxins, Chapter 14, Lesson 74
Activity Series	Unit 5: Fire, Chapter 21, Lesson 112
Standard Reduction Potentials	Unit 5: Fire, Chapter 21, Lesson 112

Some Key Formulas

Density	$D = \dfrac{m}{V}$
Ideal gas law	$PV = nRT$
Heat	$q = C_p m \Delta T$
Acidity	$\text{pH} = -\log [\text{H}^+]$
Basicity	$\text{pOH} = -\log [\text{OH}^-]$

Glossary

A

absolute zero The temperature defined as 0 K on the Kelvin scale and $-273.15\,°C$ on the Celsius scale. Considered to be the lowest possible temperature that matter can reach. (p. 279)

cero absoluto La temperatura definida como 0 K en la escala Kelvin y $-273.15\,°C$ en la escala Celsius. Se considera como la temperatura más baja que puede alcanzar la materia. (p. 279)

absorption Of light, not being transmitted because certain colors are removed by an object. Absorption transfers energy back to the object. (p. 579)

absorción De la luz: cuando esta no se transmite porque ciertos colores han sido eliminados (por un objeto). La absorción transfiere energía al objeto. (p. 579)

acid A substance that adds hydrogen ions, H^+, to an aqueous solution; a substance that donates a proton to another substance in solution. (p. 430)

ácido Sustancia que cede iones de hidrógeno, H^+, a una solución acuosa; una sustancia que dona un protón a otra sustancia en la solución. (p. 430)

actinides A series of elements that follow actinium in Period 7 of the periodic table and that are typically placed separately at the bottom of the periodic table. (p. 44)

actínidos Serie de elementos que están después del actinio en el séptimo período de la tabla periódica y que usualmente, aparecen en la parte de abajo de la tabla. (p. 44)

activation energy The minimum amount of energy required to initiate a chemical process or reaction. (p. 537)

energía de activación La mínima cantidad de energía que se necesita para iniciar una reacción o un proceso químico. (p. 537)

activity series A table showing elements in order of their chemical activity, with the most easily oxidized at the top of the list. (p. 564)

serie de actividad Una tabla que muestra elementos ordenados de acuerdo con la actividad química de cada uno de dichos elementos, empezando por aquellos que se oxidan con mayor facilidad. (p. 564)

actual yield The amount of a product obtained when a reaction is run (as opposed to the theoretical yield). (p. 473)

rendimiento real La cantidad de un producto que se obtiene cuando se ejecuta una reacción (al contrario del rendimiento teórico). (p. 473)

air mass A large volume of air that has consistent temperature and water content. (p. 288)

masa de aire Gran volumen de aire cuya temperatura y contenido de agua son constantes. (p. 288)

alkali metals The elements in Group 1A on the periodic table, except for hydrogen. (p. 44)

metales alcalinos Los elementos del grupo 1A de la tabla periódica, con excepción del hidrógeno. (p. 44)

alkaline earth metals The elements in Group 2A on the periodic table. (p. 44)

metales alcalinotérreos Los elementos del grupo 2A de la tabla periódica. (p. 44)

alpha decay A nuclear reaction in which an atom emits an alpha particle consisting of two protons and two neutrons. Alpha decay decreases the atomic number of an atom by 2 and the mass number by 4. (p. 72)

alpha particle A particle made of two protons and two neutrons, equivalent to the nucleus of a helium atom. (p. 72)

amino acid A molecule that contains both an amine ($-NH_2$) and a carboxylic acid ($-COOH$) functional group. (p. 247)

anion An ion that has a negative charge. (p. 96)

aqueous solution A solution in which water is the dissolving medium or solvent. (p. 25)

Arrhenius theory of acids and bases An acid-base theory that defines an acid as a substance that adds hydrogen ions, H^+, to an aqueous solution, and a base as a substance that adds hydroxide ions, OH^-, to an aqueous solution. (p. 434)

atmosphere (atm) A unit of measurement for gas pressure. One atmosphere is equivalent to 14.7 pounds of pressure per square inch or a barometric reading of 760 millimeters of mercury. (p. 301)

atmospheric pressure Pressure exerted by the weight, or force, of the air pressing down on a surface. Air pressure is a result of air molecules colliding with the surfaces of objects on Earth. At sea level, atmospheric pressure equals approximately 1 atm or 14.7 lb/in². (p. 301)

atom The smallest unit of an element that retains the chemical properties of that element and can exist as a separate particle. (p. 51)

atomic mass The mass of a single atom (or isotope) of an element. (p. 38)

atomic mass unit (amu) The unit used for expressing atomic mass. 1 amu = 1.66×10^{-24} g, the mass of one hydrogen atom. This is 1/12 the mass of a carbon-12 atom. (p. 38)

desintegración alfa Reacción nuclear en la que un átomo emite partículas alfa que contienen dos protones y dos neutrones. La desintegración alfa disminuye el número atómico de un átomo en 2 y el número de masa en 4. (p. 72)

partícula alfa Partícula que se compone de dos protones y dos neutrones, lo que equivale al núcleo de un átomo de helio. (p. 72)

aminoácido Molécula que contiene un grupo funcional amino ($-NH_2$) y un ácido carboxílico ($-COOH$). (p. 247)

anión Ion que tiene una carga negativa. (p. 96)

solución acuosa Solución en la que el agua es el medio de disolución o el disolvente. (p. 25)

teoría de ácidos y bases de Arrhenius Teoría ácido-base que define un ácido como una sustancia que cede iones de hidrógeno, H^+, a una solución acuosa; y una base como una sustancia que cede iones de hidróxido, OH^-, a una solución acuosa. (p. 434)

atmósfera estándar (atm) Unidad de medida de la presión de los gases. Una atmósfera estándar equivale a 14.7 libras de presión por pulgada cuadrada o a la medida barométrica de 760 milimetros de mercurio. (p. 301)

presión atmosférica La presión ejercida por el peso o fuerza del aire sobre una superficie. La presión del aire es el resultado del choque de las moléculas de aire con la superficie de los objetos que están sobre la Tierra. Al nivel del mar, la presión atmosférica es igual a 1 atm o 14.7 lb/pulg², aproximadamente. (p. 301)

átomo La unidad más pequeña de un elemento que mantiene sus propiedades químicas y que puede existir como una partícula independiente. (p. 51)

masa atómica La masa de un solo átomo (o isótopo) de un elemento. (p. 38)

unidad de masa atómica (uma) La unidad que se usa para expresar la masa atómica. 1 uma = 1.66×10^{-24} g, la masa de un átomo de hidrógeno. Esto equivale a 1/12 de la masa de un átomo de carbono-12. (p. 38)

atomic number The consecutive whole numbers associated with the elements on the periodic table. The atomic number is equal to the number of protons in the atomic nucleus of an element. (p. 41)

atomic theory A theory that states that all matter is made up of individual particles called atoms. (p. 52)

average atomic mass The weighted average of the mass of the isotopes of an element. (p. 62)

Avogadro's law A scientific law stating that equal volumes of gases at the same temperature and pressure contain equal numbers of particles. (p. 333)

Avogadro's number A number equal to 6.02×10^{23}. It is the number of particles (atoms, ions, molecules, or formula units) present in one mole of a substance. (p. 332)

B

ball-and-stick model A three-dimensional representation of a molecule that uses color-coded balls to represent atoms and sticks to represent bonds. (p. 186)

base A substance that adds hydroxide ions OH^-, to an aqueous solution, or a substance that accepts protons, H^+ ions, from another substance in solution. (p. 430)

bent shape The nonlinear shape around a bonded atom with two lone pairs of electrons. (p. 194)

beta decay A nuclear reaction in which a neutron changes into a proton and an electron, and the atom emits a beta particle, which is the electron. Beta decay increases the atomic number of the atom without changing the mass. (p. 72)

beta particle An electron emitted from the nucleus of an atom during beta decay. (p. 72)

boiling point (boiling temperature) The temperature at which both liquid and gas phases of a single substance are present and in equilibrium; the temperature at which equilibrium is established between a liquid and its vapor at a pressure of 1 atm. (p. 274)

número atómico Los números enteros consecutivos asociados con los elementos de la tabla periódica. El número atómico es igual a la cantidad de protones que hay en el núcleo de un elemento. (p. 41)

teoría atómica Teoría que afirma que toda la materia se compone de partículas individuales llamadas átomos. (p. 52)

masa atómica promedio El promedio ponderado de la masa de los isótopos de un elemento. (p. 62)

ley de Avogadro Ley científica que establece que volúmenes iguales de gases que tienen la misma temperatura y la misma presión contienen la misma cantidad de partículas. (p. 333)

número de Avogadro Número igual a 6.02×10^{23}. Es el número de partículas (átomos, iones, moléculas o unidades de fórmula) que existen en un mol de una sustancia. (p. 332)

modelo de esferas y varillas Representación tridimensional de una molécula que usa esferas de colores específicos para representar átomos y varillas para representar enlaces. (p. 186)

base Sustancia que cede iones de hidróxido, OH^-, a una solución acuosa o una sustancia que recibe protones, iones H^+, de otra sustancia en una solución. (p. 430)

forma curva La forma curvilínea alrededor de un átomo enlazado con dos pares aislados de electrones. (p. 194)

desintegración beta Reacción nuclear en la cual un neutrón pasa a ser un protón y un electrón, y el átomo emite una partícula beta, que es el electrón. La desintegración beta aumenta el número atómico de un átomo sin alterar su masa. (p. 72)

partícula beta Electrón emitido por el núcleo de un átomo durante una desintegración beta. (p. 72)

punto de ebullición (temperatura de ebullición) La temperatura a la cual los estados líquido y gaseoso de una misma sustancia están presentes y en equilibrio; la temperatura a la cual se establece el equilibrio entre un líquido y su vapor a una presión de 1 atm. (p. 274)

bond energy The amount of energy that is required to break a specific chemical bond. (p. 531)

bonded pair A pair of electrons that are shared in a covalent bond between two atoms. (p. 163)

Boyle's law The scientific law that states that the volume of a given sample of gas at a given temperature is inversely proportional to its pressure. (p. 306)

Brønsted-Lowry theory of acids and bases An acid-base theory that defines an acid as a substance that donates a proton to another substance, and a base as a substance that accepts a proton from another substance. (p. 434)

energía de enlace La cantidad de energía necesaria para romper un enlace químico. (p. 531)

par enlazado Par de electrones que están siendo compartidos en un enlace covalente entre dos átomos. (p. 163)

ley de Boyle La ley científica que establece que el volumen de una cantidad dada de un gas a una temperatura dada es inversamente proporcional a la presión. (p. 306)

teoría de ácidos y bases de Brønsted-Lowry Teoría ácido-base que define un ácido como una sustancia que dona protones a otra, y una base como una sustancia que recibe un protón de otra. (p. 434)

C

calorie (cal) A unit of measurement for thermal energy. The amount of heat required to raise the temperature of 1 gram of water by 1 Celsius degree. A kilocalorie, 1000 calories, is equal to one food Calorie. (p. 497)

calorimetry A procedure used to measure the heat transfer that occurs as a result of chemical reactions or physical changes. (p. 517)

catalyst A substance that accelerates a chemical reaction but is itself not permanently consumed or altered by the reaction. (p. 181)

cation An ion that has a positive charge. (p. 96)

Charles's law The law that states that the volume of a given sample of gas is proportional to its Kelvin temperature if the pressure is unchanged. (p. 284)

chemical bond An attraction between atoms that holds them together. (p. 129)

chemical change See **chemical reaction.**

chemical equation A representation of a chemical reaction written with chemical symbols and formulas. (p. 179)

chemical equilibrium A dynamic state in which opposing processes occur at the same time and the rate of the forward process is equal to the rate of the reverse process. (p. 620)

caloría (cal) Unidad de medida de la energía térmica. La cantidad de calor que se necesita para elevar la temperatura de 1 gramo de agua en 1 grado centígrado. Una kilocaloría, 1000 calorías, es igual a una caloría alimenticia. (p. 497)

calorimetría Procedimiento usado para medir la transferencia de calor que se produce como resultado de las reacciones químicas o de los cambios físicos. (p. 517)

catalizador Sustancia que acelera una reacción química pero que no se consume o se transforma de manera permanente por dicha reacción. (p. 181)

catión Ion que tiene una carga positiva. (p. 96)

ley de Charles La ley que establece que el volumen de una cantidad dada de un gas es proporcional a su temperatura en grados Kelvin si la presión no cambia. (p. 284)

enlace químico Fuerza de atracción entre átomos que los mantiene unidos. (p. 129)

cambio químico Ver **reacción química.**

ecuación química Representación de una reacción química escrita con símbolos químicos y fórmulas. (p. 179)

equilibrio químico Estado dinámico en el que se producen simultáneamente reacciones opuestas y la velocidad de la reacción hacia delante es igual a la velocidad de la reacción inversa. (p. 620)

chemical formula A combination of element symbols and numbers representing the composition of a chemical compound. (p. 24)

chemical reaction (chemical change) A transformation that alters the composition of one or more substances such that one or more new substances with new properties are produced. (p. 28)

chemical symbol A one- or two-letter representation of an element. The first letter is always uppercase. If there is a second letter, it is lowercase. (p. 24)

chemistry The study of substances, their properties, and how they can be transformed; the study of matter and how it can be changed. (p. 8)

coefficients The numbers in front of the chemical formulas of the reactants and products in a balanced chemical equation. They indicate the correct ratio in which the reactants combine to form the products. (p. 372)

combination reaction A reaction in which two or more reactants combine to form a single product. (p. 375)

combined gas law The law that describes the proportional relationship among the pressure, temperature, and volume of a gas. It states that the value of PV/T will be constant for a given sample of any gas. (p. 318)

combustion An exothermic chemical reaction between a fuel and oxygen, often producing flames; burning. (p. 513)

complete ionic equation A chemical equation that shows all of the soluble ionic compounds as independent ions. (p. 459)

compound A pure substance that is a chemical combination of two or more elements in a specific ratio. (p. 24)

concentration A measure of the amount of solute dissolved per unit of volume of solution, often expressed as moles of solute per liter of solution, mol/L. (p. 410)

fórmula química Combinación de símbolos de elementos y números que representan la composición de un compuesto químico. (p. 24)

reacción química (cambio químico) Transformación que altera la composición de una o más sustancias de tal manera que se producen una o más sustancias nuevas con propiedades nuevas. (p. 28)

símbolo químico Representación de un elemento por medio de una o dos letras. La primera letra es siempre mayúscula y la segunda, si la hay, es minúscula. (p. 24)

química El estudio de las sustancias, sus propiedades y cómo se pueden transformar; el estudio de la materia y cómo se puede transformar. (p. 8)

coeficientes Los números delante de las fórmulas químicas de los reactantes y los productos en una ecuación química balanceada. Indican la velocidad de reacción exacta a la que los reactantes se combinan para formar productos. (p. 372)

reacción de combinación Reacción en la que dos o más reactantes se combinan para formar un solo producto. (p. 375)

ley combinada de los gases La ley que describe la relación proporcional entre la presión, la temperatura y el volumen de un gas. Establece que el valor de PV/T será constante para una cantidad dada de un gas. (p. 318)

combustión Reacción química exotérmica entre un combustible y oxígeno que suele producir llamas; fuego. (p. 513)

ecuación iónica completa Ecuación química que muestra todos los compuestos iónicos solubles como iones independientes. (p. 459)

compuesto Sustancia pura que es una combinación química de dos o más elementos en una proporción determinada. (p. 24)

concentración Medida de la cantidad de soluto diluido por unidad de volumen de solución, usualmente expresada en moles de soluto por litro de solución, mol/L. (p. 410)

conductivity A property that describes how well a substance transmits electricity, heat, or sound. (p. 125)

core electrons All electrons in an atom that are not valence electrons. (p. 93)

covalent bonding A type of chemical bonding in which one or more pairs of valence electrons are shared between the atoms. (p. 130)

conductividad Propiedad que describe la capacidad de una sustancia de transmitir electricidad, calor o sonido. (p. 125)

electrones internos Todos los electrones de un átomo que no son electrones de valencia. (p. 93)

enlace covalente Enlace químico entre átomos en el que uno o más pares de electrones de valencia son compartidos por dichos átomos. (p. 130)

D

daughter isotope An isotope that is formed as a result of a nuclear reaction. (p. 78)

decomposition reaction A chemical change in which a single substance is broken down into two or more simpler substances. (p. 376)

density The measure of the mass of a substance per unit of volume, often expressed as grams per milliliter, g/mL, or grams per cubic centimeter, g/cm^3. (p. 17)

diatomic molecule A molecule consisting of two atoms. (p. 228)

dilution The process of adding solvent to a solution to lower the concentration of solute. (p. 442)

dipole A molecule or covalent bond with a nonsymmetrical distribution of electrical charge that makes the molecule or bond polar. (p. 224)

dissociate To break apart to form ions in solution. (p. 433)

dissolve To disperse a substance homogeneously into another substance at the molecular, ionic, or atomic level. (p. 125)

double bond A covalent bond where four electrons are shared between two atoms. (p. 166)

double exchange reaction (double displacement reaction) A chemical change in which both reactants break apart and then recombine to form two new products; chemical change where there is an exchange of ions between reactants to form new products. (p. 377)

isótopo descendiente Isótopo que se forma como resultado de una reacción nuclear. (p. 78)

reacción de descomposición Cambio químico en el que una sola sustancia se descompone en dos o más sustancias más simples. (p. 376)

densidad La medida de la masa de una sustancia por unidad de volumen, la cual se expresa usualmente en gramos por mililitro, g/mL, o gramos por centímetro cúbico, g/cm^3. (p. 17)

molécula diatómica Molécula compuesta por dos átomos. (p. 228)

dilución El proceso de añadir solvente a una solución para bajar la concentración del soluto. (p. 442)

dipolo Molécula o enlace covalente con una distribución asimétrica de la carga eléctrica que hace que dicha molécula o enlace sea polar. (p. 224)

disociar Se romper para formar iones en solución. (p. 433)

disolver Dispersar un sustancia de manera homogénea en otra sustancia a nivel molecular, iónico o atómico. (p. 125)

doble enlace Enlace covalente en el que dos átomos comparten cuatro electrones. (p. 166)

reacción de doble intercambio (reacción de doble desplazamiento) Cambio químico en el que los dos reactantes se rompen y se vuelven a combinar para formar dos nuevos productos; cambio químico en el que hay un intercambio de iones entre los reactantes para formar productos nuevos. (p. 377)

E

electrochemical cell A device used for generating electrical energy from chemical reactions. The electrical current is caused by substances releasing and accepting electrons in oxidation-reduction reactions. (p. 570)

electrolyte solution A solution that contains ions and can conduct electricity. (p. 569)

electromagnetic radiation A large range of waves, some visible, some invisible, all traveling at the speed of light, in a vacuum, and at widely varying wavelengths and frequencies. (p. 586)

electron An elementary particle with a negative charge that is located outside of the nucleus of an atom. It has a mass of about 1/1838 amu. (p. 54)

electron configuration A notation for keeping track of where the electrons in an atom are distributed among the shells and subshells in an atom. (p. 117)

electron domain The space occupied by bonded pairs or lone pairs of valence electrons in a molecule. Electron domains affect the overall shape of a molecule. (p. 192)

electron domain theory A scientific theory that states that every electron domain is located as far away as possible from every other electron domain in a molecule. (p. 193)

electronegativity A measure of the ability of an atom in a molecular substance to attract electrons to itself. (p. 224)

electroplating The process by which a material is coated with a thin layer of metal using electricity passed through a suitable ionic solution. (p. 136)

element A unique substance that cannot be broken down into simpler substances through physical or chemical processes. Elements serve as the building materials of all matter. (p. 24)

emission In general, transmission of energy from a source. Of light, a transmission of energy from the Sun or other bright body. (p. 576)

celda electroquímica Dispositivo usado para generar energía con reacciones químicas. La corriente eléctrica se genera mediante sustancias que ceden y reciben electrones en reacciones de oxidación-reducción. (p. 570)

solución de electrolitos Solución que contiene iones y que es conductora de la electricidad. (p. 569)

radiación electromagnética Una amplia gama de ondas, algunas visibles, otras invisibles, todas viajando a la velocidad de la luz (en el vacío) y a muy diversas longitudes de onda y frecuencias. (p. 586)

electrón Partícula elemental con carga negativa que está fuera del núcleo de un átomo. Tiene una masa de 1/1838 uma, aproximadamente. (p. 54)

configuración electrónica Notación que sirve para llevar la cuenta de cómo están distribuidos los electrones de un átomo en las capas y las subcapas electrónicas. (p. 117)

dominio del electrón Espacio que ocupan los pares de electrones de valencia enlazados o aislados en una molécula. El dominio de los electrones afecta la forma de una molécula. (p. 192)

teoría del dominio del electrón Teoría científica que establece que el dominio de cada electrón se encuentra lo más lejos posible del dominio de otro electrón en una molécula. (p. 193)

electronegatividad Medida de la capacidad de atraer electrones que tiene un átomo en una sustancia molecular. (p. 224)

galvanostegia Proceso mediante el cual se cubre un material con una capa delgada de metal haciendo pasar electricidad a través de una solución iónica especial. (p. 136)

elemento Sustancia única que no se puede descomponer en sustancias más simples a través de procesos físicos o químicos. Los elementos son los componentes que constituyen toda la materia. (p. 24)

emisión Se refiere en términos generales a la transmisión de energía a partir de una fuente. De luz: transmisión de energía del sol o de otro cuerpo incandescente. (p. 576)

endothermic Describes a process in which heat transfers from the surroundings to the system; heat-absorbing. (p. 487)

energy A measure of the capacity of an object or system to do work or produce heat. (p. 483)

entropy The tendency of energy or matter in a system to disperse; the energy of a system that cannot be used for external work. (p. 492)

equilibrium constant K A measure of whether starting substances or products are favored in an equilibrium mixture. A large value for K indicates a large amount of product. A small value for K indicates very little product. (p. 627)

equilibrium-constant equation A specific mathematical relationship between K and the concentrations of starting substances and products in an equilibrium mixture. (p. 631)

equivalence point The point in an acid-base titration when the acid and base have completely neutralized each other. (p. 451)

evaporation The phase change from a liquid to a gas. (p. 297)

exothermic Describes a process in which heat is transferred from a system to the surroundings; heat-releasing. (p. 484)

extensive property A characteristic, such as volume or mass, that is specific to the amount of matter and therefore changes if the quantity of the substance changes. (p. 19)

F

first law of thermodynamics The scientific law that states that energy is conserved; therefore it cannot be created or destroyed. (p. 492)

fission (nuclear) The splitting apart of an atomic nucleus into two smaller nuclei, accompanied by a release of energy. (p. 74)

endotérmico Describe un proceso en el que se transfiere calor del entorno al interior del sistema; absorción de calor. (p. 487)

energía Medida de la capacidad que tiene un objeto o un sistema de hacer trabajo o producir energía. (p. 483)

entropía La tendencia de la energía o la materia de un sistema a dispersarse; la energía de un sistema que no puede usarse para trabajo externo. (p. 492)

constante de equilibrio K Medida que indica si las sustancias iniciales o los productos finales son favorecidos en una mezcla en equilibrio. Un valor grande de K indica una gran cantidad de producto final. Un valor pequeño de K indica muy poco producto final. (p. 627)

ecuación de la constante de equilibrio Relación matemática específica entre K y las concentraciones de las sustancias iniciales y los productos finales en una mezcla en equilibrio. (p. 631)

punto de equivalencia Momento, en una valoración química ácido-base, en que el ácido y la base se han neutralizado totalmente entre sí. (p. 451)

evaporación El cambio de estado de líquido a gas. (p. 297)

exotérmico Describe un proceso en el que el calor se transfiere del interior del sistema al entorno; que emite calor. (p. 484)

propiedad extensiva Característica, como el volumen o la masa, que es específica a una cantidad de materia y por tanto, cambia si la cantidad de la sustancia cambia. (p. 19)

primera ley de la termodinámica La ley científica que establece que la energía se conserva y por tanto no se puede crear ni destruir. (p. 492)

fisión (nuclear) La ruptura del núcleo de un átomo en dos núcleos más pequeños, que viene acompañada de liberación de energía. (p. 74)

flame test A laboratory procedure used to determine the presence of certain metal atoms in a chemical sample by heating the sample in a flame and observing the resulting flame color. (p. 85)

formula unit The simplest chemical formula that can be used to represent network covalent or ionic compounds that shows the elements present in the smallest whole number ratio. (p. 372)

frequency The number of waves that pass by a certain point in space per second, abbreviated as *f*. The frequency of light is measured in hertz, Hz. (p. 582)

functional group A structural feature of a molecule; consists of a specific arrangement of atoms, responsible for certain properties of the compound. (p. 170)

fusion (nuclear) The joining of two atomic nuclei to form a larger nucleus, accompanied by a release of energy. (p. 75)

G

gamma ray Electromagnetic radiation with the shortest wavelengths, less than 10^{-11} m. Emitted when a nucleus decays or during a nuclear explosion. Used in irradiation of food and in some cancer treatments. (p. 74)

Gay-Lussac's law The scientific law that states that the pressure of a given amount of gas is directly proportional to temperature, if the gas volume does not change. (p. 310)

group A vertical column on the periodic table, also called a family. Elements in a group have similar properties. (p. 44)

H

half-cell One-half of an electrochemical cell, the site of either oxidation or reduction. (p. 570)

half-life The amount of time required for one-half of the radioactive atoms in a sample to decay. (p. 73)

ensayo a la llama Procedimiento de laboratorio que se usa para determinar la presencia de átomos de un determinado metal en una muestra química, introduciendo una muestra en una llama y observando la coloración que adquiere la llama. (p. 85)

unidad de fórmula La fórmula química más simple que se puede usar para representar una red de compuestos iónicos o covalentes, y que muestra la proporción simplificada de los elementos presentes con números enteros. (p. 372)

frecuencia Número de ondas que pasan por un punto determinado en el espacio en un segundo, su abreviatura es *f*. La frecuencia de la luz se mide en hercios, Hz. (p. 582)

grupo funcional Característica estructural de una molécula que consiste en una distribución específica de los átomos, y a la que se deben ciertas propiedades de un compuesto. (p. 170)

fusión (nuclear) La unión de dos núcleos atómicos para formar un núcleo más grande, que viene acompañada de liberación de energía. (p. 75)

rayo gamma Radiación electromagnética con las longitudes de onda más cortas de menos de 10^{-11} m. Se emiten cuando un núcleo decae o durante una explosión nuclear. Se utiliza en la irradiación de alimentos y en algunos tratamientos contra el cáncer. (p. 74)

ley de Gay-Lussac La ley científica que establece que la presión de una cantidad dada de un gas es directamente proporcional a la temperatura, si su volumen no cambia. (p. 310)

grupo Columna vertical en la tabla periódica, también denominada familia. Los elementos de un grupo tienen propiedades similares. (p. 44)

semicelda La mitad de una celda electroquímica, en la que puede producirse oxidación o reducción. (p. 570)

vida media El tiempo que se necesita para que la mitad de los átomos radiactivos de una muestra se desintegren. (p. 73)

half-reaction A chemical equation that represents either the oxidation or the reduction part of an oxidation-reduction reaction. (p. 570)

halogens The elements in Group 7A on the periodic table. (p. 44)

heat A transfer of energy between two substances, from a hotter body to a colder body. (p. 486)

heat of combustion The energy released as heat when a compound undergoes complete combustion with oxygen, usually expressed per mole of the fuel. (p. 524)

heat of formation The energy transferred as heat during a chemical reaction when a compound is formed from its constituent elements, expressed per mole of product. (p. 554)

heat of fusion The amount of heat transfer required to change a substance from a solid to a liquid, usually expressed per mole or gram. (p. 506)

heat of reaction The amount of energy transferred as heat during a chemical reaction. (p. 524)

heat of vaporization The amount of heat transfer required to change a substance from a liquid to a gas, usually expressed per mole or gram. (p. 506)

heterogeneous mixture A mixture whose composition is not uniform throughout. (p. 410)

homogeneous mixture A mixture whose composition is uniform throughout. (p. 410)

HONC 1234 rule A rule that states that in most molecules, hydrogen makes 1 bond, oxygen makes 2 bonds, nitrogen makes 3 bonds, and carbon makes 4 bonds. (p. 156)

humidity The concentration of the water vapor in the air at any given time. (p. 340)

semirreacción Ecuación química que puede representar la oxidación o la reducción de una reacción de oxidación-reducción. (p. 570)

halógenos Los elementos del grupo 7A de la tabla periódica. (p. 44)

calor Una transferencia de energía entre dos sustancias, de un cuerpo más caliente a uno más frío. (p. 486)

calor de combustión La energía liberada en forma de calor cuando un compuesto pasa por una combustión completa con oxígeno. Usualmente se expresa por mol de combustible. (p. 524)

calor de formación La energía que se transfiere en forma de calor durante una reacción química cuando se produce un compuesto a partir de sus elementos constituyentes. Se expresa por mol de producto. (p. 554)

calor de fusión La cantidad de calor que debe ser transferida para que una sustancia pase de ser un sólido a ser un líquido. Usualmente se expresa por mol o gramo. (p. 506)

calor de reacción La cantidad de energía que se transfiere en forma de calor durante una reacción química. (p. 524)

calor de vaporización La cantidad de calor que debe ser transferida para que una sustancia pase de ser un líquido a ser un gas. Usualmente se expresa por mol o gramo. (p. 506)

mezcla heterogénea Mezcla cuya composición no es uniforme. (p. 410)

mezcla homogénea Mezcla cuya composición es uniforme. (p. 410)

regla de HONC 1234 Regla que establece que en la mayoría de moléculas, el hidrógeno hace 1 enlace, el oxígeno hace 2 enlaces, el nitrógeno hace 3 enlaces y el carbono hace 4 enlaces. (p. 156)

humedad La concentración de vapor de agua en el aire en un momento determinado. (p. 340)

hydrogen bond An intermolecular attraction between a hydrogen atom in a molecule and an electronegative atom in another molecule (especially nitrogen, oxygen, or fluorine). The hydrogen atoms in water molecules form hydrogen bonds with the oxygen atoms in other water molecules. (p. 503)

hypothesis A proposed explanation for an observation or scientific problem, which can be tested by further investigation. (p. 7)

I

ideal gas law The scientific law that relates volume, pressure, temperature, and the number of moles of a gas sample: $PV = nRT$, where R is the universal gas constant. (p. 337)

indicator (acid-base) A chemical compound that indicates the relative acidity or basicity of a solution through its characteristic color changes. (p. 429)

infrared radiation Electromagnetic radiation with longer wavelengths than those of visible light, from 10^{-3} m to 10^{-6} m. Causes molecules to vibrate, which human beings sense as heat. Abbreviated as IR radiation. (p. 587)

insoluble Unable to be dissolved in another substance. (p. 125)

intensive property A characteristic, such as boiling point or density, that does not depend on the size or amount of matter and can be used to identify matter. (p. 19)

intermolecular force A force of attraction that occurs between molecules. (p. 216)

inversely proportional Related in such a way that when one quantity increases, the other decreases in a mathematically predictable way. The variables x and y are inversely proportional to each other if $y = k/x$, where k is the proportionality constant. (p. 306)

ion An atom or group of bonded atoms that has a positive or negative charge. (p. 95)

enlace de hidrógeno Atracción intermolecular entre un átomo de hidrógeno en una molécula y un átomo electronegativo en otra (especialmente de nitrógeno, oxígeno o flúor). Los átomos de hidrógeno en las moléculas de agua forma enlaces con átomos de oxígeno en otras moléculas de agua. (p. 503)

hipótesis Una explicación dada a una observación o a un problema científico, la cual se puede probar con más investigación. (p. 7)

ley de los gases ideales La ley científica que relaciona el volumen, la presión, la temperatura y el número de moles de una muestra de gas: $PV = nRT$, cuando R es la constante universal de los gases ideales. (p. 337)

indicador (ácido-base) Compuesto químico que sirve para mostrar la acidez o la basicidad relativa de una solución a través de cambios de color característicos. (p. 429)

radiación infrarroja Radiación electromagnética con longitudes de onda más largas que las de la luz visible, entre 10^{-3} m y 10^{-6} m. Causa la vibración de las moléculas, lo que los seres humanos perciben como calor. Su abreviatura es IR radiación. (p. 587)

insoluble Que no se puede disolver en otra sustancia. (p. 125)

propiedad intensiva Característica como el punto de ebullición o la densidad que no depende del tamaño o la cantidad de materia y que puede ser usada para identificar dicha materia. (p. 19)

fuerza intermolecular Fuerza de atracción que se produce entre moléculas. (p. 216)

inversamente proporcional Relacionadas de tal manera que cuando una variable aumenta, la otra disminuye de forma que se puede predecir matemáticamente. Las variables x y y son inversamente proporcionales si $y = k/x$, donde k es la constante de proporcionalidad. (p. 306)

ion Átomo, o grupo de átomos enlazados, que tiene carga positiva o negativa. (p. 95)

ionic bonding A type of chemical bonding that is the result of the transfer of electrons from one atom to another, typically between metal and nonmetal atoms. (p. 130)

ionic compound A compound that consists of positively charged metal cations and negatively charged nonmetal anions formed when valence electrons are transferred. (p. 98)

isomers Compounds with the same molecular formula but different structural formulas. Isomers differ in molecular structure and in chemical and physical properties. (p. 152)

isotopes Atoms of the same element that have different numbers of neutrons. These atoms have the same atomic number but different mass numbers. (p. 62)

J

joule (J) A unit of measurement of energy. One calorie is equal to 4.184 joules. (p. 524)

K

Kelvin scale A temperature scale with units in kelvins, K, that sets the zero point at −273.15 °C, which is also known as absolute zero. Kelvin units are equivalent in scale to Celsius units. (p. 279)

kinetic energy The energy of motion. (p. 488)

kinetic theory of gases The scientific theory that states that gases are composed of tiny particles in continuous, random, straight-line motion and collide with each other and the walls of the container. (p. 280)

L

lanthanides A series of elements that follow lanthanum in Period 6 of the periodic table; they are typically placed separately at the bottom of the periodic table. (p. 44)

law of conservation of mass The scientific law that states that mass cannot be gained or lost in a chemical reaction and that matter cannot be created or destroyed. (p. 34)

enlace iónico Atracción entre átomos que es el resultado de la transferencia de electrones de un átomo a otro, y que es común entre átomos de metales y de no metales. (p. 130)

compuesto iónico Compuesto que contiene cationes de metales con carga positiva y aniones de no metales con carga negativa que se forman al transferirse los electrones de valencia. (p. 98)

isómeros Compuestos que tienen una misma fórmula molecular pero fórmulas estructurales diferentes. Los isómeros son distintos en su estructura molecúlar y en sus propiedades químicas y físicas. (p. 152)

isótopos Átomos de un mismo elemento que tienen distinta cantidad de neutrones. Estos átomos tienen un mismo número atómico pero diferente número de masa. (p. 62)

julio (J) Unidad de medida de la energía. Una caloría es igual a 4.184 julios. (p. 524)

escala Kelvin Escala de temperatura cuyas unidades son los kelvins, K, con el punto cero en −273.15 °C, temperatura también conocida como cero absoluto. Las unidades Kelvin son equivalentes en escala a las unidades Celsius. (p. 279)

energía cinética La energía del movimiento. (p. 488)

teoría cinética de los gases La teoría científica que establece que los gases están compuestos por pequeñas partículas en movimiento continuo, aleatorio y en línea recta, y que se chocan entre sí y con las paredes que las contienen. (p. 280)

lantánidos Conjunto de elementos que están después del lantano en el sexto período de la tabla periódica y que usualmente aparecen en la parte de abajo de la tabla. (p. 44)

ley de conservación de la masa La ley científica que establece que en una reacción química no se gana ni se pierde masa y que la materia no se puede crear ni destruir. (p. 34)

Le Châtelier's principle An experimental observation that, when a stress is put on a system at equilibrium, the system responds by reducing the effect of the change. (p. 643)

Lewis dot structure A diagram that shows a molecule's structure by using dots to represent the valence electrons. (p. 162)

Lewis dot symbol A diagram that uses dots to represent the valence electrons of a single atom. (p. 162)

light detector A device that senses electromagnetic radiation, such as a radio or a photovoltaic cell. (p. 593)

light ray A model of light that shows the straight-line path that light travels and how a beam of light interacts with matter. (p. 577)

light wave Ripples of light energy created by forces that oscillate up and down, radiating in space from a source, such as the Sun. (p. 582)

limiting reactant The reactant that runs out first in a chemical reaction. It is the reactant that limits the amount of product that can be produced in the reaction. (p. 464)

linear shape A straight-line shape found in small molecules. (p. 197)

lone pair A pair of unshared valence electrons that are not involved in bonding in a molecule. (p. 163)

M

main group elements The elements in Groups 1A to 7A on the periodic table. (p. 44)

mass A measure of the quantity of matter in an object. (p. 9)

mass number The sum of the number of protons and neutrons in the nucleus of an atom. (p. 62)

matter Anything that has substance and takes up space; anything that has mass and volume. (p. 9)

principio de Le Châtelier Observación experimental que, cuando se somete un sistema en equilibrio a cierto nivel de perturbación externa, el sistema responde a fin de reducir el efecto de la perturbación. (p. 643)

estructura de puntos de Lewis Diagrama de la estructura de una molécula en el que se usan puntos para representar los electrones de valencia. (p. 162)

símbolo de puntos de Lewis Diagrama en el que se usan puntos para representar los electrones de valencia de un solo átomo. (p. 162)

detector de luz Dispositivo que detecta la radiación electromagnética, p. ej. una radio o una célula fotovoltaica. (p. 593)

rayo de luz Modelo de la luz que muestra la ruta en línea recta trazada por la luz y cómo un rayo de luz interactúa con la materia. (p. 577)

onda de luz Ondas de energía lumínica creadas por fuerzas que oscilan hacia arriba y hacia abajo; estas ondas se irradian en el espacio a partir de una fuente tal como el Sol. (p. 582)

reactante limitador El reactante que se acaba primero en una reacción química. Es el que limita la cantidad de producto que se puede producir en una reacción. (p. 464)

forma recta Forma de línea recta que aparece en moléculas pequeñas. (p. 197)

par aislado Par de electrones de valencia que no están siendo compartidos y que no están involucrados en el enlace de una molécula. (p. 163)

elementos del grupo principal Los elementos de los grupos 1A a 7A de la tabla periódica. (p. 44)

masa Medida de la cantidad de materia en un objeto. (p. 9)

número de masa La suma del número de protones y neutrones en el núcleo de un átomo. (p. 62)

materia Todo lo que tiene sustancia y ocupa un lugar en el espacio; todo lo que tiene masa y volumen. (p. 9)

melting point (melting temperature) The temperature at which both solid and liquid phases of a single substance can be present and in equilibrium. (p. 274)

meniscus The curvature of the top of a liquid in a container, which is the result of intermolecular attractions between the liquid and the container. (p. 12)

metal An element that is generally shiny and malleable and an excellent conductor of heat and electricity. Metals are located to the left of the stairstep line on the periodic table. (p. 45)

metallic bonding A type of bonding between metal atoms in which the valence electrons are free to move throughout the substance. (p. 130)

metalloid An element that has properties of both metals and nonmetals. Metalloids are located along the stair-step line of the periodic table. (p. 46)

microwave radiation Electromagnetic radiation with wavelengths longer than visible light and infrared radiation from 1 m to 10^{-3} m. Absorption of these waves cooks food. (p. 587)

mirror-image isomer Molecules whose structures are mirror images of each other and cannot be superimposed on one another. (p. 243)

mixture A blend of two or more substances that are not chemically combined. (p. 24)

model A simplified representation of a real object or process that facilitates understanding or explanation of that object or process. (p. 52)

molarity The concentration of dissolved substances in a solution, expressed in moles of solute per liter of solution. (p. 410)

molar mass The mass in grams of one mole of a substance. (p. 391)

mole A counting unit used to keep track of large numbers of particles. One mole represents 6.02×10^{23} items. (p. 332)

mole ratio The ratio of the moles of one reactant or product to the moles of another reactant or product in a balanced chemical equation. (p. 463)

punto de fusión (temperatura de fusión) La temperatura a la cual los estados sólido y líquido de un misma sustancia están presentes y en equilibrio. (p. 274)

menisco La curvatura que se forma en la parte superior de un líquido que está dentro de un envase y que es el resultado de las atracciones intermoleculares entre el líquido y el envase. (p. 12)

metal Elemento que generalmente es brillante y maleable y un excelente conductor del calor y de la electricidad. Los metales están a la izquierda de la línea escalonada en la tabla periódica. (p. 45)

enlace metálico Enlace entre átomos de un metal en el que los electrones de valencia se mueven libremente por la sustancia. (p. 130)

metaloide Elemento que tiene propiedades tanto de los metales como de los no metales. Los metaloides están sobre la línea escalonada de la tabla periódica. (p. 46)

radiación de microondas Radiación electromagnética con longitudes de onda más largas que la luz visible y la radiación infrarroja, de 1 m a 10^{-3} m. La absorción de estas ondas cocina los alimentos. (p. 587)

isómero especular Moléculas cuyas estructuras se reflejan entre sí y no se pueden superponer. (p. 243)

mezcla Combinación de dos o más sustancias que no están combinadas químicamente. (p. 24)

modelo Representación simplificada de un objeto o proceso real que facilita la comprensión o explicación de dicho objeto o proceso. (p. 52)

molaridad La concentración de sustancias disueltas en una solución. Se expresa en moles de soluto por litro de solución. (p. 410)

masa molar La masa en gramos de un mol de una sustancia. (p. 391)

mol Unidad de conteo usada para llevar la cuenta de grandes cantidades de partículas. Un mol representa 6.02×10^{23} partículas. (p. 332)

razón molar La proporción entre los moles de un reactante o producto y los moles de otro reactante o producto en una ecuación química balanceada. (p. 463)

molecular covalent bonding A type of chemical bonding characterized by the sharing of valence electrons between atoms, resulting in individual units called molecules. (p. 130)

molecular formula The chemical formula of a molecular substance, showing the identity of the atoms in each molecule and the ratios of those atoms to one another. (p. 148)

molecule A group of atoms that are covalently bonded together. (p. 131)

monatomic ion An ion that consists of only one atom. (p. 109)

N

net ionic equation A chemical equation that is written without including spectator ions. (p. 460)

network covalent bonding A type of chemical bonding characterized by the sharing of valence electrons throughout the entire solid sample. (p. 130)

neutralization reaction A chemical reaction in which an acid and base react to form a salt and water. (p. 448)

neutron A particle that is located in the nucleus of an atom and does not have an electric charge. The mass of a neutron is almost exactly equal to that of a proton, about 1 amu. (p. 54)

noble gases The elements in Group 8A on the periodic table. Noble gases are known for not being reactive. (p. 44)

nonmetal An element that does not exhibit metallic properties. Nonmetals are often gases or brittle solids at room temperature. Nonmetals are poor conductors of heat and electricity and are located to the right of the stair-step line on the periodic table. (p. 45)

nonpolar molecule A molecule that is not attracted to an electrical charge. A molecule is nonpolar if each atom shares electrons equally or there is no net dipole in the molecule. (p. 215)

enlace covalente molecular Tipo de enlace químico caracterizado por átomos que comparten electrones de valencia y forman unidades individuales llamadas moléculas. (p. 130)

fórmula molecular La fórmula química de una sustancia molecular que muestra la identidad de los átomos de cada molécula y la proporción de dichos átomos con respecto a los demás. (p. 148)

molécula Grupo de átomos que están unidos por enlaces covalentes. (p. 131)

ion monoatómico Ion que contiene un solo átomo. (p. 109)

ecuación iónica total Ecuación química en la que no se incluyen los iones espectadores. (p. 460)

enlace covalente encadenado Tipo de enlace químico caracterizado porque se comparten los electrones de valencia en toda la muestra sólida. (p. 130)

reacción de neutralización Reacción química en la que un ácido y una base reaccionan para formar sal y agua. (p. 448)

neutrón Partícula que está en el núcleo de un átomo y que no tiene carga eléctrica. La masa de un neutrón es casi igual a la de un protón, 1 uma, aproximadamente. (p. 54)

gases nobles Los elementos del grupo 8A de la tabla periódica. Los gases nobles se conocen por no ser reactivos. (p. 44)

no metal Elemento que no tiene propiedades metálicas. Los no metales suelen ser gases o sólidos quebradizos cuando están a temperatura ambiente. Los no metales no son buenos conductores del calor ni de la electricidad, y están a la derecha de la línea escalonada en la tabla periódica. (p. 45)

molécula apolar Molécula que no es atraída por una carga eléctrica. Una molécula es apolar si cada átomo comparte electrones equitativamente o si no hay bipolaridad neta en la molécula. (p. 215)

nuclear chain reaction A nuclear reaction in which neutrons emitted from the nucleus during fission strike surrounding nuclei, causing them to split apart as well. (p. 79)

nuclear equation A representation of a nuclear reaction written with isotope symbols. (p. 78)

nuclear reaction A process that changes the energy, composition, or structure of an atom's nucleus. (p. 71)

nucleus The dense, positively charged structure composed of protons and neutrons that is found in the center of an atom. (p. 54)

number density The number of gas particles per unit volume usually expressed in moles per liter or moles per cubic centimeter. (p. 328)

O

octet rule Nonmetal atoms combine by sharing electrons so that each atom has a total of eight valence electrons. After bonding, each atom resembles a noble gas in its electron arrangements. (p. 165)

opaque Not capable of transmitting light and, therefore, an object or surface not possible to see through. (p. 578)

oxidation A chemical reaction in which an atom or ion loses electrons. Oxidation is always accompanied by reduction. (p. 552)

oxidation-reduction reaction A chemical reaction that involves electron transfer. (p. 560)

P

parent isotope A radioactive isotope that undergoes decay. (p. 78)

partial charge A less than full charge on part of a molecule, created by the unequal sharing of electrons. Partial charges are represented with the symbol delta ($\delta+$ for partial positive charge and $\delta-$ for partial negative charge). (p. 215)

partial pressure The pressure exerted by one gas in a mixture of nonreacting gases. The partial pressures of all the gases add up to the total pressure exerted by that mixture of gases. (p. 341)

reacción nuclear en cadena Reacción nuclear en la que los neutrones emitidos desde el núcleo durante la fisión chocan con los núcleos alrededor haciendo que estos también se dividan. (p. 79)

ecuación nuclear Representación de una reacción nuclear que se expresa con símbolos de isótopos. (p. 78)

reacción nuclear Proceso que transforma la energía, la composición o la estructura del núcleo de un átomo. (p. 71)

núcleo Estructura densa de carga positiva que está compuesta de protones y neutrones, y que se encuentra en el centro de un átomo. (p. 54)

densidad numérica El número de partículas de gas por unidad de volumen que usualmente se expresa en moles por litro o moles por centímetro cúbico. (p. 328)

regla del octeto Los átomos de no metales se combinan con otros átomos hasta que cada uno queda rodeado por ocho electrones de valencia. Después de enlazarse, cada átomo se asemeja a un gas noble en la distribución de sus electrones. (p. 165)

opaco Que no es capaz de transmitir la luz. En el caso de superficies u objetos opacos, no es posible ver a través de ellos. (p. 578)

oxidación Reacción química en la que un átomo o un ion pierde electrones. La oxidación siempre viene acompañada de reducción. (p. 552)

reacción de oxidación-reducción Reacción química que involucra transferencia de electrones. (p. 560)

isótopo precursor Isótopo radiactivo que sufre desintegración. (p. 78)

carga parcial Una carga incompleta en un sector de una molécula que se crea porque se comparten electrones de manera desigual. Las cargas parciales se representan con el símbolo delta ($\delta+$ para cargas positivas y $\delta-$ para cargas negativas). (p. 215)

presión parcial La presión ejercida por un gas en una mezcla de gases no reactivos. La suma de la presión parcial de todos los gases es igual a la presión total ejercida por esa mezcla de gases. (p. 341)

peptide bond The bond between two amino acids; also called an amide bond. (p. 249)

percent error A calculation used to find the accuracy of a measurement. The lower the percent error, the more accurate the measurement. (p. 388)

percent yield A calculation that expresses the success of a chemical process in terms of product yield; the ratio of the actual yield to the theoretical yield expressed as a percentage. (p. 473)

period The elements in a horizontal row on the periodic table. (p. 44)

periodic table of the elements A table with elements organized in order of increasing atomic number and grouped such that elements with similar properties are in vertical columns. (p. 39)

pH scale A logarithmic scale describing the concentration of hydrogen ions, H^+, in solution. $pH = -\log [H^+]$. (p. 430)

phase The physical form of matter such as the solid, liquid, or gaseous state. (p. 25)

phase change A transition between solid, liquid, or gaseous states of matter. (p. 260)

photon A particle of light; one of many separate particles in a beam of light. (p. 590)

physical change A change that alters the form of a substance but does not change the chemical identity of a substance. (p. 259)

pigment A substance that is colored because of the light it reflects. Pigments reflect only a certain color instead of all the colors in white light from the source. (p. 609)

Planck's constant A mathematical relationship of the frequency of the electromagnetic wave to its energy. Given the symbol, h, and defined as $h = 6.63 \times 10^{-34}$ J · s. (p. 589)

polar molecule A molecule that has a negatively charged end and a positively charged end due to electronegativity differences between the atoms and/or the asymmetry of its structure. (p. 215)

enlace péptido El enlace entre dos aminoácidos; también se llama enlace amino. (p. 249)

error porcentual Cálculo que se usa para hallar la precisión de una medición. Entre más bajo es el error porcentual, más precisa es la medición. (p. 388)

rendimiento porcentual Cálculo que expresa el éxito de un proceso químico en términos del rendimiento del producto; la proporción entre el rendimiento real y el rendimiento teórico, expresada como un porcentaje. (p. 473)

período Los elementos en una línea horizontal de la tabla periódica. (p. 44)

tabla periódica de los elementos Tabla con todos los elementos organizados en orden ascendente de acuerdo a su número atómico y agrupados de tal manera que los elementos con propiedades similares están en columnas verticales. (p. 39)

escala de pH Escala logarítmica que describe la concentración de iones de hidrógeno, H^+, en una solución. $pH = -\log [H^+]$. (p. 430)

fase La forma física de la materia como el estado sólido, líquido o gaseoso. (p. 25)

cambio de fase Transición entre los estados sólido, líquido y gaseoso de la materia. (p. 260)

fotón Partícula de luz, una de las muchas partículas individuales que viajan en un rayo de luz. (p. 590)

cambio físico Cambio que altera la forma de una sustancia pero que no transforma la identidad química de la sustancia. (p. 259)

pigmento Sustancia que cambia el color de la luz reflejada. Los pigmentos reflejan solamente un determinado color en lugar de la luz blanca emitida por la fuente. (p. 609)

constante de Planck Relación matemática entre la frecuencia de la onda electromagnética y su energía. Su símbolo es h y se define como $h = 6.63 \times 10^{-34}$ J · s. (p. 589)

molécula polar Molécula que tiene un lado con carga negativa y otro con carga positiva debido a las diferencias de electronegatividad entre los átomos y/o a que su estructura es asimétrica. (p. 215)

polyatomic ion An ion that consists of two or more atoms covalently bonded. (p. 109)

potential energy Energy that is stored within a system and can be converted into other types of energy. This energy may be associated with the composition of a substance or its location. (p. 536)

precipitate A solid produced in a chemical reaction between two solutions. (p. 457)

precipitation reaction A chemical reaction that results in the formation of a solid substance (a precipitate) that separates out of a solution because it is not very soluble. (p. 457)

pressure Force applied over a specific area. Force per unit area. (p. 299)

product A substance produced as the result of a chemical reaction. (p. 180)

property A characteristic or quality of a substance. (p. 7)

proportional Related such that when one quantity increases, the other also increases. Two variables are proportional when you can multiply one variable by a constant to obtain the other. The variable y is proportional to the variable x if $y = kx$ where k is the proportionality constant. (p. 265)

proportionality constant The number that relates two variables that are proportional to one another. It is often represented by k. (p. 265)

protein A large molecule made up of chains of amino acids bonded together. Typical protein molecules consist of more than 100 amino acids. (p. 249)

proton A positively charged particle located in the nucleus of an atom. The mass of a proton is almost exactly equal to that of a neutron, about 1 amu. (p. 54)

pyramidal shape The shape assumed by other bonded atoms around an atom with one lone pair of electrons. (p. 193)

R

radiation Emission of energy as light either in the visible or invisible part of the spectrum. (p. 577)

ion poliatómico Ion que contiene dos o más átomos con enlaces covalentes. (p. 109)

energía potencial La energía que está almacenada en un sistema y que puede convertirse en otros tipos de energía. Esta energía se puede asociar con la composición de la sustancia o con su ubicación. (p. 536)

precipitado Sólido que se produce como resultado de una reacción química entre dos soluciones. (p. 457)

reacción de precipitación Reacción química que resulta en la formación de una sustancia sólida (un precipitado) que se separa de la solución porque no es muy soluble. (p. 457)

presión Fuerza aplicada sobre un área específica. Fuerza por unidad de área. (p. 299)

producto Sustancia que se produce como resultado de una reacción química. (p. 180)

propiedad Característica o cualidad de una sustancia. (p. 7)

proporcional Relacionadas de tal manera que cuando una cantidad aumenta, la otra también lo hace. Dos variables son proporcionales cuando una se puede multiplicar por una constante para obtener la otra. La variable y es proporcional a la variable x si $y = kx$ donde k es la constante de proporcionalidad. (p. 265)

constante de proporcionalidad El número que relaciona dos variables que son proporcionales entre sí. Generalmente se representa con la letra k. (p. 265)

proteína Molécula grande hecha de cadenas de aminoácidos enlazados. Las moléculas de proteínas suelen tener más de 100 aminoácidos. (p. 249)

protón Partícula de carga positiva localizada en el núcleo de un átomo. La masa de un protón es casi igual a la de un neutrón, 1 uma, aproximadamente. (p. 54)

forma piramidal Forma que adquieren otros átomos enlazados alrededor de un átomo con un par aislado de electrones. (p. 193)

radiación Emisión de energía en forma de luz, ya sea en la parte visible o invisible del espectro. (p. 577)

radio wave Electromagnetic radiation with very long wavelengths from 10^5 m to 1 m. Emitted at a television or radio station and used in communications satellites. (p. 587)

radioactive decay Spontaneous disintegration of an atomic nucleus accompanied by the emission of particles and radiation. A radioactive substance will decay with a specific half-life. (p. 71)

radioactive isotope Any isotope that has an unstable nucleus and decays over time. (p. 67)

reactants The starting materials in a chemical reaction that are transformed into products during the reaction. (p. 180)

reactivity The tendency of an element or compound to combine chemically with other substances, as well as the ease or speed of the reaction. (p. 37)

receptor site theory The currently accepted model explaining how specific molecules are detected by the nose. Molecules fit into receptor sites that correspond to the overall shape of the molecule. This stimulates a response in the body. (p. 208)

redox reaction Another name for an *oxidation-reduction* reaction. (p. 560)

reduction A chemical reaction in which an atom or ion gains electrons. Reduction is always accompanied by oxidation. (p. 559)

reflection The bouncing of light off a surface, allowing it to be seen. (p. 578)

relative humidity The amount of water vapor in the air compared to the maximum amount of water vapor possible for a specific temperature, expressed as a percentage. (p. 341)

reversible process A process that can proceed in both the forward direction to produce products and the opposite direction (reverse direction) to produce starting substances. (p. 613)

ondas de radio Radiación electromagnética con longitudes de onda muy largas, entre 10^5 m y 1 m. Emitidas por los canales de televisión o las estaciones de radio y se utilizan en los satélites de comunicaciones. (p. 587)

desintegración radiactiva Desintegración espontánea del núcleo de un átomo que viene acompañada por la emisión de partículas y radiación. Una sustancia radiactiva se desintegrará en una vida media específica. (p. 71)

isótopo radiactivo Cualquier isótopo que tiene un núcleo inestable y que se desintegra con el tiempo. (p. 67)

reactantes Los materiales iniciales de una reacción química que se transforman en productos durante dicha reacción. (p. 180)

reactividad La tendencia de un elemento o compuesto a combinarse químicamente con otras sustancias, así como la facilidad o la velocidad de la reacción. (p. 37)

teoría de los sitios receptores El modelo aceptado en la actualidad que explica cómo detectar moléculas específicas mediante la nariz. Las moléculas caben dentro de sitios receptores que corresponden a la forma general de cada molécula. Esto estimula una respuesta en el cuerpo. (p. 208)

reacción redox Otro nombre para la reacción de *oxidación-reducción* (p. 560)

reducción Reacción química en la que un átomo o un ion gana electrones. La reducción siempre viene acompañada de oxidación. (p. 559)

reflexión Rebote de la luz contra una superficie, lo que permite que dicha superficie pueda ser vista. (p. 578)

humedad relativa La cantidad de vapor de agua en el aire comparada con la máxima cantidad de vapor de agua posible para una temperatura específica. Se expresa como un porcentaje. (p. 341)

reacción reversible Reacción que puede proceder tanto en la dirección hacia delante para producir productos finales y en las direcciones opuestas (dirección inversa) para producir sustancias iniciales. (p. 613)

rule of zero charge The rule that states that in an ionic compound, the positive charges on the metal cations and the negative charges on the nonmetal anions add up to zero. (p. 99)

regla de la carga cero La regla que establece que en un compuesto iónico, la suma total de las cargas positivas de los cationes de metales y de las cargas negativas de los aniones de no metales, es cero. (p. 99)

S

saturated solution A solution that contains the maximum amount of solute that can be dissolved in a given amount of solvent at a particular temperature. (p. 411)

solución saturada Solución que contiene la cantidad máxima de soluto que se puede disolver en una cantidad dada de solvente a un temperatura determinada. (p. 411)

scientific notation A shorthand notation used for writing numbers that are very large or very small. In this notation, the number is expressed as a decimal number with one digit to the left of the decimal point, multiplied by an integer power of 10. For example, the number 890,000 is written as 8.9×10^5 in scientific notation. (p. 391)

notación científica Notación abreviada que se usa para escribir números muy grandes o muy pequeños. En esta notación, el número se expresa como un número decimal con un dígito a la izquierda del punto decimal, multiplicado por una potencia entera de 10. Por ejemplo, el número 890,000 se escribe 8.9×10^5 en notación científica. (p. 391)

second law of thermodynamics The scientific law that states that energy tends to disperse or spread out. Thermal energy is always spontaneously transferred from a hotter object to a cooler object. (p. 492)

segunda ley de la termodinámica La ley científica que establece que la energía tiende a dispersarse o diseminarse. La energía térmica siempre se transfiere espontáneamente de un objeto más caliente a uno más frío. (p. 492)

single bond A covalent bond where two electrons are shared between two atoms. (p. 156)

enlace sencillo Enlace covalente en el que dos átomos comparten dos electrones. (p. 156)

single exchange reaction (single displacement reaction) A chemical change in which an element is displaced from a compound by a more reactive element. (p. 377)

reacción de intercambio simple (reacción de desplazamiento simple) Cambio químico en el que un elemento es desplazado de un compuesto por un elemento más reactivo. (p. 377)

soluble Capable of being dissolved into another substance. (p. 125)

soluble Que tiene la capacidad de disolverse en otra sustancia. (p. 125)

solute A substance dissolved in a solvent to form a solution. (p. 409)

soluto Sustancia disuelta en un solvente para formar una solución. (p. 409)

solution A homogeneous mixture of two or more substances. (p. 409)

solución Mezcla homogénea de dos o más sustancias. (p. 409)

solvent A substance in which another substance is dissolved, forming a solution. (p. 409)

disolvente Sustancia en la que otra sustancia se disuelve, para formar una solución. (p. 409)

space-filling model A three-dimensional representation of a molecule with no space between bonded atoms, as distinct from a ball-and-stick model. (p. 200)

modelo de espacio relleno Representación tridimensional de una molécula en la que no hay espacio entre los átomos enlazados, a diferencia del modelo de esferas y varillas. (p. 200)

specific heat capacity The amount of heat required to raise the temperature of 1 gram of a substance by 1 °C. (p. 501)

spectator ion An ion that does not directly participate in a chemical reaction. Spectator ions appear on both sides of a complete ionic equation. (p. 459)

spectrometer A device for capturing electromagnetic radiation to study properties of atoms, molecules, and compounds. (p. 596)

spectroscopy A method of using electromagnetic radiation to study properties of atoms, molecules, and compounds as they interact and as they emit light. (p. 597)

speed of light The speed that light travels in a vacuum. Symbolized as c, the speed of light is 3×10^8 m/s. (p. 583)

standard temperature and pressure (STP) A standard set of conditions at which gases can be measured and compared. Standard pressure is 1 atm and standard temperature is 273 K (0 °C). (p. 332)

stoichiometry The quantitative relationship between amounts (usually moles) of reactants and products in a chemical reaction. (p. 468)

strong acid (or base) An acid (or a base) that dissociates completely in solution. (p. 435)

structural formula A two-dimensional drawing or diagram that shows how the atoms in a molecule are connected. Each line represents a covalent bond. (p. 152)

sublimation The process of changing phase from a solid to a gas without passing through the liquid phase. (p. 295)

surroundings Everything in the universe outside the system being investigated. (p. 487)

synthesis The creation of specific compounds by chemists, through controlled chemical reactions. (p. 176)

system The part of the universe being investigated. (p. 486)

capacidad calorífica específica La cantidad de calor que se necesita para aumentar la temperatura de 1 gramo de una sustancia en 1 °C. (p. 501)

ion espectador Ion que no participa directamente en una reacción química. Los iones espectadores aparecen a ambos lados de una ecuación iónica completa. (p. 459)

espectrómetro Dispositivo que se utiliza para captar la radiación electromagnética con el propósito de estudiar las propiedades de átomos, moléculas y compuestos. (p. 596)

espectroscopia Método que utiliza la radiación electromagnética para estudiar las propiedades de átomos, moléculas y compuestos a medida que interactúan y emiten luz. (p. 597)

velocidad de la luz Velocidad a la que viaja la luz en el vacío. Se simboliza como c, la velocidad de la luz es 3×10^8 m/s. (p. 583)

temperatura y presión estándar (TPE) Conjunto estándar de condiciones en las que los gases pueden ser medidos y comparados. La presión estándar es 1 atm y la temperatura estándar es 273 K (0 °C). (p. 332)

estequiometría La relación cuantitativa entre la cantidad (usualmente en moles) de reactantes y productos en una reacción química. (p. 468)

ácido (o base) fuerte Ácido (o base) que se disocia completamente en una solución. (p. 435)

fórmula estructural Dibujo o diagrama de dos dimensiones que muestra cómo están conectados los átomos en una molécula. Cada línea representa un enlace covalente. (p. 152)

sublimación El proceso de pasar de una fase sólida a una fase gaseosa sin pasar por la fase líquida. (p. 295)

entorno Todo el universo que está por fuera del sistema que se está investigando. (p. 487)

síntesis La creación de compuestos específicos por parte de científicos, a través de reacciones químicas controladas. (p. 176)

sistema La parte del universo que se está investigando. (p. 486)

T

temperature A measure of the average kinetic energy of the atoms and molecules in a sample of matter. (p. 280)

tetrahedral shape The shape defined by the symmetrical distribution of four bonded pairs of electrons around a central atom. (p. 192)

theoretical yield The maximum amount of product that could be produced in a chemical reaction when a limiting reactant is entirely consumed. The value is calculated based on a balanced chemical equation. (p. 473)

thermal energy The total kinetic energy associated with the mass and motions of the particles in a sample of matter measured as heat energy. (p. 495)

thermal equilibrium When two objects in contact with one another reach the same temperature, they are in thermal equilibrium. (p. 492)

titration An analytical procedure used to determine the concentration of an acid or a base. A measured volume of an acid or a base of known concentration is reacted with a sample in the presence of an indicator. (p. 451)

toxicity The degree to which a substance can harm an organism. Toxicity depends on the toxin and the dose in which it is received. (p. 381)

transition elements The elements in Groups 1B to 8B on the periodic table. (p. 44)

transmission Of light, passing through certain kinds of objects. (p. 578)

transparent Of an object, capable of transmitting light, meaning that it is possible to see through the object. (p. 578)

trigonal planar shape A flat triangular shape that is found in small molecules with three electron domains surrounding a central atom. (p. 196)

triple bond A covalent bond in which three electron pairs are shared between two atoms. (p. 167)

temperatura Medida del promedio de energía cinética de los átomos y las moléculas en una muestra de materia. (p. 280)

forma tetraédrica La forma definida por la distribución simétrica de cuatro pares de electrones enlazados alrededor de un átomo central. (p. 192)

rendimiento teórico La máxima cantidad de producto que puede producirse en una reacción química cuando el reactante limitador se ha consumido completamente. El valor se calcula basándose en una ecuación química balanceada. (p. 473)

energía térmica La energía cinética total asociada con la masa y el movimiento de partículas en una muestra de materia medida como energía calorífica. (p. 495)

equilibrio térmico Cuando dos objetos que están en contacto alcanzan la misma temperatura, están en equilibrio térmico. (p. 492)

valoración química Procedimiento analítico usado para determinar la concentración de un ácido o de una base. En presencia de un indicador, se hace reaccionar un ácido o una base de volumen y concentración conocidos, con una muestra de la sustancia que se quiere analizar. (p. 451)

toxicidad El grado de daño que una sustancia puede hacerle a un organismo. La toxicidad depende de la toxina y de la dosis suministrada. (p. 381)

elementos de transición Los elementos de los grupos 1B a 8B de la tabla periódica. (p. 44)

transmisión De la luz: cuando esta pasa a través de ciertos objetos. (p. 578)

transparente En relación a un objeto: cuando este es capaz de transmitir la luz, lo que significa que es posible ver a través del objeto. (p. 578)

forma plana triangular Forma plana triangular que aparece en las moléculas pequeñas con dominio de tres electrones alrededor del átomo central. (p. 196)

enlace triple Enlace covalente en el que dos átomos comparten tres pares de electrones. (p. 167)

U

ultraviolet radiation Electromagnetic radiation with shorter wavelengths than those of visible light, from 4×10^{-7} m to 10^{-8} m. Causes sunburn and breaking of chemical bonds. Abbreviated as UV radiation. (p. 588)

universal gas constant, *R* A number that relates the volume, temperature, pressure and number of moles of gas in the ideal gas law. The value of *R* is dependent on the units used. One value of *R* is 0.08206 L · atm/mol · K. (p. 337)

V

valence electrons The electrons located in the outermost electron shell of an atom, which participate in chemical bonding. (p. 90)

valence shell The outermost electron shell in an atom. (p. 90)

visible light Electromagnetic radiation with wavelengths that human beings can see from about 10^{-6} m to 4×10^{-7} m. The wavelength increases from violet to blue to green to yellow to orange to red. (p. 588)

voltage The electrical potential of an electrochemical cell expressed in volts. (p. 571)

volume The amount of space a sample of matter occupies. (p. 9)

W

water displacement (method) A method for measuring the volume of a solid object by immersing it in water. The volume of the object is equal to the amount of water displaced by the object when fully submerged. (p. 14)

wavelength The distance between two peaks or two troughs of a wave, usually measured in meters. The symbol for wavelength is λ (lambda). (p. 582)

weak acid (or base) An acid (or a base) that does not dissociate completely in solution. (p. 435)

weather The day-to-day atmospheric conditions such as temperature, cloudiness, and rainfall, affecting a specific place. (p. 259)

radiación ultravioleta Radiación electromagnética con longitudes de onda más cortas que las de la luz visible, entre 4×10^{-7} m y 10^{-8} m. Produce la ruptura de las uniones químicas y quemaduras solares. Su abreviatura es UV radiación. (p. 588)

constante universal de los gases ideales, *R* Número que relaciona el volumen, la temperatura, la presión y el número de moles de un gas en la ley de los gases ideales. El valor de *R* depende de las unidades de medida. Un valor de *R* es 0.08206 L · atm/mol · K. (p. 337)

electrón de valencia Los electrones ubicados en la capa electrónica exterior de un átomo, que participan de un enlace químico. (p. 90)

capa de valencia La capa electrónica exterior de un átomo. (p. 90)

luz visible Radiación electromagnética con longitudes de onda que los seres humanos pueden ver, entre 10^{-6} m y 4×10^{-7} m. La longitud de onda aumenta del violeta al azul al verde al amarillo al naranja al rojo. (p. 588)

voltaje El potencial eléctrico de una celda electroquímica expresado en voltios. (p. 571)

volumen La cantidad de espacio que ocupa una muestra de materia. (p. 9)

desplazamiento de agua (método) Método para medir el volumen de un objeto sólido sumergiéndolo en agua. El volumen del objeto es igual a la cantidad de agua desplazada cuando el objeto se sumerge completamente. (p. 14)

longitud de onda Distancia entre dos picos o dos valles de una onda, por lo general se mide en metros. El símbolo de longitud de onda es λ (lambda). (p. 582)

ácido (o base) débil Ácido (o base) que no se disocia completamente en una solución. (p. 435)

tiempo atmosférico Las condiciones atmosféricas diarias como la temperatura, la nubosidad y la precipitación, que afectan un lugar específico. (p. 259)

white light The mixture of all wavelengths of visible radiation (colors of the rainbow) normally seen as white rays. (p. 578)

work Describes what happens when a force causes an object to move in the direction of the force applied. Work can be calculated as force $\cdot$ distance (p. 544)

X

x-ray Electromagnetic radiation with wavelengths shorter than those of visible light and ultraviolet radiation, from 10^{-8} m to 10^{-11} m. Causes electrons to be ejected from atoms and molecules. Used mainly in medical care to create images of bones and teeth. (p. 588)

luz blanca Mezcla de todas las longitudes de onda del espectro visible (los colores del arco iris). Comúnmente se manifiesta como rayos blancos. (p. 578)

trabajo Describe qué pasa cuando una fuerza hace que un objeto se mueva en la dirección en la que se aplica dicha fuerza. El trabajo se puede calcular como fuerza $\cdot$ distancia (p. 544)

rayo X Radiación electromagnética con longitudes de onda más cortas que las de la luz visible y la radiación ultravioleta, entre 10^{-8} m y 10^{-11} m. Produce la expulsión de los electrones de los átomos y las moléculas. Se utiliza principalmente en medicina para crear imágenes de los huesos y los dientes. (p. 588)

Index

catalysts
  ammonia from, 371
  in automobiles, 181
  in chemical reactions, 180–181
  enzymes as, 541
  rate of reaction and, 541–542
cations, 96
  acids and, 628
  in electroplating, 136
  H^+, 438
  of iron, 559
  of metals, 99, 104
  of transition metal compounds, 113–114
cave paintings, with ionic compounds, 112
CCIFe. See chlorotrifluoromethane
cellulose, combustion of, 512
Celsius, Anders, 275
Celsius temperature scale, 273, 275
cerebellum, 622
cerebral palsy, 382
cerium, 67
CFCs. See chlorofluorocarbons
Chadwick, James, 53
chain reactions, 79–80
chalk, 131
charge. See also partial charge; rule of zero charge
  conductivity of, 125–126
  of electrons, 193, 215
  in electroplating, 136
  of ions, 95, 96–97
  in iron, 559
  of oxygen, 215
  of transition metal compounds, 113–114
Charles, Jacques, 282
Charles's law, 282–285
chelation therapy, 460
chemical bonds. See bonds
chemical burns, 429
chemical change. See chemical reactions
chemical equations
  for aluminum hydroxide, 468
  balanced, 371–374, 467
  chemical formulas and, 179, 181
  coefficients in, 372, 373–374, 463
  for combustion, 513, 524
  conservation of mass and, 374
  for ester functional group, 179–180
  formula units in, 372
  information in, 361
  matter in, 180
  predicting change from, 358–360
  reversible processes in, 613–614
  for toxins, 355–357
chemical equilibrium, 608–637
  forward processes in, 624
  irreversible processes and, 622
  products and, 607–637
  reverse processes in, 624
  reversible processes and, 622–624
  sleep and, 627
  of sodium chloride, 620
  starting substances and, 607–637
chemical formulas
  for acetic acid, 433
  for acids, 433–434
  for ammonia, 434
  for bases, 434
  chemical equations and, 179, 181
  for compounds, 24, 38
  elements and, 22
  for hydrochloric acid, 433
  for ionic compounds, 99, 104–106
  for magnesium hydroxide, 434
  moles and, 405
  for nitric acid, 433

for phosphoric acid, 433
physical change and, 364
for polyatomic ions, 110
for sodium carbonate, 434
for sodium hydroxide, 434
valence electrons and, 102
chemical names, 23–25
  alcohols and, 205
  alkanes and, 205
  amines and, 205
  carboxylic acid and, 205
  on household product labels, 170
  for ionic compounds, 106–107
  in periodic table of elements, 36, 43
  for polyatomic ions, 109–110
  for smell, 148–149, 205
chemical reactions. See also specific types
  bonds and, 528
  catalysts in, 180–181
  in chemistry, 8
  combustion as, 513
  compounds from, 181
  conservation of mass in, 367, 368
  in copper cycle, 28
  defined, 364
  dissolving and, 364–365
  in electroplating, 136
  elements and, 39
  energy exchange diagrams for,
    533–534
  energy from, 524, 545
  in ester functional group, 176
  fire as, 481
  functional groups and, 181
  gases from, 368
  heat of reaction and, 524
  kinetic energy and, 488
  medicines from, 467
  physical change versus, 363–365
  products in, 180
  rate of reaction of, 540–542
  reactants in, 180, 530
  reverse processes for, 535–538
  reversible processes in, 613–614
  of rust, 29
  solids from, 368
  substance from, 39
  with sugar, 360
  tools for, 3
  with toxins, 353, 354–379
chemical symbols
  for aqueous solutions, 25
  for carbon, 23
  for color, 611–613
  for copper, 32
  for elements, 24
  for indicators, 611–613
  for oxygen, 23
  in periodic table of elements, 36, 43
chemicals
  contamination of, 5
  tasting of, 4
  touching, 4
chemistry
  chemical reaction in, 8
  defined, 8
  hypothesis in, 7
  introduction to, 7–8
  language of, 23
  matter in, 8
  roots of, 7–8
  of smell, 145–256
  tools of, 3–4
chemistry lab
  calorimetry in, 519

chemical contamination in, 5
cleaning in, 5
clothing in, 5
double-checking chemicals are correct in, 5
matter and, 3–6
safety with, 4–5
waste containers in, 5
chlorine
  in acids, 433
  anions, 438
  carbon dioxide and, 376
  electronegativity scale for, 228
  with lithium, 88
  molar mass of, 394–395
  as nonmetal, 105
  reactivity of, 38
  with sodium, 88, 94, 98, 364
chlorofluorocarbons (CFCs), 232
chlorofluoromethanol, 243
chlorotrifluoromethane, 232–233
chocolate, 210
chromium
  activity series of, 565
  cancer from, 356
  electroplating of, 137, 563
  hydrochloric acid and, 355–356
  with molybdenum, 132
  as toxin, 355
  as transition element, 44
chromium chloride, 563
chromium oxide, 113
chromium sulfate, 563
circulatory shock, 644
cis isomers, 245
cis-3-hexenal, 236
citral, 186–187
citric acid
  cabbage juice and, 430
  pH of, 633–634
  as weak acid, 435
citronella, 205
citronellol
  molecular shape of, 205
  space-filling model for, 200–201
Clean Water Act, 424
cleaning, in chemistry lab, 5
clothing, in chemistry lab, 5
clouds
  with cold fronts, 190
  formation of, 324
  low pressure and, 323
  as matter, 10
  with warm fronts, 190, 291
  water in, 297
  water vapor condensation and, 339
  weather and, 259
  in weather maps, 260, 261
coastal flooding, in hurricanes, 344
cobalt
  electron configuration for, 119
  in stained-glass windows, 579
cobalt chloride
  color of, 641
  equilibrium mixtures of, 641
cobalt oxide, 113
coefficients
  in chemical equations, 372, 373–374, 463
  for H^+, 438
  mole ratios and, 463
coffee, LD_{50} of, 382
coinage metals, 41, 47
cold, sensation of, 491–492
cold fronts, 289–291
collision theory, rate of reaction and,
  540–541

covalent bonds, 129–131, 234. *See also* network
    covalent bonds
  in acids, 433
  of carbon, 131
  in diamond, 131
  electrons and, 160
  gases and, 235
  ionic bonds and, 161
  in methane, 530
  nonmetals and, 161
  nonpolar, 221, 225
  oxidation and, 553
  in oxygen, 530
  periodic table of elements and, 166
  polar, 221, 225
  properties of, 131–132
  smell and, 161–164, 235
  structural formulas and, 153
  valence electrons and, 161
crucible, for heating, 4
crystals
  ionic compounds from, 99
  of sugar, 361
  transition metal compounds in, 114
cubes
  density of, 17
  mass of, 15
  volume of, 15
cubic centimeters
  mass and, 18
  for measuring volume, 11, 13
  milliliters and, 14
cubic inches, 13
cubic meters, 11, 13
Curie, Marie, 72
cylinders, 282. *See also* graduated cylinders

D

d subshell, 117
Dalton, John, 51, 52, 53
daughter isotopes, 78–79
Dead Sea, 423
decomposition
  of aluminum oxide, 555–556
  of sugar, 359–360
  of sulfur trioxide, 644–645
  of water, 535
decomposition reactions
  oxygen from, 376
  of potassium chlorate, 376
  with toxins, 375–376
Delaney Clause of Food, Drug, and Cosmetic
    Act, 404
Democritus, 51
density
  of air, 288–289
  balance for, 18
  of copper, 17
  of cubes, 17
  defined, 17–18
  of diamond, 21
  of gases, 295–297
  of gold, 17, 19
  of ice, 270
  of lead, 17
  of liquids, 267–270
  mass and, 20
  mathematical formula for, 18
  matter and, 19, 268
  of pennies, 21, 34
  phase and, 267–268, 297–298
  shape and, 19
  of snow, 268–269, 270
  of solids, 267–270
  of solutions, 422–423

  of substance, 19
  temperature and, 289, 294
  volume and, 18, 20
  of water, 270, 297–298
  weather and, 288–291
deserts, temperature of, 501
dextrose, solution, 419
diamond
  covalent bonds in, 131
  density of, 21
  hardness of, 23
  properties of, 23
diaphragm, 315
diatomic molecules
  electronegativity scale and, 228
  as nonpolar molecules, 231
dichloromethane, 243
dilution
  of acid, 442–444
  of bases, 443
  of hydrogen chloride, 443
  with water, 442–444
dimethyl ether, 158
dipeptide, 249
dipoles
  of carbon dioxide, 222
  electronegativity and, 224
  of formaldehyde, 230
  of phosphine, 231
  from polar bonds, 222
dissociation
  of acids, 433, 628
  of formic acid, 628
  of hydrochloric acid, 628
  of phenolphthalein, 647
  of strong acids, 627–628
  of weak acids, 627–628, 646
dissolving. *See also* solutions
  of ammonia, 434–435
  chemical reactions and, 364–365
  of formic acid, 623, 628
  as forward process, 620
  of hydrochloric acid, 623
  ions from, 365, 627–628
  of metals, 127
  of nonmetals, 127
  of oxygen, 125
  physical change of, 367
  reversible processes with, 614–615
  of strong acids, 627–628
  of substances, 123–128
  of sugar, 123, 358–359
  of weak acids, 627–628
D-limonene, 245
dose. *See also* lethal dose
  for acetaminophen, 398–399
  for acetylsalicylic acid, 398–399
  for aspirin, 398–399
  for ibuprofen, 398–399
  of medicines, 398–399
  of substances, 380
  of toxins, 380, 382
double bonds, 156
  of ammonia, 196
  of carbon, 166
  electron domains and, 196
  of formaldehyde, 196
  linear shape and, 197
  of methyl methanoate, 166
  molecular shape and, 196–197
  octet rule and, 166–167
  of oxygen, 166
double exchange reactions (double
    replacement), 377
  redox and, 561

drinking water, 420
  standards for, 424
dry ice, 295–296, 297
dyes, 410

E

ears, equilibrium and, 622
earth science connection
  on stalagmites and stalactites, 459
  on winds, 323
eggs
  rotting, smell of, 214
  silver and, 561
electrical conductivity. *See* conductivity
electricity
  generation of, 544
  from light, 593
electrochemical cells. *See also* batteries
  activity series and, 570
  electrolytes in, 569
  half-cell in, 570
  half-reaction and, 570, 571
  magnesium in, 571
  metals in, 569–572
  reactants in, 571
  silver in, 569–570, 572
  tin in, 571
  voltage in, 571, 572
  zinc in, 569–570, 572
electrolytes
  in electrochemical cells, 569
  in energy drinks, 100
electromagnetic radiation, 586–592
  frequency of, 586
  light detectors for, 593
  light energy as, 588
  names for, 587–588
  photon model of light and, 590–591
  Planck's constant and, 589
  speed of light of, 587
  wavelength of, 586
  waves in, 586–587
electron cloud model, for atoms, 53
electron configurations, 116–122
  of beryllium, 117
  of boron, 117
  for cobalt, 119
  of helium, 117
  of hydrogen, 117
  of lithium, 117
  noble gases and, 120–121
  periodic table of elements and, 118–119
  of selenium, 120
  subshells and, 116–118, 119
electron domains, 191–195
  of ammonia, 193
  double bonds and, 196
  electrons in, 197
  of methane, 192–193
  theory of, 193
  of water, 194
electron shells
  of hydrogen, 89
  of lithium, 88–89
  maximum electrons in, 116
  of potassium, 89
  of sodium, 88–89
  valence electrons in, 116, 122
electronegativity
  bonds and, 220–222, 225–226, 228
  dipoles and, 224
  of formaldehyde, 230–231
  ionic bonds and, 225
  partial charges and, 220
  of polar molecules, 219, 223–226, 230–231

electronegativity scale, 227–229
  bonds and, 228
  for chlorine, 228
  diatomic molecules and, 228
  for potassium, 228
electronegativity value, 227
electronic balance
  lower weight limit for, 386–387
  for measuring mass, 11
electrons. *See also* valence electrons
  activity series and, 565–566
  in atom, 52, 54, 82
  in beta decay, 73
  bonded pairs and, 193
  charge of, 193, 215
  compounds and, 82
  core, 90–91
  covalent bonds and, 160
  in electron domain, 197
  excited, 85–86
  noble gases and, 97
  nucleus and, 86
  oxidation and, 552–553
  periodic table of elements and, 90
  probability with, 53
  reactivity and, 93
  unpaired, 162
electroplating
  chemical reaction in, 136
  of chromium, 137, 563
  of copper, 136
  of gold, 137
  of metals, 134–138
  of silver, 565
elements. *See also* periodic table of elements
  atoms and, 51
  chemical formulas and, 22
  chemical reactions and, 39
  chemical symbols for, 24
  compounds from, 24
  in copper cycle, 33–34
  formation of, 77–80
  from fusion, 80
  isotopes of, 65–67
  light signatures of, 86
  in matter, 22, 24
  from nuclear reactions, 80
  properties of, 37–40, 122
  radioactive isotopes of, 72
  reactivity of, 37–38, 46
  stable isotopes of, 72, 77
  from stars, 80
emission, of light, 576–577
endothermic chemical reactions, 487–488
  in equilibrium mixtures, 642
  heat of reaction for, 524
  potential energy for, 536
  reversing, 535
energy. *See also* bond energy; conservation of energy; kinetic energy; light energy; potential energy
  activation, 537–538
  alternative sources for, 545
  boiling point and, 506
  bonds and, 528–533
  calorimetry and, 515–522
  from chemical reactions, 524, 545
  control of, 550–574
  defined, 483
  electrochemical cells and, 569–572
  exothermic chemical reactions and, 484–485
  from fire, 481–602
  frequency and, 589
  heat and, 9, 486–487
  heat of formation and, 554–557

measurement of, 511–527
  melting and, 506
  observing, 482–509
  oxidation and, 551–562
  particle view of, 495
  radiation and, 575–599
  rate of reaction and, 540–542
  redox and, 559–562
  reversing chemical reactions, 535–538
  sound as, 9
  substances and, 483
  thermal, 495
  understanding, 528–548
  work and, 544–547
energy drinks, electrolytes in, 100
energy exchange diagrams, 533–534
engineering connection, on oxygen in
    submarines, 376
entropy, 492
  work and, 547
environmental connection
  on acid rain, 452
  on alternative energy sources, 545
  on anthocyanin, 610
  on battery disposal, 572
  on biofuels, 524
  on carbon dioxide on Mars, 262
  on chlorotrifluoromethane, 232
  on chromium and cancer, 356
  on composting, 369
  on flower colors, 612
  on geraniol, 200
  on ice, 270
  on lightning, 483
  on nitrogen dioxide, 33
  on ozone layers, 596
  on recycling, 368
  on recycling plastic, 369
  on snowpack density, 268
  on temperature changes in deserts, 501
  on water pollution, 424
Environmental Protection Agency (EPA), 420, 424
enzymes, as catalysts, 541
EPA. *See* Environmental Protection Agency
epilepsy, 106
equilibrium, 619–624. *See also* chemical
    equilibrium; thermal equilibrium
  atoms and, 619
  calculations, 631–635
  cerebellum and, 622
  sense of, 622
  sodium chloride at, 619–620
equilibrium constants, 626–629
  for acetic acid, 633
  for benzoic acid, 633
  concentration and, 647, 650
  for formic acid, 633
  for hydrochloric acid, 633
  for hydrogen sulfide, 633
  for hypochlorous acid, 633
  for lactic acid, 632
  for nitric acid, 633
  for nitrous acid, 633
  products and, 626–627
  starting substances and, 626–627
  types of, 649–650
  for weak acids, 647
equilibrium mixtures
  of cobalt chloride, 641
  concentration of, 640, 641–642
  endothermic chemical reactions in, 642
  forward processes in, 608, 639, 640
  of gases, 642–643
  hydrochloric acid in, 651
  indicators and, 640

Le Châtelier's principle and, 640–645
  pressure of, 640, 642–643
  reverse processes in, 608, 639, 640, 642
  temperature of, 640, 642
equilibrium-constant equation
  general form of, 634–635
  for liquids, 635
  products and, 635
  for solids, 635
  starting substances and, 635
  temperature and, 635
  for weak acids, 631–634
equivalence point, of solutions, 451
Erlenmeyer flask, 3
ester functional group, 170, 172
  analysis of, 179–181
  chemical equations for, 179–180
  chemical reactions in, 176
  in household products, 180
  smell and, 175
  synthesis of, 175–177
esters
  from acids and alcohols, 181
  bonds for, 181
  chemical names and, 205
ethane, 157
  heat of combustion for, 525
ethanol
  boiling point for, 507
  butyric acid and, 179
  combustion of, 521–522
  gasoline with, 158
  heat of combustion for, 524, 525
  heat of fusion for, 507
  heat of vaporization for, 507
  melting point for, 507
  phase change of, 507
  specific heat capacity for, 501
  three-dimensional structure of, 191
  in water, 364
ethene, 197
ethers, 171
ethyl acetate, 186
ethyl butyrate, 153
  from butyric acid and ethanol, 179
ethyl formate, 172
ethylthiol, 151, 152
ethyne, 198
evaporation
  of bromine, 363
  of liquids, 298
  of refrigerants, 487
  of sweat, 643
  of water, 297, 339, 621
exchange reactions, 376–377
  with toxins, 377
excited electrons, 85–86
exercise
  blood pH and, 644
  carbonic acid from, 640, 644
  sweating and, 643
exothermic chemical reactions, 484–485, 487
  aluminum oxide and, 556
  combustion as, 530
  heat from, 588
  heat of reaction for, 524
  kinetic energy and, 488
  light from, 576
  oxidation as, 554
  potential energy for, 536, 537
extensive properties, for substances, 19
extreme physical change, weather and, 343–346
eyes
  equilibrium and, 622
  melatonin and, 627

mass of, 13, 14–15
melting point for, 507
as metal of antiquity, 38
neutrons in, 81
nuclear reactions and, 81
pennies into, 21
phase change of, 507
properties of, 2, 7, 17
protons in, 81
stable isotopes of, 77
from supernovas, 78, 80
volume of, 14
graduated cylinders
measurements in, 11
meniscus in, 11, 12
for volume, 3, 11
water in, 12
gram-mole conversions, stoichiometry and,
467–468
grams, 11
Calories in, 516–517
mass in, 398
moles and, 390–391, 400
grass, smell of, 236
Great Fire of London, 483
Great Lakes Water Quality Agreement, 424
groups. *See also specific groups*
in periodic table of elements, 44

H

H⁺. *See* hydrogen ions
hafnium-144, 67
half-cell, in electrochemical cells, 570
half-life
of carbon-14, 73
in nuclear reactions, 73–74
half-reaction
of copper, 571
electrochemical cells and, 570, 571
of gold, 571
of lead, 571
of lithium, 571
of magnesium, 571
of nickel, 571
of silver, 571
voltage and, 571
halogens, 43, 44
handedness
of amino acids, 248
of molecules, 242–243
hardness
of diamond, 23
of water, 418
health connection
on altitude sickness, 337
on antacids, 430, 448
on antioxidants, 551
on aspartame, 403
on Delaney Clause of Food, Drug, and
Cosmetic Act, 404
on food additives and cancer, 404
on food Calorie intake, 516
on high-fructose corn syrup, 416
on histamine, 626
on lead paint, 395
on receptor sites and medicines, 208
on sense of smell of dogs, 148
on sour taste, 634
on trans isomers, 245
on transition elements, 44
heat. *See also* temperature; specific heat capacity
from combustion, 523–525, 552
conduction of, 490
convection of, 490
defined, 486

endothermic chemical reactions and, 487–488
energy and, 9, 486–487
exothermic chemical reactions and, 484,
487, 588
first law of thermodynamics and, 492
kinetic energy and, 488
from oxidation, 552
phase change and, 505–508
radiation and, 490, 576–577
second law of thermodynamics and, 492
sensation of, 491–492
specific heat capacity, 500–503
substances and, 500
surroundings and, 486–487, 491, 536
system and, 486, 491
work and, 545
heat of combustion, 523–525
for butane, 525
for butanol, 525
for carbon, 525
for ethane, 525
for ethanol, 524, 525
for glucose, 525
for hexane, 525
for hydrogen, 525
for methane, 525
for methanol, 524, 525
for octane, 525
heat of formation
for aluminum, 556, 557
for copper, 557
energy and, 554–557
fire and, 554–557
for gold, 557
for iron, 555, 557
for lead, 557
for magnesium, 555, 557
metals and, 554–557
mole and, 554
oxidation and, 554–557
for silver, 557
heat of fusion
for aluminum, 507
for ethanol, 507
for gold, 507
for ice, 507
for mercury, 507
for phase change, 506
heat of reaction
bond energy and, 535
calorimetry for, 535
chemical reactions and, 524
for endothermic chemical reactions, 524
for exothermic chemical reactions, 524
heat of transformation
for copper, 554–555
for zinc, 554–555
heat of vaporization
for aluminum, 507
for ethanol, 507
for gold, 507
for mercury, 507
for phase change, 506
heat transfer, 490–498
calorimetry and, 515–522
measurement of, 496–497
substances and, 506
temperature and, 506
heating
beakers for, 4
Charles's law and, 282–285
crucible for, 4
hot plate for, 4
matter and, 9
molecular size and, 236–237

physical change from, 259
of potassium chlorate, 364
safety with, 4
of sugar, 359–360
by Sun, 259
tools for, 4
heavy metals, 460, 464
Heisenberg, Werner, 53
helium
alpha particles and, 72, 79
balloons, 331, 333–334
compounds of, 93
electron configuration of, 117
molar mass of, 395
naturally occurring isotopes of, 65
in Sun, 77
valence electrons of, 94
helium-3, 66
helium-4, 65
hematite, 559
hemoglobin, 616
heptanoic acid, 198
Hero of Alexandria, 546
heterogeneous mixture, 410
hexane, 525
hexanoic acid, 153, 169
hexylamine, 151
high altitude, 317
high blood pressure, 381
high pressure, 291, 321–324
high-fructose corn syrup, 416
Hindenburg (airship), 285
histamine, 626
history connection
on absolute zero, 278
on air pressure, 299
on barometer, 329
on calorimeters, 517
on Charles's law, 282
on copper, 29
on Curie, 72
on elements, 38
on fireworks, 85
on Great Fire of London, 483
on *Hindenburg*, 285
on hot air balloon, 283
on ionic compounds, 112
on Kevlar, 127
on litmus, 646
on malachite, 554
on matches, 513
on metals of antiquity, 135
on Montgolfier brothers, 283
on Pauling, 228
on pennies, 34
on photographic flash, 512
on photography, 468
on salt, 104
on smell, 256
on steam engine, 546
on thermometers, 274
on weather glass, 328
homogeneous mixture, solution as, 410
HONC 1234 rule
Lewis dot symbols and, 162
for smell, 156–157, 160
hot air balloon, 283, 294
number density and, 331
hot plate, for heating, 4
hot springs, sulfur in, 22
household products
bases in, 429
labels for, chemical names on, 170
soap as, 631
sodium hydroxide in, 434

isotopes
  atomic number of, 67
  atoms and, 61–64
  average atomic mass of, 62–63
  of carbon, 61–62, 72
  of copper, 63–64
  daughter, 78–79
  of elements, 65–67
  human-made, 66
  mass of, 62, 63
  natural abundance of, 63
  naturally occurring, 65–67
  neutrons in, 66
  nucleus and, 61–62
  parent, 78–79
  radioactive, 65–68, 72
  stable, 65–68, 72, 77
  of uranium, 79

J
jet stream
  air pressure and, 324
  on air pressure map, 323
  in weather maps, 260, 261
joules, 524
  in electromagnetic radiation, 589

K
Kelvin temperature scale, 278–280
  Charles's law and, 284
  Gay-Lussac's law and, 310
ketones, 172
  chemical names and, 205
Kevlar, 129
kidney stones, 457
kilograms, 11
kilojoules, 524
  in mole, 525
kinetic energy
  average, 496
  chemical reactions and, 488
  defined, 488, 536
kinetic theory of gases, 280
  air pressure and, 314–315, 328
krypton, 94
Kwolek, Stephanie, 129

L
lactic acid
  in circulatory shock, 644
  equilibrium constants for, 632
language
  of chemistry, 23
  color and, 585
  oil and water and, 216
lanthanides
  electronegativity value of, 227
  in periodic table of elements, 42–43, 44, 45
Laplace, Pierre, 517
Lavoisier, Antoine, 517
law of conservation of mass. *See* conservation of mass
LD_{50}. *See* lethal dose
Le Châtelier, Henry, 643
Le Châtelier's principle
  application of, 646–652
  equilibrium mixtures and, 640–645
  indicators and, 646–648
lead
  activity series of, 566, 570
  chelation therapy for, 460
  copper and, 566
  density of, 17
  half-reaction of, 571
  heat of formation for, 557

as metal of antiquity, 38
  molar mass of, 394–395
  paint, 395
  toxicity of, 396
  as toxin, 394–396
lead iodide, 457–458
lemon juice, 430–431
lethal dose (LD_{50}), 381–382
  of acetylsalicylic acid, 382
  of arsenic, 382
  of artificial sweeteners, 403–404
  of aspirin, 382
  body weight and, 383, 405
  of caffeine, 382, 383
  of coffee, 382
  of fructose, 403–404
  of glucose, 382
  of salt, 382
  of snake venom, 382
  of sodium chloride, 382
  of sweeteners, 403–404
Lewis, Gilbert Newton, 162
Lewis dot structure, 162
  of carbon, 165
  of carbon dioxide, 167
  double bonds and, 166
  of fluorine, 165
  of nitrogen, 165
  octet rule and, 165
  of oxygen, 165
  of phosphine, 195
  of phosphorus trichloride, 163
Lewis dot symbols
  bonded pairs and, 162
  HONC 1234 rule and, 162
  lone pairs and, 162
  smell and, 161–164
lichens, litmus from, 647
light. *See also* ultraviolet radiation/light; visible light
  absorption of, 579
  from combustion, 552
  electricity from, 593
  emission of, 576–577
  from exothermic chemical reactions, 484, 576
  from nuclear reactions, 576
  from oxidation, 552
  photon model of, 590–591
  rays, 577
  reflection of, 578
  sources of, 576–577
  speed of, 583, 584, 587
  transmission of, 578
  wave model of, 581–583
  waves, 581–585
  white, 577, 579
light detectors
  for electromagnetic radiation, 593
  spectroscopy and, 593–596
light energy, 576–579
  as electromagnetic radiation, 588
light signatures
  of elements, 86
  of xenon, 93
lightning
  fire from, 483
  from volcanoes, 577
limestone
  calcium carbonate as, 608
  stalagmites and stalactites from, 617
  water through, 608, 617
limiting reactant (limiting reagent), toxins and, 464, 471–473
limonene, 242, 245
linear shape, double bonds and, 197

liquid thermometers, 272
liquids
  as aqueous solutions, 25
  density of, 267–270
  equilibrium-constant equation for, 635
  evaporation of, 298
  mass of, 9
  measurement of, 263–266
  molecules in, 279–280
  in periodic table of elements, 42, 45
  as solutes, 409
  as solvent, 409
  sugar as, 359
  volume of, 9, 11
  water as, 25
liter-atmospheres, 545
literature connection, on myths about fire, 483
liters, 11
  molarity and, 410
lithium
  batteries, 88, 564
  chlorine with, 88
  electron configuration of, 117
  electron shells of, 88–89
  flame color of, 85
  half-reaction of, 571
litmus
  paper, 430
  from photosynthetic algae, 647
L-limonene, 245
lock-and-key model. *See* receptor sites
logarithms, for H^+, 438
lone pairs
  ball-and-stick model and, 194
  Lewis dot symbols and, 162
  of nitrogen, 167
low pressure, 291, 321–324
  clouds and, 323
  precipitation and, 323
  in tropical depressions, 344
Lowry, Thomas, 434–435
Lundström, Johan, 513
lye. *See* sodium hydroxide

M
M. *See* molar
0.15 M solution, of salt, 418
macromolecules, 249
magnesium
  acid rain and, 452
  activity series of, 564–565, 570
  combustion of, 512, 513
  in electrochemical cells, 571
  half-reaction of, 571
  heat of formation for, 555, 557
  iron nitrate and, 561–562
  oxidation of, 551–553, 555
  reactivity of, 38
  redox of, 561–562
  tin and, 571
magnesium chloride, 100
  percent yield of, 474
  for tofu, 472
magnesium hydroxide, 110
  chemical formula for, 434
  hydrogen chloride and, 449
  in milk of magnesia, 449
magnesium nitrate, 561–562
main group elements
  in acids, 433
  ionic compounds with, 104
  on periodic table of elements, 42, 44
malachite, 554
maltose, 617
manganese dioxide, 113

defined, 363–364
of dissolving, 367
fire and, 481
reversible processes in, 613–614
of substances, 363
of water, 363
weather and, 259, 343–346
physics connection
on air pressure, 300
on charge of electrons, 215
on sound waves, 584
pigments
anthocyanin, 609–610
transition metal compounds as, 112–113
pistons, 282
Planck's constant, 589
plants, nitrogen and, 63
plastic
mass of, 14–15
recycling of, 369
smell of, 236
types of, 14
plum pudding model, for atoms, 52
plutonium-241, 73
poisons. *See* toxins
polar bonds, 222
polar covalent bonds
of carbon, 221
of fluorine, 221
of hydrogen, 221
of oxygen, 221
partial charges of, 225
polar molecules
ammonia as, 232
asymmetry of, 230, 231
electronegativity of, 219, 223–226, 230–231
melting points of, 222
nonpolar molecules and, 216–217
receptor sites and, 232
smell and, 215, 222, 230–233
solubility of, 222
water as, 232
pollution
of air, 33, 451
of water, 424
polyatomic ions, 109–111
chemical formulas for, 110
chemical names for, 109–110
compounds from, 110
rule of zero charge for, 110
polyethylene terephthalate (PETE), 14
ethene and, 107
polystyrene, 14
ethene and, 107
polyvinyl chloride (PVC), 14
potassium
activity series of, 565, 570
electron shells of, 89
electronegativity scale for, 228
flame color of, 85
as metal, 105
potassium bromide, 106
potassium chlorate
decomposition reactions of, 376
heating of, 364
sugar and, 360
potassium chloride, 364
solution of, 423
potassium hydroxide
limiting reactants and, 471–472
neutralization reactions with, 448
potassium sulfate, 449
potential energy
defined, 536
for endothermic chemical reactions, 536

for exothermic chemical reactions, 536, 537
of reactants, 537
pounds, 11
"powder" snow, 269
precipitates, 457
toxicity of, 460
precipitation
from humidity and temperature, 340
low pressure and, 323
measurement of, 263–266
phase of, 267–268
as reverse process, 620
reversible processes with, 614–615
of stalagmites and stalactites, 459
weather and, 259
in weather maps, 260, 261
precipitation reactions
of calcium phosphate, 465
for copper, 462
of lead iodide, 457–458
redox and, 561
of silver hydroxide, 464
of sodium hydroxide, 462–463
solubility, 458–460
for toxins, 457–460
pressure. *See also* air pressure
ammonia and, 643
atmospheric, 301–302, 341
combined gas law and, 318
of equilibrium mixtures, 640, 642–643
of gases, 294, 299, 306, 318
high blood pressure, 381
matter and, 294–302
rate of reaction and, 541
reaction rate and, 541
versus temperature, 319
volume and, 306
work and, 545
pressure cookers, 310
pressure sensors, equilibrium and, 622
probability, with electrons, 53
products
chemical equilibrium and, 607–637
in chemical reactions, 180
equilibrium constants and, 626–627
equilibrium-constant equation and, 635
Le Châtelier's principle and, 640–645
starting substances and, 607
Prometheus, 483
promethium, 66
propane
as alkane, 171
gas tank for, 310–311
properties
of acids, 429
of bases, 429
bonds and, 131–132
of covalent bonds, 131–132
of diamond, 23
of elements, 37–40, 122
extensive, 19
of gold, 2, 7, 17
of hurricanes, 343–344
intensive, 20
of ionic bonds, 131–132
of metallic bonds, 131–132
of mirror-image isomers, 244–245
of network covalent bonds, 131–132
periodic table of elements and, 45, 88
of substances, 19, 20, 124–128, 131–132
propionic acid, 171
proportional relationship, 265
inverse proportionality, 306
of mass and volume, 268
of temperature and volume, 283

proportionality constant, 265
of gases, 283
propyl acetate, 151
protein molecules
from amino acids, 241, 249
receptor sites and, 209, 249
proteins
in human body, 249–250
smell and, 249–250
proton model for atoms, 53
protons
in alpha particles, 72
atomic mass and, 56–57
atomic number and, 56
bases and, 434
in beta decay, 73
in gold, 81
in isotopes, 62
in mass number, 62
in nucleus, 53, 54
Rutherford and, 58
psychology connection, on memory and
smell, 204
pulegone, 200–201
putrid smell, 169
of butyric acid, 175
carboxyl functional group and, 171
into sweet smell, 177
PVC. *See* polyvinyl chloride
pyramidal shape
ammonia with, 193
phosphine with, 195

Q
quadruple bonds, 167

R
R groups. *See* amino acids
radiation. *See also* electromagnetic radiation;
infrared radiation; microwave radiation;
ultraviolet radiation/light
energy and, 575–599
from fire, 575–602
heat and, 490, 576–577
in nuclear reactions, 74
spectroscopy and, 591–598
thermal equilibrium and, 577
radio waves, 587
wavelength and frequency of, 589
radioactive decay, 67
alpha, 72, 78
beta, 72–73
nuclear reactions and, 71–73
subatomic particles and, 71–73
radioactive isotopes, 65–68
of argon, 67
of cerium, 67
of elements, 72
of hydrogen, 67
radon, 79
rain, 297
acid rain, 364, 451, 452
with cold fronts, 190
gauges, 263–265
in hurricanes, 343, 345
snow to, 268–269
with warm fronts, 190
water vapor condensation and, 339
rate of reaction
catalysts and, 541–542
chemical reactions of, 540–542
collision theory and, 540–541
concentration and, 541
energy and, 540–542
fire and, 540–542

LESSON 2

3. Time periods when alchemy was practiced: China (early C.E.), India (B.C.E. to Middle Ages), Middle East (early C.E. to recent), Greece (late B.C.E. to early C.E.), Spain (Middle Ages), England (Middle Ages), Egypt (B.C.E.).

4. A good answer will include: goals of alchemists in their particular region (which substances they were attempting to transform or obtain) ● contributions that alchemists in the region made to science ● a list of trustworthy sources for the information given in the answer.

5. *Possible answer:* Sodium hydroxide, or lye, is a strong base used in manufacturing paper, soaps, and detergents, and in other industrial chemical processes. In the home, sodium hydroxide is a component of many drain cleaners. Because it is a strong base, and therefore a very reactive substance, sodium hydroxide is an important industrial chemical used in many chemical manufacturing processes. It tends to break down organic materials, such as hair and grease, which makes it useful as a drain cleaner.

6. *Possible answer (any 10 examples total):* Changes that involve chemistry involve an alteration in the appearance of matter. Examples include metal rusting, cookies baking, ice cubes forming, wood burning, water boiling, and sugar dissolving. Both physical changes (such as changes of state) and chemical changes (such as objects burning or rusting) involve chemistry. ● Changes that do not involve chemistry only involve matter moving to a different location. Examples include the Sun going down, objects falling, hands moving on a clock, going for a walk, throwing a ball, and riding a bicycle.

LESSON 4

7. *Possible answers:* **a.** Pour the pancake mix into a graduated cylinder or a measuring cup and read the markings. **b.** Use a measuring spoon or measuring cup. **c.** Measure the dimensions of the box and multiply the length by the width by the height. **d.** Measure the thickness and radius of the penny and calculate its volume using the formula for the volume of a cylinder or measure the amount of liquid it displaces when it is submerged. **e.** Pour the lemonade into a graduated cylinder or a measuring

cup and read the markings. The key concept of the question is that the volume of fluids can be measured directly, while the volume of a solid material can be best measured by its dimensions or displacement.

8. *Possible answer:* The drawing should clearly show that the masses of the two objects are the same. This can be done using a balance, as shown, or by using labels. For example, a large cube could be labeled "polystyrene foam" and a small cube labeled "wood."

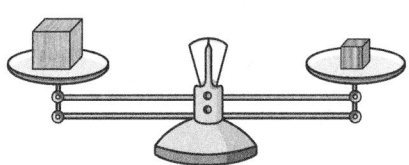

9. *Possible answer:* The drawing should clearly show that the masses of the two objects are different. This can be done using a balance, as shown, or by using labels. For example, two beakers could be shown, one labeled "sand" and the second labeled "water."

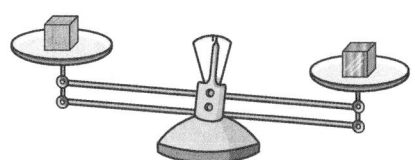

10. a. 150 mL. **b.** Because the rock is completely submerged in the water, the volume of the rock is equal to the volume of water that it displaces. The displaced volume is equal to the change in the water level: 200 mL − 150 mL = 50 mL. One mL is equal to one cm³, so the volume of the rock is 50 mL, or 50 cm³.

11. *Possible answer:* The balance or scale that was used to measure the two different cubes may not be able to measure accurately to a hundredth of a gram. The two cubes could have exactly the same mass because the measurement error is greater than 0.03 g. The exercise specifies that the cubes are exactly the same material, so they should have the same mass since they are the same volume. This eliminates the possibility that the substance is wood, plastic, or some other substance where the density may vary. The only remaining possibility is measurement error due to human error or limits to the instrument's precision.

LESSON 11

4. Both a proton and a nucleus are attracted to a negative charge because

they have a positive charge. Because an electron has a negative charge, it will be repelled by a negative charge. Because neither a neutron nor an atom has a charge, they will be neither attracted nor repelled by a negative charge.

5. *Possible answer:* The two types of atoms could have different amounts of positive fluid and different numbers of electrons, as shown in this illustration. Each atom would still have a net charge of zero.

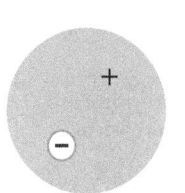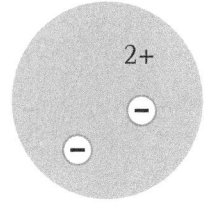

Hydrogen   Helium

6. *Possible answer:* The electrons and protons can exist together in a single mass as shown below. The number of electrons and protons in each atom is equal. The atom is held together by the attraction between the positive protons and the negative electrons.

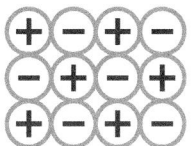

7.

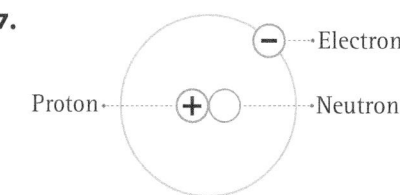

Proton   Electron   Neutron

8. 100,000

9. *Possible answers:* **a.** In the nuclear model, all of the electrons orbit the nucleus at the same distance. In the solar system model, the electrons orbit the nucleus at specific distances. **b.** In both models, negatively charged electrons revolve around a positively charged nucleus in circular orbits. Most of the atom is empty space. **c.** To change the solar system model to a more three-dimensional model, change some of the electron orbits so that they are orbiting the nucleus in different planes, similar to the way the nuclear model is drawn.

10. *Possible answers:* ● Every compound is made of two or more elements in specific proportions that never change. Because these proportions are always the same, there

must be a basic unit of each element.
● In Rutherford's experiment, alpha particles were deflected by particles in a sheet of gold foil. The gold must be made of concentrated bits of matter called *atoms* that contain mostly empty space. ● Images of atoms have been made with powerful scanning tunneling microscopes.

11. No, the Greeks were not correct. It is now known that atoms are made of smaller particles such as electrons, protons, and neutrons. Each of these particles has a mass and a volume, so they are matter.

12. *Possible answer:* Although the atomic theory has existed for over 200 years, it has changed with the addition of new discoveries. Each new discovery leads to a new model that accounts for the new information about how atoms function. For example, the discovery of protons made it necessary to revise the nuclear model to account for small particles with a single unit of positive charge.

LESSON 17

8. No, the form of the sodium chloride does not affect the results of the flame test because the atoms do not emit light until they become very hot. The water will evaporate from the aqueous solution before the sodium chloride becomes hot enough to emit light.

9. a. yellow-orange **b.** green **c.** pink-lilac **d.** pink-lilac **e.** green. The flame color produced by each compound can be determined by matching the metallic element in the compound with the flame color produced by compounds with that element.

10. C

11. D

LESSON 22

1. A polyatomic ion is an ion that consists of two or more elements.

2. An ionic compound has a polyatomic ion if there are more than two elements in the compound.

3. a. ammonium chloride
b. potassium sulfate **c.** aluminum hydroxide **d.** magnesium carbonate

4. a. Li_2SO_4 **b.** KOH **c.** $Mg(NO_3)_2$
d. $(NH_4)_2SO_4$

5. The cyanide ion has a charge of -1.

6. Each phosphate anion has a charge of -3.

7. b. KSO_4 is not a possible compound because it has a net charge of -1. The

potassium ion has a charge of $+1$ and the sulfate ion has a charge of -2.
$(+1) + (-2) = -1$

LESSON 26

3. a. Because zinc consists entirely of metal atoms, it is held together by metallic bonds. **b.** Propane is made entirely of nonmetal atoms, so it is held together by molecular covalent bonds. **c.** Calcium carbonate consists of the metal calcium and the polyatomic ion carbonate. Therefore, the compound is an ionic compound that is held together by ionic bonds.

4. *Possible answer:* a. Hair gel would probably consist of molecular covalent bonds. Hair gel is soft and malleable, dissolves in water, and is easily separated into smaller portions, so it is likely to be molecular. **b.** A silver bracelet has metallic bonds. A silver bracelet is made entirely of metal, so its atoms will be held together by metallic bonds. It is hard and bendable and conducts electricity. **c.** Motor oil would probably consist of molecular covalent bonds. Motor oil is a viscous liquid, and liquids that are not salts in an aqueous solution are usually molecular. **d.** Baking soda is the only one of these substances that could have ionic bonds. It is a solid crystalline material that dissolves easily in water. If it contains metal and nonmetal atoms, it is probably held together by ionic bonds.

5. NO_2 is a gas that is made up of nonmetal atoms, so it has molecular covalent bonds.

6. E

7. *Possible answer:* Drop the mixture into a beaker of water. The ionic bonds of sodium chloride break apart in water, causing the solid to dissolve. Carbon, held together by network covalent bonds, does not dissolve in water, and so can be filtered from the solution.

8. Copper metal can be shaped into a wire easily because of its metallic bonds. Metals are malleable and ductile. Copper chloride, as an ionic compound, is brittle and cannot be shaped into a wire. The ions in copper chloride are held in position relative to one another by strong ionic bonds.

9. Carbon is a solid because many carbon atoms are held together in a large array by network covalent bonds, such as in diamond. Gases are single atoms or molecules held together by molecular covalent bonds.

10. *Possible answers:* a. Calcium, Ca, is held together by metallic bonding so it will not dissolve in water. **b.** Sodium nitrate, $NaNO_3$, is an ionic compound, so it will dissolve in water. **c.** Silicon, Si, is a solid held together in a large array by network covalent bonds, similar to carbon. It will not dissolve in water. **d.** Methane, CH_4, is a liquid held together by molecular covalent bonds, so it will probably dissolve in water. **e.** Copper sulfate, $CaSO_4$, is an ionic compound so it will dissolve in water.

LESSON 38

1. If a molecule has three electron domains, its shape will be trigonal planar.

2. A molecule with three atoms can be linear or bent. If it has two electron domains, the molecule will be linear to keep the two domains as far apart as possible. If it has three or four electron domains, the molecule will be bent to keep the domains as far apart as possible.

3. Cl and CO_2 are linear and H_2O is bent.

4. Cl and CO_2 are both linear.

5. ● A two-atom molecule can only be linear because two points define a line. ● A three-atom molecule will be linear if the central atom has two electron domains. It will be bent if the central atom has three or four domains. ● A four-atom molecule will be linear or bent if the atoms form a chain. It will be linear if the two central atoms only have two electron domains, and it will be bent if either central atom has more than two domains. If the four-atom molecule has one central atom bonded to the other three atoms, the molecule will be trigonal planar or pyramid-shaped. It will be trigonal planar if the central molecule has three domains, and it will be pyramidal if the central molecule has four domains. ● A five-atom molecule can form a linear or bent chain, as described above. It can be tetrahedral if it has a central atom with four electron domains. Otherwise, the molecule could form one of the four-atom shapes (pyramidal or trigonal planar), with the fifth atom bonded to any atom on the pyramid or one of the atoms in the triangle.

6. CF_4 is tetrahedral, NF_3 is pyramidal, H_2Se is bent, and H_2CS is trigonal planar.

7. a.

H H H H
H:C:C:C:C:H
H H H H

b. 13

c. Possible answer: a crooked zigzag shape

d. Each carbon atom is surrounded by four electron domains. Therefore, the shape of the molecule is a series of linked tetrahedrons.

LESSON 41

1. According to the receptor site theory, the nose is lined with sites that match the shapes of molecules that have smells. When a molecule fits into this site, it stimulates nerves to send a message to the brain. The brain interprets the message as a smell.

2. Some of the molecules of a liquid or solid material must change phase to become a gas before they can travel from the material to your nose and stimulate the receptor site.

3. Possible answer: The compound would most likely have a minty smell because it appears to be a frying-pan–shaped molecule with a hydroxyl group (–OH). The nose detects the minty smell when the molecule shown fits into a receptor site in the nose that detects minty smells. To happen, some of the muscle ointment must change phase and become a gas.

4. C

5. Possible answer: Using the receptor site theory, it is likely that the smells from the two materials contain some molecules that are identical. Because the overall smell of the material is a combination of the smells of all the molecules detected by receptor sites, different materials that have very similar smells probably have some molecules with a smell in common.

6. Possible answer: A dog probably has a better sense of smell than a human because the dog has more receptor sites in its nose than humans have.

7. Possible answer: Once all of the receptor sites for a particular type of smell have been filled by matching molecules, additional molecules have no way to stimulate a signal to the brain. When this happens, you will no longer be able to detect that particular smell.

8. You can detect some smells faster than others because the molecules that cause the smell enter the gas phase and travel to your nose faster than the molecules of other smells. If a substance is not already a gas or producing a gas, it may take longer to smell.

LESSON 44

3. a. In order of increasing polarity: Li—F, Na—F, K—F and Rb—F (same polarity), Cs—F. For an alkali metal bonding with fluorine, the polarity increases from the top of the group to the bottom. **b.** In order of increasing polarity: P—S, N—F, Al—N, Mg—O, K—Cl. The polarity of the molecules increases as the distance between the atoms on the periodic table increases.

4. a. $\delta-$ H—B $\delta+$ **b.** $\delta+$ H—C $\delta-$ **c.** $\delta+$ H—N $\delta-$ **d.** $\delta+$ H—O $\delta-$ **e.** $\delta+$ H—F $\delta-$. The element with the higher electronegativity value will attract electrons more strongly and will gain a partial negative change. Hydrogen has a lower electronegativity than the other five elements (except for boron). Because electronegativity increases from left to right on the periodic table, the polarity of the bonds also increases from left to right.

5. No, hydrogen will have only a partial positive charge when it is bonded to an atom that has a greater electronegativity. If the hydrogen atom is bonded to an atom with a lesser electronegativity, such as boron, the hydrogen atom will have a partial negative charge. If the hydrogen atom is bonded to another hydrogen atom, the bond will be nonpolar.

6. Possible answer: Be—F, 2.41; Na—O, 2.51; Ca—Cl, 2.16. Any combination of elements in which the difference in electronegativities is 2.1 or greater is an acceptable answer.

7. Possible answer: O—F, 0.52; C—H, 0.45; S—F, 1.79. Any combination of elements in which the difference in electronegativities is less than 2.1 but greater than zero is an acceptable answer. Answers should be limited to nonmetal elements.

8. Possible answer: In a polar covalent bond, the electrons are shared between two atoms but are more strongly attracted to one of the atoms than to the other. In a nonpolar covalent bond, the electrons are shared equally by the two atoms. In an ionic bond, an electron is transferred from the atom with the lesser electronegativity to the atom with the greater electronegativity.

9. To say that bonding is on a continuum means that the type of bonding changes gradually as the difference in electronegativity between atoms increases. There is no sharp distinction between polar covalent and ionic bonds.

LESSON 45

1. To determine if a molecule is polar, use a Lewis dot structure or structural diagram to figure out the shape of the molecule. Then look at whether the individual bonds are polar. If the molecule is asymmetrical and has polar bonds, it will be a polar molecule.

2. Possible answer: One theory of why small nonpolar molecules do not have a smell is that these molecules do not dissolve in the mucus inside the nose. The mucous membrane in the nose is mostly water. Another possible answer is that polar molecules are attracted to the polar molecules at receptor sites and nonpolar molecules are not attracted as much.

3. a.

H
:Se:H

H_2Se is a bent molecule. Because its bonds are polar and it is asymmetrical, it is likely to have a smell.

b. H:H

H_2 is a linear molecule. Its bond is nonpolar. Because it is symmetrical and has a nonpolar bond, it is not likely to have a smell.

c. :Ar:

Because Ar consists of a single atom, it is not polar and is not likely to have a smell.

d.

:F:
:O:H

HOF is a bent molecule. Its bonds are polar and it is asymmetrical, so it is likely to have a smell.

e.

H
:F:C:Cl
:F:

$CHClF_2$ is a tetrahedral molecule and has polar bonds. Its shape is symmetrical, but because the atoms at each point in the tetrahedron are different, the bonds are not symmetrical. It is likely to have a smell.

f.

H
C::O
H

CH_2O is a trigonal planar molecule. Its bonds are polar and it is asymmetrical, so it is likely to have a smell.

4. a. 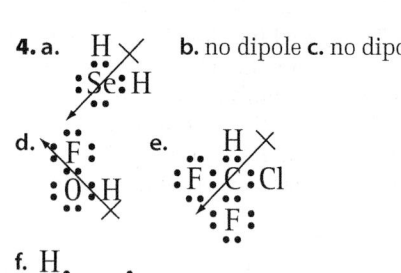 **b.** no dipole **c.** no dipole

Each dipole should point from the positive side of the molecule to the negative side.

5. Possible answer: Water does not have a smell because there are no receptors in the nose that are sensitive to water molecules. If the nose had receptor sites that were sensitive to water, they would always be filled by the water in the mucous membrane.

6. Possible answer: Smelling the air would not be useful because the molecules of the air are constantly passing through the nose and would always fill the matching receptor sites. Smelling unusual air or foul air is more useful because it warns us to find fresh air.

7. Possible answer: Although methane can be dangerous, it is odorless and hard to detect. When dimethyl sulfide, a strong-smelling polar molecule, is added to natural gas, leaks are more easily detected and can be fixed before the methane level becomes dangerous. Because the sulfur atom is not enclosed, other molecules can detect the partial negative charge on the sulfur atom.

LESSON 48

4. Possible answer: In an aqueous environment, the protein will fold such that the hydrophobic amino acids are on the inside and the hydrophilic amino acids are on the outside. Aspartic acid has an R group that is polar, because it is small and asymmetric and, therefore, hydrophilic. The aspartic acid will tend to be on the outside of the protein molecule. Glycine has an R group that is nonpolar because it is simply a hydrogen atom and, therefore, hydrophobic. The glycine will tend to be on the inside of the protein molecule.

5. Answers will vary: Humans need 20 different amino acids to make proteins. Ten of these amino acids are produced by the body. Proteins from meat, dairy products, and other foodstuffs provide all of the other amino acids that humans need.

6. Possible answer: Forming a peptide bond is similar to forming an ester

from an acid and an alcohol because in both cases the carboxyl group of an acid bonds to another functional group with two hydrogen atoms and an oxygen atom breaking off in the process. The difference is that the carboxyl group bonds to an amine in the formation of a peptide bond and an alcohol in the formation of an ester.

LESSON 50

1. Possible answer (any 2): ● Use the graph of volume versus height for a rain gauge with a base area of 2.0 cm³. Find 3 cm on the *x*-axis, move up to the graph, and then read the corresponding *y*-value. ● Use a proportionality constant. Examining the data in the rain gauge table shows that the value of the volume is always twice the value of the height. ● Use the formula volume = base · height.

2. Possible answer: Meteorologists prefer to measure rain in inches or centimeters because a height measurement will be the same for any type of gauge, and a volume measurement will depend on the gauge itself.

3. Possible answer: Both the volume and the height of water in the rain gauge will keep track of the increase because they both increase by the same proportion as additional rain falls into the gauge.

4. The beaker and a graduated cylinder will not give the same volume of rain unless they have the same diameter. They will give the same height of rain even if they have a different diameter.

5. Possible answer: The three containers would have different volumes of water in them after the storm because their bases have different areas. The greatest volume of water will be in the washtub and the least will be in the graduated cylinder. However, the height of water in each container will be the same.

6. a.

Centimeters and Inches

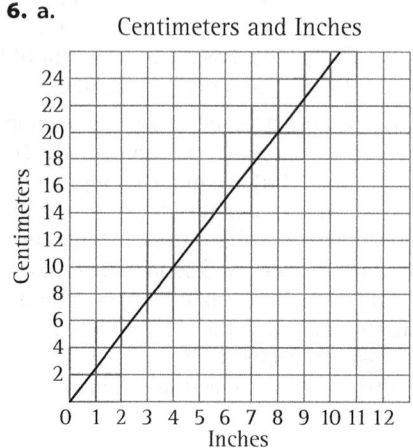

b. 30.5 cm, **c.** 3.9 in, **d.** 0.4 in. When using the graph to make conversions, answers should not be more precise than a tenth of one unit. The answer for part b can also be found by multiplying 12(2.54) = 30.48. For parts c and d, divide by 2.54.

7. a. The storm that dropped the most rain was storm number 4. The height of water in the rain gauge was greatest for that storm. The height of water in the gauge should be used to compare the storms because it is the standard way to measure the amount of rainfall. **b.** For each storm, the volume of water (measured in cubic centimeters) in the gauge is about 2.5 times the height of water (measured in centimeters) in the gauge.

c.

Volume vs. Height of a Rain Gauge

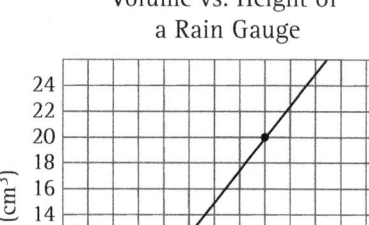

d. The points are not all in a straight line because the values for the volume and for the height are rounded to the nearest tenth. **e.** For a height of 6 cm, the volume in the gauge should be about 15 cm³. Multiply 6 by the proportionality constant 2.5 to find the volume in cubic centimeters.

LESSON 56

1. The density of a gas is much lower than the density of a solid.

2.

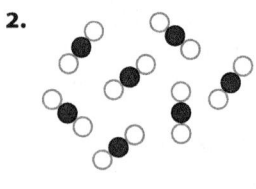

● carbon
○ oxygen

3. Possible answer: Sublimation is when a substance changes from the solid phase directly to the gas phase without becoming a liquid. Evaporation is when a substance changes from the liquid phase to the gas phase.

4. *Possible answers (any 3):* ● Fill a plastic bag with gas. You can tell the gas exists because the bag remains puffed out when you press on it. ● Move your hand rapidly back and forth in the air and you can feel the matter through which your hand is moving. This matter is a gas. ● When you connect a Bunsen burner to a gas line, turn it on, and create a spark, the gas ignites. This proves that the gas is present. ● Reduce the temperature of a trapped amount of air underneath a jar. Water vapor forms on the inside of the jar, proving that the vapor existed previously as a gas.

5. *Possible answer:* Ice floats in water, so it must have a slightly lower density than water.

6.

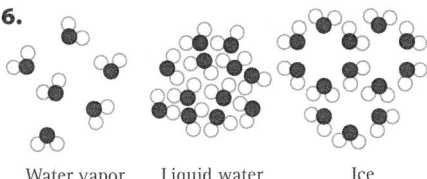

Water vapor   Liquid water   Ice

● oxygen
○ hydrogen

7. 4.1 mL

8. 6.7 g

9. 12 g

10. No, the volume of 15 grams of carbon dioxide is 7,900 mL, which is slightly less than 8 L.

LESSON 57

1. *Possible answers (any 3):* ● Water will not enter an upside-down, submerged cup because the pressure of the air inside keeps water from entering it. ● Squeezing a sealed plastic bag full of air causes the bag to pop open because of the pressure of air from inside. ● It is impossible to inflate a balloon that is inside a bottle because the air in the bottle keeps the balloon from inflating. ● When pressure is reduced at the top of a straw inserted into a drink, air pressure on the surface of the drink forces fluid up the straw.

2. Air pressure is caused by the impact of gas molecules against an object or surface.

3. Because the air inside the balloon exerts pressure on the inside wall but there is no pressure balancing it on the outside wall, the balloon will expand and most likely pop.

4. Answers will vary. *Possible answer:* Estimate the area of the back of the hand in square inches and then multiply the area in square inches by 14.7 lb/in². Reasonable methods for finding the area of a hand are to estimate the length and width of the hand with the fingers together, or to draw an outline of the hand on graph paper. For example, if the hand area is 20 in², then the pressure on the back of the hand is 20 in² · 14.7 lb/in² = 290 lb (answer given to two significant figures).

5. *Possible answer:* The air inside your ear is at the same pressure as the air around you on the ground. When the plane goes up into the atmosphere, the pressure outside your ear decreases. The pain that you experience is a result of the higher air pressure inside your ear pushing outward. Sometimes your ear pops when the air escapes from the inside of your ear, and the pain is relieved. Chewing gum or yawning helps to equalize the pressure inside the ear with that outside. The key point is that the air pressure is different inside the ear and outside the ear.

6. 33 lb/in²

7. 7.35 lb/in². Using 20 in² as the area of a hand, 20 in² · 7.35 lb/in² = 150 lb (answer given to two significant figures).

8. *Possible answer:* Gases such as helium and oxygen are transported at high pressure in tanks made of steel or other strong materials. Sometimes, a combination of the high pressure and decreased temperature of the tank causes the gas to enter the liquid phase. The gas is transported at pressure to allow a larger quantity of the substance to be transported at one time.

9. When a car tire rolls over your foot, it exerts a pressure on your foot about equal to the pressure of the air inside the tire. This pressure takes into account the weight of the car and is equal at every point outside of the tire. So, the entire weight of the car is not felt by your toes, just a small portion of the weight. A pressure of 30 lb/in² is usually not enough to break toes.

LESSON 59

2. In a flexible container, the volume of the gas can change if the pressure or temperature of the gas inside or outside the container changes. In a rigid container, the volume of the gas inside is constant, so only the temperature and pressure of the gas can change.

3. 720 L

4. a. The pressure and volume of the balloon change but the temperature remains the same. **b.** Yes, the pressure on the inside of the balloon can change so that the pressures are equal because the balloon is a flexible container. **c.** The volume occupied by the air inside the balloon will increase because the pressure has decreased, and pressure and volume are inversely proportional. **d.** 320 mL

5. a. The volume of air inside the bottle stays the same because the bottle is a rigid container. **b.** Yes, the air inside the bottle will lose energy until it has the same temperature as its surroundings because glass does not insulate against temperature changes. **c.** Pressure and temperature are directly proportional. The pressure of the air inside the bottle will decrease as its temperature decreases. **d.** 0.93 atm

LESSON 60

1. According to the kinetic theory of gases, increasing the gas volume decreases the gas pressure, because the molecules of gas have more room in which to travel and therefore strike the walls of the container less often.

2. According to the kinetic theory of gases, decreasing the gas temperature decreases the gas pressure, because molecules move more slowly at a lower temperature and therefore strike the walls of the container less often.

3. *Possible answer:* An increase in temperature probably caused the change in pressure, because pressure and temperature are proportional and the volume of a gas cylinder does not change. Other possible answers are that gas was added to the cylinder to increase the pressure or that a decrease in volume increased the pressure of the gas because the rigid container was deformed in some way.

4. Heating a gas in a sealed container can be dangerous because, as the temperature increases, the pressure will also increase. If the pressure increases too much, the container might rupture.

5. The volume of a gas cannot decrease to zero because the particles of the gas themselves occupy space.

6. *Possible answer:* As the airplane rises, the pressure outside the ear decreases. The discomfort comes from the air molecules inside the inner ear hitting the enclosing membrane more often than the air molecules outside the inner ear. This causes the membrane to flex outward until the pressure is equalized.

7. a. Because lungs act as flexible containers, a change in pressure affects their volume. The volume of your lungs at 3.5 atm is much lower than the volume at sea level. **b.** Holding your breath when you ascend quickly is dangerous because the pressure decreases rapidly as you rise. If you hold your breath, your lungs will act as sealed containers, making the pressure inside the lungs much greater than the pressure outside the lungs, possibly causing your lungs to rupture.

8. *Possible answer:* According to Boyle's law, pressure and volume are inversely proportional when the temperature is constant. As the helium balloon floats up into the sky, its volume will increase as the pressure around the balloon decreases. Because the balloon is a flexible container, the pressure inside the balloon will be the same as the pressure outside the balloon. Assuming the temperature is the same, pressure and volume are inversely proportional, so a decrease in pressure causes an increase in volume of the balloon.

9. a.

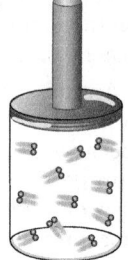

b. The pressure inside the cylinder decreases because the volume has increased. As a result, the particles strike the walls of the cylinder less frequently and exert less pressure. **c.** Boyle's law.

10. *Possible answer:* When steam heats in a boiler, its pressure increases. Injecting this heated, high-pressure steam into a cylinder causes the cylinder to push against the piston. As the volume increases, the pressure decreases, and the piston's weight is able to push back against the steam until the piston drops again. This increases the pressure on the

steam until the steam is able to push the piston back up again. The piston is connected to an axle that turns as the piston oscillates. So, heating the steam eventually leads to the mechanical energy in the axle.

LESSON 63

3. A manometer is a U-shaped tube with one end open to the atmosphere and the other end closed. The air presses down on a fluid in the open end to generate a pressure reading on the closed end. A barometer works using the same principle, but instead of a tube of fluid, a tube with one open end is inverted and placed in a dish of fluid. The air presses down on the fluid in the dish to generate a pressure reading in the inverted tube.

4. Increasing the number of gas particles in a container increases the gas pressure in the container. As the number of molecules of gas increases, the number of collisions between gas particles and the walls of the container also increases. Similarly, if the number of gas particles in a container decreases, the gas pressure also decreases.

5. *Possible answers:* ● Increase the amount of air in the tire. ● Increase the temperature of the air in the tire. ● Press on the outside of the tire to decrease its volume.

6. a. B, A, C; gas pressure is proportional to the number density, not the absolute number of particles in the sample. Sample B has a lower gas pressure than sample A because it has twice the volume of sample A but the same number of particles. Sample A has a lower pressure than sample C because it has the same volume but fewer particles.

b.

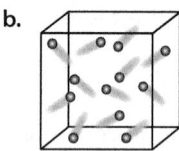

The most likely correct answers that would be drawn are ● An air sample with 10 particles shown in a volume that is in between the volumes of sample A and sample B. ● An air sample with the same volume as sample A but with 6–9 particles shown. ● An air sample with the same volume as sample B but with 11–19 particles shown.

7. When a high-pressure system moves into a region, the air pressure increases. When the air pressure on the open end of the barometer increases, it pushes on the mercury harder than it did before. Mercury is pushed higher into the closed end of the barometer until the downward pressure caused by the weight of the column of mercury is equal to the atmospheric pressure.

8. The liquid in the barometer does not spill out because the pressure of the air in the atmosphere pushes on the surface of the mercury and forces it into the tube.

LESSON 68

1. Reactants are the substances that are present before a chemical reaction, and products are the substances that are formed during the reaction.

2. *Possible answer:* Chemical reactions are important for life because the functions inside an organism depend on chemical changes. For example, digestion breaks down food to extract energy and materials for the body.

3. *Possible answer:* Toxic substances cause harm to living organisms. Answers will vary, but the key point is that toxic substances cause harm, and benign substances can become toxic if the dosage is high enough.

4. a. An aqueous solution of bleach is added to an aqueous solution of ammonia to produce a sodium hydroxide solution and chloramine gas. **b.** The three aqueous solutions all are clear. Therefore, the only thing that can be observed is the formation of a gas after mixing. Mixing the two reactants will result in the formation of bubbles.

5. a. Solid mercury chloride is added to a solution of EDTA to produce a solution containing a compound of mercury with EDTA and hydrochloric acid. **b.** When the solid mixes with the clear liquid, the solid will disappear, and a mixture of two liquids remains.

6. *Possible answers (any 3):* interfering with chemical functions in the body, ● irritating or damaging tissues, ● interfering with other compounds that the body needs, ● upsetting the balance of chemicals in the body such as making fluids acidic, ● restricting the movement of liquids or gases in the body.

LESSON 72

3. a. *Possible answer:* Make three loaves of banana bread. **b. *Possible answer:*** This is related to balancing an equation because you also have to increase the amount of the other ingredients in the same proportion as you change the amount of bread you make. This is the same as multiplying a chemical equation by a counting number.

4. a. $2K(s) + I_2(s) \rightarrow 2KI(s)$

   b. $Mg(s) + Br_2(s) \rightarrow MgBr_2(s)$

   c. $KBr(aq) + AgNO_3(aq) \rightarrow KNO_3(aq) + AgBr(s)$

   d. $2KClO_3(s) \rightarrow 2KCl(s) + 3O_2(g)$

   e. $2C_2H_6(g) + 7O_2(g) \rightarrow 4CO_2(g) + 6H_2O(l)$

   f. $4Al(s) + 3O_2(g) \rightarrow 2Al_2O_3(s)$

   g. $P_4(s) + 6H_2(g) \rightarrow 4PH_3(g)$

LESSON 85

3. In a strong acid, all of the molecules of the solute dissociate into ions, while in a weak acid only some of the molecules dissociate.

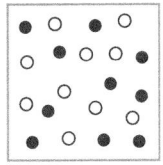 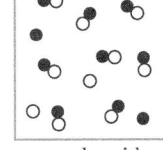

strong acid   weak acid

4. a. acid; H^+ and I^- **b.** acid; H^+ and $COOH^-$ **c.** base; Rb^+ and OH^- **d.** acid; H^+ and OCl^- **e.** acid; H^+, H^+, and SeO_4^{2-} **f.** base; PH_4^+ and OH^- **g.** acid; H^+ and ClO_4^- **h.** base; Ca^+, OH^-, OH^-

5. a. You would need 10 Br^- ions because each of the 10 hydrogen ions is formed by the dissociation of one molecule of HBr made up of one hydrogen atom and one bromine atom.

b.

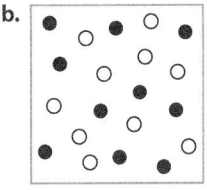

H⁺   Br⁻

6. a. You would need 20 OH^- ions because each of the 10 magnesium ions is formed by the dissociation of one molecule made up of one magnesium atom and two hydroxide groups.

b.

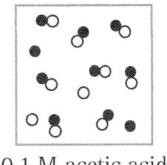

○   ●
Mg^{2+} OH^-

7. Washing soda, Na_2CO_3, forms a basic solution because it accepts a proton from water, thereby forming a hydroxide ion in solution.

8.

0.1 M HCl	0.1 M acetic acid

●H^+ ○Cl^- ●H^+ ○CH_3COOH

LESSON 94

3. *Possible answers (any 5):* I drank an *energy* drink before working out. ● To grow, all plants require *energy* from the Sun. ● Power lines transfer electrical *energy* from power plants to your home. ● The electrons in an atom are contained in different *energy* levels. ● The *energy* to run a car's engine comes from gasoline. ● I'll eat a big meal the night before the big race to make sure I have enough *energy*. ● The newer model of refrigerator is more *energy* efficient than the older model.

4. *Possible answers (any 3):* Forest fires recycle materials in the environment. ● Fire is used to cook foods. ● Fire is used to heat homes and for warmth. ● Fire is used to extract energy from fossil fuels.

5. *Possible answer:* Most of a fire is hot gases that become part of the atmosphere. Ashes are leftover material that was not converted to light and heat by a fire. The heat from the ashes transfers into the air and other materials around the site of the fire as it cools. At first, the ashes are hot because of the reactions that are taking place around them. Once the fire dies down, the ashes transfer heat to the cooler air and eventually reach equilibrium with the surrounding environment.

6. *Possible answer:* In a gas heater, natural gas is heated until a reaction occurs with oxygen to produce energy, carbon dioxide, and water. The energy heats the air around the heater, and then the hot air spreads to the rest of a room or through ducts to other rooms in the home. The combustion reaction of methane gas is $CH_4(g) + 2O_2(g) \rightarrow CO_2(g) + 2H_2O(g) +$ heat.

7. *Possible answer:* A coal-burning power plant heats coal until it bursts into flame. The result is a release of energy, carbon dioxide, nitrogen, and water. The heat that is released is used to turn water into steam. The steam is then forced into a steam turbine. The turbine's blades turn as the steam passes through them. A generator then converts the mechanical energy of the blades into electrical energy.

LESSON 95

4. a. copper sulfate, $CuSO_4$, water, H_2O, and hydrated copper sulfate, $CuSO_4 \cdot 5H_2O$. **b. *Possible answer:*** the beaker, the thermometer, air, my hand. **c.** The copper sulfate is at a lower temperature than the hydrated copper sulfate. The thermometer shows that the temperature increased after the water and the copper sulfate were mixed. **d.** The beaker will feel hot until it has a chance to reach thermal equilibrium with the surroundings. **e.** The process is exothermic because the temperature increases during the reaction. Energy was transferred from the reaction into the surrounding environment. **f.** A chemical reaction must have occurred for a change in energy to occur. The reaction must have been the source of the heat that caused the temperature reading to change on the thermometer.

5. a. barium hydroxide, $BaOH \cdot 8H_2O$, ammonium nitrate, NH_4NO_3, ammonia, NH_3, water, H_2O, and barium nitrate, $Ba(NO_3)_2$. **b. *Possible answer:*** the beaker, the pool of water outside the beaker, air, my hand. **c.** The $Ba(NO_3)_2$ is at a lower temperature than the NH_4NO_3 because the water freezing outside the beaker shows the temperature decreased after the reactants were mixed. **d.** The beaker will feel cold. An endothermic reaction will result in a transfer of heat from the surroundings, including the beaker, to the system. The beaker will feel cold to the touch until it has a chance to reach equilibrium. **e.** The process is endothermic because the temperature decreases during the reaction. Energy was transferred from the surrounding environment into the reaction. **f.** A chemical reaction must have occurred for a change in energy to occur. The reaction must have been the cause of the water outside the beaker turning into ice.

LESSON 98

8. a. The methanol will be at a higher temperature after 5 minutes because it has a lower specific heat capacity and therefore does not have to absorb as much thermal energy to increase in temperature. **b.** The final temperature for the water would be 23.50 °C, and the final temperature for the methanol would be 23.86 °C.

9. Possible answer: Water would make a poor thermal insulator for a home. Water has a higher specific heat capacity than many materials in the table on page 501, but materials that have a higher specific heat capacity have probably been developed. Also, water moves heat from place to place through convection, and insulation should not conduct heat from the outside a home to the inside or vice versa.

10. Possible answer: Energy is released as water changes state from a liquid to a solid. Water sprayed on oranges would freeze before the inside of the orange, and the orange would absorb energy as the water froze. The layer of ice acts as an insulator to protect the orange from freezing when the air temperature drops further. The key point to remember to answer this question is that freezing is an exothermic process. The energy released can be used to keep the orange warm. Ice is also a good insulator, because it reflects most incoming sunlight and because it has a high specific heat capacity.

LESSON 99

6. a. 4500 cal (to two significant figures) **b.** 81,000 cal (to two significant figures)

7. 24.4 g

8. a. 20,000 cal (to two significant figures) **b.** 14,200 cal (to two significant figures)

9. The energy needed to melt 25.0 g of ice is given by:

q = (heat of fusion) $\cdot$ m
$= (79.9 \text{ cal/g})(25.0 \text{ g})$
$= 2000 \text{ cal.}$

The energy needed to heat 25.0 g of water from 0 °C to 20 °C is given by

$q = (25.0 \text{ g})(1.00 \text{ cal/g} \cdot °C)(20 °C - 0 °C)$
$= 500 \text{ cal.}$

Therefore, the total energy absorbed by the 25.0 g ice cube is 2000 cal + 500 cal = 2500 cal. The energy released by 100 g of water that is cooled from 45 °C to 20 °C is given by

$q = (100 \text{ g})(1.00 \text{ cal/g} \cdot °C)(45 °C - 20 °C)$
$= 2500 \text{ cal.}$

Because the energy released is equal to the energy absorbed, energy is conserved.

LESSON 100

1. If a compound is ionic, or if a substance is water or carbon dioxide, it is unlikely to combust.

2. Possible answers (any 3):
● Oxygen is a reactant. ● The reaction produces heat. ● The reaction produces light. ● Energy is released.

3. a. Sodium sulfate, Na_2SO_4, will not combust because it is an ionic compound. **b.** Copper, Cu, will combust because it is a metallic element. **c.** Magnesium oxide, MgO, will not combust because it is an ionic compound. **d.** Sodium, Na, will combust because it is a metallic element. **e.** C_2H_6O will combust because it is a molecular substance containing carbon and hydrogen. **f.** Argon, Ar, will not combust because it is a noble gas.

4. a. $2C_2H_6(g) + 7O_2(g) \rightarrow$ $4CO_2(g) + 6H_2O(l)$
b. $C_2H_{12}O_6(s) + 6O_2(g) \rightarrow$ $6CO_2(g) + 6H_2O(l)$
c. $C_2H_5OH(l) + 3O_2(g) \rightarrow$ $2CO_2(g) + 3H_2O(l)$
d. $C_{21}H_{24}N_2O_4(s) + 27O_2(g) \rightarrow$ $21CO_2(g) + 12H_2O(l) + 2NO_2(g)$
e. $2C_2H_5S_2H(s) + 9O_2(g) \rightarrow$ $4CO_2(g) + 6H_2O(l) + 2SO_2(g)$. The first chemical formula (the reactant that is not oxygen) in each equation is the fuel.

5. a. carbon dioxide, CO_2, and water, H_2O:

$C_7H_6O(l) + 8O_2(g) \rightarrow$ $7CO_2(g) + 3H_2O(l)$

b. carbon dioxide, CO_2, and water, H_2O:

$CH_3COCH_3(l) + 4O_2(g) \rightarrow$ $3CO_2(g) + 3H_2O(l)$

c. carbon dioxide, CO_2, and water, H_2O:

$2C_2H_2O_4(l) + O_2(g) \rightarrow$ $4CO_2(g) + 2H_2O(l)$

d. calcium oxide, CaO:

$2Ca(s) + O_2(g) \rightarrow 2CaO(s)$

e. lithium oxide, Li_2O:

$4Li(s) + O_2(g) \rightarrow 2Li_2O(s)$

f. silicon oxide, SiO_2:

$Si(s) + O_2(g) \rightarrow SiO_2(s)$

LESSON 103

5. $2H_2(g) + O_2(g) \rightarrow 2H_2O(l)$

$C(s) + O_2(g) \rightarrow CO_2(g)$

$CH_4(g) + 2O_2(g) \rightarrow CO_2(g) + 2H_2O(l)$

$2C_2H_6(g) + 7O_2(g) \rightarrow 4CO_2(g) + 6H_2O(l)$

$2C_4H_{10}(g) + 13O_2(g) \rightarrow$ $8CO_2(g) + 10H_2O(l)$

$2C_6H_{14}(l) + 19O_2(g) \rightarrow$ $12CO_2(g) + 14H_2O(l)$

$2C_8H_{18}(l) + 25O_2(g) \rightarrow$ $16CO_2(g) + 18H_2O(l)$

$2CH_4O(l) + 3O_2(g) \rightarrow$ $2CO_2(g) + 4H_2O(l)$

$2C_2H_6O(l) + 6O_2(g) \rightarrow$ $4CO_2(g) + 6H_2O(l)$

$2C_4H_{10}O(l) + 12O_2(g) \rightarrow$ $8CO_2(g) + 10H_2O(l)$

$C_6H_{12}O_6(l) + 6O_2(g) \rightarrow$ $6CO_2(g) + 6H_2O(l)$

Possible answer: The larger the hydrocarbon, the greater the mole ratio of the oxygen required for combustion of the hydrocarbon.

6. Possible answer: Glucose combustion provides a large energy per mole because glucose is a large molecule with six carbon atoms that are available to combine with the oxygen gas. Glucose provides a small energy per gram because it already contains six oxygen atoms, so the proportion of the molecule's mass that is carbon is smaller than in alkanes and alcohols.

7. Answers will vary, but at least three reasons should be given. Reasons that hydrogen is difficult to use as a fuel in automobiles include ● the difficulty in handling a gaseous fuel ● inability to collect hydrogen on a large scale ● lack of a hydrogen distribution network ● highly explosive and easily combustible nature of hydrogen in large quantities ● lack of trained technicians to repair hydrogen-fueled cars.

LESSON 105

4. a. The diagram for the formation of methane shows an endothermic reaction. **b.** The substances with the lowest potential energy are plotted lowest on an energy diagram, carbon dioxide and water. **c.** The conservation of energy requires the kinetic energy to increase when the potential energy decreases. **d.** The reverse reaction requires a constant input of energy because the products have a higher potential energy than the reactants. For this reason, energy must be added continuously to make the reaction continue for other molecules of the reactants. **e.** The heat of the reaction is -824 kJ/mol, slightly more energy than given on the diagram.

5. a. and **b.**

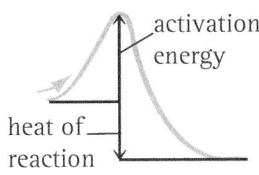

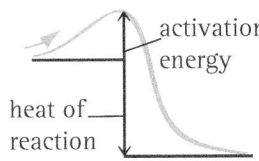

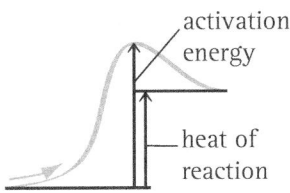

c. Exothermic reactions have a negative heat of reaction. The middle diagram has the longest heat of reaction arrow, which points downward, so it is the most exothermic reaction. **d.** All of the energy curves begin with an upward curve, so some energy input is needed at the beginning of all the reactions.

LESSON 108

1. *Possible answer:* Oxidation of a metal is similar to combustion of a covalent compound because, in both reactions, electrons are transferred to an oxygen atom and both types of reactions are exothermic. Oxidation of a metal is different from combustion of a covalent compound because it does not produce flames and can occur slowly. Oxidation of a metal produces a metal oxide, while combustion usually produces carbon dioxide and water.

2. *Possible answer:* When magnesium metal is oxidized, it reacts with oxygen to form a magnesium oxide, MgO. Each magnesium atom donates two electrons to an oxygen atom. The reaction produces heat and a bright light, but no flames.

3. a. $2Ca(s) + O_2(g) \rightarrow$ $2CaO(s)$; Ca^{2+}, O^{2-}

$4Cr(s) + 3O_2(g) \rightarrow 2Cr_2O_3(s)$; Cr^{3+}, O^{2-}

$4Li(s) + O_2(g) \rightarrow 2Li_2O(s)$; Li^+, O^{2-}

b. Calcium, Ca, is oxidized. Chromium, Cr, is oxidized. Lithium, Li, is oxidized.

4.

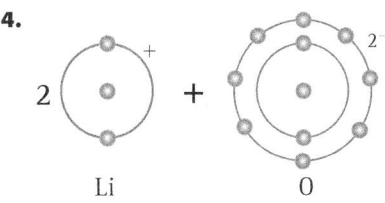

Lithium ions have a charge of $+1$ and oxygen ions have a charge of -2. Lithium ions have one filled shell and oxygen ions have two filled shells.

5.

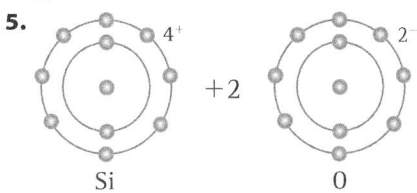

The chemical equation for the oxidation of silicon is $Si(s) + O_2(g) \rightarrow SiO_2(s)$. Silicon ions have a charge of -2. Both silicon ions and oxygen ions have two filled shells.

LESSON 112

5. *Possible answer:* The electrochemical cell with calcium and iron will have the greatest voltage because those two elements are farthest apart on the activity series table.

calcium and iron:

$Fe^{3+} + 3e^- \rightarrow Fe$ -0.04 volt (reduction)

$Ca \rightarrow Ca^{2+} + 2e^-$ -2.87 volt (oxidation)

-0.04 volt $- (-2.87$ volt$) = 2.83$ volt

copper and tin:

$Cu^{2+} + 2e^- \rightarrow Cu$ 0.34 volt (reduction)

$Sn \rightarrow Sn^{2+} + 2e^-$ 0.15 volt (oxidation)

0.34 volt $- (0.15$ volt$) = 0.19$ volt

silver and aluminum:

$Ag^+ + e^- \rightarrow Ag$ 0.80 volt (reduction)

$Al \rightarrow Al^{3+} + 3e^-$ -1.66 volt (oxidation)

0.80 volt $- (-1.66$ volt$) = 2.46$ volt

copper and lead:

$Cu^{2+} + 2e^- \rightarrow Cu$ 0.34 volt (reduction)

$Pb \rightarrow Pb^{2+} + 2e^-$ -0.13 volt (oxidation)

0.34 volt $- (-0.13$ volt$) = 0.47$ volt

So the prediction was correct.

6. *Possible answer:* The pros of combustion as a source of energy include the easy availability of fuel and the relatively large amount of energy combustion that reactions generate. The cons include the potential danger of hazardous flames and high heat.

7. *Possible answer:* The pros of electrochemistry as a source of energy include the low temperature of operation and the ease of transport, storage, and construction of different types of batteries for different applications. The cons include the high cost of materials and the need for a large amount of materials to produce large amounts of energy.

8. A good report will include: ● the type of battery researched (such as carbon, alkaline, and lithium cell batteries) ● the substances required to construct the half-cells, as well as the battery's container ● a diagram of the inside of the battery, showing the half-cells and their components ● the half-reactions that produce the voltage difference.

9. *Possible answer:* The 12-volt battery in a car is actually six 2.1-volt batteries connected in one circuit. This means that the total voltage potential of the entire battery is about 12 volts. Most car batteries are lead-acid batteries. A lead plate and lead oxide plate are immersed in a solution of sulfuric acid. When the battery is connected to a circuit, the plates react with the acid to form lead sulfate.

10. *Possible answer:* A dead battery means that all of the lead and lead oxide available in the battery has been converted to lead sulfate. Jump-starting a car means using another car's battery to start the engine. Jump-starting a car does not require that car's battery to work at all. However, once the car is started, the alternator is able to run. The alternator converts the mechanical energy of the engine back into electrical energy, which is then fed into the car battery in reverse. This provides the energy to reverse the redox reaction in the battery and convert the lead sulfate back into lead and lead oxide. The next time the car has to be started, the battery will then have enough stored chemical energy to start the car.

Periodic Table of the Elements

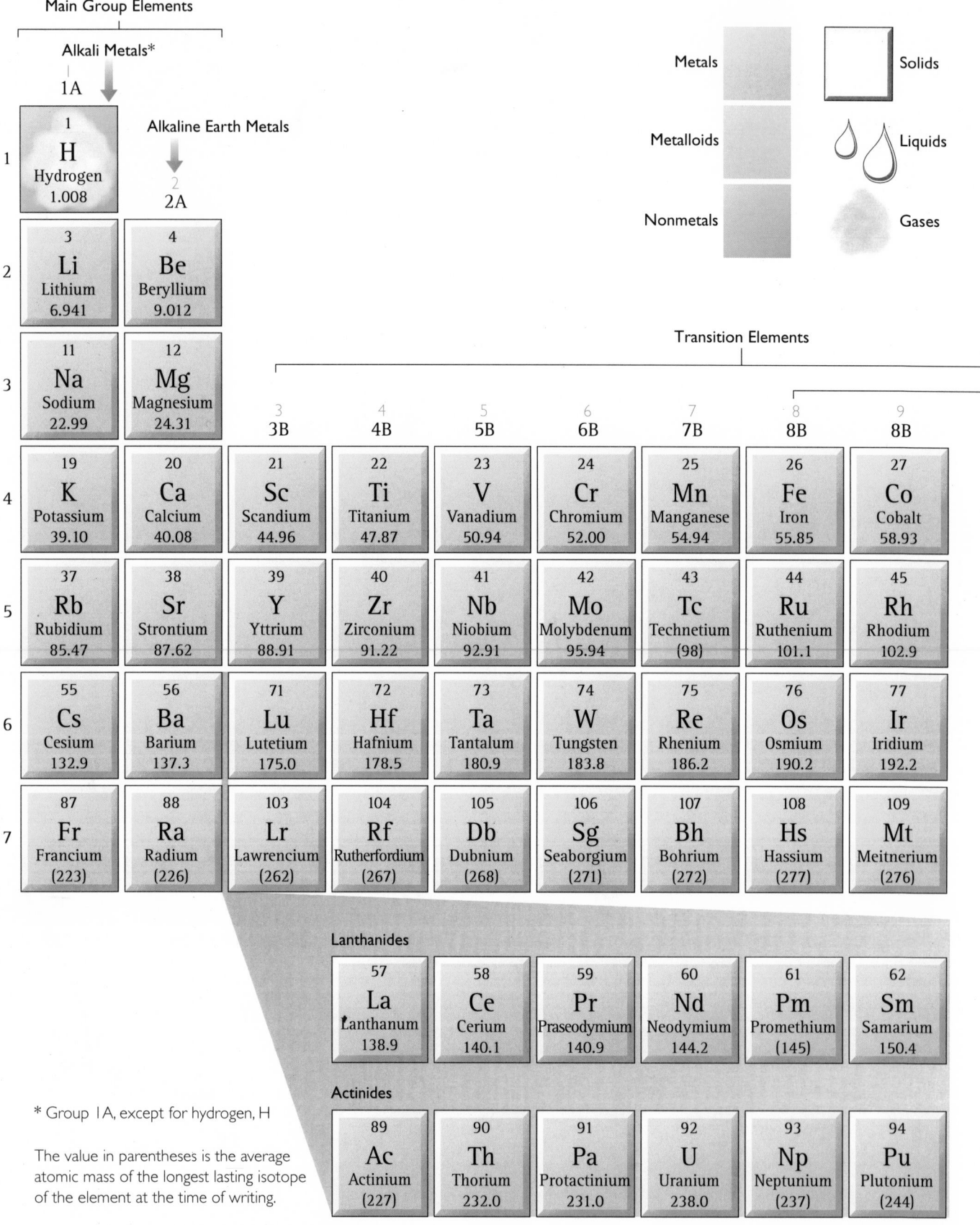

Main Group Elements

Alkali Metals*

1A

Alkaline Earth Metals

2
2A

Metals

Metalloids

Nonmetals

Solids

Liquids

Gases

Transition Elements

| 3 3B | 4 4B | 5 5B | 6 6B | 7 7B | 8 8B | 9 8B |

1
H
Hydrogen
1.008

| 3 Li Lithium 6.941 | 4 Be Beryllium 9.012 |

| 11 Na Sodium 22.99 | 12 Mg Magnesium 24.31 |

| 19 K Potassium 39.10 | 20 Ca Calcium 40.08 | 21 Sc Scandium 44.96 | 22 Ti Titanium 47.87 | 23 V Vanadium 50.94 | 24 Cr Chromium 52.00 | 25 Mn Manganese 54.94 | 26 Fe Iron 55.85 | 27 Co Cobalt 58.93 |

| 37 Rb Rubidium 85.47 | 38 Sr Strontium 87.62 | 39 Y Yttrium 88.91 | 40 Zr Zirconium 91.22 | 41 Nb Niobium 92.91 | 42 Mo Molybdenum 95.94 | 43 Tc Technetium (98) | 44 Ru Ruthenium 101.1 | 45 Rh Rhodium 102.9 |

| 55 Cs Cesium 132.9 | 56 Ba Barium 137.3 | 71 Lu Lutetium 175.0 | 72 Hf Hafnium 178.5 | 73 Ta Tantalum 180.9 | 74 W Tungsten 183.8 | 75 Re Rhenium 186.2 | 76 Os Osmium 190.2 | 77 Ir Iridium 192.2 |

| 87 Fr Francium (223) | 88 Ra Radium (226) | 103 Lr Lawrencium (262) | 104 Rf Rutherfordium (267) | 105 Db Dubnium (268) | 106 Sg Seaborgium (271) | 107 Bh Bohrium (272) | 108 Hs Hassium (277) | 109 Mt Meitnerium (276) |

Lanthanides

| 57 La Lanthanum 138.9 | 58 Ce Cerium 140.1 | 59 Pr Praseodymium 140.9 | 60 Nd Neodymium 144.2 | 61 Pm Promethium (145) | 62 Sm Samarium 150.4 |

Actinides

| 89 Ac Actinium (227) | 90 Th Thorium 232.0 | 91 Pa Protactinium 231.0 | 92 U Uranium 238.0 | 93 Np Neptunium (237) | 94 Pu Plutonium (244) |

* Group 1A, except for hydrogen, H

The value in parentheses is the average atomic mass of the longest lasting isotope of the element at the time of writing.